Brief Contents

Contents

1 Diversity Amid Globalization 1

2 The Changing Global Environment 48

10   Central Asia 418

11   East Asia 454

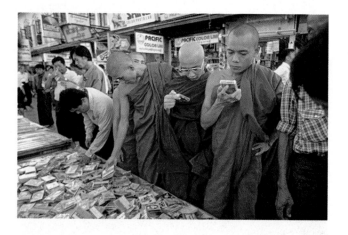

14 Australia and Oceania 592

Preface

Places fascinate geographers. From an early age, we accumulate maps and dream about faraway corners of the world. Those of us lucky enough to make a living in geography also want to understand why the world works the way it does, how its unique regions have taken shape, and how those regions are increasingly interconnected. Those fundamental curiosities brought us together to create something new and different: an interpretation of world regional geography that was deeply appreciative of global diversity and that looked with a fresh and penetrating eye at the aspects of modern life that tie us all together.

The result has been our own odyssey of exploration, a pooling of expertise and enthusiasm that we hope can offer students a perspective that will stay with them long after they leave the classroom. As an introduction to the field of geography, this book provides a view of the world that will serve students well in the early twenty-first century. This view weds environment, people, and place; ponders how those relationships play out in particular regions; and assesses how powerful processes of globalization are reshaping those relationships in new and often unanticipated ways. *Diversity Amid Globalization* is the product of a shared vision that we believe demonstrates the essential and invaluable role that geography can play in all our lives.

Objective and Approach

Diversity Amid Globalization is an issues-oriented textbook for college and university world regional geography classes that explicitly recognizes the geographic changes accompanying today's rapid globalization. With this focus, we join the many who argue that globalization is the most fundamental reorganization of the planet's socioeconomic, cultural, and geopolitical structure since the Industrial Revolution. The explicit recognition of this premise provides the point of departure for this book. As geographers, we think it essential for students to understand two interactive tensions. First, they need to appreciate and critically ponder the consequences of converging environmental, cultural, political, and economic systems through forces of globalization. Second, they need to deepen their understanding of the creation and persistence of geographic diversity and difference. The interaction and tension between these opposing forces of homogenization and diversification forms a current running throughout the following chapters and is reflected in our title, *Diversity Amid Globalization*.

Chapter Organization

As are all other world regional geography textbooks, *Diversity Amid Globalization* is structured to explain and describe the major world regions of Asia, Africa, the Americas, and so on. These 12 regional chapters, however, depart somewhat from traditional world regional textbooks. Instead of filling them with descriptions of individual countries, we place most of that important material in readily accessible ancillaries, specifically, in the textbook Web site and in the instructor's manual. This leaves us free to develop five important thematic sections as the organizational basis for each regional chapter. We begin with "Environmental Geography," which discusses the physical geography of each region as well as current environmental issues. Next, we assess "Population and Settlement" geography, in which demography, land use, and settlement (including cities) are discussed. We also provide a section on "Cultural Coherence and Diversity," which examines the geography of language and religion, yet also explores current cultural tensions resulting from the interplay of globalization and diversity. The section on each region's "Geopolitical Framework" then treats the dynamic political geography of the region, including micro-regionalism, separatism, ethnic conflicts, global terrorism, and supranational organizations. Finally, we conclude each regional treatment with a section titled "Economic and Social Development" in which we analyze each region's economic framework as well as its social geography, including gender issues.

This regional treatment follows two substantive introductory chapters that provide the conceptual and theoretical framework of human and physical geography necessary to understand our dynamic world. In the first chapter, students are introduced to the notion of globalization and are asked to ponder the costs and benefits of the globalization process, a critical perspective that is becoming increasingly common and important to understand. Following this, the geographical foundation for each of the five thematic sections is examined. This discussion draws heavily on the major concepts fundamental to an introductory university geography course. The second chapter, "The Changing Global Environment," presents the themes and concepts of global physical geography, including landforms and geology, climate, hydrology, and biogeography.

Chapter Features

Within each regional chapter, several unique features complement the thematic pedagogy of our approach:

- *Comparable Maps.* Of the 14 or 15 maps found in each regional chapter, 7 are constructed with the same theme and similar data and on the same base map so that readers can easily draw comparisons between different regions. Thus, in every regional chapter readers will find an introductory regional place-name and feature map, a map of the physical geography of the region, a climate map, an environmental issues map, a population density map, a language geography map, and a map showing the geopolitical issues of the region. In addition, each regional chapter also has 7 or 8 other maps illustrating such major themes as ethnic tensions, social development, and linkages to the global economy.

- *Comparable Regional Data Sets.* Again, to facilitate comparison between regions, as well as to provide important insight on the characteristics of each region, each chapter contains three tables. The first provides population data, including natural increase and total fertility rates for each country within the region. The second presents economic data for each country within the region on such issues as gross national income (GNI), GNI per capita, GNI with purchasing power parity (PPP), and average annual economic growth. The last standardized table provides indicators of social development and the status of women within the region with data on life expectancy at birth, child mortality, illiteracy, and female participation in the labor force.

- *Sidebars.* Within each chapter, sidebars complement text material. These sidebars have several goals. Each chapter begins with a sidebar titled "Setting the Boundaries," which gives a good overview of the region, discusses its role in the world today, often raising questions about its geographic coherence by accentuating its diversity, and looks at the vexatious issue of regional boundaries. Subsequent sidebars take on different themes; while some probe deeper into environmental issues with case studies, others elaborate on some form of popular culture, such as food or music, common to the region. Many chapters include sidebars that discuss ethnic and political tensions. Regional chapters also contain sidebars titled "Local Voices" in which the authors step aside to let local peoples—through song, literature, or the Internet—talk about their region. Finally, several chapters contain new sidebars that explicitly discuss the geopolitical implications of the war on global terrorism for that particular region.

- *Review and Research Questions.* There are two sets of review and research questions at the end of each regional chapter. The first set helps readers review basic terminology and concepts. The second set, "Thinking Geographically," asks readers to draw together more abstract themes and problems by doing further research on an issue. Resources for answering these critical thinking exercises are found in the chapter bibliographies, in the instructor's manual, and on the textbook's Web site.

- *Regional Films and Novels.* We provide a short list of films and novels that we believe capture the sense of place and flavor of the different world regions. Our experience is that these two art forms can be very important in conveying to students a more comprehensive feel for a region by putting into prose or visual form the landscape, the people, and the issues that bring meaning to land and life.

New to the Second Edition

- *Expanded Treatment of Globalization.* In the last several years, issues of globalization have moved to the center of the world stage and, in contrast to the situation several years ago, the concept of globalization is now commonly used—and misused—in all walks of life. Because this term is becoming more central to the study of world regional geography, we have expanded our initial discussion in Chapter 1 to present varying perspectives and viewpoints on globalization. Each regional chapter also has new information on globalization and its many-faceted implications.

- *New Sections on Global Terrorism.* One particularly negative aspect of globalization, global terrorism, moved to the foreground after the September 11, 2001, terrorist attacks on the United States. Because the geopolitical implications of global terrorism and the reactions against it are increasingly important for the study of world regional geography, we have added a substantial discussion of global terrorism to Chapter 1. Further discussions of the topic have been added to a number of the regional chapters, both within the text itself and in new sidebars.

- *Expanded Discussion of Environmental Geography.* To make the study of each region's physical geography more compelling, approachable, and relevant, we have added explicit linkages, often based on particular case studies, connecting pressing environmental issues, physical geography, and globalization.

- *New and Refined Maps.* New maps have been added to each chapter in order to complement text material and illustrate geographic concepts. All maps, moreover, have been updated with the most recent information and, where necessary, substantially revised. New population density maps in each regional chapter, for example, draw upon the most recent population data and incorporate state-of-the-art cartography.

- *Expanded Coverage of Central Asia.* With the recent war in Afghanistan, the world region of Central Asia has recently been placed squarely on the map of international leaders, policymakers, and scholars. Unlike almost all other world regional geography textbooks, *Diversity Amid Globalization* treats Central Asia as a region in its own right, rather than partitioning it among neighboring areas. In this second edition, the textbook takes an even more detailed look at this increasingly crucial region of the world.

- *Author Field Trips.* Six author field trips are included in the accompanying CD. These trips highlight localities that have been profoundly influenced through globalization. In selecting these places, the authors have drawn upon their own areas of research to illustrate many of the themes discussed in the regional chapters. We feel that this will be an exciting addition to the classroom that will help students visualize concepts and promote discussion.
- *Most-Current Data and Information.* Of course, all text, tables, and maps have been updated with the most current information.

Supplements: The Teaching and Learning Package

We have been pleased to work with Prentice Hall to produce a cutting-edge supplements package that will enhance teaching and learning. Not only does this package contain the traditional supplements that students and professors have come to expect from authors and publishers, but it also folds together new digital technologies that complement and enhance the learning experience.

For the Instructor

- **D.I.G.I.T. (Digital Image Gallery for Interactive Teaching)** (0-13-009778-0). All of the maps and figures from the text and many of the photographs are available digitally as high-quality JPEGs on a CD-ROM. These files are ideal for professors who use PowerPoint or a comparable presentation software in their classes or who create text-specific Web sites for their students. In addition, the CD-ROM contains editable PowerPoint presentations that will make it easier for instructors to create their own custom PowerPoint presentations for their classes.
- **Slides** (0-13-009773-X) and **Transparencies** (0-13-009771-3). Two hundred fifty full-color slides and transparencies provide all of the maps from the text, enlarged for excellent classroom visibility.
- **Test Item File** (0-13-009760-8). The test item file provides instructors with a wide variety of test questions for use in exams.
- **TestGEN-EQ** (0-13-009768-3). The test manager for this text employs TestGEN-EQ software. TestGEN-EQ is a computerized test generator that lets the instructor view and edit test-bank questions, transfer questions to tests, and print tests in a variety of customized formats. Included in each package is the QuizMaster-EQ program that lets instructors administer tests on a computer network, record student scores, and print diagnostic reports.
- **Instructor's Manual** (0-13-009776-4). Intended as a resource for both new and experienced instructors, the Instructor's Manual includes a variety of lecture outlines, additional source materials, teaching tips, advice about how to integrate visual supplements (including Web-based resources), and various other ideas for the classroom.
- **ABC Videos** (0-13-009774-8). This unique video series contains segments from award-winning shows such as *"World News Tonight," "Nightline,"* and *"Good Morning America."* Selected from the archives of ABC News, each video includes a written summary that ties the segment to particular sections of the text, making it easier for the instructor to enhance the classroom presentation with timely and relevant video programs.
- **Course Management.** Prentice Hall is proud to partner with many of today's leading course-management-system providers. These partnerships make it possible to combine market-leading online content with the powerful course management tools Blackboard, WebCT, and Prentice Hall's proprietary course management system, CourseCompass. Please visit the demonstration site, http://www.prenhall.com/demo, for more information, or contact your local Prentice Hall representative, who can provide a live demonstration of these exciting tools.

For the Student

- **Student Study Guide** (0-13-009772-1). The Study Guide includes additional learning objectives, a complete chapter outline, critical thinking exercises, problems, and short essay questions using actual figures from the text, and a self-test with an answer key in the back.
- *Diversity Amid Globalization WWW Site* (http://www.prenhall.com/rowntree). The Web site, tied chapter-by-chapter to the text, provides students with three main resources for further study:
  - **On-line Study Guide:** The on-line study guide provides immediate feedback to study questions, access to current geographical issues, and links to interesting and relevant sites on the Web. Additionally, instructors can create a customized syllabus that links directly to the Web site.
  - **Virtual Field Trips:** Integrated with the book's Web site, each field trip gives students the on-the-ground feeling of visiting a place. The field trips include information about the history, places, people, and current events of locales in each world region.
  - **Country-by-Country Data:** Also integrated with the Web site, this material summarizes the vital statistics for each country within a region.
- **Science on the Internet: A Student's Guide** by Andrew T. Stull (0-13-021308-X). The perfect tool to help students harness the power of the *Diversity Amid Globalization* Web site and the World Wide Web, this resource gives clear step-by-step instructions to access regularly updated resources, as well as navigation strategies and an overview of the Web.
- *GeoTutor* CD-ROM by Charles A. Stansfield and Jerry Westby. *GeoTutor* helps students develop basic geographic skills and allows them to explore the political, cultural, economic, and physical geography of the world through a series of interactive mapping exercises. Additionally, *GeoTutor* contains a full digital reference atlas of the world. If the instructor wishes, GeoTutor can be included *free* with *Diversity Amid Globalization*. Please contact your local Prentice Hall representative for details.

- **Map Workbook to accompany *Diversity Amid Globalization*** (0-13-009777-2). This workbook, which can be used in conjunction with either the textbook or an atlas, features the base maps from the book, printed in black and white. The workbook contains a political and a physical base map for every region in the book, along with a list of key map items for the region at hand. The map workbook is available free when packaged with *Diversity Amid Globalization*. Please contact your local Prentice Hall representative for details.
- **Rand McNally Atlas of World Geography.** This atlas includes 126 pages of up-to-date regional maps and 20 pages of illustrated world information tables. If the instructor wishes, the atlas is available free when packaged with *Diversity Amid Globalization*. Please contact your local Prentice Hall representative for details.
- **Prentice Hall/*The New York Times* Themes of the Times: Geography.** This unique newspaper-format supplement features recent articles about geography from the pages of *The New York Times*. Available *free* from your local Prentice Hall representative, this supplement encourages students to make connections between the classroom and the world around them.

Acknowledgments

We have many people to thank for their help in the conceptualization, writing, rewriting, and production of *Diversity Amid Globalization*. First, we'd like to thank the hundreds of students in our world regional geography classes who have inspired us with their energy, engagement, and curiosity; challenged us with their critical insights; and demanded a textbook that better meets their need to understand the diverse people and places of our dynamic world.

Next, we are deeply indebted to many professional geographers and educators for their assistance, advice, inspiration, encouragement, and constructive criticism as we labored through the different stages of this book. Among the many who provided invaluable comments on various drafts of *Diversity Amid Globalization* are:

Dan Arreola, Arizona State University
Bernard BakamaNume, Texas A&M University
Max Beavers, Samford University
James Bell, University of Colorado
William H. Berentsen, University of Connecticut
Kevin Blake, Kansas State University
Craig Campbell, Youngstown State University
Elizabeth Chacko, George Washington University
David B. Cole, University of Northern Colorado
Malcolm Comeaux, Arizona State University
Catherine Cooper, George Washington University
Jeremy Crampton, George Mason University
James Curtis, California State University, Long Beach
Dydia DeLyser, Louisiana State University
Jerome Dobson, University of Kansas
Caroline Doherty, Northern Arizona University
Vernon Domingo, Bridgewater State College

Joe Dymond, Louisiana State University
Jane Ehemann, Shippensburg University
Doug Fuller, George Washington University
Gary Gaile, University of Colorado
Steven Hoelscher, University of Texas, Austin
Eva Humbeck, Arizona State University
Richard H. Kesel, Louisiana State University
Rob Kremer, Front Range Community College
Robert C. Larson, Indiana State University
Alan A. Lew, Northern Arizona University
Catherine Lockwood, Chadron State College
Max Lu, Kansas State University
James Miller, Clemson University
Bob Mings, Arizona State University
Sherry D. Morea-Oakes, University of Colorado, Denver
Anne E. Mosher, Syracuse University
Tim Oakes, University of Colorado
Nancy Obermeyer, Indiana State University
Karl Offen, Oklahoma University
Jean Palmer-Moloney, Hartwick College
Bimal K. Paul, Kansas State University
Michael P. Peterson, University of Nebraska–Omaha
Richard Pillsbury, Georgia State University
David Rain, United States Census Bureau
Scott M. Robeson, Indiana State University
Susan C. Slowey, Blinn College
Philip W. Suckling, University of Northern Iowa
Curtis Thomson, University of Idaho
Suzanne Traub-Metlay, Front Range Community College
Nina Veregge, University of Colorado
Gerald R. Webster, University of Alabama
Emily Young, University of Arizona
Bin Zhon, Southern Illinois University at Edwardsville
Dr. Henry J. Zintambila, Illinois State University

In addition, we wish to thank the many publishing professionals who have been involved with this project. It has been a privilege to work with you. We thank Paul F. Corey, President of Prentice Hall's Engineering, Science, and Math division, for his early—and continued—support for this book project; Geosciences Editor and good friend Dan Kaveney, for his daily engagement, professional guidance, enduring patience, unyielding discipline, high standards, and warm companionship; Chris Rapp, Media Editor, for his creative and always stimulating additions and insights to our digital program; Amanda Griffith, Associate Editor, for gracefully taking care of the thousands of tasks necessary to this project; Marketing Manager Christine Henry for her effective sales and promotion work (and always welcome company); Editorial Assistant Margaret Ziegler, for gracefully meeting the incessant needs of four demanding authors who rarely remembered that she was also simultaneously doing the same for many other Geoscience authors; Developmental Editors Barbara Muller and Diane Culhane, for their clear articulation of what this project needed to be successful, their guidance on how that could be attained, and their firm insistence that we meet those standards; production editor Tim Flem, for his daily miracles and steady hand on the tiller so that somehow thousands of pages of manuscript were turned into a finished book; copyeditors Roberta Dempsey and Sybil Sosin, for their eagle eyes and graceful hands; and photo researchers Linda Sykes and Jerry Marshall, for their creative solutions to indulging four geographers in our quest for outstanding pictures from strange parts of the world. To all of you, your professionalism is truly inspirational and very, very much appreciated.

Finally, the authors want to thank that special group of friends and family who were there when we needed you most—early in the morning and late at night; in foreign countries and familiar places; when we were on the verge of crying, yet needed to laugh; for your love, patience, companionship, inspiration, solace, understanding, and enthusiasm: Elizabeth Chacko, Margaret Conkey, Rob Crandall, Maureen Hays, Rowan Rowntree, Paul Starrs, Karen Wigen, and Linda, Tom, and Katie Wyckoff. Words cannot thank you enough.

Les Rowntree

Martin Lewis

Marie Price

William Wyckoff

About the Authors

Les Rowntree teaches both Geography and Environmental Studies at San Jose State University in California, where he recently completed a term as the Chair of the interdisciplinary Department of Environmental Studies. As an environmental geographer, Dr. Rowntree's teaching and research interests focus on international environmental issues, the human dimensions of global change, biodiversity and conservation, and human-caused landscape transformation. He sees world regional geography as a way to engage and inform students by giving them the conceptual tools needed to assess global issues critically in their larger context. Recently Dr. Rowntree has done research in Morocco, Mexico, Australia, and Europe, as well as in his native California. Along with publishing in various geographic and environmental journals, Dr. Rowntree is also writing a book on the natural history of California's Bay Area and Central Coast.

Martin Lewis is a lecturer in International Affairs at Stanford University. He has conducted extensive research on environmental geography in the Philippines and on the intellectual history of global geography. His publications include *Wagering the Land: Ritual, Capital, and Environmental Degradation in the Cordillera of Northern Luzon, 1900-1986* (1992), and with Karen Wigen, *The Myth of Continents: A Critique of Metageography* (1997).

Marie Price is an Associate Professor of Geography and International Affairs at George Washington University. A Latin American specialist, Marie has conducted research in Belize, Mexico, Venezuela, Cuba, and Bolivia. She has also traveled widely throughout Latin America and Sub-Saharan Africa. Her studies have explored human migration, natural resource use, environmental conservation, and regional development. Dr. Price brings to *Diversity Amid Globalization* a special interest in regions as dynamic spatial constructs that are shaped over time through both global and local forces. Her publications include articles in the *Annals of the Association of American Geographers*, *Geographical Review*, *Journal of Historical Geography*, *CLAG Yearbook*, *Studies in Comparative International Development*, and *Focus*.

William Wyckoff is a geographer in the Department of Earth Sciences at Montana State University specializing in the cultural and historical geography of North America. He has written and co-edited several books on North American settlement geography, including *The Developer's Frontier: The Making of the Western New York Landscape* (1988), *The Mountainous West: Explorations in Historical Geography* (1995) and *Colorado: The Making of a Western American Landscape 1860-1940* (1999). In 1990, he received the Burlington Northern Corporation's Award for Outstanding Teaching. A World Regional Geography instructor for 18 years, Dr. Wyckoff hopes that the fresh approach taken in *Diversity Amid Globalization* will more effectively highlight the tensions evident in the world today as global change impacts particular places and people in dramatic and often unpredictable ways.

Philosophy & Approach

Diversity Amid Globalization explicitly acknowledges the geographic changes accompanying today's rapid rate of globalization, emphasizing both the homogenizing and diversifying forces inherent to the globalization process. The text challenges students to make critical comparisons between the regions of the world and to understand more fully the interconnections that bind these regions together. Examples of the sorts of topics used to accomplish this goal include:

- The rise of Islamic fundamentalism in SW Asia.
- Aboriginal groups using high-technology tools to forge common political survival strategies.
- The economic and political integration of the European Union, contrasted with micronationalism and the factionalism in Europe.
- Ethnic diversification in the face of strong participation in the global assembly line in SE Asia.
- The globalization and localization of beer consumption and production in the United States and Canada.

Thematic Structure

To facilitate students' abilities to draw comparisons and contrasts between regions, each of the twelve regional chapters employs a thematic structure. This structure organizes each chapter into five thematic sections:

- **Environmental Geography**
- **Population and Settlement**
- **Cultural Coherence and Diversity**
- **Geopolitical Framework**
- **Economic and Social Development**

This structure allows a more penetrating treatment of themes and concepts than the traditional country-by-country approach, and facilitates comparisons of specific topics between regions.

Local Voices

Each regional chapter provides a sidebar entitled **Local Voices,** in which the authors step aside to let local people talk or write about their concerns. These sidebars highlight regional issues and topics such as food, music, environmental problems, cultural tensions, and work conditions that will interest students and provide further insight into the region.

For example, the chapter on Latin America contains viewpoints from indigenous people, discussion of the cocaine trade, and depictions of how urban squatters build their communities.

LOCAL VOICES Life in Belize City

The following passage describes Belize City through the eyes of a young girl, Beka Lamb. The novel *Beka Lamb* is a Belizean coming-of-age story set in the capital city in the 1960s, when Belize was still a British colony.

At times like these, Beka was glad that her home was one of those built high enough so that she could look for some distance over the rusty zinc rooftops of the town. Many of the weathered wooden houses, built fairly close together, tilted slightly as often as not, on top of pinewood posts of varying heights. In the streets, by night and by day, vendors sold, according to the season, peanuts, peppered oranges, craboos, roasted pumpkin seeds and coconut sweets under lamp-posts that also seemed at times to lean. For a while, after heavy rainstorms, water flooded streets and yards at least to the ankles.

A severe hurricane early in the twentieth century, and several smaller storms since that time had helped to give parts of the town the appearance of a temporary camp. But this was misleading, for Belizeans loved their town which lay below the level of the sea and only through force of circumstances, moved to other parts of the country. It was a town, not unlike small towns everywhere perhaps, where each person, within his neighborhood, was an individual with well known characteristics. Anonymity, though not unheard of, was rare. Indeed, a Belizean without a known legend was the most talked about character of all.

It was a relatively tolerant town where at least six races with their roots in other districts of the country, in Africa, the West Indies, Central America, Europe, North America, Asia, and other places, lived in a kind of harmony. In three centuries miscegenation, like logwood, had produced all shades of black and brown, not grey or purple or violet, but certainly there were a few people in

town known as red ibos. Creole, regarded as a language to be proud of by most people in the country, served as a means of communication amongst the races. Still, in the town and in the country, as people will do everywhere, each race held varying degrees of prejudice concerning the others.

The town didn't demand too much of its citizens, except that in good fortune they be not boastful, not proud, and above all, not critical in any unsympathetic way of the town and country.

Source: Zee Edgell, *Beka Lamb*. London: Heinemann, 1982, pp. 11–12.

▲ **Figure 5.1.1 Belize City cottage** Residents of Belize City build their wooden cottages on stilts as protection against flooding. Shuttered cottages are typical throughout the British Caribbean. (Rob Crandall/Rob Crandall, Photographer)

Thematic Structure

Comparable, **standardized maps and data tables** in each chapter aid students' understanding of a specific world region, and help them make comparisons between regions in both cartographic and tabular form.

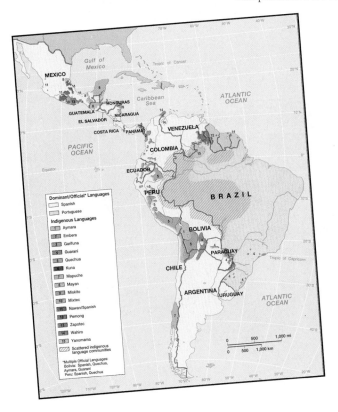

Of the fourteen or fifteen **maps** in each regional chapter, no less than seven are *organized around the same theme and designed to convey comparable material*. These are:

- a **chapter-opening map** with countries and place names
- a **physical map**, showing landforms, hydrology, and tectonic boundaries
- a **climate map**, with climograph call-outs giving temperature and precipitation data for specific cities
- a **"transformation of the Earth"** map with call-outs to environmental issues and solutions within the region
- a **population map** for the region
- a **map of regional languages**
- a **geopolitical map** with call-outs to current issues and tensions

New and refined maps in each chapter.

Each chapter contains several new maps that complement text material and geographic concepts. Along with the new maps, all previous maps have been updated with the most recent data and when appropriate, revised with a changed cartographic presentation. An example would be the new population density maps found in each regional chapter. Not only do these maps draw upon the most recent population data, but they also incorporate state-of-the-art cartography to display these data.

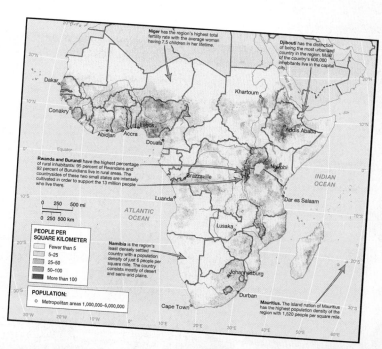

New Features of the second edition of *Diversity Amid Globalization*

NEW

Expanded treatment of globalization in Chapter 1.

In the last several years, globalization has moved to the center of the world stage and, in contrast to the situation several years ago, the concept of globalization is now commonly used (or misused) in all walks of life. Because this term is now even more central to the study of world regional geography, we have expanded our initial discussion in Chapter 1 to present varying perspectives and viewpoints on globalization. As well, each regional chapter has new information on globalization and its many-faceted implications.

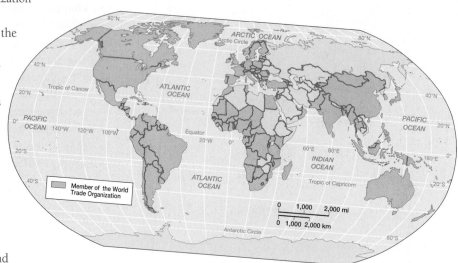

NEW

New sections on global terrorism.

Along with globalization, global terrorism also moved to center stage after the September 11, 2001 attacks on the United States. Because the geopolitical implications of global terrorism and the reactions against it are profound for the study of world regional geography, we have added an introductory discussion of the topic to Chapter 1 and the topic is discussed where relevant in the regional chapters. These regional discussions are found both within the text itself and also in new sidebars.

GEOGRAPHY IN THE MAKING The Shifting Geopolitics of Southwest Asia

What geopolitical role should the United States play in Southwest Asia? That simple question continues to confound policy makers in the United States as well as residents of that ever-troubled part of the world. In particular, the 9/11/01 terrorist attacks renewed the focus on that constantly shifting geopolitical relationship. Osama bin Laden and 12 of the 19 airline hijackers had roots in Saudi Arabia; and others involved had links to Egypt, Yemen, and nearby countries. Thus, since the United States embarked on its "War on Terror," its geopolitical presence in the region has inevitably grown (Figure 7.3.1). But many questions remain. How does the United States identify and deal with hostile elements ("unilateralism"), or depend on sharing the responsibilities with other countries or international organizations ("multilateralism")?

Specifically, three interrelated settings exemplify America's ongoing geopolitical challenges. First, U.S.–Iraqi relations, soured for more than a decade since the Persian Gulf War, continue to be a focal point for conflict. Adding fuel to the fire has been the U.S. belief that Saddam Hussein has played an active role in fostering recent anti-American terrorism. The future remains unclear, however. Indeed, some observers suggest that a post-Saddam Iraq could evolve into a moderate, Western-leaning country and a strong ally of the United States.

A second question mark is oil-rich Saudi Arabia. Unlike Iraq, that nation is an ally of the United States. But the terrorist links to Saudi Arabia have renewed friction between the two countries. Domestic challenges to the all-powerful al-Saud royal family also appear to be on the rise. Both Islamic extremists and Western-leaning progressives have expressed their dissatisfaction with the conservative, traditional Saudi government. American observers fret that a sudden change in Saudi Arabian politics would have global ramifications, particularly if oil exports are disrupted.

Finally, recent events in Israel and the occupied West Bank continue to make global headlines. Palestinian suicide bombings and Israeli reprisals and occupations have further enflamed a region already set on edge after the 9/11/01 attacks. What is the proper role for the United States in tactically improving short-term relationships between Israel and its Palestinian populations and crafting a longer-term strategic policy aimed at achieving a permanent peace in the region? All of these issues will command a great deal of attention within the United States. How the U.S. responds to Southwest Asia's shifting geopolitics seems destined to shape the region for decades to come.

▲ Figure 7.3.1 The USS *Enterprise* in the Persian Gulf Aircraft carriers such as the USS *Enterprise* demonstrate the continuing assertion of geopolitical power by the United States across strategic portions of Southwest Asia and North Africa. (*Office of Naval Research, U.S. Navy*)

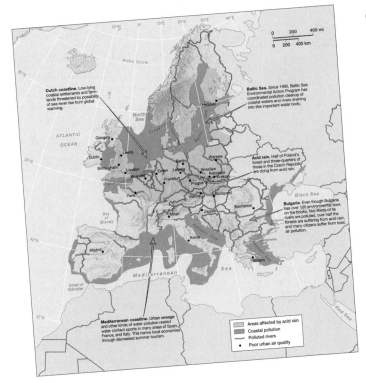

 NEW ### Expanded discussion of environmental geography.

To make the study of each region's physical geography more interesting and relevant, we have made more explicit linkages between that material and the regional and global environmental issues.

NEW ### Author Field Trip CD-ROM.

Packaged with each copy of the text is an Author Field Trips CD-ROM. These virtual field trips take students on pictorial narrated walkthroughs of six places around the world. Also included on the CD-ROM are physical and political outline maps, and the complete glossary from the book. Field trips include:

- Boom and Bust in Colorado
- Bolivian Transmigration
- Reinventing Berlin for the 21st Century
- Moroccan Crossroads: Tradition, Diversity, Globalization
- Kerala in Global Context
- Western Australia in the Global Economy

TEACHING/LEARNING PACKAGE

Testing and Teaching Aids

Test Item File (0-13-009760-8): Includes over 100 multiple choice, true/false and essay questions per chapter.

 D.I.G.I.T. (Digital Image Gallery for Interactive Teaching) (0-13-009778-0): Contains all of the maps and tables from the textbook as well as some of the photos and all of the photos from the Author Field Trip CD-ROM. Includes fully customizable PowerPoint for each chapter of the book, and Microsoft Word files of both the Instructor's Manual and Test Bank. This is a powerful resource for anyone who is using electronic visuals in the classroom. Create your own PowerPoint using the maps from the text or customize PowerPoint with your own materials. Either way, this resource will streamline the preparation of your lecture.

Transparencies (0-13-009771-3): Includes over 200 full-color images from the text, all enlarged for excellent classroom visibility. Every map and table in the book is included in the transparencies.

Slides (0-13-009773-X): Includes over 200 full-color images from the text, all enlarged for excellent classroom visibility. Every map and table in the book is included in the transparencies.

ABC Videos (0-13-009774-8): Selected from the archives of ABC news, these timely and relevant video segments offer students insight into issues effecting the regions covered in the book.

Instructor's Manual (0-13-009776-4): Includes Learning Objectives, Chapter Outlines, Activity Suggestions, Discussion Topics, Mapping Exercises, and Country-by-Country Profiles.

 TestGen-EQ CD-ROM (0-13-009768-3): TestGen-EQ is a computerized test generator that lets you view and edit testbank questions, transfer questions to tests, and print the test in a variety of customized formats. Included in each package is the QuizMaster-EQ program that lets you administer tests on a computer network, record student scores, and print diagnostic reports. Both TestGen-EQ and QuizMaster-EQ are free to adopters of *Diversity Amid Globalization*, 2/E.

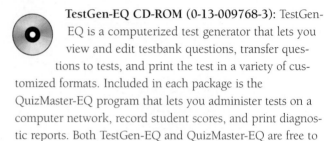 **On-Line Course Management Systems:** Prentice Hall offers content specific to *Diversity Amid Globalization*, 2/E, preloaded in the CourseCompass, BlackBoard, and WebCT course management system platforms. Each of these platforms lets you easily post your syllabus, communicate with students on-line or off-line, administer quizzes, and record student results and track their progress. Please call your local Prentice Hall representative for details on how to adopt and implement any of these options. To demo Prentice Hall's Course Management materials go to www.prenhall.com/demo.

Optional Student Aids/Package Options

 Study Guide
Includes Learning Objectives, Chapter Outlines, Practice Quizzes, and a supplemental chapter on how to read maps, charts, and tables appearing in the text. Discounted when packaged with *Diversity Amid Globalization* 2/E. To order this package use ISBN **0-13-078077-4**.

Mapping Workbook
Political and physical outline maps for each chapter, based on the maps in the text, ready for student use in identification exercises. Available at no extra charge when packaged with *Diversity Amid Globalization* 2/E. To order this package use ISBN **0-13-078067-7**.

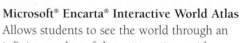 **Rand McNally Atlas of World Geography**
Available packaged for FREE with *Diversity Amid Globalization* 2/E. To order this package use ISBN **0-13-078065-0**.

Microsoft® Encarta® Interactive World Atlas
Allows students to see the world through an infinite number of dynamic options with user-customized views of more than 20 different map styles. Available at a nominal cost when packaged with *Diversity Amid Globalization* 2/E. To order this package use ISBN **0-13-078066-9**.

Student Aids

Author Field Trip CD-ROM

Packaged with every copy of the book automatically, these author field trips take students on pictorial narrated walk-throughs of six places around the world. Also included on the CD-ROM are physical and political outline maps, and the complete glossary from the book. Field trips include:

- Boom and Bust in Colorado
- Bolivian Transmigration
- Reinventing Berlin for the 21st Century
- Moroccan Crossroads: Tradition, Diversity, Globalization
- Kerala in Global Context
- Western Australia in the Global Economy

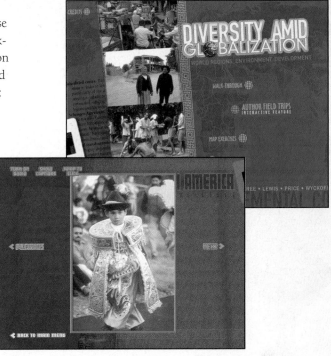

Companion Website: www.prenhall.com/rowntree

This online study guide to accompany *Diversity Amid Globalization,* provides an assortment of material to help students test themselves and explore geography on the Internet. Some items included on the Companion Website are:

- Objectives
- Multiple Choice
- Vocabulary
- Thinking Geographically

- Web Essays
- Exploring the Region
- Destinations
- Country by Country Profiles

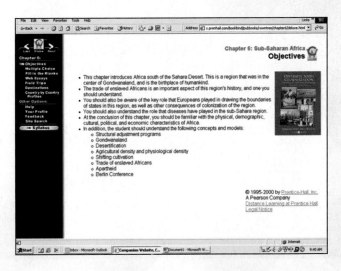

Diversity Amid Globalization

The most important challenges facing the world in the twenty-first century are associated with **globalization,** the increasing interconnectedness of people and places through converging processes of economic, political, and cultural change. Once-distant regions are now increasingly linked together through commerce, communications, and travel. Although early forms of globalization have been ongoing for several hundred years, the bonds of planetary integration are now strengthening at a pace never before witnessed. Many observers argue that contemporary globalization is the most fundamental reorganization of the planet's socioeconomic structure since the Industrial Revolution. While few dispute the widespread changes brought about by globalization, not everyone agrees on what the implications are or whether the benefits outweigh the costs.

Although economic activities may be the prime mover behind globalization, the consequences affect all aspects of land and life in the new millennium. Cultural patterns, political arrangements, and social development are all undergoing profound change. Because natural resources are now global commodities, the planet's physical environment is also implicated. Local ecosystems are altered by financial decisions made thousands of miles away, and the cumulative effect of these far-ranging activities has profound and possibly detrimental implications for the world's climates, oceans, waterways, and forests (Figure 1.1).

These immense, widespread global changes make understanding our world both a challenging and a thoroughly necessary task. Our future depends on comprehending globalization in its varied manifestations since our lives are now deeply intertwined with this worldwide phenomenon. Although understanding globalization cuts across many academic disciplines, world regional geography is an effective starting point because of its focus on regions, environment, geopolitics, culture, and economic and social development. To understand globalization, it is necessary to have comprehensive knowledge of the world in which it is played out (Figure 1.2). This book seeks to impart such knowledge by carefully outlining the basic patterns of global geography and showing how they are being constantly reorganized by global interconnections. For the same reason, the themes and structure of the book are organized around the tensions and issues resulting from globalization.

◀ **Figure 1.1 World geography** Viewed from space, both the diversity and the similarities of Earth's physical fabric are apparent. The challenge of world regional geography is to move closer so that human activities, as well as the physical environment, are examined. *(European Space Agency/Science Photo Library/Photo Researchers, Inc.)*

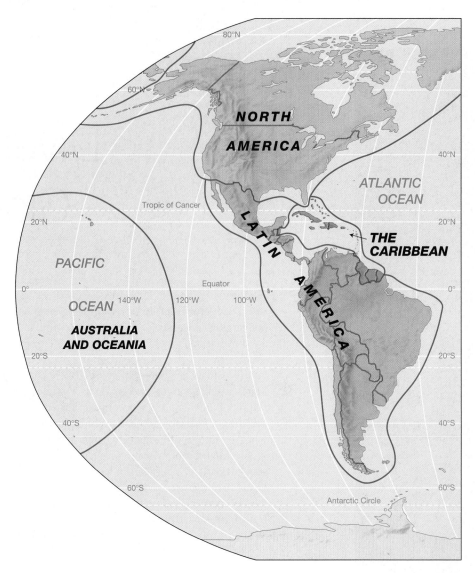

▶ **Figure 1.2 World regions**   These regions are the basis for the 12 regional chapters in this book. Countries or areas within countries that are treated in more than one chapter are designated on the map with a striped pattern. For example, western China is discussed in both Chapter 10, Central Asia, and Chapter 11, East Asia. Also, 3 countries on the South American continent are discussed as part of the Caribbean region because of their close cultural similarities with the island region.

Diversity Amid Globalization: A Geography for the Twenty-first Century

This chapter introduces the framework for studying world regional geography by first examining the varied aspects of globalization in contemporary life. While the economic implications of globalization dominate most discussions, the cultural, environmental, and geopolitical expressions must also be examined. Following this is an overview of the major concepts of global geography: human–environment interaction; areal differentiation and integration; regions; landscapes; and global-to-local scales. The third and last section introduces the five organizational themes found in each regional chapter of this textbook.

Converging Currents of Globalization

Most scholars agree that the major component of globalization is the economic reorganization of the world. Although different forms of a world economy have been in existence

for centuries, a well-integrated and thoroughly global economy is primarily a product of the last several decades. The attributes of this system are now familiar, yet bear repeating—global communication systems that link all regions on the planet instantaneously (Figure 1.3); global transportation systems capable of moving goods quickly by air, sea, and land; transnational conglomerate corporate strategies that have created global corporations more economically powerful than many sovereign nations; new and more-flexible forms of capital accumulation; international financial institutions that facilitate 24-hour trading; global agreements that promote free trade; market economies that have replaced state-controlled economies; privatized firms and services formerly operated by governments; a plethora of planetary goods and services that have arisen to fulfill consumer demand (real or imaginary); and, of course, an army of international workers, managers, and executives who have given this economic juggernaut a unique human geography (see "Is this Nebraska or India? The Globalization of Back Office Jobs").

As a result of this global reorganization, economic growth in some areas of the world has been unprecedented over recent

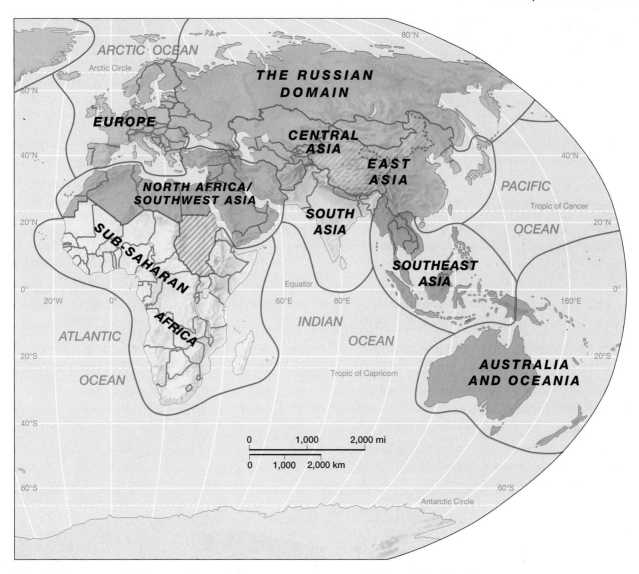

◀ **Figure 1.3 Global connections** The impacts of globalization, often through global TV, are everywhere, even in remote villages in developing countries. Here, in a small village in southwestern India, a rural family earns a few dollars a week by renting out viewing time on its globally linked television set. *(Rob Crandall/Rob Crandall, Photographer)*

Nebraska or India? The Globalization of Back Office Jobs

When you call an 800 number, instead of connecting with a service representative in Omaha, Nebraska, or somewhere else in the Midwest, you may actually be speaking to someone in India. Further, that person may have taken on a new identity to mask the international connection. Increasingly, globalization, with its instantaneous electronic communication, facilitates the subcontracting of "back office" jobs to India, where a large, relatively inexpensive, well-educated, English-speaking labor force willingly takes on these tasks.

"Hi, my name is Susan Sanders and I'm from Chicago," says C. R. Suman, 22, who is in fact a native of Bangalore, where she fields calls from customers of a large U.S. telecommunications company. While Ms. Suman's fluent English would pass muster in the bleachers of Wrigley Field, she has conjured up a fictional American life in case her callers ask personal questions as she resolves problems with their phone bill. "We watch a lot of *Friends* and *Ally McBeal* to learn the right phrases," she explains.

The point of this pretense is to convince Americans who dial toll-free numbers that the person on the other end of the line works nearby, not 8,300 miles away in a country that many Americans think has no phones at all. Indeed, call centers are a booming business in India as North American and British companies set up supermarket-size phone banks to handle the daily barrage of customer inquiries.

Call centers, though, are only the low end of a much larger industry of Indian software designers, accountants, Web site designers, and animation artists who work on projects for foreign companies from their offices in the booming cities of Bangalore, Bombay, and Hyderabad. In fact, some experts predict that India is on its way to becoming the back office of the world. One consultant predicts that back office assignments will generate 800,000 new jobs and $17 billion in revenue for India by the year 2008.

The back office business may help cushion India from the economic slowdown in the United States, for routine work like processing insurance claims or settling credit card bills goes on no matter what the economic climate. Indeed, as companies look for ways to cut costs, more of them may send work to India, where wages often run half those in the United States.

While salaries for call center jobs pay around $2,000 a year in India, these jobs are coveted. "In the U.S., these jobs are taken by housewives or kids who haven't decided what they want to do with their lives," said K. Ghanesh, the founder of Customer Asset. "Here they are career jobs for college graduates."

In the last two years, India has installed reliable high-capacity telephone lines in most of its major cities that allow people in that country to communicate with customers in the United States, by phone or over the Internet, with no discernible difference from a calling center in Nebraska. This improved telephone network has essentially erased the advantage of other countries—notably Ireland—that have offered back office services longer.

Adapted from "Hi, I'm in Bangalore (but I Dare Not Tell)," by Mark Landler, *New York Times*, March 21, 2001, p 1.

---

decades. Additionally, international corporations, along with their managers and executives, have amassed vast amounts of wealth, profiting from the new opportunities unleashed by globalization. However, not everyone has profited from economic globalization, nor have all world regions shared equally in the benefits. In Latin America, for example, the percentage of the population classified as poor increased from 33 percent in 1980 to 50 percent in 1990, largely as a result of economic restructuring. Although that figure fell to 36 percent in the late 1990s, these data suggest that as some regions and social groups move ahead and prosper, others fall behind.

Furthermore, these economic changes also trigger fundamental cultural change. Often accompanying globalization is the spread of a global consumer culture that threatens to reduce local diversity. This frequently sets up deep and serious social tensions between traditional cultures and new, external globalizing currents. Global TV, movies, and videos promote images of Western style and culture that are imitated by millions throughout the world. NBA T-shirts, sneakers, and caps are now found in small villages and large world cities alike. Fast-food franchises are changing—some would say corrupting—traditional diets with the explosive growth of McDonald's, Burger King, and Kentucky Fried Chicken outlets in most of the world's cities, including Beijing, Moscow, Singapore, and Nairobi (Figure 1.4). While these stylistic changes

▲ **Figure 1.4 Hybrid world culture**   Globalization is creating unique cultural expressions that often merge, or hybridize, the old with the new. Here, in Bangkok, a Thai woman eats lunch at a well-known hamburger chain under a poster for a uniquely Thai traditional dish, a tart papaya salad, which the hamburger chain is marketing as "Thai Spicy McSalad Shaker." *(AP/Wide World Photos)*

may seem innocuous to North Americans because of their familiarity, they suggest the more profound structural cultural changes the world is experiencing through globalization.

Although much media attention is given to the rapid spread of Western consumer culture, we should not overlook the nonmaterial culture that is also becoming more dispersed and homogenized through globalization. Language is an obvious example. Many a Western tourist in Russia or Asia has been startled by locals speaking an English made up largely of Hollywood or pop-song phrases. But far more than clichés of speech are involved, for social values, ideas, and even fundamental organizational structures are also being dispersed globally. Expectations about human rights, the role of women in society, and the intervention of nongovernmental organizations are also expressions of globalization with far-reaching implications for cultural change.

It would be a mistake, however, to view cultural globalization as a one-way flow that spreads from the United States and other Western countries into all corners of the world. In actuality, even when forms of American popular culture spread abroad, they are typically melded with local cultural traditions in a process known as hybridization. The resulting cultural "hybridities," such as world-beat music, can themselves reverberate across the planet, creating yet another dimension of globalization. It is also important to realize that U.S. culture is now much more strongly impacted by ideas and forms from the rest of the world than earlier (Figure 1.5). This is visible in the growing internationalization of American food, in the fact that more people across large portions of the United States speak Spanish than English in their homes, and even in the spread of Japanese comic book culture—Pokemon and the like—among American children.

There are also profound geopolitical facets to globalization. To many, an essential dimension of globalization is that it is little hindered by territorial, national, or jurisdictional restrictions. Globalization is just that—global—and its processes transcend traditional boundaries. For example, the creation of the United Nations following World War II was a step toward creating an international governmental structure in which all nations could find representation. Unfortunately, the simultaneous emergence of the Soviet Union as a military and political superpower led to a rigid division into Cold War blocs that inhibited further geopolitical integration. With the peaceful end of the Cold War in the late 1980s and early 1990s, the former communist countries of eastern Europe and the Soviet Union were opened almost immediately to global trade and cultural exchange. These political developments coincided with the economic and technological changes we now see as the recent wave of globalization. Some observers argue that economic imperialism has replaced the ideological divisions of the Cold War; hence, today, economic activity and politics are more intertwined than ever. They offer as an example governmental interaction with the World Trade Organization (WTO) and other trading blocs in which national interests are often overshadowed with concerns about expanded global economic activity (Figure 1.6). And while the Cold War may be over, regional military alliances and pacts are still very much a reality, again illustrating the close linkage between geopolitics and globalization. A prime example are the negotiations between the United States, Europe, and Russia over NATO expansion into former communist eastern Europe.

Beyond geopolitics, the expansion of a globalized economy is also creating and exacerbating environmental problems throughout the world as **transnational firms** disrupt local ecosystems in their incessant search for natural resources and manufacturing sites. Landscapes and resources that were previously used by only small groups of local peoples are now thought of as global commodities to be exploited and traded on the world marketplace. As a result, indigenous peoples are often deprived of their traditional resource base and displaced into marginal environments. On a larger scale, economic globalization is aggravating worldwide environmental problems, such as climate change, air pollution, water pollution, and deforestation. And yet it is only through global cooperation, such as the UN treaties on biodiversity protection or the Kyoto protocol on global warming, that these problems can be addressed.

Globalization has a clear demographic dimension as well. Although international migration is nothing new, increasing numbers of people from all parts of the world are crossing national boundaries, often permanently. Migration from Latin America and Asia has drastically changed the demographic configuration of the United States, just as migration from Africa and Asia has transformed western Europe. Countries such as Japan and South Korea that have long been perceived as ethnically homogenous now have substantial immigrant populations. Even a number of relatively poor countries, such as Nigeria and the Ivory Coast, encounter large numbers of immigrants coming from even poorer countries, such as Burkina Faso. Although international immigration is still curtailed by the laws of every country—much more so, in fact,

▲ **Figure 1.5 Global culture in the United States**   The hybridization of Vietnamese and American cultures dominates in the Eden Center shopping mall in Falls Church, Virginia, west of Washington, D.C. Often, transplanted foreign cultures deliberately attempt to create an earlier and idealized snapshot of their native culture. Here, the flag of the former South Vietnamese republic flies alongside the Stars and Stripes.   *(Rob Crandall/Rob Crandall, Photographer)*

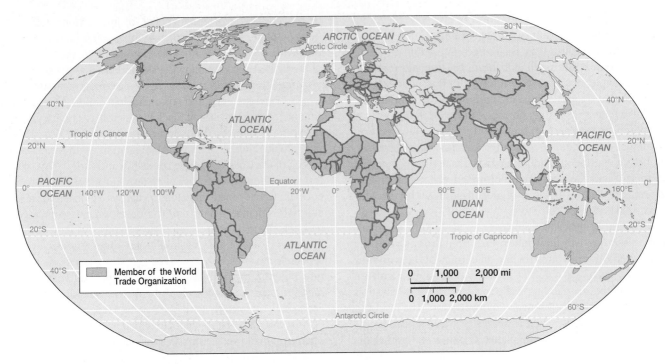

▲ **Figure 1.6 World Trade Organization**   One of the most powerful institutions of economic globalization is the World Trade Organization (WTO), which was created in 1995 to oversee trade agreements, encourage open markets, enforce trade rules, and settle disputes. With the inclusion of China in late 2001, the WTO consisted of 144 member countries. In addition to these member countries, 31 states, including Russia, Saudi Arabia, and Vietnam, have "observers" status, which obligates them to begin full membership negotiations within five years.

than the movement of goods or capital—it is still rapidly mounting, propelled by the uneven economic development associated with globalization.

Finally, there is also a significant criminal element to contemporary globalization, including terrorism (discussed later in this chapter), drugs, pornography, and prostitution. Illegal narcotics, for example, are most definitely a global commodity (Figure 1.7). Some of the most seemingly remote parts of the world, such as the mountains of northern Burma, are thoroughly integrated into the circuits of global exchange through the production of opium, and hence through the heroin trade. Even many areas that do not directly produce drugs are involved in their global sale and transshipment. Nigerians often occupy prominent positions in the international drug trade, as do members of the Russian Mafia. Many Caribbean countries have seen their economies become reoriented to drug transshipments and the laundering of drug money. Prostitution, pornography, and gambling (legal in some areas, illegal in others) have also emerged as highly profitable global businesses. Over the past decade, for example, parts of eastern Europe have become major sources of both pornography and prostitution, finding a lucrative but morally questionable niche in the new global economy.

Advocates and Critics of Globalization

Globalization, especially in its economic form, is one of the most contentious issues of the day (Figure 1.8). Supporters generally believe that it results in greater economic efficiency that will eventually result in rising prosperity for the entire world. Critics think that it will largely benefit those who are already prosperous, leaving most of the world poorer than before while reducing cultural and ecological diversity. Economic globalization is generally applauded by corporate leaders and economists, and it has substantial support among the leaders of both the Republican and Democratic parties in the United States. Opposition to economic globalization is widespread in the labor and environmental movements, and among many student groups. Hostility toward globalization is sometimes deeply felt, as has been made evident in massive antiglobalization protests in Seattle, Quebec City, and Genoa, Italy.

The Pro-globalization Stance   Advocates argue that globalization is a logical and inevitable expression of contemporary international capitalism that will benefit all nations and all peoples by increasing global commerce and wealth. These new riches, they say, will eventually trickle down to enrich even the poorest of peoples in all the world's different regions. Economic globalization can work such wonders, they contend, by enhancing competition, allowing the flow of capital to poor areas, and encouraging the spread of beneficial new technologies and ideas. As countries reduce their barriers to trade, inefficient local industries will be forced to become more efficient in order to compete with the new flood of imports, thereby enhancing overall national productivity. Every country and region of the world, moreover, ought to be able

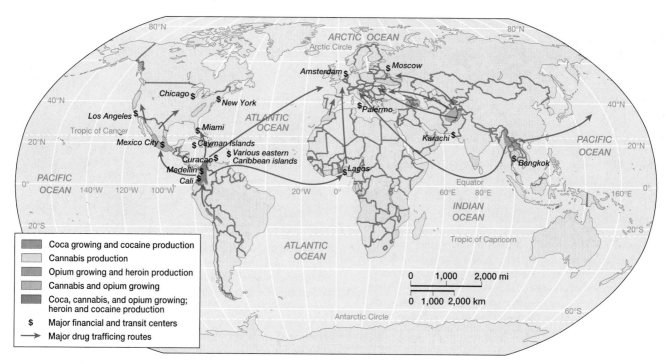

▲ **Figure 1.7 The global drug trade**   The cultivation, processing, and transshipment of coca (cocaine), opium (heroin), and cannabis are global issues. The most important cultivation centers are Colombia, Mexico, Afghanistan, and northern Southeast Asia, whereas the major drug financing centers are mostly located in the Caribbean, the United States, and Europe. Additionally, Nigeria and Russia also have significant roles in the global transshipment of illegal drugs.

to concentrate on those activities for which it is best suited in the global economy. Enhancing such geographical specialization, the pro-globalizers argue, creates a more efficient world economy. Such economic restructuring is made increasingly possible by the free flow of capital, which ought to flow smoothly to those areas that have the greatest opportunities. By making access to capital more readily available throughout the world, many economists argue, globalization should eventually result in a certain global **economic convergence,** meaning that the world's poorer countries will gradually catch up with the more advanced economies.

Thomas Friedman, one of the most influential proponents of economic globalization, argues that the need to attract capital from abroad forces countries to adopt sound economic policies. Friedman describes the great power that the global "electronic herd" of bond traders, currency speculators, and fund managers enjoy as they direct money to, or withhold it from, developing economies (Figure 1.9). If a country is to please the "herd" and thus get the capital it needs to modernize and compete internationally, Friedman argues, a country must adopt a policy of fiscal responsibility (which he calls the "golden straightjacket") to avoid wasteful expenditures.

To the committed pro-globalizer, even the global spread of **sweatshops**—crude factories in which workers sew clothing, assemble sneakers, or perform similar labor-intensive tasks for extremely low wages—is to be applauded. People go to work in sweatshops, they argue, because the alternatives in the local

▲ **Figure 1.8 Protests against globalization**   Meetings of international groups such as the World Trade Organization, World Bank, and International Monetary Fund commonly draw large numbers of protesters against globalization. This demonstration took place at a Washington, D.C., meeting of the World Bank and IMF.   *(Rob Crandall/Rob Crandall, Photographer)*

▲ Figure 1.9 The global "electronic herd" One facet of economic globalization is the rapid and dynamic movement of capital as bond traders, currency speculators, and fund managers direct money into or out of the economies of developing countries. In this picture, traders and clerks work on the Euro-dollar Futures floor of the Chicago Mercantile Exchange. *(Tim Boyle/Getty Images, Inc.)*

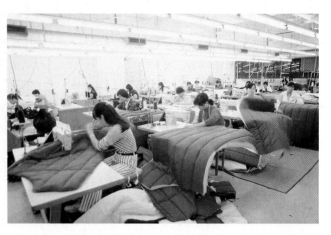

▲ Figure 1.10 Global sweatshops One of the most debated aspects of economic globalization is the crude factories, or sweatshops, in which workers sew clothing, assemble sneakers, stitch together soccer balls, or perform similar labor-intensive tasks for low wages. While pro-globalizers and free-trade advocates argue that these sweatshops provide better opportunities than the local economy, critics of globalization point to long hours, low wages, uncertain employment, and unhealthy working conditions as the seamy and unethical side of economic globalization. *(Macduff Everton/The Image Works)*

economy are even worse (Figure 1.10). Once countries achieve full employment through sweatshops, the argument goes, they can gradually improve conditions and move into more sophisticated, better-paying industries. Proponents of economic globalization also commonly argue that multinational firms based in North America or Europe often offer better pay and safer working conditions than do local firms, and thus contribute to worker well-being. The more extreme pro-globalizers go so far as to argue that poor countries ought to take advantage of their generally lax environmental laws in order to attract highly polluting industries from the wealthy countries, since acquiring such industries would enhance their overall economic positions.

The pro-globalizers in general strongly support the large multinational organizations that facilitate the flow of goods and capital across international boundaries. Three such organizations are particularly important: the World Bank, the International Monetary Fund (IMF), and the World Trade Organization (WTO). The primary function of the World Bank is to make loans to poor countries so that they can invest in infrastructure and generally build more modern economic foundations. The IMF is more concerned with making short-term loans to countries that are in financial difficulties—those having trouble, for example, making interest payments on the loans that they had previously taken. The WTO, a much smaller organization that the other two, is explicitly concerned with lowering trade barriers between countries to enhance economic globalization. It also tries to mediate between countries and trading blocks that are engaged in trade disputes, although not always with success.

To support their claims, pro-globalizers often argue that countries that have been highly open to the global economy have generally had much more economic success than those that have isolated themselves, seeking self-sufficiency. The world's most isolated countries, such Burma and North Korea, have often been economic disasters, with little growth and rampant poverty, whereas those that have opened themselves to global forces, such as Singapore and Thailand, have in the same period seen rapid growth and a substantial reduction of poverty.

Critics of Globalization Virtually all of the claims of the pro-globalizers are denied strongly by the anti-globalizers. Opponents often begin by arguing that globalization is not a "natural" process. Instead, it is the product of an explicit economic policy promoted by free-trade advocates, capitalist countries (mainly the United States, but also Japan and those of western Europe), financial interests, international investors, and multinational firms. In addition, the processes of globalization so evident today are much more pervasive than those of the past, such as during the period of European colonialism. Thus, while global economic and political linkages have been around for centuries, their modern expression is unprecedented.

Since the globalization of the world economy appears to be creating greater inequity between rich and poor, the "trickle-down" model of developmental benefits for all people in all regions has yet to be demonstrated. Evidence for this inequity comes from the fact that the percentage of poor people in most world regions is actually increasing (Figure 1.11). A recent United Nations report notes that, in terms of social conditions, 60 countries are worse off now than 30 years ago and that there is a clear trajectory in the direction of increasing inequity within the world. On a global scale, the richest 20 percent of the world's people consume 86 percent of the world's

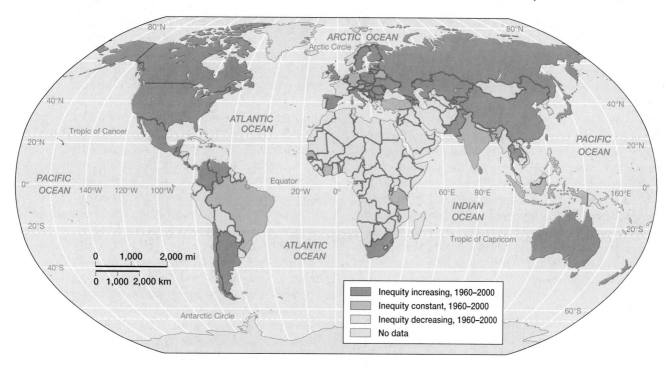

▲ **Figure 1.11 Global economic inequity**   One of the most problematic issues associated with economic globalization is the trajectory of increased income inequity. Put differently, the accusation that globalization has allowed the
"rich to get richer while the poor stay poor" seems to be true despite the claim by pro-globalizers that everyone
would benefit from enhanced global economic activity. This map shows the results of a United Nations study, "Inequality, Growth and Poverty in the Era of Liberalization and Globalization" (2001), that shows inequity increasing
in 48 of the 73 countries studied. Further, though the UN study found inequity was unchanged in 16 countries, it
does not mean there is not a high degree of income inequity. For example, India and Brazil are mapped as "constant" yet have very high levels of inequity.

resources, while the poorest 80 percent use only 14 percent
of global resources. The growing inequality of this age of globalization is apparent on both global and national scales. Globally, the wealthiest countries have in general grown much
richer over the past two decades, while the poorest have become more impoverished. Nationally, even in successful countries such as the United States, the wealthiest 10 percent of the
population have reaped almost all of the gains that globalization has offered, whereas the poorest 10 percent have seen
their income decline in recent decades.

Opponents also commonly contend that globalization promotes free-market, export-oriented economies at the expense
of localized, sustainable activities. World forests, for example, are increasingly cut for export timber rather than serving
local needs. As part of their economic structural adjustment
package, the World Bank and the International Monetary
Fund encourage developing countries to expand their resource exports so they will have more hard currency to make
payments on their foreign debts. This strategy, however, usually leads to overexploitation of local resources. Opponents
also note that the IMF often requires developing countries to
adopt programs of fiscal austerity that entail a substantial reduction in funds for basic public services, including education, health, and food subsidies. By adopting such policies,
critics warn, poor countries may end up with even more impoverished populations than before.

Anti-globalizers also dispute the empirical evidence on national development offered by the pro-globalizers. Highly successful developing countries such as South Korea, Taiwan,
and Malaysia, they argue, have indeed been engaged with the
world market, but they have generally done so on their own
terms, rather than those preferred by the IMF and other advocates of full-fledged economic globalization. These countries have actually protected many of their domestic industries
from foreign competition and have, at various times, controlled the flow of capital. The "free-market" economic model
commonly promoted for developing countries is not even the
one that Western industrial countries used for their own economic development. In Germany, France, and even to some
extent the United States, governments have historically played
a strong role in directing investment, managing trade, and
subsidizing chosen sectors of the economy.

Those who challenge globalization also worry that the entire system—with its near instantaneous transfers of vast sums
of money over nearly the entire world on a daily basis—is in-
herently unstable. The noted critic John Gray, for example,
thinks that the same "electronic herd" that Thomas Friedman
applauds is a dangerous force because it is so liable to "stam-
pedes." These international managers of capital tend to panic
when they think their funds are at risk; and when they do so,
the entire intricately linked global financial system can quickly become destabilized, potentially leading to a crisis of global

proportions. Even when the "herd" spots opportunity, trouble may still ensue. As vast sums of money flow into a developing country, they may create a speculatively inflated "**bubble economy**" that cannot be sustained.

Such a bubble economy emerged in Thailand and many other parts of Southeast Asia in the mid-1990s. So much money flooded Thailand that property values in the capital city of Bangkok came to exceed those of most U.S. cities—clearly an economically inefficient arrangement when one considers how much less productive the Thai economy is than that of the United States. When the bubble burst in 1997, a severe depression crippled most of Southeast Asia, and this depression has never really ended in Indonesia, the biggest country of the region (Figure 1.12). This Asian crisis did not undermine the entire world economy, as many anti-globalizers feared it would, but some evidence suggests that it very nearly did so.

A Middle Position? A number of experts, not surprisingly, argue that both the anti-globalization and the pro-globalization stances are exaggerated. Those in the middle ground

▲ **Figure 1.12 Economic recession in Thailand** One of the unfortunate symbols of Thailand's 1997 economic financial crisis are more than 300 unfinished high-rise buildings that now litter Bangkok's downtown skyline. Before the economic collapse, Thailand's "bubble economy" led to widespread speculation in downtown property values and new buildings. When the bubble burst, bankrupt developers walked away from their high-rise office construction, leaving these abandoned steel skeletons. *(AFP/Corbis)*

tend to argue that economic globalization is indeed unavoidable; even the anti-globalization movement, they point out, is made possible by the globalizing power of the Internet and is, therefore, itself an expression of globalization. They further contend that while globalization holds both promises and pitfalls, it can be managed, at both the national and international levels, to reduce economic inequalities and protect the natural environment. Such scholars stress the need for strong yet efficient governments, strengthened yet also substantially reformulated international institutions (such as the UN, World Bank, and IMF), and globalized networks of environmental, labor, and human rights groups.

One of the most influential of the middle-ground globalization scholars is Dani Rodrik, who argues that openness to the global economy can indeed be highly beneficial, but that countries must "make openness work" by investing in education and maintaining social cohesion. He concludes, "The world market is a source of disruption and upheaval as much as it is an opportunity for profit and economic growth. Without the complementary institutions at home—in the areas of governance, judiciary, civil and political liberties, social insurance, and education—one gets too much of the former and too little of the latter" (1999, p. 96).

The debate about globalization is one of the most important issues of the day, and one of the most complicated. While this book does not pretend to resolve the controversy, it does encourage readers to reflect on these critical points as they apply to different world regions and locales.

Diversity in a Globalizing World

The ever-increasing globalization of the contemporary world has led many observers to foresee a future world far more uniform and homogeneous than that of today. Optimists imagine a universal global culture uniting all humankind into a single community untroubled by war, ethnic strife, or resource shortage—a global utopia of sorts. More common, however, is the view that the world is becoming blandly homogeneous as different places, peoples, and environments lose their distinctive character and become indistinguishable from their neighbors. While diversity may be the hardest thing for a society to live with, it may also be the most dangerous thing to live without. Nationality, ethnicity, cultural distinctiveness—all are the legitimate legacy of humanity. If this diversity is blurred, denied, or repressed through global homogenization, humanity loses one of its desirable defining traits.

But even if globalization is generating a certain degree of homogenization, the world is still a fantastically diverse place (Figure 1.13). One still finds marked differences in culture (including language, religion, architecture, urban form, foods, and many other attributes of daily life), economy, and politics—and in the natural environmental as well. Such diversity is so vast that it cannot be readily extinguished even by the most powerful forces of globalization. And it is important to realize that globalization itself often provokes a strong reaction on the part of local people, making them all the more determined to maintain what is distinctive about their own way of life. Thus, globalization is really understandable only

▲ **Figure 1.13 Local cultures** This family in Ghana reminds us that although few places are beyond the reach of globalization, there are, nevertheless, still many unique local landscapes, economies, and cultures. For example, most of the material culture in this photograph is local in origin and comes from long-standing cultural tradition.   *(Caroline Penn/Panos Pictures)*

if one also examines the diversity that continues to characterize the world.

Our concern with geographic diversity takes many forms and goes beyond merely celebrating traditional cultures and unique places. People have many ways of making a living throughout the world. Different kinds of agriculture produce different food resources, and different uses of natural resources produce different landscapes. As a globalized economy becomes increasingly fixated on mass-produced goods, it is important to recognize this broader spectrum of economic diversity. Furthermore, a stark reality of today's economic landscape is unevenness: while some places and people prosper, others suffer from unrelenting poverty. Unfortunately, this also is a form of diversity amid globalization. As the gap between rich and poor becomes ever greater throughout the world, this critical dimension of human geography cannot be overlooked.

The politics of diversity also demands increasing attention these days as we labor to understand worldwide tensions over terrorism, ethnic separateness, regional autonomy, and political independence. Groups of people throughout the world seek self-rule of territory they can call their own. Today most wars are fought *within* countries, not between them. Each conflict is unique, understandable only in the light of the specific cultural and political environments in which it occurs. But as a group the conflicts also illustrate processes that are found across much of the globe. We must therefore seek a balance between describing the commonalities of the world's regions and explaining the distinctiveness of places, environments, and landscapes.

In summary, globalization can be defined as the increasing interconnectedness of people and places through converging processes of economic, political, and cultural change. It is a dominant feature of the contemporary world because of its pervasive influences and uneven benefits. Although economic restructuring is a prime mover, globalization is not simply the growth and expansion of international trade. It also represents converging and homogenizing forces that many seek to resist. Since many environmental, cultural, political, economic, and social differences predate contemporary globalization, they could be thought of as the extant fabric of a highly diverse world. An equally important theme, thus, is how geographic diversity comes in conflict with globalization; how globalization in different places and at different times is suppressed, renegotiated, hybridized, protected, preserved, or extinguished. Because of the importance of this theme, diversity and globalization should be examined as inseparable—often in conflict, yet also often complementary.

Geography Matters: Environments, Regions, Landscapes

Geography is one of the most fundamental sciences, a discipline awakened and informed by a long-standing human curiosity about our surroundings and the world environment. The term *geography* has its roots in the Greek words for "describing the Earth," and this discipline has been advanced since classical times by all cultures and civilizations. With the inherent satisfaction of knowing about different environments come pragmatic benefits for physical exploration, resource use, world commerce, and travel. In some ways geography can be compared to history: while historians describe and explain what has happened over time, geographers describe and explain Earth's spatial dimensions, showing how the world differs from place to place.

Given the broad scope of the geographical charge, it is no surprise that geographers have many different approaches to examining the world based upon the conceptual emphasis guiding their study. At the most basic level, geography can be broken into two complementary pursuits, physical and human geography. Physical geography examines climate, landforms, soils, vegetation, and hydrology, while human geography concentrates on the spatial analysis of economic, social, and cultural systems. For example, a physical geographer in the Amazon Basin of Brazil might be interested primarily in the ecological diversity of the tropical rain forest or the ways the destruction of that dense vegetation changes the local climate and hydrology. The human geographer, on the other hand, might focus on the social and economic factors explaining the migration of settlers into the rain forest or the tensions and conflicts over land and resources between these migrants and indigenous peoples.

Another basic division is that between focusing on a specific topic or theme, such as climatology or cultural geography, and synthesizing various topics as they apply to a specific region or area of the world, such as Latin America or Europe. The former approach is referred to as *thematic* or *systematic geography,* while the latter is called *regional geography.* These two perspectives are complementary and by no means mutually exclusive. This textbook, for example, draws upon a regional scheme for its overall architecture, dividing the globe into 12 separate world regions, yet it organizes each chapter thematically, examining topics such as environment, population and settlement, cultural differentiation, geopolitics, and economic development. In doing so, each chapter brings together both physical and human geography.

Human-Environment Interaction

A fundamental component of geographic inquiry is the examination of the basic yet highly complex relationship between humans and their environment. Though this analysis may take many different forms, three overarching questions form the foundation:

1. Where and under what conditions does nature control, constrain, or disrupt human activities, and what kinds of societies are most vulnerable to these environmental influences (Figure 1.14)?
2. How have human activities affected natural systems such as the world's climate, vegetation, rivers, and oceans, and what are the consequences of these actions?
3. Is there evidence that humans can live in balance with nature so that human use of natural resources might be sustainable over a longer period of time, or is environmental damage an inevitable consequence of human settlement?

While there are no simple answers to these questions, the study of world regional geography offers valuable insights from different cultural and regional perspectives. The first question, for example, can frame discussions of how human vulnerability to flooding, earthquakes, and famine differs among various societies. The second question illuminates the ever-important issue of what kind of human activities cause the most (or the least) environmental damage. This comparative perspective can be helpful for generating solutions to existing problems, as well as constraining actions known to be environmentally harmful. Finally, the third question drives discussions about resource and environmental planning for the near- and long-term future.

▲ **Figure 1.14 Natural hazards**   One pressing question in geographic inquiry is how people and societies are affected by natural hazards. Often, as a population increases in an area, more and more people become susceptible to natural hazards, such as hurricanes, floods, and drought. This picture is of a flood in Bangladesh.   *(James Blair/NGS Image Collection)*

Areal Differentiation and Integration

At a fundamental level, geography can be considered the spatial science, charged with the study of Earth's space or surface area, the uniqueness of places, and the similarities between them. One component of that responsibility is explaining the differences that distinguish one piece of the world from another. The geographical term for this is **areal differentiation** (*areal* means "pertaining to area"). Why is one part of Earth humid and verdant while another just a few hundred kilometers away is an arid desert (Figure 1.15)? This is a question of spatial or areal differentiation, the systematic study of the differences between parts of Earth's surface.

Geographers are also interested in the connections between areas and how they are linked through interactions. This concern is one of **areal integration,** or the study of how areas interact with each other. How and why are the economies of Singapore and the United States closely intertwined even though the two countries are situated in entirely different

◄ **Figure 1.15 Areal differentiation**
The California–Mexico border, with different agricultural landscapes on opposite sides of the border, reminds us of the many variables that shape cultural landscapes into different patterns. The larger red pattern to the upper left is the agricultural fields of the United States, while to the bottom right, the much smaller fields of Mexico show a different landscape signature. Explaining these differences (and the similarities) is a central focus of geography. *(Earth Satellite Corporation/SPL/Photo Researchers, Inc.)*

physical, cultural, and political environments? Such questions of areal integration are becoming increasingly important because of the world linkages inherent in globalization.

Regions

The human intellect seems driven to make sense of the universe by lumping phenomena together into categories of similarity. Biology has its taxa of living objects, history its eras and periods of time, geology its epochs of Earth history. Geography, too, makes sense of the world by compressing and synthesizing vast amounts of information into spatial categories based on similar traits. The resulting areal units are referred to as **regions** (see "Defining the Region: The Metageography of World Regions").

Sometimes the unifying threads of a region are physical, such as climate and vegetation, resulting in a regional designation such as the Sahara Desert or a tropical rain forest. Other times the threads are more complex, combining economic and cultural traits, as in the use of the term *Midwest* for the central United States. Human beings compress large amounts of information into stereotypes; often the geographic region is just that, a spatial stereotype for a section of Earth that has some special signature or characteristic that sets it apart from other places (Figure 1.16).

Some caution is advised. First, no regions are homogeneous throughout. While there may be a single characteristic that unites an area, there will also be diversity and differences within it. A second caution is urged regarding borders between regions. Except in regard to political borders, such as those separating countries, states, or provinces, rarely are boundaries quite as abrupt as they appear on a map. Moun-

tain ranges usually rise gradually from lowland areas, and deserts transition only gradually into grasslands; manufacturing districts lessen in intensity toward the periphery; and linguistic or other cultural regions typically blend gradually into one another. Usually those traits used to define a geographic region are most explicit and clear in the core area. In moving away from the core toward the boundaries, the defining traits usually grow less and less apparent, weakening in intensity toward the region's periphery.

Space into Place: The Cultural Landscape

Human beings conceptually transform space into distinct places that are unique and heavily loaded with meaning and symbolism. This diverse fabric of "placefulness" is of great interest to geographers because it tells us much about the human condition as it varies across the surface of Earth. Places can tell us how humans interact with nature and between themselves; where there are tensions and where there is peace; where people are rich and where they are poor. Consequently, the systematic study of places is a critical element of geography.

A common tool for the analysis of place is the concept of the **cultural landscape,** which is, simply stated, the visible, material expression of human settlement, past and present. The cultural landscape is the tangible expression of the human habitat. It visually reflects the most basic needs of humans: shelter, food, work. Furthermore, the cultural landscape acts to bring people together (or keep them apart), since it is a marker of cultural values, attitudes, and symbols. Because cultures vary greatly around the world, so do cultural landscapes (Figure 1.17).

DEFINING THE REGION   Metageography of World Regions

One of the perplexing issues of world regional geography is the definition and demarcation of the regions themselves. For example, historically Europe has been separated from Asia by the Straits of Bosporus, the narrow waterway that divides Turkey into European Turkey to the west and Asiatic Turkey to the east. Since the country is unified by language, religion, and culture, any traveler to Istanbul can see the fallacy of this traditional division.

The problem of dividing the world into a small number of regional units has been apparent since the earliest days of geography, associated with Greek civilization. The Greeks conceptualized the world as divided into a few continents from their experiences as seafaring explorers. To them, it made sense to divide the great landmasses of Europe, Asia, and Africa based upon seas and waterways; thus, the Bosporus and the Red Sea became the landmarks separating those continent-based world regions. European explorers didn't argue with this threefold division even though, once again, a large country—Russia—had to be separated into European Russia west of the Ural Mountains and Asiatic Russia to the east.

In spite of this problem, European explorers continued to divide the world into continents: North and South America were referred to as separate continents, although they were clearly joined by the narrow isthmus of Panama (or pan-America, as some call it); Australia was added as a continent, even though many argued it was simply a large island. But if Australia was a continent, shouldn't Madagascar also be a continent? Not according to European explorers, who were quite comfortable with considering it a part of the African continent. Finally, Antarctica was given continental status even though it was totally uninhabited.

This 7-part "myth of the continents" seemed primarily useful to allow Europeans to differentiate themselves from other cultures, specifically Asian and African cultures. It seemed much easier to draw cultural and ethnic boundaries if they were placed on distinct continents, separated from Europe by mountains, seas, and waterways. However, this scheme makes little sense in today's world. There is, for example, no such thing as Asian culture. That is, no common traits link the people of East Asia with those of Southwest Asia.

World War II and the subsequent aid programs focused greater emphasis on cultural and geopolitical regions than on continents. Asia was subdivided into a handful of regions roughly comparable to historically defined civilizations; thus, East Asia, centered on China and Japan, was differentiated from South Asia, anchored by India. While this postwar scheme represented a major advance over the continental scheme, it was not flawless, nor was any one regional division accepted by all. Though progress was made by moving away from the myth of continents, there is still a good deal of quibbling about where to draw the boundaries for different world regions. For readers interested in that discussion, each of our regional chapters contains a sidebar titled "Setting the Boundaries" that examines in more detail this problematic issue of regional definition and boundary-drawing.

Most of the French countryside, for example, with its modestly sized agricultural fields enclosed by hedgerows and small vineyards adjacent to stone houses clustered together in villages, looks very different from a typical farm landscape of the midwestern United States, with its fields covering several square miles and people separated from each other by large distances, comfortable in isolated wood frame houses far from the nearest town. From the contrasting look of these two landscapes, we can infer much about the social and economic systems underlying the different societies. While these two landscapes may be relatively familiar and easily imagined in our minds, all landscapes convey much important information about land, life, and the environment around the world.

Increasingly, however, we see the uniqueness of places being eroded by the homogeneous landscapes of globalization—shopping malls, fast-food outlets, business towers, theme parks, industrial complexes. Although these landscapes are familiar and often taken for granted, understanding the forces behind their spread is important because they tell us much about the expansion of global economies and cultures. While a modern shopping mall in Hanoi, Vietnam, may seem familiar to someone from North America, this new landscape conveys a profound message about change and culture in the twenty-first century as yet another component of globalized world culture is implanted into a once remote and distinctive city.

Scales: Global to Local

There is a sense of scale to all systematic inquiry, whatever the discipline. In biology, for example, some scientists study the smaller units of cells, genes, or molecules, while others take a larger view, examining plants, animals, or ecosystems. Similarly, some historians may focus on a specific individual at a specific point in time, while others take a broader view of international events over many centuries. Geographers also work at different scales. While some may concentrate on analysis of a local landscape—perhaps a single village in southern India—others will focus on a broader regional picture, examining, for example, all of southern India. Others do research on a still larger global scale, perhaps studying emerging trade networks between southern India's center of information technology in Bangalore and North America's Silicon Valley, or investigating how India's monsoon might be connected to and affected by the Pacific Ocean's El Niño. But even though geographers may be working at different scales, they never lose sight of the interactivity and connectivity be-

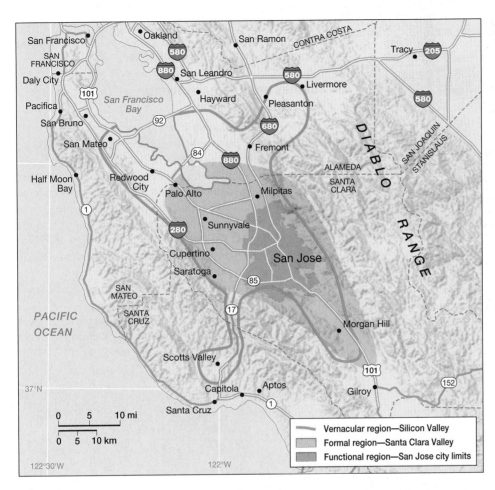

▲ **Figure 1.16 Geographic regions** This map illustrates three different kinds of regions. A vernacular region is a region with vague, cognitive borders used by the public to refer to a general area. Another example of a vernacular region would be "the Midwest." Usually, these venacular regions are a combination of many traits, real or imagined. In contrast, formal regions have distinct boundaries that are constructed by using one specific trait. In the case of the Santa Clara Valley, that trait is the topographic boundary of the valley. Last, a functional region is based on a certain activity or organization, such as the civic government of San Jose, which is then delimited by the city limits.

Map legend:
Vernacular region—Silicon Valley
Formal region—Santa Clara Valley
Functional region—San Jose city limits

tween local, regional, and global scales, of ways that the village in southern India might be linked to world trade patterns, or the late arrival of the monsoon (delayed, perhaps, by El Niño) could affect agriculture and food supplies in different parts of India.

This ability to move between different scales—global, regional, local—is critical for understanding contemporary world regional geography because of the way globalization links all peoples and places. Few villages today, however remote, are unconnected to the modern world. Global economies draw upon local crops and resources, and, conversely, fluctuations in world commodity prices affect the well-being of local people. Global TV and videos introduce foreign styles, ideas, mannerisms, and expectations into small towns and remote settlements in formerly isolated parts of the world. Global tourism in its different faces brings strangers (some welcome, some not) to far-flung localities, often corrupting and compromising the very uniqueness the tourists sought. Further, very few places and people are isolated from global politics and tensions, from the influences of superpowers and supranational organizations.

In this book, we use the phrase "global to local" to capture the flexibility in the scale of analysis necessary to understanding placing regions, places, landscapes, and people in their globalized context. Since our premise is that no place in today's world is completely isolated from the larger currents,

▲ **Figure 1.17 The cultural landscape** A major geographic tool is landscape analysis, which examines how humans shape the environment into distinctive forms that give places their special—and spatial—identities. This village is in Yunnan province, China. *(Mike Yamashita/Woodfin Camp & Associates)*

▲ **Figure 1.18 Global-to-local connections**   Today, most localities are linked to the world economy as both consumers and producers. As a result, fluctuations in the global economy have repercussions that ripple into the smallest communities. In this photo, women in a small village in China produce clothing for world markets.   *(James Montgomery/Bruce Coleman Inc.)*

we need this perspective to examine the complicated linkages connecting people and places with the larger, dynamic world (Figure 1.18).

Themes and Issues in World Regional Geography

Geography is the academic discipline that describes Earth and explains the patterns on its surface. Although this task involves different and often contrasting approaches and perspectives, the themes and concepts held in common give coherence to the field, offering an effective perspective on how globalization affects different parts of the world. A general background on global environmental geography is found in Chapter 2, "The Changing Global Environment." That chapter outlines the global environmental elements fundamental to human settlement—climate, topography, vegetation, and hydrology—and discusses the linkages between environmental issues and globalization.

Following the two introductory, thematic chapters, the book adopts a regional perspective, grouping all of Earth's countries

▶ **Figure 1.19 World population**
This world population map shows the differing densities of population in the regions of the world. East Asia stands out as the most populated region, with high densities in Japan, Korea, and eastern China. The second most populated region is South Asia, dominated by India, which is second only to China in population. In North Africa and Southwest Asia, population clusters are often linked to the availability of water for irrigated agriculture, as is apparent with the population cluster along the Nile River. Higher population densities in Europe, North America, and other countries are usually associated with large cities, their extensive suburbs, and nearby economic activities.

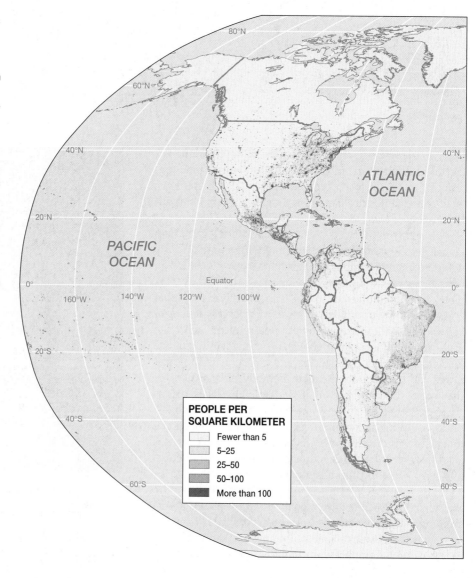

PEOPLE PER
SQUARE KILOMETER
Fewer than 5
5–25
25–50
50–100
More than 100

into the framework of world regions. We begin with the region most familiar to most of our readers, North America, then move to Latin America, the Caribbean, Africa, the Middle East, Europe, Russia, and the different regions of Asia, before concluding with Australia and Oceania. Each of the 12 regional chapters employs the same 5-part thematic structure: environmental geography; population and settlement; cultural coherence and diversity; geopolitical framework; and economic and social development. Let us now examine each of these major themes.

Population and Settlement: People on the Land

The human population is far larger than it ever was in the past, and there is considerable debate about whether continued growth will benefit or harm the human condition. There is concern not only with aggregate numbers, but also with geographic patterns of human settlement across the planet. While some parts of the world are densely populated, with intertwined towns and cities, other places remain almost empty (Figure 1.19).

Global population growth is one of the most vexing issues facing the world. With more than 6 billion humans on Earth, we currently add another 86 million each year, or about 10,000 new people each hour. About 90 percent of this population growth takes place in the developing world, especially in Africa, Latin America, and South and East Asia. Because of rapid growth in developing countries, perplexing questions dominate discussions of all global issues. Can these countries absorb the rapidly increasing population and still achieve the necessary economic and social development to ensure some level of well-being and stability for the total population? What role should the developed countries of North America, Europe, and East Asia play in those population issues now that their own populations have nearly stabilized? While population is a complicated and contentious topic, several points help focus the issue:

1. Very different rates of population growth are found in various regions of the world. While some countries are growing rapidly, others have essentially no natural increase; instead, any population increase comes from in-migration.

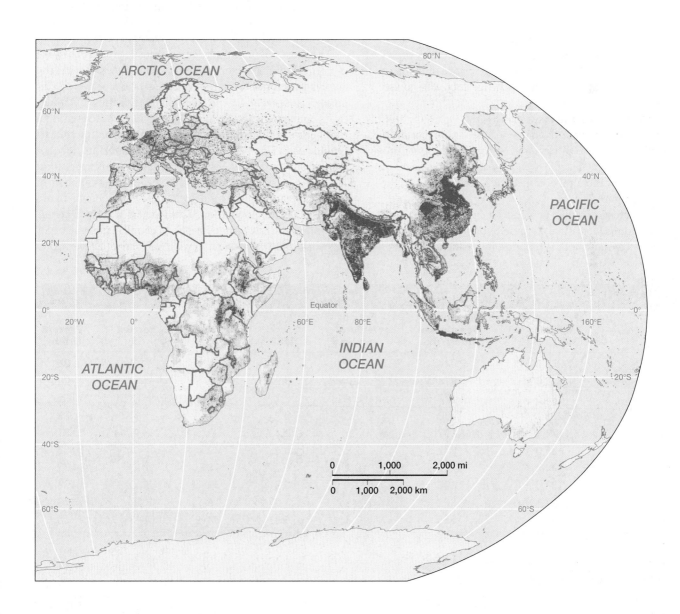

▲ Figure 1.20 Family planning policies   Many countries in the developing world have concluded that unrestrained population growth may keep them from realizing their developmental goals, and therefore have put family planning policies in place. This poster in Vietnam urges smaller families.   *(Chris Stowers/ Panos Pictures)*

is driven by a desire for a better life in the richer regions of the developed world, we should not lose sight of the millions of migrants who are refugees from civil strife, persecution, and environmental disasters.

4. The greatest migration in human history (at least in aggregate numbers) is now going on as people move from rural to urban environments. If current rates continue, more than half the world's population will soon live in cities. Once again, it is the developing countries of Africa, Latin America, and Asia that are experiencing the most dramatic changes caused by urbanization.

Population Growth and Change

Because of the centrality of population growth, each regional chapter contains a table with pertinent data for every country within that region. These tables present the following information for each country: total population, density of population per square mile, rate of natural increase, total fertility rate, percentages of the population under 15 years of age and over 65, and percentage of the population living in cities (Table 1.1). Although the statistics in these tables might seem daunting at first glance, this information is crucial to understanding the population geography of each country.

The most common population statistic used is the **rate of natural increase** (abbreviated as RNI), which depicts the annual growth rate for a country or region as a percentage increase. This statistic is produced by subtracting a population's number of deaths in a given year from its annual number of births to find the net number added through natural increase. Gains or losses through migration are not considered in this figure. Instead of using raw numbers for a large population, demographers divide the gross numbers of births or deaths by the total population, producing a figure per 1,000 of the population. This is referred to as the *crude birthrate* or the *crude death rate*. For example, current data for the world as a whole show a crude birthrate of 22 per 1,000

India is an example of the former, Europe of the latter. Other countries, such as those of the former Soviet Union, have both low birthrates and negligible rates of immigration; unless their circumstances change, they will soon see major population losses.

2. Population planning takes many forms, from the rigid one-child policies of China to the "more children, please" programs of western Europe. While some programs attempt to achieve their goals through coercion and force, others rely on incentives and social cooperation (Figure 1.20).

3. Not all attention should be focused on natural growth, since migration is increasingly a cause of population change in this globalized world. While most international migration

TABLE 1.1 *Demographic Indicators of the Ten Largest Countries*							
Country	Population (Millions, 2001)	Population Density, per square mile	Rate of Natural Increase	TFR[a]	Percent < 15[b]	Percent > 65	Percent Urban
China	1,273.30	344	0.9	1.8	23	7	36
India	1,033.00	814	1.7	3.2	36	4	28
United States	284.5	77	0.6	2.1	21	13	75
Indonesia	206.1	280	1.7	2.7	31	4	39
Brazil	171.8	52	1.5	2.4	30	5	81
Pakistan	145.0	472	2.8	5.6	42	4	33
Russia	144.4	22	−0.7	1.2	18	13	73
Bangladesh	133.5	2,401	2	3.3	40	3	21
Japan	127.1	872	0.2	1.3	15	17	78
Nigeria	126.6	355	2.8	5.8	44	3	36

[a]Total fertility rate

[b]Percent of population younger than 15 years of age

Source: Population Reference Bureau. World Population Data Sheet, 2001.

and a crude death rate of 9 per 1,000. Thus, the natural growth rate is 13 births per 1,000. This is expressed as the rate of natural increase simply by converting that figure to a percentage. As a result, the current RNI for the world is 1.3 percent per year. Because birthrates vary greatly between countries (and even regions of the world), rates of natural increase also vary greatly. In Africa, for example, many countries have crude rates of more than 40 births per 1,000 people, with several close to 50 per 1,000. Since their death rates are generally less than 20 per 1,000, we find RNIs around 3 percent per year, which are among the highest rates of natural growth found anywhere in the world.

Though the crude birthrate gives some insight to current conditions in a country, demographers also use the **total fertility rate,** or TFR. The TFR is a synthetic rate that measures the fertility of a statistically fictitious, yet average, group of women moving through their childbearing years. If women marry early and have many children over a long span of years, the TFR will be a high number; conversely, if data show that women have few children, the number will be correspondingly low. From population data collected in the second half of the 1990s, the current TFR for the world is 2.8 children; this is the average number of children borne by a statistically average woman. For the same period the TFR for Africa is 5.2, compared to slow-growth or no-growth Europe with 1.4.

One of the best indicators of the momentum (or lack) for continued population growth is the youthfulness of a given population, since this figure shows the proportion of a population about to enter the prime reproductive years. The common conceptualization for this measure is the percentage of a population under the age of 15. Equally important, this figure gives some indication of the health and nutritional needs of a society providing for a youthful population, which is most vulnerable to famine, malnutrition, and disease. As a global average, 30 percent of the population is younger than age 15, but in fast-growing Africa, that figure is 43 percent, with several African countries approaching 50 percent (Figure 1.21). This suggests strongly that rapid population growth in Africa will continue for at least another generation despite the tragedy of the AIDS epidemic. In contrast, Europe averages 18 percent and North America 21 percent. At the other end of the age spectrum is the percentage of a population over the age of 65. These data suggest the need for social welfare services such as health care and social security, which are usually provided through taxation of those of working age. Japan and European countries, for example, have a relatively high proportion of people over age 65 and a very small percentage of young people approaching their reproductive years. In combination, these data raise concerns about whether these countries can continue to support social security and other public services necessary for the elderly.

The Demographic Transition

Population growth rates change over time, as is well documented by the historical record. More specifically, historically in Europe, North America, and Japan, growth declined as countries became increasingly industrialized and urbanized. From this historical data, demographers generated the **demographic transition** model, a four-stage conceptualization that tracks changes in birthrates and death rates through time as a population urbanizes. While the demographic transition

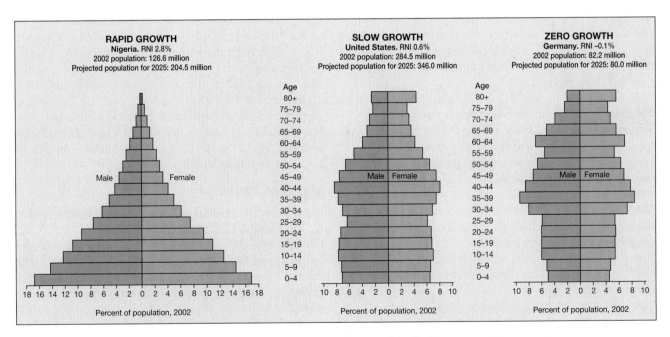

▲ Figure 1.21 Population pyramids   The term *population pyramid* comes from the form assumed by a rapidly growing country such as Nigeria when data for age and sex are plotted graphically as a percentage of total population. Rapid natural growth will probably continue in this country because of the high percentage of the population entering fertility years. In contrast, slow-growth and zero-growth countries have a very narrow base, which indicates relatively few people in the childbearing ages.

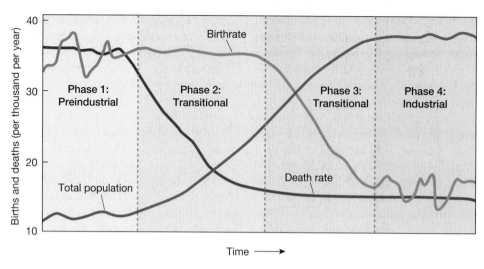

▲ Figure 1.22 Demographic transition   As a country goes through industrialization, its population moves through the four phases in this diagram, referred to as the *demographic transition*. In the first phase, population growth is low because high birthrates are offset by high death rates. Rapid growth takes place in Phase 2 as death rates decline. Stage 3 is characterized by a decline in birthrates. The transition ends with low growth once again in Phase 4, resulting from a relative balance between low birthrates and death rates.   *(From Knox and Marston, 2001,* Human Geography, *Second Edition, Upper Saddle River, NJ: Prentice Hall)*

model seems valid for describing historical population change, there is considerable debate about whether it offers an accurate tool for predicting change in countries currently undergoing industrialization and urbanization. If the model is an accurate predictor of population growth, some population workers and politicians argue that the best form of birth control might be financial aid that furthers economic development. However, a large number of skeptics question the application of the demographic transition model to all developing countries. They draw their findings from countries where low birthrates are not linked to economic development, such as in the former communist countries of eastern Europe. Because of these contrasting views, there is a spirited debate in international aid circles about the best course of action regarding family planning. Often the demographic transition model is at the center of this discussion (Figure 1.22).

In the demographic transition model, Phase 1 is characterized by both high birthrates and death rates, leading to a very slow rate of natural increase. Historically, this stage is associated with Europe's preindustrial stage, a period that also predated common public health measures such as sewage treatment, the scientific understanding of disease transmission, and the most fundamental aspects of modern medicine. Little wonder that death rates were high and life expectancy short. Unfortunately, these conditions are still common in some parts of the world.

In Phase 2, death rates fall dramatically while birthrates remain high. This produces a rapid rise in the RNI. Again, both historically and contemporaneously, this decrease in death rates is usually associated with the onset of public health measures and modern medicine. One of the assumptions of the demographic transition model is that these services become increasingly available after some degree of economic development takes place and, further, that they are increasingly accessible to the growing number of urban dwellers. However, even as death rates fall, it takes time for people to respond with lower birthrates, which begin in Phase 3. This, then, is the transitional stage as people apparently become aware of the advantages of smaller families in an urban and industrial setting. In Phase 4, a very low RNI results from a combination of low birthrates and very low death rates. As a result, there is very little natural population increase.

For many professionals, the demographic transition model is a reasonable predictor of where and when we can expect population growth rates to slow in the developing world. However, it also has detractors who see too many complicating variables and differences between countries for universal application in the twenty-first century. Minimally, the demographic transition concept is a useful tool for organizing information about the vast differences in birthrates and death rates that we find around the world (Figure 1.23).

Migration Patterns

Never before in human history have so many people been on the move and lived outside of their countries of birth. Today about 125 million, or 2 percent, of the total world population can be considered migrants of some sort. Estimates suggest that this figure may increase by some 4 million annually. Much of this international migration is directly linked to the new globalized economy. About half of the migrants live either in the developed world or in those developing countries with vibrant industrial, mining, or petroleum extraction economies. In the oil-rich countries of Kuwait and Saudi Arabia, for example, the labor force is primarily composed of foreign migrants. In terms of total numbers, fully a third of the world's migrants live in 7 industrial countries: Japan, Germany, France,

▲ **Figure 1.23 Fertility and mortality**   Birthrates and death rates vary widely around the world. Fertility rates result from an array of variables, including state family-planning programs and the level of a woman's education. This family is in Bangladesh, a country that is working hard to reduce its birthrate through education programs.   *(Trygve Bolstad/Panos Pictures)*

Canada, the United States, Italy, and the United Kingdom. Since industrial countries usually have very low birthrates, immigration accounts for a large proportion of total population growth. Roughly one-third of the annual growth in the United States, for example, comes from in-migration.

But not all migrants move for economic reasons. War, persecution, famine, and environmental destruction cause people to flee to safe havens elsewhere. Accurate data on refugees are often difficult to obtain, but UN officials estimate that some 27 million people (or about 20 percent of the total migrant population) should be considered refugees. More than half of these are in Africa and western Asia (Figure 1.24). While the causes of migration are often complicated, three interactive concepts are helpful in understanding this process. First, *push* forces, such as civil strife, environmental degradation, or unemployment, drive people from their homelands. Second, *pull* forces, such as better economic opportunity or health servic-

▲ **Figure 1.24 Refugees**   About 27 million people, or 20 percent of the world's migrant population, are refugees from ethnic warfare and civil strife. Most of these refugees are in Africa and western Asia. These women wait in line for water at a refugee camp in southern Sudan.   *(Liz Gilbert/Corbis/Sygma)*

es, attract migrants to certain locations, whether within or beyond national boundaries. Connecting the two are the informational *networks* of families, friends, and sometimes labor contractors who provide information on the mechanics of migration, transportation details, or news of housing and jobs.

Settlement Geography

Geography has long been concerned with the different patterns of human settlement in the landscape. For example, in some regions, such as Europe and South Asia, it is common for people to cluster in villages and towns; this contrasts with the dispersed settlement pattern of rural North America, where people tend to scatter across the countryside in individual farms and homesteads. Additionally, the density, pattern, and dynamic of settlement convey much information about regional issues, both historically and currently. In many regions, such as Africa, Latin America, and Asia, it was common for European colonial powers to relocate indigenous people into new settlements to suit their own purposes. Plantation labor needs, for example, were often met by moving native peoples from villages to agglomerated settlements. And, of course, there was the massive disruption in Africa of traditional settlement patterns by the slave trade.

Today settlement geography gives insight into the profound changes resulting from globalization: whether people are leaving the countryside or staying; whether they are moving to small towns or large cities; whether jobs from new plants and factories are decentralized into the countryside or agglomerated around large cities; whether mechanization is replacing farmworkers who then must seek other options in other places; whether civil and ethnic strife ravage the countryside, forcing people to flee as refugees; whether the rural environment is viable and productive or degraded. This is not to suggest that changes in settlement patterns should be viewed as only problematic. While the unfortunate aspects cannot be overlooked, there is also much to learn of a positive nature from how people arrange themselves on the land. Settlement patterns convey information about social organizations, economic structures, cultural values, political systems, population growth and change, and the meaning and emotion that people attach to the environment. When taken together, a settlement pattern can express a sense of a place, the uniqueness of a landscape, and the nature of the home territory of a specific group of human beings in this complex world (Figure 1.25).

An Urban World   Cities are the focal points of the contemporary, globalizing world, the fast-paced centers of deep and widespread economic, political, and cultural change. Because of this vitality and the options cities offer to impoverished and uprooted rural peoples, they are also magnets for migration. The scale and rate of growth of some world cities is staggering: estimates are that both Mexico City and São Paulo (Brazil), cities of more than 20 million, are adding nearly 10,000 new people each week. Urban planners predict that these two cities will actually double in size within the next 15 years. Assuming that predictions about the migration rate

(a)

(b)

▲ Figure 1.25 Settlement landscapes   City and village landscapes differ widely because of the interplay between contemporary and historical forces, socioeconomic and cultural. (a) A small Peruvian village in the Andes that is populated by indigenous people and expresses their local traditions. (b) A town in Venezuela, where the Spanish colonial presence is still found in the plaza, street pattern, and building architecture.   (Rob Crandall/Rob Crandall, Photographer)

to cities are correct, the world is quickly approaching the point at which more than half of its people will be urban dwellers. Evidence for this comes from data on the **urbanized population,** which is that percentage of a country's population living in cities. Currently, 46 percent of the world's population lives in cities. Many demographers predict, however, that the world will be 60 percent urbanized by the year 2025.

Tables in this book's regional chapters include data on the urbanization rate for each country. For example, more than

75 percent of the populations of Europe, Japan, Australia, and the United States live in cities. Generally speaking, most countries with such high rates of urbanization are also highly industrialized, since manufacturing tends to cluster around urban centers (Figure 1.26). In contrast, the urbanized rate for developing countries is usually less than 50 percent, with figures closer to 40 percent not uncommon. Roughly 30 percent of Sub-Saharan Africa's people, for example, live in cities. These data imply a low level of industrial activity. Urbanization fig-

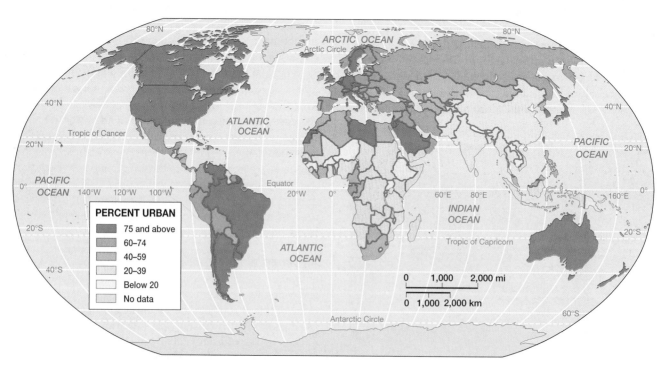

▲ Figure 1.26 World urbanization   In general, developed countries have a higher percentage of population living in cities as a result of the clustering of industry and business in urban places. This can be seen by contrasting Europe and North America with Africa. However, the Chinese government attempts to restrict the migration of people to cities to prevent overcrowding. Thus, China's urbanization is lower than expected.

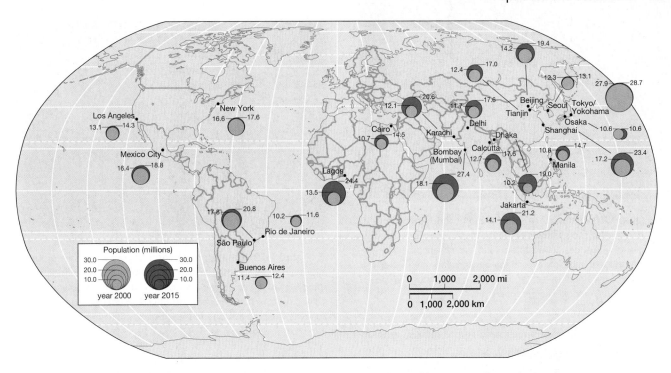

▲ **Figure 1.27 Growth of world cities** This map shows the largest cities in the world (over 10 million), along with predicted growth by the year 2015. Note that the greatest population gains are expected in the large cities of the developing world, such as Lagos, Nigeria; Karachi, Pakistan; and Bombay (Mumbai). In contrast, large cities in the developed world are predicted to grow slowly over the next decades. Tokyo, for example, will add fewer than a million people.

ures also show where there is high potential for urban migration. If the urbanized population is relatively small, such as in Zimbabwe (Africa), where only some 32 percent of the population lives in cities, the probability of high rates of urban migration in the next decades is great. Accompanying the movement of people from countryside to cities will likely be profound changes in family structure, national policies, and market and manufacturing economics. Such changes will be accompanied by widespread transformations of the landscape and natural environment as people move from rural to urban settlements.

Conceptualizing the City Given the important role of cities in world regional geography, it seems appropriate to offer a brief conceptual framework for thinking about urban settlements.

1. Cities are interdependent with other cities and towns, both regionally and globally, and are linked in an urban system of interactive settlements. As some cities expand or come to specialize in a specific function, such as banking or manufacturing, the relationships and interactions within that urban system may change. The largest city within the city system may have **urban primacy,** a term used to describe a city that is disproportionately large and dominates economic, political, and cultural activities within the country (Figure 1.27). In many countries, one large primate city usually stands out in that country's urban system. Examples include Bangkok, Mexico City, Cairo, Paris, and Buenos Aires.

2. Although cities are highly complicated settlements, there are often shared traits of **urban structure.** This term refers to the distribution and patterning of land use within the

city. Land use is linked to specific functions, such as finance, governmental activities, housing, retailing, and industry. Almost all large cities, for example, contain a central business district (CDB) in which banks and business headquarters are located. Because these functions are not randomly distributed, but instead follow well-understood location principles, it is possible to generate models of urban structure that convey the commonalties between cities within a specific region. An example is the model of Latin American cities discussed in Chapter 4.

3. Closely related to urban structure is the landscape of the city, the **urban form** or physical arrangement of buildings, streets, parks, and architecture that gives each city its unique sense of place. For example, Cairo clearly looks and feels very different from New Delhi or Caracas (Figure 1.28). The reasons for these differences in urban landscapes range from different histories, economies, and planning policies to ideologies and cultural values. In the regional chapters, we explain why certain cities look and feel as they do.

4. Because urban migration has taken place so rapidly in many countries, resulting in explosive growth, many cities have reached a state of **overurbanization,** in which the urban population grows more quickly than support services such as housing, transportation, waste disposal, and water supply. Unfortunately, this a fairly common situation in the developing world. The hundreds of thousands of "street people" of Calcutta are legend, although the phenomenon of homelessness is by no means restricted to South Asia. **Squatter settlements,** illegal developments of makeshift housing on land neither owned nor rented by

▲ Figure 1.28 Urban form This photo of downtown Caracas, Venezuela, shows how city landscapes express both local and global influences. In the foreground are low-income "ranchitos" where local migrants live; in the background, the high-rise buildings of the financial center, linked to global trade. *(Rob Crandall/Rob Crandall, Photographer)*

▲ Figure 1.29 Squatter settlements Because of the massive migration of people to world cities, adequate housing for the rapidly growing population becomes a daunting problem. Often, migrant housing needs are filled with illegal squatter settlements, such as this one in New Delhi. *(Porterfield/Chickering/Photo Researchers, Inc.)*

their inhabitants, are widespread in the rapidly growing cities of the developing world. Such settlements are often built on steep hillside slopes or on river floodplains that expose the occupants to the dangers of landslide and floods. Squatter settlements are also often found in the open space of public parks or along roadways where they are regularly destroyed by government authorities, only to be quickly rebuilt by migrants who have no other alternatives (Figure 1.29).

Even for those who have found more permanent housing, water and sewer connections are often rudimentary at best. Water service can be sporadic and restricted to a few hours a day, and open sewers are a disturbingly common feature of many cities. The World Bank estimated recently that only 40 percent of the people in less-developed cities lived in homes that are connected to sewers. Although difficult to measure accurately in developing countries, urban unemployment and underemployment are of epidemic proportion, often encompassing up to 50 percent of the potential workforce. Without regularly paid work, urban migrants survive by begging, running errands, being paid to stand in lines, street vending, or scavenging public dumps. Because cities are experiencing unprecedented and seemingly relentless growth, this multitude of problems is a good reason for concern about the quality of life of the world's recent urban migrants.

Cultural Coherence and Diversity: The Geography of Tradition and Change

Culture can be thought of as the weaving that keeps the world's diverse social fabric together. A cursory reading of the daily newspaper, however, raises questions about whether the world is literally coming unraveled since the frequency and intensity of cultural conflict seem ever-increasing. With the recent rise of global communication systems (satellite TV,

movies, video, etc.), moreover, stereotypic Western culture is spreading at a rapid pace. While this is willingly accepted by many throughout the world, others resist what they consider a new form of cultural imperialism through protests; censorship; restrictions on Western film, TV, and music; and even terrorism.

In response to the spread of a common world culture, many people across the globe are revisiting their traditional and historical identities as ethnic or folk groups. This is not simply a matter of quaint customs and costumes; instead, it often leads to overt conflict with neighboring groups with different identities and self-interests. Over much of the world, we are seeing a new interest in small-group autonomy and self-determination. This phenomenon is particularly pronounced in the former Soviet Union, where ethnic identities were suppressed under communism for most of this century; but it is also encountered in other large nation-states—including those in western Europe—where new voices are demanding regional autonomy, separate territories, and control over natural resources. Religious conflicts, too, seem to be on the increase. Some of this is coincidental with the return of religious choice to former Soviet lands, yet other tensions result from missionary activity by Christian and Islamic sects. Religious fundamentalism is on the rise as a reaction to globalizing culture, and this, too, often breeds cultural tensions.

The geography of cultural cohesion and tension, then, entails an examination of tradition and change, of language and religion, of group belonging and identity, and of those complex and varied currents that underlie and pervade twenty-first-century ethnic factionalism and separatism (Figure 1.30).

Culture in a Globalizing World

Given the diversity of cultures around the world, coupled with the dynamic changes associated with globalization, traditional definitions of *culture* must be stretched somewhat to provide

▲ **Figure 1.30 Ethnic tensions**   Contemporary cultural change is characterized by both convergence and homogeneity as people are pulled together by globalized culture, or by factionalism and separatism, often fueled by ethnic tensions, such as the religious strife shown here in India between Hindu and Muslim peoples.   *(Robert Nickelsberg/Getty Images, Inc.)*

a conceptual framework applicable to the new millennium. Though social scientists offer varied definitions of culture, there is enough overlap and agreement between these definitions to provide a starting point. **Culture** is learned, not innate, and is shared, not individual, behavior; it is held in common by a group of people, empowering them with what, for lack of a better term, could be called a "way of life."

Additionally, culture has both abstract and material dimensions: speech, religion, ideology, livelihood, and value systems, but also technology, housing, foods, and music. These varied expressions of culture are germane to the study of world regional geography because they tell us much about the way people interact with their environment. Finally, especially given the widespread influences of globalization, it is best to think of culture as dynamic rather than static. That is, culture is a process, not a condition, something that is constantly adapting to new circumstances. As a result, there are always tensions between the conservative, traditional elements of a culture and newer forces promoting change and modernity (Figure 1.31).

Given that group culture is expressed in many ways and that each group has different internal characteristics, there is value to creating a hypothetical spectrum of cultures ranging from the most tradition-bound to those more contemporary and fluid. The term *folk culture* is used to describe a rural, largely self-sufficient, and relatively homogeneous group of people with a conservative, traditional way of life. Here families or clans keep the social fabric together, and individualism is generally weak and unrewarded. Although there is typically little division of labor in rural cultures, gender roles are usually strong. The cultural landscape of folk cultures—house architecture, field types, crops, public monuments, and symbols—offers clues to how these people interact with the environment. Folk cultures were far more prevalent historically than today, since industrial society has eroded and changed what may have historically been a common social fabric. North American examples of folk cultures would in-

clude certain religious communities such as the Amish or the remnants of Appalachian Mountain culture. In Europe, folk culture can still be found in parts of rural Ireland, eastern Europe, or the remote mountain valleys of the Alps.

More fluid than a folk culture with defining traits based upon common ancestry, race, religion, or language is an *ethnic culture,* in which there is usually a strong sense of ethnic tradition controlled by clear lines of authority through family, clan, tribe, or church. Ethnic group identity is often constructed from the shared feeling that its members are a minority within a larger cultural context. Many parts of Latin America contain good examples of ethnic cultures, as do Asia and Africa.

In contrast to folk and ethnic cultures, *popular culture* is primarily urban-based, encompasses great heterogeneity, and is rapidly changing and dynamic. Relationships between people are often ephemeral, perhaps even shallow, with a much weaker family structure than in traditional society. Labor division is high, with distinct specialization of jobs; gender roles may be less rigidly defined; and material goods are mass-produced in international factories. The quintessential landscapes of popular culture include the mall, the franchise strip mall, the parking lot, and the suburban housing tract. Although many disparage popular culture because of its faddishness and superficiality, it probably has more adherents around the globe than folk or ethnic cultures.

As a subset or variant of popular culture, today one finds an emerging *world culture* that is empowered by globalization. Transnational economies, electronic communication, and international politics produce citizens of indeterminate nationality and home base who travel widely, work in different countries, and, as a result, generate tastes and cultural values that are an amalgamation of international stimuli. Its members are equally at home in Hong Kong, New York, Singapore, and any of the world's other mega-cities, all of which have interchangeable landscapes of offices, hotels, restaurants, and discos. Increasingly, it seems, the one real home base for this world culture is the Internet, where instantaneous electronic communication has replaced traditional forms of social interaction.

These four categories offer a scheme to help comprehend the wide spectrum of cultural expressions found in the contemporary world. For many human beings, membership in one culture or another is beyond choice. Until recently, social scientists argued that a defining characteristic of folk and ethnic cultures was that individuals were born into them; people could choose to leave or disassociate through time, but their early membership was decided at birth. However, this seems an increasingly questionable position today, given the fluidity of cultural opportunities and the possibilities for multiple cultural associations. It seems reasonable that a person might choose to exercise a narrow cultural identity at one point in time or in a particular place, while participating in popular or world cultures at other times and places. As an example, one could point to the proclivity of some Amish young people to interact with popular culture while still taking their primary identity from their folk culture. This sort of dual cultural membership is becoming an increasingly common characteristic of contemporary globalized culture.

(a)

(b)

(c)

(d)

▲ Figure 1.31 Folk, ethnic, popular, and global culture   A broad spectrum of cultural groups characterize the world's contemporary cultural geography, ranging from traditional folk cultures to the ubiquitous popular and global cultures. Representing global culture (a) is a Korean woman in downtown Washington, D.C. with laptop and cell phone, while the young couple (b) in Sofia, Bulgaria, express the universal style of popular culture. At the opposite end of the spectrum are the ethnic cultures of (c) Farifuna women in Honduras and (d) the Chonguinada in Peru. *(Rob Crandall/Rob Crandall, Photographer)*

When Cultures Collide

Cultural change often takes place within the context of international tensions. Sometimes one cultural system will replace another; at other times resistance will stave off change. More commonly, however, a newer, hybrid form of culture results that is an amalgamation of two cultural traditions. Historically, colonialism was the most important perpetuator of these cultural collisions; today, globalization in its varied forms can be thought of as the major vehicle of cultural collision and change.

Cultural imperialism is the active promotion of one cultural system at the expense of another. Although there are still many expressions of cultural imperialism today, the most severe examples occurred in the colonial period when European cultures spread worldwide, sometimes overwhelming, eroding, and even replacing indigenous cultures. During this period, Spanish culture spread widely in Latin America; French culture diffused into parts of Africa; and British culture entered India. Of course, North America was also highly influenced by these three cultures. New languages were mandated, new education systems implanted, and new administrative institutions took the place of the old. Foreign dress styles, diets, gestures, and organizations were added to existing cultural systems. Many vestiges of colonial culture are still evident today. In India the makeover was so complete among a few social groups that wags are fond of saying, with only slight exaggeration, that "the last true Englishman will be an Indian." Today's cultural imperialism is seldom linked to an explicit colonizing force, but more often comes as a fellow traveler with globalization. Though many expressions of cultural imperialism carry a Western (even U.S.) tone—such as McDonald's, MTV, Marlboro cigarettes, and even the use of English as the dominant language of the Internet—these facets result more from a search for new consumer markets than from deliberate efforts to spread modern American culture throughout the world.

The reaction against cultural imperialism, **cultural nationalism,** is the process of protecting and defending a cultural system against diluting or offensive cultural expressions while at the same time actively promoting national and local cultural values. Often, cultural nationalism takes the form of explicit legislation or official censorship that simply outlaws the unwanted cultural traits. Examples of legislated cultural nationalism are common. France has long fought the Anglicization of its language by banning "Franglais," the use of English words such as "Le weekend," in official French. More recently, France has sought to protect its national music and film industries by legislating that radio DJs play a certain percentage (40 percent at this writing) of French songs and artists each broadcast day. France is also erecting tax and tariff barriers to Hollywood's products. Many Muslim countries censor Western cultural influences by restricting and censoring international TV, which they consider the source of many undesirable cultural influences. Most Asian countries, as well, are increasingly protective of their cultural values, and many are demanding changes to tone down the sexual content of MTV and other international TV networks.

As noted above, the most common product of cultural collision is the blending of forces to form a new, synergistic form of culture. This is called **cultural syncretism or hybridization** (Figure 1.32). To characterize India's culture as British, for example, would be to grossly oversimplify and exaggerate England's colonial influence. Indians have adapted many British traits to their own circumstances, infusing them with their own meanings. India's use of English, for example, has produced a unique form of "Indlish" that often befuddles visitors to South Asia. Nor should we forget how India has added many words to our own English vocabulary—khaki, pajamas, veranda, bungalow. In sum, both the Anglo and Indian cultures have been changed by the British colonial presence in South Asia.

Language and Culture in Global Context

Language and culture are so intertwined that, in the minds of many, language is the characteristic that best differentiates and defines cultural groups. Furthermore, since language is the primary means for communication, it obviously folds together many other aspects of cultural identity, such as politics, religion, commerce, folkways, and customs. Language is fundamental to cultural cohesiveness (Figure 1.33). It not only brings people together but also sets them apart; it can be an

▲ **Figure 1.32 Cultural syncretism**   When two cultures are brought together, a new culture is often created, as was the case in India as a result of the fusion of British colonial culture and local South Asian influences. This photograph shows British colonial architecture and modern buses in Bombay (Mumbai), India, coupled with local cabs, scooters and ox carts.   *(Rob Crandall/ Rob Crandall, Photographer)*

important component of national or ethnic identity and a way of creating and maintaining boundaries necessary to the reinforcement of regional identity and character (see "Local Voices: Who Are They, and Why Should We Listen?").

Because most languages have common historical (and even prehistoric) roots with other languages, linguists have grouped the thousands of languages found throughout the world into a handful of *language families*. These are simply a first-order grouping of languages into large units based upon common ancestral speech. For example, about half of the world's people speak languages of the Indo-European family, which includes not only European languages such as English and Spanish, but also Hindi and Bengali, which are widespread languages of South Asia. This grouping demonstrates that these diverse languages evolved from common linguistic roots. Within language families are smaller units that also give clues to the common history and geography of peoples and cultures. *Language branches and groups* (also called *subfamilies*) are closely related subsets within a family in which there are usually similar sounds, words, and grammar. Well known are the similarities between German and English, or between French and Spanish. Because of those similarities, these languages are placed into the same linguistic grouping (Figure 1.34).

Even individual languages often have very distinctive forms associated with specific regions. Such different forms of speech are called *dialects*. Although dialects of the same language have their own unique pronunciation and grammar (think of the distinctive differences, for example, between British, North American, and Australian English), they are—sometimes with considerable effort—mutually intelligible. Additionally, when people from different cultural groups cannot communicate directly in their native languages, they often agree upon a third

▶ **Figure 1.33 World languages** Most languages of the world belong to a handful of major language families. About 50 percent of the world's population speaks a language belonging to the Indo-European language family, which includes languages common to Europe, but also major languages in South Asia, such as Hindi. They are in the same family because of their linguistic similarities. The next largest family is the Sino-Tibetan family, which includes languages spoken in China, the world's most populous country. *(Adapted from Rubenstein, 2002, The Cultural Landscape: An Introduction to Human Geography, 7th ed., Upper Saddle River, NJ: Prentice Hall)*

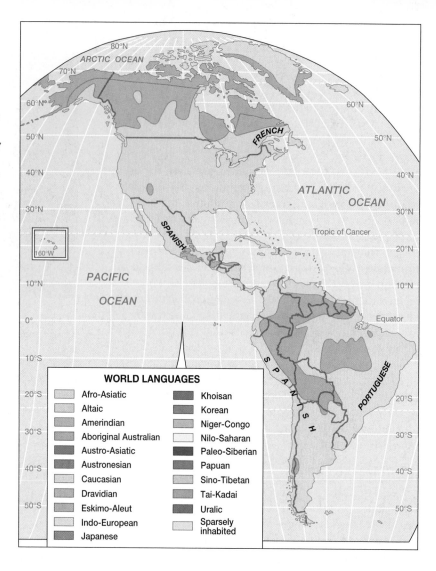

WORLD LANGUAGES

Afro-Asiatic	Khoisan
Altaic	Korean
Amerindian	Niger-Congo
Aboriginal Australian	Nilo-Saharan
Austro-Asiatic	Paleo-Siberian
Austronesian	Papuan
Caucasian	Sino-Tibetan
Dravidian	Tai-Kadai
Eskimo-Aleut	Uralic
Indo-European	Sparsely inhabited
Japanese	

LOCAL VOICES   Who Are They, and Why Should We Listen?

One of the best ways to learn about a foreign place is by listening to the people who live there: by listening to them talk about their country, their village, their neighborhood, their interests and concerns; about the problematic issues they face and the solutions they propose. Hearing these "local voices" is particularly important these days because the ubiquitous and loud voices of globalization, with satellite TV, global music videos, Hollywood movies, and 1970s TV reruns, threaten to mask the diversity of local expression.

While the messages of globalization are important and should not be dismissed, we wish to temper them somewhat by offering an opportunity to listen to the locals. Thus, most chapters have a sidebar titled "Local Voices" that offers brief exposure to diverse individuals and groups speaking about their homelands. Sometimes these voices come from rich works of regional fiction; other sidebars draw upon a group of local voices, perhaps students from a foreign university finding their voice through an Internet Web site. Still other examples are issue-oriented, as we listen to an environmental group present its views on the problems of pollution and nature protection in the homeland, or a separatist faction argue its case for territorial autonomy. Some local voices represent mainstream ideas, while others are less often heard—voices from the margins of globalization, so to speak.

The common theme of these local voices is that they add to our understanding of global geography by taking us down to ground level and providing us with local perspectives on the diverse fabric of places, people, environments, and issues that constitute world regional geography.

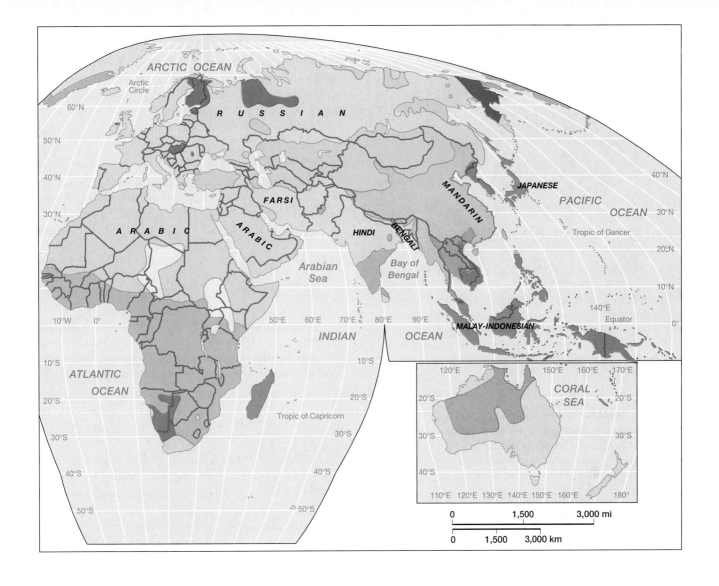

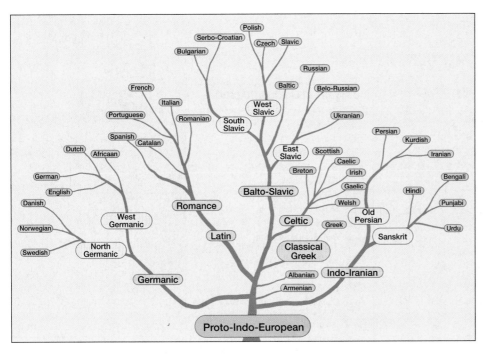

Figure 1.34 Language groups and branches The notion of a language family, such as the one shown here for Indo-European, is that similar historical roots have imparted linguistic similarities to each group and language branch. Although the exact time and place of the origin of proto-Indo-European languages have yet to be agreed upon and thus remain a fascinating geographical and linguistic issue, there is little question for linguists that the Indo-Iranian languages of South Asia and Germanic languages of Europe and North America share a common root. The relationships between later language groups and branches are much clearer and can be usually historically documented.

▲ Figure 1.35 English as global language   A traffic sign in Saudi Arabia reminds us of how some familiar English phrases have become almost universal. In many ways, English has become a global lingua franca for diverse activities such as international business, aviation traffic, communications, popular music, and road signs.   *(Ray Ellis/Photo Researchers, Inc.)*

language to serve as a common tongue, or **lingua franca**. Swahili has long served that purpose in polyglot eastern Africa, and French was historically the lingua franca of international politics and diplomacy. Today English is increasingly the common language of international business (Figure 1.35).

A Geography of World Religions

Another extremely important and often defining trait of cultural groups is religion. Indeed, with the erosion of atheistic communism in the former Soviet Union and the search for differentiating cultural identities in this era of a totalizing global culture, religion is becoming increasingly important in defining cultural identity (Figure 1.36). Recent ethnic violence and unrest in far-flung places such as the Balkans, Afghanistan, and Indonesia illustrate the point. Additionally, widespread missionary activity with its resulting conversions has both spread certain religions and aggravated religious tensions between certain groups.

Universalizing religions, such as Christianity, Islam, and Buddhism, attempt to appeal to all peoples regardless of lo-

▶ Figure 1.36 Major religious traditions
This map shows the geographical mosaic of major religious traditions found throughout the world. For most people, religious tradition is a major component of cultural and ethnic identity. While Christians of different sorts account for about 34 percent of the world's population, this religious tradition is highly fragmented. Within Christianity, there are about twice as many Roman Catholics as Protestants. Adherents to Islam account for about 20 percent of the world's population. The Sunni branch within Islam is three times larger than the Shiite. Hindus account for about 14 percent of the global population.

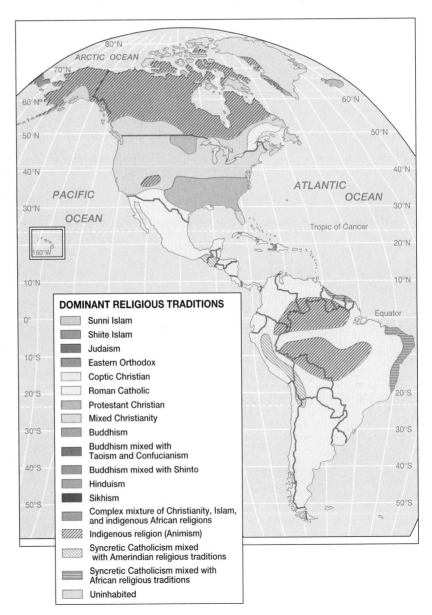

DOMINANT RELIGIOUS TRADITIONS
- Sunni Islam
- Shiite Islam
- Judaism
- Eastern Orthodox
- Coptic Christian
- Roman Catholic
- Protestant Christian
- Mixed Christianity
- Buddhism
- Buddhism mixed with Taoism and Confucianism
- Buddhism mixed with Shinto
- Hinduism
- Sikhism
- Complex mixture of Christianity, Islam, and indigenous African religions
- Indigenous religion (Animism)
- Syncretic Catholicism mixed with Amerindian religious traditions
- Syncretic Catholicism mixed with African religious traditions
- Uninhabited

cation or culture; these religions usually have a proselytizing or missionary program that actively seeks new converts. In contrast are the **ethnic religions,** which remain identified closely with a specific ethnic, tribal, or national group. Judaism and Hinduism, for example, are usually regarded as ethnic religions because they normally do not actively seek new converts. Christianity, a universalizing religion, is the world's largest in both areal extent and number of adherents. Although fragmented into separate branches and churches, Christianity as a whole claims almost 2 billion adherents, or about a third of the world's population. The largest number of Christians can be found in Europe, Africa, Latin America, and North America. Islam, which has spread from its origins on the Arabian Peninsula as far east as Indonesia and the Philippines, has about 1.2 billion members. While not as severely fragmented as Christianity, Islam should not be thought of as a homogeneous religion because it also has split into separate groups. The major branches are Shiite Islam, which comprises about 11 percent of the total Islamic population and represents a majority in Iran and southern Iraq, and the mainstream

Sunni Islam, which is found from the Arab-speaking lands of North Africa to Indonesia. Both of these forms of Islam are experiencing a fundamentalist revival in which proponents are interested in maintaining a purity of faith distanced from Western influences (Figure 1.37).

Hinduism, which is closely linked to India, has about 750 million adherents. Outsiders often regard Hinduism as polytheistic, since Hindus worship many deities. Most Hindus argue, however, that all of their faith's gods are merely representations of different aspects of a single divine, cosmic unity. Historically, Hinduism is linked to the caste system with its segregation of peoples based upon ancestry and occupation. But because the caste system is now anathema to India's official democratic ideology, connections between religion and caste are now much less explicit than in the past. Buddhism, which derived from Hinduism 2,500 years ago as a reform movement, is widespread in Asia, extending from Sri Lanka to Japan and from Mongolia to Vietnam. In its spread, Buddhism came to coexist with other faiths in certain areas, making it difficult to accurately estimate the number of its

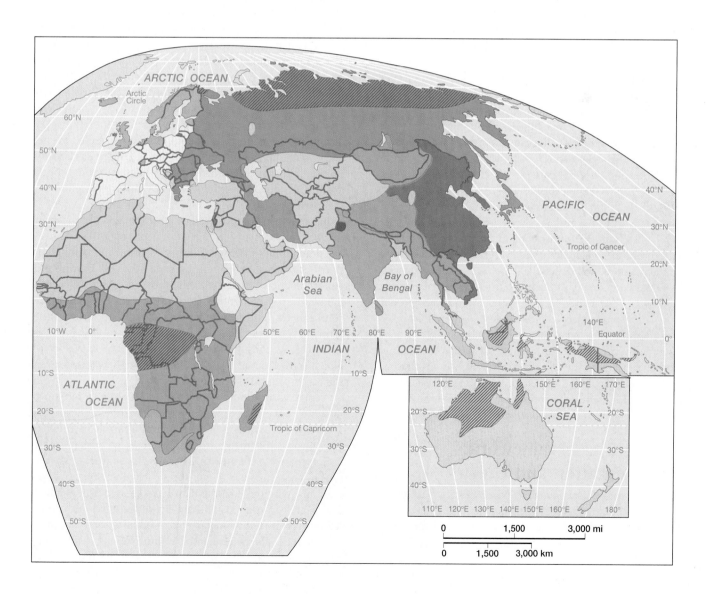

▲ **Figure 1.37 Religious landscapes** Minarets, from which Muslims are called to prayer, surround this mosque in Istanbul, Turkey, creating an instantly recognizable landscape linked to Islam. A similar landscape feature would be the towering steeple of a Roman Catholic cathedral, which would be associated with that religion regardless of location. *(Rob Crandall, Photographer)*

▲ **Figure 1.38 End of the Cold War** With the fall of Soviet communism, many nations rushed to remove the symbols of their former governments to forget the past and make room for a new future. In this photo taken in 1990, a monument to Lenin is toppled in Bucharest, Romania. *(Bi/Getty Images, Inc.)*

adherents. Estimates of the total Buddhist population thus range from 350 million to 900 million people.

Judaism, the parent religion of Christianity, is also closely related to Islam. Although tensions are often high between Jews and Muslims because of the Israeli-Palestinian conflict, these two religions, along with Christianity, actually share historical and theological roots in the Hebrew prophets and leaders. Judaism now numbers about 18 million adherents, having lost perhaps a third of its total population during the systematic extermination of Jews by the Nazis during World War II.

Finally, in some parts of the world, religious practice has declined for different reasons, giving way to secularization, in which people consider themselves either nonreligious or outright atheistic. Though secularization is difficult to measure, social scientists estimate that about 1 billion people fit into this category worldwide. Perhaps the best example of secularization comes from the former communist lands of Russia and eastern Europe, where there was overt hostility between government and church. Since the fall of Soviet communism in 1989, however, many of these countries have experienced a modest religious revival. However, secularization has probably grown more pronounced in recent years in western Europe. Although historically and to some extent still culturally a Roman Catholic country, France now reportedly has more people attending Muslim mosques on Fridays than Christian churches on Sundays. Japan and the other countries of East Asia are also noted for their high degree of secularization.

Geopolitical Framework: Fragmentation and Unity

The term *geopolitics* is used to describe and explain the close link between geography and political activity. More specifically, geopolitics focuses on the interactivity between power, territory, and space at all scales from the local to the global. There is little question that one of the dominant characteristics of the last several decades has been the rapidity, scope, and character of political change in various regions of the world.

With the dissolution of the Soviet Union in 1991 came opportunities for self-determination and independence in eastern Europe and Central Asia. As a result, there have been fundamental changes in economic, political, and even cultural alignments and alliances. While religious freedom helps drive national identities in some new Central Asian republics, eastern Europe is primarily concerned with economic and political links to western Europe. Russia itself wavers perilously between different geopolitical pathways. Phrases and terms common to an earlier generation may disappear; for example, the Cold War between the two superpowers, the United States and Soviet Union, has already become a historical artifact (Figure 1.38). Although optimism at the end of the Cold War was captured briefly with the term "the New World Order," hopes for a new global stability were also short-lived as it became quickly apparent that the superpowers' protective concern over their respective domains left many countries susceptible to internal tensions once that protection was withdrawn. Ethnic and cultural factions, giddy with the prospect of self-rule, promoted diverse and contradictory agendas that brought many countries to the brink of anarchy. Continued strife in the former Yugoslavia, the Caucasus, and Central Asia serve as examples. As a result, factionalism, ethnic tensions, and regional separatism are becoming increasingly widespread in the early twenty-first century.

While wide-ranging international conflicts remain a nagging concern of diplomats and governments (particularly in South Asia, where the proliferation of nuclear weapons has inflamed historical animosities), perhaps more common are strife and tension within—rather than between—nation-states. Civil unrest, tribal tensions, terrorism, and religious factionalism have created a new fabric and scale of political tension. So widespread and problematic is this tension in some parts of the world that scholars increasingly describe the nation-state as a possibly archaic political structure that may soon be replaced by tribal and ethnic micro-states or even warlord fiefdoms. The war in Afghanistan illustrates the point; ethnic group

Because the world we live in is constantly changing, so is world regional geography. Current events, be they tragic or uplifting, assure that there is always a geography in the making as complicated and intertwined global processes unfold. Sometimes the pace and nature of change is gradual and takes a familiar course; other times, global change comes more quickly and is largely unanticipated. Such was the case with the end of the Cold War and the geopolitical and economic devolution of the Soviet Union in the early 1990s.

Today, largely as a result of the September 11, 2001 terrorist attacks on the United States, we may be entering an episode of rapid world change. Or perhaps not. Opinions differ. Robert Kaplan, an astute commentator of world affairs, writes: "As Americans, we have a natural tendency to believe that world events over the next few years will unfold from September 11. But in truth, much of the world will evolve without regard to September 11. Civil wars will continue, diseases will break out, and local economic crises will run their course. The challenge is to anticipate how these other processes will intersect with the war on terrorism." After this qualification, Kaplan then goes on to discuss how the war on terrorism might have the unintended consequence of disturbing regional politics in the Middle East to a degree unknown since the end of World War I and, additionally, how a new "concert of powers" between Russia, China, and the United States could provide a framework for cooperation against militant Islam in the Asian steppes. Further, Kaplan opines that Sub-Saharan Africa may experience the implosion of larger countries such as Nigeria, Kenya, and Côte d'Ivoire and, accompanying the loss of central authority, these countries could well become havens for global terrorist groups.

Even though the world and its geography are dynamic and continue to change on a daily basis, a world regional geography book must, at some point, meet its deadline by going to press. Such was the case with this book at mid-year 2002. Within this text, we have discussed world events (including the war on terrorism) as they were at press time. In addition, to the integration of current events into the text itself, most regional chapters contain new sidebars entitled "Geography in the Making." Here, we discuss breaking news and current events pertinent to world regional geography. We encourage readers to look at this timely material. Additionally, we encourage students of world regional geography to link material gleaned from daily newspapers and newscasts to the contents of this book to fully appreciate—and understand—our changing world and how geography is always in the making.

Quote from Robert D. Kaplan, "The World in 2005," *The Atlantic Monthly*, March 2002, page 54.

and tribal membership was an important factor in the shifting allegiances among both Taliban and Northern Alliance members.

Global Terrorism

In many ways, the September 2001 terrorist attacks on the United States underscore the need to expand and redefine our conceptualization of globalization and geopolitics. While previous terrorist acts were usually connected to nationalist or regional geopolitical aspirations to achieve independence or autonomy, that was not the case with the attacks on the World Trade Center and the Pentagon. Unlike IRA terrorist bombings in Great Britain or those of the extremist wing of Basque nationalists in Spain, the September 2001 terrorism transcended conventional geopolitics as a relatively small group of religious fanatics attacked the symbols and superstructure of Western culture, power, and finance. Moreover, these attacks were a stark reminder of the increasingly close and now unpredictable interconnections between political activity, cultural identity, and the economic linkages that bind our contemporary world.

In geopolitical terms, global terrorism demands a new way of looking at the world to understand the attacks and to assess the profound implications of that terrorism. More to the point, many experts argue that global terrorism is both a product as well as an expression of globalization. Unlike earlier geopolitical conflicts, the geography of global terrorism is not defined by a war between well-established political states. Instead, the Al Qaeda terrorists appear to belong to a web of small, well-organized cells located in many different countries, linked in a decentralized network that provides guidance, financing, and political autonomy. This network has used the tools and means of globalization to its advantage. Members communicate instantaneously via cell phones and the Internet. Transnational members travel between countries quickly and frequently. The network's nefarious activities are financed through a complicated array of holding companies and subsidiaries that traffic in a range of goods, such as honey, diamonds, and opium. To recruit members and political support, the network feeds upon the unrest and inequities (real and imagined) resulting from economic globalization; and global terrorism attacks targets that symbolize the modern global values and activities it opposes. Even though the September 2001 attacks focused terror on the United States, the casualties and resultant damage were truly international. For example, citizens from more than 80 countries were killed in the World Trade Center tragedy alone. Additionally, the attacks may have pushed parts of the global economy from temporary economic slowdown to full-on recession. The devastating effects on U.S. and international airline companies as well as the loss of tourist revenue to Muslim countries such as Morocco and Egypt serve as examples.

The military and political responses to global terrorism also demand an expanded conceptualization of geopolitics. In 2001, for example, in Afghanistan, the military muscle of a nation-state superpower—the United States—was directed not at another nation-state but, rather, at a confederation of religious extremists, the Taliban and Al Qaeda. Military strategists refer to this kind of fighting as **asymmetrical warfare,** a term that aptly describes the differences between a superpower's military technology and strategy and the lower level technology and guerilla tactics used by Al Qaeda and the Taliban. Most super-power military strategists agree that the war on terrorism will be fought with political and battlefield asymmetry. Further, because the global context for the war on terrorism is a complex mosaic of religious, cultural, political, and economic undercurrents, there is debate in geopolitical circles about whether the United States can—or should—pursue its agenda as a member of a coalition of explicitly committed countries (Britain and Russia, for example), or whether it will be done essentially alone, or unilaterally. Many geopolitical thinkers posit that resolution of this coalition-unilateral issue will be decided by how the United States defines its goals for the war on terrorism. A global coalition seems likely if the United States sets a specific and limited goal such as the destruction of Al Qaeda. However, the broader goals discussed in Washington, namely the displacement of hostile governments in Iraq or North Korea or rebuilding nation-states in the Middle East, may have limited support from the world community.

Nation-states

A conventional starting point for examining political geography is the concept of the nation-state. The hyphen links two diverse concepts, the term *state,* which is a political entity with territorial boundaries recognized by other countries and internally governed by an organizational structure, with the term *nation,* which refers to a large group of people who share numerous cultural elements, such as language, religion, tradition, or simple cultural identity, and, more importantly, view themselves as forming a single political community. The **nation-state,** then, is ideally a relatively homogeneous cultural group with its own fully independent political territory. Nation-states, however, do not always fit neatly into the boundaries of actual states. In fact, perfect nation-state congruence is relatively rare. While some large cultural groups may consider themselves to be nations lacking recognized, self-governed territory (such as the Kurdish people of Southwest Asia), many, if not most, independent states include cultural and ethnic groups within their established boundaries that seek autonomy and self-rule (Figure 1.39). In Spain, for example, both the Catalans and the Basques think of themselves as forming separate, non-Spanish nations, and thus seek political autonomy from the central government. Uniform cultural (or national) homogeneity within any political unit is actually rare. On a world scale, out of the more than 200 different countries that now make up the global geopolitical fabric, only several dozen would qualify as true nation-states.

Although the nation-state may be considered the ideal political model, multinational states containing different cultural and ethnic groups are much more numerous than homogeneous ones. Micronationalism, moreover, or group identity with the goal of self-rule within an existing nation-state, has been on the rise for the last half century and remains a considerable source of geopolitical tension throughout the world.

▶ **Figure 1.39 A nation without a state**   Not all nations or large cultural groups control their own political territories or states. For example, the Kurdish people of Southwest Asia traditionally occupy a large cultural territory that is currently in four different political states: Turkey, Iraq, Syria, and Iran. As a result of this political fragmentation, the Kurds are considered minorities, and each of the these four political units attempts in its own way to limit (and even eliminate) Kurdish unity and power so as to weaken claims for an independent Kurdish state.

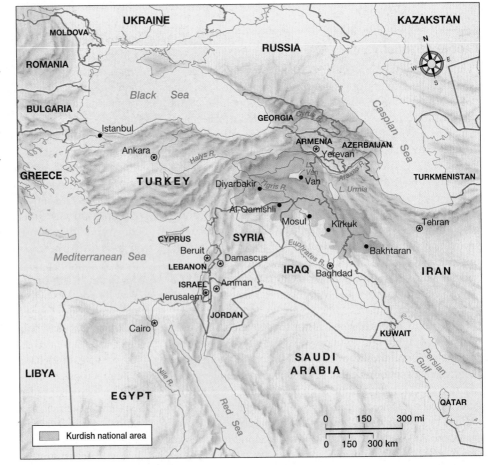

Kurdish national area

▲ **Figure 1.40 Basque separatism** The outspokenness and militancy of the Spanish Basques are just one expression of the ethnic separatism that is still common in contemporary Europe and, for that matter, the world. *(Alberto Arzoz/Panos Pictures)*

Centrifugal and Centripetal Forces

Cultural and political forces acting to weaken or divide an existing state are called **centrifugal forces**, since they pull away from the center. Many have already been mentioned: linguistic minority status, ethnic separatism, territorial autonomy, and disparities in income and well-being. Separatist tendencies in French-speaking Quebec (Canada) or the Basque region in Spain are good examples (Figure 1.40). Counteracting these dissipating forces are those that promote political unity and reinforce the state structure. These **centripetal forces** could include a shared sense of history, a need for military security, an overarching economic structure, or simply the advantages that come from a larger political apparatus that builds and maintains the infrastructure of highways, airports, and schools.

The overriding question in much of the world is whether centrifugal forces, which often remain modest and controlled, will increase in strength and furor so that they overcome the centripetal forces, leading to new, smaller independent units. Also important is whether this process takes place through violent struggle or peaceful means. Given the current widespread nature of these internal tensions, there is much evidence to suggest that separatist struggles will continue to dominate regional geopolitics for decades to come.

Boundaries and Frontiers

A key component of a state's definition is that its territory is demarcated by borders recognized by a large number of other states. Although these boundaries may be one of the most recognizable features of a world map, the issue of agreed-upon borders is often complicated and contentious. Border wars were frequent in the twentieth century, and if the current trend toward an increasing number of independent states continues, boundary disputes may become an even more important facet of the geopolitical landscape.

A cursory glance at the world political map reminds us that boundaries come in many forms. Some are drawn along physical features such as mountain ridges, rivers, or coastlines. Al-

though historically it was thought that using clearly marked landscape features would lessen border tensions, this is not always the case. The Rio Grande between Texas and Mexico is notoriously porous to illegal boundary crossings. Similarly, the sinuous coastline of southern Spain offers North Africans ample opportunity for illegal entry into Europe. **Ethnographic boundaries,** which follow cultural traits such as language or religion, have long been a model for international politics, since they complement the ideal of the nation-state. Many boundaries in Europe were adjusted ethnographically after World War I. Similarly, attempts were made to define the new states of Bosnia and Serbia with ethnographic boundaries after the dissolution of Yugoslavia in the early 1990s. However, there is not always congruence between ethnic territory and political boundaries (Figure 1.41).

In contrast, **geometric boundaries** are perfectly straight lines drawn without regard for physical or cultural features that usually follow a parallel of latitude or a meridian of longitude. Historically, geometric boundaries were often drawn by colonial powers as a convenient way of quickly demarcating spheres of influence; the use of the 49th parallel to draw the western part of the boundary between Canada and the United States is an excellent example. Other illustrations can be found in the arid lands of North Africa. However, when space is divided by geometric borders without consideration for the diverse cultural fabric on the landscape, border wars and even civil wars often result.

Colonialism and Decolonialization

One of the overarching themes in world regional geography is the waxing and waning of European colonial power over much of the world. **Colonialism** refers to the formal establishment of rule over a foreign population. A colony has no independent standing in the world community, but instead is seen only as an appendage of the colonial power. Generally speaking, the main period of colonialization by European states was from 1500 through the mid-1900s, though even today a few colonies remain (Figure 1.42).

Decolonialization refers to the process of a colony's gaining (or regaining) control over its territory and establishing a separate, independent government. As was the case with the Revolutionary War in the United States, this process often begins as a violent struggle. As wars of independence became increasingly prevalent in the mid-twentieth century, some colonial powers recognized the inevitable and began working toward peaceful disengagement from their colonies. In the late 1950s and early 1960s, for example, Britain and France granted independence to most of their former African colonies, often after a period of warfare or civil unrest. This process was nearly completed in 1997 when Hong Kong was peacefully restored to China by the United Kingdom. But decades and even centuries of colonial rule are not easily erased; hence, the impress of colonialism is still found in the new nations' governments, education, agriculture, and economies. While some countries may enjoy special status with, and receive continued aid from, their former colonial masters, others remain disabled and disadvantaged because of a much-reduced resource base.

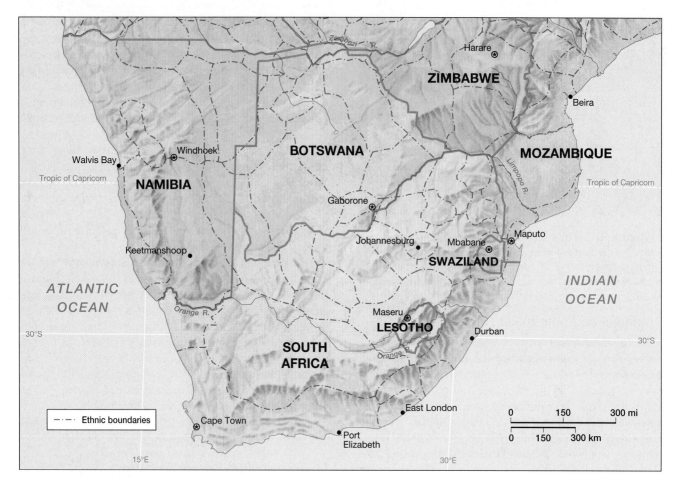

▲ **Figure 1.41 Ethnic and state boundaries**   This map shows the lack of congruence between ethnic group territory and modern political borders in southern Africa. Instead of adhering to enthographic boundaries, several of these political boundaries were drawn geometrically (such as the Nambia–Botswana border) or along physical features such as rivers and mountains (the South Africa–Botswana border).

Some scholars regard the continuing economic ties between certain imperial powers and their former colonies as a form of exploitative neocolonialism. On the other hand, some remaining colonies, such as the Dutch Antilles in the Caribbean, have made it clear that they have no wish for independence, finding both economic and political advantages in continued dependency. Because the consequences of colonialism differ greatly from place to place, the final accounting is far from complete (Figure 1.43).

International and Supranational Organizations

As another attribute of globalization, international organizations increasingly link two or more states for the express purpose of military, economic, political, or environmental cooperation. At the top of this organizational chart, of course, is the United Nations (UN), which has lately become a major peacekeeping force on the international scene. Virtually all independent countries belong to the UN, although a few, such as Switzerland, do not. Other well-known organizations are the Organization of Petroleum Exporting Countries (OPEC), which includes 12 oil-producing countries around the world; the North Atlantic Treaty

Organization (NATO), a military alliance linking the United States and Canada to western and central Europe (including Turkey); and the Association of South-East Asian Nations (ASEAN), a regional forum that encompasses all of Southeast Asia. Regional trade alliances are also a common part of the global landscape, with the North American Free Trade Association (NAFTA) linking three North American countries—Canada, the United States, and Mexico—into a free-trade area.

Different from an international organization is one that is *supranational*. This is defined as an organization of states linked by a common goal, within which individual state power may be reduced to achieve the organization's goals. Put differently, the higher goals of the supranational organization may often necessitate a reduction in state sovereignty by individual members (Figure 1.44). Perhaps the best example is that of the European Union, or EU, which began as an economic union, and which has expanded its powers so that it now holds elections, has a parliament, promotes a common monetary currency, and works toward reducing border formalities between member states. These EU goals remain controversial, however, as discussed in Chapter 8.

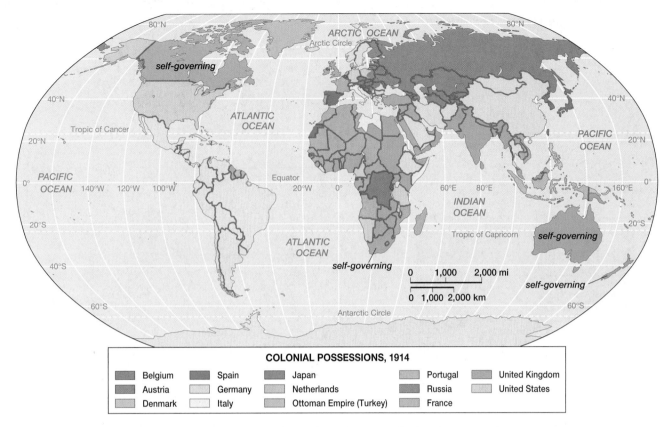

COLONIAL POSSESSIONS, 1914

Belgium	Spain	Japan	Portugal	United Kingdom
Austria	Germany	Netherlands	Russia	United States
Denmark	Italy	Ottoman Empire (Turkey)	France	

▲ **Figure 1.42 The colonial world, 1914** This world map shows the extent of colonial power and territory just prior to World War I. At that time, most of Africa was under colonial control, as were Southwest Asia, South Asia, and Southeast Asia. Australia and Canada were very closely aligned with England. Also note that in Asia, Japan controlled colonial territory in Korea and northeastern China, which was known as Manchuria at that time.

Economic and Social Development: The Geography of Wealth and Poverty

The pace of global economic change and development has accelerated dramatically in the last decade. As a result, we now talk about global assembly lines, transnational corporations, commodity chains, and international electronic offices. Few regions of the world are untouched by these changes. The overarching question, however, is whether the positive changes of economic globalization outweigh the negative. While answers are varied, sometimes elusive, and often inconclusive, an important first step in world regional geography is to link economic change to social development.

Economic development is commonly accepted as desirable because it generally brings increased prosperity to individuals, regions, and nations, which also, by conventional thinking at least, usually translates into social improvements such as better health care, improved education systems, and

◀ **Figure 1.43 Colonial vestiges in Vietnam** The red star flag of communist Vietnam flies in front of the Hotel de Ville in Ho Chi Minh City (formerly Saigon). This juxtaposition of the contemporary government's symbol and an artifact of the French colonial period captures the process of decolonialization and independence. *(Catherine Karnow/Woodfin Camp & Associates)*

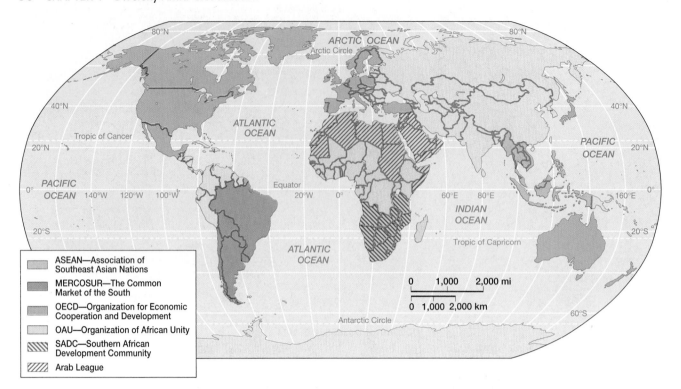

▲ **Figure 1.44 International and supranational organizations**   Countries increasingly are joining international or supranational organizations that have specific goals and agendas. Some international organizations exist for mutual economic and trade advantages, which is the purpose of MERCOSUR in Latin America. Others exist to work out political problems, which is the case for ASEAN in Southeast Asia. Supranational organizations, such as the OAU and the Arab League, often require members to surrender some sovereignty to pursue certain cultural, economic, and political objectives.

more enlightened labor practices. The operative assumption is that, as the economy of an area develops, so will its social infrastructure. One of the more troubling expressions of recent economic growth, however, has been the geographic unevenness of prosperity and infrastructural improvement. While some regions prosper, others languish and, in fact, fall behind. As a result, the gap between rich and poor regions has increased. This geographic unevenness in development, prosperity, and social infrastructure is a characteristic signature of the early twenty-first century. According to the World Bank, more than 1.3 billion people live on less than a dollar a day, about 60 percent of them in Sub-Saharan Africa and South Asia.

Regional inequities are also problematic because of their inseparable interaction with political, environmental, and social concerns. Political instability and civil strife within a nation, for example, are often driven by the economic disparity between a poor periphery and an affluent, industrial core. In those backwaters, poverty, social tensions, and environmental degradation often drive civil unrest that ripples through the rest of the country. The World Bank's *World Development Reports* document these inequities. In China's Gansu province, for example, per capita income is almost 50 percent lower than the national average, and partly because of poor environmental conditions, most of Gansu's population lives in poverty while many other regions of China prosper. Other examples of internal unevenness are common: Italy, Poland,

Mexico, Kenya, Brazil, and India all have striking geographical disparities in wealth. This theme of the unevenness of social and economic development is a major story line in the regional chapters of this book.

More- and Less-Developed Countries

Until recently, economic development was centered in North America, Japan, and Europe, while most of the rest of the world remained gripped in poverty. This uneven distribution of power led scholars to devise a **core-periphery model** of the world economy. According to this model, the U.S. and Canada, western Europe, and Japan constitute the global economic core of the North, while most of the areas to the south make up a less-developed global periphery. Although an oversimplification of sorts, this core-periphery dichotomy does contain some truth. All the G-7 countries—the exclusive club of the world's richest nations made up of the United States, Canada, France, England, Germany, Italy, and Japan—are located in the northern hemisphere core (for certain meetings, the G-7 includes Russia, at which time the group becomes the G-8) Another assumption of this core-periphery model is that the developed core achieved its wealth primarily by exploiting the periphery, either through historical colonial relationships or through more recent economic imperialism (Figure 1.45).

Today, for example, much is made of "North–South tensions," a phrase implying that the rich and powerful countries

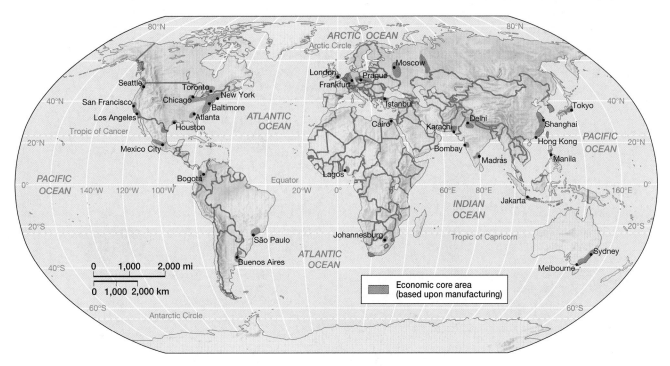

▲ **Figure 1.45 World economic core areas**   Within most world regions, there are centers of economic activity where manufacturing and business are clustered. Often these are the more prosperous and most urbanized areas within the region. Outlying areas within the region may lie in the shadow of these robust economic core areas, suffering from the consequences of underdevelopment. This economic diversity, both on a global scale and within regions (and even within states), has given rise to the concept of core-periphery interactions. One assumption of this model is that economic cores prosper only by exploiting poorer periphery territories.

of the Northern Hemisphere are at odds with poor and less powerful countries to the south. This dichotomy is based upon the historical arrangement of northern powers—mainly European—dominating and exploiting southern colonies in Latin America, Sub-Saharan Africa, and Southeast Asia. Over the past several decades, however, the global economy has grown much more complicated. A few former colonies of the south, most notably Singapore, have become very wealthy. Meanwhile, the northern countries of the former Soviet Union have experienced an extremely rapid economic and political decline, and are therefore now sometimes placed within the zone of global poverty. Australia and New Zealand, moreover, never fit into the neat North–South division. For these reasons, many global experts conclude that the term *North–South* is antiquated and should be avoided.

Another term, the *Third World,* is often used to refer to the developing world. This phrase carries connotations of a low level of economic development, unstable political organizations, and a rudimentary social infrastructure. Originally, the term was a product of Cold War jargon and was used to describe those countries that were independent and not allied with either the capitalist and democratic First World or communist Second World superpowers. Since the Second World of the Soviet Union and its eastern European allies no longer exists, the concept of a Third World must be thrown into question. Also significant is the fact that a number of so-called

Third World countries, such South Korea, Taiwan, and Singapore, have experienced such rapid growth over the past several decades that they are no longer underdeveloped. In this book, therefore, we avoid terms such as *Third World* and *First World* and use relational terms that capture the complex spectrum of economic and social development, namely *more-developed country* (MDC), and *less-developed country* (LDC). A world map of income per person shows the spatial patterns of rich, moderate, and poor countries (Figure 1.46).

Indicators of Economic Development

The terms *development* and *growth* are often used interchangeably when referring to international economic activities. There is, however, value in keeping them separate. *Development* has both qualitative and quantitative dimensions. A dictionary definition uses phrases such as "expanding or realizing potential; bringing gradually to a fuller or better state." When we talk about economic development, then, we usually imply structural changes, such as a shift from agricultural to manufacturing activity, with accompanying changes in the uses of labor, capital, and technology. Along with these changes are assumed improvements in standard of living, education, and even political organization. The structural changes experienced by Southeast Asian countries such as Thailand or Malaysia in the last several decades illustrate this process.

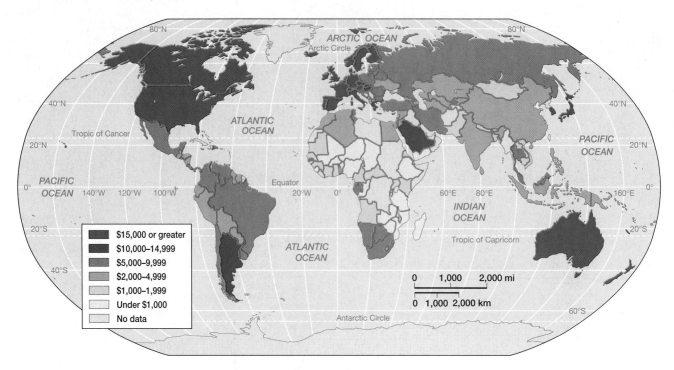

▲ **Figure 1.46 World GNI per capita** Gross national income, or GNI (formerly referred to as GNP), is one of the best measures to show the geographic pattern of rich and poor countries. Whereas 75 percent of the world's population fall into the lower categories, only 10 percent are in the highest category. *(Data from the* World Bank Atlas, *2001)*

Growth, in contrast, is simply the increase in size of a system. The agricultural or industrial output of a country may grow, as it has for India in the last decade, and this growth may—or may not—have positive implications for development. Many expanding economies have experienced increased poverty. When something grows, it gets bigger; when it develops (sticking with the dictionary meaning, at least), it improves. Critics of the world economy are often heard to say

that we need less growth and more development. This critique should not be dismissed as unrealistic or naïve.

In this book, each of the regional chapters includes a table of economic development indicators (Table 1.2). A few introductory comments are necessary to put these data in proper perspective. The traditional measure for the size of a country's economy is the value of all final goods and services produced within its borders (gross domestic product, or

TABLE 1.2 *Economic Indicators of the Ten Largest Countries*

Country	Total GNI (Millions of $U.S., 1999)	GNI per Capita ($U.S., 1999)	GNI per Capita, PPP* ($Intl, 1999)	Average Annual Growth % GDP per Capita, 1990–1999
China	979,894	780	3,550	9.5
India	441,834	440	2,230	4.1
United States	8,879,500	31,910	31,910	2
Indonesia	125,043	600	2,660	3
Brazil	730,424	4,350	6,840	1.5
Pakistan	62,915	470	1,860	1.3
Russia	328,995	2,250	6,990	−5.9
Bangladesh	47,071	370	1,530	3.1
Japan	4,054,545	32,030	25,170	1.1
Nigeria	31,600	260	770	−0.5

*Purchasing power parity

Source: The World Bank Atlas, *2001.*

GDP) plus the net income from abroad. These two factors constitute the **gross national income**, or **GNI** (formerly referred to as gross national product, or GNP). However, although widely used, GNI can be an incomplete and even misleading economic indicator because it ignores nonmarket economic activity, such as bartering or household work, and also because it does not take into account ecological degradation or depletion of natural resources, which may constrain future economic growth. For example, if a country clear-cut its forest, which could limit future growth, this act would nevertheless increase the GNI for that year. Further, GNI does not reveal the distribution of wealth within a country, the cost of living, or the social organizations or infrastructure supporting the economy. Diverting educational funds to purchase military weapons might increase a country's GNI in the short run, but the economy would likely suffer in the future because of its less-educated population. For this reason, it is important to couple GNI data with an array of other economic and social indicators. A first step is to divide the GNI by the country's population, which produces a **GNI per capita** figure. For example, according to World Bank data, the annual GNI for the United States in 1999 was about $8.8 billion. Dividing that figure by the population of around 280 million results in a GNI per capita of $31,910. Because of the ratio between population and GNI, similar figures will result for countries that have both a smaller GNI and a smaller population. Japan, for example, has a GNI about half that of the United States at $4.0 billion. However, because that country also has a smaller population (127 million), its GNI per capita is slightly higher at $32,030.

An important qualification to these GNI per capita data is the concept of adjustment through **purchasing power parity (PPP)**, which gives us a comparable figure for a standard "market basket" of goods and services purchased with the local currency (Figure 1.47). Until recently, GNI data were usually based on the market exchange rate for a particular country's national currency as compared to the U.S. dollar, even though the GNI data might be inflated or undervalued depending on the strength or weakness of that currency. If the Japanese yen were to fall against the dollar overnight as a result of currency speculation, for example, the official figure for Japan's GNI would correspondingly drop, despite the fact that Japan had experienced no real decline in economic output. Because of these problems, the PPP adjustment can provide a more accurate sense of the local cost of living, and these data can also be used to adjust GNI for currency inflation. For example, when Japan's GNI per capita is adjusted for PPP, which takes out the inflationary factor, the figure is $25,170.

Lastly, the **economic growth rate** (the annual rate of expansion for GNI) can also be distorted by inflated currencies, but can be standardized by PPP. To illustrate, China's economic growth rate averaged nearly 9 percent per year during the 1980s and early 1990s based upon conventional GNI data. However, when adjusted for PPP-based currency, this annual growth rate is closer to 5 percent. Increasingly, world organizations are using PPP-based measures of economic activity for making international comparisons. In this book, whenever possible, we use PPP-adjusted data in the economic development tables in the regional chapters.

Indicators of Social Development

Although economic growth is a major component of development, there is equal interest in the conditions and quality of human life in this rapidly changing world. While the standard assumption is that economic development and growth will spill over into the social infrastructure so that there will also be a concomitant improvement in public health, working conditions, gender equality, and education, this assumption must be supported with objective data. For that reason, we include several measures of social development in the regional chapters (Table 1.3).

◀ Figure 1.47 Village food market The concept of purchasing power parity (PPP), which is a common market basket of goods, is used to generate a sense of the true cost of living for different countries. This outdoor produce market is in Chichicastenango, Guatemala. *(Rob Crandall/Rob Crandall, Photographer)*

TABLE 1.3 *Social Indicators and Status of Women of the Ten Largest Countries*

Country	Life Expectancy at Birth		Under Age 5 Mortality (per 1,000)		Percent Illiteracy (Ages 15 and over)		Female Labor Force Participation (% of total, 1999)
	Male	Female	1980	1999	Male	Female	
China	69	73	65	37	9	25	45
India	60	61	177	90	32	56	32
United States	74	80	15	8	0	0	46
Indonesia	65	70	125	52	9	19	41
Brazil	65	72	81	40	15	15	35
Pakistan	60	61	161	126	41	70	28
Russia	59	72	—	20	0	1	49
Bangladesh	59	59	211	89	48	71	42
Japan	77	84	11	4	0	0	41
Nigeria	52	53	196	151	29	46	36

Sources: Population Reference Bureau, Data Sheet, *2001, Life Expectancy (M/F); The World Bank,* World Development Indicators, *2001, Under Age 5 Mortality Rate (1980/99) and Percent Illiteracy (ages 15 and over); The World Bank Atlas, 2001, Female Participation in Labor Force.*

One indicator of social development is *life expectancy,* the average length of life expected at birth for a hypothetical male or female based upon national death statistics. A large number of social factors influence life expectancy, such as availability of health services, nutrition, prevalence (or absence) of disease, accident rates, sanitation, and homicide rates. Additionally, life expectancy is affected by biological factors, which may explain why women usually live longer than men. When social and biological factors are folded together, life expectancy data provide important insights into local conditions. For the few countries in which men live longer than women, for example, one can deduce that the social position of women remains quite low.

Life expectancy figures vary widely between countries in different world regions, but they have generally been improving over time. For the world as a whole, life expectancy in 1975 was 58 years, whereas today it is 66. Over the same period, Africa's figure has gone from 46 to 55, although in some African countries longevity has dropped recently because of the AIDS epidemic. In Russia, life expectancy has also fallen lately—especially for men—with the deterioration of economic conditions. Unfortunately, national life expectancy data often mask differences within a country, such as the wide disparity between rich and poor, differences between ethnic groups, or even discrepancies between specific regions. Within the United States, for example, the gap in life expectancy between African Americans and whites is 7 years, although this is an improvement over the 15-year gap recorded in 1900. Local-scale data also tell important stories. For example, partly because of the high homicide rates for African American males in major urban areas such as New York City, Detroit, and Chicago, their calculated life expectancy is shorter than that of men in Bangladesh, one of the world's poorest countries.

Under age 5 mortality, which is the measure of the number of children who die per 1,000 persons, is another important indicator of social conditions. This figure gives insight into

conditions such as food availability, health services, and public sanitation. In the first 5 years of life, a child moves from the personal protection and nurturing of his or her mother into a larger social environment; these first 5 years are a time of high vulnerability to nutritional deprivation, infection, disease, accidents, and other human tragedies. Child mortality data are given for two points in time, 1980 and 1999, to indicate whether there has been an improvement in child mortality rates over that period.

Adult illiteracy rates, provided for both men and women, provide two kinds of information relevant to the social development of a country. In general, high rates of illiteracy are usually products of low levels of national investment in education, which does not bode well for participation in a globalizing world. Second, gender disparities can also be evident in these data. In some countries, we see a striking discrepancy between men and women, often with much higher rates of illiteracy for females than for males. This suggests that there are strong cultural, social, or economic constraints on the education of girls and women. The implications of low female literacy rates are many. For example, there is a demonstrated connection between acceptance of family planning and female education; usually a higher birthrate accompanies high female illiteracy. The role that women play in a country's economy is also closely linked to their levels of education and literacy (Figure 1.48).

In many countries, women play a prominent role in economic life; thus, data on the percentage of *females in the labor force* are important (Figure 1.49). However, three points must qualify these data. First, much of women's work has traditionally been undervalued and hidden from the national census because it is considered part of the unpaid sector of household and subsistence. Therefore, the long hours women spend gathering firewood, drawing water, cooking, tending gardens and fields, and raising children are not tabulated as "work" or "employment" and do not appear in the labor force

▲ **Figure 1.48 Women and literacy**   Gender disparities are often found in national illiteracy rates. When there is a large difference between the percentage of males and females characterized as illiterate, often this can be explained by cultural, social, or economic constraints placed on the education of girls and women.   *(Ron Giling/Panos Pictures)*

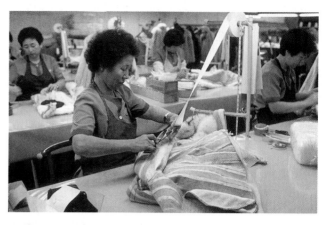

▲ **Figure 1.49 Women in the workforce**   Women workers have become an important component of the modern globalized economy. However, women are often paid less than men, and usually don't have the job security to prevent them from being the first to be fired or laid off as the global economy fluctuates. *(Nathan Benn/Woodfin Camp & Associates)*

statistics (men also perform work in the unpaid domestic sector, but usually to a lesser degree than women). Second, many women work in the "informal" economic sector, earning subsistence wages selling vegetables or handicrafts in village markets, cleaning houses, or doing other forms of temporary labor. National employment data rarely include such important forms of work. Third, even when employed in the formal

economic sector, such as on a factory assembly line, women are usually paid less than men and are also disproportionately concentrated in jobs with little security or chance of advancement (see "Women in the Workforce: The First to Go as Asian Economies Shrink").

Even though women may comprise the majority of workers in many developing countries, the quality and conditions

WOMEN IN THE WORK FORCE   The First to Go as Asian Economies Shrink

As globalization led to booming Asian economies in the 1990s, more and more women were brought into the workforce. Some women found jobs as assembly workers in new offshore factories; others were given managerial responsibilities in jobs previously reserved for males; and still others were hired as office ornaments—"flowers of the workplace," as they're called in Japanese.

Later, though, with fiscal crises, shrinking economies, and global recession battering many Asian countries, women were the first to be fired. As a result, the Asian financial crisis was particularly disastrous for women. "The impact on women and girls is just catastrophic," said the American envoy to the Asian Development Bank in the Philippines. From the high-rise office suites of Japan to the peasant villages of Thailand, the bottom line is that resources have also been allocated disproportionately to men and boys. During the good times of the Asian boom, there were plenty of leftovers for women; but in lean times, it is the women who pay the most.

Much of this discrimination comes from the revered status of the male in Asian cultures, with companies and firms attempting to minimize the pain of layoffs in a male-dominated family structure. "In a crisis, first of all we would have to fire the women," said the owner of a Japanese construction company. "We would retain men because they are the pillar of household earnings. . . . I would never want to dismiss the main income-

earner because that might destroy a whole family." When the Japanese economy first slowed in the 1990s, the job market for women was referred to as the *hyogaki* or "ice age"; now Japanese women call it the *cho-hyogaki* or "super ice age."

Even those females not yet in the workforce are affected. Young girls are pulled out of school in rural Indonesia because families can't afford the $2 charge for uniforms. "People say it's better for girls to stay at home, so that they can save money for the boys," said a 17-year-old girl on the island of Sumba. Indonesia girls are 6 times more likely than boys to drop out of school before completing the fourth grade. Once they drop out, rarely do they go back, and many of these dropout girls are then pressured or forced into prostitution.

The problem is not limited to rural villages. Even in the cities of South Korea and Japan, families are pulling their daughters out of private cram schools in order to allocate the tuition money to the boys who will carry on the family name. Is that fair? Many Asian women appear resigned to the situation. "It would be best if everyone could get opportunities," said one South Korean woman, "but I think it's right that a son gets the most attention."

Source: Adapted from "With Asia's Economies Shrinking, Women Are Being Squeezed Out," by Nicholas D. Kristof. *New York Times*, June 11, 1998.

of their work often mimic the early and unpleasant days of European industrialization, when people worked long hours under dangerous conditions for low wages with little job security. While it took Europe almost a century to enact legislation to improve these working conditions, it seems reasonable to expect that the economic, political, and social currents of the early twenty-first century might move faster to produce humane working conditions.

The Vision of Sustainable Development

As the environmental and social costs of globalized economic development become increasingly apparent, there is widespread discussion of the notion of **sustainable development** by international aid workers, development planners, environmentalists, and community groups. This can be defined simply as an agenda of economic change and growth that seeks a balance with issues of environmental protection and social equity so that the short-term needs of contemporary society do not compromise the future. Some reduce the definition to its essence by referring to sustainable development as "intergenerational equity." For example, instead of wholesale harvesting of Indonesian rain forests as a way of amortizing current foreign debt, a sustainable development plan would emphasize the paced, longer-term use of this valuable resource so that the economic benefits might be spread out in a more equitable and self-sustaining manner. Another illustration would be a country's investment in female education, which, even if such an investment diverted money from contemporary social programs, would probably have positive future consequences for both economic and social development.

⊕ Conclusion

As the world becomes increasingly interdependent, globalization is driving a fundamental reorganization of economies and cultures through such vehicles as trade agreements, supranational organizations, military alliances, and cultural exchanges. At one level, the world seems to be converging and becoming more homogeneous because of globalization. Nonetheless, there is still great diversity. While some areas prosper, others stagnate; as global TV promotes a common world culture, small group identity becomes increasingly important; as nation-states become world players through supranational agreements, separatist groups seek autonomy and independence. This tension and interplay between globalization and diversity—and between the global, the regional, and the local—gives world regional geography its primary focus.

Our description and analysis of each world region are organized around 5 issue-oriented themes: environmental geography; population and settlement; cultural coherence and diversity; geopolitical framework; and economic and social development. Although the specific issues differ from region to region, certain general themes link the chapters. Environmental issues, be they at the global or local scale, affect all world regions. Examples include global climate change, water pollution, air pollution, forest destruction, and the protection of biodiversity.

In most regions of the developing world, population and settlement issues revolve around four issues: rapid population growth, family planning (or its absence), migration to new centers of economic activity (both within and outside the region), and rapid urbanization (combined with the concern about whether cities can keep pace with the ever-increasing demand for jobs, housing, transportation, and public facilities).

A major theme in our treatment of global cultural geography is the tension between the forces of cultural homogenization resulting from globalization and the countercurrents strengthening small-scale cultural and ethnic identity. Throughout the world, small groups are setting themselves apart from larger national cultures with renewed interest in ethnic traits, languages, religion, territory, and shared histories. This is not merely a matter of colorful folklore and revitalized local customs; in many parts of the world, cultural diversity is translated into politics with outspoken calls for regional autonomy or separatism.

Because of cultural factionalism, the geopolitical issues of many of the world's regions are dominated by matters of global terrorism, ethnic strife and territorial disputes within nation-states, border tensions with neighbors of different cultural traditions (Pakistan and India are examples), and new forms of military alliances and agreements to deal with the ever-changing nature of national security. The ideological differences between former Cold War global superpowers now seems eclipsed by more localized internal and regional tensions, which often involve the reaction, response, alignment, and alliances of external countries and organizations.

Last, the theme of economic and social development is dominated by one issue: the increasing disparity between the rich and the poor, between countries and regions that already have wealth—and are getting even richer through globalization—and those that remain impoverished. Economic inequities are found on every scale from global to local, including pockets of poverty within the world's richest countries. On a global scale, the core of developed countries is surrounded by a periphery of less-developed nations aspiring to the same success; on the smaller scale of regions and individual nations, vibrant centers of economic development are linked to the globalized world, while more remote backwater areas languish and stagnate. Often the same blatant inequities in social development, schools, health care, and working conditions accompany these disparities in wealth.

Understanding this complex world is a challenging, yet necessary, task. Think of this book as a beginning rather than an end, as a way to gain skills in using the conceptual tools of geography to engender critical thinking about the complicated issues and themes of world regional geography replete with diversity amid globalization.

⊕ Key Terms

areal differentiation (page 12)
areal integration (page 12)
asymmetrical warfare (page 33)
bubble economy (page 10)
centrifugal forces (page 35)
centripetal forces (page 35)
colonialism (page 35)
core-periphery model
    (page 38)
cultural imperialism (page 26)
cultural landscape (page 13)
cultural nationalism (page 27)
cultural syncretism or
    hybridization (page 27)

culture (page 25)
decolonialization (page 35)
demographic transition
    (page 19)
economic convergence
    (page 7)
economic growth rate
    (page 41)
ethnic religion (page 31)
ethnographic boundaries
    (page 35)
geometric boundaries
    (page 35)
globalization (page 1)

gross national income (GNI)
    (page 41)
gross national income (GNI)
    per capita (page 41)
lingua franca (page 30)
nation-state (page 34)
natural increase (page 18)
overurbanization (page 23)
purchasing power parity
    (PPP) (page 41)
region (page 13)
squatter settlements (page 23)
sustainable development
    (page 44)

sweatshop (page 8)
total fertility rate (TFR)
    (page 19)
transnational firm (page 5)
universalizing religion
    (page 30)
urban form (page 23)
urban primacy (page 23)
urban structure (page 23)
urbanized population
    (page 22)

⊕ Questions for Summary and Review

1. Why is globalization a signature of twenty-first century, and what has facilitated this?

2. What role did the end of the Cold War play in furthering globalization?

3. Summarize the major assets and liabilities of globalization as discussed in this chapter.

4. What geographic concept(s) best capture(s) the appearance of places and the "look of the land"?

5. Give examples of how changing scale from global to local might change an analysis of areal differentiation and integration for a specific area.

6. If the rate of natural increase (RNI) for the world remains consistent at 1.3 percent per year, what will the planet's population be in 10 years?

7. Explain the four stages of the demographic transition. How do birthrates and death rates differ in each stage? At what point is RNI highest?

8. Why is there an association between urbanized population rates and economic development in different countries of the world? Under what conditions would this association not hold?

9. What are the expressions and implications of overurbanization for cities in the developing world?

10. Why is it difficult to define *culture* in this globalized world? What conceptual and geographic solutions are helpful in solving this problem?

11. What are the differences between language families, branches, and groups? Give examples by noting their geographic distribution in the world.

12. Using the concepts presented in the section on "Geopolitical Framework," explain what happened to the "New World Order."

13. Using a world map or atlas, come up with five examples each of ethnographic and geometric boundaries. Is there any clue about problems associated with these different borders?

14. What are the shortcomings of GNI? How are these rectified?

15. Why are adult illiteracy rates a good indicator of social development?

16. What are the strengths and shortcomings of sustainable development in a globalized world?

⊕ Thinking Geographically

1. Select an economic, political, or cultural activity in your city, and discuss how it has been influenced by globalization.

2. Choose a specific country (or region) of the world to examine the benefits and liabilities of globalization. Remember to look at different facets of globalization, such as the environment and cultural cohesion and conflict, as well as the economic effects on different segments of the population.

3. What natural hazards create the most problems in your local area? What groups are most (and least) vulnerable to these problems? How have these hazards and vulnerabilities changed over the last several decades?

4. Using the concepts of areal integration and city systems, discuss the global links between your city and other urban areas, both nationally and internationally.

5. How would you characterize your location in terms of regions? That is, in what physical, economic, and cultural regions are you located?

6. How important is migration, both nationally and internationally, to your locality? Using the concepts of push and pull forces, along with that of informational networks, discuss why people leave and arrive, how they get jobs, and how many find housing in your area.

7. Compare the issues and problems with urban growth in your or a nearby city to that of a similarly sized city in a foreign country. Think, too, about how urban structure and form might be similar and different in these two cities.

8. Discuss the cultural geography of your area, noting the distribution, landscapes, and interaction associated with different cultural groups.

9. Drawing upon information in current newspapers, magazines, TV, and the Internet, apply the concepts of cultural imperialism, nationalism, and syncretism to a region or place experiencing cultural tensions.

10. Choose a large nation-state of interest and elucidate the different centrifugal and centripetal forces within that country. Based upon your findings, speculate on conditions in that country in 10 years.

11. Using the concept of ethnographic borders as a focal point, critique the way boundaries have been redrawn recently in the former Yugoslavia. Note those areas where the new boundaries seem to work well, contrasted with areas where there are still ethnic tensions.

12. Apply the core-periphery concept to a country of your choice by delimiting what you think to be cores of economic development and contrasting them with peripheries of much lower development. If possible, collect and analyze data that test your findings.

13. Using the tables of social indicators in the regional chapters of this book, identify traits shared by countries in which there is a high percentage of female illiteracy. What general conclusions result from your inquiry?

14. Apply and critique the definition of sustainable development to these projects: (a) a large dam and hydroelectric project in the Amazon; (b) clear-cut timber harvesting in British Columbia, with logs sold to Japanese lumber firms; (c) intensive usage of chemical fertilizer and irrigation water on alfalfa agriculture in the Colorado high plains; (d) agroecology in the Amazon rain forest.

⊕ Bibliography

Agnew, John A. 1998. *Geopolitics: Re-Visioning World Politics.* London and New York: Routledge.

Anderson, Benedict. 1983. *Imagined Communities: Reflections on the Origin and Spread of Nationalism.* London: Verso.

Anderson, Sarah, and Cavanagh, John. 1999. *Field Guide to the Global Economy.* New York: New Press.

Barber, Benjamin R. 1995. *Jihad vs. McWorld.* New York: Times Books.

Burtless, Gary, Lawrence, Robert Z., Litan, Robert E., and Shapiro, Robert J. 1998. *Globaphobia: Confronting Fears About Open Trade.* Washington DC: Brookings Institution and Progressive Policy Institute.

Buttimer, Anne. 1998. "Close to Home: Making Sustainability Work at the Local Level." *Environment* 40(3), 12–40.

Castells, Manuel. 1996. *The Rise of Network Society.* Cambridge, MA: Blackwell Publishers.

Cohen, Saul. 1991. "Global Geopolitical Change in the Post-Cold-War Era." *Annals of the Association of American Geographers* 81, 551–80.

Cosgrove, Denis. 1994. "Contested Global Visions: One-World, Whole Earth, and the Apollo Space Photographs." *Annals of the Association of American Geographers* 84, 270–94.

Cox, Kevin R., ed. 1997. *Spaces of Globalization: Reasserting the Power of the Local.* New York: Guilford Press.

Dickinson, Robert E. 1969. *The Makers of Modern Geography.* New York: Frederick A. Praeger.

Friedman, Thomas L.1999. *The Lexus and the Olive Tree.* New York: Farrar, Straus & Giroux.

Fukuyama, Francis. 1992. *The End of History and the Last Man.* New York: Avon.

Gallopin, Gilberto C., and Raskin, Paul. 1998. "Windows on the Future: Global Scenarios and Sustainability." *Environment*, 40(3) 6–31.

Garten, Jeffrey E., ed. 2000. *World View: Global Strategies for the New Economy.* Cambridge, MA: Harvard Business School Press.

Gilbert, A. 1988. "The New Regional Geography in English and French-Speaking Countries." *Progress in Human Geography* 12, 208–28.

Gilpin, Robert. 2000. *The Challenge of Global Capitalism: The World Economy in the 21st Century.* Princeton, NJ: Princeton University Press.

Gray, John. 1998. *False Dawn: The Delusions of Global Capitalism.* New York: The New Press.

Gupta, A., and Ferguson, J. 1992. "Beyond 'Culture': Space, Identity, and the Politics of Difference." *Cultural Anthropology* 7, 6–23.

Huntington, Samuel P. 1996. *The Clash of Civilizations and the Remaking of World Order.* New York: Simon & Schuster.

Jameson, Fredric, and Miyoshi, Masao, eds. 1998. *The Cultures of Globalization.* Durham, NC: Duke University Press.

Johnston, R. J., Taylor, Peter J., and Watts, Michael, eds. 1995. *Geographies of Global Change: Remapping the World in the Late 20th Century.* Cambridge, MA: Blackwell.

Kane, Hal. 1995. *The Hour of Departure: Forces That Create Refugees and Migrants.* Washington, DC: Worldwatch Institute.

Kaplan, Robert D. 2000. *The Coming Anarchy: Shattering the Dreams of the Post Cold War.* New York: Random House.

Katzner, Kenneth. 1995. *The Languages of the World.* London: Routledge.

Knox, Paul, and Marston, Sallie. 1997. *Human Geography: Places and Regions in Global Context.* Upper Saddle River, NJ: Prentice Hall.

Knox, Paul, and Taylor, Peter J., eds. 1995. *World Cities in a World System.* Cambridge: Cambridge University Press.

Korten, David C. 2001. *When Corporations Rule the World.* Bloomfield, CT: Kumarian Press.

Lewis, Martin W. 1991. "Elusive Societies: A Regional-Cartographical Approach to the Study of Human Relatedness." *Annals of the Association of American Geographers* 81, 605–26.

Lewis, Martin W., and Wigen, Karen. 1997. *The Myth of Continents: A Critique of Metageography.* Berkeley: University of California Press.

Luttwak, Edward. 1999. *Turbo-Capitalism: Winners and Losers in the Global Economy.* New York: HarperCollins.

Mander, Jerry, and Goldsmith, Edward, eds. 1996. *The Case Against the Global Economy and for a Turn Toward the Local.* San Francisco: Sierra Club Books.

Martin, Philip, and Widgren, Jonas. 1996. *International Migration: A Global Challenge.* Washington, DC: Population Reference Bureau.

McFalls, Joseph, Jr. 1995. *Population: A Lively Introduction.* Washington, DC: Population Reference Bureau.

Mikesell, Marvin. 1983. "The Myth of the Nation-State." *Journal of Geography* 82, 257–60.

Murphy, Alexander B. 1991. "Regions as Social Constructs: The Gap Between Theory and Practice." *Progress in Human Geography* 15, 22–35.

Riley, Nancy. 1997. *Gender, Power, and Population Change.* Washington, DC: Population Reference Bureau.

Robertson, Roland, and Khondker, H. H. 1998. "Discourses of Globalization: Preliminary Considerations." *International Sociology* 13, 25–40.

Rodrik, Dani. 1997. *Has Globalization Gone Too Far?* Washington, DC: Institute for International Economics.

Rodrik, Dani. 1999. *The New Global Economy and Developing Counties: Making Openness Work.* Washington DC: Overseas Development Council.

Rosenau, James N. 1997. "The Complexities and Contradictions of Globalization." *Current History,* 96, 360–64.

Sassen, Saskia. 1998. *Globalization and Its Discontents: Essays on the New Mobility of People and Money.* New York: New Press.

Schaeffer, Robert K. 1997. *Understanding Globalization: The Social Consequences of Political, Economic, and Environmental Change.* Lanham, MD: Rowman & Littlefield.

Taylor, Peter J. 1994. "The State as Container: Territoriality in the Modern World System." *Progress in Human Geography* 18, 151–62.

Thrift, Nigel. 1993. "For a New Regional Geography 3." *Progress in Human Geography* 17, 92–100.

The World Bank. 2001. *The World Bank Atlas.* Washington, DC: International Bank for Reconstruction and Development/The World Bank.

World Resources Institute. 2001. *World Resources: A Guide to the Global Environment. 2001-02.* New York: Oxford University Press.

The Changing Global Environment

The human imprint is everywhere on Earth, from the highest mountains to the deepest ocean depths; from dry deserts to lush tropical forests; from frozen Arctic ice caps to the cloudless atmosphere (Figure 2.1). Hundreds of spent oxygen canisters clutter the heights of Mt. Everest (Figure 2.2); radioactive debris accumulates in deep offshore waters of the North Atlantic; deserts in North America bloom with irrigated cotton, while tropical forests in Brazil are laid waste by logging; agricultural pesticides appear in Arctic ice caps; and carbon dioxide from burnt fossil fuels collects in the skies. While many changes to the global environment are intentional and have improved the human habitat, other environmental changes are inadvertent and are often harmful to human welfare. Because of the importance of environmental issues, a study of the changing global environment is central to the study of world regional geography.

Environmental issues are also intertwined with globalization and diversity (see "Globalization, Trade Agreements, and the Environment"). The destruction of tropical rain forests, for example, is a response to international demand for wood products and beef. Similarly, global warming through human-caused climate change is closely linked to patterns of world industry and commerce, as is the diversion of streams and rivers for irrigated agriculture. With 6 billion people on Earth, there is a long list of the ways humans interact with—and change—the natural environment.

The purpose of this chapter is to provide an overview of Earth's environmental systems—geology, climate, hydrology, and vegetation—to set the scene for better understanding the environmental geography of the 12 world regions taken up in the following chapters. This chapter concludes with a discussion of world food supply issues, since that topic brings together many aspects of environmental geography.

Geology and Human Settlement: A Restless Earth

Geology shapes the fundamental form of Earth's surface by giving distinctive character to world landscapes through the physical fabric of mountains, hills, valleys, and plains. The geological environment is also critical to a wide spectrum of human activities and concerns, such as the relationship between soil fertility and agriculture or the distribution of mineral resources such as iron and coal. Additionally, the geologic environment presents humans with challenges and hazards in the form of devastating earthquakes, unstable landscapes, and explosive volcanoes. Clearly, a basic understanding of the physical processes that shape Earth's surface is crucial to comprehending human settlement in different parts of the world.

◀ **Figure 2.1 Earth's landscapes**   Jewel of the Hawaiian Islands, Kauai features an abundance of diverse natural landscapes. However, as is true in most of the world, Kauai's mosaic of varied landforms, tropical vegetation, and spectacular river valleys has been much modified by humans. *(Kevin O. Mooney/Odyssey Productions)*

▲ Figure 2.2 Trash on Mt. Everest Few places on Earth have not been affected by humans in the early twenty-first century. Even isolated Mt. Everest is littered with base camp garbage and discarded oxygen canisters at 17,000 feet. *(Mountain Light Photography, Inc.)*

Plate Tectonics

The starting point for understanding the dynamic geology of Earth is **plate tectonics**, a geophysical theory that postulates that Earth is made up of a large number of geological plates that move slowly across its surface. This theory explains and describes both the inner workings of our planet and many surface landscape features. Additionally, plate tectonics gives clues about the world distribution of earthquakes and volcanoes.

Earth's interior is separated into three major zones with very different physical characteristics. These three parts are the core, the mantle, and the outer crust. Tectonic plate theory is built on the assumption that a significant heat exchange takes place deep within Earth. This heat exchange results from the cooling of the inner core. As a result, plastic-like molten material is circulated through the mantle area (Figure 2.3).

This heat exchange occurs in numerous **convection cells**, large areas of very slow-moving molten rock within Earth. Much like a stove provides heat for boiling water within a kettle, radioactive decay deep within Earth's core drives the convection cells. As the molten material reaches Earth's surface, it cools and becomes more dense. This causes the material to sink back into the mantle. Although geophysicists assume that convection cells move material at different rates, the rate of movement is generally in a few inches or centimeters per year, a rate roughly similar to the growth of human fingernails.

The convection cells drag the **tectonic plates** in different directions, sometimes forcing them to collide, sometimes spreading them apart. On top of the slowly moving tectonic plates sit parts of the major continents and the major ocean basins. The map of world tectonic plates (Figure 2.4) shows that these plates vary significantly in size. It is important to note that the continents are not identical to the tectonic plates. In fact, rarely is there a close match between the two. More often, continents straddle several tectonic plates, a fact that explains many of the world's mountainous landscapes. For example, western North America straddles two tectonic plates, the Pacific Plate and the North American Plate, on what is called a **convergent plate boundary.** This means that the two plates are converging, or being forced together by convection cells deep within Earth. The infamous San Andreas Fault, which runs north and south through coastal California, is the actual plate boundary. Major mountain ranges have resulted from the tectonic convergence or collision of these two plates. The volcanic Cascade Range of Oregon and Washington and the massive Sierra Nevada of California are two examples. Similarly, in Latin America, the Andes Mountains are also products of two colliding plates, the eastward-moving Nazca Plate and the westward-moving South American Plate.

Often in these collision zones one tectonic plate dives below another, creating a **subduction zone** characterized by deep trenches where the ocean floor has been pulled downward by tectonic movement. This is true off the western coast of South America and also near the Philippines, where the Mariana Trench forms the deepest ocean depths in the world at 35,000 feet (10,700 meters).

In other parts of the world, tectonic plates move in opposite directions, forming a **divergent plate boundary.** As plates diverge, magma often flows from Earth's interior, creating mountain ranges with active volcanoes. In the North Atlantic, Iceland is formed on a divergent plate boundary that bisects the Atlantic Ocean. In other places, divergent boundaries form deep depressions, or **rift valleys,** such as that occupied by the Red Sea between northern Africa and Saudi Arabia. To the west, in Africa, a splinter of this boundary has created the extensive African rift, a part of which is the famous Olduvai Gorge, where tectonic activity has preserved the earliest traces of our human ancestors in volcanic soils (Figure 2.5).

Geologic evidence suggests that some 250 million years ago all the world's plates were tightly consolidated into a supercontinent centered on present-day Africa. Through time, this large area, now called *Pangaea,* was broken up as convection cells moved the tectonic plates apart. A hint of this former continent can be seen in the jigsaw puzzle fit of South America to Africa and of North America to Europe.

Globalization, Trade Agreements, and the Environment

Economic globalization depends to a great extent on unfettered world trade, and to facilitate international commerce, many countries have entered into free agreements and pacts under the World Trade Organization (WTO). An example is the North American Free Trade Agreement, or NAFTA, which binds together Canada, Mexico, and the United States. Although such trade agreements may be an economic boom, their effect on the environment can be troublesome. More specifically, national and local control of environmental protection can be sacrificed to global trade. Several illustrations follow.

- In 1997 the U.S. Environmental Protection Agency (EPA) weakened its Clean Air Act regulations to comply with a World Trade Organization ruling barring U.S. limits on contaminants in foreign gasoline. Venezuela claimed that the EPA limits acted as an unfair trade barrier against their petroleum.
- In October 1998 the World Trade Organization ruled against the U.S. ban on shrimp imports from nations whose fishing fleets do not use devices to keep endangered sea turtles out of the shrimp nets. As a result of this ruling, the Clinton administration had to revise the U.S. Endangered Species Act to comply with the WTO ruling.
- In response to a recent WTO ruling, U.S. consumers will no longer find "dolphin-safe" tuna on supermarket shelves. This is because Congress had to weaken the U.S. Marine Mammal Protection Act to comply with the WTO decision that the United States was restricting free trade by discriminating against countries that use mile-long nets to catch tuna. These nets also snare and kill thousands of dolphins each year.
- In June 1999 the Canadian-based Methanex Corporation filed a $970 million lawsuit against the state of California under NAFTA. This lawsuit was a reaction to California's decision to ban MTBE, the gasoline additive that is blamed for polluting the state's groundwater and lakes. It was similar to a 1998 lawsuit by a U.S. company that forced Canada to overturn its ban on another gasoline additive. In both cases, the companies pleaded loss of business from the bans.

The MTBE case is "just what we predicted would happen under NAFTA and what we predict will happen under the WTO," said Congressman George Miller of California. "It's happened with dolphin-safe tuna, and it could happen with lots of other laws. This is the New World Order's assault on democracy. . . . Local [environmental] legislation can be nullified because a secret traded tribunal says so."

Source: Adapted from Robert Collier and Glen Martin, "U.S. Laws Diluted by Trade Pacts," *San Francisco Chronicle*, July 24, 1999.

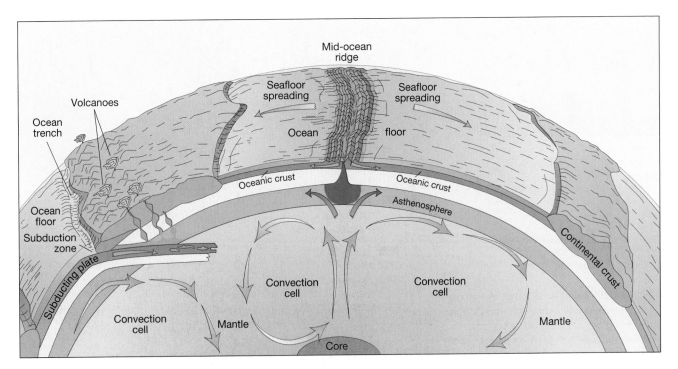

▲ **Figure 2.3 Plate tectonics**   Plate tectonic theory suggests that vast convection cells in Earth's mantle drive molten rock toward the surface along mid-ocean ridges, creating new crust. Elsewhere, continental plates collide or are subducted beneath the surface in deep ocean trenches.   *(Adapted from McKnight and Hess, 2002, Physical Geography: A Landscape Appreciation, 7th Ed., Upper Saddle River, NJ: Prentice Hall)*

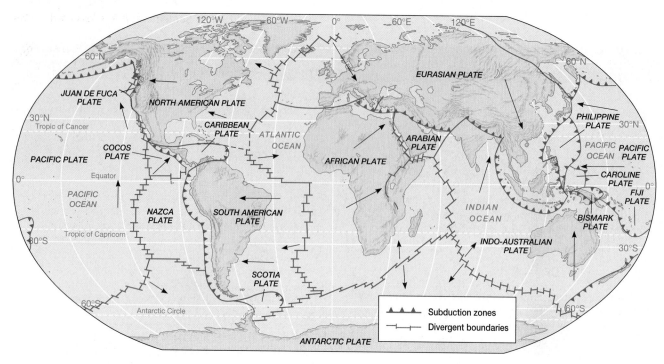

▲ **Figure 2.4 Global tectonic plates** Akin to a fractured jigsaw puzzle, Earth's tectonic plates vary greatly in size and shape. Where plates converge and collide, active volcano and earthquake zones frequently appear, creating new mountains and significant environmental hazards. In other areas one plate may dive beneath another, creating ocean trenches more than 30,000 feet deep, such as east of the Philippines. *(Adapted from McKnight and Hess, 2002, Physical Geography: A Landscape Appreciation, 7th Ed., Upper Saddle River, NJ: Prentice Hall)*

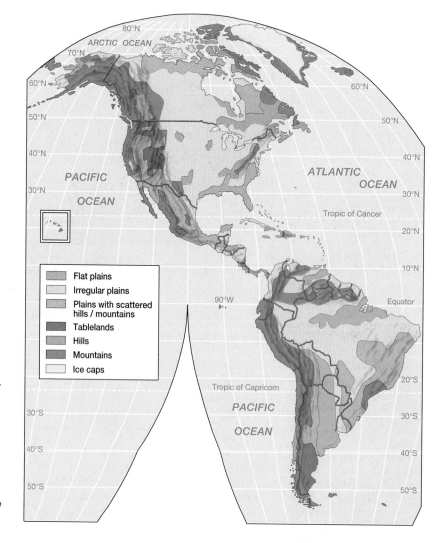

▶ **Figure 2.6 Global landforms** A vast mosaic of mountains and plains, the world's major landform regions are the product of many dynamic processes acting on the planet's complex surface for millions of years. While many mountain ranges follow major tectonic boundaries, other uplifted areas are the products of local and regional forces. *(Adapted from McKnight and Hess, 2002, Physical Geography: A Landscape Appreciation, 7th Ed., Upper Saddle River, NJ: Prentice Hall)*

While tectonic plate theory explains many of the world's large mountain ranges, it does not account for all highlands. For example, both the Himalaya and Alpine mountain ranges have been created by the colliding forces of tectonic plates, as have the westernmost mountains of North America. However, a glance at the world map of landforms (Figure 2.6) shows that there are numerous mountain ranges far removed from plate boundaries. In North America, the Rocky Mountains serve as an illustration, as do the Ural Mountains of Russia. This reminds us that local geologic forces also play an important role in shaping the landscape. Many geoscientists acknowledge that plate tectonic theory does not explain all the world's landforms and that many geophysical questions remain unanswered.

Geologic Hazards: Earthquakes and Volcanoes

Although the human toll from geologic hazards is not nearly as great as that from such other natural disasters as floods and hurricanes, earthquakes and volcanoes nevertheless can have a major impact on human settlement and activities. In August 1999, more than 14,000 people died when a strong earthquake struck western Turkey (Figure 2.7). In January 1995, Kobe, Japan, suffered $100 million in damage from a serious earthquake that killed more than 5,000 people. In

▲ **Figure 2.5 African rift valley**   Where tectonic plates pull apart from one another, large structurally defined rift valleys often create dramatic landforms. These settings in East Africa preserve remains of our early human ancestors.   *[Altitude (Y. Arthus-B.)/Peter Arnold, Inc.]*

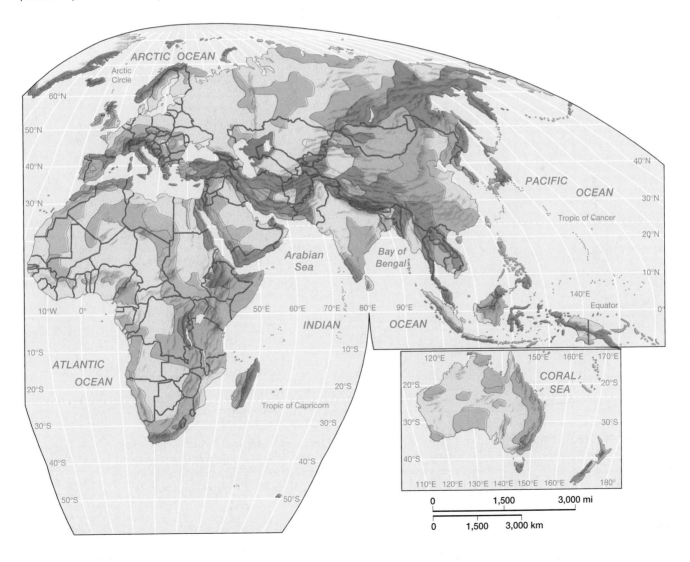

▲ **Figure 2.7 Earthquake in Turkey** A mosque stands amid the rubble of damaged and collapsed buildings in the western town of Golcuk, about 60 miles east of Istanbul. As shown on the map of earthquake epicenters (Figure 2.8), this area of the eastern Mediterranean experiences a large number of severe and damaging earthquakes. *(Enric Marti/AP/Wide World Photos)*

1976 a quarter-million people were killed in rural China when a strong earthquake hit Tangshan. Generally, more than 100 earthquakes cause considerable damage to human settlements in different parts of the world each year.

Predicting earthquakes is much more difficult than warning people about hurricanes, tidal waves (tsunamis), or even volcanic explosions, and no scientific breakthrough is ex-

pected in the next decade that will make earthquake prediction more accurate. Because of this, many cities in earthquake-prone regions emphasize building codes for more resistant structures, land-use planning that discourages building in hazardous areas, and better preparation for post-quake search and rescue. Unfortunately, these measures are expensive; hence, usually only wealthier countries have been able to reduce the toll from seismic disasters. As a result, damage from earthquakes is usually higher in countries that are less able to cope with such a disaster. This disparity is illustrated by the higher loss of life and more widespread damage from similar-magnitude earthquakes in, say, Iran or Mexico than in Japan or California.

However, this does not mean that developed countries can dismiss the threat of major earthquakes. Professionals estimate that when the inevitable "big one" strikes California, the loss of life could be between 30,000 and 50,000 people, depending on the proximity of the earthquake to a major urban area such as Los Angeles or San Francisco. Despite planning and emergency preparedness, the threat of major destruction and a disastrous loss of life from a catastrophic earthquake is a nagging reality along North America's West Coast from San Diego to Seattle, and even inland to Salt Lake City.

Volcanic eruptions are also found along most tectonic plate boundaries and can cause major destruction (Figure 2.8). In 1985, for example, 23,000 deaths resulted in Colombia, South America, from a volcanic eruption. In some cases eruptions can be predicted days in advance, which usually pro-

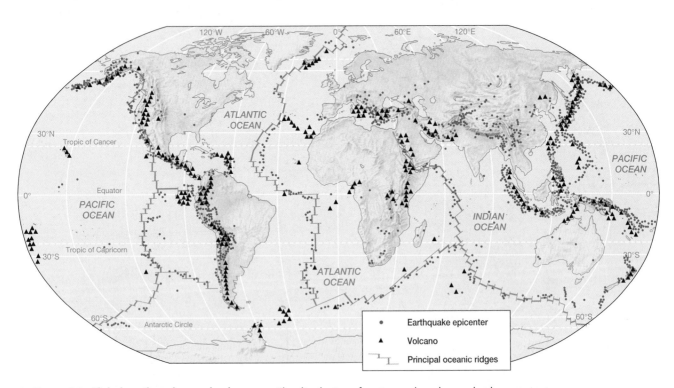

▲ **Figure 2.8 Global earthquakes and volcanoes** The distribution of major earthquakes and volcanoes is strongly associated with tectonic plate boundaries around the world. The circum-Pacific zone of activity from the western Americas to East Asia is particularly active. Large populations found near zones of tectonic activity will continue to pay a price for living in these high-risk locations. *(Adapted from McKnight and Hess, 2002, Physical Geography: A Landscape Appreciation, 7th Ed., Upper Saddle River, NJ: Prentice Hall)*

vides enough time for evacuation. During the 1991 eruption of Mt. Pinatubo in the Philippines, 60,000 people were evacuated, although 800 did die in the disaster. Because of this predictability, the loss of life from volcanoes is generally a fraction of that from earthquakes. In the twentieth century, an estimated 75,000 people were killed by volcanic eruptions, compared with the approximately 1.5 million who died in earthquakes (Figure 2.9).

Unlike earthquakes, volcanoes provide some benefits to people in certain areas. In Iceland, New Zealand, and Italy, geothermal activity produces energy to heat houses and power factories. In other parts of the world, such as the islands of Indonesia, volcanic ash has enriched soil fertility for food crops. Additionally, local economies benefit from tourists attracted by scenic volcanoes in such places as Hawaii, Japan, and the Pacific Northwest.

Global Climates:
An Uncertain Forecast

Human settlement and food production are closely linked to patterns of local weather and climate. Where it is dry, such as in the arid parts of Southwest Asia, life and landscape differ considerably from the wet tropical areas of Southeast Asia. People in different parts of the world adapt to weather and climate in widely varying ways, depending on their culture,

▲ **Figure 2.9 Plymouth, Montserrat** Recent volcanic eruptions on the Caribbean island of Montserrat ravaged the island's small population and reminded residents of the costly and unpredictable consequences of living in a tectonically active landscape. *(John McConnico/AP/Wide World Photos)*

economy, and technology. Although some desert areas of California are covered with high-value irrigated agriculture that produces vegetables for the global marketplace, most of the world's arid regions support very little agriculture and barely participate in global commerce. Moreover, when drought hits one portion of the world (such as Russia's grain belt or Africa's Sahel), the socioeconomic repercussions are felt throughout the world. In many ways, climate links us in our globalized economy, providing opportunities for some, hardships for others, and challenges for all in the struggle to supply the world with food (see "Natural or Unnatural Disasters?").

Natural or Unnatural Disasters?

Hurricane Mitch swept through Central America in December 1998, causing widespread damage and destruction. By the time it went back to sea, more than 10,000 people had been killed, with thousands more listed as missing. Damage was estimated at more than $8.5 billion. Throughout Central America, towns and cities were flooded, roads and bridges destroyed, and agricultural fields wiped out. Was this a natural disaster, or one made by humans?

Although Mitch was unusually violent, with days of rain and high winds, the areas that bore the brunt of the hurricane were not necessarily the areas that suffered the most damage. That is, the killer floods and mudslides that did major damage did not follow the path of the storm itself. "The areas affected the worst were those that are heavily deforested," said George Pilz, head of the natural resources department at the Panamerican Agricultural College at Zamorano, Honduras. In the deforested areas, floods surged unobstructed, soil became waterlogged, and denuded hills collapsed with mudslides. "In areas with forests, there was less damage," Pilz said.

Until recently, tropical forests still blanketed much of the country. However, in the 1990s, huge melon farms and cattle ranches covered the fertile lowlands, along with tropical fruit plantations owned by U.S. companies. With the best land devoted to export markets and global trade, hundreds of thousands of Honduran subsistence farmers were pushed off the lowlands into the steep hillsides covered by tropical forests. Clearing land for subsistence crops meant clearing

the forests. With the forests gone, soil nutrients were lost after a few years, so the farmers had to move deeper into the forested mountains. After a decade of tropical forest clearance by these peasant farmers, much of the upland forest was gone. With its demise, the landscape was poised for destruction by tropical storms. Not only did these farmers suffer most from Hurricane Mitch with loss of life and subsistence, but the cleared forestlands in the mountains also contributed to the widespread flooding that devastated the lowlands.

Disasters such as Hurricane Mitch are not "natural" disasters. Unfortunately, they are not uncommon. In China, recent flooding on the Yangtze River caused more than 4,000 deaths, affected over 200 million people, and cost the country $36 billion in damages. Experts blame deforestation in the extensive Yangtze watershed for the flooding, for in the last few decades 85 percent of the forest cover has been cleared by logging and agriculture. Furthermore, these "unnatural" disasters will continue—and probably increase—throughout the developing world until governments address the short-term management of their natural resources and the deep social inequities that usually determine the way land is used.

Sources: Adapted from Barbara Goldoflas, "Unnatural Disasters," *San Francisco Chronicle*, January 11, 1999, and Janet Abramovitz, "Averting Unnatural Disasters," *State of the World 2001*, New York: W.W. Norton and Company.

Understanding the complex meteorological processes that influence our global climates presents quite an intellectual challenge. Nevertheless, it is highly rewarding because it provides insight into some of the reasons the human condition varies so widely around the world.

Climatic Controls

Although weather and climate differ tremendously around the world, an accepted set of atmospheric processes control and influence meteorological conditions. More specifically, there are five main factors that must be understood—solar energy, latitude, interaction between land and water, world pressure systems, and global wind patterns.

Solar Energy   Both the surface of Earth and the atmosphere immediately above it are heated by energy from the sun as our planet revolves around that large star. Solar energy is one of the most important variables that produce different world climates, for it explains the great differences between the tropical climates near the equator and the cold climates closer to the poles. Most incoming solar energy, or **insolation,** is absorbed by Earth's land and water surfaces. These, in turn, heat the lower atmosphere through the process of re-radiation. This re-radiated energy is trapped by clouds and water mois-

ture in the air adjacent to Earth, providing a warm envelope that makes life possible (Figure 2.10). Because there is some similarity between this process and the way a garden greenhouse traps sunlight to make that structure's interior warmer than the outside, this natural process of atmospheric heating is known as the **greenhouse effect.** Were it not for this process, Earth would be far too cold for human habitation; some scientists suggest our climate would be much like that on the planet Mars.

Latitude   Because of the curvature of Earth, the highest amounts of insolation are received in the equatorial region, which is the area between the equator and 23.5° north and south latitude. Poleward of this equatorial region, Earth receives far less insolation. As a result, not only are the tropics much warmer than the middle or high latitudes, but there is also a buildup of heat energy that must be redistributed through other processes, namely global wind systems, ocean currents, and even massive tropical storms such as typhoons and hurricanes (Figure 2.11).

Interaction Between Land and Water   Because land and water differ in their abilities to absorb and re-radiate insolation, the global arrangement of oceans and continents is a

▶ **Figure 2.10 Solar radiation**
Essential for Earth's survival, incoming solar radiation, or insolation, warms the planet as heat is absorbed by the surface. This surface heat is then radiated back to the lower atmosphere with infrared radiation, where most heat is trapped by the greenhouse effect.

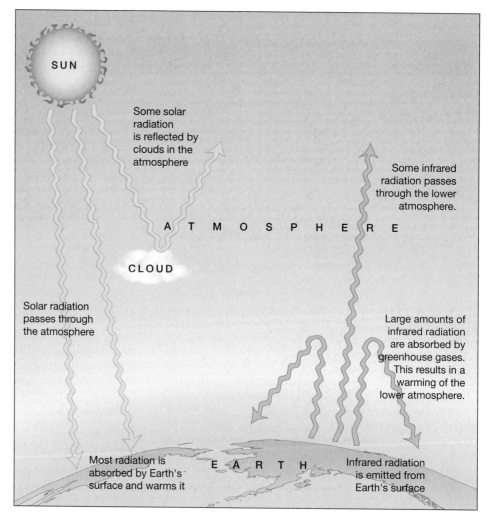

SUN

Some solar radiation is reflected by clouds in the atmosphere

Some infrared radiation passes through the lower atmosphere.

A T M O S P H E R E

CLOUD

Solar radiation passes through the atmosphere

Large amounts of infrared radiation are absorbed by greenhouse gases. This results in a warming of the lower atmosphere.

Most radiation is absorbed by Earth's surface and warms it

E A R T H

Infrared radiation is emitted from Earth's surface

▲ Figure 2.11 Cumulus clouds   A vivid signature of Earth's atmosphere, these massive cumulus clouds display the transfer of heat and moisture in the tropics. As heat energy builds up in low latitudes, ocean currents and wind and weather systems redistribute it globally.   *(Jules Bucher/Photo Researchers, Inc.)*

major influence on world climates. Land areas heat and cool faster than do bodies of water, which explains why temperature extremes such as hot summers and cold winters are always found away from coasts. Conversely, since water bodies retain solar heat longer than land areas, oceanic or maritime climates usually have moderate temperatures without the seasonal extremes found inland. The climatic characteristics of land and water differ so much that geographers use the term **continentality** to describe inland climates with hot summers and cold, snowy winters such as those found in interior North America and Russia. In contrast, **maritime climates,** those close to the ocean, generally have cool, cloudy summers (Oregon, Washington, and British Columbia are good examples), with winters that are cold but lack the subfreezing temperatures of interior locations.

Global Pressure Systems   The uneven heating of Earth due to latitudinal differences and the arrangement of oceans and continents produces a regular pattern of high- and low-pressure cells that, in turn, drive the world's wind and storm systems (Figure 2.12). For example, the interaction between high- and low-pressure systems in the North Pacific produces the storms that are driven onto the North American continent in both winter and summer. The same processes in the North Atlantic produce winter and summer weather for Europe. Farther south in the subtropical zones, large oceanic cells of high pressure cause different conditions. These high-pressure cells expand during the warm summer months because of the subsidence of warm air from the equatorial regions. As they enlarge, the cells produce the warm, rainless

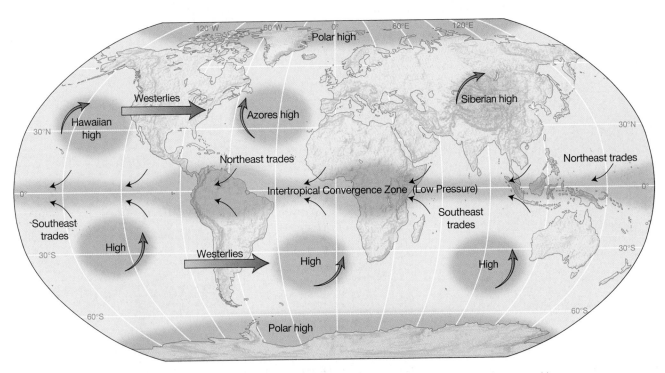

▲ Figure 2.12 Global pressure systems   Predictable patterns of low and high pressure result from the unequal heating of Earth's surface and the positioning of the continents. The intertropical convergence zone often causes unsettled weather near the equator, while fast-flowing westerlies move storm systems across the middle latitudes.

summers of the Mediterranean area of Europe and California. In the equatorial zone itself, summer weather spawns the strong tropical storms known as typhoons in Asia and hurricanes in North America and the Caribbean. The map of global pressure systems (Figure 2.12) shows that this same alternating pattern of high- and low-pressure systems is also found in the Southern Hemisphere.

Global Wind Patterns The pressure systems also produce global wind systems. It is important to remember that air flows from high pressure to low (just as water flows from high elevations to low); thus, winds flow away from high-pressure and into low-pressure cells. This explains the monsoon in India, for example, which arrives in June as moisture-laden air masses flow from the warm Indian Ocean over land into the low-pressure area above northern India and Tibet. In the winter the opposite is true: as high pressure builds over these same areas, winds flow outward from cold Tibet and the snowy Himalayas toward the low pressure over the warm Indian Ocean. (More detail on the monsoon is found in Chapter 12, South Asia.)

World Climate Regions

The interaction of the meteorological processes just discussed produces the world's weather and climate. Before going further, it is important to note the difference between these two terms. *Weather* is the short-term day-to-day (or even hourly) expression of atmospheric processes; our weather can be rainy, cloudy, sunny, hot, windy, calm, or stormy all within a short time period. Measures of this ever-changing weather are taken at regular intervals each day, often hourly. Data are compiled on temperature, pressure, precipitation, humidity, and so on. Over a period of time, statistical averages from these daily observations provide a quantitative picture of common or usual conditions. From this long-term view, a sense of a regional *climate* is generated. Usually at least 30 years of daily weather data are required before climatologists and geographers construct a picture of an area's climate. In summary, weather is the short-term expression of highly variable meteorological processes, whereas climate is the long-term, average conditions.

Since weather observations have been taken for far longer than 30 years in most parts of the world, it is possible to use these data to generate a picture of the climate for thousands

▶ **Figure 2.13 World climate regions** A standard scheme, called the *Köppen system* after the Austrian geographer who devised it in the early twentieth century, is used to describe the world's diverse climates. A combination of letters refer to the general climate type, along with its precipitation and temperature characteristics. More specifically, the *A* climates are tropical, the *B* climates are dry, the *C* climates are generally moderate and are found in the middle latitudes, and the *D* climates are associated with continental and high-latitude locations.

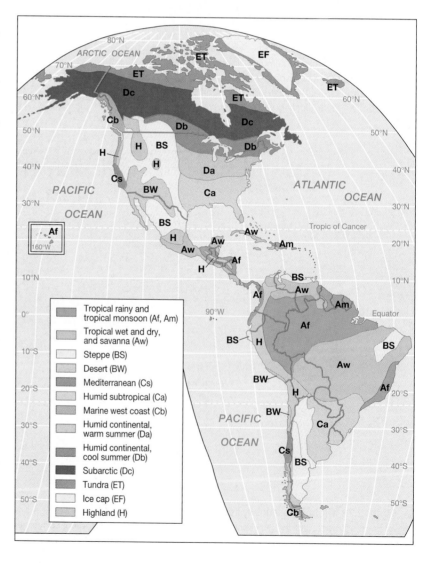

of places all over the globe. Where similar conditions prevail over a larger area, boundaries are drawn cartographically around that area, which is called a **climate region,** so it can be folded into a larger classification of world climates. Knowing the climate type for a given part of the world not only conveys a clear sense of average rainfall and temperatures, but also translates to larger inferences about human activities and settlement. If an area is categorized as desert, then we infer that rainfall is so limited that any agricultural activities require irrigation. Without supplemental watering, no crops can be grown, so people must turn to other strategies to obtain food. In contrast, if we see an area characterized by the climatic designation for tropical monsoon, we know there is enough rainfall for agriculture.

A standard scheme of climate types is used throughout this book, and each regional chapter contains a map showing the different climate regions in some detail. Figure 2.13 shows these climatic regions generalized to a world scale. The regional climate maps also contain **climographs,** graphs of average high and low temperatures and precipitation for an entire year. In Figure 2.13 the climate of Cape Town, South Africa, is presented as an example. Climographs for specific

cities and locales are used to illustrate different climatic regions of the world in the various chapters of this book. Two lines for temperature data are presented on each climograph: the upper one plots average high temperatures for each month, while the lower shows average low temperatures. These monthly averages convey a good sense of what typical days might be like during different seasons. Besides temperatures, climographs also contain bar graphs depicting average monthly precipitation. Not only is the total amount of rain and snowfall important, but the seasonality of this precipitation provides valuable information for making inferences about agriculture and food production. In many midlatitude regions, most moisture falls as rainfall during the warm summer months, which is good for agriculture because it comes during the growing season. However, in other parts of the world, precipitation falls primarily as snowfall during the cold winter months, when it cannot be used directly by crops but instead adds to soil moisture (Figure 2.14). In a few areas, such as in California and the Mediterranean region of Europe and northern Africa, the summers are dry, with rain falling during the winters. Only irrigated agriculture can thrive during the summer season.

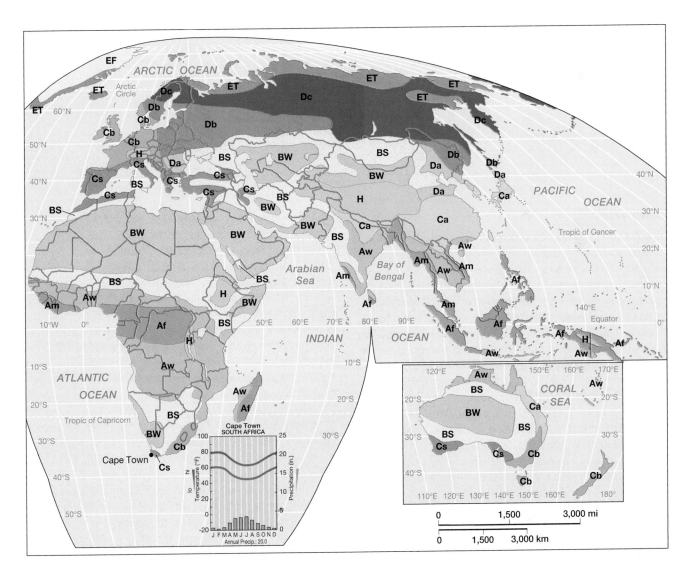

▶ **Figure 2.14 Global precipitation** This map shows the range of annual precipitation totals for the world. As a general rule, the highest amounts are found in the tropics, particularly in places where mountains wring moisture from the tropical clouds. To illustrate, on the island of Kauai, Hawaii, annual totals of more than 400 inches are found in the uplands. Directly north and south of these wet tropics lie expansive deserts and steppes with low rainfall. These are caused by the subtropical high-pressure cells that dominate weather in these regions. Poleward of the arid lands lie the midlatitudes, where precipitation falls as both rain and snow and varies highly depending on storm tracks, coastal location, and topography. *(Adapted from Clawson and Fisher, 2001, World Regional Geography, 7th Ed., Upper Saddle River, NJ: Prentice Hall)*

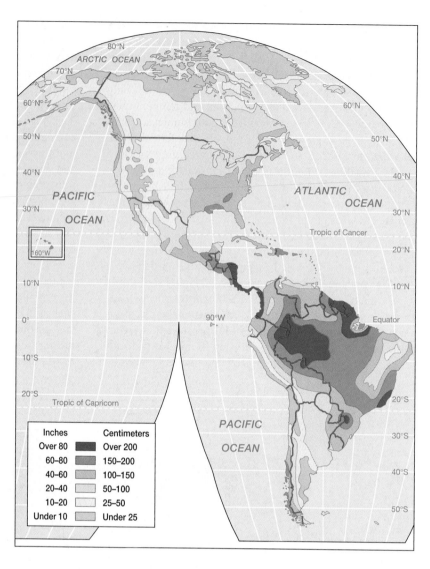

Inches	Centimeters
Over 80	Over 200
60–80	150–200
40–60	100–150
20–40	50–100
10–20	25–50
Under 10	Under 25

Global Warming

Human activities connected with economic development and industrialization are changing the world's climate in ways that may have significant and problematic consequences. More specifically, **anthropogenic,** or human-caused, emissions into the lower atmosphere are increasing the natural greenhouse effect so that worldwide global warming is taking place. This warming, in turn, may change rainfall patterns; increase aridity in some areas; melt polar ice caps, causing an increase in sea levels; and possibly lead to a greater intensity in tropical storms. Moreover, because of climate change from global warming, the world may experience a dramatic change in food production regions. While some prime agricultural areas, such as the lower Midwest of North America, may lose out because of increased dryness, other areas, such as the Russian steppes, may actually gain a more favorable climate for agriculture. Though it is too early to tell if and when this will happen, the implications for world food supplies could be profound.

Causes of Global Warming As noted earlier, a natural greenhouse effect provides us with a warm atmospheric envelope that supports human life. This comes from the trap-

ping of incoming and outgoing (or re-radiated) solar radiation by moisture and an array of natural greenhouse gases in the atmospheric layer close to Earth. Although natural greenhouse gases have varied somewhat over long periods of geologic time, they seem to have been relatively stable until recently. However, within the last 130 years, coinciding with the Industrial Revolution in Europe and North America, the composition and amount of these greenhouse gases have changed dramatically. This has primarily resulted from the burning of fossil fuels associated with industrialization. Four major greenhouse gases account for the bulk of this atmospheric change:

1. *Carbon dioxide (CO$_2$)* accounts for more than half of the human-generated greenhouse gases. The increase in atmospheric CO$_2$ is primarily a result of burning fossil fuels (coal and petroleum); the burning of wood is also a major contributor to the problem. To illustrate this increase, in 1860 atmospheric CO$_2$ was measured at 280 parts per million (ppm). Today it is more than 360 ppm and is expected to exceed 450 ppm by 2050. Computer models predict that CO$_2$ could reach 970 ppm by the end of the

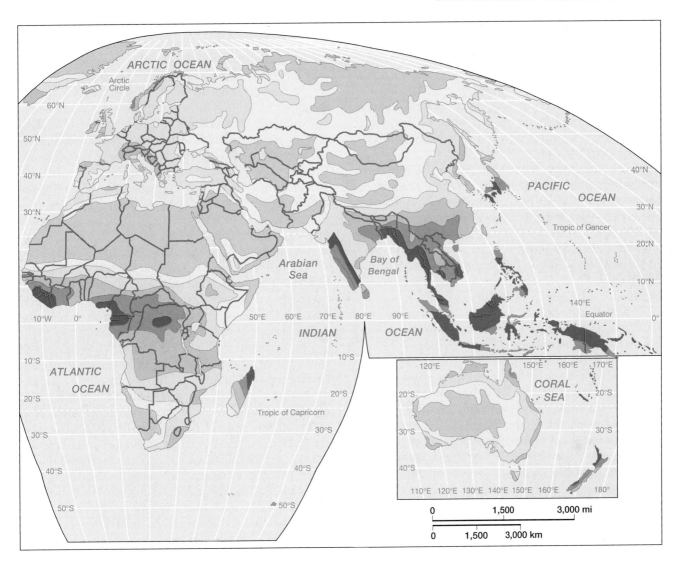

century, which would be a 250 percent increase over pre-industrial levels.

2. *Chlorofluorocarbons (CFCs)*, which make up nearly 25 percent of the human-generated greenhouse gases, come mainly from widespread usage of aerosol sprays and refrigeration, including air conditioning. Although CFCs have been banned in North America and most of Europe, they are still increasing in the atmosphere at the rate of 4 percent each year. These gases are highly stable and reside in the atmosphere for a long time, perhaps as long as 100 years. As a result, their role in global warming is highly significant. Recent research has shown that a molecule of CFC absorbs a thousand times more infrared radiation from Earth than a molecule of CO_2.

3. *Methane (CH_4)* has increased 151 percent since 1750 as a result of vegetation burning associated with rainforest clearing, anaerobic activity in flooded rice fields, cattle and sheep digestion, and leakage of pipelines and refineries connected with natural gas production. Currently, CH_4 accounts for about 15 percent of anthropogenic greenhouse gases.

4. *Nitrous oxide (N_2O)* is responsible for just over 5 percent of human-caused greenhouse gases. It results primarily from the widespread usage of chemical fertilizers in modern agriculture.

Effects of Global Warming   The complexity of the global climate system leaves some uncertainty about how the world's climate may change as a result of human-caused greenhouse gases. Increasingly, though, the high-powered computer models used by scientists to refine our understanding of climate change are reaching consensus on the possible effects of global warming.

Unless countries of the world drastically reduce their emission of greenhouse gases in the next few years, computer models predict that average global temperatures will increase 2 to 4 °F (1 to 2 °C) by 2030. While this may not seem dramatic at first glance, it is about the same magnitude of change as the cooling that caused the Ice Age glaciers to cover much of Europe and North America 30,000 years ago. Further, this temperature increase is projected to double by 2100.

Such a change in climate could cause a shift in major agricultural areas. For example, the Wheat Belt in the United States might receive less rainfall and become warmer and drier, endangering grain production as we know it today. While more northern countries, such as Canada and Russia, might experience a longer growing season because of global warming, the soils in these two areas are not nearly as fertile as in the United States. As a result, scientists are predicting a decrease in the world's grain production by 2030. Further, the southern areas of the United States and the Mediterranean region of Europe can expect a warmer and drier climate that will demand even more irrigation for crops.

Warmer global temperatures will cause rising sea levels from the thermal expansion of the oceans and the melting of polar ice sheets. Although projections differ on the predicted sea level rise—from several inches to almost 3 feet (1 meter)—even the smallest increase will endanger low-lying island nations throughout the world and coastal areas in Europe, Asia, and North America. Island nations in the Pacific and the Indian Ocean are particularly concerned, since they may be flooded out of existence.

Globalization and Climate Change: The International Debate on Limiting Greenhouse Gases

By the early 1990s, the United Nations recognized that climate change was "a common concern of humankind" and that all nations have a "common but differentiated responsibility" for combating global warming. At the Rio de Janiero Earth summit in 1992, the first legal instrument addressing global warming was formulated. Signatories were bound by international law to reduce their greenhouse gas emissions by agreed-upon target dates. Within a year, 167 countries had signed. Unfortunately, those countries emitting the most greenhouse gases, namely, the United States, Japan, India, and China, did not come close to meeting the emission reductions agreed on in Rio.

Consequently, another attempt to develop an international treaty was made at a December 1997 meeting in Kyoto, Japan. From these discussions emerged the current vehicle for international action on climate change—the Kyoto Protocol. In the protocol, 38 industrialized countries agree to reduce their emissions of major greenhouse gases below 1990 levels. The United States, Japan, and Canada, for example, agree to a reduction of about 7 percent below 1990 levels

by the year 2012. If the Kyoto Protocol is ratified by countries emitting 55 percent of the world's greenhouse gases, it will become international law. However, in subsequent meetings, it became clear that obtaining international ratification will be extremely difficult. There are several reasons for this difficulty.

Within the United States, which contributes the most greenhouse emissions, the debate over global warming and the Kyoto Protocol remains contentious, partisan, and unresolved (Table 2.1). While many politicians see the need for greenhouse gas controls, others emphatically resist any action because they fear that controls will constrain business, slow the economy, and increase the cost of living for U.S. citizens. Early in 2001, for example, while agreeing that global warming was a serious issue, President George W. Bush went on record opposing the Kyoto Protocol. He was concerned that restrictions on atmospheric emissions could harm the U.S. economy, and because large developing countries, namely China and India, were not yet bound to specific greenhouse gas reductions.

The European Union (EU), in contrast to the United States, is one of the strongest advocates of international emission regulations and cutbacks. This is primarily because Europe is considerably more energy efficient than the United States and could thus more easily meet reduced emission quotas. Another reason for Europe's assertiveness is that western European countries are grouped together in the European Union, the supranational organization representing 15 states, which gives them an advantage in calculating their emissions. As a group the EU can better meet requirements for a cutback because since 1990 the British coal industry has effectively closed down and the unification of Germany has resulted in reduced emissions due to the closure of industries in the former East Germany. These events offset higher emissions since 1990 in other EU countries such as Italy and France. As a result, EU emissions overall are lower today than in 1990.

One of the most vexing tensions is between the developed nations and less-developed countries. Unrestricted emissions in the industrial world (North America, Europe, Japan) cre-

---

TABLE 2.1 *Carbon dioxide emissions (selected examples)*

Country	CO$_2$ emissions 1999*	Percent global total	Projected CO$_2$ emissions, 2010*	Percent global total	Average annual growth
United States	1,511	24.8	1,809	23	1.4
Western Europe	940	15.4	1,040	13.2	0.9
China	669	10.9	1,131	14.4	4.5
Russia	607	10	712	9	1.7
India	242	4	351	4.5	3.3
Global total	6,091	65.1	7,835	64.1	2.3

*Million metric tons carbon equivalent

Source: *International Energy Annual 1999* and *Annual Energy Outlook 2001*, Department of Energy, Washington, DC.

▲ **Figure 2.15 Bangladesh brick factory** Increasing industrialization in the developing world poses new threats to global environmental health. Often free of stringent pollution controls, manufacturing operations in these countries offer new economic opportunities, but at a high environmental price. *(S. Noorani/ Woodfin Camp & Associates)*

ated the global warming problem, and these countries contribute more than half of the total human-caused greenhouse gases today. Many international experts argue that these same developed countries should be required not only to take stringent steps to curb their own emissions, but also should subsidize and underwrite emission controls in developing countries. Understandably, the less-developed countries are reluctant to sign an emission control agreement that will constrain their economic future when, as they argue, they have contributed very little to the global warming problem up to now (Figure 2.15). Because of issues such as these, a timely solution to global warming is elusive. While the technology exists for controlling most greenhouse gas emissions, a strong commitment from the major global polluters appears to be lacking.

Water on Earth: A Scarce and Polluted Resource

Fresh water is one of the most critical—and scarcest—resources on Earth. It is needed not only to sustain human life itself, but also to support agricultural systems, industry, transportation, and even recreation. However, water is unevenly distributed about the world; some regions have abundant water resources, while others have serious shortages. Currently about 40 percent of the world's population lives in arid or semiarid regions, such as Southwest Asia, North Africa, and Central Asia.

The Global Water Budget

A world map shows that more than 70 percent of the surface area of the world is covered by oceans. As a result, 97 percent of the total global water supply is salt water, with only 3 per-

cent fresh water. Of the minuscule amount of fresh water on Earth, almost 70 percent is locked up in polar ice caps and mountain glaciers. Groundwater supplies account for almost 30 percent of the world's fresh water, which leaves less than 1 percent accessible from surface rivers and lakes.

Another way of conceptualizing this limited amount of fresh water is to think of the total global water supply as 100 liters, or 26 gallons. Of that amount, only 3 liters (0.8 gallon) would be fresh water; and of that small supply, only a mere 0.003 liters, or about half a teaspoon, would be available to humans. Given this limited amount of available water coupled with growing demands for its use, political and economic tensions over efficient water usage are pronounced, and are bound to increase.

International planners use the concept of **water stress** to describe and predict where water resource problems will be greatest (Figure 2.16). Water stress data are generated from the amount of fresh water available on a per capita basis in different parts of the world. By dividing known population growth rates into the amount of water available, one can develop a picture of future problem areas.

Africa stands out as a region of high water stress. More specifically, it is predicted that three-quarters of the population will experience water shortages by the year 2025. Other problem areas will be northern China, India, much of Southwest Asia, Mexico, and even parts of Russia. Additionally, water supply problems throughout the world are aggravated by the great amount of water that is polluted. Most commonly this pollution results from urban sewage that contaminates surface streams and rivers. In India, for example, less than a third of all sewage receives any treatment; in many developing countries the figure is closer to 25 percent. Additionally, the world's groundwater supplies, which currently provide water for about 2 billion people along with vast acreages of agricultural land, are also showing signs of pollution. This means that if the world is to solve its water shortages, solutions will have to come from combined efforts to increase water supplies while at the same time cleaning up and protecting polluted rivers and lakes.

Flooding

Ironically, given the severe global water shortages, floods cause the most deaths of all natural disasters—far more than earthquakes, hurricanes, or volcanic eruptions. Floods are responsible for almost 50 percent of deaths from natural disasters. Further, as the world's population increases, this figure seems to also be increasing (Figure 2.17). This can be attributed to several causes. As the population of a densely settled area increases from natural growth or from in-migration, people are often forced to settle in flood-prone areas, such as river floodplains and deltas. In Bangladesh, for example, an ever-expanding population has forced people to settle in low-lying delta lands subject to serious flooding during the monsoon season. Another factor causing flooding is deforestation. Without forest cover to absorb and slow rainfall runoff, flooding often increases downstream.

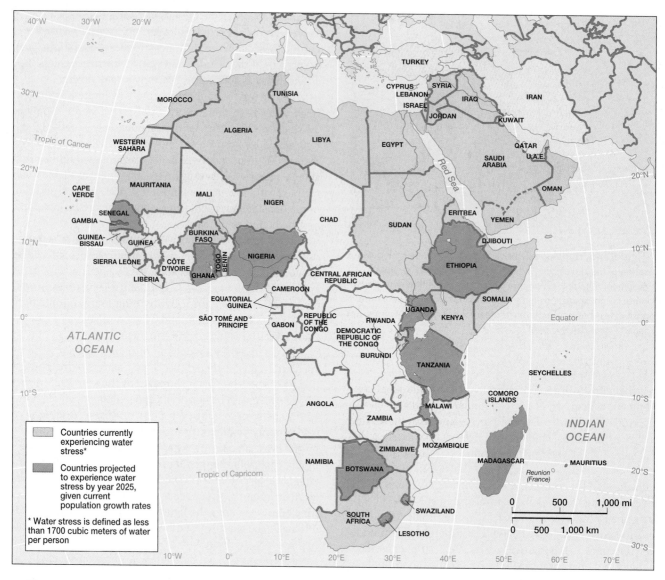

▲ **Figure 2.16 Water stress in Africa and Southwest Asia**   International resource planners use the concept of water stress to target areas where populations currently suffer from shortages of water, as well as areas where shortages might occur in the future. Projections to the year 2025 are based on current water availability combined with present-day population growth rates. As can be seen on this map, many countries in Africa and Southwest Asia already do or soon will experience severe water stress.

Human Impacts on Plants and Animals: The Globalization of Nature

One aspect of Earth's uniqueness compared to other planets in our solar system is the rich diversity of plants and animals that covers its continents. Geographers and biologists think of the cloak of vegetation as the "green glue" that binds together life, land, and atmosphere because it is both a product of and an influence on climate, geology, and hydrology. Humans are very much a part of this interaction. Not only are we evolutionary products of a specific **bioregion**, or as-semblage of local plants and animals (most probably the tropical savanna of Africa), but our subsequent human pre-history was deeply intertwined with the domestication of certain plants and animals. From this process of domesti-cation has come agriculture. Further, humans have changed the natural pattern of plants and animals dramatically by plowing grasslands, burning woodlands, cutting forests, and hunting animals. The pace of such change has accelerated in recent decades, and these actions have led to a crisis in the biological world as forests are devastated, plants and ani-mals exterminated, and watersheds denuded. The very vi-tality of our life-giving biosphere is threatened in many parts of the world.

PRESERVING NATURE   The Case of Antarctica

Thanks to international cooperation, Earth's fifth largest continent—Antarctica—is dedicated to scientific research and environmental protection. From the first Antarctica Treaty in 1959 to the most current agreements on environmental protection, the global community is committed to preserving this huge world region from the ravages of economic exploitation.

Larger than Australia and almost one and a half times the size of the United States, Antarctica is the windiest, coldest, highest (based upon average elevation), and driest continent on Earth. Despite the fact that 98 percent of the continent is covered with ice, only about 2 inches (50 cm) of precipitation falls each year. While winter temperatures in the heart of Antarctica have reached –128 °F (–89 °C), balmy summer temperatures of 50 °F (15 °C) are not uncommon on the coast. In the interior, the Vinson Massif, with elevations of more than 15,000 feet (4,500 meters), rivals the world's highest peaks.

The first Antarctic Treaty, signed by a handful of counties in 1959, prohibited the establishment of military bases and the dumping of radioactive wastes. Additionally, all forms of economic exploitation—particularly mining—were banned. More than 40 countries are signatories to the most current version of the treaty, which focuses on environmental protection. These countries include superpowers, such as the United States, Great Britain, France, and Russia, neighboring nations (primarily Argentina, New Zealand, and Australia), and a number of developing countries, including China, India, Peru, and Bulgaria.

As a result of this treaty, there are 27 permanent research stations on the continent representing 16 national governments. During the Antarctic summer about 4,000 scientists are in residence; this drops to 1,000 during the fierce winter. These scientists work on a wide range of topics relevant to contemporary environmental issues. To name just a few, studies are currently under way on the ozone hole's effects on plants and animals; the impact of global warming on the massive Antarctic ice shelves; reconstructing past climates through analysis of deep ice cores; and the effects of global warming on ocean life.

Despite this international protection, there are still serious problems in Antarctica. For example, the effects of tourism on the environment are a concern. Tourist visits (mainly on cruise ships) have increased to more than 13,000 people per year. More pressing still are problems associated with commercial fishing and oceanic warming as they affect Antarctic animals such as penguins and whales. Recent evidence suggests strongly that penguins are highly stressed because of reduced fish populations in nearby waters. Whether this results from overfishing or global warming is not clear. There is no question, though, that illegal fishing is a serious issue in the vast Antarctic waters. To illustrate, 9,000 tons of Patagonian toothfish (which is marketed as Chilean sea bass) were poached recently by fishing boats from the distant island nation of Mauritius, which until recently was better known for sugarcane than for illegal fishing. Perhaps this serves as warming that even Antarctica isn't beyond the reach of economic globalization.

▲ **Figure 2.17 Flooding in the Philippines**   Growing world populations create more opportunities for natural disasters. Increasingly, human settlements such as Makati City (near Manila) in the Philippines are being constructed in flood-prone areas, and this trend is likely to continue in the future.   *(Bullit Marquez/ AP/Wide World Photos)*

Many of these problems can be explained by the globalization of nature and of local ecologies. Until the last half-century, for example, tropical forests were primarily homes for small populations of indigenous peoples who made modest demands on local environments for their sustenance and subsistence. Today, however, these same tropical forests are capital for multinational corporations that clear-cut forests for international trade in wood products or search out plants and animals to meet the needs of far-removed populations. For example, Japanese lumber companies cut South American rain forests; German pharmaceutical corporations harvest medicinal plants in Africa; North American bears are killed by poachers who sell their gallbladders on the Asian black market as elixirs and sexual stimulants. The list of human impacts on nature is a long one.

Biomes and Bioregions

Biome is the biogeographical term used to describe a grouping of the world's flora and fauna into a large ecological province or region. In this book we use the terms *biome* and *bioregion* interchangeably. Biomes are closely connected with climate regions, since the major characteristics of a climate

▶ **Figure 2.18 World bioregions**   Although global vegetation has been greatly modified by clearing land for agriculture and settlements and cutting forests for lumber and paper pulp, there is still a recognizable pattern to the world's bioregions, ranging from tropical forests to arctic tundra. Each bioregion has its own array of ecosystems containing plants, animals, and insects. These species constitute the biodiversity so necessary for robust gene pools. Put differently, biodiversity can be thought of as the genetic library that keeps life going on Earth.   *(Adapted from Clawson and Fisher, 2001,* World Regional Geography, *7th Ed., Upper Saddle River, NJ: Prentice Hall)*

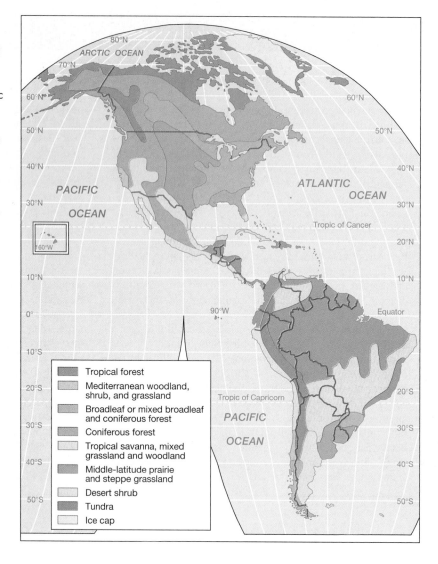

Tropical forest

Mediterranean woodland, shrub, and grassland

Broadleaf or mixed broadleaf and coniferous forest

Coniferous forest

Tropical savanna, mixed grassland and woodland

Middle-latitude prairie and steppe grassland

Desert shrub

Tundra

Ice cap

region—temperature, precipitation, and seasonality—are also the major factors influencing the distribution of natural vegetation and animals (Figure 2.18).

In the preindustrial world, there was a very close correlation between climate and vegetation. That linkage is less clear now because of the ways humans have altered the environment. Desert areas now bloom with agricultural crops because of irrigation; once-extensive forests are gone, replaced in many cases with pastures or farm fields; plants and animals earlier restricted to particular environments are now domesticated and thrive worldwide. Nevertheless, even with these human-induced changes, large tracts of nature's original pattern can still be found in the landscape. A brief overview of the more important bioregions follows.

Tropical Forests and Savannas

Most tropical forests are found in the equatorial climate zones of high average annual temperatures, long days of sunlight throughout the year, and heavy amounts of rainfall. This bioregion covers about 7 percent of the world's land area (roughly the size of the contiguous United States) in Central and South America, Sub-Saharan Africa, Southeast Asia, Australia, and on many tropical Pacific islands. More than half of the known plant and animal species live in the tropical forest bioregion, making it the most diverse of all biomes.

The dense tropical forest vegetation is usually arrayed in three distinct levels that are adapted to decreasing amounts of sunlight closer to the forest floor. The tallest trees, around 200 feet (61 meters) high, receive open sunlight; the middle level (around 100 feet, or 31 meters) gets filtered sunlight; the third level is the forest floor, where plants can survive with very little direct sunlight (Figure 2.19). Even though much organic material accumulates on the forest floor in the form of falling leaves, tropical forest soils tend to be very low in stored nutrients. The nutrients are stored instead in the living plants. As a result, tropical forest soils are not well suited for intensive agriculture.

If the tropical rainfall is seasonal and there is a distinct dry season, the dense tropical rain forest is replaced by a

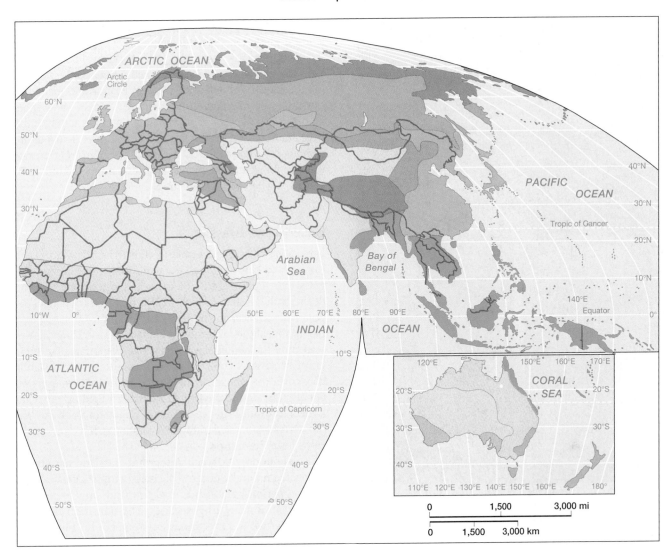

more open forest where trees drop their leaves during the hot, dry season. At the poleward margins of the equatorial zone, where there are even longer dry seasons, the tropical forest is missing completely and is replaced with tropical grassland or savanna characterized by widely scattered trees and lush grassland during the wet season. This is the environment of large mammals and also the ancestral home of early humans.

Deforestation in the Tropics

For a number of reasons, tropical forests are now being devastated at an unprecedented rate, creating a crisis that tests our political, economic, and ethical systems. Although deforestation rates differ from region to region, each year an area of tropical forest about the size of Wisconsin or Pennsylvania is denuded. Spatially, almost half of this activity is in the Amazon Basin of South America. However, actual deforestation appears to be occurring faster in Southeast Asia, where some estimates suggest logging three times faster than in the Ama-

zon (see "Case Study: Illegal Logging in Cambodia"). If those estimates are accurate, Southeast Asia could be completely stripped of forests within 15 years. Behind this widespread cutting of tropical forests lies the recent globalization of commerce in international wood products. By all accounts, Japan was the first to globalize its timber business by reaching far beyond its national boundaries to purchase timber in North and South America and Southeast Asia. This was a two-pronged strategy for meeting the increased demand for wood products in Japan while at the same time protecting its own rather limited forests for recreational use. Currently, about one-half of all tropical forest timber is destined for Japan. Unfortunately, much of it is used for throwaway items such as chopsticks and newspapers.

Another factor contributing to the rapid destruction of tropical rain forests is the world's seemingly insatiable appetite for beef. Cattle species originally bred to survive the hot weather of India are now raised on grassland pastures created by cutting tropical forests throughout the tropical world. Unfortunately, because tropical forest soils are poor in

▲ **Figure 2.19 Tropical rain forest** As fragile as they are diverse, tropical rain forest environments feature a complex, multilayered canopy of vegetation. Plants on the forest floor are well adapted to receiving very little direct sunlight. *(Gary Braasch/ Woodfin Camp & Associates)*

porarily deflect the pressures of land hunger by opening up land for migration and settlement.

Often these three processes work together. First, international logging companies are granted timber concessions to log the forest. Following this, cutover lands are opened to settlement by migrants from other areas. At the same time, deforested lands are made available for cattle ranching by both domestic and foreign firms. When all three demands are combined—wood products, beef, and land hunger—the tropical rain forest and the plants, animals, and indigenous tribal groups who dwell within are the losers.

Deserts and Grasslands

Large areas of arid and semiarid climate lie poleward of the tropics, and here are found the world's extensive deserts and grasslands. Fully a third of Earth's land area qualifies as true desert, with less than 10 inches (25 centimeters) of rainfall per year. In areas receiving more than 10 inches of rainfall, grassy plants appear, often forming a verdant cover during the wet season; in the higher elevations where evapotranspiration is lower because of cooler temperatures, extensive grasslands may cover the landscape. In North America, the midsection of both Canada and the United States is covered by grassland known as **prairie.** In other parts of the world, such as Central Asia, Russia, and Southwest Asia, shorter, less dense grasslands form the **steppe.**

The boundary between desert and grassland has always fluctuated naturally because of changes in the climate. During wet periods, grasslands might expand, only to contract once again during drier decades. The transition zone between the two is a precarious environment for humans, as the United States found out during the 1930s when the semiarid grasslands of the western prairie lands turned into the notorious "Dust Bowl." At that time, thousands of farmers saw their fields devastated by wind erosion and drought—a calamity that led to an exodus from these once-productive farmlands. Farming marginal lands may actually worsen the situation, leading to **desertification,** or a spread of desert-like conditions (Figure 2.20). This has happened on a large scale throughout the world—in Africa, Australia, and South Asia, to name just a few regions. In fact, in the last several decades an area estimated to be about the size of Brazil has become desertified through poor cropping practices, overgrazing, and the buildup of salt in soils from irrigation. In northern China an area the size of Denmark became desertified between 1950 and 1980 with the expansion of farming into marginal lands. Historically this region averaged three sandstorms a year; today 25 such storms are common each year. Although some scientists say the case is somewhat overstated, the United Nations recently estimated that about 60 percent of the world's rangelands are threatened by desertification. If such an amount of rangeland were to become desert, according to the UN, the agricultural livelihood of some 1.2 billion people would be threatened.

nutrients, cattle ranching is not a sustainable activity in these new grassland areas. After a few years, soil nutrients become exhausted, making it necessary to move ranching activities into newly cleared forest land. As a result, more forest is cut; more pasture created; more soil destroyed; and the process goes on at the expense of forest lands.

A third factor explaining tropical forest destruction is that these forest areas are often the last settlement frontiers for the rapidly growing population of the developing world. Brazil has used much of its interior Amazon rain forest for settlement in order to alleviate population pressure along its densely settled eastern coast, where rural lands are controlled by powerful land owners. Brazil had a choice: either address the troublesome issue of land reform and break up the coastal estates or open up the interior rain forests. It chose the path of least resistance, allowing settlers to clear and homestead the tropical forest lands of the interior. Many countries do the same, looking at the vast tracts of tropical rain forest as a safety valve of sorts, a landscape that can be used to tem-

CASE STUDY   Illegal Logging in Cambodia

On the road south from Cambodia's capital city of Phnom Penh, Route 4 crosses a spectacular mountain pass into a wide valley flanked by mountains covered with tropical forests. Until recently, the road was dotted with sawmills that processed logs brought down from the mountains. To the west are legal timber concessions; to the east, Bokor National Park, which was supposedly off-limits to logging. Yet these forests are now laid bare.

For now the sawmills along Route 4 are quiet as the Cambodian government responds to international pressure to crack down on illegal logging, but the damage has been done. Half of Cambodia's forests have been cut in the last several decades, most during an unbroken period of civil unrest. Government and military officials collaborated to strip the forests in an atmosphere of unbridled profit-taking. Much of the illegal logging was done in front of military and police officials who plundered national parks and nature preserves, immune to penalty because of high-level political connections and powerful weaponry.

The World Bank estimated that if the government had not stepped in with strong reform measures, Cambodia's forests would be completely worthless by the year 2003. "Cambodia is a textbook example of bad forestry management. You name it, they've done it," said one World Bank official. "The political environment for the last five years has been very difficult, and the worst victim was the forestry sector."

Cambodia is one of the world's poorest countries and is highly dependent on foreign aid. Earlier, international donors made it clear to the Cambodian government that their aid would stop unless real progress was made in forestry reform. In response, Prime Minister Hun Sen implemented a plan to curb deforestation that included cancellation of major timber concessions, increased revenues from foreign logging firms, and strong prohibitions against logging national parks and nature preserves. These new regulations will be enforced by Cambodian army troops. Government officials claim that 95 percent of illegal logging has now been stopped and more than 800 small, unlicensed sawmills have been shut down.

Illegal tropical forest logging is not unique to Cambodia. Unfortunately, illicit timber cutting is widespread throughout the tropics from Africa to Latin America and throughout Southeast Asia. All too common is illegal collaboration among government, military, and police officials that allows timber cutting in national parks and nature reserves. This is done overtly by international logging companies or, more commonly, by small local timber cutters who then sell the illegal logs to giant international firms.

Source: Adapted from Mark Lioi, "Cambodia Slows Its Destruction of Jungles," *San Francisco Chronicle*, July 28, 1999.

▲ Figure 2.20 Desertification Climatic fluctuations and human misuse combine to produce desertification in certain localities, where marginal lands are overcropped or grazed heavily, resulting in expansion of nearby deserts. Globally, many rangelands remain threatened by desertification. (Mark Edwards/Still Pictures/Peter Arnold, Inc.)

Temperate Forests

The large tracts of forests found in middle and high latitudes are called *temperate forests*. Their vegetation is different from the low-latitude forests found in the equatorial regions. In temperate forests, two major tree types dominate. One type is softwood coniferous or evergreen trees, such as pine, spruce, and firs, that are found in higher elevations and higher latitudes. The second category comprises deciduous trees, which drop their leaves during the winter. Examples are elm, maple, and beech. Because these trees are hardwood, hence harder to mill, they are generally less favored by the timber industry than softwood species.

In North America, conifers dominate the mountainous West, Alaska, and Canada's western mountains, while deciduous trees are found on the eastern seaboard of the United States north to New England. There the two tree types intermix before giving way to the softwood forests of Maine and the Maritime provinces of eastern Canada. In the coniferous forests of western North America, the struggle between timber harvesting and environmental concerns remains controversial. While timber interests argue that they must meet high demand for lumber and other wood products through increased cutting, environmental concerns,

▲ **Figure 2.21 Clear-cut forest** Commercial logging in Washington's Olympic Peninsula has dramatically reshaped this landscape. Throughout the Pacific Northwest, environmental lobbies have successfully restricted logging to protect habitat for endangered species and recreation. *(Calvin Larsen/Photo Researchers, Inc.)*

such as the protection of habitat for endangered species (for example, the spotted owl), have caused the government to place large tracts of forest off limits to commercial logging (Figure 2.21). Further complicating the future of western forests are global market forces. Many Japanese timber firms pay premium prices for logs cut from U.S. and Canadian forests, thus outbidding domestic firms for these scarce resources. Since these trees are often cut from public lands, this facet of globalization raises an interesting question about the appropriate use of public forests that are maintained by tax money.

In Europe, hardwoods were the natural tree cover in western countries such as France and Great Britain before these once-extensive forests were cleared away for agriculture. In the higher latitudes of Norway and Sweden, coniferous species prevail in the remaining forests. These conifers also populate extensive forests across Germany and eastern Europe, and through Russia into Siberia, creating an almost unbroken land-

scape of dense forests. The Siberian forest remains a resource that could become a major source of income for financially strapped Russia. Some argue that if the Siberian forests are put on the market for global trade, it will reduce logging pressure on North America's western forests, which, in turn, could make it easier to enact and enforce comprehensive environmental protection in the United States. This illustrates once again how the multifaceted forces of globalization are intertwined with the fate of local ecosystems.

Food Resources: Environment, Diversity, Globalization

If the human population continues to grow at expected rates, food production must double by 2025 just to provide each person in the world with a basic subsistence diet. Every minute of each day, 170 people are born who need food; during the same minute, about 10 acres of existing cropland are lost because of environmental problems such as soil erosion and desertification. Many experts argue that food scarcity will be the defining issue of the next several decades, just as ideological tensions between superpowers defined the recent past. Because of the close ties between world food problems, the environment, and globalization, we conclude this chapter by introducing the concepts underlying world food resource issues that will be discussed in subsequent chapters.

Industrial and Traditional Agriculture

Food is produced in a great variety of ways around the world, and these different production strategies and techniques yield widely varying results. A useful, though necessarily simplified starting point for assessing these strategies is the differentiation between *industrial* and *traditional* food production systems. Industrial agriculture is practiced on about 25 percent of the world's croplands. It is characterized by the use of large amounts of fossil fuel, both to power farm machinery and to provide the industrial basis for chemical fertilizers and pesticides (Figure 2.22). Additionally, irrigation water is generally used to a high degree. Because of the high level of mechanization, labor requirements are low. In the United States, for example, less than 2 percent of the total workforce is employed in agriculture.

In contrast, traditional agriculture is practiced by almost half the people on Earth. Besides being labor-intensive, traditional agriculture is characterized by less mechanization, less fertilizer, infrequent applications of pesticides, and, as a result, generally lower yields. Until recently, traditional agriculture was synonymous with **subsistence agriculture,** defined as farming that produces only enough crops or livestock for a farm family's survival. However, because of the increasing influence of market economies through globalization, many traditional farmers now also grow cash crops that can be traded on world markets to supplement their subsistence needs (Figure 2.23).

The amount of capital (which includes labor) necessary for an agricultural system provides two further evaluative ref-

▲ **Figure 2.22 Industrial farming in Iowa**   The harvesting of corn in Iowa reveals the imprint of technology on modern agriculture. In such settings, large capital investments in machinery, fuel, and agrochemicals produce high yields while keeping labor costs low.   *(Andy Levin/Photo Researchers, Inc.)*

▲ **Figure 2.23 Subsistence farming in Burkina Faso**   These West African women are harvesting sorghum. Such traditional agricultural operations remain pivotal in feeding people in the less-developed world. While this kind of agriculture requires low inputs of technology, it is often quite labor-intensive.   *(Mark Edwards/Still Pictures/Peter Arnold, Inc.)*

erence points: *intensive* and *extensive* farming. Intensive farming can be found in both industrial and traditional systems. Intensive industrial farming requires high inputs of capital to support fossil fuels, machinery, and fertilizers and generally produces high yields. Intensive traditional farming primarily requires high inputs of labor. Padi rice agriculture in Southeast Asia is an example. Extensive farming is used where labor and capital inputs are considerably lower, such as in open-range livestock ranching in North America or nomadic herding in North Africa. In both cases, large amounts of grazing land are needed to support herd animals. Another example of extensive agriculture is traditional wheat farming in places such as Turkey and Iran, where animals rather than machinery are used for plowing and harvesting; thus, the capital input for fossil fuels or tractors is nonexistent.

The Green Revolution

During the last 40 years of the twentieth century, the world's population doubled. Even more remarkably, during the same period global food production also doubled to keep pace with this population explosion. This increase in food production came primarily from the expansion of intensive, industrial agriculture into areas that previously produced subsistence crops through extensive and traditional means.

More specifically, since 1950 the increases in global food production have come from three interconnected processes that are known as the first stage of the **Green Revolution.** These three processes are, first, the change from traditional mixed crops to monocrops, or single fields, of genetically altered, high-yield rice, wheat, and corn seeds; second, intensive applications of water, fertilizers, and pesticides; and, third, further increases in the intensity of agriculture by reduction in the fallow or field-resting time between seasonal crops.

Since the 1970s, a second stage of the Green Revolution has evolved. This has emphasized new strains of fast-growing wheat and rice specifically bred for tropical and subtropical climates. When combined with irrigation, fertilizers, and pesticides, these new varieties allow farmers to grow two or even three crops a year on a parcel that previously supported just one. Using these methods, India actually doubled its food production between 1970 and 1992.

However, many argue that these agricultural revolutions also carry high environmental and social costs. Since these crops draw heavily on fossil fuels, there has been a 400 percent increase in the agricultural use of fossil fuels during the last several decades. As a result, Green Revolution agriculture now consumes almost 10 percent of the world's annual oil output. Cheap oil prices have facilitated much of this increased agricultural usage of fossil fuel, but if oil prices rise, one must wonder how this will affect food prices in the less-developed world. The environmental costs of the Green Revolution include damage to habitat and wildlife from diversion of natural rivers and streams to agriculture; pollution of rivers and water sources by pesticides and chemical fertilizers applied in heavy amounts to fields; and increased regional air pollution from factories and chemical plants that produce these agricultural chemicals.

There is also evidence that the Green Revolution can bring high social costs and disruption to some areas. Because the financial costs to farmers participating in the Green Revolution are higher than with traditional farming, those farmers must have access to capital or bank loans. Money is needed for hybrid seeds, fertilizers, pesticides, and even new machinery. For those with high social standing, a family support system, or good credit, the rewards can be great, but for those without access to loans or family support, the toll can be heavy. Usually traditional farmers cannot compete against those raising Green Revolution crops in the regional market-place. As a result, they often struggle at a poverty level while Green Revolution farmers prosper. In some wheat-growing areas of India, the social and economic distance between the well-off Green Revolution farmers and poor, traditional farmers has become a major source of economic, social, and political tension. What was once an area where all shared a common plight has now become a highly stratified society of rich and poor. The social costs can be condoned only because of the pressing need for food in this rapidly growing country.

Problems and Projections

Even though agriculture has been able to keep up with population growth in the last decades, few are completely confident that it will again in the twenty-first century. While fuller discussion of food and agriculture issues is found in the regional chapters of this book, four key points offer a starting point for understanding the issues.

1. While overall food production remains an important global issue, it is in fact local and regional problems that often keep people from obtaining food. To many experts, the issue is less global food production and more the widespread poverty and civil unrest at local levels that keep people from growing, buying, or receiving adequate food supplies. If the total global agricultural output were somehow shared equally among all people in the world, each person would have approximately four times his or her basic daily needs. Given this view, some say that there is already enough food to feed the world's population. The real issue, they argue, is distribution and purchasing power.

2. Political problems are usually more responsible for food shortages and famines than are natural events, such as drought and flooding. Distribution of world food supplies is often highly politicized, starting at the global level and continuing right on down to the local. Food aid goes to political friends and allies, while enemies go without. During the Cold War, both the Soviet Union and the United States provided food aid to their ideological allies, but never to those of the other camp. Put differently, the abundance of food available in the world was selectively distributed depending on political outlook and allegiance. While the Cold War is over, we still see food being used as a political weapon to support allies and deprive opposition groups of sustenance. This occurs frequently in refugee camps in Pakistan, Africa, and the Balkans, for example.

3. Globalization is causing dietary preferences to change worldwide, and the implications of this change could be profound. Currently two-thirds of the world population is primarily vegetarian, eating only small portions of meat because it is so expensive. However, because of recent economic booms in some developing countries, an increasing number of people are now eating meat, having moved up the food chain from a subsistence diet based on local resources to a global diet containing a greater pro-

▲ **Figure 2.24 Cattle ranching in Brazil** Growing global demands for meat are reshaping the world's agricultural landscapes. Cattle ranching in western Brazil's rain forest reflects these changing market conditions, but the practice is also introducing new environmental problems. *(Martin Wendler/Peter Arnold, Inc.)*

portion of meat (Figure 2.24). In many cases this change in diet comes not just from a new economic prowess, but also from changing cultural tastes and values as people are exposed to new products through economic and cultural globalization. There are, however, constraints on the world's ability to supply meat because of the agricultural support system needed to produce market meat. Some food experts say that the global food production system could sustain only half the current world population if everyone ate the standard meat-rich diet of North America, Europe, and Japan.

4. Most food supply experts agree that the two world regions of greatest concern are Africa and South Asia. Until 1970 Africa was self-sufficient in food, but since that time there have been serious disruptions and even breakdowns in the food supply system. The causes for these problems in Africa are twofold—rapid population increase and civil disruption from tribal warfare. As a result, one of every four people faces food shortages in Sub-Saharan Africa. More detail on these issues is found in Chapter 6. South Asia's future is also problematic. The United Nations predicts that by the year 2010 almost 200 million people in South Africa will suffer from chronic undernourishment. Further discussion of these problems is found in Chapter 12.

There is some good news in this otherwise bleak picture. Because population growth rates are generally declining in the industrializing areas of East Asia and food production there is still increasing, the UN predicts that the percentage of undernourished people in that region will actually drop by almost 5 percent in the next 10 years. Similarly, gains are being made in Latin America because of lower population growth and higher agricultural production. Predictions are that the proportion of chronically hungry in Latin America will fall to about 6 percent in 2010, which is half the rate of 1990.

⊕ Conclusion

Environmental geography is fundamental to the study of world regional geography because of its close links to issues of globalization and diversity. For example, understanding the natural diversity of the world and the uniqueness of its local landscapes necessitates understanding the processes shaping a region's geology, climate, vegetation, and hydrology. The interaction of those natural forces forges the character of a place. Additionally, these processes provide humans with the fabric of our habitat in terms of environmental resources.

Global environmental change has always been a part of the human backdrop. Sometimes this change results from purely natural processes, such as the unpredictable climatic fluctuations that produce a drought year or an unusually wet season. Increasingly, however, global environmental change is being driven by human activities. These changes take place at all scales, from the global to the local. While some environmental changes, such as the logging of an Alaskan rain forest, are expected byproducts of world and national economic activity, others are unanticipated, inadvertent, and accidental. For example, decades ago only a few scientists saw the large-scale consequences of global climate change resulting from the atmospheric by-products of fossil fuel consumption. Today, however, the problems of global warming are recognized by most.

Globalization is both a help and a hindrance to world environmental problems. From a positive perspective, some would argue that the world's nation-states are increasingly willing to sign international agreements to solve environmental problems. Examples are treaties on whaling, ocean pollution, fisheries, and the protection of wildlife species. Because of these supranational agreements, much progress has been made in some areas of environmental protection.

A conflicting view argues that globalization has aggravated global environmental problems. This is the case, say the critics, because of environmental damage resulting from superheated economic activity that exploits the widest possible range of international resources, regardless of the consequences. Additionally, since unrestricted world free trade is a handmaiden of globalization, trade agreements often conflict with national and local environmental protection. Demonstrations and protests at World Trade Organization (WTO) meetings around the globe attest to this concern. The conflict between globalization and the environment is an important theme discussed in the chapters that follow. Each begins with a section on the environmental geography of the region. Additionally, each contains a map that focuses on environmental issues. On these maps, callouts highlight the most pressing environmental controversies, and if a noteworthy solution has taken place, this also is mentioned. Besides the environmental issues maps, each chapter provides cartographic displays of the physical and environmental geography of the region, along with a separate map of climates. Taken together, these resources should provide readers with a good sense of the environmental geography for each of the 12 world regions.

⊕ Key Terms

anthropogenic (page 60)
biome (page 65)
bioregion (page 64)
climate region (page 59)
climograph (page 59)
continentality (page 57)
convection cells (page 50)

convergent plate boundary (page 50)
desertification (page 68)
divergent plate boundary (page 50)
greenhouse effect (page 56)
Green Revolution (page 71)

insolation (page 56)
maritime climate (page 57)
plate tectonics (page 50)
prairie (page 68)
rift valley (page 50)
steppe (page 68)
subduction zone (page 50)

subsistence agriculture (page 70)
tectonic plates (page 50)
water stress (page 63)

⊕ Questions for Summary and Review

1. Explain how convection cells are linked to tectonic plate theory.
2. Describe several kinds of tectonic plate boundaries, along with the landforms and landscapes usually associated with these boundaries.
3. Describe the natural greenhouse effect.
4. How do continental and maritime climates differ? What causes this difference?
5. What major human activities cause global climate change and warming? More specifically, how have greenhouse gases changed because of these activities?
6. Describe the major unresolved issues in achieving an international agreement to limit global warming.
7. Which natural disaster takes the highest human toll—earthquakes, volcanic eruptions, or flooding? Why?
8. Describe the ecological characteristics of a tropical rain forest.
9. What are the different causes of tropical forest deforestation? How are they linked to globalization?
10. What is *desertification*? Why, and where, is it a problem?
11. How does industrial agriculture differ from traditional agriculture?
12. Describe the characteristics of the Green Revolution.
13. Explain the different variables important to providing the world's population with adequate food. How might these change in the next 10 years?

⊕ Thinking Geographically

1. What are the most threatening natural hazards in your region—earthquakes, tornados, hurricanes, floods, drought? Contact local agencies or visit the public library to find copies of disaster preparedness plans for your community. After reading these, discuss them with your class to determine whether they are adequate. If not, suggest how they might be improved.

2. What climate region do you live in? What are the major weather problems faced by people in your area? How do they adjust to these problems? Has your climate changed over the last decades? You can gain some insight into this by plotting annual temperature and precipitation data over the last 30 years to see if there is a trend toward warmer or drier conditions.

3. Visit some of the many Web sites on the Internet that discuss global warming. Once you get an overview of the different positions, concentrate on one or two of the most contentious issues, such as the debate within the United States between environmentalists and business interests, or the difference in opinion between developed and developing countries. Make sure you also look at the Web sites for the small island nations threatened by flooding. If possible, set up a debate in your class that will bring out the different issues, perspectives, and agendas held by vested interests and different countries.

4. How has the vegetation in your area been changed by human activities in the last 100 years? Has this change led to the extinction of any plants or animals or placed them on the endangered list? If so, what is being done to protect them or restore their habitat?

5. Study the globalization of wood products by acquainting yourself with the source areas for different items, such as building lumber, paper, furniture, or other items found in a local import store. More specifically, does the lumber used in construction in your area come from Canada or the United States? What items in local stores come from tropical forests? In what parts of the world—Asia, Africa, or Latin America—are these forests located?

6. Conduct a detailed analysis of the food issues in a foreign country of your choice. Consider the following points. In general, has food supply kept up with population growth? Has the country suffered recently from food shortages or famine? If so, were these from natural causes, such as drought or floods, or from distribution problems resulting from civil disruption? What segments of the population have food security and which segments have recurring problems obtaining adequate food? How is this inequity best explained? Finally, have food preferences or diets changed recently? If so, how and why?

⊕ Bibliography

Alexandatos, Nikos, ed. 1995. *World Agriculture: Toward 2010. An FAO Study.* Chichester, UK: John Wiley.

Allen, J. C., and Barnes, D. F. 1985. "The Causes of Deforestation in Developing Countries." *Annals, Association of American Geographers* 75, 163–84.

American Association for the Advancement of Science. 2000. *AAAS Atlas of Population and Environment.* Berkeley: University of California Press.

Blaikie, Piers, and Brookfield, Harold. 1995. *Land Degradation and Society.* London: Methuen.

Blaikie, Piers, et al. 1994. *At Risk: Natural Hazards, People's Vulnerability, and Disasters.* London: Routledge.

Brown, Lester. 1996. *Tough Choices: Facing the Challenge of Food Scarcity.* New York: W. W. Norton.

Brown, Lester; Flavin, Christopher; and French, Hilary. 2001. *State of the World.* New York: W. W. Norton.

Bush, Mark. 2000. *Ecology of a Changing Planet.* 2nd ed. Upper Saddle River, NJ: Prentice Hall.

Christopherson, Robert. 2002. *Geosystems: An Introduction to Physical Geography.* 4th ed. Upper Saddle River, NJ: Prentice Hall.

Gelbspan, Ross. 1997. *The Heat Is On: The High Stakes Battle Over Earth's Threatened Climate.* Reading, MA: Addison Wesley.

Gleick, Peter, ed. 1993. *Water in Crisis: A Guide to the World's Fresh Water Resources.* New York: Oxford University Press.

Global Environment Facility. 1998. *Valuing the Global Environment: Actions and Investments for a 21st Century.* Washington, DC: Global Environment Facility.

Goudie, Andrew. 1994. *The Human Impact on the Natural Environment.* 4th ed. Cambridge, MA: The MIT Press.

Hecht, Susanna, and Cockburn, Alexander. 1990. *The Fate of the Forest: Developers, Destroyers, and Defenders of the Amazon.* New York: HarperCollins.

Hertsgaard, Mark. 1998. *Earth Odyssey: Around the World in Search of Our Environmental Future.* New York: Broadway Books.

Hidore, John. 1996. *Global Environmental Change: Its Nature and Impact.* Upper Saddle River, NJ: Prentice Hall.

IPCC. 2001. *Summary for Policymakers: Climate Change 2001— A Report of Working Group 1 of the Intergovernmental Panel on Climate Change.* New York: United Nations Environmental Programme (UNEP).

Kump, Lee; Kustins, James; and Crane, Robert. 1999. *The Earth System.* Upper Saddle River, NJ: Prentice Hall.

Livernash, Robert, and Rodenburg, Eric. 1998. "Population Change, Resources, and the Environment." *Population Bulletin* 53(1).

Manning, Richard. 2000. *Food's Frontier: The Next Green Revolution.* New York: North Point Press.

Marchak, Patricia. 1995. *Logging the Globe.* Montreal: McGill-Queens University Press.

McNeill, J. R. 2000. *Something New Under the Sun: An Environmental History of the Twentieth-Century World.* New York: W. W. Norton.

Moser, Susanne. 1996. *Human Driving Forces and Their Impacts on Land Use/Land Cover.* Washington, DC: Association of American Geographers.

Ojima, D. S., et al. 1995. "The Global Impact of Land Use Changes." *Bioscience* 44(5), 300–3.

Ott, Hermann. 1998. "The Kyoto Protocol: Unfinished Business." *Environment* 40(6), 16ff.

Postel, Sandra. 1996. *Dividing the Waters: Food Security, Ecosystem Health, and the New Politics of Scarcity.* Worldwatch Paper 132. Washington, DC: Worldwatch Institute.

Sandalow, David, and Bowles, Ian. 2001. "Fundamentals of Treaty-Making on Climate Change." *Science* 292, 1839–40.

Schaeffer, Robert. 1997. *Understanding Globalization: The Social Consequences of Political, Economic, and Environmental Change.* Lanham, MD: Rowman & Littlefield.

Soroos, Marvin. 1998. "The Thin Blue Line: Preserving the Atmosphere as a Global Common." *Environment* 40(2), 6ff.

Sponsel, Leslie; Headland, Thomas; and Bailey, Robert. 1996. *Tropical Deforestation: The Human Dimension.* New York: Columbia University Press.

Turner, B. L., et al. 1990. *The Earth as Transformed by Human Action.* Cambridge, UK: Cambridge University Press.

World Resources Institute. 1999. *World Resources, 1999–2000.* New York: Oxford University Press.

Wright, Richard, and Nebel, Bernard. 2002. *Environmental Science: Towards a Sustainable Future.* 8th ed. Upper Saddle River, NJ: Prentice Hall.

RUSSIA

ARCTIC OCEAN

GREENLAND (DENMARK)

Baffin Bay

Prudhoe Bay

ALASKA (U.S.)
Fairbanks

Anchorage

YUKON

Mackenzie R.

NORTHWEST TERRITORIES

N U N A V U T

Iqaluit

Whitehorse

Juneau

Yellowknife

Slave R.

NEWFOUNDLAND AND LABRADOR

Prince Edw Isla

QUEBEC

BRITISH COLUMBIA

ALBERTA

SASKATCHEWAN

Peace R.

Athabasca R.

Fraser R.

Churchill

MANITOBA

C A N A D A

Hudson Bay

NORTH AMERICA
Political Map
⊗ ● Over 1,000,000
✪ • 500,000–1,000,000 (selected cities)
★ • Selected smaller cities
(National capitals shown in red)

Edmonton

Vancouver

Calgary

Regina

Winnipeg

ONTARIO

St. Lawrence R.

NEW BRUNSWI
Charlottetc
St. John

Hali

Quebec

Montreal

Ottawa

Toronto

VT.
MAINE

N.H.
Boston

Seattle

WASHINGTON

Portland

Columbia R.

OREGON

MONTANA

IDAHO

NORTH DAKOTA

SOUTH DAKOTA

MINN.

WIS.

Minneapolis–St. Paul

Mississippi R.

Milwaukee

Detroit

MICH.

Albany

N.Y.

Rochester

Buffalo

Cleveland

PENN.

MA

R.I.

Provider
Hartford

CT.

New York

PACIFIC OCEAN

Sacramento

Reno

NEVADA

Salt Lake City

UTAH

WYOMING

Cheyenne

NEBRASKA

Denver

COLORADO

IOWA

Chicago

ILL.

IND.

OHIO

Indianapolis

Dayton

Columbus

Pittsburgh

Philadelphia

Baltimore (MD.)

DEL.

Washington, D.C.

N.J.

San Francisco

CALIFORNIA

Las Vegas

Salton Sea

Los Angeles

San Diego

ARIZONA

Phoenix

Tucson

NEW MEXICO

Albuquerque

U N I T E D   S T A T E S

Kansas City

KANSAS

MO.

St. Louis

Missouri R.

Cincinnati

Louisville

KY.

Richmond

VA.

W.VA.

Norfolk

Ohio R.

Oklahoma City

OKLAHOMA

Memphis

TENN.

Nashville

N.C.

Charlotte

S.C.

ARK.

MISS.

ALA.

Atlanta

Birmingham

GEORGIA

Dallas–Ft. Worth

TEXAS

LA.

New Orleans

Houston

San Antonio

Rio Grande

Mississippi R.

Jacksonville

Orlando

Tampa–St. Petersburg

FLORIDA

Miami

Gulf of Mexico

MEXICO

22°N

Honolulu

PACIFIC OCEAN

HAWAII

20°N

0 75 150 mi
0 75 150 km

160°W   158°W   156°W

130°W   120°W   90°W

0 250 500 mi
0 250 500 km

80°N

50°N

40°N

30°N

North America

Elevation in meters

4000+
2000–4000
500–2000
200–500
0–200
Sea Level
Below sea
level

Island of
Newfoundland
St. John's

St. Pierre and
Miquelon (Fr.)

Cape Breton I.

Nova
Scotia

ATLANTIC
OCEAN

50°N

40°N

30°N

Tropic of Cancer

20°N

70°W   60°W

ICELAND

UNITED
KINGDOM

IRELAND

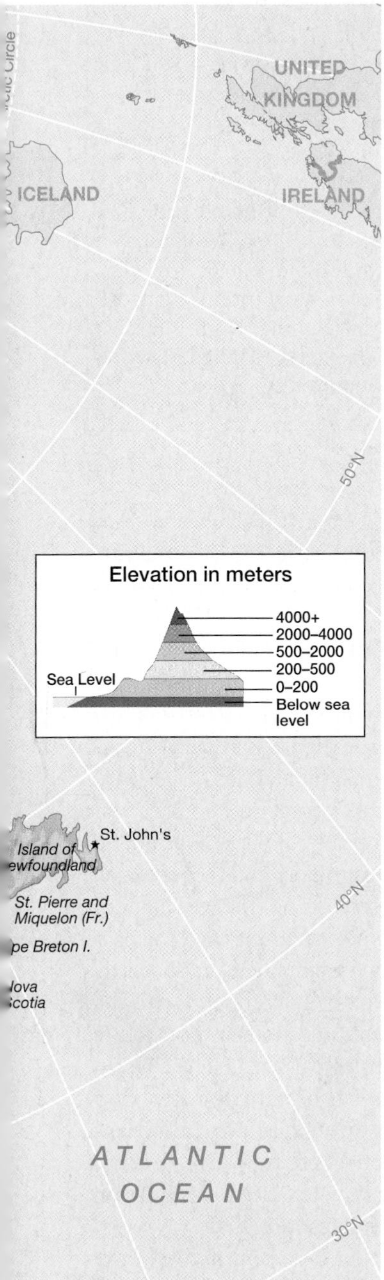

North America encompasses the United States and Canada, a culturally diverse and resource-rich region that has seen unparalleled human modification and economic development over the past two centuries (Figure 3.1; see also "Setting the Boundaries"). The result is one of the world's most affluent regions, where two highly urbanized and mobile societies are oriented around the processes of globalization and the highest rates of resource consumption on Earth. Indeed, the realm superbly exemplifies a **postindustrial economy** in which human geographies are shaped by modern technology, by innovative financial and information services, and by a popular culture that dominates both North America and the world beyond.

Globalization has fundamentally reshaped North America. Stroll down any busy street in Toronto, Tucson, or Toledo, and ponder the landscape (Figure 3.2). Count the ways in which international products, foods, culture, and economic connections shape the everyday scene. The simple visual lesson is revealing: today the North American region stands at the center of the globalization process, and the results are transforming the cultural and economic geographies of the region. From farmers to computer workers, most North Americans are employed in occupations that are either directly or indirectly linked to the global economy. Sizable foreign-born populations in each nation also provide direct links to every part of the world. Tourism brings in millions of additional foreign visitors and billions of dollars that are spent everywhere from Las Vegas to Disney World. In more subtle ways, North Americans embrace globalization in their everyday lives. They consume ethnic foods, tune in to international sporting events on television, enjoy the sounds of salsa and Senegalese music, surf the Internet from one continent to the next, and invest their pensions in global mutual funds.

Globalization is a two-way street, and North American capital, culture, and power are ubiquitous. By any measure of multinational corporate investment and global trade, the region plays a dominant role that far outweighs its population of 315 million residents. North American automobiles, consumer goods, information technology, and investment capital circle the globe. In addition, North American foods and popular culture are diffusing globally at a rapid pace. A new McDonald's restaurant opens every few hours somewhere on the planet. North American music, cinema, and fashion have also spread rapidly around the world. It is no accident that more than two-thirds of the world's Internet hosts are based in North America.

◀ **Figure 3.1 North America** North America plays a pivotal role in globalization. The region also contains one of the world's most highly urbanized and culturally diverse populations. With 315 million people and extensive economic development, North America is also one of the largest consumers of natural resources on the planet.

SETTING THE BOUNDARIES

The United States and Canada are commonly referred to as "North America," but that regional terminology can sometimes be confusing. In some geography textbooks the realm is called "Anglo America" because of its close and abiding connections with Britain and its Anglo-Saxon cultural traditions. The increasingly visible cultural diversity of the realm, however, has discouraged the widespread use of the term in more recent years. While more culturally neutral, the term "North America" also has its problems. As a physical feature, the North American continent commonly includes Mexico, Central America, and often the Caribbean. Culturally, however, the United States–Mexico border seems a better dividing line, although the growing Hispanic presence in the southwest United States, as well as ever-closer economic links across the border, make problematic even that regional division. In some geography texts North America's boundaries extend northeastward to include Greenland, a colony of Denmark that is home to more than 55,000 indigenous people. To the west, North America reaches to Alaska's Aleutian Islands and to the state of Hawaii in the mid-Pacific Ocean. While the future may find Mexico even more intimately tied to its northern neighbors, our coverage of the "North American" realm concentrates on Canada and the United States, two of the world's largest and most affluent nation-states.

Globalization has also brought political risk and uncertainty to the region as the sudden terrorist attacks on the World Trade Center and Pentagon demonstrated in September 2001. The attacks were a tragic reminder that the global reach of U.S. power and influence makes the country a highly vulnerable target, and that its traditions of freedom and openness offer opportunities for terrorism. Indeed, the United States remains the last global superpower, and such status seems destined to keep the country center stage in times of global tensions, whether they are in the Middle East, East Asia, or South Asia. It is clear that the United States stands at the epicenter of the globalization process, and it should not be a surprise that its largest metropolitan area, New York City (20.2 million people), was singled out as a prime objective of the 2001 attacks.

North of the United States, Canada is the other political unit within the region. While slightly larger in area than the United States [3.83 million square miles (9.97 million square kilometers) versus 3.68 million square miles (9.36 million square kilometers)], Canada's population is only about 10 percent that of the United States.

Widespread abundance and affluence characterize North America. The region is extraordinarily rich in natural resources, such as navigable waterways, good farmland, fossil fuels, and industrial metals. Combine that continental good fortune with the acquisitive nature of its European colonizers and an accelerating pace of technological innovation, and the results are reflected everywhere on the modern scene. Indeed, contemporary North America displays both the bounty and the price of the development process. On one hand, the realm shares the benefits of modern agriculture, globally competitive industries, excellent transport and communications infrastructure, and two of the most highly urbanized societies in the world. The cost of abundance, however, has been high: native populations were all but eliminated by encroaching Europeans, forests were logged, grasslands converted into farms, precious soils eroded, numerous species threatened with extinction, great rivers diverted, and natural resources often wasted. Today, although home to only about 6 percent of the world's population, the region consumes 30 percent of the world's commercial energy budget and produces carbon dioxide emissions at a per person rate almost 10 times that of Asia.

Nevertheless, economic growth has vastly improved the standard of living for many North Americans, who enjoy high rates of consumption and varied urban amenities that are the envy of the less-developed world. Satellite dishes, sushi, and shopping malls are within easy reach of most North American residents. Amid this material abundance, however, there are persisting disparities in income and in the quality of life. Poor rural and inner-city populations still struggle to match the affluence of their wealthier neighbors, and these social geographies of poverty have been slow to disappear, even given the unprecedented economic growth of the second half of the twentieth century.

▲ **Figure 3.2 Toronto's cultural landscape**   Toronto's varied ethnic population is celebrated during the Caribana Parade through the city. Powerful forces of globalization have reshaped the cultural and economic geographies of dozens of North American cities.   *(Canadian Tourism Commission)*

North America's unique cultural character also defines the realm. The cultural glue that holds this region together involves a common process of colonization, a heritage of Anglo dominance, and a shared set of civic beliefs in representative democracy and individual freedom. But the history of the region has also juxtaposed Native Americans, Europeans, Africans, and Asians in fresh ways, and the results are two societies unlike any other. Bicultural Canada is shaped by both its French and English antecedents, and major political tensions are associated with the geographic concentration of its French-speaking citizens within the province of Quebec. Adding to Canada's cultural variety is a diverse mix of other European, African, and Asian immigrants who arrived in the country after 1900. South of the Canadian border, cultural pluralism characterizes an ethnically complex United States. The country witnessed early and enduring English and African contributions that were later supplemented by other European, Latin American, and Asian influences. Adding to the mix is a popular culture that today exerts a powerful homogenizing influence on North American society.

Environmental Geography: A Threatened Land of Plenty

North America's physical geography is incredibly diverse. Its varied environmental settings include the rich lowlands of the humid Midwest, the deserts of the American Southwest, and the cool rain forests of the British Columbia coast (Figure 3.3). It has also witnessed unprecedented changes over the past 400 years. California's largest lake, the Salton Sea, illustrates the dynamic and often unpredictable nature of those changes. The 35-mile-long lake, located in the southeast corner of the state, resulted from a simple accident in 1905. An irrigation diversion project on the nearby Colorado River went astray, and for 18 months the river's entire drainage flooded into the low-lying Salton Basin. The huge inland "sea" transformed the environment and the economy of the entire region, with consequences that persist a century later. A great variety of birds were drawn to the area, and in 1930 a large wildlife refuge was created to protect them. More than 350 types of birds now frequent the refuge. Fish were also introduced into the artificially created environment, and by the 1950s a large sportfishing industry had attracted thousands of tourists to the desert setting. Nearby irrigated agriculture in the Imperial Valley also benefited because water from farm operations seeped into the lake instead of waterlogging the fields. Problems have also surfaced. The Salton Sea is getting larger (from agricultural overflows), saltier (from high rates of evaporation, leaving dissolved salts behind), and dirtier (agricultural and municipal pollution drain into the area). Periodically, large die-offs of bird and fish populations (from diseases, toxic salts, and algae blooms) have decimated the regional environment. The lake has also raised international tensions: a major borderlands issue in the area involves controlling the northward flow of Mexicali's (a city in nearby Mexico) industrial waste into the Salton Sea. Indeed, the setting is a splendid example of how quickly people can transform the environment and how those sudden changes can have far-reaching and long-lasting consequences.

North Americans have modified their physical setting in many ways. Processes of globalization and accelerated urban and economic growth have transformed North America's landforms, soils, vegetation, and climate. Indeed, problems such as acid rain, nuclear waste storage, groundwater depletion, and toxic chemical spills are all manifestations of a way of life unimaginable only a century ago. The result is an extensively modified landscape that has proven highly productive but also faces daunting environmental challenges in the twenty-first century. Understanding the basic characteristics of the region's physical geography is essential to appreciating the scope and speed of the human transformations and to anticipating the unfolding environmental dramas that will confront this generation and those that follow.

A Diverse Physical Setting

The North American landscape is dominated by vast interior lowlands bordered by more mountainous topography (Figure 3.3) in the western portion of the region. Extensive coastal plains stretch from southern New York to Texas and include a sizable portion of the lower Mississippi Valley. These plains are flat, sometimes poorly drained, and prone to coastal and river flooding. The present coastline is an intricate landscape of drowned river valleys, bays, swamps, and low barrier islands (Figure 3.4). Decent farmland, good accessibility, and early European settlement contributed to the area's extensive development and later urbanization.

The nearby Piedmont consists of rolling hills and low mountains that are much older and less easily eroded than the lowlands. West and north of the Piedmont are the Appalachian Highlands, an internally complex zone of higher and rougher country that typically crests between 3,000 and 6,000 feet (915 and 1,829 meters, respectively). From Pennsylvania south, folded layers of sedimentary rock, often rich in coal, create a diverse setting of ridges, valleys, and plateaus. North of Pennsylvania, the underlying rock is crystalline, and surface forms are more irregular. Here the highlands reach the Atlantic and include New England and the Canadian Maritimes. The result is the famed and highly indented coastline that extends from southern Maine to the rugged, fjordlike landscapes of western Newfoundland. Far to the southwest, Missouri's Ozark Mountains and the Ouachita Plateau of northern Arkansas resemble portions of the southern Appalachians.

Much of the North American interior is a vast and easily accessible lowland. These mainly depositional plains have accumulated thick sediments, and they extend from west central Canada to the coastal lowlands near the Gulf of Mexico. Eastward, they include the southern Great Lakes and the lower Ohio River Valley. To the west, the Great Plains are formed from younger sediments that have been eroded from the Rocky Mountains. Glacial forces, particularly north of the Ohio and Missouri rivers, also have actively carved and reshaped the landscapes of this lowland zone.

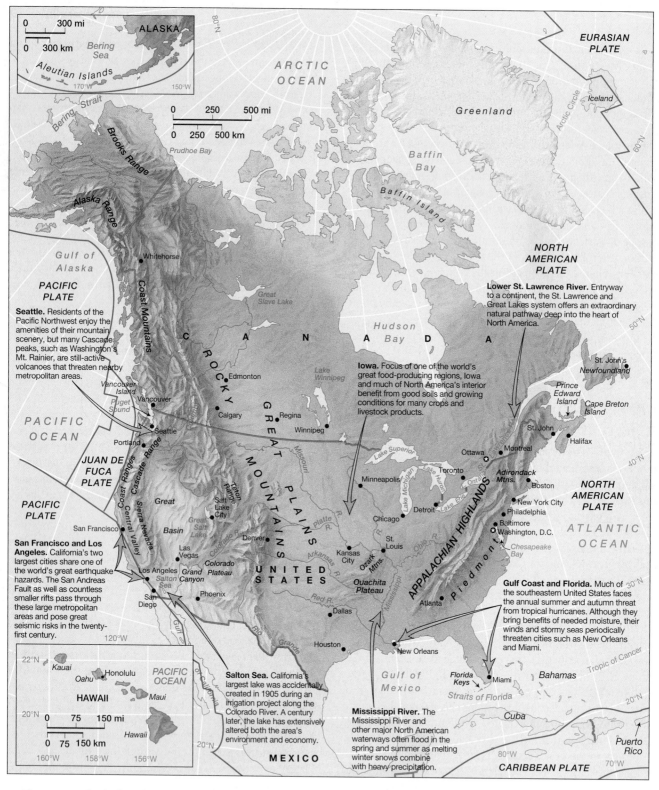

Scale bars (Alaska inset):
0 — 300 mi
0 — 300 km

Scale bars (main map):
0 — 250 — 500 mi
0 — 250 — 500 km

Map labels:

ALASKA
Bering Sea
Aleutian Islands
170°W — 150°W

Bering Strait
Brooks Range
Prudhoe Bay
ARCTIC OCEAN
Greenland
Iceland
EURASIAN PLATE
Arctic Circle
80°N — 60°N

Alaska Range
Gulf of Alaska
PACIFIC PLATE
Whitehorse
Baffin Bay
Baffin Island
NORTH AMERICAN PLATE

Coast Mountains
Great Slave Lake
Hudson Bay
C A N A D A
50°N

Seattle. Residents of the Pacific Northwest enjoy the amenities of their mountain scenery, but many Cascade peaks, such as Washington's Mt. Rainier, are still-active volcanoes that threaten nearby metropolitan areas.

Lower St. Lawrence River. Entryway to a continent, the St. Lawrence and Great Lakes system offers an extraordinary natural pathway deep into the heart of North America.

Vancouver Island
Puget Sound
Vancouver
Seattle
PACIFIC OCEAN
JUAN DE FUCA PLATE
PACIFIC PLATE
Portland
Cascade Range
Coast Ranges
Edmonton
Calgary
Regina
Winnipeg
Lake Winnipeg
R O C K Y
G R E A T   P L A I N S
M O U N T A I N S
St. John's
Newfoundland
Prince Edward Island
Cape Breton Island
St. John
Montreal
Halifax

Iowa. Focus of one of the world's great food-producing regions, Iowa and much of North America's interior benefit from good soils and growing conditions for many crops and livestock products.

Sierra Nevada
Great Basin
Teton Range
Salt Lake City
Great Salt Lake
San Francisco
Central Valley
Las Vegas
Colorado Plateau
Denver
Colorado River
Platte R.
Missouri
Minneapolis
Chicago
Lake Superior
Lake Michigan
Lake Huron
Lake Erie
Lake Ontario
Ottawa
Toronto
Detroit
Adirondack Mtns.
Boston
New York City
Philadelphia
Baltimore
Washington, D.C.
Chesapeake Bay
NORTH AMERICAN PLATE
ATLANTIC OCEAN
40°N

San Francisco and Los Angeles. California's two largest cities share one of the world's great earthquake hazards. The San Andreas Fault as well as countless smaller rifts pass through these large metropolitan areas and pose great seismic risks in the twenty-first century.

Los Angeles
Salton Sea
San Diego
Grand Canyon
Phoenix
U N I T E D   S T A T E S
Kansas City
St. Louis
Ozark Mtns.
Ouachita Plateau
Arkansas R.
Ohio R.
APPALACHIAN HIGHLANDS
Piedmont
Atlanta
Dallas
Red R.
Mississippi
Houston
Rio Grande
New Orleans
Gulf of Mexico
120°W

Gulf Coast and Florida. Much of the southeastern United States faces the annual summer and autumn threat from tropical hurricanes. Although they bring benefits of needed moisture, their winds and stormy seas periodically threaten cities such as New Orleans and Miami.

Salton Sea. California's largest lake was accidentally created in 1905 during an irrigation project along the Colorado River. A century later, the lake has extensively altered both the area's environment and economy.

Mississippi River. The Mississippi River and other major North American waterways often flood in the spring and summer as melting winter snows combine with heavy precipitation.

Florida Keys
Miami
Straits of Florida
Bahamas
Tropic of Cancer
Cuba
Puerto Rico
CARIBBEAN PLATE
30°N — 20°N
80°W — 70°W

Hawaii inset:
22°N
Kauai
Oahu
Honolulu
Maui
PACIFIC OCEAN
HAWAII
20°N
Hawaii
0 — 75 — 150 mi
0 — 75 — 150 km
160°W — 158°W — 156°W

Gulf of California
MEXICO
20°N

▲ **Figure 3.3 Physical geography of North America** North America's diverse physical setting includes both Arctic tundra and tropical forests. Stretching from northern Canada to the Hawaiian Islands, the region reveals varied climates, vegetation, and landforms. While its physical setting provides great opportunities and resources, North America also faces a daunting assortment of natural hazards, including earthquakes, volcanoes, and hurricanes.

The western third of North America features a staggering variety of landforms that resist easy generalization. Active mountain-building (including large earthquakes and volcanic eruptions), alpine glaciation, and often spectacular processes of erosion produce a regional topography quite unlike that of eastern North America. The Rocky Mountains are a series of uplifts, many more than 10,000 feet (3,048 meters) in height, that stretch from Alaska's Brooks Range to northern New Mexico's Sangre de Cristo Mountains. The Rockies form a dramatic backdrop to cities such as Calgary and Denver,

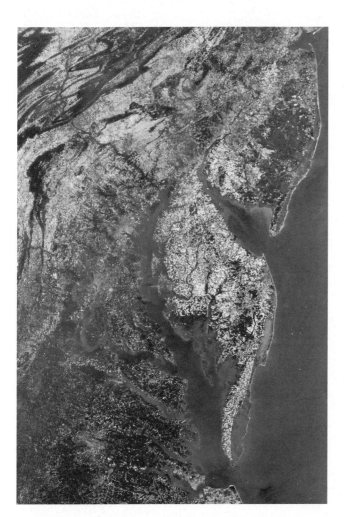

▲ Figure 3.4 Satellite image of the Chesapeake Bay   This view of the Middle Atlantic Coast reveals the intricate shoreline of the Chesapeake Bay (lower center). The coastal sector is characterized by drowned river valleys, barrier islands, and sandy beaches. The Piedmont zone and Appalachian Highlands appear to the northwest.   *(Earth Satellite Corporation/Science Photo Library/Photo Researchers, Inc.)*

▲ Figure 3.5 Landforms of the Colorado Plateau   Monument Valley's colorful sedimentary rocks are part of the unique regional character of the Colorado Plateau. Such eroded uplifts are common across northern Arizona and southern Utah. The region's aridity and lack of vegetation help to highlight the dramatic colors of these buttes and mesas.   *(Rob Crandall/Rob Crandall, Photographer)*

and their twin riches of metals and spectacular scenery have drawn European and American settlers for the last 150 years. Perhaps their greatest resource today is the moisture they capture as accumulated snowfall. Thirsty residents nearby depend on that seasonal surplus. The human geographies of the Platte, Rio Grande, Columbia, and Colorado river basins are increasingly shaped by the challenges of getting water where it is needed for everything from hay fields and vegetable gardens to hydroelectric facilities and swimming pools.

West of the Rockies, the Colorado Plateau includes uplifted blocks of highly colorful sedimentary rock eroded into spectacular buttes and mesas (Figure 3.5). Nevada's vast and sparsely settled basin and range country features numerous north–south-trending mountain ranges interrupted by structural basins with no outlet to the sea. To the north, lava plateaus and low mountains are interspersed with river valleys to create a complex tangle of landforms from eastern Oregon to the Yukon. North America's western border is marked

by the mountainous and rain-drenched coasts of southeast Alaska and British Columbia; the Coast Ranges of Washington, Oregon, and California; the lowlands of the Puget Sound, Willamette Valley, and Central Valley; and the complex uplifts of the Cascade Range and Sierra Nevada. Much of the geological variety of the region is related to the crustal contact zone between the North American and Pacific tectonic plates. Mountain-building is so dynamic and frequent that it is often chronicled in the daily newspaper headlines. Major California earthquakes and smoking Cascade volcanoes offer an ongoing display of landscape changes that have direct and sometimes painful consequences for regional residents (Figure 3.3). Nature has also bestowed its riches on the province: spectacular harbors include developments at Vancouver, Seattle, and San Francisco. Where water is available, some of North America's most productive farmland lies in California's Central Valley and in the Willamette Valley south of Portland.

Patterns of Climate and Vegetation

North America's climates and vegetation are as varied as its landforms. Not surprisingly, the region's size, latitudinal range, and varied terrain have contributed to a diversity of temperature and precipitation patterns (Figure 3.6). These, in turn, have shaped patterns of vegetation. Much of North America south of the Great Lakes is characterized by a relatively long growing season, 30 to 60 inches (76.2 to 152.4 centimeters) of precipitation annually, and a deciduous broadleaf forest that has been extensively modified by commercial agriculture. From the Great Lakes north, the coniferous evergreen or **boreal forest** dominates the cooler continental interior, and these conditions extend to sparsely settled portions of eastern

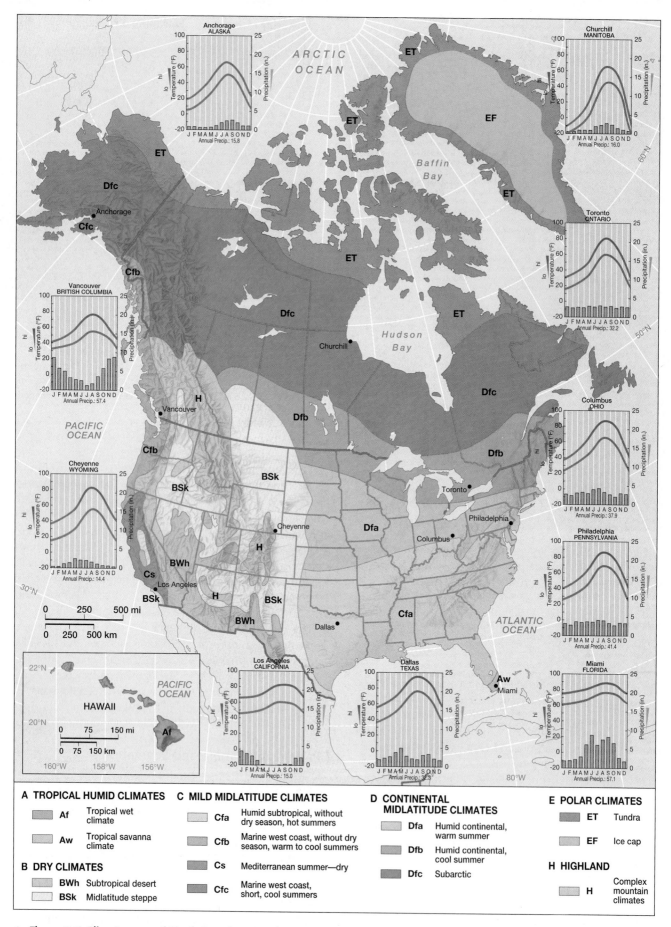

▲ **Figure 3.6 Climate map of North America** North American climates include everything from tropical savanna (Aw) to tundra (ET) environments. Most of the region's best farmland and densest settlements lie in the mild (C) or continental (D) midlatitude climate zones.

Quebec. Near Hudson Bay and across harsher northern tracts, trees give way to **tundra**, a mixture of low shrubs, grasses, and flowering herbs that grow briefly in the short growing seasons of the high latitudes. The drier continental climates found from west Texas to Alberta feature tremendous seasonal ranges in temperature, frequent winds (including tornadoes), and unpredictable precipitation that averages between 10 and 30 inches (25.4 and 76.2 centimeters) annually. Indeed, the region straddles the transition zone between humid and arid portions of the North American interior, a boundary broadly defined as the 20-inch line of annual precipitation that runs near the 100th meridian. Much of this subhumid region is blessed with fertile soils and was originally clothed in **prairie** vegetation dominated by tall grasslands in the East and by short grasses and scrub vegetation in the West.

Western North American climates and vegetation are greatly complicated by the Pacific Ocean and the region's complex landforms. Marine west coast climates dominate north of San Francisco, while a dry summer Mediterranean climate holds sway across central and southern California. Vegetation varies from coastal British Columbia's coniferous forests to the mixed oak woodlands and chaparral scrub found south of San Francisco. Variations related to elevation complicate these broad patterns, and cities also modify local climates and patterns of native vegetation. Farther east, the intermontane interior and the Rocky Mountains experience the typical seasonal variations of the middle latitudes, but the patterns are modified by the effects of topography. Vegetation is strongly influenced by altitude. A single mountain front in the Rockies, for example, might include sagebrush and pinyon pine on its lower slopes; a denser cover of ponderosa pine, western fir, and quaking aspen above; and an alpine zone of meadows, tundra, and snowfields on the ridge crest.

Natural Processes and Natural Hazards

North Americans live in a dynamic, sometimes unpredictable natural setting. Over the longer term, North America's present-day climates and vegetation are merely a snapshot in time, reflecting dynamic natural processes. Climate cycles have brought glacial advances and retreats, the most recent dating to about 15,000 years ago, when ice sheets expanded across Canada and much of the eastern United States north of the Missouri and Ohio rivers. Although these glaciers melted, the current temperate period could be followed by another era of extraordinary change in which climate and vegetation belts are rapidly redistributed. If global warming continues, perhaps fueled by natural as well as human causes, many of North America's cities, including New York, Miami, and Houston, could be submerged by rising world sea levels.

Shorter term, North Americans continue to face other natural hazards. As human populations have grown in once sparsely settled areas of North America, they have inevitably encountered a growing list of natural environmental threats. For example, farmers found they had to adjust to periodic drought cycles on the Great Plains and prairies. Although the

Dust Bowl era of the 1920s and 1930s was the most famous of these dry spells, there is no reason to suspect that the cyclical pattern will not continue in the future. Elsewhere, western North Americans from Anchorage to Los Angeles have discovered the seismic rigors of living within a zone of active earthquakes, tsunamis, and volcanoes.

Ironically, many of North America's most dramatic natural hazards have been magnified by the economic affluence of the region's population. In the past 50 years, millions of North Americans have been attracted to scenic, high-amenity areas that are especially vulnerable to environmental disasters. For example, thousands of miles of North America's spectacular coastline are now densely settled. Resort hotels, retirement condominiums, and exclusive private estates now ring much of the continent and its nearby islands. The result is that natural events such as tropical hurricanes, coastal mudslides, heavy winter surf, and beach erosion create many more hazards than they did a century or two ago. Future global warming may further raise global sea levels, increasing these hazards even more dramatically in the twenty-first century.

Another example is the wildfire hazards in the mountains and foothills of western North America. Fire has always been a part of the natural world here, but recent drought episodes, again made more severe by global warming, have combined with vastly increased numbers of suburban and rural residents drawn to the amenities of such settings. The West suffered one of its worst fire seasons in 90 years during the summer of 2000, with more than 5.5 million acres burned and $1 billion in economic losses (Figure 3.7). In the future, as populations grow and drought cycles continue to visit the region, such costly conflagrations are likely to be repeated.

▲ **Figure 3.7 Western fires**   The summer of 2000 witnessed some of the West's worst fires in 90 years. This scene in western Montana may be repeated in the future as global warming and periodic droughts combine with growing populations in the West's forested mountain zones. *(BLM/Alaska Fire Service)*

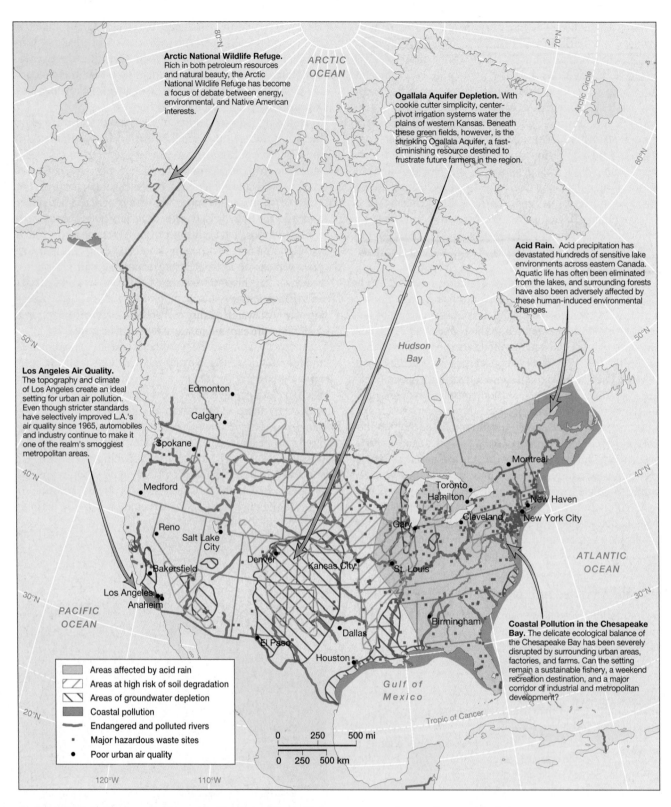

Arctic National Wildlife Refuge. Rich in both petroleum resources and natural beauty, the Arctic National Wildlife Refuge has become a focus of debate between energy, environmental, and Native American interests.

Ogallala Aquifer Depletion. With cookie cutter simplicity, center-pivot irrigation systems water the plains of western Kansas. Beneath these green fields, however, is the shrinking Ogallala Aquifer, a fast-diminishing resource destined to frustrate future farmers in the region.

Acid Rain. Acid precipitation has devastated hundreds of sensitive lake environments across eastern Canada. Aquatic life has often been eliminated from the lakes, and surrounding forests have also been adversely affected by these human-induced environmental changes.

Los Angeles Air Quality. The topography and climate of Los Angeles create an ideal setting for urban air pollution. Even though stricter standards have selectively improved L.A.'s air quality since 1965, automobiles and industry continue to make it one of the realm's smoggiest metropolitan areas.

Coastal Pollution in the Chesapeake Bay. The delicate ecological balance of the Chesapeake Bay has been severely disrupted by surrounding urban areas, factories, and farms. Can the setting remain a sustainable fishery, a weekend recreation destination, and a major corridor of industrial and metropolitan development?

Legend:
- Areas affected by acid rain
- Areas at high risk of soil degradation
- Areas of groundwater depletion
- Coastal pollution
- Endangered and polluted rivers
- Major hazardous waste sites
- Poor urban air quality

0   250   500 mi
0   250   500 km

▲ **Figure 3.8 Environmental issues in North America** Many environmental issues threaten North America. Acid rain damage is widespread in regions downwind from industrial source areas. Elsewhere, widespread water pollution, cities with high levels of air pollution, and zones of accelerating groundwater depletion pose health dangers and economic costs to residents of the region. Since 1970, however, both Americans and Canadians have become increasingly responsive to the dangers posed by these environmental challenges.

84

The Costs of Human Modification

The story of North America's dynamic environment becomes even more complex when we consider how extensive human modifications have transformed the region's soils and vegetative cover, water resources, and atmosphere (Figure 3.8). The speed and magnitude of that transformation are remarkable, largely taking place over the past 200 years and leaving few areas of North America free from human imprint. Some changes are beneficial, including agricultural adaptations that easily feed the region's population. Other impacts have been less welcome, the unintended consequences of activities with lasting and destructive effects on the natural environment and its human occupants. Reacting to these issues, environmental movements in Canada and the United States have increasingly shaped national policies over the past 30 years.

Transforming Soils and Vegetation   Particularly after Europeans arrived on the scene, countless new species of plants and animals invaded the continent, fundamentally reshaping its ecology to include exotics such as wheat, cattle, and horses. Forest cover was removed from millions of acres. Similarly, grasslands were plowed under and replaced with grain and forage crops not native to the region. Widespread soil erosion was greatly accelerated by careless cropping and ranching practices, and many areas of the Great Plains and South suffered lasting damage. With twentieth-century urbanization, many earlier agricultural transformations were swept aside to make way for suburbs and shopping centers. From the nonnative lawns of southern California to the wheat fields of Montana and Alberta, North Americans today are living in rural and urban environments that bear little resemblance to settings across the region two centuries ago.

Managing Water   North Americans consume huge amounts of water. While conservation efforts and better technology have slightly reduced per capita rates of water use over the past 25 years, city dwellers still use an average of more than 170 gallons daily. When society's overall demands are figured to include total agricultural and industrial needs, every North American consumes more than 1,500 gallons of water per day! Approximately 45 percent of the water used in the United States is employed in manufacturing and energy production, 40 percent in agriculture, and the remainder for home and business use. While water is essential to North America's survival and level of affluence, managing it effectively and efficiently is a growing environmental challenge (see "Environment: Sink or Swim? Life on the Modern Mississippi"). It is a fundamental problem of geography: water resources are not equally spread across North America, nor are demands for water evenly distributed across time and space. The result is a mammoth problem of moving water from one place to another, while at the same time maintaining or enhancing its quality so that it can be safely used where it is needed.

Many North American localities are threatened by water shortages. Metropolitan areas in eastern Canada and the United States struggle with outdated municipal water supply systems. Beneath the Great Plains, the waters of the Ogallala Aquifer are also being depleted. The largest in North America, this aquifer is a huge reserve of precious groundwater created during the last Ice Age, and today it irrigates about 20 percent of all U.S. cropland. Center-pivot irrigation systems are steadily tapping into the supply; water tables across much of the region have fallen more than 100 feet (30 meters) in the past 50 years; and the the costs of pumping are rising steadily. Farther west, California's elaborate system of water management is a reminder of that state's ever-growing demands (Figure 3.9). Aqueducts move the precious resource from an already overutilized Colorado River system into southern California's agricultural and metropolitan regions. Both the eastern (Los Angeles Aqueduct) and western (Central Valley Project) slopes of the Sierra Nevada are also tapped to satisfy the thirsts of Golden State homeowners, farmers, and industrialists.

Issues of water quality are equally problematic. North Americans are exposed to water pollution in varied ways, and

▲ **Figure 3.9 California aqueduct**   Thirsty Californians have extensively modified the geography of water in their state. Large aqueducts have dramatically reconfigured the distribution of this precious resource, promoting tremendous agricultural and metropolitan expansion within western North America.   *(Alexander Lowry/Photo Researchers, Inc.)*

Back in Mark Twain's time, life on the Mississippi River had challenges that ranged from shifting channels to unsavory waterfront characters. Today, the Mississippi poses a new set of problems for residents along America's greatest river. Record spring and early summer rains in 1993 produced one of the continent's greatest floods of the previous 200 years. More than 14 million acres (6 million hectares) were inundated in the deluge. Almost $20 billion in crop and property damage resulted as the swirling waters of the Mississippi and Missouri rivers overwhelmed levees and forced more than 50,000 people from their homes (Figure 3.1.1). Since 1993, additional flooding has inundated many stretches of the river, suggesting that future floods will be a common occurrence along the Mississippi.

What the floods dramatically demonstrate are the unanticipated consequences incurred as people interact with their environment. Since Mark Twain's time, billions of federal dollars have been spent to control the Mississippi River: its channel was narrowed and deepened to handle commercial barge traffic, and thousands of miles of levees were constructed to hold back the river's flow. At the same time, floodplain acreage filled with farms, suburban housing tracts, and shopping malls. Federally subsidized crop and flood insurance further emboldened residents to challenge the odds.

Even more daunting are the possible effects of global climate change on flood probabilities in the valley or along North America's thousands of miles of coastline. Indeed, global warming might increase the chances for more cataclysmic floods on the Mississippi, and rising sea levels may offer the same troubling scenario for increasingly developed coastal settings. Should billions of dollars be spent on development in such places, only to see it periodically washed away? Who should pay for insuring such activities? What risks *are* worth taking in these localities? Although Huck Finn never worried about such things, modern life along the Mississippi demands that residents confront such issues and that they ponder solutions that can help them stay above water in the future.

▶ **Figure 3.1.1 Satellite views of the confluence of the Missouri and Mississippi rivers** The upper image, showing the setting in the drought year of 1988, contrasts dramatically with the lower image, taken during the peak of the 1993 floods as millions of acres in the region were inundated. St. Louis is just downriver from the confluence. *(Courtesy of Spaceimaging.com)*

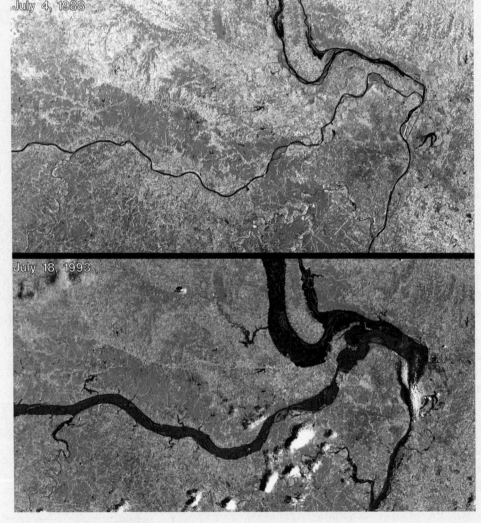

July 4, 1988

July 18, 1993

even mandates, such as the U.S. Clean Water Act or Canada's Green Plan, cannot stop ongoing abuses. Mining operations and industrial users such as chemical, paper, and steel plants generate toxic wastewater or metals that enter surface and groundwater supplies. Accidental chemical and petroleum spills are another ongoing threat. For example, in the fall of 2000, an oil tanker leaked more than 560,000 gallons of crude oil into the environmentally sensitive waters of the lower Mississippi River in Louisiana; this was the largest North American oil spill since Alaska's *Exxon Valdez* disaster in 1989. Metropolitan areas also produce vast amounts of raw sewage that annually costs billions to repurify. Overall, the United States generates hazardous wastes at a much higher per capita level than either Canada or western Europe. America's most toxic localities include the petroleum-rich Texas and Louisiana Gulf coasts and the older industrial centers of the Northeast and Midwest. Toxic threats in the West are found near industrial sites, in former mining regions from Arizona to Montana, and in nuclear fuel and chemical warfare storage areas such as Hanford, Washington; Rocky Flats (near Denver), Colorado; and Tooele (near Salt Lake City), Utah. In addition, the widespread use of fertilizers and pesticides in modern North American agriculture raises land productivity, but many of these chemicals are carried into natural runoff. In the U.S. Southwest, another critical water problem is salinization, a process in which irrigation water leaves behind a toxic buildup of salts in the soil.

Altering the Atmosphere   North Americans humanize the very air they breathe; in doing so, they change local and regional climates as well as the chemical composition of the atmosphere. Urban heat islands, particularly on clear, calm nights, generate nighttime temperatures some 9 to 14 °F (5 to 8 °C) warmer than nearby rural areas. Most significantly, air pollution affects large numbers of plants, animals, and people in a myriad of localities. At the local level, industries, utilities, and automobiles contribute carbon monoxide, sulfur, nitrogen oxides, hydrocarbons, and particulates to the urban atmosphere. While some of the region's worst offenders are U.S. cities such as Houston and Los Angeles, Canadian cities such as Toronto, Hamilton, and Edmonton also experience significant problems of air quality.

On a broader scale, North America is plagued by **acid rain,** industrially produced sulfur dioxide and nitrogen oxides in the atmosphere that damage forests, poison lakes, and kill fish. Many acid rain producers are located in the Midwest and southern Ontario, where industrial plants, power-generating facilities, and motor vehicles contribute emissions. Prevailing winds transport the pollutants and deposit damaging acidic precipitation across the Ohio Valley, Appalachia, the northeastern United States, and eastern Canada. Fortunately, tougher controls and technological improvements have reduced pollution levels, and utility companies have reduced their sulfur emissions by 40 to 50 percent since 1980. Still, much damage has already been done to sensitive forests and lakes in New York's Adirondack Mountains, southern Quebec, and the Maritimes.

The Price of Affluence

Globalization has brought many benefits to North America, but with the accompanying urbanization, industrialization, and heightened consumption, the realm is also paying an environmental price for its affluence. Energy consumption within the region, for example, remains extremely high, imposing a growing list of environmental and economic costs both within North America and beyond. North Americans consume energy at a per capita rate almost double that of the Japanese and more than 16 times that of India's population. California's electricity crisis in 2001 suggests that difficult choices may need to be made between further boosting energy supplies and guaranteeing environmental quality. For all its economic growth and material wealth, the twentieth century in North America will also be remembered for its toxic waste dumps, frequently unbreathable air, and wildlands lost to development.

Still, many environmental initiatives in the United States and Canada have addressed local and regional problems. For example, the improved water quality of the Great Lakes over the past 30 years is an achievement to which both nations contributed and that benefits both. Tougher air quality standards have selectively reduced emissions in many North American cities. Though forces still oppose them, preservation groups in both countries have secured new tracts of endangered forest, alpine, and wetland environments, including acreage in the Rocky Mountains and in the Alaskan and Canadian North. Whatever the outcome of future policy debates, North America's humanized environment will continue to reflect both the costs and rewards of living in one of the world's most developed regions.

Population and Settlement: Refashioning a Continental Landscape

The North American landscape is the product of four centuries of extraordinary human change. During that period, Europeans, Africans, and Asians converged upon the realm, displaced a continental expanse of Native American peoples, and created a new geography of human settlement. Today, more than 315 million people live in the region, and they are some of the world's most affluent and highly mobile populations. The contemporary North American scene dramatically displays how its population has refashioned the settlement landscape to meet the needs of a modern, postindustrial society. Above all, urbanization shapes contemporary North American settlement, and more than 75 percent of the region's population now lives in urban areas, a striking contrast to 1850, when only about one in five (20 percent) people lived in an urban area.

Modern Spatial and Demographic Patterns

Metropolitan clusters dominate North America's population geography, producing strikingly uneven patterns of settlement across the region (Figure 3.10). The largest number of

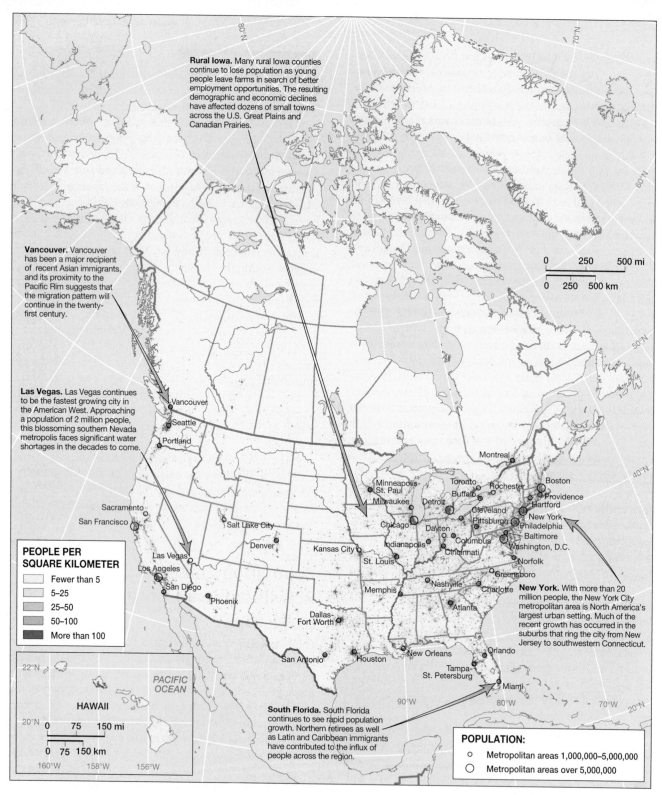

Rural Iowa. Many rural Iowa counties continue to lose population as young people leave farms in search of better employment opportunities. The resulting demographic and economic declines have affected dozens of small towns across the U.S. Great Plains and Canadian Prairies.

Vancouver. Vancouver has been a major recipient of recent Asian immigrants, and its proximity to the Pacific Rim suggests that the migration pattern will continue in the twenty-first century.

Las Vegas. Las Vegas continues to be the fastest growing city in the American West. Approaching a population of 2 million people, this blossoming southern Nevada metropolis faces significant water shortages in the decades to come.

New York. With more than 20 million people, the New York City metropolitan area is North America's largest urban setting. Much of the recent growth has occurred in the suburbs that ring the city from New Jersey to southwestern Connecticut.

South Florida. South Florida continues to see rapid population growth. Northern retirees as well as Latin and Caribbean immigrants have contributed to the influx of people across the region.

PEOPLE PER SQUARE KILOMETER
- Fewer than 5
- 5–25
- 25–50
- 50–100
- More than 100

HAWAII

PACIFIC OCEAN

22°N
20°N
0   75   150 mi
0   75   150 km
160°W   158°W   156°W

POPULATION:
- ○ Metropolitan areas 1,000,000–5,000,000
- ◯ Metropolitan areas over 5,000,000

▲ **Figure 3.10 Population map of North America** North America's geography of population reveals a strikingly clustered pattern of large cities interspersed with more sparsely settled zones. Notable concentrations are found on the eastern seaboard between Boston and Washington, D.C., along the shores of the Great Lakes, and across the Sunbelt from Florida to California.

cities and the densest collection of rural settlements are found south of central Canada and east of the 100th meridian. Within this broad region, Canada's "Main Street" corridor contains most of that nation's urban population, led by the cities of Toronto (4.2 million) and Montreal (3.3 million). The feder-

al capital of Ottawa (1 million) and the industrial center of Hamilton (620,000) are also within the Canadian urban corridor. **Megalopolis,** the largest settlement agglomeration in the United States, includes Washington, D.C. (4.7 million), Baltimore (2.5 million), Philadelphia (6 million), New York

City (20.2 million), and Boston (3.3 million). Beyond these two national core areas, other sprawling urban centers cluster around the southern Great Lakes, various parts of the South, and along the Pacific Coast.

North America's population has increased greatly since the beginning of European colonization. Before 1900, high rates of natural increase produced large families. In addition, waves of foreign immigration swelled settlement, a pattern that continues today. In Canada, a population of less than 300,000 Indians and Europeans in the 1760s grew to an impressive 3.2 million a century later. For the United States, a late colonial (1770) total of around 2.5 million increased more than tenfold to more than 30 million by 1860. Both countries saw even higher rates of immigration in the late nineteenth and twentieth centuries, although birthrates gradually fell after 1900. After World War II, birthrates rose once again in both countries, resulting in the "baby boom" generation born between 1946 and 1965. Today, however, as in much of the developed world, rates of natural increase in North America are below 1 percent annually (Table 3.1), and the overall population is growing older. Still, the region continues to attract immigrants: more than 31 million foreign-born migrants now live in North America. These growing numbers, along with higher birthrates among immigrant populations, have led demographic experts to increase long-term population projections for the twenty-first century. Indeed, predictions that the region's population will reach 375 million (338 million in the United States; 37 million in Canada) by 2025 may prove conservative.

Occupying the Land

Europeans began occupying North America about 400 years ago. It is important to remember that Europeans were not peopling an empty land. North America was populated for at least 12,000 years by peoples as culturally diverse as those who came to conquer them. Native Americans were broadly distributed across the realm and made diverse adaptations to its many natural environments. Their precontact numbers are impossible to reconstruct precisely, although cultural geographers estimate A.D. 1500 populations at 3.2 million for the continental United States and another 1.2 million for Canada, Alaska, Hawaii, and Greenland. European diseases and disruption decimated these Native American populations as

contacts increased. Native societies could hardly have anticipated the magnitude of coming changes after 1600 as the European world expanded its reach. The continental sweep of settlement that followed took shape in three stages, and the twentieth-century results fundamentally reordered North America's human geography.

The first stage of this dramatic new settlement geography began with a series of European colonial footholds, mostly within the coastal regions of eastern North America (Figure 3.11). Established between 1600 and 1750, these regionally distinct societies were anchored on the north by the French settlement of the St. Lawrence Valley and nearby areas of the Canadian Maritimes. English Puritans dominated nearby southern New England, imposing their own brand of cultural orthodoxy by the 1640s. Farther south, both the Dutch colony of New Netherlands (later English-controlled New York) and the English Quaker colony of Pennsylvania attracted a varied collection of farmers, merchants, and tradesmen. Bicultural European and African settlements concentrated in the plantation South, with the largest colonies in English-controlled Virginia and South Carolina. Additional French settlements concentrated along the Gulf of Mexico (New Orleans was founded in 1718), and there was an early Spanish presence in the Southwest (Santa Fe was founded in 1610) and Florida (St. Augustine was founded in 1565).

The second stage in the Europeanization of the North American landscape took place between 1750 and 1850, and it was highlighted by the infilling of much of the better agricultural land within the eastern half of the continent (Figure 3.11). Restrictive English colonial policies failed to deter frontier settlement in the upper Ohio and Tennessee valleys. Following the American Revolution (1776) and a series of Indian conflicts, pioneers surged across the Appalachians. They found much of the Interior Lowlands region almost ideal for agricultural settlement. As a result, most of the Midwest and interior South were occupied by 1850. Expansion in early Canada was more modest. Few major changes in settlement came as the region shifted from French to English control in 1763. More important, much of southern Ontario, or Upper Canada, was opened to widespread development in 1791.

The third stage in North America's settlement expansion accelerated after 1850 and continued until just after 1910 (Figure 3.11). During this period, most of the region's remaining

TABLE 3.1 Demographic Indicators

Country	Population (Millions, 2001)	Population Density, per square mile	Rate of Natural Increase	TFR[a]	Percent <15[b]	Percent >65	Percent Urban
Canada	31	8	0.3	1.4	19	13	78
United States	284.5	77	0.6	2.1	21	13	75
Total	315.5						

[a]Total fertility rate
[b]Percent of population younger than 15 years of age
Source: Population Reference Bureau. World Population Data Sheet, 2001.

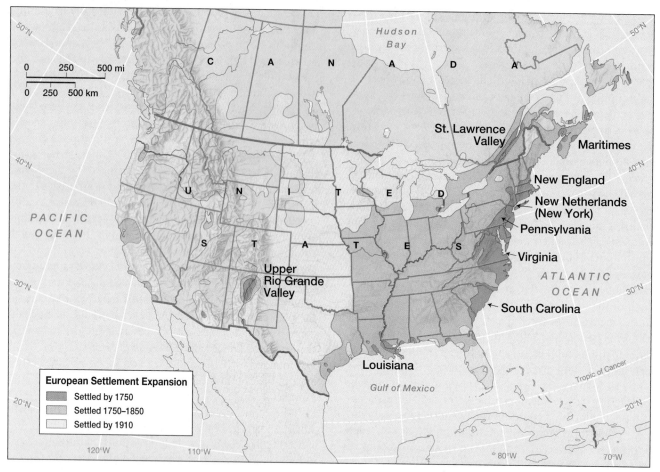

▲ **Figure 3.11 European settlement expansion**   Sizable portions of North America's East Coast and the St. Lawrence Valley were occupied by Europeans before 1750. The most remarkable surge of settlement, however, occurred during the next century as vast areas of land were opened to European dominance and Native American populations were dramatically disrupted.

agricultural lands were settled by a mix of native-born and immigrant farmers. Often, pioneers tested their luck and ingenuity in unfamiliar environmental surroundings. Farmers were challenged and sometimes defeated by aridity, mountainous terrain, and short northern growing seasons. In the American West, settlers were attracted by opportunities in California, the Oregon country, Mormon Utah, and the Great Plains. In Canada, few flocked to the far north, but thousands occupied southern portions of Manitoba, Saskatchewan, and Alberta. Mineral rushes led to initial development in areas such as Colorado, Montana, and British Columbia's Fraser Valley.

Incredibly, in a mere 160 years, much of the North American landscape was domesticated as expanding populations sought new land to occupy and as a globe-encircling capitalist economy demanded resources to fuel its growth. It was one of the largest and most rapid human transformations of the landscape in the history of the human population. This European-led advance forever reshaped North America in its own image, and in the process also changed the larger globe in lasting ways by creating a "New World" destined to reshape the Old.

North Americans on the Move

From the mythic days of Davy Crockett and Daniel Boone to the twentieth-century sojourns of John Steinbeck and Jack Kerouac, North Americans have been on the move. Indeed, almost one in every five Americans moves annually, suggesting that residents of the region are quite willing to change addresses in order to improve their income or their quality of life. Although interregional population flows are complex in both the United States and Canada, several trends dominate the picture.

Westward-moving Populations   The most persistent regional migration trend in North America has been the tendency for people to move west. Indeed, the dominant thrust in the past two centuries (Figure 3.11) has been to follow the setting sun, and many North Americans continue that pattern to the present. By 1990, more than half of the population of the United States lived west of the Mississippi River, a dramatic shift from colonial times. Between 1990 and 2000, some of the fastest-growing states were in the American West (including California, Texas, Arizona, and Nevada), as well as in the western Canadian provinces of Alberta and British Columbia.

Much of the extraordinary growth in the Mountain states has been fueled by new job creation in high-technology industries and services, as well as by the region's scenic and recreational amenities (Figure 3.12). Larger metropolitan areas such as Salt Lake City, Phoenix, and Las Vegas see their numbers continue to swell. In addition, many smaller nonmetropolitan centers, such as St. George, Utah; Coeur d'Alene, Idaho; and Kalispell, Montana, have also attracted a flood of new migrants, including many outward-bound Californians. The sustained move to the Interior West has many implications: the demand for water in the arid region continues to grow; the mix of natives and recent migrants creates cultural tensions; and the area's growing political power (it gained four seats in Congress in 2000) and economic presence is redefining its traditionally peripheral role in national affairs. Indeed, these growth centers are full participants in the globalization process. For example, Denver's international exports more than doubled in the 1990s.

Black Exodus from the South   African Americans displayed a somewhat different pattern of interregional migration. Originally, black slave populations were concentrated in the plantation South. In fact, at the end of the American colonial period, African Americans constituted the majority of the population in southern states such as South Carolina. Even after the legal emancipation of blacks in the 1860s, most remained economically bound to the rural South, where they worked as sharecroppers, often for their former owners. Conditions changed, however, in the twentieth century. Many African Americans migrated because of declining demands for labor in the agricultural South and growing industrial opportunities in the North and West. Two waves of migration, one between 1910 and 1920 and the other from 1940 to 1960, propelled blacks into many new American settings. Overwhelmingly, migrants ended up in cities where job opportunities beckoned. Boston, New York, and Philadelphia became key destinations for blacks from Georgia and the Carolinas,

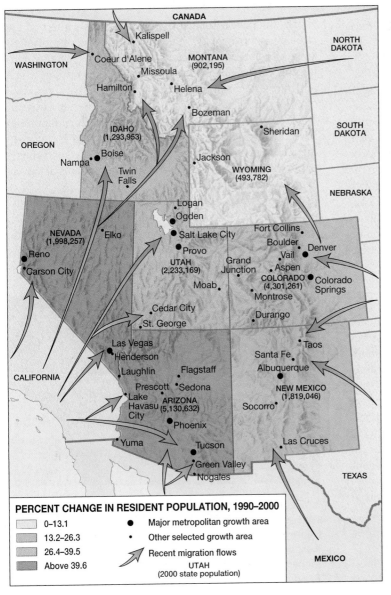

◀ **Figure 3.12 Intermountain West growth, 1990–2000**   Nevada (66 percent growth) and Arizona (40 percent growth) were the two fastest-growing states between 1990 and 2000. Many people are flocking to the region's larger cities to find employment, often in growing high-technology industries. Other amenity-bound migrants and retirees are attracted to the region's smaller towns and recreational areas.   *(Modified from the U.S. Census, 2000, Percent Change in Resident Population for the 50 States, 1990 to 2000)*

PERCENT CHANGE IN RESIDENT POPULATION, 1990–2000

0–13.1	● Major metropolitan growth area
13.2–26.3	• Other selected growth area
26.4–39.5	↗ Recent migration flows
Above 39.6	UTAH (2000 state population)

while Midwest cities such as Detroit and Chicago attracted many blacks from Alabama, Mississippi, and Louisiana. Particularly in the later migration wave, Los Angeles and San Francisco/Oakland drew many blacks to the West. Since 1970, however, more blacks have moved from North to South. Indeed, Sun Belt jobs and federal civil rights guarantees now attract many northern urban blacks to growing southern cities. The net result is still a profound change from 1900: at the beginning of the century, more than 90 percent of African Americans lived in the South, while today only about half of the nation's 35 million blacks reside within the region.

Rural-to-Urban Migration   Another persistent trend in North American migration has taken people from the country to the city. Two centuries ago, only 5 percent of North Americans lived in urban areas (cities of more than 2,500 people), whereas today more than 75 percent of the North American population is urban. Shifting economic opportunities account for much of the transformation: as mechanization on the farm reduced the demand for labor, many young people left for new employment opportunities in the city. Well-paying manufacturing jobs and a growing service economy proved to be powerful magnets for many rural residents. Historically, rates of urbanization jumped first in the Northeast as industrialization spread through the region between 1800 and 1850. The scene of rapid urban expansion shifted westward to the Great Lakes region later in the nineteenth century as industrial growth transformed cities such as Toronto and Chicago. In the twentieth century, people were drawn by many urban employment opportunities in more peripheral zones, resulting in large migrations to cities in the South and West. Most of the twentieth-century urban growth took place in cities of fewer than 1 million people. The consequences of this rural-to-urban shift transcend mere job relocation. Larger processes of modernization and globalization were greatly facilitated by the overwhelmingly urban orientation of North America's population.

Growth of the Sun Belt South   Twentieth-century moves to the American South are clearly related to other dominant trends in North American migration, yet the pattern deserves closer inspection. Particularly after 1970, southern states from the Carolinas to Texas grew much more rapidly than states in the Northeast and Midwest. During the 1990s, Georgia, Florida, Texas, and North Carolina each grew by more than 20 percent, while Ohio and New York eked out 5 percent gains. The South's buoyant economy, modest living costs, adoption of air conditioning, attractive recreational opportunities, and appeal to snow-weary retirees all contributed to its growth. Movements have been selective, however; many rural agricultural and mountain counties within the South have seen few new residents, while amenity-rich coastal settings and job-generating metropolitan areas witnessed spectacular growth. Atlanta's bustling metropolitan area (3.8 million), for example, is now larger than that of Boston (3.3 million) (Figure 3.13).

The Counterurbanization Trend   During the 1970s, certain nonmetropolitan areas in North America began to see significant population gains, including many rural settings that had previously lost population. Selectively, that pattern of **counterurbanization**, in which people leave large cities and move to smaller towns and rural areas, continues today. Some participants in counterurbanization are part of the growing retiree population in both Canada and the United States, but a substantial number are younger, so-called *lifestyle migrants*. They find or create employment in affordable smaller cities and rural settings that are rich in amenities and often removed from the perceived problems of urban America. In fact, recent nonmetropolitan population growth has exceeded metropolitan growth in most western states. Other smaller communities outside the West, such as Mason City, Iowa; Mankato, Minnesota; and Traverse City, Michigan, also are seen as desirable destinations for migrants interested in downsizing from their metropolitan roots.

Settlement Geographies: The Decentralized Metropolis

North America's settlement landscape bears witness to the population movements, shifting regional economic fortunes, and technological innovations that shaped twentieth-century geographical change across the realm. The ways in which settlements are organized on the land—the actual appearance of cities, suburbs, and farms—as well as the very ways in which North Americans socially construct their communities have changed greatly in the past century. Today's cloverleaf interchanges, sprawling suburbs, outlet malls, and theme parks would have struck most 1900-era residents as utterly extraordinary.

Settlement landscapes of North American cities boldly display the consequences of **urban decentralization**, in which metropolitan areas sprawl in all directions and suburbs take on many of the characteristics of traditional downtowns. Although both Canadian and U.S. cities have experienced decentraliza-

▲ **Figure 3.13 Downtown Atlanta, Georgia**   Sunbelt cities such as Atlanta have been transformed by rapid job creation. Healthy growth in office space, specialty retailing, and entertainment districts has fueled downtown Atlanta's expansion and reshaped the look of the central city skyline.   *(Bill Bachmann/Photo Researchers, Inc.)*

tion, the impact has been particularly profound in the United States, where inner-city problems, poor public transportation, widespread automobile ownership, and fewer regional-scale planning initiatives have encouraged many middle-class urban residents to move beyond the central city. Even beyond North America, observers note a globalization of urban sprawl: many Asian, European, and Latin American cities are taking on attributes of their North American counterparts as they experience similar technological and economic shifts. Indeed, much as they have in Seattle and Albuquerque, suburban Wal-Marts, semiconductor industrial parks, and shopping malls may become increasingly familiar sights on the peripheries of Kuala Lumpur or Mexico City.

Historical Evolution of the City in the United States

Changing transportation technologies decisively shaped the evolution of the city in the United States (Figure 3.14). The pedestrian/horsecar city (pre-1888) was compact, essentially limiting urban growth to a three- or four-mile-diameter ring around downtown conveniently accessible by foot or horse-powered trolley cars. The invention of the electric trolley in 1888 propelled the urbanized landscape farther into new "streetcar suburbs," often 5 or 10 miles from the city center. Indeed, the electric streetcar city (1888–1920) offered commuters affordable mass transit at the dizzying speed of 15 or 20 miles an hour. A star-shaped urban pattern resulted, with growth extending outward along and near the streetcar lines.

The biggest technological revolution came after 1920 with the widespread adoption of the automobile. The recreational automobile city (1920–1945) promoted the infilling of areas beyond the reach of the streetcar and added even more distant suburban rings in the surrounding countryside. Essentially, it allowed many middle-income residents, particularly whites, to leave the central city in favor of lower density and less ethnically complex suburban settings. Following World War II, the outer city (1945 to the present) promoted even more dramatic decentralization along commuter routes as built-up areas appeared 40 to 60 miles from downtown. The perception that inner-city crime, congestion, and racial tensions were worsening after 1960 accelerated the movement of whites toward the presumed peace of the suburbs.

Urban decentralization also reconfigured land-use patterns in the city, producing metropolitan areas today that are strikingly different from their counterparts of the early twentieth century. In the city of the early twentieth century, idealized in the **concentric zone model,** urban land uses were neatly organized in rings around a highly focused central business district (CBD) that contained much of the city's retailing and office functions. Residential districts beyond the CBD were added as the city expanded, with higher-income groups seeking more desirable, peripheral locations. The model nicely summarized the urban geography of many cities in the United States during the 1920s, and elements of the model are still relevant today, particularly in examining how eras of economic expansion add rings of residential construction around the periphery of a city.

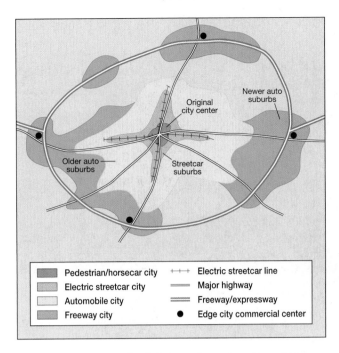

▲ **Figure 3.14 Growth of the American city** Many U.S. cities became increasingly decentralized as they moved through the pedestrian/horsecar, electric streetcar, automobile, and freeway eras. Each era left a distinctive mark on metropolitan America, including the recent growth of edge cities on the urban periphery.

Today's **urban realms model** highlights new suburban growth characterized by a mix of peripheral retailing, industrial parks, office complexes, and entertainment facilities. These areas of activity, often called "edge cities," have fewer functional connections with the central city than they have with other suburban centers. For most suburban residents of an edge city, jobs, friends, and entertainment are located in other peripheral realms, rather than in the old downtown. Tysons Corner, Virginia, is a superb example of the edge-city landscape on the dynamic periphery of a North American metropolis (Figure 3.15).

The Consequences of Sprawl

The rapid evolution of the North American city continues to transform the urban landscape and those who live in it. As suburbanization accelerated in the 1960s and 1970s, many inner cities, especially in the Northeast and Midwest, suffered absolute losses in population, increased levels of crime and social disruption, and a shrinking tax base that often brought them to the brink of bankruptcy. The inner city, particularly in the United States, still faces daunting educational challenges, homeless populations, and immense planning problems. Poverty rates average almost three times those of nearby suburbs. Unemployment rates, while declining somewhat in the 1990s, remain above the national average. Inner-city populations often lack the training, family stability, and social networks that can aid in obtaining employment. Above all, central cities in the United States remain places of racial tension, the product of decades of discrimination, segregation, and poverty. While there are

▲ **Figure 3.15 Tysons Corner, Virginia** North America's edge-city landscape is superbly illustrated by Tysons Corner, Virginia. Far from a traditional metropolitan downtown, this sprawling complex of suburban offices and commercial activities reveals how and where many North Americans will live their lives in the twenty-first century. *(Rob Crandall/Rob Crandall, Photographer)*

▲ **Figure 3.16 Baltimore's Harborplace** Inner-city revitalization is exemplified by the dramatic transformation of Baltimore Harbor. The elaborate development offers visitors food, entertainment, and specialty shopping in an upscale setting that caters to a cosmopolitan urban clientele. *(Rob Crandall/Rob Crandall, Photographer)*

growing numbers of middle-class African Americans and Hispanics, many exit the central city for the suburbs, further isolating the urban underclass that remains behind.

Amid these challenges, inner-city landscapes are also enjoying a selective renaissance. Referred to as **gentrification**, the process involves the displacement of lower-income residents of central-city neighborhoods with higher-income residents, the rehabilitation of deteriorated inner-city landscapes, and the construction of new shopping complexes, entertainment attractions, or convention centers in selected downtown locations. The older and more architecturally diverse housing stock of the central city is also a draw, serving as specialty shops and restaurants for a cosmopolitan urban clientele and offering residential opportunities for upscale singles who wish to live near downtown. Seattle's Pioneer Square, Toronto's Yorkville district, and Baltimore's Harborplace exemplify how such new public and private investments shape the central city (Figure 3.16).

The suburbs are also changing. Construction of new corporate office centers, fashion malls, and industrial facilities has created true "suburban downtowns" in suburbs that are no longer simply bedroom communities for central-city workers. Indeed, such localities have been growing players in the continent's globalization process. Many of North America's key internationally connected corporate offices (IBM, Microsoft), industrial facilities (Boeing, Sun Microsystems, Oracle), and entertainment complexes (Disneyland, Walt Disney World, and the Las Vegas Strip) are now in such settings, and they are intimately tied to global information, technology, capital, and migration flows (see "The Local and the Global: Sprawling Las Vegas").

The edge-city lifestyle has also transformed the United States into a continent of suburban commuters in which people live in one suburb and work in another. The average daily commute now exceeds 30 miles per person in cities such as Atlanta, Georgia, and Birmingham, Alabama. As urban settlements of the twenty-first century grow in area but decline in density, it will become more difficult to build roads, schools, and other public infrastructure in cost-effective ways. Another consequence of edge-city sprawl is the rapid loss of surrounding rural land. Particularly on the outer suburban fringe, the boundary between city and country is blurring; as urbanites reach toward the amenities and privacy of the rural landscape, modern technologies and metropolitan growth bring the countryside inexorably closer to the city.

Settlement Geographies: Rural North America

Rural North American landscapes trace their origins to initial European settlement. Although many New World communities replicated elements of the Old World, North Americans favored more dispersed rural settlement patterns than their European antecedents. An abundance of available acreage, liberal land disposal laws, and perhaps a cultural predilection for independence and privacy all contributed to a pattern of dispersed settlement in North America in which settlers lived on their own farmsteads, often at some distance from their nearest neighbors.

In portions of the United States settled after 1785, the federal government surveyed and sold much of the rural landscape. Surveys were organized around the simple, rectangular

The spectacular growth of Las Vegas, Nevada, illustrates the tangible local imprint of global processes. As North America's fastest growing major metropolitan area, Las Vegas reveals what happens when a locality becomes increasingly tied to a dynamically growing segment of the global economy. Las Vegas is a quintessential global niche city. It has emerged as a national and international destination resort, entertainment, and retirement center. It now hosts 4,000 conventions, meetings, and trade shows annually, and its 125,000 hotel rooms (the largest total in North America) play host to more than 32 million visitors a year, many of them from Europe and Asia. The city's huge hotels, entertainment complexes, shopping areas, and golf courses are now globally recognized amenities. More generally, the city has profited from North America's mushrooming affluence, much of it created through the region's growing participation in the global economy.

What have been the local consequences of this globally driven expansion? Most important, the region's population soared from 64,000 people in 1960 to 1.5 million in 2001. Predictions are for more than 3 million residents by 2020. The accompanying suburban expansion has been a low-density assortment of single-family homes, retirement complexes, and decentralized shopping malls. The local environmental impacts of this space-extensive growth have been

dramatic. The metropolis has rapidly sprawled into the nearby desert environment both east and west of the central city (Figure 3.2.1). The city has also made ever-increasing demands on rapidly dwindling regional water supplies, mostly for residential use. Surface water resources have been completely allocated, and the region's groundwater supplies are being quickly depleted. Meanwhile, as housing subdivisions multiply, nearby wildlife species, such as the desert tortoise, are threatened.

Social and cultural changes have also accelerated. Local governments are furiously building schools to keep up with the 7 percent annual growth rate in enrollments. The area's 238 schools in 2000 are expected to grow to more than 320 by 2008. The city's cultural geography is becoming as diverse as nearby California's, with thousands of new Hispanic and Asian workers attracted to the area's strong job market. In addition, retirees from nearby California and from the world beyond have flocked to the region in huge numbers. Chronic problems of crime, drug abuse, and gang violence have grown along with the population. What the Las Vegas Valley will look like in 50 years is anyone's guess. What is certain is that North American and global affluence has forever reshaped the city's settlement geography, cultural landscapes, and environmental setting.

▲ **Figure 3.2.1 Sprawling Las Vegas**   These satellite images from 1990 and 2000 reveal the extraordinary pace of change in Henderson, Nevada, the most rapidly growing suburb of Las Vegas. Henderson is located 8 miles southeast of the famous Las Vegas Strip.   *(Landiscor)*

pattern of the federal government's township-and-range survey system, which offered a convenient method of dividing and disposing of the public domain in six-mile-square townships. In a repeating pattern across the U.S. interior, grid towns were laid out amid a seemingly endless geometry of straight lines that neatly squared off the surrounding rural landscape (Figure 3.17). Canada developed a similar system

of regular surveys that stamped much of southern Ontario and the western provinces with a strikingly rectilinear character. Exceptions to the predictable pattern were found in former American colonies such as Virginia, where irregular surveys dominated, and in Canada's St. Lawrence Valley, in which distinctive long lot lines lay perpendicular to the river to give more farmers access to the water.

▶ **Figure 3.17 Iowa settlement patterns** The regular rectangular look of this Iowa town and the nearby rural setting reveals the North American penchant for simplicity and efficiency. In the United States, the township-and-range survey system stamped such predictable patterns across vast portions of the North American interior. *(Craig Aurness/Corbis)*

Commercialized agriculture and technological changes further reorganized the settlement landscape. Railroads opened new corridors of development, provided access to markets for commercial crops, and facilitated the establishment of towns that served surrounding rural populations. By 1900, several transcontinental lines spanned North America, radically transforming the farm economy and the pace of rural life. After 1920, however, even more profound changes accompanied the arrival of the automobile, farm mechanization, and better rural road networks. The need for farm labor declined with mechanization, and many smaller market centers became unnecessary as farmers equipped with automobiles and trucks could travel farther and faster to larger, more diverse towns. Other technological innovations, including center-pivot irrigation systems, better yielding seeds, and new fertilizers and insecticides, also made their mark on the rural settlement landscape.

Today, many rural North American settings face population declines as they adjust to the changing conditions of modern agriculture (see "Virtual Field Trip: Boom and Bust in Colorado"). Both U.S. and Canadian farm populations fell by more than two-thirds during the last half of the twentieth century. Typically, larger but fewer farms dot the modern rural scene, and many young people leave the land to obtain employment elsewhere. The population drain in settings such as rural Iowa, southern Saskatchewan, and eastern Montana also affects towns within these regions. Fewer shoppers visit local stores; doctors and dentists retire and move; and schools and churches are forced to close or consolidate. Indeed, the very fabric of rural and small-town communities is threatened. The visual legacies of abandonment provide poignant reminders of painful adjustments: weed-choked driveways, empty farmhouses, roofless barns, and the empty marquees

of small-town movie houses tell the story more powerfully than any census or government report.

Other rural settings show signs of economic health and population growth. No single explanation suffices for these burgeoning hinterlands. Some localities are merely feeling the fringe effects of encroaching edge cities. Suddenly, almost overnight, such places are overrun with suburbanites who snap up farmhouses once beyond the reach of the city. Other growing rural settings lie beyond direct metropolitan influence but are seeing new populations who seek amenity-rich environments removed from city pressures. These impulses toward counterurbanization are shaping the settlement landscape from British Columbia's Vancouver Island to Michigan's Upper Peninsula. Newly subdivided land parcels, a plethora of real estate offices, the appearance of resort and golf complexes, and the telltale in-migration of espresso bars all signal the changes afoot in such surroundings.

Cultural Coherence and Diversity: Shifting Patterns of Pluralism

North America's cultural geography is both globally dominant and internally pluralistic. On one hand, history and technology have produced a contemporary North American cultural force that is second to none in the world. Many people outside the United States speak of cultural imperialism when they describe the increasing global dominance of American popular culture, which they often see as threatening the vitality of other cultural values. Yet North America is also a mosaic of different peoples who retain part of their traditional cultural identities. In fact, North Americans celebrate their pluralistic folk roots and acknowledge the realm's multicultural character.

The Roots of a Cultural Identity

Powerful forces forged a common dominant culture within the realm. Historically, both Canada and the United States were strongly tied to Great Britain, and these links transcended colonial politics. With some exceptions, most early Europeans who played key cultural, economic, and political roles in the region were from the British Isles. Key Anglo legal institutions and social mores solidified the common set of core values that many North Americans shared with Britain and, eventually, with one another. Traditional beliefs, defined within the limits of the Anglo world view, emphasized representative government, separation of church and state, liberal individualism, privacy, pragmatism, and social mobility. The American Revolution (1776) and the Confederation of Canada (1867), while distancing the realm from direct British authority, were nevertheless assertions of cultural continuity that combined the Anglo world view with converging national cultural identities.

As settlement spread and economic linkages forged truly continental connections in the late nineteenth century, these core values offered a set of experiences shared by many North Americans, both native- and foreign-born. From those traditional roots, particularly within the United States, consumer culture blossomed after 1920, producing a common, if more diffuse, set of shared experiences that included mass advertising, movies, national radio and television programming, and an increasingly secular society oriented toward convenience and consumption. Since World War II, these modern technologies and values, with the help of growing political and multinational corporate imperatives, have cumulatively shaped the world in North America's cultural image.

But North America's cultural coherence coexists with pluralism, the persistence and assertion of distinctive cultural identities. Closely related is the concept of **ethnicity,** in which a group of people with a common background and history identify with one another, often as a minority group within a larger society. Indeed, the roots of these impulses are deep and sometimes divisive, and they have created ongoing tensions within the two countries that often have a powerful geographical expression. For Canada, the early and enduring French colonization of Quebec complicates its contemporary cultural geography. Canadians face the challenge of creating a truly bicultural society where issues of language and political representation are central concerns. Within the United States, given its unique immigration history, a greater diversity of ethnic groups exists, and key geographical divisions of social space are often local as well as regional. The result can be a mosaic of Anglos, African Americans, Hispanics, and Asians often coexisting within a single city yet retaining their cultural identities and claims to distinctive areas of local territorial dominance.

Peopling North America

North America is a region of immigrants. Quite literally, global-scale migrations made North America possible. Decisively displacing Native Americans in most portions of the realm, immigrant populations created a new cultural geography of ethnic groups, languages, and religions. Early migrants often had considerable cultural influence, even though their numbers were minuscule compared to the flood of later arrivals. Over time, varied immigrant groups and their changing destinations produced a culturally heterogeneous landscape that continues to evolve today. Also varying between groups was the pace and degree of **cultural assimilation,** the process in which immigrants were absorbed by the larger host society.

Migration to the United States   In the United States, variations in the number and source regions of migrants produced five distinctive chapters in the country's history (Figure 3.18). In Phase 1 (prior to 1820), English and African influences dominated. Other Europeans, particularly Irish, Dutch, French, and Germans, were also important, but it was the English who played the pivotal role. Slaves, mostly from West Africa, contributed additional cultural influences in the South. Northwest Europe served as the main source region of immigrants between 1820 and 1870 (Phase 2). The emphasis, however, shifted away from English dominance to include greater numbers of Irish and Germans. The pace of immigration also increased greatly: only 130,000 migrants arrived in the 1820s versus more than 2.8 million in the 1850s.

As Figure 3.18 shows, immigration reached a much higher peak around 1900, when almost 1 million foreigners entered the United States *annually*. During Phase 3 (1870–1920), the source regions of this great wave of migrants also shifted. Northwest Europeans declined in importance while southern (particularly Italians) and eastern (Poles, Russians, Austro-Hungarians) Europeans dominated. Political strife and poor economies in Europe sharply contrasted with news of available land, expanding industrialization, and well-paying jobs in America. Between 1880 and 1920, Scandinavians also swept through the northern interior, where they settled on still-available farmland. By 1910, almost 14 percent of the nation was foreign-born. Very few of these immigrants, however, targeted the job-poor American South, creating a cultural divergence that persists to the present.

Twentieth-century migrations also profoundly shaped the nation's cultural geography. Between 1920 and 1970 (Phase 4), more immigrants came from neighboring Canada and Latin America, but overall totals plunged, a function of more restrictive federal immigration policies (the Quota Act of 1921 and the National Origins Act of 1924), the Great Depression, and the disruption caused by World War II. Since 1970 (Phase 5), the realm has witnessed a sharp reversal in numbers, however, and now arrivals match those of the early twentieth century (Figure 3.18). The current surge was made possible by economic and political instability abroad, a growing postwar American economy, and a loosening of immigration laws (the Immigration Acts of 1965 and 1990, and the Immigration Reform and Control Act of 1986). Illegal immigration also rose after 1970, although several federal laws passed in 1996 greatly increased the number of U.S. border patrol agents and made it more difficult for immigrants, both legal and illegal, to receive federal welfare benefits. Most migrants since 1970 originated in Latin America or Asia.

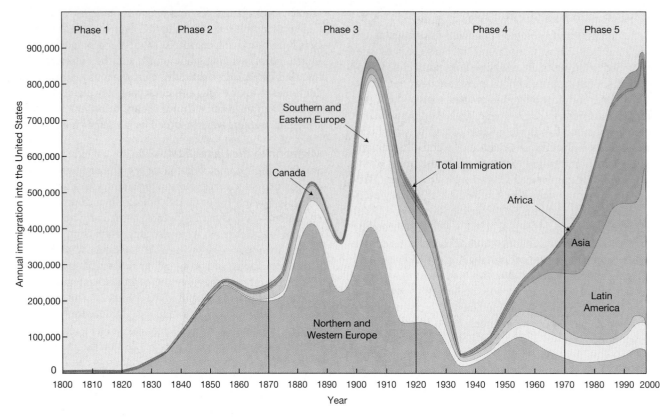

▲ **Figure 3.18 United States immigration, by year and group**   Annual immigration rates peaked around 1900, declined in the early twentieth century, and then surged again, particularly since 1970. The source areas of these migrants have also shifted. Note the decreased role Europeans currently play versus the growing importance of Asians and Latin Americans.   *(Modified from Rubenstein, 2002,* An Introduction to Human Geography, *Upper Saddle River, NJ: Prentice Hall)*

Today, about one-fourth of U.S. immigrants arrive from Mexico, and Mexicans make up more than 60 percent of the nation's Hispanic population. Although most Mexican migrants live in California or Texas, they are increasingly moving to other areas. Recently, states such as Wisconsin, Georgia, and Arkansas have witnessed dramatic increases in their immigrant Hispanic populations, a trend likely to continue in the early twenty-first century. The cultural implications of these Hispanic migrations are profound: today the United States is the fifth largest Spanish-speaking nation on Earth.

In percentage terms, Asian newcomers are the fastest-growing immigrant group, and they account for almost 4 percent of the population. Asian migrants often move to large cities. In fact, more than 40 percent of the nation's Asian population lives in the Los Angeles (1.7 million), San Francisco (1.2 million), and New York City (1.2 million) metropolitan areas. Beyond these key gateway cities, Asians are also moving to growing communities in Washington, D.C.; Chicago; Seattle; and Houston. Filipinos, Vietnamese, Chinese, and Indians dominate the flow, but Korean and Japanese communities are also well established, particularly in the West Coast states and Hawaii. One of the distinctive attributes of the Asian population nationally is that their median household incomes average 10 percent *higher* than those of white households. Still, growing numbers of poorer immigrants, particularly from

East and Southeast Asia, have also arrived in the United States, both legally and illegally, since 1980.

The future cultural geography of the United States will be dramatically redefined by these recent immigration patterns (Figure 3.19). Indeed, the increasing ethnic diversity of the country is simply one more manifestation of the globalization process and the powerful pull of North America's economy and political stability. By 2070, Asians may total over 10 percent of the U.S. population, and almost one American in three will be Hispanic. Indeed, it is likely that the U.S. non-Hispanic white population will achieve minority status by that date.

The Canadian Pattern   The peopling of Canada broadly parallels the U.S. story, but there are some important differences. Early French arrivals concentrated in the St. Lawrence Valley and constituted a well-defined cultural nucleus of 60,000 residents by the mid-eighteenth century. After 1765, however, many new migrants came from Britain, Ireland, and the United States. Canada then experienced the same surge and reorientation in migration flows seen in the United States around 1900. Between 1900 and 1920, more than 3 million foreigners ventured to Canada, an immigration rate far higher than for the United States, given Canada's much smaller population. Eastern Europeans, Italians, Ukrainians, and Russians were prominent participants in these later movements.

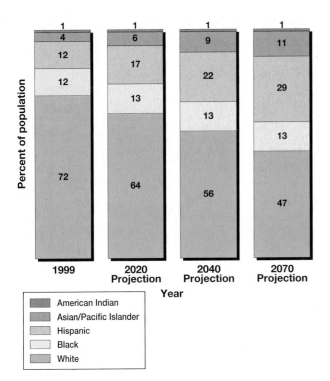

▲ **Figure 3.19 Projected U.S. ethnic composition, 1999 to 2070** By late in the century, almost one in three Americans will be Hispanic, and non-Hispanic whites will achieve minority status amid an increasingly diverse U.S. population. *(Modified from the U.S. Census,* Census of Population, National Projections, *2020-2070)*

Some settled in Ontario cities such as Toronto, while others pioneered on the Canadian prairies. Today, more than 60 percent of Canada's recent immigrants are Asians. As in the United States, more liberal immigration laws since the 1960s encouraged movement to Canada, and its 16 percent foreign-born population is among the highest in the developed world. Just as in the United States, rapidly shifting immigrant populations are changing the look of many Canadian cities. In Toronto, more than 1.7 million immigrants reveal a slight bias toward European backgrounds, while the west-coast metropolis of Vancouver is clearly dominated by Asian, particularly Chinese, populations (Figure 3.20).

Culture and Place in North America

Cultural and ethnic identity are often strongly tied to place. North America's cultural diversity is expressed geographically in two ways. First, similar people congregate near one another and derive meaning from the territories they occupy in common. Second, culture marks the visible scene: the everyday landscape is filled with the artifacts, habits, language, and values of different groups. Boston's Italian North End simply looks and smells different than nearby Chinatown, and rural French Quebec is a world away from a Hopi village in Arizona. In exploring the relationship between culture and place in North America, both large, regional-scale cultural homelands as well as smaller, more local-scale ethnic neighborhoods illustrate the connection.

Persisting Cultural Homelands French-Canadian Quebec superbly exemplifies the cultural homeland: it is a culturally distinctive nucleus of settlement in a well-defined geographical area, and its ethnicity has survived over time, stamping the cultural landscape with an enduring personality (Figure 3.21). More than 80 percent of the population of Quebec speaks French, and language remains the cultural glue that holds the homeland together. Indeed, policies adopted after 1976 strengthened the French language within the province by requiring French instruction in the schools and by mandating national bilingual programming by the Canadian Broadcasting Corporation (CBC). Also fostering unity are the group's minority status within a larger Anglo-dominated Canada (only about 25 percent of Canadians are French) and the feeling that the French-Canadians are second-class citizens within their own country. Ironically, many Quebeçois feel that the greatest cultural threat may come not from Anglo-Canadians but rather from recent immigrants to the province. Southern Europeans or Asians in Montreal, for example, show little desire to learn French, preferring instead to put their children in English-speaking private schools.

Another well-defined cultural homeland is the Hispanic Borderlands (Figure 3.21). It is similar in geographical magnitude to French-Canadian Quebec, significantly larger in total population, but more diffuse in its cultural and political articulation. Historical roots of the homeland are deep, extending back to the sixteenth century, when Spaniards opened the region to the European world. The homeland's core is in northern New Mexico, including Santa Fe and much of the surrounding rural hinterlands. A rich legacy of Spanish place names, earth-toned Catholic churches, and traditional Hispanic settlements dot the rolling highlands of northern New Mexico and southern Colorado. From California to Texas, other historical sites, place names, missions, and presidios also reflect the Hispanic legacy.

Unlike Quebec, however, massive twentieth-century migrations from Latin America brought an entirely new wave of Hispanic settlement to the Southwest. More than 32 million Hispanics now live in the United States, with more than half in California, Texas, New Mexico, and Arizona combined. Indeed, by 2015, Hispanics will likely outnumber non-Hispanic whites in California. Within the homeland, Hispanics have created a distinctive Borderlands culture that mixes many elements of Latin and North America. These newer migrants augment the rural Hispanic presence in agricultural settings such as the lower Rio Grande Valley in Texas and the Imperial and Central valleys in California. Cities such as San Antonio and Los Angeles also play leading roles in articulating the Hispanic presence within the Southwest. Regionally distinctive Latin foods and music add internal cultural variety to the region. New York City, Chicago, and Miami serve as key points of Hispanic cultural influence beyond the homeland.

African Americans also retain a cultural homeland, but it has diminished in intensity because of outmigration (Figure 3.21). Reaching from coastal Virginia and North Carolina to east Texas, a zone of enduring African American population (the Black Belt) remains a legacy of the Cotton South, when

▶ Figure 3.20 Vancouver's immigrant population, by place of birth, 1996   While Vancouver is home to many Europeans, particularly from Great Britain, most of the city's immigrants have come from East, Southeast, and Southern Asia. About equal numbers of immigrant residents have moved to Vancouver from the United States, Latin America/Caribbean, Africa, and Australia/Oceania.   *(Data from Statistics Canada, 1996 Census; photograph by Michael Collier/DRK Photo)*

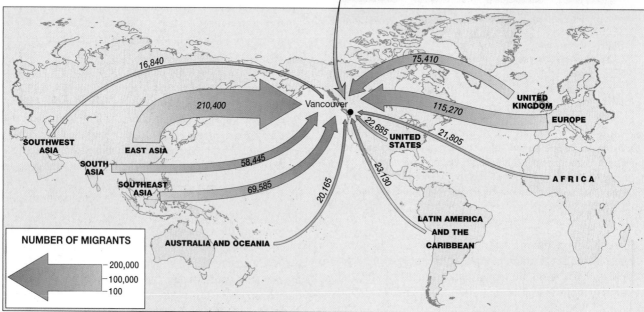

a vast majority of American blacks resided within the region. Today, while many blacks have left for cities, dozens of rural counties in the region still have large black majorities. More broadly, the South is home to many black folk traditions, including such music as black spirituals and the blues, which have now become popular far beyond their rural origins. Regrettably, even though the rural neighborhoods of the black homeland differ greatly in density and appearance from African American urban neighborhoods in the North, poverty plagues both types of communities.

A second rural homeland in the South is Acadiana, a zone of persisting Cajun culture in southwestern Louisiana (Figure 3.21). This homeland was created in the eighteenth century when French settlers were expelled from eastern Canada and relocated to Louisiana. Nationally popularized today through their food and music, the Cajuns retain an enduring attachment to the bayous and swamps of southern Louisiana.

Native American populations are also strongly tied to their homelands. Indeed, many native peoples maintain intimate

relationships with their surroundings, weaving elements of the natural environment together with their material and spiritual lives. Almost 4 million Indians, Inuits, and Aleuts live in North America, and they claim allegiance to more than 1,100 tribal bands. Particularly in the American West and the Canadian and Alaskan North, native peoples control sizable reservations, including the 16-million-acre (6.5-million-hectare) Navajo Reservation in the Southwest, as well as self-governing Nunavut in the Canadian North (Figure 3.21). Although these homelands preserve traditional ties to the land, they have also been settings for pervasive poverty and increasing cultural tensions (Figure 3.22). Within the United States, many Native American groups have taken advantage of the special legal status of their reservations and have built gambling casinos and tourist facilities that bring in much-needed capital, but also challenge traditional life ways. In both Canada and Alaska, friction has also developed among native peoples, private natural resource interests, and various governmental agencies.

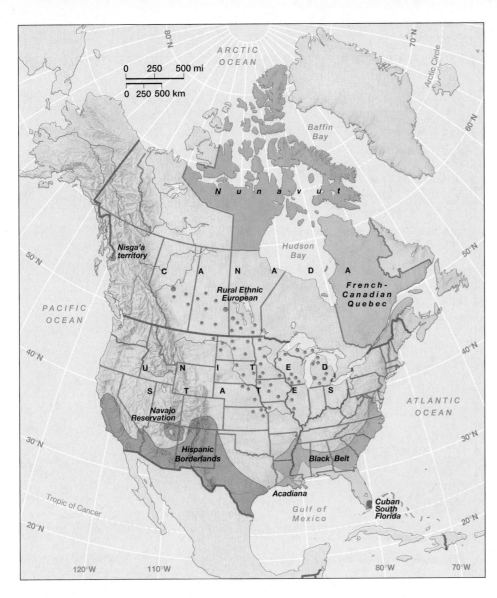

◀ Figure 3.21 Selected cultural regions of North America From northern Canada's Nunavut to the Southwest's Hispanic Borderlands, different North American cultural groups strongly identify with traditional local and regional homelands. Such patterns stamp the modern landscape with an enduring personality. (Modified from Jordan, Domosh, and Rowntree, 1998, The Human Mosaic, Upper Saddle River, NJ: Prentice Hall)

▲ Figure 3.22 Native American poverty Navajo youngsters enjoy a game of basketball on the reservation. Poor housing, low incomes, and persistent unemployment plague many Native American settings across the rural West. (Jim Noelker/The Image Works)

A Mosaic of Ethnic Neighborhoods North America's cultural mosaic is also enlivened by smaller-scale ethnic signatures that shape both rural and urban landscapes. For example, distinctive rural communities that range from Amish settlements in Pennsylvania to Ukrainian neighborhoods in southern Saskatchewan add cultural variety. When much of the agricultural interior was settled, immigrants often established close-knit communities. Among others, German, Scandinavian, Slavic, Dutch, and Finnish neighborhoods took shape, held together by common origins, languages, and religions. Although many of these ties weakened over time, rural landscapes of Wisconsin, Minnesota, the Dakotas, and the Canadian prairies still display many of these cultural imprints. Folk architecture, distinctive settlement patterns, ethnic place names, and the simple elegance of rural churches selectively survive as signatures of cultural diversity upon the visible scene of rural North America.

Ethnic neighborhoods also enrich the urban landscape and reflect both global-scale and internal North American migration patterns. Complex social and economic processes are clearly at work. Employment opportunities historically fueled population growth in North American cities, but the cultural

▶ **Figure 3.23 Ethnicity in Los Angeles** Varied economic opportunities have attracted a wide variety of migrants to North American cities, producing varying ethnic mosaics of distinctive neighborhoods and communities. In Los Angeles, several cycles of economic expansion have drawn a diverse collection of residents from around the globe. *(Reprinted from Rubenstein, 2002, An Introduction to Human Geography, Upper Saddle River, NJ: Prentice Hall)*

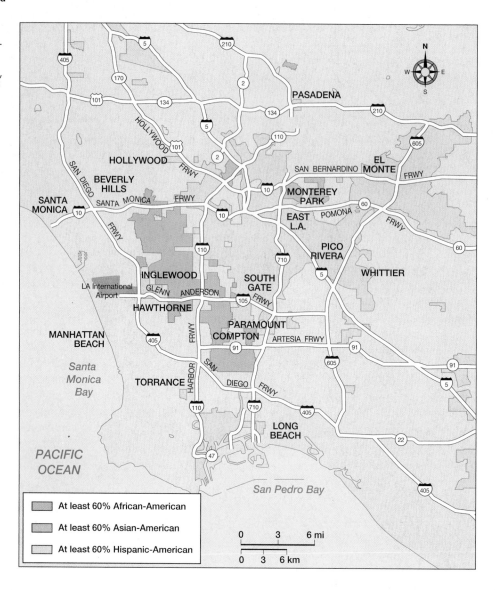

makeup of the incoming labor force varied, depending on the timing of the economic expansion and the relative accessibility of an urban area to different cultural groups. The ethnic geography of Los Angeles exemplifies the interplay of the economic and cultural forces at work (Figure 3.23). Since most of its economic expansion took place during the twentieth century, its ethnic geography reflects the movements of more recent migrants. African American communities on the city's south side (Compton and Inglewood) represent the legacy of black population movements out of the South. Hispanic (East Los Angeles) and Asian (Alhambra and Monterey Park) neighborhoods are a reminder that almost 40 percent of the city's population is foreign-born. These shifting cultural patterns also have significant political impacts, and many local and state representatives mirror the varied ethnic backgrounds of their constituents.

Particularly in the United States, ethnic concentrations of nonwhite populations increased in many cities during the twentieth century as whites exited for the perceived safety of the suburbs. In terms of central-city population, African Americans make up 75 percent of Detroit and more than 60 percent of Atlanta, while Los Angeles is now more than 40 percent

Hispanic. Often these central-city ethnic neighborhoods are further isolated from the urban mainstream by high levels of poverty and unemployment. Ethnic concentrations are also growing in the suburbs of some U.S. cities: southern California's Monterey Park has been called the "first suburban Chinatown," and growing numbers of middle-class African Americans and Hispanics shape suburban neighborhoods in metropolitan settings such as Atlanta and San Antonio.

Patterns of North American Religion

Many religious traditions also shape North America's human geography. Reflecting its colonial roots, Protestantism is dominant within the United States, accounting for about 60 percent of the population. In some settings, hybrid American religions sprang from broadly Protestant roots. By far the most successful modern articulation of this impulse are the Latter-day Saints (Mormons), regionally concentrated in Utah (77 percent of the population) and Idaho (27 percent of the population), and claiming 5 million North American members. Within Canada, almost 40 percent of the population is

Protestant, with the United Church of Canada and the Anglican Church claiming the largest numbers of adherents. Roman Catholicism remains important in regions that received large numbers of Catholic immigrants. French-Canadian Quebec is a bastion of Catholic tradition and makes Canada's population (44 percent) distinctly more Catholic than that of the United States (25 percent). Still, major regional concentrations of American Catholics persist, particularly in the urban Northeast and the Hispanic Southwest.

Millions of other North Americans practice religions outside of the Protestant and Catholic traditions or are nonbelievers. Orthodox Christians congregate in the urban Northeast, where many Greek, Russian, and Serbian Orthodox communities were established between 1890 and 1920. The telltale domes of Ukrainian Orthodox churches still dot the Canadian prairies of Alberta, Saskatchewan, and Manitoba. More than 7 million Jews live in North America, concentrated in East and West Coast cities. In the United States, the rapidly growing Nation of Islam (Black Muslims) also has a strong urban orientation, reflecting its appeal to many economically dispossessed African Americans. There has been additional recent growth in other North American Islamic (6 million), Buddhist (1 million), and Hindu (1 million) populations where sizable numbers of Asians have settled. While only about 8 percent of people in the United States classify themselves as nonbelievers, a recent survey showed that 30 percent of the population claimed to have a largely secular lifestyle in which religion was rarely practiced.

The Globalization of American Culture

Simply put, North America's cultural geography is becoming more global at the same time that global cultural geographies are becoming more North American (influenced particularly by the United States). As a dominant player in globalization, the United States is inevitably changing in the process, but to what extent? Similarly, how is the globalization of American culture transforming the larger world? Are basic values shifting with globalization; is a new "global culture" being forged; or are most of the technological innovations and elements of popular culture being incorporated into existing traditions? There are no simple answers to these questions, but they should be central concerns to human geographers and all citizens of the twenty-first century.

North Americans: Living Globally
More than ever before, North Americans in their everyday lives are exposed to people from beyond the region. In the United States, about 70,000 foreigners arrive daily, mostly as tourists. Annually, more than 22 million tourists visit the country, while 4 million business visitors spend time within its borders. At American colleges and universities, 500,000 international students add a global flavor to the ordinary curriculum. Canada experiences a similar saturation of such global influences, with the U.S. presence particularly dominant.

Globalization presents cultural challenges for North Americans. In the United States, one key issue revolves around the English language, which some have described as the key "social glue" that holds the nation together. Since 1980, the continuing flood of non-English-speaking immigrants into the country has sharpened the debate over the role English should play in American culture. Many in the United States argue that English should be the country's only officially recognized language in order to compel immigrants to learn it and thus speed their assimilation into the host culture. On the other hand, immigrant groups suggest that they need to maintain their traditional languages both to function within their ethnic communities and to preserve their cultural heritage. In reaction to the immigrants, a growing number of states, especially in the South and West, have recognized English as the "official language." Programs in bilingual education in several states, including Florida and California, have been challenged so that immigrants drop their native language and learn English in school. At the same time, foreign-language media outlets flourish in many multicultural markets around the country. New York City's WSKQ, known as "Mega," is a leading Spanish-language radio station, and Los Angeles is home to Korean, Cantonese, and Japanese television broadcasts. Satellite dishes and cable programming have also given birth to multilingual offerings such as the Filipino Channel, Native American Nations, and TV Asia.

North Americans are going global in other ways. By 2000, more than half the region's people were Internet users, thus opening the door for far-reaching journeys in cyberspace. North Americans also travel much more widely than ever before. Residents of the United States took more than 50 million international flights in 2000, almost four times the rate 20 years earlier. The popularity of ethnic restaurants has peppered the realm with a bewildering variety of Cuban, Ethiopian, Basque, and Pakistani eateries. Chinese and Italian foods dominate the taste buds of all North Americans, with Mexican food a rapidly growing choice in the United States. The growing affinity for foreign beverages mirrors the pattern; imported beer sales in the United States are growing 14 percent annually, with the top producers selling 159 million cases during 1997 (Figure 3.24). Americans also have increased their consumption of foreign red wines and have rapidly Europeanized their coffee-drinking habits. In fashion, Gucci, Armani, and Benetton are household words for millions who keep their eyes on European styles. While British pop music has been an accepted part of North American culture for four decades, the beat of German techno bands, Gaelic instrumentals, and Latin rhythms also have become an increasingly seamless part of daily life within the realm. Indeed, from acupuncture and massage therapy to soccer and New Age religions, North Americans are tirelessly borrowing, adapting, and absorbing the larger world around them.

The Global Diffusion of U.S. Culture
In a parallel fashion, the lives of billions of people beyond the region have been forever changed by U.S. culture. Although the economic and military power of the United States was notable by 1900, it was not until after World War II that the country's popular culture reshaped global human geographies in fundamental

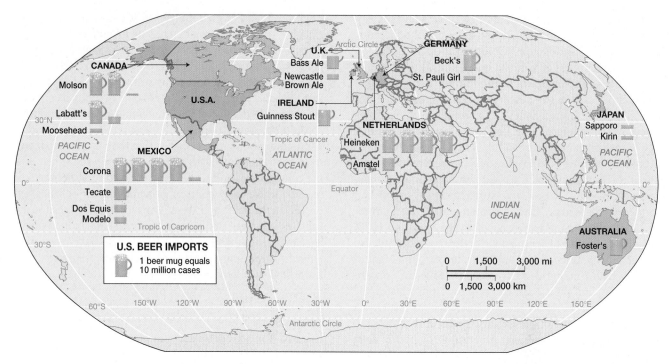

▲ **Figure 3.24 Annual beer imports to the United States**   Whether they are aware of it or not, North Americans are increasingly eating and drinking globally. Rising beer imports, including many upscale foreign labels, exemplify the pattern. The nation's beer drinkers know no bounds to their thirsts, drawing diversely from Asian, Australian, European, and Latin American producers.   *(Data from* Forbes, *October 5, 1998)*

ways. The Marshall Plan and Peace Corps initiatives exemplified the growing presence of the United States on the world stage even as European colonialism waned. Rapid improvements in global transportation and information technologies, much of them engineered in the United States, also brought the world more surely under the region's spell. Perhaps most critical was the marriage between growing global demands for consumer goods and the rise of the multinational corporation, which was superbly structured to meet and cultivate those needs. Today, many U.S. companies dominate such transactions, and with new industries, products, and services have come subtle but fundamental shifts in attitudes and values that challenge older ways of life. While the results are not a simple Americanization of traditional cultures or a single, synthesized global culture shaped in the U.S. image, millions of people, particularly the young, are strongly attracted by that image's emphasis on individualism, wealth, freedom, youth, and mobility. The wide popularity of English-language teaching programs in places from China to Cuba is testimony to America's growing power. Indeed, English is already an official language in more than 75 countries around the world.

Global information flows illustrate the country's cultural influence on the world. Companies such as AOL-Time Warner and Walt Disney increasingly dominate multiple entertainment media. The emergence of English as the de facto global language has obviously been an advantage to a variety of U.S. media outlets. U.S. book and magazine publishing is thriving, and there is an expanding international appetite for everything from technical journals to romance novels and science fiction.

Cosmopolitan magazine produces dozens of foreign-language editions, and its international circulation tops 4 million copies, almost twice its readership in the United States. The United States also dominates television; as television sets, cable networks, and satellite dishes spread, so do U.S. sitcoms, CNN, and MTV. Even the Asia-wide STAR TV network, which broadcasts from Hong Kong, carries a preponderance of North American and English-language programming. Movie screens and VCRs tell a similar story. In western Europe, the United States claims 70 percent of the film market, and more than half of Japan's movie entertainment is dubbed fresh from Hollywood. The country also dominates cyberspace, both in terms of the number of host sites on the Internet as well as technological innovation.

The United States shapes the popular cultural landscape of every corner of the globe. Global corporate advertising, distribution networks, and mass consumption bring Cokes and Big Macs to Moscow and Beijing, golf courses to Thai jungles, Mickey and Minnie Mouse to Tokyo and Paris, and Avon cosmetics to millions of beauty-conscious Chinese (Figure 3.25). Western-style business suits have become the professional uniform of choice, while T-shirts and jeans offer standardized global comfort on days away from work. In the built landscape, central city skylines become indistinguishable from one another, suburban apartment blocks take on a global sameness, and one airport hotel looks the same as another eight time zones away.

The country's cultural control has not gone unchallenged. For example, Canadian government agencies routinely chastise their radio, television, and film industries for letting in too much U.S. cultural influence. As an antidote to their overbearing southern neighbor, the government requires certain levels of Canadian content in much of the media programming. Even so, more than 70 percent of actual television viewing time of English-speaking Canadians is spent watching foreign programs, and the overwhelming majority of these are American. The French have also been critical of U.S. dominance in such media as the Internet. Public subsidies to France's Centre National de la Cinematographie are designed to foster filmmaking for a national audience deluged with English-language productions. Elsewhere, Afghanistan and Iran banned satellite dishes and many U.S. films, although illegal copies of top box-office hits often found their way through national borders. In one way or another, U.S. cultural influences manage to reach beyond the nation's borders, a reminder that the region's cultural values will continue to have an ever greater impact on the larger world beyond.

▲ **Figure 3.25 Tokyo Disneyland** While some traditional cultures resist American influences, many people around the world have embraced globalization, especially when it is brought by franchised ambassadors such as Mickey and Minnie Mouse. Tokyo Disneyland is one of a growing number of Magic Kingdoms around the planet. *(AP/Wide World Photos)*

Geopolitical Framework: Patterns of Dominance and Division

In disarmingly simple fashion, North America's political geography brings together two of the world's largest states. The creation of these states, however, was neither simple nor preordained, but rather the complex outcome of historical processes that might have created quite a different North American map. Once established, the states have coexisted in a close relationship of mutual economic and political interdependence. President John F. Kennedy summarized the links in a speech to the Canadian parliament in 1962: "Geography has made us neighbors, history has made us friends, economics has made us partners and necessity has made us allies." That cozy continental relationship has not been without its tensions, and some persist today. In addition, both nations have had to deal with fundamental internal political complexities that have not only tested the limits of their federal structures, but also challenged their very existence as states.

Creating Political Space

The United States and Canada sprang from very different political roots. The United States broke cleanly and violently from Great Britain. The American Revolution created a powerful sense of nationalism that sped the process of spatial expansion and produced a political rhetoric oriented around the country's preordained role as a continental, indeed global, power. Canada, by contrast, was a country of convenience, born from a peaceful, incremental separation from Britain and then assembled as a collection of distinctive regional societies that only gradually acknowledged their common political destiny.

Uniting the States The creation of the United States replaced one continental division of political space with another, and that process of annihilation and invention proceeded in ways that no one could have anticipated. Turning back the clock to the early eighteenth century reveals a political geography very much in the making. Beyond scattered frontiers of European settlement lay vast domains of Native American-controlled political space. Although boundaries were not formally surveyed or mapped, native peoples carved up the continent in an elaborate geography of homelands, allied territories, and enemy terrain.

With amazing rapidity, Europe and then the United States imposed their own political boundaries across the realm. The 13 English colonies, sensing their common destiny after 1750, finally united two decades later and clashed violently with their colonial parent. By the 1790s, the young nation's political claims had reached the Mississippi River; the new republic was busily coercing land cessions from native peoples; and the Ordinance of 1787 had provided a template for western territory and state formation that served as the model of expansion for the next century. Soon the Louisiana Purchase (1803) nearly doubled the national domain, creating a new political geography that was as vast as it was unexplored. By mid-century, Texas had been annexed; treaties with Britain had secured the Pacific Northwest; and an aggressive war with

Mexico had captured much of the Southwest. The territorial acquisition of Alaska (1867) and Hawaii (1898) eventually rounded out the present political domain of 50 states.

Assembling the Provinces

Canada was created under quite different circumstances. The modern pattern of provinces assembled in a slow and uncertain fashion. After the American Revolution, England's remaining territorial claims in the region came under the control of colonial administrators in British North America. The Quebec Act of 1774 allowed for continued French settlement in the St. Lawrence Valley and provided the initial template for governing the region. Soon, however, Anglo settlers near Lake Ontario and Lake Erie pressed for more local colonial representation. The result was the Constitutional Act of 1791, which divided the colony into Upper Canada (Ontario) and Lower Canada (Quebec). Frustration with that system led to the Act of Union in 1840, thereby reuniting the two Canadas. In 1867 the British North America Act united the provinces of Ontario, Quebec, Nova Scotia, and New Brunswick in an independent Canadian Confederation. The peaceful separation from the mother country also guaranteed to Quebec special legal and cultural privileges that presaged some of the modern power struggles within the country.

Once created, the Canadian Confederation grew in piecemeal fashion, more out of geographical convenience than from any compelling nationalism at work to unite the northern portion of the continent. Within a decade, the Northwest Territories (1870), Manitoba (1870), British Columbia (1871), and Prince Edward Island (1873) joined Canada, and the continental dimensions of the country took shape. Soon, the Yukon Territory (1898) separated from the Northwest Territories; Alberta and Saskatchewan gained provincial status (1905); and Manitoba, Ontario, and Quebec were enlarged (1912) north to Hudson Bay. Newfoundland finally joined in 1949. The recent addition of Nunavut Territory (1999), carved from the Northwest Territories, represents the latest change in Canada's political geography.

Continental Neighbors

Geopolitical relationships between Canada and the United States have always been intimate: their common 5,525-mile (8,900-kilometer) boundary compels both nations to pay close attention to one another. More than that, their status as continental neighbors has generated tremendous interaction, trade, and mutual cooperation, while at the same time offering potential for conflict and controversy.

Canadians, for good reason, have worried about being in the political shadow of the United States. Historically, the War of 1812 included U.S. invasions of British North American territory, a military and geopolitical offense not soon forgotten. Canada's independence in 1867 was probably linked in no small way to fear of its southern neighbor. U.S. rhetoric in the post-Civil War era bristled with the language of Manifest Destiny, and open talk of annexing Canadian lands may have hastened British actions to unite the region.

During the twentieth century, however, political cooperation outweighed lingering suspicions, and the two countries saw both their political and economic destinies increasingly intertwined. There was also the recognition that the inevitable conflicts that did occur would need to be swiftly adjudicated if the two countries were to benefit from their proximity and shared interests. In 1909, the Boundary Waters Treaty created the International Joint Commission, an early step in the common regulation of cross-boundary issues involving water resources, transportation, and environmental quality. That tradition of cooperation was illustrated by the St. Lawrence Seaway project in 1959, opening the Great Lakes region to better global trade connections. The two nations joined in cleaning up Great Lakes pollution and in making plans to reduce acid rain in eastern North America. Internationally, both cooperated in creating a common North American defense and in serving in the North Atlantic Treaty Organization (NATO).

It has been in the realm of trade relations that the close political ties between these neighbors have mattered most. The United States receives 80 percent of Canada's exports and supplies more than two-thirds of its imports. Conversely, Canada is the United States' most important trading partner, accounting for roughly 20 percent of its exports and imports. These historically close ties, fed by advantages of accessibility and nourished through shared economic growth, have been further strengthened by an increasing number of trade agreements between the countries. One landmark reached in 1989 was the signing of the bilateral Free Trade Agreement (FTA). Five years later, the larger **North American Free Trade Agreement (NAFTA)** extended the alliance to Mexico and laid out a 15-year plan for drastically reducing all barriers to trade and capital investment among the three nations. Paralleling the success of the European Union (EU), NAFTA has forged the world's largest trading bloc, including more than 400 million consumers and a huge free-trade zone that stretches from beyond the Arctic Circle to Latin America.

Political conflicts still divide the two countries. Environmental issues produce cross-border tensions. Many U.S. waterways begin in Canada, while Canada receives much of its water from upstream settings in the United States. Inevitable conflicts result when environmental degradation in one nation affects the other. For example, Montana's North Flathead River flows out of British Columbia, where Canadian logging and mining operations periodically threaten fisheries and recreational lands south of the border. Similarly, industrial pollution in the United States often becomes Canada's problem when fouled waters flow through the Great Lakes system. Agricultural and natural resource competition also engenders periodic fits of controversy between the two neighbors. For example, problems recently developed when Canadian wheat and potato growers were accused of dumping their products into U.S. markets, thus depressing prices and profits for U.S. farmers. Similar issues have arisen in the logging industry, with American timber companies accusing their Canadian counterparts of being subsidized by provincial governments willing to sell off public resources cheap-

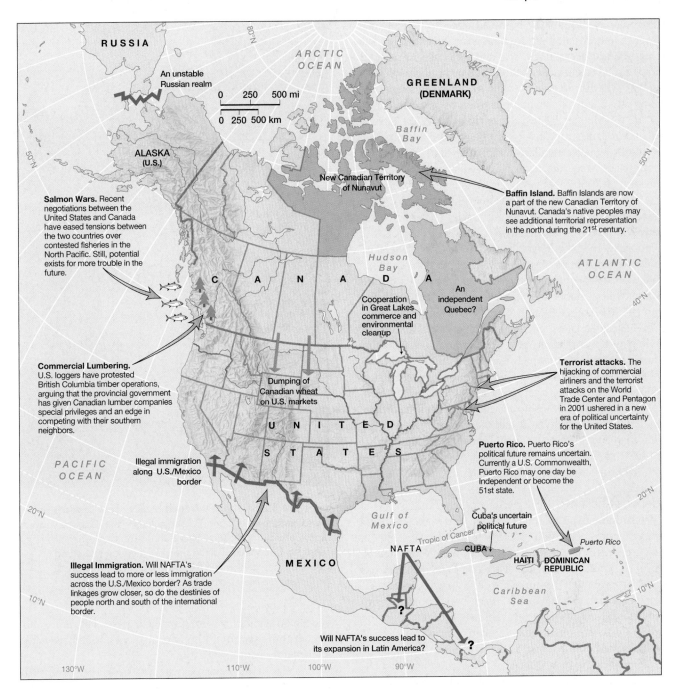

▲ Figure 3.26 Geopolitical issues in North America   Although Canada and the United States share a long and peaceful border, many political issues still divide the two countries. In addition, internal political conflicts, particularly in bicultural Canada, cause tensions. Given its global prominence and the openness of its society, the United States also remains vulnerable to acts of global terrorism.

ly. In the 1990s "salmon wars" flared along the Pacific Coast, where U.S. and Canadian fishermen compete keenly for fish claimed by both sides. By 2000 conditions improved when the two countries signed an agreement on fishing rights, but other border disputes still linger (Figure 3.26). Occasional foreign policy disagreements also divide the pair, including long-standing political differences over Cuba. Still, these neighbors coexist in a remarkably harmonious marriage, a

political conjoining born from both the heart and the pocketbook, and one that is likely to be sustained in the twenty-first century.

The Legacy of Federalism

The United States and Canada are **federal states** in that both nations allocate considerable political power to units of government beneath the national level. Other nations, such as

France, have traditionally been **unitary states**, in which power is centralized at the national level. Federalism leaves many political decisions to local and regional governments and often allows distinctive cultural and political groups to be recognized as distinct entities within a country. Although both nations have federal constitutions, their origins and evolution are very different. The U.S. Constitution (1787), created out of a violent struggle with the powerful British nation, specifically limited centralized authority, giving all unspecified powers to the states or the people. In contrast, the Canadian Constitution (1867), which created a federal parliamentary state, was an act of the British Parliament. Originally, it reserved most powers to central authorities and maintained many political links between Canada and the British Crown.

Ironically, the evolution of the United States as a federal republic produced an increasingly powerful central government, while Canada's geopolitical balance of power shifted toward more provincial autonomy and a relatively weak national government. For example, the federal government largely controls U.S. public lands, but in Canada provincial authorities retain power over public Crown lands. In addition, Canadian central authorities traditionally faced powerful regional political identities, particularly in Quebec. The close connection to Britain did little to foster Canadian nationalism. Indeed, separate Canadian citizenship apart from that of Britain came about only after World War II, and the modern Canadian flag is less than 50 years old. It was only with the Constitution Act of 1982 that Canada formally transferred all legal authority from the British to the Canadian Parliament.

Quebec's Challenge   The greatest challenge facing Canada is determining what role Quebec will play in the country's future (Figure 3.26). Will it secede from Canada, remain a province, or seek another form of political autonomy from the rest of the country? Quebec's distinctive French society has deep historical roots (Figure 3.27). Over time, economic disparities between the Anglo and French populations have reinforced differences between the two groups, with the French-Canadians often suffering when compared with their wealthier neighbors in Ontario. Beginning in the 1960s, a separatist political party in Quebec increasingly voiced French-Canadian concerns. When the party won provincial elections in 1976, it quickly moved to declare French the official language of Quebec and to schedule a provincial referendum on remaining within Canada. Although only 41 percent of the electorate voted for separation in 1980, the referendum precipitated the drafting of a new federal constitution two years later. Many French-Canadians were not impressed, however, fearing that their distinctive cultural identity might be threatened. They refused to sign the document, leading federal lawmakers in Ottawa to propose another solution, the Meech Lake Accord of 1990. The Accord included stronger guarantees of Quebec's special status as a "distinct society." Manitoba and Newfoundland proceeded to derail the agreement, arguing it gave Quebec autonomy that was not being extended to other provinces. After a national compromise failed (the Charlottetown Agreement), Quebec again held a refer-

▲ **Figure 3.27 Canada's French legacy**   Four centuries of French settlement in Canada, particularly within Quebec, leave many cultural and political issues unanswered. Most French-speaking Canadians want to preserve their language and culture. Can they do it within the present political framework? *(R. Sidney/The Image Works)*

endum on separating from Canada in 1995. This time, only a razor-thin majority (50.6 percent) voted to remain within the larger country.

Canada's future remains clouded by the Quebec issue. Federalism has given the larger nation the flexibility to accommodate a "distinct society," but that same federal structure has proven cumbersome in bringing the country together in a lasting agreement that all parties can adopt. In 2000 the federal parliament ruled (the Clarity Act) that Quebec or any Canadian province could secede only if a "clear" majority of its population voted to do so. Many questions remain. How would a sovereign Quebec function? How would Quebec's sizable non-French population react? Would the rest of Canada accede to its wishes? Would other Canadian provinces ponder independence? Final answers to these questions and a solution to Canada's political challenges are not likely soon.

Native Peoples and National Politics   Another challenge to federal political power has come from North American Indian and Inuit populations, in both Canada and the United

LOCAL VOICES   Nunavut Meets the Internet

Baffin Island resident Adamee Itorcheak is wired to the world, and he is helping fellow citizens of Canada's Nunavut Territory surf the Web as well. Itorcheak lives in Iqaluit on the southern portion of Baffin Island; he runs Nunanet Communications, the only Internet service provider within a thousand miles of the village. Globalization first arrived in Iqaluit during World War II when a refueling airstrip was built on the edge of town. Gradually thereafter, regular mail and passenger air service, telephone connections, and satellite television linked the villagers with the larger world. Although some native Inuits have greeted cyberspace with suspicion, there has also been a cultural heritage of borrowing that portends well for the future of Itorcheak's high-technology gamble. The native word for e-mail roughly translates to "letters through the artificial brain," and Baffin Islanders increasingly see the virtues of such quick communications in a land where travel is often difficult.

Itorcheak's Nunanet service has hundreds of customers and serves a varied clientele, including businessmen, seal hunting guides, and educators. Itorcheak had to teach his first few hundred customers how to use the Internet, but now he is busily supplementing his service provider business as a consultant to local entrepreneurs and government agencies. His customers monitor wholesale fish prices in Japan, set up dogsled excursions for incoming tourists, participate in political chat rooms, and disseminate information on preserving native languages and customs to other Arctic communities.

Whether Nunanet survives or not, the Internet has a firm foothold in Nunavut, another reminder of the role this technology is playing in the diffusion of information in the twenty-first century. Its diverse uses also illustrate how the complex process of globalization transforms particular places in varied ways. Thanks to Itorcheak, Baffin Island will never be quite the same; conversely, the new connection brings the newly wired Arctic a bit closer to the outer world.

Source: Adapted from "Across Tundra and Cultures, Entrepreneur Wires Arctic," *Wall Street Journal,* October 19, 1998.

States. Within the United States, the renewed assertion of Native American political power began in the 1960s and marked a decisive turn away from earlier policies of assimilation. Since passage of the Indian Self-Determination and Education Assistance Act of 1975, the trend has been toward increased Native American autonomy. The Indian Gaming Regulatory Act (1988) offered potential economic independence for many tribes. In the western American interior, where Indians control roughly 20 percent of the land, tribes are solidifying their hold on resources, reacquiring former reservation acreage, and participating in political interest groups such as the Native American Fish and Wildlife Society and the Council of Energy Resource Tribes. In Alaska, native peoples acquired title to 44 million acres (18 million hectares) of land in 1971 under the Alaska Native Claims Settlement Act.

In Canada even more ambitious challenges to a weaker centralized government have yielded dramatic results. As natives pressed their claims for land and political power in the 1970s, Canada established the Native Claims Office (1975) and began negotiating settlements with various groups, particularly within the country's vast northern interior. Agreements with native peoples in Quebec, Yukon, and British Columbia turned over millions of acres of land to aboriginal control and increased native participation in managing remaining public lands. By far, the most ambitious agreement has been to create the new Canadian territory of Nunavut out of the eastern portion of the Northwest Territories in 1999 (Figure 3.26). Nunavut is home to 25,000 people (85 percent Inuit) and is the largest territorial/provincial unit within Canada. Its creation represents a new level of native self-government in North America (see "Local Voices: Nunavut Meets the Internet"). Recently, agreements between the federal Parliament and British Columbia tribes

have initiated a similar move towards more native self-government in that western province. The Nisga'a tribe, for example, now controls a 770-square-mile (1,992-square-kilometer) portion of the Nass River Valley near the Alaska border (Figure 3.21). Growing discord among regional business groups, however, threatens to slow the move toward more native control. In British Columbia, critics of the native policy have argued that declining investments in natural resource industries are related to growing concerns about who really owns the province. Indeed, native groups on both sides of the border appear determined to regain more control of a continent they once entirely possessed.

A Global Reach

The geopolitical reach of the United States, in particular, has taken its influence far beyond the bounds of the realm. The Monroe Doctrine (1824) asserted that U.S. interests were hemispheric and transcended national boundaries, but it was not until after 1895 that the United States accelerated its global expansion. Two principal settings served as early laboratories for political imperatives in the United States. In the Pacific, the United States claimed the Philippines as a prize of the Spanish-American War (1898), and further annexations of Guam (1898) and the Hawaiian Islands (1898) presaged the country's twentieth-century dominance of the region. In Central America and the Caribbean, the growing role of the American military between 1898 and 1916 shaped politics in Cuba, Puerto Rico, Panama, Nicaragua, Haiti, Mexico, and elsewhere. Further, the country's role in World War I raised its stakes in European affairs.

While the 1920s and 1930s briefly returned the United States to isolationist policies, World War II and its aftermath

GEOGRAPHY IN THE MAKING   The War on Terror and Daily Life in the United States

After the September 11, 2001 terrorist attacks in the United States, many observers have contemplated the subtle yet pervasive changes shaping U.S. society. Naturally, those directly affected by the attacks saw their lives transformed forever on that September morning, and the urban landscapes of New York City and Washington, D.C., will never be quite the same. But much broader shifts in everyday life also have been generated by the attacks and by the U.S.-led "War on Terror" that has followed.

Consider the changing nature of travel in the United States. Billions of dollars have been spent on security in the airline and air-transport industries, travelers put up with long delays, and a new federal workforce has been created to monitor luggage at airports around the country. The nation's infrastructure and public spaces have also been cast in a new light. Everyday American landscape features such as bridges, dams, oil pipelines, theme parks, and shopping malls are now seen from the perspective of possible terrorist attacks. In addition, the health-care system has come into the spotlight, particularly after the anthrax incidents that followed so closely after the terrorist attacks of September 11. New medical initiatives target responses to biological warfare, and experts scramble to improve vaccines for smallpox and potential biohazards.

How will recent events shape longer-term geographies of immigration into the United States? No one knows for sure, but more than 300,000 foreigners were ordered out of the country after the attacks for various reasons, and new restrictions on student visas, international travel, and immigrant workers inevitably will shape patterns of both legal and illegal immigration to the country.

Finally, the October 2001 passage of the USA Patriot Act raised real questions about personal freedom and privacy in the United States. Under the act, the government has been granted sweeping new powers to gather information on both immigrants and U.S. citizens. Others question how many restrictions need to be placed on public documents and Internet access in order to fight terrorism. There are no easy answers to these questions, but the "War on Terror" seems destined to subtly redefine the daily lives of every American.

Source: Adapted from "How Sept. 11 Changed America," *Wall Street Journal,* March 8, 2002.

---

forever redefined the country's role in world affairs. Victorious in both the Atlantic and Pacific theaters, postwar America emerged from the conflict as the world's dominant political power. Quickly, however, a resurgent Soviet Union challenged the United States, and the Cold War began in the late 1940s. In response, the Truman Doctrine promised aid to struggling postwar economies and actively challenged communist expansion in Europe and elsewhere. The United States also fashioned multinational political and military agreements, such as those establishing the North Atlantic Treaty Organization (NATO) and the Organization of American States (OAS), which were designed to cast a broad umbrella of U.S. protection across much of the noncommunist world. Violent conflicts in Korea (1950–1953) and Vietnam (1961–1973) pitted U.S. political interests against communist attempts to extend their Asian dominance beyond the Soviet Union and China. Tensions also ran high in Europe as the Berlin Wall crisis (1961) and nuclear weapons deployments by NATO- and Soviet-backed forces brought the world closer to another global war. The Cuban missile crisis (1962) reminded Americans that traditional political boundaries provided little defense in a world uneasily brought closer together by technologies of potential mass destruction.

Even as the Cold War gradually receded during the late 1980s, the global political reach of the United States continued to expand. A few examples suggest the pattern. Interventionist policies in Central America favored regimes friendly to the United States. President Carter's successful Middle East Peace Treaty between Israel and Egypt (1979) guaranteed a continuing diplomatic and military presence in the eastern Mediterranean. When Iraq's Saddam Hussein threatened Persian Gulf oil supplies in 1990, the United States led a United Nations coalition to contain the aggression. In the late 1990s, Serbian aggression within Kosovo prompted an American and NATO-led intervention, which included major air attacks on the Serbian capital of Belgrade (1999) and a peacekeeping presence (with the United Nations) in the disputed area of Kosovo.

While the United States remains the world's key superpower, the early twenty-first century has dawned beneath a cloud of geopolitical uncertainty. As the commercial aircraft hijackings and terrorist attacks of late 2001 suggest, the United States will be challenged by the changing calculus and unpredictable technologies of global terrorism as anti-American groups attempt to destabilize the region (Figure 3.26, Figure 3.28). As U.S. leaders have made clear, the sweeping responses necessary to combat such threats are unlike earlier regional or global conflicts, a reminder that the geopolitical realities of globalization have brought both opportunities and risks into the lives of every American.

▲ Figure 3.28 Attack on the Pentagon   Symbol of U.S. political power, the Pentagon was forever changed on September 11, 2001, as terrorists crashed a commercial airliner into the heart of the nation's military control center. The act demonstrated that even the world's greatest superpower was vulnerable to acts of terrorist violence and was a sobering reminder that globalization often comes with a price.   *(Rob Crandall/Rob Crandall, Photographer)*

Economic and Social Development: Geographies of Abundance and Affluence

North America possesses the world's most powerful economy and its most affluent population. Its 315 million people consume huge quantities of global resources, but also produce some of the world's most sought-after manufactured goods and services. North America's size, geographic diversity, and resource abundance have all contributed to the realm's global dominance in economic affairs. More than that, however, the region's human capital—the skills and diversity of its population—has enabled North Americans to achieve high levels of economic development (Table 3.2). Even amid this continental affluence, however, many North Americans struggle to escape from poverty. Indeed, since 1950, the social and economic gap between rich and poor has widened across the region, and these disparities will continue to challenge both Canada and the United States in the twenty-first century.

An Abundant Resource Base

North America is blessed with a varied storehouse of natural resources. The realm's climatic and biotic variety, its soils and terrain, and its abundant energy, metals, and forest resources have provided diverse raw materials for development. Indeed, the direct extraction of natural resources still makes up 3 percent of the U.S. economy and more than 6 percent of the Canadian economy. Some of these North American resources are then exported to global markets, while other raw materials are imported to the region.

Opportunities for Agriculture   North Americans have created one of the most productive food-producing systems in the world (Figure 3.29). Farmers within the realm practice highly commercialized, mechanized, and specialized agriculture that emphasizes the importance of efficient transportation systems, global markets, and large capital investments in farm machinery. Agricultural productivity climbed rapidly during the twentieth century, and crop yields rose at the same time so that demands for labor declined. Today, agriculture employs only a small percentage of the labor force in both the United States (2.6 percent) and Canada (3.7 percent). At the same time, farm consolidations have sharply reduced the number of operating units while average farm sizes have steadily risen. For example, in the United States, there are fewer than 2 million farms currently in operation, representing a 20 percent decline since 1975. On the other hand, average farm size has increased from 420 acres in 1975 to more than 470 acres in 2000. As a result, farm populations have plummeted in many communities from Illinois to Alberta. Still, agriculture remains a critically important part of the North American economy, producing between $200 billion and $300 billion annually in sales. In addition, agriculture remains a dominant land use across much of the realm. Although suburbanization has taken its toll, almost half of the United States is classified as cropland, pasture, or potentially productive range lands. Given Canada's northern latitude, only about 7 percent of its land is potentially agricultural, but its extensive forests have made it the world's largest exporter of timber, pulp, and newsprint.

The geography of North American farming represents the marriage between diverse environments, varied continental and global markets for food, and historical patterns of settlement and agricultural evolution. In the Northeast, dairy operations and truck farms specializing in fresh produce

TABLE 3.2 *Economic Indicators*

Country	Total GNI (Millions of $U.S., 1999)	GNI per Capita ($U.S., 1999)	GNI per Capita, PPP* ($Intl, 1999)	Average Annual Growth % GDP per Capita, 1990–1999
Canada	$614,003	$20,140	$25,440	1.7%
United States	$8,879,500	$31,910	$31,910	2%

*Purchasing power parity
Source: The World Bank Atlas, *2001.*

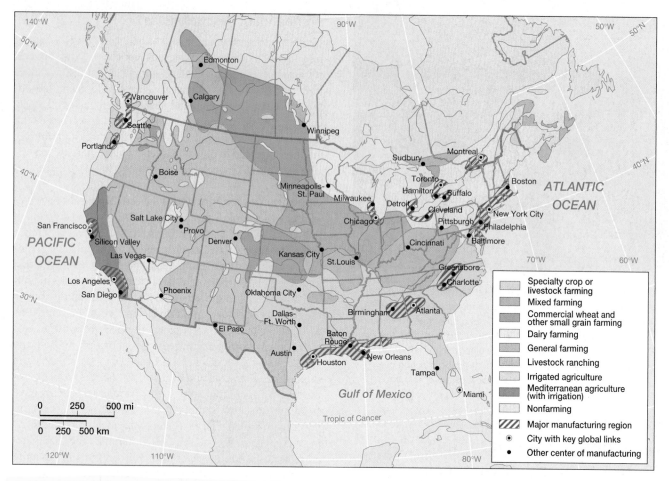

▲ **Figure 3.29 Major economic activities of North America**   Varied environmental settings, settlement histories, and economic conditions have produced the modern map of North American economic activities. A growing array of major metropolitan areas are playing increasingly visible roles in the global economy.   *(Modified from Clawson and Fisher, 1998,* World Regional Geography, *Upper Saddle River, NJ: Prentice Hall)*

capitalize on the proximity of the region to major cities in Megalopolis and southern Canada. Corn and soybeans dominate croplands across the U.S. Midwest and western Ontario, where a tradition of mixed farming combines feed grains with commercially raised livestock, particularly cattle and hogs. Indeed, more than 15 percent of the agricultural wealth of the United States sprouts from the three heartland states of Illinois, Iowa, and Nebraska. North of the Corn Belt, cooler, shorter growing seasons support more dairy operations across Wisconsin and central Michigan. To the south, only remnants of the old Cotton Belt remain, largely replaced by a diverse geography of subtropical specialty crops; poultry, catfish, and livestock production; and commercial forestry (Figure 3.30). West of the 100th meridian, extensive, highly mechanized commercial grain-growing operations stretch from Kansas to Saskatchewan and Alberta. Depending on surface and groundwater resources, irrigated agriculture across western North America also offers opportunities for more intensive operations. Indeed, California's agricultural output, nourished by large agribusiness operations in the irrigated Central Valley, accounts for more than 10 percent of the overall farm economy of the United States.

Industrial Raw Materials   North Americans produce and consume huge quantities of other natural resources. While the region is well endowed with a variety of energy and metals resources, the scale and diversity of the North American economy have led to the need to import additional raw materials. Petroleum use in the United States exemplifies the pattern. The country produces about 12 percent of the world's oil, but consumes more than 25 percent. As a result, the United States imports more than half of its oil, much of it coming from the Americas (Venezuela, Mexico, and Canada), the Middle East (Saudi Arabia), and Africa (Nigeria). Within the realm, the major areas of oil and gas production are the Gulf Coast of Texas and Louisiana; the Central Interior, including west Texas, Oklahoma, and eastern Kansas; Alaska's North Slope; and Central Canada (especially Alberta). The most abundant fossil fuel in the United States is coal, but its relative importance in the overall energy economy declined in the twentieth century as industrial technologies changed and environmental concerns grew. Still, the country's 400-year supply of coal reserves (23 percent of the world's total) will be an important energy resource, both for domestic consumption as well as for export. The nation's

▲ **Figure 3.30 Specialty agriculture in the American South** Catfish ponds near Greenville, Mississippi, offer a new look to the South's rural landscape. Today, the old Cotton Belt has largely been replaced by specialty operations engaged in subtropical cropping, commercial forestry, and poultry, catfish, and livestock production.   *(C. C. Lockwood/DRK Photo)*

leading coal-producing region is Appalachia, but demand has been increasing for the environmentally cleaner low-sulfur coals of the western interior (Great Plains, Intermountain West, and Alberta). North America also remains a major producer of metals resources, although global competition, rising extraction costs, and environmental concerns pose challenges for this sector of the economy. Still, the realm is endowed with more than 20 percent of the world's copper, lead, and zinc reserves, and it accounts for more than 20 percent of global gold, silver, and nickel production.

Creating a Continental Economy

The timing of European settlement in North America proved pivotal in its rapid economic transformation. The region's abundant resources came under the control of varied European powers armed with new technologies that reshaped the landscape and reorganized its economic geography. By the nineteenth century, North Americans themselves were actively contributing to those technological changes, and as new resources were unlocked in the interior and new immigrant populations flooded the continent, the region took full advantage of the marriage between its storehouse of raw materials and the Industrial Revolution. In the twentieth century, although natural resources remained important, new industrial innovations and greatly expanded service employment diversified the economic base and extended its global reach.

Connectivity and Economic Growth   Dramatic improvements in North America's transportation and communication systems laid the foundation for urbanization, industrialization, and the commercialization of agriculture.

Indeed, the region's economic miracle was a function of its **connectivity,** or how well its different locations became linked with one another through an improved transportation and communication infrastructure. Those links greatly facilitated the potential for interaction between locations and dramatically reduced the cost of distance. Before 1830, North American connectivity gradually improved with the help of better roads, but it still took a week to travel from New York City to Ohio. For commercial traffic, canal construction also facilitated the movement of bulky goods. More than 1,000 miles (1,600 kilometers) of canals crisscrossed the eastern United States, including the Erie Canal (completed in 1825), which linked New York City with the North American interior. Other canals connected the Great Lakes and Ohio River systems, further integrating the Midwest into the Northeast economic core.

Tremendous technological breakthroughs revolutionized North America's economic geography between 1830 and 1920. Railroads cemented together whole new economic relationships. By 1860 more than 30,000 miles (48,387 kilometers) of track had been laid in the United States, and the network grew to more than 250,000 miles (403,226 kilometers) by 1910. Farmers in the Midwest and Plains found ready markets for their products in cities hundreds of miles away. Industrialists collected raw materials from vast distances, processed them, and shipped manufactured goods to their final destinations. After 1869, travelers could board a train in Chicago and conveniently cross the continent in a few days. The telegraph brought similar changes to information: long-distance messages flowed across eastern North America by the late 1840s, and 20 years later undersea cables firmly linked the realm to Europe, another milestone in the larger process of globalization.

North America's integrated transportation and communications systems were redefined after 1920 as automobiles, mechanized farm equipment, paved highways, commercial air links, national radio broadcasts, and dependable transcontinental telephone service further reduced the cost of distance across the region. Cheap mass-produced automobiles reached many middle-class North Americans by the 1920s, and the subsequent construction of major North American highways by 1970 produced the world's largest integrated road system. After World War II, continental connectivity also benefited from the St. Lawrence Seaway between the Atlantic Ocean and Great Lakes (1959), vast improvements in jet airline connections, and the increasing prevalence of television. Perhaps most profoundly, the region has taken the lead in the global information age, integrating computer, satellite, and Internet technologies in a web of connections that facilitates the flow of knowledge both within the realm and beyond.

The Sectoral Transformation   Changes in employment structure signaled North America's economic modernization just as surely as its increasingly interconnected society. The **sectoral transformation** refers to the evolution of a nation's labor force from one highly dependent on the *primary* sector (natural resource extraction) to one with more employment in the *secondary* (manufacturing or industrial), *tertiary* (services), and *quaternary* (information processing) sectors. The sectoral transformation thus reflects the way innovation changes employment opportunities in a society. For example, with agricultural mechanization, lower demands for primary-sector workers are replaced by new opportunities in the expanding industrial sector. In the twentieth century, new services (trade, retailing) and information-based activities (education, data processing, research) created employment opportunities in these postindustrial sectors of the economy.

Both the United States and Canada have witnessed dramatic changes in employment that reveal the formative impacts of technological innovation and economic restructuring. In the late nineteenth century, primary-sector employment in agriculture and mining dominated the occupational structures of the two countries. Rapid industrialization after 1870 contributed to steady growth in the secondary sector, and a century later it accounted for about 30 percent of the workforce. Modern technology has had the opposite effect on the primary sector, however, as machines replaced labor, and employment in the sector has fallen to less than 4 percent of the workforce. In the later twentieth century, both nations also ex-

perienced relative declines in manufacturing employment compared with the rapid growth of the tertiary and quaternary sectors. Today these two latter sectors employ more than 70 percent of the labor force in both Canada and the United States. In addition, the service sector generates more than 75 percent of the region's annual economic output. These recent trends also reveal the tangible imprint of globalization on the North American labor force. Much of North America's sustained growth in the tertiary and quaternary sectors is directly tied to the ability of those industries to export innovations in services, financial management, and information processing to a worldwide clientele.

Eastern Canada's Maritime provinces illustrate the challenges of successfully engineering the sectoral transformation in more peripheral areas of North America (Figure 3.31). The region's long dependence on a small number of extractive activities, such as fishing, logging, and marginal agriculture, has encouraged many of its young people to leave and has boosted unemployment levels well above the national norm. As its residents struggle to find reliable alternatives to its traditional economic base, many have seen their wages slip to only 70 or 80 percent of the national average. Canada's federal government has pledged to spend millions to support new busi-

▶ **Figure 3.31 Eastern Canadian unemployment**   Eastern Canada's Maritime provinces continue to suffer from high unemployment as the regional economy struggles to shift away from dependence on primary sector activities. Regional weekly earnings also remain well below the national average.

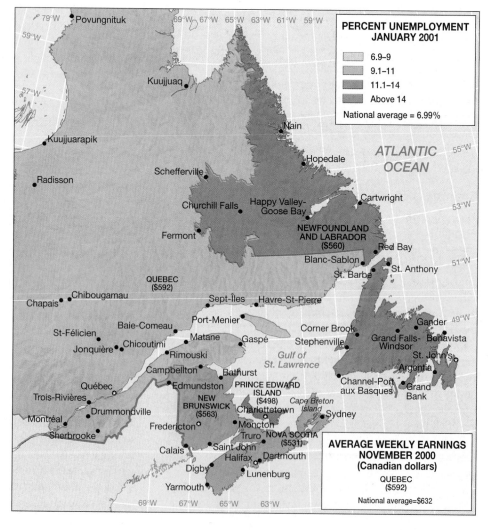

nesses and attract more high-wage jobs to the Maritimes, but the economy of the region is likely to lag as long as its reliance on the waning primary sector continues.

Regional Economic Patterns   North America's industries reveal important regional patterns of concentration. **Location factors** are the varied influences that explain *why* an economic activity is located where it is, and North America's economic geography suggests that these factors change over time and that many influences, both within and beyond the realm, shape patterns of economic activity. Patterns of industrial location illustrate the concept (Figure 3.29). The historical manufacturing core includes Megalopolis (Boston, New York, Philadelphia, and Baltimore), southern Ontario (Toronto and Hamilton), and much of the industrial Midwest. Several location factors account for the core's nineteenth-century development and the ongoing importance of its diverse industrial economy. Its proximity to *natural resources* (farmland, coal, and iron ore); increasing *connectivity* (canals and railroad networks, highways, air traffic hubs, and telecommunications centers); a ready supply of *productive labor;* and a growing national, then global, *market demand* for its industrial goods spurred sustained *capital investment* within the core. Traditionally, the core has dominated in the production of steel, automobiles, machine tools, and agricultural equipment, as well as played a pivotal role in producer services such as banking and insurance.

In the last half of the twentieth century, however, industrial and service-sector growth has gravitated to peripheral centers in the South and West. Many older industrial centers in the core experienced drastic job losses in manufacturing, thus reducing its overall dominance within the realm. Conversely, cities of the South's Piedmont manufacturing belt (Greensboro to Birmingham) have seen industrial expansion after 1960, partly because lower labor costs and Sun Belt amenities attracted new investment. The Gulf Coast industrial region is strongly tied to nearby fossil fuels that provide raw materials for its many energy refining and petrochemical industries (Figure

▲ **Figure 3.32 Gulf Coast petroleum refining**   Petroleum-related manufacturing has transformed many Gulf Coast settings. Much of Houston's twentieth-century growth has been fueled by the dramatic expansion of oil-related industries. The port of Houston remains a major center of North America's refining and petrochemical operations.   *(Walter Frerck/Odyssey Productions)*

3.32). The varied West Coast industrial region stretches discontinuously from Vancouver, British Columbia, to San Diego, California (and beyond into northern Mexico), and it reveals the increasing importance of Pacific Basin trade. Large aerospace operations in the West also suggest the role of *government spending* as a location factor. Silicon Valley is now North America's leading region of manufacturing exports. Its proximity to Stanford, Berkeley, and other universities demonstrates the importance of *access to innovation and research* for many fast-changing high-technology industries and the advantages of *agglomeration economies,* in which many companies with similar and often integrated manufacturing operations locate near one another (Figure 3.33). Many smaller industrial centers also thrive throughout North America, suggesting that some industries

◀ **Figure 3.33 Silicon Valley**   The high-technology industrial landscape of California's Silicon Valley contrasts sharply with the look of traditional manufacturing centers. Here similar industries form complex links, benefiting from their proximity to one another and to nearby universities such as Stanford and Berkeley.   *(George Hall/ Woodfin Camp & Associates)*

can be relatively footloose and succeed in many localities. Places such as Provo, Utah, and Austin, Texas, that specialize in high-technology industries demonstrate the growing role of *lifestyle amenities* in shaping industrial location decisions, both for entrepreneurs and for the skilled workers who need to be attracted to such opportunities.

North America and the Global Economy

Together with Europe and Japan, North America plays a pivotal role in the global economy. In prosperous times, the realm benefits from global economic growth, but in periods of international instability, globalization means that the region is more vulnerable to economic downturns. These interconnections transcend the abstract statistics of trade flows and foreign investment. Increasingly, North American workers and localities find their futures directly tied to export markets in Latin America, the rise and fall of Asian imports, or the pattern of global investments in U.S. stock and bond markets. No community remains untouched by these links, and the consequences will shape the lives of every North American in the twenty-first century.

The region is home to a growing number of truly "global cities" that serve as key connecting points and decision-making centers in the world economy (Figure 3.29). While New York City is the largest, smaller metropolitan areas such as Miami, Toronto, Chicago, and Vancouver are emerging as pivotal urban players on the global stage (see "Latinized Miami: Emergent Global City"). The global status of these urban centers has had profound local consequences. Life changes as thousands of new jobs are created; suburbs grow with a rush of new and ethnically diverse migrants; and a new, more cosmopolitan culture replaces older regional traditions.

The United States, with Canada's firm support, played a formative role in creating much of this new global economy and in shaping many of its key institutions. In 1944 allied nations met at Bretton Woods, New Hampshire, to discuss economic affairs. Under U.S. leadership, the group set up the International Monetary Fund (IMF) and the World Bank and gave these global organizations the responsibility for defending the world's monetary system and making key postwar investments in infrastructure. The United States also spurred the creation (1948) of the General Agreement on Tariffs and Trade (GATT). Renamed the World Trade Organization (WTO) in 1995, its 140 member states are dedicated to reducing global barriers to trade. In addition, the United States and Canada participate in the **Group of Seven (G-7)**, a collection of powerful countries (including Japan, Germany, Great Britain, France, Italy, and sometimes Russia) that confers regularly on key global economic and political issues.

Patterns of Trade   Global geographies of trade reveal North America's prominence in both the sale and purchase of goods and services within the international economy. Both countries import diverse products from many global sources. Dominated overwhelmingly by the United States, Canada imports large quantities of manufactured parts, vehicles, computers, and foodstuffs. Imports to the United States continue to grow, cre-

ating a persistent global trade deficit for the country. Canada, Japan, Mexico, China, and world oil exporters supply the United States with a diversity of raw materials, low-cost consumer goods, and high-quality vehicles and electronics products.

Outgoing trade flows suggest what North Americans produce most cheaply and efficiently. Canada's exports include large quantities of raw materials (grain, energy, metals, and wood products), but manufactured goods are becoming increasingly important, particularly in its pivotal trade with the United States. Since 1994, trade initiatives with the Pacific Rim have offered Canadians new opportunities for export growth. The United States also enjoys many lucrative global economic ties, and its geography of exports reveals particularly strong links to other portions of the more developed world (Figure 3.34). By 2000, international sales of U.S. software and entertainment products totaled more than $60 billion annually, more than any other industry. Sales of automobiles, aircraft, computer and telecommunications equipment, financial and tourism services, and food products also contributed to the nation's flow of exports. The destinations for these products and services spanned the globe, with Canada, Japan, Mexico, and western Europe serving as major purchasers.

Patterns of Investment in North America   Patterns of capital investment and corporate power place the North American realm at the center of global money flows and economic influence. Given its relative stability, the region attracts huge inflows of foreign capital, both as investments in North American stocks and bonds and as foreign direct investment (FDI) by international companies. For Canada, the wealth and proximity of the United States have meant that 80 percent of foreign-owned corporations in the country are based in the United States. These investments include a large U.S. presence in Canada's automobile, chemical, electronics, and natural resource industries. In the United States, sustained economic growth and supportive government policies have encouraged large foreign investments, particularly since the late 1970s. Today, the United States is the largest destination of foreign investment in the world. For example, Japanese carmakers build U.S. assembly plants; foreign banks set up U.S. operations; and many parcels of the country's most prestigious real estate are purchased by eager Japanese and European buyers.

Doing Business Globally   The growing impact of U.S. investments in foreign stock markets and the dominance of U.S.-controlled multinational companies suggest how flows of outbound capital are transforming the way business is done throughout the world. Since 1980, aging U.S. baby boomers have poured billions of pension fund and investment dollars into Japanese, European, and "emerging" stock markets. U.S. investments in foreign countries also flow through direct investments made by multinational corporations based in the United States. In 2000, thirteen of the world's 20 largest public companies were based in the United States. Coca-Cola, Intel, and Procter & Gamble are

Latinized Miami: Emergent Global City

Walk through Miami's bustling international airport, and you are immediately aware of the unique role this city plays in linking the economic and cultural worlds of Latin America with those of North America. Spanish-speaking travelers and international flight announcements dominate airport conversation as planes arrive from and depart to cities throughout Central and South America. Indeed, this busy city of 3.7 million has seen much of its post-1960 economic growth fueled by its increasingly close ties with the Latin American world.

Much of the recent Latin connection came with Castro's takeover of Cuba. Between 1960 and 1980, hundreds of thousands of refugees left Cuba and went to Miami's Cuban community. Many brought capital and entrepreneurial skills, establishing hundreds of Cuban businesses in the city. By the late 1970s, Latin America's economic growth demanded closer ties with the United States. What better place to serve as a gateway than Miami? In 1977 Miami's Chamber of Commerce launched the Greater Miami Foreign-Trade Zone, Inc., an organization to foster economic links with Latin America. Two years later, the Florida International Bankers Association (FIBA) was formed to facilitate international banking in the city. Since then, other business groups, such as the Brazilian-American Chamber of Commerce of Florida and the Ecuadorian American Chamber of Commerce of Greater Miami, have tied their home countries to the southern Florida economy.

As a result, modern Miami and its nearby suburbs, such as Coral Gables, have become the leading North American corporate and trade center for Latin America. Indeed, some have termed it the "capital of the Americas." Today, Miami handles more than 50 percent of all U.S. trade with Central America and the Caribbean, and it is involved in more than a third of all U.S. trade with Latin America. Airplane connections, money flows, and even the illegal drug trade demon-

▲ **Figure 3.3.1 A Latin American bank office in Miami** Many major financial institutions in Miami have key connections to Latin America, making the city one of the region's primary contact points with the global economy. Language, landscapes, and money flows confirm the Latin link, and Miami's economy and society are now forever wed to their hemispheric neighbors. *(Robert Frerck/Odyssey Productions)*

strate Miami's centrality. Newspapers such as the *Colombian Post, Venezuela Al Dia,* and the *Diario Las Americas* keep local residents and Latin visitors in touch with news from the home country. Agencies such as Elite International Realty (with offices in Caracas, Lima, Rio de Janeiro, and São Paulo) cater to the needs of an exclusive Latin American clientele who plan to spend time in the city. As journalist Joel Garreau noted in the 1980s, Miami has become a new sort of financial and cultural capital for Latin America, and in the process it has attained the status of a global city and one of North America's most foreign metropolitan areas.

household words from Bangladesh to Bolivia. The economic influence of these corporations exceeds that of many nation-states. General Electric is the world's largest public company. Valued at $350 billion, it is "bringing good things to life" across the world at an annual sales rate of more than $130 billion. Indeed, GE now generates more than 40 percent of its sales from foreign customers. The success of these complex multinational entities is increasingly related to their penetration of foreign markets. Their global orientation is explicit: the Internet version of IBM's annual report, for example, is available to readers in Chinese, English, French, German, Italian, Japanese, and Spanish!

The geographical organization of multinational corporations parallels their global financial reach. U.S. companies such as AT&T, IBM, and Motorola have equipment assembly plants in France, China, and Thailand; service contracts in Mexico and Saudi Arabia; and joint ventures with dozens of other firms in places such as Japan and the Dominican Republic. The processes of globalization in the automobile and

electronics industries are also creating cars and computers literally constructed from parts and by workers from every corner of the world.

Persisting Social Issues

Profound economic and social problems shape the human geography of North America. Even with its continental wealth, great differences persist between rich and poor. High median household incomes in the United States and Canada fail to reveal the differences in wealth within the two countries. Broader measures of social well-being suggest disparities in health care and education. In addition, both nations face enduring gender inequity and aging, which remain key social and political issues for the future. One consequence of globalization is that many of these economic and social challenges are increasingly defined beyond the realm. Poverty in the rural American South may be related to low Asian wage rates, for example, and a viral outbreak in Hong Kong might be only a plane flight away from suburban Vancouver.

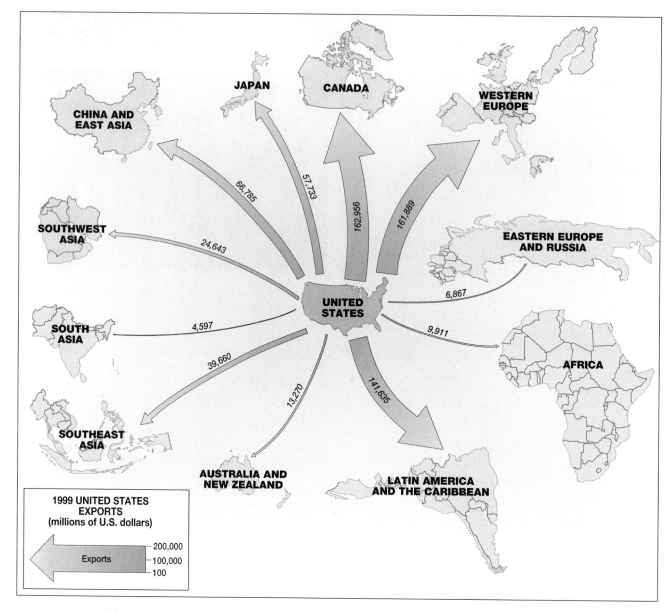

▲ Figure 3.34 Global Links: U.S. exports, 1999   A large flow of goods links the United States with nearby Canada and Latin America, particularly Mexico. Asian trade has also grown greatly in the past 30 years, and the European Union's recent economic consolidation may create new twenty-first-century trade opportunities with that region of the world.   (Data from Direction of Trade Statistics Yearbook, International Monetary Fund, 2000)

Wealth and Poverty   The North American landscape vividly displays contrasting signatures of wealth and poverty. Elite suburbs, gated and guarded neighborhoods, upscale shopping malls, and posh resorts are all geographical expressions of the spatial exclusivity that characterizes many wealthier North American communities, particularly within the United States (Figure 3.35). On the other hand, signatures of poverty are displayed in a great range of local and regional settings. Substandard housing, abandoned property, aging infrastructure, and unemployed workers are visual reminders of the gap between rich and poor within the realm (Figure 3.36). Specifically, within the United States, black household incomes remain only 64 percent of the national average, while

Hispanic incomes fare slightly better at 72 percent of the national average.

The distribution of wealth and poverty varies widely across the United States and Canada. Many of America's wealthiest communities are suburbs on the edge of large metropolitan areas. Just outside New York City, Fairfield, Connecticut, is one of the nation's wealthiest counties; similar settings are found in suburban New Jersey, Maryland, and Virginia, as well as on the peripheries of dozens of southern and western cities. Resort and retirement communities are havens for the rich as well: Palm Beach, Florida, and Aspen, Colorado, have some of the nation's most desirable real estate and costliest housing. In terms of average income, the Northeast and West remain

◀ **Figure 3.35 Gated America**   Wealthier North Americans have had a tremendous impact on the region's cultural landscapes, and their influence far outweighs their relatively modest numbers. Many gated communities in North American suburbs and resorts display a desire for privacy, safety, and social exclusivity.   *(James Patelli)*

America's richest regions, although the South is now within 8 percent of the national norm. In Canada, Ontario and British Columbia are the country's wealthiest provinces, with Vancouver's high house prices vying with those of San Francisco.

Poverty levels have declined in both countries since 1980, but poor populations are still clustered in a variety of geographical settings. By 2000, poverty rates in the United States had fallen to about 12 percent of the population; a similar measure of low-income Canadians declined to just below 20 percent. The problems of the rural poor remain major regional social issues in the Canadian Maritimes, Appalachia, the Deep South, and the Southwest. In the United States, ethnicity is often linked with rural poverty, particularly in the case of southern blacks, Hispanics in the Borderlands, and Native Americans on reservation lands in the Southwest. Most poor people in the United States, however, live in central city locations, and the links between ethnicity and poverty are even stronger in these communities. Nationally, 24 percent of the country's African Americans and 23 percent of its Hispanic populations live below the poverty line, and the great majority reside in central city ghettos or barrios. In the context of the information economy, there is also a pronounced

◀ **Figure 3.36 Inner-city neighborhood** Many poor inner-city neighborhoods, such as this community in New York City, suffer from substandard housing and aging infrastructure. Ethnic bonds often supply the social support necessary to ease the challenges of high unemployment and troubled family life. *(Erica Lananer/Black Star)*

digital divide in North America that suggests that the region's poor and underprivileged groups have significantly less access to Internet communications than the wealthy. Blacks, for example, are far less likely to be linked to the Web, and, within the United States, households making more than $75,000 annually are twenty times more likely to use the Internet than Americans living below the poverty level.

Twenty-first Century Challenges   Measures of social well-being in North America compare favorably with those of most other world regions (Table 3.3). Still, many economic and social challenges confront the realm. As workers, North Americans compete globally, and the declining percentage of stable, high-paying jobs with long-term security and benefits suggests that companies and employees are scrambling to adjust to the uncertainties of the world economy.

Education is also a major public policy issue in Canada and the United States. Although political parties differ in their approach, most public officials agree that more investment in education can only improve North America's chances for competing successfully in the global marketplace. In the United States, dropout rates average about 13 percent for 18- to 24-year-olds, but rates are much higher in many poor urban neighborhoods and rural areas. In addition, American whites are two or three times more likely than blacks or Hispanics to hold a college degree. Another challenge, particularly in the United States, is debating an effective national education policy in the face of a very strong tradition of local control of education.

Since World War II, both nations have seen great changes in the role that women play in society, but the "gender gap" is yet to be closed when it comes to salary issues, working conditions, and political power. Women widely participate in the workforce in both countries (Table 3.3) and are as educated as men, but they still earn only about 75 cents for every dollar that men earn. Although the gap is shrinking, corporate America's "glass ceiling" still makes it extraordinarily difficult for women to advance to top managerial positions and pay scales. Even more daunting is the fact that women head the vast majority of poorer single-parent families in the United States; more than 40 percent of these women are unwed mothers. Canadian women, particularly single mothers who work full-time, are also greatly disadvantaged, averaging only about 65 percent of the salaries of Canadian men. In fact, more than half the single women who head families in Canada are classified as low-income workers. In addition, political power remains largely in male hands, even though women make up the majority of the population and electorate. While Canadian women have voted since 1918 and U.S. women since 1920, females remain in the minority in the Canadian Parliament and the U.S. Congress in the early twenty-first century.

Health care and aging are also key concerns within a world realm of graying baby boomers. A recent report on aging in the United States predicted that 20 percent of the nation's population will be older than 65 by 2050 and that the most elderly senior citizenry (age 85+) are the fastest-growing part of the population. With fewer young people to support their parents and grandparents, officials debate the merits of reforming social security programs. Whatever the outcome of such debates, the geographical consequences of aging are already abundantly clear. Whole sections of the United States—from Florida to southern Arizona—have become increasingly oriented around retirement (Figure 3.37). Communities cater to seniors with special assisted-living arrangements, health-care facilities, and recreational opportunities. Florida and California have the largest number of senior citizens; farm-belt states struggling to hold on to their younger residents also have higher percentages of the aged in their populations.

Although Canada and the United States have distinctive health-care delivery systems, they share many of the health-related benefits and costs of living within the realm. As long average life spans and low childhood mortality rates suggest (Table 3.3), the two countries reap many rewards from modern health-care systems that offer the latest in high technology. Still, North Americans voice concern over escalating health-care costs and uneven access to premium care. While U.S. residents spend over 14 percent of GDP on health care (Canadians spend slightly less), there are 45 million Americans without any health-care insurance, 30 percent more than in 1990.

The rising incidence of chronic diseases associated with aging (heart disease, cancer, and stroke are the three leading causes of death) will continue to pressure both health-care systems. Another cost has been the care and treatment of the realm's 800,000 AIDS victims. The price of the disease will still be broadly borne in the twenty-first century, but particularly among poorer black and Hispanic populations, in which rates of new infection are still on the rise.

TABLE 3.3 *Social Indicators and Status of Women*

Country	Life Expectancy at Birth		Under Age 5 Mortality (per 1,000)		Percent Illiteracy (Ages 15 and over)		Female Labor Force Participation (% of total, 1999)
	Male	Female	1980	1999	Male	Female	
Canada	76	81	13	6	3	3	46
United States	74	80	15	8	3	3	46

Sources: Population Reference Bureau, Data Sheet, 2001, Life Expectancy (M/F); The World Bank, World Development Indicators, 2001, Under Age 5 Mortality Rate (1980/99); The World Bank Atlas, 2001, Female Participation in Labor Force; CIA World Factbook, 2001, Percent Illiteracy (ages 15 and over).

▲ **Figure 3.37 Tomorrow's baby boom landscape?**   This Arizona retirement community, boldly stamped on the desert landscape, suggests the twenty-first-century destination for many aging baby boomers. As North Americans grow older, entire subregions may be devoted to age-dependent demands for housing, recreation, and health services.   *(Cont/Frank Fournier/ Woodfin Camp & Associates)*

The realm also reveals striking variations in the geographies of disease. Certain populations are much more likely to bear the risks and costs of particular health-care problems. Gay men in larger North American cities have borne much of the AIDS burden since the early 1980s. Elsewhere, a combination of poor diet, poor medical care, and genetic factors may contribute to the existence of the "stroke belt" in the eastern Carolinas and Georgia, where death rates from the disease are double the national norm. Although cancer is a leading cause of death in both countries, certain regions and populations appear prone to variations of the disease. For example, soaring rates of bladder and lung cancer in southern Louisiana appear strongly linked to the region's petrochemical industry and to problems of toxic waste and pollution.

⊕ Conclusion

North Americans have reaped the natural abundance of their realm, and in the process they have transformed the environment, created a highly affluent society, and extended their global economic, cultural, and political reach. In a remarkably short time period, a unique mix of varied cultural groups contributed to the settlement of a huge and resource-rich continent that is now the world's most urbanized realm. Along the way, North Americans produced two societies that are closely intertwined, yet face distinctive national political and cultural issues. For both, the twenty-first century brings uncertainties. In Canada, the nation's very existence remains problematic as it works through the persisting realities of its bicultural character and both the costs and benefits of its proximity to its ever-present,

sometimes overbearing continental neighbor. For the United States, social problems linked to ethnic diversity and enduring poverty remain central concerns, along with the environmental price tag associated with high rates of resource consumption. In addition, the pivotal role of the United States in the international economy, in the global balance of geopolitcal power, and in the ongoing, but sometimes problematic expansion of popular culture makes the country on ongoing target for groups focused on destabilizing the key player on the world stage that epitomizes globalization more than any other. How the United States responds to those challenges will directly shape the role it assumes on the world stage as well as the way the processes of globalization play out in the first decade of the twenty-first century.

⊕ Key Terms

acid rain (page 87)
boreal forest (page 81)
concentric zone model (page 93)
connectivity (page 113)
counterurbanization (page 92)
cultural assimilation (page 97)
digital divide (page 120)
ethnicity (page 97)
federal state (page 107)
gentrification (page 94)
Group of 7 (G-7) (page 116)
location factors (page 115)
Megalopolis (page 88)
North America Free Trade Agreement (NAFTA) (page 106)
postindustrial economy (page 77)
prairie (page 83)
sectoral transformation (page 113)
tundra (page 83)
unitary state (page 108)
urban decentralization (page 92)
urban realms model (page 93)

⊕ Questions for Summary and Review

1. Describe North America's major landform regions, and suggest ways in which their physical setting has shaped patterns of human settlement.

2. How are patterns of commercial agriculture related to underlying climatic patterns in North America? What key economic and technological trends have shaped agriculture?

3. What were some of the dominant North American migration flows during the twentieth century?

4. Describe the principal patterns of land use within the modern U.S. metropolis. Include a discussion of (a) the central city and (b) the suburbs/edge city. How have forces of globalization shaped North American cities?

5. How have (a) railroads, (b) the township-and-range survey system, and (c) freeways shaped North America's settlement geography?

6. What are the distinctive eras of immigration in U.S. history, and how do they compare with those of Canada?

7. How do the political origins of the United States and Canada differ?

8. Cite examples of globalization that illustrate the impact of North American cultural, political, and economic influence elsewhere in the world.

9. Briefly summarize North America's endowment of natural resources, and describe where the resources can be found within the realm.

10. What is the sectoral transformation, and how does it help us understand economic change in North America?

11. Cite five types of location factors, and illustrate each with examples from your local economy.

Thinking Geographically

1. Compare and contrast the African American experience in the Black Belt and in urban America. Assess the unique challenges faced by each group.

2. Explain how "natural hazards" can be "culturally" defined. In other words, what role do humans play in shaping the distribution of hazards?

3. Summarize and map the ethnic background and migration history of your own family. Discuss how these patterns parallel or depart from larger North American trends.

4. Discuss the strengths and weaknesses of federalism, and cite examples from both Canada and the United States.

5. "An independent Quebec is a far more viable and coherent nation-state than a divided Canada." Do you agree or disagree? Why?

6. The environmental price for North American development has often been steep. Suggest why it may or may not be worth the price, and defend your answer.

7. Who will America's leading trading partner be in 2050? Suggest why this may be the case.

8. Assume you are a leader of a small African nation. Describe the chief benefits and costs of North American globalization on your part of the world.

9. Discuss why the United States has become a target for political terrorism. How do terrorist groups justify such actions in cultural, political, and economic terms? How does the relative openness of U.S. society complicate the American response to terrorism?

Regional Novels and Films

Novels

Willa Cather, *My Antonia* (1918, Houghton Mifflin)

Ivan Doig, *This House of Sky* (1978, Harcourt Brace Jovanovich)

William Faulkner, *The Bear* (1964, Random House)

F. Scott Fitzgerald, *The Great Gatsby* (1925, Scribner)

Frederick Philip Grove, *Settlers of the Marsh* (1925, Ryerson Press)

William Kennedy, *Ironweed* (1983, Viking Press)

Jack Kerouac, *On the Road* (1957, New American Library)

E. Annie Proulx, *The Shipping News* (1993, Scribner)

Ole Rolvaag, *Giants in the Earth* (1927, Harper and Brothers)

John Steinbeck, *The Grapes of Wrath* (1939, Viking Press)

Wallace Stegner, *Angle of Repose* (1971, Doubleday)

George Stewart, *Storm* (1941, Random House)

Mark Twain, *The Adventures of Huckleberry Finn* (1885)

Nathaniel West, *The Day of the Locust* (1939, Random House)

Tom Wolfe, *Bonfire of the Vanities* (1987, Farrar, Straus, Giroux)

Films

American Beauty (1999, U.S.)

Avalon (1990, U.S.)

Black Robe (1991, Canada)

Chinatown (1974, U.S.)

Dances with Wolves (1990, U.S.)

The Deer Hunter (1978, U.S.)

Fargo (1996, U.S.)

The Godfather (1972, U.S.)

Gone with the Wind (1939, U.S.)

L.A. Confidential (1997, U.S.)

Little Big Man (1970, U.S.)

Pulp Fiction (1994, U.S.)

To Kill a Mockingbird (1962, U.S.)

Traffic (2000, U.S.)

West Side Story (1961, U.S.)

⊕ Bibliography

Adams, John S. 1970. "Residential Structure of Midwestern Cities." *Annals of the Association of American Geographers* 60, 37–62.

Allen, James P., and Turner, Eugene J. 1987. *We the People: An Atlas of America's Ethnic Diversity.* New York: Macmillan.

"American Pop Penetrates Worldwide." 1998. *The Washington Post,* October 25.

Andrews, Alice, and Fonseca, James. 1996. *The Atlas of American Society.* New York: New York University Press.

Annals of the American Academy of Political and Social Science 551 (May). 1997. Special issue on globalization and American cities.

Birdsall, Stephen S., and Florin, John W. 1992. *Regional Landscapes of the United States and Canada.* New York: John Wiley & Sons.

Borchert, John R. 1967. "American Metropolitan Evolution." *Geographical Review* 57, 301–32.

Braus, Patricia. 1998. "Strokes and the South." *American Demographics* 20(5), 26–29.

Butzer, Karl W., ed. 1992. "The Americas Before and After 1492: Current Geographical Research." *Annals of the Association of American Geographers* 82, 345–68.

Castells, Manuel. 1996. *The Information Age. Economy, Society and Culture. Volume 1: The Rise of the Network Society.* Cambridge: Blackwell.

Clay, Grady. 1994. *Real Places: An Unconventional Guide to America's Generic Landscape.* Chicago: University of Chicago Press.

Conzen, Michael P., ed. 1990. *The Making of the American Landscape.* Boston: Unwin Hyman.

Ford, Larry R. 1994. *Cities and Buildings. Skyscrapers, Skid Rows, and Suburbs.* Baltimore: Johns Hopkins University Press.

Garreau, Joel. 1981. *The Nine Nations of North America.* Boston: Houghton Mifflin.

Garreau, Joel. 1991. *Edge City: Life on the New Frontier.* New York: Doubleday.

Gentilcore, R. Louis, ed. 1990. *Historical Atlas of Canada. Volume 2: The Land Transformed.* Toronto: University of Toronto Press.

Getis, Arthur, and Getis, Judith, eds. 1995. *The United States and Canada: The Land and the People.* Dubuque, IA: William C. Brown.

Harris, R. Cole, ed. 1987. *Historical Atlas of Canada. Volume 1: From the Beginning to 1800.* Toronto: University of Toronto Press.

Harris, R. Cole, and Warkentin, John. 1974. *Canada Before Confederation.* New York: Oxford University Press.

Heubusch, Kevin. 1998. "Small Is Beautiful." *American Demographics* 20(1), 43–49.

Hugill, Peter J. 1993. *World Trade Since 1431: Geography, Technology, and Capitalism.* Baltimore: Johns Hopkins University Press.

Kacapyr, Elia. 1998. "The Well-Being of American Women." *American Demographics,* 20(8), 30–32.

Kerr, Donald, and Holdsworth, Deryck, eds. 1993. *Historical Atlas of Canada. Volume 3: Addressing the Twentieth Century.* Toronto: University of Toronto Press.

Knox, Paul L., and Marston, Sallie A. 2001. *Human Geography: Places and Regions in Global Context.* Upper Saddle River, NJ: Prentice Hall.

Martin, Philip, and Midgeley, Elizabeth. 1999. *Immigration to the United States, Population Bulletin* 54(2), 1–44.

McIlwraith, Thomas F., and Muller, Edward K., eds. 2001. *North America: The Historical Geography of a Changing Continent.* Lanham, MD: Rowman and Littlefield.

McKnight, Tom L. 1997. *Regional Geography of the United States and Canada,* 2nd ed. Upper Saddle River, NJ: Prentice Hall.

Meinig, Donald W. 1986. *The Shaping of America: A Geographical Perspective on 500 Years of History. Volume 1, Atlantic America, 1492–1800.* New Haven: Yale University Press.

Meinig, Donald W. 1993. *The Shaping of America: A Geographical Perspective on 500 Years of History. Volume 2, Continental America, 1800–1867.* New Haven: Yale University Press.

Meinig, Donald W. 1998. *The Shaping of America: A Geographical Perspective on 500 Years of History. Volume 3, Transcontinental America, 1850-1915.* New Haven: Yale University Press.

National Geographic Society. 1985. *Atlas of North America: Space Age Portrait of a Continent.* Washington, DC: National Geographic Society.

Opdycke, Sandra. 2000. *The Routledge Historical Atlas of Women in America.* London: Routledge.

Riche, Martha. 2000. "America's Diversity and Growth: Signposts for the 21st Century." *Population Bulletin* 55(2), 1–43.

Riebsame, William E., ed. 1997. *Atlas of the New West: Portrait of a Changing Region.* New York: W. W. Norton.

Shinagawa, Larry, and Yang, Michael. 1998. *Atlas of American Diversity.* Lanham, MD: Rowman and Littlefield.

Shortridge, James R. 1991. "The Concept of the Place-Defining Novel in American Popular Culture." *Professional Geographer* 43, 280–91.

Statistics Canada. 2001.

United States Bureau of the Census. Census of 2000.

Wheeler, James; Aoyama, Yuko; and Warf, Barney, eds. 2000. *Cities in the Telecommunications Age: The Fracturing of Geographies.* London: Routledge.

Yeates, Maurice H. 1990. *The North American City,* 4th ed. New York: Harper & Row.

Zelinsky, Wilbur. 1992. *The Cultural Geography of the United States: A Revised Edition.* Englewood Cliffs, NJ: Prentice Hall.

UNITED STATES

70°W   60°W   50°W   40°W

Tropic of Cancer

30°N

Tijuana

Ciudad
Juárez

Monterrey

20°N

MEXICO

Guadalajara

México City ⊛   Veracruz
Puebla

ATLANTIC
OCEAN

Caribbean
Sea

HONDURAS

Guatemala City   Tegucigalpa

GUATEMALA

San Salvador   NICARAGUA

EL SALVADOR   Managua

San José ⊛
COSTA RICA
PANAMA   Panama

Barranquilla   Maracaibo
Cartagena   Valencia   Caracas
Panama

VENEZUELA   Ciudad
Bolívar

Medellín   Orinoco R.

Cali   Bogotá

COLOMBIA

Equator   Quito

ECUADOR   Negro R.

Galápagos Is.
(ECUADOR)

Guayaquil

Iquitos   Manaus   Amazon R.   Belém

São Luis

Forta

Trujillo   PERU   Natal

0°

Callao   BRAZIL   Recife

Lima

10°S   Cuzco   São Francisco R.   Salvad

Arequipa   BOLIVIA   Brasília

La Paz   Santa Cruz   Goiânia

Arica   Sucre
Potosí

Tropic of Capricorn

20°S   Belo Horizonte

CHILE   PARAGUAY   São Paulo   Rio de Janeiro

Asunción   Santos

Curitiba

0   500   1,000 mi

Córdoba   Pôrto Alegre

Valparaíso   Rosario

0   500   1,000 km

Santiago ⊛

Concepción   ARGENTINA   URUGUAY

Buenos Aires   Montevideo
La Plata

30°S

Bahía Blanca

LATIN AMERICA
Political Map

⊛ ●   Over 1,000,000

○ •   500,000–1,000,000
(selected cities)

★ •   Selected smaller cities

40°S

Falkland Is.
(U.K.)

50°S

Puntas Arenas

120°W   100°W   80°W   60°W   40°W   20°W

10°N

0°

10°S

20°S

30°S

PACIFIC
OCEAN

CHAPTER 4

Latin America

Beginning with Mexico and extending to the tip of South America, Latin America's regional coherence stems largely from its shared colonial history, rather than from the disparate levels of development seen today. More than 500 years ago, the Iberian countries of Spain and Portugal began their conquest of the Americas. Iberia's mark is still visible throughout the area: officially, two-thirds of the inhabitants speak Spanish, and the rest speak Portuguese. Iberian architecture and town design add homogeneity to the colonial landscape. The vast majority of the population nominally practices Catholicism, although mainline Protestant denominations and Pentecostal groups have made inroads since the 1960s. These European traits blended with those of different Amerindian peoples. The Indian presence remains especially strong in Bolivia, Peru, Ecuador, Guatemala, and southern Mexico, where large and diverse indigenous populations maintain their native languages, dress, and traditions. After the initial conquest, other cultural groups were added to this mix of indigenous and Iberian peoples. The legacy of slavery imparted a strong African presence, primarily on the coasts of Colombia and Venezuela and throughout Brazil. In the nineteenth and twentieth centuries new waves of settlers came from Spain, Italy, Germany, Japan, and Lebanon. The end result is one of the world's most racially mixed regions. The modern states of Latin America are multiethnic, with distinct indigenous and immigrant profiles and very different rates of social and economic development (Figure 4.1; see also "Setting the Boundaries").

Through colonialism, immigration, and trade, the forces of globalization have been embedded in the Latin American landscape. The early Spanish Empire concentrated on extracting precious metals, sending galleons laden with silver and gold across the Atlantic. The Portuguese became prominent producers of dyewoods, sugar products, gold, and later coffee. In the late nineteenth and early twentieth centuries, exports to North America and Europe fueled the region's economy. Most countries specialized in one or two products: bananas and coffee, meats and wool, wheat and corn, petroleum and copper. Such a primary export tradition, according to Latin American economists, led to an unhealthy economic dependence. They argued in the 1960s that Latin American economies were too specialized and faced unequal terms of trade that inhibited overall development. Since the 1960s the countries of the region have industrialized and diversified their production, but they continue to be major producers of primary goods for North America, Europe, and Japan. Changes

◀ **Figure 4.1 Latin America** Roughly equal in size to North America, Latin America supports a larger population and far greater ecological diversity. The 17 countries included in this region share a history of Iberian colonization. Seventy-four percent of the region's 490 million people live in cities, making it the most urbanized region of the developing world. It is noted for its production of primary exports; however, rates of economic development vary greatly between states.

SETTING THE BOUNDARIES

The concept of Latin America as a distinct region has been popularly accepted for nearly a century. The boundaries of this region are straightforward, beginning at the Rio Grande (called the Rio Bravo in Mexico) and ending at Tierra del Fuego. French geographers are credited with coining the term "Latin America" in the nineteenth century to distinguish the Spanish- and Portuguese-speaking republics of the Americas plus Haiti from the English-speaking territories. There is nothing particularly "Latin" about the area, other than the predominance of romance languages. The term stuck because it was vague enough to be inclusive of different colonial histories while also offering a clear cultural boundary from Anglo-America, the region referred to as North America in this text.

Latin America is one of the world regions that North Americans are most likely to visit. The trend, of course, is to visit the northern fringe of this region. Tourism is robust, especially along Mexico's northern border and coastal resorts such as Cancun. Unfortunately, there is a tendency to visit one area in the region and generalize for all of it. Although it is historically sound to think of Latin America as a major world region, there are extreme variations in the physical environment, levels of social and economic development, and the influences of indigenous societies.

Since the region is so large, geographers often divide Latin America. The continent of South America is typically distinguished from Middle America (which includes Central America, Mexico, and the Caribbean). The term "Middle America" was created to identify an area culturally distinct from North America but physically part of it, since most of Mexico and all of Cuba rest on the North American Plate. Such a division has its merits, but it also separates countries with very similar histories (such as Mexico and Peru) while joining countries that have very little in common (such as El Salvador and Jamaica). In this text the Americas are divided slightly differently. Latin America will consist of the Spanish- and Portuguese-speaking countries of Central and South America, including Mexico. Chapter 5 will examine the Caribbean, consisting of the islands of the Antilles, the Guianas, and Belize. So divided, the important Indian and Iberian influences of mainland Latin America will be emphasized. Similarly, the Caribbean's unique colonial and demographic history will be discussed in Chapter 5.

in the economic landscape, however, are reflected in increased levels of manufacturing for internal consumption and export. Intraregional trade within Latin America, as well as the prospect of a **Free Trade Area of the Americas (FTAA)** that would include all of Latin America, the Caribbean (minus Cuba), and North America in a hemispheric free-trade zone by 2005, are indicators of heightened economic integration.

Despite the region's growing industrial capacity, extractive industries will continue to prevail, in part, because of the area's impressive natural resources. Latin America is home to Earth's largest rain forest, the greatest river by volume, and massive reserves of natural gas, oil, and copper. With its vast territory, its tropical location, and its relatively low population density (Latin America has half the population of India in nearly seven times the area), the region is also recognized as one of the world's great reserves of biological diversity. How this diversity will be managed in the face of global demand for natural resources is an increasingly important question for the countries of this region (Figure 4.2). Roughly equal in area to North America, Latin America has a much larger and faster-growing population of 490 million people. Its most populous state, Brazil, has more than 170 million people, making it the fifth largest country in the world. The next largest state, Mexico, has a population of 100 million. Throughout the region, population growth rates have slowed dramatically in the last 40 years, and the typical Latin American woman now has three children, compared with six in the 1960s. Unlike most areas of the developing world today, Latin America is decidedly urban. Prior to World War II, most people lived in rural settings and worked as farmers. Today three-quarters of Latin Americans are city dwellers. Even more startling is the number of **megacities.** São Paulo, Mexico City, Buenos Aires, and Rio de Janeiro all have more than 10 million inhabitants. In addition, there are more than 40 cities of at least 1 million residents.

The urban nature of life in Latin America provides a landscape of contrasts. Modern automotive and high-tech industries, high-rise apartments, and stately residential suburbs are juxtaposed with shantytowns, street vendors, and artisans

▲ **Figure 4.2 Tropical wilderness**   The biological diversity of Latin America—home to the world's largest rain forest—is increasingly seen as a genetic and economic asset. These forested mesas (called tepuis) in southern Venezuela are representative of the wild lands that many conservationists seek to protect. *(Rob Crandall)*

who fill the ranks of the urban poor. Collectively, many Latin American states fall into the middle-income category and support a significant middle class. But national debt, political scandal, currency devaluation, and triple-digit inflation have triggered grave economic hardships throughout the region, especially during the 1980s. Today, neoliberal policies that encourage foreign investment, export production, and privatization have been adopted by many states. These policies exemplify the current impact of economic globalization on Latin America. The results are mixed, with some states experiencing impressive economic growth but increased disparity between rich and poor.

Environmental Geography: Neotropical Diversity

Much of the region is characterized by its tropicality. Travel posters of Latin America showcase verdant forests and brightly colored parrots. The diversity and uniqueness of the **neotropics** (tropical ecosystems of the Western Hemisphere) have long been attractive to naturalists eager to understand their unique flora and fauna. It is no accident that Charles Darwin's insights on evolution were inspired by his two-year journey in tropical America. Even today, scientists throughout the region work to understand complex ecosystems, discern new species, conserve genetic resources, and interpret the impact of human settlement, especially in neotropical forests. Not all of the region is tropical. Important population centers extend below the Tropic of Capricorn, most notably Buenos Aires and Santiago. Much of northern Mexico, including the city of Monterrey, is north of the Tropic of Cancer. Yet it is Latin America's tropical climate and vegetation that prevail in popular images of the region.

The movement of tectonic plates explains much of the region's basic topography. As the South and North American plates slowly drift westward, the Nazca, Cocos, and Pacific plates are subducted below them (Figure 4.3). In this contact zone, deep oceanic trenches, such as the Humboldt Trench off the coast of Peru and Chile, form. The submerged plates have folded and uplifted the mainland's surface, creating the geologically young western mountains, such as the Andes and the Volcanic Axis of Central America. This area is also vulnerable to powerful earthquakes that threaten people and damage property. In January 2001, for example, a major earthquake struck El Salvador, killing nearly 900 people and destroying more than 100,000 homes. In contrast, the Atlantic side is characterized by humid lowlands interspersed with upland plateaus called **shields**. The Brazilian shield is the largest, followed by the Patagonian and Guiana shields. Across these lowlands meander some of the great rivers of the world, including the Amazon. An exception to the pattern of Atlantic lowlands is the Central Plateau of Mexico, whose eastern escarpment forms the Sierra Madre Oriental, which averages 7,000 to 8,000 feet in height (2,100 to 2,400 meters). On the North American Plate, the Sierra Madre Oriental is geologically similar to the Appalachian Mountains in the eastern United States.

Environmental Issues Facing Latin America

Given the territory's immense size and relatively low population density, Latin America has not experienced the same levels of environmental degradation witnessed in East Asia and Europe. The worst environmental problems are found in cities, their surrounding rivers and coasts, and intensely farmed zones. Concerns over deforestation, loss of biodiversity, degradation of farmland, and the livability of urban areas are the major environmental issues for the region (Figure 4.4). Yet vast areas remain relatively untouched, supporting an incredible diversity of plant and animal life. Throughout the region, national parks offer some protection to unique communities of plants and animals. A growing environmental movement in countries such as Costa Rica and Brazil has yielded both popular and political support for "green" initiatives. In short, Latin Americans enter the twenty-first century with a real opportunity to avoid many of the environmental missteps seen in other regions of the world. At the same time, economic pressures brought on by global market forces are driving governments to aggressively exploit their natural resources (minerals, fossil fuels, forests, soils). One challenge lies in managing the region's immense natural resources and balancing the economic benefits of extraction with the ecological soundness of conservation. Another major challenge is to make the large urban centers of Latin America more environmentally sound for the millions of people living in them (see "Local Voices: The Environmentalist Message of the Kogi").

The Valley of Mexico At the southern end of the great Central Plateau of Mexico lies the Valley of Mexico, the cradle of Aztec civilization and the site of Mexico City, a metropolitan area of approximately 18 million people. The severe problems facing this vital ecosystem underscore how environment, population, technology, and politics are interwoven in the daunting effort to make Mexico City livable. This once-idyllic high-altitude basin with year-round mild temperatures, fertile soils, and ample water surrounded by snow-capped volcanoes was an early center of indigenous settlement and plant domestication, and the site of one of Spain's most important colonial cities. Given these features, it is no wonder that Mexico City has retained its primacy into the twenty-first century, even though the environmental setting that made it attractive centuries ago has also created modern challenges. Some of the most pressing problems are air quality, adequate water, and subsidence (soil sinkage) caused by overdrawing the valley's aquifer.

The legendary and perpetual smog of Mexico City is so bad that most modern-day visitors have no idea that mountains surround them because they are rarely visible. Air quality has been a major issue for Mexico City since the 1960s, driven in part by the city's phenomenal rate of growth. (Between 1950 and 1980 the city's annual rate of growth was 4.8 percent.) It is hard to imagine a better setting for creating air pollution. The city sits in a bowl 7,400 feet (2,250 m) above sea level, where thermal inversion layers regularly trap a toxic concoction of exhaust, industrial smoke, garbage, and

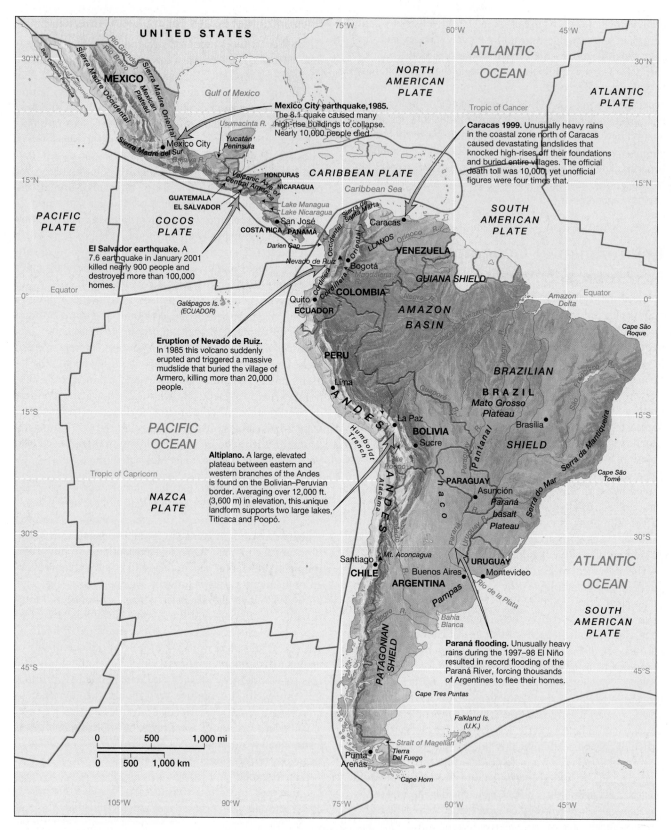

UNITED STATES

MEXICO

Sierra Madre Occidental

Sierra Madre Oriental

Rio Grande / Rio Bravo

Baja California Peninsula

Gulf of California

Mexican Plateau

Gulf of Mexico

Usumacinta R.

Yucatán Peninsula

Mexico City earthquake, 1985. The 8.1 quake caused many high-rise buildings to collapse. Nearly 10,000 people died.

Mexico City

Grijalva R.

Sierra Madre del Sur

Volcanic Axis of Central America

HONDURAS

GUATEMALA
EL SALVADOR

NICARAGUA

Lake Managua
Lake Nicaragua

San José

COSTA RICA PANAMA

CARIBBEAN PLATE

Caribbean Sea

PACIFIC PLATE

COCOS PLATE

El Salvador earthquake. A 7.6 earthquake in January 2001 killed nearly 900 people and destroyed more than 100,000 homes.

Darien Gap

Nevado de Ruiz

Cordillera Occidental

Cordillera Oriental

Sierra de Santa Marta

Caracas

LLANOS

Orinoco R.

VENEZUELA

Bogotá

Magdalena R.

COLOMBIA

Quito

ECUADOR

Galápagos Is. (ECUADOR)

Eruption of Nevado de Ruiz. In 1985 this volcano suddenly erupted and triggered a massive mudslide that buried the village of Armero, killing more than 20,000 people.

Caracas 1999. Unusually heavy rains in the coastal zone north of Caracas caused devastating landslides that knocked high-rises off their foundations and buried entire villages. The official death toll was 10,000, yet unofficial figures were four times that.

SOUTH AMERICAN PLATE

GUIANA SHIELD

Amazon Delta

AMAZON BASIN

Negro R.

Amazon R.

Cape São Roque

BRAZILIAN

B R A Z I L

Mato Grosso Plateau

São Francisco R.

SHIELD

Brasília

PERU

Lima

Guaporé R.

ANDES

Humboldt Trench

Atacama

La Paz

BOLIVIA

Sucre

Lake Poopó

Pantanal

Paraguay R.

PARAGUAY

Asunción

Chaco

Paraná basalt Plateau

Serra da Mantiqueira

Cape São Tomé

Serra do Mar

Altiplano. A large, elevated plateau between eastern and western branches of the Andes is found on the Bolivian–Peruvian border. Averaging over 12,000 ft. (3,600 m) in elevation, this unique landform supports two large lakes, Titicaca and Poopó.

PACIFIC OCEAN

NAZCA PLATE

Mt. Aconcagua

Santiago

CHILE

Buenos Aires

Paraná R.

Uruguay R.

URUGUAY

Montevideo

ARGENTINA

Pampas

Río de la Plata

Bahía Blanca

Negro R.

ATLANTIC OCEAN

SOUTH AMERICAN PLATE

Paraná flooding. Unusually heavy rains during the 1997–98 El Niño resulted in record flooding of the Paraná River, forcing thousands of Argentines to flee their homes.

PATAGONIAN SHIELD

Cape Tres Puntas

Falkland Is. (U.K.)

Strait of Magellan

Tierra Del Fuego

Punta Arenas

Cape Horn

ATLANTIC OCEAN

ATLANTIC PLATE

NORTH AMERICAN PLATE

Tropic of Cancer

Equator

Tropic of Capricorn

0   500   1,000 mi
0   500   1,000 km

▲ **Figure 4.3 Physical geography of Latin America**   Centered in the tropics but extending into the midlatitudes, Latin American landforms include mountains, shields, highland plateaus, vast river basins, and grassy plains. Due to the movement of continental plates, earthquakes and volcanic eruptions are always threatening, especially on the Pacific coast, where the region's highest mountains are found. The lower landforms on the Atlantic side include the Amazon, Plata, and Orinoco river basins, as well as the highly eroded surfaces of the Brazilian, Guiana, and Patagonian shields.

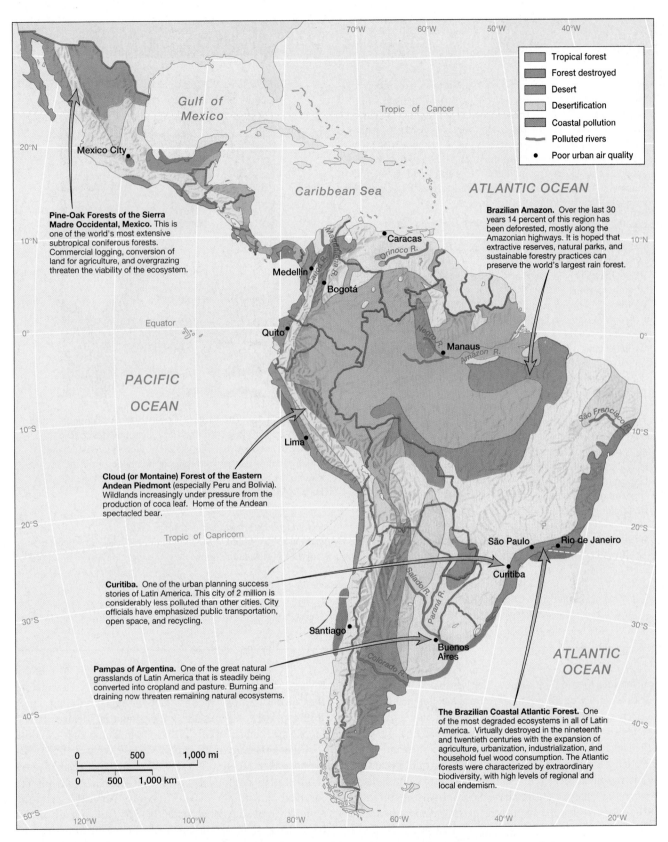

Legend:
- Tropical forest
- Forest destroyed
- Desert
- Desertification
- Coastal pollution
- Polluted rivers
- Poor urban air quality

Pine-Oak Forests of the Sierra Madre Occidental, Mexico. This is one of the world's most extensive subtropical coniferous forests. Commercial logging, conversion of land for agriculture, and overgrazing threaten the viability of the ecosystem.

Brazilian Amazon. Over the last 30 years 14 percent of this region has been deforested, mostly along the Amazonian highways. It is hoped that extractive reserves, natural parks, and sustainable forestry practices can preserve the world's largest rain forest.

Cloud (or Montaine) Forest of the Eastern Andean Piedmont (especially Peru and Bolivia). Wildlands increasingly under pressure from the production of coca leaf. Home of the Andean spectacled bear.

Curitiba. One of the urban planning success stories of Latin America. This city of 2 million is considerably less polluted than other cities. City officials have emphasized public transportation, open space, and recycling.

Pampas of Argentina. One of the great natural grasslands of Latin America that is steadily being converted into cropland and pasture. Burning and draining now threaten remaining natural ecosystems.

The Brazilian Coastal Atlantic Forest. One of the most degraded ecosystems in all of Latin America. Virtually destroyed in the nineteenth and twentieth centuries with the expansion of agriculture, urbanization, industrialization, and household fuel wood consumption. The Atlantic forests were characterized by extraordinary biodiversity, with high levels of regional and local endemism.

▲ Figure 4.4 **Environmental issues in Latin America** Tropical forest destruction, desertification, water pollution, and poor urban air quality are some of the pressing environmental problems facing Latin America. Yet vast areas of tropical forest are still present, supporting a wealth of genetic and biological diversity. *(Adapted from* DK World Atlas, *1997, pp. 7, 55. London: DK Publishing)*

LOCAL VOICES   The Environmentalist Message of the Kogi

The Kogi live in the Sierra de Santa Marta in Colombia. A small indigenous group that has struggled to survive for five centuries, they continue to perceive themselves as the "Elder Brothers" and the rest of the world as "Younger Brothers." One of the fundamental beliefs of the Kogi is that they are the guardians of the environment and that Younger Brothers are recklessly destroying Earth. Kogi spiritual leaders, the Mamas, expressed their environmentalist agenda in front of a BBC film crew in the 1980s.

We are the Elder Brothers
We have not forgotten the old ways
How could I say that I do not know how to dance?
We still know how to dance.
We have forgotten nothing.
We know how to call the rain.
If it rains too hard we know how to stop it.
We call the summer.
We know how to bless the world and make it flourish.

But now they are killing the Mother.
The Younger Brother, all he thinks about is plunder.

The Mother looks after him too, but he does not think.

He is cutting into her flesh.
He is cutting into her arms.
He is cutting off her breasts.
He takes out her heart.
He is killing the heart of the world.

When the final darkness falls everything will stop.
The fires, the benches, the stones, everything.
All the world will suffer.

If that happened and all we Mamas died,
*   and there was no one doing our work,*
well, the rain wouldn't fall from the sky.

▲ **Figure 4.1.1 The Kogi of Colombia**   A Kogi man walks with his *poporo* (a gourd filled with lime powder) and a satchel of coca leaves. The chewing of coca leaf with lime, which reduces hunger and fatigue, has been practiced by indigenous peoples in the Andes for centuries.   *(Victor Englebert/Englebert Photography, Inc.)*

It would get hotter and hotter from the sky,
and the trees wouldn't grow,
and the crops wouldn't grow.

Or am I wrong, and they would grow anyway?

Source: Alan Ereira, *The Elder Brothers.* New York: Alfred Knopf, 1992, pp. 113–14.

fecal matter (Figure 4.5). During the worst pollution emergencies when a winter inversion layer traps pollutants, schoolchildren are required to stay indoors, and high-polluting vehicles are barred from the streets. Steps were finally taken in the late 1980s to reduce emissions from factories and cars. Unleaded gas is now widely available for the 3 million cars in the metropolitan area, and some of the worst polluting factories in the Valley of Mexico have closed. In fact, a study in 2001 suggested that the high levels of lead, carbon dioxide, and sulfur dioxide in the atmosphere were starting to decline. Still, the health costs of breathing such contaminated air are real, as elevated death rates due to heart disease, influenza, and pneumonia suggest. By the 1990s, Mexico City's growth rate had slowed to less than 1 percent a year. Apparently, people were fleeing the pollution as well as the crowding, high costs, and crime associated with the Federal District. Small

towns and cities up to 100 miles (160 kilometers) away from the downtown are now considered within commuting range. If this sprawl continues, it is anticipated that Mexico's megalopolis may contain approximately 50 million people by 2050.

One of Mexico City's most relentless environmental problems is water. Mexican President Vicente Fox declared water (both scarcity and quality) a national security issue, not just for the capital but for the entire country. Ironically, it was the abundance of water that made this site attractive for settlement. Large shallow lakes once filled the valley, but over the centuries most were drained to expand agricultural land. As surface water became scarce, wells were dug to tap the basin's massive freshwater aquifer. Today approximately 70 percent of the water used in the metropolitan area is drawn from the valley's aquifer. There is troubling evidence that the aquifer is being overdrawn and at risk of contamination, especially in

its impact is similar throughout the city. Building foundations are destroyed, and water and sewer lines rupture. Repair brigades race through city streets, patching some 40,000 ruptures in the water lines every year. Cherished landmarks, such as the national cathedral in the central plaza, have cracks, list to one side, and are supported by scaffolding. Because of subsidence, groundwater is no longer pumped from the city center, so that sinking has slowed to 1 inch (2.5 centimeters) per year. In the periphery, where the aquifer is actively tapped, some areas are sinking as much as 20 inches (50 centimeters) per year.

These serious urban environmental problems are made worse by poverty and governmental inaction. During the 1970s and 1980s, politicians and industrialists denied that there were problems. For most of the century, one-party rule in Mexico reduced the likelihood of meaningful environmental reforms. When steps were finally taken to introduce unleaded gasoline and catalytic converters, open dumps, aging cars and minibuses, and unregulated factories continued to spew pollutants into the air. Mexico City's poorest citizens suffer the most from urban contamination. Imagine being one of the 2 million people in Cuidad Nezahualcoyotl in the northeast corner of the metropolitan area. Bordering the slum is a vast, dusty lake bed into which much of the city's garbage is dumped. A canal running through the foul-smelling dump carries black ooze through the neighborhoods. Hundreds of thousands of residents have no indoor plumbing and use the dump and the canal as a substitute. Understandably, respiratory diseases and other infections are rampant in this neglected corner of the city, and children seem to suffer the worst effects. Air quality is so poor that breathing is equivalent to smoking two packs of cigarettes a day. For the vast majority of Latin American urban poor, the lack of reliable water and clean air is the most pressing environmental problem.

Deforestation   Perhaps the environmental issue most commonly associated with Latin America is deforestation. The Amazon Basin and portions of the eastern lowlands of Central America and Mexico still maintain unique and impressive stands of tropical forest. Other areas, such as the Atlantic coastal forests of Brazil and the Pacific forests of Central America, have nearly disappeared as a result of agriculture, settlement, and ranching. In the midlatitudes, the ecologically unique evergreen rain forest of southern Chile (the Valdivian forest) is being cleared to export wood chips to Asian markets (Figure 4.6). The coniferous forests of northern Mexico are also falling, in part because of a bonanza for commercial logging stimulated by the North American Free Trade Association (NAFTA) agreement.

In terms of biological diversity, however, the loss of tropical rain forests is the most critical. Tropical rain forests account for only 6 percent of Earth's landmass, but at least 50 percent of the world's species are found in this biome. Moreover, the Amazon contains the largest undisturbed stretches of rain forest in the world. Unlike Southeast Asian forests, where hardwood extraction drives forest clearance, Latin American forests are usually seen as an agricultural frontier that state governments

▲ **Figure 4.5 Air pollution in Mexico City** Mexico City is notorious for its poor air quality, and its high elevation and immense size make management of air quality difficult. While lead levels have declined with the introduction of unleaded gasoline, respiratory ailments are on the rise. *(Tom Owen Edmunds/Getty Images, Inc.)*

areas where unlined drainage canals can leak pollutants into the surrounding soil, which then leach into the aquifer. To reduce reliance on the aquifer, the city now pumps water nearly a mile uphill from more than 100 miles (160 kilometers) away. Still there are shortages, as the city has one of the world's leakiest distribution systems and one-third of the water pumped into the system leaks out. Remedies for many of these water problems are costly, but efforts to charge industrial consumers higher rates, implement modern conservation efforts, and recharge the most depleted areas of the aquifer are being tested.

A lesser known but equally vexing problem is that Mexico City is sinking. As the metropolis grows and pumps more water from its aquifer, subsidence worsens. Mexico City sank 30 feet (9 meters) during the twentieth century. By comparison, Venice, an Italian city knows for its subsidence problems, sank only 9 inches (23 centimeters) during the same period. Subsidence is a huge problem that will not go away because the city is still so reliant on ground water. Although the amount of subsidence varies across the metropolitan area,

▲ **Figure 4.6 Chilean wood chips**   A mountain of wood chips awaits shipment to Japanese paper mills from the southern Chilean port of Punta Arenas. The exploitation of wood products from both native and plantation forests supports Chile's booming export economy.   *(Rob Crandall/Rob Crandall Photography)*

limited resources for expansion into the region. During the 1990s forest loss was averaging 0.5 percent annually. However, there is growing interest in harvesting tropical hardwoods from the Brazilian Amazon; Southeast Asian logging firms have sought concessions from the Brazilian government. At the same time, tropical forestry experts argue that only through sustainable management can forest extraction support the region's economy while maintaining its resource base.

The conversion of tropical forest into pasture, called **grassification,** is another practice that has contributed to deforestation. Particularly in southern Mexico, Central America, and the Brazilian Amazon, a hodgepodge of development policies from the 1960s through the 1980s encouraged deforestation to make room for cattle. The preference for ranching as a status-conferring occupation seems to be a carryover from Iberia. The image of the *vaquero* (cowboy) looms large in the region's history. Even poor farmers appreciate the value of having livestock. Like a savings account, cattle can be quickly sold for cash. Although there are many natural grasslands such as the Llanos, the Chaco, and the Pampas suitable for grazing, it was the rush to convert forest into pasture that made ranching a scourge on the land. Ultimately a cattle bust occurred in the 1980s when world prices collapsed. Even in cases where domestic demand for beef increased, ranching in remote tropical frontiers was seldom economically self-sustaining. All told, nearly 300 million acres of tropical forest were destroyed in the 1980s, and much of this went to pasture.

Degradation of Farmlands   The pressure to modernize agriculture has produced a series of environmental problems. As peasants were encouraged to adopt new hybrid varieties of corn, beans, and potatoes, an erosion of genetic diversity occurred. Efforts to preserve dozens of native domesticates are under way at agricultural research centers in the central Andes and Mexico. It is feared, nonetheless, that many useful native plants are already lost. Modern agriculture also depends on chemical fertilizers and pesticides that eventually run off into surface and groundwater supplies. Consequently, many rural areas suffer from contamination of local water supplies. More troublesome still is the direct exposure of farmworkers to toxic agricultural chemicals. Mild exposure to pesticides and fertilizers typically occurs due to mishandling, resulting in rashes and burns. Yet there are areas, such as Sinaloa, Mexico, where the widespread application of chemicals parallels a rise in serious birth defects.

Soil erosion and fertility decline occur in all agricultural areas. Certain soil types in Latin America are particularly vulnerable to erosion, most notably the volcanic soils and the reddish oxisols found in the humid lowlands. The productivity of the Paraná basalt plateau in Brazil, for example, has declined over the decades due to the ease with which these volcanic soils erode and the failure to apply soil conservation methods. The oxisols of the tropical lowlands, by contrast, can quickly degrade into a baked claypan surface when the natural cover is removed, making permanent agriculture nearly impossible. Ironically, the consolidation of the large-scale modern farms in the basins and valleys of the highlands tends

divide in an attempt to appease landless peasants or reward political cronies. Thus, forests usually fall to the ax and fire, with settlers and politicians carving them up to create permanent settlements, slash-and-burn plots, or large cattle ranches. In addition, some tropical forest clearance has been motivated by the search for gold (Brazil, Venezuela, and Costa Rica) and the production of coca leaf for cocaine (Peru, Bolivia, and Colombia).

Brazil has incurred more criticism than other countries for its Amazon forest policies. Through these policies, some 15 percent of the Brazilian Amazon has been deforested. In states such as Rondônia, where settlers streamed in along a popular road known as BR364, close to 60 percent of the state has been deforested (Figure 4.7). What most alarmed environmentalists and forest dwellers (Indians and rubber tappers) was the dramatic increase in the rate of clearing, especially in the 1980s. Rates of deforestation have slowed in response to both pressure from environmental, indigenous, and development agencies and Brazil's economic crisis, which

Figure 4.7 Frontier settlement in Rondônia, Brazil Colonization along Amazonian highways, such as BR364, sparked a wave of forest clearing in the 1980s. Paired satellite images from Rondônia, Brazil, in 1975 (left) and 1992 (right) reveal the extent of forest clearing that occurred along this major highway and its side roads. *(EROS Data Center, U.S. Geological Survey)*

to push peasant subsistence farmers into marginal areas. On these hillside farms, gullies and slides reduce productivity (Figure 4.8). Lastly, the sprawl of Latin American cities consumes both arable land and water, eliminating some of the region's best farmland.

Urban Environmental Challenges For most Latin Americans, air pollution, inadequate water, and garbage removal are the pressing environmental problems of everyday life. Consequently, many environmental activists from the region focus their efforts on making urban environments cleaner by introducing "green" legislation and rousing popular action. In this most urbanized region of the developing world, city dwellers do have better access to water, sewers, and electricity than their counterparts in Asia and Africa. Moreover, the density of urban settlement seems to encourage the widespread use of mass transportation—both public and private bus and van routes make getting around most cities fairly easy. Yet the inevitable environmental problems that come from dense urban settings ultimately require expensive infrastructural remedies. The money for such projects is never enough, thanks to currency devaluation, inflation, and foreign debt. Since many urban dwellers tend to reside in unplanned squatter settlements, retroactively servicing these communities with utilities is difficult and costly. These settlements are

Figure 4.8 Degraded Bolivian farmlands The farmlands of Cochabamba have historically been Bolivia's breadbasket. Agricultural productivity has declined in recent decades due to increased aridity and severe soil erosion. In this photo, soil erosion has produced deep gullies that divide fields and remove valuable topsoil. *(Rob Crandall/Rob Crandall Photography)*

especially vulnerable to natural hazards, as witnessed by the catastrophic landslides in Caracas, Venezuela, in December 1999. Unusually heavy rains on the coastal mountains north of the city triggered massive mudflows and landslides that buried entire communities. Death toll estimates range from 10,000 to 40,000 people, making it one of Latin America's worst natural disasters in the twentieth century.

Industrialization is also a major cause of pollution. Factories, electricity generation, and transportation all contribute to urban pollution. A lax attitude about enforcing environmental laws tends to be the norm. The consequences are, in the worst cases, a serious threat to people and the environment. In the 1980s the Brazilian industrial center of Cubatão, near São Paulo, became synonymous with urban environmental nightmares. For years people complained of headaches and nausea from the belching factory smokestacks, but their complaints were not taken seriously. In 1984 a leak occurred in a gasoline pipeline that ran through one of the poorest squatter settlements. The smell of leaking gas went unnoticed because of the omnipresent stench of industrial pollutants in the valley. When the gas was finally ignited, as many as 200 people were incinerated in the resulting explosion and fire. A year later, a break in an ammonia pipeline forced the evacuation of 6,000 people and the hospitalization of 65. While industry downplayed these disasters as part of the risk of doing business, traumatized residents mobilized to address the worst abuses. The events in Cubatão, more than the destruction of rain forest, are credited with invigorating the environmental movement in Brazil.

Not far from Cubatão is the capital of Paraná state, Curitiba. Curitiba is the celebrated "green city" of Brazil because of some relatively simple yet innovative planning decisions. More than 2 million people inhabit this industrial and commercial center for Paraná state, yet it is significantly less polluted than similar-size cities. Since the location was vulnerable to flooding, city planners built drainage canals and set aside the remaining natural drainage areas as parks in the 1960s, well before explosive growth would have made such a policy difficult. This action added green space and reduced the negative impacts of flooding. Next, public transportation became a top priority. An innovative and clean bus system that featured rapid loading and unloading of passengers made Curitiba a model for transportation in the developing world. The city's Free University for the Environment offers short courses at no cost to anyone interested in learning simple measures to improve the urban environment. And lastly, a low-tech but effective recycling program has greatly reduced solid waste. Cities such as Curitiba demonstrate that designing with nature makes sense both ecologically and economically. To comprehend the diversity of environments Latin Americans must grapple with when making development decisions, an understanding of the region's physical geography is essential.

Western Mountains and Eastern Shields

Historically, the most important areas of settlement in tropical Latin America were not along the region's major rivers, but across its shields, plateaus, and fertile intermontane basins. In these localities the combination of arable land, benign climate, and sufficient rainfall produced the region's most productive agricultural areas and its densest settlement. The Mexican Plateau, for example, is a massive upland area ringed by the Sierra Madre mountains. The southern end of the plateau is where the Valley of Mexico is located. Similarly, the elevated and well-watered basins of Brazil's southern mountains provide an ideal setting for agriculture. These especially fertile areas are able to support high population densities, so it is not surprising that the two largest cities, Mexico City and São Paulo, emerged in these settings. The Latin American highlands also lend a special character to the region. Luxuriant tropical valleys nestled below snow-covered mountains hint at the diversity of ecosystems found in close proximity to each other. The most dramatic of these highland areas, the Andes, runs like a spine down the length of the South American continent (refer again to Figure 4.3).

The Andes   Beginning in northwestern Venezuela and ending at Tierra del Fuego, the Andes are relatively young mountains that extend nearly 5,000 miles (8,000 kilometers). Created by the collision of oceanic and continental plates, the mountains are a series of folded and faulted sedimentary rocks with intrusions of crystalline and volcanic rock. The Andes are still forming, so active volcanism and regular earthquakes are common in this zone. The end result is an ecologically and geologically diverse mountain chain with some 30 peaks higher than 20,000 feet (6,000 meters). Due to the violent and complex origins of this mountain chain, many rich veins of precious metals and minerals are found here. The initial economic wealth of many Andean countries came from mining silver, gold, tin, copper, and iron.

Given the length of the Andes, the mountain chain is typically divided into northern, central, and southern components. In Colombia the northern Andes actually split into three distinct mountain ranges before merging near the border with Ecuador. High-altitude plateaus and snow-covered peaks distinguish the central Andes of Ecuador, Peru, and Bolivia. The Andes reach their greatest width here. Of special interest is the treeless high plain of Peru and Bolivia, called the **Altiplano**. The floor of this elevated plateau ranges from 11,800 feet (3,600 meters) to 13,000 feet (4,000 meters) in altitude, and it has limited usefulness for grazing. Two high-altitude lakes, Titicaca on the Peruvian and Bolivian border and the smaller Poopó in Bolivia, are located in the Altiplano, as well as many mining sites (Figure 4.9). The southern Andes are shared by Chile and Argentina. The highest peaks of the Andes are found in the southern Andes, including the highest peak in the Western Hemisphere, Aconcagua, at almost 23,000 feet (6,958 meters). South of Santiago, Chile, the mountains are lower and the chain less compact. The impact of glaciation is most evident in the southernmost extension of the Andes.

The Uplands of Mexico and Central America   The Mexican Plateau and the Volcanic Axis of Central America are the most important elevated lands in terms of settlement. Most

▲ **Figure 4.9 Bolivian Altiplano**   The Altiplano is an elevated plateau straddling the Bolivian and Peruvian Andes. This stark, windswept land is inhabited mostly by Amerindians. Considered one of the poorer areas of the Andes, mining and pastoral activities dominate.   *(Loren McIntyre/Woodfin Camp & Associates)*

▲ **Figure 4.10 Guatemalan farmer**   A peasant farmer works his small plot of subsistence and cash crops. Behind him looms a volcano. The fertile volcanic soils, ample rainfall, and temperate climate of the Guatemala highlands have supported dense populations for centuries.   *(Sean Sprague/Panos Pictures)*

major cities of Mexico and Central America are found here. The Mexican Plateau is a large, tilted block that has its highest elevations in the south, about 8,000 feet (2,500 meters) around Mexico City, and its lowest, just 4,000 feet (1,200 meters), at Ciudad Juárez. The southern end of the plateau, the Mesa Central, contains a number of flat-bottomed basins interspersed with volcanic peaks. Mexico's megalopolis—a concentration of the largest population centers such as Mexico City, Guadalajara, and Puebla—is in the Mesa Central. (The Valley of Mexico, discussed earlier, is one of the basins of the Mesa Central.) Yet this productive region is also under severe stress. The once-ample ground and surface waters of the Mesa Central have been depleted. Water shortages have threatened the agricultural productivity of the area, which was long thought of as Mexico's breadbasket. Across the Mexican Plateau are rich seams of silver, copper, and zinc. The quest for silver drove much of the economic activity of colonial Mexico. Today the Mexican economy is driven by petroleum and gas production along the Gulf Coast, not by the metals of the plateau.

Along the Pacific coast of Central America lies a chain of volcanoes that stretches from Guatemala to Costa Rica. The Volcanic Axis of Central America is a handsome landscape of rolling green hills, elevated basins with sparkling lakes, and conical volcanic peaks. More than 40 volcanoes are found here, many of them still active. Their legacy is a rich volcanic soil that yields a wide variety of domestic and export crops. Most of Central America's population is also concentrated in this zone, in the capital cities or the surrounding rural villages. The bulk of the agricultural land is tied up in large holdings that produce beef, cotton, and coffee for export. Yet most of the farms are small subsistence properties that produce corn, beans, squash, and assorted fruits (Figure 4.10). A major rift (a large crustal fracture) east of the Volcanic Axis

exists in Nicaragua. In this low-lying valley are Lakes Managua and Nicaragua, the largest in Central America. Surrounded by mountains, this rift zone presently yields sorghum and cotton for export.

The Shields   As mentioned earlier, South America has three major shields—large upland areas of exposed crystalline rock that are similar to upland plateaus found in Africa and Australia. (The Guiana shield will be discussed in Chapter 5.) The Brazilian and Patagonian shields vary in elevation between 600 and 5,000 feet (200 and 1,500 meters) and are remnants of the ancient landmass of Gondwanaland. The Brazilian shield is the larger and more important in terms of natural resources and settlement. Far from a uniform land surface, the Brazilian shield covers much of Brazil from the Amazon Basin in the north to the Plata Basin in the south. It is studded with isolated low ranges and flat-topped plateaus in the north. In southeastern Brazil a series of mountains (Serra da Mantiqueira and Serra do Mar) reach elevations of

▲ **Figure 4.12 Patagonian wildlife**   Guanacos thrive on the thin steppe vegetation found throughout Patagonia. Native to South America, the numbers of guanacos fell dramatically due to hunting and competition with introduced livestock.   *(Rob Crandall/Rob Crandall, Photographer)*

▲ **Figure 4.11 Brazilian oranges**   Most estate-grown oranges in Brazil are processed into frozen concentrate and exported. São Paulo and Paraná have some of the finest soils in Brazil. In addition to oranges, coffee and soybeans are widely cultivated. *(Stephanie Maze/NGS Image Collection)*

9,000 feet (2,700 meters). In between these ranges are elevated basins that offer a mild climate and fertile soils. In one of these basins is the city of São Paulo, the largest urban conglomeration in South America. The other major population centers are on the coastal fringe of the plateau, where large protected bays made the sites of Rio de Janeiro and Salvador attractive to Portuguese colonists. Finally, the Paraná basalt plateau, located on the southern end of the Brazilian shield, is celebrated for its fertile red soils *(terra roxa)*, which yield coffee, oranges, and soybeans. The basalt plateau, much like the Deccan plateau in India, is an ancient lava flow that resulted from the breakup of Gondwanaland. So fertile is this area that the economic rise of São Paulo is attributed to the expansion of commercial agriculture, especially coffee, into this area (Figure 4.11).

The vast low-lying shield of Patagonia lies in the southern tip of South America. Often described as the end of the world, much of Patagonia is open steppe country with few settlements. Located in the midlatitudes, Patagonia captured the imagination of many nineteenth-century travelers. Because most transoceanic traffic prior to the building of the Panama Canal used the Strait of Magellan, many voyagers had the opportunity to observe Patagonia and write about it. Beginning south of Bahia Blanca and extending to Tierra del Fuego, the region to this day remains sparsely settled and hauntingly beautiful. It is treeless, covered by scrubby steppe vegetation,

and home to wildlife such as the guanaco (Figure 4.12). Sheep were introduced to Patagonia in the late nineteenth century, spurring a wool boom. More recently, offshore oil production has renewed the economic importance of Patagonia.

River Basins and Lowlands

Three great river basins drain the Atlantic lowlands of South America: the Amazon, Plata, and Orinoco (refer to Figure 4.3). Within these basins are vast interior lowlands, less than 600 feet (200 meters) in elevation, that lie over young sedimentary rock. From north to south they are the Llanos, the Amazon lowlands, the Pantanal, the Chaco, and the Pampas. With the exception of the Pampas, most of these lowlands are sparsely settled and offer limited agricultural potential except for grazing livestock. Yet the pressure to open new areas for settlement and to exploit natural resources has created pockets of intense economic activity in the lowlands. Areas such as the Amazon and the Chaco have witnessed marked increases in resource extraction and settlement since the 1970s.

Amazon Basin   The Amazon drains an area of roughly 2.4 million square miles (6.1 million square kilometers), making it the largest river system in the world by volume and area and the second longest by length. Everywhere in the basin, rainfall is more than 60 inches (150 centimeters) a year, and in many places more than 80 inches (200 centimeters). The basin's largest city, Belém, averages close to 100 inches (250 centimeters) a year. Although there is no real dry season, there are definitely drier and wetter times of year, with August and September being the driest months. In the basin, rainfall is likely most days, but showers often pass quickly, leaving bright blue skies. The immensity of this watershed and its hydrologic cycle is underscored by the fact that 20 percent of all freshwater discharged into the oceans comes from the Amazon.

The river is navigable to Iquitos, Peru, some 2,100 miles (3,600 kilometers) upstream. More than 200 tributaries drain the eastern slope of the Andes, as well as portions of the Brazilian and Guiana shields. Since the Amazon Basin draws from nine countries, it would seem that this watershed would be an ideal network to integrate the northern half of South America. Ironically, compared with the other great rivers of the world, settlement in the basin continues to be sparse, in large part due to the poor quality of the forest soils. The best soils are found in the floodplain, where natural levees reach heights of 20 feet (6 meters) (Figure 4.13). Thousands of years of alluvium deposited on these levees have made them extremely fertile, as well as safe from normal flooding. Consequently, most of the older settlements, such as Manaus, are found on the levees. Active colonization of the Brazilian portion of the Amazon since the 1960s has made the population soar. According to the 2000 census, 12 million people live in the Brazilian Amazon, which is only 7 percent of the country's total population. Still, the population in the Amazonian states is increasing at nearly 4 percent a year.

Plata Basin   The region's second largest watershed begins in the tropics and discharges into the Atlantic in the midlatitudes. Three major rivers make up this system: the Paraná, the Paraguay, and the Uruguay. The Paraguay River and its tributaries drain the eastern Andes of Bolivia, the Brazilian shield, and the Chaco. The Paraná primarily drains the Brazilian uplands before the Paraguay River joins it in northern Argentina. The Paraná and the considerably smaller Uruguay empty into the Rio de la Plata estuary, which begins north of Buenos Aires.

Unlike the Amazon, much of the Plata Basin is now economically productive through large-scale mechanized agriculture, especially soybean production. Arid areas such as the Chaco and inundated lowlands such as the Pantanal support livestock. The Plata Basin contains several major dams, including the region's largest hydroelectric plant, the Itaipú on the Paraná, which generates all of Paraguay's electricity and much of southern Brazil's. As agricultural output in this watershed grows, sections of the Paraná River are being canalized and dredged to enhance the region's fluvial transportation network. In the late 1990s plans were made to canalize the Paraguay River so that eastern Bolivia could have a water route to the Atlantic. The costs of such a project would be enormous, and if it is carried out, much of the Pantanal (a unique Brazilian wetland) will be drained.

Orinoco Basin   The third largest river basin by area is the Orinoco in the northern part of South America. The Orinoco River meanders through much of southern Venezuela and part of eastern Colombia, giving character to a tropical grassland called the *Llanos*. Although it is just one-sixth the size of the Amazon watershed, its discharge roughly equals that of the Mississippi River. Like the Amazon, this basin is home to very few individuals; 90 percent of Venezuela's population lives north of the basin. With the exception of the industrial developments between Ciudad Guayana and Ciudad Bolívar, cities are rare. Much of the Orinoco drains the Llanos, which are inundated by several feet of water during the rainy season. Since the colonial era, these grasslands have supported large cattle ranches. Cattle are still important, but the Llanos have become a dynamic area of petroleum production for both Colombia and Venezuela.

Grijalva-Usumacinta   In Mexico and Central America the largest watershed by volume, the Grijalva-Usumacinta Basin, also flows through a sparsely populated tropical forest zone in Mexico and Guatemala. In the Mexican state of Tabasco, the Usumacinta joins the Grijalva and flows into the Bay of

◀ **Figure 4.13 Amazonian floodplains**   More water flows through the Amazon Basin than through any other river system. Historically, settlement has been along the river's natural levees and floodplains (called *varzeas*). Agricultural interest in the fertile varzeas is growing as farmers seek to develop sustainable agricultural practices within this tropical ecosystem. *(Victor Englebert/Englebert Photography, Inc.)*

▶ **Figure 4.14 Satellite image of Mexican deforestation** This view from space shows the clearing of Mexican forests adjacent to the Guatemalan border. Frontier settlement programs brought a wave of Mexican farmers and ranchers into this area in the 1970s and 1980s. Environmentalists and Indian-rights groups fear that Guatemala's forests may soon meet a similar fate. *(Earth Satellite Corporation/ SPL/Photo Researchers, Inc.)*

Campeche, accounting for nearly half of Mexico's freshwater river flow. Hydroelectric dams have been built or are planned, and an active agricultural frontier exists. This frontier is contested, however, by indigenous and environmental groups that are fighting to slow the incorporation of this area into the modern economy. Political interest in the basin has intensified over the years because the watershed may be critical for satisfying the water and energy demands of Mexico (Figure 4.14).

Climate

In tropical Latin America average monthly temperatures in localities such as Managua, Quito, or Manaus show little variation (Figure 4.15). Precipitation patterns, however, do vary and create distinct wet and dry seasons. In Managua, for example, January is typically a dry month, and October is a wet one. The tropical lowlands of Latin America, especially east of the Andes, are usually classified as tropical humid climates that support forest or savanna, depending on the amount of rainfall. The region's desert climates are found along the Pacific coasts of Peru and Chile, Patagonia, northern Mexico, and the Bahia of Brazil. Thus, a city such as Lima, Peru, which is clearly in the tropics, averages only 1.5 inches (4 centimeters) of rainfall a year due to the hyperaridity of the Peruvian coast.

Midlatitude climates, with hot summers and cold winters, prevail in Argentina, Uruguay, and parts of Paraguay and Chile. Of course, the midlatitude temperature shifts in the Southern Hemisphere are the inverse of those in the Northern Hemisphere (cold Julys and warm Januarys). Chile's climate is a mirror image of the west coast of Mexico and the United States, with the Atacama Desert in the north (like Baja California), a Mediterranean dry summer around Santiago (similar to Los Angeles), and a marine west coast climate with no dry season south of Concepción (like the coasts of Oregon and Washington). In the mountain ranges, complex climate patterns result from changes in elevation. To appreciate how humans adapt to tropical mountain ecosystems, one must understand the concept of **altitudinal zonation,** the relationship between cooler temperatures at higher elevations and changes in vegetation.

Altitudinal Zonation   First described in the scientific literature by Prussian naturalist Alexander von Humboldt in the early 1800s, altitudinal zonation has had practical applications that have been intimately understood by all the region's native inhabitants. Humboldt systematically recorded declines in temperature as he ascended to higher elevations, a phenomenon referred to today as the **environmental lapse rate** (averaging a temperature decline of 3.5 °F for every 1,000 feet in elevation or 6.5 °C for every 1,000 meters). He also noted changes in vegetation by elevation, demonstrating that plant communities common to the midlatitudes could thrive in the tropics at higher elevations. These different altitudinal zones

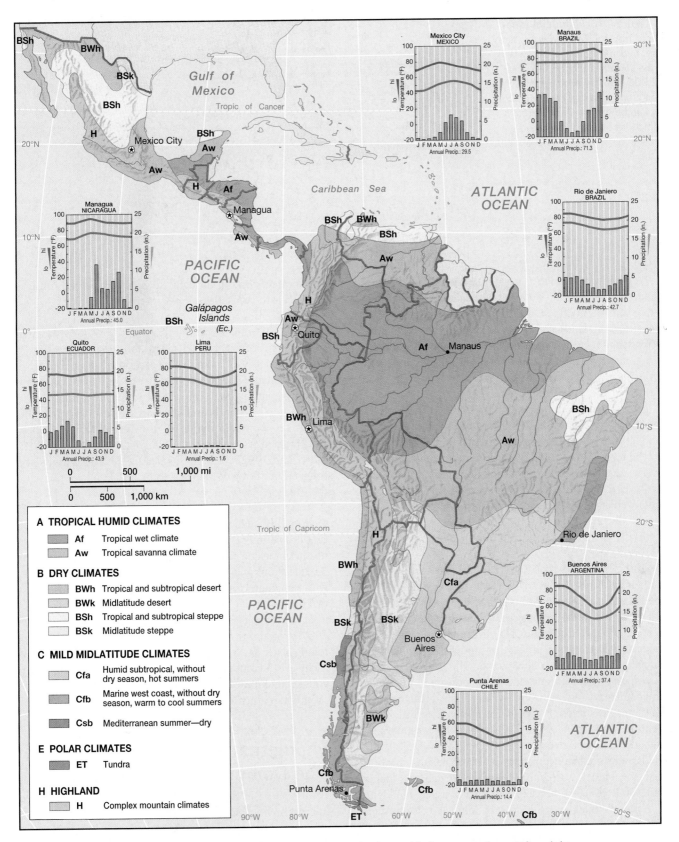

▲ Figure 4.15 Climate map of Latin America Latin America includes the world's largest rain forest (Af) and driest desert (BWh), as well as nearly every other climate classification. Latitude, elevation, and rainfall play important roles in determining the region's climates. Note the contrast in rainfall patterns between humid Quito and arid Lima. *(Temperature and precipitation data from Pearce and Smith, 1984.* The World Weather Guide. *London: Hutchinson)*

▶ **Figure 4.16 Altitudinal zonation** Tropical highland areas support a complex array of ecosystems. In the tierra fría zone (6,000 to 12,000 feet), for example, midlatitude crops such as wheat and barley can be grown. This diagram depicts the range of crops and wildlife found at different elevations in the Andes.

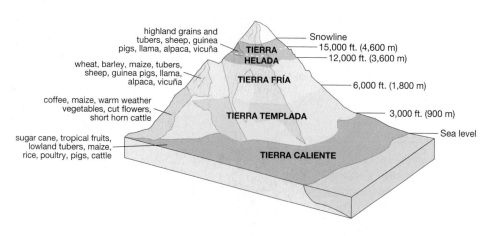

are commonly referred to as the *tierra caliente* (hot land) from sea level to 3,000 feet (900 meters); the *tierra templada* (temperate land) at 3,000 to 6,000 feet (900 to 1,800 meters); the *tierra fría* (cold land) at 6,000 to 12,000 feet (1,800 to 3,600 meters); and the *tierra helada* (frozen land) above 12,000 feet (3,600 meters). Exploitation of these zones allows agriculturists, especially in the uplands, access to a great diversity of domesticated and wild plants (Figure 4.16).

The concept of altitudinal zonation is most relevant for the Andes, the highlands of Central America, and the Mexican Plateau. Traditional Andean farmers, for example, might use the high pastures of the Altiplano for grazing llamas and alpacas, the tierra fría for potato and quinoa production, and the lower temperate zone to produce corn. All the great precontact civilizations, especially the Incas and the Aztecs, systematically extracted resources from these zones, thus ensuring a diverse and abundant resource base.

El Niño   El Niño is probably the most talked-about weather phenomenon in Latin America and the world. Its name (a reference to the Christ child) comes from a warm Pacific current that usually arrives along coastal Ecuador and Peru in December, around Christmastime. Every decade or so, an abnormally large current arrives that produces torrential rains, signaling the arrival of an El Niño year. The 1997–98 El Niño was especially bad. Devastating floods occurred in Peru and Ecuador. Heavy May rains in Paraguay and Argentina caused the Paraná River to rise 26 feet (8 meters) above normal. Flooding drove some 350,000 people from their homes in Peru. In Argentina 150,000 fled. At least 900 people were killed by floods and storms in Latin America, the single most devastating being a hurricane that hit Acapulco, Mexico, and took at least 200 lives. The other, less talked about aspect of El Niño is drought. While the Pacific coast of South and North America experienced record rainfall in the 1997–98 El Niño, Colombia, Venezuela, northern Brazil, Central America, and Mexico battled drought. In addition to crop and livestock losses, estimated to be in the billions of dollars, hundreds of brush and forest fires left their mark. The amount of smoke produced by forest fires in northern Mexico in the spring of 1998 was so great that it caused haze in the southeastern United States.

The indirect costs of drought, such as fire-charred hillsides vulnerable to landslides, are impossible to calculate.

Drought   Drought conditions seem to regularly threaten certain areas of Latin America. In northeastern Brazil from Fortaleza to Salvador lies an area of relatively high rural population density with a notoriously precarious environment. When the intertropical convergence zone (ITCZ), described in Chapter 2, fails to oscillate here, devastating droughts occur. The worst was in the 1870s, when a half million people perished from drought. Droughts in the 1970s, 1983, 1991, and 1992, while not as deadly, destroyed crops and triggered outflows of migrants to Brazil's southern cities and the Amazon.

Today, few Latin American droughts cause the fatalities seen in Sub-Saharan Africa, yet too little or late rainfall can produce serious economic hardships. The 1996 drought in northern Mexico caused basic grain production to decline by 3 million tons and half a million cattle to be sent prematurely to slaughterhouses. Similarly, droughts in the densely settled highlands of Central America and the Andes often result in agricultural losses and disruptions in electric power. In 1992 several cities in Colombia instituted the practice of rolling blackouts (often for several hours a day) to ration electricity use. A combination of below-average rainfall and reservoir sedimentation had drastically reduced the country's production of hydroelectricity.

Population and Settlement: The Dominance of Cities

Great river basin civilizations like those in Asia never existed in Latin America. In fact, the great rivers of the region are surprisingly underutilized as areas of settlement or corridors for transportation. While the major population clusters of Central America and Mexico are in the interior plateaus and valleys, the interior lowlands of South America are relatively empty. Historically, the highlands supported most of the region's population during the pre-Hispanic and colonial eras. In the twentieth century population growth and immigration to the Atlantic lowlands of Argentina and Brazil, along with continued growth of Andean coastal cities such as Guayaquil, Barranquilla, and Maracaibo, have reduced the demographic importance of the highlands. Major highland cities such as Mexico City, Guatemala City, Bogotá, and La Paz still dominate their national economies, but the majority of large cities are on or near the coasts (Figure 4.17).

Like the rest of the developing world, Latin America experienced dramatic population growth in the 1960s and 1970s. In 1950 its population totaled 150 million people, which equaled the population of the United States at that time. By 1995 the population had tripled to 450 million; in comparison, the United States will probably not reach 300 million until 2010. Latin America outpaced the United States because its infant mortality rate declined and life expectancy soared, while its birthrate consistently outpaced that of the United States. In 1950 Brazilian life expectancy was only 43 years; by the 1980s it was 63. In fact, most countries in the region experienced a 15- to 20-year improvement in life expectancy

between 1950 and 1980, which pushed up growth rates. Four countries account for 75 percent of the region's population: Brazil with more than 170 million, Mexico with 100 million, Colombia with 43 million, and Argentina with 38 million (Table 4.1).

During the 1980s, population growth rates in Latin America suddenly began to slow, and by the 1990s most countries reported rates of less than 2 percent. This relatively sudden shift surprised demographers who had predicted in 1985 that the region's population would reach 750 million in 2025. Today's projection is for only 650 million. One of the reasons for this fertility decline is the shift to urban living, which tends to limit family size.

The Latin American City

A quick glance at the population map of Latin America shows a concentration of people in cities. One of the most significant demographic shifts has been the movement out of rural areas

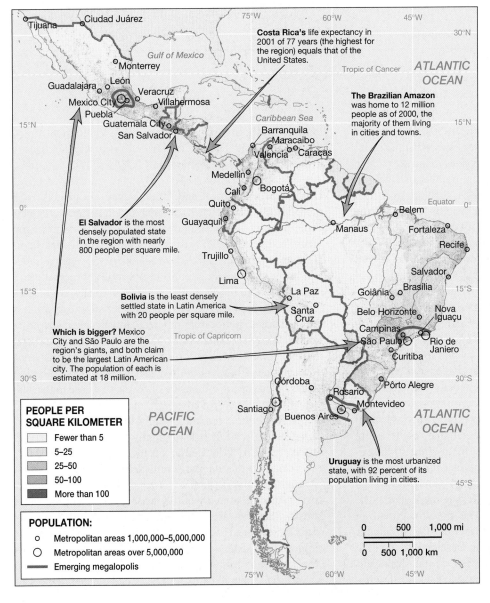

◀ **Figure 4.17 Population map of Latin America** The concentration of population in urban and coastal settlements is evident in this map. Population density in central and southern Mexico, as well as Central America, is quite high. In South America, the majority of people live on or near the coasts, leaving the interior of the continent lightly populated.

Costa Rica's life expectancy in 2001 of 77 years (the highest for the region) equals that of the United States.

The Brazilian Amazon was home to 12 million people as of 2000, the majority of them living in cities and towns.

El Salvador is the most densely populated state in the region with nearly 800 people per square mile.

Bolivia is the least densely settled state in Latin America with 20 people per square mile.

Which is bigger? Mexico City and São Paulo are the region's giants, and both claim to be the largest Latin American city. The population of each is estimated at 18 million.

Uruguay is the most urbanized state, with 92 percent of its population living in cities.

PEOPLE PER SQUARE KILOMETER
- Fewer than 5
- 5–25
- 25–50
- 50–100
- More than 100

POPULATION:
- ○ Metropolitan areas 1,000,000–5,000,000
- ◯ Metropolitan areas over 5,000,000
- ━━ Emerging megalopolis

TABLE 4.1 *Demographic Indicators*

Country	Population (Millions, 2001)	Population Density, per square mile	Rate of Natural Increase	TFR[a]	Percent <15[b]	Percent >65	Percent Urban
Argentina	37.5	35	1.1	2.6	28	10	90
Bolivia	8.5	20	2.4	4.2	40	4	63
Brazil	171.8	52	1.5	2.4	30	5	81
Chile	15.4	53	1.3	2.3	28	7	86
Colombia	43.1	98	1.8	2.6	32	5	71
Costa Rica	3.7	188	1.8	2.6	32	5	45
Ecuador	12.9	118	2.2	3.3	34	4	62
El Salvador	6.4	788	2.3	3.5	36	5	58
Guatemala	13.0	309	2.9	4.8	44	3	39
Honduras	6.7	155	2.8	4.4	43	4	46
Mexico	99.6	132	1.9	2.8	34	5	74
Nicaragua	5.2	104	3.0	4.3	43	3	57
Panama	2.9	100	2.1	2.6	31	6	56
Paraguay	5.7	36	2.7	4.3	40	5	52
Peru	26.1	53	1.8	2.9	34	5	72
Uruguay	3.4	49	0.7	2.3	24	13	92
Venezuela	24.6	70	2.0	2.9	34	5	87

[a]Total fertility rate

[b]Percent of population younger than 15 years of age

Source: Population Reference Bureau. World Population Data Sheet, *2001.*

to cities, which began in earnest in the 1950s. In 1950 just one-quarter of the region's population was urban; the rest lived in small villages and the countryside. Today the pattern is reversed, with three-quarters of the population living in cities. In the most urbanized countries, such as Argentina, Chile, Uruguay, and Venezuela, more than 85 percent of the population live in cities (Table 4.1). This preference for urban life is attributed to cultural as well as economic factors. Under Iberian rule, residence in a city conferred status and offered opportunity. Initially, only Europeans were allowed to live in the colonial cities, but this exclusivity was not strictly enforced. Over the centuries colonial cities became the hubs for transportation and communication, underscoring their primary role in structuring regional economies.

Latin American cities are noted for high levels of **urban primacy,** a condition in which a country has a primate city three to four times larger than any other city in the country. Examples of primate cities are Lima, Caracas, Guatemala City, Santiago, Buenos Aires, and Mexico City. Primacy is often viewed as a liability, since too many national resources are concentrated into one urban center. In an effort to decentralize, some governments have intentionally built new cities far from existing primate cities (for example, Ciudad Guayana in Venezuela and Brasília in Brazil). Despite these efforts, the tendency toward primacy remains. Moreover, the growth of urbanized regions has inspired the label of Megalopolis for three areas in Latin America. Emerging megalopolises include Mexico City–Puebla–Toluca–Cuernavaca on the Mesa Central, the Niterói–Rio de Janeiro–Santos–São Paulo–Campinas

axis in southern Brasil, and the Rosario–Buenos Aires–Montevideo–San Nicolás corridor in Argentina and Uruguay's lower Rio Plata Basin (see Figure 4.17).

Urban Form   Latin American cities have a distinct urban morphology that reflects both their colonial origins and their contemporary growth (Figure 4.18). Usually a clear central business district (CBD) exists in the old colonial core. Radiating out from the central business district is older middle- and lower-class housing found in the zones of maturity and *in situ* accretion. In this model, residential quality declines as one moves from the center to the periphery. The exception is the elite spine, a newer commercial and business strip that extends from the colonial core to newer parts of the city. Along the spine one finds superior services, roads, and transportation. The city's best residential zones, as well as shopping malls, are usually on either side of the spine. Close to the elite residential sector, a limited area of middle-class tract housing is typically found. Most major urban centers also have a *periférico* (a ring road or beltway highway) that circumscribes the city. Industry is located in isolated areas of the inner city and in larger industrial parks outside the ring road.

Straddling the periférico is a zone of peripheral squatter settlements where many of the urban poor live in the worst housing. Services and infrastructure are extremely limited: roads are unpaved; water is often trucked in; and sewer systems are nonexistent. The dense ring of squatter settlements (variously called *ranchos, favelas, barrios jovenes,* or *pueblos nuevos*) that encircle Latin American cities reflect the speed

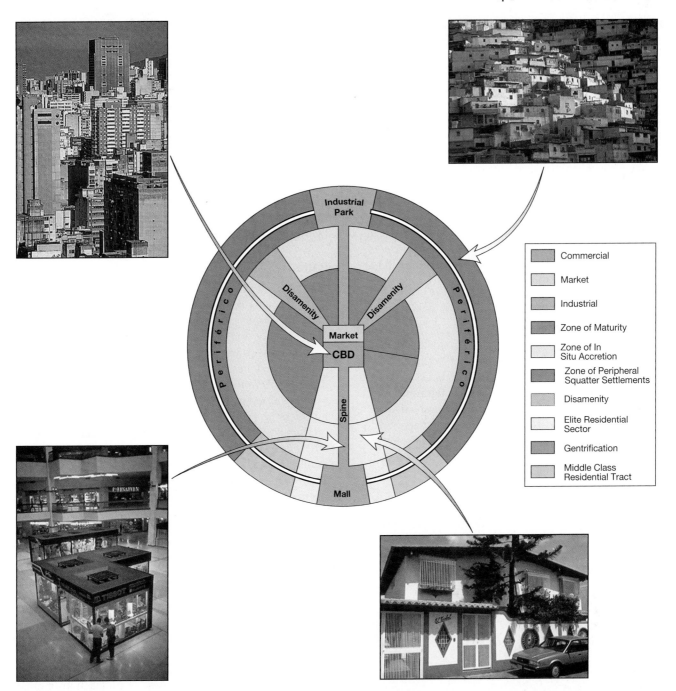

▲ Figure 4.18 Latin American city model   This urban model highlights the growth of Latin American cities and the class divisions within them. While the central business district (CDB), elite spine, and residential sectors may have excellent access to services and utilities, life in the zone of peripheral squatter settlements is much more difficult. In many Latin American cities, one-third of the population resides in squatter settlements.   *[Model reprinted from Ford, 1996. "New and Improved Model of Latin American City Structure." Geographical Review 86(3), 437-40. Photos by Rob Crandall/Rob Crandall, Photographer]*

and intensity with which these zones were created. In some cities more than one-third the population lives in these self-built homes of marginal or poor quality. These kinds of dwellings are recognizable throughout the developing world, yet the practice of building one's home on the "urban frontier" has a longer history in Latin America than in most Asian and African cities. The combination of a rapid inflow of migrants (at times reaching 1,000 people per day), the inability of governments to meet pressing housing needs, and the eventual official recognition of many of these neighborhoods with land titles and utilities meant that this housing strategy was rarely discouraged. Each successful colonization encouraged more.

Among the inhabitants of these neighborhoods, the **informal sector** is a fundamental force that houses, services, and employs them. Definitions of the informal sector are much debated. The term usually refers to the economic sector that relies on self-employed, low-wage jobs (such as street vending, shoe shining, and artisan manufacturing) that are unregulated and untaxed. Some scholars include illegal activities such as drug smuggling, sale of contraband items such as illegally copied videos and tapes, and prostitution as part of the informal sector. One of the most interesting expressions of informality is the housing in the squatter settlements. In Lima, Peru, an estimated 40 percent of the population lives in self-built housing, often of very poor quality. Typically these settlements begin as illegal invasions of open spaces that are carefully planned and timed to avoid the risk of eviction by city authorities. If the hastily built communities go unchallenged, squatters steadily improve their houses (Figure 4.19). The creation of these landscapes reflects a conscious and organized effort on the part of the urban poor to make a place for themselves in Latin American cities.

Rural-to-Urban Migration   Since the 1950s, as conditions in rural areas deteriorated due to the consolidation of lands, mechanization of agriculture, and increased population pressure, peasants began to pour into the cities of Latin America in a process referred to as **rural-to-urban migration.** As life in the countryside worsened, many rural households would send family members to the cities for employment as domestics, construction workers, artisans, and vendors. Once in the cities, rural migrants generally found conditions better, especially access to education, health care, electricity, and clean water. It was not poverty alone that drove people out of rural areas, but individual choice and an urban preference. Migrants believed in, and often realized, greater opportunities in cities, especially the capital cities. Those who came were usually young (in their twenties) and better educated than those who stayed behind. Women slightly outnumbered men

▲ **Figure 4.19 Lima squatter settlement**   In arid Lima, squatters initially build their homes using straw mats. As settlements become established, residents invest in adobe and cinder block to improve their homes. Life on the urban frontier is harsh. Water is trucked in; electricity is irregular; and travel toward the city center is costly and slow.   *(Rob Crandall/Rob Crandall, Photographer)*

in this migrant stream. The move itself was made easier by extended kin networks formed by earlier migrants who settled in discrete areas of the city and aided new arrivals. The migrants maintained their links to their rural communities by periodically sending remittances and making return visits.

As better roads and transport evolved, the connections between rural and urban areas were reinforced. One indication is a pattern of seasonal returns to the countryside. In Caracas, for example, the city empties out during the Christmas and Easter holidays as thousands of city dwellers jam into buses and cars to visit relatives in the countryside. Although Latin America is the most urbanized region in the developing world, urban residents often maintain close ties with extended family living in rural areas.

Patterns of Rural Settlement

Throughout Latin America a distinct rural lifestyle exists, especially among peasant subsistence farmers. While the majority of people live in cities, approximately 130 million people do not. In Brazil alone at least 30 million people live in rural areas. Interestingly, the absolute number of people living in rural areas today is roughly equal to the number in the 1960s. Yet rural life has definitely changed. The links between rural and urban areas are much improved, making rural folks less isolated. In addition to village-based subsistence production, in most rural areas highly mechanized capital-intensive farming occurs. Much like the region's cities, the rural landscape is divided by extremes of poverty and wealth. The root of social and economic tension in the countryside is the uneven distribution of arable land.

Rural Landholdings   The control of land in Latin America was the basis for political and economic power. Historically, colonial authorities granted large tracts of land to the colonists, who were also promised the service of Indian laborers as part of the **encomienda** system. These large estates typically took up the best lands along the valley bottoms and coastal plains. The owners were often absentee landlords, spending most of their time in the city and relying on a mixture of hired, tributary, and slave labor to run their rural operations. Passed down from one generation to the next, many estates can trace their ownership back several centuries. The allocation of large blocks of land also denied peasants access to land, so they were forced to labor for the estates. This entrenched practice of maintaining large estates is called **latifundia.**

Although the pattern of estate ownership is well documented, peasants have always farmed small plots for their subsistence. The practice of **minifundia** can lead to permanent or shifting cultivation. Small farmers typically plant a mixture of crops for subsistence as well as for trade. Peasant farmers in Colombia or Costa Rica, for example, grow corn, fruits, and various vegetables alongside coffee bushes that produce beans for export. Strains on the minifundia system occur when demographic pressures create land scarcity or political elites reallocate land for their needs.

Much of the turmoil in twentieth-century Latin America surrounded the question of land, with peasants demanding its redistribution through the process of **agrarian reform.** Governments have addressed these concerns in different ways. The Mexican Revolution in 1910 yielded a system of communally held lands called *ejidos*. In the 1950s Bolivia crafted agrarian reform policies that led to the expropriation of estate lands and their reallocation to small farmers. As part of the Sandinista revolution in Nicaragua in 1979, lands were expropriated from the political elite and converted into collective farms. Each of these programs met with resistance and proved to be politically costly. Eventually the path chosen by most governments was to make frontier lands available to land-hungry peasants. The opening of tropical frontiers, especially in South America, was a widely practiced strategy that changed national settlement patterns and began waves of rural-to-rural and even urban-to-rural migration.

Agricultural Frontiers   The creation of agricultural frontiers served several purposes: providing peasants with land, tapping unused resources, and shoring up political bound-

aries. Frontier colonization efforts in South America are some of the more noteworthy cases. In addition to settlement along Brazil's Trans-Amazon highway, Peru developed its Carretera Marginal (Marginal Highway) in the 1960s in an effort to lure colonists into the cloud and rain forests of eastern Peru. In Bolivia, Colombia, and Venezuela, agricultural frontier schemes in the lowland tropical plains attracted peasant farmers and large-scale investors. Mexico sent colonists, some displaced by dam construction, into the forests of Tehuantepec. Guatemala developed its northern Petén region. El Salvador had no frontier left, but many desperately poor Salvadorans poured into the neighboring states of Honduras and Belize in search of land. In short, although the dominant demographic trend has been a rural-to-urban movement, an important rural-to-rural flow has changed previously virgin areas into agricultural communities (Figure 4.20).

The opening of the Brazilian Amazon for settlement was the most ambitious frontier colonization scheme in the region. In the 1960s Brazil began its frontier expansion by constructing

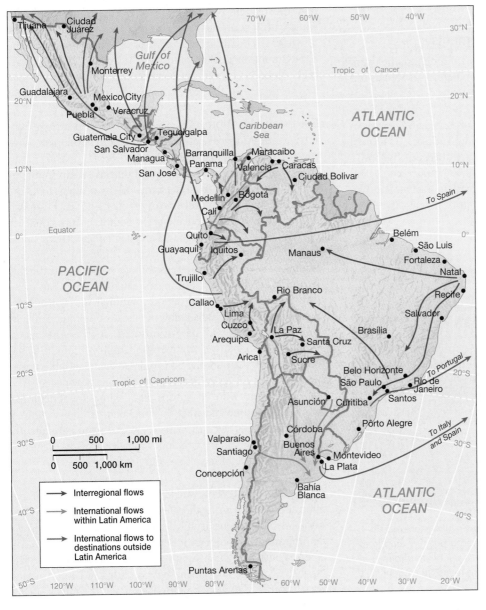

◀ **Figure 4.20 Principal Latin American migration flows**   Internal and international migrations have opened frontier zones and created transnational communities. Over the past two decades, the flow of Latin Americans to the United States has grown. In the 2000 U.S. Census, 35 million people of Hispanic ancestry were counted. Most of these people were either born in or have ancestral ties to Latin America. *(Adapted from Clawson, 2000. Latin America and the Caribbean: Lands and People, 2nd ed., Boston: McGraw-Hill.)*

several new Amazonian highways, a new capital (Brasília), and state-sponsored mining operations. It was the Brazilian military who directed the opening of the Amazon to provide an outlet for landless peasants and to extract the region's many resources. Yet the generals' plans did not deliver as intended. Throughout the basin, thin forest soils were incapable of supporting permanent agricultural colonies and in the worst cases degraded into baked claylike surfaces devoid of vegetation. Government-promised land titles, agricultural subsidies, and credit were slow to reach small farmers, even if they were fortunate enough to be given land on *terra roxa,* a nutrient-rich purple clay soil. Instead, a disproportionate amount of money went to subsidizing large cattle ranches through tax breaks and improvement deals where "improved" land meant cleared land. Despite a concerted effort by Brazilian officials to settle the entire region, most commercial activities and residents are concentrated in a few sites. Four times more people lived in the Amazon in the 1990s than in the 1960s. The 12 million people living in the Brazilian Amazon in 2000 is projected to reach 15 million by 2010; thus, increased human modification of the Brazilian Amazon is inevitable (Figure 4.21).

Population Growth and Movements

The high growth rates in Latin America throughout the twentieth century are attributed to natural increase as well as immigration. The 1960s and 1970s were decades of tremendous growth resulting from high fertility rates and increasing life expectancy. In the 1960s, for example, a typical Latin American woman had six or seven children. By the 1980s the TFR (total fertility rate) was half this in some countries. As Table 4.1 shows, the 2001 TFR was 2.4 for Brazil, 2.6 for Colombia, and 2.8 for Mexico. A number of factors explain this: more urban families, which tend to be smaller than rural ones; increased

participation of women in the workforce; higher education levels of women; state support of family planning; and better access to birth control. The exceptions to this trend are the poor and more rural countries, such as Guatemala and Bolivia, where the average woman has four or five children. Cultural factors may also be at work, as Amerindian peoples in the region tend to have more children.

Even with family sizes shrinking and nearing replacement rates in Uruguay and Chile, there is built-in potential for continued growth because of the relative demographic youth of these countries. The average percentage of the population below the age of 15 is 32 percent. In North America the similar cohort is 21 percent of the population, and in western Europe it is just 17 percent. This means that a proportionally larger segment of the population has yet to enter into its childbearing years.

Waves of immigrants into Latin America and migrant streams within Latin America have influenced population size and patterns of settlement. Beginning in the late nineteenth century, new immigrants from Europe and Asia added to the region's size and ethnic complexity. Important population shifts within countries have also occurred in recent decades, as witnessed by the growth of Mexican border towns and the demographic expansion of the Bolivian plains. In an increasingly globalized economy, even more Latin Americans live and work outside the region, especially in the United States.

European Migration After gaining their independence from Iberia, Latin America's new leaders sought to develop their territories through immigration. Firmly believing in the dictum "to govern is to populate," many countries set up immigration offices in Europe to attract hardworking peasants to till the soils and "whiten" the **mestizo** (people of mixed European and Indian ancestry) population. The Southern Cone countries of Argentina, Chile, Uruguay, Paraguay and southern Brazil were the most successful in attracting European immigrants from the 1870s until the depression of the 1930s. During this period, some 8 million Europeans arrived (more than came during the entire colonial period), with Italians, Portuguese, Spaniards, and Germans being the most numerous. Some of this immigration was state-sponsored, such as the nearly 1 million laborers (including entire families) brought to the coffee estates surrounding São Paulo at the start of the twentieth century. Other migrants came seasonally, especially the Italian peasants who left Europe in the winter for agricultural work in Argentina and were thus nicknamed "the swallows." Still others paid their own passage, intending to permanently settle and prosper in the growing commercial centers of Buenos Aires, São Paulo, Montevideo, and Santiago.

Asian Migration Less well known are the Asian immigrants who also arrived during the late nineteenth and twentieth centuries. Although considerably fewer, over time they established an important presence in the large cities of Brazil, Peru, Argentina, and Paraguay. Beginning in the mid-nineteenth century, the Chinese and Japanese who settled in Latin America were contracted to work on the coffee estates in southern Brazil

▲ **Figure 4.21 Settlers in the Amazon** Newly arrived settlers in the Brazilian Amazon have cleared a patch of land from the forest and constructed their new home out of wood and palm. Thousands of such homesteads are found throughout the Amazon. *(Brian Godfrey)*

and the sugar estates and guano mines of Peru. In the 1990s a Japanese-Peruvian, Alberto Fujimori, was president of Peru. The Japanese in Brazil are the most studied Asian immigrant group. Between 1908 and 1978 a quarter-million Japanese immigrated to Brazil; today the country is home to 1.3 million people of Japanese descent. Initially, most Japanese were landless laborers, yet by the 1940s they had accumulated enough capital so that three-quarters of the migrants had their own land in the rural areas of São Paulo and Paraná states. As a group, the Japanese have been closely associated with the expansion of soybean and orange production. Today Brazil leads the world in exports of orange juice concentrate, with most of the oranges grown on Japanese-Brazilian farms. Increasingly, second- and third-generation Japanese have taken professional and commercial jobs in Brazilian cities; many of them have married outside their ethnic group and are losing their fluency in Japanese. Even with intermarriage, they continue to maintain a strong sense of their ethnic identity, with more than two-thirds of Japanese-Brazilians speaking both Portuguese and some Japanese (Figure 4.22).

The latest Asian immigrants are from South Korea. Unlike their predecessors, most of the Korean immigrants came with enough capital to invest in small businesses and settled in cities, rather than in the countryside. According to official South Korean statistics, 120,000 Koreans emigrated to Paraguay between 1975 and 1990. While many have stayed in Paraguay, there seems to be a pattern of secondary immigration to Brazil and Argentina. Recent Korean immigrants in São Paulo have created more than 2,500 small businesses. Unofficial estimates of the number of Koreans living in Brazil range from 40,000 to 120,000. As a group they are decidedly commercial in orientation and urban in residence; their cities of choice are Asunción and Ciudad del Este in Paraguay, São Paulo in Brazil, and Buenos Aires in Argentina.

Latino Migration and Hemispheric Change   Movement within Latin America and between Latin America and North America has had a significant impact on sending and receiving communities alike. Within Latin America, international migration is shaped by shifting economic and political realities. Thus, Venezuela's oil wealth during the 1960s and 1970s attracted between 1 and 2 million Colombian immigrants who tended to work as domestics or agricultural laborers. Argentina has long been a destination for Bolivian and Paraguayan laborers. And, of course, farmers in the United States have depended on Mexican laborers for most of the twentieth century (refer to Figure 4.20).

Political turmoil also sparked waves of international migrants. Chilean intellectuals fled to neighboring countries in the 1970s when General Pinochet wrested power from the socialist government led by Salvador Allende. Nicaraguans likewise fled when the socialist Sandanistas came to power in 1979. The bloody civil wars in El Salvador and Guatemala sent waves of refugees into neighboring countries, such as Mexico and the United States. With democratization on the rise in the region, many of today's immigrants are classified as economic migrants, not political asylum seekers.

▲ **Figure 4.22 Japanese-Brazilians**   Retired Japanese-Brazilians play the board game "Go" in a city plaza. Most Brazilians of Japanese ancestry live in the southern states of São Paulo, Paraná, and Santa Catarina. The majority are descended from Japanese who immigrated to Brazil in the first half of the twentieth century.   *(Gary Payne/Getty Images, Inc.)*

Presently, Mexico is the largest country of origin of legal immigrants to the United States, followed by the Philippines, China, Korea, and Vietnam (Figure 4.23). Twenty-two million people claimed Mexican ancestry in the 2000 U.S. Census, of whom approximately 8 million were immigrants.

▲ **Figure 4.23 Mexican–U.S. border crossing**   Mexican day workers cross the border into El Paso, Texas, from Ciudad Juárez. Mexicans have long used these busy border crossings to enter the United States. Other Latino immigrants, especially from Central America, now join them.   *(Rob Crandall/Rob Crandall, Photographer)*

Odyssey to the North: Salvadorans in Washington, D.C.

Civil war and economic uncertainty forced nearly 1 million Salvadorans to flee their country in the 1980s and early 1990s. Many came to the United States, especially to Los Angeles and Washington, D.C. In fact, in the Washington metropolitan area, immigrants from El Salvador represent the largest national group of recent immigrants (Figure 4.2.1). In his novel *Odyssey to the North*, Salvadoran author Mario Bencastro weaves a story of exile and struggle around the character of Calixto—an undocumented immigrant who fled the violence of his country and ended up working in a Washington restaurant. In the following excerpt, Calixto describes life in far-off El Salvador to other immigrant kitchen workers:

Fate plays with us. I never imagined that one day I would be washing dishes in a foreign land. In my country I did everything; I began as a day laborer in a village, and after I went to the capital I worked as a shoemaker and then as a mason. This is, when I wasn't on a binge, because the desperation of unemployment and misery pushes you to alcoholism. But here in this country I'm like a new man, I rarely drink and I work in a kitchen, a job usually reserved for women in our country . . .

What's the name of the town you come from?

Ojo de Agua.

Is it big?

No, it's just a tiny village on a mountain full of snakes and iguanas.

It's not that bad; I bet it's a pretty little town.

It was. But there were battles there between the guerrillas and the army. The bombs destroyed the place and those of us who survived fled to other cities. Now it's a ghost town.

I can see you really miss your country.

Why wouldn't I? I was born and raised there, and learned such important things.

Like what? What can you learn in a place as far from civilization as that?

Well, to use a machete. And herd oxen and cows. To ride a horse. And plant crops. Harvest cotton. And hunt iguanas. Drink moonshine. And chew tobacco. All the things a man has to learn in order to survive in the country . . .

In the country you work really hard. It's not a life for wimps. That's for sure. I used to get up real early in the morning, long

▲ **Figure 4.2.1 Salvadorans in the suburbs** Day laborers wait for employment in a Maryland suburb outside Washington, D.C. War and economic hardship drove many Salvadorans from their country in the 1980s and 1990s. After Los Angeles, the second largest community of Salvadorans is in the Washington, D.C., area. *(Rob Crandall/Rob Crandall, Photographer)*

before the sun came up. My breakfast was a couple of tortillas with beans and cheap coffee.

Why did you get up so early?

Because I used to walk as much as five kilometers to get to the hacienda where I picked cotton.

Five kilometers just to get there?

That was nothing. In the country everyone walks everywhere. On the hacienda I'd work from sunup 'til sundown, for a few colones and a couple of tortillas with salt every day, and with cruel crew leaders as bosses.

The owners of the hacienda?

No, the crew leaders are poor peasants like the rest of the workers. But they think they're powerful and end up even meaner than the owners themselves.

Source: Mario Bencastro, *Odyssey to the North*. Houston: Arte Público Press, 1998, pp. 62–63.

Mexican labor migration to the United States dates back to the late 1800s when relatively unskilled labor was recruited to work in agriculture, mining, and railroads. This practice was formalized in the 1940s through the 1960s with the *Bracero* program, which granted temporary employment residence to 5 million Mexican laborers (much like the "guest workers" recruited from southern Europe and Turkey by West Germany). Today roughly 60 percent of the Hispanic population (both foreign-born and native-born) in the United States claims Mexican ancestry. Mexican immigrants are most concentrated in California and Texas, but increasingly they are found throughout the country. Although Mexicans continue to have the greatest presence among Latinos in the United States, the number of immigrants from El Salvador, Guatemala, Nicaragua, Colombia, Ecuador, and Brazil has steadily grown. The 2000 Census counted 35 million Hispanics in the United States (both foreign- and native-born). Most of this population has ancestral ties with peoples from Latin America and the Caribbean (see Chapter 5 on Caribbean migration).

Population shifts from one country to another change both demographic and cultural patterns. The cultural complexity of Latin America is attributable in part to immigration; today's emigrants from Latin America are weaving their culture into the fabric of North American and European societies. For example, Brazil, which has long been an immigrant destination, is experiencing waves of out-migration to the United States and Portugal. As noted above, many migrants maintain close contact with their home countries, a phenomenon some scholars have labeled **transnationalism.** A cultural and economic outcome of globalization, transnationalism highlights the social and economic links that form between home and host countries (see "Author Fieldtrip: Bolivian Transnational Migration" on the enclosed CD-ROM). Technological advances that make communication both faster and cheaper, as well as improved banking and courier services, allow immigrants to maintain contacts with their home countries in ways that earlier generations could not. The transnational migrants maintain a dual or hybrid identity, which also is seen as a cultural expression of globalization. Salvadorans working in Washington, D.C., for example, maintain regular contact with their rural villages in El Salvador while also developing vital immigrant social networks in their new home (see "Odyssey to the North: Salvadorans in Washington, D.C.").

Patterns of Cultural Coherence and Diversity: Repopulating a Continent

The Iberian colonial experience imposed a political and cultural coherence on Latin America that makes it distinguishable today as a world region. Yet this was not a simple transplanting of Iberia across the Atlantic. Often a syncretic process unfolded in which European and Indian traditions blended as indigenous groups were subsumed into either the Spanish or the Portuguese empires. In some areas such as southern Mexico, Guatemala, Bolivia, Ecuador, and Peru, Indian cultures have showed remarkable resilience, as evidenced by the survival of Amerindian languages. Yet the prevailing pattern is one of forced assimilation in which European religion, languages, and political organization were imposed on the surviving fragments of native society. Later other cultures, arriving as both forced and voluntary migrants, added to the region's cultural mix. Perhaps the single most important factor in the dominance of European culture in Latin America was the demographic collapse of native populations.

Demographic Collapse

It is hard the grasp the enormity of cultural change and human loss due to this cataclysmic encounter between two worlds. Throughout the region archaeological sites are poignant reminders of the complexity of precontact civilizations. Dozens of stone temples found throughout Mexico and Central America, where the Mayan and Aztec civilizations flourished, attest to the ability of these societies to thrive in the area's tropical forests and upland plateaus. In the Andes, stone terraces built by the Incas are still being used by Andean farmers; earthen platforms for village sites and raised fields for agriculture are still being discovered and mapped. Ceremonial centers such as Cuzco—the core of the great Incan empire that was nearly leveled by the Spanish—and the Incan site of Machu Picchu—unknown to most of the world until American archaeologist Hiram Bingham investigated the site in the early 1900s—are evidence of the complexity of precontact civilizations (Figure 4.24). The Spanish, too, were impressed by the sophistication and wealth they saw around them, especially in the incomparable Tenochtitlán, where Mexico City sits today. Tenochtitlán was the political and ceremonial center of the Aztecs, supporting a complex metropolitan area with some 300,000 residents. The largest city in Spain at the time was considerably smaller.

◄ **Figure 4.24 Machu Picchu** This complex, ancient city near Cusco was not known to the outside world until the early 1900s. Located above the humid Urubamba River valley, Machu Picchu is one of Peru's major tourist destinations. *(Rob Crandall/Rob Crandall, Photographer)*

Within and around the Rio Plátano Biosphere Reserve in northeastern Honduras are the communities of Miskito, Pech, Garífuna, and *ladinos* (the Honduran term for mestizo). This is the largest area of road-free rain forest in Central America. For centuries indigenous people have inhabited this area, maintaining its biological diversity while extracting resources through the traditional economic activities of hunting, gathering, slash-and-burn agriculture, and fishing.

Through participatory mapping techniques using global positioning systems and local knowledge of resource zones, indigenous people are outlining their communal land-use zones (the resource areas needed for subsistence). With technical assistance by geographer Peter Herlihy, the native groups mapped the "resource-sheds" they rely on (Figure 4.3.1). By mapping communal land-use zones, they could express their claims to this area in a powerful but nonconfrontational way. Potential problem areas between groups are readily apparent.

For example, the several ethnic groups rely on the Tinto-Ibans at the park's northern border. More important, mapping indigenous land use acknowledges the Amerindian presence in and around the park.

The biggest challenge to maintaining the biosphere's biological and cultural diversity is dealing with some 6,000 ladino settlers on the southwestern perimeter who are clearing land to create permanent agricultural colonies. One strategy is to legally secure the territorial claims of indigenous people within the biosphere and also recognize the rights of existing ladino settlements outside of the biosphere zone while discouraging further expansion into the park itself. A new participatory mapping initiative forms one component of a Honduran-German project to protect and manage the Rio Plátano Biosphere Reserve. Through this process, indigenous and ladino communities have come together to define their own land-use zoning system and management guidelines.

▶ **Figure 4.3.1 Communal land-use zones and the Rio Plátano Biosphere Reserve** *Various ethnic communities including the Garífuna, Miskito, Pech, and ladinos live within the Rio Plátano Biosphere Reserve. The first three groups are using mapping to maintain access to parklands for fishing, hunting, and farming. (Modified from "Tierra Indigenas de la Mosquitia Hondureña-1992: Zonas de Subsistencia" by Masta Mopawi, Peter Herlihy, and Andrew Leake)*

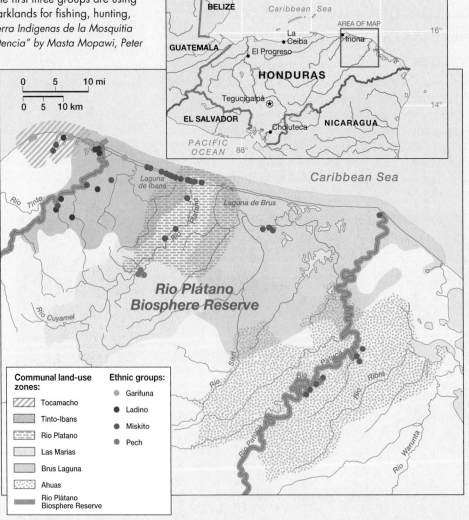

The most telling figures of the impact of European expansion are demographic. It is widely believed that the precontact Americas had 54 million inhabitants; by comparison, western Europe in 1500 had approximately 42 million. Of the 54 million, about 47 million were in what is now Latin America, and the rest were in North America and the Caribbean. There were two major population centers: one in Central Mexico with 14 million people and the other in the Central Andes (highland Peru and Bolivia) with nearly 12 million. By 1650, after a century and a half of colonization, the indigenous population was one-tenth its precontact size. The human tragedy of this population loss is difficult to comprehend. The relentless elimination of 90 percent of the indigenous population was largely caused by epidemics of influenza and smallpox, but warfare, forced labor, and starvation due to a collapse of food production systems also contributed to the death rate.

The tragedy of conquest did not end in 1650, the population low point for Amerindians, but continued throughout the colonial period and to a much lesser extent continues today. After the indigenous population began its slow recovery in the Central Andes and Central Mexico, there were still tribal bands in southern Chile (the Mapuche) and Patagonia (Araucania) that experienced the ravages of disease three centuries after Columbus landed. Even now, the isolation of some Amazonian tribes has made them vulnerable to disease. Conflicts with outsiders who invade their territories in search of land or gold still occur. In an all too familiar story, a common cold can prove deadly to forest dwellers who lack the needed immunities.

The Columbian Exchange   Historian Alfred Crosby likens the contact period between the Old World (Europe, Africa, and Asia) and the New World (the Americas) as an immense biological swap, which he terms the **Columbian exchange.** According to Crosby, Europeans benefited greatly from this exchange, and Amerindian peoples suffered the most from it. The human ecology of both sides of the Atlantic, however, was forever changed through the introduction of new diseases, peoples, plants, and animals. Take, for example, the introduction of Old World crops. The Spanish, naturally, brought their staples of wheat, olives, and grapes to plant in the Americas. Wheat did surprisingly well in the highland tropics and became a widely consumed grain over time. Grapes and olive trees did not fare as well, but eventually grapes were produced commercially in the temperate zones of South America. The Spanish grew to appreciate the domestication skills of Indian agriculturalists who had developed valuable starch crops such as corn, potatoes, and bitter manioc, as well as condiments such as hot peppers, tomatoes, pineapple, cacao, and avocados. Corn never became a popular food for Europeans, but many African peoples adopted it as a vital staple food. After initial reluctance, Europeans and Russians widely consumed the potato as a basic food. Domesticated in the highlands of Peru and Bolivia, the humble potato has an impressive ability to produce a tremendous volume of food in a very small area, even when climatic conditions are not ideal. This root crop is credited with driving Europe's rapid population increase in the eighteenth century when peasant farmers from Ireland to Russia became increasingly dependent on it as a basic food. This potato dependence also made them vulnerable to potato blight, a fungal disease that emerged in the nineteenth century and came close to unraveling Irish society.

Tropical crops transferred from Asia and Africa reconfigured the economic potential of the region. Sugarcane, an Asian transfer, became the dominant cash crop of the Caribbean and the Atlantic tropical lowlands of South America. With sugar production came the importation of millions of African slaves. Coffee, a later transfer from East Africa, emerged as one of the leading export crops throughout Central America, Colombia, Venezuela, and Brazil in the nineteenth century. Introduced African pasture grasses enhanced the forage available to livestock.

The movement of Old World animals across the Atlantic had a profound impact on the Americas. Initially these animals hastened Indian decline by introducing animal-borne diseases and by producing feral offspring that consumed everything in their paths. For the most part, native agriculturalists did not have to contend with grazing animals; and their vast gardens of corn, beans, and squash proved to be attractive fodder for the rapidly multiplying swine, cows, and horses. The utility of these animals was eventually appreciated by native survivors. Draft animals were adopted, as was the plow, which facilitated the preparation of soil for planting. Wool became a very important fiber for indigenous communities in the uplands. And slowly, pork, chicken, and eggs added protein and diversity to the staple diets of corn, potatoes, and cassava. Ironically, the horse, which was a feared and formidable weapon of the Europeans, became a tool of resistance in the hands of skilled riders who inhabited the plains of the Chaco and Patagonia. Much like native peoples of North America, these tribal groups challenged European conquest by using their horsemanship in combat or for flight. With the major exception of disease, many transfers of plants and animals ultimately benefited both worlds. Still, it is clear that the ecological and material basis for life in Latin America was completely reworked through this exchange process initiated by Columbus.

Indian Survival   Presently, Mexico, Guatemala, Ecuador, Peru, and Bolivia have the largest indigenous populations. Not surprisingly, these areas had the densest native populations at contact. Indigenous survival also occurs in isolated settings where the workings of national and global economies are slow to penetrate. The Sierra de Santa Marta in Colombia, home of the Kogi, or the Gran Sabana in Venezuela, where the Pemon live, are places where relatively small groups have managed to maintain a distinct Indian way of life despite pressures to assimilate.

In many cases Indian survival comes down to one key resource—land. Indigenous peoples who are able to maintain a territorial home, formally through land title or informally through long-term occupancy, are more likely to preserve a distinct ethnic identity. Because of this close association between identity and territory, native peoples are increasingly insisting on a recognized space within their countries. These efforts to define indigenous territory are seldom welcomed by the state.

Today the state of Panama recognizes four *comarcas* that encompass six native groups; the most successful is Comarca San Blas on the Caribbean coast, where 40,000 Kuna live. A comarca is a loosely defined territory similar to a province or a homeland. What distinguishes the comarca from an Indian reservation is that the native people have defined the territory and assert political and resource control within its boundaries. From Amazonia to the highlands of Chiapas, many native groups are demanding formal political and territorial recognition as a means to redress centuries of injustice. Whether these efforts will actually reshape political and cultural space for Latin America's Amerindians is still uncertain (see "Mapping for Indigenous Survival," on page 150). Yet there are hopeful signs of increased political participation by Amerindian peoples. In 2001 Peruvians elected President Alejandro Toledo, an Amerindian who rose from acute poverty to obtain a doctorate in economics from Stanford University. He is the first person of Amerindian ancestry to hold the presidential office in Peru.

Patterns of Ethnicity and Culture

The Indian demographic collapse enabled Spain and Portugal to refashion Latin America into a European likeness. Yet instead of a neo-Europe rising in the tropics, a complex ethnic blend evolved. Beginning with the first years of contact, unions between European sailors and Indian women began the process of racial mixing that over time became a defining feature of the region. The courts of Spain and Portugal officially discouraged racial mixing, but not much could be done about it. Spain, which had a far larger native population with which to deal than did the Portuguese in Brazil, became obsessed with the matter of race and maintaining racial purity among its colonists. An elaborate classification system was constructed to distinguish emerging racial castes. Thus, in Mexico in the eighteenth century a Spaniard and an Indian union resulted in a *mestizo* child. A child of a mestizo and a Spanish woman was a *castizo*. However, the children from a castizo woman and a Spanish man were considered Spanish in Mexico but a quarter mestizo in Peru. Likewise, *mulattoes* were the progeny of European and African unions, and *zambos* were the offspring of Africans and Indians.

After generations of intermarriage, such a classification system collapsed under the weight of its complexity, and four broad categories resulted: *blanco* (European ancestry), *mestizo* (mixed ancestry), *indio* (Indian ancestry), and *negro* (African ancestry). The blancos (or Europeans) continue to be well represented among the elites, yet the vast majority of people are of mixed racial ancestry. In Venezuela, for example, the phrase "cafe con leche" (coffee with milk) is used to describe the racial makeup of the majority of the population who share European, African, and Indian characteristics. Dia de la Raza, the region's observance of Columbus Day, recognizes the emergence of a new *mestizo* race as the legacy of European conquest. Throughout Latin America, more than other regions of the world, miscegenation (or racial mixing) is the norm, which makes the process of mapping racial or ethnic groups especially difficult.

Languages   Roughly two-thirds of Latin Americans are Spanish speakers, and one-third speak Portuguese. These colonial languages were so prevalent by the nineteenth century that they were the unquestioned languages of government and instruction for the newly independent Latin American republics. In fact, until recently many countries actively discouraged, and even repressed, Indian tongues. It took a constitutional amendment in Bolivia in the 1990s to legalize native-language instruction in primary schools and to recognize the country's multiethnic heritage (more than half the population is Indian; and Quechua, Aymara, and Guaraní are widely spoken) (Figure 4.25).

Because Spanish and Portuguese dominate, there is a tendency to neglect the influence of indigenous languages in the region. Mapping the use of indigenous languages, however, reveals important pockets of Indian resistance and survival. In the Central Andes of Peru, Bolivia, and southern Ecuador, more than 10 million people still speak Quechua and Aymara, along with Spanish. In Paraguay and lowland Bolivia there are 4 million Guaraní speakers, and in southern Mexico and Guatemala at least 6 to 8 million speak Mayan languages. Small groups of native-language speakers are found scattered throughout the sparsely settled interior of South America and the more isolated forests of Central America, but many of these languages have fewer than 10,000 speakers.

Blended Religions   Like language, the Roman Catholic faith appears to have been imposed upon the region without challenge. Most countries report 90 percent or more of their population as Catholic. Every major city has dozens of churches, and even the smallest hamlet maintains a graceful church on its central square (Figure 4.26). In some countries, such as El Salvador and Uruguay, a sizable portion of the population attend Protestant evangelical churches, but the Catholic core of this region is still intact.

Exactly what native peoples absorbed of the Christian faith is unclear. Throughout Latin America **syncretic religions**, blends of different belief systems, enabled animist practices to be folded into Christian worship. These blends took hold and endured, in part because Christian saints were easy surrogates for pre-Christian gods and because the Catholic Church tolerated local variations in worship as long as the process of conversion was under way. The Mayan practice of paying tribute to spirits of the underworld seems to be replicated today in Mexico and Guatemala via the practice of building small cave shrines to favorite Catholic saints and leaving offerings of fresh flowers and fruits. One of the most celebrated religious icons in Mexico is the Virgin of Guadeloupe, a dark-skinned virgin seen by an Indian shepherd boy who became the patron saint of Mexico.

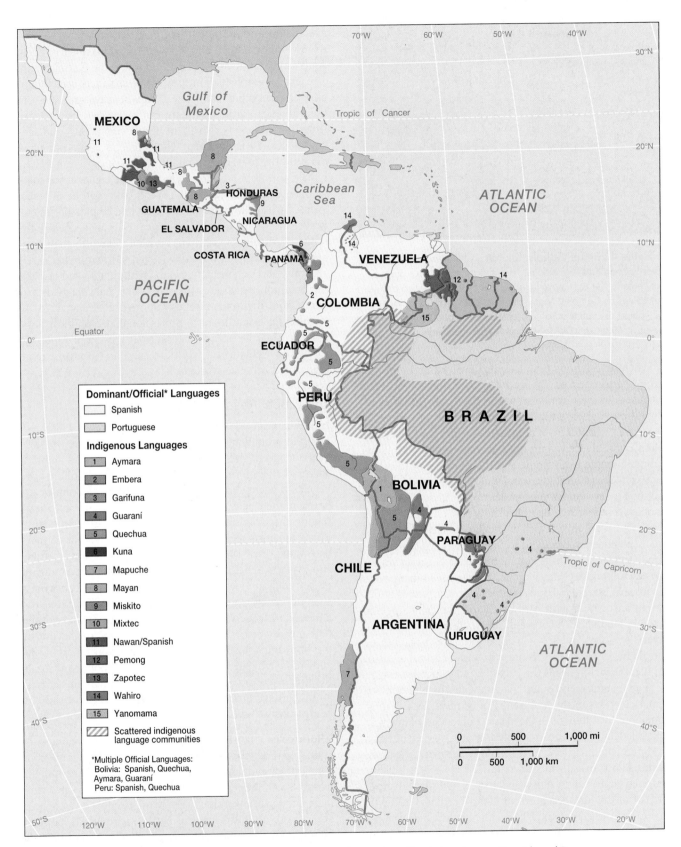

Dominant/Official* Languages

- ☐ Spanish
- ☐ Portuguese

Indigenous Languages

1	Aymara
2	Embera
3	Garifuna
4	Guaraní
5	Quechua
6	Kuna
7	Mapuche
8	Mayan
9	Miskito
10	Mixtec
11	Nawan/Spanish
12	Pemong
13	Zapotec
14	Wahiro
15	Yanomama

⬚ Scattered indigenous language communities

*Multiple Official Languages:
Bolivia: Spanish, Quechua, Aymara, Guaraní
Peru: Spanish, Quechua

▲ **Figure 4.25 Language map of Latin America**   The dominant languages of Latin America are Spanish and Portuguese. Nevertheless, there are significant areas in which indigenous languages persist and, in some cases, are recognized as official languages. Smaller language groups exist in Central America, the Amazon Basin, and southern Chile.   *(Adapted from the Atlas of the World's Languages, 1994, New York: Routledge)*

▲ **Figure 4.26 The Catholic Church**   Churches, such as the Dolores Church in Tegucigalpa, are important religious and social centers. The vast majority of people in Latin America define themselves as Catholic. Many churches built in the colonial era are valued as architectural treasures and are beautifully preserved.   *(Rob Crandall/Rob Crandall, Photographer)*

Syncretic religious practices also evolved and endured among African slaves. By far the greatest concentration of slaves was in the Caribbean, where slaves were used to replace the indigenous population, which was wiped out by disease (see Chapter 5). Within Latin America the Portuguese colony of Brazil received the most Africans—at least 4 million. In Brazil, where the volume and the duration of the slave trade were the greatest, the transfer of African-based religious and medical systems is most evident. West African-based religious systems such as Batuque, Umbanda, Candomblé, and Shango are often mixed with or ancillary to Catholicism and widely practiced in Brazil. So accurate were some of these religious transfers that it is common to have Nigerian priests journey to Brazil to learn forgotten traditions. In many parts of southern Brazil, Umbanda is as popular with people of European ancestry as with Afro-Brazilians. Typically a person becomes familiar with Umbanda after falling victim to a magician's spell by having some object of black magic buried outside his or her home. In order to regain control of his or her life, the victim needs the help of a priest or priestess.

The syncretic blend of Catholicism with African traditions is most obvious in the celebration of carnival, Brazil's most popular festival and one of the major components of Brazilian national identity. The three days of carnival known as the Reign of Momo combines Christian Lenten beliefs with pagan influences and African musical traditions epitomized by the rhythmic samba bands. Although the street festival was banned for part of the nineteenth century, Afro-Brazilians in Rio de Janeiro resurrected it in the 1880s with nightly parades, music, and dancing. Within fifty years the street festival had given rise to formalized samba schools and helped break down racial barriers. By the 1960s, carnival became an important symbol for Brazil's multiracial national identity. Today the festival—which is most associated with Rio—draws thousands of participants from all over world, although increased incidences of violent crime and robbery have tarnished its image.

Machismo and Marianismo   Cultural traits often ascribed to men (**machismo**) and women (**marianismo**) in Latin America reflect the influence of Iberia and the Catholic faith. The term *macho* (which means "male") is widely used by English speakers to characterize men who have great faith in their ability to attract women, but machismo in the Latin American context is also about honor, risk-taking, and self-confidence. Thus, someone who is macho would not allow his authority to be questioned in the home or in public. The female cultural counterpart, *marianismo,* reflects the ideal woman and refers to María (Mary), the mother of Jesus. Women who follow marianismo strive to be patient, loving, gentle, and willing to suffer in silence. They are keepers of the home, nurturers of children, and deferential to their husbands. Women are also acknowledged for their higher moral authority and, accordingly, treated with dignity and respect. The traditional standards for male and female behavior that result from these stereotypes are often challenged. Many women, for example, have professional and political positions outside the home. In recent decades Latin American women have entered the workforce, and now comprise more than one-third of all workers.

The Global Reach of Latino Culture

Latin American culture, vivid and diverse as it is, is widely recognized throughout the world. Whether it is the sultry pulse of the tango or the fanaticism with which Latinos embrace soccer as an art form, aspects of Latin American culture have been absorbed into a globalizing world culture. A dramatic example of the reach of Latino culture can be seen every Saturday on the Univision television network, based in Miami. There, a charismatic Chilean, Don Francisco, has hosted *Sabado Gigante* since 1986. This four-hour-long Spanish variety show is viewed in 28 countries and draws a weekly audience of 120 million viewers. In the arts, Latin American writers such as Gabriel García Marquez and Isabel Allende have obtained worldwide recognition. In terms of popular culture, new musical artists such as Colombia's Shakira and Brazil's hip-hop samba singer Max de Castro are gaining international audiences, while Latino superstars Jennifer Lopez and Ricky Martin are icons of world pop culture. Through music, literature, and even *telenovelas* (soap operas), Latino culture is being transmitted to an eager worldwide audience.

Telenovelas   Popular nightly soap operas are a mainstay of Latin American television. These tightly plotted series are filled with intrigue and double deals. Unlike their counterparts in the United States, they end, usually after 100 episodes. Once standard fare for the working class, many telenovelas take hold and absorb an entire nation. During particularly popular episodes, the streets are noticeably calm as millions of people tune in to catch up on the lives of their favorite heroines. Brazil, Venezuela, and Mexico each produce scores of telenovelas, but the Mexican ones are international mega-hits.

Televisa, a Mexican production agency, has aggressively marketed its inventory of soap operas to an eager global public. Mexican telenovelas are avidly watched in countries as diverse as Croatia, Russia, China, South Korea, Iran, the United States, and France, as well as throughout Latin America. Predictably scripted as Mexican Cinderella stories, these sagas of poor underclass women (often domestics) falling in love with members of the elite, battling jealous rivals, and ultimately emerging triumphant seem to resonate with fans around the world. When one Mexican telenovela star, Veronica Castro, visited Russia, her plane had trouble arriving at the gate because of the large crowds that came to greet her. In addition to their broad appeal, telenovelas are big business, perhaps Mexico's largest international export. While Hollywood and Mumbai grind out movies, much of Mexico's entertainment industry is geared toward producing this vernacular and loved art form.

National Identities   Viewed from the outside, there is considerable homogeneity to this region, yet distinct national identities and cultures flourish in Latin America. Since the early days of the republics, countries celebrated particular elements from their pasts when creating their national histories. In the case of Brazil, the country's interracial characteristics were highlighted to proclaim a new society in which the color lines between Europeans and Africans ceased to matter. Mexico turned to its Aztec past, celebrating the architectural and cultural achievements of its predecessors while at the same time forging an assimilationist strategy that discouraged surviving indigenous culture and language.

Musical and dance traditions evolved and became emblematic of these new societies: the tango in Argentina, the vallenato and cumbia in Colombia, the mariachi in Mexico, the huaynos in Peru, and the samba in Brazil are easily distinguished styles that often become popular anthems for the nations in the region (Figure 4.27). Literature also reflects the perceived cleavages in identity found in Latin America. Writers such as Isabel Allende, Gabriel García Marquez, Mario Vargas Llosa, Carlos Fuentes, and Jorge Amado situate their stories in their native countries and, in so doing, celebrate the unique characteristics of Chileans, Colombians, Peruvians, Mexicans, and Brazilians. Distinct political cultures evolved, which at times led to expansionist policies that brought neighbors into conflict. The geopolitical dimensions of these intraregional disputes will be discussed in the following section.

Geopolitical Framework: Redrawing the Map

It is Latin America's colonial history, more than its present condition, that unifies this region. For the first 300 years after Columbus's arrival, Latin America was a territorial prize sought by various European countries but effectively settled by Spain and Portugal. By the nineteenth century the independent states of Latin America had formed, but they continued to experience foreign influence and, at times, overt political pressure, especially from the United States. At other times a more neutral Pan-American vision of American relations and hemispheric cooperation has held sway, represented by the formation of the **Organization of American States (OAS)**. The present organization was chartered in 1948, but its origins date back to 1889. Yet there is no doubt that U.S. policies toward trade, economic assistance, political development, and at times military intervention are often seen as compromising to the sovereignty of these states. Many Latin Americans view the proposed **Free Trade Area of the Americas (FTAA)** as another example of the United States' imposing its economic interests on the rest of the hemisphere.

Within Latin America there have been cycles of intraregional cooperation and antagonism. Neighboring countries have fought over territory, closed borders, imposed high tariffs, and cut off diplomatic relations. Even today there are a dozen long-standing border disputes in Latin America that

(a)

(b)

▲ **Figure 4.27 Musical traditions**   (a) Mariachi musicians pose in their performance costumes. Mariachi blends European horn and string instruments with lyrics that celebrate Mexican life. (b) A samba band marches in the streets of Rio de Janeiro. Samba is the quintessential music of carnival and draws inspiration from African rhythmic traditions.   *[(a) Werner Bertsch/Bruce Coleman, Inc.; (b) Murilo Dutra/Corbis/Stock Market]*

occasionally erupt into armed conflict. The 1990s witnessed a revival in the trade block concept with the formation of **Mercosur** (the Southern Cone Common Market, which includes Brazil, Uruguay, Argentina, and Paraguay as full members and Bolivia and Chile as associate members) and NAFTA (Mexico, the United States, and Canada). It is possible that, as these economic ties strengthen, these trade blocks could form the basis for a new alignment of political and economic interests in the region.

Iberian Conquest and Territorial Division

Because it was Christopher Columbus who claimed the Americas for Spain, the Spanish were the first active colonial agents in the Western Hemisphere. In contrast, the Portuguese presence in the Americas was the result of the **Treaty of Tordesillas,** brokered by the pope in 1493–94. By that time Portuguese navigators had charted much of the coast of Africa in an attempt to find a water route to the Spice Islands (Moluccas) in Southeast Asia. With the help of Christopher Columbus, Spain sought a western route to the Far East. When Columbus discovered the Americas, Spain and Portugal asked the pope to settle how these new territories should be divided. Without consulting other European powers, the pope divided the Atlantic world in half—the eastern half containing the African continent was awarded to Portugal, the western half with most of the Americas was given to Spain. This treaty was never recognized by the French, English, or Dutch, who also asserted territorial claims in the Americas, but it did provide the legal apparatus for the creation of a Portuguese territory in America—Brazil—which would later become the largest and most populous state in Latin America (Figure 4.28).

The Treaty of Tordesillas presented a number of interesting geographical problems. It decreed that a line would be drawn 370 leagues west of the Portuguese-controlled Cape Verde Islands, yet exactly how long a league was and which island was the starting point was not specified. Today it is estimated that the line fell somewhere between the 48th and 49th west lines of longitude, which awarded the eastern third of what is now Brazil to the Portuguese. To complicate matters further, Columbus was convinced that the West Indian islands he discovered were in Asia, off the coast of Cathay (China). The existence of the Pacific Ocean, let alone the continents of North and South America, was unknown when the treaty was signed. Thus, the treaty established a far larger Spanish colonial territory than any of the participants realized; Spain ultimately extended its claims across the Pacific to the islands of the Philippines.

Six years after the treaty was signed, Portuguese navigator Alvares Cabral inadvertently reached the coast of Brazil on a voyage to southern Africa. The Portuguese soon realized that this territory was on their side of the Tordesillas line. Initially they were unimpressed by what Brazil had to offer; there were no spices or major indigenous settlements. Quickly, however, they came to appreciate the utility of the coast as a provisioning site as well as a source for brazilwood, used to produce a valuable dye. Portuguese interest in the territory intensified in the late sixteenth century with the development of sugar estates and the expansion of the slave trade and in the seventeenth century with the discovery of gold in the Brazilian interior.

Spain, in contrast, aggressively pursued the conquest and settlement of its new American territories from the very start. After discovering little gold in the Caribbean, by the mid-sixteenth century Spain's energy was directed toward developing the silver resources of Central Mexico and the Central Andes (most notably Potosí in Bolivia). Gradually the economy diversified to include some agricultural exports, such as cacao and sugar, as well as a variety of livestock. In terms of foodstuffs, the colonies were virtually self-sufficient. Some basic manufactured items, such as crude woolen cloth and agricultural tools, were also produced, but, in general, manufacturing was forbidden in the Spanish American colonies in order to keep them dependent on Spain.

Revolution and Independence It was not until the 1800s, with a rise of revolutionary movements between 1810 and 1826, that Spanish authority on the mainland was challenged. Ultimately elites born in the Americas gained control, displacing the representatives of the crown. In Brazil the evolution from Portuguese colony to independent republic was a slower and less violent process that spanned eight decades (1808–89). In the nineteenth century Brazil was declared a separate kingdom from Portugal with its own monarch, and later became a republic.

It was the territorial division of Spanish and Portugese America into administrative units that provided the legal basis for the modern states of Latin America (Figure 4.28). The Spanish colonies were first divided into two viceroyalties (New Spain and Peru) and within these were various subdivisions that later became the basis for the modern states. (In the eighteenth century the Viceroyalty of Peru, which included all of Spanish South America, was divided to form three viceroyalties: La Plata, Peru, and New Granada.) Unlike Brazil, which evolved from a colony into a single republic, the former Spanish colonies experienced fragmentation in the nineteenth century. Prominent among revolutionary leaders was Venezuelan-born Simon Bolívar (Figure 4.29), who advocated his vision for a new and independent state of **Gran Colombia.** For a short time (1822–30) Bolívar's vision was realized, as Colombia, Venezuela, Ecuador, and Panama were combined into one political unit. Similarly, in 1823 the **United Provinces of Central America** was formed to avoid annexation by Mexico. By the 1830s this union also broke apart into the states of Guatemala, Honduras, El Salvador, Nicaragua, and Costa Rica. Today the former Spanish mainland colonies include 16 states (plus three Caribbean islands), with a total population of 340 million. If the Spanish colonial territory had remained a unified political unit, it would now have the third largest population in the world, following China and India.

Persistent Border Conflicts As the colonial administrative units turned into states, it became clear that the territories were not clearly delimited, especially the borders that stretched

▶ **Figure 4.28 Shifting political boundaries**   The evolution of political boundaries in Latin America began with the 1494 Treaty of Tordesillas, which gave much of the Americas to Spain and a slice of South America to Portugal. The larger Spanish territory was gradually divided into viceroyalties and audiencias, which formed the basis for many modern national boundaries. The 1830 borders of these newly independent states were far from fixed. Bolivia would lose its access to the coast; Peru would gain much of Ecuador's Amazon; and Mexico would be stripped of its northern territory by the United States. *(From Lombardi, Cathryn L., and John V. Lombardi. Latin American History: A Teaching Atlas. © 1993. Reprinted by permission of the University of Wisconsin Press.)*

into the sparsely populated interior of South America. This would later become a source of conflict as new states struggled to demarcate their territorial boundaries. Numerous border wars erupted in the nineteenth and twentieth centuries, and the map of Latin America has been redrawn many times. Some of the more noted conflicts were the War of the Pacific (1879–82) in which Chile expanded to the north and Bolivia lost its access to the Pacific; warfare between Mexico and the United States in the 1840s, which resulted in the present border under the Treaty of Hidalgo (1848) and Mexico's loss of

▲ **Figure 4.29 Simon Bolívar** Heroic likenesses of Simon Bolívar (the Liberator) are found throughout South America, especially in his native country of Venezuela. This statue stands in the central plaza of the Andean city of Mérida, Venezuela. *(Rob Crandall/Rob Crandall, Photographer)*

what became the southwestern United States; and the War of the Triple Alliance (1864–70), the bloodiest war of the post-colonial period, which occurred when Argentina, Brazil, and Uruguay allied themselves to defeat Paraguay in its claim to control the upper Paraná River Basin. It is estimated that the adult male population in Paraguay was reduced by nine-tenths as a result of this conflict. Sixty years later, the Chaco War (1932–35) resulted in a territorial loss for Bolivia in its eastern lowlands and a gain for Paraguay. In the 1980s Argentina lost a war with Great Britain over control of the Falkland, or Malvinas, Islands in the South Atlantic. And as recently as 1998, Peru and Ecuador skirmished over a disputed boundary in the Amazon Basin (Figure 4.30).

Outright war in the region is less common than ongoing and nagging disputes over international boundaries. Any of a dozen dormant claims erupt from time to time, given the political climate between neighbors. These include Guatemala's claim to Belize, Venezuela and Guyana's border dispute, Ecuador's claims to the Peruvian eastern lowlands, and the maritime boundary dispute between Honduras and Nicaragua. Many of these disputes are based on historical territorial claims dating back to the colonial period.

One of the more interesting geopolitical conflicts in the Southern Cone is based on territorial claims to Antarctica. Seven claimant states—Argentina, Chile, Norway, the United Kingdom, France, New Zealand, and Australia—plus five nonclaimant states (including the United States) seek resource rights to Antarctica. Even though an Antarctic Treaty has existed since 1959 stating that the landmass should be used for peaceful purposes, Chile and Argentina have incorporated their Antarctic territorial claims on national maps and even postage stamps. A Chilean president, Pinochet, spent a week touring the Chilean claim in the 1970s. Argentina held a national cabinet meeting on its portion of Antarctica in the 1970s and then sent a pregnant Argentine woman to give birth to the continent's first child. For the most part, these nationalist claims are symbolic. Recent treaties show an international inclination toward cooperation and the banning of commercial exploitation of the continent.

The Trend Toward Democracy Early in the twenty-first century, most of the 17 countries in this region will celebrate their bicentennials. Compared with most of the developing world, Latin Americans have been independent for a long time. Yet political stability is not a hallmark of the region. Among the countries in the region, some 250 constitutions have been written since independence, and military coups have been alarmingly frequent. Since the 1980s, however, the trend has been toward democratically elected governments, the opening of markets, and broader popular participation in the political process. Where dictators once outnumbered elected leaders, by the 1990s each country in the region had a democratically elected president. (Cuba, the one exception, will be discussed in Chapter 5.)

Democracy may not be enough for the millions frustrated by the slow pace of political and economic reform. In survey after survey, Latin Americans register their dissatisfaction with politicians and governments. Most of the newly elected democratic leaders are also free-market reformers who are quick to eliminate state-backed social safety nets, such as food subsidies, government jobs, and pensions. Many of the poor and middle class have grown skeptical about whether this brand of democracy could make their lives better. The political left, however, has yet to produce an alternative to privatization and market-driven policies. For now, the status quo continues, although popular frustration with falling incomes, rising violence, and chronic underemployment are a recipe for political instability as seen, for example, in the Andean countries and Central America (Figure 4.31).

Regional Organizations

At the same time democratically elected leaders struggle to address the pressing needs of their countries, political developments at the supranational and subnational levels pose new challenges to their authority. The most discussed **supranational organizations** (governing bodies that include several states) are the trade blocks. **Subnational organizations** (groups that represent areas or people within the state) often form along ethnic or ideological lines and can provoke serious internal divisions. Indigenous groups seeking territorial recognition (such as the Kuna) and insurgent groups espousing Marxist ideology (such as the FARC in Colombia) have challenged the authority of the states. Finally, the financial and political force of drug cartels, especially for smaller countries, transcends state boundaries and undermines judicial systems.

Trade Blocks Beginning in the 1960s, regional trade alliances were attempted in an effort to foster internal markets and reduce trade barriers. The Latin American Free Trade Association (LAFTA), the Central American Common Market (CACM), and the Andean Group have existed for decades, but their ability to influence economic trade and growth is

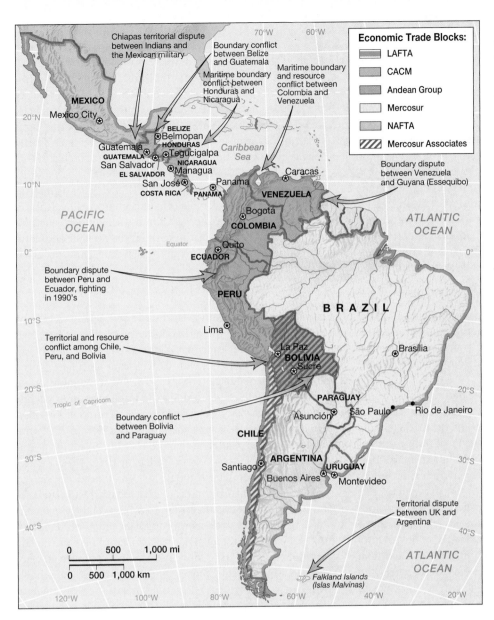

◀ Figure 4.30 Border disputes and trade blocks in Latin America Of the five economic trade blocks depicted, Mercosur and NAFTA are the most dynamic. In fact, several members of the Andean Group have expressed interest in joining Mercosur, while members of the Central American Common Market would like to be included in NAFTA. Most of Latin America's disputed boundaries are not being actively challenged. Yet these long-standing but dormant border disputes often contribute to tense relations between neighbors. (Data from Child, 1985. Geopolitics and Conflict in South America. New York: Praeger Press; and Allcock, 1992. Border and Territorial Disputes, 3rd ed. Harlow, Essex, UK: Longman Group)

▲ Figure 4.31 Protesting political violence Women protestors demand justice for the alleged killing of "disappeared" labor organizers in Honduras. Political violence at the hands of state, paramilitary, and guerrilla organizations is a tragic undercurrent affecting the lives of many Latin Americans. (Rob Crandall/Rob Crandall, Photographer)

limited at best. In the 1990s Mercosur and NAFTA emerged as supranational structures that could influence development (see Figure 4.30). For Latin America, the lessons of Mercosur and NAFTA are causing politicians to rethink the value of regional trade.

Mercosur was formed in 1991 with Brazil and Argentina, the two largest economies in South America, and the smaller states of Uruguay and Paraguay as members. Since its formation, trade among these countries has grown tremendously, so much so that Chile and Bolivia have now joined the group as associate members. This is significant in two ways: it reflects the growth of these economies and the willingness to put aside old rivalries (especially long-standing antagonisms between Argentina and Brazil) for the economic benefits of cooperation. As of 2000, Mercosur represented a market of 240 million people with a combined GNP of $1,100 billion. Intraregional trade within the group (excluding Bolivia and Chile) was estimated at nearly $19 billion in 1999. The size and productivity of this market have not gone unnoticed. Mercosur's leaders negotiate directly with the European Union to develop separate trade agreements. With Chile on board, expansion into Asian-Pacific markets is also likely.

If enthusiasm for Mercosur lasts, transportation and communication between Southern Cone countries will certainly improve. Plans abound to fundamentally rework the flow of goods and communication in this area. As privatization of the telephone companies has benefited telecommunication, major joint engineering projects are being considered. Studies are under way to improve navigation along the Paraná and Paraguay rivers, which are already important arteries for the transport of grains. Other schemes include building a bridge across the Plata River estuary to form a more direct link between the capital cities of Uruguay and Argentina, a tunnel through the Andes between Chile and Argentina, and a regional network of natural gas pipelines that would link Bolivia, Argentina, and southern Brazil. In short, the elaboration of Mercosur could change the way individuals in the member countries relate to one another and think about themselves. Much like the European Union, which has fostered a sense of European identity, Mercosur may over time shape a Southern Cone identity.

NAFTA took effect in 1994 as a free trade area that would gradually eliminate tariffs and ease the movement of goods among the member countries (Mexico, the United States, and Canada). NAFTA has increased intraregional trade (estimated at more than $700 billion in 1999), but there is considerable controversy about costs to the environment and to employment (see Chapter 3). NAFTA did prove, however, that a free trade area combining industrialized and developing states was possible. Moreover, it gave further impetus to the hemispheric vision of the Free Trade Area of the Americas (FTAA), which was first proposed in 1994 at the Miami Summit of the Americas. Thirty-four states from Canada, the United States, and all of Latin America (except Cuba, which was not invited) have agreed to conclude negotiations by 2005. FTAA embodies the ideals of neoliberalism, which holds that increased trade and economic integration will improve the standard of living for people in the Americas. If an agreement is reached, FTAA will be the largest free trade area in the world, including more than 800 million people. Yet there are many detractors, as witnessed by organized protests in Quebec, Canada, and Porto Alegre, Brazil, in 2001.

Insurgencies, Drug Traffickers, and Protest

Guerrilla groups such as the Shining Path (*Sendero Luminoso*) in Peru and the FARC (Revolutionary Armed Forces of Colombia) in Colombia have controlled large territories of their countries through the use of patronage, extortion, kidnapping, and violence. In the 1980s the Shining Path terrorized the rural highlands of Peru and caused citywide blackouts in Lima. Thousands of peasants fled the highlands, especially Ayacucho, to live as virtual refugees in Lima. With the capture of the Shining Path's leader in 1992, its destructive force was greatly diminished. In Colombia, however, two major well-armed insurgency groups exist: the FARC and the ELN (National Liberation Army). Their tactics of kidnapping, extortion, blowing up pipelines, and closing public roads are legendary. The two groups were enriched by the drug

trade in the 1990s, and there are now a recognized FARC-controlled zone in the southern Llanos and a proposed ELN-zone in the central Magdalena river basin. The level of violence in Colombia has escalated further with the rise of paramilitary groups—well-armed vigilante groups that seek to "cleanse" areas of insurgency sympathizers. The paramilitary groups are blamed for the rise in politically motivated murders, some 6,000 in 2000. Colombia also has the distinction of the highest murder rate in the world, ten times that of the United States. Perhaps as many as 2 million Colombians have been internally displaced by violence since the late 1980s, most fleeing rural areas for towns and cities. More alarming still is the number of Colombians who are leaving the country. Between 1996 and 2001, nearly 1 million Colombians fled to Venezuela, the United States, Ecuador, and Central America.

The drug trade is often seen as the root of many of the region's problems, but this illegal trade also generates billions of dollars for Latin Americans (from the small coca farmer in Bolivia to the cartel leader in Colombia). The drug trade began in earnest in the 1970s with Colombia at its center. By the 1980s the Medellín Cartel was a powerful and wealthy crime syndicate that used narco-dollars to bribe or murder anyone who got in its way. The organization was profoundly weakened by the death of its leader, Pablo Escobar, in 1993, but by then the Cali Cartel was poised to take control. Its power was reduced by arrests of key leaders in 1995. No one cartel dominates the drug trade today; instead, it has decentralized into dozens of smaller syndicates that are productive and creative. Initially, most Latin governments cared little about controlling the drug trade as it brought in much-needed hard currency. Within the region, drug consumption was scarcely a problem. Some drug lords even became popular folk heroes, lavishly spending money on housing, parks, and schools for their communities. The social costs of the drug trade to Latin America became evident by the 1980s when the region was crippled by a badly damaged judicial system. By paying off police, the military, judges, and politicians, the drug syndicates wield incredible political power that threatens the civil fabric of the states in which they work. Years of counter-narcotics work have done little to reduce the overall flow of drugs to North America and Europe, but the programs have changed the areas of production (see "Plan Colombia and the Hemispheric Drug War").

At the grassroots level, scores of organizations have formed to protest the consequences of globalization. It can be argued that the Zapatista rebellion in southern Mexico was in part a reaction to globalization. The rebellion began on January 1, 1994, in Chiapas, the day NAFTA took effect. Although Zapatistia supporters—largely Amerindian peasants—are mostly interested in local issues, such as land and basic services, their movement reflects a general concern about how increased foreign trade and investment hurt rural peasants. In South America, the city of Cochabamba, Bolivia, was engulfed in protest in 1999 as some 600,000 peasants rallied in the town plaza to denounce the privatization of the water supply. The government had sold the right to distribute water to a multinational corporation led by Spain. As a result of the sale, water prices soared even as overall service declined. A Bolivian union organizer, Oscar Olivera, headed a popular coalition that oversaw the return of the waterworks to local control.

Economic and Social Development: Dependent Economic Growth

Most Latin American economies fit into the broad middle-income category set by the World Bank. Clearly part of the developing world, their people are much better off than those in Sub-Saharan Africa, South Asia, and China. Still the economic contrasts are sharp both between states and within them (Table 4.2). Generally, the Southern Cone states (including southern Brazil and excluding Paraguay) are the richest. Argentina leads the way with a per capita GNI of $7,550, followed by Uruguay at $6,220. Brazil and Mexico, the largest countries in Latin America, have per capita GNIs around $4,400. The poorest countries in the region lie in Central America and the Andes. In 2001 Nicaragua, Honduras, and Bolivia had per capita income figures of less than $1,000. Even with its middle-income status, extreme poverty is evident throughout the region; nearly one-third of all Latin Americans live on less than $2 a day.

Development policies of the 1960s and 1970s accelerated industrialization and infrastructural development, but they also fostered debt and rural displacement. All sectors of the economy were radically transformed. Agricultural production increased with the application of "green revolution" technology and mechanization. State-run industries reduced the need for imported goods, and the service sector ballooned as a result of new government and private jobs. Protectionist tariffs ensured that domestically manufactured goods would be consumed over imports. Yet this rush to modernize produced victims. In rural areas poverty, landlessness, and inadequate systems of credit stymied the productivity of small farmers. As rural people were pushed off the land, there were not enough jobs in industry to absorb them, so they created their own niche in the urban informal economy. In the end, most countries made the transition from predominantly rural and agrarian economies dependent on one or two commodities to more economically diversified and urbanized countries with mixed levels of industrialization.

Development Strategies

Since the 1950s Latin America has experimented with various development strategies, from closed economies reliant on **import substitution** (policies that foster domestic industry by imposing inflated tariffs on all imports) to state-run nationalized industries and various attempts at agrarian reform. By the 1960s Brazil, Mexico, and Argentina all seemed poised to enter the ranks of the industrialized world. Multilateral agencies such as the World Bank and the Inter-American Development Bank loaned money for big development projects: continental highways, dams, mechanized agriculture, and power plants. Yet the dream of the 1960s became a nightmare by the 1980s. Argentina's economy, which was larger than

TABLE 4.2 *Economic Indicators*				
Country	Total GNI (Millions of $U.S., 1999)	GNI per Capita ($U.S., 1999)	GNI per Capita, PPP* ($Intl, 1999)	Average Annual Growth % GDP per Capita, 1990–1999
Argentina	$276,097	$7,550	$11,940	3.6
Bolivia	8,092	990	2,300	1.8
Brazil	730,424	4,350	6,840	1.5
Chile	69,602	4,630	8,410	5.6
Colombia	90,007	2,170	5,580	1.4
Costa Rica	12,828	3,570	7,880	3.0
Ecuador	16,841	1,360	2,820	0
El Salvador	11,806	1,920	4,260	2.8
Guatemala	18,625	1,680	3,630	1.5
Honduras	4,829	760	2,270	0.3
Mexico	428,877	4,440	8,070	1.0
Nicaragua	2,012	410	2,060	0.4
Panama	8,657	3,080	5,450	2.4
Paraguay	8,374	1,560	4,380	−0.2
Peru	53,705	2,130	4,480	3.2
Uruguay	20,604	6,220	8,750	3.0
Venezuela	87,313	3,680	5,420	−0.5

*Purchasing power parity

Source: The World Bank Atlas, *2001.*

Plan Colombia and the Hemispheric Drug War

In 1989 President George H. W. Bush declared war on cocaine production in the Andes. At that time, the major production centers were in the cloud forests of Peru and Bolivia, and the producers were peasants working small plots. During the 1990s coca leaf production (the main ingredient for cocaine) shifted to the rain forests of Colombia, and coca was being grown on large, modern estates. Plan Colombia, introduced by President Clinton in 2000, upped the stakes in the drug war with a commitment of $1.3 billion to Colombia alone. Since 1990 the United States has spent between $3 billion and $4 billion to reduce Andean coca and cocaine production. Proponents of this production-oriented strategy are claiming victory, as Bolivian and Peruvian production has tumbled (see Figure 4.4.1). Yet the overall area of coca cultivation has fallen only slightly, from 209,000 hectares in 1996 to 185,000 in 2000.

The battle against the coca plant begs for geographical analysis regarding the ecological requirements of the crop, where it can be produced, what the distribution flows are, and the ecological consequences of both coca production and eradication. Coca is a perennial shrub native to the eastern slopes of the Andes. It requires a tropical climate and ample rain. Coca leaf was domesticated by Andean agriculturalists because of its ability to suppress hunger and fatigue. Chewing coca leaf also reduces nausea and headache brought on by exposure to high altitudes. And coca is present in Coca-Cola, although a decocainized leaf is used for flavoring. A higher-elevation variety is still legally grown in Bolivia and Peru for local consumption as a chewable leaf, but a newer, lowland variety has become the shrub of choice in Colombia.

The Central Andean states of Peru and Bolivia, with U.S. support, used several strategies, including crop substitution, alternative development, interdiction of coca-carrying aircraft, and hand eradication of coca fields to cut production (Figure 4.4.2). Success in Bolivia and Peru led coca producers to relocate into Colombia, closer to the processing laboratories where cocaine is made. (To complicate matters, a Colombia insurgency group, FARC, grew rich from the surge of coca farms in its home states of Putomayo and Caqueta, thus becoming better armed and increasing

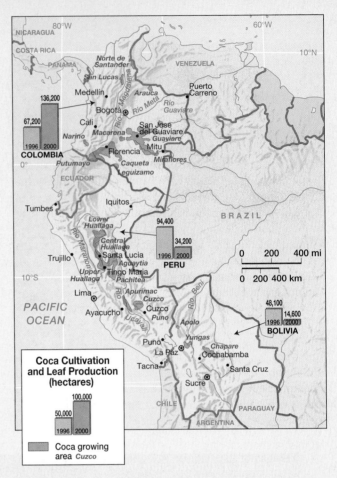

▲ Figure 4.4.1 Coca-growing areas in South America Although the oldest coca-growing regions are in Bolivia and Peru, Colombian traffickers turned the processing and distribution of cocaine into an international narcotics trade. By the late 1990s the bulk of coca production had shifted out of Bolivia and Peru and to Colombia. The U.S. government estimated nearly 400,000 acres (185,000 hectares) of coca were in production in 2000. (Source: U.S. Government. "Latin American Narcotics Cultivation and Production Estimates, 2000.")

Japan's in 1960, ranked seventeenth in 1999, while Japan's was ranked second. Oil wealth helped foster a large middle class in Mexico that was badly shaken by the debt crisis in the early 1980s. And Brazil, the eighth largest economy in the world, is the developing world's largest debtor. Brazil also maintains the region's worst income disparity: 10 percent of the country's richest people control nearly half the country's wealth, while the bottom 40 percent control only 10 percent.

Industrialization Since the 1960s most government development policies have emphasized manufacturing. Various strategies have been employed, from **growth poles** (planned industrial centers) and nationalized industries to import substitution. The results have been mixed. Today at least 25 to 30 percent of the male labor force in Mexico, Argentina, Brazil, Chile, Colombia, Peru, Uruguay, and Venezuela is employed in industry (including mining, construction, and energy). Yet this is far short of the hoped-for levels of industrial manufacturing, especially when the size of urban populations is taken into account. What is more, the most industrialized areas tend to be around the capitals or in planned growth poles such as Ciudad Guayana in Venezuela and the Mexican border cities of Ciudad Juárez and Tijuana.

▲ Figure 4.4.2 Coca eradication   Hand-eradication of coca has been practiced throughout the Bolivian Chapare and the Upper Huallaga Valley of Peru. In this photo an eradication line is destroying a coca field near Tinga Maria, Peru.

work is constantly evolving. When drug enforcement efforts close one route, a new one soon takes it place.

Cocaine production also requires supplies: pesticides, fertilizers, and processing chemicals to turn leaf into paste and then cocaine. This imposes some geographical limits on production, as coca needs to be grown near roads or waterways where the necessary materials can be brought in. Consequently, the most remote rainforest areas, where roads or river traffic is not available, are unlikely settings for large-scale cocaine production.

Less appreciated are the environmental consequences of coca production. Geographer Ken Young estimates that in Peru alone as many as 2.5 million acres (1 million hectares) of tropical forest were degraded in the 1980s by coca cultivation. Delicate tropical ecosystems and their waterways are being contaminated by the tons of chemicals used to produce cocaine. Unfortunately, serious ecological damage may also be occurring as a result of eradication efforts. The preferred method in southern Colombia is aerial spraying (with U.S helicopters), but fumigation has not been that successful, and it causes serious health problems, destruction of legal crops, increased social tension, and damage to both terrestrial and aquatic ecosystems. The World Wildlife Fund has compared the potential environmental impact of fumigation to that of Agent Orange in Vietnam.

It is the profitability of the cocaine trade that drives it. As long as there is demand, the supply will be met. Still confined to South America, coca can grow in other tropical areas of Southeast Asia, the Pacific, and Africa. The poppy industry, the base for heroin, offers an instructive comparison. Once poppies were almost an exclusive crop of Southeast Asia, but today Central Asia as well as Colombia and Mexico are major producers of poppy for the heroin trade. Given the global nature of the drug trade, it is likely that a victory in Colombia against coca cultivation will result in another "coca war" elsewhere. This doesn't mean that nothing should be done; it only underscores the inherent problems associated with a supply-oriented drug policy.

the level of violence in the region.) The variety of coca being grown in Colombia is hardier, with a larger potential growing area than traditional varieties. This is why Ecuador, Brazil, and Venezuela, neighbors of Colombia, fear that Plan Colombia may ultimately push coca growers into the frontier forests of their states.

The coca trade relies upon an intricate distribution network of both supplies and product in order to thrive. The Colombians are credited with organizing the production of cocaine. They developed a variety of shipping techniques, which include flying light aircraft over the jungle, stuffing sacks of cocaine into carnation boxes bound for New York, and driving carloads of cocaine across the Mexican border. This net-

There are cases in which industry has flourished in non-capital cities and without direct state support. Monterrey, Mexico; Medellín, Colombia; and São Paulo, Brazil, all developed important industrial sectors initially from local investment. Long before Medellín (2 million people) was associated with cocaine, it was a major center of textile production, more industrialized than the larger capital of Bogotá. A popular image of Medellín's inhabitants—the South American Yankees—surfaced in the 1950s. Celebrated as hardworking and entrepreneurial to the core, residents developed a strong sense of regional pride so that being from the department of Antioquia (where Medellín is capital) meant more

than being Colombian. Similar perceptions exist for Monterrey (3 million people), a city that is not well known outside Mexico, but is perceived as innovative, resourceful, and solidly middle class within Mexico.

The industrial giant of Latin America is metropolitan São Paulo in Brazil. Rio de Janeiro has greater name recognition and was the capital before Brasília was built, but it does not have the economic muscle of São Paulo. This city of 18 million, which vies with Mexico City for the title of Latin America's largest, began to industrialize in the early 1900s when the city's coffee merchants started to diversify their investments. Since then, a combination of private and state-owned industries have agglomerated around São Paulo. Within a 60-mile radius of the city center, automobiles, aircraft, chemicals, processed foods, and construction materials are produced. There are also heavy industry and industrial parks. With the port of Santos nearby and the city of Rio de Janeiro a few hours away, São Paulo is the uncontested financial center of Brazil. What stuns most first-time visitors to São Paulo is the forest of high-rises that greets them, a tropical version of Manhattan, only larger.

Maquiladoras and Foreign Investment

The Mexican assembly plants that line the border with the United States, called **maquiladoras,** are characteristic of manufacturing systems in an increasingly globalized economy. More than 4,000 maquiladoras exist, employing 1.3 million people who assemble automobiles, consumer electronics, and apparel. Between 1994 and 2000, 3 out of every 10 new jobs in Mexico were in the maquiladoras, which account for nearly half of Mexico's exports. As part of a border development program that began in the 1960s, the Mexican government allowed the duty-free import of machinery, components, and supplies from the United States to be used for manufacturing goods for export back to the United States. Initially, all products had to be exported, but changes in the law in 1994 now allow up to half of the goods to be sold in Mexico. The program was slow to develop, but it took off in the 1980s as foreign companies realized tremendous profits from the inexpensive Mexican labor. An autoworker in a General Motors plant in Mexico averages $10 a day, while the same worker in the United States earns more than $200 a day in wages and benefits.

Considerable controversy surrounds this form of industrialization on both sides of the border. Organized labor in the United States complains that well-paying manufacturing jobs are being lost to low-cost competitors, while environmentalists decry serious industrial pollution resulting from lax government regulation. Mexicans worry that these plants are poorly integrated with the rest of the economy and that many of the workers are young unmarried women who are easily exploited. With NAFTA, foreign-owned manufacturing plants are no longer restricted to the border zone, which may result in slower rates of industrialization for the border towns. Foreign-owned plants are increasingly being constructed near the population centers of Monterrey, Puebla, and Veracruz. Mexican workers and foreign corporations, however, continue to locate in the border zone because there are unique advantages to being positioned next to the U.S. border (see "Profile of Ciudad Juárez").

Mexico's competitive advantage is its location along the U.S. border and membership in NAFTA, but other Latin American states are attracting foreign companies through tax incentives and low labor costs. Assembly plants in Honduras, Guatemala, and El Salvador are drawing foreign investors, especially in the apparel industry. A recent report from El Salvador claims that not one of its 229 apparel factories has a union. Making goods for American labels such as the Gap, Liz Claiborne, and Nike, many Salvadoran garment workers complain they do not make a living wage, work 80-hour weeks, and face mandatory pregnancy tests. The situation in

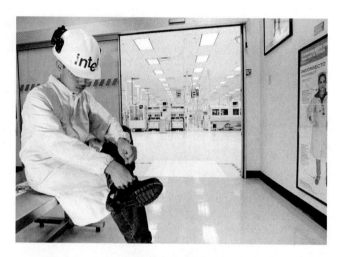

▲ **Figure 4.32 High-tech in Costa Rica** A Costa Rican worker straps on protective clothing before entering the manufacturing area of Intel's plant in Belen, not far from the capital city of San José. In 2000 Costa Rica earned more foreign exchange from exporting computer chips than it did from coffee, its traditional export. *(AP/Wide World Photos)*

Costa Rica, which is now a major chip manufacturer for Intel and exported over $2 billion in chips in 1999, is quite different (Figure 4.32). With a well-educated population, low crime rate, and stable political scene, Costa Rica is now attracting other high-tech firms. Hopeful officials claim that Costa Rica is transitioning from a banana republic (bananas and coffee were the country's long-standing exports) to a high-tech manufacturing center. As a result, the Costa Rican economy averaged 3 percent annual growth from 1990 to 1999.

The Entrenched Informal Sector

Even in prosperous San José, Costa Rica, a short drive to the urban periphery shows large neighborhoods of self-built housing filled with street traders and family-run workshops. Such activities make up the informal sector, the provision of goods and services without the benefit of government regulation, registration, or taxation. Most people in the informal economy are self-employed and receive no wages or benefits except the profits they clear. The most common informal activities are housing construction (in many cities half of all residents live in self-built housing), manufacturing in small workshops, street vending, transportation services (messenger services, bicycle delivery, and collective taxis), garbage picking, street performing, and even line-waiting (Figure 4.33). These activities are legal. Illegal informal activities also exist: drug trafficking, prostitution, and money laundering, for example. The vast majority of people who are reliant on informal livelihoods produce legal goods and services.

No one is sure how big this economy is, in part because it is difficult to separate formal activities from informal ones. One estimate in 1998 showed that nearly 60 percent of the total non-agricultural employment was in the informal sector. Visit Lima, Belém, Guatemala City, or Guayaquil and it is easy

Ciudad Juárez is big, with over 1 million people, and much larger than El Paso, Texas, its sister city across the border. Maquiladora-based manufacturing profoundly changed the look, size, and rhythm of the city. The factories are east of the old town, as are most of the middle- and upper-class houses. To the west, squatters have settled in shantytowns that take up three-fifths of the city's land area. Every morning buses grind their way through the west side of town, slowing picking up factory workers. The trip can take up to an hour each way. The workers—young men and women—are neatly dressed. Their day of labor will earn them anywhere from $5 to $10.

Founded in 1659 as Paso del Norte on the floodplain of the Rio Grande, the city was renamed in 1888 in honor of Mexican President Benito Juárez. Yet it is the sprawling, shabby newness of the place, and not its history, that impresses the visitor. Figures tell only part of the story. It is a city of young workers with a high birthrate, but no one really knows how many people live there. Estimates for the year 2000 range from 1.3 to 2.4 million, although most experts favor the lower figures. Since the 1980s several hundred acres of housing have been added to the city each year. In the west, the roads are mostly dirt, and the houses are built of cinder blocks, adobe, and plywood. Access to electricity is spotty, and water is usually delivered by truck. The east side of town is another world. In between the assembly plants are the nicest suburbs, with sidewalks, garages, and neatly landscaped yards. Drive down the commercial strip and stop at McDonald's, Burger King, or the 7-Eleven.

Most geographers find border zones intriguing cultural landscapes that invite the question of how a line in the sand can create two distinct faces. At this particular bend of the Rio Grande, there are three faces: El Paso to the north and the dual faces of Ciudad Juárez's west and east sides.

(a)

(b)

▲ Figure 4.5.1 Ciudad Juárez The core of the city sits along the Rio Grande/Bravo, while the periphery radiates out for several miles. The industrial zone, filled with maquilas, is east of downtown. [Modified from The Mexican Border Cities (1993), p. 39, by Daniel Arreola and James Curtis. Tucson, AZ: University of Arizona Press.]

▲ Figure 4.5.2 Elite and shanty housing in Ciudad Juárez (a) The majority of the city's newcomers reside in self-built homes like these on the urban periphery. (b) In contrast, small elite suburbs are interspersed with assembly plants on the east side of the city. (Photos by Rob Crandall/Rob Crandall, Photographer)

to get the impression that the informal economy *is* the economy. From self-help housing that dominates the landscape to hundreds of street vendors that crowd the sidewalks, it is impossible to avoid. There are advantages in the informal sector—hours are flexible, children can work with their parents, and there are no bosses. Peruvian economist Hernando de Soto even argues that this most dynamic sector of the economy should be encouraged and offered formal lines of credit. As important as this sector may be, however, widespread dependence on it signals Latin America's poverty, not its

▲ Figure 4.33 Peruvian street vendors   A street vendor selling produce in Huancayo, Peru. Street vending plays a critical
role in the distribution of goods and the generation of income.
Contrary to popular opinion, it is often regulated by city governments and by the vendors themselves.   *(Rob Crandall/Rob Crandall, Photographer)*

wealth. It reflects the inability of the formal economies of the
region, especially in industry, to absorb labor. For millions of
urban dwellers, the thought of finding formal employment
that offers benefits, safety, and a living wage is still a dream.
With nowhere else to go, the ranks of the informally employed continue to grow.

Primary Export Dependency

Historically, Latin America's abundant natural resources were
its wealth. In the colonial period silver, gold, and sugar generated tremendous riches for the colonists. With independence in the nineteenth century, a series of export booms
introduced commodities such as bananas, coffee, cacao,
grains, tin, rubber, copper, wool, and petroleum to an expanding world market. One of the legacies of this export-led
development was a tendency to specialize in one or two major
commodities, a pattern that continued into the 1950s. During that decade Costa Rica earned 90 percent of its export
earnings from bananas and coffee; Nicaragua earned 70 percent from coffee and cotton; 85 percent of Chilean export income came from copper; half of Uruguay's export income
came from wood. Even Brazil generated 60 percent of its export earnings from coffee in 1955; by 1998 coffee accounted
for less than 5 percent of the country's exports, yet Brazil remained the world leader in coffee production.

Agricultural Production   Since the 1960s, the trend in Latin
America has been to diversify and to mechanize agriculture.
Nowhere is this more evident than in the Plata Basin, which
includes southern Brazil, Uruguay, northern Argentina,
Paraguay, and eastern Bolivia. Soybeans, used for oil and animal feed, transformed these lowlands in the 1980s and early
1990s. Added to this are acres of rice, cotton, and orange
groves, as well as the more traditional plantings of wheat and

sugar. The speed with which the plains were converted into
fields alarmed environmentalists. Throughout the 1990s, efforts to conserve some of the ecosystem increased.

Similar large-scale agricultural frontiers exist along the
piedmont zone of the Venezuelan Llanos (mostly grains) and
the Pacific slope of Central America (cotton and some tropical fruits). In northern Mexico water supplied from dams
along the Sierra Madre Occidental has turned the valleys in
Sinaloa into intensive producers of fruits and vegetables for
consumers in the United States. The relatively mild winters in
northern Mexico allow growers to produce strawberries and
tomatoes during the winter months.

In each of these cases, the agricultural sector is capital-
intensive and dynamic. By using machinery, high-yielding hybrids, chemical fertilizers, and pesticides, many corporate
farms are extremely productive and profitable. What these operations fail to do is employ many rural people, which is especially problematic in countries where a third or more of the
population depends on agriculture for its livelihood. As industrialized agriculture becomes the norm in Latin America,
subsistence peasant producers are further marginalized. The
overall trend is that agricultural production is increasing while
proportionally fewer people are employed by it. In Peru
45 percent of the male labor force worked in agriculture in
1980; by 1998 the figure had plummeted to just 7 percent. In
absolute terms, however, the number of people living in rural
areas is about the same as it was in 1960 (roughly 130 million).
The major difference over the last 40 years is that many of
these people are worse off as traditional rural support networks break down and small farmers are forced onto marginal lands that are vulnerable to drought and erosion. Peasant
farmers who are able to produce a surplus of corn or wheat see
its value undercut by cheaper imported grains.

Mining and Forestry   The exploitation of silver, zinc, copper, iron ore, bauxite, gold, oil, and gas is the economic mainstay for many countries in the region. The oil-rich nations of
Venezuela, Mexico, and Ecuador are able to meet their own
fuel needs and to earn vital state revenues from oil exports.
Venezuela is most dependent on revenues from oil, earning up
to 90 percent of its foreign exchange from crude petroleum
and petroleum products. In 1998 Venezuela also exported
more oil to the United States than any other country, including oil-rich Saudi Arabia (Figure 4.34). Vast oil reserves also
exist in the Llanos of Colombia, yet a costly and vulnerable
pipeline that connects the oil fields to the coast is a regular target of guerrilla groups. By the late 1990s Colombian oil production had improved, but it is far from reliable.

Like agriculture, mining has become more mechanized
and less labor intensive. Even Bolivia, a country dependent on
tin production, cut 70 percent of its miners from state payrolls
in the 1990s. The measure was part of a broad-based austerity program, yet it suggests that the majority of the miners
were not needed. Similarly, the vast copper mines of northern Chile are producing record amounts of copper with fewer
miners. Gold mining, in contrast, continues to be labor intensive, offering employment for thousands of prospectors.

▲ **Figure 4.34 Oil production**   A portable drilling platform on Lake Maracaibo, Venezuela. Oil production accelerated the pace of development for countries like Venezuela, Mexico, and Ecuador, but these economies struggled in the 1990s as oil prices declined.   *(Rob Crandall/Rob Crandall, Photographer)*

▲ **Figure 4.35 Amazonian gold**   Serra Pelada, in the Brazilian state of Pará, is one of the most productive gold mines in Latin America.   *(Peter Frey/Getty Images, Inc.)*

Gold rushes are occurring in remote tropical regions of Venezuela, Brazil, Colombia, and Costa Rica. The famous Serra Pelada mine in Brazil, an anthill of workers with ladders, shovels, and burlap sacks filled with gold-bearing soil, is the most dramatic example of labor-intensive mining (Figure 4.35). Generally, gold miners tend to work in small groups. Moreover, many gold strikes occur illegally on Indian lands or within the borders of national parks (as in the case of Costa Rica). Invariably, if a large enough strike occurs, the state steps in to regulate mining operations and to get its share of the profits.

Logging is another important, and controversial, extractive activity. Ironically, many of the forest areas cleared for cattle were not systematically harvested. More often than not, all but the most valuable trees were burned. Logging concessions are commonly awarded to domestic and foreign timber companies, which export boards and wood pulp. These one-time arrangements are seen as a quick means for foreign exchange, particularly if prized hardwoods like mahogany are found. Logging can mean a short-term infusion of cash into a local economy. In 1995, $31 million in forest products was exported from the Bolivian state of Santa Cruz (a relative newcomer to the tropical hardwood market). Yet rarely do long-term conservation strategies exist, making this system of extraction unsustainable. There is growing interest in certification programs that designate when a wood product has been produced sustainably. This is due to consumer demand for certified wood, mostly in Europe. Unfortunately, such programs are small and the lure of profit usually overwhelms the impulse to conserve for future generations.

Several countries rely on plantation forests of introduced species of pines, teak, and eucalyptus to supply domestic fuelwood, pulp, and board lumber. These plantation forests grow single species and fall far short of the complex ecosystems occurring in natural forests. Still, growing trees for paper or fuel reduces the pressure on other forested areas. Leaders in plantation forestry are Brazil, Venezuela, Chile, and Argentina. Considered Latin America's economic star in the 1990s, Chile relied on timber and wood chips to boost its export earnings. Thousands of hectares of exotics (eucalyptus and pine) have been planted, systematically harvested, cut into boards, or chipped for wood pulp (see Figure 4.6). Japanese capital is heavily involved in this sector of the Chilean economy. The recent expansion of the wood chip business, however, led to a dramatic increase in the logging of native forests. By 1992, more than half of all wood chips were from native species.

Latin America in the Global Economy

In order to conceptualize Latin America's place in the world economy, **dependency theory** was advanced in the 1960s by scholars from the region. The premise of the theory is that expansion of European capitalism created the region's underdevelopment. For the developed "cores" of the world to prosper, the "peripheries" became dependent and impoverished. Dependent economies, as those in Latin America, were export-oriented and vulnerable to fluctuations in the global market. Even when they experienced economic growth, it was subordinate to the economic demands of the core (North America and Europe).

Economists who embraced this interpretation of Latin America's history were convinced that economic development could occur only through self-sufficiency, growth of internal markets, agrarian reform, and greater income equality. In short, they argued for vigorous state intervention and an uncoupling from the economic cores. Policies such as import substitution industrialization, and nationalization of key industries were partially influenced by this view. Dependency theory has its detractors. In its simplest form it becomes a means to blame forces external to Latin America for the region's problems. Implicit in dependency theory is also the notion that the path to development taken by Europe and North America cannot be easily replicated. This was a radical idea for its time.

A look at trade in the Western Hemisphere reveals both increased integration within Latin America as well as the dominance of the North American market (Figure 4.36). In terms of regional integration, the emergence of Mercosur in 1991 stimulated new levels of trade, especially between Brazil and Argentina. The success of Mercosur, along with the expansion of trade within the Andean Community, resulted in a 16 percent increase in intraregional trade in Latin America during the 1990s. Even with this increase, the level of intraregional trade among the Mercosur countries ($18.3 billion) is a small fraction of the intraregional NAFTA trade ($700 billion). As shown in Figure 4.36, the level of intraregional trade within Mercosur, Central America, or the Andean Community is much less than each group's trade with North America. What these data do not show is the disproportionate flow of exports to the United States. In the late 1990s, over 80 percent of all

Mexican exports and roughly 40 percent of all Central American and Andean exports were bound to the United States. By comparison, only 15 percent of the total Mercosur trade went to the United States. (Mercosur countries have proportionally more trade with Europe and Asia.) Generally, the increase in intraregional trade is widely recognized as a positive sign of greater economic independence for Latin America. Nevertheless, the historic pattern of dependence on North America still exists as the United States continues to be the major trading partner for most states in the region.

From a U.S. perspective, the FTAA agreement would only formalize an established pattern of trade. Not all Latin Americans see it that way. Brazilian President Fernando Cardoso speaks eloquently of the asymmetries of globalization and how they impact Brazil. The financial crises in Asia and Russia in the late 1990s caused a drastic reduction in the amount of direct foreign investment in Latin America. This, in turn, caused Brazil to devalue its currency in 1999. Devaluation resulted in cheaper Brazilian products, which caused trade instability with one of Brazil's Mercosur partners, Argentina. This worsened Argentina's already unstable economy, and in the summer of 2001 Argentina was forced to refinance its debt. Without major changes, argues Cardoso, the FTAA agreement would be an example "protectionism of the fittest." Free trade in the Amer

▶ **Figure 4.36 Global linkages: hemispheric trade** Latin American trade has well-established links with North America and Europe. In the past two decades, intraregional trade has surged. Many credit the growth in intraregional trade to the success of Mercosur and NAFTA. *(Source: The Economist, April 21, 2001)*

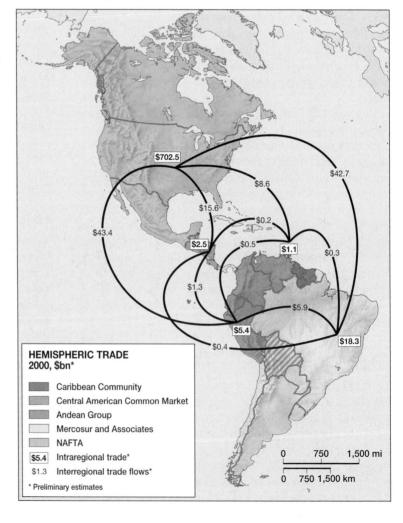

icas—with additional safety nets—may have the potential to increase both economic and social development in the hemisphere, which is what its supporters argue. Yet the staunchest critics of FTAA see it as a new form of U.S. imperialism (see "Geography in the Making: Argentina Is Bust").

Neoliberalism as Globalization
By the 1990s governments and the World Bank had become champions of neoliberalism as a sure path to economic development. **Neoliberal policies** stress privatization, export production, direct foreign investment, and few restrictions on imports. They epitomize the forces of globalization by turning away from policies that emphasize state intervention and self-sufficiency. Most Latin American political leaders are embracing neoliberalism and the benefits that come with it, such as increased trade and more favorable terms for debt repayment. Yet there are signs of discontent with neoliberalism throughout the region. The usually tranquil city of Porto Alegre, Brazil, hosted the World Social Forum in 2001 as an antiglobalization counterpoint to the World Economic Forum's meeting in Switzerland. This leftist international forum staged rallies denouncing neoliberalism and the FTAA and advocating reduction of poverty and inequality as Latin America's top priorities.

Chile is an outspoken champion of neoliberalism. Its growth rate between 1990 and 1999 averaged 5.6 percent, the region's healthiest. In 1995 alone the Chilean economy grew 10.4 percent, placing it in the same league as the Asian tigers. Consequently, it is the most studied and watched country in Latin America. By the numbers, Chile's 15.4 million people are doing well, but the country's accomplishments are not readily transferable. To begin with, the radical move to privatize state-owned business and open the economy occurred under an oppressive military dictatorship that did not tolerate opposition. Thankfully, such dictatorships are rare in Latin America today, but in a democracy the pace of reform must be slower. Much of Chile's export-led growth has been based on primary products: fruits, seafood, copper, and wood. Although many of these are renewable, Chile will need to develop more value-added goods if it hopes to be labeled "developed." Furthermore, its relatively small and homogeneous population in a resource-rich land does not have the same ethnic divisions that hinder so many states in Latin America. Experiments in neoliberalism are still new, so all the social and environmental costs associated with this economic free-for-all are not known.

Dollarization
As financial crises spread through Latin America in the late 1990s, governments began to consider the economic benefits of **dollarization,** a process by which a country adopts—in whole or in part—the U.S. dollar as its

GEOGRAPHY IN THE MAKING   "Argentina Is Bust"

It was obvious that Argentina's economy was in peril by 1999. Several attempts were made to restructure loans, but the inevitable collapse came in early 2002 when Argentina defaulted on its $155 billion public debt, the largest such default by any country in history. At the same time, the Argentine peso was uncoupled from its one-to-one exchange rate with the U.S. dollar, triggering a painful devaluation that squeezed many middle class workers below the poverty line. Unemployment surged to more than 18 percent. The economic crisis triggered a political crisis, bringing down the presidency of Fernando de la Rua in December 2001. As the world watched, the Argentine legislature nominated various interim presidents, finally selecting Eduardo Duhalde to lead the country until elections could be held in 2003. In a speech to the Congress that selected him on January 1, 2002, President Duhalde declared, "Argentina is bust."

The stunning descent of Argentina is remarkable considering that just a few years earlier it was heralded as a model for free-market, neo-liberal reform. Between 1991–1997, Argentina's economy—the third largest in Latin America—grew at an average annual rate of 6.1%, the highest in Latin America. Argentines were the wealthiest Latin Americans in 1999 with a GNI of $7,550 per capita. With devaluation in 2002, the GNI per capita plummeted to an estimated $3,500.

During the 1990s most utilities were privatized, and private capital poured into the country. At this time the Argentine currency board was created, which pegged the peso to the U.S. dollar and for a time secured fiscal stability. The problem with this strategy was that as the dollar rose in value compared to other currencies in the 1990s, so did the peso. By the late 1990s Argentine exports were less competitive, and this negatively impacted employment. Argentines used their overvalued currency to become major importers, leading to a trade imbalance, especially with Brazil. With economic ruin in sight, those who had capital moved it abroad. In 2002, the state tried to stem capital flight by imposing a $1,000 a month withdrawal limit on all bank accounts. All payment and pricing systems quickly became chaotic. Companies could not pay international suppliers because accounts were partially frozen. Imports, from luxury items to prescription drugs, were nearly impossible to get in the first half of 2002.

Some politicians blame the IMF for imposing neo-liberal policies and insisting on privatization and open markets. Others blame the rigidity of the currency board and its inability to respond to external shocks in the global economy. Still others point to decentralized state governments that overspent and an ineffective national tax collection system. While scholars and policymakers debate the reasons for this default, the results have been devastating for Latin America's most developed country.

official currency. In a totally dollarized economy, the U.S. dollar becomes the only medium of exchange and the country's national currency ceases to exist. This was the radical step taken by Ecuador in 2000 to address the dual problems of currency devaluation and hyperinflation. El Salvador was considering total dollarization in 2001 as a means of reducing the cost of borrowing money. Dollarization is not a new idea; back in 1904 Panama dollarized its economy the year after it had gained independence from Colombia. Until 2000, however, Panama was the only fully dollarized state in Latin America.

A more common strategy in Latin America is limited dollarization, in which U.S. dollars circulate and are used alongside the country's national currency. Limited dollarization exists in many countries around the world, but most notably in Latin America. Since the economies of Latin America are prone to currency devaluation and hyperinflation, limited dollarization is a type of insurance. Many banks in Latin America, for example, allow customers to maintain accounts in dollars to avoid the problem of capital flight should a local currency be devalued. Other countries keep their national currency but peg its value one-for-one to the dollar; this was the innovative strategy adopted by Argentina in 1991. Dollarization, partial or full, tends to reduce inflation, eliminate fears of currency devaluation, and reduce the cost of trade by eliminating currency conversion costs.

There are drawbacks to dollarization. The obvious one is that a country no longer has control of its monetary policy, making it reliant on the decisions of the U.S. Federal Reserve. Foreign governments do not have to ask permission to dollarize their economies. At the same time, the United States insists that all its monetary policies be based exclusively on domestic considerations, regardless of the impact such decisions may have on foreign countries. The political impact of eliminating a national currency is real. The case of Ecuador is instructive. In 1999, when President Mahuad announced his plan to dollarize the economy to head off hyperinflation, he was quickly forced out of office by a coalition of military and Indian activists. When Vice-President Naboa became president and the economic situation worsened, the country's political leadership went ahead with dollarization. In short, dollarization may help in a time of economic duress, but it is a potent indicator of the prominence of one currency—the dollar—in an increasingly global economy.

Social Development

Marked improvements in life expectancy, child survival, and educational attainment have occurred in Latin America the past three decades. One telling indicator is the mortality rate for children younger than 5 years old, which was cut by 50 percent or more in most Latin America countries between 1980 and 1999 (Table 4.3). Brazil in 1980 had 81 deaths per 1,000 children under age 5; in 1999 the number was 40. Nicaragua dropped from 143 deaths per 1,000 in 1980 to 43 in 1999. This indicator is important because when children

TABLE 4.3 Social Indicators and Status of Women

Country	Life Expectancy at Birth		Under Age 5 Mortality (per 1,000)		Percent Illiteracy (Ages 15 and over)		Female Labor Force Participation (% of total, 1999)
	Male	Female	1980	1999	Male	Female	
Argentina	70	77	38	22	3	3	33
Bolivia	60	64	170	83	8	21	38
Brazil	65	72	81	40	15	15	35
Chile	72	78	35	12	4	5	33
Colombia	68	74	58	28	9	9	38
Costa Rica	75	79	29	14	5	5	31
Ecuador	68	73	101	35	7	11	28
El Salvador	67	73	120	36	19	24	36
Guatemala	63	68	121	52	24	40	28
Honduras	64	68	103	46	26	26	31
Mexico	73	78	74	36	7	11	33
Nicaragua	66	70	143	43	33	30	35
Panama	72	76	36	25	8	9	35
Paraguay	71	76	61	27	6	8	30
Peru	66	71	126	48	6	15	31
Uruguay	70	78	42	17	3	2	42
Venezuela	70	76	42	23	5	9	34

Sources: Population Reference Bureau, Data Sheet, *2001, Life Expectancy (M/F); The World Bank*, World Development Indicators, *2001, Under Age 5 Mortality Rate (1980/99); The World Bank Atlas, 2001, Female Participation in Labor Force; CIA World Factbook, 2001, Percent Illiteracy (ages 15 and over).*

younger than 5 years of age are surviving, it suggests that basic nutritional and health-care needs are being met. One can also infer that resources are being used to sustain women and their children. By comparison, the United States has a death rate of 8 per 1,000 children under age 5 and Japan's rate is 4. Most other developing countries have much higher under-5 death rates than Latin America; India's is 90 per 1,000 and Kenya's is 118. Despite economic downturns, the region's social networks have been able to mitigate the negative effects on children.

Grassroots and nongovernmental organizations (NGOs) play a fundamental role in contributing to social well-being. Initiated by international humanitarian organizations, church organizations, and community activists, these popular groups provide many services that state and local governments cannot. Catholic Relief Services and Caritas, for example, work with the rural poor throughout the region to improve their water supplies, health care, and education. Other groups lobby local governments to build schools or recognize squatters' claims. Grassroots organizations also develop cooperatives that market anything from sweaters to cheeses. Cooperative organizations are able to realize some economies of scale as well as improve access to credit. In the 1980s Lima, Peru, witnessed the mobilization of 2,000 popular kitchens run by poor women to feed poor families with local and foreign donations. At their peak in the late 1980s, it was estimated that 1 million meals a day were served. Had this not occurred, the human cost of Peru's economic collapse would have been far worse.

Other important gauges for social development are life expectancy, literacy, and access to improved water sources. In aggregate, 80 percent of the people in the region have access to an adequate amount of water from an improved source; literacy rates are over 85 percent; and life expectancy (men and women) is 71 years. Masked by this aggregate data are extreme variations between rural and urban areas, between regions, and along racial and gender lines. No matter how difficult city life may be, the poorest people in Latin America are in rural areas. The move to cities for many people is a development strategy.

A running joke in Latin America is that the economy is doing well but the people are not. Even today when Chile or Costa Rica is heralded for making impressive economic gains, fear persists that a large segment of the population is excluded. The 1980s were dubbed the "lost decade" for Latin America due to the debt crisis, inflation, and slow-to-negative growth rates. During that period, several countries witnessed setbacks in both their economic and social indicators—most notably Peru, Bolivia, Nicaragua, Venezuela, Mexico, and Brazil. Many economic indicators improved in the 1990s. Still, Mexico and Brazil, the two largest countries with more than half the region's population, struggle with inflation, currency devaluation, and a per capita GDP that is lower than it was in 1980 constant dollars.

Within Mexico and Brazil tremendous internal differences exist in socioeconomic indicators (Figure 4.37). The northeast part of Brazil lags behind the rest of the country in every social indicator. The country has a literacy rate of 85 percent, but in the northeast it is only 60 percent. Moreover, 70 percent of city residents in the northeast are literate, but only 40 percent of rural ones are. In Mexico the levels of poverty are highest in the Indian states of Chiapas and Oaxaca. In contrast, Mexico City, the state of Nuevo Leon (Monterrey is the capital), and Quintana Roo (home to Mexico's largest resort, Cancun) have a per capita GDP of more than $6,000. All countries have spatial inequities regarding income and availability of services, but the contrasts tend to be sharper in the developing world. In the cases of Mexico and Brazil, it is hard to ignore ethnicity and race when trying to explain these patterns.

Race and Inequality   There is much to admire about race relations in Latin America. The complex racial and ethnic mix that was created in Latin America fostered tolerance for diversity. That said, Indians and blacks are disproportionately represented among the poor of the region. More than ever, racial discrimination is a major political issue in Brazil. Reports of organized killings of street children, most of them Afro-Brazilian, make headlines. For decades, Brazil espoused its vision of a color-blind racial democracy. True, residential segregation by race is rare in Brazil and interracial marriage is common, but certain patterns of social and economic inequity seem best explained by race.

There are major problems with trying to assess racial inequities in Brazil. The Brazilian census asks few racial questions, and all are based on self-classification. In 1987 only 11 percent of the population called itself black and 53 percent white. Some Brazilian sociologists, however, claim that more than half the population is of African ancestry, making Brazil the second largest "African state" after Nigeria. Racial classification is always highly subjective and relative, but there are patterns that support the existence of racism. Evidence from northeastern Brazil, where Afro-Brazilians are the majority, shows death rates approaching those of some of the world's poorest countries. Throughout Brazil, blacks suffer higher rates of homelessness, landlessness, illiteracy, and unemployment. There is even a movement in the northeastern state of Bahia to extend the same legal protection given Indians to hundreds of Afro-Brazilian villages that were once settled by runaway slaves.

In areas of Latin America where Indian cultures are strong, one also finds low socioeconomic indicators. In most countries areas where native languages are widely spoken invariably correspond with areas of persistent poverty. In Mexico the Indian south lags behind the booming north and Mexico City. Prejudice is embedded in the language. To call someone an *indio* (Indian) is an insult in Mexico. In Bolivia, women who dress in the Indian style of full, pleated skirts and bowler hats are called *cholas,* a descriptive term referring to the rural mestizo population that suggests backwardness and even cowardice. No one of high social standing, regardless of skin color, would ever be called a *chola* or *cholo.*

It is difficult to separate status divisions based on class from those based on race. From the days of conquest, being European meant an immediate elevation in status over the

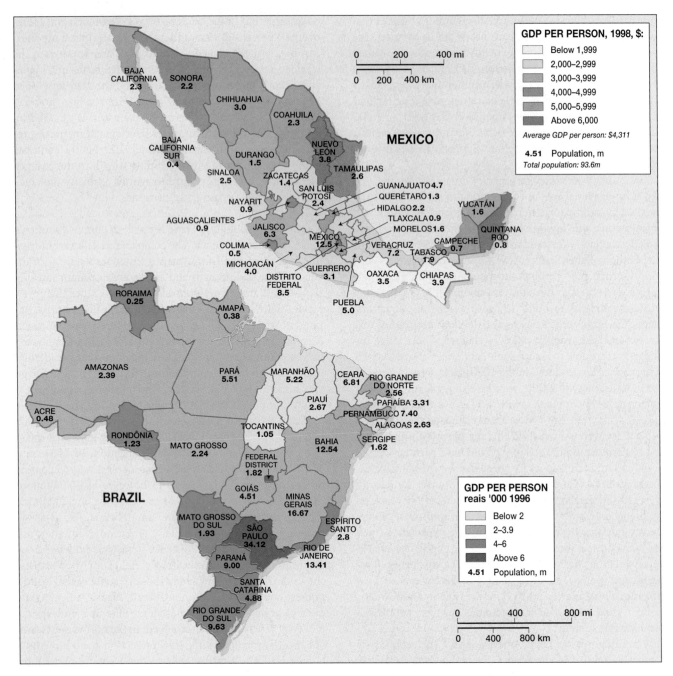

GDP PER PERSON, 1998, $:

	Below 1,999
	2,000–2,999
	3,000–3,999
	4,000–4,999
	5,000–5,999
	Above 6,000

Average GDP per person: $4,311

4.51 Population, m
Total population: 93.6m

**GDP PER PERSON
reais '000 1996**

	Below 2
	2–3.9
	4–6
	Above 6

4.51 Population, m

▲ **Figure 4.37 Mapping poverty and prosperity**   Mexico and Brazil are the largest states in the region, and yet the levels of poverty vary greatly within each country. In Mexico, the southern states of Chiapas and Oaxaca have the lowest GDP per capita. In Brazil, the northeastern states of Paraíba, Alagoas, Tocantins, Piauí, Ceará, and Maranhão have the country's lowest per capita GDP.   *(©1999 and 2000 The Economist Newspaper Group, Inc. Reprinted with permission. Further reproduction prohibited.)*

Indian, African, and mestizo populations. Class awareness is very strong. Race does not necessarily determine one's economic standing, but it certainly influences it. Most people recognize the power of the elite and envy their lifestyle.

An emerging middle class also exists that is formally employed, aspires to own a home and car, and strives to give its children a university education. The vast majority of people, however, are the working poor who struggle to meet basic food, shelter, clothing, and transport needs. These class dif-

ferences express themselves in the landscape. Go to any large city and find handsome suburbs, country clubs, and trendy shopping centers. High-rise luxury apartment buildings with beautiful terraces offer all the modern amenities, including maids' quarters. The elite and the middle class even show a preference for decentralized suburban living and dependence on automobiles, as do North Americans. Yet near these same residences are shantytowns where urban squatters build their own homes, create their own economy, and eke out a living.

The Status of Women   Many contradictions exist with regard to the status of women in Latin America. The gender stereotypes of machismo and marianismo are breaking down in the younger generations. Many Latina women work outside the home. In most countries the formal figures hover between 30 and 40 percent of the workforce, not far off from many European countries, but lower than in the United States (Table 4.3). Legally speaking, women can vote, own property, and sign for loans, although they are less likely to do so than men, which reflects the patriarchal tendencies in the society. Even though Latin America is predominantly Catholic, divorce is legal and family planning is promoted. In most countries, however, abortion remains illegal.

Overall, access to education in Latin America is good compared to other developing regions, and thus illiteracy rates tend to be low. The rates of adult illiteracy are slightly higher for women than for men, but usually by only a few percentage points (Table 4.3). The highest levels of illiteracy are found in Central America. In Guatemala 40 percent of adult women are illiterate; the rate for men is lower but still well above the regional average at 26 percent. Once again, Costa Rica stands out in Central America with its high levels of literacy; only 5 percent of its adult population is illiterate. Throughout higher education in Latin America, male and female students are equally represented today. Consequently, in the fields of education, medicine, and law, women are regularly employed.

The biggest changes for women are the trends toward smaller families, urban living, and educational parity with men. These factors have greatly improved the participation of women in the labor force. Since 1970 in 11 of the 17 Latin American countries, the average family size has dropped by at least two children. In the most populous states of Brazil and Mexico, the average woman has three fewer children than in 1970 (Figure 4.38). And in nearly every country, women's participation in the labor force has climbed by 10 to 15 percent in the last three decades. In the countryside, however, serious inequities remain. Rural women are less likely to be educated and tend to have larger families. In addition, they are often left to care for their families alone as husbands leave seasonally in search of employment. In most cases, the conditions and prejudices facing rural women have been slow to improve.

Women are increasingly playing an active role in politics. In 1990 Nicaragua elected the first woman president in Latin America, Violeta Chamorro, the owner of an opposition newspaper. Nine years later, Panamanians voted Mireya Moscoso into power. A Mayan woman from Guatemala, Rigoberta Manchú, won the Nobel Peace Prize in 1992 for denouncing human rights abuses in her native country. Women are also active organizers and participants in cooperatives, microenterprises, and unions. In a relatively short period, urban women have won a formal place in the economy and a political voice. Moreover, evidence suggests that this trend will continue, reflecting an improved status for women in the region.

Conclusion

Latin America and the Caribbean were the first world regions to be fully colonized by Europe. In the process, perhaps 90 percent of the native population died from disease, cruelty, and forced resettlement. The slow demographic recovery of native peoples and the continual arrival of Europeans and Africans resulted in an unprecedented level of racial and cultural mixing. It took nearly 400 years for the populations of Latin America and the Caribbean to reach 54 million again, their precontact level. During this long period, European culture, technology, and political systems were transplanted and modified. Indigenous peoples integrated livestock and wheat into their agricultural practices, yet they held true to their preference for native corn, potatoes, and cassava. In short, a syncretic process unfolded in which many indigenous customs were preserved beneath the veneer of Iberian ones. Over time, a blending of indigenous and Iberian societies, with some African influences, gave distinction to this part of the world. The music, literature, and artistry of Latin America are widely acknowledged.

Compared with Asia, this is still a region rich in natural resources and relatively lightly populated. Yet as population continues to grow along with economic expectations, there is considerable concern that much of this natural endowment could be squandered for short-term gains. In the midst of a boom in natural resource extraction, popular concern for the state of the environment is growing. Not only was Brazil the site of the 1992 United Nations Earth Summit, but hundreds of locally based en-

vironmental groups have formed to try to protect forests, grasslands, indigenous peoples, and freshwater supplies. This brand of environmentalism is pragmatic; it recognizes the need for economic development, but hopes to improve urban environmental quality and sustainable resource use.

In Latin America the trend toward modernization began in the 1950s, and the pace of change has been rapid. Unlike other developing areas, most Latin Americans live in cities. This shift started early and reflects a cultural bias toward urban living with roots in the colonial past. Not everyone who came to the city found employment, however. Thus, the dynamics of the informal sector were set in place. Even though population growth rates have declined, the overall makeup of the population is young. Serious challenges lie ahead in educating and finding employment for the cohort younger than age 15. In this matter some countries are faring better than others. Industrial employment in Mexico and the Southern Cone is up, yet in many parts of the Central Andes subsistence farming is widely practiced.

The 1990s were better years for much of Latin America if measured by real annual growth. What is deceptive about economic growth is that the percentage of people living in poverty (as well as the total number of people) was higher in 2000 than in 1980. So too was the widening gap between rich and poor in the region. Latin America has the world's highest inequality when it comes to the distribution of wealth. Brazil's measures of income distribution are nearly the same as South Africa's when it

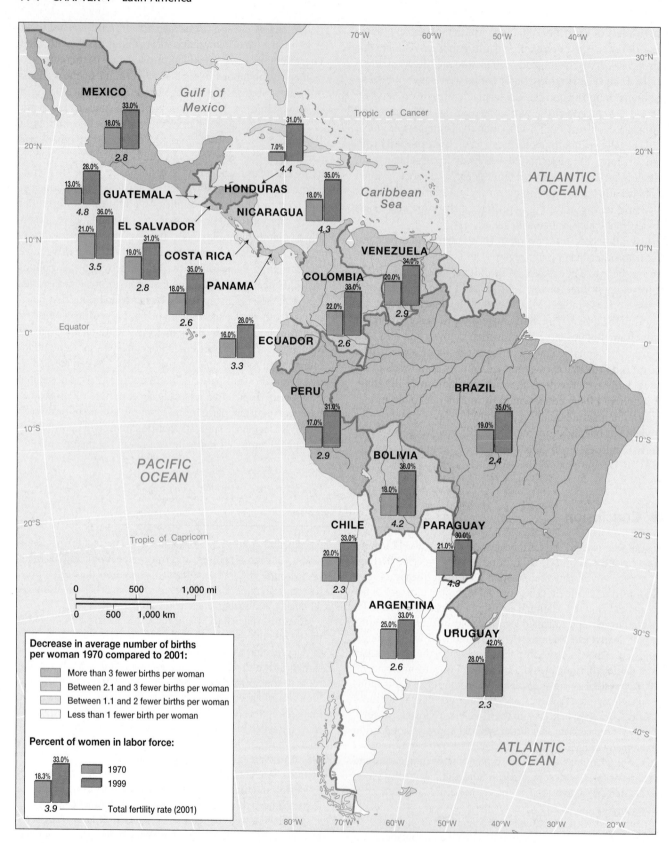

▲ Figure 4.38 Changing status of women   This map shows that 11 states have seen total fertility rates decline by 2 or more since 1970. Over the same period, the number of women in the labor force has grown dramatically (see bar graphs).   *(Sources: Population Reference Bureau,* World Population Data Sheet, *2001, for current TFR; United Nations World Fertility Patterns, 1997, for 1970 TFR; World Bank Atlas, 2001, Female Labor Force Participation; U.S. Bureau of the Census, International Data Base for 1970 Female Labor Force)*

was still under apartheid (see Chapter 6). Moreover, a number of important economies (namely, Brazil, Argentina, Colombia, and Venezuela) are faltering in the new millennium as they wrestle with problems such as currency devaluation, debt burdens, inflation, and growing unemployment. How new waves of economic turmoil will influence attitudes toward neoliberalism, globalization, and hemispheric trade remains to be seen. It does seem, however, that new political actors are emerging in Latin America—from union organizers to environmentalists—who are challenging the old ways of doing things.

⊕ Key Terms

agrarian reform (page 145)
Altiplano (page 134)
altitudinal zonation (page 138)
Columbian Exchange (page 151)
dependency theory (page 167)
dollarization (page 169)
El Niño (page 140)
encomienda (page 144)
environmental lapse rate (page 138)

Free Trade Area of the Americas (FTAA) (page 126)
Gran Colombia (page 156)
grassification (page 132)
growth poles (page 162)
import substitution (page 161)
informal sector (page 143)
latifundia (page 144)
machismo (page 154)
maquiladora (page 164)

marianismo (page 154)
megacity (page 126)
Mercosur (page 156)
mestizo (page 146)
minifundia (page 144)
neoliberal policies (page 169)
neotropics (page 127)
Organization of American States (OAS) (page 155)
rural-to-urban migration (page 144)

shields (page 127)
subnational organizations (page 158)
supranational organizations (page 158)
syncretic religions (page 152)
transnationalism (page 149)
Treaty of Tordesillas (page 156)
United Provinces of Central America (page 156)
urban primacy (page 142)

⊕ Questions for Summary and Review

1. Why is this region called *Latin America?*

2. Discuss the different environmental problems facing Mexico City. Are megacities such as Mexico City viable?

3. What is El Niño, and how does it impact Latin America and other parts of the world?

4. What is altitudinal zonation, and how is it relevant to agricultural practices?

5. What was the Columbian Exchange? Discuss ways in which it still occurs today.

6. What factors contributed to the rapid and early rate of urbanization in Latin America?

7. What are the origins of latifundia? Discuss the relationship between latifundia and minifundia.

8. Besides Iberia, what are the source countries for Latin American immigrants? What impact have these different groups had on the region?

9. Give two examples of syncretic religions. What processes create them?

10. Discuss the origins of the current political boundaries in Latin America. Where are boundary disputes occurring and why?

11. How are trade blocks reshaping international relations in Latin America?

12. Is the dependency theory useful in understanding Latin America's position in the world economy? Can the theory be applied to other regions in the developing world?

13. What is dollarization, and how is it being used in Latin America?

14. What social and demographic shifts account for improvements in the lives of women in Latin America?

15. Discuss the influence of race when measuring levels of well-being in Mexico and Brazil.

⊕ Thinking Geographically

1. Discuss the processes driving tropical deforestation in Latin America and how they compare with deforestation in other areas of the world.

2. Agrarian reform has been advocated throughout Latin America as a means to reduce social and economic inequalities. Has it worked in the region? How have land distribution programs fared? Where else in the world have agrarian reforms been attempted?

3. How is neoliberalism influencing the way Latin America interacts with the rest of the world? What are the social and en-

vironmental costs of neoliberalism? Is this a model for understanding the impact of globalization in the developing world?

4. After examining the language map, what conclusions can you draw about the patterns of Indian survival in Latin America?

5. Given the dominance of cities in this region, what particular urban environmental problems face cities in the developing world? How might Latin America's megacities use their size and density to reduce the environmental problems associated with urbanization?

6. Indigenous homelands or reservations can be a source of cultural repression or cultural survival. Compare the creation of comarcas in Panama or communal land-use zones in Honduras with Indian reservations in the United States. What different political processes are behind the creation of these indigenous territories?

7. Discuss the social, environmental, and economic consequences behind the modernization of agriculture. How does Latin America's experience with modern agricultural systems compare with North America's?

⊕ Regional Novels and Films

Novels

Isabel Allende, *House of the Spirits* (1982, Barcelona Plaza & Janes)

Jorge Amado, *Dona Flor and Her Two Husbands* (1967, Losada)

Jorge Amado, *Gabriela, Clove and Cinnamon* (1958, Biblioteca Ayacucha)

Mario Bencastro, *Odyssey to the North* (1998, Arte Público)

Jorge Luis Borges, *Labyrinths* (1962, New Directions)

Carlos Fuentes, *The Old Gringo* (1985, Fondo de Cultura Economica)

Gabriel García Marquez, *One Hundred Years of Solitude* (1967, Biblioteca Ayacucha)

Gabriel García Marquez, *Love in the Time of Cholera* (1988, Penguin)

Mario Vargas Llosa, *The Real Life of Alejandro Mayta* (1986, Farrar, Straus and Giroux)

Films

Black Orpheus (1958, Brazil)

Bolivian Blues (2000, U.S.)

Children of Rio (1990, Brazil)

Coffee: A Sack Full of Power (1997, France)

El Norte (1983, U.S.)

Geraldo Off-Line (2000, U.S.)

Like Water for Chocolate (1993, U.S.)

Lines of Blood: The Drug War of Colombia (1992, U.S.)

The Posse (2000, U.S.)

Traffic (2000, U.S.)

Transnational Fiesta (1992, U.S.)

Trinkets & Beads (1996, U.S.)

⊕ Bibliography

Arreola, Daniel D., and Curtis, James R. 1993. *The Mexican Border Cities, Landscape Anatomy and Place Personality*. Tucson, AZ: University of Arizona Press.

Bebbington, Anthony. 2000. "Reencountering Development: Livelihood Transitions and Place Transformations in the Andes." *Annals of the Association of American Geographers* 90(3), 495–520.

Brown, Ed. 1996. "Articulating Opposition in Latin America: The Consolidation of Neoliberalism and the Search for Radical Alternatives." *Political Geography* 15(2), 169–92.

Butzer, Karl W., guest ed. 1992. "The Americas Before and After 1492: Current Geographical Research." *Annals of the Association of American Geographers* 82(3), 343–568.

Caviedes, César, and Knapp, Gregory. 1995. *South America*. Englewood Cliffs, NJ: Prentice-Hall.

Clapp, Roger Alex. 1998. "Waiting for the Forest Law: Resource-Led Development and Environmental Politics in Chile." *Latin American Research Review* 33(2), 3–36.

Clawson, David L. 2000. *Latin America and the Caribbean, Lands and Peoples*, 2nd ed. Boston: McGraw-Hill.

Crosby, Alfred. 1972. *The Columbian Exchange, Biological and Cultural Consequences of 1492*. Westport, CT: Greenwood Press.

Dean, Warren. 1995. *With Broadax and Firebrand: The Destruction of the Brazilian Atlantic Forest*. Berkeley, CA: University of California Press.

Denevan, William M. 1992. *The Native Population of the Americas in 1492*, 2nd ed. Madison, WI: University of Wisconsin Press.

de Soto, Hernando. 1989. *The Other Path: The Invisible Revolution in the Third World*. New York: Harper & Row.

Dodds, Klaus-John. 1993. "Geopolitics, Cartography and the State in South America." *Political Geography* 12, 361–81.

Ezcurra, Exequiel, and Mazari-Hiriart, Marisa. 1996. "Are Megacities Viable? A Cautionary Tale from Mexico City." *Environment* 38(1), 6–15, 26–35.

Gilbert, Alan. 1998. *The Latin American City*, 2nd ed. New York: Monthly Review Press.

Godfrey, Brian. 1992. "Modernizing the Brazilian City." *Geographical Review* 81(1), 18–34.

Hays-Mitchell, Maureen. 1994. "Streetvending in Peruvian Cities: The Spatio-Temporal Behavior of Ambulantes." *The Professional Geographer* 46(4), 425–38.

Hecht, Susanna, and Cockburn, Alexander. 1989. *The Fate of the Forest: Developers, Destroyers and Defenders of the Amazon*. London: Verson Press.

Herlihy, Peter H. 1989. "Panama's Quiet Revolution: Comarca Homelands and Indian Rights." *Cultural Survival Quarterly* 13(3), 17–24.

Korzeniewicz, Roberto Patricio, and Smith, William C. 2000. "Poverty, Inequality and Growing in Latin America: Searching for the High Road to Globalization" *Latin American Research Review* 35(3), 7–54.

Lovell, W. George. 1992. "Heavy Shadows and Black Night: Disease and Depopulation in Colonial Spanish America." *Annals of the Association of American Geographers* 82(3), 426–43.

Mahone, James E., Jr. 1999. "Economic Crisis in Latin America: Global Contagion, Local Pain." *Current History* 98(626), 105–10.

Mörner, Magnus. 1985. *The Andean Past, Land, Societies and Conflicts.* New York: Columbia University Press.

Place, Susan E., ed. 1993. *Tropical Rainforests, Latin American Nature and Society in Transition.* Wilmington, DE: Scholarly Resources, Inc.

Preston, David. 1996. *Latin American Development, Geographical Perspectives,* 2nd ed. Essex, England: Longman Scientific & Technical.

Price, Marie. 1994. "Ecopolitics and Environmental Nongovernmental Organizations in Latin America." *Geographical Review* 84(1), 42–58.

Rabinovitch, Jonas. 1997. "A Success Story of Urban Planning: Curitiba." In *Cities Fit for People*, edited by Üner Kirdar. New York: United Nations.

Sánchez-Albornoz, Nicolás. 1974. *The Population of Latin America: A History.* Berkeley, CA: University of California Press.

Slater, David. 1993. "The Geopolitical Imagination and the Enframing of Development Theory." *Transactions* 18, 419–37.

Sluyter, Andrew. 2001. "Colonialism and Landscape in the Americas: Material Conceptual Transformations and Continuing Consequences." *Annals of the Association of American Geographers* 91(2), 410–28.

Smith, Nigel J. H. 1996. *The Enchanted Amazon Rain Forest: Stories from a Vanishing World.* Gainesville, FL: University Press of Florida.

Tenenbaum, Barbara, ed. 1996. *Encyclopedia of Latin American History and Culture.* New York: Charles Scribner's Sons.

Thrupp, Lori Ann. 1995. *Bittersweet Harvests for Global Supermarkets: Challenges in Latin America's Agricultural Export Boom.* Washington, DC: World Resources Institute.

Voeks, Robert. 1997. *Sacred Leaves of Candomblé: African Magic, Medicine and Religion in Brazil.* Austin: University of Texas Press.

West, Robert, and Augelli, John P. 1989. *Middle America, Its Lands and Peoples,* 3rd ed. Englewood Cliffs, NJ: Prentice Hall.

Williams, Robert G. 1994. *States and Social Evolution, Coffee and the Rise of National Governments in Central America.* Chapel Hill, NC: University of North Carolina Press.

Winn, Peter. 1992. *Americas: The Changing Face of Latin America and the Caribbean.* Berkeley, CA: University of California Press.

Young, Emily. 1999. "Local People and Conservation in Mexico's El Vizcaíno Biosphere Reserve." *Geographical Review* 89(3), 364–90.

Young, Kenneth R. 1996. "Threats to Biological Diversity Caused by Coca/Cocaine Deforestation in Peru." *Environmental Conservation* 23(1), 7–15.

Zimmerer, Karl. 1996. *Changing Fortunes: Biodiversity and Peasant Livelihood in the Peruvian Andes.* Berkeley, CA: University of California Press.

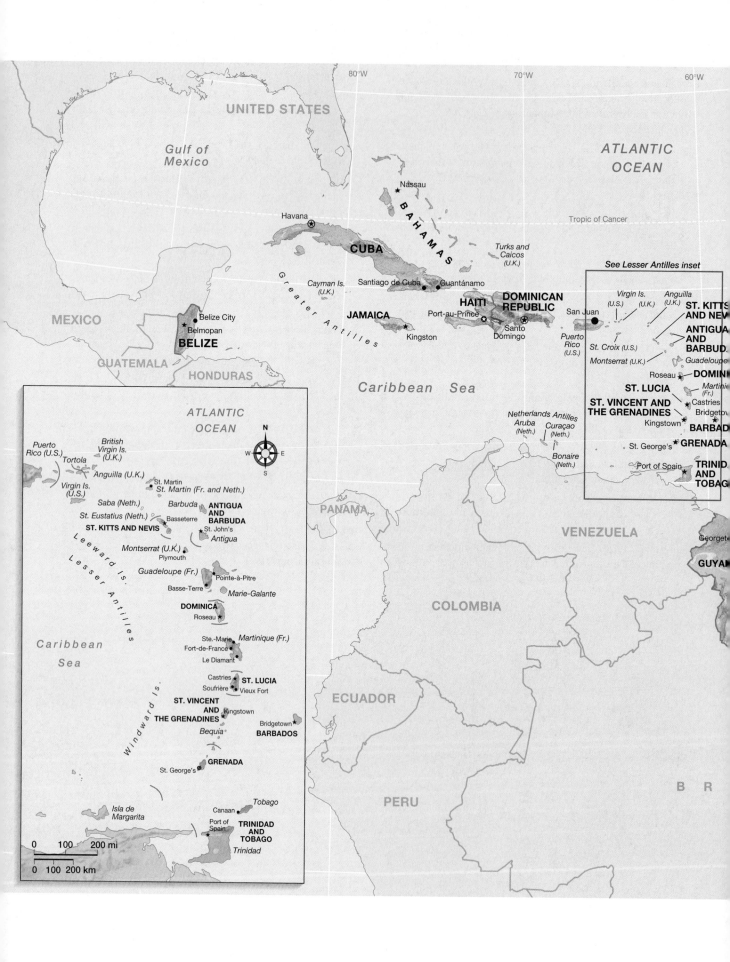

UNITED STATES

Gulf of
Mexico

ATLANTIC
OCEAN

Nassau

Havana

B A H A M A S

Tropic of Cancer

CUBA

Turks and
Caicos
(U.K.)

See Lesser Antilles inset

Cayman Is.
(U.K.)

Santiago de Cuba   Guantánamo

*Virgin Is.
(U.S.)*   *Anguilla
(U.K.)*   ST. KITTS
AND NEV

MEXICO

Belize City
Belmopan
BELIZE

JAMAICA

Kingston

HAITI
Port-au-Prince

DOMINICAN
REPUBLIC

Santo
Domingo

San Juan

*Puerto
Rico
(U.S.)*

St. Croix (U.S.)

Montserrat (U.K.)

ANTIGUA
AND
BARBUD.

Guadeloupe

Roseau   DOMINI

ST. LUCIA   *Martini
(Fr.)*

ST. VINCENT AND
THE GRENADINES   Castries
Bridgetow

Kingstown   BARBAD

GRENADA

GUATEMALA

HONDURAS

Greater Antilles

Caribbean   Sea

*Netherlands Antilles
Aruba
(Neth.)*   *Curaçao
(Neth.)*

*Bonaire
(Neth.)*

PANAMA

St. George's

Port of Spain   TRINID
AND
TOBAG

VENEZUELA

Georget

GUYA

COLOMBIA

ECUADOR

B   R

PERU

ATLANTIC
OCEAN

N
W   E
S

*Puerto
Rico (U.S.)*

*British
Virgin Is.
(U.K.)*

Tortola

Anguilla (U.K.)

*Virgin Is.
(U.S.)*

St. Martin
St. Martin (Fr. and Neth.)

Saba (Neth.)   *Barbuda*   ANTIGUA
AND
BARBUDA

St. Eustatius (Neth.)   Basseterre

ST. KITTS AND NEVIS   St. John's
Antigua

Leeward Is.

Lesser Antilles

Montserrat (U.K.)
Plymouth

Guadeloupe (Fr.)
Basse-Terre   Pointe-à-Pitre

Marie-Galante

*Caribbean
Sea*

DOMINICA
Roseau

Ste.-Marie   *Martinique (Fr.)*
Fort-de-France
Le Diamant

Castries   ST. LUCIA
Soufrière   Vieux Fort

ST. VINCENT
AND
THE GRENADINES   Kingstown

Windward Is.

Bridgetown
BARBADOS

Bequia

GRENADA
St. George's

*Isla de
Margarita*

Tobago

Canaan

Port of
Spain   TRINIDAD
AND
TOBAGO

Trinidad

0   100   200 mi

0   100   200 km

CHAPTER 5

The Caribbean

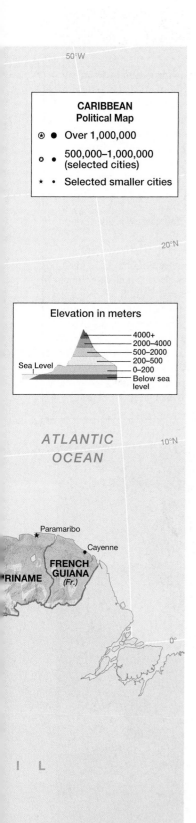

**CARIBBEAN
Political Map**

⊗ ● Over 1,000,000

✿ ● 500,000–1,000,000
(selected cities)

★ • Selected smaller cities

Elevation in meters

4000+
2000–4000
500–2000
200–500
0–200
Below sea level

Sea Level

ATLANTIC
OCEAN

20°N

10°N

0°

50°W

Paramaribo

Cayenne

FRENCH
GUIANA
(Fr.)

RINAME

I L

| 0 | 200 | 400 mi |
| 0 | 200 | 400 km |

The Caribbean was the first region of the Americas to be extensively explored and colonized by Europeans. Yet its modern regional identity is ambiguous, often merged with Latin America but also viewed as apart from it (see "Setting the Boundaries"). Today the region is home to 38 million inhabitants scattered across 25 countries and dependent territories. They range from the small British dependency of the Turks and Caicos, with 12,000 people, to the island of Hispaniola, with nearly 16 million. In addition to the Caribbean Islands, Belize of Central America and the three Guianas—Guyana, Suriname, and French Guiana—of South America are included as part of the Caribbean. For historical and cultural reasons, the peoples of these mainland states identify with the island nations and are thus included in this chapter (Figure 5.1).

Historically, the Caribbean was a battleground between rival European powers competing for territorial control of these tropical lands. In the early 1900s, the United States took over as the dominant geopolitical force in the region, regularly sending troops and maintaining what some have called a neocolonial presence. As in many developing areas, external control of the Caribbean produced highly dependent and inequitable economies. The plantation was the dominant production system and sugar the leading commodity. Over time other commodities, such as bananas, citrus, coffee, spices, timber, nickel, and bauxite, began to have economic importance, as did the international tourist industry. Increasingly, governments have sought to diversify their economies by expanding nontraditional exports (such as flowers, nuts, and processed fruits), offshore banking, and manufacturing to reduce the region's dependence on select primary exports and tourism.

Generally, when one thinks of the Caribbean, images of white sandy beaches and turquoise tropical waters come to mind. Tucked between the tropic of Cancer and the equator, with year-round temperatures averaging in the high 70s, the hundreds of islands and picturesque waters of the Caribbean have often inspired comparisons to paradise. Columbus began the tradition by describing the islands of the New World as the most marvelous, beautiful, and fertile lands he had ever known, filled with flocks of parrots, exotic plants, and friendly natives. Writers today are still lured by the sea, sands, and swaying palms of the Caribbean. Since the 1960s, the Caribbean has earned much of its international reputation as a playground for northern vacationers. There is, of course, another Caribbean that is far poorer and

◀ **Figure 5.1 The Caribbean** Named for one of its former native inhabitants—the Carib Indians—the modern Caribbean is a culturally complex and economically dependent world region. Settled by various colonial powers, most of the region's inhabitants are a mix of African, European, and South Asian peoples. Twenty-five countries and dependent territories make up the Caribbean, yet most of the population resides on the four islands of the Greater Antilles: Cuba, Hispaniola, Jamaica, and Puerto Rico. Home to 38 million people, the Caribbean is demographically larger than Australia and Oceania.

SETTING THE BOUNDARIES

This culturally diverse area of the world does not fit neatly into current world regional schemes. In fact, in most textbooks it is scarcely mentioned, being folded into Latin America or appearing as part of Middle America (along with Mexico and Central America). By recognizing the Caribbean as a distinct world region, however, one can see clearly the long-term processes of colonization and globalization that have created it.

For much of the colonial period, the islands themselves were referred to as "the Indies" or "the Spanish Main," and the sea surrounding them was called *Mar del Norte* (North Sea). It was not until the late eighteenth century that an English cartographer first applied the name "Caribbean" to the sea (a reference to the Carib Indians who once inhabited the islands). It took another 200 years for the term *Caribbean* to be generally applied to this region of islands and rimland.

The rationale for treating the Caribbean as a distinct area lies within its particular cultural and economic history. Culturally, the islands and portions of the littoral can be distinguished from the largely Iberian-influenced mainland of Latin America because of their more diverse European colonial history and much stronger African imprint. Demographically, the native population of the Caribbean was virtually elimi-

nated within the first 50 years following Columbus's arrival. Thus, unlike Latin America and North America, the Amerindians of the insular Caribbean have virtually no legacy, save a few place names. In terms of economic production, the dominance of export-oriented plantation agriculture explains many of the social, economic, and environmental patterns in the region. African slaves were thrust upon the scene as replacements for lost native labor. As Caribbean forests and savannas became sugarcane fields, a pattern was set in which exotic species of plants replaced native ones. The legacy of plantation life is still visible in the grossly uneven distribution of land and resources.

Today, organizations such as the Caribbean Common Market exist to foster the common interests of island and mainland states. Yet intraregional alliances often split along colonial lines, so that former Spanish, British, French, and Dutch colonies at times have more affinity for each other than for the region as a whole. All of the Caribbean states with the exception of Cuba are eager to join the proposed Free-Trade Area of the Americas. Given the peripheral economic status of the Caribbean and its location as the crossroads of the Americas, it is likely that the region will experience increased economic integration while maintaining its distinct cultural identity.

---

economically more dependent than the one portrayed on travel posters. Haiti, by many measures the poorest country in the Western Hemisphere, has the third largest population in the region with roughly 7 million people. The two largest countries—Cuba (11 million) and the Dominican Republic (8 million)—also suffer from serious economic problems.

The majority of Caribbean people are poor, living in the shadow of North America's vast wealth. The concept of **isolated proximity** has been used to explain the region's unusual and contradictory position in the world. The *isolation* of the Caribbean sustains the area's cultural diversity (Figure 5.2) but also explains its limited economic opportunities. Caribbean writers note that this isolation fosters a strong sense of place and an inward orientation by the people. Yet the relative *proximity* of the Caribbean to North America (and, to a lesser extent, Europe) ensures its transnational connections and economic dependence. For example, each year Dominican workers in the United States send more than $1 billion to family and friends in the Dominican Republic who rely on this money for their sustenance. History has shown the inhabitants of the Caribbean that their fate is not under their own control. Through the years, the Caribbean has evolved as a distinct but economically peripheral world region. This status expresses itself today as workers flee the region in search of employment, while foreign companies are attracted to the Caribbean for its cheap labor. The economic well-being of most Caribbean countries is precarious. Despite such uncertainty, an enduring cultural richness and attachment to place is witnessed here that may explain a growing counter-current of returnees to their homelands.

▲ **Figure 5.2 Carnival dancer**   A woman in elaborate costume celebrates Carnival in Port of Spain, Trinidad. Linking a Christian pre-Lenten celebration with African musical rhythms, Carnival has become an important symbol of Caribbean identity. *(Rob Crandall/Rob Crandall, Photographer)*

Environmental Geography: Paradise Undone

The levels of environmental degradation and poverty found in Haiti are unmatched in the Western Hemisphere. Haiti's story, albeit extreme, illustrates how social and economic inequities contribute to unsound natural resource choices, which in turn lead to more poverty. What was once considered France's richest colony—its tropical jewel—now has a per capita income of $460. Haiti's life expectancy (49 years) is the lowest in the Americas, while its levels of child malnutrition and infant mortality are the highest in the hemisphere. The explanations for Haiti's misfortune are complex, but political economy, demography, and natural resources are critical factors.

In the late eighteenth century, Haiti's plantation economy yielded vast wealth for several thousand Europeans (mostly French) who exploited the labor of half a million African slaves. During the colonial period the lowlands were cleared to make way for cane fields, but most of the mountains remained forested. The brutality and inequities of Haiti's plantation economy led to the world's first successful slave uprising, followed by Haitian independence in 1804. Even though Haiti was the second state in the Americas to achieve independence, social and economic inequities persisted. The last thing former slaves were interested in was producing sugarcane. A long period of economic isolation and internal conflict ensued that left most Haitians to fend for themselves as subsistence farmers where they planted food crops along with some coffee and cacao. In 1915 the United States sent in the marines to quell political unrest and safeguard U.S. interests in the region. U.S. troops and advisors remained for nearly 20 years, during which time they rewrote the country's laws so that foreign companies could own land. The U.S. occupation resulted in improvements in Haitian infrastructure, but it also caused increased land pressure as foreign sugar companies purchased the best lands, forcing the majority of rural peasants onto the marginal soils of the hillsides.

By the 1970s, a destructive cycle of environmental and economic impoverishment was established that Haiti still confronts. Political corruption and favoritism during the father and son Duvalier dictatorships (1957–86) benefited foreign and local elites, but saw the conditions of the rural poor worsen. More than two-thirds of Haiti's people are peasants who work small hillside plots and seasonally labor on large estates. As the population grew, people sought more land. They cleared the remaining hillsides, subdivided their plots into smaller units, and abandoned the practice of fallowing land in an effort to eke out an annual subsistence. When the heavy tropical rains came, the exposed and easily eroded mountain soils were washed away, leaving rocky and semiarable surfaces for the next season. Downstream, soil collected in irrigation ditches and dams, which negatively affected agriculture, electricity production, and water supply throughout the country. The state of the environment was further aggravated by dependence upon wood fuel. Because of their poverty and inadequate infrastructure, most Haitians use charcoal (made from trees) to cook meals and heat water. By the late 1990s it was estimated that only

▲ **Figure 5.3 Deforestation on Hispaniola** The border between Haiti and the Dominican Republic is easily seen in this aerial photo. Haitians, the poorest people in the Western Hemisphere, rely almost exclusively on firewood and charcoal for their energy needs. With so many trees removed, soils are more vulnerable to erosion, and agricultural yields tend to decline. Successful reforestation efforts have yet to be implemented. *(James Blair/NGS Image Collection)*

3 percent of Haiti remained forested. The deterioration of the resource base is evident from the air; aerial photos reveal a sharp boundary between a denuded Haiti and a forested Dominican Republic (Figure 5.3).

Although most agree that Haiti needs a major reforestation program to help restore some ecological balance, more than a decade of political turmoil has made it impossible to enact such a policy. Throughout the 1990s, military and prodemocracy forces battled for control. Eventually U.S. and UN troops were called upon in 1994 to help restore the peace. The years of political uncertainty led to greater economic isolation and environmental inaction. As the political, economic, and environmental situation worsened, many Haitians sought work on the sugar estates of the Dominican Republic and French Guiana. Others fled the region entirely, seeking work, and in some cases asylum, in North America and Europe. The levels of environmental degradation, political instability, lack of infrastructure, and population pressure have inhibited most efforts to curb the cycle of social and environmental impoverishment facing Haiti.

Environmental Issues

Ecologically speaking, it is difficult to image an area that has been so completely reworked through colonization and global trade as the Caribbean. For nearly five centuries the destruction of forests and the unrelenting cultivation of soils resulted in the extinction of many endemic Caribbean plants and animals, including various shrubs and trees, songbirds, large mammals, and monkeys. As beautiful as the Caribbean is, severe depletion of biological resources has occurred, which helps explain some of the present economic and social instability of the region (Figure 5.4). Most of the environmental problems in the region are associated with agricultural practices, soil erosion, excessive reliance upon biofuels, and water and air pollution associated with sprawling and impoverished cities.

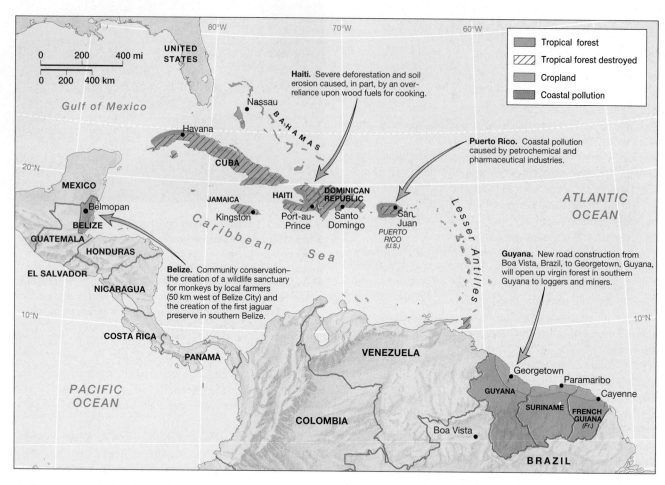

▲ **Figure 5.4 Environmental issues in the Caribbean** It is hard to imagine a region in which the environment has been so completely transformed. Most of the island forests were removed long ago for agriculture or fuel wood, and soil erosion is a chronic problem. Coastal pollution is serious around the largest cities and industrial zones. The forest cover of the rimland states, however, is largely intact and is attracting the interest of environmentalists. As tourism becomes increasingly important for the Caribbean, efforts to protect the beaches and reefs along with the fauna and flora are growing. *(Adapted from DK World Atlas, 1997, pp. 7, 55. London: DK Publishing)*

Agriculture's Legacy of Deforestation Much of this region was covered in tropical rain forests and deciduous forests prior to the arrival of Europeans. The great clearing of the Caribbean's forests began in earnest on the smaller islands of the eastern Caribbean in the seventeenth century and spread westward. The island forests fell not only to make room for sugarcane, but also to provide the fuel necessary to turn the cane juice into sugar, as well as to provide lumber for housing, fences, and ships. Primarily, however, tropical forests were removed because they were seen as unproductive; the European colonists valued cleared land. Yet unlike in the mid-latitudes, the newly exposed tropical soils easily eroded and ceased to be productive after several harvests, a situation that led to two distinct land-use strategies. On the larger islands of Cuba and Hispaniola and on the mainland, new lands were constantly cleared and older ones abandoned or fallowed (left untilled for several seasons) in an effort to keep up sugar production. On the smaller islands, such as Barbados and Antigua, where land was limited, labor-intensive efforts to conserve soil and maintain fertility were employed. In either

case, the island forests were replaced by a landscape devoted to crops for world markets.

While Haiti has lost most of its forest cover, on Jamaica and the Dominican Republic nearly 30 percent of the land is still forested. On Puerto Rico roughly one-quarter of the land has forest cover, whereas on Cuba the figure has dropped to 20 percent. Cuba has experienced a surge in charcoal production brought on by the economic and energy crises that began in 1990. Since much of the country's electricity is generated by imported fuel, locals have turned to Cuba's forests to make charcoal for domestic energy needs. A recent governmental report acknowledged that up to 60 percent of the national territory is being desertified by mechanized agriculture and a surge in charcoal production.

Managing the Rimland Forests The Caribbean **rimland** is the coastal zone of the mainland, beginning with Belize and extending along the coast of Central America to northern South America. In general, the biological diversity and stability of the rimland states are less threatened than in the rest of the Caribbean. Thus, current conservation efforts could produce important results. Even though much of Belize was selectively logged for mahogany in the nineteenth and twentieth centuries, healthy forest cover still supports a diversity of mammals, birds, reptiles, and plants. Public awareness of the negative consequences of deforestation is also greater now. In the 1980s the Coca Cola Corporation purchased thousands of acres around Dangriga, Belize, with the intention of turning forest and savanna into orange groves for juice concentrate. International protests, especially from U.S. environmentalists, forced the Atlanta-based multinational corporation to cancel its plans. Limited and locally owned orange production occurred instead. Many protected areas, most notably the first jaguar reserve in the Americas, have been established in Belize in the last decade.

The relatively pristine interior forests of Guyana are becoming a battleground between conservationists and developers. During the 1990s, the Guyanese government made the wood processing industry a priority by encouraging private investment and negotiating with companies from Malaysia and China about forest concessions and saw-mill operations. A new dry-season highway traverses the length of the country, connecting Boa Vista, Brazil, with Georgetown, Guyana. Not only does this road improve trade between Guyana and Brazil, giving Brazil access to the North Atlantic, but it also provides access to the forest and mineral resources (chiefly gold) in southern Guyana. While governments in both states are encouraged by the economic possibilities of this road, a coalition of conservationists (local and foreign) and indigenous peoples is attempting to establish a vast national park in part of this region where no commercial extraction could occur.

Failures in Urban Infrastructure The growth of Caribbean cities has produced localized environmental strains, especially water contamination and waste disposal. The urban poor are the most vulnerable to health problems associated with overly strained or nonexistent water and sewage services. According to 2001 World Bank estimates, only half of Haiti's urban population has access to improved water sources, typically through shared neighborhood faucets. The figures are better for Jamaica and the Dominican Republic, where more than 80 percent of city dwellers have access to improved water supplies. In rural areas, however, many people take their chances with surface water or collected rainwater. The expense of improving basic urban infrastructure far exceeds the capabilities of most island economies. Several dams already exist on the larger islands to supply water, while some of the smaller islands rely on expensive desalination plants. Still, existing freshwater supplies fall far short of domestic needs. As tourism, offshore manufacturing, and a growing urban population demand more water and produce more waste, Caribbean countries will be forced to make hard decisions about their infrastructure.

In addition to being a public health concern, water contamination poses serious economic problems for states dependent on tourism. Governments are caught between a desire to fix the problem before tourists notice and a tendency not to discuss it at all. Doing nothing, however, may not be an option. In Grenada, for example, ocean-dumped sewage from the capital, St. George, is killing the sea fans that, among other things, protect the island's most important tourist beach from erosion. In the more industrialized Puerto Rico, it is estimated that half the country's coastline is unfit for swimming, mostly due to contamination from sewage.

Throughout the Caribbean it is increasingly recognized that protecting the environment is not a luxury but a question of economic livelihood. Various environmental groups lobby their governments to create new laws or comply with existing ones. Although many see increased environmental awareness as a hopeful sign, both urban and rural ecosystems suffer from serious degradation. To appreciate the environmental complexity of the region, an understanding of the physical geography of the islands and rimland is essential.

The Sea, Islands, and Rimland

It is the Caribbean Sea itself, that body of water enclosed between the *Antillean* islands (the arc of islands that begins with Cuba and ends with Trinidad) and the mainland of Central and South America, that links the states of the region (Figure 5.5). Historically the sea connected people through its trade routes and sustained them with its marine resources of fish, green turtle, manatee, lobster, and crab. While the sea is noted for its clarity and biological diversity, the quantities of any one species are not great, so it has never supported large commercial fishing. The surface temperature of the sea ranges from 73° to 84 °F (23° to 29 °C), over which forms a warm tropical marine air mass that influences daily weather patterns. This warm water and tropical setting continue to be a key resource for the region as millions of tourists visit the Caribbean each year (Figure 5.6).

The arc of islands that stretches across the sea is its most distinguishing feature. The Antillean islands are divided into two groups: the Greater and Lesser Antilles. The rimland (the Caribbean coastal zone of the mainland) includes Belize and the Guianas, as well as the Caribbean littoral of Central and South America. In contrast to the islands, the rimland has low population densities and tremendous biological diversity.

Greater Antilles The four large islands of Cuba, Jamaica, Hispaniola (shared by Haiti and the Dominican Republic), and Puerto Rico make up the **Greater Antilles.** On these islands are found the bulk of the region's population, arable lands, and large mountain ranges. Given the popular interest in the Caribbean coasts, it surprises many people that Pico Duarte in the Cordillera Central of the Dominican Republic is more than 10,000 feet tall (3,000 meters), Jamaica's Blue Mountains top 7,000 feet (2,100 meters), and Cuba's Sierra Maestra is more than 6,000 feet tall (1,800 meters). The mountains of the Greater Antilles were of little economic interest since plantation owners preferred the coastal plains and valleys. Yet the mountains were an important refuge for runaway slaves and subsistence farmers, and thus figure prominently in the cultural history of the region.

The best farmlands are found in the central and western valleys of Cuba, where a limestone base contributes to the

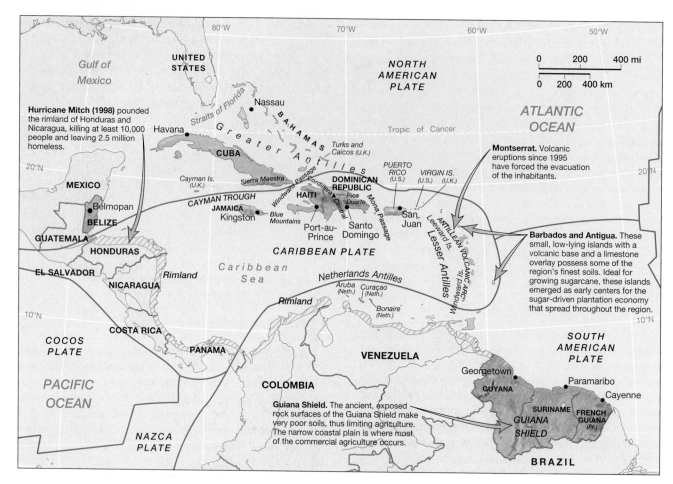

▲ **Figure 5.5 Physical geography of the Caribbean** Centered on the Caribbean Sea, the region falls neatly within the neotropical belt north of the equator. In addition to a warm year-round climate and beautiful waters, the Greater Antilles have several important mountain ranges, whereas some of the Lesser Antilles have active volcanoes. The Guianas, on the South American Plate, contain some of the oldest rock surfaces found on Earth. Cuba and Belize are on the North American Plate; these countries have flat limestone formations similar to Florida.

▲ **Figure 5.6 Caribbean Sea** Noted for its calm turquoise waters, steady breezes, and treacherous shallows, the Caribbean Sea has both sheltered and challenged sailors for centuries. This aerial photograph shows the southern Caribbean islands of Los Roques, off the Venezuelan coast. *(Rob Crandall/Rob Crandall, Photographer)*

formation of a fertile red clay soil (locally called *matanzas*) and a gray or black soil type called *rendzinas* (also found in Antigua, Barbados, and lowland Jamaica). The rendzinas soils consist of a gravelly loam with a high organic content ideal for sugar production and are actively exploited wherever found. Surprisingly, given the area's agricultural orientation, many of the soils are nutrient poor, heavily leached, and acidic. These *ferralitic* soils are found in the wetter areas where crystalline base rock exists (as in parts of Hispaniola, the Guianas, and Belize). They are characterized by heavy accumulations of red and yellow clays and offer little potential for permanent intensive agriculture.

Lesser Antilles The **Lesser Antilles** form a double arc of small islands stretching from the Virgin Islands to Trinidad. Smaller in size and population than the Greater Antilles, early on they were important footholds for rival European colonial powers. The islands from St. Kitts to Grenada form the inner arc of the Lesser Antilles. These mountainous islands, with peaks ranging from 4,000 to 5,000 feet (1,200 to 1,500 meters), have volcanic origins. In this subduction zone, the heavier North and South American plates go underneath the

Caribbean Plate, producing volcanic activity. Erosion of the island peaks and the accumulation of ash from eruptions have created small pockets of arable soils, although the steepness of the terrain places limits on agricultural development. The latest round of volcanic activity began in July 1995 on Montserrat. A series of volcanic eruptions of ash and rock have taken several lives and forced most of the island's 10,000 inhabitants to nearby islands, and even to London. Residents of Plymouth, the capital, were forced to evacuate in 1996 (Figure 5.7).

Just east of this volcanic arc are the low-lying islands of Barbados, Antigua, Barbuda, and the eastern half of Guadeloupe. Covered in limestone that overlays volcanic rock, these lands were much more inviting for agriculture. Trinidad and Tobago are on the South American Plate and consist of sedimentary rather than volcanic rock. These islands include alluvial soils and, more important, sedimentary basins that contain oil reserves.

Other island groups in the Caribbean are the Dutch islands of Aruba, Bonaire, and Curaçao (known as the ABC islands) off the coast of Venezuela; the British dependencies of the Cayman Islands and the Turks and Caicos on either side of Cuba; and the Bahamian archipelago northeast of Cuba. The Bahamas encompass hundreds of barren islands and islets as well as 20 inhabited ones. While technically the Bahamas and the Turks and Caicos are in the Atlantic Ocean and not the Caribbean Sea, convention places them as part of the Caribbean region.

Rimland States   This chapter includes the rimland states of Belize and the Guianas (the other rimland states are discussed in Chapter 4). Much of low-lying Belize is limestone. Sugarcane dominates in the drier north, while citrus is produced in the wetter central portion of the state. The Guianas, however, are characterized by the rolling hills of the Guiana Shield. The shield's crystalline rock explains the area's overall poor soil quality. Most agriculture in the Guianas occurs on the narrow coastal plain, where sugar and rice are produced. Unlike the rest of the Caribbean, these rimland territories still contain significant amounts of forest cover. In Suriname alone, 96 percent of the territory is forest or woodland. Timber continues to be an important export for these rimland states. Metal extraction (bauxite and gold) is also vital to the economies of Guyana and Suriname. French Guiana, which is an overseas territory of France, relies mostly on French subsidies but exports shrimp and timber. It is also home to the French space center at Kourou.

Climate and Vegetation

Coconut palms framed by a blue sky evoke the postcard image of the Caribbean. In this tropical region, it is warm all year and rainfall is abundant. Much of the Antillean islands and rimland receive more than 80 inches (200 centimeters) of rainfall annually and can support tropical forests. Amid the forests are pockets of naturally occurring grasslands in parts of Cuba, Hispaniola, and southern Guyana. Distinctly dry areas exist, such as the rain shadow basin in western His-

▲ **Figure 5.7 Volcanic destruction**   The eruptions of the Soufrière Hills volcano on the island of Montserrat sent mudflows and volcanic ash into the capital city of Plymouth, partially submerging a clock tower and a telephone booth in the town center. Damage was widespread on the island, and many of the 10,000 residents of this British colony had to be evacuated. Eruptions began on July 18, 1995, and are ongoing.   *(Rob Huibers/Panos Pictures)*

paniola. As explained earlier, much of the natural vegetation on the islands has been removed to accommodate agriculture and fuel needs, and only small forest fragments remain today. Tropical ecosystems in the rimland are largely intact.

As in many tropical lowlands, seasonality in the Caribbean is defined by changes in rainfall more than temperature. Although some rain falls throughout the year, the rainy season is from July to October. In Belize City and Havana, October is the wettest month; while in Bridgetown and San Juan the wettest month is November (Figure 5.8). This is when the Atlantic high-pressure cell is farthest north and easterly winds generate moisture-laden and unstable atmospheric conditions that sometimes yield hurricanes. During the slightly cooler months of December through March, rainfall declines. This time of year corresponds with the peak tourist season.

In the Guianas a different rainfall cycle is evident. These territories, on average, receive more rain than the Antillean islands. In Cayenne, French Guiana, an average of 126 inches (320 centimeters) fall each year. Unlike the Antilles, the Guianas experience a brief dry period in late summer (September to October). Also, January tends to be a wet period for the mainland, while it is a dry time for the islands. This difference is due to the influence of the intertropical convergence zone (ITCZ) on equatorial South America. The ITCZ is a circulation of equatorial air masses and winds. Within the zone, converging trade winds are forced upward by equatorial warming, creating active cloud bands and precipitation. The ITCZ shifts to the north in winter and the south in summer, causing variations in rainfall. Climatically, the Guianas are also distinguishable from the rest of the region because they are not affected by hurricanes.

Hurricanes   Each year several **hurricanes** form, pounding the Caribbean as well as Central and North America with heavy rains and fierce winds. Beginning in July, westward-moving low-pressure disturbances form off the coast of West

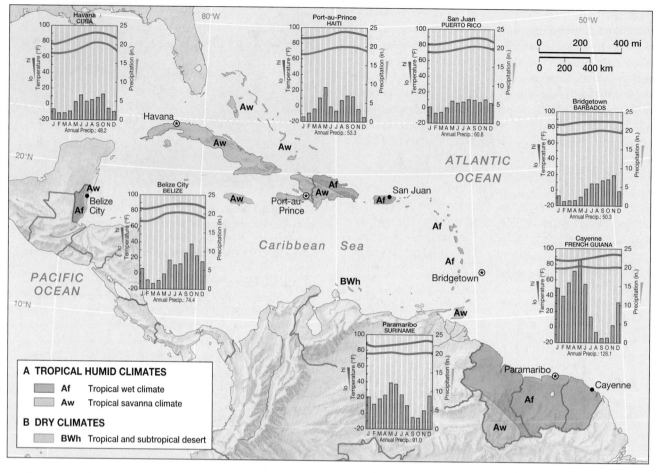

▲ **Figure 5.8 Climate map of the Caribbean**   Most of the region is classified as having either a tropical wet (Af) or tropical savanna (Aw) climate. Temperature varies little across the region, with highs slightly above 80 degrees and lows around 70 degrees. Important differences in total rainfall and the timing of the dry season distinguish different places. In the Guianas, for example, the dry season is September through October, whereas the drier months for the islands are December through March.   *(Temperature and precipitation data from* The World Weather Guide *(1984) by E. A. Pearce and C. G. Smith. London: Hutchinson.)*

Africa, picking up moisture and speed as they move across the Atlantic. The air masses are usually no more than 100 miles across, but to achieve hurricane status they must reach velocities of more than 75 miles per hour. Hurricanes may take several paths through the region, but they typically enter through the Lesser Antilles. They then arc north or northwest and collide with the Greater Antilles, Central America, Mexico, or southern North America before moving to the northeast and dissipating in the Atlantic Ocean. The hurricane zone (Figure 5.9) lies just north of the equator on both the Pacific and Atlantic sides of the Americas. Typically a half dozen to a dozen hurricanes form each season and move through the region, causing limited damage.

There are, of course, exceptions, and most long-time residents of the Caribbean have felt the full force of at least one major storm in their lifetime. The destruction caused by these storms is not just from the high winds, but also from the heavy downpours that can cause severe flooding and deadly coastal tidal surges. In 1998 the torrential rains of Hurricane Mitch, the most deadly tropical storm in a century, resulted in the death

of at least 10,000 people in Honduras, Nicaragua, and El Salvador. Mud slides and flooding ravaged structures and roads, leaving an estimated 2.5 million people homeless. While Hurricane Mitch largely bypassed the Antilles, in 1988 Hurricane Gilbert took 260 lives and pounded Jamaica and the Yucatán before slamming into Texas. Eighty percent of the houses in Jamaica lost their roofs. Plantations of coconut palms on the Yucatán were leveled like matchsticks. Hurricane Hugo came the following year, leaving nearly everyone in Montserrat homeless, wiping out the infrastructure and tourist economy of St. Croix, and damaging Puerto Rico to such an extent that troops from the mainland were sent in to help restore order.

Modern tracking equipment has improved hurricane forecasting and reduced the number of fatalities, primarily by evacuation of threatened areas. Forecasting cannot reduce the damage to crops, forests, or infrastructure. A badly timed storm can destroy a banana harvest or shut down resorts for a season or more. The sheer force of a storm can radically transform the landscape as well, turning a palm-strewn sandy beach into a treeless rocky shore covered with debris.

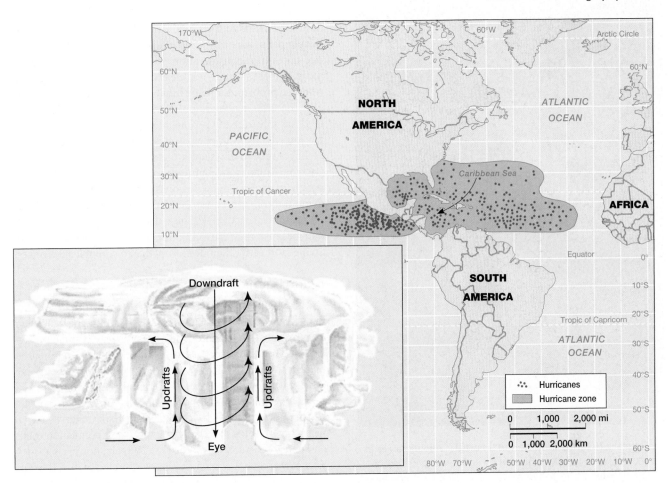

▲ **Figure 5.9 Hurricanes**   From July until November each year, people of the Caribbean know that several hurricanes will hit, some with deadly consequences. These tropical disturbances bring heavy rains and winds (often more than 100 miles per hour). They rip roofs off houses, destroy crops, and flood settlements. As Hurricane Mitch showed in 1998, the worst storms kill thousands of people and devastate a country's infrastructure.   *(Modified from McKnight, 1996, Physical Geography, Upper Saddle River, NJ: Prentice Hall)*

Forests, Savannas, and Mangroves   Most of the region is a tropical, wet climate that can support forests. Today tropical forests are almost exclusively found in the rimland, the Guianas and Belize, and even these areas are increasingly under pressure. As for the rest of the Caribbean, small pockets of forest remain broken by fields and savannas. On the island of Puerto Rico, for example, the small El Yunque rain forest offers a glimpse of the Caribbean's forested past. Protected by the Spanish in the nineteenth century, today El Yunque is the smallest national forest in the U.S. system, but also one of the most diverse (Figure 5.10).

The palm savannas are important biomes in the Caribbean as well, found mostly in the tropical savanna (Aw) zones. The palm savannas of Hispaniola and Cuba are quite fertile and easily adapted to agriculture. In central Cuba these natural grasslands studded with palms have the best soils in the entire country and are mostly planted in sugarcane and citrus. The savanna soils of southern Guyana, however, are acidic with little agricultural potential.

For centuries the coastal mangrove swamps were largely left alone. This wet environment is poorly suited to human settlement, although it is a vital nursery for young crustaceans and fish. Found throughout the Caribbean, especially on the calmer leeward shores, the tangled stands of woody mangrove are regularly cleared to create open beaches. Mangroves also decline when disturbances upstream increase the silt load in the water. The removal of mangrove, besides eliminating a vital marine habitant, exposes coasts to increased erosion.

Arid Zones   Semiarid vegetation of thorn-scrub brush and even cactus exists in the rain shadow of the Antillean mountains and in the generally drier leeward sides of the islands, where comparatively less rain falls (20 to 40 inches annually). These scrublands have limited agricultural potential unless irrigation is introduced or a drought-resistant crop such as sisal or agave is planted. Often subsistence grazing of goats occurs in this ecological zone. Aruba, Bonaire, and Curaçao, as well as Anguilla and the Cayman Islands, typify the shallow soils and sparse vegetation found in these arid areas. With the climate too dry to support agriculture, the colonial economy hobbled along by producing salt and raising goats. It wasn't until the 1960s that the dry climate, white sands, and tropical reefs facilitated the transformation of these islands into world-class resorts.

▲ **Figure 5.10 El Yunque National Forest** A tourist takes in the scene of a waterfall surrounded by lush forest in El Yunque National Forest. An important tourist destination for Puerto Rico, this small national forest is a protected remnant of the extensive tropical forests that once existed across the Antilles. *(Robert Fried/Stock Boston)*

Population and Settlement: Densely Settled Islands and Rimland Frontiers

In the Caribbean, the population density is generally quite high and, as in neighboring Latin America, increasingly urban. Eighty-six percent of the region's population is concentrated on the four islands of the Greater Antilles (Figure 5.11). Add to this Trinidad's 1.3 million and Guyana's 700,000, and most of the population of the Caribbean is accounted for by six countries and one U.S. territory (Puerto Rico). Of these, Puerto Rico has the highest population density, with 1,139 people per square mile (439 people per square kilometer), whereas Jamaica has 624 people per square mile (241 people per square kilometer) and Hispaniola averages over 500 per square mile (roughly 200 people per square kilometer).

In absolute terms, few people inhabit the Lesser Antilles; nevertheless, some of these microstates are densely settled. The small island of Barbados is an extreme example. With only 166 square miles (430 square kilometers) of territory, it has 1,620 people per square mile (625 people per square kilometer). Population densities on St. Vincent, Martinique, and the Netherlands Antilles, while not as high, are still over 700 people per square mile (270 people per square kilometer). If one takes into consideration the scarcity of arable land on some of these islands, it is clear that access to land is a basic resource problem for many inhabitants of the Caribbean. The growth in the region's population coupled with its scarcity of land has forced many people into the cities or abroad. It also has forced many Caribbean states to be net importers of food.

In contrast to the islands, the mainland territories of Belize and the Guianas are lightly populated; Guyana averages 8 people per square mile (3 people per square kilometer), Suriname only 7, and Belize 29 (2.6 and 11 people per square kilometer, respectively). These areas are sparsely settled, in part because the relatively poor quality and accessibility of arable land made them less attractive to colonial enterprises.

Demographic Trends

During the years of slave-based sugar production, mortality rates were extremely high due to disease, inhumane treatment, and malnutrition. Consequently, the only way population levels could be maintained was through the continual importation of African slaves. With the end of slavery in the mid to late nineteenth century and the gradual improvement of health and sanitary conditions on the islands, natural population increase began to occur. In the 1950s and 1960s many states achieved peak growth rates of 3.0 or higher, causing population totals and densities to soar. Over the past 20 years, however, growth rates have come down or stabilized. As noted earlier, the current population of the Caribbean is 38 million. However, the population is now growing at an annual rate of 1.3 percent, and projected population in 2025 is 48 million (Table 5.1).

Fertility Decline The most significant demographic trend in the Caribbean is that of fertility decline (Figure 5.12). Cuba and Barbados have the region's lowest rates of natural increase (Table 5.1). In socialist Cuba the education of women combined with the availability of birth control and abortion means that the average woman has 1.6 children (compared to 2.1 in the United States). In capitalist Barbados similar results have been achieved. Population pressures in Barbados have also been eased by a steady out-migration of young Barbadians overseas, especially to Great Britain. The decision to emigrate along with a preference for smaller families have contributed to slower growth rates throughout the region. Even states with relatively high TFR rates, such as Haiti, have seen a decline in family size. Haiti's total fertility rate fell from 6.0 in 1980 to 4.7 in 2001.

The Rise of HIV/AIDS The rate of HIV/AIDS infection in the Caribbean is more than three times that in North America, making the disease an important regional issue. Although nowhere near the infection rates in Sub-Saharan Africa (see Chapter 6), slightly over 2 percent of the Caribbean population between the ages of 15 and 49 has HIV/AIDS. The highest rates are in some of the most populated islands. In Haiti, one of the earliest locations where AIDS was detected, more than 5 percent of the population between the ages of 15 and

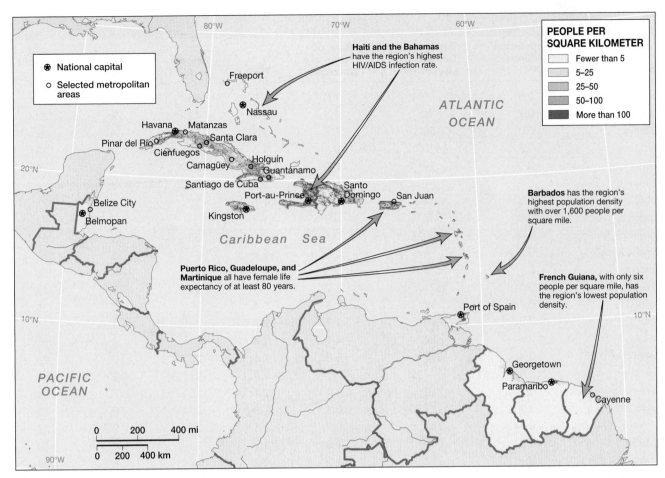

▲ **Figure 5.11 Population of the Caribbean**   The major population centers are on the islands of the Greater Antilles. The pattern here, like the rest of Latin America, is a tendency toward greater urbanism. The largest city of the region is Santo Domingo, followed by Havana. In comparison, the rimland states are very lightly settled. All of Suriname has just 400,000 people, and Belize has fewer than 300,000.

49 is infected with the virus. The Dominican Republic infection rate has climbed to nearly 3 percent; Guyana also had a 3 percent infection rate as of 1999. AIDS is already the largest single cause of death among young men in the English-speaking Caribbean.

There is a relationship between HIV/AIDS transmission, international tourism, and prostitution. While many Caribbean countries do not keep statistics on infection rates, those that do have seen infection rates on the rise. The Bahamas, for example, has an estimated HIV/AIDS infection rate among the 15- to 49-year-old population of 4 percent. Officials there have introduced drug treatments to help prevent mother-to-child transmission. Cuba, which witnessed a surge in both tourism and prostitution in the 1990s, claims to have an infection rate of only 0.2 percent among its 15- to 49-year-old population. Education programs and an effective screening and reporting system for the disease have apparently kept the infection rate down in Cuba.

Emigration   Driven by the region's limited economic opportunities, a pattern of emigration to other Caribbean islands, North America, and Europe began in the 1950s. For more than 50 years a **Caribbean diaspora**—the economic flight of Caribbean peoples across the globe—has defined existence and identity for much of the region (Figure 5.13). Barbadians generally choose England, most settling in the London suburb of Brixton with other Caribbean immigrants. In contrast, one out of every three Surinamese has moved to the Netherlands, with most residing in Amsterdam. As for Puerto Ricans, only slightly more live on the island than reside on the U.S. mainland. In the 1980s roughly 10 percent of Jamaica's population legally immigrated to North America (some 200,000 to the United States and 35,000 to Canada). Cubans have made the city of Miami their destination of choice since the 1960s. Today they are the majority of that city's population.

Intraregional movements are also important. While perhaps one-fifth of all Haitians do not live in their country of birth, the most common destination is the neighboring Dominican Republic, followed by the United States, Canada, and French Guiana. Dominicans also leave; the vast majority come to the United States, settling in New York. Others, however, simply cross the Mona Passage and settle in Puerto Rico. The economic implications of this labor-related migration are significant and will be discussed later.

TABLE 5.1 *Demographic Indicators*

Country	Population (Millions, 2001)	Population Density, per square mile	Rate of Natural Increase	TFR[a]	Percent < 15[b]	Percent > 65	Percent Urban
Anguilla	0.01	—	1.0	1.8	26	7	—
Antigua and Barbuda	0.1	394	1.6	2.4	28	8	37
Bahamas	0.3	58	1.5	2.4	31	6	84
Barbados	0.3	1620	0.5	1.6	23	9	38
Belize	0.3	29	1.9	3.2	41	5	49
Cayman	0.03	—	0.9	2.1	22	8	—
Cuba	11.3	264	0.6	1.6	22	10	75
Dominica	0.1	262	0.8	1.8	33	9	71
Dominican Republic	8.6	456	2.1	3.1	35	5	61
French Guiana	0.2	6	2.3	3.4	31	5	79
Grenada	0.1	678	1.3	2.4	38	4	34
Guadeloupe	0.5	691	1.2	1.9	25	9	48
Guyana	0.7	8	1.3	2.5	31	5	36
Haiti	7.0	650	1.7	4.7	43	4	35
Jamaica	2.6	624	1.5	2.4	31	7	50
Martinique	0.4	897	0.8	1.8	23	10	93
Montserrat	0.01	—	1.0	1.9	24	12	—
Netherlands Antilles	0.2	722	1.1	2.1	26	7	70
Puerto Rico	3.9	1139	0.8	1.9	25	10	71
St. Kitts and Nevis	0.04	281	0.9	2.5	31	9	43
St. Lucia	0.2	656	1.3	2.1	33	6	30
St. Vincent and the Grenadines	0.1	757	1.2	2.2	32	6	44
Suriname	0.4	7	1.9	3	33	5	69
Trinidad and Tobago	1.3	656	0.7	1.7	26	7	72
Turks and Caicos	0.2	—	2.1	3.3	33	4	—

[a]Total fertility rate

[b]Percent of population younger than 15 years of age

Source: Population Reference Bureau. World Population Data Sheet, 2001. *Data for dependent territories from CIA World Factbook, 2000.*

Most migrants, with the exception of the Cubans, are part of a **circular migration** flow. In this type of migration, a man or woman typically leaves children behind with relatives in order to work hard, save money, and return home. Other times a **chain migration** begins, in which one family member at a time is brought over. In some cases large numbers of residents from a Caribbean town or district send migrants to a particular locality in North America. Thus, chain migration can account for the formation of immigrant enclaves.

The Rural–Urban Continuum

Initially, plantation agriculture and subsistence farming shaped Caribbean settlement patterns. Low-lying arable lands were dedicated to export agriculture and controlled by the colonial elite. Only small amounts of land were set aside for subsistence production. Over time, villages of freed or runaway slaves were established, especially in remote areas of the interior. Still the vast majority of people lived on estates as own-

ers, managers, or slaves. The cities that formed existed to serve the administrative and social needs of the colonizers, but most were small, containing a small fraction of a colony's population. The colonists who linked the Caribbean to the world economy saw no need to develop major urban centers.

Even today the structure of Caribbean communities reflects the plantation legacy. Many of the region's subsistence farmers are ancestors of former slaves who continue to work their small plots and seek seasonal wage-labor on estates. The social and economic patterns generated by slavery still mark the landscape. Rural communities tend to be loosely organized; labor is transient; and small farms are scattered wherever pockets of available land are found. Since men tend to leave home for seasonal labor, matriarchal family structures and female-headed households are common.

The construction of **houseyards** in the Lesser Antilles typifies the blending of rural subsistence, economic survival, and a matriarchal social structure. These small enclosed properties of a half acre or less include a number of dwellings, small

▲ **Figure 5.12 Fewer children**   A mother from the Bahamas with her two children. The average Caribbean woman has far fewer children now than 30 years ago. Higher levels of education, improved availability of contraception, an increase in urban living, and the large percentage of women in the labor force contribute to slower population growth rates in many countries. *(Tony Arruza/Tony Arruza Photography)*

livestock, fruit trees, herb gardens, and a protected play and work space. Typically the houseyard is owned by a woman, and often her extended family of married children lives there. It is not unusual for a woman to be associated with one yard for most of her life, while a man's connection to this domestic space is more tenuous (Figure 5.14).

Caribbean Cities   Since the 1960s, the mechanization of agriculture, offshore industrialization, and rapid population growth have caused a surge in rural-to-urban migration. Cities have grown accordingly, and today 60 percent of the region is classified as urban. Of the large countries, Cuba is the most urban (75 percent) and Haiti the least (35 percent). Caribbean cities are not large by world standards, as only four have 1 million or more residents: Santo Domingo (2.6 million), Havana (2.2 million), Port-au-Prince (1.5 million), and San Juan (1.0 million). All but Port-au-Prince were laid out by the Spanish.

Like their counterparts in Latin America, the Spanish Caribbean cities were laid out on a grid with a central plaza. Vulnerable to raids by rival European powers and pirates, these cities were usually walled and extensively fortified. The oldest continually occupied European city in the Americas is Santo Domingo in the Dominican Republic, settled in 1496, and today the largest city in the region. Havana emerged as the most important colonial city in the region, serving as a port for all incoming and outgoing Spanish galleons. Strategically situated on Cuba's north coast at a narrow opening to a natural deep-water harbor, Havana became an essential city for the Spanish empire. Consequently, Havana possesses a handsome collection of colonial architecture, especially from the eighteenth and nineteenth centuries (Figure 5.15). Only recently did Santo Domingo edge out Havana as the Caribbean's largest city.

Other colonial powers left their mark on the region's cities. For example, Paramaribo, the capital of Suriname, has been

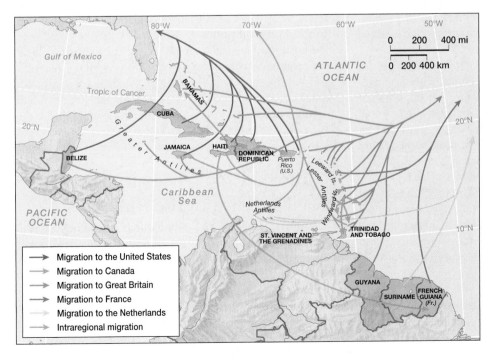

◀ **Figure 5.13 Caribbean diaspora**   Emigration has long been a way of life for Caribbean peoples. With relatively high education levels but limited professional opportunities, migrants from the region head to North America, Great Britain, France, and the Netherlands. Intraregional migrations between Haiti and the Dominican Republic or the Dominican Republic and Puerto Rico also occur. *(Data from Barry Levin, 1987,* Caribbean Exodus, *Praeger Publishers)*

▶ **Figure 5.14 Miss Joy's house-yard** Caribbean houseyards are an adaption to conditions of land scarcity, poverty, out-migration, and matriarchal family structures. Compound life offers a social network, some subsistence, and a relatively safe and inexpensive way to live. (Source: Lydia Pulsipher, 1993, "Changing Roles in the Life Cycles of Women in Traditional West Indian Houseyards," pp. 50–64, in Women and Change in the Caribbean, Janet Momsen, editor)

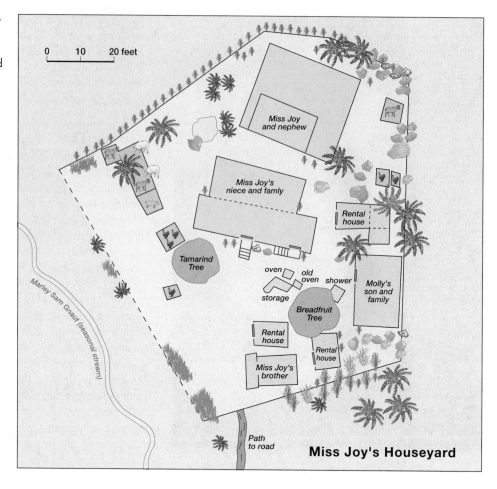

described as a tropical, tulipless extension of Holland. In the British colonies a preference for wooden whitewashed cottages with shutters was evident. Yet the British and French colonial cities tended to be unplanned afterthoughts; these port cities were built to serve the rural estates, not the other way around. Most of them have grown dramatically over the last 40 years. No longer small ports for agricultural exports, increasingly these cities are oriented to welcoming cruise ships and sun-seeking tourists.

Caribbean cities and towns do have their charms and reflect a melange of cultural influences. Throughout the region, houses are often simple structures (made of wood, brick, or stucco), raised off the ground a few feet to avoid flooding, and painted in soft pastels. Most people still get around by foot, bicycle, or public transportation; neighborhoods are filled with small shops and services that are within easy walking distance. Streets are narrow, and the pace of life is markedly slower than in North America and Europe. Even when space is tight in town, most settlements are close to the sea and its cooling breezes. An afternoon or evening stroll along the waterfront is a common activity. Flowering shrubs and swaying palms add to the tropical ambiance (see "Local Voices: Life in Belize City").

Housing The sudden surge in urbanization in the Caribbean is best explained by an erosion of rural jobs, rather than a rise in urban opportunities. Thousands poured into the cities as economic migrants, erecting shantytowns and filling the ranks of the informal sector. Squatter settlements in Port-au-Prince and Santo Domingo are especially bad, with residents living in wretched housing without the benefit of sewers and running water. Electricity is usually pirated from nearby power lines.

The one place that dramatically breaks from this pattern is Cuba. Forged in a socialist mode, Cubans are housed in uniform government-built apartment blocks like those seen throughout Russia and eastern Europe (Figure 5.16). These unimaginative complexes contained hundreds of identical one- and two-bedroom apartments where basic modern amenities (plumbing, electricity, and sewage) are provided. Compared to other large cities of the developing world, the paucity of squatter settlements makes Havana unusual.

Cultural Coherence and Diversity: A Neo-Africa in the Americas

Linguistic, religious, and ethnic differences abound in the Caribbean. A score of former European colonies, millions of descendents of ethnically distinct Africans and indentured workers from India and China, and isolated Amerindian communities on the mainland challenge any notion of cultural coherence.

Within such diversity, common historical and cultural processes provide the rudimentary glue for the region. Euro-

▲ **Figure 5.15 Old Havana**   A Cuban family strolls down a narrow cobble-stoned street near the Cathedral in Old Havana, Cuba. The best examples of eighteenth- and nineteenth-century colonial architecture in the Caribbean are found in Havana. In 1982 UNESCO declared Old Havana a World Heritage Site, and new funds became available to aid its restoration.   *(Rob Crandall/Rob Crandall, Photographer)*

▲ **Figure 5.16 Government housing**   Cubans in a state-built apartment block on the outskirts of greater Havana. Under socialism, thousands of standardized apartment blocks were built to ensure that all Cubans had access to basic modern housing. *(Rob Crandall/Rob Crandall, Photographer)*

pean colonies, with their plantation-based economies, reproduced similar social structures throughout the region. The imprint of more than 7 million African slaves, creating a neo-Africa in the Americas, is the focus of vigorous scholarly research to show the important linkages between the Caribbean and the wider Atlantic world. Last, in a process called **creolization**, African and European cultures were blended in the Caribbean. Through this mixing, European languages were transformed into vibrant local dialects, and at times entirely new languages were created (Papiamento and French Creole). This melding also produced the rich and diverse musical traditions heard throughout the world: reggae, salsa, merengue, and calypso.

The Cultural Imprint of Colonialism

The formation of European colonies in the Caribbean destroyed indigenous societies and imposed completely different social systems and cultures. Plantation-based agriculture dependent on forced and indentured labor was the clearest expression of how the colonists ordered space and society. Unlike colonies that relied on indigenous labor, Caribbean plantations depended on millions of foreign laborers.

The arrival of Columbus in 1492 triggered a devastating chain of events that depopulated the region within 50 years. A combination of Spanish brutality, enslavement, warfare, and disease reduced the densely settled islands, which supported up to 3 million Caribs and Arawaks, into an uninhabited territory ready for the colonizer's hand. The demographic collapse of Amerindian populations occurred throughout the Americas (see Chapter 4), but the death rates were highest in the Caribbean. Only fragments of Amerindian communities survive, mostly on the rimland.

By the mid sixteenth century, as rival European states vied for Caribbean territory, the lands they fought for were virtually uninhabited. In many ways this simplified their task, as they did not have to acknowledge indigenous land claims or work amid Amerindian societies. Instead, the Caribbean territories were reorganized to serve a plantation-based production system. The critical missing element was labor. Once slave labor from Africa, and later indentured labor from Asia, was secured, the small Caribbean colonies became surprisingly profitable. Much of Caribbean culture and society today can be traced to the same processes that created plantation America.

Plantation America   The term **plantation America** was coined by anthropologist Charles Wagley to designate a cultural region that extends from midway up the coast of Brazil through the Guianas and the Caribbean into the southeastern United States. Ruled by a European elite dependent on an African labor force, this society was primarily coastal and produced agricultural exports. Other characteristics included a reliance on **mono-crop production** (a single commodity, such as sugar) under a plantation system that concentrated land in the hands of elite families. Such a system engendered rigid class lines, as well as the formation of a multiracial society in which people with lighter skin were privileged. The term *plantation America* is not meant to describe a race-based division of the Americas, but rather a production system that engendered specific ecological, social, and economic relations (Figure 5.17).

The following passage describes Belize City through the eyes of a young girl, Beka Lamb. The novel *Beka Lamb* is a Belizean coming-of-age story set in the capital city in the 1960s, when Belize was still a British colony.

At times like these, Beka was glad that her home was one of those built high enough so that she could look for some distance over the rusty zinc rooftops of the town. Many of the weathered wooden houses, built fairly close together, tilted slightly as often as not, on top of pinewood posts of varying heights. In the streets, by night and by day, vendors sold, according to the season, peanuts, peppered oranges, craboos, roasted pumpkin seeds and coconut sweets under lamp-posts that also seemed at times to lean. For a while, after heavy rainstorms, water flooded streets and yards at least to the ankles.

A severe hurricane early in the twentieth century, and several smaller storms since that time had helped to give parts of the town the appearance of a temporary camp. But this was misleading, for Belizeans loved their town which lay below the level of the sea and only through force of circumstances, moved to other parts of the country. It was a town, not unlike small towns everywhere perhaps, where each person, within his neighborhood, was an individual with well known characteristics. Anonymity, though not unheard of, was rare. Indeed, a Belizean without a known legend was the most talked about character of all.

It was a relatively tolerant town where at least six races with their roots in other districts of the country, in Africa, the West Indies, Central America, Europe, North America, Asia, and other places, lived in a kind of harmony. In three centuries miscegenation, like logwood, had produced all shades of black and brown, not grey or purple or violet, but certainly there were a few people in

town known as red ibos. Creole, regarded as a language to be proud of by most people in the country, served as a means of communication amongst the races. Still, in the town and in the country, as people will do everywhere, each race held varying degrees of prejudice concerning the others.

The town didn't demand too much of its citizens, except that in good fortune they be not boastful, not proud, and above all, not critical in any unsympathetic way of the town and country.

Source: Zee Edgell, *Beka Lamb*. London: Heinemann, 1982, pp. 11–12.

▲ **Figure 5.1.1 Belize City cottage**   Residents of Belize City build their wooden cottages on stilts as protection against flooding. Shuttered cottages are typical throughout the British Caribbean.   *(Rob Crandall/Rob Crandall, Photographer)*

▶ **Figure 5.17 Sugar plantation**   A historical illustration (1823) of slaves harvesting sugarcane on a plantation in Antigua. Sugar production was profitable but arduous work. Several million Africans were enslaved and forcibly relocated into the region.   *(Michael Holford/Michael Holford Photographs)*

Asian Immigration   Before detailing the pervasive influence of Africans on the Caribbean, the lesser-known Asian presence deserves mention. By the mid nineteenth century, most colonial governments in the Caribbean had begun to free their slaves. Fearful of labor shortages, they sought **indentured labor** (workers contracted to labor on estates for a set period of time, often several years) from South and Southeast Asia.

The legacy of these indentured arrangements is clearest in Suriname, Guyana, and Trinidad and Tobago. In Suriname, a former Dutch colony, more than one-third of the population is of South Asian descent, and 16 percent are Javanese (from Indonesia). Guyana and Trinidad were British colonies, and most of their contract labor came from India. Today half of Guyana's population and 40 percent of Trinidad and Tobago's claim South Asian ancestry. Hindu temples are found in the cities and villages, and many families speak Hindi in the home (Figure 5.18). Such racial diversity has led to tension. During Guyana's general election in 2001, tensions between Afro- and Indo-Guyanese were heightened, and international election observers were called in to monitor the vote. In the end an Indo-Guyanese candidate was peacefully elected, but many fear that Guyana is teetering on the brink of serious ethnic conflict. Most of the former English colonies have Chinese populations of not more than 2 percent. Some Caribbean countries, such as the Dominican Republic, recruited Japanese immigrants as late as the 1950s. Once these East Asian immigrants fulfilled their agricultural contracts, they often became merchants and small-business owners, positions they still hold in Caribbean society.

▲ **Figure 5.18 Afro- and Indo-Guayanese politicians**
During a 2001 campaign rally, the Indo-Guayanese President, Bharrat Jagdeo and the Afro-Guayanese Prime Minister, Samuel Hinds, stand side by side. In Guyana, peoples of Indian (South Asian) and African ancestry make up the majority of the population. *(AFB/CORBIS)*

Creating a Neo-Africa

African slaves were first introduced to the Americas in the sixteenth century, partly in response to the demographic collapse of the Amerindians. The flow of slaves continued into the nineteenth century. This forced migration of Africans to the Americas was only part of a much more complex **African diaspora**—the forced removal of Africans from their native area. The slave trade also crossed the Sahara to include North Africa and linked East Africa with a slave trade in the Middle East (see Chapter 6). The best documented slave route was the transatlantic one; at least 10 million Africans landed in the Americas, and it is estimated that another 2 million died en route. More than half of these slaves were sent to the Caribbean (Figure 5.19).

This influx of slaves, combined with the extermination of nearly all the native inhabitants, recast the Caribbean as the area with the greatest concentration of African transfers in the Americas. The African source areas extended from Senegal to Angola, and slave purchasers intentionally mixed tribal groups in order to dilute ethnic identities. Consequently, intact transfer of religion and languages into the Caribbean did not occur; instead languages, customs, and beliefs were blended.

Maroon Societies Communities of runaway slaves—termed **maroons**—offer the most compelling examples of African cultural diffusion across the Atlantic. Clandestine settlements of escaped slaves existed wherever slavery was practiced. Called maroons in English, *palenques* in Spanish, and *quilombos* in Portuguese, many of these settlements were short-lived, but others have endured and allowed for the survival of African traditions, especially farming practices, house designs, community organization, and language maintenance.

The maroons of Jamaica formed dozens of isolated and independent communities in the forested mountains of the island's interior. A constant source of aggravation for the white plantation owners, these small communities endured and are

well documented. The most striking characteristic of these settlements (with names like Me No Seen and You No Come) was their ability to maintain some African language and religious practices, especially the belief in Obi (an African god) and Obeah (black magic). With the abolition of slavery in Jamaica, the political necessity for maroon villages ceased to exist, and maroons were gradually assimilated. Today villages that began as maroon settlements might look like any other rural community in Jamaica, except that there is a local appreciation for the origin of these communities.

The so-called Bush Negros of Suriname still manifest clear links to West Africa. Whereas other maroon societies gradually assimilated into their local populations, to this day the Bush Negros maintain a distinct identity. These runaways fled the Dutch coastal plantations in the seventeenth and eighteenth centuries, forming riverine settlements amid the interior rain forest. Six distinct Bush Negro tribes formed, ranging in size from a few hundred to 20,000. Clear manifestations of West African cultural traditions persist, including religious practices, crafts, patterns of social organization, agricultural systems, and even dress (Figure 5.20). Living relatively undisturbed for 200 years, these rainforest inhabitants fashioned a rich ritual life for themselves involving oracles, spirit possession, and witch doctors.

Bush Negros were familiar with Surinamese society through trade with the coastal settlements and male out-migration for seasonal work, but the engagement was on their terms. This changed in the 1960s when the ambitions of a developing state collided with Bush Negro traditions. Nearly half of Saramaka territory (one of the largest tribes) was flooded with the construction of a hydroelectric dam, and 6,000 Saramaka were forced to resettle. This incident compelled the Bush Negros to modify their isolationist ideology and become politically engaged with the state in order to defend their territorial claims. It also fostered a desire to affirm their cultural roots. In

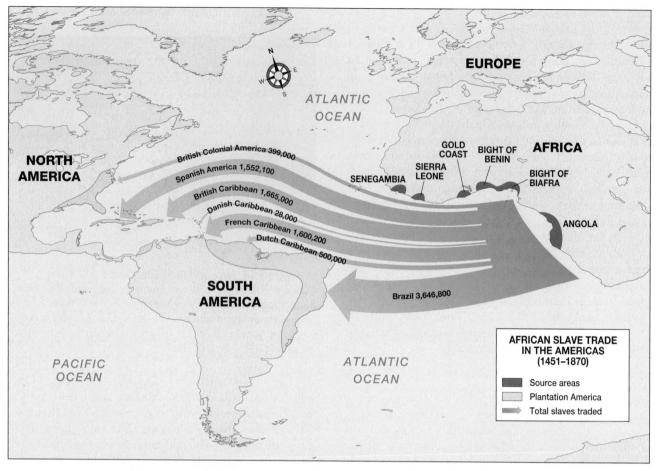

▲ Figure 5.19 Transatlantic slave trade At least 10 million Africans landed in the Americas during the four centuries in which the Atlantic slave trade operated. Most of the slaves came from West Africa, especially the Gold Coast (now Ghana) and the Bight of Biafra (now Nigeria). Angola, in southern Africa, was also an important source area. (*Data based on Philip D. Curtin, 1969*, The Atlantic Slave Trade, A Census. *Madison: University of Wisconsin Press, p. 268.*)

▼ Figure 5.20 Bush Negros A man weaves a basket in the Bush Negro settlement of Bolopasi, Suriname. The so-called Bush Negros fled Dutch plantations in the eighteenth and nineteenth centuries and settled in the interior rain forests of Suriname, where they continue many West African cultural practices. (*Martha Cooper/The Viesti Collection, Inc.*)

recognition of the cultural continuities across the Atlantic, tribal elders made a historic trip to West Africa in the 1970s to meet with various heads of state.

African Religions Linked to maroon societies, but more widely diffused, is the transfer of African religious and magical systems to the Caribbean. These patterns are another reflection of neo-Africa in the Americas and are most closely associated with northeastern Brazil and the Caribbean. In Chapter 4 we discussed how millions of Brazilians practice the African-based religions of Umbanda, Macuba, and Candomblé, along with Catholicism. Likewise, Afro-religious traditions in the Caribbean have evolved into unique forms that have clear ties to West Africa. The most widely practiced are Voodoo (also Vodoun) in Haiti, Santería in Cuba, and Obeah in Jamaica. These religions have their own priesthood and unique patterns of worship. Their impact is considerable; the father and son dictators of Haiti, the Duvaliers, were known to hire Voodoo priests to scare off government opposition. Moreover, as Figure 5.21 shows, many of these religions have diffused from their areas of origin. Santería is practiced in Florida and New York by some Cuban immigrants. Likewise, belief in Obeah diffused when Jamaicans migrated to Panama and Los Angeles.

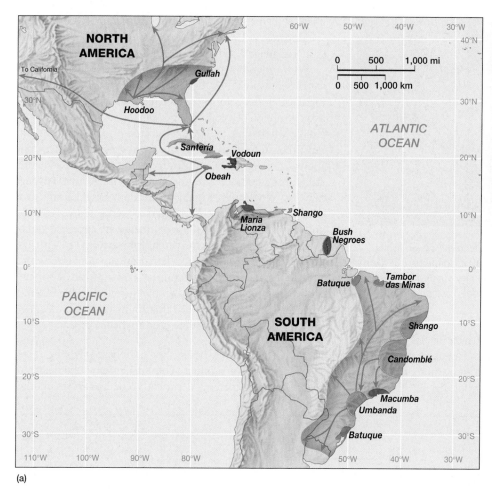

(a)

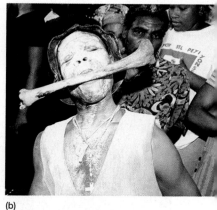

(b)

◀ **Figure 5.21 African religious influences** (a) African religious practices are found in the Americas, where large concentrations of slaves existed. Practitioners of such religions as Voodoo (Vodoun), Santería, Obeah, and Shango often mix their beliefs with Christianity. (b) A voodoo practitioner in a trance during a ceremony honoring the God of the Dead in a Port-au-Prince cemetery. [Source: (a) Robert Voeks, 1993, "African Medicine and Magic in the Americas," Geographical Review, 83(1), 66-78; (b) AFB/CORBIS].

Creolization and Caribbean Identity

Creolization refers to the blending of African, European, and even some Amerindian cultural elements into the unique sociocultural systems found in the Caribbean. Early anthropologists tended to shun the Caribbean as "cultureless," since an indigenous society had been replaced with fragments of European and African traditions. Yet the Creole identities that have formed over time are complex; they illustrate the dynamics of identity formation and social change that challenge static interpretations of culture. Today Caribbean writers (V. S. Naipaul, Derek Walcott, and Jamaica Kinkaid), musicians (Bob Marley, Celia Cruz, and Juan Luís Guerra), and artists (Trinidadian costume designer Peter Minshall) are internationally regarded. Collectively these artists are representative of their individual islands and of Caribbean culture as a whole.

The story of the Garifuna people illustrates creolization at work. Settled along the Caribbean rimland from the southern coast of Belize to the northern coast of Honduras, the Garifuna (formerly called the *Black Carib*) are descendants of African slaves who speak an Amerindian language. Unions between Africans and Carib Indians on the island of St. Vincent produced an ethnic group that was predominantly African but spoke an Indian language. In the late eighteenth century Britain forcibly resettled some 5,000 Garifuna from St. Vin-

cent to the Bay Islands in the Gulf of Honduras. Over time the Garifuna settled along the Caribbean coast of Central America, living in isolated fishing communities from Honduras to Belize (Figure 5.22). In addition to maintaining an Indian language, the Garifuna are the only group in Central America who regularly eat bitter manioc—a root crop common in lowland tropical South America. It is assumed that they acquired their taste for manioc from their exposure to Carib culture. The Afro-Indian blend that the Garifuna manifest is unique, but the process of creolization is recognizable throughout the Caribbean, especially in language and music.

Language   The dominant languages in the region are European: Spanish (24 million speakers), French (8 million), English (6 million), and about half a million Dutch (Figure 5.23). Yet these figures tell only part of the story. In Cuba, the Dominican Republic, and Puerto Rico, Spanish is the official language, and it is universally spoken. As for the other countries, colloquial variants of the official language exist, especially in spoken form, that can be difficult for a nonnative speaker to understand. In some cases completely new languages emerge; in the ABC islands, Papiamento (a trading language that blends Dutch, Spanish, Portuguese, English, and African languages) is the *lingua franca,* with usage of

▲ **Figure 5.22 Garifuna girl** A Garifuna girl sits on fishing nets in Batalla Village on the north coast of Honduras. The Garifuna are mostly of African ancestry, but they speak an Amerindian language. *(Rob Crandall/Rob Crandall, Photographer)*

guage of the street, the home, and oral tradition. Most Haitians speak *patois*, but only the formally educated know French (see "Language and Identity: Caribbean English").

With independence in the 1960s, Creole languages became politically and culturally charged with national meaning. During the colonial years Creole was negatively viewed as a corruption of standard European forms. Those who spoke Creole where considered uneducated or backward. As linguists began to study these languages, they found that while the vocabulary came from Europe, the syntax or semantic structure had other origins, notably from the African language families. While most formal education is taught using standard language forms, the richness of vernacular expression and its ability to instill a sense of identity are appreciated. Locals rely on their ability to switch from standard to vernacular forms of speech. Thus, a Jamaican can converse with a tourist in standard English and then switch to a Creole variant when a friend walks by, effectively excluding the outsider from the conversation. While this ability to switch is evident in many cultures, it is widely used in the Caribbean.

Dutch declining. Similarly, French Creole or *patois* in Haiti has constitutional status as a distinct language. In practice, French is used in higher education, government, and the courts, but *patois* (with clear African influences) is the lan-

Music The rhythmic beats of the Caribbean might be the region's best-known product. This small area is the hearth of reggae, calypso, merengue, rumba, zouk, and scores of other

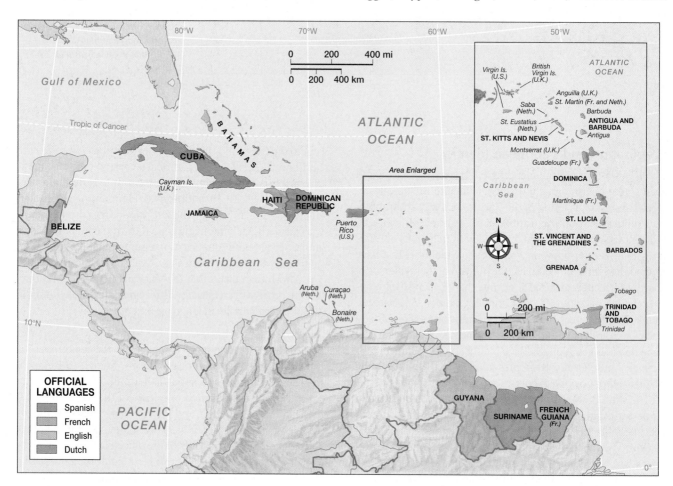

▲ **Figure 5.23 Caribbean language map** Since this region has no significant Amerindian population (except on the mainland), the dominant languages are European: Spanish (24 million), French (8 million), English (6 million), and Dutch (0.5 million). However, many of these languages have been creolized, making it difficult for outsiders to understand them.

LANGUAGE AND IDENTITY   Caribbean English

Caribbean English reflects a diverse blend of influences. Lexicographers study the various loan words from African, Spanish, French, Hindi, and Amerindian languages to try to explain the origins of certain terms. Expressions unique to the region are also based on a shared history and often are politically charged. In Belize and Jamaica a *charley-price* is a very large rat. The term is credited to Sir Charles Price, an eighteenth-century politician and planter who is believed to have introduced the rat to Jamaica.

Besides unique terms, patterns of speech distinguish spoken Caribbean English. For example, repeating words makes the intensity of expression evident: *big big big, stupid stupid, fraidy-fraidy*. Or *he talk-talk till I get weary*. Typically these English Caribbean phrasings are part of a spoken vernacular that readily separates locals from outsiders; it is seldom written.

The following terms and gestures reflect the blending of African and European words that make up Caribbean Creole.

backra: A white person of authority, usually a man. A loan word traced to the Efik language in eastern Nigeria. The Efik were established middlemen in the slave trade and probably originated the term. *Backra land* refers to England. *Example:* Le[t] me go and do de backra wo[r]k.

jook: To poke, stab, or wound; to prick or pierce the skin; or to give an infection. This word has phonic correlates in various African languages, including Hausa, Fulani, Mende, and Tsonga. However, European sailors might have reinforced the spread of this word. *Example:* He start jooking me under me arm with his hands and asking if I know him.

joukoutoo: An insignificant, unschooled, or of-no-account person. Used in Grenada, this is probably from French Creole. *Example:* But you dam[ned] rude! Joukoutoo want to wash yo[ur] mout[h] [u]pon me? (But you're damned rude! Even you want to say rude things about me?)

mamaguy: To fool; to trick or deceive, especially by flattery. Used in Trinidad, Tobago, and Grenada. It probably is borrowed from a Venezuelan cock-fighting term (*mamar gallo*) that refers to a fighting cock that poses but does not fight. *Example:* After you mamaguy me and take my money.

suck-teeth: The action or sound of sucking your teeth; to make an insulting sound with saliva. A salivary sound with similar significance is found throughout Sub-Saharan Africa, especially West Africa. *Example:* Give a suck-teeth; *or* He looked up in surprise and answered with a long-drawn suck-teeth.

Source: Richard Allsopp, ed., *Dictionary of Caribbean English Usage.* Oxford: Oxford University Press, 1996.

---

musical forms. The roots of modern Caribbean music reflect a combination of African rhythms with European forms of melody and verse. These diverse influences coupled with a long period of relative isolation sparked distinct local sounds. As circulation among Caribbean inhabitants increased, especially during the twentieth century, musical traditions were grafted onto each other, but characteristic sounds remained.

The famed steel pan drums of Trinidad were created from oil drums discarded from a U.S. military base there in the 1940s. The bottoms of the cans are pounded with a sledge hammer to create a concave surface that produces different tones. During Carnival, racks of steel pans are pushed through the streets by dancers while the panmen sound off. So skilled are these musicians that they even perform classical music, and government agencies encourage troubled teens to learn steel pan (Figure 5.24).

The eclectic sound and the ingenious rhythms make Caribbean music so popular. Yet it is more than good dancing music; the music is closely tied with Afro-Caribbean religions and is a popular form of political protest. In Haiti, *ra-ra* music mixes percussion instruments, saxophones, and bamboo trumpets, while weaving in funk and reggae basslines. The songs are always performed in French Creole and typically celebrate Haiti's African ancestry and the use of Voodoo. The lyrics address difficult issues, such as political oppression or poverty. Consequently, ra-ra groups and other musicians have been banned from performing and even forced

▲ **Figure 5.24 Carnival drummer** A steel pan drummer performs while his drum cart is pushed through the streets during Carnival. Steel drums originated from discarded oil cans that local peoples fashioned into drums by hammering the tops into a concave surface. Many steel drum bands perform internationally, playing everything from calypso to classical music. *(Rob Crandall/Rob Crandall, Photographer)*

into exile—most notably, folk singer Manno Charlemange, who later returned to Haiti and was elected mayor of Port-au-Prince in the 1990s.

The music of the late Bob Marley also takes a political stand. Jamaican-born Marley sang of his life in the Kingston ghetto of Trenchtown. He was a devout Rastafarian who believed that Jah was the living force, that New World Africans should look to Africa for a prince to emerge (determined to be Haile Selassie of Ethiopia), and that *ganja* (marijuana) should be consumed regularly. It was Marley's political voice as peacemaker, however, that touched so many lives. His first hit, "Simmer Down," was written to quell street violence that had erupted in Kingstown in 1964. Other songs, such as "Stand-up" and "No Woman No Cry," had a message of social unity and freedom from oppression that resonated in the 1970s. Commercial success never dulled Marley's political edge. He was wildly popular in Africa, and one of his last concerts before his death in 1981 was in Zimbabwe to mark its independence.

Geopolitical Framework: Colonialism, Neocolonialism, and Independence

Caribbean colonial history is a patchwork of rival powers dueling over profitable tropical territories. By the seventeenth century, the Caribbean had become an important proving ground for European colonial ambitions. Spain's grip on the region was tentative, and rivals felt confident that they could win territory by gradually moving from the eastern edge of the sea to the west. Many territories, especially islands in the Lesser Antilles, changed hands several times (Figure 5.25). Trinidad, for example, went from Spanish to limited French control, and finally became an English colony in the eighteenth century. In a few instances, contested colonial holdings produced contemporary border disputes. Only recently did the Guatemalan government give up its claim to Belize, arguing that the British ignored Spanish claims to the area and illegally acquired it. To this day, most Guatemalan-made maps show Belize as part of Guatemala. Also, there are several long-standing border disputes among the Guianas. The most notable is Venezuela's claim to the western half of Guyana.

Europeans viewed the Caribbean as a strategic and profitable region in which to produce sugar, rum, and spices. Geopolitically, rival European powers also felt that their presence in the Caribbean checked Spanish hegemony there. Yet Europe's geopolitical dominance in the Caribbean began to wane by the mid nineteenth century just as the U.S. presence increased. Inspired by the **Monroe Doctrine,** which claimed that the United States would not tolerate European military involvement in the Western Hemisphere, the U.S. government made it clear that it considered the Caribbean to be within its sphere of influence. This view was underscored during the Spanish-American War in 1898. Even through several English, Dutch, and French colonies persisted after this date, the United States indirectly (and sometimes direct-

ly) asserted its control over the region, ushering in a period of **neocolonialism.** In an increasingly global age, however, even neocolonial ties are not a given. The Caribbean has not attracted the level of private foreign investment witnessed by other regions. Moreover, as the Caribbean's strategic significance in a post-Cold War era fades and old colonial ties fray, the leaders of the region openly worry about becoming even more peripheral.

Life in the "American Backyard"

To this day, the United States maintains a proprietary attitude toward the Caribbean, referring to it as "the American backyard." Initial foreign policy objectives were to free it from European tyranny and foster democratic governance. Yet time and again, American political and economic ambitions betrayed those goals. President Theodore Roosevelt made his priorities clear with imperialistic policies that emphasized the construction of the Panama Canal and the maintenance of open sea-lanes. The United States later offered benign-sounding development packages such as the Good Neighbor Policy (1930s), the Alliance for Progress (1960s), the Caribbean Basin Initiative (1980s), and most recently the proposed Free Trade Area of the Americas (2005). The Caribbean view of these initiatives has been guarded at best. Rather than feeling liberated, many residents believe that one kind of political dependence was being traded for another: colonialism for neocolonialism.

In the early 1900s the role of the United States in the Caribbean was overtly military and political. The Spanish-American War (1898) secured Cuba's freedom from Spain and also resulted in Spain's ceding the Philippines, Puerto Rico, and Guam to the United States; the latter two are still U.S. territories. The U.S. government also purchased the Danish Virgin Islands in 1917, renaming them the U.S. Virgin Islands and developing the harbor of St. Thomas. French, English, and Dutch colonies were tolerated as long as these allies recognized the supremacy of the United States in the region. Avowedly against colonialism, the United States had become much like an imperial force.

One of the requirements of an empire is the ability to impose one's will, by force if necessary. When a Caribbean state refused to abide by U.S. trade rules, U.S. Navy vessels would embargo its ports. Marines landed and U.S.-backed governments were installed throughout the Caribbean basin. These were not short-term engagements; U.S. troops occupied the Dominican Republic from 1916 to 1924, Haiti from 1913 to 1934, and Cuba from 1906 to 1909 and 1917 to 1922 (Figure 5.26). Even today several important military bases are in the region, including Guantánamo in eastern Cuba. There is greater reluctance to commit troops in the area now, but as recently as 1994 President Bill Clinton sent troops to Haiti to suppress political violence and prevent a mass exodus of Florida-bound refugees.

Many critics of U.S. policy in the Caribbean complain that business interests overshadow democratic principles in the determinination of foreign policy. U.S. banana companies settled the coastal plain of the Caribbean rimland and operated

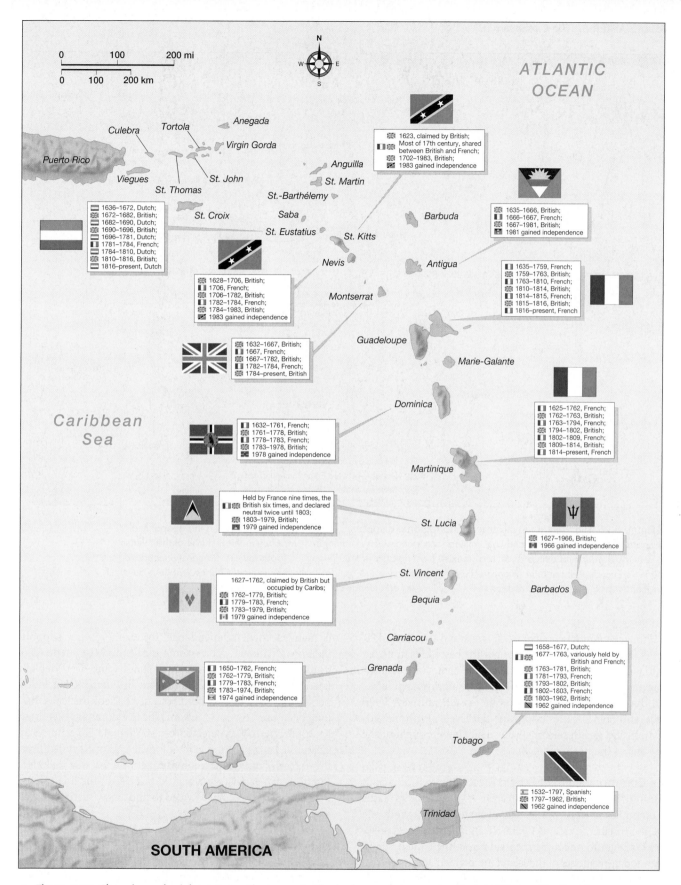

ATLANTIC OCEAN

Puerto Rico

Culebra
Tortola
Anegada
Virgin Gorda
Vieques
St. John
St. Thomas
Anguilla
St. Martin
St.-Barthélemy
St. Croix
Saba
St. Eustatius
St. Kitts
Nevis
Montserrat

Barbuda

Antigua

1623, claimed by British;
Most of 17th century, shared between British and French;
1702–1983, British;
1983 gained independence

1636–1672, Dutch;
1672–1682, British;
1682–1690, Dutch;
1690–1696, British;
1696–1781, Dutch;
1781–1784, French;
1784–1810, Dutch;
1810–1816, British;
1816–present, Dutch

1628–1706, British;
1706, French;
1706–1782, British;
1782–1784, French;
1784–1983, British;
1983 gained independence

1635–1666, British;
1666–1667, French;
1667–1981, British;
1981 gained independence

1635–1759, French;
1759–1763, British;
1763–1810, French;
1810–1814, British;
1814–1815, French;
1815–1816, British;
1816–present, French

1632–1667, British;
1667, French;
1667–1782, British;
1782–1784, French;
1784–present, British

Guadeloupe

Marie-Galante

Caribbean Sea

Dominica

1632–1761, French;
1761–1778, British;
1778–1783, French;
1783–1978, British;
1978 gained independence

1625–1762, French;
1762–1763, British;
1763–1794, French;
1794–1802, British;
1802–1809, French;
1809–1814, British;
1814–present, French

Martinique

Held by France nine times, the British six times, and declared neutral twice until 1803;
1803–1979, British;
1979 gained independence

St. Lucia

1627–1966, British;
1966 gained independence

Barbados

St. Vincent

1627–1762, claimed by British but occupied by Caribs;
1762–1779, British;
1779–1783, French;
1783–1979, British;
1979 gained independence

Bequia

Carriacou

1658–1677, Dutch;
1677–1763, variously held by British and French;
1763–1781, British;
1781–1793, French;
1793–1802, British;
1802–1803, French;
1803–1962, British;
1962 gained independence

1650–1762, French;
1762–1779, British;
1779–1783, French;
1783–1974, British;
1974 gained independence

Grenada

Tobago

1532–1797, Spanish;
1797–1962, British;
1962 gained independence

Trinidad

SOUTH AMERICA

▲ **Figure 5.25 Changing colonial masters** The Lesser Antilles had several changes in colonial affiliation. The French and British traded islands such as Tobago, Grenada, Dominica, and Guadeloupe several times. Many of these territories gained their independence in the 1960s through the 1980s. *(Data source: Bonham Richardson, 1992, The Caribbean in the Wider World, 1492-1992, Cambridge University Press, p. 56, reprinted with the permission of Cambridge University Press; and D.P. Henige, 1970, Colonial Governors from the Fifteenth Century to the Present, University of Wisconsin Press)*

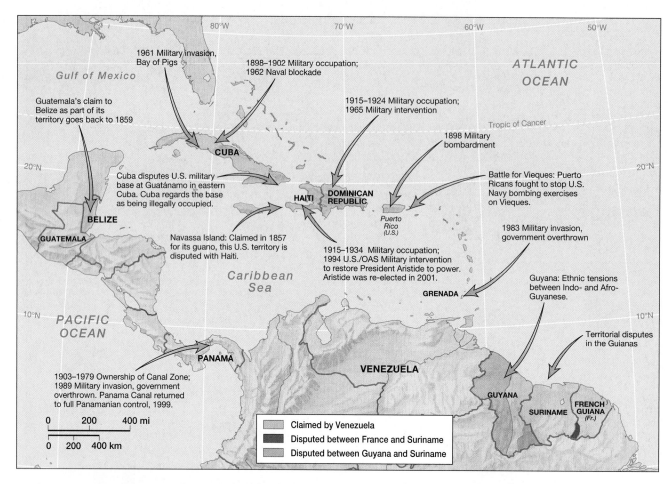

▲ **Figure 5.26 U.S. military involvement and regional disputes** The Caribbean was labeled the geopolitical backyard of the United States, and U.S. military occupation was a common occurrence in the first half of the twentieth century. Border and ethnic conflicts also exist, most notably in the Guianas. *(Data sources: Barbara Tenenbaum, ed., 1996 Encyclopedia of Latin American History and Culture, vol. 5, p. 296, with permission of Charles Scribner's Sons; and John Allcock, 1992, Border and Territorial Disputes, 3rd ed., Harlow, Essex, UK: Longman Group)*

as if they were independent states. Sugar and rum manufacturers from the United States bought the best lands in Cuba, Haiti, and Puerto Rico. Meanwhile, truly democratic institutions remained weak, and there was little improvement in social development. True, exports increased, railroads were built, and port facilities improved; but levels of income, education, and health remained abysmally low throughout the first half of the twentieth century.

The Commonwealth of Puerto Rico Puerto Rico is both within the Caribbean and apart from it because of its status as a commonwealth of the United States. Throughout the twentieth century, various Puerto Rican independence movements sought to uncouple the island from the United States. Even today, residents of the island are divided about their island's political future. At the same time, Puerto Rico depends on U.S. investment and welfare programs; U.S. food stamps are a major source of income for many Puerto Rican families. Commonwealth status also means that Puerto Ricans can freely move between the island and the U.S. mainland, a right they actively assert. In other ways Puerto Ricans symbolical-

ly manifest their independence; for example, they support their own "national" sports teams and send a Miss Puerto Rico to international beauty pageants. The dispute over Vieques Island has also showcased a current of Puerto Rican independence. The small island off of Puerto Rico's east coast has long been used by the U.S. Navy for live bombing exercises. In 2000 Puerto Rico elected its first female governor, Sila Calderon, on a platform calling for the end of Navy-led live-fire exercises on Vieques. After years of protest that included Puerto Rican activists' placing themselves in the line of fire, President George W. Bush agreed to find a new space for naval exercises by 2003.

Puerto Rico led the Caribbean in the transition from an agrarian economy to an industrial one, beginning in the 1950s. For some U.S. officials, Puerto Rico became the model for the rest of the region. Puerto Rican President Muñoz Marín championed an industrialization program called "Operation Bootstrap." Through tax incentives and cheap labor, hundreds of U.S. textile and apparel firms relocated to Puerto Rico. Over the next two decades, 140,000 industrial jobs were added, resulting in a marked increase in per capita GNP. In the

1970s, when Puerto Rico faced stiff competition from Asian apparel manufacturers, the government encouraged petrochemical and pharmaceutical plants to relocate to the island (Figure 5.27). By the 1990s, Puerto Rico was one of the most industrialized localities in the region, with a significantly higher PPP than its neighbors (Table 5.2). Yet it still showed many signs of underdevelopment, including rampant out-migration, low rates of educational attainment, and widespread poverty and crime.

Cuba and Regional Politics   The most profound challenge to U.S. authority in the region came from Cuba and its superpower ally, the former Soviet Union. Historically, Havana was the dominant city of the region. After Spain lost its mainland territories in the 1820s, it invested heavily in the sugar and tobacco industries in Cuba and Puerto Rico. Cubans formally challenged Spain's colonial rule in 1895. Over a three-year period, 400,000 Cubans died in their fight for independence.

▲ **Figure 5.27 Puerto Rican factory**   A pharmaceutical plant near San Juan, Puerto Rico. Operation Bootstrap helped launch the rapid industrialization of Puerto Rico. The island is the Caribbean's most industrialized location.   *(Robert Frerck/Odyssey Productions)*

TABLE 5.2 *Economic Indicators*

Country	Total GNI (Millions of $U.S., 1999)	GNI per Capita ($U.S., 1999)	GNI per Capita, PPP* ($Intl, 1999)	Average Annual Growth % GDP per Capita, 1990–1999
Anguilla	—	—	7,900	—
Antigua and Barbuda	606	8,990	9,870	2.7
Bahamas	—	—	15,500	−0.1
Barbados	2,294	8,600	14,010	1.5
Belize	673	2,730	4,750	0.7
Cayman	—	—	24,500	—
Cuba	—	—	1,700	—
Dominica	238	3,260	5,040	1.8
Dominican Republic	16,130	1,920	5,210	3.9
French Guiana	—	—	6,000	—
Guadeloupe	—	—	9,000	—
Grenada	334	3,440	6,330	2.2
Guyana	651	760	3,330	5.2
Haiti	3,584	460	1,470	−3.4
Jamaica	6,311	2,430	3,390	−0.6
Martinique	—	—	10,700	—
Montserrat	—	—	—	—
Netherlands Antilles	—	—	11,800	—
Puerto Rico	—	—	9,800	1.9
St. Kitts and Nevis	259	6,330	10,400	4.9
St. Lucia	590	3,820	5,200	0.9
St. Vincent and the Grenadines	301	2,640	4,990	2.6
Suriname	—	—	3,780	3.3
Trinidad and Tobago	6,142	4,750	7,690	2.0
Turks and Caicos	—	—	7,700	—

*Purchasing power parity

Source: The World Bank Atlas, *2001. Data for dependent territories from* CIA World Factbook, *2000.*

Even though the United States is credited with winning what is commonly known as the Spanish-American War in 1898, U.S. forces were relative latecomers to this struggle. In the end, many Cubans resented the presence of the United States on their island because it undermined their hard-fought independence.

In the 1950s another revolutionary effort began in Cuba, led by Fidel Castro against the pro-American Batista government. Cuba's economic productivity had soared but its people were still poor, uneducated, and increasingly angry. The contrast between the lives of average cane workers and the foreign elite was stark and inescapable. In the 1950s Havana emerged as the top Caribbean tourist destination for Americans, further demonstrating the gap between the classes. Beach resorts, casinos, and several hundred brothels cemented Havana's reputation as "Sin City." Yet the vast majority of rural workers felt betrayed by six decades of American neocolonialism. In defiance, they lent their support in 1959 to a charismatic, cigar-smoking former law student named Fidel.

Castro tapped a deep vein of Cuban resentment against U.S. policy. After Castro's government nationalized American industries and took ownership of all foreign-owned properties, the United States responded by refusing to buy Cuban sugar and ultimately ending diplomatic relations with the state. Various U.S. trade embargoes against Cuba have existed for over four decades. What sealed Cuba's fate as a geopolitical enemy was its establishment of diplomatic relations with the USSR in 1960 during the height of the Cold War. With the Soviet Union financially and militarily backing Castro, a direct U.S. invasion of Cuba was too risky. The fall of 1962 produced one of the most dangerous episodes of the Cold War when Soviet missiles were discovered on Cuban soil. Ultimately, the Soviet Union removed its weapons; in return, the United States promised not to invade Cuba.

The Cuban missile crisis signaled that American dominance in the Caribbean was not universally accepted. Cuba by itself was not a threat to the United States, but the geopolitical implications of its turn toward the Soviet Union could not be ignored. Moreover, Castro wanted to extend his socialist vision abroad, supporting at different times revolutionaries in Colombia, Venezuela, Guyana, Suriname, Bolivia, and across the Atlantic in Angola. Other Caribbean states watched the Cuba socialist experiment with interest, especially its remarkable strides in literacy and public health. By the mid-1970s Cuba had the lowest infant mortality rate in all of the Caribbean and Latin America.

Cuban-style socialism did not readily transfer to other countries, however, because the United States was determined to prevent it, and Soviet support of Cuba proved too expensive. Even with the end of the Cold War, Cuba maintains an identity apart from its regional neighbors and in defiance of the United States. By the 1990s, every other independent country in the region had a democratically elected leader. And as the geopolitical landscape of the region is again realigned with the Free Trade Area of the Americas treaty, Cuba has not been part of negotiations (see "Cuba: A Caribbean Exception").

Independence and Integration

Given the repressive colonial history of the Caribbean, it is no wonder that the struggle for political independence began more than 200 years ago. Haiti was the second colony in the Americas to gain independence in 1804; the United States was the first in 1776. However, the achievement of political independence by many states in the region has not guaranteed economic independence. Many Caribbean states struggle to meet the basic needs of their people. Surprisingly, today some Caribbean territories maintain their colonial status as an economic asset. For example, the French territories of Martinique, Guadeloupe, and French Guiana are overseas departments of France; residents have full French citizenship and social welfare benefits.

Perhaps the hardest task facing the Caribbean is promoting regional integration. Scattered islands, a divided rimland, different languages, and limited economic resources inhibit the formation of a meaningful regional trade block. More common is cooperation between groups of islands with a shared colonial background.

Independence Movements   Haiti's revolutionary war began in 1791 and ended in 1804. Spanish, French, and British forces were involved, as well as factions within Haiti that had formed along racial lines. In the end, a full-blown race war ensued. The island's population was halved through casualties and emigration; ultimately, the former slaves became the rulers. Independence, however, did not allow this crown of the French Caribbean to prosper. "Plantation America" watched in horror as Haitian slaves used guerrilla tactics to gain their freedom. Fearing that other colonies might follow Haiti's lead, plantation owners in other countries were on guard for the slightest hint of revolt. For its part, Haiti did not become a leader in liberation. Mired in economic and political problems, it was shunned by the European powers and never embraced by the states of the Spanish mainland when they became independent in the 1820s.

Several revolutionary pulses followed in the nineteenth century. In the Greater Antilles, the Dominican Republic finally gained independence in 1844 after wresting control of the territory from Spain and Haiti. Cuba and Puerto Rico were freed from Spanish colonialism in 1898, but their independence was compromised by greater U.S. involvement. The British colonies also faced revolts, especially in the 1930s, yet it was not until the 1960s that independent states emerged from the English Caribbean. First the larger colonies of Jamaica, Trinidad and Tobago, Guyana, and Barbados gained their independence. Other British colonies followed throughout the 1970s and early 1980s: the Bahamas in 1973, Grenada in 1974, Dominica in 1978, St. Vincent and the Grenadines in 1979, St. Lucia in 1979, Antigua and Barbuda in 1981, Belize in 1981, and St. Kitts and Nevis in 1983. Suriname, the only Dutch colony on the rimland, became an autonomous territory in 1954 but remained part of the Kingdom of the Netherlands until 1975, when it declared itself an independent republic.

Cuba: A Caribbean Exception

Cuba has proven to be a regional exception. Socialism fell in the former Soviet Union and Eastern Europe, but not in Cuba. However, the economic crisis that ensued in 1991 forced Fidel Castro to loosen government controls of the economy. Whereas having U.S. dollars was once a punishable offense, today Cubans freely carry and exchange them. More important, Cuban expatriates are now able to send cash remittances to their relatives on the island. The streets of Havana buzz with small private enterprises, including flea markets, street vendors, and privately owned restaurants, all of which are licensed and sanctioned by the government. The state opened the door to foreign investment with joint ventures in tourism, mining, industry, and food processing. And perhaps the most radical change of all, agricultural reforms broke up the state-run farms and encouraged smaller cooperatives and privately owned farms to respond to market incentives.

In 1991 Cuba faced the reality of the "zero option": no subsidized food or fuel from the former Soviet Union, and no guaranteed market for Cuban sugar and citrus. Overnight, the economic rules changed, and Cubans were asked to ride bikes, pick fruit, endure blackouts, wait in lines, create small businesses, and rediscover basic market forces.

The earliest crisis was one of fuel. Dependent on subsidized Soviet imports of petroleum, suddenly Cuba had a chronic fuel shortage that affected transportation and electricity production. The most striking impact was the emptiness of Cuban streets. In the early 1990s auto traffic in Havana nearly ceased; the few buses that ran were crammed beyond capacity. The solution became bicycles. Nearly 2 million of them were imported from or built with the help of China. Fuel and parts to run agricultural equipment were also scarce; thus, oxen, carts, and manual harvesting were resurrected. Many factories were reduced to shorter hours as they lacked the energy or parts to run. Brownouts and scheduled blackouts also became normal in the early 1990s because of fuel shortages. Today fuel is more widely available, but energy costs have escalated.

The second crisis was food. Cuba's overly specialized agricultural sector produced sugar and citrus, relying on generous imports of staples from the Soviet bloc. In the worst years of the crisis, food was rationed on a monthly basis according to family size. Cuba's sugar harvest was also cut in half due to lack of fuel and working farm equipment. By 1995, however, the sugar harvest began to recover, although the government now earns significantly less from its principal export. To increase and diversify agricultural production, the state redistributed two-thirds of its land, primarily to newly created cooperatives and thousands of small private farms. After producers deliver their contracted amounts to the state, they can sell their surplus on the open market. Local farmer's markets now exist, and prices are negotiated, not set. While overall production levels are far below the 1980s, a greater diversity of foods for domestic consumption is available.

Desperate for hard currency, Cuba reluctantly turned to tourism to earn foreign exchange. Initially the government tightly regulated tourism, trying to isolate foreign tourists from the daily lives of Cubans and in some cases forbidding Cubans from entering tourist hotels and restaurants. This policy was deeply resented by Cubans, who prefer to interact with tourists. Tourism is now a way of life. In fact, a brisk informal sector catering to tourists has emerged; it includes cigar sales, family-run restaurants, and prostitution. In 1999 more than 1.5 million tourists came to Cuba, mostly from Canada, Europe, and Latin American countries. Some 2 million tourists were expected in 2001.

Cuba has managed to cope without its superpower patron and despite antagonism from the United States. Although U.S. firms cannot invest in the country, Cuba now has joint ventures with more than 300 companies, many of them from Spain, Canada, and Italy. Just how Cuba's brand of tropical socialism will evolve is anyone's guess, but so far Cuba has been able to remake itself.

▲ **Figure 5.2.1 Bicycle power** Cubans riding bicycles participate in a patriotic parade at Playa Larga, Cuba. The scarcity of fuel has led to a dramatic increase in bicycle transport. *(Rob Crandall/Rob Crandall, Photographer)*

▲ **Figure 5.2.2 Plowing with oxen** A Cuban farmer plows his privately owned plot in preparation for planting corn. Oxen returned to the fields, and for the first time in socialist Cuba farmers could sell their surplus produce on the open market. *(Rob Crandall/Rob Crandall, Photographer)*

Present-day Colonies　Britain still maintains several crown colonies in the region: the Cayman Islands, the Turks and Caicos, Anguilla, and Montserrat. The combined population of these islands is just 71,000 people, yet their standard of living is high, due in part to their specialization in the recently developed industry of offshore banking. French Guiana, Martinique, and Guadeloupe are each departments of France and thus, technically speaking, not colonies. Together they total 900,000 people. The Dutch islands in the Caribbean are considered autonomous countries that are part of the Kingdom of the Netherlands. Curaçao, Bonaire, St. Martin, Saba, and St. Eustatius comprise the federation of the Netherlands Antilles. Aruba left the federation in 1986 and governs without its influence. Together the population of the Dutch islands is a quarter million people.

Regional Integration　Experimentation with regional trade associations as a means to improve the economic competitiveness of the Caribbean began in the 1960s. The goal of regional cooperation was to improve employment rates, increase intraregional trade, and ultimately reduce external dependence. The countries of the English Caribbean took the lead in this development strategy. In 1963 Guyana proposed an economic integration plan with Barbados and Antigua. In 1972 the integration process intensified with the formation of the **Caribbean Community and Common Market (CARICOM).** Representing the former English colonies, CARICOM proposed an ambitious regional industrialization plan and the creation of the Caribbean Development Bank to assist the poorer states. CARICOM also oversees the University of the West Indies, with campuses in Trinidad, Jamaica, and Barbados. As important as this trade group is as an institutional symbol of collective identity, it has produced limited improvements in intraregional trade. There are 13 full member states—all of the English Caribbean and French-speaking Haiti. Other dependencies, such as Anguilla, Turks and Caicos, and the British Virgin Islands, are associate members. Still, the predominance of English-speaking territories in CARICOM underscores the deep linguistic fractures in the Caribbean. The dream of regional integration as a way to produce a more stable and self-sufficient Caribbean has never been realized. One scholar of the region argues that a limiting factor is a small-islandist ideology. Islanders tend to keep their backs to the sea, oblivious to the needs of neighbors. At times such isolationism results in suspicion, distrust, and even hostility toward nearby states. There is a desire to remain inward-looking, but economic necessity dictates engagement with partners outside the region. And so this peculiar status of isolated proximity unfolds in the Caribbean, expressing itself in uneven social and economic development trends.

Economic and Social Development: From Cane Fields to Cruise Ships

Collectively, the population of the Caribbean, albeit poor by U.S. standards, is economically better off than most of Sub-Saharan Africa, South Asia, and China. Despite periods of economic stagnation in the Caribbean, social gains in education, health, and life expectancy are significant. Historically the Caribbean's links to the world economy were through tropical agricultural exports, yet several specialized industries, such as tourism, offshore banking, and assembly plants, have challenged the dominance of agriculture. These industries grew because of the region's proximity to North America and Europe, the availability of cheap labor, and the implementation of policies that created a nearly tax-free environment for foreign-owned companies. Unfortunately, growth in these sectors does not employ all the region's displaced rural laborers, so the lure of jobs in North America and Europe is still strong.

From Fields to Factories and Resorts

Agriculture used to dominate the economic life of the Caribbean. Decades of turbulent commodity prices and decline in preferential trade agreements with former colonial masters have produced more hardship than prosperity. Ecologically, the soils are overworked, and there are no frontier areas to expand production, save for areas of the rimland. Moreover, agricultural prices have not kept pace with rising production costs, so wages and profits remain low. With the exception of a few mineral-rich territories, such as Trinidad, Guyana, Suriname, and Jamaica, most countries have systematically tried to diversify their economies, relying less on their soils and more on manufacturing and services.

Comparing export figures over time demonstrates the shift away from mono-crop dependence. In 1955 Haiti earned more than 70 percent of its foreign exchange through the export of coffee; by 1990 coffee accounted for only 11 percent of its export earnings. Similarly, in 1955 the Dominican Republic earned close to 60 percent of its foreign exchange through sugar, but 35 years later sugar earned less than 20 percent of the country's foreign exchange, and pig iron exports nearly equaled that of sugar. The one exception to this trend is Cuba, which earned approximately 80 percent of its foreign exchange through sugar production from the 1950s to 1990. Cuba, however, was forced to diversify in the 1990s when Russia could no longer guarantee the price supports that kept sugar a lucrative commodity.

Sugar and Coffee　The economic history of the Caribbean cannot be separated from the production of sugarcane. Even relatively small territories such as Antigua and Barbados yielded fabulous profits because there was no limit to the demand for sugar in the eighteenth century. Once considered a luxury crop, it became a popular necessity for European and North American laborers by the 1750s. It sweetened tea and coffee and made jams a popular spread for stale bread. In short, it made the meager and bland diets of ordinary people tolerable, and it also boosted caloric intake. Distilled into rum, sugar produced a popular intoxicant. Though it is hard to imagine today, consumption of a pint of rum a day was not uncommon in the 1800s.

Sugarcane is still grown throughout the region for domestic consumption and export. Its economic importance has declined, however, mostly due to increased competition from sugar beets grown in the midlatitudes. The Caribbean and Brazil are the world's major sugar exporters. Up until 1990, Cuba alone accounted for more than 60 percent of the value of world sugar exports. Cuba's dominance in sugar exports had more to do with its subsidized and guaranteed markets in eastern Europe and Russia than with unprecedented productivity. Since 1990 the Cuban sugar harvest has been reduced by half.

Coffee is planted in the mountains of the Greater Antilles. Haiti has been the most dependent on coffee, relying on peasant sharecroppers to tend the plants and harvest the beans. For other countries coffee is a valued specialty commodity. Beans harvested in the Blue Mountains of Jamaica, for example, fetch two to three times the going price for similar highland coffee grown in Colombia. Puerto Rico and Cuba are also trying to develop a niche in the gourmet coffee market. An important production distinction with coffee, in contrast to sugar, is that it is mostly grown on small farms and sold to buyers or delivered to cooperatives. Typically, farmers plant other crops between the coffee bushes so that they can meet their subsistence needs as well as produce a cash crop. Even with this self-provisioning system, peasants often seasonally abandon their farms for work elsewhere as laborers. The instability of coffee prices, which were especially low in 2000 and 2001, makes the economics of growing coffee hard to predict.

The Banana Wars   The major banana exporters are in Latin America, not the Caribbean. In fact, the success of banana plantations is mixed in this region, as banana plants are especially vulnerable to hurricanes. Still, several small states in the Lesser Antilles, most notably Dominica, St. Vincent, and St. Lucia, have become dependent on bananas, gaining as much as 60 percent of their export earnings from the yellow fruit. Bananas have not made people rich, yet their production for export has fostered greater economic and social development. In the eastern Caribbean, where most bananas are grown on small farms of five acres, the landowners are the laborers and thus earn two to four times more than banana plantation workers in Ecuador and Central America. Moreover, for these small states, banana exports are their link with the global economy. Pressures on the European Union to drop the preferential treatment given to banana growers from the former colonies has undermined the economic viability of this crop in the Caribbean (Figure 5.28). Where once a market and a minimum price were guaranteed, under the new global order neither is certain.

The case of the eastern Caribbean underscores what happens to the losers in globalization. In 1996 the United States, Ecuador (the world's leading banana exporter), Mexico, Guatemala, and Honduras took the EU's restrictive banana trade agreement to the World Trade Organization court. The agreement was denounced as unfair, and the EU was told to eliminate it by 1998. To make matters worse, consumer tastes had changed, giving preference to a uniformly large and unblemished yellow banana typical of the Latin American plantations but not always grown in places like St. Lucia. The growers of the eastern Caribbean, to survive in this newly competitive environment, will have to produce a more stan-

▲ **Figure 5.28 Banana farm**   Small banana farms struggle to be economically viable in a globalizing marketplace. At one time, small island producers such as St. Lucia, shown here, received preferential access to the European market. As this preferential status is dropped, many banana-dependent Caribbean nations fear for their livelihood.   *(Bob Krist/Corbis)*

dardized fruit and increase their yield per acre. While island governments are trying to aid in this transition, local farmers have experimented with new crops, such as okra, tomatoes, avocados, and marijuana. A few well-tended marijuana plants will earn 30 times more per pound than bananas. Just what will happen to family-run banana farms is hard to tell, but many Caribbean growers fear that the days of the banana economy are numbered.

The banana story also reflects another major shift for the Caribbean, the decline in the economic importance of agriculture. True, there are still fields of swaying sugarcane, groves of shaded coffee, and plots of bananas and tobacco. What has changed is the amount of labor and land needed to produce similar yields. Beginning in the early 1900s and intensifying in the 1950s, sugar production was mechanized, making many rural laborers unnecessary. Coffee and bananas are still harvested by hand, but they grow in limited areas. In addition, the green revolution meant that the use of chemical fertilizers, pesticides, and high-yielding varieties of crops enabled yields to increase even when the area under cultivation remained the same or declined. The combination of mechanization, green revolution technology, and increased global competition has forced many rural laborers to find employment elsewhere.

Assembly-plant Industrialization   One regional strategy to deal with the unemployed was to invite foreign investors to set up assembly plants and thus create jobs. This was first tried successfully in Puerto Rico in the 1950s and was copied throughout the region. In Puerto Rico's "Operation Bootstrap," island leaders encouraged U.S. investment by offering cheap labor, local tax breaks, and, most important, federal tax exemptions (something only Puerto Rico can do because of its special status as a commonwealth of the United States). Initially the program was a tremendous success, so that by 1970 nearly 40 percent of the island's gross domestic product (GDP) came from manufacturing. Today about 13 percent of the Puerto Rican labor force is employed in manufacturing, and this sector accounts for nearly half of the island's GDP. Yet competition from other states with even lower wages and the U.S. Congress's decision in 1996 to phase out many of the tax exemptions may threaten Puerto Rico's ability to maintain its specialized industrial base.

Through the creation of **free trade zones (FTZs)**—duty free and tax-exempt industrial parks for foreign corporations—legalization of foreign ownership, pursuit of direct foreign investment, and cheap labor, the Caribbean is an increasingly attractive location to assemble goods for North American consumers. Jamaica began developing its light manufacturing sector in the 1980s and has since seen manufacturing rise to account for 15 percent of its GDP. Most production takes place in export-oriented free trade zones that receive some government support, offer tax breaks, and have no import restrictions. These zones exist in the island's major cities: Kingston, Spanish Town, and Montego Bay. Beginning with food, drink, and tobacco products, they have expanded to include textiles and garments. Investors in the free trade zones are principally from the United States, Taiwan, and Hong Kong. As part of the Caribbean Basin Initiative, the United States agreed to purchase most of the garments manufactured in Jamaica.

The Dominican Republic offers another example of the globalizing trends in manufacturing. Like Jamaica, it took advantage of tax incentives and guaranteed access to the U.S. market offered through the Caribbean Basin Initiative. The number of operational free trade zones in the Dominican Republic in 2001 was 16, including an FTZ on the island's north shore, near the Haitian border, optimistically named "Hong Kong of the Caribbean." Plans exist to double the number of FTZs. Firms from the United States and Canada are the most frequent investors in these zones, followed by Dominican, South Korean, and Taiwanese firms. Traditional manufacturing on the island was tied to sugar refining, whereas up to 60 percent of production in the FTZs is in garments and textiles (Figure 5.29). Manufacturing is the largest economic sector in the Dominican Republic, accounting for nearly one-fifth of the country's GDP.

In each of these cases, the growth in manufacturing depended on national and international development policies that supported export-led growth through direct foreign investment. Certainly, new jobs are being created and national economies are diversifying in the process, yet critics believe that foreign investors gain more than the host countries. Since most goods are assembled from imported materials, there is little integration with national suppliers. Often higher than local averages, wages are still miserable compared to the developed world—sometimes just two to three dollars a day. Moreover, as other developing countries compete with the Caribbean for the establishment of FTZs, this strategy may become less significant over time.

Offshore Banking   The rise of offshore banking in the Caribbean is most closely associated with the Bahamas, which began this industry back in the 1920s. **Offshore banking** centers appeal to foreign banks and corporations by offering

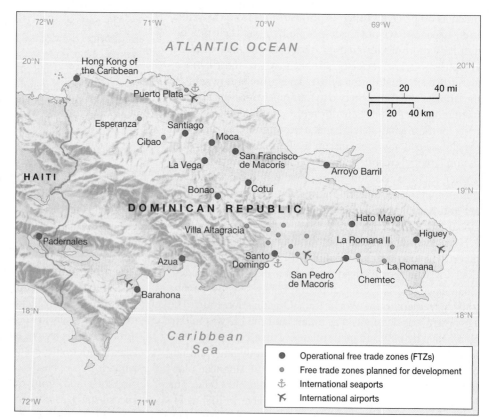

▶ **Figure 5.29 Free trade zones in the Dominican Republic**   Another sign of globalization is the proliferation of duty-free and tax-exempt industrial parks in the Caribbean. Currently 16 FTZs are operational, with foreign investors from the United States, Canada, South Korea, and Taiwan. *(Source: Modified from Barney Warf, 1995, "Information Services in the Dominican Republic," Yearbook, Conference of Latin Americanist Geographers, p. 15)*

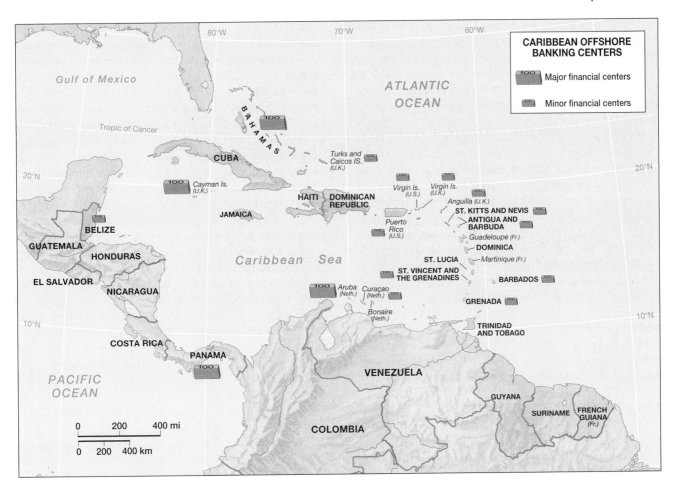

▲ Figure 5.30 Caribbean offshore banking centers Another development strategy used by the smaller islands is to offer specialized banking services to international corporations. These services are both confidential and tax-exempt. The Bahamas, Cayman Islands, and Panama dominate offshore banking in the region. Many other countries, however, are beginning to offer similar services. [Source: Susan Roberts, 1995, "Small Place, Big Money: The Cayman Islands and the International Financial System," Economic Geography, 71(3), 237–256]

specialized services that are confidential and tax-exempt. Localities that provide offshore banking make money through registration fees, not taxes. The Bahamas were so successful in developing this sector that by 1976 the country was the third largest banking center in the world. Its dominance began to decline due to competitors from the Caribbean, Hong Kong, and Singapore and because there was no longer a tax advantage in booking large international loans offshore. Concerns about corruption and laundering of drug money also hurt the islands' financial status in the 1980s, and major reforms were introduced to remove illegal funds. By 1998 the Bahamas' ranking in terms of global financial centers had dropped to 15th. Still, offshore banking remains an important part of the Bahamian economy. In the 1990s, the Cayman Islands emerged as the region's leader in financial services. With a population of 30,000, this crown colony of Britain has some 50,000 registered companies and the highest per capita PPP of the region at $24,500. In 2001 it was estimated that Cayman banks had more than $800 billion on deposit; those deposits were estimated to be growing at $100 billion a year.

Each of the offshore banking centers in the Caribbean (Figure 5.30) tries to develop special financial niches to attract clients, such as banking, functional operations, insurance, or trusts. The Caribbean is an attractive locality for such services because of its proximity to the United States (home of many of the registered firms), client demand for these services in different localities, and the steady improvement in telecommunications that make this industry possible. On the supply side, the resource-poor islands of the region see financial services as a way to bring hard currency to state coffers. Envious of the economic success of the Bahamas and the Cayman Islands, countries such as Antigua, Aruba, Barbados, and Belize hope that greater prosperity will come from establishing close ties to international finance.

The growth of offshore financial centers and the manufacturing zones discussed earlier are indicative of the forces of globalization. Ironically, banking does not employ that many people and can quickly unravel with the slightest sign of political instability. Offshore banking also attracts money tied to the drug trade. A major transshipment area for cocaine bound for North America and Europe, the Caribbean is also a major money laundering center. This brings its own problems to the region, such as increases in drug consumption, corruption of local officials, and even drug-related murders. International efforts in the 1990s sought to curb money laundering by threatening the suspension of privacy provisions whenever criminal activities are suspected. After the terrorist attacks on the United States in September 2001, the Bush Administration had renewed interest in promoting know-your-customer laws as a means of tracking down the offshore assets of terrorist organizations. The administrative burden of reporting clients, however, remains with bank personnel, who may be reluctant to turn down potentially lucrative business. Thus the financial services industry continues to evolve in the Caribbean, transforming many island economies and underscoring this region's ties to the global economy. What is less certain is whether such development trends will greatly improve local earnings and standards of living.

Tourism   Environmental, locational, and economic factors converge to buttress tourism in this part of the world. The earliest visitors to this tropical sea admired its clear and sparkling turquoise waters. By the nineteenth century, elite

North Americans were fleeing winter to enjoy the restorative warmth of the Caribbean during the dry season. Entrepreneurs later realized that the Caribbean dry season's complement of the Northern Hemisphere's winter was ideal for beach-oriented resorts. By the twentieth century tourism was well established with both destination resorts and cruise lines. By the 1950s the leader in tourism was Cuba, and the Bahamas was a distant second. Castro's rise to power, however, eliminated this sector of the island's economy for nearly three decades and opened the door for other islands to step in.

Five islands hosted 70 percent of the 14 million international tourists who came to the Caribbean in 1999: Puerto Rico, the Bahamas, the Dominican Republic, Jamaica, and Cuba (Figure 5.31). Puerto Rico saw its tourist sector begin to grow with commonwealth status in 1952. San Juan is now the largest home port for cruise lines and the second largest cruise-ship port in the world in terms of total visitors. The cruise business, in combination with stayover visitors, totaled over 3 million arrivals in 1999. The Bahamas attributes most of its economic development and high per capita income to tourism. With more than 1.5 million stayover visitors in 1999 and 1.7 million cruise ship passengers, the Bahamas is another major hub for tourism in the region. Some 30 percent

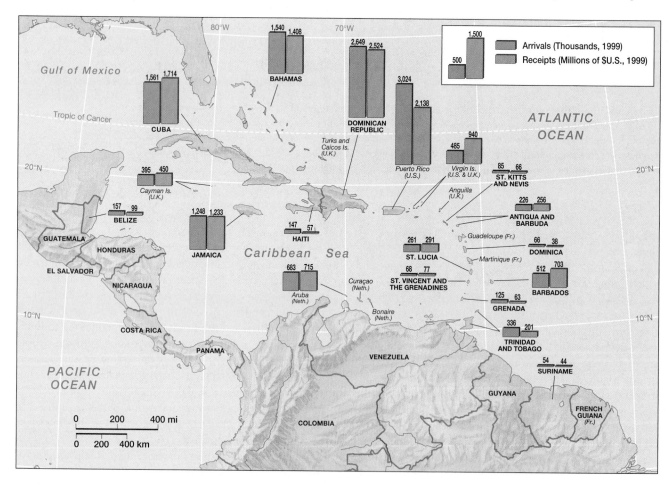

▲ **Figure 5.31 Global linkages: International tourism in the Caribbean**   Tourism directly links the Caribbean to the global economy. Each year 14 million tourists come to the islands, mostly from North America, Latin America, and Europe. The most popular destinations are the Dominican Republic, Puerto Rico, Cuba, the Bahamas, and Jamaica.   *(Source: World Bank Atlas, 2001)*

▲ **Figure 5.32 Caribbean cruise ships**   Cruise ships in port at Nassau, the Bahamas. Tourism is vital to many Caribbean states, but most cruise ships are owned by companies outside the region and offer relatively little employment for Caribbean workers.   *(Galen Rowell/Corbis)*

▲ **Figure 5.33 Ecotourism**   A scuba diver explores the Belizean reef. Many Caribbean and Latin American countries are betting on ecotourism as a means of generating income and better managing their environments. In Belize, for example, part of the reef was declared a marine park in hope of protecting species diversity and attracting more tourists.   *(Doug Perrine/Innerspace Visions)*

of the Bahamian population is employed in tourism, and tourism represents nearly half the country's GDP (Figure 5.32). Since the Bahamas is closer to the United States than most Caribbean destinations, the majority of its international clientele are American (about 80 percent), followed by Europeans and Canadians.

The Dominican Republic also sees more than 2.6 million visitors, many of them Dominican nationals who live overseas. Since 1980 tourist receipts have increased twenty-fold, making tourism the leading foreign-exchange earner with over $2.5 billion in receipts. The Dominican Republic earned more tourism dollars in 1999 than any other Caribbean country. Jamaica has become similarly dependent on tourism for hard currency, earning over $1 billion in tourist receipts in 1999. The vast majority of tourists to Jamaica are from the United States and the United Kingdom. While drawing 1.2 million stayover arrivals in 1999, Jamaica was a port-of-call for another half million cruise ship passengers.

After years of neglect, Cuba is reviving tourism in an attempt to earn badly needed hard currency. Tourism represented less than 1 percent of the national economy in the early 1980s. By 1999 more than 1.5 million tourists (mostly Canadians and Europeans) yielded gross receipts approaching $2 billion. Conspicuous in their absence are travelers from the United States, forbidden to travel to Cuba because of the U.S.-imposed sanctions. The Helms-Burton Law of 1996 even tried to reduce the flow of foreign investment by threatening to sue major foreign investors in Cuba. While some investors delayed their plans because of this law, others, such as a leading Spanish hotel group, are busy building the island's tourist capacity in the hope that the U.S. ban will someday be lifted. It is estimated that about 2 million people will visit Cuba in 2001, either as stayover guests or on cruise ships.

As important as tourism is for the larger islands, it is often the principal source of income for smaller ones. The Virgin Islands, Barbados, Turks and Caicos, and, recently, Belize all greatly depend on international tourists. To show how quickly this sector can grow, consider this example: Belize began promoting tourism in the early 1980s when just 30,000 arrivals came a year. Close to North America and English-speaking, it specialized in a more rustic ecotourism that showcased its interior tropical forests and coastal barrier reef (Figure 5.33). By 1994 the number of tourists topped 300,000, and tourism was credited for employing one-fifth of the workforce. Five years later, however, international arrivals were halved and tourist receipts equaled one-seventh of Belize's GNP, which shows the unpredictable nature of the tourist industry.

For more than four decades, tourism has been one of the Caribbean's mainstays. Yet this regional industry has grown more slowly in recent years compared to other tourist destinations in the Middle East, southern Europe, and even Central America. It seems that Americans are favoring domestic destinations, such as Hawaii, Florida, and Las Vegas, or are going to more "exotic" localities, such as Costa Rica. European tourists also seem to be sticking closer to home or venturing to new locations, such as Dubai on the Persian Gulf or Goa in India. A destination's reputation can suddenly deteriorate as a result of social or natural forces. In the 1980s a wave of crimes against tourists in Jamaica caused a drop in visitors. It took a locally owned hotel chain to market all-inclusive trips on private beaches to restore the island's image as a safe place to visit. Hurricane Hugo virtually shut down St. Croix to visitors for an entire season, but islanders managed to rebuild most of their infrastructure by the following year.

Tourism-led growth has detractors for other reasons. It is subject to the overall health of the world economy and current political affairs. Thus if North America experiences a recession or international tourism declines due to heightened fears of terrorism, the flow of tourist dollars to the Caribbean dries up. Where tourism is on the rise, local resentment may build as residents confront the disparity between their own lives and those of the tourists. There is also a serious problem of **capital leakage,** which is the huge gap between gross receipts and the total tourist dollars that remain in the Caribbean. Since many guests stay in hotel chains with corporate headquarters outside the region, leakage of profits is inevitable. On the plus side, tourism tends to promote stronger environmental laws and regulation. Countries quickly learn that their physical environment is the foundation for success. And while tourism does have its costs (higher energy and water consumption and demand for more imports), it is environmentally less destructive than traditional export agriculture and currently more profitable.

Social Development

While the record for economic growth in the region is inconsistent, measures of social development are stronger, in part due to Cuba's accomplishments in health care and education. With the exception of Haiti's abysmally low life expectancy of 49 years, most Caribbean peoples have an average life expectancy of 70 years or more (Table 5.3). Levels of adult illiteracy are very low, especially in Cuba, the Bahamas, Trinidad, Grenada, Guyana, the Cayman Islands, Barbados, and the Netherlands Antilles, where only 2 to 3 percent of the population cannot read. Higher educational attainment and outmigration have contributed to a marked decline in natural increase rates over the last 30 years, so that the average Caribbean woman has two or three children. Despite real social gains, many inhabitants are chronically underemployed, poorly housed, and dependent on foreign remittances or subsistence agriculture. For rich and poor alike, the temptation to leave the region in search of better opportunities is always present.

Education   Many Caribbean states have excelled in educating their citizens. As indicated in Table 5.3, literacy is the norm, and the expectation is for most people to receive at least a high school degree. In many respects, Cuba's educational accomplishments are the most impressive, given the size of the coun-

try and its high illiteracy rates in the 1960s. Today 80 percent of the population receives secondary education, and illiteracy among adults is just 3 to 4 percent. Cuba also excels in training medical personnel and advancing biomedical technology. Hispaniola is the obvious contrast to Cuba's success. Less than 15 percent of Haitians and about half of the Dominican Republic's citizens acquire secondary education. While the Dominican Republic has made strides to improve adult literacy (83 percent of adults are literate), over half of all Haitian adults are illiterate. Political stability and economic growth have helped the Dominican Republic better its social conditions over the past decade. In fact, many Haitians have fled to the Dominican Republic because conditions, although far from ideal, are much better than in their homeland. Three former English colonies (Belize, Jamaica, and St. Lucia) have much higher rates of adult illiteracy than the regional average. Roughly one-third of adults in St. Lucia and Belize are illiterate, while nearly half of Jamaican adults are illiterate.

Education is expensive for these nations, but it is considered essential for development. Ironically, many states express frustration about training professionals for the benefit of developed countries in a phenomenon called **brain drain**. In the early 1980s the prime minister of Jamaica complained that 60 percent of his country's newly trained workers left for the United States, Canada, and Britain, representing a subsidy to these economies far greater than the foreign aid Jamaica received from them. During the economically depressed 1980s, Barbados, Guyana, and even the Dominican Republic and Haiti lost up to 20 percent of their most educated citizens. Brain drain occurs throughout the developing world, especially between former colonies and the mother countries. Yet given the small population of many Caribbean territories, each professional person lost to emigration can negatively impact local health care, education, and enterprise. For the time being, stronger economic performance in the Caribbean has slowed the outflow of professionals.

Status of Women   The matriarchal basis of Caribbean households is often heralded as a distinguishing characteristic of the region. The rural custom of men leaving home for seasonal employment tends to nurture strong and self-sufficient female networks. Women typically run the local street markets. With men absent for long periods of time, women tend to make household and community decisions. While giving women local power, this position does not always confer status. In rural areas female status is often undermined by the relative exclusion of women from the cash economy; men earn wages while women provide subsistence.

As Caribbean society urbanizes, more women are being employed in assembly plants (the garment industry, in particular, prefers to hire women), in data-entry firms, and in tourism. With new employment opportunities, female labor force participation has surged; in countries such as the Bahamas, Barbados, Jamaica, and Martinique more than 40 percent of the workforce is female (Table 5.3). Increasingly, women are the principal earners of cash, which challenges established gender roles. One impact of greater female participation in the formal economy seems to be smaller families.

TABLE 5.3 *Social Indicators and Status of Women*

Country	Life Expectancy at Birth		Under Age 5 Mortality (per 1,000)		Percent Illiteracy (Ages 15 and over)		Female Labor Force Participation (% of total, 1999)
	Male	Female	1980	1999	Male	Female	
Anguilla	73	79	—	—	5	5	—
Antigua and Barbuda	68	72	—	—	10	12	—
Bahamas	70	75	—	—	2	2	47
Barbados	70	75	—	—	2	3	46
Belize	70	74	—	—	30	30	24
Cayman	76	81	—	—	2	2	—
Cuba	73	77	22	8	3	4	39
Dominica	70	76	—	—	6	6	—
Dominican Republic	67	71	92	47	17	17	30
French Guiana	72	79	—	—	16	18	—
Grenada	63	66	—	—	2	2	—
Guadeloupe	73	80	—	—	10	10	—
Guyana	62	68	—	—	1	2	—
Haiti	47	51	—	—	52	58	34
Jamaica	70	73	200	118	49	53	43
Martinique	76	82	39	24	18	10	46
Montserrat	76	80	—	—	3	3	—
Netherlands Antilles	72	76	—	—	2	1	—
Puerto Rico	71	80	—	—	10	12	—
St. Kitts and Nevis	66	71	—	—	7	6	37
St. Lucia	70	73	—	—	35	31	—
St. Vincent and the Grenadines	70	73	—	—	4	4	—
Suriname	68	74	—	—	5	9	—
Trinidad and Tobago	68	73	—	—	1	3	33
Turks and Caicos	71	76	40	20	5	8	34

Sources: Population Reference Bureau, Data Sheet, *2001, Life Expectancy (M/F); The World Bank, World Development Indicators, 2001, Under Age 5 Mortality Rate (1980/99); The World Bank Atlas, 2001, Female Participation in Labor Force; CIA World Factbook, 2001, Percent Illiteracy (ages 15 and over).*

Long before offshore assembly plants and resorts began hiring women, Cuba's educational and labor policies yielded the most educated and professional women in the Caribbean. Here female doctors outnumber their male counterparts, and women have achieved near parity in the sciences, which is not true for many developed countries. Yet economic hardship may undermine these achievements as professionals see their wages and purchasing power erode. Cuba has not experienced the same level of brain drain as its neighbors, mostly because of restrictive emigration policies. Yet when the United States held a lottery for 6,000 visas in 1986, more than 400,000 Cubans applied.

Labor-Related Migration Inequitable economies and limited resources conspire to force many residents to migrate. Besides the obvious economic rationale to migrate, this strategy impacts community and household structures as well.

Historically, labor circulated within the Caribbean in pursuit of the sugar harvest—Haitians went to the Dominican Republic, residents of the Lesser Antilles journeyed to Trinidad. During the construction of the Panama Canal in the early 1900s, thousands of Jamaicans and Barbadians journeyed to the Canal Zone, and many remained.

After World War II, better transportation and political developments in the Caribbean produced a surge of migrants to North America. This trend began with Puerto Ricans going to New York in the early 1950s and intensified in the 1960s with the arrival of nearly half a million Cubans. Since then, large numbers of Dominicans, Haitians, Jamaicans, Trinidadians, and Guyanese have also migrated to North America, typically settling in Miami, New York, Los Angeles, and Toronto.

Crucial in this exchange of labor from north to south is the flow of cash **remittances** (monies sent back home). Immigrants are expected to send something back, especially when

LOCAL TO GLOBAL   The Caribbean Diaspora

The diaspora of the second half of the twentieth century has made Caribbeans an international people. In New York, Boston, Toronto, Montreal, Miami, London, Amsterdam, and Paris, immigrant communities regard their home as a distant island that they may never have seen, or barely remember, or remember in a way that has not existed for decades. Even the governments of those distant places recognize these unfamiliar exiles as a part of the greater nationality. Although the migratory experience has been painful for many Caribbeans, it has become a recognized part of the Caribbean way of life and a fixed element in the Caribbean economic system. The following story personalizes the immigrant experience as an immigrant from Barbados to Canada describes his encounter with North America.

Going to Canada was a physical shock for me. It was the first time that I had experienced temperatures below about 70 degrees, or maybe on a rare occasion I'd seen it get down to 65 here. Anyway, I went straight from balmy Barbados into 30- and 40-degree temperatures, and since I arrived in October each week would get colder and colder, until in winter the thermometer sometimes gets down to 40 below. My friends always think you are overdressed, but you are really just trying to get warm.

Most of the people I met didn't know where Barbados was. Many assumed it was part of Jamaica. But that began to change *when the weather report on the radio started giving the temperatures in Barbados. I think it was the Barbados tourist board that got the radio station to do that, and it worked well. When people didn't know where Barbados was I'd show them my passport, which I always had with me. On my passport cover was a map of the West Indian islands, starting with the tip of Florida and coming right down the chain to Trinidad and Tobago. So I could say, "Okay, here is Florida, here is Jamaica, and there is Barbados." They would always say, "Oh, it's really far south . . ."*

Then, in terms of speaking, it took you at least two months before anybody could fully understand what you were saying, in spite of the fact that we were all speaking English. The Canadians were always saying, "Pardon me," and I was forever repeating myself. I got tired of doing that, and I found it was easier to change my speech . . .

Another thing that freaked me out was that on Friday evenings everybody would leave class and go line up to get in a bar. It's below freezing, and there in the street is a line waiting to get into a bar, waiting to sit down and drink beer. I could not believe it.

Source: George Gmelch. 1992. *Double Passage: The Lives of Caribbean Migrants Abroad and Back Home*, pp. 248–250. Ann Arbor: University of Michigan Press.

immediate family members are left behind. Collectively, remittances add up; it is estimated that nearly $1 billion is annually sent to the Dominican Republic by immigrants in the United States, making remittance income the country's second leading industry. Governments and individuals alike depend on these transnational family networks. Families carefully select the member most likely to succeed abroad in the hopes that money will flow back and a base for future immigrants will be established. A Caribbean nation in which no one left would soon face a crisis. Economic opportunities are limited, and the populations have expanded far faster than the economies. Many of the people who have left would have been jobless at home or would have forced someone else to be jobless.

A common strategy for many immigrants is to work abroad and retire in the Caribbean with a pension and savings that ensure a comfortable old age. Returnees often bring cash, new skills, and increased expectations. There is anecdotal evidence that this cohort of migrants can introduce positive economic and political changes in particular localities. Still, it is unlikely that migrants and their remittances can challenge the larger structural forces that make migration attractive in the first place. Their impact is experienced primarily at the household level and is too fragmented to represent a national development force (see "Local to Global: The Caribbean Diaspora").

⊕ Conclusion

As a world region, the Caribbean is more integrated into the global economy than most areas of the developing world, albeit as a dependent economic periphery. Even though this is the smallest world region discussed, it offers some of the best examples of the long-term effects of globalization. This region was fashioned five centuries ago by Europeans who eliminated the native population and imported millions of Africans to support a plantation-based economy that supplied raw materials to Europe and North America. Consequently, most of the human and economic resources of the region went toward enriching other parts of the world, especially Europe and later North America. In a blatant example of colonial exploitation, little was invested in the social and economic development of these colonial localities. Today the region includes 20 independent countries, although each maintains strong international links. Rather than harvesting sugarcane, Caribbean laborers are more likely to sew garments, assemble telephones, manage offshore banks, and cater to the needs of the 14 million tourists who visit each year. The economies of the region have diversified, although they still depend on foreign investment, exports, and tourism. Perhaps more significant are the region's strides in social development, especially in education, health, and the status of women, which distinguish it from other developing areas.

There is a definite transnational quality to life in this region as social and economic influences from Latin America, Africa, North America, and Europe are woven into the lifestyle. Rap music blares from boom boxes in the streets of Havana and Kingston, while African-based religions are intertwined with Catholicism in the countryside. Cricket is a popular sport in the eastern Caribbean, while baseball dominates in Cuba and the Dominican Republic. Similarly, as Caribbean residents emigrate in search of opportunities, they introduce their own traditions to host countries. In Toronto, for example, Caribbean immigrants hold Canada's largest Carnival. The input of so many cultures has triggered an eclectic identity for the Caribbean, most notably through its literature, music, and arts.

The Caribbean also maintains important ties with Europe and Africa. European colonies still exist in the region and maintain special trade relations with their former colonizers. Even independent states look to their former rulers for help; for example,

since the fall of the Soviet Union, Spain is one of Cuba's largest foreign investors. In contrast, Caribbean links with Africa are more cultural and political than economic. The cultural transfers of music, religion, and dance from West Africa to the Caribbean are well documented. Moreover, Afro-Caribbean scholars have supported the development of a Pan-African consciousness in their writings.

With the end of the Cold War, many of the microstates in the region fear that their lack of strategic significance may result in neglect by the United States and Europe and may prejudice their ability to participate in world trade. It seems that the intervention of the United States in the past few years has been driven more by a fear of thousands of refugees seeking sanctuary in Florida than by the ideologies and geopolitics of the past. Certainly it is difficult to untangle the Caribbean from the larger regions that surround it. Perhaps it is best to view it as the crossroads of the Americas. Whether through offshore banking, money laundering, the drug trade, or commercial airline traffic, the Caribbean is a vibrant intersection for the Western Hemisphere.

⊕ Key Terms

African diaspora (page 195)
brain drain (page 212)
capital leakage (page 212)
Caribbean Community and
   Common Market
   (CARICOM) (page 206)
Caribbean diaspora
   (page 189)

chain migration (page 190)
circular migration (page 190)
creolization (page 193)
free trade zones (page 208)
Greater Antilles (page 183)
houseyards (page 190)
hurricanes (page 185)

indentured labor (page 194)
isolated proximity (page 180)
Lesser Antilles (page 184)
maroons (page 195)
mono-crop production
   (page 193)
Monroe Doctrine (page 200)

neocolonialism (page 200)
offshore banking (page 208)
plantation America (page 193)
remittances (page 213)
rimland (page 182)

⊕ Questions for Summary and Review

1. What characteristics define the Caribbean as a world region?

2. What environmental, economic, and locational factors contributed to the growth of tourism in the Caribbean?

3. What is the typical path of a hurricane in the Caribbean, and when are these meteorological disturbances likely to occur?

4. What significance does sugarcane play in the region's history?

5. What factors contributed to the deforestation of the Caribbean islands? Why are the rimland forests, in comparison, still intact?

6. What happened to the Amerindian population in the Caribbean?

7. How is AIDS impacting this region, and what is being done about it?

8. How does the experience of the Garifuna people illustrate the process of creolization?

9. What cultural traditions transferred from Africa to the Caribbean via the slave trade?

10. How is the cultural history of the Caribbean reflected in its music? What musical styles are associated with the region?

11. What geopolitical role did Cuba play during the Cold War? What is Cuba's current geopolitical significance?

12. Which countries rely on mining to bring in revenue? What is mined in this region?

13. Where are the Asian populations in the Caribbean? What explains their presence?

14. Which territories in the region are still colonies? Why have they not sought independence?

15. Where is "plantation America," and what characteristics define it?

16. What are the banana wars and how do they illustrate the tensions brought on by globalization?

⊕ Thinking Geographically

1. After looking at the map of offshore banking in the Caribbean, what conclusions can you draw about where this occurs?

2. Contrast the historical African diaspora to the contemporary Caribbean one. What patterns are formed by these two distinct population movements? What social and economic forces are behind them? Are they comparable?

3. What advantages and disadvantages exist when pursuing tourism as a development strategy?

4. What advantages might Caribbean free trade zones retain over their competitors in Southeast and East Asia? What disadvantages might they face?

5. Why have U.S. actions in the region been considered neo-colonial? Are there other regions in the world where the United States exerts neocolonial tendencies?

6. Compare and contrast the industrial development of Cuba with Puerto Rico. How were these two islands integrated into the world economy during the twentieth century?

7. Why does chain migration occur between the Caribbean and North America or Europe? How does a migrant benefit from being part of a chain migration? How does chain migration affect sending communities?

8. In the twenty-first century, will agricultural exports continue to be important for Caribbean economies?

9. How might the September 2001 terrorist attacks on the United States impact the regional economy of the Caribbean?

⊕ Regional Novels and Films

Novels

Reinaldo Arenas, *Farewell to the Sea: A Novel of Cuba* (1994, Penguin)

Aimé Césaire, *Notebook on a Return to the Native Land* (2001, Wesleyan University Press)

Patrick Chamoiseau, *Texaco* (1988, Vintage International)

Edwidge Danticat, *The Farming of Bones* (1998, Soho)

Edwidge Danticat, ed., *The Butterfly's Way: Voices from the Haitian Dyaspora in the United States* (2001, Soho)

Zee Edgell, *Beka Lamb* (1982, Heinemann)

Cristina Garcia, *Dreaming in Cuba* (1992, Random House)

Jamaica Kincaid, *At the Bottom of the River* (1983, Farrar, Straus & Giroux)

Jamaica Kincaid, *A Small Place* (1988, Penguin)

William Luis, ed., *Dance Between Two Cultures: Latino Caribbean Literature Written in the United States* (1997, Vanderbilt)

V. S. Naipaul, *The Middle Passage* (1963, Macmillan)

V. S. Naipaul, *Miguel Street* (1959, Vanguard Press)

Derek Walcott, *Omeros* (1990, Farrar, Straus & Giroux)

Derek Walcott, *Sea Grapes* (1976, Farrar, Straus & Giroux)

Derek Walcott, *The Star-Apple Kingdom* (1979, Farrar, Straus & Giroux);

Alec Waugh, *Island in the Sun* (1955, Farrar, Straus, and Cudahy)

Films

Bitter Cane (1986, Haiti)

Buena Vista Social Club (1999, Cuba)

Cuba—In the Shadow of Doubt (1987, Cuba)

The Emperor's Birthday: The Rastafarians Celebrate (1997, Jamaica)

The Harder They Come (1972, Jamaica)

Inside Castro's Cuba (1995, Cuba)

The King Does Not Lie: The Initiation of a Shango Priest (1993, Trinidad)

Sugar Cane Alley (1983, Martinique)

⊕ Bibliography

Allsopp, Richard, ed. 1996. *Dictionary of Caribbean English Usage*. Oxford: Oxford University Press.

Babcock, Elizabeth, and Conway, Dennis. 2000. "Why International Migration Has Important Consequences for the Development of Belize." *Yearbook, Conference of Latin Americanist Geographers*, 26, 71–86.

DeSouza, Roger Mark. 1998. "The Spell of the Cascadura: West Indian Return Migration," in *Globalization and Neo-Liberalism: The Caribbean Context*, edited by Thomas Klak. Lanham, MD: Rowman and Littlefield.

Europa. 2001. *South America Central America and the Caribbean*, 9th ed. London: Europa Publications Limited.

"Expelled from Eden." *The Economist*, December 20, 1997, pp. 3–38.

Gmelch, George. 1992. *Double Passage: The Lives of Caribbean Migrants Abroad and Back Home*. Ann Arbor: University of Michigan Press.

Gordon, Joy. 1997. "Cuba's Entrepreneurial Socialism." *Atlantic Monthly*, January, 18–30.

Green, Cecilia. 1998. "The Asian Connection: The US Caribbean Apparel Circuit and a New Model of Industrial Relations." *Latin American Research Review,* 33(3), 7–47.

Horst, Oscar, and Asagiri, Katsuhiro. 2000. "The Odyssey of Japanese Colonists in the Dominican Republic." *Geographical Review,* 90(3), 335–58.

Klak, Thomas, ed. 1998. *Globalization and Neoliberalism: The Caribbean Context.* Lanham, MD: Rowman and Littlefield.

Levinson, David. 1998. *Ethnic Groups Worldwide: A Ready Reference Handbook.* Phoenix, AZ: Oryx Press.

Levitt, Peggy. 2001. *The Transnational Villagers.* Berkeley: University of California Press.

Lowell, B. Lindsay, and de la Garza, Rodolfo. 2000. "The Development Role of Remittances in U.S. Latino Communities and in Latin American Countries." Report from the Inter-American Dialogue and The Tomás Rivera Policy Institute.

Mintz, Sidney. 1985. *Sweetness and Power: The Place of Sugar in Modern History.* New York: Viking Penguin.

Mintz, Sidney. 1989. *Caribbean Transformations.* New York: Columbia University Press.

Mintz, Sidney, and Price, Sally. 1985. *Caribbean Contours.* Baltimore: Johns Hopkins University Press.

Momsen, Janet H., ed. 1993. *Women and Change in the Caribbean.* London: James Currey Limited.

Mullings, Beverly. 1999. "Sides of the Same Coin? Coping and Resistance Among Jamaican Data-Entry Operators." *Annals of the Association of American Geographers,* 89(2), 290–311.

Okpewho, Isidore; Davies, Carole Boyce; and Mazrui, Ali A. 1999. *The African Disapora: African Origins and New World Identities.* Bloomington, IN: Indiana Univeristy Press.

Pattullo, Polly. 1996. *Last Resorts: The Cost of Tourism in the Caribbean.* New York: Monthly Review Press.

Price, Richard. 1979. *Maroon Societies: Rebel Slave Communities in the Americas.* Baltimore: Johns Hopkins University Press.

Pulsipher, Lydia. 1993. "Changing Roles in the Life Cycles of Women in Traditional West Indian Houseyards," in *Women and Change in the Caribbean* (pp. 50–64), edited by Janet Momsen. Bloomington, IN: Indiana University Press.

Richardson, Bonham C. 1992. *The Caribbean in the Wider World, 1492–1992.* Cambridge, England: Cambridge University Press.

Roberts, Susan. 1995. "Small Place, Big Money: The Cayman Islands and the International Financial System." *Economic Geography,* 71(3), 237–56.

Sale, Kirkpatrick. 1990. *The Conquest of Paradise: Christopher Columbus and the Columbian Legacy.* New York: Knopf.

Sauer, Carl Ortwin. 1969. *The Early Spanish Main.* Berkeley, CA: University of California Press.

Scarpaci, Joseph. 2000. "Reshaping Habana Vieja: Revitalization, Historic Preservation, and Restructuring in the Socialist City." *Urban Geography,* 21(8), 724–44.

Segre, Roberto; Coyula, Mario; and Scarpaci, Joseph L. 1997. *Havana: Two Faces of the Antillean Metropolis.* New York: John Wiley & Sons.

Tenenbaum, Barbara, ed. 1996. *Encyclopedia of Latin American History and Culture.* New York: Charles Scribner's Sons.

Thornton, John. 1992. *Africa and Africans in the Making of the Atlantic World, 1400–1680.* Cambridge, England: Cambridge University Press.

Tulchin, Joseph, and Espach, Ralph, eds. 2000. *Security in the Caribbean Basin: The Challnege of Regional Cooperation.* Boulder, CO: Lynne Rienner Publishers.

U.S. Committee for Refugees. 2001. *World Refugee Survey.*

Voeks, Robert. 1993. "African Medicine and Magic in the Americas." *Geographical Review,* 83(1), 66–78.

Wagley, Charles. 1970. "Plantation-America: A Culture Sphere," in *Caribbean Studies: A Symposium* (pp. 3–13), edited by Vera Rubin. Seattle: University of Washington Press.

Warf, Barney. 1995. "Information Services in the Dominican Republic." *Yearbook, Conference of Latin Americanist Geographers,* 21, 13–23.

Watts, David. 1987. *The West Indies: Patterns of Development, Culture and Environmental Change Since 1492.* Cambridge, England: Cambridge University Press.

Wiley, James. 1996. "The European Union's Single Market and Latin America's Banana Exporting Countries." *Yearbook, Conference of Latin Americanist Geographers,* 22, 31–40.

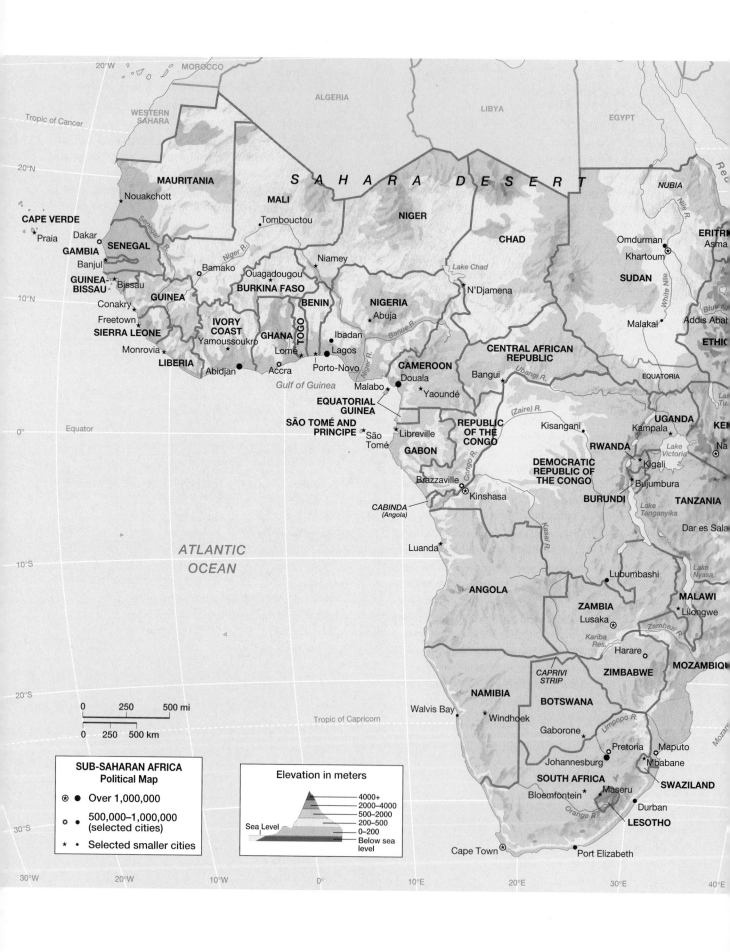

20°W · MOROCCO

Tropic of Cancer

WESTERN
SAHARA

ALGERIA

LIBYA

EGYPT

20°N

MAURITANIA

Nouakchott ★

MALI

S A H A R A   D E S E R T

NUBIA

NIGER

CHAD

Nile R.

Red

ERITREA

Asma

CAPE VERDE

Praia

Dakar ★

Tombouctou

Niger R.

Omdurman
Khartoum ⊛

SENEGAL

GAMBIA

Banjul

GUINEA-
BISSAU

Bissau ★

Conakry ★

Freetown ★

SIERRA LEONE

Monrovia ★

LIBERIA

Bamako ★

Niamey ★

Ouagadougou ★

BURKINA FASO

BENIN

IVORY
COAST

Yamoussoukro ★

GHANA

Lomé ⊙

TOGO

Abidjan ★

Accra ⊙

Ibadan ●

Lagos ●

Porto-Novo ⊙

Abuja ★

NIGERIA

Lake Chad

N'Djamena ★

Benue R.

SUDAN

Malakal ★

White Nile

Blue Nile

Addis Abab

ETHIO

10°N

CENTRAL AFRICAN
REPUBLIC

Bangui ⊙

EQUATORIA

Gulf of Guinea

CAMEROON

Douala ●

Malabo ★

Yaoundé ★

Ubangi R.

Equator 0°

EQUATORIAL
GUINEA

SÃO TOMÉ AND
PRINCIPE

São
Tomé

Libreville ★

GABON

(Zaire) R.

REPUBLIC
OF THE
CONGO

Kisangani ★

DEMOCRATIC
REPUBLIC OF
THE CONGO

UGANDA

Kampala ⊛

RWANDA

Kigali ★

Bujumbura ★

Lake
Victoria

KE

Na

Brazzaville ⊛

Congo R.

Kinshasa ⊛

BURUNDI

Lake
Tanganyika

TANZANIA

Dar es Sala

CABINDA
(Angola)

Kasai R.

10°S

Luanda ●

ATLANTIC
OCEAN

ANGOLA

Lubumbashi ★

Lake
Nyasa

MALAWI

Lilongwe ★

ZAMBIA

Lusaka ⊛

Zambezi R.

Kariba
Res.

Harare ⊙

MOZAMBIQU

ZIMBABWE

CAPRIVI
STRIP

20°S

NAMIBIA

Walvis Bay ★

Windhoek ★

BOTSWANA

Gaborone ★

Limpopo R.

Maputo ⊙

Mbabane ★

SWAZILAND

Tropic of Capricorn

250   500 mi

0   250   500 km

SUB-SAHARAN AFRICA
Political Map

⊛ ● Over 1,000,000

⊙ ● 500,000–1,000,000
      (selected cities)

★ • Selected smaller cities

Elevation in meters

4000+
2000–4000
500–2000
200–500
0–200
Below sea
level

Sea Level

Johannesburg ●

Pretoria ●

SOUTH AFRICA

Bloemfontein ★

Maseru ★

Durban ●

Orange R.

LESOTHO

Moz

Cape Town ⊛

Port Elizabeth ★

30°S

30°W   20°W   10°W   0°   10°E   20°E   30°E   40°E

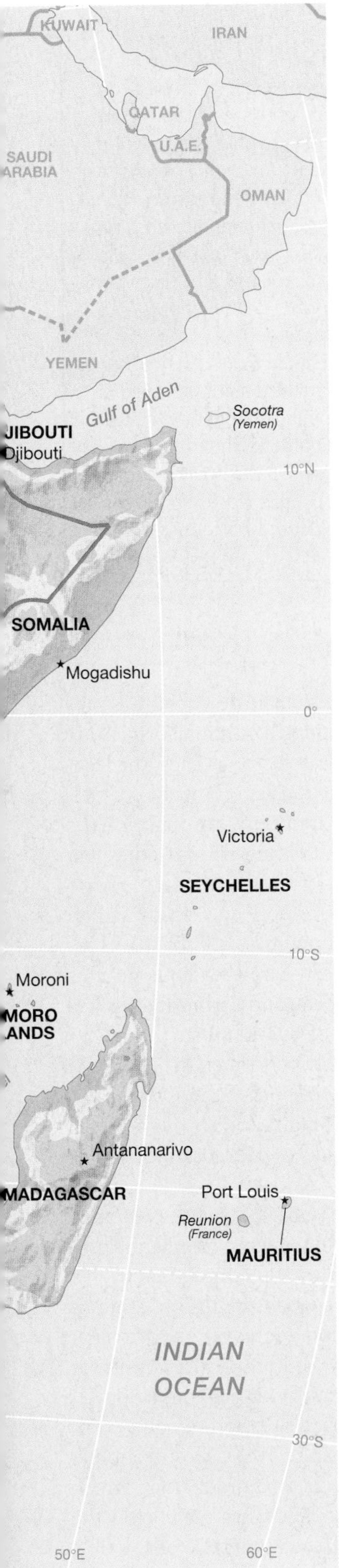

CHAPTER 6

Sub-Saharan Africa

Compared with Latin America and the Caribbean, Africa south of the Sahara is poorer and more rural, and its population is very young. More than 670 million people reside in this region, which includes 48 states and one territory. Demographically, this is the world's fastest-growing region (2.5 percent rate of natural increase); in most countries, nearly half the population (45 percent) is younger than 15 years old. Income levels are low (a per capita GNI of $520 in 1998), and more than one-third of the region's population lives on less than $1 per day. Life expectancy is only 51 years. For many people, this part of the world has become synonymous with poverty, disease, violence, and refugees. Overlooked in the all-too-frequent negative headlines are initiatives by private, local, and state groups to improve the region's quality of life. Many states have reduced infant mortality, expanded basic education, and increased food production in the past two decades despite the profound economic and political crises that beleaguer this region.

Sub-Saharan Africa—that portion of the African continent lying south of the Sahara Desert—is a commonly accepted world region (Figure 6.1). It is historically referred to as "Black Africa" because of the skin color of its indigenous inhabitants, but such a race-based division falls apart under closer inspection. The coherence of this region has less to do with skin color than with similar livelihood systems and a shared colonial experience. No common religion, language, philosophy, or political system ever united the area. Instead, diffuse cultural bonds developed from a variety of lifestyles and idea systems that evolved here. The impact of outsiders also helped to determine the region's identity. Slave traders from Europe, North Africa, and Southwest Asia treated Africans as chattel; up until the mid-1800s millions of Africans were taken from the region and sold into slavery. In the late 1800s the entire African continent was divided by European colonial powers, imposing political boundaries that remain to this day. In the postcolonial period, which began in the 1960s, Sub-Saharan African countries faced many of the same economic and political challenges, creating still more commonality (see "Setting the Boundaries").

The cultural complexity of the region is underappreciated. It is not uncommon for 20 or more languages to be spoken in one large state. Consequently, most Africans understand and speak several languages. Ethnic identities do not conform to the political divisions of Africa, sometimes resulting in bloody ethnic warfare, as witnessed in Rwanda in the mid-1990s. Nevertheless, throughout the region peaceful coexistence between

◀ **Figure 6.1 Sub-Saharan Africa** Africa south of the Sahara includes 48 states and one territory. This vast region of rain forest, tropical savanna, and desert is home to more than 670 million people. Much of the region consists of broad plateaus ranging from 500 to 2,000 meters in elevation. Although the population is growing rapidly, the overall population density of Sub-Saharan Africa is low. Considered one of the least developed regions of the world, it remains an area rich in natural resources.

SETTING THE BOUNDARIES

Sub-Saharan Africa includes 43 mainland states plus the island nations of Madagascar, Cape Verde, São Tomé and Principe, Seychelles, Mauritius, and the French territory of Reunion. When setting this particular regional boundary, the major question one faces is how to treat North Africa.

Many scholars argue for Africa's continental integrity as a culture region. The Sahara has never formed a complete barrier between the Mediterranean north and the remainder of the African landmass. Moreover, the Nile River forms a corridor several thousand miles long of continuous settlement linking North Africa directly to the center of the continent. There is no obvious place to divide the watershed between northern and Sub-Saharan Africa. Political units, such as the Organization of African Unity, are modern examples of the continent's indivisibility.

The lack of a clear divide across Africa does not invalidate its division into two world regions. North Africa is generally considered more closely linked, both culturally and physically, to Southwest Asia. Arabic is the dominant language and Islam the dominant religion of North Africa. Consequently, North Africans feel more closely connected to the Arab hearth in Southwest Asia than to the Sub-Saharan world.

The decision to view Sub-Saharan Africa as a world region still presents the problem of where to divide it from Africa north of the Sahara. Political boundaries were preserved when drawing the line so that the Mediterranean states of North Africa will be discussed with Southwest Asia. Sudan will be discussed in both Chapters 6 and 7 because it shares characteristics common to both regions. Sudan is Africa's largest state in terms of area; it is one-fourth the size of the United States. In the more populous and powerful north, Muslim leaders have forged an Islamic state that is culturally and politically oriented toward North Africa and Southwest Asia. Southern Sudan, however, has more in common with the animist and Christian groups of the Sub-Saharan region. Moreover, peoples in the south continue to fight for their independence from Khartoum.

---

distinct ethnic groups is the norm. The cultural impress of European colonizers cannot be ignored: European languages, religions, educational systems, and political ideologies were adopted and modified. Yet the daily rhythms of life are far removed from the industrial world. Most Africans still engage in subsistence and cash-crop agriculture. Women in particular are charged with tending crops and procuring household necessities (Figure 6.2). Male roles revolve around tending livestock and the public life of the village and market. As is happening elsewhere in the developing world, cities are growing rapidly, although just 30 percent of the population is classified as urban.

▲ **Figure 6.2 Women in the fields** Women from the Merina tribe in Madagascar plant paddy rice. The majority of Sub-Saharan people engage in subsistence agriculture. Women are typically responsible for producing food crops, while men are more likely to tend livestock. *(Carl D. Walsh/Aurora & Quanta Productions)*

The influence of African peoples outside the region is profound, especially when one considers that human origins are traceable to this part of the world. In historic times, the legacy of the slave trade resulted in the transfer of African peoples, religious systems, and musical traditions throughout the Western Hemisphere. Many of the rhythms and melodies that make up the core of popular Western music are traceable to West African musical traditions. American jazz, Brazilian samba, and Cuban rumba would not exist but for the influence of Africans. Even today, African-based religious systems are widely practiced in the Caribbean and Latin America, especially in Brazil. The popularity of "world music" has created a large international audience for performers such as Papa Wemba from the Democratic Republic of the Congo and South Africa's Ladysmith Black Mambazo. Through music and the arts, Sub-Saharan Africa has exerted a significant influence on global culture.

The African economy, however, is marginal when compared with the rest of the world. According to the World Bank, Sub-Saharan Africa's economic output in 1999 amounted to just 1 percent of global output, even though the region contains 11 percent of the world's population. Moreover, the gross national product of just one country, South Africa, accounts for nearly half (44 percent) of the region's total economic output. Even the region's share of world primary exports (such as copper, cocoa, and coffee) has fallen since 1980. Most troubling of all, Sub-Saharan Africa's per capita food production, which made gains in the 1960s and early 1970s, has been below the region's 1961 output level for the last 25 years. No one doubts that there is undercounting of subsistence activities in the region, but rapid population growth has greatly increased food demands, and it is hard to ignore this troubling decline.

Many scholars feel that Sub-Saharan Africa has benefited little from its integration (both forced and voluntary) into the global economy. Slavery, colonialism, and export-oriented mining and agriculture served the needs of consumers outside the region but undermined domestic production for those within. Foreign assistance in the postindependence years initially improved agricultural and industrial output, but also led to mounting foreign debt and corruption, which over time undercut the region's economic gains. One consequence of mounting debt was the implementation of **structural adjustment programs** by the World Bank and the International Monetary Fund in the 1980s. These controversial economic measures are designed to reduce government spending and encourage private sector initiatives. Yet they typically trigger drastic cutbacks in government-supported services and food subsidies, which disproportionately affect the poor. The idea of debt forgiveness for Africa's poorest states, championed by the unlikely duo of economist Jeffrey Sachs and rock star Bono from U2, is gradually gaining acceptance as a strategy to reduce human suffering in the region.

Given the region's history, most scholars argue that foreign involvement in Sub-Saharan Africa has led to greater underdevelopment than development. Ironically, many of these same scholars worry that global pessimism about the region has produced a pattern of neglect. Private capital investment in Sub-Saharan Africa lags far behind investment rates in Latin America, the Caribbean, and East and Southeast Asia. It is hard to imagine how the region can economically recover without new private investment.

Small political and social movements within the region offer valuable lessons on the importance of meeting Africa's basic needs first. Thousands of nongovernmental organizations have formed to support women's groups, cooperatives, schools, and health centers. By addressing basic needs and empowering community members to solve their own problems, organizations such as the Hadasko Women's Farmers Association of Ghana, the Samanko women's group in Mali, and the Green Belt Movement in Kenya have become progressive agents for local change. Although there are serious human rights issues throughout the region, the near-bloodless transition from white rule to black rule in South Africa in 1994 underscores the politics of promise for Sub-Saharan Africa. To recognize the scale of what is possible for this region, familiarity with its vast yet capricious resource base is necessary.

Environmental Geography: The Plateau Continent

The largest landmass straddling the equator, Sub-Saharan Africa is vast in scale and its physical environment is remarkably beautiful. Called the plateau continent, the African interior is dominated by extensive uplifted areas that resulted from the breakup of **Gondwanaland**, an ancient megacontinent that included Africa, South America, Antarctica, Australia, Madagascar, and Saudi Arabia. Some 250 million years ago, it began to split apart through the forces of continental drift. As this process unfolded, the African landmass experienced a series of continental uplifts that left much of the area with vast elevated plateaus. The highest areas are found on the eastern edge of the continent, where the Great Rift Valley forms a complex upland area of lakes, volcanoes, and deep valleys. In contrast, lowlands prevail in West Africa, although smaller elevated areas exist, such as the Guinea and the Andamawa highlands.

The landscape of Sub-Saharan Africa offers a palette of intense colors: deep red soils studded with plantings of subsistence crops; the blue tropical sky; golden savannas that ripple with the movement of animal herds; dark rivers meandering through towering rain forests; and sun-drenched deserts. Amid this beauty, however, one finds relatively poor soils, endemic disease, and vulnerability to drought. Large areas for potential agricultural development still exist, however, especially in southern Africa; and throughout the continent vast water resources, biodiversity, and mineral wealth abound.

Africa's Environmental Issues

The prevailing perception of Africa south of the Sahara is one of environmental scarcity and degradation, no doubt fostered by televised images of drought-ravaged regions and starving children. Single explanations such as rapid population growth or colonial exploitation cannot fully capture the complexity of African's environmental issues or the ingenious ways that people have adapted to living in marginal ecosystems. Because much of Sub-Saharan Africa's population is rural and poor, earning its livelihood directly from the land, sudden environmental changes are keenly felt and can cause mass migrations, famine, and even death. As Figure 6.3 illustrates, deforestation and **desertification**, the expansion of desertlike conditions as a result of human-induced degradation, are commonplace. Sub-Saharan African is also vulnerable to drought, most notably in the Horn of Africa, parts of southern Africa, and the Sahel. As Sub-Saharan cities grow in size and importance, urban environments increasingly face problems of air and water pollution as well as sewage and waste disposal. At the same time, wildlife tourism is an increasingly important source of foreign exchange for many African states. Throughout the region, national parks have been created in an effort to strike a balance between humans' and animals' competing demands for land.

The Sahel and Desertification In the 1970s the **Sahel** became an emblem for the dangers of unchecked population growth and human-induced environmental degradation when a relatively wet period came to an abrupt end and six years of drought (1968–1974) ravaged the land. The Sahel is a zone of ecological transition between the Sahara to the north and wetter savannas and forest in the south. Life depends on a delicate balance of limited rain, drought-resistant plants, and a pattern of animal **transhumance** (the movement of animals between wet-season and dry-season pasture). During the drought, rivers in the area diminished and desertlike conditions began to move south. Unfortunately,

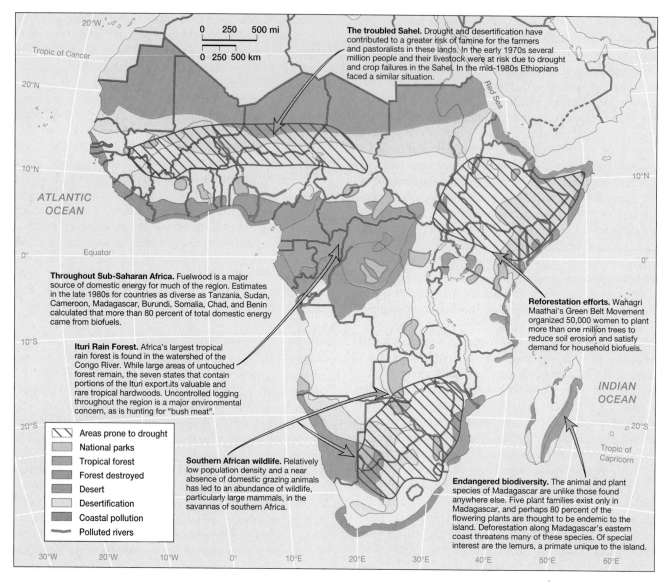

The troubled Sahel. Drought and desertification have contributed to a greater risk of famine for the farmers and pastoralists in these lands. In the early 1970s several million people and their livestock were at risk due to drought and crop failures in the Sahel. In the mid-1980s Ethiopians faced a similar situation.

Throughout Sub-Saharan Africa. Fuelwood is a major source of domestic energy for much of the region. Estimates in the late 1980s for countries as diverse as Tanzania, Sudan, Cameroon, Madagascar, Burundi, Somalia, Chad, and Benin calculated that more than 80 percent of total domestic energy came from biofuels.

Ituri Rain Forest. Africa's largest tropical rain forest is found in the watershed of the Congo River. While large areas of untouched forest remain, the seven states that contain portions of the Ituri export its valuable and rare tropical hardwoods. Uncontrolled logging throughout the region is a major environmental concern, as is hunting for "bush meat".

Reforestation efforts. Wahagri Maathai's Green Belt Movement organized 50,000 women to plant more than one million trees to reduce soil erosion and satisfy demand for household biofuels.

Southern African wildlife. Relatively low population density and a near absence of domestic grazing animals has led to an abundance of wildlife, particularly large mammals, in the savannas of southern Africa.

Endangered biodiversity. The animal and plant species of Madagascar are unlike those found anywhere else. Five plant families exist only in Madagascar, and perhaps 80 percent of the flowering plants are thought to be endemic to the island. Deforestation along Madagascar's eastern coast threatens many of these species. Of special interest are the lemurs, a primate unique to the island.

Legend:
- Areas prone to drought
- National parks
- Tropical forest
- Forest destroyed
- Desert
- Desertification
- Coastal pollution
- Polluted rivers

▲ Figure 6.3 Environmental issues in Sub-Saharan Africa   Given the immense size of Sub-Saharan Africa, it diffi-cult to generalize about environmental problems. Dependence on trees for fuel places strains on forests and wooded savannas throughout the region. In semiarid regions, such as the Sahel, population pressures and land-use practices seem to have exacerbated desertification. Yet Sub-Saharan Africa also supports the most impressive array of wildlife, especially large mammals, on Earth. The rapid growth of cities in the past three decades has yielded new concerns about water quality and pollution.   (Adapted from DK World Atlas, 1997, p. 75. London: DK Publishing)

tens of millions of people inhabited this area, and farmers and pastoralists whose livelihoods had come to depend on the more abundant precipitation of the relatively wet period were temporarily forced out.

Considerable disagreement persists over the root causes of the Sahelian drought. Was it a matter of too many humans degrading their environment, ill-conceived settlement schemes encouraged by European colonizers, or a failure to understand cycles in the global atmospheric system? Certainly human activity in the region greatly increased in the mid-twentieth century, making the case for human-induced de-sertification more plausible. But it is worth noting that parts of the Sahel were important areas of settlement long before European colonization. What appears to be desert wasteland

in April or May is transformed into a verdant garden of mil-let, sorghum, and peanuts after the drenching rains of June. Relatively free of the tropical diseases found in the wetter zones to the south, Sahelian soils are also quite fertile, which helps to explain why people continue to live there despite the unreliable rainfall patterns of recent decades.

The main culprits in the case for desertification are the ex-pansion of agriculture and overgrazing, leading to the loss of natural vegetation and declines in soil fertility. Through most of the Sahel, French colonial authorities forced villagers to grow peanuts as an export crop, a policy continued by the newly independent states of the region. Unfortunately, peanuts tend to deplete several key soil nutrients, which means that peanut farms are often abandoned after a few years as cultivators move on to fresh sites. A number of years are required for natural vegetation to become reestablished and begin to hold the soil in place and absorb rainfall. Since peanuts grow underground, moreover, the soil must be overturned at harvest time. This typically occurs at the onset of the dry season, when dry winds from the Sahara can carry away the fine dirt of the newly harvested field. With the loss of topsoil, the ground no longer absorbs the precipitation that does fall—often in drenching torrents—during the brief rainy season. The end result of this grim process is the spread of desertlike conditions regardless of actual changes in precipitation.

Overgrazing has also been implicated in Sahelian desertification, as is true in other regions of the world. Livestock is a traditional product of the region, and animal production dramatically expanded after World War II as the population swelled and the need for export earnings grew. In many areas, natural pasture was greatly reduced by intensified grazing pressure, leading to increased wind erosion. National and international developmental agencies, hoping to increase production further, began to dig deep wells in areas that had previously been unused by herders through most of the year. The new supplies of water, in turn, allowed year-round grazing in places that, over time, could not withstand it (Figure 6.4). Large barren circles around each new well began to appear even on satellite images. Some scholars questioned whether recovery of such degraded areas was possible even if "normal" precipitation patterns were to return (see "From the Field: Mobility and Livelihood in the Sahel").

Pessimists view the process of Sahelian degradation as virtually irreversible and argue that the region can no longer support the millions of people who presently reside there; a few foresee a future of mass starvation. Optimists contend that the land can quickly recover if appropriate practices are followed, and a few maintain that, given proper adjustments, larger populations could subsist in the area even if dry conditions continue. Debates also surround the issue of climatic fluctuation itself. Some climatologists believe that the Sahel now faces a prolonged dry period, while others argue that wetter conditions could soon return. A few suggest that, in the warmer global climate of the future, precipitation should increase markedly in the area. Regardless of the causes, many of the areas experiencing desertification are the same ones most vulnerable to drought and famine.

Deforestation   Although Sub-Saharan Africa still contains extensive forests, much of the region has relatively little tree cover. Unlike tropical America, forest clearance in the wet and dry savanna is of greater local concern than the limited commercial logging of the rain forest. North of the equator, for example, only a few wooded areas remain in a landscape dominated by grasslands, savanna (grassland with scattered trees and shrubs), and cropland. Highland Ethiopia was once covered with lush forests, but these have long since been reduced to a few remnant patches. Loss of woody vegetation has resulted in extensive hardship, especially for women and children who must spend many hours a day scrounging for wood (Figure 6.5). Deforestation, especially in the woodlands of the savannas, aggravates problems of soil erosion, moisture loss, and shortages of **biofuels** (wood and charcoal used for household energy needs, especially cooking).

In the southern tropical savanna where human population has historically been lighter, extensive tracts of dry woodland remain intact. Zambia and sections of Mozambique, Angola, and Tanzania are still extensively wooded, although the actual tree coverage is not dense. The trees of the dry forest have little commercial value compared to some of the species found in the rain forest, but they provide vital subsistence resources for local people as well as wildlife habitat. Even in these more abundant areas, biofuel scarcity is common around the larger towns and villages. In some countries,

▲ **Figure 6.4 Desertification in the Sahel**   In southern Niger, pastoralists rely upon wells to water their herds of goats and cattle. The relationship between livestock, farming, and the Sahelian ecosystem is complex. Many researchers believe that increased grazing pressure brought on by more livestock and wells may have made the Sahel more vulnerable to desertification.   *(Victor Englebert/Englebert Photography, Inc.)*

▲ **Figure 6.5 Gathering fuel**   A wagon loaded with firewood in rural Mali. Villagers and city dwellers alike rely upon wood for household energy needs, placing an enormous strain on Africa's woodlands. In areas with the greatest scarcity, women and children may spend many hours each day gathering fuel. *(Betty Press/Woodfin Camp & Associates)*

After a two-year stint in the 1980s in Niger with the Peace Corps, geographer David Rain returned to the Sahelian city of Maradi, Niger, in the 1990s to study the significance of seasonal migration in the lives of Sahelian farmers and pastoralists. The following exert is from his book, *Eaters of the Dry Season* (Westview, 1999). The title comes from a Huasa phrase, *masu cin rani*, which means those "who look for food" or "who are able to eat during the dry season." During the dry season (October through May), thousands of people move to nearby or distant towns in order to sustain themselves.

The African Sahel has been portrayed as a portent of a global environmental future, and it is an oft-cited example of environmental ruin. Prior to the devestating drought that occurred in the late 1960s, the Sahel was virtually unkonwn to the Western world. While the conditions there were capturing the world's attention and begging for its compassion, an entity called "the Sahel" was created, first by French anthropologists and international aid workers. . . .

What visitors witnessed during those times of drought were massive out-migrations of entire villages and clans, the decimation of livestock herds, and widespread human suffering. What the visitors did not see occurred later, as farmers and herders returned to their homes after the rains began again. The resilience of the people was not in evidence, and the relatively low loss of life from the multiyear drought likewise escaped the attention of most observers and commentators. This resilience, which is bred into the backbone of the cultures of the Sahel, has permitted the continued settlement of a region inhospitable to many activities. . . .

In recounting depredations old and new in comparative perspective, it helps to take the long view. People who have over their history withstood the ravages of slave raiding, epidemics, and attacks by predators should perhaps be insulted that "we" think a half-century of inept European colonial occupation could destroy the fabric of society and stop civilization in its tracks. Where is the evidence that today's ravages are presenting a challenge against which the people are not equipped? The secret of the masu cin rani (dry-season migration) is that "they" are far better at surviving the challenges of the twenty-first century than "we" in the West believe them to be. Who survives, however—often the oldest and the most influential as opposed to the children—might indeed be an ethical point of contention in the hypothetical debate between Sahelians and the West.

By extending this observation to activities to promote economic development, it is clear that policies to root peasants and to force them to invest in places instead of people are not in the peasants' interests. . . . The implication of a value system that defines "places" as constellations of people and commitments, instead of as inventories of physical contents, is that there is no apparent advantage to investing in villages to make them more permanent, unless such investments are perceived to be in the interests of the inhabitants. Instead of trying to deny or negate complicated social webs, concerned parties should acknowledge them and then work with them to make them more resilient. Community and livelihoods should be emphasized over permanence of settlement. (Rain, 1999: 8–9, 236–37)

▶ **Figure 6.1.1 Map of Sahelian migration flows**   An urban network exists in the Sahel in which migrants move seasonally from north to south in lines that are perpendicular to rainfall gradients. These same routes were once used by the long-distance caravans and are still used by pastoralists moving livestock. *(Adapted from David Rain. 1999. Eaters of the Dry Season, p. 146, Boulder: Westview Press.)*

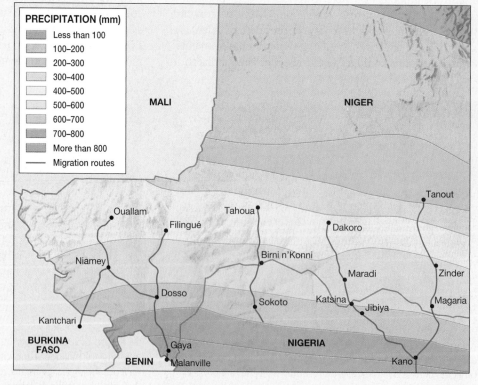

village women have organized into community-based non-governmental organizations (NGOs) to plant trees and create greenbelts to meet future fuel needs. One of the most successful efforts is in Kenya under the leadership of Wangari Maathai. Maathai's Green Belt Movement has more than 50,000 members, mostly women, organized into 2,000 local community groups. Since the group's inception in 1977, millions of trees have been successfully planted. In those areas, village women now spend less time collecting fuel, and local environments have improved. Kenya's success has drawn interest from other African countries, spurring a Pan-African Green Belt Movement largely organized through nongovernmental organizations interested in biofuel generation, protection of the environment, and the empowerment of women.

Destruction of rain forest for logging is most evident in the southern fringes of Central Africa's Ituri forest (again see Figure 6.3). Given the vastness of this forest and the relatively small number of people living there, however, it is less threatened than other forest areas. Two smaller rain forests, one along the Atlantic coast from Sierra Leone to western Ghana and the other along the eastern coast of the island of Madagascar, have nearly disappeared. These rain forests have been severely degraded by commercial logging and agricultural clearance. At present rates, several countries in West Africa will be completely deforested early in this century. Madagascar's eastern rain forests, as well as its western dry forests, have suffered serious degradation in the past three decades (Figure 6.6). Deforestation in Madagascar is especially worrisome because the island forms a unique environment with a large number of endemic species. Many species of fauna (including birds, reptiles, and lemurs) are now endangered, and several face imminent extinction.

Wildlife Conservation   Sub-Saharan Africa is justly famous for its wildlife. In no other region of the world can one find such abundance and diversity, especially of large mammals. The survival of wildlife here reflects, to some extent, the historically low human population density and the fact that sleeping sickness (transferred by the tsetse fly) and other diseases have kept people and their livestock out of many areas. In addition, many African peoples have developed various means of successfully coexisting with wildlife.

But as is true elsewhere in the world, wildlife is quickly declining in much of Sub-Saharan Africa. West Africa now has few areas of prime habitat, and what remains is rapidly shrinking. The most noted wildlife reserves are in East Africa; in Kenya and Tanzania these reserves are major tourist attractions and are thus economically important. Even there, however, population pressure, political instability, and poverty make the maintenance of large wildlife reserves difficult. Poaching is a major problem, particularly for rhinoceroses and elephants; the price of a single horn or tusk in distant markets represents several years' wages for most Africans. Ivory is avidly sought in East Asia, especially in Japan. In China powdered rhino horn is used as a traditional medicine, whereas in Yemen rhino horn is prized for dagger handles. During the 1980s the region's elephant population fell by more than half, to 600,000, in part due to the ivory trade. Rhino populations are far smaller, and the black rhino is considered an endangered species.

The most secure wildlife reserves now seem to be located in southern Africa (Figure 6.7). In fact, elephant populations are considered to be too large for the land to sustain in countries such as Zimbabwe. South Africa's wildlife reserves are so well managed that even white rhino populations are said to be too large. Some wildlife experts here contend that herds should be culled to prevent overgrazing, and that the ivory and rhino horn should to be legally sold in the international market in order to generate revenue for further conservation.

▲ **Figure 6.6 Deforestation in Madagascar**   Forest clearing for agriculture and fuel has denuded much of the countryside in Madagascar. Serious problems with soil erosion exist throughout the country, limiting agricultural productivity. During the rainy season, the discharge of topsoil into the Indian Ocean is visible from space. *(Bios/M. Gunther/Peter Arnold, Inc.)*

▲ **Figure 6.7 South African wildlife**   An African elephant traversing the savannas of Kruger National Park in South Africa, one of the region's oldest wildlife parks. South Africa, Botswana, Zimbabwe, and Zambia have the largest elephant populations. Other countries have seen a troubling decline in elephant numbers. *(Rob Crandall/Rob Crandall, Photographer)*

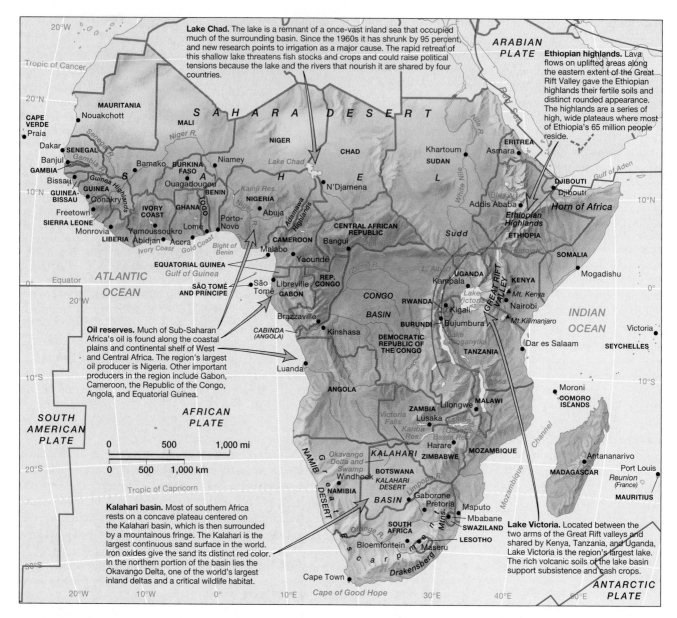

▲ **Figure 6.8 Physical geography of Sub-Saharan Africa** The so-called plateau continent was gradually uplifted with the breakup of Gondwanaland 250 million years ago. While much of the landmass is elevated, there are relatively few mountains except for those associated with the Rift Valley in eastern Africa, where a complex chain of mountains, lakes, and valleys exist. Four major river basins are associated with the region: the Congo, Nile, Niger, and Zambezi. The Congo flows through the largest rain forest in the world after the Amazon. The Nile, the lifeline for Sudan and Egypt, is the world's longest river.

Only such a market-oriented approach, they argue, will give African countries a long-term incentive to preserve habitat. Many environmentalists, not surprisingly, disagree strongly.

In 1989 a worldwide ban on the legal ivory trade was imposed as part of the Convention on International Trade in Endangered Species (CITES). While several African states, such as Kenya, lobbied hard for the ban, others, such as Zimbabwe, Namibia, and Botswana, complained that their herds were growing and the sale of ivory had helped to pay for conservation efforts. Conservationists feared that lifting the ban would bring on a new wave of poaching and illegal trade. In the late 1990s the ban was lifted so some southern African states could sell down their substantial inventories of ivory

and continue with limited sales. Botswana, for example, planned to export up to 25 tons of raw ivory to Japan in 1999. The long-term repercussions for elephant survival are not yet known, although political leaders from Kenya and India have voiced strong opposition to all ivory sales. The ivory controversy shows how global markets and international policies impact elephant survival in Sub-Saharan Africa. At the same time, differences within the region make it difficult to come up with a consistent conservation strategy. It is these basic differences in the region's physical geography that must be understood in order to comprehend the dynamic human–environmental relationships that shape African livelihoods.

Plateaus and Basins

A series of plateaus and elevated basins dominate the African interior and explain much of the region's uniqueness. Generally, elevations increase toward the south and east of the continent. Most of southern and eastern Africa lies well above 2,000 feet (600 meters), and sizable areas sit above 5,000 feet (1,500 meters). This is typically referred to as High Africa; Low Africa includes West Africa and much of Central Africa (Figure 6.8). The higher plateaus of Kenya, Zimbabwe, and Angola are noted for their cooler climates and relatively abundant moisture. Steep escarpments form where plateaus abruptly end, as illustrated by the majestic Victoria Falls on the Zambezi River (Figure 6.9). Much of southern Africa is rimmed by a landform called the **Great Escarpment,** which begins in southwestern Angola and ends in northeastern South Africa, creating a formidable impediment to coastal settlement. South Africa's Drakensberg Mountains (with elevations reaching 10,000 feet, or 3,100 meters) rise up from the Great Escarpment. Because of this landform, coastal plains tend to be narrow; few natural harbors exist; and river navigation is impeded by a series of falls. Fairly wide coastal lowlands are found in parts of western Africa, Mozambique, and Somalia, but even these lack extensive areas of fertile alluvial soil.

Though Sub-Saharan Africa is an elevated landmass, it has few significant mountain ranges. The one extensive area of mountainous topography is in Ethiopia, which lies in the northern portion of the Rift Valley zone. Yet even there the dominant features are high plateaus intercut with deep valleys rather than actual mountain ranges. Receiving heavy rains in the wet season, the Ethiopian Plateau forms the headwaters of several important rivers, most notably the Blue Nile, which joins the White Nile at Khartoum, Sudan.

A discontinuous series of volcanic mountains, some of them quite high, are associated with the southern half of the Rift Valley. Kilimanjaro at 19,000 feet (5,900 meters) is the continent's tallest mountain, and nearby Mount Kenya (17,000 feet, or 5,200 meters) is the second tallest. Both are snow-covered peaks that rise from the grassy upland plains. The physiographically complex Rift Valley reveals the slow but inexorable progress of geological forces (Figure 6.10). Eastern Africa is slowly being torn away from the rest of the continent, and within some tens of millions of years it will form a separate landmass. Such motion has already produced a great gash across the uplands of eastern Africa, much of which is occupied by elongated and extremely deep lakes (most notably Nyasa, Malawi, and Tanganyika). In central eastern Africa this rift zone splits into two separate valleys, each of which is flanked by volcanic uplands. Between the eastern and western rifts lies a bowl-shaped depression, the center of which is filled by Lake Victoria—Africa's largest body of water. Not surprisingly, some of the densest areas of settlement are found amid the fertile and well-watered soils that border the Rift Valley.

Watersheds   Africa south of the Sahara conspicuously lacks the broad, alluvial lowlands that influence patterns of settlement throughout other regions. The four major river systems

▲ **Figure 6.9 Victoria Falls**   The Zambezi River descends over Victoria Falls. A fault zone in the African plateau explains the existence of a 360-foot (110-meter) drop. The Zambezi has never been important for navigation, but it is a vital supply of hydroelectricity for Zimbabwe, Zambia, and Mozambique.   *(Rob Crandall/Rob Crandall, Photographer)*

are the Congo, Niger, Nile and Zambezi. Smaller rivers, such as the Orange in South Africa; the Senegal, which divides Mauritania and Senegal; and the Limpopo in Mozambique, are locally important but drain much smaller areas. Ironically, most people think of Africa south of the Sahara as suffering from water scarcity and tend to discount the size and importance of the watersheds (or catchment areas) that these river systems drain.

▲ **Figure 6.10 The Rift Valley**   An aerial view of a portion of the Rift Valley in Kenya. This is one of the continent's most dramatic landforms, extending nearly 5,000 miles from Lake Nyasa to the Red Sea. The Rift Valley zone is a series of faults that have created volcanoes, escarpments, elongated lakes, and valleys. *(Altitude/Y. Arthus-B./Peter Arnold, Inc.)*

The Congo River (or Zaire) is the largest watershed in terms of drainage and discharge in the region. It is second only to South America's Amazon River in terms of annual discharge. The Congo flows across a relatively flat basin that lies more than 1,000 feet (300 meters) above sea level, meandering through Africa's largest tropical forest, the Ituri. Entry from the Atlantic into the Congo Basin is prevented by a series of rapids and falls, making the Congo River only partially navigable. Despite these limitations, the Congo River has been the major corridor for travel within the Republic of the Congo and the Democratic Republic of the Congo (formerly Zaire); the capitals of both countries, Brazzaville and Kinshasa, rest on opposite sides of the river (Figure 6.11). Political turmoil of the past decade, however, has greatly reduced commerce on this vital waterway.

The Nile River, the world's longest, is discussed in Chapter 7 as it is the lifeblood for Egypt and Sudan. Yet this river originates in the highlands of the Rift Valley zone, and it is an important physical expression of the linkages between North and Sub-Saharan Africa. The Nile begins in the lakes of the rift zone (Victoria and Edward) before descending into a vast wetland in southern Sudan known as the Sudd. In the Sudd the river divides into two main channels before reuniting in Malaka, Sudan. Agricultural development projects in the 1970s greatly increased the agricultural potential of the Sudd, especially its peanut crop. Unfortunately, the past two decades of civil war in Sudan (between northern Muslims and southern Christians) ravaged this area, turning farmers and herders into refugees and undermining the productive capacity of this important ecosystem.

Like the Nile, the Niger River is the critical source of water for the arid countries of Mali and Niger. Beginning in the humid Guinea highlands, the Niger flows first to the northeast and then spreads out to form a huge inland delta in Mali before making a great bend southward at the margins of the Sahara near Gao. On the banks of the Niger River are the capitals of Mali (Bamako) and Niger (Niamey), as well as the historic city of Tombouctou (Timbuktu). Tombouctou flourished as a center of trans-Saharan trade in the sixteenth century, with a population of nearly 100,000. (Its population is now one-fifth that, reflecting the decline of long-distance caravan trade.) After flowing through the desert north, the Niger River returns to the humid lowlands of Nigeria, where the Kainji Reservoir temporarily blocks its flow to produce electricity for Africa's most populous state.

The considerably smaller Zambezi River begins in Angola and flows east, spilling over an escarpment at the incomparable Victoria Falls, and finally reaching Mozambique and the Indian Ocean. More than other rivers in the region, the Zambezi is a major supplier of commercial energy. Sub-Saharan Africa's two largest hydroelectric installations, the Kariba on the border of Zambia and Zimbabwe and the Cabora Bassa in Mozambique, are on this river. Flooding in the lower Zambezi and Limpopo rivers devastated Mozambique in 2000, afflicting nearly 2 million people with failed crops, property damage, and disease. It seems that the worst flooding in 50 years was aggravated when countries upstream released additional waters from their locks to keep dams from bursting.

▲ **Figure 6.11 The Congo River** *Heavily laden barges ferry cargo and people between the cities of Kinshasa and Kisangani in the Democratic Republic of the Congo. The Congo is Sub-Saharan Africa's largest river by volume. Meandering through a relatively flat basin, the river is an important transportation corridor. (Georg Gerster/NGS Image Collection)*

Even though the navigational offerings of the Zambezi are limited, this did not stop German colonialists from demanding access to the river as they formed the territory of Southwest Africa in the 1880s, which eventually became Namibia. The resulting Caprivi Strip is a narrow belt of Namibian territory that separates Botswana from Zambia. It is one of the oddest geopolitical boundaries of the continent, but it does give arid Namibia access to the Zambezi.

Soils Scholars concerned about Africa's growing population often pay close attention to the poverty of its soils. With a few major exceptions, Sub-Saharan Africa's soils are relatively infertile and thus cannot easily support the intensive agriculture needed to feed large populations. Generally speaking, fertile soils are young soils, those deposited in recent geological time by rivers, volcanoes, glaciers, or windstorms. In older soils—especially those located in moist tropical environments—natural processes tend to wash out most plant nutrients over time. Over most of Sub-Saharan Africa, the agents of soil renewal have largely been absent; the region has few alluvial lowlands where rivers periodically deposit fertile silt, and it did not experience significant glaciation in the last ice age, as did North America.

Portions of Sub-Saharan Africa are, however, noted for their natural soil fertility, and not surprisingly these areas support denser settlement. Some of the most fertile soils are in the Rift Valley, enhanced by the volcanic activity associated with the area. The population densities of rural Rwanda and Burundi, for example, are partially explained by the highly productive soils. The same can be said for highland Ethiopia, which supports the region's second largest population of more than 65 million people. The Lake Victoria lowlands and central highlands of Kenya are also noted for their sizable populations and productive agricultural bases.

In the drier grasslands and semidesert areas one finds a soil type called *alfisols*. High in aluminum and iron, these red soils are heavily leached but have more organic matter, and thus greater fertility, than comparable soils found in wetter zones. This helps to explain the tendency of farmers to plant in drier areas, such as the Sahel, even though they risk exposure to drought. With irrigation, many agronomists feel that the southern African countries of Zambia and Zimbabwe could greatly increase their commercial grain production.

Knowing areas of soil fertility can help explain patterns of settlement and can even suggest potential areas for development, but soil fertility alone does not determine where Africans live. The Nigerian case is instructive in this regard. Although some of Nigeria's soils are fertile, many are mediocre or poor. Nigerian peoples long ago developed methods of enhancing soil fertility, allowing them to move from shifting cultivation to permanent-field agriculture and thus sustain impressive population density. In recent decades Nigeria has been able to feed its people in large part by importing food from abroad, paying for it by exporting oil. This strategy was not possible several generations ago, however, when Nigeria was even then the most populous country of Sub-Saharan Africa.

Climate and Vegetation

Sub-Saharan Africa lies in the tropical latitudes. Beginning just north of the tropic of Cancer, crossing the equator, and extending past the tropic of Capricorn in the south, it is the largest tropical landmass on the planet. Only the far south of the continent extends into the subtropical and temperate belts. Much of the region averages high temperatures from 70 °F to 80 °F (22° to 28 °C) year-round. The seasonality and amount of rainfall, more than temperature, determine the different vegetation belts that characterize the region. As Figure 6.12 shows, Addis Ababa, Ethiopia, and Walvis Bay, Namibia, have similar average temperatures, but the former is in the moist highlands and receives nearly 50 inches (127 centimeters) of rainfall annually, while Walvis Bay rests on the Namibian Desert and receives less than 1 inch (2.5 centimeters).

To understand the relationship between climate and vegetation in Sub-Saharan Africa, refer to Figure 6.12. Imagine a series of concentric vegetation belts that begin in the western equatorial zone as forest (Am and Af), followed by woodlands and grasslands (Aw), semidesert (BSh), and finally desert (BWh). The montane zones of East Africa, Cameroon, Guinea, and South Africa (especially along the Drakensberg Range) exhibit the forces of vertical zonation discussed in Chapter 4. Capturing more rainfall than the surrounding lowlands, these mountains often support unique forests and woodlands with high rates of endemic species. The only midlatitude climates of the region are found in South Africa. The southwestern corner of South Africa contains a small zone of Mediterranean climate that is comparable to that of California and noted for its pro-

duction of fine wine (Csb). The eastern coast of South Africa has a moist subtropical climate, not unlike that of Florida (Cfa).

Tropical Forests   The core of Sub-Saharan Africa is remarkably moist. The world's second-largest expanse of humid equatorial rain forest, the Ituri, lies in the Congo Basin, extending from the Atlantic Coast of Gabon two-thirds of the way across the continent, including the northern portions of the Republic of the Congo and the Democratic Republic of the Congo (Zaire). The conditions here are constantly warm to hot and precipitation falls year-round (see graph for Kisangani, Figure 6.12).

Commercial logging and agricultural clearing have degraded the western and southern fringes of this vast forest, but much of it is still intact. Considering the high rates of tropical deforestation in Southeast Asia and Latin America, the Central African case is a pleasant exception. Certainly the area's low population base of subsistence farmers does not strain the resource base. Moreover, Gabon and the Republic of the Congo derive substantial foreign exchange from oil exports, making forest extraction less necessary. The political chaos engulfing the Democratic Republic of the Congo has made large-scale logging a difficult proposition. In the future, however, it seems likely that Central Africa's rain forests could suffer the same kind of degradation experienced in other equatorial areas. As deforestation proceeds elsewhere in the world, the trees of equatorial Africa become increasingly valuable and, hence, more vulnerable.

Savannas   Wrapped around the Central African rain forest belt in a great arc lie Africa's vast tropical wet and dry savannas. Savannas are dominated by a mixture of trees and tall grasses in the wetter zones immediately adjacent to the forest belt and shorter grasses with fewer trees in the drier zones (Figure 6.13). North of the equatorial belt, rain generally falls only from May to October. The farther north one travels, the less the total rainfall and the longer the dry season. Climatic conditions south of the equator are similar, only reversed, with the wet season occurring between October and May and precipitation generally decreasing toward the south (see Lusaka in Figure 6.12). A larger area of wet savanna exists south of the equator, with substantial woodlands in southern portions of the Democratic Republic of the Congo, Zambia, and eastern Angola. These savannas are also a critical habitat for the region's large fauna. Elephants, zebras, rhinoceroses, and lions are found in the wooded grasslands, although their numbers have declined in most areas since the 1980s.

Deserts   The vast extent of tropical Africa is bracketed by several deserts. The Sahara, the world's largest desert and one of its driest, spans the landmass from the Atlantic coast of Mauritania all the way to the Red Sea coast of Sudan. A narrow belt of desert extends to the south and east of the Sahara, wrapping around the **Horn of Africa** (the northeastern corner that includes Somalia, Ethiopia, Djibouti, and Eritrea) and pushing as far as eastern and northern Kenya. An even drier zone is found in southwestern Africa. In the

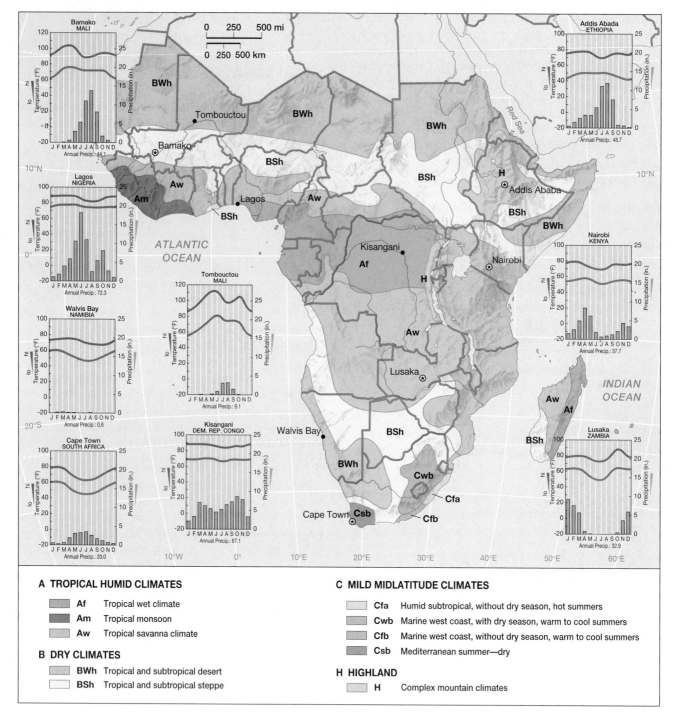

▲ **Figure 6.12 Climate map of Sub-Saharan Africa** Much of the region lies within the tropical humid and tropical dry climatic zones; thus, the seasonal temperature changes are not great. Precipitation, however, varies significantly from month to month. Compare the distinct rainy seasons in Lusaka and Lagos: Lagos is wettest in June and Lusaka receives most of its rain in January. Although there are important tropical forests in West and Central Africa, much the territory is tropical savanna. *[Temperature and precipitation data from* The World Weather Guide *(1984) by E. A. Pearce and C. G. Smith. London: Hutchinson.]*

Namib Desert of coastal Namibia, rainfall is a rare event, although temperatures are usually mild (see Figure 6.12 for Walvis Bay). Inland from the Namib lies the Kalahari Desert. Most of the Kalahari is not dry enough to be classified as a true desert, since it receives slightly more than 10 inches

(25 centimeters) of rain a year (Figure 6.14). Its rainy season, however, is brief. Most of the precipitation is immediately absorbed by the underlying sands. Surface water is thus scarce, giving the Kalahari a desertlike aspect for most of the year.

▲ Figure 6.14 Kalahari Desert   Sand dunes are one of the attractions found in Kalahari Gemsbok Park in South Africa. Centered in Botswana, the Kalahari extends into parts of Namibia and South Africa. Sub-Saharan Africa has many semiarid areas but few true deserts.   *(Clem Haagner/Photo Researchers, Inc.)*

▲ Figure 6.13 African savannas   A line of wildebeests marches across the tree-studded savannas found in Masai Mara National Park, Kenya. The savannas, an essential habitat for Africa's wildlife, are steadily being converted into agricultural and pastoral lands for human needs.   *(Nik Wheeler)*

Population and Settlement: Young and Restless

Sub-Saharan's population is growing quickly. While the global population is projected to increase by nearly 50 percent by 2050, the projected population growth for Africa south of the Sahara is 130 percent over the same time period. It is also a very young population, with 45 percent of the people younger than age 15, compared to just 18 percent for more-developed countries (Table 6.1). Only 3 percent of the region's population is over 65 years of age, whereas in Europe 15 percent is over 65. Families tend to be large, with an average woman having five or six children. Yet high child and maternal mortality rates also exist, reflecting disturbingly low access to basic health services. The most troubling indicator for the region is its stubbornly low life expectancy, which dropped from 52 years in 1995 to 48 years in 1998, in part due to the AIDS epidemic, and was estimated to be at 51 years in 2001. Life expectancy in other developing nations is much better; India's

is 10 years higher (61 years) and China's 20 years (71 years). The growth of cities is also a major trend; in 1980 it was estimated that 23 percent lived in cities, now the figure is approximately 30 percent.

Behind these demographic facts lie complex differences in settlement patterns, livelihoods, belief systems, and access to health care. Although the region is experiencing rapid population growth, Sub-Saharan Africa is not densely populated. The entire region holds some 670 million persons—roughly half the population that is crowded into the much smaller land area of South Asia. In fact, the overall population density of the region (72 people per square mile) is similar to that of the United States (77 people per square mile). Just six states account for half of the region's population: Nigeria, Ethiopia, the Democratic Republic of Congo, South Africa, Tanzania, and Sudan (see Table 6.1). True, there are states with very high population densities: Rwanda has more than 700 per square mile and the island nation of Mauritius has a population density of over 1,500. Yet there are many sparsely inhabited states. Chad has just 18 people per square mile (7 people per square kilometer) and Namibia has only 6 people per square mile (2 people per square kilometer). Such states do not consider population growth a serious issue. Many of the governments of the more densely settled territories, however, began to seriously promote family planning policies in the 1980s.

Crude population density is an imperfect indicator of whether a country is overpopulated. Geographers are often more interested in **physiological density**, which is the number of people per unit of arable land. The physiological density in Chad, where only 3 percent of the land is arable, is

TABLE 6.1 Demographic Indicators

Country	Population (Millions, 2001)	Population Density, per square mile	Rate of Natural Increase	TFR[a]	Percent < 15[b]	Percent > 65	Percent Urban
Angola	12.3	26	2.4	6.9	48	3	32
Benin	6.6	152	3	6.3	48	2	39
Botswana	1.6	7	1	3.9	41	4	49
Burkina Faso	12.3	116	3	6.8	48	3	15
Burundi	6.2	579	2.5	6.5	48	3	8
Cameroon	15.8	86	2.7	5.2	43	3	48
Cape Verde	0.4	287	3	4	43	7	53
Central African Republic	3.6	15	2	5.1	44	4	39
Chad	8.7	18	3.3	6.6	48	3	21
Comoros	0.6	692	3.5	6.8	46	5	29
Congo	3.1	24	3	6.3	43	3	41
Dem. Rep. of Congo	53.6	59	3.1	7	48	3	29
Djibouti	0.6	71	2.7	6.1	43	3	83
Equatorial Guinea	0.5	43	3.1	5.9	44	4	37
Eritrea	4.3	95	3	6	43	3	16
Ethiopia	65.4	153	2.9	5.9	44	3	15
Gabon	1.2	12	1.6	4.3	40	6	73
Gambia	1.4	323	3	5.9	46	3	37
Ghana	19.9	216	2.2	4.3	43	3	37
Guinea	7.6	80	2.3	5.5	44	3	26
Guinea-Bissau	1.2	88	2.2	5.8	44	3	22
Ivory Coast	16.4	132	2	5.2	42	2	46
Kenya	29.8	133	2	4.4	44	3	20
Lesotho	2.2	186	2	4.3	40	5	16
Liberia	3.2	75	3.1	6.6	43	3	45
Madagascar	16.4	71	3	5.8	45	3	22
Malawi	10.5	231	2.3	6.4	47	3	20
Mali	11.0	23	3	7	47	3	26
Mauritania	2.7	7	2.8	6	44	2	54
Mauritius	1.2	1,520	1	2	26	6	43
Mozambique	19.4	63	2.1	5.6	44	3	28
Namibia	1.8	6	1.9	5	43	4	27
Niger	10.4	21	2.9	7.5	50	2	17
Nigeria	126.6	355	2.8	5.8	44	3	36
Reunion	0.7	744	1.5	2.3	27	7	73
Rwanda	7.3	719	1.8	5.8	44	3	5
São Tomé and Principe	0.2	445	3.5	6.2	48	4	44
Senegal	9.7	127	2.8	5.7	44	3	43
Seychelles	0.1	449	1.1	2	29	8	63
Sierra Leone	5.4	196	2.6	6.3	45	3	37
Somalia	7.5	30	3	7.3	44	3	28
South Africa	43.6	92	1.2	2.9	34	5	54
Sudan	31.8	33	2.4	4.9	43	3	27
Swaziland	1.1	165	2	5.9	46	3	25
Tanzania	36.2	99	2.8	5.6	45	3	22
Togo	5.2	235	2.9	5.8	47	2	31
Uganda	24.0	258	2.9	6.9	51	2	15
Zambia	9.8	34	2.3	6.1	45	2	38
Zimbabwe	11.4	75	0.9	4	44	3	32

[a]Total fertility rate

[b]Percent of population younger than 15 years of age

Source: Population Reference Bureau. World Population Data Sheet, *2001.*

much higher than its crude population density. Perhaps a more telling indicator of population pressure and potential food shortages is **agricultural density**, the number of farmers per unit of arable land. Since the majority of people in Sub-Saharan Africa earn their livings from agriculture, agricultural density indicates the number of people who directly depend on each arable square mile. The agricultural density of many Sub-Saharan countries is 10 times greater than their crude population density.

Population Trends and Demographic Debates

It is the combination of growth in particular areas of Sub-Saharan Africa (high agricultural densities and rates of natural increase) and reversals in some economic and social indicators that make demographers concerned about the region's overall well-being. The Sahel, for example, is not crowded by European or Asian standards, but it may already contain too many people for the land to support given the unpredictability of its limited rainfall.

This assertion remains controversial, however—as does the entire field of African demography. Some believe that the region could support many more people than it presently does; these people even argue that Sub-Saharan Africa as a whole is still underpopulated and therefore will benefit from continued high population growth. Doomsayers, however, argue that the region is a demographic time bomb, and that unless fertility is quickly reduced, Sub-Saharan Africa will face massive famines in the near future. The majority of African states officially support lowering rates of natural increase and are slowly promoting modern contraception practices both to reduce family size and to protect people from sexually transmitted diseases.

Family Size   A preference for large families is the basis for the region's demographic growth. In the 1960s many areas in the developing world had comparable total fertility rates (TFR) of 6.0 or higher, but by the mid-1990s only people in Sub-Saharan Africa and Southwest Asia continued to have such large families. For Southwest Asia the dominance of Islam is used to explain high fertility rates. In Sub-Saharan Africa a combination of cultural practices, rural lifestyles, and economic realities encourage large families (Figure 6.15).

Throughout the region large families are considered prestigious and guarantee a family's lineage and status. Even now most women marry young, typically when they are teenagers, which increases their opportunity to have children. Demographers often point to the limited formal education available to women as another factor contributing to high fertility. Ethnic rivalries may also encourage pronatal practices. In ethnically divided states, such as Kenya and Nigeria, resources are often divided according to ethnic affiliation. An ethnic group's decision to reduce its numbers could, over time, weaken its political influence and thus be seen as unsound. Religious affiliation has little bearing on the region's fertility rates; Muslim, Christian, and animist communities all have similarly high birthrates.

Attitudes about family size, however, are shaped by high child mortality rates. In the 1960s, for example, it was not

▲ **Figure 6.15 Large families**   A Gambian father poses with his 10 children. Large families are still common in Sub-Saharan Africa. Growth rates soared in the 1970s and 1980s, but they began to slow slightly in the 1990s. In southern Africa total fertility rates have dropped to 3 and 4 children per woman. The overall average for the region is 5.6 children per woman. *(Mark Boulton/Photo Researchers, Inc.)*

uncommon for a woman to lose several children before they reached the age of five. As child health improved in the 1970s and 1980s, women continued to have the same number of children, but more lived, which pushed fertility rates even higher—in some countries reaching averages of eight children per woman.

The everyday realities of rural life make large families an asset. Children are an important source of labor; from tending crops and livestock to gathering fuelwood, they add more to the household economy than they take. Also, for the poorest places in the developing world, such as Sub-Saharan Africa, children are seen as social security. When the parent's health falters, they expect their grown children to care for them.

Government policies toward family size have shifted dramatically in the past three decades. During the 1970s population growth was not perceived as a problem by many African governments; in fact, many equated limiting population size with a neocolonial attempt to slow regional development. By the 1980s a shift in national policies occurred. For the first time, government officials argued that smaller families and slower population growth were needed for social and economic development. Following the United Nations International Conference on Population and Development in Cairo, Egypt, in 1994, organizers announced the following ambitious goals: to bring the natural increase rate down to 2.0 by the year 2020 and to increase the rate of contraceptive use to 40 percent. They may reach their goal; as of 2001 the regional rate of natural increase was down to 2.5, and nearly 20 percent of married women used some method of contraception.

Other factors are converging to slow the growth rate. As African states slowly become more urban, there is a corresponding decline in family size—a pattern seen throughout the world. Tragically, declines in natural increase are also occurring as a result of AIDS.

The Impact of AIDS on Africa   In April 1999 the president of Zimbabwe, Robert Mugabe, announced that 1,200 Zimbabweans were dying each week from AIDS. This was the first public acknowledgment of the gravity of the epidemic that is reversing the social and economic gains Zimbabwe has made since independence. If it were not for AIDS (acquired immunodeficiency syndrome), life expectancy in Zimbabwe would be in the high 60s. Instead, it plummeted to 40 years in 2001, and one-quarter of the population aged 15 to 49 years is infected with HIV (the human immunodeficiency virus that causes AIDS) or with AIDS.

Southern Africa is ground zero for the AIDS epidemic that is ravaging the region (Figure 6.16). In South Africa, the most populous state in southern Africa, nearly 5 million people (one in five people aged 15–49) are infected with HIV/AIDS. The rate of infection in neighboring Botswana is a staggering 36 per-

cent for the same age group. Although the rate of infection is highest in southern Africa, East Africa and West Africa are seeing increases. In Nigeria it is estimated that 5 percent of the 15–49 age group has HIV or AIDS. For Kenya the figure is 14 percent. By comparison, only 0.6 percent of the same age group in North America is infected. As of 2000, two-thirds of the HIV/AIDS cases in the world were found in Sub-Saharan Africa, and it was estimated that the epidemic had already claimed 17 million lives in the region. The virus is thought to have originated in the forests of the Congo, possibly crossing over from chimpanzees to humans sometime in the 1950s. Yet it was not until the late 1980s that the impact of the disease was widely felt in some of the more populated parts of the region.

Infection rates are such that demographers anticipate a slower population growth rate for the entire region. Sub-Saharan Africa will still grow, but population estimates for

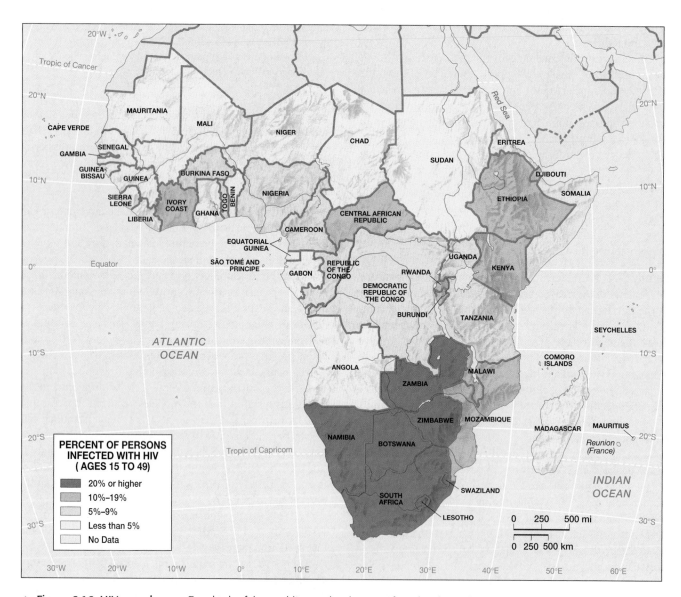

▲ **Figure 6.16 HIV prevalence**   Two-thirds of the world's people who are infected with HIV/AIDS are in Sub-Saharan Africa. As of 2000, more than 17 million lives have been lost in the region due to AIDS. Infection rates at the end of 1999 were highest in southern Africa, especially Botswana (35.8 percent) and Zimbabwe (25.1 percent). [Data Source: Population Reference Bureau, World Population Data Sheet, 2001]

2025 have been trimmed by as much as 200 million. Sadly, the social and economic implications of this epidemic are hard to measure. AIDS typically hits the portion of the population that is most active economically. Time lost to care for sick family members and the outlay of workers' compensation benefits could reduce economic productivity and overwhelm public services in hard-hit areas. Infection rates among newborns are high, and many areas struggle to care for children orphaned by AIDS.

Unlike the developed world, where expensive and potent drug therapies have prolonged the lives of people with HIV/AIDS, Sub-Saharan Africa does not have the money for this option. The government of South Africa insisted on its right to use cheap generic drugs to prolong the lives of those with HIV/AIDS, even though this violated patent laws. The world's major drug companies promptly sued South Africa, but backed down in 2001 in the face of growing international pressure (and the realization that pharmaceutical companies in developing countries such as India would happily provide generic alternatives). Through a UN-led agreement, generic drugs are now available, but a year's supply of a three-drug cocktail costs close to $1,000, which is still beyond the economic reach of most Africans.

For now, the surest way to stem the epidemic is through prevention, mostly through educating people about how the virus is spread and convincing them to change their sexual behavior. In Uganda, state agencies, along with NGOs, began a national no-nonsense campaign for AIDS awareness in the 1980s that focused on the schools. As a result of explicit materials, role-playing games, and frank discussion, the prevalence of HIV among women in prenatal clinics had declined by the late 1990s. Senegal, which has an infection rate of only 1.8 percent, also mounted a vigorous campaign to educate its citizens. One indicator of its success is a tenfold increase in condom purchases from 1988 to 1997. Yet many of the hardest-hit areas, such as Botswana, South Africa, and Zimbabwe, were reluctant to launch serious education efforts until the 1990s. The consequences of this inaction have been deadly (see "Mapping Risky Behavior: Fighting AIDS in Tanzania").

Patterns of Settlement and Land Use

Because of the dominance of rural settlements in Sub-Saharan Africa, people are widely scattered throughout the region (Figure 6.17). Population concentrations are the highest in West Africa, highland East Africa, and the eastern half of South Africa. The first two areas have some of the region's best soils, and indigenous systems of permanent agriculture developed there. In the last case, an urbanized economy based on mining, as well as the forced concentration of black South Africans into eastern homelands, contributed to the region's overall density.

West Africa is more heavily populated than most of Sub-Saharan Africa, although the actual distribution pattern is patchy. Density in the far west, from Senegal to Liberia, is moderate; this area is characterized by broad lowlands with decent soils, and in many areas the cultivation of wet rice has

enhanced agricultural productivity. Greater concentrations of people are found along the Gulf of Guinea, from southern Ghana through southern Nigeria, and again in northern Nigeria along the southern fringe of the Sahel. Nigeria is moderately to densely settled through most of its extensive territory; with 127 million inhabitants, it stands as the demographic core of Sub-Saharan Africa. The next largest country, Ethiopia, has only half of Nigeria's population.

The population centers in East Africa are highland Ethiopia, centered around Addis Ababa, the Lake Victoria basin, and the Kenyan portion of the Rift Valley. On the high plateaus of Ethiopia, agriculture is based on temperate crops such as wheat and barley as well as varieties of lentils, peas, and potatoes. In addition to subsistence production, coffee is an important export that was first domesticated by farmers in this area hundreds of years ago. In the volcanic highland zones of Kenya, Tanzania, Uganda, Rwanda, and Burundi, subsistence agriculture relies more on maize and root crops than on wheat. Export crops such as tea, coffee, and cotton are also produced.

As more Africans move to cities, patterns of settlement are evolving into clusters of higher concentration. Localities that were once small administrative centers for colonial elites mushroomed into major cities. The region even has its own mega-city; Lagos topped 10 million in the 1990s. Throughout the continent, African cities are growing faster than rural areas. But before examining the Sub-Saharan urban scene, a more detailed discussion of rural subsistence is needed.

Agricultural Subsistence   The staple crops over most of Sub-Saharan Africa are millet, sorghum, and corn (maize), as well as a variety of tubers and root crops such as yams. Irrigated rice is widely grown in West Africa and Madagascar. Wheat and barley are grown in parts of South Africa and Ethiopia. Intermixed with subsistence foods are a variety of export crops—coffee, tea, rubber, bananas, cocoa, cotton, and peanuts—that are grown in distinct ecological zones and often in some of the best soils.

In areas that support annual cropping, population densities are greater. In parts of humid West Africa, for example, the yam became the king of subsistence crops. The Ibos' mastery of yam production allowed them to procure more food and live in denser permanent settlements in the southeastern corner of present day Nigeria. Much of traditional Ibo culture is tied to the arduous tasks of clearing the fields, tending the delicate plants, and celebrating the harvest.

Over much of the continent, African agriculture remains relatively unproductive, and population densities tend to be low. Amid the poorer tropical soils, cropping usually entails shifting cultivation (or **swidden**). This process involves burning the natural vegetation to release fertilizing ash and planting crops such as maize, beans, sweet potatoes, banana, papaya, manioc, yams, melon, and squash. Each plot is temporarily abandoned once its source of nutrients has been exhausted. Swidden cultivation is often a very finely tuned adaptation to local environmental conditions, but it is unable to support high population densities.

Mapping Risky Behavior: Fighting AIDS In Tanzania

On the shores of Lake Victoria, hundreds of Tanzanian communities have discovered a potential key to slowing the spread of AIDS: mapping where risky behavior occurs and crafting local laws to discourage such behavior from occurring. In the past few years, local governments and village councils have cracked down on traditional social and sexual practices that have facilitated the spread of HIV/AIDS. First-time violators may be fined a chicken or beaten once with a club by village police. The punishment for repeat offenders is more serious: two strokes with a club, surrendering a goat, or in the most extreme cases, expulsion from the village.

The new laws resulted from a Dutch-funded research project. The goal was to draw maps of 900 villages in the Magu district and locate places where AIDS transmission would likely occur, such as dance clubs or abandoned huts. The Tanzanian-European researchers asked separate groups of men and women to draw maps identifying the places in their village where they thought AIDS might be transmitted. The men usually marked the local bar, the guest house, and the dance club. Women often added the school, the well, and the forest, suggesting the prevalence of rape in isolated places and sexual abuse of schoolgirls by teachers.

When the maps where completed, men and women sat down in teams and compared their findings. These gatherings produced candid discussions about sexual practices, something that is rare in Africa. Moreover, the teams were encouraged to suggest ways to change local laws so that safe behavior could be encouraged. For example, on one village map the shoreline of Lake Victoria was marked—an acknowledgment of what happened when villagers sent their daughters to buy fish without money to pay for it. The resulting new law prohibits sending children to purchase dinner without sufficient funds. In another village, the traditional

healer's hut was a red zone because he reused blades to make incisions. Now the healer must follow a one-patient, one-blade regimen or risk being fined a cow.

Other new ordinances outlaw dancing after dark or serving women alcohol after 6 P.M. Evening trips to the well or abandoned huts are forbidden. The village of Nyakaboja even made it a crime to flirt, fining the offender one chicken. The villagers formed school AIDS committees to educate students and draft rules for safe standards of conduct.

One village AIDS chairman observed that "we're penalizing people less often because almost everyone is behaving better." So far it is not clear that these new rules of conduct are reducing the HIV infection rate. There are some encouraging signs. A recent survey of Magu district residents reported that 38 percent of the residents see safer behavior in their villages. Teen pregnancy rates are down, which means that youths are having less sex or using condoms. Either would serve to reduce infection rates.

At the national level, the AIDS epidemic is forcing dramatic changes in Tanzanian customs and gender roles. For example, for the polygamous Sukuma people, tradition mandates giving a widow, her children, and her property to her late husband's younger brother. The idea is that the woman and her children would be protected. But if a husband dies of AIDS, the widow may put the younger brother and his other wives at risk. In 1998 a national law was passed that allows widows to keep their family, land, and property. These days few widows are marrying their brothers-in-law.

Adapted from "Tanzanian Villages Try New Anti-HIV Tool: Local Militia," by Michael M. Phillips. *The Wall Street Journal Europe,* January 12–13, 2001.

Madagascar's agricultural patterns are unique and thus deserve special mention. Some 1,500 years ago, settlers from Indonesia landed on the island. These seafarers had been blown off course, or perhaps were merely sailing into the unknown, hoping to find new land suitable for settlement. At any rate, once they landed on Madagascar, they lost contact with Southeast Asia, and within a few hundred years they had settled throughout the island. Their main focus of occupation was the central highlands, where they built irrigated rice fields and created a cultural landscape reminiscent of their ancestral home. Somewhat later another wave of immigrants began to arrive, this time from the African mainland, introducing the shifting cultivation techniques commonly used there. The ancestry of Madagascar's 16 million residents seems to be evenly divided between African and Southeast Asian stock. The bulk of the population continues to reside in the eastern highlands, but deforestation and serious erosion problems are undermining the productivity of traditional subsistence practices.

Plantation Agriculture Plantation agriculture, designed to produce crops for export, is critical to the economies of many states. If African countries are to import the modern goods and energy resources they require, they must sell their own products on the world market. Since the region has few competitive industries, the bulk of its exports are primary products derived from farming, mining, and forestry.

A number of African countries rely heavily on one or two export crops. Coffee, for example, is vital for Ethiopia, Kenya, Rwanda, Burundi, and Tanzania. Peanuts have long been the primary foreign exchange earner in the Sahel, while cotton is tremendously important for Sudan and the Central African Republic. Ghana and the Ivory Coast have long been the world's main suppliers of cocoa (the source of chocolate); Liberia produces plantation rubber; and many farmers in Nigeria specialize in palm oil (Figure 6.18). The export of such products can bring good money when commodity prices are high, but when prices collapse, as they periodically do, economic devastation may follow.

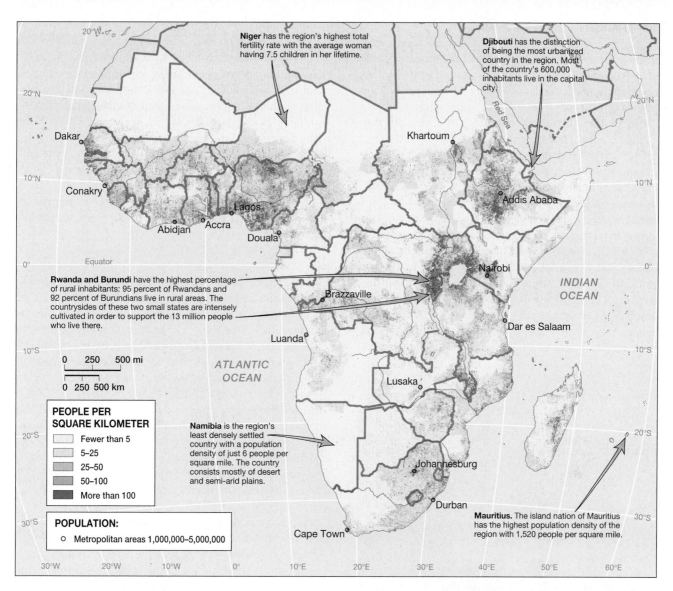

Niger has the region's highest total fertility rate with the average woman having 7.5 children in her lifetime.

Djibouti has the distinction of being the most urbanized country in the region. Most of the country's 600,000 inhabitants live in the capital city.

Rwanda and Burundi have the highest percentage of rural inhabitants: 95 percent of Rwandans and 92 percent of Burundians live in rural areas. The countrysides of these two small states are intensely cultivated in order to support the 13 million people who live there.

Namibia is the region's least densely settled country with a population density of just 6 people per square mile. The country consists mostly of desert and semi-arid plains.

Mauritius. The island nation of Mauritius has the highest population density of the region with 1,520 people per square mile.

PEOPLE PER SQUARE KILOMETER

- Fewer than 5
- 5–25
- 25–50
- 50–100
- More than 100

POPULATION:

- ○ Metropolitan areas 1,000,000–5,000,000

▲ Figure 6.17 Population distribution   The majority of people in Sub-Saharan Africa live in rural areas. Some of these rural zones, however, are densely settled, such as West Africa and the East African highlands. Major urban centers, especially in South Africa and Nigeria, support millions. Overall, the region has few large cities with more than 1 million people.

Herding and Livestock   Animal husbandry is extremely important in Sub-Saharan Africa, particularly in the semiarid zones. Camels and goats are the principal animals in the Sahara and its southern fringe, but farther south cattle are primary. Many African peoples have traditionally specialized in cattle raising and are often tied into symbiotic relationships with neighboring farmers. Such **pastoralists** typically graze their stock on the stubble of harvested fields during the dry season and then move them to drier uncultivated areas during the wet season when the pastures turn green. Farmers thus have their fields fertilized by the manure of the pastoralists' stock, while the pastoralists find good dry-season grazing. At the same time, the nomads can trade their animal products for grain and other goods of the sedentary world. Several pastoral peoples of East Africa, however, are noted for their extreme reliance on cattle and general (but never complete) independence from agriculture. The Masai of the Tanzanian-Kenyan borderlands traditionally derive a large

▲ Figure 6.18 Harvesting palm oil   A man harvests palm oil near Ibadan, Nigeria. Palm oil is a major export for western Nigeria. Other important Sub-Saharan plantation crops include cocoa, rubber, tea, coffee, cotton, and peanuts.   (M. & E. Bernheim/Woodfin Camp & Associates)

237

percentage of their nutrition from drinking a mixture of milk and blood (Figure 6.19). The blood is obtained by periodically tapping the animal's jugular veins, a procedure that evidently causes little harm.

Large expanses of Sub-Saharan Africa have been off limits to cattle because of infestations of **tsetse flies,** which spread sleeping sickness to cattle, humans, and some wildlife. Where wild animals, which harbor the disease but are immune to it, were present in large numbers, especially in environments containing brush or woodland (which are necessary for tsetse fly survival), cattle simply could not be raised. Some evidence suggests that tsetse fly infestations dramatically increased in the late 1800s, greatly harming African societies dependent on livestock but benefiting wildlife populations. In colonial Uganda, for example, where the burning of brush was outlawed, cases of sleeping sickness surged. At present, tsetse fly eradication programs are reducing the threat, and cattle raising is spreading into areas that were previously forbidden. This process is beneficial for African peoples, but it obvious-

ly bodes ill for the continued survival of large numbers of wild animals. When people and their stock move into new areas in sizable numbers, wildlife almost inevitably declines.

As the tsetse fly story shows, Sub-Saharan Africa presents a difficult environment for raising livestock because of the virulence of its animal diseases. In the tropical rain forests of Central Africa, cattle have never survived well and the only domestic animal that thrives is the goat. Raising horses, moreover, has historically been feasible only in the Sahel and in South Africa. As we shall see later, the disease environment of tropical Africa has presented a variety of problems for humans as well.

Urban Life

Although Sub-Saharan Africa is considered the least urbanized region in the developing world, most Sub-Saharan cities are growing at twice the national growth rates. If present trends continue, half of the region's population may well be living in cities by 2025. One of the consequences of this surge in city living is urban sprawl. Rural-to-urban migration, industrialization, and refugee flows are forcing the cities of the region to absorb more people and use more resources. As in Latin America, the tendency is toward urban primacy, the condition in which one major city is dominant and at least three times larger than the next largest city. Kinshasa, the capital of the Democratic Republic of the Congo, is an example. In the 1960s less than half a million people resided there; by the late 1990s it was a city of more than 4 million, dwarfing the country's other cities.

Sub-Saharan Africa's largest city, Lagos, has some 12 million inhabitants (Figure 6.20). In 1960 it was a city of only 1 million. Unable to keep up with the surge of rural migrants, Lagos's streets are clogged with traffic; for those living on the city's periphery, three- and four-hour commutes (one way) are common. City officials struggle to build enough roads and

▲ Figure 6.19 Masai pastoralists   A Masai man holds a cow steady while a woman collects blood being drained from the animal's neck. Pastoral groups such as the Masai live in the drier areas of Sub-Sahara Africa. The Masai live in Kenya and Tanzania. Other pastoral groups are found in the Sahel and the Horn of Africa.   *(© Kennan Ward/Corbis)*

▲ Figure 6.20 Downtown Lagos, Nigeria   Lagos is home to 12 million people, making it the region's largest city. A classic primate city, its streets teem with buses, collective taxis, thousands of pedestrians, and street vendors. Infrastructure has not kept up with the city's rapid growth. The demand for roads and utilities is far greater than the city's government is able to provide.   *(Daniel Lainé/Corbis)*

provide electricity, water, sewage service, and employment for all of these people. In many cases the informal sector (unregulated services and trade) provides urban employment and services. Crime is another major problem for Lagos. The chances of being attacked on the streets or robbed in one's home are quite high, even though most windows are barred and houses are fenced. To deal with this problem, Lagos has the highest density of police in Nigeria, but the weak social bonds between urban migrants, widespread poverty, and the gap between rich and poor seem to encourage lawlessness.

European colonialism greatly influenced urban form and development in the region. Africans, however, had an urban tradition prior to the colonial era, although a very small percentage of the population lived in cities. Ancient cities, such as Axum in Ethiopia, thrived 2,000 years ago. Similarly, in the Sahel, prominent trans-Saharan trade centers, such as Timbuktu (Tombouctou) and Gao, have existed for more than a millennium. In East Africa, an urban mercantile culture emerged that was rooted in Islam and the Swahili language. The prominent cities of Zanzibar, Tanzania, and Mombasa, Kenya, flourished by supporting a trade network that linked the East African highlands with the Persian Gulf. The stone ruins of Great Zimbabwe in southern Africa are a testimony to the achievements of stone working, metallurgy, and religion achieved by Bantu groups in the fourteenth century. West Africa, however, had the most developed precolonial urban network, reflecting both indigenous and Islamic traditions. It also supports some of the region's largest cities today.

West African Urban Traditions   The West African coastline is dotted with cities, from Dakar, Senegal, in the far north to Lagos, Nigeria, in the east. Over one-third of Nigerians live in cities, and it has half a dozen metropolitan areas with populations of more than 1 million. Historically, the Yoruba cities in southwestern Nigeria are the best documented. Developed in the twelfth century, cities such as Ibadan were walled and gated, with a palace encircled by large rectangular courtyards at the city center. An important center of trade

for an extensive hinterland, Ibadan was also a religious and political center. Lagos was also a Yoruba settlement. Founded on a coastal island on the Bight of Benin, most of the modern city has spread onto the nearby mainland. Its coastal setting and natural harbor made this relatively small indigenous city attractive to colonial powers. When the British took control in the mid-nineteenth century, the city's size and importance grew.

Most West African cities are hybrids, combining Islamic, European, and national elements such as mosques, Victorian architecture, and streets named after independence leaders. Accra is the capital city of Ghana and home to more than 1 million people. Originally settled by the Ga people in the sixteenth century, it had become a colonial administrative center by the late 1800s. As Figure 6.21 suggests, the modern city is largely divided along income lines. The wealthier inhabitants live to the north and east of the central business district in an area first constructed for the European elite. The lowest-income zone is to the west, behind the major industrial area. Like many West African cities, interspersed amid the low-income neighborhoods are *zongos* (so-called stranger communities filled with ethnic groups). In Accra, Hausa traders from the north created zongos so they could assist each other and practice their Islamic faith. More recently, Accra has experienced sprawl on its perimeter, largely for the development of upper-income suburbs.

Urban Industrial South Africa   The major cities of southern Africa, unlike those of West Africa, are colonial in origin. Most of these cities grew as administrative or mining centers such as Lusaka, Zambia, or Harare, Zimbabwe. The nation of South Africa is one of the most urbanized states in the entire region, and it is certainly the most industrialized. The foundations of South Africa's urban economy rest largely on its incredibly rich mineral resources (diamonds, gold, chromium, platinum, tin, uranium, coal, iron ore, and manganese). Eight metropolitan areas have more than 1 million people; the largest of them are Johannesburg, Durban, and Cape Town.

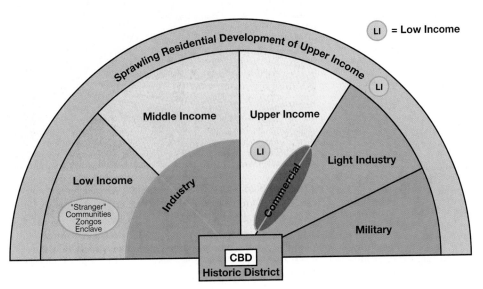

◀ Figure 6.21 Urban model of Accra, Ghana   In the late nineteenth century Accra was a small colonial administrative center, with most of its European population living in the historic district. Like most African cities, Accra has grown tremendously since independence. It exhibits a sectoral pattern of land use with upper-income, middle-income, and lower-income groups living in distinct areas. Pockets of low-income groups (marked LI) are found in upper-income zones. *(Adapted from Aryeetey-Attoh, 1997, Geography of Sub-Saharan Africa, Upper Saddle River, NJ: Prentice Hall)*

▶ **Figure 6.22 Racial segregation in Cape Town** Under apartheid, most land in Cape Town was designated for white use. Coloureds, blacks, and Indians (South Asians) were crammed into far smaller areas on the less desirable flat and sandy soils east of Table Mountain. *(Source: A. J. Christopher, 1994, The Atlas of Apartheid, p. 156, London: Routledge)*

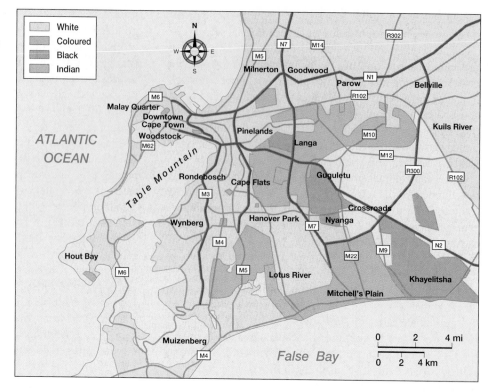

The form of South African cities continues to be imprinted by the legacy of **apartheid** (an official policy of racial segregation that shaped social relations in South Africa for nearly 50 years). Even though apartheid was abolished in 1994, it is still evident in the landscape. Under apartheid rules, cities such as Cape Town were divided into residential areas according to racial categories: white, **coloured** (a South African term describing people of mixed African and European ancestry), Indian (South Asian), and African (black). Whites occupied the largest and most desirable portions of the city, especially the scenic areas below Table Mountain (Figure 6.22). Blacks were crowded into the least desired areas, forming squatter settlements called *townships* in places such as Gugulethu (Figure 6.23). Today blacks, coloureds, and Indians are legally allowed to live anywhere they want. Yet the economic disparity between racial groups as well as entrenched animosity hinder residential integration. Post-apartheid settlements, such as Cape Town's Delpht South, promote residential integration of black and coloured households, something that was forbidden during the apartheid years.

Cultural Coherence and Diversity: Unity Through Adversity

No world region is culturally homogeneous, but most have been partially unified in the past by widespread systems of belief and communication. Traditional African religions, however, were largely limited to local areas, and the religions that did become widespread, Islam and Christianity, are primarily associated with other world regions. Certainly a few indigenous religious ideas and practices were shared over large expanses of Sub-Saharan Africa, but no institutionalized form of religion ever came close to unifying the region. A handful of African trade languages similarly have long been understood over vast territories (Swahili in East Africa, Mandingo and Hausa in West Africa), but none span the entire Sub-Saharan region. Sub-Saharan Africa also lacks a history of widespread political union or even of an indigenous system of political relations. Powerful African kingdoms and empires existed in past centuries, but all were limited to distinct subregions of the landmass.

▲ **Figure 6.23 Cape Town township** Gugulethu is a densely settled black township in the sandy flats east of downtown Cape Town. One of the older black townships, it was considered an undesirable area by whites because of fierce winter winds that whip through the area. Residents stack heavy rocks onto their metal roofs to keep them from blowing away. *(Rob Crandall/ Rob Crandall, Photographer)*

The lack of traditional cultural and political coherence across Sub-Saharan Africa is not surprising if one considers the region's vast scale. Sub-Saharan Africa is more than four times larger than Europe or South Asia. Had foreign imperialism not impinged on the region, it is quite possible that West Africa and southern Africa would have developed into their own distinct world regions.

An African identity south of the Sahara was forged through a common history of slavery and colonialism as well as struggles for independence and development. More telling, the people of the region often define themselves as African, especially to the outside world. That Sub-Saharan Africa is poor, no one will argue. And yet the cultural expressions of its people—its music, dance, and art—are joyous. Africans share an extraordinary resilience and optimism that visitors to the region often comment on. The cultural diversity of the region is obvious, yet there seems to be a unity drawn from surviving adversity.

Language Patterns

In most Sub-Saharan countries, as in other former colonies, the persistence of multiple languages reflects the layers of ethnic, colonial, and national identity. Indigenous languages, many from the Bantu subfamily, are often localized to relatively small rural areas. More widely spoken African trade languages, such as Swahili or Hausa, serve as a lingua franca over broader areas. Overlaying indigenous languages are Indo-European and Afro-Asiatic ones (French and English; Arabic and Somali). Figure 6.24 illustrates the complex pattern of language families and major languages found in Africa today. Contrast the larger map with the inset that shows current "official" languages. A comparison of the two shows that most African countries are multilingual, which can be a source of tension within states. In Nigeria, for example, the official language is English, yet there are 25 million Hausa speakers, 24 million Yoruba, 20 million Igbo (or Ibo), 11 million Ful (or Fulani), and 6 million Efik, as well as speakers of dozens of other languages.

African Language Groups Three of the six language groups mapped in Figure 6.24 are unique to the region (Niger-Congo, Nilo-Saharan, and Khoisan), while the other three (Afro-Asiatic, Austronesian, and Indo-European) are more closely associated with other parts of the world. Afro-Asiatic languages, especially Arabic, dominate North Africa and are understood in Islamic areas of Sub-Saharan Africa as well. Amharic in Ethiopia and Somali of Somalia are also Afro-Asiatic languages. The Malayo-Polynesian language family is limited to the island of Madagascar, which many believe was first settled by seafarers from Indonesia some 1,500 years ago. Indo-European languages, especially French, English, Portuguese, and Afrikaans, are a legacy of colonialism and widely used today.

Of the three language groups found exclusively in the region, Khoisan includes only a few languages spoken in a limited area. Khoisan speakers, who probably inhabited all of southeastern Africa several thousand years ago, are now confined to the arid lands of the Kalahari (Figure 6.24). Speakers of the Nilo-Saharan languages seem to have originated in

what is now southern Sudan and spread westward along the Sahel corridor and then to the south, into the heart of East Africa. Dinka and Nuer in Sudan, Songhai in Mali, and Turkana in Kenya are part of this group.

The Niger-Congo language group is by far the most important one in the region. This linguistic group originated in West Africa and includes Mandingo, Yoruba, Ful(ani), and Igbo, among others. Around 3,000 years ago a people of the Niger-Congo stock began to expand out of western Africa into the equatorial zone. This group, called the Bantu, commenced one of the most far-ranging migrations in human history, which introduced agriculture into large areas of central and southern Africa. One Bantu group migrated east across the fringes of the rain forest to settle in the Lake Victoria basin in East Africa, where they formed an eastern Bantu core that later pushed south all the way to South Africa (Figure 6.25). Another group moved south, into the rain forest proper. The equatorial rainforest belt immediately adjacent to the original Bantu homeland had been very sparsely settled by the ancestors of the modern pygmies (a distinct people noted for their short stature and hunting skills). Pygmy groups, having entered into close trading relations with the Bantu newcomers, eventually came to speak Bantu languages as well. While several pygmy populations have persisted to the present, their original languages disappeared long ago.

Once the Bantu migrants had advanced beyond the rain forest into the savannas and woodlands, their agricultural techniques proved highly successful and their influence expanded (Figure 6.25). Sometime around A.D. 650, Bantu-speaking peoples reached South Africa. Over the centuries the various languages and dialects of the many Bantu-speaking groups, which were often separated from each other by considerable distances, gradually diverged from each other. Today there are several hundred distinct languages in the Bantu subfamily of the great Niger-Congo group. All Bantu languages, however, remain closely related to each other, and a speaker of one can generally learn any other without undue difficulty.

Most individual Sub-Saharan languages are limited to relatively small areas and are significant only at the local scale. One language in the Bantu subfamily, Swahili, eventually became the most widely spoken Sub-Saharan language. Swahili originated as a trade language on the East African coast, where a number of merchant colonies from Arabia were established around A.D. 1100. A hybrid society grew up in a narrow coastal band of modern Kenya and Tanzania, one speaking a language of Bantu structure enriched with many Arabic words. While Swahili became the primary language only in the narrow coastal belt, it spread far into the interior as the language of trade. After independence was achieved, both Kenya and Tanzania adopted Swahili as an official language. Swahili, with some 47 million speakers, is the lingua franca of East Africa. It has generated a fairly extensive literature, and is often studied in other regions of the world.

Language and Identity Ethnic identity as well as linguistic affiliation have historically been highly unstable over much of the region. The tendency was for new groups to form when

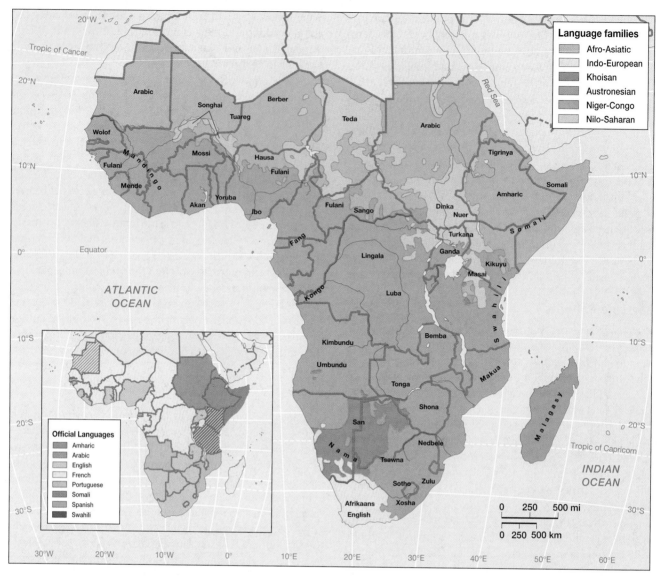

▲ Figure 6.24 African language groups and official languages   Mapping language is a complex task for Sub-Saharan Africa. There are languages with millions of speakers, such as Swahili, and there are languages spoken by a few hundred people living in isolated areas. Six language families are represented in the region. Among these families are scores of individual languages (see the labels on the map). Since most modern states have many indigenous languages, the colonial language often became the "official" language because it was less controversial than picking from among several indigenous languages. English and French are the most common official languages in the region (see inset).

people threatened by war fled to less-settled areas, where they often mixed with refugees from other places. In such circumstances, new languages arise quickly and divisions between linguistically distinct groups are blurred. Nevertheless, distinct **tribes** formed that consisted of a group of families or clans with a common kinship, language, and definable territory. The impetus to formalize tribal boundaries came from European colonial administrators who were eager to establish a fixed indigenous social order to better control native peoples. In this process a cultural map of Sub-Saharan Africa evolved, albeit flawed. Some tribes were artificially divided; meaningless names were applied; and territorial boundaries were often misinterpreted.

Social boundaries between different ethnic and linguistic groups have become more stable in recent years, and a number of individual languages have emerged as particularly important vehicles of communication at the national scale. Wolof in Senegal; Mandingo and other closely related Mande languages in Mali; Mossi in Burkina Faso; Yoruba, Hausa, and Igbo in Nigeria; Kikuyu in central Kenya; and Zulu, Xosha, and Sotho in South Africa are all nationally significant languages spoken by millions of persons (Figure 6.26). None, however, has the status of being the official language of any country. Indeed, there are only a handful of Sub-Saharan countries in which any single language has a clear majority status, which is partially explained by the arbitrary territori-

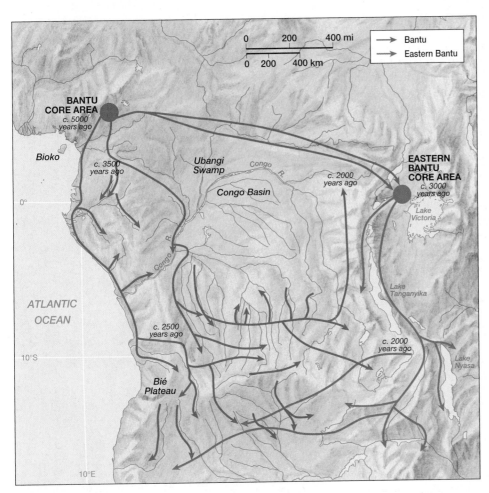

◀ Figure 6.25 Bantu migrations
Bantu languages, a subfamily of the Niger-Congo language family, are widely spoken throughout Sub-Saharan Africa. The out-migration of Bantu tribes from an original core in West Africa and a secondary core in East Africa helps to explain the diffusion of Bantu languages, which include Zulu, Swahili, Bemba, Shona, Lingala, and Kikuyu, among others. *(Source: James Newman, 1995, The Peopling of Africa, p. 141, New Haven: Yale University Press)*

al boundaries superimposed by Europeans. The more linguistically homogeneous countries of the region include Somalia (where virtually everyone speaks Somali) and the very small states of Rwanda, Burundi, Swaziland, and Lesotho.

European Languages   In the colonial period, European countries used their own languages for administrative purposes in their African empires. Education in the colonial period also stressed literacy in the language of the imperial power. In the postindependence period, most Sub-Saharan African countries have continued to use the languages of their former colonizers for government and higher education. Few of these new states had a clear majority language that they could employ, and picking any minority tongue would have aroused the opposition of other peoples. The one exception is Ethiopia, which maintained its independence during the colonial era. The official language is Amharic, although other indigenous languages are used, especially in the southwestern corner of the country.

Two vast blocks of European languages exist in Africa today: Francophone Africa, encompassing the former colonies of France and Belgium, where French serves the main language of administration; and Anglophone Africa, where the use of English prevails (see inset, Figure 6.24). Early Dutch settlement in South Africa resulted in the use of Afrikaans (a

▲ Figure 6.26 Multilingual South Africa   A South African sign warns pedestrians not to cross a highway in three languages: English, Afrikaans, and Zulu. Throughout Sub-Saharan Africa, many people are multilingual because of the diversity of languages that exist in each state.   *(Rob Crandall/Rob Crandall, Photographer)*

Dutch-based language) by several million South Africans. In Mauritania and Sudan, Arabic serves as the main language, although in the case of Sudan, this has resulted in severe internal tensions.

Religion

Indigenous African religions are generally classified as animist. This is a somewhat misleading catchall term used to classify all local faiths that do not fit into one of the handful of "world religions." Most animist religions are centered on the worship of nature and ancestral spirits, but the internal diversity within the animist tradition is vast. Classifying a religion as animist says more about what it is not than what it actually is.

With such warnings in mind, one can still say that much of Sub-Saharan Africa has an animist religious tradition. If one goes back far enough, the entire region was animist—but that is, of course, true for the rest of the world as well. Both Christianity and Islam actually entered the region early in their histories, but they advanced slowly for many centuries. Since the beginning of the twentieth century, both religions have spread rapidly—more rapidly, in fact, than in any other part of the world. But tens of millions of Africans still follow animist beliefs, and many others combine animist practices and ideas with their observances of Christianity and Islam.

The Introduction and Spread of Christianity   Christianity came first to northeast Africa. Kingdoms in both Ethiopia and central Sudan were converted by A.D. 300—the earliest conversions outside of the Roman Empire. The peoples of northern and central Ethiopia adopted the Coptic form of Christianity and have thus historically looked to Egypt's Christian minority for their religious leadership (Figure 6.27). At present, roughly half of the population of both Ethiopia and Eritrea profess Coptic Christianity; most of the rest are Muslim, but there are still some animist communities, especially in Ethiopia's western lowlands.

European settlers and missionaries introduced Christianity to other parts of Sub-Saharan Africa beginning in the 1600s. The Dutch, who began to colonize South Africa at this time, brought their Calvinist Protestant faith. Today most of their descendants, the Afrikaners as well as many coloureds, still practice this rather puritanical faith (other coloureds are Muslims). Later European immigrants to South Africa brought Anglicanism and other Protestant creeds, as well as Catholicism. A substantial Jewish community also emerged, concentrated in the Johannesburg area. Most black South Africans eventually converted to one or another form of Christianity as well. In fact, churches in South Africa were instrumental in the long fight against white racial supremacy. Religious leaders such as Bishop Desmond Tutu were outspoken critics of the injustices of apartheid and worked to bring down the system.

Elsewhere in Africa, Christianity came with European missionaries, most of whom arrived after the mid-1800s. As was true in the rest of the world, missionaries had little success where Islam had proceeded them, but they eventually made numerous conversions in animist areas. As a general rule, Protestant Christianity prevails in areas of former British colonization, while Catholicism is more important where France, Belgium, and Portugal had staked their empires. In the post-colonial era, African Christianity has diversified, at times taking on a life of its own independent from foreign

▲ **Figure 6.27 Eritrean Christians at prayer**   Coptic Christians gather in front of St. Mary's Church in Asmara, Eritrea, on Good Friday, a holy day for Christians. Half of the populations in Eritrea and Ethiopia belong to the Coptic Church, which has ties with the Christian minority in Egypt.   *(Ed Kashi/Corbis)*

missionary efforts. Still active in the region are various Pentecostal, Evangelical, and Mormon missionary groups, mostly from the United States. Yet many areas have also seen the emergence of syncretic faiths, in which Christianity is complexly intertwined with traditional belief systems. It is difficult to map the distribution of Christianity in Africa, however, since it has spread irregularly across the entire non-Islamic portion of the region.

The Introduction and Spread of Islam   Islam began to advance into Sub-Saharan Africa 1,000 years ago (Figure 6.28). Berber traders from North Africa and the Sahara introduced the religion to the Sahel, and by 1050 the Kingdom of Tokolor in modern Senegal emerged as the first Sub-Saharan Muslim state. Somewhat later, the ruling class of the powerful Mande-speaking mercantile empires of Ghana and Mali converted as well. In the fourteenth century the emperor of

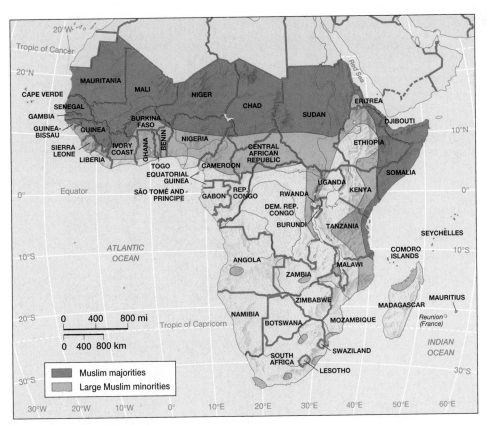

◀ **Figure 6.28 Extent of Islam** Muslim majorities prevail in the Sahelian states that border North Africa, as well as Somalia and Djibouti. Throughout West and East Africa there are also large Muslim minorities. Except in Sudan, religion has not been a source of political tension in the region. *(Source: Claude S. Phillips, 1984, The African Political Dictionary, p. 196, Santa Barbara, CA: Clio Press Ltd.)*

Mali astounded the Muslim world when he and his huge entourage made the pilgrimage to Mecca, bringing with them so much gold that they set off an inflationary spiral throughout Southwest Asia.

Mande-speaking traders, whose networks spanned the area from the Sahel to the Gulf of Guinea, gradually introduced the religion to other areas of West Africa. There, however, many peoples remained committed to animism, and Islam made slow and fitful progress. Even in the Sahel, syncretic forms of Islam prevailed through the 1700s. In the early 1800s, however, the pastoral Fulani people launched a series of successful holy wars designed to shear away animist practices and to establish pure Islam. Today orthodox Islam prevails through most of the Sahel. Farther south Muslims are mixed with Christians and animists, but their numbers continue to grow and their practices tend to be orthodox as well (Figure 6.29).

Interaction Between Religious Traditions The southward spread of Islam from the Sahel, coupled with the northward dissemination of Christianity from the port cities, has generated a complex religious frontier across much of West Africa. In Nigeria the Hausa are firmly Muslim, while the southeastern Igbo are largely Christians. The Yoruba of the southwest are divided between Christians and Muslims. In the more remote parts of Nigeria, moreover, animist traditions remain vital. But despite this religious diversity, there has been little overt religious animosity in Nigeria—or elsewhere in West Africa, for that matter. This may change; in 2000 seven Nigerian states imposed Muslim sharia laws,

which triggered much violence, especially in the northern city of Kaduna. Observers described it as the worst violence since Nigeria's civil war in the late 1960s. Still, most of West Africa's regional conflicts continue to be framed in ethnic and linguistic terms, rather than in religious. For example, recent ethnic tensions between Hausa and Yoruba citizens in Lagos resulted in more than 100 killings in October 2000.

▲ **Figure 6.29 West African mosque** A mosque rises above the houses in the town of Mankono, Ivory Coast. Islam entered this part of West Africa more than 600 years ago, but conversions were limited, with many Ivorians retaining their animist religious practices. Today Christian, Muslim, and animist faiths are practiced in the Ivory Coast, which reflects a pattern of religious tolerance seen throughout much of Sub-Saharan Africa. *(Victor Englebert/Englebert Photography, Inc.)*

CULTURAL DIFFUSION   South Asians and Hinduism in Africa

East Africa and South Asia, resting on opposite sides of the Arabian Sea, share a long history of interaction. In earlier centuries, the primary connections were between Ethiopia and India, with Ethiopia supplying substantial numbers of mercenaries and military slaves for the traditional kingdoms of South Asia. Few Indians, however, seem to have sailed to Africa. This was to change radically with British conquest in both areas. The British rulers in East and South Africa often found that they had inadequate labor at their disposal. In the east their main concern was railway building; in the south it was sugar growing in Natal, South Africa. Both required large amounts of labor, and since the British territories in South Asia had a labor surplus, the practice of recruiting Indian labor to Africa added another cultural layer to the region's ethnic complexity. In West India the region of Gujarat soon became the main recruiting ground. Both Hindu and Muslim Gujaratis migrated to Africa in the hundreds of thousands.

Many Indians returned to their homeland once their labor contracts were fulfilled. Others, however, elected to stay, usually turning their efforts to business and trade. Within a few decades the Indian communities of countries such as Kenya and Uganda had emerged as business leaders. Many Indians also prospered in South Africa, but this community suffered, along with nonwhite groups, under the apartheid system. It is notable that the founder of modern India, Mohandas Gandhi, began his career as a lawyer representing the Indian community in South Africa against a repressive, racist state.

After African nations achieved their independence in the 1960s, Indian merchants sometimes found themselves treated as economic scapegoats. When Uganda's economy began to falter in the 1970s, the country's ruthless dictator, Idi Amin, blamed the Indians, claiming that they were foreign economic predators unconcerned about the well-being of indigenous Africans. As a result, he decided to expel the entire community, many of whom fled to the United States. When the South Asians were deported, the country lost much of the business and professional expertise that had kept it afloat. After suffering a long interval of economic devastation and political chaos, Uganda emerged in the 1990s with a competent and reformist government that set the country on the road to eco-

nomic reconstruction. One of the new government's first acts was to extend an invitation to the country's former South Asian residents. A number have returned, and Uganda now has one of the most dynamic economies in Sub-Saharan Africa.

▲ **Figure 6.2.1 South Asian merchant**   A South Asian merchant sells toys to a South African couple in Johannesburg. South Asians were brought to South and East Africa as indentured laborers under British rule. Those who stayed continue to practice their religion (either Hinduism or Islam) and typically work in the service sector as merchants.   *(Don L. Boroughs)*

Religious conflict has historically been far more acute in northeastern Africa, where Muslims and Christians have struggled against each other for centuries. Islam came early to the coastal areas of the Horn, and soon it had virtually isolated the Ethiopian highlands from the rest of the Christian world. In due time large areas on the plateau itself were brought into the Muslim sphere as well, and in the early 1500s it seemed that Islam might prevail throughout the Ethiopian highlands. Animist invasions from the lowlands in the 1600s then put both groups on the defensive, but by the 1800s the Christians had acquired modern weapons and had gained the upper hand. In more recent years, religious conflict in the Horn has remained muted and has been largely replaced by ethnic and ideological struggles. The new country of Eritrea,

which is roughly half Christian and half Muslim, has in the past few years emerged as something of a model of peaceful coexistence between members of different faiths.

Sudan, on the other hand, is currently the scene of an intense conflict that is both religious and ethnic in origin. Here Islam was introduced in the 1300s by an invasion of Arabic-speaking pastoralists who extinguished the indigenous Coptic Christian kingdoms of the area. Within a few hundred years, central and northern Sudan were thoroughly Islamized. The southern Equatoria province of Sudan, however, where tropical diseases and extensive wetlands prevented Arab incursions, remained animist. During the British colonial era, many of the Dinka, Nuer, and other peoples of this area converted to Christianity. In the postcolonial period, the Arabic-

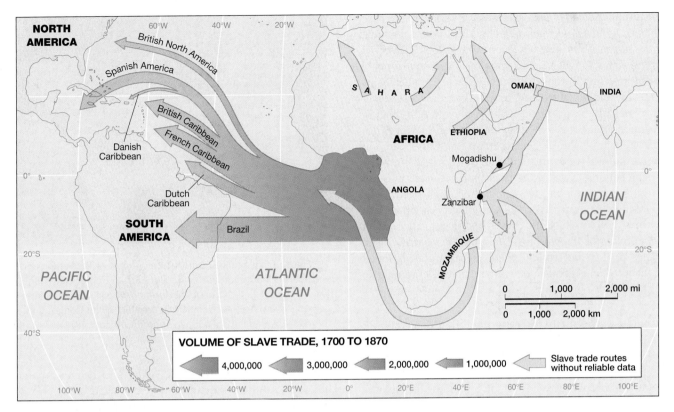

▲ Figure 6.30 African slave trade The slave trade had a devastating impact on Sub-Saharan societies. From ship logs, it is estimated that 12 million Africans were shipped to the Americas to work as slaves on sugar, cotton, and rice plantations; the majority went to Brazil and the Caribbean. Yet other slave routes existed, although the data are less reliable. Africans from south of the Sahara were used as slaves in North Africa. Others were traded across the Indian Ocean into Southwest Asia and South Asia.

speaking Muslims of the north and center quickly emerged as the country's dominant group, and in the 1970s they began to build an Islamic state. Experiencing both religious discrimination and economic exploitation, the peoples of the south subsequently launched a massive rebellion. Human rights groups report that some southerners are being enslaved by people in northern and central Sudan. Fighting since the 1980s has been intense, with the government generally controlling the main towns and roads and the rebels maintaining power in the countryside. The war has periodically prevented the distribution of food, resulting in horrific famines and the massive flight of refugees.

The final zone of widespread Islamization in Sub-Saharan Africa, the eastern coast, exhibits yet another historical-geographical pattern. Islam was introduced by merchants from southern Arabia around A.D. 1000. The Swahili-speaking community of the eastern coastal plain has continued to observe a relatively orthodox form of Islam ever since. Unlike West Africa, however, in East Africa the religion did not spread to any great extent. A few peoples of central Tanzania eventually accepted Islam, but the hinterland has remained largely animist, with some areas later experiencing numerous conversions to Christianity (see "Cultural Diffusion: South Asians and Hinduism in Africa").

Sub-Saharan Africa is a land of profound religious vitality. Both Christianity and Islam are spreading rapidly, but animism continues to hold widespread appeal. Many new and syncretic forms of religious expression are also emerging. With such a diversity of faiths, it is fortunate that religion has seldom been the cause of overt conflict. Religious vitality and tolerance are distinguishing features of Sub-Saharan Africa that deserve greater study. So, too, does the question of the region's influence on global culture, especially its musical traditions.

Globalization and African Culture

The slave triangle that linked Africa to the Americas and Europe set in process patterns of cultural diffusion that transferred African peoples and practices across the Atlantic. Tragically, slavery undermined the demographic and political strength of African societies, especially in West Africa, from where the most slaves were taken. An estimated 12 million Africans were shipped to the Americas as slaves from the 1500s until 1870. As Figure 6.30 shows, slavery impacted the entire region, sending Africans not just to the Americas, but to Europe, North Africa, and Southwest Asia. The vast majority, however, toiled on plantations across the Americas.

Out of this tragic diaspora came a melding of African cultures with Amerindian and European ones. Rumba, jazz, bossa nova, the blues, and even rock and roll have African rhythms at their core. Brazil, the largest country in Latin America, is claimed to be the second largest "African state" (after Nigeria) because of its huge Afro-Brazilian population.

Cultural exchanges are never one-way. Foreign language, religion, and dress were absorbed by Africans who remained in the region. With independence in the 1960s and 1970s, several states sought to rediscover their ancestral roots by openly rejecting European cultural elements. Ironically, the search for traditional religions found African scholars traveling to the Caribbean and Brazil to consult with Afro-religious practitioners about ceremonial elements lost to West Africa. Musical styles from the United States, the Caribbean, and Latin America were imitated and transformed into new African sounds. Kwaito, the latest musical craze in South Africa, sounds a lot like gangsta rap from the United States. A closer listen, however, reveals an incorporation of local rhythms, lyrics in Zulu and Xhosa, and themes about life in the post-apartheid townships. The complexity of cultural exchange, politics, and world markets can be illustrated through an examination of contemporary music in the Democratic Republic of the Congo and Nigeria.

Congo's Authenticity Movement

The Democratic Republic of the Congo (formerly Zaire) became a major center for African music in the 1970s, with Kinshasa as its hub. Despite the myriad of political and economic problems facing this country, its people have a reputation for leading the region in popular music and dance forms. This was not an accident, but rather part of then-President Mobutu's "authenticity movement," which began in the 1970s. Mobutu, who ruled the country as president-for-life from 1965 until 1997, desired a distinct Congolese musical voice. Unlike other leaders, he invested in this vision by subsidizing musical groups and sponsoring state competitions. A leader in the authenticity movement was singer and guitarist Franco Luambo Makiadi, known throughout Africa as Franco. Franco's OK Jazz band re-Africanized the Afro-Cuban rumba sound by borrowing elements from Congolese folk music, and it became wildly popular throughout the region. Patriotic and moralistic, Franco even wrote national anthems that workers were required to sing once a week. Although supported by the state, Franco's relationship with the government was not trouble-free; his records were sometimes banned when he became too critical of the state.

The authenticity movement also supported musicians working on contemporary dance music. Soukous (both a dance step and a music style) became an international sensation in the 1980s, especially in Western Europe and Japan (Figure 6.31). Papa Wemba, a star Soukou performer, made regular trips to Japan in the 1980s and 1990s. Before the political turmoil of the late 1990s, Japanese tourists were frequent visitors in Kinshasa's hip music clubs. A few Japanese groups even played Soukous, performing in typical dress and singing in Lingala (the local language). Interestingly, since the state financially backed its top musicians, the music rarely acquired a political edge, despite the mounting unrest in the country. Maintaining an apolitical line and producing an irresistible beat, Soukous became one of the Congo's biggest exports, both within the region and beyond it.

▲ **Figure 6.31 Soukou performers**   Tabu Ley Rochereau and members of his band perform Soukous for an appreciative audience in Lafayette, Louisiana. This form of African pop music originated in Kinshasa, the Democratic Republic of the Congo, and is widely played throughout Sub-Saharan Africa.   *(Philip Gould/ Corbis)*

Music as Political Conscience

Nigeria is the musical center of West Africa, with a well-developed and cosmopolitan recording industry. Modern Nigerian styles such as juju, highlife, and Afro-beat are influenced by jazz, rock, reggae, and gospel, but they are driven by an easily recognizable African sound. Two of the country's musical leaders are Sunny Ade and Fela Kuti. Both are from the Yoruba ethnic group, but their styles are strikingly different.

Sunny Ade became an international pop star in the 1980s and was dubbed the King of Juju Music (a combination of intricate lead guitar melodies supported by the percussive beat of talking drums). Because Sunny Ade sang in Yoruba, his music was not readily accessible to Western audiences, who gradually lost interest after a few years. Within Nigeria juju is still a popular sound, filling the buses and streets with its lilting, tropical rhythms.

Singer Fela Kuti, on the other hand, became a voice of political conscience for Nigerians struggling for true democracy. From an elite family and educated in England, Kuti borrowed from jazz, traditional, and popular music to produce the Afrobeat sound in the 1970s. Yet it was his searing and angry lyrics that attracted the most attention. Acutely critical of the military government, he sang of police harassment, the inequities of the international economic order, and even Lagos's infamous traffic. Singing in English and Yoruba, his message was transmitted to a larger audience, and he became a target of state harassment. At times self-exiled in Ghana, in the 1980s Kuti was jailed briefly by the Nigerian government, but widespread protests eventually led to his early release. Though Fela Kuti's protest music is unpopular with the state, it has been copied by other groups, making music an important form of political expression in Nigeria as well as in other Sub-Saharan states. Sadly, Kuti died in Lagos in 1997 from complications of AIDS; he was 58 years old. His son, however, is now an important recording artist in Nigeria.

Sub-Saharan Africa has shared many musical and spiritual traditions with the world, and the creative vitality of its people continues to be recognized. In contrast to the fairly peaceful coexistence of diverse artistic and religious traditions in Sub-Saharan Africa, ethnic tensions have often boiled over into violent disputes. To understand the roots of ethnic conflict in Sub-Saharan Africa, it is necessary first to examine the region's political history, paying particular attention to the legacy of European colonization.

Geopolitical Framework: Legacies of Colonialism and Conflict

The duration of human settlement in Sub-Saharan Africa is unmatched by any other region. After all, humankind originated there, evidently evolving from a rather apelike *Australopithicus* all the way to modern *Homo sapiens*. Over the millennia, many diverse ethnic groups formed that defy simple classification. Conflicts among these groups existed, with certain groups (the Bantu, for example) overwhelming others (the Khoisan). But cooperation and coexistence among different peoples were also evident.

With the arrival of Europeans, patterns of human relatedness and ethnic relations were changed forever. As Europeans rushed to carve up the continent to serve their imperial ambitions, they instituted various policies that heightened ethnic tensions and promoted hostility. Many of the region's modern conflicts can trace their roots back to the colonial era, especially the arbitrary drawing of political boundaries. Others are attributed to struggles over national identity and political control among different ethnic groups.

Indigenous Kingdoms and European Encounters

The first significant state to emerge in Sub-Saharan Africa was Nubia, which controlled a large territory in central and northern Sudan some 3,000 years ago (Figure 6.32); 1,000 years later the Kingdom of Axum arose in northern Ethiopia and Eritrea. Both of these states were strongly influenced by political models derived from Egypt and Arabia. The first wholly indigenous African states were founded in the Sahel around A.D. 700. Kingdoms such as Ghana, Mali, Songhai, and Kanem-Bornu grew rich by exporting gold to the Mediterranean and importing salt from the Sahara, and they maintained power over lands to the south by monopolizing horse breeding and mastering cavalry warfare.

Over the next several centuries, a variety of other states emerged in West Africa. Some were large but diffuse empires organized through elaborate hierarchies of local kings and chiefs; others were centralized states focused on small centers of power. The Yoruba of southwestern Nigeria, for example, developed a city-state form of government, and it is not coincidental that their homeland is still one of the most urbanized and densely populated parts of Africa. The most powerful Sub-Saharan states continued to be located in the Sahel until the 1600s, when European coastal trade undercut the lucrative trans-Saharan networks. Subsequently, the focus of power moved to the Gulf of Guinea, where well-organized states took advantage of the lucrative opportunities presented by the slave trade.

Early European Encounters Prior to European colonization, Sub-Saharan Africa presented a complex mosaic of kingdoms, states, and tribal societies. With the intensification of the trans-Atlantic slave trade in the sixteenth and seventeenth centuries, various indigenous African states (such as Dahomey and Ashanti) increased their military and economic power. Certain states in West Africa were well positioned to profit from the slave trade; by selling slaves to Europeans, they could obtain the firearms that would give them further military advantages over their enemies. Thus, the slave trade seems to have accelerated the gradual process of state formation almost everywhere in the region. Smaller tribal groups were usually the main victims of slave raiders, since they typically lacked effective military force and were easily captured by well-armed rivals.

Unlike the relatively rapid colonization of the Americas, Europeans needed centuries to gain effective control of Sub-Saharan Africa. Portuguese traders arrived along the coast of West Africa in the 1400s, and by the 1500s they were well established in East Africa as well. Initially the Portuguese made large profits, converted a few local rulers to Christianity, established several fortified trading posts, and acquired dominion over the Swahili trading cities of the east. They stretched themselves too thin, however, and had little ultimate success in any of their endeavors. Only where a sizable population of mixed African and Portuguese descent emerged, as along the coasts of modern Angola and Mozambique, could Portugal maintain power. Along the Swahili, or eastern, coast they were eventually expelled by Arabs from Oman, who subsequently established their own mercantile empire in the area.

The Disease Factor One of the main reasons for the Portuguese failure was the disease environment of Africa. With no resistance to malaria and other tropical diseases, roughly half of all Europeans who remained on the African mainland died within a year. Protected both by their formidable armies and by the diseases of their native lands, African states were able to maintain an upper hand over European traders and adventurers well into the 1800s. Unlike the Americas, where European conquest was facilitated by the introduction of Old World diseases that devastated native populations (see Chapters 4 and 5), in Sub-Saharan Africa endemic disease limited European settlement until the mid-nineteenth century.

The hazards of malaria and other tropical diseases such as sleeping sickness were compensated by the lure of profit, and soon other European traders followed the Portuguese. By the 1600s Dutch, British, and French firms dominated the lucrative export of slaves, gold, and ivory from the Gulf of Guinea. The Dutch also established a settler colony in South Africa, safely outside of the tropical disease zone, to supply their ships bound for Indonesia. For the next 200 years European traders

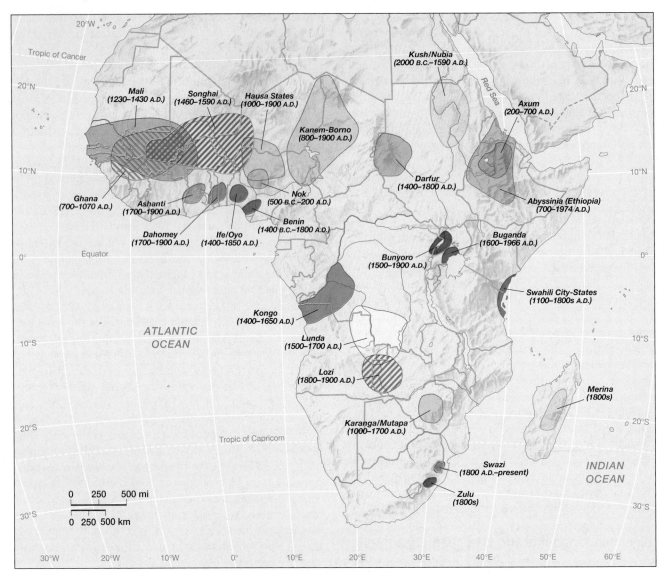

▲ **Figure 6.32 Early Sub-Saharan states and empires** Lost in the current political boundaries of Sub-Saharan Africa are the many African states and empires that existed long before Europeans advanced their territorial claims in the region. Most African kingdoms ceased to exist by 1900, but several, such as Buganda (in Uganda) and Abyssinia (Ethiopia), existed well into the mid-twentieth century. *(Sources: Samuel Aryeetey-Attoh, 1997,* Geography of Sub-Saharan Africa, *p. 63, Upper Saddle River, NJ: Prentice Hall; and Robert Stock, 1995,* Africa South of the Sahara, *p. 62, New York: Guilford Press)*

came and went, occasionally building fortified coastal posts, but they almost never ventured inland and seldom had any real influence on African rulers. By exporting millions of slaves, however, they had a profoundly negative impact on African society.

In the 1850s European doctors discovered that a daily dose of quinine would offer protection against malaria, radically changing the balance of power in Africa. Explorers immediately began to penetrate the interior of the continent, while merchants and expeditionary forces began to move inland from the coast. The first imperial claims soon followed. The French quickly grabbed power along the easily navigat-

ed Senegal River, while the British established protectorates over the indigenous states of the Gold Coast (modern-day Ghana).

European Colonization

In the 1880s European colonization of the region quickly accelerated, leading to the so-called scramble for Africa. By this time, after the invention of the machine gun, no African state could long resist European force. The exact reasons for the abrupt division of Africa among the colonial powers remain controversial, but several developments seem to have been crucial. One was the British seizure in 1882 of Egypt, a territory that the French had long coveted. In compensation, the infuriated French began to seize additional lands in West Africa, equatorial Africa, and Madagascar.

Another precipitating factor was the desire of several new European countries to join the game of empire-building. Since Asia was either occupied by established European powers or controlled by still-formidable indigenous empires, Africa emerged the main arena of rivalry and expansion. Even though Belgium had been a country only since 1830, its king quickly began to carve out a personal empire along the Congo River, using particularly brutal techniques. The German government—which itself dated back only to 1871—began to claim territories wherever German missionaries were active, and it soon staked out the colonies of Togo, Cameroon, Namibia, and Tanganyika (modern Tanzania minus the island of Zanzibar). The Italians eyed the Horn of Africa, while Spain acquired a small coastal foothold in equatorial West Africa. Alarmed by such activity, the Portuguese began to push inland from their coastal possessions in Angola and Mozambique.

Also in the early 1800s, two small territories were established in West Africa so that freed and runaway slaves would have a place to return to in Africa. The territory that was to become Liberia was set up by the American Colonization Society in 1822 to settle African American slaves. By 1847 it was the independent and free state of Liberia. Sierra Leone served a similar function for ex-slaves from the British Caribbean, but it remained a protectorate of Britain until the 1960s. Despite the good intentions behind the creation of these territories, they too were colonies. Liberia was imposed on existing indigenous groups who viewed their new "African" leaders with contempt.

The Berlin Conference As the scramble intensified, tensions among the participating countries mounted. Rather than risk war, 13 countries convened in Berlin at the invitation of the German chancellor Bismarck in 1884 in a gathering known as the **Berlin Conference.** During the conference, which no African leaders attended, rules were established as to what constituted "effective control" of a territory, and Sub-Saharan Africa was carved up and traded like properties in a game of Monopoly® (Figure 6.33). Exact boundaries in the interior, which was still poorly known, were not determined, and a decade of "orderly" competition remained as imperial armies marched inland. While European arms were by the 1880s far superior to anything found in Africa, several indigenous states did mount effective resistance campaigns. In central Sudan an Islamic-inspired force held out against the British until 1900, and as late as 1914 the Darfur region of western Sudan maintained tenuous independence.

Eventually European forces prevailed everywhere, with one major exception: Ethiopia. The Italians had conquered the Red Sea coast and the far northern highlands (modern Eritrea) by 1890, and they quickly set their sights on the large Ethiopian kingdom, called Abyssinia, which had itself been vigorously expanding for several decades. In 1896, however, Abyssinia vanquished the invading Italian army, earning the respect of and recognition by the European powers. In the 1930s fascist Italy launched a major invasion of the country, now renamed Ethiopia, to redeem its earlier defeat, and with the help of poison gas and aerial bombardment, it quickly prevailed. By 1942 Ethiopia had regained its freedom.

Although Germany was a principal instigator of the scramble for Africa, it lost its own colonies after suffering defeat in World War I. Britain and France then partitioned most of Germany's African empire between themselves. Figure 6.33 shows the colonial status of the region in 1913, prior to Germany's territorial loss. The French held most of West Africa, but the British controlled populous Nigeria and several other coastal territories. The French also colonized Gabon in western equatorial Africa, Madagascar, and the small but strategic enclave of Djibouti at the southern end of the Red Sea. The British holdings were larger still, covering a continuous swath of territory in the east from Sudan to South Africa. Belgium and Portugal formed the other main colonial powers. The government of Belgium had taken direct control over the personal domain of King Leopold II and had extended it to the south, eventually reaching the mineral-rich copper belt. Portugal, the weakest European power, controlled huge territories in southwestern and southeastern Africa (Angola and Mozambique). Britain's drive to the north from South Africa, organized by the imperial dreamer and diamond magnate, Cecil Rhodes, thwarted the Portuguese effort to bridge the continent.

Establishment of South Africa While the Europeans were cementing their rule over Africa after World War I, South Africa was inching toward political freedom, at least for its white population. South Africa was one of the oldest colonies in Sub-Saharan Africa, and it became the first to obtain its political independence from Europe in 1910. Its economy is the most productive and influential of the region, yet the legacy of apartheid made it an international pariah, especially within Africa. Territorial divisions in South Africa make for a classic study of applied political geography, in which an elite uses its control of space to ensure power. The evolution of South Africa's internal political boundaries merits extended consideration.

The original body of Dutch settlers in South Africa had grown slowly, expanding to the north and east of its original nucleus around Cape Town. As a farming and pastoral people largely isolated from the European world, these Afrikaners, or Boers, developed an extremely conservative cultural outlook marked by an intensifying belief in their racial superiority over the indigenous population. In 1806 the British, then the world's unchallenged maritime power, seized the Cape district from the Dutch. Relations between the British rulers and their new Afrikaner subjects quickly soured, in part because the British attempted to restrict the enslavement of Africans as part of a larger abolitionist movement. In the 1830s the bulk of the Afrikaner community opted to leave British territory and strike out for new lands. After being rebuffed by the powerful Zulu state in the Natal area, they settled on the northeastern plateau, known as the high veldt. By the 1850s they had established two "republics" in the area: the South African Republic (commonly called the Transvaal) and the Orange Free State (Figure 6.34).

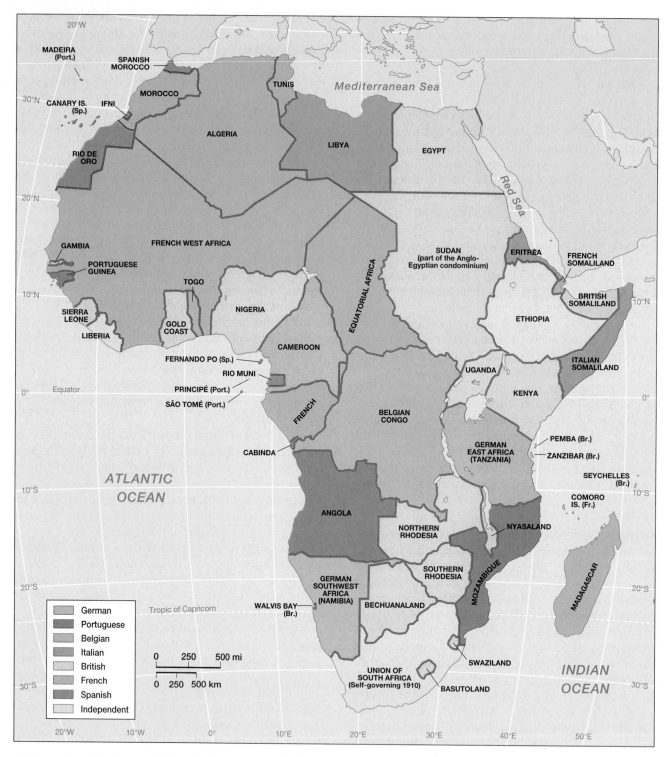

▲ Figure 6.33 European colonization in 1913   Before 1880 few areas of Africa were under direct European control. By 1884, when the Berlin Conference began, Africa was carved up and traded between European powers. France and Britain controlled the most territory, but Germany, Portugal, Belgium, Spain, and Italy all had their claims. By 1913 the entire continent, except Ethiopia, Liberia, and South Africa, were under European colonial control. (Source: The Times Atlas of World History. 1989. Hammond Inc.)

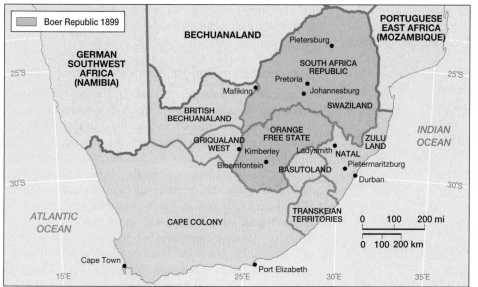

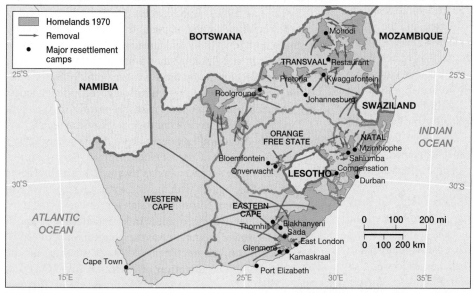

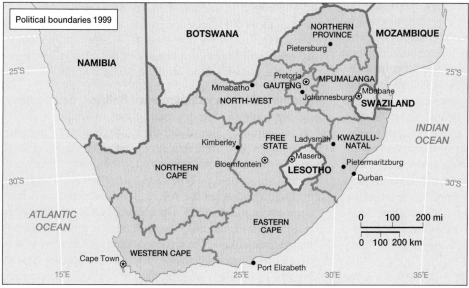

◀ Figure 6.34 The evolution of South African political boundaries   South Africa's internal political boundaries have been redrawn several times in the past century. To appease the Afrikaans (Dutch) settlers, a Boer republic was established in the late nineteenth century with British approval. It was later reabsorbed into the Union of South Africa when gold and diamonds were discovered. During the apartheid years, homelands were created with the intent of having all South African blacks live in these ethnic-based rural settings. In order to accomplish this, 3 million black South Africans were forcibly relocated. Lastly, in the new South Africa, homelands were eliminated and several new provinces were created.   *(Source: A. J. Christopher, 1994, The Atlas of Apartheid, pp. 16, 83, London: Routledge)*

The Boers were able to occupy the high veldt relatively easily because the area had been largely depopulated by the Zulu wars of the preceding decade. The Zulus had previously been a relatively minor group until a Zulu king introduced centralized rule and new military techniques in the 1820s. From then on Zulu armies were virtually invincible, and rather than suffer defeat at their hands, many other people chose to flee, some migrating as far north as Central Africa. Once relocated on the high veldt, the Afrikaners found themselves threatened by Zulu armies. After a series of inconclusive wars, the British intervened in 1878 on behalf of the Dutch settlers. The first British army sent into Zululand was annihilated; the second, equipped with machine guns, prevailed. By 1900 the British had incorporated the Zulu into South Africa's Natal province.

The British had more difficulty subduing the Afrikaners. By the late 1800s it had become clear that the two Boer republics were sitting above one of the world's greatest troves of mineral wealth, inciting British imperial ambition. English-speaking people were increasingly drawn to the area, and they grew resentful of Afrikaner power. The result was the Boer War, which turned into a protracted and brutal guerrilla struggle between the British Army and mobile Afrikaner bands. By 1905 the Boers relented, after which the British joined the two former republics to Cape Province and Natal to form the Union of South Africa. Five years later South Africa was given its independence from Britain. Britons and other Europeans continued to settle there, but the Afrikaners remained the majority white group. Black and coloured South Africans, however, greatly outnumbered whites.

It wasn't until 1948, when the Afrikaner's National Party gained control of the government, that they introduced the policy of "separateness" known as *apartheid*. British South Africans had enacted a series of laws that were prejudicial to nonwhite groups, but it was under Afrikaner leadership that racial separation become more formalized and systematic. Operating on three scales—petite, meso-, and grand—apartheid managed social interaction by controlling space. Petite apartheid, like Jim Crow laws in the United States, created separate service entrances for government buildings, bus stops, and restrooms based on skin color. Meso-apartheid divided the city into residential sectors by race, giving whites the best and largest urban zones. Finally, grand apartheid was the construction of black **homelands** by ethnic group. Technically blacks were to become citizens of the nominally independent homelands, such as KwaZulu and Transkei (see the 1970 map in Figure 6.34). Likened to the reservations created for Native Americans in the United States, homelands were rural, overcrowded, and on marginal land. Moreover, to ensure the notion that every black had a homeland, some 3 million blacks were forcibly relocated into homelands during apartheid, and residence outside of the homelands was strictly regulated.

Granted its independence in 1910, South Africa was the first state in the region freed from colonial rule. Yet because of its formalized system of discrimination and racism, it was hardly a symbol of liberty. Ironically, at the same time that the Afrikaners tightened their political and social control over the nonwhite population, the rest of the continent was preparing for political independence from Europe.

Decolonization and Independence

Decolonization of the region happened rather quickly and peacefully beginning in 1957. Independence movements, however, had sprung up throughout the continent, some dating back to the early 1900s. Workers' unions and independent newspapers became voices for African discontent and the hope for freedom. Black intellectuals, who had typically studied abroad, were influenced by the ideas of the **Pan-African Movement** led by W. E. B. Du Bois and Marcus Garvey in the United States. Founded in 1900, the movement's slogan of "Africa for Africans" encouraged a trans-Atlantic liberation effort. Nevertheless, Europe's hold on Africa remained secure through the 1940s and early 1950s, even as other colonies in South and Southeast Asia gained their independence.

By the late 1950s, Britain, France, and Belgium decided that they could no longer maintain their African empires and began to withdraw. (Italy had already lost its colonies during World War II, and Britain gained Somalia and Eritrea.) Once started, the decolonization process moved rapidly. By the mid-1960s, virtually the entire region had achieved independence. In most cases the transition was relatively peaceful and smooth.

Dynamic African leaders put their mark on the region during the early decades after independence. Men such as Kenya's Jomo Kenyatta, Ivory Coast's Felix Houphuët-Boigney, Julius Nyerere of Tanzania, and Ghana's Kwame Nkrumah became powerful father figures who molded their new nations (Figure 6.35). President Nkrumah's vision for Africa was the most expansive. After helping to secure independence for Ghana in 1957, his ultimate aspiration was the political unity of Africa. While his dream was never realized, it set the stage for the founding of the **Organization of African Unity (OAU)** in 1963. The OAU is a continent-wide organization. Its main role has been to mediate disputes between neighbors, although its record is mixed. Certainly in the 1970s and 1980s it was a constant voice of opposition to South Africa's minority rule, and it intervened in some of the more violent independence movements in southern Africa.

Southern Africa's Independence Battles
Independence did not come easily to southern Africa. In Southern Rhodesia (modern-day Zimbabwe) the problem was the presence of some 250,000 white residents, most of whom owned large farms. Unwilling to see power pass to the country's black majority, then some 6 million strong, these settlers unilaterally declared themselves the rulers of an independent, white-supremacist state in 1965. The black population continued to resist, however, and in 1978 the Rhodesian government was forced to capitulate. The renamed country of Zimbabwe was henceforth ruled by the black majority, although the remaining whites still form an economically privileged community. Since the mid-1990s, disputes over government land reform (splitting up the large commercial farms mostly owned by whites and giving the land to black farmers) has resulted in

▲ **Figure 6.35 A monument to Kwame Nkrumah**   Charismatic independence leader Kwame Nkrumah is remembered with this monument in Accra, Ghana. Nkrumah led Ghana to an early independence in 1957; he was also a founder of the Organization of African Unity (OAU).   *(Victor Englebert/Englebert Photography, Inc.)*

▲ **Figure 6.36 Displaced Angolans**   Since 1999 an estimated 3 million Angolans have been internally displaced from their villages due to fighting between government troops and UNITA. Technically speaking, they are not refugees, as they have not left their country, but they live in refugee-like conditions.   *(Malcolm Linton/Getty Images, Inc.)*

serious racial and political tensions. High unemployment, low economic growth, political opposion to President Mugabe's party, and the ravages of AIDS have dimmed the near-term hopes for Zimbabwe's future.

In the former Portuguese colonies, independence came violently. Unlike the other imperial powers, Portugal refused to relinquish its colonies in the 1960s. As a result, the people of Angola and Mozambique turned to armed resistance. The most powerful rebel movements adopted a socialist orientation and received support from the Soviet Union and Cuba. A new Portuguese government came to power in 1974, however, and it withdrew abruptly from its African colonies. At this point Marxist regimes quickly came to power in both Angola and Mozambique. The United States, and especially South Africa, responded to this perceived threat by supplying arms to rebel groups that opposed the new governments. As was common in much of Sub-Saharan Africa, the resulting struggles turned out to be more firmly grounded in ethnic loyalty than in Cold War ideology. Nevertheless, many of Africa's postindependence political struggles were caught up

in the Cold War, which tended to increase the supply of arms to the region.

Fighting dragged on for several decades in Angola and Mozambique. The countryside in both states is now so heavily riddled with land mines that it can hardly be used. With the end of the Cold War, however, outsiders lost their interest in perpetuating these conflicts, and sustained efforts have been under way to broker a peace. At present Mozambique is at peace, but Angola's attempts at reconciliation during the 1990s were fitful. As of 2001 the military strength of UNITA, a rebel group, has weakened considerably, but it still wages war against the government of Angola. It is estimated that from 1999 to 2001, more than 3 million Angolans were displaced from their homes by fighting. The peace in Mozambique has held, and new investment is coming into the country. Yet the years of warfare ravaged the country's infrastructure and crippled its agricultural production. In the mid-1990s its per capita GNI was just $80 per person (25 cents a day). By 2000 the GNI had climbed nearly threefold to $220 per person. Although a poor country by any measure, at least there are measurable signs of improvement in Mozambique (Figure 6.36).

Apartheid's Demise in South Africa   While fighting continued in the former Portuguese zone, South Africa underwent a remarkable transformation. Through the 1980s, its government had remained firmly committed to white supremacy. Under apartheid only whites enjoyed real political freedom, while blacks were denied even citizenship in their own country—technically, they were citizens of homelands. Yet since labor was needed in South African cities and mines, nonwhites were allowed to reside in segregated neighborhoods on the outskirts of cities, called **townships.** Some 20 miles southwest of Johannesburg, for example, is the famous black township of Soweto. Home to Nelson Mandela and Desmond Tutu, it was the principal home for many blacks working in Johannesburg and a center for political resistance to apartheid. In the post-apartheid era, Soweto is a mixture of lean-tos and middle-class housing that still attracts mostly black migrants from former homelands and neighboring countries. Planners estimate that nearly 2 million people reside in Soweto.

Opposition to apartheid began in the 1960s, intensifying and becoming more violent by the 1980s. Blacks led the opposition, but coloureds and Asians (who suffered severe, but less extreme, discrimination) also opposed the Afrikaner government. As international pressure mounted, white South Africans found themselves ostracized. Many corporations refused to do business there, and South African athletes (regardless of color) were banned from most international competitions, such as the Olympics and World Cup Soccer. Increasing numbers of whites also opposed the apartheid system, and many businesspeople began to believe that apartheid threatened to undermine their economic endeavors.

The first major change came in 1990, when South Africa withdrew from Namibia, which it had controlled as a protectorate since the end of World War I. South Africa now stood alone as the single white-dominated state in Africa. A few years later the leaders of the Afrikaner-dominated political party decided they could no longer resist the pressure for change. In 1994 free elections were held in which Nelson Mandela, a black leader who had been imprisoned for 27 years by the old regime, emerged as the new president. Black and white leaders pledged to put the past behind them and work together to build a new, multiracial South Africa. The homelands themselves were the first to be eliminated from the political map of the new South Africa (see the 1999 map in Figure 6.34). Residential segregation is officially illegal, but neighborhoods are still sharply divided along race lines. In 1999 peaceful elections were held and a new black president, Thabo Mbeki, replaced Nelson Mandela as the country's leader. And two years later the city of Durban, South Africa, hosted the UN-led World Conference Against Racism.

Unfortunately, South Africa's racial and economic problems were more difficult to eliminate. Many whites remained opposed to the new regime, and a number of Afrikaners indicated a desire to create a new, smaller, white-dominated state. Distrust also grew between blacks and coloureds and between blacks and South Asians; no longer did these groups face a common enemy and a similar set of political disadvantages. Divisions within the black community also grew more pronounced. Many Zulus felt more loyalty to their own ethnic group than to South Africa, and members of the rather conservative Zulu Inkatha Party engaged in violent conflicts with other black political groups. Violent crime, moreover, increased in many parts of the country. Under the new system a black middle class was rapidly emerging, but most blacks remained extremely poor (and most whites remained quite prosperous). Since the political change was not matched by a significant economic transformation, the hopes of many people were frustrated. Such a situation may not be conducive to political stability.

Enduring Political Conflict

Although most Sub-Saharan countries made a relatively peaceful transition to independence, virtually all of them immediately faced a difficult set of institutional and political problems. In several cases the old authorities had done virtually nothing to prepare their colonies for independence. Lacking an institutional framework for independent government, countries such as the Democratic Republic of the Congo confronted a chaotic situation from the beginning. Only a handful of Congolese had received higher education, let alone been trained for administrative posts. The indigenous African political framework had been essentially destroyed by colonization, and in most cases very little had been built in its place.

Even more problematic, in the long run, was the political-geographical structure of the newly independent states. Civil servants could always be trained and administrative systems built, but little could be done to rework the region's basic political map. The fundamental problem was the fact that the European colonial powers had essentially ignored indigenous cultural and political geographies, both in dividing Africa among themselves and in creating administrative subdivisions within their own imperial territories.

The Tyranny of the Map   All over Africa, different ethnic groups found themselves forced into the same state with peoples of disparate linguistic and religious backgrounds, many of whom had recently been their enemies. At the same time, a number of the larger ethnic groups of the region found their territories split between two or more countries. The Hausa people of West Africa, for example, were divided between Niger (formerly French) and Nigeria (formerly British), each of which they had to share with several former enemy groups.

Given the imposed political boundaries, it is no wonder that many African countries struggle to generate a common sense of national identity or establish stable political institutions. **Tribalism,** or loyalty to the ethnic group rather than to the state, has emerged as the bane of African political life. Especially in rural areas, tribal identities usually supersede national ones. In urban areas, however, the reverse is increasingly true. The few African countries that approach the ideal nation-state condition, one in which a single people are united under a single state, are the small, landlocked states of southern Africa (Lesotho and Swaziland) and Somalia. Even for these states the situation is far from ideal, since many members of the nationalities reside outside of the national boundaries (Sothos and Swazis in South Africa and Somalis in Kenya, Ethiopia, and Djibouti). Multiethnic Tanzania, however, is an example of a Sub-Saharan state that has forged a national identity, in part because of its socialistic policies in the 1970s.

Since virtually all of Africa's countries inherited an inappropriate set of colonial borders, one might assume that they would have been better off scrapping the system altogether and drawing a new political map based on indigenous identities. Such a strategy was impossible, as all the leaders of the newly independent states realized. Any new territorial divisions would have created winners and losers, and thus would have proved tremendously contentious.

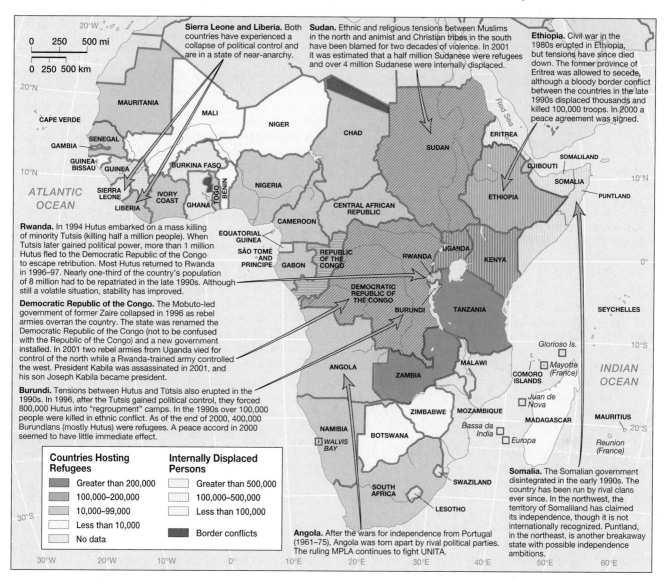

Sierra Leone and Liberia. Both countries have experienced a collapse of political control and are in a state of near-anarchy.

Sudan. Ethnic and religious tensions between Muslims in the north and animist and Christian tribes in the south have been blamed for two decades of violence. In 2001 it was estimated that a half million Sudanese were refugees and over 4 million Sudanese were internally displaced.

Ethiopia. Civil war in the 1980s erupted in Ethiopia, but tensions have since died down. The former province of Eritrea was allowed to secede, although a bloody border conflict between the countries in the late 1990s displaced thousands and killed 100,000 troops. In 2000 a peace agreement was signed.

Rwanda. In 1994 Hutus embarked on a mass killing of minority Tutsis (killing half a million people). When Tutsis later gained political power, more than 1 million Hutus fled to the Democratic Republic of the Congo to escape retribution. Most Hutus returned to Rwanda in 1996–97. Nearly one-third of the country's population of 8 million had to be repatriated in the late 1990s. Although still a volatile situation, stability has improved.

Democratic Republic of the Congo. The Mobuto-led government of former Zaire collapsed in 1996 as rebel armies overran the country. The state was renamed the Democratic Republic of the Congo (not to be confused with the Republic of the Congo) and a new government installed. In 2001 two rebel armies from Uganda vied for control of the north while a Rwanda-trained army controlled the west. President Kabila was assassinated in 2001, and his son Joseph Kabila became president.

Burundi. Tensions between Hutus and Tutsis also erupted in the 1990s. In 1996, after the Tutsis gained political control, they forced 800,000 Hutus into "regroupment" camps. In the 1990s over 100,000 people were killed in ethnic conflict. As of the end of 2000, 400,000 Burundians (mostly Hutu) were refugees. A peace accord in 2000 seemed to have little immediate effect.

Somalia. The Somalian government disintegrated in the early 1990s. The country has been run by rival clans ever since. In the northwest, the territory of Somaliland has claimed its independence, though it is not internationally recognized. Puntland, in the northeast, is another breakaway state with possible independence ambitions.

Angola. After the wars for independence from Portugal (1961–75), Angola was torn apart by rival political parties. The ruling MPLA continues to fight UNITA.

Countries Hosting Refugees
- Greater than 200,000
- 100,000–200,000
- 10,000–99,000
- Less than 10,000
- No data

Internally Displaced Persons
- Greater than 500,000
- 100,000–500,000
- Less than 100,000
- Border conflicts

▲ **Figure 6.37 Postcolonial conflicts**   The independence period has witnessed many ethnic conflicts as well as civil wars between competing political groups. In comparison to other world regions, Sub-Saharan Africa has had only one successful secessionist movement, which led to the creation of Eritrea. One consequence of political turmoil has been several million refugees and internally displaced persons.   *(Data sources: U.S. Committee for Refugees, 2001,* World Refugee Survey; *and John Allcock et al., 1992,* Border and Territorial Disputes, *Third Ed., Longman Current Affairs)*

Moreover, since ethnicity in Sub-Saharan Africa was traditionally fluid, and since many groups were territorially interspersed among their neighbors, it would have been difficult to generate a clear-cut system of division. Finally, most African ethnic groups were considered too small to form viable countries. With such complications in mind, the new African leaders, meeting in 1963 to form the Organization of African Unity, agreed that colonial boundaries should remain. The violation of this principle, they argued, would lead to pointless wars between states and interminable civil struggles within them.

Despite the determination of Africa's leaders to build their new nations within existing boundaries, challenges to the states began soon after independence. Figure 6.37 maps the numerous postcolonial political conflicts that have disabled

parts of Africa. Ethnic and secessionist conflicts, as well as military governments, have undermined African hopes for peaceful democracy. The human cost of this turmoil is several million refugees. **Refugees** are people who flee their state because of a well-founded fear of persecution based on race, ethnicity, religion, or political orientation. More than 3 million Africans were considered refugees in 2000. Added to this figure are at least another 13 million internally displaced persons. **Internally displaced persons** have fled from conflict but still reside in their country of origin. Sudan had the largest number of internally displaced people (4 million), followed by Angola (estimates range from 2 to 4 million) and the Democratic Republic of the Congo (2 million) at the end of 2000. These populations are hard to assist because they are not technically considered

LOCAL VOICES   A Sudanese "Lost Boy"

For nearly two decades a bloody civil war has devastated the southern provinces of Sudan. Fighting between the Sudanese army and the main rebel group, the SPLA (Sudan People's Liberation Army), continues with no resolution in sight. As of 2000 the conflict has resulted in 4 million internally displaced people and more than half a million Sudanese refugees scattered throughout Africa and beyond. The following testimonial is that of a young Dinka man who spent most of his life (14 years) on the run and in refugee camps in Ethiopia and Kenya before being given asylum in the United States. Thousands of children and teens, called the lost boys, wander the fringes of Sudan, displaced by war and in search of a home.

My name is William Majek Deng. I am Sudanese by nationality. I was born in southern Sudan in the upper Nile, in Bor. My hometown is Mading Bor. The community lived a solid and stable life. It was a splendid, noble, and grand life in all its dimensions. In my native home, life was beautiful and authentic.

I was born into a family of ten: parents, seven boys and a daughter. My family was an extended one, comprised of aunts, uncles, grandparents, cousins, nieces and nephews. My father was a pastoralist. He had a large farm and many herds of cattle. My father was also a prominent person in my community, the Dinka community, which is one of the Nilotic groups and also the largest tribe in Sudan.

However, the world is not static. The genuine life was interrupted by a civil war that broke out in 1983. This civil war intensified and spread in the whole of Sudan. It was marked by unreasonable looting of properties, demolition of shelters, abduction of children, and later on when all the livelihood was swept away, deliberate killings were launched on us—the southerners. We were attacked both aerially through bombing, and on the ground through shooting. Thousands of lives were lost. It was a genocide. Life became valueless and meaningless. I lost my father and many relatives. That was 1986. As that abomination raged on, life became unbearable, and a massive displacement of people occurred as a result of the ceaseless killings.

In 1987 things completely fell apart in the south . . . . I was separated from my brothers, mum, sister, and relatives. I found myself among the mass of people, mainly children, heading to an

unknown destination. We wandered in despair and agony in the Sudan looking for peace. The only relative who was with me during those doom days was my uncle's son. He used to carry me on his shoulder some distance, and I walked the rest on my own. We headed eastward of the Sudan. On our way, we were bothered by mosquitoes and terrified by wild animals. We experienced a burning thirst in some distances and, of course, starvation. It was a long journey through hell.

After three months of wandering, Deng crossed into Ethiopia, where he lived for nearly four years in a UN-run refugee camp. Then the Mengistu regime fell in 1991 and the new Ethiopian government expelled Sudanese refugees from their camps. As they returned to Sudan, the refugees were bombed and shot at by the Sudanese and Ethiopian military. Many were captured and taken to Khartoum, Sudan. Deng escaped with his life and, after several months of wandering the borderlands of Sudan and Ethiopia, he was picked up by UN relief workers and taken to another refugee camp in Kakuma, Kenya, where he lived for eight years. He describes camp life in northern Kenya:

Our arrival to Kakuma's hostile environment in 1992 was not an easy one. Kakuma was an abyss for natural disasters, whirlwinds, and blazing equatorial sun. The only rain was a dusting. It blew or rained day and night. I learned that forces without care must be endured.

The camp was composed of different refugees from different nationalities within Africa. The obvious one were Somalis, Ethiopians, Congolese, Burundians, and of course the Sudanese, who were predominant. In fact, Kakuma was a multinational, multiethnic, multicultural, and multireligious camp. That huge number of people were confined in the camp. As the number grew larger and larger, the shortages of food, water, and living space and all the health problems became major crises . . . . We had to live like prisoners.

In February 2001 Deng was one of forty-five youths given asylum and resettled in the United States. He lives in Houston, Texas.

Source: U.S. Committee for Refugees, *World Refugee Survey 2001*, pp. 108–111.

refugees, making it difficult for humanitarian nongovernmental organizations and the United Nations to help them (see "Local Voices: A Sudanese 'Lost Boy'").

Over the past decade the number of internally displaced people has risen, while the total number of refugees has declined. This decline is due, in part, to the reluctance of neighboring states to take on the burden of hosting refugees. Tanzania had more than half a million refugees (mostly from Burundi) and Guinea 400,000 (from Sierra Leone and Liberia) in 2000. With international aid on the decline, it is understandable that poor African states that struggle to serve their own people are disinclined to respond to the needs of refugees.

Ethnic Conflicts   In 1994 the world was shocked by the genocide in Rwanda among the Hutus and Tutsis. Rwanda and its neighbor Burundi inherited a potentially explosive ethnic mix from the precolonial period. Both countries had been indigenous kingdoms in which a Tutsi aristocracy ruled a Hutu peasantry. During the colonial period, the practice of giving one group privileges over the other intensified ethnic animosities. The Hutus gained political power in Rwanda after independence. In 1994 the Hutu president died in a plane crash under somewhat mysterious conditions, which Hutu leaders interpreted as the work of Tutsi rebels. In an orchestrated wave of genocide, half a million Tutsis were slaughtered over the next several months. Rebel Tutsi soldiers soon thereafter managed to oust the government and gain power, after which more than a million Hutus, fearing reprisals, fled the country (again, see Figure 6.37).

Images of a million refugees living in camps on the Rwandan border filled the international media. Efforts to repatriate the refugees were stymied for two years. Ultimately, political turmoil in the Democratic Republic of the Congo (a major host country) caused a massive and sudden repatriation in 1996. In four days, a half million Rwandans left the Congolese camps and walked back to their homes. The Rwandan case underscores the cruel reality of how ethnic conflict begets further turmoil.

A revamped Tutsi-led Rwandan army attacked the mostly Hutu refugee camps in the Democratic Republic of the Congo in 1996, forcing the repatriation of Hutus back to Rwanda. The new army proved more capable than anyone expected. It headed west, picking up allies from the Democratic Republic of the Congo as well as Uganda and eventually took over the capital, Kinshasa, installing Laurent Kabila as president in 1997. Under Kabila's rocky leadership, which ended with his assassination in 2001, two countries invaded the Democratic Republic of the Congo, Rwanda, and Uganda, and new rebel groups formed. By 2001 the Kinshasa-based government controlled the western and southern portions of the country, while two rebel groups from Uganda and one from Rwanda controlled the northern and eastern portions. As civil unrest continues, the number of displaced people has soared while economic output has plummeted. Meanwhile, the Democratic Republic of the Congo's nine neighboring states fear spreading destabilization and a potential outflow of refugees.

Liberia and Sierra Leone both saw their states virtually collapse in the 1990s as a result of rebellions, ethnic conflict, and economic decay. During Liberia's seven-year civil war (1989–96), virtually all the country's 3 million people fled their homes for periods ranging from a few weeks to years. In 1997 a former rebel leader, Charles Taylor, became president, and Liberians began to repatriate. Unfortunately, new waves of violence in 2000 led to international condemnation of Taylor's government for supporting rebel groups in Sierra Leone and trafficking in illicit diamonds from rebel-controlled mines. The current situation in Sierra Leone is even more unstable. When a UN peacekeeping mission failed in 1999, renewed fighting broke out. Some 400,000 Sierra Leoneans were refugees at the end of 2000, and perhaps a million were internally displaced. Fueling the conflict is a lucrative international diamond trade controlled by the Revolutionary United Front (RUF) (Figure 6.38). In the fall of 2001, news reports from West Africa showed how this diamond trade is tied to global terrorism. Representatives from Osama bin Laden's Al Qaeda network are one of the major buyers of RUF's diamonds. Al Qaeda uses diamonds to protect its assets and make millions. As a commodity, diamonds are easily hidden, hold their value, and remain nearly untraceable, which makes them ideal for sheltering money (see "Geography in the Making: Al Qaeda in Africa").

Some 400,000 Somalis were living as refugees in Ethiopia, Kenya, Yemen, and Djibouti at the end of 2000. Somalia had

politically disintegrated a few years earlier. Its collapse was somewhat ironic, since it is one of the very few ethnically homogenous states of Sub-Saharan Africa. Its people, however, have long been divided into fiercely independent clans, and it has no tradition of central authority. **Clans** are social units that are branches of a tribe or ethnic group larger than a family. As clan warfare intensified, food could no longer be easily produced or distributed, and a massive famine followed. In 1992 United Nations forces, led by the United States, arrived to impose order and distribute food. While they were rather successful at the latter task, they were utterly ineffective at the former, and they were soon forced to withdraw. Since the 1990s Somalia has been informally divided into several clan territories, and sporadic fighting continues. In the northern half of the country, the territory of Somaliland has declared its independence and elected a president. However, as of 2001 no country has recognized this territory as a new state.

Secessionist Movements Knowing how problematic African political boundaries are, one would expect territories to secede and form new states. The Shaba (or Katanga) province in what was then the state of Zaire tried to leave its national territory soon after independence. The rebellion was crushed a couple of years after it started with the help of France and Belgium. Zaire was unwilling to give up its copper-rich territory, and the former colonialists had economic interests in the territory worth defending. Similarly, the Igbo in oil-rich southeastern Nigeria proclaimed an independent state of Biafra in 1967. After a short but brutal war during which Biafra was essentially starved into submission, Nigeria was reunited.

Only one territory in the region has successfully seceded. In 1993 Eritrea gained independence from Ethiopia after two decades of civil conflict. What is striking about this territorial secession is that Ethiopia forfeited its access to the Red Sea,

▲ **Figure 6.38 Diamond miners in Sierra Leone** Workers pan for diamonds in a government-controlled diamond mine near Kenema, Sierra Leone. Diamond panners earn about 25 cents and two cups of rice a day unless they find a diamond. The battle to control Sierra Leone's diamond mines has led to escalated violence between the government and RUF rebels. *(Chris Hondros/Getty Images, Inc.)*

GEOGRAPHY IN THE MAKING   Al Qaeda in Africa

Long before the September 11th terrorist attacks on the United States made *Al Qaeda* a household term, the power of Al Qaeda was expressly felt in Africa. African states have been used by Al Qaeda as bases of operation, sources for income, and settings to execute acts of terror. In an era of globalized terrorism, it seems worthwhile to ask why Al Qaeda spread into Africa.

Al Qaeda originated during the Soviet invasion of distant Afghanistan in 1979. Afghan rebel fighters (called *mujahedin*) were joined by Muslim supporters from across the Islamic world, including many Africans. The resistance to the Soviets was headquartered in Peshawar, Pakistan, and led by a Palestinian academic named Abdallah Azzam. Over the decade of fighting, Osama bin Laden provided much of the financial backing to this operation that became known as Al Qaeda (solid base).

When Soviet forces left Afghanistan in 1989, Al Qaeda redefined its mission to establish a purer form of Islam in Arabia. After the Gulf War in 1991, the group took a decidedly anti-American turn. That same year bin Laden relocated his large construction businesses and his terrorist network to Khartoum, Sudan, where he stayed until 1996. Sudan borders Egypt and is across the Red Sea from Saudi Arabia—countries where Al Qaeda hoped to exert its influence. The Sudanese government's conservative interpretation of the Quran and imposition of shari'a matched Al Qaeda's own desire for Islamic purity. The cash-strapped Sudanese government also welcomed bin Laden's many private investments in shipping, trade, road and bridge construction, agriculture, leather production, and furniture manufacture.

During the years in Sudan, Al Qaeda grew to include cells throughout North and East Africa by supporting radicalized Muslims in these areas. Africa also provided situations where Americans and U.S. institutions were vulnerable to attack. It is now believed that bin Laden-trained guerrillas were responsible for the deaths of 18 service men in Somalia in 1993. After years of planning, Al Qaeda operatives in 1998 simultaneously bombed the U.S. embassies in Kenya and Tanzania, killing over 200 Africans and 12 Americans.

A robust trade in illegal diamonds in West Africa also provided an important source of revenue for Al Qaeda. Millions of dollars were earned from 1998–2001 from the sale of diamonds mined by Liberian rebels loyal to Charles Taylor working in Sierra Leone. These "conflict diamonds" are an ideal way to generate income and launder capital. Diamonds mined in Sierra Leone were smuggled into Liberia and then sold to Belgian diamond dealers for considerable profits.

The elaboration of such a terrorist network is a fluid geography in the making. In the case of Al Qaeda, its substantial resources were systematically used in economically weak and often politically unstable states in Africa. In 2001, one-third of the 35 states with Al Qaeda cells were located in Africa.

---

making it landlocked. Yet the creation of Eritrea still did not bring about peace. After years of fighting, the transition to Eritrean independence began remarkably well. Unfortunately, border disputes between the two countries erupted in 1998, resulting in some 100,000 troop deaths. In 2000 a peace accord was reached and the fighting stopped. If Ethiopia can accept Eritrea's independence, it might be possible for other areas torn by ethnic warfare to do so. The bloody civil war in southern Sudan, which has produced a half million refugees and 4 million internally displaced persons, might be resolved by granting independence to the province of Equatoria. Still, any thorough transformation of Africa's political map should not be expected.

Big Man Politics   Out of political turmoil, "big man" politics emerged. Presidents, both military and civilian, grabbed the reigns of power and refused to let go. Military governments, one-party states, and presidents-for-life became the norm. So, too, were violent changes in government, which occurred in virtually every state, and often several times. Strongmen such as General Abacha in Nigeria (1993–1998), president-for-life Mobuto Sese-Seko (1965–1997), and Liberia's current president Charles Taylor (1997–present) developed reputations as ruthless leaders who would not permit open political debate. Many see the fragility and corruption of Africa's political institutions as the biggest impediment to its development. Furthermore, the reliance on bullets over ballots encourages disproportionate spending on the military. In countries such as Sudan, Mozambique, and Liberia, and even relatively peaceful ones such as Tanzania, military expenditures per capita far outstrip spending on health care.

Fortunately, the 1990s saw a growth in the number of states with multiple political parties and free elections. Uganda, for example, previously suffered from one of Africa's most repressive regimes and was burdened by a legacy of bitter ethnic estrangement and capricious rule. Today it has a stable government, a growing economy, and reduced ethnic tensions. In many other Sub-Saharan countries as well, authoritarian governments have recently been giving way to more liberal regimes promising increased freedom and democratic rule. Looking exclusively at debacles such as Liberia, Burundi, or Somalia—as the U.S. press tends to do—makes it difficult to appreciate the political progress that has been occurring in much of Sub-Saharan Africa.

Economic and Social Development: The Struggle to Rebuild

By almost any measure, Sub-Saharan Africa is the poorest and least developed world region. Only a few African countries, such as Gabon, South Africa, Seychelles, Botswana, and Mauritius, are counted among the middle-income economies of the world (with per capita GNIs more than $3,000) (Table 6.2).

TABLE 6.2 *Economic Indicators*

Country	Total GNI (Millions of $U.S., 1999)	GNI per Capita ($U.S., 1999)	GNI per Capita, PPP* ($Intl, 1999)	Average Annual Growth % GDP per Capita, 1990–1999
Angola	3,276	270	1,100	−2.8
Benin	2,320	380	920	1.8
Botswana	5,139	3,240	6,540	1.8
Burkina Faso	2,602	240	960	1.4
Burundi	823	120	570	−5.0
Cameroon	8,798	600	1,490	−1.5
Cape Verde	569	1,330	4,450	3.2
Central African Republic	1,035	290	1,150	−0.3
Chad	1,555	210	840	−0.9
Comoros	189	350	1,430	−3.1
Congo	1,571	550	540	−3.3
Dem. Rep. of Congo	—	—	—	−8.1
Djibouti	511	790	—	−5.1
Equatorial Guinea	516	1,170	3,910	16.3
Eritrea	779	200	1,040	2.2
Ethiopia	6,524	100	620	2.4
Gabon	3,987	3,300	5,280	0.6
Gambia	415	330	1,550	−0.6
Ghana	7,451	400	1,850	1.6
Guinea	3,556	490	1,870	1.5
Guinea-Bissau	194	160	630	−1.9
Ivory Coast	10,387	670	1,540	0.6
Kenya	10,696	360	1,010	−0.3
Lesotho	1,158	550	2,350	2.1
Liberia	—	—	—	—
Madagascar	3,712	250	790	−1.2
Malawi	1,961	180	570	0.9
Mali	2,577	240	740	1.1
Mauritania	1,001	390	1,550	1.3
Mauritius	4,157	3,540	8,950	3.9
Mozambique	3,804	220	810	3.8
Namibia	3,211	1,890	5,580	0.8
Niger	1,974	190	740	−1.0
Nigeria	31,600	260	770	−0.5
Reunion	—	—	—	—
Rwanda	2,041	250	880	−3.0
São Tomé and Principe	40	270	—	−0.9
Senegal	4,685	500	1,400	0.6
Seychelles	520	6,500	—	1.3
Sierra Leone	653	130	440	−7.0
Somalia	—	—	—	—
South Africa	133,569	3,170	8,710	−0.2
Sudan	9,435	330	—	—
Swaziland	1,379	1,350	4,380	−0.2
Tanzania	8,515	260	500	−0.1
Togo	1,398	310	1,380	−0.5
Uganda	6,794	320	1,160	4.0
Zambia	3,222	330	720	−2.4
Zimbabwe	6,302	530	2,690	0.6

*Purchasing power parity

Source: The World Bank Atlas, 2001.

The rest are ranked at the bottom of most lists comparing per capita GNI or purchasing power parity. Using social indicators such as life expectancy or adult illiteracy, the region also fares poorly. Interestingly, female labor force participation far exceeds rates in other parts of the world (Table 6.3). Yet the overall status of African women is mixed, and labor force participation is not necessarily an indicator of equality.

The most troubling indicators show that the region's economic base actually declined during the 1990s. Table 6.2 shows most countries' low to negative growth rates between 1990 and 1999. The impact of falling economic output was made worse by rapid population growth. (The region's population growth rate of 2.5 percent in 2001 was considerably higher than Latin America's 1.7 percent rate.) This means that the national economies of Africa needed to grow briskly just to maintain their living standards; instead, they declined. The economic and debt crisis that ensued prompted the introduction of structural adjustment programs. Promoted by the IMF and the World Bank, structural adjustment programs typically reduce government spending, cut food subsidies, and encourage private-sector initiatives. Yet these same policies spurred immediate hardships for the poor, especially women and children, and ignited social protest, most notably in cities.

Over the past few years there have been some signs of an impending economic turnaround. The economy of the region as a whole seems to have experienced some growth beginning in 1995, but this growth has been at a slow pace. The late 1990s witnessed African economies growing at a slightly higher rate than their population, something that has not happened in many years. A few African countries, moreover, have done much better over the past decade, most notably Uganda and Equatorial Guinea. Economic and social development will likely improve by a combination of foreign investment, debt relief, state reform, and private initiatives led by individuals and NGOs. How much Africa should push for greater integration into the world economy is hotly debated, as are the roots of the region's poverty.

Roots of African Poverty

In the past outside observers often attributed Africa's poverty to its environment. Favored explanations included the infertility of its soils, the erratic patterns of its rainfall, the paucity of its navigable rivers, and the virulence of its tropical diseases. Most contemporary scholars, however, argue that such handicaps are not prevalent throughout the region, and that even where they do exist they can be—and have often been—overcome by human labor and ingenuity. The favored explanations for African poverty now look much more to historical and institutional factors than to environmental circumstances.

Numerous scholars have singled out the slave trade for its debilitating effect on Sub-Saharan African economic life. Large areas of the region were effectively depopulated, and many people were forced to flee into poor, inaccessible refuges. Colonization was another blow to Africa's economy. European powers invested little in infrastructure, education, or public health and were instead interested mainly in extracting mineral and agricultural resources for their own benefit. Several plantation and mining zones did achieve a degree of prosperity under colonial regimes, but internally dynamic national economies failed to develop. In almost all cases, the rudimentary transport and communications systems were designed to link administration centers and zones of extraction directly to the colonial powers, rather than to their own hinterlands or neighboring areas. On achieving independence, Sub-Saharan African countries thus faced economic challenges that were as daunting as their political problems.

Failed Development Policies   The first decade or so of independence was a time of relative prosperity and optimism for many African countries. Most of them relied heavily on the export of mineral and agricultural products, and through the 1970s commodity prices generally remained high. Some foreign capital was attracted to the region, and in many cases the European economic presence actually increased after decolonization. (This is one reason why critics refer to the period as one of "neocolonialism.")

The relatively buoyant economies of the 1960s and early 1970s were to disappear in the 1980s as most commodity prices began to decline. Sizable foreign debt began to weigh down many Sub-Saharan countries. By the end of the 1980s, most of the region entered a virtual economic tailspin. By the early 1990s, the region's foreign debt was around $200 billion. Although low compared to other developing regions (such as Latin America), as a percentage of its economic output, Sub-Saharan Africa's debt was the highest in the world.

Many economists contend that Sub-Saharan African governments enacted counterproductive economic policies and thus brought some of their misery on themselves. Keen to build their own economies and reduce their dependency on the former colonial powers, most African countries followed a course of economic nationalism. Thus, they set about building steel mills and other forms of heavy industry in which they were simply not competitive (Figure 6.39). In many cases inefficient and often corrupt government ministries took over large segments of the economy. Local currencies were also often maintained at artificially elevated levels, which benefited the elite who consume imported products but undercut exports. Some former French colonies kept their currencies pegged to the franc, which almost precluded successful export strategies.

Food Policies   The largest blunders made by Sub-Saharan leaders were in agricultural and food policies. The main objective was to retain a cheap supply of staple foods in the urban areas. A modern industrial economy could emerge, or so it was argued, only if manufacturing wages remained low, which in turn was feasible only if food could remain inexpensive. Another reason for keeping food inexpensive was that it would help maintain political stability in urban areas. The main problem with this policy was that the majority of Africans were farmers, who could not make money from their crops because prices were kept artificially low. They thus opted to grow food mainly for subsistence, rather than to

TABLE 6.3 Social Indicators and Status of Women

Country	Life Expectancy at Birth		Under Age 5 Mortality (per 1,000)		Percent Illiteracy (Ages 15 and over)		Female Labor Force Participation (% of total, 1999)
	Male	Female	1980	1999	Male	Female	
Angola	37	39	261	208	—	—	46
Benin	49	51	214	145	45	76	48
Botswana	41	42	94	95	26	21	45
Burkina Faso	47	47	—	210	67	87	47
Burundi	46	47	193	176	44	61	49
Cameroon	55	56	173	154	19	31	38
Cape Verde	65	72	—	—	—	—	39
Central African Republic	43	46	—	151	41	67	—
Chad	48	52	235	189	50	68	45
Comoros	54	59	—	—	—	—	42
Congo	47	52	125	144	13	27	43
Dem. Rep. of Congo	45	50	210	161	28	51	43
Djibouti	44	48	—	—	—	—	—
Equatorial Guinea	48	52	—	—	—	—	36
Eritrea	53	57	—	105	33	61	47
Ethiopia	51	53	213	180	57	68	41
Gabon	51	54	194	133	—	—	45
Gambia	51	54	216	110	57	72	45
Ghana	56	59	157	109	21	39	51
Guinea	43	47	280	167	—	—	47
Guinea-Bissau	44	46	290	214	42	82	40
Ivory Coast	45	47	—	—	—	—	—
Kenya	48	49	115	118	12	25	46
Lesotho	52	55	168	141	28	7	37
Liberia	49	52	—	—	—	—	40
Madagascar	52	56	216	149	27	41	45
Malawi	39	40	44	39	26	55	49
Mali	45	47	—	223	53	67	46
Mauritania	49	52	175	142	48	69	44
Mauritius	67	74	40	23	12	19	32
Mozambique	69	76	223	203	41	72	48
Namibia	47	45	114	108	18	20	41
Niger	41	41	317	252	77	92	44
Nigeria	52	53	196	151	29	46	36
Reunion	70	79	—	—	—	—	—
Rwanda	39	40	—	203	27	41	49
São Tomé and Principe	63	66	—	—	—	—	—
Senegal	51	54	—	124	54	73	43
Seychelles	67	73	—	—	—	—	—
Sierra Leone	42	47	336	283	—	—	37
Somalia	45	48	—	—	—	—	43
South Africa	52	54	91	76	14	16	38
Sudan	55	57	145	109	31	55	29
Swaziland	40	41	—	—	—	—	38
Tanzania	52	54	176	152	16	34	49
Togo	53	58	188	143	26	60	40
Uganda	42	43	180	162	23	45	48
Zambia	37	38	149	187	15	30	45
Zimbabwe	41	39	108	118	8	16	44

Sources: Population Reference Bureau, Data Sheet, *2001, Life Expectancy (M/F); The World Bank,* World Development Indicators, *2001, Under Age 5 Mortality Rate (1980/99); The World Bank Atlas, 2001, Female Participation in Labor Force;* CIA World Factbook, *2001, Percent Illiteracy (ages 15 and over).*

▲ **Figure 6.39 Industrialization** Heavy industry, such as this chemical plant in Kafue, Zambia, has failed to deliver Sub-Saharan Africa from poverty. In the worst cases, these industrial enterprises were unable to produce competitive products for world and domestic markets. *(Marc & Evelyne Bernheim/Woodfin Camp & Associates)*

sell—often at a loss—to national marketing boards. At the same time, they were encouraged to shift into export crops, such as coffee, cocoa, peanuts, and cotton. The end result was the failure to meet staple food needs at a time when the population was growing explosively. In the 1980s famine occurred in 22 African states, partly as a result of failed agricultural policies combined with drought.

It is also important to realize that, although Sub-Saharan Africa is poorer than South Asia in terms of per capita GNI, its people are generally better nourished and often better housed. Starvation has followed war and drought in Africa, but on a day-to-day level food intake is adequate. Such "nutritional prosperity" can be explained in part by the region's low population density and by the fact that many people still rely largely on subsistence production, which does not figure into official economic statistics.

Many African countries could vastly increase their production of staple foods, according to prevailing economic wisdom, if cultivators could be offered adequate incentives. Most countries in the region began to reform their agricultural policies in the 1980s, but this has been difficult to do because, as explained above, the urban poor have grown dependent on low prices of staple goods.

Corruption Although prevalent through most of the world, corruption also seems to have been particularly rampant in several African countries. Civil servants are sometimes not paid a living wage, and are thus virtually forced to solicit bribes. According to a recent poll of international businesspeople, Nigeria ranks as the world's most corrupt country. (Skeptical observers, however, point out that several Asian nations with highly successful economies, such as China, are also noted for high levels of corruption, so that corruption alone may not be the problem.)

With millions of dollars in loans and aid pouring into the region, officials at various levels were tempted to skim from the

top. Some African states, such as the former Zaire, were dubbed kleptocracies. A **kleptocracy** is a state in which corruption is so institutionalized that politicians and government bureaucrats siphon off a huge percentage of the country's wealth. Again, President Mobutu of the Democratic Republic of the Congo was a legendary kleptocrat. While his country was saddled with an enormous foreign debt, he reportedly skimmed several billion dollars and deposited them in Belgian banks.

Links to the World Economy

In terms of trade, Sub-Saharan Africa's connection with the world is limited. The level of overall trade is low both within the region and outside it. Most exports are bound for the European Union, especially the former colonial powers. The United States is the second most common destination. Exports to Asia are surprisingly few—mostly oil and tropical hardwoods. The import pattern mirrors exports. Despite four decades of independence, the majority of African countries turn to Europe for imports. Increasingly, however, cheaper goods from Asia are making their way across the Indian Ocean to African consumers.

By most measures of connectivity, Sub-Saharan Africa lags behind other developing regions. Telephone lines are scarce; in Nigeria there are only 4 lines per 1,000 persons. The regional average is slightly better, at 14 lines per 1,000, but still far behind North Africa and Southwest Asia. For every 1,000 people in the region there are 40 television sets, whereas the number of televisions in high-income countries averages 700 per 1,000. One hopeful change is the expansion of mobile telephones in Africa. No longer reliant on expensive fixed telephone lines for communication, multinational providers are now competing for mobile-phone customers. South Africa alone has more than 8 million mobile phone subscribers (Figure 6.40).

▲ **Figure 6.40 Mobile phones for Africa** A shopkeeper has steady customers for mobile phones in Abidjan, Ivory Coast. Mobile telephones are quickly becoming the preferred instrument of communication in Sub-Saharan Africa. They do not require expensive fixed lines and multinational firms are competing for the region's business. *(Andre Ramasore/Galbe.com)*

Africa also lacks the infrastructure and goods to facilitate more intraregional trade. Only southern Africa has a telecommunications network of any note. Of the few roads that exist, especially in West and East Africa, the majority are not paved. Nothing comparable to MERCOSUR (the trade bloc in South America) is found in Sub-Saharan Africa. This, too, is a poignant legacy of colonialism, and one that keeps Africans from nurturing economic linkages with each other.

Aid Versus Investment   As a poor region, Sub-Saharan Africa is linked to the global economy more through the flow of financial aid and loans than through the flow of goods. As Figure 6.41 reveals, for several states aid accounts for 10 percent of the GNI. In the extreme case of Malawi, economic assistance in 1999 accounted for one-quarter of the GNI. Most of this aid comes from a handful of developed countries (see the insert in Figure 6.41). France has provided the most aid, closely fol-

lowed by the United States and then Japan, Germany, and the United Kingdom. By contrast, net flows of private capital are extremely low. Countries such as South Africa, Equatorial Guinea, Gabon, and Angola are the exceptions because of their oil or mineral wealth. The majority of states in the region attract less than $10 per capita in private foreign investment, whereas the figure for Latin America is 10 times greater.

The reasons foreign investors shun Africa are fairly obvious. The region is generally perceived to be too poor and unstable to merit much attention, and most foreign investors eager to take advantage of low wages put their money in Asia or Latin America. Sub-Saharan Africa, some economists contend, is thus starved for capital. Other scholars, however, see foreign investment as more of a trap—and one that the region would be wise to avoid. Recently in a number of African countries a small boom has occurred in mineral exploration and production, attracting considerable overseas interest. Optimists interpret this

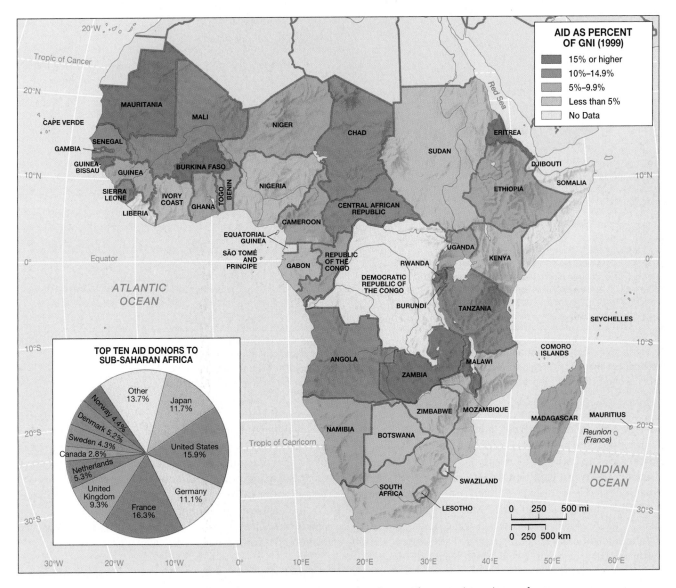

▲ Figure 6.41 Global linkages: aid dependency   Many states in Sub-Saharan Africa are dependent on foreign aid as their primary link to the global economy. This figure maps aid as a percentage of GNI, which ranges from less than 1 percent to more than 25 percent in Guinea-Bissau and Malawi. (Source: World Development Indicators, 2001)

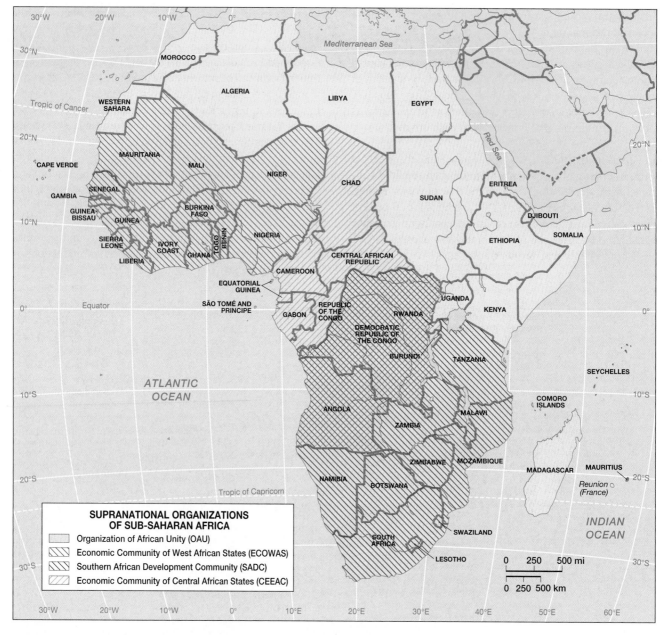

▲ Figure 6.42 Supranational organizations of Sub-Saharan Africa   Political affiliations in Sub-Saharan Africa are both continental and regional. The OAU includes all African countries. Smaller organizations, such as SADC, CEEAC, and ECOWAS, represent regional affiliations. Of these, SADC shows the most economic promise.

as a sign of impending African economic renewal, but others counter that in the past mineral extraction has failed to lead to broad-based and lasting economic gains.

Finally, the World Bank and the IMF proposed in 1996 to reduce debt levels for heavily indebted poor countries, many of which are in Sub-Saharan Africa. Most Sub-Saharan states are indebted to official creditors such as the World Bank, not to commercial banks (as is the case in Latin America and Southeast Asia). Under this new World Bank/IMF program, substantial debt reduction will be given to Sub-Saharan countries that are determined to have "unsustainable" debt burdens. Mauritania, for example, spends six times more money on repaying its debts than it does on health care. States qual-

ify for different levels of debt relief provided they present a poverty reduction strategy. Uganda was the first state to qualify for the program, using the money it saved on debt repayment to expand primary schooling. Other countries that will benefit from debt reduction are Tanzania, Mozambique, Ghana, Cameroon, Mauritania, Mali, Senegal, Burkina Faso, and Benin. With luck, more African countries will be liberated from burdensome foreign debt payments.

Future Paths   Sub-Saharan Africa's future economic path is highly uncertain. Will the various countries of the region opt for integration into the global system, or will they instead attempt to build more autonomous and less capitalistic economic systems? The latter option was tried in the first few decades after independence, but the former seems to be gaining momentum. Many African countries have begun to privatize sectors of their economies, a process that often leads to faster economic growth but also tends to heighten income disparities and negative environmental impacts. Trade groups, such as the Southern African Development Community, have the potential to build up regional economies and infrastructure. Neither African socialism nor global capitalism offers an easy fix for the region's problems, but both hold out promises for future improvement.

Economic Differentiation Within Africa

As in most regions, considerable disparity in levels of economic and social development persist. In many respects, the small island nations of Mauritius and the Seychelles have little in common with the mainland. With a high per capita GNI, life expectancies averaging in the low 70s, and economies built on tourism, they could more easily fit into the Caribbean were it not for their Indian Ocean location. In contrast, Chad has a per capita GNI of $210, and more than half the adult population is illiterate.

Given the scale of the African continent, it is not surprising that groups of states formed trade blocs to facilitate intraregional exchange and development. The two most active regional organizations are the Southern African Development Community (SADC) and the Economic Community of West African States (ECOWAS). Both were founded in the 1970s but became more prominent in the 1990s (Figure 6.42). SADC and ECOWAS are anchored by the region's two largest economies: South Africa and Nigeria. The Economic Community of Central African States (CEEAC) has been hampered in its efforts to foster economic integration, in part because of the political instability of the area. Other prosperous states are those that exploit oil resources. The following section contrasts Africa's better-off states with its poorer ones.

South Africa   South Africa is the unchallenged economic powerhouse of the region. The GNI of Nigeria, the next largest economy, is just one-fourth that of South Africa. Only South Africa has a well-developed and well-balanced industrial economy. It also boasts a healthy agricultural sector and, more important, it stands as one of the world's mining superpowers. South Africa remains unchallenged in gold production and is a leader in many other minerals as well. But while South Africa is undeniably a wealthy country by African standards and while its white minority is prosperous by any standard, it is also a country beset by severe and widespread poverty. In the townships lying on the outskirts of the major cities and in the rural districts of the former homelands, employment opportunities remain limited and living standards marginal. Despite the end of apartheid, South Africa continues to suffer from one of the worst distributions of income in the world.

Oil and Mineral Producers   Another group of relatively well-off Sub-Saharan countries benefits from substantial oil and mineral reserves and small populations (Figure 6.43).

The prime example is Gabon, a country of noted oil wealth that is inhabited by 1.2 million people. Its neighbors, the Republic of the Congo and Cameroon, also benefit from oil, but both have experienced economic declines in recent years, with that of Cameroon being particularly dramatic. Equatorial Guinea, a former Spanish colony, began producing significant amounts of oil in 1998. American investment is behind much of the offshore drilling, most notably by Exxon Mobil. As this small country of just half a million people begins to earn substantial oil revenues, observers hope that the lives of average citizens will improve. Farther south, Namibia and Botswana also have the advantage of small populations and abundant mineral resources, especially diamonds. Both countries have also enjoyed sound government over the past few years and have experienced significant economic growth.

The Leaders of ECOWAS   Nigeria has the largest oil reserves in the region, but its huge population has kept its per capita GNI at a low $260. Oil money has also helped make Nigeria notoriously inefficient. A small minority of its population has grown fantastically wealthy more by manipulating the system than by engaging in productive activities. Most Nigerians, however, remain trapped in poverty. Oil money also led to the explosive growth of the former capital of Lagos, which by the 1980s had become one of the most expensive—and least livable—cities in the world. As a result, the Nigerian government opted to build a new capital city in Abuja, located near the country's center, a move that has proved tremendously expensive.

Ivory Coast and Senegal, formerly the core territories of the French Sub-Saharan empire, still function as commercial centers, but they also suffered economic downturns in the 1980s. In the mid-1990s, the Ivorian economy again began to grow. Boosters within the country call it an emerging

▲ Figure 6.43 Nigerian oil exports   An oil tanker loads up at Shell-owned Bonny Oil Terminal in the Niger Delta area of Nigeria. Currently, Shell Petroluem of Nigeria is being sued in New York Court for taking land for oil development without paying adequate compensation to the Ogoni tribe who live in southeastern Nigeria. Nigeria is the region's largest oil exporter and a member of OPEC.   (Chris Hondros/Getty Images, Inc.)

"African elephant" (comparing it to the successful "economic tigers" of eastern Asia). Yet its real annual growth throughout the 1990s was less than 1 percent. Ghana, a former British colony and ECOWAS member, also began an economic recovery in the 1990s. In 2001 it negotiated with the IMF and World Bank for debt relief to reduce its nearly $6 billion foreign debt.

East Africa   Long the commercial and communications center of East Africa, Kenya experienced economic decline and political tension throughout the 1990s. Despite a per capita GNI of just $360, Kenya boasts good infrastructure by African standards, and about 1 million foreign tourists come each year to marvel at its wildlife. Agricultural exports of coffee, sisal, and tea dominate the economy. As for social indicators, Kenyan women are having three fewer babies now than in the late 1970s (down from an average of 8 to 4.4 children per woman), and those children are better educated. If Kenya can avoid political implosion, it could lead East Africa into better economic integration with the southern and western parts of the continent.

In the 1960s and 1970s Tanzania was considered something of a showcase for Africa development, and it made some advances in health and education. However, its ambitious plans for building a distinctly African form of socialism, termed *Ujaama* in Swahili, largely failed. Peasants resisted attempts to form agricultural cooperatives, and large industrial projects came to naught. Such failures proved bitterly disappointing not only for Tanzania's leadership, but also for developmental specialists in the United States and Europe who saw great promise in this unique and well-meaning experiment. For many years Tanzania was the world's largest per capita recipient of foreign aid, but the funds that it received seem to have had little lasting impact. Its per capita income is just $260.

The Poorest States   The poorest parts of Sub-Saharan Africa include the Sahel, the Horn, and several war-torn states. Sahelian countries such as Mali, Burkina Faso, Niger, and Chad have suffered from prolonged droughts and serious environmental degradation, and they have poorly developed infrastructures and commercial networks. More so than the rest of Africa, their economies remain dominated by agriculture, with very little urban or industrial development (the largest city in Chad, N'Djamena, has only 280,000 inhabitants). In the Horn of Africa, Ethiopia, Eritrea, and Somalia have all been the victims of drought as well as war. As of 1999, Ethiopia had the distinction of being the region's poorest country on a per capital basis ($100 per year). Currently there are no estimates for Somalia's GNI.

Conflict has destroyed the economies of several African states, such as Burundi, Rwanda, Sierra Leone, Liberia, Angola, and the Democratic Republic of the Congo. Yet conflict alone cannot explain why more than half the states in the region average a GNI per capita of less than $1 a day. Many of these impoverished states, such as Malawi, Guinea Bissau, Gambia, and Zambia, will be eligible for debt relief. Still, with limited infrastructure and economic output, it is hard to imagine the conditions in these countries improving any time soon.

Measuring Social Development

By world standards, measures of social development in Africa are extremely low. Yet unlike Africa's economic indicators, at least there are some positive trends in social development. Rates of child survival, which improved dramatically in the 1960s, have continued to get better since 1980 (Table 6.3). Still, compared to the rest of the world, Sub-Saharan Africa's child mortality rates are high. Levels of adult illiteracy vary greatly in the region, but several states, such as Kenya, Zimbabwe, and the Republic of the Congo, have invested in education (Figure 6.44). Troubling gaps do exist between male

▶ **Figure 6.44 Educating children**   An all-girls school in Kenya reflects an important investment in the country's social development. Education of girls, in particular, is seen as a way to raise the status of women, lower birthrates, and improve economic opportunities. *(Betty Press/Woodfin Camp & Associates)*

and female literacy figures. In most states, illiteracy is higher among women than men, and in some cases, such as Ghana and Eritrea, adult illiteracy among women is twice as high as that among men.

Life Expectancy   Sub-Saharan Africa's figures on life expectancy are, overall, the world's lowest. Only the poorest Asian countries, such as Afghanistan, Nepal, and Laos, stand at the average African level. Angola's miserable figures of 37 years for men and 39 for women reflect, in part, years of conflict and political disorder. Despite these dismal figures, some progress has been made in enhancing life expectancy in Sub-Saharan Africa. While the region's average life expectancy of 51 years may seem incredibly short, it must be remembered that infant and childhood mortality figures depress these numbers; average life expectancy for adults is substantially higher. States such as Kenya, Botswana, and Zimbabwe established relatively high life expectancy figures in the 1980s, but the growing AIDS epidemic eroded their accomplishments.

The causes of low life expectancy are generally related to extreme poverty, environmental hazards (such as drought), and various environmental and infectious disesases (malaria, schistosomiasis, cholera, AIDS, and measles). Often these factors work in combination. Malaria, for example, kills a half million African children each year. The death rate is exacerbated by poverty, undernourished children being the most vulnerable to the effects of high fevers. Cholera outbreaks occur in crowded slums and villages where food or water is contaminated by the feces of infected persons. Again, unsanitary practices result from lack of basic infrastructure. Tragically, diseases that are preventable, such as measles, occur when people have no access to or cannot afford vaccines.

Health Issues   A paucity of doctors and health facilities, especially in rural areas, also helps to explain Sub-Saharan Africa's high mortality levels. The more successful countries in the region have discovered that inexpensive rural clinics dispensing rudimentary treatments, such as oral rehydration therapy for infants with severe diarrhea, can substantially improve survival rates. Another problem is the severity of the African disease environment itself. Malaria is making a comeback in many areas as the disease-causing organisms develop resistance to common drugs and the mosquito carriers develop resistance to insecticides. While most Africans have partial immunity to malaria, it remains a major killer, and many who survive infections remain debilitated.

Many other tropical diseases also continue to plague the region. Schistosomiasis is carried by freshwater snails and can cause chronic diarrhea and cramping. The construction of dams, reservoirs, and irrigation ditches has greatly increased the environment for these snails, yet basic eradication programs are often not implemented. Consequently, 200 million Africans are affected by schistosomiasis, which can lead to dehydration and susceptibility to other diseases. In some West African areas, up to 10 percent of villagers have lost their sight from river-blindness, a disease spread by a small fly. Fortunately, effective pest control is reducing the incidence of river-blindness in the region. Collectively, national and international health agencies, along with local NGOs, have improved the availability of basic health care in Sub-Saharan Africa, but much work remains to be done.

Women and Development

Development gains cannot be made in Africa unless the economic contributions of African women are recognized. Officially, women are the invisible contributors to local and national economies. In agriculture, women account for 75 percent of the labor that produces more than half the food consumed in the region. Tending subsistence plots, taking in extra laundry, and selling surplus produce in local markets all contribute to household income. Yet since many of these activities are considered informal economic activities, they are not counted. For many of Africa's poorest people, however, the informal sector is the economy. Some research suggests that it can account for 30 to 50 percent of the gross domestic product in certain states. Within this sector, women dominate.

Status of Women   The social position of women is difficult to gauge for Sub-Saharan Africa. Women traders in West Africa, for example, have considerable political and economic power. By such measures as female labor force participation, many Sub-Saharan African countries show relative gender equality. And women in most Sub-Saharan societies do not suffer the kinds of traditional social liabilities encountered in much of South Asia, Southwest Asia, and North Africa (see "Women and Development: The Nanas-Benz of Togo").

By other measures, however, such as the prevalence of polygamy, the practice of the "bride-price," and the denial of property inheritance, African women do suffer discrimination. Perhaps the most controversial issue regarding women's status is the practice of female circumcision, or genital mutilation. In Sudan, Ethiopia, Somalia, and Eritrea, as well as parts of West Africa, almost 80 percent of girls are subjected to this practice, which is extremely painful and can have serious health consequences. Yet because the practice is considered traditional, most African states are unwilling to ban it.

Regardless of their social position, most African women still live in remote villages where educational and wage-earning opportunities remain limited and bearing numerous children remains a major economic contribution to the family. As educational levels increase and urban society expands—and as reduced infant mortality provides greater security—one can expect fertility in the region to gradually decrease. Governments can greatly quicken the process by providing birth control information and cheap contraceptives—and by investing more money in health and educational efforts aimed at women. As the economic importance of women receives greater attention from national and international organizations, more programs are being directed exclusively toward them.

Building from Within   Major shifts in the way development agencies view women and women view themselves have the potential to transform the region. All across the continent, support groups and networks have formed, raising women's

WOMEN AND DEVELOPMENT   The Nanas-Benz of Togo

The *Nanas* (a Mina word for chief) are women of power and prestige touring the streets of Lomé, Togo, in their Mercedes-Benz automobiles. Rooted in a West African tradition of textile manufacturing and vending, the Nanas have cornered the lucrative market for dress fabric in this part of Africa. Togo's cloth traders have earned a reputation for being astute, well-connected, and successful businesswomen. Even though many of the founding traders were illiterate, this has not stopped them from acquiring loans, making investments, and conducting million-dollar transactions. Honing their skills in the markets of West Africa and not in the business schools of the West, the largest cloth traders do thousands of dollars of trade each day with Europe and Asia. In Togo they rely on a network of formal and informal markets, limited bookkeeping, and mostly cash payments. The story of their success and influence is legendary.

The Nanas formed the Association of Cloth Merchants, an organization for women who sell cloth, in the 1960s. Prior to that, Ghana was the undisputed center of the African cloth trade, but Togo's Nanas took advantage of political instability in Ghana during the 1960s and secured exclusive import contracts with top textile firms in Europe. By the 1980s the Association of Cloth Merchants represented some 3,000 retailers and wholesalers. The Nanas-Benz, the most successful of the group, are so named for their preference for Mercedes-Benz cars. A story is told about the president of Togo needing a fleet of cars for a state occasion in 1977. The Nanas were contacted and in short order the president had 30 Mercedes-Benzes at his disposal.

As the Nanas prospered, so too did their business holdings, expanding to restaurants, hair salons, bakeries, and real estate operations that employ many people. Always on the lookout for new opportunities, their latest ventures include building luxurious rental properties for Lomé-based embassies and corporations.

Active politically, the Nanas-Benz support the lowering of import tariffs and encourage women to vote. Their political and religious activism has allowed them to enter spheres usually reserved for men. As economically and politically powerful women, they symbolize what is possible for Africa in the modern global economy.

Source: Adapted from "Africa's Mercedes Ladies," *World Press Review,* January 1985.

consciousness, offering women micro-credit loans for small businesses, and harnessing their economic power. From farm-labor groups to women's market associations, investment in the organization of women has paid off. In Kenya, for example, hundreds of women's groups organize tree plantings to prevent soil erosion and ensure future fuel supplies.

Whether inspired by feminism, African socialism, or the free market, community organizations have made a difference in meeting basic needs in sustainable ways. No doubt the majority of the groups fall short of all of their objectives. Yet for many people, especially women, the message of creating local networks to solve community problems is an empowering one.

Conclusion

By simply looking at the numbers, it is easy to grow despondent about Africa's future. Population growth has been outstripping economic expansion for some years, and the living standards of most of the region's inhabitants have steadily declined. Many Sub-Saharan cities are now growing rapidly, but few jobs await the rural migrants or the youngsters born and raised in the urban environment. In many respects, however, opportunities in the villages are even more limited. Frustration is often particularly acute for young people who have struggled to obtain an education, only to find that jobs are scarce.

Some pessimists think that Sub-Saharan Africa could experience an almost complete political and economic breakdown within the next few decades. Large cohorts of unemployed and disillusioned young people, they remind us, often lead to political instability. In the worst-case situation—such as currently exists in Liberia—young men may see joining a predatory armed band as their only option. As various armed forces vie for power, an entire country can be quickly devastated—as occurred in the 1990s in the Democratic Republic of the Congo. Ethnic conflicts and the spread

of AIDS exacerbate the region's many problems. Meanwhile, desertification and other forms of environmental degradation, which are themselves linked to runaway population growth, threaten the very survival of tens of millions of Africans, as well as the economic and political foundation of the entire region. For the confirmed pessimists, Rwanda—densely populated and torn by an ethnic conflict of genocidal proportions—foretells Africa's future.

The optimists' vision of Africa's future, however, can be equally compelling. Most of the region remains lightly populated, and large expanses of land could be productively farmed. Countries such as the Central African Republic, Gabon, and Angola could keep growing for many generations before crowding becomes a serious problem. In other words, most of the region still enjoys a crucial breathing space in which to address its problems.

Many recent developments give cause for hope. These include initial signs of a fertility decline, improvements in health and education, and the tentative movement away from authoritarianism and toward democracy. Equally important has been the seemingly successful transition to majority rule and a multiracial society in South Africa. Now that peace seems to have come at long last to Mozambique, southern Africa as a whole has emerged as a promising area. As the Southern African Development Community, a group of 14 countries pledged to economic cooperation, matures, it could quickly improve infrastructure and attract more foreign investment. Positive examples include the new Trans-Kalahari highway, which links Johannesburg, South Africa, with Windhoek, Namibia. Also planned is an improved highway between Johannesburg, South Africa, and Maputo, Mozambique, as well as new investment in Maputo's war-ravaged port.

In focusing on Africa's traumas, the U.S. media miss many of the region's success stories. We have heard much about Somalia, Liberia, and Rwanda in recent years, but very little about Botswana. Yet the story of Botswana is in many respects just as instructive. It has a stable, democratic government, and it enjoys a strong and growing economic base that averaged 1.8 percent real annual growth in the 1990s. Botswana's levels of social development (under age five mortality and adult literacy) are among the best in the region. The country still faces a number of significant problems—most notably the AIDS epidemic—but thus far it has demonstrated an impressive ability to address them in a constructive manner. Let us hope Botswana, rather than Rwanda, becomes the best emblem for Sub-Saharan Africa at the beginning of the new millennium.

⊕ Key Terms

agricultural density (page 233)
apartheid (page 240)
Berlin Conference (page 251)
biofuels (page 223)
clan (page 259)
coloureds (page 240)
desertification (page 221)
Gondwanaland (page 221)

Great Escarpment (page 227)
homelands (page 254)
Horn of Africa (page 229)
internally displaced persons (page 257)
kleptocracy (page 264)
Organization of African Unity (OAU) (page 254)

Pan-African Movement (page 254)
pastoralists (page 237)
physiological density (page 231)
refugee (page 257)
Sahel (page 221)

structural adjustment programs (page 221)
swidden (page 235)
township (page 255)
transhumance (page 221)
tribalism (page 256)
tribe (page 242)
tsetse fly (page 238)

⊕ Questions for Summary and Review

1. Discuss the relationship between soil types and patterns of settlement in Sub-Saharan Africa.

2. How might water resources be exploited to help in Africa's development?

3. Is desertification a natural or a human-induced process? Where does it occur?

4. What are the demographic, social, and economic consequences of AIDS in Sub-Saharan Africa?

5. Explain the factors that contribute to high population growth rates in the region.

6. What was the political significance of the 1884 Berlin Conference for Africa?

7. Why have there been relatively few boundary changes since African nations won their independence?

8. How are private and grassroots organizations shaping African development?

9. Explain the history and policies that make South Africa distinct from other states in the region.

10. How might the food policies of African governments have undermined the region's ability to feed itself?

11. What are the region's key economic resources that may facilitate its integration into the global economy?

12. Discuss why private foreign investors have generally avoided Sub-Saharan Africa.

⊕ Thinking Geographically

1. What factors might explain why European conquest and settlement occurred much earlier in tropical America than in tropical Africa?

2. What are some of the cultural and political ties that unite Africa with the Americas? With Southwest Asia? With South Asia?

3. Discuss how increasing urbanization in the twenty-first century might affect the overall structure of the population of Sub-Saharan Africa.

4. More than any other region, Sub-Saharan Africa is noted for its wildlife, especially large mammals. What environmental and historical processes explain the existence of so much

fauna? Why are there relatively fewer large mammals in other world regions?

5. Compare and contrast the role of tribalism in Sub-Saharan Africa with that of nationalism in Europe.

6. Historically, how was Sub-Saharan Africa integrated into the global economy? Was its role similar to or different from that of other developing regions?

7. Sub-Saharan Africa is a particularly problematic world region because of its relationship to North Africa. Should the Sahara be considered a cultural divide between these two regions? Why or why not?

⊕ Regional Novels and Films

Novels

Chinua Achebe, *Things Fall Apart,* (1959, Fawcett Crest)

J. M. Coetzee, *Waiting for the Barbarians* (1982, Penguin)

Nadine Gordimer, *Burger's Daughter* (1979, Viking)

Nadine Gordimer, *Conservationist* (1975, Viking)

Nadine Gordimer, *Crimes of Conscience* (1991, Heinemann)

Nadine Gordimer, *Lifetimes Under Apartheid* (1986, Knopf)

Barbara Kingsolver, *The Poisonwood Bible* (1998, Harper)

Doris Lessing, *African Laughter: Four Visits to Zimbabwe* (1992, Harper)

Alan Paton, *Cry, the Beloved Country* (1948, Scribner's)

Ngugi Wa Thiong'o, *A Grain of Wheat* (1967, Heinemann)

Ngugi Wa Thiong'o, *Petals of Blood* (1977, Heinemann)

Zoë Wicomb, *You Can't Get Lost in Cape Town* (1987, Pantheon)

Films

Black Girl (1966, Senegal)

Borom Sarret (1966, Senegal)

Breaker Morant (1980, Australia)

Burden on the Land (1992, U.S.)

Ceddo (1978, Senegal)

Emitai (1978, Senegal)

General Idi Amin Dada (1975, Uganda)

The Gods Must Be Crazy (1981, Botswana/South Africa)

Haramuya (1995, Burkina Faso)

Lumumba (2000, France)

Tilai (1990, Burkina Faso)

With These Hands (1987, U.S.)

Yaaba (1990, Burkina Faso)

⊕ Bibliography

Abraham, Willie E. 1962. *The Mind of Africa*. Chicago: The University of Chicago Press.

Aryeetey-Attoh, Samuel, ed. 1997. *Geography of Sub-Saharan Africa*. Upper Saddle River, NJ: Prentice Hall.

Bell, Morag. 1986. *Contemporary Africa: Development, Culture and the State*. White Plains, NY: Longman.

Berg, Robert J., and Whitaker, Jennifer S. 1986. *Strategies for African Development*. Berkeley: University of California Press.

Binns, Tony. 1994. *Tropical Africa*. London: Routledge.

Broughton, Simon; Ellington, Mark; Muddyman, David; and Trillo, Richard, eds. 1994. *World Music*. London: The Rough Guides.

Christopher, A. J. 1984. *Colonial Africa: A Historical Geography*. Totowa, NJ: Barnes & Noble.

Crush, Jonathan, ed. 1995. *Power of Development*. London: Routledge.

Curtin, Philip D. 1969. *The Atlantic Slave Trade*. Madison: University of Wisconsin Press.

Drake, St. Clair. 1987. *Black Folk Here and There*. Los Angeles: Center for Afro-American Studies, University of California, Los Angeles.

Elliott, Michael. 2001. "Special Report: Inside Al Qaeda." *Time Magazine,* November 12, 2001.

Fage, John D. A. 1995. *A History of Africa*. London: Routledge.

Farah, Douglas. 2001. "Al Qaeda Cash Tied to Diamond Trade," *The Washington Post,* November 2, 2001.

Fortes, Meyer, and Evans-Pritchard, Edward E. 1940. *African Political Systems*. London: Oxford University Press.

Franke, Richard W., and Chasin, Barbara H. 1980. *Seeds of Famine: Ecological Destruction and the Development Dilemma in the West African Sahel*. Totowa, NJ: Rowman and Allanheld.

Gaile, Gary, and Ferguson, Alan. 1996. "Success in African Social Development: Some Positive Indicators." *Third World Quarterly* 17(3), 557–72.

Gleave, M. B., ed. 1992. *Tropical African Development: Geographical Perspectives*. New York: Wiley.

Goliber, Thomas. 1997. "Population and Reproductive Health in Sub-Saharan Africa." *Population Bulletin* 52(4). Washington, DC: Population Reference Bureau.

Griffiths, Ienan L. L. 1984. *An Atlas of African Affairs*. London: Methuen.

Grinker, Roy Richard. 1994. *Houses in the Rainforest: Ethnicity and Inequality Among Farmers and Foragers in Central Africa*. Berkeley: University of California Press.

Grove, Alfred T. 1994. *The Changing Geography of Africa*. Oxford: Oxford University Press.

Kwamena-Poh, Michael; Tosh, John; Waller, Richard; and Tidy, Michael. 1982. *African History in Maps*. London: Longman.

Maier, Karl. 2000. *The House Has Fallen: Midnight in Nigeria*. New York: Perseus Books Group.

Manning, Patrick. 1990. *Slavery and African Life: Occidental, Oriental, and African Slave Trades*. Cambridge: Cambridge University Press.

Marcus, Harold G. 1994. *A History of Ethiopia*. Berkeley: University of California Press.

Mortimore, Michael. 1989. *Adapting to Drought: Farmers, Famines and Desertification in West Africa*. Cambridge: Cambridge University Press.

Mountjoy, Alan B., and Embleton, Clifford. 1967. *Africa: A New Geographical Survey*. New York: Praeger.

Mudimbe, Vumbi Y. 1988. *The Invention of Africa: Gnosis, Philosophy, and the Order of Knowledge.* Bloomington, IN: Indiana University Press.

Murdock, George P. 1959. *Africa: Its People and Their Culture History.* New York: McGraw-Hill.

Ndegwa, Stephen N. 1996. *The Two Faces of Civil Society: NGOs and Politics in Africa.* West Hartford, CT: Kumarian Press.

Newman, James L. 1995. *The Peopling of Africa: A Geographical Interpretation.* New Haven: Yale University Press.

Oliver, Roland, and Crowder, Michael, eds. 1981. *The Cambridge Encyclopedia of Africa.* Cambridge: Cambridge University Press.

Peil, Margaret. 1991. *Lagos: The City Is the People.* Boston: G.K. Hall & Co.

Rain, David. 1999. *Eaters of the Dry Season: Circular Labor Migration in the West African Sahel.* Boulder, CO: Westview Press.

Ramsay, F. Jeffress, ed. 2001. *Global Studies: Africa,* 9th ed. Guilford, CT: Dushkin/McGraw-Hill.

Reader, John. 1997. *Africa: A Biography of the Continent.* London: Penguin Books.

Senior, M., and Okunrotifa, P. 1983. *A Regional Geography of Africa.* London: Longman.

Stock, Robert. 1995. *Africa South of the Sahara: A Geographical Interpretation.* New York: Guilford Press.

Stren, Richard, and White, Rodney, eds. 1989. *African Cities in Crisis: Managing Rapid Urban Growth.* Boulder, CO: Westview.

Taylor, Fraser, and Mackenzie, Fiona. 1992. *Development from Within: Survival in Rural Africa.* London: Routledge.

Thornton, John. 1992. *Africa and Africans in the Making of the Atlantic World, 1400–1680.* Cambridge: Cambridge University Press.

The UNESCO General History of Africa. 1981–present. (Various authors). Berkeley: University of California Press.

U.S. Committee for Refugees. 2001. *World Refugee Survey.* Washington, DC.

Vansina, Jan. 1990. *Paths in the Rainforest: Toward a History of Political Tradition in Equatorial Africa.* Madison: University of Wisconsin Press.

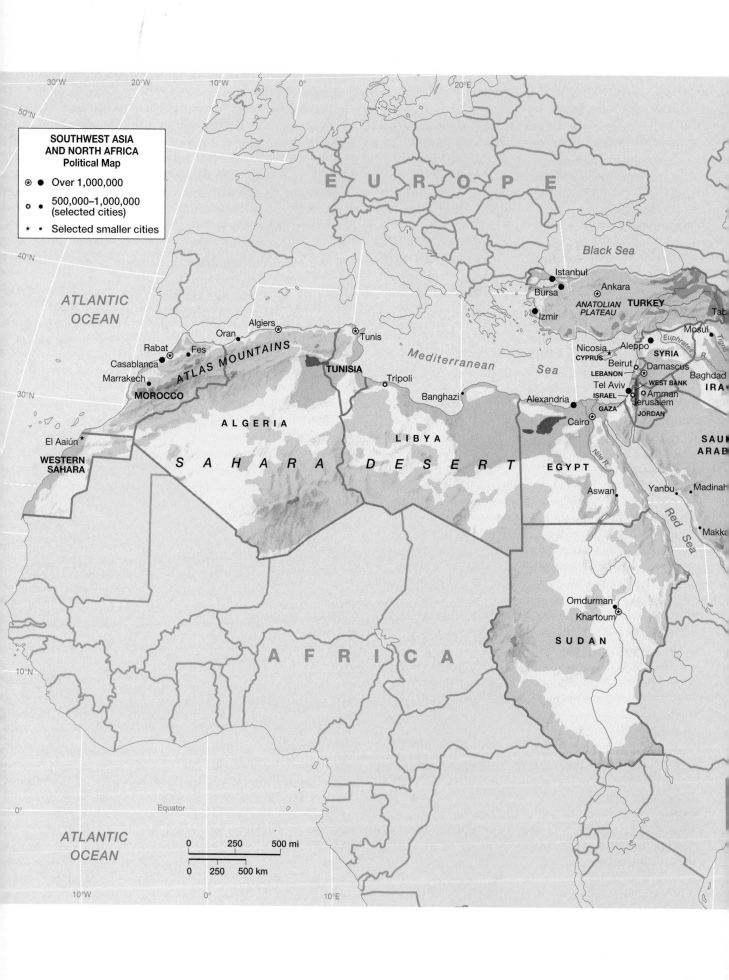

SOUTHWEST ASIA
AND NORTH AFRICA
Political Map

⊗ ● Over 1,000,000

✧ ✦ 500,000–1,000,000
(selected cities)

★ • Selected smaller cities

EUROPE

ATLANTIC
OCEAN

Black Sea

Istanbul
Bursa
Ankara
Izmir
ANATOLIAN
PLATEAU
TURKEY
Tab

Mosul
Euphrates R.
Tigris R.

Algiers
Oran
Tunis
Nicosia
CYPRUS
Aleppo
SYRIA

Rabat
Fes
Beirut
LEBANON
Damascus
Baghdad

Casablanca
ATLAS MOUNTAINS
Mediterranean   Sea
Tel Aviv
WEST BANK
IRA

Marrakech
TUNISIA
Tripoli
ISRAEL
Amman
JORDAN
Jerusalem

MOROCCO
Banghazi
Alexandria
GAZA
SAU
ARAB

ALGERIA
LIBYA
Cairo

El Aaiún
SAHARA   DESERT
EGYPT
Nile R.

WESTERN
SAHARA
Aswan
Yanbu
Madinah

Red Sea
Makka

Omdurman
Khartoum

SUDAN

AFRICA

ATLANTIC
OCEAN

Equator

0   250   500 mi

0   250   500 km

7

Southwest Asia and North Africa

The map on the left shows elevation in meters for the region.

ASIA

Tehran

Isfahan

IRAN

Shiraz

uwait City

Persian Gulf

BAHRAIN
Manama
QATAR
Doha
OMAN
Abu Dhabi
UNITED ARAB EMIRATES
Muscat

ARABIAN PENINSULA

OMAN   20°N

YEMEN

Socotra
(Yemen)

10°N

INDIAN OCEAN

Elevation in meters

4000+
2000–4000
500–2000
200–500
0–200
Below sea level

Sea Level

50°E

Climate, culture, and oil all help define the complex Southwest Asia and North Africa world region (see "Setting the Boundaries"). Straddling the historic meeting ground between Europe, Asia, and Africa, the region sprawls across thousands of miles of parched deserts, rugged plateaus, and oasis-like river valleys. It extends 4,000 miles (6,400 kilometers) between Morocco's Atlantic coastline and Iran's eastern boundary with Pakistan. More than two dozen nations are included within its borders, with the largest populations found in Egypt, Turkey, and Iran (Figure 7.1). Generally, its arid climates make it a part of the dry world, although the region's diverse physical geography causes precipitation to vary considerably. Culturally, diverse languages, religions, and ethnic identities have molded land and life within the region for centuries, strongly wedding people and place in ways that have had profound social and political implications. One traditional zone of conflict is the Middle East, where Jewish, Christian, and Islamic peoples have yet to resolve long-standing cultural tensions and political differences, particularly as they relate to the state of Israel and the pivotal Palestinian issue. In addition, Southwest Asia and North Africa's extraordinary petroleum resources place the area in the global economic spotlight. The strategic value of oil has combined with ongoing ethnic and religious conflicts to produce one of the world's least-stable political settings, one prone to geopolitical conflicts both within and between the countries of the region.

Sudan's recently changing oil fortunes illustrate these connections between the global economy, petroleum resources, ethnic differences, and political conflicts. In late 1999 oil began flowing to the Red Sea from southern Sudan's Greater Nile Oil Project, where reserves may exceed 2 billion barrels. Numerous international oil companies from North America, western Europe, and Southeast Asia have made huge investments in the area, which promise to bring much-needed capital to the poor North African country. Unfortunately, the oil fields are located in the midst of a 50-year-old civil war within Sudan that has killed more than 2 million people since 1983 and displaced another 4.5 million. As a result, Muslim-dominated government troops from the north have been aggressively stepping up their attacks on dozens of non-Muslim African villages in the oil district, killing or displacing tens of thousands of locals in the hope of securing the area's valuable resource base. While the government's forced depopulation scheme has raised protests worldwide, the oil still flows. The situation offers a poignant example of how global economic forces

◀ **Figure 7.1 Southwest Asia and North Africa** This vast region extends from the shores of the Atlantic Ocean to the Caspian Sea. Within its boundaries, striking cultural differences and globally important petroleum reserves have contributed to recent political tensions. The region's economic geography reveals great differences in wealth, as it includes some of the world's richest and poorest nations. In addition, its fragile arid environments and growing populations pose enduring twenty-first century challenges.

SETTING THE BOUNDARIES

"Southwest Asia and North Africa" is both a cumbersome term and a complex region. Often the same area is simply called the "Middle East," but some experts would exclude the western parts of North Africa as well as Turkey and Iran from such a region. In addition, the Middle East carries with it a peculiarly European vantage point—Lebanon is in the "middle of the east" only from the perspective of the western Europeans who colonized the region and still shape the names we give the world today. Instead, "Southwest Asia and North Africa" offers a useful and straightforward way to describe the general limits of the region. Largely arid climates, an Islamic religious heritage, abundant fossil fuel reserves, and persistent political instability are dominant in many areas, but internal variations on these themes make ready generalizations difficult.

There are also problems with simply defining the geographical limits of the region. On the northwest, a small piece of Turkey actually sits west of the Bosporus, generally considered to be the dividing line between Europe and Asia (Fig-

ure 7.1). Nearby Cyprus, located in the eastern Mediterranean, also resists easy regional definition since the population is divided between ethnic Greeks and Turks. We include it in Southwest Asia and North Africa because of its proximity to the Turkish and Syrian coasts. To the northeast, the largely Islamic peoples of Central Asia share many religious and linguistic ties with Turkey and Iran, but we have chosen to treat these groups in a separate Central Asia chapter (see Chapter 10).

African borders are also problematic. The conventional regional division of "North Africa" from "Sub-Saharan Africa" often cuts directly through the middle of modern Mauritania, Mali, Niger, Chad, and Sudan. We place all of these transitional countries in Chapter 6, Sub-Saharan Africa, but also include some discussion of Sudan within our Southwest Asia and North Africa material because of its status as an "Islamic republic," a political and cultural designation that ties it strongly to the Muslim world.

---

have redefined everyday life for so many people in the region, often in unanticipated ways.

Indeed, in the longer sweep of history, no world region better exemplifies the theme of globalization than Southwest Asia and North Africa. A key global **culture hearth**, the region witnessed many cultural innovations that subsequently diffused widely to other portions of the world (Figure 7.2). As a seedbed of agriculture, several great civilizations, and three major world religions, the region has been a pivotal human crossroads for thousands of years. Great trade routes have connected North Africa with the Mediterranean and Sub-Saharan Africa. Southwest Asia has also had historic ties to Europe, the Indian subcontinent, and the vastness of Central Asia. As a result, innovations within the region have often spread far beyond its bounds. Ponder the global importance of wheat and cattle, both domesticated in Southwest Asia. Consider, as well, the far-reaching impacts of urban-based civilizations that originated within its bounds and then formed models for city-building that spread to Europe and beyond. In addition, religions born within the region (Judaism, Christianity, and Islam) have shaped many other parts of the world.

Particularly within the past century, globalization has also operated in the opposite direction: the region's strategic importance has made it increasingly vulnerable to outside influences. The twentieth-century development of the petroleum industry, largely initiated by U.S. and European investment, has had enormous, but selective, consequences for economic development. Global demand for oil and natural gas has powered rapid industrial change within the region, defining its pivotal role in world trade. Many key members of **OPEC (Organization of Petroleum Exporting Countries)** are found within the region, and these countries profoundly influence

▲ **Figure 7.2 Aleppo, Syria** For centuries, the northern Syrian city of Aleppo has served as a cultural and economic crossroads. Today, the city has more than 1.5 million residents and reveals traditional mosques and minarets next to newer apartments and office buildings. *(Jean-Leo Dugast/Panos Pictures)*

global prices and production targets for petroleum. Still, the regional patterns of petroleum resources are complex. Oil-rich nations such as Saudi Arabia, Kuwait, and Iran have been fundamentally transformed, while petroleum-poor neighbors such as Jordan, Lebanon, and Tunisia have seen less dramatic impacts. Politically, petroleum resources have also elevated the region's strategic importance, adding to many traditional cultural tensions existing within the realm. As a result, the area was a central theater of conflict in the Cold War between the United States and the Soviet Union, and international tensions remain focused there.

More recently, **Islamic fundamentalism** has advocated a return to more traditional practices within the religion and has challenged the encroachment of global popular culture. Fundamentalists have also advocated merging civil and religious authority and rejecting many characteristics of modern, Western-style consumer culture. While often demonized in the Western press, fundamentalism can be seen as a cultural reaction to the disruptive role that forces of globalization have played across the region. Recently, it also has provided a dynamic political element within the region that has had consequences far beyond its boundaries. Indeed, the early twenty-first century will no doubt see the continuation of these cultural and political tensions as traditional values are juxtaposed with the omnipresent influence of the modern world (Figure 7.3).

The realities of the regional setting provide additional challenges for the populations of Southwest Asia and North Africa. The availability of water in this largely dry portion of the world has shaped both the physical and human geographies of the region. More fundamental to regional residents than their petroleum riches, water remains a daily and omnipresent need. Biologically, the region's plants and animals must adapt

▲ **Figure 7.3 Modern Istanbul** Turkey's largest city has been a major urban place for more than 2,000 years. The city was a center for Christianity before Islam arrived in the fifteenth century. Modern Western influences now add cultural diversity to this scene on Independence Avenue. *(Rob Crandall/Rob Crandall, Photographer)*

to the pervasive aridity of long dry seasons and short, often unpredictable rainy periods. Similarly, the geography of human settlement is wedded to water. Whether it comes from precipitation, underground aquifers, or rivers, water has shaped patterns of human settlement and has placed severe limits on agricultural development across vast portions of the region. Just as important, however, have been the human transformations that have dramatically reconfigured the region's environmental setting. With remarkable ingenuity, residents have captured water in varied ways that include underground drainage systems, elaborate deep-water wells, life-giving canal networks, and huge dam projects that expand irrigable acreage. However, the region's growing population of 400 million people will stress these available resources even further in the future and act as an additional catalyst for potential economic and political instability.

Environmental Geography: Life in a Fragile World

In the popular imagination, much of Southwest Asia and North Africa is a land of shifting sand dunes, searing heat, and scattered oases. Although examples of those stereotypes certainly can be found across the region, the actual physical setting, in terms of both landforms and climate, is considerably more complex (Figure 7.4). In reality, the regional terrain varies greatly, with rocky plateaus and mountain ranges more common than sandy deserts. Even the climate, although dominated by aridity, varies remarkably from the dry heart of North Africa's Sahara Desert to the well-watered highlands of northern Morocco, coastal Turkey, and western Iran. One theme is pervasive, however: a lengthy legacy of human settlement has left its mark on a fragile environment, and the entire region will be faced with increasingly daunting ecological problems in the decades ahead (see "Environmental History: Socotra's Splendid Isolation").

An ominous example of the interplay between people, politics, and environment in the region unfolded during the Persian Gulf War in the early 1990s. Iraq decided to use environmental devastation as a tool of warfare. Certainly there is a long human legacy of "environmental warfare," including the use of fire to flush out opponents or the more modern use of defoliants to reveal enemy movements through dense vegetation. Still, Iraq's aggressive use of such strategies in the conflict does not bode well for the future environmental health of a region filled with fragile ecological niches, including densely settled river valleys and vulnerable desert settings.

Most dramatically, the Iraqis used the very oil they coveted as an environmental weapon, particularly once they saw that they could not win the military conflict against UN-supported forces. They set fire to more than 700 wells in Kuwait, and many burned for months after the Iraqi army withdrew. In addition, 11 million barrels of petroleum were released into the desert, forming vast pools of damaging oil. Iraqis also targeted the Gulf, and inflows of crude oil devastated many shoreline settings. More than a decade later, the regional environment, on both land and sea, is only slowly recovering.

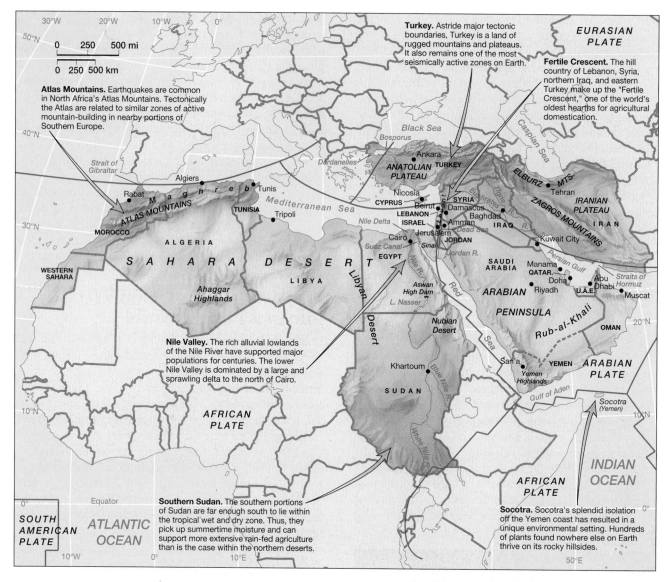

▲ Figure 7.4 Physical geography of Southwest Asia and North Africa   Vast deserts stretch across much of the region and dominate the physical setting. Major mountains and highland zones complicate the scene by altering local climate and vegetation patterns and offering distinctive environments for human settlement. In addition, converging tectonic plates from North Africa to Iran provide opportunities for active mountain-building, and earthquake hazards are high across much of the region.

Following the war, Iraq encountered increased civil resistance at home. While Kurds lobbied for independence in the north, the Marsh Arabs, an ethnic minority in the far south, also resisted Iraqi rule. Their traditional homelands lay in marshes near the mouths of the Tigris and Euphrates rivers. The area provided a living through its farmlands and fisheries, as well as a denser vegetative cover that offered protection from excessive Iraqi interference. In 1991, however, Iraqi engineers constructed a causeway that drained much of the marsh country, devastating the region's ecology and forcing many Marsh Arabs from their homes. While the Iraqis claimed it was part of a "reclamation" scheme, the real purpose was clear. The result has been to cause great environmental damage and to force many Marsh Arabs out of their protective settlements.

Regional Landforms

A quick tour of the region reveals a surprising diversity of environmental settings and landforms. In North Africa, the **Maghreb** region (meaning "western island") includes the nations of Morocco, Algeria, and Tunisia and is dominated near the Mediterranean coastline by the Atlas Mountains. The rugged flanks of the Atlas rise like a series of islands above the narrow coastal plains to the north and the vast stretches of the

ENVIRONMENTAL HISTORY   Socotra's Splendid Isolation

Socotra's stony slopes rise out out of the shimmering waters of the Indian Ocean about 230 miles (368 kilometers) southeast of Yemen. The island's unique natural and cultural history have recently caught the world's attention. Separated for millions of years from the mainland of the Arabian Peninsula, Socotra's environment evolved in splendid isolation. Over 30 percent of the island's 850 plants are found nowhere else on Earth. Exotic dragon's blood trees dot many of the island's dry and rocky hillsides (Figure 7.1.1), and dozens of other species have only recently been catalogued by botanists. Offshore, equally rare coral reefs and unusual fish populations have also evolved as a part of the island's unique environmental setting in the wet-dry tropics of the northwestern Indian Ocean.

The record of human settlement adds to Socotra's exceptional character. First occupied several thousand years ago, the island became an early and exotic outpost of Christian missionaries, was later visited by Marco Polo, and then was briefly colonized by curious Portuguese. Today, most of the island's 80,000 inhabitants (politically part of Yemen) are Muslims engaged in fishing and livestock herding.

Socotra's unique environmental heritage now hangs in the balance. Global participants in the 1992 environmental summit at Rio de Janeiro established a $5 million international biodiversity project on Socotra, and Edinburgh's Royal Botanic Garden has conducted extensive studies on the island. Some of the island's unique plants have already been commercially developed as cosmetics and herbal remedies, and in 1998 a secretive French expedition, financed by a large pharmaceutical company, gathered hundreds of rare plants for laboratory testing. At the same time, the Yemeni government has invited British Gas to explore for petroleum just offshore, and it has pondered a grand scheme for developing a series of luxury tourist hotels that would forever change the island's character. Only time will tell how Socotra, which is often cited as the Galápagos of the Indian Ocean, will fare in the twenty-first century. A recently developed mixed-use zoning plan and a newly paved airstrip suggest that some development is inevitable. Hopefully, there will be still be room for both people and the dragon's blood trees in this remote and imperiled paradise.

Source: Adapted from "An Imperiled Arabian Paradise," *The Economist*, August 5, 2000.

▲ **Figure 7.1.1 Socotra's dragon's blood tree**   Unique to Socotra, the rare dragon's blood tree reflects the island's environmental isolation. Evolving as a part of the island's ecosystem, the tree survives in the region's dry tropical climate. *(Chris Hellier/Corbis)*

---

lower Saharan deserts to the south (Figure 7.5). The Atlas Mountains are related to Europe's nearby alpine system and are a complex series of folded highlands. They reach heights of more than 13,000 feet (3,965 meters) in central Morocco and sweep eastward to dominate the physical settings of northern Algeria and Tunisia. South and east of the Atlas Mountains, interior North Africa varies between rocky plateaus and extensive lowlands. Even here, however, the visual scene includes diverse desert topography: the bare and steeply sloping mountains of southern Algeria contrast markedly with the sandier reaches of the Libyan Desert and the river-carved lowlands of the Nile as it flows northward through Sudan and Egypt.

Southwest Asia is generally more mountainous than North Africa. In the **Levant,** or eastern Mediterranean region, mountains rise within 20 miles (32 kilometers) of the sea, and the highlands of Lebanon reach heights of more than 10,000 feet (3,048 meters). Farther south, the Arabian Peninsula forms a massive tilted plateau, with western highlands higher than 5,000 feet (1,524 meters) gradually sloping eastward to extensive lowlands in the Persian Gulf area. The Persian Gulf separates the Arabian Peninsula from nearby Iran, reaching its narrowest point at the strategically pivotal Straits of Hormuz, through which pass vast amounts of the region's petroleum exports. Both the southwest (Yemen) and southeast (Oman) corners of the peninsula are marked by particularly rugged highlands whose steep slopes and narrow canyons depart strikingly from the drifting dune fields that dominate other portions of the Arabian interior.

North and east of the Arabian Peninsula lie the two great upland areas of Southwest Asia: the Iranian and Anatolian plateaus (*Anatolia* refers to the large peninsula of Turkey, sometimes called Asia Minor) (Figures 7.4 and 7.6). Both of these plateaus average between 3,000 and 5,000 feet (915 to 1,524 meters) in elevation and are associated with tectonically active mountain uplifts and fault zones that are very prone to earthquakes (Figure 7.4). One dramatic quake in western Turkey in August 1999 measured 7.8 on the Richter

▲ **Figure 7.5 Atlas Mountains**   Residents of Morocco's Atlas Mountains adapt to the region's steep slopes, and their Mediterranean-style agriculture takes advantage of higher rates of annual precipitation than are found in the arid deserts below.   *(Sean Sprague/Stock Boston)*

▲ **Figure 7.6 Anatolian Plateau**   Turkey's vast interior is a mix of high mountains and elevated plateaus. The Anatolian Plateau provides a complex environment in which traditional agriculture includes midlatitude crops and livestock adapted to the region's semiarid setting.   *(Adam Woolfitt/Woodfin Camp & Associates)*

ments. Although much smaller, the distinctive Jordan River Valley is also a notable lowland region that straddles the strategic borderlands of Israel, Jordan, and Syria and drains southward to the Dead Sea.

Patterns of Climate

Although often termed the "dry world," a closer look at Southwest Asia and North Africa reveals a more complex climatic pattern (Figure 7.7). Both latitude and altitude come into play. Aridity dominates large portions of the region. A nearly continuous belt of desert lands stretches eastward from the Atlantic coast of southern Morocco across the continent of Africa, through the Arabian Peninsula, and into central and eastern Iran. Throughout this vast dry zone, plant and animal life have adapted to extreme conditions. Deep or extensive root systems and rapid life cycles allow desert plants to benefit from the limited moisture they receive. Similarly, animals adjust by efficiently storing water, hunting nocturnally, or migrating seasonally to avoid the worst of the dry cycle. Camels are justly famous as beasts of burden, going a week or more between water holes.

Some of the driest conditions are found across North Africa. Away from the Atlas Mountains, the Sahara Desert dominates much of the region. Much of the central Sahara receives less than 1 inch (2.5 centimeters) of rain a year, and can thus support only the most meager forms of vegetation. Large expanses of sand dune and rocky waste have virtually no plant life at all. Saharan summers feature extremely hot days and warm evenings (the world's record high, 136 °F [58 °C] was recorded in Libya), but winters are generally pleasant, even cool at

scale, killed more than 17,000 people, and left 350,000 residents homeless. The Elburz Mountains of northern Iran reach more than 18,000 feet (5,485 meters) in height. They offer a dramatic contrast to Caspian Sea lowlands immediately to the north. Portions of these plateaus also feature interior drainage, with streams flowing from bordering mountains into basins, only to evaporate in extensive salt flats or saline lakes.

Smaller lowlands characterize other portions of Southwest Asia. Narrow coastal strips are common in the Levant, along both the southern (Mediterranean) and northern (Black Sea) Turkish coastlines, and north of the Iranian Elburz near the Caspian Sea. Iraq contains the most extensive alluvial lowlands in Southwest Asia, dominated by the Tigris and Euphrates rivers, which flow southeast to empty into the Persian Gulf. Neighboring portions of Kuwait and southwestern Iran are also part of these river valley settings that once formed part of a larger sea before being filled in by river-borne sedi-

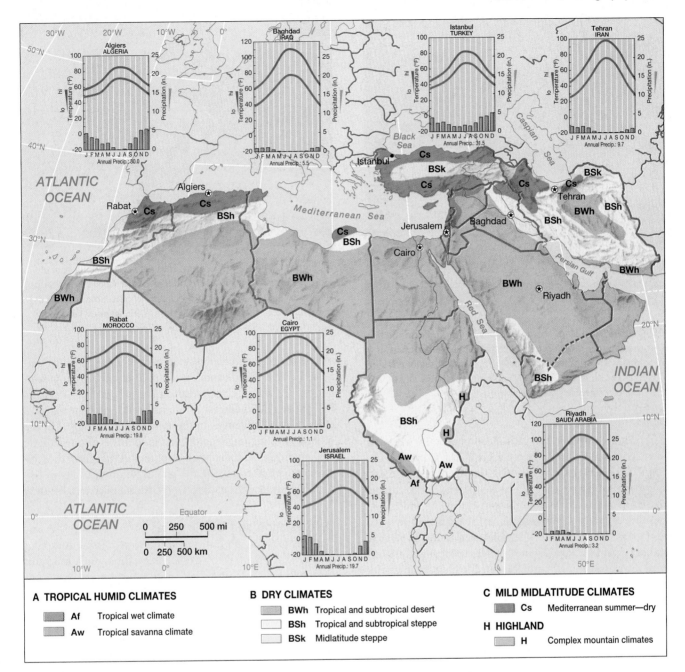

▲ Figure 7.7 Climate map of Southwest Asia and North Africa   Dry climates dominate from western Morocco to eastern Iran. Within these zones, persistent subtropical high-pressure systems offer only limited opportunities for precipitation. Elsewhere, mild midlatitude climates with wet winters are found near the Mediterranean Basin and Black Sea. To the south, tropical savanna climates provide summer moisture to southern Sudan.

night. Only in the tropical latitudes of southern Sudan do precipitation levels rise significantly. Here, summer rains produce between 20 and 50 inches (51 to 127 centimeters) of precipitation, and tropical savannas and woodlands replace the desert vegetation to the north.

Across the Red Sea, deserts also dominate Southwest Asia. Most of the Arabian Desert is not quite as dry as the Sahara proper, although the Rub-al-Khali along Saudi Arabia's southern border is one of the world's most desolate areas. On the fringe of the summer-monsoon belt, the Yemen Highlands along the southwestern edge of the Arabian Peninsula receive much more rain than the rest of the region and thus form a more favorable site for human habitation. To the northwest, precipitation also increases slightly in the central Tigris and Euphrates river valleys, with Baghdad, Iraq, averaging just over 5 inches (13 centimeters) of rain annually. Another major arid zone lies across Iran, a country divided into a series of minor mountain ranges and desert basins (Figure 7.8).

▲ **Figure 7.8 Arid Iran** Large portions of Iran are arid. Settlements are strongly oriented around zones of higher moisture or opportunities for irrigated agriculture. This view of remote desert lands to the north of the Zagros Mountains is near the city of Shiraz. *(Isabella Tree/Hutchison Library)*

▲ **Figure 7.9 Algerian orange harvest** The Mediterranean moisture in northern Algeria produces an agricultural landscape similar to that of southern Spain or Italy. Winter rains create a scene that contrasts sharply with deserts found elsewhere in the region. *(Dott. Giorgio Gualco/Bruce Coleman Inc.)*

Elsewhere, altitude and latitude dramatically alter the desert environment and produce a surprising amount of climatic variety. For example, the Atlas Mountains and the nearby lowlands of northern Morocco, Algeria, and Tunisia experience a distinctly Mediterranean climate in which hot, dry summers alternate with cooler, relatively wet winters. In these areas the landscape resembles that found in nearby southern Spain or Italy (Figure 7.9). Both Rabat and Algiers reveal the classic climatic profile of the Mediterranean world. Annual precipitation averages between 20 and 30 inches (51 to 76 centimeters); most moisture comes in winter; and summers are hot, dry, and dusty.

A second zone of Mediterranean climate extends along the Levant coastline into the nearby mountains and northward across sizable portions of northern Syria, Turkey, and northwestern Iran. The mountains of Lebanon, for example, receive heavy winter snows, and coastal Beirut gets more annual rainfall than either London or Seattle. Farther south, the Israeli capital of Jerusalem is somewhat drier, but still averages more than 20 inches (51 centimeters) of precipitation. Western Turkey also benefits from seasonal Mediterranean moisture, with Istanbul's climate resembling that of the Po River valley in northeastern Italy. Eastward, precipitation decreases gradually across the Anatolian Plateau, and Mediterranean woodlands are replaced by semiarid grassland (or steppe) vegetation.

Legacies of a Vulnerable Landscape

The environmental history of Southwest Asia and North Africa illustrates both the shortsighted and ingenious legacies of its human occupants. Littered with examples of despoiled vegetation, blighted soils, and depleted water resources, the region reveals the hazards of lengthy human settlement in a marginal land (Figure 7.10). Indeed, the pace of population growth and technological change during the past century suggests that the region's vulnerable environment is destined to face even greater challenges in the twenty-first century. On the other hand, the long record of its residents is also testimony to their remarkable ability to adapt to and transform their regional environment and in the process to produce a truly humanized landscape that differs greatly from the natural setting.

Deforestation and Overgrazing Deforestation is an ancient problem in Southwest Asia and North Africa. Although much of the region is too dry for trees, the more humid and elevated lands that ring the Mediterranean once supported heavy forests. Included in these woodlands are the cedars of Lebanon, despoiled in ancient times and now reduced to a few scattered groves that survive in a largely denuded landscape. Even 4,000 years ago, monarchs in both Egypt (along the Nile River) and Mesopotamia (the area between the Tigris and Euphrates rivers) were forced to launch major expeditions to obtain lumber needed for city building and ship construction. Elsewhere, growing demands for agricultural land pressured upland forests, which were removed and replaced with grain fields, orchards, and pastures.

Human activities have conspired with natural conditions to reduce most of the region's forests to grass and scrub. Mediterranean forests often grow slowly, are highly vulnerable to fire, and usually fare poorly if subjected to heavy grazing. Browsing by goats in particular has often been blamed for much of the region's forest loss, but other livestock have also had an impact on the vegetative cover, particularly in steeply sloping semiarid settings that are slow to recover from grazing. The lack of trees has been costly. Lumber, pulp, and other forest products generally have to be imported. Deforestation has also resulted in a millennia-long deterioration of the region's water supplies and in accelerated soil erosion.

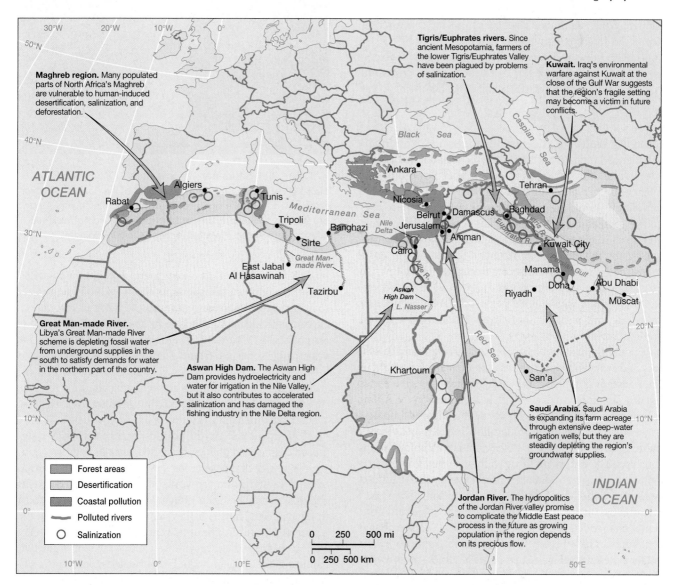

▲ **Figure 7.10 Environmental issues in Southwest Asia and North Africa** Growing populations, pressures for economic development, and pervasive aridity combine to create environmental hazards across the region. Long human occupance has contributed to deforestation, irrigation-induced salinization, and expanding desertification. Saudi Arabia's deep-water wells, Egypt's Aswan High Dam, and Libya's Great Man-made River are all recent technological attempts to expand settlement, but they may carry a high long-term environmental price tag.

Some forests survive in the mountains of northern and southern Turkey, and northern Iran also retains considerable tree cover. Scattered forests can be found in western Iran, the Levant, and the Atlas Mountains. Moreover, several governments have launched reforestation drives and forest preservation efforts in recent years. For example, both Israel and Syria have expanded their coverage of wooded lands since the 1980s, and more than 5 percent of Lebanon's total area is now part of the Chouf Cedar Reserve formed in 1996 to protect old-growth cedars.

Salinization Salinization, or the buildup of toxic salts in the soil, is another ancient environmental issue in a region where irrigation has been practiced for centuries (Figure 7.10). The accumulation of salt in the topsoil is a common

problem wherever desert lands are subjected to extensive irrigation. All freshwater contains a small amount of dissolved salt, and when water is diverted from streams into fields, salt remains in the soil after the moisture has been absorbed by the plants and evaporated by the sun. In humid climates, accumulated salts are washed away by saturating rains, but in arid climates this rarely occurs. Where irrigation is practiced, salt concentrations build up over time, leading to lower crop yields and eventually to land abandonment.

Hundreds of thousand of acres of once-fertile farmland within the region have been destroyed or degraded by salinization. The problem has been particularly acute in Iraq, where centuries of canal irrigation along the Tigris and Euphrates rivers have seriously degraded land quality. Similar conditions plague central Iran, Egypt, and irrigated portions of the

Maghreb. While salinization dates back to ancient times, recent attempts to intensify agriculture, made necessary by the region's rapidly growing population, have worsened the problem. Agronomists are attempting to breed salt-resistant crops and to devise inexpensive ways to desalinize soil, but their success so far has been minimal.

Managing Water   Residents of the region are continually challenged by many other problems related to managing water in one of the driest portions of Earth. For thousands of years, the pervasive attitude has been that any available technology should be used to bring water to where it is needed. As technological change has accelerated, the scale and impact of water management schemes have had a growing effect on the region's environment. Sometimes the costs are justified if large new areas are brought into productive and sustainable agricultural use. In other cases, however, the long-term environmental price will far outweigh any immediate and perhaps short-lived gains in agricultural production.

Regional populations have been modifying drainage systems and water flows for thousands of years. The Iranian **qanat system** of tapping into groundwater through a series of gently sloping tunnels was widely replicated on the Arabian Peninsula and in North Africa. With simple technology, farmers directed the underground flow to fields and villages where it could be efficiently utilized. Other ingenious adaptations such as waterwheels, carefully managed flood irrigation, and increasingly elaborate canal systems also redefined the local distribution of water across the region.

In the past half century, however, the scope of environmental change has been greatly magnified. Water management schemes have become huge engineering projects, major budget expenditures, and even political issues, both within and between countries of the region. One remarkable example is Egypt's Aswan High Dam, completed in 1970 on the Nile River south of Cairo (Figure 7.10). Many benefits came with the dam's completion. Most important, greatly increased storage capacity in the upstream reservoir promoted more year-round cropping and an expansion of cultivated lands along the Nile, critical changes in a country that continues to experience rapid population growth. The dam also generates large amounts of clean electricity for the region. But the environmental costs have been high. The dam has changed methods of irrigation along the Nile, greatly increasing salinization. While fresh sediments and soil nutrients once annually washed across the valley floor, more controlled irrigation has meant greatly increased inputs of costly fertilizers, as well as the infilling of Lake Nasser behind the dam with accumulating sediments. Other problems with the dam include an increased incidence of *schistosomiasis* (a debilitating parasitic disease spread by waterborne snails in irrigation canals) and the collapse of the Mediterranean fishing industry near the Nile Delta, an area previously nourished by the river's silt.

Elsewhere, **fossil water,** or water supplies stored underground during earlier and wetter climatic periods, has also been put to use by modern technology. Libya's "Great Man-

▲ **Figure 7.11 Saudi Arabian irrigation**   One of Saudi Arabia's largest deep-water irrigation sites is at the oasis of Al Hofuf east of Riyadh. While significantly expanding the country's food production, such efforts are rapidly depleting underground supplies of fossil water.   *(Minosa/Scorpio/Corbis/Sygma)*

made River" scheme taps underground water in the southern part of the country, transports it 600 miles (965 kilometers) to northern coastal zones, and uses the precious resource to expand agricultural production in one of North Africa's driest countries (Figure 7.10). Similarly, Saudi Arabia has invested huge sums to develop deep-water wells, allowing it to greatly expand its food output (Figure 7.11). Unfortunately, these underground supplies are being depleted much more rapidly than they are recharged, thus limiting the long-term sustainability of such ventures.

Most dramatically, **hydropolitics,** or the interplay of water resource issues and politics, has raised tensions between countries that share drainage basins. For example, Sudan's plans to expand its irrigation networks along the upper Nile and Ethiopia's Blue Nile Dam project are both causes of concern in Egypt. To the north, Turkey's growing development of the upper Tigris and Euphrates rivers (the Southeast Anatolian Project) has raised issues with Iraq and Syria, who argue that capturing "their" water might be considered a provocative political act. Hydropolitics has also played into negotiations between Israel, the Palestinians, and other neighboring states, particularly in the valuable Jordan River drainage, which runs through the center of the area's most hotly disputed lands (Figure 7.10). Israelis fear Palestinian and Syrian pollution; nearby Jordanians argue for more water from Syria; and all regional residents must deal with the uncomfortable reality that, regardless of their political differences, they must drink from the same limited supplies of freshwater (Figure 7.12).

Population and Settlement: Patterns in an Arid Land

The geography of human population across Southwest Asia and North Africa demonstrates the intimate tie between water and life in this part of the world. The pattern is complex: large areas of the population map remain almost devoid of permanent settlement, while more moisture-favored lands suffer increasingly from problems of crowding and overpopulation (Figure 7.13). Almost everywhere across the region, human occupants have uniquely adapted themselves to living within arid or semiarid settings, and these cultural and technological transformations have been enduringly stamped upon the visual scene. In addition, rapid increases in population, levels of urbanization, and contacts with the outside world have left their mark and suggest that future patterns of population growth and economic development will continue reshaping the settlement landscape in dramatic and unanticipated ways.

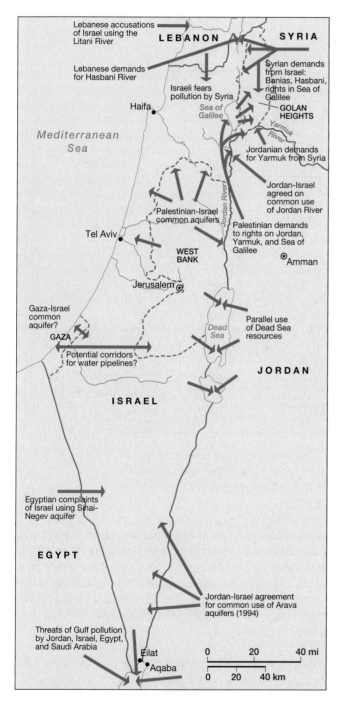

▲ Figure 7.12 Hydropolitics in the Jordan River Basin
Many water-related issues complicate the geopolitical setting in the Middle East. The Jordan River system has been a particular focus of conflict.   *(Modified from Arnon Soffer, Rivers of Fire: The Conflict over Water in the Middle East, 1999, p. 180. Reprinted by permission of Rowman and Littlefield Publishers, Inc.)*

The Geography of Population

Today, more than 400 million people live in Southwest Asia and North Africa (Table 7.1). The distribution of that population is strikingly varied: in countries such as Egypt, large zones of almost empty desert land stand in sharp contrast to crowded, well-watered locations, such as those along the Nile River. While overall population densities in such countries appear modest, **physiological densities,** a statistic that relates the number of people to the amount of arable land, are among the highest on Earth. Patterns of urban geography are also highly uneven: although less than two-thirds of the overall population is urban, many nations are overwhelmingly dominated by huge and sprawling cities that concentrate urban functions and produce the same problems of urban crowding found elsewhere in the developing world. Rates of recent urban growth have also been phenomenal: Cairo, a modest-size city of 3.5 million people in the 1960s, has quadrupled in population in the past 40 years.

Across North Africa, two dominant clusters of settlement, both shaped by the availability of water, account for most of the region's population (Figure 7.13). In the Maghreb, the moister slopes of the Atlas Mountains and nearby better-watered coastal districts have accommodated denser populations for centuries. Today, concentrations of both rural and urban settlement extend from south of Casablanca in Morocco to Algiers and Tunis on the shores of the southern Mediterranean. Indeed, most of the populations of Morocco, Algeria, and Tunisia crowd into this crescent of more-favored country, a stark contrast to the almost empty lands south and east of the Atlas Mountains. Casablanca and Algiers are the largest cities in the Maghreb. They have rapidly growing metropolitan populations of around 4 million residents each. Farther east, much of Libya and western Egypt is very thinly settled. Egypt's Nile Valley, however, is home to the other great North African population cluster. The vast majority of Egypt's 70 million people live within 10 miles of the river. Optimistic politicians and planners hope to create a second corridor of denser settlement in Egypt's "New Valley" west of the Nile in the twenty-first century by diverting water from Lake Nasser into the Western Desert. The 20-year project remains in its early stages.

Southwest Asian populations also reflect the realities of living in a dry world: the region's residents are clustered in favored coastal zones, moister highland settings, and desert localities where water is available from nearby rivers or subsurface aquifers. High population densities are found in better-watered portions of the eastern Mediterranean (Israel, Lebanon, and Syria) and Turkey. Nearby Iran is home to more than 65 million residents, but population densities vary considerably from thinly occupied deserts in the east to more concentrated settlements near the Caspian Sea and across the more humid highlands of the northwest. Turkey's Istanbul (formerly Constantinople) and Iran's Tehran are Southwest Asia's largest urban areas, and both have grown in recent years as rural populations gravitate toward the economic opportunities of these cities. Elsewhere, sizable populations are scattered through the Tigris and Euphrates valley, the Yemen Highlands, and near oases where groundwater can be tapped to support agricultural or industrial activities.

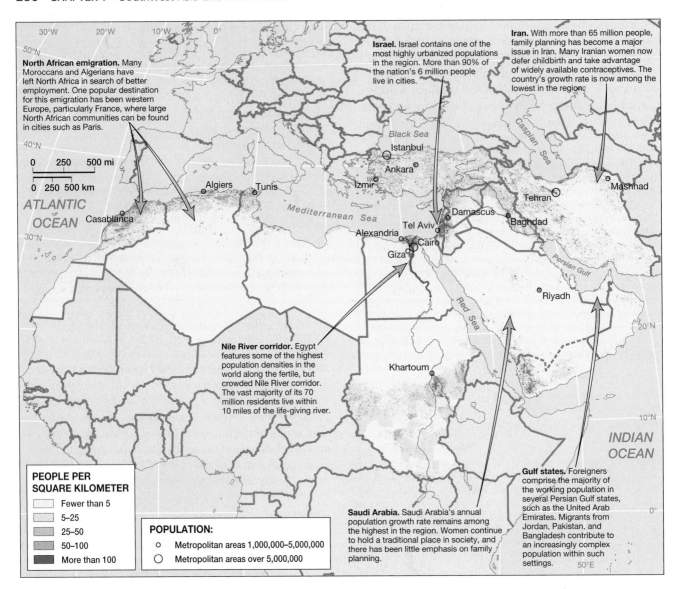

North African emigration. Many Moroccans and Algerians have left North Africa in search of better employment. One popular destination for this emigration has been western Europe, particularly France, where large North African communities can be found in cities such as Paris.

Israel. Israel contains one of the most highly urbanized populations in the region. More than 90% of the nation's 6 million people live in cities.

Iran. With more than 65 million people, family planning has become a major issue in Iran. Many Iranian women now defer childbirth and take advantage of widely available contraceptives. The country's growth rate is now among the lowest in the region.

Nile River corridor. Egypt features some of the highest population densities in the world along the fertile, but crowded Nile River corridor. The vast majority of its 70 million residents live within 10 miles of the life-giving river.

Saudi Arabia. Saudi Arabia's annual population growth rate remains among the highest in the region. Women continue to hold a traditional place in society, and there has been little emphasis on family planning.

Gulf states. Foreigners comprise the majority of the working population in several Persian Gulf states, such as the United Arab Emirates. Migrants from Jordan, Pakistan, and Bangladesh contribute to an increasingly complex population within such settings.

PEOPLE PER SQUARE KILOMETER
- Fewer than 5
- 5–25
- 25–50
- 50–100
- More than 100

POPULATION:
- ○ Metropolitan areas 1,000,000–5,000,000
- ◯ Metropolitan areas over 5,000,000

▲ **Figure 7.13 Population map of Southwest Asia and North Africa**   The striking contrasts between large, sparsely occupied desert zones and much more densely settled regions where water is available are clearly evident. The Nile Valley and the Maghreb region contain most of North Africa's people, while Southwest Asian populations cluster in the highlands and along the better-watered shores of the Mediterranean.

Water and Life: Rural Settlement Patterns

Water and life are intimately linked across the rural settlement landscapes of Southwest Asia and North Africa. Indeed, the diverse environments of Southwest Asia, in particular, are home to one of the world's earliest hearths of **domestication**, where plants and animals were purposefully selected and bred for their desirable characteristics. Beginning around 10,000 years ago, increased experimentation with wild varieties of wheat and barley initiated a long and complex series of cultural innovations that subsequently included domesticated animals, such as cattle, sheep, and goats. Much of the early agricultural activity focused on the **Fertile Crescent**, an ecologically diverse zone that stretches from the Levant inland through the fertile hill country of northern Syria into Iraq (Figure 7.4). Between 5,000 and 6,000 years ago, better knowledge of irrigation techniques and increasingly centralized political states

promoted the diffusion of agriculture into nearby valleys such as the Tigris and Euphrates (Mesopotamia) and North Africa's Nile Valley. Since then, different peoples of the region have adapted to its environmental diversity and limitations in distinctive ways. In the process, they have practiced forms of agriculture appropriate to their settings and have left their own unique imprints upon the landscape (Figure 7.14). Throughout the region, the signatures of rural settlement reflect the lasting interrelationship between water and life, an uneasy marriage between a limited natural resource and a dynamic, unpredictable, and growing human population.

Pastoral Nomadism   Pastoral nomadism, most common in the drier portions of the region, is a traditional form of subsistence agriculture in which practitioners depend on the seasonal movement of livestock for a large part of their livelihood.

TABLE 7.1 *Demographic Indicators*

Country	Population (Millions, 2001)	Population Density, per square mile	Rate of Natural Increase	TFR[a]	Percent < 15[b]	Percent > 65	Percent Urban
Algeria	31.0	34	1.9	3.1	39	4	49
Bahrain	0.7	2,688	1.9	2.8	31	2	88
Cyprus	0.9	247	0.6	1.8	23	10	66
Egypt	69.8	181	2.1	3.5	36	4	43
Gaza and West Bank	3.3	1,365	3.7	5.9	47	4	
Iran	66.1	108	1.2	2.6	36	5	64
Iraq	23.6	139	2.7	5.3	42	3	68
Israel	6.4	791	1.6	3	29	10	91
Jordan	5.2	150	2.2	3.6	40	5	79
Kuwait	2.3	297	1.8	4.2	26	1	100
Lebanon	4.3	1,061	1.7	2.5	29	7	88
Libya	5.2	8	2.4	3.9	37	4	86
Morocco	29.2	169	2.0	3.4	33	5	55
Oman	2.4	29	3.5	6.1	41	2	72
Qatar	0.6	139	2.7	3.9	27	2	91
Saudi Arabia	21.1	25	2.9	5.7	43	2	83
Sudan	31.8	33	2.4	4.9	43	3	27
Syria	17.1	231	2.6	4.1	41	3	50
Tunisia	9.7	154	1.3	2.3	31	6	62
Turkey	66.3	221	1.5	2.5	30	6	66
United Arab Emirates	3.3	103	1.4	3.5	26	1	84
Western Sahara	0.3	3	2.9	6.8			95
Yemen	18.0	88	3.3	7.2	48	3	26
Total	418.6						

[a]Total fertility rate

[b]Percent of population younger than 15 years of age

Source: Population Reference Bureau. World Population Data Sheet, *2001.*

An offshoot of sedentary agriculture, pastoral nomadism is an appropriate adaptation to life in arid settings where inadequate moisture and forage make permanent settlement impossible. Arabian Bedouins, North African Berbers, and Iranian Bakhtiaris provide surviving examples of nomadism within the realm. Today, however, with fewer than 10 million nomads remaining, the lifestyle is in decline, victim of more constricting political borders, a reduced demand for traditional beasts of burden, competing land uses, and selective overgrazing. In addition, government resettlement programs in Saudi Arabia, Syria, Egypt, and elsewhere are actively promoting a more settled lifestyle for many nomadic groups.

The settlement landscape of pastoral nomads reflects their need for mobility and flexibility as they seasonally move camels, sheep, and goats from place to place. Typically, nomads use lightweight and easily transportable tents, and tribes often split up into smaller kin-related clans to move about the landscape easily and take advantage of limited water and pasture resources. Near highland zones such as the Atlas Mountains or the Anatolian Plateau, nomads practice **transhumance** by seasonally moving their livestock to cooler, greener high country pastures in the summer and then re-

turning them to valley and lowland settings for fall and winter grazing. Elsewhere, broad horizontal movements often incorporate huge territories of desert to support small groups of a few dozen families. Where moisture conditions allow, some nomads divide further and often put women and children to work on small agricultural plots that supplement the tribe's dependence on livestock. In addition, nomads trade with sedentary agricultural populations, a symbiotic relationship in which they exchange meat, milk, hides, and wool for cereal and orchard crops available at desert oases.

Oasis Life   Permanent oasis settlements dot the arid landscape where high groundwater levels or modern deep-water wells provide reliable moisture in otherwise arid locales (Figures 7.11 and 7.15). Tightly clustered, often walled villages, their sun-baked mud houses blending into the surrounding scene, sit adjacent to small, but intensely utilized fields where underground water is carefully applied to tree and cereal crops. In more recently created oases, concrete blocks and prefabricated housing add a contemporary look. Surrounded by large zones of desert, these green islands of rural activity stand out in sharp contrast to the sand- and rock-strewn landscape.

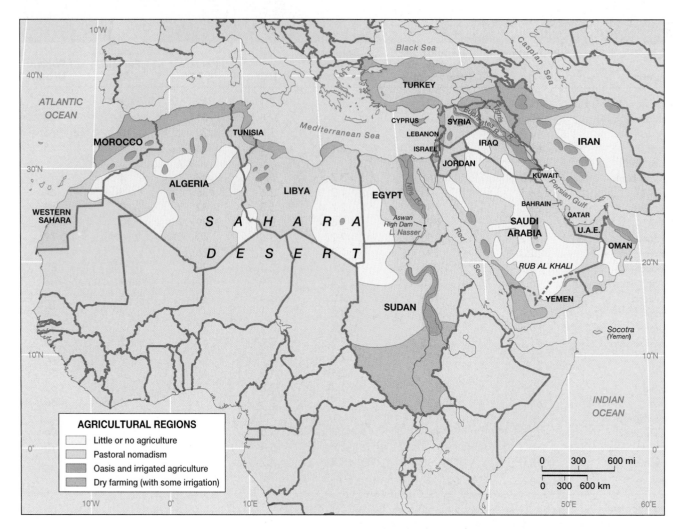

▲ Figure 7.14 Agricultural regions of Southwest Asia and North Africa   Important agricultural zones include oases and irrigated farming where water is available. Elsewhere, dry farming supplemented with irrigation is practiced in midlatitude settings.   *(Modified from Clawson and Fisher, 1998,* World Regional Geography, *Upper Saddle River, NJ: Prentice Hall; and Bergman and Renwick, 1999,* Introduction to Geography, *Upper Saddle River, NJ: Prentice Hall)*

Traditional oasis settlements are composed of close-knit families who work their own irrigated plots or, more commonly, toil for absentee landowners. Although oases are usually small, eastern Saudi Arabia's Hofuf oasis covers more than 30,000 acres (12,150 hectares). While some crops are raised for local consumption, commercial trade has always played a role in such settings. Although oases are less pivotal today, centuries ago camel caravans and pastoral nomads frequently stopped at such settlements, trading their goods for the fruits of oasis life. In the past century, the expanding world demand for products such as figs and dates has included even these remote locations in the global economy, and many products end up on the tables of hungry Europeans or North Americans. From central Algeria and Libya eastward to the dry reaches of the Arabian Peninsula, oasis life continues for some residents of the region. New drilling and pumping technologies, particularly in Saudi Arabia, have added to the size and number of oasis settlements. But oasis life across the region faces major challenges: population growth, groundwater depletion, and the pressures of global cultural change threaten the economic and social integrity of these settlements.

Settlement Along Exotic Rivers   For centuries, the region's densest rural settlement has been tied to its great river valleys and their seasonal floods of water and enriching nutrients. In such settings, **exotic rivers** transport precious water and nutrients from distant, more humid lands into drier regions, where the resources are utilized for irrigated farming. The Nile and the Tigris and Euphrates rivers are the largest regional examples of such activity, with both systems also characterized by large, densely settled deltas. Other linear irrigated settlements can be found near the Jordan River in Israel and Jordan, along short streams originating in North Africa's Atlas Mountains, and on the more arid peripheries of the Anatolian and Iranian plateaus. These settings, while capable of supporting sizable rural populations, are also among the most vulnerable to overuse, par-

▲ **Figure 7.15 Oasis agriculture** Date palms dominate this Tunisian oasis. Often, these commercial farm products find their way to consumers in distant European and North American markets. *(Alain Le Garsmeur/Panos Pictures)*

ticularly if irrigation results in salinization. Farmlands were abandoned centuries ago in ancient Mesopotamia (modern Iraq) as toxic salts lowered crop yields. These hazards of rural settlement in dry lands persist, and the growing buildup of salts in the modern Nile Valley suggests that such problems will continue.

Farming in such localities supports much higher population densities than is the case with pastoral nomadism or traditional desert oases. Fields are small, intensely utilized, and connected with closely managed irrigation systems designed to store and move water efficiently through the settlement. In Egypt, farmers of the Nile Valley live in densely settled, clustered villages near their fields; work much of their land with the same tools and technologies as their ancestors did; and grow a mix of cotton, rice, wheat, and forage crops. The scale and persistence of their agricultural imprint have become one of the world's most visible and dramatic signatures of the human transformation of Earth's surface (Figure 7.16). But rural life is also changing in such settings. New dam- and canal-building schemes in Egypt, Israel, Syria, Turkey, and elsewhere are increasing the storage capacity of river systems, which allows for more year-round agricultural activity. Higher-yielding rice and wheat varieties and more mechanized agricultural methods have also raised food output, particularly in places such as Egypt and Israel. Some of the most efficient farms in the region are associated with Israeli **kibbutzes,** collectively worked settlements that produce grain, vegetable, and orchard crops irrigated by waters from the Jordan River and from the country's elaborate feeder canals. For example, Israel's National Water Carrier system takes water from the Sea of Galilee and moves it south and west, where intensively worked agricultural operations produce key food crops for nearby Tel Aviv and Jerusalem.

The Challenge of Dryland Agriculture

Mediterranean climates in portions of the region permit varied forms of dryland agriculture that depend largely on seasonal moisture to support farming. These zones include the better-watered valleys and coastal lowlands of the northern Maghreb, lands along the

▲ **Figure 7.16 Nile Valley** This satellite image of the Nile Valley dramatically reveals the impact of water on the North African desert. Cairo lies at the southern end of the delta, where it begins to widen toward the Mediterranean Sea. *(NASA/MODIS)*

shore of the eastern Mediterranean, and favored uplands across the Anatolian and Iranian plateaus. A different regional variant of dry farming is also present across Yemen's terraced highlands in the moister corners of the southern Arabian Peninsula, and a smaller zone appears in Oman just east of the Persian Gulf.

Often, a multiplicity of crops and livestock surrounds the Mediterranean villages of the region. Drought-resistant tree crops diversify the scene, with olive groves, almond trees, and citrus orchards producing output for both local consumption and commercial sale. Elsewhere, favored locations support grape vineyards, while more marginal settings can be used to grow wheat and barley, or to raise forage crops to feed cattle, sheep, and goats. Vulnerability to drought and the availability of more sophisticated water management strategies are leading some Mediterranean farmers to utilize more irrigated cropping, thus improving production of cotton, wheat, citrus fruits, and tobacco across portions of the region. More mechanization, crop specialization, and fertilizer use are also transforming such agricultural settings, following a pattern set earlier in nearby areas of southern Europe. One commercial adaptation of growing regional and global importance is Morocco's flourishing hashish crop. More than 200,000 acres (80,000 hectares) of cannabis are cultivated in the hill country near Ketama in northern Morocco, generating over $2 billion annually in illegal exports (mostly to Europe).

Many-Layered Landscapes: The Urban Imprint

Cities have also played a pivotal role in the region's human geography. Indeed, some of the world's oldest urban places are located in the region. Today, enduring political, religious, and economic ties wed the city and countryside. This lengthy and intimate relationship has shaped the urban landscape just as surely as cities have transformed the rural scene.

A Long Urban Legacy Cities in the region have traditionally played important functional roles as centers of political and religious authority, as well as key focal points of local and long-distance trade. Urbanization in Mesopotamia (modern Iraq) began by 3500 B.C., and cities such as Eridu and Ur reached populations of 25,000 to 35,000 residents. Similar centers appeared in Egypt by 3000 B.C., with Memphis and Thebes assuming major importance amid the dense populations of the middle Nile Valley. These ancient cities were key centers of political and religious control. Temples, palaces, tombs, and public buildings dominated the urban landscapes of such settlements, and surrounding walls (particularly in Mesopotamia) offered protection from outside invasion. By 2000 B.C., however, a different kind of city was emerging, particularly along the shores of the eastern Mediterranean and at the junction points of important caravan routes. Centers such as Beirut, Tyre, and Sidon, all in modern Lebanon, as well as Damascus in nearby Syria, exemplified the growing role of trade in creating the urban landscapes of selected cities. Expanding port facilities, warehouse districts, and commercial thoroughfares suggested how trade and commerce shaped these urban settlements, and many of these early Middle Eastern trading towns have survived to the present.

Islam also left an enduring urban signature because cities traditionally served as centers of Islamic religious power and education. By the eighth century, Baghdad had emerged as a religious center, followed soon thereafter by the appearance of Cairo as a seat of religious authority and expansion. Urban settlements from North Africa to Turkey felt the influences of Islam. Indeed, the Moors carried its characteristic signature to Spain, where it enduringly shaped centers such as Córdoba and Málaga. Its impact upon the settlement landscape merged with older urban traditions across the region and established a characteristic Islamic cityscape that exists to this day (Figure 7.17). Its traditional attributes include a walled urban core, or **medina**, dominated by the central mosque and its associated religious, educational, and administrative functions. A nearby bazaar, or "suq," functions as a marketplace where products from city and countryside are traded. Housing districts feature an intricate maze of narrow, twisting streets that maximize shade and accentuate the privacy of residents, particularly women. Houses have small windows, frequently are situated on dead-end streets, and typically open inward to private courtyards often shared by extended families with similar ethnic or occupational backgrounds.

More recently, European colonialism added another layer of urban landscape features in selected cities. Particularly in North Africa, coastal and administrative centers during the late nineteenth century sprouted dozens of architectural reminders of British and French urban traditions (see "Settlement Patterns: The Changing Urban Landscape of Fes, Morocco"). Victorian building blocks, French mansard roofs,

▲ **Figure 7.17 The Islamic landscape** Iranian mullahs discuss the religious questions of the day beneath the minarets of the Moussavi Mosque in Qom. Islam has left a widespread mark on the region's cultural landscapes. *(S. Franklin/Corbis/Sygma)*

suburban housing districts, and wide European-style commercial boulevards complicated the settlement landscapes of dozens of cities, both old and new, within the region. Centers such as Algiers (French) and Cairo (British) vividly displayed the effects of colonial control, and many of these signatures remain a legacy on the modern scene.

Signatures of Globalization Since 1950, dramatic new forces have transformed the urban settlement landscape. Cities have become the key gateways to the global economy. As the region has been opened to new investment, industrialization, and tourism, the urban landscape reflects the fundamental changes taking place. Expanded airports, commercial and financial districts, industrial parks, and luxury tourist facilities all mark the impress of the global economy.

Further, as urban centers become focal points of economic growth, surrounding rural populations are drawn to the new employment opportunities, thus fueling rapid population increases. The results are both impressive and problematic. Many traditional urban centers, such as Algiers and Istanbul, have more than doubled in size in recent years. Booming demand for homes has produced scores of ugly, cramped high-rise apartment houses in some government-planned neighborhoods, while elsewhere sprawling squatter settlements provide little in the way of quality housing or municipal services.

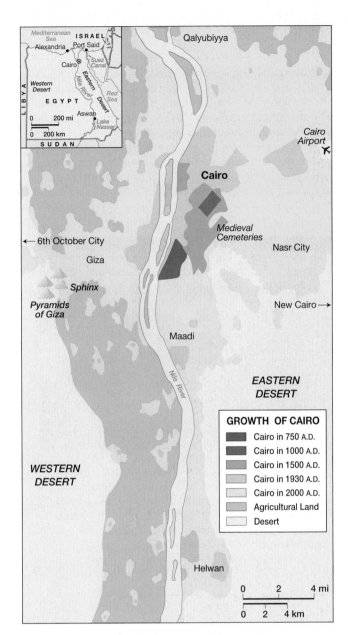

▲ **Figure 7.18 The growth of Cairo**   Today a city of more than 15 million people, Cairo has grown immensely in area since its Islamic origins more than 12 centuries ago. Just since 1930, the city has sprawled in all directions from its roots along the Nile, and now the metropolitan area includes a growing number of industrial and residential suburbs.   *(Modified from Max Rodenbeck,* Cairo: The City Victorious, *1999, p. x. Reprinted by permission of Alfred A. Knopf, Inc.)*

Crowded Cairo, now with 15 million people, exemplifies the pattern of urban expansion. The city has become much more densely settled than it once was, as well as larger in size (Figure 7.18). Its legendary "City of the Dead" neighborhood, now home to almost 1 million residents, is an urban cemetery intermingled with homes and apartment houses, a vivid acknowledgment of the premium put on living space. Many central-city neighborhoods have become so crowded and congested that wealthier urbanites have left, moving to outlying suburbs where large villas and lower population densities

▲ **Figure 7.19 Abu Dhabi cityscape**   The capital city of the oil-rich United Arab Emirates vividly reveals the impact of Western wealth and architecture on the modern Arab world.   *(M. Attar/ Corbis/Sygma)*

beckon. Other suburbs are sites for major industrial expansion, often financed through foreign investment. In some cases, entirely new industrial centers are being established, part of Egypt's ambitious New-Town Program.

Undoubtedly, the oil-rich states of the Persian Gulf display the most extraordinary changes in the urban landscape (Figure 7.19). Before the twentieth century, urban traditions were relatively weak in the area, and even as late as 1950 only 18 percent of Saudi Arabia's population lived in cities. All that changed, however, as the global economy's demand for petroleum mushroomed. Today, the Saudi Arabian population is more urban (83 percent) than many industrialized nations, including the United States, and the capital city of Riyadh has grown to 3 million people. Particularly after 1970, other cities, such as Abu Dhabi (United Arab Emirates), Doha (Qatar), and Kuwait City (Kuwait), also bore the bold signatures of modern Western urban design, futuristic architecture, and new transportation infrastructure. In addition, investments in petrochemical industries have fueled the creation of new urban centers, such as Jubail along Saudi Arabia's Persian Gulf coastline. The result is an urban settlement landscape where traditional and global influences curiously intermingle, producing cityscapes where domed mosques, mirrored bank buildings, and oil refineries uneasily coexist beneath the dusty skies and desert sun.

A Region on the Move

While pastoral nomads have crisscrossed the region for ages, entirely new patterns of migration have been sparked by the dynamic global economy and by recent political events. Several

The settlement patterns of Fes, Morocco, dramatically display the contrasting signatures of Islamic and Western influences on the modern scene. The northern Moroccan city, home to about a million people and growing rapidly because of in-migration, is a fascinating amalgam of the traditional and modern worlds, and a reminder of how the urban landscape can become a vast visual accumulation of differing cultural, economic, and technological influences.

The old city lies to the east, a classic walled medina typical of the urban Islamic world (Figure 7.2.1). The traditional city is primarily made up of tightly-knit neighborhoods of three-story residences, a sprinkling of palaces of the formerly well-to-do, endless small shops clustered together in handicraft zones, and mosques, both large and small. Additionally, there is the *kasbah*, a fortified quarter formerly used by local officials for defense. The narrow, curved streets form a labyrinth of cul-de-sacs that visitors find inpenetrable (Figure 7.2.2). Only with the help of a local guide can strangers find their way about the medina. Along the exterior walls, a series of city gates allow passage into the protected inner city. This intricate, densely interconnected urban landscape evolved centuries ago, in harmony with the needs of Islamic leaders who valued both its functional and symbolic characteristics.

The Treaty of Fes (1912), however, formally turned over control of Morocco to the French. Colonialism's signature quickly appeared on the urban landscape: Fes expanded to the west in a series of formal, straight boulevards and radiating avenues that echoed the baroque-style planning traditions of post-Renaissance France. In addition, the French built a rail station and hospital to serve the needs of the colonial regime. A new university was also constructed nearby. Since then, even more contemporary influences have reshaped the urban scene, with office buildings and apartment houses reflecting the needs of an ever-expanding urban population.

Thus, contemporary Fes remains a complex accumulation of the past and present, a record of changing cultural influences visible on the map and in the everyday landscapes of its city streets. While the medina remains the heart of post-colonial Fes, this historic city is also plagued by overcrowding, deteriorating housing, and inadequate sanitation. As local families move from the medina to modern apartments on the city's outskirts, their houses are filled quickly with migrants from rural parts of Morocco hoping for a new start in the now-cramped quarters of the old medina.

▲ **Figure 7.2.1 The Old City of Fes, Morocco**   These narrow streets in the old city of Fes are lined with shops and commercial stalls. The winding labyrinth of alleyways and cul-de-sacs forms an almost impenetrable maze to visitors, and it exemplifies the complex urban landscapes of the traditional Islamic city. *(Dave Bartruff/Corbis)*

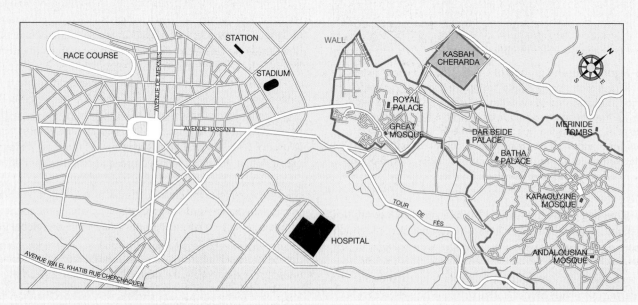

▲ **Figure 7.2.2 Fes, Morocco**   The tiny neighborhoods and twisting lanes of the old walled city reveal features of the traditional Islamic urban center. To the west, however, the rectangular street patterns, open spaces, and broad avenues suggest colonial European influences.

major migration streams have profoundly reshaped the human geography of the region. First, the rural-to-urban shift seen so widely in many parts of the less-developed world is fundamentally reworking population patterns across Southwest Asia and North Africa. The Saudi Arabian example is echoed in many other countries within the region. Second, large numbers of workers are migrating within the region to areas with growing job opportunities. Labor-poor countries such as Saudi Arabia, Kuwait, and the United Arab Emirates (UAE), for example, have attracted thousands of Jordanians, Palestinians, Yemenis, and Iranians to construction and service jobs within their borders. Indeed, the demand for workers has been so strong that large numbers of South Asians, Southeast Asians, and East Africans, particularly from Muslim nations, have flooded into the region, fundamentally redefining its cultural geography. Today, for instance, more than 70 percent of Saudi Arabia's labor force are foreigners who annually export more than $18 billion in wages.

A third major pattern of movement involves residents' migrating to job opportunities elsewhere in the world. Because of its strong economy and its proximity, Europe has been a particularly powerful draw. More than 1.5 million Turkish guest workers live in Germany. Both Algeria and Morocco also have seen large out-migrations to western Europe, particularly France. Only eight miles separates Africa from Europe, and illegal northbound boat traffic has reached unprecedented levels. Thousands more have journeyed to North America, particularly more educated professionals who seek high-paying jobs.

Political forces have also encouraged migrations. Thousands of wealthier residents, for example, left Lebanon and Iran during the political turmoil of the 1980s and are today living in cities such as Toronto, Los Angeles, and Paris. Elsewhere, internal movements in countries such as Sudan, Cyprus, and Iraq have often been provoked by political instability. The large influx of Jewish immigrants to Israel has also accelerated in the past 15 years, particularly after the breakup of the former Soviet Union.

Shifting Demographic Patterns

While high population growth remains a critical issue throughout the region, the demographic picture is shifting. Uniformly high rates of population growth in the 1950s and 1960s have been replaced by more varied regional patterns, and many nations are seeing birthrates fall fairly rapidly. For example, women in Tunisia, Iran, and Turkey are now averaging fewer than three births, representing a large decline in the total fertility rate (Table 7.1). Indeed, some experts now predict that fertility rates on the European and North African sides of the Mediterranean will almost converge (at just over two births per woman) by 2030. Many factors help explain the changes. More urban, consumer-oriented populations have opted for fewer children. Many Arab women are now delaying marriage into the middle 20s and even early 30s. Even in more traditional Bahrain, the average marriage age of women has risen from 15 to 23 years of age since the early 1970s. Family planning initiatives are expanding in many

countries. For example, programs in Tunisia, Egypt, and even Iran have greatly increased access to contraceptive pills, IUDs, and condoms since 1980.

Still, demographic challenges loom. Areas such as the West Bank, Gaza, and Libya experience some of the highest rates of natural increase on the planet. In some localities, persisting patterns of poverty and traditional ways of rural life contribute to the large rates of population increase, and even in more urban and industrialized Saudi Arabia, annual growth rates remain around 3 percent. Such growth rates result from the combination of high birthrates and very low death rates.

Elsewhere, even as new births are declining, already existing populations pose problems. For most countries in the region, more than 35 percent of the present population is younger than 15 years old, promising another large generation of people in need of food, jobs, and housing. In Egypt, for example, even though birthrates are likely to decline, the labor market will need to absorb more than 500,000 new workers annually over the next 10 to 15 years just to keep up with the country's large, youthful population. As that population ages in the mid-twenty-first century, it will continue to demand jobs, housing, and social services (Figure 7.20). Given recent trends in job growth and migration, the region's large cities probably will bear the brunt of future population increases. Growing populations will also impose increasingly daunting demands on the region's already limited water resources.

Cultural Coherence and Diversity: Signatures of Complexity

While Southwest Asia and North Africa clearly define the heart of the Islamic and Arab worlds, a surprising degree of cultural diversity characterizes the region. Muslims practice their religion in varied ways, often disagreeing profoundly on basic religious tenets as well as on how much of the modern world and its mass consumer culture should be incorporated into their daily lives. In addition, diverse religious minorities complicate the region's contemporary cultural geography. Linguistically, Arabic languages form an important cultural core historically centered on the region. Still, many non-Arab peoples, including Persians, Kurds, and Turks, also populate important homelands. Understanding these varied patterns of cultural geography is essential to comprehending many of the region's political tensions, as well as appreciating why many of its residents resist processes of globalization.

Patterns of Religion

Religion permeates the lives of most people within the region. Its centrality stands in sharp contrast to largely secular cultural impulses that dominate life in many other parts of the world. Whether it is the quiet ritual of morning prayers or profound discussions about contemporary political and social issues, religion suffuses the daily routine of most regional residents from Casablanca to Tehran. The geographies of religion—their points

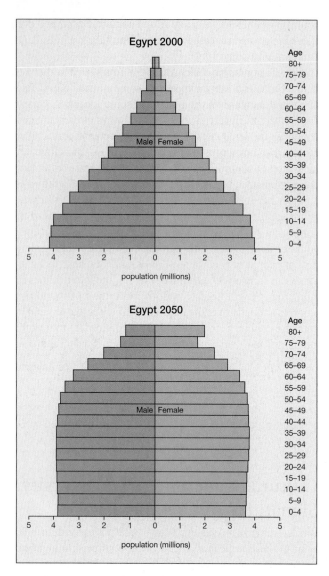

Egypt 2000

Egypt 2050

▲ Figure 7.20 Egypt's changing population   While Egypt's annual population growth rates are slowing, its 2000 population pyramid reveals a large number of young people already in the population. Future (2050) implications are clear: Egypt's population will grow larger and older in the years to come.   (U.S. Census Bureau, International Data Base)

of origin, paths of diffusion, and patterns of modern regional articulation—are essential elements in defining varied cultural and political zones of both conflict and coexistence. In some settings religious differences have had the power to pit one neighbor against another, while elsewhere ancient religious rivalries appear to be slowly healing.

Hearth of the Judeo-Christian Tradition   Both Jews and Christians trace their religious roots to the eastern Mediterranean, and while neither group is numerically dominant across the area, each plays a pivotal cultural role. The roots of Judaism lie deep in the past: Abraham, an early patriarch in the Jewish tradition, lived some 4,000 years ago and led his people from Mesopotamia to Canaan (modern-day Israel),

near the shores of the Mediterranean. From Jewish history, recounted in the Old Testament of the Holy Bible, springs a rich religious heritage focused on a belief in one God (or **monotheism**), a strong code of ethical conduct, and a powerful ethnic identity that continues to the present. Around A.D. 70, during the time of the Roman Empire, most Jews were forced to leave the eastern Mediterranean after they challenged Roman authority. The resulting forced migration, or diaspora, of the Jews took them to the far corners of Europe and North Africa. Only in the past century have many of the world's far-flung Jewish populations returned to the religion's hearth area, a process that accelerated greatly with the formation of the Jewish state of Israel in 1948.

Christianity also emerged in the vicinity of modern-day Israel and has left a lasting legacy across the region. An outgrowth of Judaism, Christianity was based on the teachings of Jesus and his disciples, who lived and traveled in the eastern Mediterranean about 2,000 years ago. While many Christian traditions became associated with European history, some forms of early Christianity remained potent near the religion's original hearth. To the south, one stream of Christian influences associated with the Coptic Church diffused into northern Africa, shaping the cultural geographies of places such as Egypt and Ethiopia. In the Levant, another group of early Christians, known as the Maronites, retained a separate cultural identity that survives today.

The Emergence of Islam   Islam originated in Southwest Asia in A.D. 622, forming yet another cultural hearth of global significance. While Muslims can be found today from North America to the southern Philippines, the Islamic world is still centered on its Southwest Asian origins. Most Southwest Asian and North African peoples still follow its religious precepts and moral doctrines. Muhammad, the founder of Islam, was born in Makkah (Mecca) in A.D. 570 and taught in nearby Medinah (Medina) (Figure 7.21). In many respects, his creed parallels the Judeo-Christian tradition. Muslims believe that both Moses and Jesus were true prophets and that both the Hebrew Bible (or Old Testament) and the Christian New Testament, while incomplete, are basically accurate. Ultimately, however, Muslims hold that the **Quran** (or Koran), a book of revelations received by Muhammad from Allah (God), represents God's highest religious and moral revelations to mankind.

The basic tenets of Islam offer an elaborate blueprint for leading an ethical and religious life. Islam literally means "submission to the will of God," and the creed rests on five essential pillars: (1) repeating the basic creed ("There is no God but God, and Muhammad is his prophet"); (2) praying facing Makkah five times daily; (3) giving charitable contributions; (4) fasting between sunup and sundown during the month of Ramadan; and (5) making at least one religious pilgrimage, or **Hajj**, to Muhammad's birthplace of Makkah (Figure 7.22). Islam is a more austere religion than most forms of Christianity, and its modes of worship and forms of organization are generally less ornate. Muslims regard religious images as idolatrous, and the strictest interpretations even forbid the depiction of the human form. Followers of Islam are prohibited

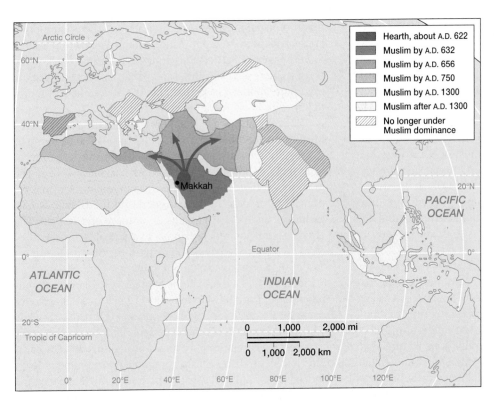

◀ **Figure 7.21 Diffusion of Islam** The rapid expansion of Islam that followed its birth is shown here. From Spain to Southeast Asia, Islam's legacy remains strongest nearest its Southwest Asian hearth. In some settings, its influence has ebbed or has come into conflict with other religions, such as Christianity, Judaism, and Hinduism. *(Modified from Rubenstein, 2002, An Introduction to Human Geography, Upper Saddle River, NJ: Prentice Hall)*

from drinking alcohol and are instructed to lead moderate lives, avoiding excess. Many Islamic fundamentalists still argue for a **theocratic state,** such as modern-day Iran, in which religious leaders (ayatollahs) guide policy.

A major religious schism divided Islam early on, and the differences endure to the present. The breakup occurred almost immediately after the death of Muhammad in A.D. 632. Key questions surrounded the succession of religious power. One group, now called the **Shiites,** favored passing power on within Muhammad's own family, specifically to Ali, his son-in-law. Most Muslims, later known as **Sunnis,** advocated passing power down through the established clergy. This group emerged victorious. Ali was martyred, and his Shiite supporters went underground. Ever since, Sunni Islam has formed the mainstream branch of the religion, to which Shiite Islam has presented a recurring, and sometimes powerful, challenge. The Shiites argue that a successor to Ali will someday return to reestablish the pure, original form of Islam. The Shiites are also more hierarchically organized than the Sunnis. Iran's ayatollahs, for example, wield overriding religious and political power.

In a very short period of time, Islam diffused widely from its Arabian hearth, often following camel caravan routes and Arab military campaigns as it expanded its geographical range and converted thousands to its creed. By Muhammad's death in A.D. 632, the peoples of the Arabian peninsula were united under its banner. Shortly thereafter, the Persian Empire fell to Muslim forces and the Eastern Roman (or Byzantine) Empire lost most of its territory to Islamic influences. By A.D. 750, Arab armies had swept across North Africa, conquered most of Spain and Portugal, and established footholds in Central and South Asia. At first only the Arab conquerors followed

Islam, but the diverse inhabitants of Southwest Asia and North Africa were gradually absorbed into the religion, albeit with many distinct local variants. By the thirteenth century, most people in the region were Muslims, while older creeds such as Christianity and Judaism became minority faiths or disappeared altogether.

Between 1200 and 1500, Islamic influences expanded in some areas and contracted in others. The Iberian Peninsula (Spain and Portugal) returned to Christendom in 1492, although many Moorish (Islamic) cultural and architectural features remained behind and still shape the region today.

▲ **Figure 7.22 Makkah** Millions of the faithful make the pilgrimage to Makkah, where they visit the holy al-Ka'ba, the black, cubelike structure in the center of the picture. In Muslim theology, Abraham and Ishmael constructed the al-Ka'ba, and the site was later honored by Muhammad. *(Elkoussy/Corbis/Sygma)*

At the same time, Muslims expanded their influence southward and eastward into Africa. In addition, Muslim Turks largely replaced Christian Greek influences in Southwest Asia after 1100. One group of Turks moved into the Anatolian Plateau and finally conquered the last vestiges of the Byzantine Empire in 1453. These Turks soon created the vast **Ottoman Empire** (named after one of its leaders, Osman), which included southeastern Europe (including modern-day Albania, Bosnia, and Kosovo) and most of Southwest Asia and North Africa. The legacy of the Ottoman Empire was considerable: it offered a new, distinctly Turkish interpretation of Islam, and it provided a dynamic and centralized expression of Muslim political power within the region until the Empire's disintegration in the late nineteenth and early twentieth centuries.

Modern Religious Diversity   Today, Muslims form the majority population in all of the countries of Southwest Asia and North Africa except Israel, where Judaism is the dominant religion, and Cyprus, where Greek Orthodox outnumber Turkish Muslims (Figure 7.23). Still, divisions within Islam have created key cultural differences within the region. While most (73 percent) of the region is dominated by Sunni Muslims, the Shiites (23 percent) remain an important element in the contemporary cultural mix. Particularly dominant in Iran, southern Iraq, Lebanon, Sudan, and Bahrain, the Shiites also form a major religious minority in many other countries, including Algeria, Egypt, and Yemen. Strongly associated with the recent flowering of Islamic fundamentalism, the Shiites also have benefited from rapid growth rates because their

brand of Islam is particularly appealing to the poorer, powerless, and more rural populations of the region. While some Sunnis have been attracted to fundamentalism as well, many reject its more radical cultural and political precepts and argue for a more modern Islam that incorporates some accommodation with Western values and traditions.

While the Sunni–Shiite split is the great divide within the Muslim world, other variations of Islam can also be found in the region. One division, for example, separates the mystically inclined form of Islam—known as *Sufism*—from the more scripturally oriented mainstream tradition. Sufism is especially prominent in the peripheries of the Islamic world, including the Atlas Mountains of Morocco and Algeria. The Druze of Lebanon practice another variant of Islam, and they form a cohesive religious minority in the Shouf Mountains east of the Lebanese capital of Beirut.

Southwest Asia is also home to many non-Islamic communities. Israel has a Muslim minority (14 percent) that is dominated by that nation's Jewish population (80 percent), while Christians comprise another 2 percent of the total. Even within Israel's Jewish community, increasing cultural differences divide Jewish fundamentalists from more reform-minded Jews. Indeed, this cultural diversity within Judaism has shaped the way Israel has dealt with political and social issues involving Muslims within and beyond its borders. In neighboring Lebanon, there was a slight Christian (Maronite and Orthodox) majority as recently as 1950. Christian out-migration and differential birthrates, however, have created a nation that today is 60 to 70 percent Muslim. Christians also form approximately 10 percent of Syria's population, while Iraqi Christians, con-

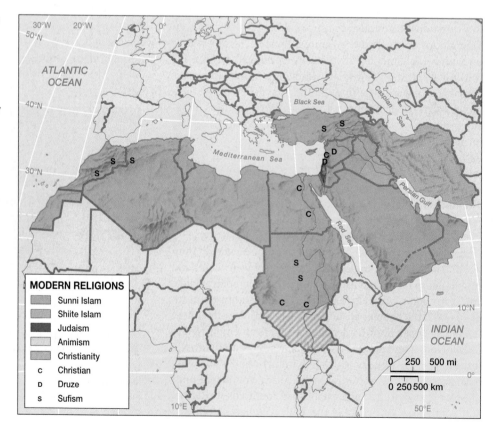

▶ **Figure 7.23 Modern religions** Islam persists as the dominant religion across the region. Most Muslims are tied to the Sunni branch, while Shiites are found in places such as Iran and southern Iraq. In some locales, however, Christianity and Judaism remain important. African animism is found in southern portions of Sudan. *(Modified from Rubenstein, 2002, An Introduction to Human Geography, Upper Saddle River, NJ: Prentice Hall)*

MODERN RELIGIONS
- Sunni Islam
- Shiite Islam
- Judaism
- Animism
- Christianity
- c Christian
- D Druze
- s Sufism

▲ **Figure 7.24 Old Jerusalem** Jerusalem's historic center reflects its varied religious legacy. Sacred sites for Jews, Christians, and Muslims all cluster within the Old City. The Western Wall, a remnant of the ancient Jewish temple, stands at the base of the Dome of the Rock and Islam's Al Aqsa Mosque. *(Bernard Boutrit/Woodfin Camp & Associates)*

centrated mostly in the rugged northern uplands, make up about 3 percent of its population.

Within the Middle East, the city of Jerusalem (now the Israeli capital) holds special religious significance for several groups and also stands at the core of the region's political problems (Figure 7.24). Indeed, the sacred space of this ancient Middle Eastern city remains deeply scarred and divided as different groups argue for more control of contested neighborhoods and nearby suburbs. Historically, Jews particularly revere the city's old Western Wall (the site of a Roman-era temple); Christians honor the Church of the Holy Sepulchre (the burial site of Jesus); and Muslims hold sacred religious sites in the city's eastern quarter (including the place from which the prophet Mohammad reputedly ascended to heaven). Further complicating the traditional religious divisions of the city, Israel's victory in the 1967 war emboldened it to establish many new Jewish settlements within the city's redefined and expanded eastern suburbs. The move angered Islamic Palestinian residents of the area, who saw the new settlements encroaching on their homeland.

Important Jewish and Christian communities also have left a long legacy across North Africa. Roman Catholicism was once dominant in much of the Maghreb, but it disappeared several hundred years after the Muslim conquest. The Maghreb's Jewish population, on the other hand, remained prominent until the post-World War II period, when most of the region's Jews migrated to the new state of Israel. In Egypt, Coptic Christianity has maintained a stable presence through the centuries and today includes approximately 7 percent of the country's population. In earlier years the Coptic community had a secure place in Egyptian society, and numerous Copts held high-level posts in government and business. Today, however, Egypt's Christians are being increasingly marginalized, and some of their communities have been put under pressure, even subjected to physical attack, by extremist Islamic elements. Other Christian communities are located in southern Sudan, but unlike those of Egypt, these are mostly recent converts from traditional African religions.

Geographies of Language

Although the region is often referred to as the "Arab World," linguistic complexity creates many important cultural divisions across Southwest Asia and North Africa (Figure 7.25). These varied geographies of language offer insights into regional patterns of ethnic identity, potential cultural conflicts that exist at linguistic borders, and the instability that often characterizes geopolitical relationships within and between states.

Semites and Berbers Afro-Asiatic languages dominate much of the region. Within that family, Arabic-speaking Semitic peoples can be found from Morocco to Saudi Arabia. Before the expansion of Islam, Arabic was limited to the Arabian Peninsula. Today, however, Arabic is spoken from the Persian Gulf to the Atlantic and reaches southward into Sudan, where it borders the Nilo-Saharan speaking peoples of Sub-Saharan Africa. As the language has diffused, it has slowly diverged into local dialects. As a result, the everyday Arabic spoken on the streets of Fes, Morocco, is not easily intelligible to an Arabic speaker from the United Arab Emirates. The Arabic language also has a special religious significance for all Muslims because it was the sacred language in which God delivered his message to Muhammad. While most of the world's Muslims do not speak Arabic, the faithful often memorize certain prayers in the language, and many Arabic words have entered the other important languages of the Islamic world. Advanced Islamic learning, moreover, demands competence in Arabic.

Hebrew, another traditional Semitic language of Southwest Asia, was recently reintroduced into the region with the creation of Israel. This Semitic language originated in the Levant and was used by the ancient Israelites 3,000 years ago. Today its modern version survives as the sacred tongue of the Jewish people and is the official language of Israel, although the country's non-Jewish population largely speaks Arabic. In addition, many recent Jewish immigrants speak Russian, while English is widely used as a second or third language throughout the country.

While Arabic eventually spread across North Africa, several older languages survive in more remote areas. Older Afro-Asiatic tongues endure in the Atlas Mountains and in certain parts of the Sahara. Collectively known as Berber, these languages are related to each other but are not mutually intelligible. Most Berber languages have never been written, and none has generated a significant literature. Indeed, a Berber-language version of the Quran was not completed until 1999. The decline in pastoral nomadism and pressures of modernization threaten the integrity of these Berber languages. Scattered Berber-speaking communities are found as far to the east as Egypt, but Morocco is the center of this language group, where it plays an important role in shaping that nation's cultural identity.

Persians and Kurds Although Arabic spread readily through portions of Southwest Asia, much of the Iranian Plateau and nearby mountains are dominated by older Indo-European languages. Here the principal tongue remains Persian, although, since the tenth century, the language has been

▶ **Figure 7.25 Modern languages** Arabic is a Semitic Afro-Asiatic language, and it dominates the region's cultural geography. Turkish, Persian, and Kurdish, however, remain important exceptions, and such differences within the region have often had enduring political consequences. Israel's more recent reintroduction of Hebrew further complicates the region's linguistic geography. *(Modified from Rubenstein, 2002, An Introduction to Human Geography, Upper Saddle River, NJ: Prentice Hall)*

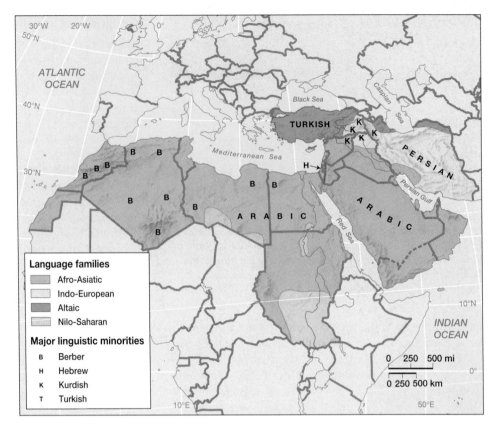

enriched with Arabic words and written in the Arabic script. Persian, like other languages, developed distinct local dialects. Today Iran's official language is called *Farsi,* which denotes the form of Persian spoken in Fars, the area around the city of Shiraz. Thus, while both Iran and neighboring Iraq are Islamic nations, their ethnic identities spring from quite different linguistic and cultural traditions.

The Kurdish speakers of northern Iraq, northwest Iran, and eastern Turkey add further complexity to the regional pattern of languages. Kurdish, also an Indo-European language, is spoken by 10 to 15 million persons in the region. Kurdish has not historically been a written language, but the Kurds do have a strong sense of shared cultural identity. Indeed, "Kurdistan" has sometimes been called the world's largest nation without its own political state since the group remains a minority in several countries of the region (Figure 7.26). Making matters worse, the Kurds have been discriminated against and attacked by surrounding groups, particularly by antagonistic governments in Iraq and Turkey.

The Turkish Imprint Turkish languages provide more variety across much of modern Turkey, in portions of far northern Iran, and on the northern third of Cyprus. The Turkish languages are a part of the larger Altaic language family that originated in Central Asia. Turkey remains the largest nation in Southwest Asia dominated by the family. Tens of millions of persons in other countries of Southwest and Central Asia speak related Altaic languages, such as Azeri, Uzbek, and Uighur. During the era of the Ottoman Empire, Turkish speakers ruled much of Southwest Asia and North Africa, but Iran

is the only other large country in the region today where the language persists. Turks also contribute to the linguistic diversity and the cultural tensions of Cyprus in the eastern Mediterranean: a Turkish minority occupies the north, while a Greek (Indo-European family) majority dominates most of the southern part of the island.

Regional Cultures in Global Context

Southwest Asian and North African peoples increasingly find themselves intertwined with cultural complexities and conflicts external to the region. Islam links the realm with a globally dispersed population. In addition, European and American cultural influences have multiplied greatly across the region since the mid-nineteenth century. Colonialism, the boom in petroleum investment, and the growing presence of Western-style popular culture have had enduring impacts on the complex regional mosaic that stretches from the Atlas Mountains to the Indian Ocean.

Islamic Internationalism While Islam is geographically and theologically divided, all Muslims recognize the fundamental unity of their religion. This religious unity extends far beyond Southwest Asia and North Africa. Islamic communities are well established in such distant places as central China, European Russia, central Africa, and the southern Philippines. Today, Muslim congregations are also expanding rapidly in the major urban areas of western Europe and North America, largely through migration but also through local conversions. Islam is thus emerging as a truly global

▲ Figure 7.26 Kurdish family As an ethnic minority, these Kurdish-speaking settlers from eastern Turkey face cultural and political discrimination in their own country, as do their Kurdish neighbors in nearby Iran, Iraq, and Syria. *(J. Egan/Hutchison Library)*

religion. Even with its global reach, however, Islam remains centered on Southwest Asia and North Africa, the site of its origins and its most holy places. As Islam expands in number of adherents and geographical scope, the religion's tradition of pilgrimage ensures that Makkah will become a city of increasing global significance in the twenty-first century. The global growth of Islamic fundamentalism also focuses attention on the region, where much of the contemporary movement recently burst upon the scene. In addition, the oil wealth accumulated by many Islamic nations will be used to sustain and promote the religion. Countries such as Saudi Arabia and Libya invest in Islamic banks and economic ventures and make donations to Islamic cultural causes, colleges, and hospitals worldwide.

Globalization and Cultural Change The region is also grappling with how its growing role in the global economy is changing traditional cultural values. Indeed, the expansion of Islamic fundamentalism is in many ways a reaction to the threat posed by external cultural influences, particularly those of western Europe and the United States. European colonialism left its own cultural legacy not only in the architectural landscapes still found in the old colonial centers, but also in the widespread use of English and French among the Western-educated elite across the region. In oil-rich countries, huge capital investments have had important cultural implications as the number of foreign workers has grown and as

more affluent young people have embraced elements of Western-style music, literature, and clothing.

Technology also contributes to cultural change. Educated Turks, Tunisians, and Egyptians have been embracing the Internet and its power to access a global wealth of information and entertainment. In Morocco, the use of cellular phones more than quadrupled in 2000. While less than 6 percent of the country's population is served by fixed phone lines, more than 85 percent is within the reach of cellular phones. Even in conservative Iran, where satellite dishes are officially banned, millions of people have access to them, beaming in multicultural programming from around the globe. While some fundamentalists in Islamic republics, such as Iran and Sudan, would like to impose a complete technological and cultural wall between themselves and the West, such actions remain problematic in today's world. Indeed, many middle-class and well-to-do Iranians privately enjoy their wealth, including the pleasures of Western-style food, drink, fashion, and popular culture.

In some settings, there are strong historical links between the realm and other world regions. Former colonial links between Algeria and France, for example, have encouraged a sizable out-migration of North Africans to western Europe, thus increasing interactions between the former colony and its mother country. Large numbers of Turkish workers have also spent considerable time in European employment, particularly in Germany. While some have remained as guest workers, many others work periodically in Europe and then return to their Turkish homeland. Indeed, Turkey's general openness to European and U.S. investment and popular culture has forever reshaped the cultural scene, particularly in the urban areas (see "The Local and the Global: The Marlboro Man Arrives in Istanbul"). In the case of Israel, its role as the center of Judaism and its historical significance as the hearth of Christianity have made it a destination of global cultural importance. Akin to the Hajj in Islam, many of the world's Jews and Christians consider a visit to their Holy Land in Israel an essential part of their religious lives. Close economic and military connections between Israel and the United States also enhance the cultural glue that binds the region to the world beyond.

Geopolitical Framework: A Region of Persisting Tensions

Geopolitical tensions remain high in Southwest Asia and North Africa (Figure 7.27). Some of the tensions relate to age-old patterns of cultural geography in which different ethnic, religious, and linguistic groups still struggle to live with one another in a rapidly changing world of new nation-states and political relationships. The region's brief but complex ties to the era of European colonialism also contribute to present difficulties, since the boundaries of many countries grew from an imposed political geography that still shapes the scene today. Geographies of wealth and poverty enter the geopolitical mix: some residents profit mightily from petroleum resources and industrial expansion, while others struggle just to feed their

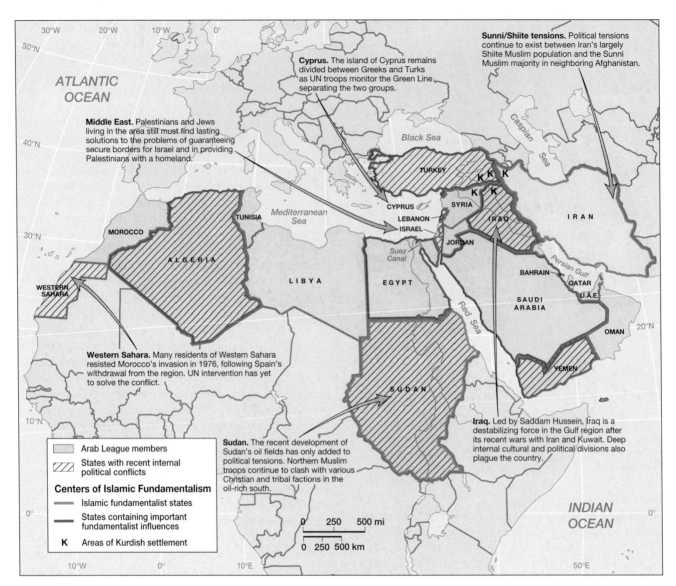

Cyprus. The island of Cyprus remains divided between Greeks and Turks as UN troops monitor the Green Line separating the two groups.

Sunni/Shiite tensions. Political tensions continue to exist between Iran's largely Shiite Muslim population and the Sunni Muslim majority in neighboring Afghanistan.

Middle East. Palestinians and Jews living in the area still must find lasting solutions to the problems of guaranteeing secure borders for Israel and in providing Palestinians with a homeland.

Western Sahara. Many residents of Western Sahara resisted Morocco's invasion in 1976, following Spain's withdrawal from the region. UN intervention has yet to solve the conflict.

Sudan. The recent development of Sudan's oil fields has only added to political tensions. Northern Muslim troops continue to clash with various Christian and tribal factions in the oil-rich south.

Iraq. Led by Saddam Hussein, Iraq is a destabilizing force in the Gulf region after its recent wars with Iran and Kuwait. Deep internal cultural and political divisions also plague the country.

Arab League members

States with recent internal political conflicts

Centers of Islamic Fundamentalism

Islamic fundamentalist states

States containing important fundamentalist influences

K Areas of Kurdish settlement

▲ Figure 7.27 **Geopolitical issues in Southwest Asia and North Africa** Political tensions continue across much of the region. While the central conflict remains oriented around Israel, its neighboring states, and the rights of resident Palestinians, other regional trouble spots periodically erupt in violence. Islamic fundamentalism challenges political stability in settings from Algeria to Sudan. Elsewhere, persisting ethnic conflicts shape daily life in Sudan, Lebanon, Turkey, Cyprus, and Iraq.

families. The result is a political climate charged with tension, a region in which the sounds of bomb blasts and gunfire have been an all too common characteristic of everyday life.

The Colonial Legacy

European colonialism arrived relatively late in Southwest Asia and North Africa, but the era left an important impact upon the region's modern political geography. The Turks were one reason for Europe's tardy participation in imposing colonial rule. Between 1550 and 1850, much of the region was dominated by the Turkish Ottoman Empire, which expanded from its Anatolian hearth to engulf much of North Africa as well as

nearby areas of the Levant, the western Arabian Peninsula, and modern-day Iraq. Ottoman rule imposed an outside imperial political order and a larger colonial economic framework that had lasting consequences. The tide began to turn, however, in the early nineteenth century as the European presence increased and Ottoman power ebbed. Still, it took a century of shifting geopolitical fortunes for Ottoman influences to be replaced with largely European colonial dominance after World War I (1918). While much of that direct European control ended by the 1950s, old colonial ties persist in a variety of economic, political, and cultural contexts. Indeed, it is still common to encounter British English on the streets of Cairo, while French can still be heard in Algiers and Beirut.

THE LOCAL AND THE GLOBAL   The Marlboro Man Arrives in Istanbul

The impact of globalization has taken many forms in Turkey, but few examples are more dramatic than the arrival of the Marlboro Man in Istanbul. Even in a land where native tobacco is revered, Philip Morris, the U.S. manufacturer of Marlboro products, has succeeded in capturing almost 25 percent of the Turkish cigarette market. Its effective use of U.S.-style marketing campaigns has especially appealed to young urban Turks, who equate the Marlboro name with a new, more progressive lifestyle. As one young man in Istanbul noted, "None of my friends smoke local cigarettes . . . that would be humiliating."

The success of Philip Morris illustrates how many U.S. and European corporations have westernized Turkish culture. In a society where more than 40 percent of the population smokes, the company saw great potential in its Marlboro strategy. It lobbied the Turkish government to open domestic markets to foreign competition. Then it invested more than $230 million in a new cigarette manufacturing facility, carefully engineering the local Marlboros to Turkish tastes.

Most powerfully, the company's marketing efforts have inundated the country with the Marlboro name and insignia. Salesmen dress as cowboys as they deliver their products in vans designed to look like traveling packs of cigarettes. Sidewalk diners around Istanbul are protected from the hot Turkish sun by umbrellas emblazoned with the Marlboro logo; fashion-conscious urbanites proudly swagger city streets in "Marlboro Classic Khakis"; and the company's advertising campaign successfully features the same independent cowboys and sweeping western vistas as it does in the United States. Marlboros are also penetrating the Turkish countryside, a subtle but omnipresent signature of increasing globalization both in terms of changing cultural values and the ways consumers spend their money.

Source: Adapted from "How Philip Morris Got Turkey Hooked on American Tobacco," *Wall Street Journal*, September 11, 1998.

Imposing European Power   French colonial ties have long been a part of the region's history. Beginning around 1800, France became committed to a colonial presence in North Africa. Later in the nineteenth century, Algeria ended up in French hands. Several million French, Italian, and other European immigrants poured into the country, taking the best lands from the Algerian people. The French government expected this territory to become an integral part of France, dominated by the growing French-speaking immigrant population (Figure 7.28). Later, France established protectorates in Tunisia (1881) and Morocco (1912), ensuring an enduring French political and cultural presence in the Maghreb. Finally, France's victory over the German–Ottoman Turk alliance in World War I produced additional territorial gains in the Levant, as France gained control of the northern zone encompassing the modern nations of Syria and Lebanon.

Great Britain's colonial fortunes also grew within the region before 1900. To control its sea-lanes to India, Britain established a series of nineteenth-century protectorates over the small coastal states of the southern Arabian Peninsula and the Persian Gulf. In this manner, such places as Kuwait, Bahrain, Qatar, the United Arab Emirates, and Aden (in southern Yemen) were loosely incorporated into the British Empire. Nearby Egypt also caught Britain's attention. Once the British-engineered **Suez Canal** linked the Mediterranean and Red seas in 1869, European banks and trading companies gained more influence over the Egyptian economy. The British took more direct control in 1883. In the process, Britain also inherited a direct stake in Sudan, since Egyptian soldiers and traders had been pushing south along the Nile for decades.

Another series of British colonial gains within the region came at the close of World War I. In Southwest Asia, British and Arab forces joined to expel the Turks during the war. To obtain Arab trust, Britain promised that an independent Arab

state would be created in the former Ottoman territories. At roughly the same time, however, Britain and France signed a secret agreement to partition the area. When the war ended, Britain opted to slight its Arab allies and honor its treaty with France, with one exception. The Saud family convinced the British that a smaller country (Saudi Arabia) should be established, focused on the desert wastes of the Arabian Peninsula. Saudi Arabia became fully independent in 1932. Elsewhere, however, Britain carried out its plan to partition lands with France. Britain divided its new territories into three entities: Palestine (now Israel) along the Mediterranean coast; Transjordan to the east of the Jordan River (now Jordan); and a third zone that later became Iraq. Iraq, in particular, was a

▲ **Figure 7.28 Algiers**   European colonial influences still abound in the old French capital of Algiers in northern Algeria. The city's modern bustle belies the increasing political and religious tensions that have torn the country apart since 1992. *(Francoise Perri/Woodfin Camp & Associates)*

contrived territory that combined three dissimilar former Ottoman provinces. It included the centers of Basra in the south (an Arabic-speaking Shiite area), Baghdad in the center (an Arabic-speaking Sunni area), and Mosul in the north (a Kurd-dominated zone).

Other settings within the region felt more marginal colonial impacts. Libya, for example, was long regarded by Europeans as a desert wasteland. Italy, never a dominant colonial power, expelled Turkish forces from the coastal districts by 1911, but they did not subdue the Saharan oases until the 1930s. Spain also carved out a territorial stake within the region, gaining control over southern Morocco (now Western Sahara) in 1912. To the east, Persia and Turkey were never directly occupied by European powers, but both played into the calculus of the global geopolitics of the time. In Persia, the British and Russians agreed to establish mutual spheres of economic influence in the region (the British in the south, the Russians in the north), while respecting Persian independence. In 1935, Persia's modernizing ruler, Reza Shah, changed the country's name to Iran.

In nearby Turkey the old core of the Ottoman Empire was almost partitioned by European powers following World War I. After several years of fighting, however, the Turks expelled the French from southern Turkey and the Greeks from western Turkey. The key to the successful Turkish resistance was the spread of a new modern, nationalist ideology under the leadership of Kemal Ataturk. Ataturk decided to emulate the European countries and establish a culturally unified and resolutely secular state. He was successful, and Turkey was quickly able to stand up to European power.

Decolonization and Independence   European colonial powers began their withdrawal from several Southwest Asian and North African colonies before World War II. By the 1950s, most of the countries in the region were independent, although many maintained political and economic ties with their former colonial rulers. In North Africa, Britain finally withdrew its troops from Sudan and Egypt in 1956. Libya (1951), Tunisia (1956), and Morocco (1956) achieved independence peacefully during the same era, but the French colony of Algeria became a major problem. Since several million French citizens resided there, France had no intention of simply withdrawing. A bloody war for independence began in 1954, and France agreed to an independent Algeria in 1962, but the two nations continued to share a close—if not always harmonious—relationship thereafter.

Southwest Asia also lost its colonial status between 1930 and 1960, although many of the imposed colonial-era boundaries continue to shape the regional geopolitical setting. While Iraq became independent from Britain in 1932, its later instability in part resulted from its artificial borders, which never recognized much of its cultural diversity. Similarly, the French division of its Levant territories into the two independent states of Syria and Lebanon (1946) greatly angered local Arab populations and set the stage for future political instability in the region. As a favor to its small Maronite Christian majority, France carved out a separate Lebanese state from largely Arab Syria, even guaranteeing the Maronites constitutional control of the national government. The action created a culturally divided Lebanon as well as a Syrian state that repeatedly has asserted its influence over its Lebanese neighbors.

Modern Geopolitical Issues

The geopolitical instability in Southwest Asia and North Africa will persist in the twenty-first century. It remains difficult to predict political boundaries that seem destined to shift through negotiated settlements or political conflict. Several key problems loom within the region. The Arab–Israeli conflict zone in the eastern Mediterranean will continue as a major trouble spot. Islamic fundamentalism will also spark ongoing cultural and political tensions wherever the movement achieves prominence. Internal political conflicts are likely where multiple cultural groups claim sometimes overlapping territories within a single country. Finally, future tensions between states are likely where neighboring nations disagree over cultural issues, natural resources, or political boundaries.

The Arab–Israeli Conflict   The 1948 creation of the Jewish state of Israel produced an enduring zone of cultural and political tensions within the eastern Mediterranean. Jewish migration to the area accelerated after the British took Palestine from the defeated Ottoman Empire after World War I. In 1917 Britain issued the Balfour Declaration, considered by some to be a pledge to encourage the "establishment of Palestine as a home for the Jewish people." After World War II, the British agreed to withdraw from the area, and the United Nations divided the region into two states, one to be predominantly Jewish, the other primarily Muslim (Figure 7.29). Indigenous Arab Palestinians rejected the partition, and war erupted as soon as the British departed. Jewish forces proved victorious, and by 1949 Israel had gained a larger share of land than it had originally been allotted. The remainder of Palestine, including the West Bank and the Gaza Strip, passed to Jordan and Egypt, respectively. Hundreds of thousands of Palestinian refugees fled from Israel to neighboring countries, such as Egypt, Jordan, and Lebanon, where many of them remained in makeshift camps. Under these difficult conditions, the Palestinians nurtured the idea of creating their own state in the land that had become Israel. The Israelis, not surprisingly, remained adamant that the country was theirs.

Israel's relations with neighboring countries were bitter from the beginning. The supporters of Arab unity and Muslim solidarity sympathized with the Palestinians, while their antipathy toward Israel grew. In the 1950s and 1960s, leaders of Egypt, Syria, Iraq, and Jordan viewed Israel as an unwanted Western neocolonial presence. Indeed, Israel's presence fostered pan-Arab solidarity that for a time promoted the creation of a region-wide Arab state, a political impulse that continues today in the form of the Arab League. The Israelis saw things very differently: many were born in the area, and many others migrated from other portions of Southwest Asia and North Africa.

Israel and its Arabic-speaking neighbors fought major wars in 1956, 1967, and 1973. In territorial terms, the Six-Day War of 1967 was the most important conflict. In this strug-

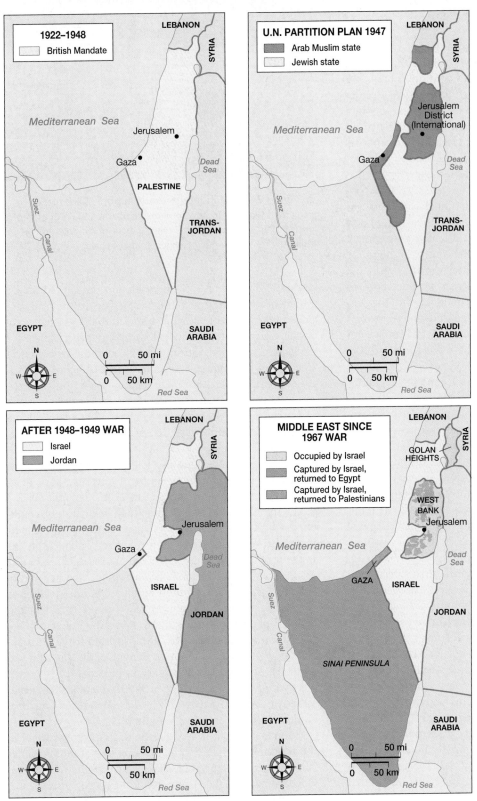

◀ Figure 7.29 Evolution of Israel   Modern Israel's complex evolution began with an earlier British colonial presence and a United Nations partition plan in the late 1940s. Thereafter, multiple wars with nearby Arab states produced Israeli territorial victories in settings such as Gaza, the West Bank, and the Golan Heights. Each of these regions continues to figure importantly in the country's recent relations with nearby states and with resident Palestinian populations. (Modified from Rubenstein, 2002, An Introduction to Human Geography, Upper Saddle River, NJ: Prentice Hall)

gle against Egypt, Syria, and Jordan, Israel gained substantial new territories in the Sinai Peninsula, the Gaza Strip, the West Bank, and the Golan Heights. Israel annexed the eastern part of the formerly divided city of Jerusalem, arousing particular bitterness among the Palestinians, since Jerusalem is a sacred city in the Muslim tradition (it contains the Dome of the Rock and the holy Al-Aqsa Mosque). The center is sacred in Judaism as well (it contains the Temple Mount and the holy Western Wall of the ancient Jewish temple), and Israel remains adamant in its claims to the entire city. A peace treaty with Egypt later resulted in the return of the Sinai Peninsula in 1982, but tensions then focused on the other occupied territories that remained under Israeli control. To strengthen its geopolitical claims, Israel also built additional Jewish settlements on the West Bank and in the Golan Heights, further angering Palestinian residents and nearby Syrians.

A cycle of violent actions and equally violent reprisals accelerated with the Palestinian uprising in 1987 that became known as the "Intifada." The conflict continued unabated until 1993, when the two sides began to negotiate a settlement. Preliminary agreements called for the construction of a quasi-independent Palestinian state in the Gaza Strip and across much of the West Bank. Radical factions on both sides denounced the settlement as a sellout, and political polarization among both the Palestinians and the Israelis heightened tensions across the region. Israel also has harshly resisted ongoing Palestinian terrorism, while Palestinian leaders such as Yasser Arafat have criticized continuing construction of new Jewish settlements in the West Bank. A tentative agreement late in 1998 strengthened the potential control of the ruling **Palestinian Authority (PA)** in the Gaza Strip and West Bank (Figure 7.30). A new cycle of heightened violence erupted late in 2000, however, as the Israelis continued with the construction of new settlements in occupied lands and resisted PA moves toward more autonomy. In addition, Palestinian unemployment and poverty rates also have soared since 2000, furthering tensions between Arabs and Jews.

Problems also continue between Israel and its neighbors. The Israeli occupation of the Golan Heights remains the key point of conflict with Syria. Syrian authorities argue that peace in the region will be possible only after the territory is returned to Syrian control. The government of Israel, however, counters that it is vulnerable to attack from any potentially hostile state that might occupy this strategic highland. The region's relatively abundant water supplies add to Israel's determination to retain it. To the north, Israel continues to impose its influence in southern Lebanon, although many of its occupying troops were withdrawn in 1999 and 2000. Farther east, Iraq remains a potential enemy. Indeed, during the Persian Gulf War, Israel was among the nearby countries Iraq subjected to missile attacks. Tensions have been reduced between Israel and nearby Jordan and Egypt, but shifting political winds among the Israelis make future relations hard to predict. One thing is certain: geographical issues will remain at the center of the regional conflict. Palestinians hope for a land they can call their own, and Israelis continue their search for more secure borders that guarantee their own political integrity in a region where they are surrounded by potentially hostile neighbors. Ultimately, the sacred geography of Jerusalem, a mere 220 acres of land within the Old City, stands at the center of the conflict. Imaginative compromises in defining that political space will need to recognize its special value to both its Jewish and Palestinian residents.

Politics of Fundamentalism Islamic fundamentalism is another catalyst for continuing regional geopolitical tensions. Fundamentalism dramatically appeared on the political scene in 1978 and 1979 as Shiite Muslim clerics in Iran overthrew the shah, an authoritarian, pro-Western ruler friendly to U.S. political and economic interests. The Ayatollah Khomeni, a once-exiled religious leader, took power in 1979 and proclaimed an Islamic republic in which religious officials ruled both clerical and political affairs. Building on considerable

▲ **Figure 7.30 Rafah International Airport** The Palestinian-controlled Gaza Strip now features an international airport that adds further legitimacy to the ruling power of the Palestinian Authority (PA). *(Nadav Neuhaus/Corbis/Sygma)*

domestic distrust of both the shah and the United States, Khomeni fomented a revolutionary fervor that not only engulfed his own country, but also spread to many other states in the region. Fundamentalists throughout the realm have challenged existing political regimes, advocated more traditional roles for women, and criticized the influx of heretical ideas from the West. Khomeni died in 1989, and since the mid-1990s somewhat more moderate leaders have come to power in Iran. They are slowly redefining that nation's brand of fundamentalism in ways that may permit more domestic democratic reforms.

Sudan's Sunni Muslims also followed the fundamentalist path. An Islamic-led military coup in 1989 overthrew a democratic government and imposed an Islamic republic (Figure 7.31). Immediately, the Sudanese fundamentalists strengthened official ties with Iran. Further, elements of the new regime became associated with revolutionary political groups in the region, provoking U.S. military reprisals in 1998. The regime also imposed Islamic law across the country and in the process antagonized both moderate Sunni Muslims as well as the nation's large non-Muslim (mostly Christian and animist) population in the south. Persisting differences between northern and southern groups within the country reflect larger global tensions on the margins of the Islamic world and have produced intermittent but highly disruptive civil wars within the country. The recent development of oil reserves in southern Sudan will likely only intensify regional rivalries within the war-torn country.

Fundamentalism is a potent political force elsewhere in the region. Most notably, the growing role of fundamentalism in Algeria has further destabilized a country already divided by increasingly extreme political factions. Once a stable and moderate state within the region, Algeria has been plunged into an escalating cycle of internal violence and terrorism that has killed tens of thousands since 1992. Elections in late 1991 made it clear that Islamic fundamentalists were on the verge of assuming power in the country. Other Algerian nationalists and the military feared the fundamentalists and they negated the election results and suspended democracy in the country. Bombings, kidnappings, and assassinations have since been a regular feature of Algerian life. Since 1997, murderous attacks against unarmed villagers

▲ Figure 7.31 Khartoum, Sudan The politics of fundamentalism spread to the streets of Khartoum in 1989 as a military regime overthrew a democratic government. Today, signs of the city's religious heritage mingle with office buildings and commercial streets. *(J. Hartley/Panos Pictures)*

have been carried out, some of them very close to the capital city of Algiers. In addition, portions of Algeria's Berber population (approximately 20 percent of the total) have periodically erupted in violence, protesting an increasingly hostile national government.

The Algerian debacle has heightened concerns over fundamentalism in other countries within the region, particularly Egypt, Turkey, and Saudi Arabia. In Egypt, various fundamentalist factions have attacked government officials and tourists. Although Turkey has long been regarded as a moderate, more secular state, the rising tide of fundamentalism has challenged the political status quo and clouded the country's potential links with the European Union. Even in Saudi Arabia, a bastion of conservative Sunni traditions and mostly pro-U.S. politics, fundamentalists threaten to undermine the authority of the ruling Saudi family.

Conflicts Within States As the Algerian and Sudanese situations suggest, the region contains many settings in which domestic cultural and political differences have led to civil war. While each country has a unique geopolitical history, broader regional and global influences also play pivotal roles in these settings. Unpredictable relations with the West, the shifting price of oil, and longer-term demographic and economic pressures suggest that these internal geopolitical conflicts will continue in the future.

Several examples illustrate the challenges faced by nations within the region. Lebanon has seen some of the most destructive fighting over the past several decades. From its birth

in the 1940s, Lebanon faced potential discord among its many distinct Christian and Muslim communities. In addition, the Israeli–Palestinian struggle spilled over into neighboring Lebanon as Palestinian refugees in southern Lebanon used the area as a base for attacks in Israel. Israel responded with armed reprisals. The fighting soon spread among the Lebanese, and by 1975 civil war raged within the country. Fighting generally pitted Christians against Muslims, but the existence of numerous sects and factions on both sides made the actual struggle far more complicated. By the late 1970s, the capital city of Beirut was a scarred and divided war zone. Both Syria and Israel sent in armies, and an international peacekeeping force, backed by the United States, vainly attempted to pacify Beirut. Although the conflict flared throughout the 1980s, the nation's various armed factions finally reached a peace agreement in 1995. Syria still maintains a sizable military force in the country, but the fighting has subsided and rebuilding has begun.

Troubled Iraq is another nation born during the colonial era that has yet to escape the consequences of its externally imposed geopolitical origins. Simmering beneath the bluster of Saddam Hussein's Iraqi regime have been enduring cultural differences within the country that periodically produce internal political conflicts. When the country was carved out of the British Empire in 1932, it contained the cultural seeds of its later troubles. One problem zone centers on the lower Tigris and Euphrates valley south of Baghdad, where most of the country's 11 million Shiites live. Indeed, the region focused around the city of Basra contains some of the holiest Shiite sites in the world. Its Marsh Arab population has periodically resisted control from Baghdad, and it briefly threatened independence just after the Persian Gulf War in the early 1990s.

Conditions also remain grim in northern Iraq. For decades, the culturally distinctive Kurds have attempted to create a Kurdish homeland politically separate from Iraq, Turkey, and other nearby nations. The Kurdish population engaged in periodic revolts and raised its flags of rebellion in the 1980s, only to face the full wrath of the Iraqi military. Entire villages were bombed with poison gases. After the Persian Gulf War, the United Nations forbade Iraqi air forces from operating north of the 36th parallel, an area encompassing the main Kurdish districts. As a result, northern Iraq's political identity remains ambiguous: officially it remains part of Iraq, but local forces under UN supervision control some portions of the area.

Divided Cyprus in the eastern Mediterranean offers a classic example of a country torn in two by virtue of its complex historical geography. The island's ancient connections with the Greek world were disrupted in the sixteenth century when Ottoman Turks occupied the area. While Cyprus passed into British hands three centuries later, an important Turkish minority population was well established on the island. When Cyprus gained independence in 1960, the country's national leaders hoped that the island's mixed Greek Orthodox and Islamic Turkish populations could find common ground in their newly won freedom. Unfortunately, divisiveness won out in the 1970s, and civil war erupted. The conflict intensified when Turkey sent troops to the island in 1974, and war with neighboring Greece became a real possibility. One dramatic geographical consequence of the conflict was the spatial polarization of the population: Turks concentrated on the northern portion of the island, while Greeks settled to the south. As a result, UN peacekeepers established a **Green Line** that separates the two groups, a dramatic symbol of failed political compromise that runs through the heart of the country's divided capital of Nicosia (Figure 7.32). In 1983 the Turks (20 percent of the population) declared their own Turkish Republic of Northern Cyprus, an orphan state that has gone unrecognized by the rest of the world (with the exception of Turkey) and that has languished economically as the Greek Cypriots enjoy more rapid economic development.

Conflicts Between States Political tensions within the region often involve multiple states, and these trouble spots are not limited to the Arab–Israeli conflict. The sources of these conflicts vary. Natural resources—including contested claims for oil, water, and farmland—also play a role. Elsewhere, persisting cultural differences still fire contemporary antagonisms, including the enduring problems introduced by boundaries imposed during the colonial era.

Varied North African settings threaten the region's political stability. One trouble spot is Western Sahara, just south of Morocco. The former Spanish colony was invaded and annexed by Morocco in the late 1970s, a move resisted by most of the local population. Although home to only about 200,000 people, the marginal desert land contains substantial mineral (especially phosphate) deposits. Algerian support for the Western Saharans has also periodically strained relations with neighboring Morocco, and UN intervention has yet to end the conflict. To the east, Libya's leader, Colonel Muammar al-Qaddafi, has intermittently fomented regional tensions since he took power in 1969. Libya has financed violent political movements directed against Israel, western Europe, and the United States. Its relationships with nearby Egypt and Chad have also been strained, although Qaddafi has recently campaigned for greater African unity. Internal conflicts within Sudan's fundamentalist Islamic state have spilled into adjacent portions of Ethiopia, Eritrea, Chad, and Uganda as refugee populations have fled the country. Suspected Sudanese involvement in the 1995 assassination attempt on Egypt's President Mubarak strained relations between those two nations.

In Southwest Asia, regional geopolitical tensions persist, reflecting the recent legacy of bloodshed in the Iran–Iraq (1980–88) and Persian Gulf (1990–91) wars. Iraq, led by Saddam Hussein, remains one focus of continuing instability. In 1980 Hussein invaded oil-rich but politically weakened Iran to gain a better foothold in the Persian Gulf. Eight years of bloody fighting resulted in a stalemate, and the conflict left Iraq's finances in disarray. After the war, Iraq received little economic support from wealthy Persian Gulf states even though they opposed the radicalism of the Iranian regime. It tried to convince nearby Kuwait to hand over several strategically located islands in the area, but Kuwait refused. In response, Iraq invaded and overran Kuwait in 1990, claiming it as an Iraqi province. A U.S.-led UN coalition, receiving sub-

▲ **Figure 7.32 Nicosia, Cyprus** The United Nations-enforced Green Line separating Greeks and Turks on the island runs through the major city of Nicosia. The result is an unusual urban landscape that resembles scenes in once-divided East and West Berlin. *(David Rubinger/TimePix)*

stantial support from Saudi Arabia, expelled Iraq from Kuwait in early 1991. In many respects, however, the war in Iraq continued. The country remained under UN-imposed economic sanctions, devastating the Iraqi economy. Large "no-fly" zones were imposed upon Iraqi air space north of 36° and south of 33° latitude in the northern and southern thirds of the country, limiting Iraqi territorial control over those portions of the country. In the late 1990s, tensions again increased as UN inspection teams withdrew from the country and British and U.S. warplanes bombed strategic military and communications sites.

Elsewhere in Southwest Asia, other cultural divides and contested resource issues threaten to ignite larger-scale political conflicts. For example, Turkey's geopolitical tensions lie both to the west and to the east. The country's relations with Greece remain troubled over the issue of Cyprus as well as conflicts in the Balkans. Turkish Kurds remain a destabilizing element in the eastern part of the country that also spills over into relations with neighboring states containing their own Kurdish minorities. Indeed, Turkey and Syria came close to war late in the late 1990s over questions concerning Kurdish rebels, conflicting water rights in the upper Euphrates River, and territorial disputes that date back to the late 1930s. On the eastern borders of the region, Iran faces uncertainties from post-Taliban Afghanistan, a central Asian state that continues to experience political instability.

An Uncertain Political Future

Few areas of the world pose more geopolitical questions than Southwest Asia and North Africa. Twenty years from now, the region's political map could look quite different than it does today. The region's strategic global importance increased greatly after World War II, propelled into the international spotlight by the creation of Israel, the tremendous growth in the world's petroleum economy, Cold War tensions between the United States and the Soviet Union, and the more recent rise of Islamic fundamentalism. Now that the Cold War between global superpowers has ended, the region is experiencing a

What geopolitical role should the United States play in Southwest Asia? That simple question continues to confound policy makers in the United States as well as residents of that ever-troubled part of the world. In particular, the 9/11/01 terrorist attacks renewed the focus on that constantly shifting geopolitical relationship. Osama bin Laden and 12 of the 19 airline hijackers had roots in Saudi Arabia; and others involved had links to Egypt, Yemen, and nearby countries. Thus, since the United States embarked on its "War on Terror," its geopolitical presence in the region has inevitably grown (Figure 7.3.1). But many questions remain. How does the United States identify and deal with hostile elements? Should the United States go it alone in these efforts ("unilateralism"), or depend on sharing the responsibilities with other countries or international organizations ("multilateralism")?

Specifically, three interrelated settings exemplify America's ongoing geopolitical challenges. First, U.S.–Iraqi relations, soured for more than a decade since the Persian Gulf War, continue to be a focal point for conflict. Adding fuel to the fire has been the U.S. belief that Saddam Hussein has played an active role in fostering recent anti-American terrorism. The future remains unclear, however. Indeed, some observers suggest that a post-Saddam Iraq could evolve into a moderate, Western-leaning country and a strong ally of the United States.

A second question mark is oil-rich Saudi Arabia. Unlike Iraq, that nation is an ally of the United States. But the terrorist links to Saudi Arabia have renewed friction between the two countries. Domestic challenges to the all-powerful al-Saud royal family also appear to be on the rise. Both Islamic extremists and Western-leaning progressives have expressed their dissatisfaction with the conservative, traditional Saudi government. American observers fret that a sudden change in Saudi Arabian politics would have global ramifications, particularly if oil exports are disrupted.

Finally, recent events in Israel and the occupied West Bank continue to make global headlines. Palestinian suicide bombings and Israeli reprisals and occupations have further enflamed a region already set on edge after the 9/11/01 attacks. What is the proper role for the United States in tactically improving short-term relationships between Israel and its Palestinian populations and crafting a longer-term strategic policy aimed at achieving a permanent peace in the region? All of these issues will command a great deal of attention within the United States. How the U.S. responds to Southwest Asia's shifting geopolitics seems destined to shape the region for decades to come.

▲ Figure 7.3.1 The USS *Enterprise* in the Persian Gulf Aircraft carriers such as the USS *Enterprise* demonstrate the continuing assertion of geopolitical power by the United States across strategic portions of Southwest Asia and North Africa. *(Office of Naval Research, U.S. Navy)*

geopolitical reorientation. Allies of the former Soviet Union, such as Libya, Syria, and Iraq, have often had to turn to other sources for military equipment and expertise.

Political relations between the United States and the various countries of the region remain complex. Israel and Turkey are close U.S. allies and recipients of large amounts of military aid, just as Iran, Iraq, Syria, and Libya remain firmly opposed to the United States. Relations with the oil-rich states of the Arabian Peninsula are more ambiguous. Most countries in the region, especially Saudi Arabia, have been major purchasers of U.S. military hardware, and all have relied to some extent on U.S. forces for their protection. The United States has eagerly supplied such protection because of its reliance on the oil resources of the region, a mutual dependency made evident in the Gulf War of the early 1990s. In the future, however, it is uncertain whether the Arab-speaking world will readily join forces with the United States, as many people in the region view the United States as an intrusive superpower traditionally too closely allied with Israel.

The likelihood of future political conflicts in the region is high, as is the probability of drawing in outside military participants from Europe, the United States, or even South Asia. Most countries in the region spend a disproportionate amount of money on defense. Traditions of peaceful democratic political regimes are few and far between, and they often run counter to fundamentalist leanings. On the other hand, many authoritarian governments within the region face opposition from groups both within and beyond their borders. The realm's petroleum riches also promise to keep it in the global spotlight. In the long run, a lasting settlement of the Palestinian issue and the securing of stable borders for Israel seem essential before the region can move toward greater political stability.

Economic and Social Development: Lands of Wealth and Poverty

Southwest Asia and North Africa is a region of both incredible wealth and disheartening poverty. While a few of its countries enjoy great prosperity, due mainly to rich reserves of petroleum and natural gas, other nations within the region are among the least developed in the world (Table 7.2). Overall, recent economic growth rates have lagged those of the more-developed world. Persisting political instability has also contributed to the regional economic malaise. These

TABLE 7.2 *Economic Indicators*

Country	Total GNI (Millions of $U.S., 1999)	GNI per Capita ($U.S., 1999)	GNI per Capita, PPP* ($Intl, 1999)	Average Annual Growth % GDP per Capita, 1990–1999
Algeria	46,548	1,550	4,840	−0.5
Bahrain				0.8
Cyprus	9,086	11,950	19,080	2.8
Egypt	86,544	1,380	3,460	2.4
Gaza and West Bank	5,063	1,780		−0.2
Iran	113,729	1,810	5,520	1.9
Iraq				
Israel	99,574	16,310	18,070	2.3
Jordan	7,717	1,630	3,880	1.1
Kuwait				
Lebanon	15,796	3,700		5.7
Libya				
Morocco	33,715	1,190	3,320	0.4
Oman				0.3
Qatar				
Saudi Arabia	139,365	6,900	11,050	−1.1
Sudan	9,435	330		
Syria	15,172	970	3,450	2.7
Tunisia	19,757	2,090	5,700	2.9
Turkey	186,490	2,900	6,440	2.2
United Arab Emirates				−1.6
Western Sahara				
Yemen	6,088	360	730	−0.4

*Purchasing power parity

Source: The World Bank Atlas, *2001*.

economic stumbling blocks have had profound social consequences: investments in education, health care, and new employment opportunities have slowed considerably in many nations from the heady gains of the late 1970s and 1980s. Petroleum will no doubt figure prominently into the region's future economic relationships with the rest of the world, but many countries in the area also have focused on increasing agricultural output, investing in new industries, and promoting tourism as important ways to broaden the regional economic base.

The Geography of Fossil Fuels

The striking global geographies of oil and natural gas reveal the region's persisting importance in the world oil economy, as well as the extremely uneven distribution of these resources within the region (Figure 7.33). Saudi Arabia remains one of the major producers of petroleum in the world, and Iran, the United Arab Emirates, Libya, and Algeria also contribute significantly. The region plays an important though less dominant role in natural gas production. The distribution of fossil fuel reserves suggests that regional supplies will not be exhausted anytime soon. Overall, with only 7 percent of the world's population, the region holds a staggering 68 percent of the world's

proven oil reserves. Saudi Arabia's pivotal position, both regionally and globally, is also clear: its 21 million residents live atop 26 percent of the planet's known oil supplies.

Two major geological zones supply much of the region's output of fossil fuels. The world's largest concentration of petroleum lies within the Arabian-Iranian sedimentary basin, a geological formation that extends from northern Iraq and western Iran to Oman and the lower Persian Gulf (Figure 7.34). All of the states bordering the Persian Gulf reap the benefits of oil and gas deposits within this geological basin, and it is not surprising that the world's densest concentration of OPEC members is found in the area. A second important zone of oil and gas deposits includes eastern Algeria, northern and central Libya, and scattered developments in northern Egypt. As in the Persian Gulf region, these North African fields are tied to regional processing points and to global petroleum markets by a complex series of oil and gas pipelines and by networks of technologically sophisticated oil shipping facilities.

Even with all these riches, the geography of fossil fuels is strikingly uneven across the region. Some nations—even those with tiny populations (Bahrain, Qatar, and Kuwait, for example)—contain incredible fossil fuel reserves, especially when considered on a per capita basis. Many other countries, however, and millions of regional inhabitants, reap relatively few

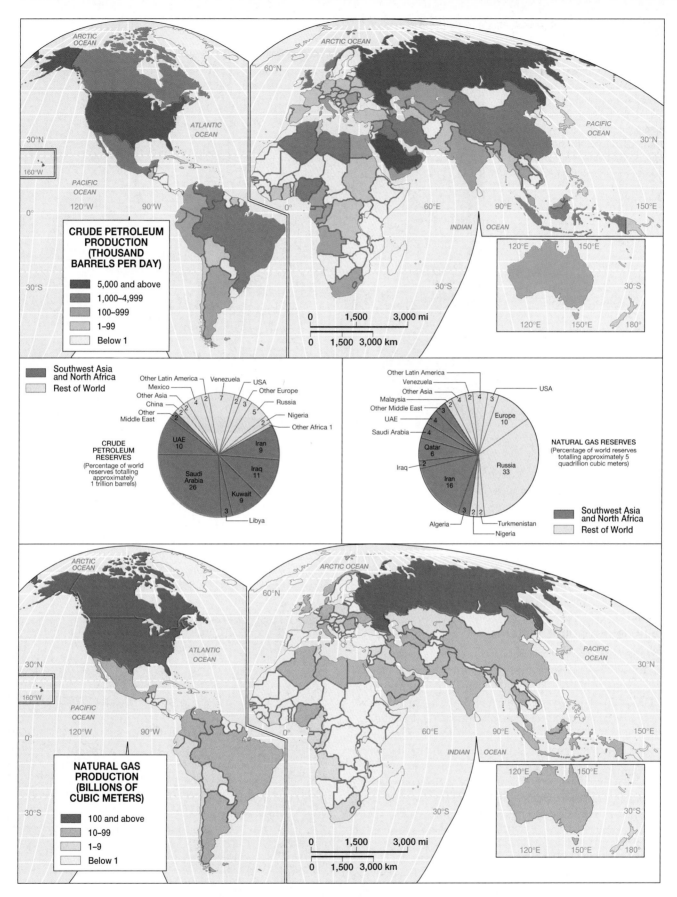

▲ Figure 7.33 Crude petroleum and natural gas production and reserves   The region plays a pivotal role in
the global geography of fossil fuels. Abundant regional reserves suggest that the pattern will continue.   *(Modified
from Rubenstein, 2002,* An Introduction to Human Geography, *Upper Saddle River, NJ: Prentice Hall)*

▶ Figure 7.34 Persian Gulf This satellite view of the Persian Gulf reveals one of the world's richest sources of petroleum. Sedimentary rocks, both on land and offshore, contain additional reserves that can sustain production for decades. *(Earth Satellite Corporation/Science Photo Library/Photo Researchers, Inc.)*

benefits from the oil and gas economy. In North Africa, for example, Morocco and Sudan possess few developed petroleum reserves, although recent discoveries will lead to increased production in both countries. Even in oil-rich Southwest Asia, the distribution of fossil fuels is amazingly fickle: Israel, Jordan, and Lebanon all lie outside favored geological zones for either petroleum or natural gas. While Turkey has some developed fields in the far southeast, it must import substantial supplies to meet the needs of its large and industrializing population.

Regional Economic Patterns

As Table 7.2 suggests, remarkable economic differences characterize the region. Some oil-rich countries have prospered greatly since the early 1970s, but in many cases fluctuating oil prices, political disruptions, and rapidly growing populations have reduced prospects for economic growth. Other nations, while poor in oil and gas reserves, have seen brighter prospects through moves toward greater economic diversification. Finally, some countries in the region exemplify the ongoing reality of persisting poverty, where rapid population growth and the basic challenges of economic development combine with political instability to produce very low standards of living.

Higher-Income Oil Exporters The most prosperous countries of Southwest Asia and North Africa owe their wealth to massive oil reserves. Nations such as Saudi Arabia, Kuwait (after recovering from the Gulf War), Qatar, Bahrain, and the United Arab Emirates benefit from fossil fuel production, as well as from their relatively small populations. Since the oil-price hikes of the 1970s, these countries have reaped billions

in revenues that have had a fundamental impact on their economies (Figure 7.35). Particularly in the case of Saudi Arabia, huge investments in transportation infrastructure, in urban commercial and financial centers, and in other petroleum-related industries have reshaped the cultural landscape. The Saudi petroleum-processing and shipping centers of Jubail (on the Persian Gulf) and Yanbu (on the Red Sea) exemplify this commitment to expand their economic base be-

▲ Figure 7.35 Saudi Arabian oil refinery Eastern Saudi Arabia's Ras Tanura Oil Refinery links the oil-rich country to the world beyond. Huge foreign and domestic investments since 1960 have dramatically transformed many other settings in the region. *(Minosa/Scorpio/Corbis/Sygma)*

yond the simple extraction of crude oil. With these new industrial hubs, the Saudis participate more broadly in the benefits of the oil economy by refining and manufacturing petroleum-related products before they leave the country. While much of the oil wealth in these nations remains concentrated among the ruling elite, petrodollars also have provided real improvements for the larger population. Billions of dollars have poured into new schools, medical facilities, low-cost housing, and modernized agriculture, fundamentally raising the standard of living in the past 40 years. In addition, the Saudis are opening their economy to more foreign investment both within and beyond the petroleum sector.

Still, problems remain, even in these centers of relative wealth. Poor people do reside in Saudi Arabia and elsewhere in the oil-rich region. For example, the Shiite minority in eastern Saudi Arabia has a standard of living far below that of urban populations elsewhere in the country. In addition, large foreign workforces are common across the oil-rich zone, and they typically are paid far less than domestic laborers. Foreign men usually work in construction and the oil industry, while foreign women mostly work as domestic servants. While foreign laborers typically live and work under tough conditions, they receive wages that are far higher than they could earn at home.

Another challenge relates to the shifting fortunes of the oil economy itself. When prices rise, as they did in the 1970s, early 1980s, and in 2000, these oil-rich nations gain an economic windfall. But dependence on oil and gas revenues clearly has a darker side: falling prices, such as those seen in the mid-1980s, the late 1990s, and in 2001 created immediate economic pain for the major Middle East producers. In fact, the UAE, Qatar, and Oman all experienced real declines in income during the 1990s. Since 1990, even Saudi Arabia has been forced to reduce government spending as global petroleum prices have fallen and its external debt has ballooned. While OPEC continues to be a significant global economic force, many non-OPEC producers also compete in the world marketplace (including Europe, Mexico, and Russia), thus diminishing OPEC's power to set prices. As a result, future economic growth in these wealthier oil-dependent states will likely be moderate, at best. While Saudi Arabia, blessed with abundant reserves of easily accessible oil, can survive with high or low prices in the future, countries such as Bahrain and Oman are faced with the additional problem of rapidly depleting their reserves over the next 20 to 30 years.

Lower-Income Oil Exporters   Other states in the region are important secondary players in the oil trade, but different political and economic variables have often hampered sustained economic growth. In North Africa, Algeria and Libya illustrate this scenario. Algerian oil and natural gas overwhelmingly dominate its exports, but the 1990s brought lower fossil-fuel prices, political instability, and increasing shortages of consumer goods. While the country contains some excellent agricultural lands in the north, the overall amount of arable land has increased little over the past 25 years, even as the country's population has grown by more than 50 percent. The pressure on the rural econ-

omy has accelerated movement to increasingly crowded cities, and many of the nation's upwardly mobile workers have left the country for employment opportunities in western Europe. Nearby Libya also remains a major exporter of oil and natural gas, but foreign investment beyond the petroleum industry has remained limited. Still, Qaddafi has undertaken ambitious efforts to expand the agricultural economy and improve water availability in his desert land, and he has recently pushed for expanding the country's tourism industry in the hope of attracting more outsiders.

Other economic and political challenges face Iraq in Southwest Asia. Iraq has been its own worst economic enemy. Sanctions imposed after the Persian Gulf War virtually halted Iraq's oil exports, crippling its economy. Since Iraq is not self-sufficient in most basic products, living standards plummeted to extremely low levels. In the late 1990s the United Nations allowed Iraq to export oil in limited quantities, but only in exchange for food and medicine. Even so, health care in the country has deteriorated greatly, and food shortages are much more common in many areas than they were 20 years ago. Iraq certainly has the potential to be a prosperous country: its population density is moderate and its resource base is substantial. Yet as long as the current regime remains in power, Iraq will likely continue to suffer.

The situation in Iran is more complex. The country is large and populous, and has a relatively diverse economy. Iran's oil reserves are huge and have seen active commercial development since 1912. The country also has a sizable industrial base, much of it built in the 20 years prior to the fundamentalist revolution of 1979. In addition, considerable capital flowed into modernizing Iran's agricultural infrastructure during the same period, bringing the benefits of increased mechanization and new irrigation projects. Still, most observers agree that today Iran is relatively poor, burdened with a stagnating if not declining standard of living. Since 1980, the country's fundamentalist leaders have downplayed the role of international trade in consumer goods and services, fearing they would import unwanted cultural influences from abroad. Making matters worse were the twin challenges of a costly and bloody war with Iraq in the 1980s followed by the struggling oil economy of the 1990s. Recently, however, the country's economic prospects have brightened somewhat with new economic links with Central Asia, a partial recovery in oil prices, and the potential for greater economic interaction with the West.

Prospering Without Oil   Some countries, while lacking petroleum resources, have nevertheless found paths to increasing economic prosperity. Israel, for example, supports the highest standard of living in the region, even with its political challenges. The nation's sparse natural resource base would hardly seem conducive to economic development. But the Israelis have invested large amounts of capital to create a highly productive agricultural and industrial base. The country is also emerging as a global center for high-tech computer and telecommunications products. Many U.S. and European companies maintain development and production centers there, and the country is increasingly becoming known for its fast-paced and highly entrepreneurial business culture, which resembles California's Silicon Valley. Tourism is another important sector of the economy, as the country attracts global visitors interested in the region's rich cultural heritage. Even so, Israel has many economic problems. Its persisting struggles with the Palestinians and with neighboring states have sapped much of its potential vitality. Defense spending absorbs a large share of total gross national income (GNI), necessitating high tax rates. Poverty among the Palestinians is also widespread, and the gap between rich and poor within the country has widened considerably.

Turkey also has a diversified economy. While its per capita income is modest even by regional standards, it has shown far greater economic dynamism in recent years than have many of the major oil producers in the region. Lacking petroleum, Turkey produces varied agricultural and industrial goods for export. Almost half of the population remains employed in agriculture, and the country's principal commercial products include cotton, tobacco, wheat, and fruit. The industrial economy has grown since 1980, including exports of textiles, food, and chemicals. Turkey remains the most important tourist destination in the region as well, attracting more than 6 million visitors annually during the 1990s. In addition, Istanbul enjoys increasing prominence as a regional finance and investment center. Many Turkish leaders hope to bolster the economy by eventually joining the European Union. Turkey already has closer links to the West than any other country of the region except Israel (it is, for example, a member of NATO), and its trade potential with Europe is great. Still, economic integration with the West will not be easy. Some European leaders are wary of extending their economic union to what they consider a politically unstable non-European state. Greece also resists the closer integration of Turkey into the European economic and political block, and conservative religious leaders in Turkey oppose stronger ties to Europe.

Other oil-poor countries in the region aspire to economic prosperity. In North Africa, Tunisia has fostered relative economic stability without the benefit of large oil reserves. Although the country possesses a few important oil fields south of Tunis, most of its workforce is in agriculture, services (particularly tourism), and manufacturing. Recent government policies have favored economic reforms, increased private investment, and encouraged participation in global markets, such as those that bring the region's citrus fruits to grocery stores in the United States. A similar situation benefits Cyprus. Although the island is divided along ethnic lines, the Greek-dominated southern portion has invested in infrastructure, education, and tourism, while favorable tax and trade policies have attracted considerable foreign capital. Even war-torn Lebanon has the potential for prosperity. At one time, Lebanon was the region's financial center and wealthiest state. Then the country was devastated by war during most of the 1970s and 1980s. A slow economic recovery, however, began in the 1990s as the political climate improved somewhat. Tourism is on the rise, and the country's new telecommunications industry signals the potential for growth in that sector of the economy.

▲ **Figure 7.36 Sudanese village** Poverty and political instability remain a part of life in many Sudanese villages. The traditionally cultivated fields around the village center contrast vividly with the skyscrapers and industrial complexes found in oil-rich portions of the region. *(Liba Taylor/Hutchison Library)*

Regional Patterns of Poverty The poorer countries of the region share the problems of much of the less-developed world. In North Africa, the nations of Sudan, Morocco, and Egypt face distinctive economic challenges. For Sudan, devastating political problems have stood in the way of progress. Civil war has resulted in major food shortages, particularly in the south. In addition, the conflicts have disrupted major agricultural improvement schemes in the Gezira district, between the White and Blue branches of the Nile River. The country's infrastructure has seen little new investment; settlement remains overwhelmingly rural; and secondary school enrollments stand at less than 25 percent of the school-age population (Figure 7.36). On the other hand, Sudan's fertile soils could support more farming, and its new oil pipeline suggests petroleum's expanding role in the economy. Still, the country's sustained economic development appears forestalled by continuing political instability.

Morocco's economic picture is brighter, but the country remains poorer than either Algeria or Tunisia. Poverty is especially widespread in the Atlas Mountains, where many Berber communities have little access to modern services or infrastructure. Elsewhere in the country, economic reforms have spurred privatization, encouraged investment by foreign banks, and slowed inflation. Still, illiteracy is widespread, and the country suffers from the **brain drain** phenomenon as some of its brightest young people leave for better jobs in western Europe. Rural population pressures and a stagnating agricultural economy have also spurred internal movements from the countryside to major cities, including Casablanca and Rabat, but often migrants are frustrated by ongoing problems of underemployment even in these rapidly expanding metropolitan areas.

Egypt's economic prospects are also unclear. On the one hand, the country experienced real economic growth during the 1990s as President Hosni Mubarak pushed for smaller government deficits and a multibillion-dollar privatization

LOCAL VOICES   The Egypt of Naguib Mahfouz

Born in 1911, Nobel Prize-winning novelist Naguib Mahfouz has written many stories about life in his native Egypt (Figure 7.4.1). His most widely read books in English are the Cairo Trilogy (*Palace Walk, Palace of Desire,* and *Sugar Street*), originally published in the 1950s, which remain an extraordinary reconstruction of Egyptian life in the early twentieth century. Critic Edward Said notes that "as a geographical place and as history, Egypt for Mahfouz has no counterpart in any other part of the world. Old beyond history, geographically distinct because of the Nile and its fertile valley, Mahfouz's Egypt is an immense accumulation of history, stretching back in time for thousands of years, and despite the astounding variety of its rulers, regimes, religions, and races, nevertheless retaining its own coherent identity."

Mahfouz's long familiarity with Cairo's many streets and neighborhoods enables him to take the reader into the city with an intimate look at the interplay between people and place. In *Palace of Desire,* young Yasin returns to his boyhood neighborhood:

> *When his feet brought him to al-Gamaliya Street, he was so choked up he felt he would die. He had not been there for eleven years . . . yet it remained exactly the way it had been when he was growing up. Nothing had changed. The street was still so narrow a handcart would almost block it when passing by. The protruding balconies of the houses almost touched each other overhead. The small shops resembled the cells of a beehive, they were so close together and crowded with patrons, so noisy and humming. The street was unpaved, with gaping holes full of mud. The boys who swarmed along the sides of the street made footprints in the dirt with their bare feet. There was the same never-ending stream of pedestrian traffic. Uncle Hasan's snack shop and Uncle Sulayman's restaurant too remained just as he had known them. . . . His heart pounded so strongly it almost deafened his ears . . .*

English translation obviously reworks the flow of language and eliminates the nuances of the native Arabic. Still, this Egyptian writer's sense of place is now widely accessible outside the Arab World since the publication of the translated trilogy in the early 1990s.

Sources: Edward W. Said, "The Cruelty of Memory," *The New York Review of Books,* November 30, 2000, and Naguib Mahfouz, *Palace of Desire* (New York: Doubleday, 1991).

▲ **Figure 7.4.1 Naguib Mahfouz** Egypt's great novelist captures a unique sense of place and time in the stories he tells. Mahfouz's Cairo Trilogy has been widely translated into English and other languages. *(Thomas Hartwell/TimePix)*

program to put government-controlled assets under more efficient management. Egypt also actively courts foreign investment in its expanding industrial sector, and the streets of Cairo are increasingly sprinkled with the flash of Rolls-Royces, Porsches, and Lamborghinis. Even so, many Egyptians still live in poverty, and the gap between rich and poor continues to widen (see "Local Voices: The Egypt of Naguib Mahfouz"). Millions till tiny plots of land along the Nile, while others scramble for living wages in low-paying service and manufacturing jobs in Cairo and Alexandria. Furthermore, the demographic clock is ticking: Egypt's 70 million people already make it the region's most populous state, and recent efforts to expand the nation's farmland have met with numerous environmental, economic, and political problems. Future prosperity hinges on the country's ability to expand its economy faster than its population.

In Southwest Asia, Yemen remains the poorest country on the Arabian Peninsula. Positioned far from most of the region's principal oil fields, Yemen's low per capita GNI puts it on par with many nations in impoverished Sub-Saharan Africa or South Asia. The largely rural country relies mostly on marginally productive subsistence agriculture, and much of its mountain and desert interior lacks effective links to the outside world. The present state emerged in 1990 with the political union of North and South Yemen, but periodic civil unrest continues between the two parts of the country. Many Yemenis work outside their country, but the nation's support of Iraq in the Persian Gulf War led to many Yemeni deportations to their impoverished homeland.

Issues of Social Development

Measures of social development vary widely across the realm and are broadly associated with overall levels of economic development. Religion and cultural traditions also shape social geographies across the region.

Varied Regional Patterns   Israel's population reaps many benefits from its relative wealth within the region. Israelis enjoy high-quality health care and education. In fact, Israel does better in many social measures than might be expected on the basis of its per capita GNI. Average life spans and rates of infant mortality parallel those of the United States, and the vast majority of the population is literate. Social conditions are appreciably better, however, for Israel's Jewish majority than for its substantial Muslim minority. Other relatively wealthy countries—particularly high-income oil exporters such as Saudi Arabia—have lower figures of social well-being than might be expected. In Saudi Arabia, for example, almost one-quarter of the population remains illiterate, and infant mortality is much higher than in Israel. The recent timing of Saudi Arabia's development helps explain the divergence, and future figures may improve substantially based on recent investments in health care and education. Not surprisingly, illiteracy rates are high and average life expectancy is low in the poorer countries of the region, such as Sudan and Yemen.

A Woman's Changing World   The role of women in the largely Islamic region also remains a major social issue. Female labor participation rates in the workforce are the lowest in the world, and large gaps typically exist between male and female literacy (Table 7.3). In the most conservative parts of the region, few women are allowed to work outside of the home. Even in parts of Turkey, where Western influences are widespread, it is rare to see women selling goods in the marketplace or driving cars in the street. More orthodox Islamic states impose legal restrictions on the activities of women. In Saudi Arabia, for example, women are not allowed to drive. In Iran, full veiling remains mandatory in more conservative parts of the country. Generally, Islamic women lead more private lives than men: much of their domestic space is shielded from the world by walls and shuttered windows, and even their public appearances are filtered through the use of the face veil or chador (full-body veil).

Yet even in a fundamentalist state such as Iran, women's roles are changing. Young working women in Tehran, for example, are much more likely to be wearing Western-style fashions than was the case 10 years ago (Figure 7.37). Educational opportunities are also increasingly open to girls across much of the region. In Sudan and Saudi Arabia, a growing number of women pursue high-level careers. Education may be segregated, but it is available. Libya's Qaddafi has singled out the modernization of women as a high priority, and today more women than men graduate from the nation's university system. Women also have a more visible social position in Israel, except in fundamentalist Jewish communities where conservative social mores emphasize more traditional domestic roles.

TABLE 7.3 *Social Indicators and Status of Women*

Country	Life Expectancy at Birth		Under Age 5 Mortality (per 1,000)		Percent Illiteracy (Ages 15 and over)		Female Labor Force Participation (% of total, 1999)
	Male	Female	1980	1999	Male	Female	
Algeria	68	70	139	39	23	44	27
Bahrain	70	75	—	—	—	—	20
Cyprus	75	79	—	—	—	—	39
Egypt	65	68	175	61	34	57	30
Gaza and West Bank	70	74	—	26	—	—	—
Iran	69	71	126	33	17	31	27
Iraq	58	60	95	128	35	55	19
Israel	76	80	19	8	2	6	41
Jordan	69	71	49	31	6	17	24
Kuwait	72	73	35	13	16	21	31
Lebanon	68	73		32	8	20	29
Libya	73	77	80	28	10	33	23
Morocco	67	71	152	62	39	65	35
Oman	69	73	95	24	21	40	16
Qatar	69	74	—	—	—	—	15
Saudi Arabia	66	69	85	25	17	34	15
Sudan	55	57	145	109	31	55	29
Syria	70	70	73	30	12	41	27
Tunisia	70	74	100	30	20	41	31
Turkey	67	71	133	45	7	24	37
United Arab Emirates	71	76		9	26	22	14
Western Sahara	—	—	—	—	—	—	—
Yemen	57	61	198	97	33	76	28

Sources: Population Reference Bureau, Data Sheet, 2001, Life Expectancy (M/F); The World Bank, World Development Indicators, 2001, Under Age 5 Mortality Rate (1980/99); The World Bank Atlas, 2001, Female Participation in Labor Force; CIA World Factbook, 2001, Percent Illiteracy (ages 15 and over).

▲ Figure 7.37 The changing role of women   Traditional and Western clothing styles mingle on the streets of Tehran, Iran, suggesting slow but inexorable shifts in the roles played by women in many Islamic societies within the region.   *(Corbis)*

Global Economic Relationships

Southwest Asia and North Africa share close economic ties with the world. While oil and gas remain critical commodities that cement these international linkages, the growth of manufacturing and tourism are redefining the region's role in the world. New intraregional and interregional ties promise to further rework the ways in which this dynamic part of the world functions in the twenty-first century's global economy.

OPEC's Changing Fortunes   OPEC has not gone away, and the region's enduring role in oil and gas production ensures that fossil fuels will continue to be a major international export for many nations in Southwest Asia and North Africa (Figure 7.33). While OPEC can no longer control oil and gas prices globally, it still influences the cost and availability of these pivotal products within the developed and less-developed worlds. Western Europe, the United States, Japan, and many less-industrialized countries depend on the region's fossil fuels. In the case of Saudi Arabia, for example, crude oil shipments make up more than 70 percent of its exports, with refined oil and petrochemicals constituting another 20 percent. One recent trend evident in many major oil-producing countries (such as Saudi Arabia) is their increasing willingness to form larger production and refining partnerships with foreign cor-

porations, a pattern that will accelerate the economic integration of the region with the rest of the world.

Another energy-related aspect of globalization relates to the flow of investment capital out of the region, a process long dominated by petrodollars. As oil prices increased tenfold between 1970 and 1980, many of those investments ended up in foreign bank accounts, stock markets, and real estate. Although the region's role in the global oil economy has declined some since 1980, recent estimates still value Gulf Arab foreign investments at more than $800 billion. These investments influence everything from the cost of office buildings in downtown Singapore to the yields on Brazilian bonds.

Beyond the key OPEC producers, other countries within the region are less dependent on oil-related exports. Turkey, for example, ships textiles, food products, and manufactured goods to its principal trading partners, Germany, Russia, Italy, and the United States. Similarly, Israeli exports emphasize the country's highly skilled workforce: products such as cut diamonds, electronics, and machinery parts are exported to the United States, western Europe, and Japan.

Regional and International Linkages   Future interconnections with the global economy may depend increasingly on cooperative economic initiatives far beyond OPEC. Relations with the European Union (EU) are critical. Since 1996, Turkey has enjoyed closer economic ties with the EU, but recent attempts at full membership in the organization have failed. Other so-called Euro-Med agreements also have been signed between the EU and Morocco, Tunisia, Jordan, and Israel. Supporters argue that these agreements will bring more export-oriented industries to the region. Most Arab countries, however, are wary of too much European dominance. They formed a regional political organization known as the Arab League in 1945, and 18 League members established the Arab Free-Trade Area (AFTA) in 1998, designed to eliminate all intraregional trade barriers by 2008 and to spur economic cooperation within the region. Smaller organizations tie together other regional constituencies. For example, the six-member Gulf Cooperation Council provides major oil producers in the area with a forum in which to discuss shared security and trade-related issues, and the five-member Union of the Arab Maghreb is dedicated to promoting economic integration and a free trade zone across North Africa. In addition, Saudi Arabia has played a particularly pivotal role in regional economic development through organizations such as the Islamic Development Bank and the Arab Fund for Economic and Social Development.

The Geography of Tourism   Tourists are another link to the global economy. Traditional magnets such as ancient historical sites and globally significant religious localities (for a multitude of faiths) draw millions of visitors annually. As the developed world becomes wealthier, there is also a growing global demand for recreational spots that can offer beaches, sunshine, and novel entertainment. Indeed, many miles of the Mediterranean, Black, and Red Sea coastlines are now lined with the upscale, but often ticky-tacky, landscapes of

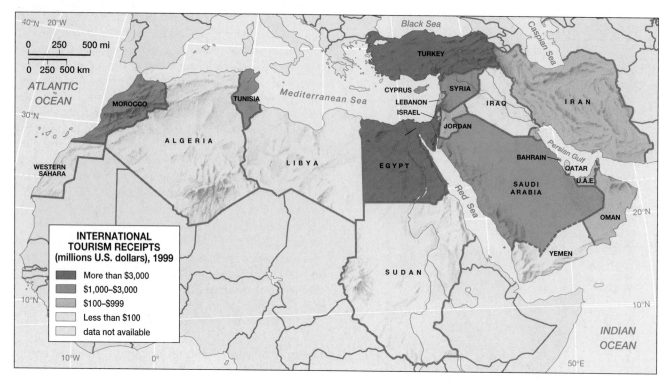

▲ **Figure 7.38 International tourism receipts**   Dramatic differences in tourism expenditures reveal the varying importance of that industry in connecting Southwest Asia and North Africa to the world beyond. The multibillion-dollar links provided by tourism for Turkey, Egypt, and Israel contrast sharply with tourist income in politically unstable settings such as Algeria, Sudan, and Iraq, where little tourism exists.   *(Data from World Bank Atlas, 2001)*

resort hotels and condominiums dedicated to serving the traveler's needs. More adventurous travelers seek ecotourist activities such as snorkeling in Naama Bay on Egypt's Sinai Coast or four-wheeling among the Berbers in the Moroccan backcountry. Endangered wildlife also beckon photographers and poachers hoping to catch a glimpse of a South Arabian grey wolf, Cyprian mouflon (wild sheep), Nubian ibex, or a darting Persian squirrel.

All of this activity means big business to many countries in the region, and the economic impacts of tourism seem likely to grow during the twenty-first century. Tourism is already a huge part of the regional economy in settings such as Turkey, Israel, and Egypt (Figure 7.38). Elsewhere, smaller inflows of tourist dollars still have an extraordinary impact on the economic geographies of settings such as northern Morocco and Tunisia, coastal Cyprus (in the south), and the holy Saudi Arabian cities of Makkah and Medinah. In some cases tourism remains much less developed, but the small existing inflows

in such places as Libya, Jordan, and Iran may be the beginning of a much larger economic trend in future years, depending on the region's political stability.

The heady growth of tourism, while further wedding the region to the global economy, has come at a considerable price. In addition to the visual blight of high-rise hotels, real environmental damage is increasingly the product of the growing human presence in many fragile regional settings. Various archeological and sacred sites have also been impacted across the region. More broadly, the economic realities of the tourist economy have produced a local underclass of poorly paid service workers, a growing sense of social distance between the haves and have-nots in tourist settings, and an amenity, consumption-oriented lifestyle that contrasts sharply with many traditional ways of life. Still, these changes appear inevitable for a region that, even in uncertain political times, seems to irresistibly attract tourists from every corner of the globe.

⊕ Conclusion

Positioned at the meeting ground of Earth's largest landmasses, the Southwest Asia and North Africa region has played a pivotal role in world history and in processes of globalization that bind the planet ever more tightly together. In ancient times, the region's inhabitants were early contributors to the Agricultural Revolution, a long process of plant and animal domestication destined to reshape the world's cultural landscapes. The realm also served as a crucible for urban civilization, offering in the process a fundamentally new way of human settlement that continues to reshape the distribution of global populations today. Three of the world's great religions—Judaism, Christianity, and Islam—emerged beneath its desert skies. More recently, from the sweeping expansion of Ottoman rule to the unpredictable machinations of OPEC oil ministers, regional forces have continued to reach beyond its boundaries, reconfiguring global political, economic, and cultural relationships.

Despite this rich legacy of global influence and power, the peoples of Southwest Asia and North Africa are struggling at the beginning of the twenty-first century. Indeed, most of the countries within the region suffer from significant economic stagnation and political uncertainties. Twentieth-century population growth across the region was dramatic. At the same time, it has been difficult and costly to expand the region's limited supplies of agricultural land and water resources. The results, apparent from the eroded soils of the Atlas Mountains to overworked garden plots along the Nile, are a classic illustration of the environmental price paid when population growth outstrips the ability of the land to support it. Political conflicts have also disrupted economic development across the region. Civil wars, conflicts between states, and regional tensions have worked against initiatives for greater cooperation and trade. Perhaps most important, the region must deal with the intrinsic contradictions between modernity and the more fundamentalist interpretations of Islam. One thing is certain: future cultural change will be guided by a complex response to Western influences, a mix of fascination and suspicion that will no doubt yield its own unique regional cultural amalgam.

The region still commands global attention early in the twenty-first century. It remains the hearth of Christendom, the spatial and spiritual core of Islam, and the political and territorial salvation of modern Judaism. Muslims worldwide are influenced by its cultural and political evolution, and the Arab–Israeli conflict will continue to be a critical element in global geopolitics. Indeed, broader regional hopes for peace seem limited until the diverse peoples of the eastern Mediterranean reach a political rapprochement that satisfies both the Palestinians and their Jewish neighbors. In addition, the accidents of geology and the thirsts of the global economy dictate that the region will remain prominent in world petroleum markets. Also likely are moves toward economic diversification and integration, initiatives that gradually will draw the region closer to Europe and other participants in the global economy. Even so, Southwest Asia and North Africa will retain its distinctive regional identity, a character defined by its setting, the rich cultural legacy of its history, the selective abundance of its natural resources, and its stubbornly persistent political problems.

⊕ Key Terms

brain drain (page 312)
culture hearth (page 276)
domestication (page 286)
exotic rivers (page 288)
Fertile Crescent (page 286)
fossil water (page 284)
Green Line (page 306)
Hajj (page 294)
hydropolitics (page 284)

Islamic fundamentalism
  (page 277)
kibbutzes (page 289)
Levant (page 279)
Maghreb (page 278)
medina (page 290)
monotheism (page 294)

OPEC (Organization of
  Petroleum Exporting
  Countries) (page 276)
Ottoman Empire (page 296)
Palestinian Authority (PA)
  (page 304)
pastoral nomadism (page 286)
physiological densities
  (page 285)

qanat system (page 284)
Quran (page 294)
Shiites (page 295)
Suez Canal (page 301)
Sunnis (page 295)
theocratic state (page 295)
transhumance (page 287)

⊕ Questions for Summary and Review

1. Why do Southwest Asia and North Africa form a useful world region? What are some of the problems associated with defining the region?

2. Describe the climatic changes you might experience as you traveled from the eastern Mediterranean coast to the highlands of Yemen. What are some of the key climatic variables that explain these variations?

3. Discuss five important human modifications of the Southwest Asian and North African environment, and assess whether these changes have benefited the region.

4. Discuss how pastoral nomadism, oasis agriculture, and dryland wheat farming represent distinctive adaptations to the regional environments of Southwest Asia and North Africa. How does each of these rural lifeways create distinctive patterns of settlement?

5. Compare the modern maps of religion and language for the region, and identify three major examples where Islam dominates non-Arabic-speaking areas. Explain why that is the case.

6. Describe the role played by the French and British in shaping the modern political map of Southwest Asia and North Africa. Provide specific examples of their lasting legacy.

7. Outline three regional examples where Islamic fundamentalism has redefined the domestic geopolitical setting since the late 1970s.

8. Explain why internal political tensions in both Lebanon and Cyprus are related to neighboring states.

9. Describe the basic geography of oil reserves across the region, and compare the pattern with the geography of natural gas reserves.

10. What strategies for economic development have recently been employed by nations such as Turkey, Israel, Egypt, and Morocco? How successful have they been, and how do they relate to the theme of globalization?

⊕ Thinking Geographically

1. How might a major project for transferring water from Turkey to the Arabian Peninsula affect the development of Saudi Arabia? What would be some of the potential political and ecological ramifications of such a project?

2. Why are birthrates declining in selected countries within the region? Despite the cultural differences with North America, what common processes seem to be at work in both regions that have contributed to this demographic transition?

3. What economic changes could occur if Israel and the Palestinians were to reach a lasting peace? What kinds of general connections might be found between political conflict and economic conditions throughout the region?

4. Are relations between the region's Muslim societies and the United States likely to improve or worsen? What factors would you cite to defend your answer?

5. What might be some of the reasons for Southwest Asia and North Africa's general failure to match the rates of economic and industrial growth found in North America?

6. Why has the idea of Arab nationalism failed to achieve any lasting geopolitical changes in the region?

7. As a ruler of a conservative Arab state, what might be the advantages and disadvantages of opening up your country to the Internet?

⊕ Regional Novels and Films

Novels

Shmuel Agnon, *Shira* (1989, Schocken)

Hanan Al-Shaykah, *Beirut Blues: A Novel* (1995, Doubleday)

Simin Daneshvar, *Savushun: A Novel About Modern Iran* (1990, Mage)

Kahlil Gibran, *The Prophet* (1923, Knopf)

Kahlil Gibran, *Broken Wings: A Novel* (1957, Citadel)

Barbara Hodgson, *The Tattooed Map* (1995, Chronicle Books)

Ulfat Idlibi, *Sabruya: Damascus Bitter Sweet, A Novel* (1980, Interlink)

Sahar Kalifeh, *Wild Thorns* (1988, Olive Branch Press)

Yasar Kemal, *Salman the Solitary* (1998, Harvill)

Muhsin Mahdi, ed., *The Arabian Nights* (1992, Knopf)

Naguib Mahfouz, *Palace Walk* (1991, Doubleday)

Naguib Mahfouz, *Palace of Desire* (1991, Doubleday)

Naguib Mahfouz, *Sugar Street* (1992, Doubleday)

Abdelrahman Munif, *Cities of Salt: A Novel* (1987, Random House)

Mustapha Tlili, *Lion Mountain* (1988, Arcade Publishing)

Films

The Battle of Algiers (1965, Italy)

Casablanca (1942, U.S.)

The English Patient (1996, U.S.)

The House on Chelouche Street (1973, Israel)

Kadosh (1999, Israel)

Kazablan (1974, Israel)

Late Summer Blues (1987, Israel)

Lawrence of Arabia (1962, U.S.)

Lion of the Desert (1981, U.K.)

Raiders of the Lost Ark (1981, U.S.)

Sallah (1965, Israel)

The Sheltering Sky (1990, U.S.)

The Ten Commandments (1956, U.S.)

Wedding in Galilee (1987, Israel)

The White Balloon (1995, Iran)

Bibliography

Abun-Nasr, Jamil M. 1975. *A History of the Maghrib.* Cambridge, U.K.: Cambridge University Press.

Amery, Hussein, and Wolf, Aaron, eds. 2000. *Water in the Middle East: A Geography of Peace.* Austin: University of Texas Press.

Anderson, Ewan W. 2000. *The Middle East.* London: Routledge.

Barakat, Halim. 1993. *The Arab World: Society, Culture, and State.* Berkeley: University of California Press.

Blake, Gerald H. 1988. *The Cambridge Atlas of the Middle East and North Africa.* Cambridge, U.K.: Cambridge University Press.

Canfield, Robert, ed. 1991. *Turko-Persia in Historical Perspective.* Cambridge, U.K.: Cambridge University Press.

Cohen, Saul B. 1992. "Middle East Geopolitical Transformation: The Disappearance of a Shatter Belt." *Journal of Geography* 91 (January/February), 2–10.

Cole, Juan R., ed. 1992. *Comparing Muslim Societies: Knowledge and the State in a World Civilization.* Ann Arbor: University of Michigan Press.

Cressey, George B. 1960. *Crossroads: Land and Life in Southwest Asia.* Chicago: J. B. Lippincott.

English, Paul Ward. 1966. *City and Village in Iran: Settlement and Economy in the Kirman Basin.* Madison: The University of Wisconsin Press.

Esposito, John L., ed. 1999. *The Oxford History of Islam.* Oxford, U.K.: Oxford University Press.

Held, Colbert C. 1994. *Middle East Patterns: Places, Peoples, and Politics.* Boulder, CO: Westview Press.

Hillel, Daniel. 1994. *Rivers of Eden: The Struggle for Water and the Quest for Peace in the Middle East.* Oxford, U.K.: Oxford University Press.

Hourani, Albert. 1991. *A History of the Arab Peoples.* New York: Warner Books.

Jabbur, Jibrail. 1995. *The Bedouins and the Desert: Aspects of Nomadic Life in the Arab East.* Albany, NY: State University of New York Press.

Keyder, Caglar, ed. 1999. *Istanbul: Between the Global and the Local.* Lanham, MD: Rowman and Littlefield.

Kliot, Nurit. 1994. *Water Resources and Conflict in the Middle East.* London: Routledge.

Lapidus, Ira M. 1988. *A History of Islamic Societies.* Cambridge, U.K.: Cambridge University Press.

Lawrence, Bruce B. 1989. *Defenders of God: The Fundamentalist Revolt Against the Modern Age.* San Francisco: Harper & Row.

Lewis, Bernard. 1993. *The Arabs in History.* Oxford, U.K.: Oxford University Press.

Miller, Judith. 1996. *God Has Ninety-Nine Names: Reporting from a Militant Middle East.* New York: Simon & Schuster.

Mostyn, Trevor, ed. 1988. *The Cambridge Encyclopedia of the Middle East and North Africa.* New York: Cambridge University Press.

Peters, F. E. 1994. *The Hajj: The Muslim Pilgrimage to Mecca and the Holy Places.* Princeton, NJ: Princeton University Press.

Raymond, Andre. 2000. *Cairo.* Cambridge, MA: Harvard University Press.

Rodenbeck, Max. 1999. *Cairo: The City Victorious.* New York: Knopf.

Said, Edward. 2000. "The Cruelty of Memory." *The New York Review of Books,* November 30.

Sciolino, Elaine. 2000. *Persian Mirrors: The Elusive Face of Iran.* New York: Free Press.

Seal, Jeremy. 1995. *A Fez of the Heart: Travels Around Turkey in Search of a Hat.* New York: Harcourt Brace and Company.

Soffer, Arnon. 1999. *Rivers of Fire: The Conflict Over Water in the Middle East.* Lanham, MD: Rowman and Littlefield.

Starr, Joyce R., and Stoll, Daniel. 1988. *The Politics of Scarcity: Water in the Middle East.* Boulder, CO: Westview Press.

Stewart, Dona. 1996. "Cities in the Desert: The Egyptian New-Town Program." *Annals of the Association of American Geographers* 86, 459–80.

"A Survey of Egypt." 1999. *The Economist,* March 20.

Sutton, Keith. 1999. "Demographic Transition in the Maghreb." *Geography* 84; 111–18.

Swearingen, Will D., and Bencherifa, Abdellatif, eds. 1996. *The North African Environment at Risk.* Boulder, CO: Westview Press.

Tessler, Mark. 1994. *A History of the Israeli-Palestinian Conflict.* Bloomington, IN: Indiana University Press.

"War, Famine and Oil in Sudan." 2001. *The Economist,* April 14.

8

Europe

EUROPE
Political Map

⊗ ● Over 1,000,000

○ ○ 500,000–1,000,000
(selected cities)

★ • Selected smaller cities

R U S S I A

Black Sea

T U R K E Y

Elevation in meters

4000+
2000–4000
500–2000
200–500
0–200
Below sea
level

Sea Level

30°E

Europe is one of the most diverse regions in the world, encompassing a wide array of people and places in an area considerably smaller than North America. More than half a billion people reside in this region, living in 37 countries that range in size from giant Germany to microstates such as Andorra and Monaco (Figure 8.1).

The region's remarkable cultural diversity produces a geographical mosaic of different languages, religions, and landscapes. Commonly, a day's journey finds a traveler speaking two or three languages, possibly changing money several times, and sampling distinct regional food and drink. Cultural landscapes also vary widely, with diverse house types, settlement forms, and field patterns characterizing the everyday scene. Indeed, much of Europe's regional cohesion and identity comes from a shared history that has unfolded in close geographic proximity (see "Setting the Boundaries").

Though the traveler may revel in Europe's cultural and environmental diversity, these regional differences are also entangled with Europe's troubled past, a history of neighbors warring with each other because of the very same cultural and political differences. It is often said that Europe invented nationalism and its political expression, the nation-state. But nationalism has also been Europe's Achilles heel, leading it into devastating wars and destructive regional rivalries. In the twentieth century alone, Europe was the principal battleground of two world wars, followed by a 40-year **Cold War** that divided the continent and the world into two hostile, highly armed camps—Europe and the United States against the former Soviet Union.

Today, however, a spirit of cooperation prevails as Europe sets aside nationalistic agendas and works toward regional economic, political, and cultural integration through the **European Union (EU).** This supranational organization is made up of 15 countries, anchored by the western European states of Germany, France, Italy, and the United Kingdom. Because of the EU's demonstrated success, many countries want to join the organization, including most of the former Soviet satellites in eastern Europe. Undoubtedly, the geographical reach and economic policies of the EU will continue to transform the region during the twenty-first century.

Some Europeans, however, are not convinced that the EU is the answer to Europe's problems. In Norway, for example, the government has voted to join the EU on two occasions,

◄ **Figure 8.1 Europe** Stretching from Iceland in the Atlantic to the Black Sea, Europe includes 37 countries, ranging in size from large states, such as France and Germany, to the microstates of Liechtenstein, Andorra, San Marino, and Monaco. Currently the population of the region is about 523 million. Europe is highly urbanized and, for the most part, affluent, particularly the western portion. However, economic and social disparities between eastern and western Europe remain a problem.

SETTING THE BOUNDARIES

The European region is small compared to the United States. In fact, Europe from Iceland to the Black Sea would fit easily into the eastern two-thirds of North America. A more apt comparison would be Canada, since Europe, too, is a northern region; more than half of Europe lies north of the 49th parallel, the line of latitude forming the western border between the United States and Canada (Figure 8.3).

Europe currently contains 37 countries, and that number may grow in the near future. These states range in size from large countries, such as France and Germany, to microstates, such as Liechtenstein, Andorra, Monaco, and San Marino. Currently the population for these combined states totals more than 523 million people. Most of these states are fully independent, with their own sovereign governments and unique histories. However, tensions exist between and within many of these varied political units. Also at issue is regional European integration and cooperation.

The notion that Europe is a continent with clearly defined boundaries is a misconception rooted in history. The Greeks and Romans divided their worlds into the three continents of Europe, Asia, and Africa separated by the Mediterranean Sea, the Red Sea, and the Bosporus Straits. A northward extension of the Black Sea was thought to separate Europe from Asia, and only in the sixteenth century was this proven false. Instead, explorers and cartographers discovered that the "continent" of Europe was firmly attached to the western portion of Asia.

Since that time, geographers have not agreed on the eastern boundary of Europe. During the existence of the Soviet Union, some took the "continent's" border all the way east to the Ural Mountains, while others drew the line at the western boundary of the Soviet Union. However, with the disintegration of the Soviet Union in 1990, the eastern boundary of Europe became even more problematic. Now some geography textbooks extend Europe to the border with Russia, which places the two countries of Ukraine and Belarus, former Soviet republics, in eastern Europe. Though we appreciate that an argument can be made for that expanded definition of Europe, recent events, along with a bit of crystal-ball gazing into the near future, lead us to reject that approach and instead to draw our eastern border at Poland, Romania, and Moldova. To the north, the three Baltic republics of Estonia, Lithuania, and Latvia are also included in Europe. Our justification is this: these six countries are currently clearly engaged with Europe and aspire to become even more closely linked to the West in the future. This is not the case with Ukraine and Belarus, which show a decidedly eastern orientation toward Russia.

only to have its population twice reject the idea in national referenda. Opponents of the EU argue that individual countries lose control over important sovereign matters; for some people—and a few states—EU membership carries more costs than benefits. Thus, tension exists between large-scale regional cooperation and integration and smaller-scale impulses toward retaining independence and autonomy, between a new and shared sense of "Europeaness" and traditional national and regional identities (Figure 8.2).

Europe, like most world regions, is caught up in the tension between globalization and national and local diversity. Given Europe's considerable impact on the rest of the world as the hearth of Western civilization, the cradle of the Industrial Revolution, and the home of global imperialism, many would argue that Europe actually invented globalization. But while all world regions struggle with this vexing problem of trading off global convergence against national interests, Europe finds itself with an added layer of complexity as it moves into the new and untested waters of economic, political, and cultural integration. Europe is truly at a juncture: either it will succeed dramatically by moving past its historical legacy of nationalistic in-fighting, or it will fail miserably by fracturing further into small ethnic-based divisions.

▲ **Figure 8.2 Anti-EU protest** Farmers in western England burn a European Union flag during an anti-EU demonstration. These farmers are protesting the EU ban on British beef imports into continental Europe because of BSE/Mad Cow disease. Protests against EU agricultural policies are fairly common throughout Europe as farmers struggle against the submergence of local ways to European integration. *(AP/Wide World Photos)*

Environmental Geography: Human Transformation of a Diverse Landscape

Despite its small size, Europe's environmental diversity is extraordinary. Within its borders are found a startling array of landscapes from the Arctic tundra of northern Scandinavia to the barren hillsides of the Mediterranean islands, and from the explosive volcanoes of southern Italy to the glaciated seacoasts of western Norway.

Four factors explain this environmental diversity. First, the complex geology of this western extension of the Eurasian landmass has produced some of the newest, as well as the oldest, landscapes in the world. Second, Europe's latitudinal extent creates opportunities for diversity, since the region extends from the Arctic to the Mediterranean subtropics (Figure 8.3). Third, these latitudinal controls are further modified by the interaction of land and sea as the Atlantic Ocean and Black, Baltic, and Mediterranean seas shape regional climates and bioregions. Last, the long history of human settlement has transformed and modified Europe's natural landscapes in fundamental ways over thousands of years. A million years ago the first fire-bearing hunter-gatherers arrived; while their imprint on the landscape was minimal, this was not the case when Neolithic agriculturists arrived in Europe about 7,000 years ago. From that time on, human hands have left a major imprint on the European environment.

Environmental Issues, Local and Global, East and West

Because of its long history of agriculture, resource-extraction, industrial manufacturing, and urbanization, Europe has its share of serious environmental problems. Compounding the situation is the fact that pollution rarely respects political boundaries. Air pollution from England, for example, creates serious acid-rain problems in Sweden, and water pollution of the upper Rhine River by factories in Switzerland creates major problems for the Netherlands, where Rhine River water is used for municipal drinking supplies. When environmental problems cross national boundaries, solutions must come from intergovernmental cooperation (Figure 8.4).

Since the 1970s when the European Union (EU) added environmental issues to its economic and political agenda, western Europe has been increasingly effective in addressing its varied environmental problems with regional solutions. Besides the more obvious environmental problems of air and water pollution, the EU has also taken the lead on matters of recycling, waste management, reduced energy usage, and sustainable resource use. As a result, western Europe is probably the "greenest" of the major world regions. This heightened environmental sensitivity is also found in the political arena, where the Green Party regularly gets about 10 percent of the vote and in some areas, such as southern England, close to 20 percent. In Germany this environmental party recently shared national power with the Social Democrats in what was

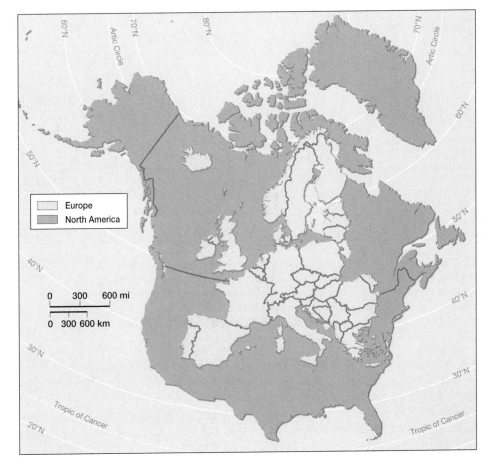

◀ Figure 8.3 Europe: Size and northerly location   Europe is about two-thirds the size of North America, as shown in this cartographic comparison. Another important characteristic is the northerly location of the region, which affects its climate, vegetation, and agriculture. Much of Europe lies at the same latitude as Canada; even the Mediterranean lands are farther north than the U.S.–Mexico border.

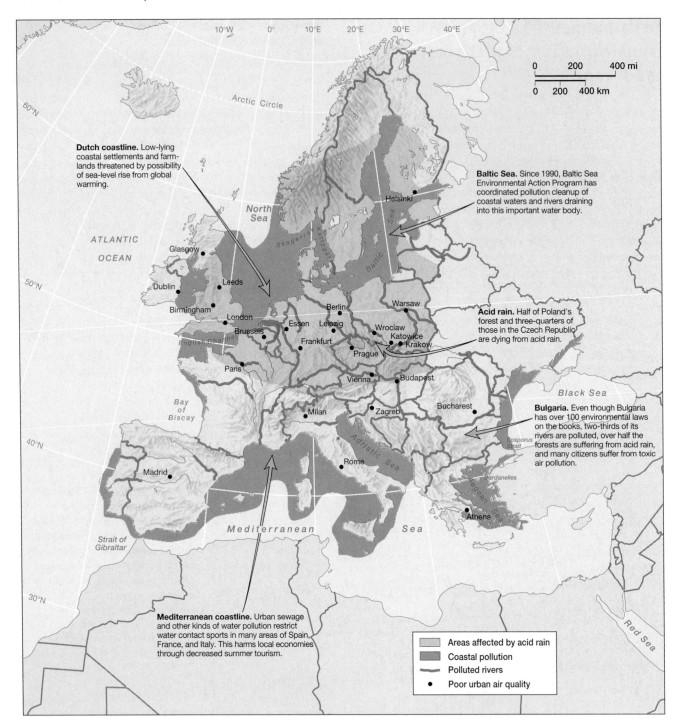

Dutch coastline. Low-lying coastal settlements and farmlands threatened by possibility of sea-level rise from global warming.

Baltic Sea. Since 1990, Baltic Sea Environmental Action Program has coordinated pollution cleanup of coastal waters and rivers draining into this important water body.

Acid rain. Half of Poland's forest and three-quarters of those in the Czech Republic are dying from acid rain.

Bulgaria. Even though Bulgaria has over 100 environmental laws on the books, two-thirds of its rivers are polluted, over half the forests are suffering from acid rain, and many citizens suffer from toxic air pollution.

Mediterranean coastline. Urban sewage and other kinds of water pollution restrict water contact sports in many areas of Spain, France, and Italy. This harms local economies through decreased summer tourism.

Legend:
- Areas affected by acid rain
- Coastal pollution
- Polluted rivers
- • Poor urban air quality

▲ **Figure 8.4 Environmental issues in Europe**   In terms of environmental protection, there is a major disparity between western and eastern Europe. While the West has worked energetically over the last 30 years to solve problems such as air and water pollution, those same problems are still widespread in the East because of the long environmental neglect coincidental with the communist period. Because Europe is composed of a fabric of relatively small nation-states, most environmental problems must be solved at the regional level, rather than by each country alone.

referred to as a "Red-Green Coalition." Voter support for the environment is also expressed on the international scene. The European Union has become an aggressive advocate and world leader for the reduction of atmospheric pollutants responsible for global climate change. More specifically, as an energy efficient group of countries, the EU has pushed other industrial- ized countries—including the United States—to reduce green- house emissions and attain the same level of energy efficien- cy. Given the Bush administration's skeptical position on greenhouse gas reductions, political tensions have increased between the EU and the United States over these internation- al environmental agreements.

While western Europe is successfully addressing environmental issues in that part of the region, the situation is decidedly grimmer in eastern Europe. During the period of Soviet economic planning (1945–90), little attention was paid to environmental issues because of an overt emphasis on short-term industrial output. Additionally, since communist economics had no way to calculate environmental costs, they were not a concern. As a result, the environment was not an issue; consequently, there were few controls on air and water pollution, environmental safety and health, and the dumping of toxic and hazardous wastes.

Unfortunately, the contemporary costs of those historical decisions have been extraordinarily high whether measured in environmental, human, or monetary terms. For example, 90 percent of Poland's rivers have no aquatic or plant life, and more than 50 percent of the country's forest trees show signs of damage from air pollution. Humans are also suffering. Fully one-third of Poland's population is expected to suffer from an environmentally induced disease such as cancer or respiratory illness. Sadly, the nation's own Polish Academy of Sciences pronounced their country the most polluted in the world.

Nearby countries in eastern Europe fare no better. In the neighboring Czech Republic, almost three-quarters of the forests are dying from air pollution and acid rain, though in all fairness it should be noted that some of this airborne pollution originated in the industrial countries to the west (Figure 8.5). Nevertheless, the air pollution is so bad in the Czech industrial heartland that human life expectancy is 11 years less than the national average. Though Bulgaria had about 100 environmental laws on its books during the communist period, they were never enforced. As a result, two-thirds of its rivers are badly polluted today, and almost half of its trees show environmental damage. Around the uranium plant in Rokovski, Bulgaria, the death rate for children is three times the national average.

▲ **Figure 8.5 Acid rain and forest death**   Acid precipitation has taken a devastating toll on eastern European forests, such as those shown here in Bohemia, the Czech Republic. In this country, three-quarters of the forests are dead or injured from acid precipitation, which is caused by industrial and auto emissions. *(Karol Kallay/Bilderberg Archiv der Fotografen)*

Ill-designed and poorly maintained Soviet-built nuclear power plants are also common in the east, threatening all of Europe with a repeat of the 1986 Chernobyl disaster in Ukraine. At that time, the meltdown of a nuclear reactor sent radioactive contamination into the atmosphere that not only polluted the immediate area, but was also carried into western and northern Europe by wind circulating around a high-pressure system. Schoolchildren in Austria were kept indoors for a week as a safety measure. Dairy supplies were irradiated by fallout, as were vegetables and other food supplies, which were subsequently destroyed. In Scandinavia, radioactive fallout was absorbed by lichens and moss, the fodder for large reindeer herds that provide a livelihood for Sami peoples of the Arctic region. For several years reindeer meat, a major food source, was unacceptably high in radioactivity, causing widespread misery and deprivation for the Sami. Another nuclear meltdown of a Chernobyl-type reactor took place in Lithuania in 1992, this time resulting in limited damage because of a different wind pattern that did not spread the radiation.

Solutions to eastern Europe's environmental problems will not come easily because of several complicating factors. First, most eastern European states are still struggling with their post-1990 economic and political transitions, and it is unclear whether these new governments will either enact or enforce the necessary environmental controls. Instead, there is widespread concern that environmental regulations will inhibit economic recovery by burdening industry with higher operating costs.

Second, there is increasing evidence that capital is not flowing into certain areas or specific industrial sites because western businesses fear they will be charged for cleaning up toxic dumps or repairing environmental damage. Until eastern European governments show a willingness to share those cleanup costs, capital will stay away from these polluted areas and move instead into less-damaged regions (Figure 8.6).

Third, pollution control is intertwined with the privatization of former state-owned industries. Communist state-owned factories that had no wastewater treatment facilities or air emission controls caused many environmental problems. To solve the problems, pollution controls must now be added. However, this will add to overhead costs as these industries are moved into the private sector.

Last, as is the case in western Europe, regional solutions that transcend political borders are necessary. Water pollution problems in the Danube River, for example, must be solved through cooperative agreements between the different states of the lower drainage basin, including Hungary, Yugoslavia, Bulgaria, and Romania. Similarly, regional air pollution problems must also be resolved through cooperation of different states.

Hope that regional cooperation can be extended into eastern Europe comes from recent action on the Baltic Sea, a large brackish sea that is a major nursery for fish species. Before 1989, the **Iron Curtain** (the ideological and political border between democracy and communism) prevented cooperative solutions to pollution problems. As a result, measures taken

▲ Figure 8.6 Toxic landscape in Romania   A troublesome legacy of Soviet communism in eastern Europe are the numerous toxic dump sites and polluted landscapes found in the region, such as this site in Romania. Not only do these environmental problems threaten public health, but the prospect of high cleanup costs also inhibits investment by Western firms.   *(Filip Horvat/ Corbis/SABA Press Photos, Inc.)*

by Denmark, Sweden, Finland, and Western Germany to control water pollution were largely offset by unchecked water pollution from East Germany, Poland, and other Soviet states. A double standard in environmental pollution existed, to the detriment of Baltic ecosystems. In 1990, however, immediately after the fall of communism in eastern Europe and with the unification of Germany, a new Baltic Sea Joint Comprehensive Environmental Action Program was founded by all states sharing a shoreline with or drainage into the Baltic. Much progress has been made since then, as new standards and enforcement regulations have been put into place. Financial resources from the World Bank, along with the European Bank for Reconstruction and Development, have helped eastern European countries, such as Poland, Estonia, Belarus, and even Russia, build sewage treatment plants, protect coastal wetlands, and clean up hazardous waste sites around the Baltic rim. This cooperative effort gives hope that other eastern European environmental problems will also be resolved in the early twenty-first century.

Landform and Landscape Regions

European landscapes can be organized into four general topographic regions. First, the European Lowland forms an arc from southwest France to the northeast plains of Poland and includes southeastern England. Second, the Alpine mountain systems extend from the Pyrenees in the west to the Balkan mountains of southeast Europe. Third, the Central Uplands

are positioned between the Alps and the European Lowland, stretching from France into eastern Europe. Last, the Western Uplands include mountains in Spain, portions of the British Isles, and the highlands of Scandinavia (Figure 8.7).

The European Lowland   This lowland (also known as the North European Plain) is the unquestionable focus of western Europe with its high population density, intensive agriculture, large cities, and major industrial regions. Though not completely flat by any means, most of this lowland lies below 500 feet (150 meters) in elevation, though it is broken in places by rolling hills, plateaus, and uplands (such as in Brittany, France), where elevations exceed 1,000 feet (300 meters). Many of Europe's major rivers, such as the Rhine, the Loire, the Thames, and the Elbe, meander across this lowland and form broad estuaries before emptying into the Atlantic. Several of Europe's great ports inhabit these strategic lowland settings, including London, Le Havre, Rotterdam, and Hamburg.

The Rhine River delta conveniently divides the unglaciated lowland to the south from the glaciated plains to the north, which were covered by a Pleistocene ice sheet until about 15,000 years ago. Because of these continental glaciers, the North European Lowland, including Netherlands, Germany, Denmark, and Poland, is far less fertile for agriculture than the unglaciated portion in Belgium and France (Figure 8.8). Rocky clay materials in Scandinavia were eroded and transported south by glaciers. As the glaciers later retreated with a warming climate, piles of glacial debris known as **moraines** were left on the plains of Germany and Poland. Elsewhere in the north, glacial meltwater created infertile outwash plains that have limited agricultural potential. Accordingly, the irregular pattern of drainage in the postglacial period left a landscape of lakes, marshlands, and bogs that severely limits agriculture in the Baltic areas, particularly when contrasted with the unglaciated and more fertile areas of the southern European Lowland.

The Alpine Mountain System   The Alpine Mountain System forms the topographic spine of Europe. It consists of a series of east–west-running mountains from the Atlantic to the Black Sea and the southeastern Mediterranean. Though these mountain ranges carry distinct regional names, such as the Pyrenees, Alps, Carpathians, Dinaric Alps, and Balkan Ranges, they have similar geologic traits. All were created more recently (about 20 million years ago) than other upland areas of Europe, and all are constructed from a complex arrangement of rock types.

The *Pyrenees* form the political border between Spain and France (including the microstate of Andorra). This rugged range extends almost 300 miles (480 kilometers), stretching from the Atlantic to the Mediterranean. Within the mountains, glaciated peaks reaching to 11,000 feet (3,350 meters) alternate with broad glacier-carved valleys. The Pyrenees have long been home to the Basque people in the western reaches as well as to distinctive Catalan-speaking minorities in the east. Insurrection and separatism produce a fascinating, but often violent, human geography in these two mountainous areas.

▲ Figure 8.7 Physical geography of Europe   Much of the mountain and upland topography of Europe is a function of the gradual northward movement of the African tectonic plate into the Eurasian Plate. These tectonic forces have created the east-west trending Alpine mountain chain. Besides these tectonic forces, Pleistocene glaciation has also shaped the European region. Until about 15,000 years ago, much of the region was covered by continental glaciers that extended south to the mouth of the Rhine River. The southern extent of this glaciated area is shown by the dotted line on the map.

The centerpiece of the larger geologic system is the prototypical Alpine range itself, the *Alps*, reaching more than 500 miles (800 kilometers) from France to eastern Austria. These impressive mountains are highest in the west, reaching more than 15,000 feet (4,575 meters) in Mt. Blanc on the French-Italian border; in Austria, to the east, few peaks exceed 10,000 feet (3,050 meters). Though easily crossed today by car or train through a system of long tunnels and valley-spanning bridges, these mountains have formed an important cultural divide between the Mediterranean lands to the south and central and western Europe in the north. The Alps also restrict the Mediterranean's distinctive climate and bioregions from spilling over into western and central Europe.

▲ **Figure 8.8 The European Lowland**   Also known as the North European Plain, this large lowland extends from southwestern France to the plains of northern Germany and into Poland. Although this landform region has some rolling hills, most of it is less than 500 feet (150 meters) in elevation.   *(P. Vauthey/ Corbis/Sygma Photo News)*

The *Appenine* Mountains are located south of the Alps; however, the two ranges are physically connected by the hilly coastline of the French and Italian Riviera. Forming the mountainous spine of Italy, the Appenines are generally lower and lack the spectacular glaciated peaks and valleys of the true Alps. Farther to the south, the Appenines take on their own distinct character with the impressive and explosive volcanoes of Mt. Vesuvius (just over 4,000 feet, or 1,200 meters) outside of Naples and the much higher (almost 11,000 feet, or 3,350 meters) Mt. Etna off Italy's toe on the island of Sicily. These two active volcanoes give clues to the geophysical heritage of the Alpine mountain chain as the tectonic meeting ground of the African and Eurasian plates. The northward movement of the African Plate against the Eurasian Plate created both the Mediterranean basin and the Alpine mountain system that forms its northern rim.

To the east, the *Carpathian* Mountains define the limits of the Alpine system in eastern Europe. They are a plow-shaped upland area that extends from eastern Austria to the Iron Gate gorge, where the borders of Romania and Yugoslavia intersect, and they create a formidable passage for Danube River traffic. About the same length as the main Alpine chain, the Carpathians are not nearly as high. The highest summits in Slovakia and southern Poland are less than 9,000 feet (2,780 meters).

In southeastern Europe, the complicated tail of the Alpine system is expressed in a series of uplands with several distinct regional names. The *Dinaric* Alps fringe the Adriatic Sea to the west in Croatia, Bosnia, and Yugoslavia. To the southeast are the *Balkan Ranges*, which trend both north–south and east–west, reaching almost to the Black Sea in Bulgaria. Finally, the *Rhodope*

Mountains occupy northern Greece and Macedonia. Elevations throughout these mountain ranges are generally less than 10,000 feet (3,050 meters).

Central Uplands   In western Europe, a much older highland region occupies an arc between the Alps and the European Lowland in France and Germany. These mountains are much lower in elevation than the Alpine system, with their highest peaks at 6,000 feet (1,830 meters). Dated to about 100 million years ago, much of this upland region is characterized by rolling landscapes of about 3,000 feet (less than 1,000 meters). Their importance to western Europe is great because they contain the raw materials for Europe's industrial areas. In both Germany and France, for example, these uplands have provided the iron and coal necessary for each country's steel industry; in the eastern reaches, mineral resources from the Bohemian highlands have also fueled major industrial areas in Germany, Poland, and the Czech Republic.

Western Highlands   The Western Highlands define the western edge of the European subcontinent, extending from Portugal in the south, through the portions of the British Isles in the northwest, to the highland backbone of Norway, Sweden, and Finland in the far north. These are Europe's oldest mountains, formed about 300 million years ago. Their unity as a landform region, however, comes more from similar topography and elevation than from a uniform geology.

As with other upland areas that traverse many separate countries, specific place-names for these mountains differ from country to country. A portion of the Western Highlands form the highland spine of England, Wales, and Scotland, where picturesque glaciated landscapes are found at modest elevations of 4,000 feet (1,220 meters) or less. These U-shaped glaciated valleys are also appear in Norway's uplands, where they produce a spectacular coastline of **fjords**, or flooded valley inlets similar to the coastlines of Alaska and New Zealand.

Though less elevated, the Fenno-Scandian Shield of Sweden and northern Finland is noteworthy because it is made up of some of the oldest rock formations in the world, dated conservatively at 600 million years. This **shield landscape** was eroded to bedrock by Pleistocene glaciers and, because of the cold climate and sparse vegetation, still has extremely thin soils that severely limit agricultural activity (Figure 8.9). As with other heavily glaciated areas, such as the Canadian Shield of North America, numerous small lakes dot the countryside, giving clues to the impressive erosional power of ice sheets.

Europe's Climates

Three principal climates characterize Europe (Figure 8.10). Along the Atlantic coast, a moderate and moist maritime climate dominates, modified by oceanic influences. Farther inland, continental climates prevail, with hotter summers and colder winters. Finally, dry-summer Mediterranean climates are found in southern Europe, from Spain to Greece. An extensive area of warm-season high pressure inhibits summer storms and rainfall in this region, creating the seemingly endless blue skies so attractive to tourists from northern Europe.

North Atlantic Drift.

◄ Figure 8.9 Northern landscapes   Northern Europe is a
harsh land characterized by landscapes with little soil, glaciated
rock expanses, sparse vegetation,
and thousands of lakes. Pleistocene
glaciers sculpted this region
20,000 years ago, and its high-latitude location maintains an Ice Age
character that includes a very short
growing season.   (Macduff Ever-
ton/Corbis)

One of the most important climate controls is that of the Atlantic Ocean. Though most of Europe is at a relatively high latitude (London, England, for example, is slightly farther north than Vancouver, British Columbia), the mild North Atlantic current, which is a continuation of the warm Atlantic Gulf Stream, moderates coastal temperatures from Norway to Portugal and even inland to the western reaches of Germany. As a result, this maritime influence gives Europe a climate 5 to 10 °F (2.8 to 5.7 °C) warmer than comparable latitudes without this oceanic effect.

In the **Marine west coast climate** region, no winter months average below freezing, though cold rain, sleet, and an occasional blizzard are common winter visitors. Summers are often cloudy and overcast with frequent drizzle and rain as moisture flows in from the ocean. Ireland, the Emerald Isle, offers an appropriate snapshot of this maritime climate.

With increasing distance from the ocean (or where a mountain chain limits the maritime influence, as in Scandinavia), landmass heating and cooling becomes a strong climatic control, producing hotter summers and colder winters. Indeed, all **continental climates** average at least one month below freezing during the winter. In Europe, the transition between maritime and continental climates takes place close to the Rhine River border of France and Germany. Farther north, although Sweden and other nearby countries are close to the moderating influence of the Baltic Sea, high latitude and the blocking effect of the Norwegian mountains produces cold winter temperatures characteristic of continental climates. Precipitation in continental climates comes as rain from summer frontal systems and local thundershowers. Usually this rainfall is sufficient for nonirrigated agriculture, though supplemental watering is increasingly common where high-value crops are grown.

The **Mediterranean climate** is characterized by a distinct dry season during the summer, which results from the warm season expansion of the Atlantic (or Azores) high-pressure area. This high pressure is produced by the global circulation of air warmed in the equatorial tropics that subsides or descends between latitudes 30° and 40°, thus inhibiting summer rainfall. This same phenomenon also produces the Mediterranean climates of California, western Australia, parts of South Africa, and Chile. While these rainless summers may attract tourists from northern Europe, the seasonal drought can be problematic for agriculture. It is no coincidence that traditional Mediterranean cultures, such as the Arab, Moorish, Greek, and Roman, have been major innovators of irrigation technology (Figure 8.11).

Of course, within these three major climate types there are significant variations. For example, in the high-latitude Scandinavian countries, where there is a transition from maritime to continental climates, short (or even nonexistent) winter days give the region its unique climatological character. Even farther north in Scandinavia, these climates are replaced by Arctic tundra climates, characterized by low precipitation, long, cold winters, and the briefest of summers.

In the Alpine mountain chain, elevation and terrain shape local variation in temperature and precipitation. Further, these mountains act as a barrier between the maritime and continental climates of the north and the Mediterranean zone to the south. Crossing the Alps in March, for example, via a high mountain road (if it is free of snow) or an underground tunnel quickly takes a traveler from a northern winter of snow and mud into the brilliant sunshine and blossoming vegetation of southern Europe.

Seas, Rivers, Ports, and Coastline

Europe remains a maritime region with strong ties to its surrounding seas. Even its landlocked countries, such as Austria and the Czech Republic, have access to the ocean through an interconnected network of navigable rivers and canals.

Europe's Ring of Seas   Five major seas encircle Europe; these water bodies are connected to each other through narrow straits with strategic importance for controlling waterborne trade and naval movement (Figure 8.7). In the north, the Baltic Sea separates Scandinavia from north-central Europe. Denmark and Sweden have long commanded the narrow Skagerrak and Kattegat straits, which connect the Baltic to the North Sea. Besides its historic role as a major fishing ground, the North Sea is now well known for its rich oil and natural gas fields mined from deep-sea drilling platforms.

The English Channel (in French, *La Manche*) separates the British Isles from continental Europe. At its narrowest point, the Dover Straits are only 20 miles (32 kilometers) wide. Though England has thought of the Channel as a protective

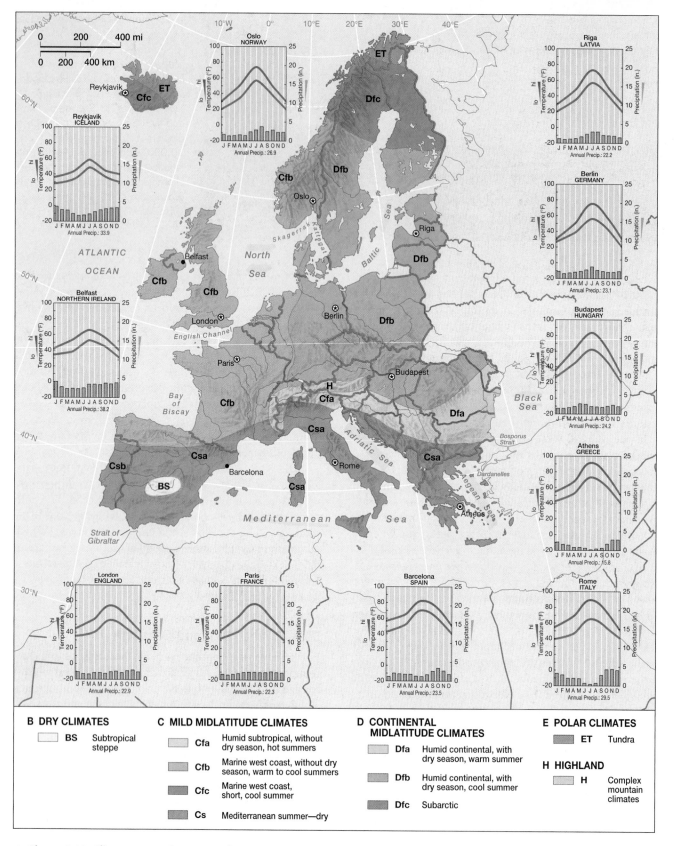

▲ **Figure 8.10 Climate map of Europe** Three major climate zones dominate Europe. Close to the Atlantic Ocean the Marine west coast climate is found, with cool seasons and steady rainfall throughout the year. Farther inland, continental climates are found. They have at least one month averaging below freezing, and they have hot summers, with a precipitation maximum falling during the summer season. The dry summer Mediterranean climate is found in southern Europe. Unlike the case in continental climates, most precipitation falls during the cool winter period in the Mediterranean region.

▲ **Figure 8.11 Mediterranean agriculture** Because of water scarcity during the hot, dry summers in the Mediterranean climate region, local farmers have evolved agricultural strategies for making the best use of soil and water resources. In this photo from Portugal, the terraces used to prevent soil erosion on steep slopes and the mixture of tree and ground crops are evident. *(Getty Images, Inc.)*

▲ **Figure 8.12 Polder landscape** Diked agricultural settlements, or polders, are situated along coastal Netherlands. Because these lands, such as the fields on the left side of this photo, are reclaimed from the sea, many are at or below sea level. The English Channel and North Sea are to the right. *(Adam Woolfitt/Woodfin Camp & Associates)*

moat, it has primarily been a symbolic barrier for it deterred neither the French Normans from the continent nor Viking raiders from the north. Since 1993, after decades of resistance, England has been connected to France through the 31-mile (50-kilometer) Eurotunnel and a high-speed rail system carrying passengers, autos, and freight.

Gibraltar guards the narrow straits between Africa and Europe at the western entrance to the Mediterranean Sea. Britain's stewardship of this passage remains an enduring symbol of its once great sea-based empire. Finally, on Europe's southeastern flanks are the Straits of Bosporus and the Dardanelles, the narrows connecting the eastern Mediterranean with the Black Sea. Disputed for centuries, these pivotal waters are now controlled by Turkey. Though these straits are often used as a physical boundary between Europe and Asia, they are easily bridged in several places to facilitate truck and train transportation within Turkey and between Europe and Southwest Asia.

Rivers and Ports Europe is also a region of navigable rivers connected by a system of canals and locks that allow inland barge travel from the Baltic and North seas to the Mediterranean, and between western Europe and the Black Sea. Many rivers on the European Lowland, such as the Loire, Seine, Rhine, Elbe, and Vistula, flow into Atlantic or Baltic waters. However, the Danube, Europe's longest river, flows east and south, rising in the Black Forest of Germany only a few miles from the Rhine River and running southeastward to the Black Sea. It offers a connecting artery between central and eastern Europe. Similarly, the Rhône headwaters rise close to those of the Rhine in Switzerland, yet it flows southward into the Mediterranean. Both the Danube and the Rhône are connected by locks and canals with the rivers of the European Lowland, making it possible for barge traffic to travel between all of Europe's fringing seas and oceans.

Major ports are found at the mouths of most western European rivers, serving as transshipment points for inland wa-terways as well as focal points for rail and truck networks. From south to north, these ports include Bordeaux at the mouth of the Garonne, Le Havre on the Seine, London on the Thames, Rotterdam (the world's largest port in terms of tonnage) at the mouth of the Rhine, Hamburg on the Elbe River, and, to the east in Poland, Szcezin on the Oder, and Gdansk on the Wisla. Of the major Mediterranean ports, only Marseilles, France, is close to the mouth of a major river, the Rhône. Other modern-day ports, such as Genoa, Naples, Venice, and Barcelona, are some distance from the delta harbors that served historic trade. Apparently, as Mediterranean forests were cut in past centuries, erosion on the hillslopes carried sediment down the rivers to the delta regions, effectively filling in the historic ports used by the Greeks and Romans.

Reclaiming the Dutch Coastline Much of the Netherlands landscape is a product of the people's long struggle to protect their agricultural lands against coastal and river flooding. Beginning around A.D. 900, dikes were built to protect these fertile but low-lying lands against periodic flooding from the nearby Rhine River, as well as from the stormy North Sea. By the twelfth century, these protected and reclaimed landscapes were known as **polders**, or diked agricultural settlements, a term that is still used today to describe coastal reclamation. While windmills had long been used to grind grain in the low countries, this wind power was also employed to pump water from low-lying wetlands. As a result, windmills became increasingly common on the Dutch landscape to drain marshes. This technology worked so well that the Dutch government embarked on a widespread coastal reclamation plan in the seventeenth century that converted an 18,000-acre (7,275-hectare) lake into agricultural land (Figure 8.12).

With steam- (and later, electric-) powered pumps, even more ambitious polder reclamation was possible, culminating in the massive Zuider Zee project of the twentieth century, in which the large bay north of Amsterdam was dammed, drained, and converted to agricultural lands over the course of a half century. This project both alleviated a major flooding hazard and opened up new lands for farming and settlement in the heart of the Netherlands. New suburbs of Amsterdam are now spilling onto these reclaimed lands.

For the last several decades, major efforts have been directed toward curbing flooding where the Rhine **distributaries,** or delta channels, flow into the English Channel. After a disastrous flood in 1953, the Dutch government conceived the Delta Project, which has constructed four large sea dikes that essentially keep the many mouths of the Rhine River within their channels. By all accounts, these projects have succeeded in stabilizing the North Sea coast and providing lowland Europe with more area in its densely settled heart for both rural and urban settlement.

Settlement and Population: Slow Growth and Rapid Migration

Population Density in the Core and Periphery

The map of Europe's population distribution (Figure 8.13) shows that, in general, population densities are higher in the historical industrial core areas of western Europe (England, the Netherlands, northern France, northern Italy, and western Germany) than in the periphery to the east and north. While this generalization overlooks important urban clusters in Mediterranean Europe, it does convey an accurate sense of a densely settled European core set apart from a more rural, agricultural periphery. Much of this distinctive population pattern is linked to areas of early industrialization, and there are many contemporary consequences of this core-periphery distribution. For example, economic subsidies from the affluent, highly urbanized core to the less affluent, agricultural periphery have been central to the EU's development policies for several decades. Further, while the urban-industrial core is characterized by extremely low natural growth rates, it is also the target area for migrants—both legal and illegal—from Europe's peripheral countries as well as from outside Europe.

Natural Growth: Beyond the Demographic Transition

Probably the most striking characteristic of Europe's population is its continued slow natural growth (Table 8.1). More to the point, in many European countries the death rate exceeds the birthrate, meaning that there is simply no growth at all. Instead, many countries are experiencing negative growth rates; were it not for in-migration from other countries and other world regions, these countries would record a decline in population over the next few decades. Italy, for example, currently has a population of 57.8 million. Yet if current natural growth holds true for 20 years and is not offset by immigration, the population will decrease to 55 million by the year 2025.

There seem to be several causes for zero population growth in western Europe. First of all, recall from Chapter 1 that the concept of the demographic transition was formulated from the historical change in European growth rates as the population moved from rural settings to more urban and industrial locations. What we see today is an extension of that model, namely, the continued expression of the low fertility–low mortality of the fourth stage of the demographic transformation. Some demographers suggest adding a fifth stage to the model—a "post-industrial" phase in which population falls below replacement levels. Evidence for this explanation comes from the highly urbanized and industrialized populations of Germany, France, and England, all of which are below zero natural population growth.

These low-birthrate countries have all attempted to increase their rate of natural growth. In Germany, for example, couples are given an outright gift of $650 if they produce a child. In Austria, the government gives a modest cash award to couples when they marry, and still more follows if they give birth. France provides monthly payments to families with small children that last until the children reach school age. Despite these incentives, most European countries still have natural growth rates below replacement levels.

Additional explanations must be sought to explain the extraordinary negative growth in the eastern European countries. As seen in Table 8.1, the most striking examples of negative population growth are found in Bulgaria, Romania, Hungary, the Czech Republic, Latvia, Estonia, and Lithuania.

During the immediate postwar years, the growth of centralized planning and industrial development under the Soviet Union led to a labor shortage in eastern Europe, as well as in the Soviet Union itself. This was aggravated by the huge losses suffered during World War II. As a result, women were needed in the workforce. So that child-rearing would not conflict with jobs, Soviet policy was to make all forms of family planning and birth control available, including abortion. As a result, small families became the norm. Another factor was the widespread housing shortage that prevailed in most Soviet countries, a result of concentrating people in cities for industrial jobs, along with priorities placed on other forms of construction. Evidence from eastern Europe suggests that people experiencing a housing crunch tend to have few (if any) children.

When eastern European governments realized these policies were leading to negative population growth, several tried to reverse the trend. Perhaps the most notorious example was Romania, where the Ceausescu dictatorship outlawed all contraception and abortion. Reportedly, security police monitored pregnancies to make sure they reached full term. This was to no avail. Instead, many women died from illegal abortions, and the infant mortality rate was the highest in Europe, probably because of parental neglect. When the communist government fell in 1989, some 14,000 unwanted children were living in orphanages under questionable conditions. Although one of the first acts of the new government was to legalize both contraception and abortion, this act simply institutionalized the tendency toward a low birthrate.

Finally, there is the intangible factor of outlook and mood in contemporary eastern Europe as it influences birthrates and family planning. Uncertainty about the future is com-

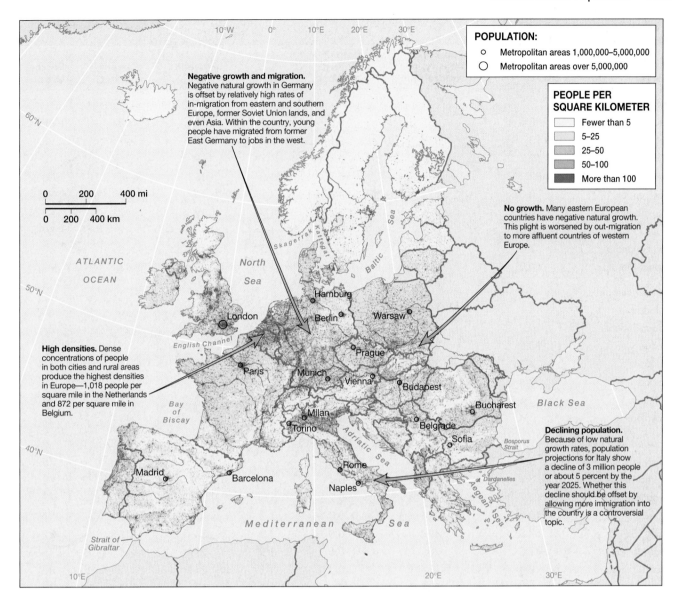

▲ **Figure 8.13 Population map of Europe**   The European region includes more than 523 million people, many of them clustered in large cities in both western and eastern Europe. As can be seen on this map, the most densely populated areas are in England, the Netherlands, Belgium, western Germany, northern France, and south across the Alps to northern Italy. Since most European countries have very little, if any, natural growth, in-migration is a major issue.

Refugees

monplace, as one quickly learns by talking with eastern Europeans. In many of these countries, jobs, wages, food, and housing are scarcer now than before 1989. Given these factors, it is understandable that the birthrate remains very low.

Migration to and Within Europe

Migration is one of the most challenging population issues facing Europe today because the region is caught in a web of conflicting policies and values (Figure 8.14). Currently there is widespread resistance to unlimited migration into Europe, ostensibly because of high unemployment in the western European industrial countries, but also because of an array of other concerns, ranging from foreign terrorism to dilution of

a country's dominant culture. Many Europeans argue that scarce jobs should go first to European citizens, not to "foreigners." However, should the economy improve, there could be a significant labor shortage that could be solved only through immigration. To illustrate, recent studies suggest that Germany would need about 400,000 in-migrants each year to provide an adequate labor force for its industry. Additionally, given the extraordinarily low rates of natural growth in most of Europe, coupled with the aging of the population, without immigration to fill the void most countries will experience severe shortfalls in the tax revenue needed to support social security and other programs demanded by an older population. Many demographers (and some politicians) believe an open-door immigration policy will be necessary to ensure

TABLE 8.1 *Demographic Indicators*

Country	Population (Millions, 2001)	Population Density, per square mile	Rate of Natural Increase	TFR[a]	Percent < 15[b]	Percent > 65	Percent Urban
Western Europe							
Austria	8.1	251	0.0	1.3	17	15	65
Belgium	10.3	872	0.1	1.6	17	17	97
France	59.2	278	0.4	1.9	19	16	74
Germany	82.2	597	−0.1	1.3	16	16	86
Liechtenstein	0.03	534	0.6	1.4	19	10	23
Luxembourg	0.4	446	0.4	1.7	19	14	88
Netherlands	16.0	1,018	0.4	1.7	19	14	62
Switzerland	7.2	453	0.2	1.5	18	15	68
United Kingdom	60.0	635	0.1	1.7	19	16	90
Eastern Europe							
Bulgaria	8.1	190	−0.5	1.2	16	16	68
Czech Republic	10.3	337	−0.2	1.1	17	14	77
Hungary	10.0	278	−0.4	1.3	17	15	64
Moldova	4.3	328	−0.1	1.4	24	9	46
Poland	38.6	310	0.0	1.4	20	12	62
Romania	22.4	243	−0.1	1.3	18	13	55
Slovakia	5.4	286	0.0	1.3	20	11	57
Southern Europe							
Albania	3.4	310	1.2	2.8	33	6	75
Bosnia-Herzegovina	3.4	173	0.4	1.6	20	8	72
Croatia	4.7	197	−0.2	1.4	20	12	77
Greece	10.9	214	0.0	1.3	15	17	81
Italy	57.8	497	0.0	1.3	14	18	82
Macedonia	2.0	205	0.5	1.9	23	10	75
Malta	0.4	3,157	0.3	1.7	21	12	80
Portugal	10.0	282	0.1	1.5	17	15	79
San Marino	0.03	1,166	0.4	1.3	15	16	89
Slovenia	2.0	256	−0.1	1.2	16	14	50
Spain	39.8	204	0.0	1.2	15	17	64
Yugoslavia (Serbia)	10.7	270	0.1	1.6	21	13	52
Northern Europe							
Denmark	5.4	322	0.2	1.7	18	15	72
Estonia	1.4	78	−0.4	1.3	18	14	69
Finland	5.2	40	0.2	1.7	18	15	60
Iceland	0.3	7	0.8	2.0	23	12	93
Ireland	3.8	142	0.6	1.9	22	11	58
Latvia	2.4	95	−0.6	1.2	18	15	69
Lithuania	3.7	147	−0.1	1.3	20	13	68
Norway	4.5	36	0.3	1.8	20	15	74
Sweden	8.9	51	0.0	1.5	19	17	84

[a]Total fertility rate

[b]Percent of population younger than 15 years of age

Source: Population Reference Bureau. World Population Data Sheet, *2001.*

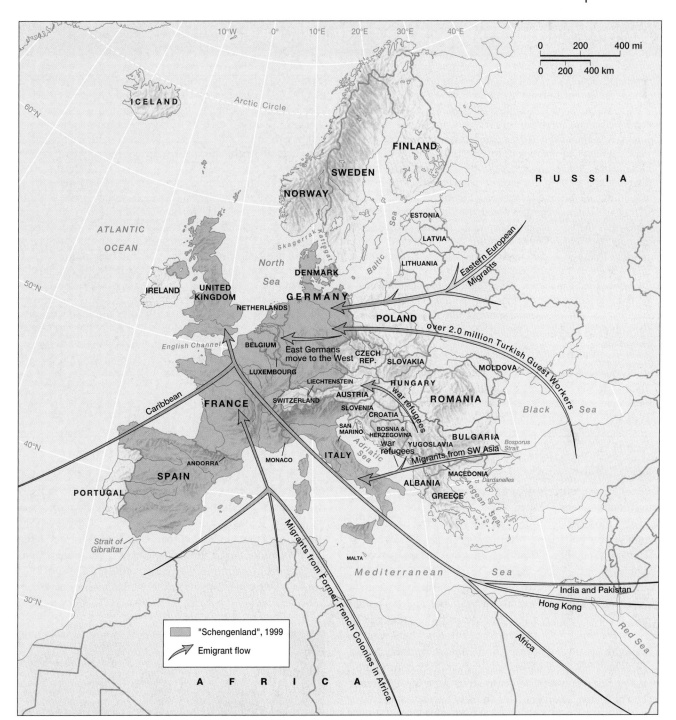

▲ **Figure 8.14 Migration into Europe**   A labor shortage during the reconstruction of Europe following World War II opened the door for a large number of guest workers to migrate into Germany. In the initial stage, these workers came from Italy and Yugoslavia; later they came from Turkey and Greece. Most recently, economic stagnation in Europe's heartland has slowed the migration of guest workers. However, immigrants continue to migrate into Europe from former colonial countries in Asia and Africa as well as from eastern Europe and the war-torn Balkans.

Europe's economic vitality in the next decade. Complicating the issue, though, is the social and political unease linked to the presence of large numbers of foreigners within European countries that are historically culturally and ethnically homogenous. In the recent past, immigration policies were decided by each individual country. Today, however, the topic has become so controversial that the European Union (EU)

has committed itself to forging a common immigration policy by the year 2004. During the 1960s postwar recovery when western Europe's economies were booming, many countries looked to migrant workers to alleviate labor shortages. Germany, for example, depended on workers from Europe's periphery, namely Italy, Yugoslavia, Greece, and Turkey, for industrial and service jobs. These *gastarbeiter*, or **guest**

IMMIGRANT GHETTO OF THE NEW EUROPE   Rinkeby, Sweden

Though Rinkeby has a Mediterranean flavor, since most of its residents are from northern Africa, Turkey, and Palestine, it is a suburb of Stockholm, only a short subway ride from the central city. This utilitarian suburb of high-rise apartments was built in the early 1970s as a government project to provide 1 million new apartments for Swedes. Today 80 percent of Rinkeby's 14,000 residents are immigrants, living in a troubled city in a country that only decades ago preached racial and ethnic tolerance to the rest of the world.

In the last two decades, Sweden—like so much of Europe—has seen its traditionally homogenous population changed by the influx of migrants, guest workers, and refugees from the world's many ethnic battles. Today immigrants make up fully 10 percent of the Swedish population. Tragically, the assimilation and acceptance of ethnic diversity that Sweden loudly wished to see in other mixed societies such as the United States have not occurred there. Instead, these immigrants live in a racially segregated, even discriminatory world.

Mazhar Goker came from Turkey in 1972 with his family. At that time they were the only immigrants on the block. Today, however, there are no native Swedes left. "When I first came here, we were exotic creatures, and people liked to look at us and feel our hair. Now they take detours not to see us."

Because of the high number of foreigners, Rinkeby has been stigmatized as a haven for welfare cheats and a center of crime. Throughout the country, ill-spoken Swedish is known as "Rinkeby Swedish," used by urban toughs and middle-class youths eager for a little street credibility. If there is any kind of trouble in Stockholm, the newspapers usually point the finger at Rinkeby.

Recently, though, there were headlines of a different sort when one of Rinkeby's schools produced students with some of the best test scores in all Sweden. Headlines across the Sunday papers read, "It's a Sensation—Rinkeby Students Get Good Grades." This was no surprise to one of the teachers. Reciting the same formula heard over and over again in Europe's immigrant ghettos, the teacher said, "The reason they are so motivated is they feel they have to be twice as good as any other Swede." In the room, 13 teenagers, students from Bangladesh, Somalia, Iran, Turkey, Chile, and the Balkans, were taking a language class that would help them become fluent in three languages—Swedish, English, and French or Spanish. For them, this is a required passport into the New Europe.

Source: Adapted from "A Swedish Dilemma: The Immigrant Ghetto," by Warren Hoge, *New York Times*, October 6, 1998.

---

workers, arrived by the thousands. As a result, large ethnic enclaves of foreign workers became a common part of the German urban landscape. Today, there are 2.5 million Turks in Germany, most of whom are there because of the open-door foreign worker policies of past decades.

However, these foreign workers are now the target of considerable animosity and resentment. Economic recessions and stagnation have plagued western Europe for the last decade, with unemployment rates approaching 25 percent for young people. As noted, many native Europeans protest that guest workers are taking jobs that could be filled by them. Complicating the issue is that many of the first-generation guest workers have now become citizens of their host country, and in many cases only ethnic distinctions separate different groups of "native" workers. This subtlety opens the door for racially based discrimination (see "Immigrant Ghetto of the New Europe: Rinkeby, Sweden").

Additionally, the region has witnessed a massive influx of migrants from former European colonies in Asia, Africa, and the Caribbean. The former colonial powers of England, France, and the Netherlands have been the major recipients of this immigration. England, for example, has inherited large numbers of former colonials from India, Pakistan, Jamaica, and, most recently, Hong Kong. Indonesians (from the former Dutch East Indies) are common in the Netherlands, while emigrants in France are often from former colonies in both northern and Sub-Saharan Africa (Figure 8.15).

More recently, political and economic troubles in eastern Europe and the former Soviet Union have generated a new

wave of migrants to Europe. Within Germany, for example, thousands of former East Germans have taken advantage of the Berlin Wall's dismantling by moving to the more prosperous and dynamic western parts of unified Germany. With the total collapse of Soviet border controls in 1990, emigrants from Poland, Bulgaria, Romania, Ukraine, and other former Soviet satellite countries poured into western Europe, looking for a better life. The major receiving countries have been

▲ Figure 8.15 Immigrant ghettos in Europe   A large number of Africans from France's former colonies have migrated to large cities where they form distinct ethnic colonies—even ghettos—in high-rise apartment buildings on the outskirts of Paris, Lyon, and Marseille.   (S. Elbaz/Corbis/Sygma)

Germany, Austria, Switzerland, Denmark, and Sweden. This flight from the post-1989 economic and political turmoil has also included refugees from war-torn regions of what was Yugoslavia, particularly from Bosnia and Kosovo. Europe's conscience was tested severely by the plight of these people, who are reminiscent of another generation's experience during and immediately after World War II.

As a result of these different migration streams, Germany has become a reluctant land of immigration that receives 400,000 newcomers each year. Currently, about 7.5 million foreigners live in Germany, making up about 9 percent of the population. This is about the same percentage as the foreign-born share of the U.S. population. However, while the U.S. celebrates its immigration heritage, Germany, along with other European countries, struggles with this new kind of cultural diversity.

The Geography of "Fortress Europe"

An important and perhaps even aggravating ingredient in European concerns about foreign migration has been the agreement between the heartland countries of western Europe to facilitate free movement across borders without passport inspections and checks. This agreement is reshaping the political geography of Europe by creating divisions between insiders and outsiders. To those on the inside it creates a long-dreamed-of "Europe without borders," while to those on the outside it presents a "Fortress Europe," a defensive perimeter hostile to migrants from Asia, Africa, and even eastern Europe.

Taking its name from the city of Schengen, Luxembourg, where the original declaration of intent was signed in 1985, the goals of the EU's **Schengen Agreement** are to gradually reduce border formalities for travelers moving between France, Germany, and the Benelux countries (Belgium, Netherlands, and Luxembourg). As other countries, such as England, Italy, Spain, and Denmark, have shown interest in joining this agreement, the emphasis on free travel between these states has been replaced with increasingly restrictive controls at the external borders of this new "Schengenland." To illustrate, a French national (or "Schengen National" in the agreement's parlance) can cross the border into Germany by auto without the usual traffic stop; instead, cars will simply slow down for a visual check by border police. However, on the Schengen perimeter—say, at the border between Germany and Poland—border checks are much more rigorous and time-consuming. Because of these procedures, critics of the Schengen Agreement refer to a "Fortress Europe," where outsiders are often subject to intimidating customs interrogation designed to weed out undesirable migrants (Figure 8.16).

Though reduced border formalities within western Europe seemed like reasonable and desirable goals in 1985, the situation is much more complicated today because of the large number of migrants from the former Soviet lands, Africa, and southwest Asia. Because of their long Mediterranean shorelines, Spain and Italy in particular carry the burden for policing Schengen's outer limits, a task not always performed to the satisfaction of their other partners. This concern about illegal immigrants from north Africa and southwest Asia was elevated by the September 2001 terrorist attacks.

▲ Figure 8.16 "Fortress Europe" border station   This border station between the Czech Republic and Germany has replaced the Iron Curtain. However, the border station now provides entry into "Schengenland," the group of western European countries that agreed to facilitate free movement by reducing border formalities. *(Linda Sykes)*

Italy, in particular, has had trouble stemming the flow of illegal immigrants. Many emigrants travel from Istanbul by ship, paying thousands of dollars for a five- or six-day journey (often under steerage conditions) that ends when they are dumped on the beaches of Italy at night to take their chances with immigration police. However, Italian law simply requires illegal immigrants to leave the country within 15 days, and illegal immigrants from Albania, Turkey, Iraq, Iran, and southeast Asia can easily obtain transit visas to more northern European countries, which are usually their destinations of choice anyway.

The Landscapes of Urban Europe

One of the major characteristics of Europe's population is its high level of urbanization. All but three countries (Albania, Portugal, and Yugoslavia) have more than half their population in cities, and several countries, such as the United Kingdom and Belgium, are more than 90 percent urbanized. Once again, the gradient from the highly urbanized European heartland to the less-urbanized periphery reinforces the distinctions between the affluent, industrial core area and the more rural, less well-developed periphery. While historical factors explain much of this difference, given that the core countries have been industrialized for almost 200 years, it also speaks to the slower pace of contemporary economic development in the peripheral countries.

Historical Momentum   While widespread urbanization is relatively recent in Europe, dating back only a century or two, the traditions of European urban life are much older. More specifically, the spread of cities into Europe can be associated with the classical empires of Greece and Rome, which makes many cities in Europe more than 2,000 years old. Great cities such as London, Paris, Frankfurt, and Vienna, to name just a few, all sprouted from Roman roots. People before the

Romans tended to settle on defensive sites, such as marshes, hilltops, or small islands, while the Romans located their early cities at river crossings (Frankfurt, Paris, London) to complement their expansive road network. These Roman locations were exceptionally well suited for the historical development of trade and the exercise of political power. For these reasons, the sites of these cities outlasted the Roman Empire and provide continuity between today's urban geography and its classical antecedents.

However, this model explains the location only of those cities within the Roman settlement area, which generally included England and Europe south of the Rhine and Danube rivers. To the north and east of these Roman lands, the founding of true cities came later, during the medieval period. Many of these northern cities, such as Berlin, Copenhagen, Hamburg, and Stockholm, arose with the expansion of regional trade that centered on the Baltic Sea. Even in these areas, most major cities were founded by 1400, thus providing Europe with a dynamic urban legacy that is still apparent in today's urban landscape.

The Past in the Present   Three historical eras dominate most European city landscapes. The medieval (roughly A.D. 900–1500), Renaissance–Baroque (1500–1800), and industrial (1800–present) periods each left characteristic marks on the European urban scene. Learning to recognize these stages of historical growth provides visitors to Europe's cities with fascinating insights into both past and present landscapes (Figure 8.17).

▲ **Figure 8.17 Urban landscapes**   This aerial view of the Marais district of Paris shows how the contemporary landscape expresses different historical periods. The landscape of narrow streets and small buildings is from the medieval period, while the broad boulevards and large public buildings are examples of later Renaissance–Baroque city planning.   *(Corbis)*

The **medieval landscape** is one of narrow, winding streets, crowded with three- or four-story masonry buildings with little setback from the street. This is a dense landscape with few open spaces, except around churches or public buildings. Here and there, public squares or parks are clues about earlier open-air marketplaces where medieval commerce was transacted. As picturesque as we find medieval-era districts today, they nevertheless present challenges to modernization because of their narrow, congested streets and antiquated housing. Often modern plumbing and heating are lacking, and rooms and hallways are small and cramped compared to our contemporary expectations. Because these medieval sectors usually lack modern facilities, they are often inhabited by low- and fixed-income residents. As a result, in many areas these quarters are dominated by the elderly, university students, and ethnic migrants.

In contrast to the cramped and dense medieval landscape, those areas of the city built during the **Renaissance–Baroque** period produce a landscape that is much more open and spacious, with expansive ceremonial buildings and squares, monuments, ornamental gardens, and wide boulevards lined with elaborate residences. During this period (1500–1800), a new aesthetic and sense of urban planning arose in Europe that resulted in the restructuring of many European cities, particularly the large capitals. These changes were primarily for the benefit of the new urban elite: the royalty, aristocrats, and successful merchants. City dwellers of lesser means remained in the medieval quarters, which became increasingly crowded and cramped as more of the city was devoted to the space-extensive pleasures of the ruling classes.

Aggravating crowding during this time was the fact that defensive structures, such as city walls and other fortifications, constrained the outward spread of growing cities. With the advent of high-powered assault artillery, European cities were forced to build an extensive system of defensive walls. Once girdled by these walls, the cities could not expand laterally; instead, densities had to increase within their confines. A common solution was to add several new stories to the medieval houses.

These Renaissance–Baroque landscapes are still very much a part of the contemporary city and are, in fact, often the most prominent landmarks sought out by visitors. For example, after its devastating fire in 1666, London was completely rebuilt in the Baroque style with showpiece parks, boulevards, and monumental buildings, such as St. Paul's Cathedral. In Paris, although major boulevard-building took place in the mid-nineteenth century, it was still very much an expression of the Baroque mentality. In Vienna, the grand boulevard of the *Ringstrasse,* or ring-street, that encircles the inner city was built in the late nineteenth century when city fortifications were declared obsolete and torn down.

Industrialization dramatically altered the landscape of European cities. Historically, factories clustered together in cities beginning in the early nineteenth century, drawn by their large markets and labor force and supplied by raw materials shipped by barge and railroad. Full-blown industrial districts of factories and worker tenements grew up around these

level of urbanization

transportation lines. In continental Europe, where many cities retained their defensive walls until the late nineteenth century, the new industrial districts were often located outside the fortifications, removed somewhat from the historic central city. In Paris, for example, when the railroad was constructed in the 1850s, it was not allowed to penetrate the city walls. As a result, terminals and train stations for the network of tracks were located beyond the original fortifications. Although the walls of Paris are long gone, this pattern of outlying train stations persists today. As noted, the historical industrial areas were also along these tracks, outside the walls in what were essentially suburban locations. This contrasts with the geography of industrial areas in North American cities, where factories and working-class housing often occupied central city locations.

In London, factories and workers located in the eastern part of the city, close to the busy docks of the Thames River. Early city planning (some would call it "social engineering") reinforced the distinction between the middle- and upper-class West End and London's impoverished blue-collar, industrial east side. This historical west–east difference has been lessened in postwar decades with massive urban renewal projects of office buildings, shopping malls, and middle-class housing. Much of the impetus for these recent projects came from the need to rebuild London's east side, which was heavily damaged by air raids in World War II. Postwar rebuilding has been a mixed blessing for some other European cities that suffered heavy war damage. In many cities, such as Berlin, Rotterdam, and Warsaw, the historic landscape was largely destroyed. In its place today is a landscape of modern office buildings, apartment houses, downtown highways, and shopping malls.

Protecting the Sense of Place   Europeans take a strong sense of cultural identity from their cities and, as a result, have legislated a number of innovative programs designed to protect their urban landscapes. Many of these planning and preservation measures have now spread to other countries, including those in North America. Many cities have unique skylines that have become signatures or symbols, and some have taken pains to protect those skylines. The best example can been seen in contrasting the skylines of Paris and London. In Paris, strong building limitations protect the inner-city skyline so that the historic bell towers of Notre Dame and the more modern Eiffel Tower remain visually prominent. As a result, the traditional skyline remains largely intact (Figure 8.18). New high-rise buildings are clustered in outlying nodal points, such as in the high-density office and residential neighborhood of La Défense just to the west of the inner city. London, in contrast, has done little to protect its skyline, and it is now dominated by modern high-rise office buildings that obscure the landmark cathedral of St. Paul's and the spires of Westminster.

At the street level in European cities, it is common to find historic preservation measures that are designed to protect and renovate medieval neighborhoods while at the same time maintaining the traditional urban fabric and sense of place.

These are often expensive projects, since restoring old buildings often costs more than demolition and replacement with modern structures. Still, most European cultures have decided that the benefits outweigh the costs. Once again, Paris was a leader in this historic preservation movement; in the 1960s it allocated public funds to protect its medieval district. By the late 1970s similar historic preservation measures were found throughout Europe; examples include Salzburg, Austria, and Regensburg, Germany.

Landscape protection is also common in the rural environments of Europe, reminding us of the link between a traditional sense of place and cultural identity. In England, Denmark, France, Austria, and Germany, for example, many rural areas and small settlements have enacted strong local ordinances to protect the visual qualities of their landscapes. In these formally designated landscape protection zones, new and renovated houses must use traditional building materials and designs so that structures fit into the existing environment. Potential visual blemishes from billboards and convenience stores are tightly regulated and, in a few extreme cases, even the appearance of agricultural field patterns, fences, and land cover is regulated by local ordinances (Figure 8.19).

Though these regulations may seem odious to individuals and cultures emphasizing individual freedoms over communal values, the fact such landscape protection measures exist in Europe demonstrates the strong attraction people have to cities, villages, and countryside that exhibit a unique sense of place. For many, these preserved landscapes are a welcome relief from the blandness of modern, globalized cities, shopping centers, and malls.

▲ **Figure 8.18 Landscape of central Paris**   For several decades Paris has attempted to preserve its historic skyline by banning high-rises in the central city, as shown in this photo. Instead, the modern structures are usually clustered in development zones outside the inner city.   *(Allard/REA/Corbis/SABA Press Photos, Inc.)*

▲ **Figure 8.19 Protecting traditional rural landscapes** In many European countries, cultural identity is linked to traditional rural landscapes. Thus, states and regions attempt to protect these landscapes from the commercial influences of globalization by banning billboards, controlling architecture so that new buildings fit in with existing structures, and limiting the size and shape of new structures. This photo of the Auvergne area of France shows a traditional French landscape. *(John Brooks/Getty Images, Inc.)*

▲ **Figure 8.20 Global culture in Europe** U.S. popular culture is received with great ambivalence in Europe. While many people embrace everything from fast food to Hollywood movies, at the same time they protest the loss of Europe's traditional regional cultures. *(David R. Frazier Photolibrary, Inc.)*

Cultural Coherence and Diversity: A Mosaic of Differences

The rich cultural geography of Europe demands our attention for several reasons. First, the highly varied and fascinating mosaic of languages, customs, religions, ways of life, and landscapes that characterize Europe has also shaped strong local and regional identities that have all too often stoked the fires of separatism and nationalism. In many cases, this cultural fabric has been torn asunder by geopolitical conflicts that arose from competing cultural differences.

Second, European cultures have played leading roles in processes of globalization. Through European colonialism and imperialism, regional languages, religion, economies, and values have transformed every corner of the globe. If you doubt this, consider cricket games in Pakistan, high tea in India, Dutch architecture in South Africa, and the millions of French-speaking inhabitants of equatorial Africa. Even without such modern technologies as satellite TV, the Internet, and Hollywood films and video, European culture spread across the world, changing the speech, religion, belief systems, dress, and habits of millions of people on every continent.

Today, though, many European countries resist the varied expressions of global culture (Figure 8.20). France, for example, struggles against both U.S.-dominated popular culture and the multicultural influences of its large migrant population. In many ways the same is true of Germany and England. The cultural geography of contemporary Europe is also complicated by the fact that culture operates at many different scales and in many different ways. While Europe fends off global culture with one hand, with the other it creates its own unique supranational culture through economic and political integration. As individual countries come together in political alliances and unions that transcend national sovereignty, local and regional cultures demand autonomy and independence. Indeed, the list of cultural contradictions is a long and fascinating one. Underlying the complexity of this dynamic cultural mosaic are the cultural fundamentals of language and religion, traits that form the basis of so many of Europe's cultural patterns.

Geographies of Language

Language has always been an important component of nationalism and group identity in Europe (Figure 8.21). This is true not only of the past, but also today as tensions between cultural convergence and divergence continue to play out. While some small ethnic groups, such as the Irish or the Bretons, work hard to preserve their local language in order to reinforce their cultural identity, millions of Europeans are busy learning multiple languages so they can communicate across cultural and national boundaries. In this age of globalization and world culture, rare is the European who does not speak at least two languages.

As their first tongue, 90 percent of Europe's population speaks Germanic, Romance, or Slavic languages belonging to western linguistic groups of the Indo-European family. Germanic and Romance speakers each number almost 200 million in the European region. There are far fewer Slavic speakers (about 80 million) when Europe's boundaries are drawn to exclude Russia, Belarus, and Ukraine.

Cultural Coherence and Diversity

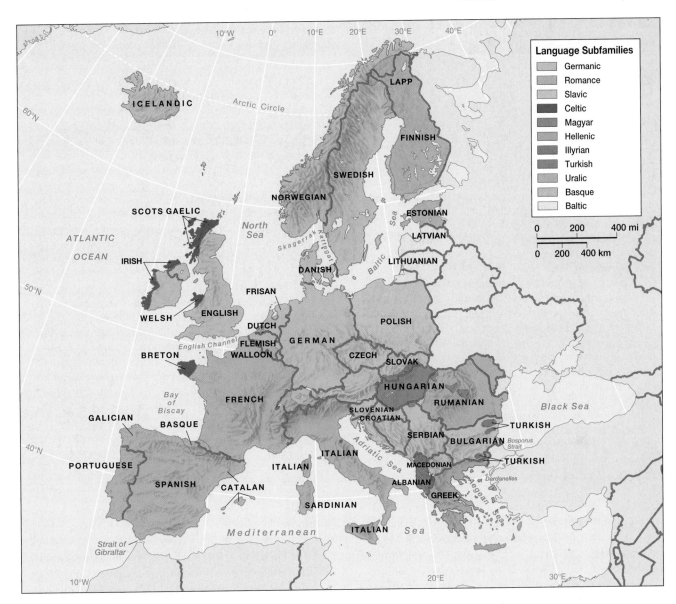

▲ **Figure 8.21 Language map of Europe**   Ninety percent of Europeans speak an Indo-European language, grouped into the major categories of Germanic, Romance, and Slavic languages. As a first language, 90 million Europeans speak German, which places it ahead of the 60 million who list English as their native language. However, given the large number of Europeans who speak fluent English as a second language, one could make the case that English is the dominant language of modern Europe.

Germanic Languages   Germanic languages dominate Europe north of the Alps. This linguistic region includes Germany, Scandinavia, two Baltic republics (Latvia and Lithuania), and the English-speaking areas of the United Kingdom. Approximately 5,000 years ago, the earliest Germanic tribes were centered along the Baltic coast of what is today Germany and Denmark. From this prehistoric cultural hearth, Germanic peoples spread south, east, and west over the course of thousands of years, and with this diffusion, regional dialects evolved into distinct languages, such as German, English, and Dutch.

Today about 90 million people speak German as their first language. This is the dominant language of Germany, Austria, Liechtenstein, Luxembourg, eastern Switzerland, and several small areas in Alpine Italy. Until recently, there were also large German-speaking minorities in Romania, Hungary, and Poland, but many of these people left eastern Europe when the Iron Curtain was lifted in 1989. As "ethnic Germans," they were given automatic citizenship in Germany itself. As is true of most languages, there are very strong regional dialects in German that set apart German-speaking Swiss, for example, from the *Letzeburgish* German spoken in Luxembourg. All dialects, though, are mutually intelligible.

English is the second-largest Germanic language, with about 60 million speakers using it as their native tongue. Additionally, uncounted millions learn English as a second language, particularly in the Netherlands and Scandinavia, where many are as fluent as native speakers (see "Language and Cultural

LANGUAGE AND CULTURAL IDENTITY   English Invades Europe

English, the language of globalization, is both savior and villain in Europe. While on the one hand English is considered by many as the language that might, in the end, bind the European continent together, there is an equally strong movement to protect national cultural identities and languages from homogenization. In a recent EU poll, 70 percent of those surveyed agreed that "everyone should speak English." However, the same percentage said their own language needed to be protected. Capturing the right balance is the issue. When is English practical? And when does it threaten national identity?

In Switzerland, for example, the English language program in grammar schools has caused an uproar. While parents are delighted that their offspring are learning English from the second grade on, critics attack the program with passion, asking why children are learning English as their first foreign language when French, German, Italian, and Romansch are the official languages of Switzerland.

Neighboring France has been perhaps the fiercest defender of its native language. During the last decade, numerous laws have been enacted to protect the French language, including a law making the French language mandatory for advertising, labeling, and instruction manuals. To combat the onslaught of English, a government commission on terminology regularly turns out French alternatives to English computer and business terms. While the government is obliged to use these "official" terms, many of them seem clunky to the French ear. Rarely do these terms catch on in common speech.

As well, similar legal action to protect national languages is under debate in several other countries, including Germany, Poland, and Romania. German today is awash in English expressions, so much so that people refer to this linguistic potpourri as "Denglisch" (pronounced dinglish). "Flirt," "baby," "power," and "sex appeal" are now part of everyday vocabulary, and the list grows longer every day. But the tide may be turning against English. A recent poll showed 53 percent of the people opposed to the use of English words in German. One conservative German politician is pushing a language purification bill with heavy fines for those caught using this bastardized tongue.

Yet English marches on, for it is now the most widely spoken second language in Europe. Some say the language has become so common that it is increasingly thought of as a basic skill, like mathematics, rather than a foreign language. As one expert said, "Learning it has less to do with wanting to talk to the British or Americans and more to do with talking to one another. In some countries, such as Sweden, the Netherlands and Denmark, you can't live a full life as a citizen without speaking English, because some domains—business, science, even intellectual discussions—are in English now."

Adapted from "In Europe, Some Fear National Languages Are Endangered," by Suzanne Daley, *The New York Times*, April 16, 2001.

Identity: English Invades Europe"). Linguistically, English is closest to the Low German spoken along the coastline of the North Sea, which reinforces the notion that an early form of English evolved in the British Isles through contact with the coastal peoples of northern Europe. However, one of the distinctive traits of English that sets it apart from German is that almost a third of the English vocabulary is comprised of Romance words brought to England during the Norman French conquest of the eleventh century.

Elsewhere in this region, Dutch (Netherlands) and Flemish (northern Belgium) account for another 20 million people, and roughly the same number of Scandinavians speak the closely related languages of Danish, Norwegian, and Swedish. Icelandic, though, is a distinct language because of its long separation from its Scandinavian roots.

Romance Languages   Romance languages, such as French, Spanish, and Italian, evolved as regional dialects from their common ancestry in the vulgar (or everyday) Latin used within the Roman Empire. Today about 60 million Europeans speak Italian as their first language. Italian is also an official language of Switzerland and is spoken on the French island of Corsica.

French is spoken in France, western Switzerland, and southern Belgium, where it is known as *Walloon*. Today there are about 55 million native French speakers in Europe. As with other languages, French also has very strong regional dialects, so strong that linguists differentiate between two forms of French in France itself, that spoken in the north (the official form because of the dominance of Paris) and the language of the south, or *langue d'oc*. This linguistic divide expresses long-standing tensions between Paris and southern France. In the last decade the strong regional consciousness of the southwest (centered on Toulouse and the Pyrenees) has led to a resurrection of its own distinct language, *Occitanian*.

Spanish also has very strong regional variations. About 25 million people speak Castillian Spanish, the country's official language, which dominates the interior and northern areas of that large country. However, the *Catalan* form, which some argue is a completely separate language, is found along the eastern coastal fringe, centered on Barcelona, Spain's major city in terms of population and the economy. This distinct language reinforces a strong sense of cultural separateness that has led to the state of Catalonia's being given autonomous status within Spain. Portuguese is spoken by another 12 million speakers in that country and in the northwestern corner of Spain. Portugal's colonial legacy is implicated in the fact that far more people speak this language in its former colony of Brazil in Latin America than in Europe.

Finally, Romanian represents the most eastern extent of the Romance language family; it is spoken by 24 million people in Romania and in parts of the new state of Moldova. Though unquestionably a Romance language, Romanian also contains many Slavic words. Further, in Moldava, the language is written in the **Cyrillic alphabet,** which is based on Greek letters and is used by Russian and other Slavic languages.

The Slavic Language Family   Slavic is the largest western subfamily of the Indo-European language family if one includes European Russia and its neighboring states. Traditionally, Slavic speakers are separated into northern and southern groups, divided by the non-Slavic speakers of Hungary and Romania.

To the north, Polish has 35 million speakers, and Czech and Slovakian about 14 million each. These numbers pale in comparison, however, with the number of northern Slav speakers in nearby Ukraine, Belarus, and Russia, where one can easily count more than 150 million. Southern Slav languages include three groups: 14 million Serbo-Croatian speakers (now considered separate languages because of the political troubles between Serbs and Croats), 11 million Bulgarian-Macedonian, and 2 million Slovenian.

The use of two alphabets further complicates the geography of Slavic languages (Figure 8.22). In countries with a strong Roman Catholic heritage, such as Poland and the Czech Republic, the Latin alphabet is used in writing. In contrast, countries with close ties to the Orthodox church use the Greek-derived Cyrillic alphabet, as is the case in Bulgaria, Macedonia, Croatia, and the Serbian areas of Yugoslavia.

Minor Indo-European Languages   One of the most interesting minor language families is the Celtic group, which includes languages native to Ireland, Scotland, and the Brittany region of France. Though Celtic languages were probably dominant in much of Europe several thousand years ago, today Celtic speakers number just over 1 million. The prehistoric cultural hearth for Celts, dating from 1500 B.C., is identified from archeological materials stretching along the northern foothills of the Alps, including portions of Austria, Switzerland, and France. One theory suggests that Celts fled from this hearth as they were caught between the northward movement of Romans and the southward advance of Germanic peoples in A.D. 200 or 300. As a result, Celts moved to the western fringes of Europe, reestablishing their culture on the Brittany peninsula as well as in the highlands of what are now Wales, Scotland, and Ireland. These fringe areas then served as the new hearth area for Celtic cultures that flourished for more than 1,000 years.

Modern Celtic languages, however, are struggling for survival. In western France, the Celtic Breton language counts fewer than half a million speakers, about one-third of its population in 1900. A major decline took place after World War II when the Catholic clergy in Celtic Brittany switched to French. In Wales, about the same number of people speak Welsh, though several groups are trying to preserve the language and culture. In the highlands and islands of Scotland the situation is even more desperate. Here the number of Gaelic speakers is thought to be less than 100,000. In Ireland the government has tried different strategies for promoting the use of Irish, but to little avail. Until recently the teaching

▲ **Figure 8.22 The alphabet of ethnic tension**   With the departure of many of Kosovo's Serbs, signs in the Cyrillic alphabet were taken off many stores, shops, and restaurants as Kosovars of Albanian ethnicity restated their claims to the region. Here the Albanian owner of a restaurant in the capital city of Pristina scrapes off Cyrillic letters following the recent war.   *(David Brauchli/AP/Wide World Photos)*

of Irish was mandatory in Irish public schools, yet that was dropped in the 1980s because it was thought to be highly impractical. Land grants and other forms of settlement subsidies were offered to Irish speakers, but this did not stem the increasing use of English. Today only about 60,000 people use Gaelic in everyday conversation.

In southeastern Europe, modern residents trace their linguistic heritage to the Hellenic language of ancient Greece. About 10 million people in Greece and Cyprus speak this tongue. Much of the vitality of Hellenic languages comes from its use in the Greek Orthodox Church. It is said that modern Greek has the same relationship to classical Greek as contemporary Italian does to Latin. Finally, in the northeastern corner of Europe, the Baltic languages include Lettish (spoken in Latvia) and Lithuanian. Together they number about 4 million speakers.

Non-Indo-European Languages   Only 5 percent of Europe's population speak non-Indo-European languages, and with the exception of the Hungarians, they are found primarily on the outer reaches of the region. *Magyar* counts some 13 million, mainly in Hungary, but this linguistic region also spills over into nearby Romania, Slovakia, and northern Yugoslavia. Magyar tradition maintains that their ancestors were a herding people who entered Europe from the Asian steppes around A.D. 900. Once settled in the central Danube basin, they created an important culture that, despite its linguistic differences, has played a major role in Europe's affairs. The Austro-Hungarian Empire, for example, dominated eastern Europe until World War I.

Far to the north, the Uralic linguistic family includes Estonian speakers, an even smaller number of Sami people (often called Lapps), and a larger number of Finns. In total, about 5 million people speak these related languages, which possibly originated in the forest and tundra lands of northern Eurasia.

On the southeastern periphery of Europe, the *Altaic* language family includes Turkish. Because of the long rule of the Ottoman Turks in southeastern Europe, significant Turkish minorities remain in countries such as Bulgaria, where, despite discrimination and attempts to expel them, they constitute almost 10 percent of the nation's population. Also important are the more than 1.5 million Turks currently in Germany, guest workers invited to that country to solve its labor problems, but now the focus of considerable discrimination and blatant racism.

Last, the Basque people on the Atlantic border of France and Spain at the western end of the Pyrenees mountain chain speak *Euskara,* a language unlike any other in Europe. Some argue that the Basques are ancestors of the original prehistoric Europeans. Perhaps more important than this uncertain past is the violent terrorism that today pits many Spanish Basque people against the Madrid government as they seek political autonomy for their unique region. Even though a large number of Basques have emigrated to North and South America, there are still more than 600,000 speakers in this part of Europe.

Geographies of Religion, Past and Present

Religion is absolutely inseparable from the geography of cultural coherence and diversity in Europe because so many of today's cultural tensions are embedded in historical events. To illustrate, strong cultural borders in the Balkans and eastern Europe are drawn based upon the eleventh-century split into Christianity's eastern and western churches; in Northern Ireland blood is still shed over the tensions between the seventeenth-century division into Catholicism and Protestantism; and much of the terrorism of ethnic cleansing in Yugoslavia comes from the historical struggle between Christianity and Islam in that part of Europe. Additionally, there is considerable tension regarding the large Muslim migrant populations in England, France, and Germany. Understanding these important contemporary issues thus involves a brief look back at the historical geography of Europe's religious complexity (Figure 8.23).

The spread of early Christianity into Roman Europe was slow until the Edict of Tolerance by the emperor Constantine in A.D. 313. After that, the new religion appears to have spread more rapidly both within and outside of the Empire. The Celtic peoples of Britain adopted Christianity by A.D. 600, and their missionaries played a central role in spreading this religion to the Germanic tribes of central Europe by 800, then even farther east and north by the tenth century. This spread appears to have taken place by **hierarchical diffusion**, with local peoples converting because of the adoption of Christianity by a tribal leader or ruler. Centuries later, this same sort of hierarchical diffusion was important during the conflict between Catholics and Protestants.

The Schism Between Western and Eastern Christianity

In southeastern Europe, early Greek missionaries spread Christianity through the Balkans and into the lower reaches of the Danube. Progress was slower than in western Europe, perhaps because of continued invasions by peoples from the Asian steppes. There were other problems as well, primarily the refusal of these Greek missionaries to accept the growing church hierarchy and control of Roman bishops.

This tension with western Christianity led in 1054 to an official split of the eastern church from Rome. This eastern church subsequently splintered into sects closely linked to specific nations and states. Today, for example, we find Greek Orthodox, Bulgarian Orthodox, and Russian Orthodox churches, all of which have different rites and rituals yet share cultural affinity. The current political ties between Russians and Serbian Yugoslavs, for example, grow from their shared language and religion.

Another factor that distinguished eastern Christianity from western was the Orthodox use of the Cyrillic alphabet instead of the Latin. Since Greek missionaries were primarily responsible for the spread of early Christianity in southeastern Europe, it is not surprising that they used an alphabet based on Greek characters. More precisely, this alphabet is attributed to the missionary work of St. Cyril in the ninth century. As a result, the division between western and eastern churches, and between the two alphabets, remains one of the most problematic cultural boundaries in Europe.

Conflicts with Islam   Both eastern and western Christian churches also struggled with incursions from Islamic empires to Europe's south and east. Even though historical Islam was reasonably tolerant of Christianity in its conquered lands, Christian Europe did not reciprocate by accepting Muslim imperialism. The first crusade to reclaim Jerusalem from the Turks took place in 1095. After the Ottoman Turks took Constantinople in 1453 and gained control over the Straits of Bosporus and the Black Sea, they moved rapidly to spread their Muslim empire throughout the Balkans and arrived at the gates of Vienna in the middle of the sixteenth century. There, Christian Europe drew its line in the sand and turned back the Turks. However, Ottoman control of southeastern Europe lasted until the empire's demise in the early twentieth century. This historical presence of Islam explains the current problematical mosaic of religions in the Balkans, with intermixed areas of Muslims, Orthodox, and Roman Catholics. Commonly, ethnic boundaries in that part of Europe are drawn along religious lines (Figure 8.24).

Muslim incursions into Spain from North Africa left an enduring Moorish imprint throughout the Iberian Peninsula in terms of architecture and technology. Irrigation and water-control techniques and strategies brought by the Muslims, for example, benefited not just the Spanish, but also those parts of the world colonized by Spain and Portugal.

The Protestant Revolt   Besides the division between western and eastern churches, the other great split within Christianity occurred between Catholics and Protestants. This division arose in Europe during the sixteenth century and has

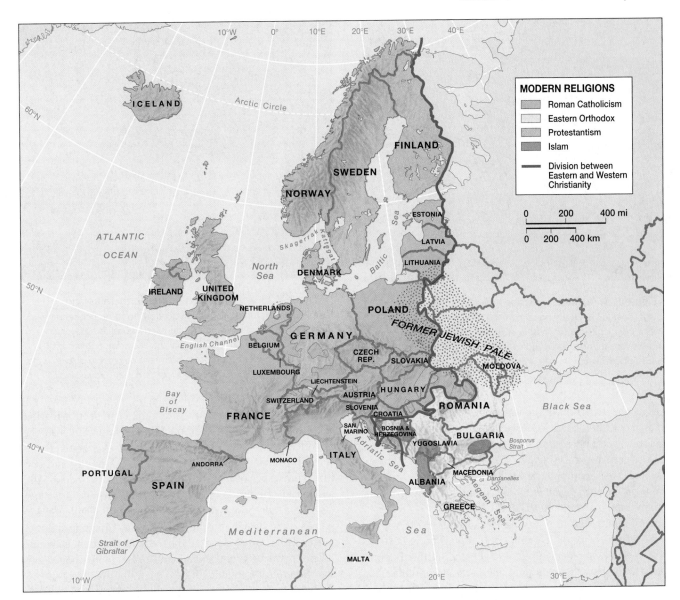

▲ **Figure 8.23 Religions of Europe**   This map shows the divide in western Europe between the Protestant north and the Roman Catholic south. Historically this distinction was much more important than it is today. Note the location of the former Jewish Pale, which was devastated by the Nazis during World War II. Today ethnic tensions with religious overtones are found primarily in the Balkans, where adherents to Roman Catholicism, Eastern Orthodoxy, and Islam are found in close proximity.

divided the region ever since. With the exception of the troubles in Northern Ireland, tensions today between these two major groups are far less damaging than in the past, when the European landscape was scarred by religious battles.

Perhaps the low point came during the Thirty Years' War (1618–48), when central and western Europe were laid waste by war, famine, and the plague. Archival evidence suggests that more than 50 percent of the affected population was either killed or died from starvation or sickness. Fields lay fallow, forests expanded, and villages were deserted as this religious struggle devastated Europe from Bohemia (now the Czech Republic) to Belgium on the Atlantic shore. Today this contested landscape still delineates the north–south boundary between Protestant and Catholic Europe.

A Geography of Judaism   Europe has long been a problematic homeland for Jews after their forced dispersal from Palestine during the Roman Empire. At that time, small Jewish settlements were found in cities throughout the Mediterranean. Later, by A.D. 900, about 20 percent of the Jewish population was clustered in the Muslim lands of the Iberian Peninsula, where Islam showed greater tolerance than Christianity had for Judaism. Furthermore, Jews played an important role in trade activities both within and outside of the Islamic lands. After the Christian reconquest of Iberia, however, Jews once more faced severe persecution and fled from Spain to more tolerant countries in western and central Europe.

One focus for migration was the area in eastern Europe that became known as the Jewish Pale. In the late Middle

▲ **Figure 8.24 Balkan landscape** The urban landscape of Mostar, Bosnia-Herzegovina, shows the different religious affiliations of the local people. Muslim mosques, with their distinctive minarets, are found side by side with Roman Catholic and Eastern Orthodox churches. *(A. Ramey/Woodfin Camp & Associates)*

▲ **Figure 8.25 Religious tensions in Northern Ireland** Despite considerable and continual efforts to forge peace between Protestants and Catholics in Northern Ireland, outbreaks of violence are still common. Here, in Belfast, sectarian fighting broke out only hours after groups from both religions met to discuss ways to end the fighting. *(Cathal McNaughton/Getty Images, Inc.)*

Ages, at the invitation of the Kingdom of Poland, which offered an edict of tolerance, Jews settled in cities and small villages in what is now eastern Poland, Belarus, western Ukraine, and northern Romania (Figure 8.23). Jews collected in this region for several centuries in the hope of establishing a true European homeland, despite the poor natural resources of this marshy, marginal agricultural landscape.

Until emigration to North America began in the 1890s, 90 percent of the world's Jewish population lived in Europe. Most were clustered in the Pale. Even though many emigrants to the United States and Canada came from this area, the Pale remained the largest aggregation of Jews in Europe until World War II. Tragically, Nazi Germany was able to use this ethnic clustering to its advantage by focusing extermination activities on this area.

In 1939, on the eve of World War II, there were 9.5 million Jews in Europe, or about 60 percent of the world's Jewish population. During the war German Nazis murdered some 6 million Jews during the horror of the Holocaust. Today fewer than 2 million Jews live in Europe. Since 1990 and the lifting of constraints on Jewish emigration from Russia, Belarus, and Ukraine, more than 100,000 Jews have emigrated to Germany, giving it the fastest-growing Jewish population outside of Israel.

The Patterns of Contemporary Religion

In Europe today there are about 250 million Roman Catholics and fewer than 100 million Protestants. Generally, Catholics are found in the southern half of the region, except for significant numbers in Ireland and Poland, while Protestants dominate in the north. Additionally, since World War II there has been a marked loss of interest in organized religion, mainly in western Europe, which has led to plummeting church attendance in many

areas. This trend is so marked that the term **secularization** is used, referring to the widespread movement away from the historically prominent organized religions of Europe.

Catholicism dominates the religious geography and cultural landscapes of Italy, Spain, France, Austria, Ireland, and southern Germany. In these areas large cathedrals, monasteries, iconic monuments to Christian saints, and religious place-names draw heavily on pre-Reformation Christian culture. Since visible beauty is part of the Catholic tradition, elaborate religious structures and monuments offer a more ornate landscape than in Protestant lands.

Protestantism is most widespread in northern Germany, the Scandinavian countries, and England, and is intermixed with Catholicism in the Netherlands, Belgium, and Switzerland. Because of its reaction against the perceived aesthetic excesses of the Catholic Church, the landscape of Protestantism is much more sedate and subdued. Large cathedrals and religious monuments in Protestant countries are associated primarily with the Church of England, which has strong historical ties to Catholicism; St. Paul's Cathedral and Westminster Abbey in London are examples.

Tragically, another sort of religious landscape has emerged in Northern Ireland, where barbed-wire fences and concrete barriers separate Protestant and Catholic neighborhoods in an attempt to reduce violence between warring populations (Figure 8.25). Religious affiliation is a major social force that influences where people live, work, attend school, shop, and so on. Of the 1.6 million inhabitants of North Ireland, 54 percent are Protestant and 42 percent Roman Catholic. The Catholics feel they have been discriminated against by the Protestant majority and have been treated as second-class citizens. As a result, they want a closer relationship with the Republic of Ireland across the border to the south. The Protestant reaction is to forge even stronger ties with the United Kingdom. Unfortunately, these differences have been expressed in prolonged violence that has led to about 4,000 deaths since the 1960s, largely resulting from paramilitary groups on both sides who promote their political and social agendas through terrorist activities.

This religious and social strife goes back to the seventeenth century when Britain attempted to colonize Ireland and dilute Irish nationalism by subsidizing the migration and settlement of Protestants from England and Scotland. This strategy was most successful in the north, as that population chose to remain part of the United Kingdom when the rest of Ireland declared its independence from Britain in 1921. However, peace between the two religious groups was short-lived in Northern Ireland. The breakdown of law and order was so bad that British troops were ordered onto the streets of Belfast in 1969, leading to a military occupation that lasted for decades. Current peace talks have centered on three vexing issues: the Protestant insistence that their desire for a continued relationship with the United Kingdom be honored; the Catholic insistence on a closer relationship with Ireland that might lead to eventual unification with that country; and third, the need for some form of political power-sharing by these two groups. While mainstream paramilitary groups have declared a series of cease-fires, splinter groups remind the population that a lasting end to violence has not yet been achieved.

The emerging geography of religion in eastern Europe remains highly differentiated and complex. In Poland, where the Catholic Church was strong and remained an active opponent of communism during the Cold War years, religion seems to be playing a prominent role in contemporary life. Three-quarters of the Polish people consider themselves practicing Catholics. This reemergence of religious commitment far exceeds church attendance in some of the historically Catholic countries of western Europe. Historically, Catholicism also dominated in Hungary, and recent census figures suggest that more than two-thirds of the population considers itself Catholic in today's post-communism environment. This is also the case in nearby Slovakia. In contrast, religious adherence in the Czech Republic is evenly divided between Catholics (40 percent) and those who consider themselves atheist (also 40 percent).

European Culture in a Global Context

Europe, like all world regions, is currently caught up in a period of profound cultural change; in fact, many would argue that the pace of cultural change in Europe is accelerated because of the crosscurrents and complicated interactions between globalization and Europe's internal agenda of political and economic integration. While newspaper pundits celebrate the "New Europe" of integration and unification, other critics refer to a more tension-filled New Europe of foreign migrants and guest workers haunted by ethnic discrimination and racism (Figure 8.26).

Globalization and Cultural Nationalism   Since World War II, Europe has been defensive about, even hostile to, cultural contamination from North America. Some countries are more reactionary than others. While a few large countries, such as England and Italy, seem to accept the onrush of American popular culture, other countries have expressed outright indignation over the corrupting impact of U.S. popular culture on speech, music, food, and fashion. France, for example, is often thought of as the poster child for the antiglobalization struggle as it fights to preserve local foods and culture against the homogenizing onslaught of fast food outlets and genetically engineered crops. Currently, the outspoken leader of French antiglobalization is José Bové, a sheep farmer who became internationally famous for trashing a McDonald's restaurant to protest a 100 percent tax placed on Roquefort cheese by the United States, which was the American response to the EU's banning U.S. hormone-treated beef.

France has taken cultural nationalism to new levels with protective legislation. Radio stations, for example, must devote at least 40 percent of their air play to French songs and musicians, and French filmmakers are subsidized and supported by government subventions in the hope of staving off a complete takeover by Hollywood. The official body of the French Academy, whose job it is to legislate proper usage of the French language, has a long list of English and American words and phrases that are banned from use in official publications and from highly visible advertisements or billboards. Though the French commonly use "le weekend" or "le software" in their everyday speech, these words will never be found in official governmental speeches or publications.

While France sets the model for a continental brand of cultural nationalism, these ideas seem to be spreading to other European countries. Germany, for example, recently started public debate over banning certain English phrases and words in commercial advertisements and "proper" speech. There is a clear irony here, given that Europe's languages have borrowed heavily from each other. Examples include the high number

▲ **Figure 8.26 Turkish store in Germany**   This Turkish store in the Kreuzberg district of Berlin, sometimes referred to as "Little Istanbul," serves not only the large Turkish population, but also German shoppers who have found this aspect of migrant culture an appealing aspect of the country's increasing ethnicity. Unfortunately, to many, other expressions of Europe's migrant cultures are more problematic.   *(Stefano Pavasi/Matrix International, Inc.)*

of French and German cognates in the English language, the similarities between Romance languages, and the shared vocabulary of Scandinavian tongues.

Migrants and Culture   Migration patterns are also influencing the cultural mix in Europe. Historically, Europe spread its cultures worldwide through aggressive colonialization and imperialism. Today, however, the region is experiencing a reverse flow as millions of migrants move into Europe, bringing their own distinct cultures from the far-flung countries of Africa, Asia, and Latin America. Unfortunately, in some areas of Europe, the products of this cultural exchange are highly problematic.

Ethnic clustering, even ghettoization, is now common in the cities and towns of western Europe. The high-density apartment buildings of suburban Paris, for example, are home to large numbers of French-speaking Africans and Arab Muslims caught in the cultural crossfire of high unemployment, poverty, and racial discrimination. Recent estimates are that there are 4.5 million Muslims in France, constituting 7.5 percent of the total population. Similarly, more than 2.5 million Muslim Turks live in Germany (Figure 8.26). Added to that are increasing numbers of emigrants from eastern Europe and the former Soviet Union. As a result, many European countries are experiencing significant culture wars that are fought on many fronts. French leaders, unsettled by the country's large Muslim migrant population, attempted to speed assimilation of female high school students into French mainstream culture by banning a key symbol of conservative Muslim life, the head scarf. This rule triggered riots, demonstrations, and counter-demonstrations. The political landscape of many European countries now contains far-right, nationalistic parties with an exclusionary and discriminatory political agenda ("France is for the French"). Many of these political parties also celebrate the cultural customs, symbols, and icons associated with an idealized, nationalistic past, a social and political construction that clearly does not include migrant foreigners. In Germany, for example, conservative politicians have ignited a firestorm by promoting the notion of a German *leitkultur*, or "guiding culture," that immigrants should subscribe to before being granted visas or citizenship. Although the full content of this guiding culture remains elusive and coded, it most certainly includes the German language, Christianity, and adherence to traditional forms of social behavior. Most noxious, perhaps, is the reaction of young German skinheads who purport to defend with terrorism and violence "true and pure" Aryan values historically associated with Nazi Germany (Figure 8.27).

Geopolitical Framework: A Dynamic Map

One of Europe's unique characteristics is its dense fabric of 37 independent states within a relatively small area. No other world region demonstrates the same kind of geopolitical fragmentation. Europe invented the nation-state. Later though, it fueled the flames of political independence and democracy worldwide, often promoting these same engaging ideals as replacements for its colonial rule in Asia, Africa, and the Americas.

But Europe's varied geopolitical landscape has been as much problem as promise. Twice in the last century Europe shed blood to redraw its political borders. Within the last decade alone, seven new states have appeared in eastern Europe, more than half through war. The map of geopolitical troubles suggests that still more political fragmentation may take place in the near future, much of it violently (Figure 8.28).

Most of Europe, however, sees a brighter geopolitical future. For many, this is based on a widespread spirit of cooperation through unification rather than continued fragmentation. Now empowered by the European Union's recent successes at political and economic integration, many Europeans are still giddy over the sudden and unanticipated end of the Cold War in 1989. Many argue that the disasters of the twentieth century were of Europe's own making. If true, this region seems determined to avoid those mistakes in the twenty-first century by giving the world a new geopolitical model for peace and prosperity. The first decade of the new century should give a clue about whether this optimism is well founded.

Geopolitical Background: From Empire to Nation-State

As might be expected, the political space of Europe has been organized in various ways through history. More important, these different forms have influenced the present geopolitical framework of the region. Broadly put, three scales of political space have been apparent: the empire, the feudal territory, and the nation-state. Though presented here chronologically, one could argue that all three of these geopolitical forms persisted in various expressions well into the twentieth century.

The Legacy of Rome   The Roman Empire (in existence from approximately 300 B.C. to A.D. 400) had a profound and lasting influence on Europe's geography, so much so that its heritage remains apparent on the landscape today. Major west-

▲ **Figure 8.27 Neo-Nazis in Germany**   Purporting to embrace "pure and true" Aryan values, these neo-Nazi Germans provoke police at a rally against foreigners. A rise in extreme forms of nationalism has become increasingly problematic in Germany, with anger often directed against Turks and other immigrants.   *(Joanna B. Pinneo/Aurora & Quanta Productions)*

ern European cities such as London, Paris, and Frankfurt trace their origins back to the Romans. These Roman settlements were linked together with an impressive highway system that remained in use long after Rome had faded. Long, straight stretches of road cutting through a rolling landscape suggest a highway's Roman origins. Bridges and aqueducts built by Roman engineers still span rivers and valleys in many parts of Europe from the Atlantic to the Black Sea.

Geopolitically, the western Roman Empire declined in the fifth century because it was unable to defend itself against the attacks of tribes from northern and central Europe who coveted the richer lands of the Mediterranean civilization. These invasions spelled the end of the spatial and civil organization that linked Rome's far-flung settlements and rural estates. As a result of this warfare in the west, political power and eco-

nomic activity shifted from Rome to Constantinople (Istanbul) in Turkey. This eastern or Byzantine Empire persisted until the fifteenth century, when it was replaced by the Islamic Ottoman Empire.

Feudal Territories   After a period of decline in western Europe following the fall of Rome (the so-called Dark Ages), a revival and expansion of trade beginning in the eleventh century led to a resurgence of urban life, artisan guilds, and merchant associations. Trade based on English wool moved across the Channel into the blossoming commercial cities of Ghent and Bruges in the Low Countries. In the Baltic Sea, the Hanseatic League of cities, anchored by Luebeck and Hamburg (Germany), linked emerging economies in Scandinavia and northern Europe. In the Mediterranean, Genoa and

▲ **Figure 8.28 Geopolitical issues in Europe**   While the major geopolitical issue of the early twenty-first century remains the integration of eastern and western Europe in the European Union, numerous issues of micro- and ethnic nationalism also engender geopolitical fragmentation. In other parts of Europe, such as Spain, France, and Great Britain, questions of local ethnic autonomy within the nation-state structure challenge central governments.

Venice dominated trade as their merchants moved readily through settlements in Europe, Africa, and even Asia.

Contrasting with the urban power of a merchant class was the rural polity based on **feudalism,** or the formal relationship between a superior and a vassal or person of lesser standing. Fundamental to this system was the obligation of the rural serf or peasant to farmland provided by the noble. The regional noble offered protection in return for agricultural labor, obedience, and military service. As a result, a complex geopolitical hierarchy of contracts linked individuals to their masters, who in turn were tied to higher nobles. This network of privileges and rights, of obligations and duty, went right to the top: even emperors and popes thought of themselves as "vassals of God." As a result of these feudal contracts, the geopolitical landscape of medieval Europe was dominated by regional aristocrats who expanded their holdings and power base through war, marriage, and even territorial exchange (Figure 8.29). Spatial and personal identities were oriented to religious and aristocratic authority, not to a shared sense of community, culture, or nation.

Rise of Nationalism and Nation-States
Europe is considered the birthplace of the nation-state, a completely new geopolitical and spatial entity fostered by ethnic and cultural nationalism. While this new ideology began in Europe, it spread far beyond the region and became the preferred modern model for peoples seeking independence from colonialism, authoritarianism, and other forms of unwanted political control. Today, as western Europe moves toward a supranational unity that plays down the power of individual nation-states, peoples in eastern and southern Europe continue to struggle, often violently, for nation-state status. Though some scholars argue that Europe is now in a "post-nation-state" phase, many ethnic groups in the Balkans beg to differ.

Between the fifteenth and eighteenth centuries, European society and politics underwent fundamental changes that undermined the feudal hierarchy. Overseas exploration and trade brought untold wealth to Europe. Urban life blossomed. Scientific discoveries brought new technologies, and also philosophies concerned with new notions of truth. Humanistic thought empowered individuals with tools for shedding superstition and challenging authority. The Reformation reacted against corrupt power in religious life; notions of democracy questioned absolute rule; and the printing press provided the means for spreading new ideas.

With this foundation, Europeans sought to construct political structures based upon much more than a single royal household or court. As European civilization became more stratified and complicated, a broad spectrum of institutions was needed to guide society, and as these institutions evolved, new territorial identities also flourished. State political structures emerged that intermingled social control with unique cultural and territorial identities. The notion of congruence between polity and space became increasingly common by the eighteenth century.

Often nation-state formation can be seen as a process of territorial accretion or growth out from a dominant core area. A congruence between a shared culture and political space

▲ **Figure 8.29 The feudal landscape**   The imposing and picturesque fortress of Carcassonne, France, reminds visitors of the feudal period in Europe's geopolitical history. The castle was sited as a defensive structure on the major transportation routes between the Mediterranean and the Atlantic in southwestern France.   *(Pierre Toutain Dorbec/Corbis/Sygma)*

results as core areas extend their own language, religion, social values, and ways of life over peripheral areas until there is an extensive common "national" culture occupying the territory within the state's political control. France is often used as an example of this process, as French culture expanded incrementally outward from the Paris Basin core area. The geopolitical expansion of other European countries, however, is not so spatially compelling, even though proponents of the core area model continue to make their case.

While the conceptual model of European nation-state formation probably is in need of revision, there is no question about the result. Over the course of several centuries, anchored by the aggregation of smaller regional polities into the "unified" nation-states of Italy and Germany in the second half of the nineteenth century, the political landscape of western Europe changed from a mosaic of small feudal fiefdoms into one of nation-states. Even though these states were ruled in various ways, they shared an important trait: their population found identity and common bonds in the fervor of nationalism, shared culture, and shared territory. This fiery spirit of nationalism, however, brought with it explosive fuel for the devastation of Europe in the twentieth century's two world wars.

Redrawing the Map of Europe Through War
Two world wars redrew the geopolitical maps of twentieth-century Europe (Figure 8.30). Because of these conflicts, empires, nation-states, and even remnant feudal territories have appeared and disappeared within the last 100 years. By the early twentieth century, Europe was divided into two opposing and highly armed camps with exaggerated national egos that tested each other for a decade before the tragic outbreak of World War I in 1914. France, Britain, and Russia were allied against the new nation-states of Italy and Germany, along

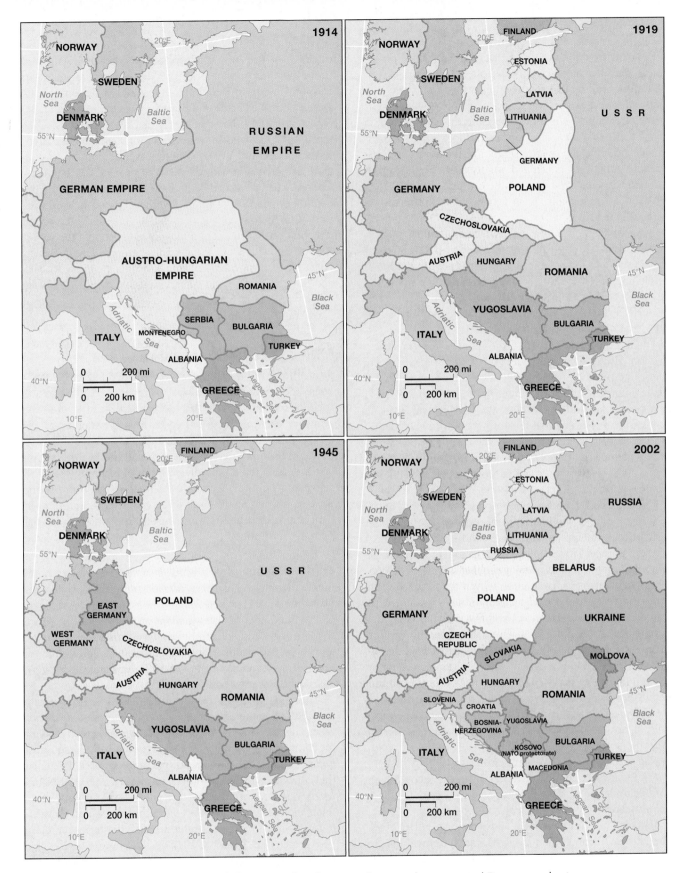

▲ Figure 8.30 A century of geopolitical change When the twentieth century began, central Europe was dominated by the German, Austro-Hungarian (or Hapsburg), and Russian empires. Following World War I, these empires were largely replaced by a mosaic of nation-states. More border changes followed World War II, largely as a result of the Soviet Union's turning that area into a buffer zone between itself and western Europe. With the demise of Soviet hegemony in 1989, further political fragmentation took place.

with the ineffectual Austro-Hungarian or Hapsburg Empire, which controlled a complex mosaic of ethnic groups in central Europe and the Balkans. Though at the time World War I was referred to as the "war to end all wars," it fell far short of solving Europe's geopolitical problems. Instead, according to many experts, it made a further European land war inevitable. Additionally, American entry into the war in 1917 ensured that nation's continued involvement in European geopolitical affairs for the rest of the century.

When Germany and Austria-Hungary surrendered in 1918, the Treaty of Versailles peace process set about redrawing the map of Europe with two goals in mind: to punish the losers through loss of territory and severe fiscal reparations, and to recognize the nationalistic aspirations of certain peoples by creating new nation-states. As a result, the new states of Czechoslovakia and Yugoslavia were born. Additionally, Poland was reestablished, as were the Baltic states of Finland, Estonia, Latvia, and Lithuania.

Though the goals were lofty, the redrawn map created a fragmented Europe in which few states were satisfied. New states were generally resentful when kinfolk were left outside the new borders and became minorities in other new states. This created an epidemic of interwar **irredentism,** or state policies for reclaiming lost territory (real or imagined) beyond borders inhabited by people of the same ethnicity. Examples include the large German population in the western portion of the new state of Czechoslovakia and the Hungarians in western Romania.

This imperfect geopolitical solution was aggravated greatly by the global economic depression of the 1930s, which brought high unemployment, food shortages, and political unrest to Europe. Three competing ideologies promoted their own solutions to Europe's pressing problems: Western democracy (and capitalism), communism from the Soviet revolution to the east, and a fascist totalitarianism promoted by Mussolini in Italy and Hitler in Germany. With industrial unemployment commonly approaching 25 percent in western Europe, public opinion fluctuated wildly between extremist solutions. In 1936 Italy and Germany once again joined forces through the Rome-Berlin "axis" agreement. Once again this alignment was countered with mutual protection treaties between France, Britain, and the Soviet Union. When an imperialist Japan signed a pact with Germany, the scene was set for a second global war.

Nazi Germany tested Western resolve in 1938 by first annexing Austria, the country of Hitler's birth, and then Czechoslovakia under the guise of providing protection for Germans located there. After signing a nonaggression pact with the Soviet Union, Hitler invaded Poland on September 1, 1939. Two days later France and Britain declared war on Germany. Within a month the Soviet Union moved into eastern Poland, the Baltic states, and Finland to reclaim territories lost through the peace treaties of World War I. Nazi Germany then moved westward and occupied Denmark, the Netherlands, Belgium, and France, and began preparations to invade England. In 1941 the war took several startling new turns. In June, Hitler broke the nonaggression pact with the Soviet Union and,

catching the Red Army by surprise, took the Baltic states and drove deep into Soviet territory. When Japan attacked Pearl Harbor, Hawaii, in December, the United States entered the war in both the Pacific and Europe.

By early 1944 the Soviet army had recovered most of its territorial losses and moved against the Germans in eastern Europe, beginning a long communist hegemony in that region. By agreement with the Western powers, the Red Army stopped when it reached Berlin in April 1945. At that time, Allied forces crossed the Rhine River and began their occupation of Germany. With Hitler's suicide, Germany signed an unconditional surrender on May 8, 1945, ending the war in Europe. But with Soviet forces firmly entrenched in the Baltics, Poland, Czechoslovakia, Bulgaria, Romania, Hungary, Austria, and eastern Germany, the military battles of World War II were quickly replaced by the geopolitical reality of an ideological Cold War that lasted more than 40 years, until 1989.

A Divided Europe, East and West

From 1945 until 1989, Europe was divided into two geopolitical and economic blocs, east and west, separated by the infamous Iron Curtain that descended shortly after the armistice of World War II. In 1961 this division took concrete expression when the Soviets built the Berlin Wall to stop Germans from fleeing to the west. East of this border, the Soviet Union imposed the heavy imprint of communism on all activities—political, economic, military, and cultural. To the west, as Europe rebuilt from the ravages of the war, new alliances and institutions were created to counter the Soviet presence in Europe.

Cold War Geography   The seeds of the Cold War are commonly thought to have been planted at the Yalta Conference of February 1945, when Britain, the Soviet Union, and the United States met to plan the shape of postwar Europe. Since the Red Army was already in eastern Europe and moving quickly on Berlin, Britain and the United States agreed that the Soviet Union would occupy eastern Europe and the Western allies would occupy parts of Germany.

The larger geopolitical issue, though, was the Soviet desire for a **buffer zone** between its own territory and western Europe. This buffer zone consisted of an extensive bloc of satellite countries that would cushion the Soviet heartland against possible attack from western Europe. Some scholars argue that this concern goes back to the Russian civil war (1918–21), when western states attempted to intervene in the hope of influencing the outcome of the communist revolution. Regardless of its historical roots, the result was that former military allies ceased to be friends in 1945. Instead, Europe fell into two hostile blocs serving the interests of the two new global superpowers, the United States and the Soviet Union.

In the east the Soviet Union took control of the Baltic states, Poland, Czechoslovakia, Hungary, Bulgaria, Romania, Albania, and Yugoslavia. The last two states later distanced themselves from the Soviet Union by taking different approaches to communism. Austria and Germany were divided into occupied sectors by the four (former) allied powers. In both cases the Soviet Union dominated the eastern portion of each country, areas that contained the capital cities of Berlin and Vienna. Both capital cities, in turn, were divided into French, British, U.S., and Soviet sectors. In 1955, with the creation of an independent and neutral Austria, the Soviets withdrew from their sector, effectively moving the Iron Curtain eastward to the Hungary–Austria border. This was not the case with Germany, however, which quickly evolved into two separate states (West Germany and East Germany) that remained apart until 1990.

Along the border between east and west, two hostile military forces faced each other for almost half a century. Both sides prepared for and expected an invasion by the other across the barbed wire of a divided Europe (Figure 8.31). In the west, NATO (the North Atlantic Treaty Organization) forces, including the United States, were encamped from West Germany south to Turkey. To the east, the Warsaw Pact forces were

anchored by the Soviets, but also included token armies from most satellite countries. Both NATO and the Warsaw Pact countries were armed with nuclear weapons, making Europe a tinderbox for what could potentially be another world war.

Berlin was the flashpoint that brought these forces close to a fighting war on two occasions. In winter 1948 the Soviets imposed a blockade on the city by denying Western powers access to Berlin across its East German military sector. This attempt to starve the city into submission was thwarted by a nonstop airlift of food and coal by NATO. In August 1961 the Soviets built the Berlin Wall to curb the flow of East Germans seeking political asylum in the west. The Wall became the concrete-and-mortar symbol of a firmly divided postwar Europe. For several days while the Wall was being built and the West agonized over destroying it, NATO and Warsaw Pact tanks and soldiers faced each other with loaded weapons at point-blank range. Though war was averted, the Wall stood for 28 years, until November 1989.

The Cold War Thaw   The symbolic end of the Cold War in Europe came on November 9, 1989, when East and West Berliners joined forces to rip apart the Wall with jackhammers and hand tools (Figure 8.32). By October 1990 East and West Germany were officially reunified into a single nation-state (see Author Fieldtrip, "Reinventing Berlin for the 21st Century"). During this period, all other Soviet satellite states, from the Baltic to the Black Sea also underwent major geopolitical changes that have resulted in a mixed bag of benefits and problems. While some, like the Czech Republic, appear to have made a successful transformation to democracy and capitalism, others, such as Romania, appear stalled in their search for new directions.

The Cold War's end came as much from a combination of problems within the Soviet Union (discussed in Chapter 9) as from rebellion in eastern Europe. By the mid-1980s, the Soviet leadership was advocating an internal economic restructuring and also recognizing the need for a more open dialogue with the West. Fiscal problems from supporting a huge military establishment, along with heavy losses from an unsuccessful war in Afghanistan, tempered the Soviet appetite for intervention.

In August 1989 Poland elected the first noncommunist government to lead an eastern European state since World War II. Following this, with just one exception, peaceful revolutions with free elections spread throughout eastern Europe as communist governments renamed themselves and broke with doctrines of the past. In Romania, though, street fighting between citizens and military resulted in the violent overthrow and execution of the communist dictator, Nicolae Ceaucescu.

In October 1989, as Europeans nervously awaited a Soviet response to developments in eastern Europe, President Mikhail Gorbachev said that the Soviet Union had no moral or political right to intervene in the domestic affairs of eastern Europe. Following this statement, Hungary opened its borders to the west, and eastern Europeans freely visited and migrated to western Europe for the first time in 50 years. Estimates are that more than 100,000 East Germans fled their country during this time.

▲ **Figure 8.31 The Iron Curtain**   The former Czechoslovakian–Austrian border was marked by a barbed-wire fence as part of an extensive border zone that included mine fields, tank barricades, watch towers, and a sanitized security zone in which all trespassers were shot on sight.   *(James Blair/NGS Image Collection)*

(a)

(b)

◀ **Figure 8.32 The Berlin Wall** In August 1961, the East German and Soviet armies built a concrete and barbed-wire structure, known simply as "the Wall," to stem the flow of East Germans leaving the Soviet zone. It was the most visible symbol of the Cold War until November 1989, when Berliners physically dismantled it after the Soviet Union renounced its control over eastern Europe. *[(a) Bettmann/Corbis; (b) Anthony Suau/Getty Images, Inc.]*

As a result of the Cold War thaw, the map of Europe began changing once again. Germany reunified in 1990. Elsewhere, separatism and ethnic nationalism, long suppressed by the Soviets, were unleashed in southeastern Europe. Yugoslavia fragmented into the independent states of Slovenia, Croatia, and Bosnia. On January 1, 1993, Czechoslovakia was replaced by two separate states, the Czech Republic and Slovakia.

A New Geopolitical Stability? While the final shape of post-Cold War Europe is still far from clear, there is little question that geopolitical tensions have been reduced between former adversaries. Major steps have been taken to foster stability in the historically treacherous region of eastern Europe. In early 1999, three former Warsaw Pact enemies—Poland, the Czech Republic, and Hungary—became members of NATO. Russia resisted this encroachment into its former sphere of influence and buffer zone, so much so that NATO offered Moscow a special relationship of its own within the alliance. In a historic document, NATO affirmed that it no longer viewed Russia as an enemy and, had no intention of deploying nuclear weapons on the territory of its new members, and, further, would not station permanent combat forces in those countries. While Russia grudgingly accepted this initial expansion of NATO, it said further plans for NATO expansion to bordering countries such as Estonia, Latvia, Lithuania, or Moldova would be considered a hostile act. However, after the September 2001 terrorist attacks and the resulting Afghanistan war, Russia's position on NATO expansion, along with its closer interactions with Europe and the United States, appears to be more flexible (see "Geography in the Making: Europe and the War on Terrorism").

Yugoslavia and the Balkans: Europe's Geopolitical Nightmare

Yugoslavia tests the mettle of the new, unified Europe. More specifically, both NATO and the EU are haunted by their inability to solve the complexities of ethnic conflict, micronationalism, and state disintegration. In 1918 the world responded to Balkan factionalism by creating the new state of Yugoslavia. However, by the last decade of the twentieth century, it was painfully apparent that this was not a solution at all. Again and again, ethnic fault lines in the Balkans have threatened global stability.

That the problems in Yugoslavia are global—not merely local or regional aberrations—needs to be emphasized. Indeed, a complicating geopolitical dimension of the Balkan issue is that these struggles and tensions play out on four different and sometimes contradictory scales. First, there are the local and regional tensions within Yugoslavia itself, which consists of the provinces of Serbia, Montenegro, and Kosovo. Second, there are the tensions between Yugoslavia and its neighboring states of Macedonia, Albania, Bosnia, Croatia, and Slovenia. Third, there are relations with the rest of Europe, with NATO and the EU, that range from military and peacekeeping actions to EU offers of asylum to refugees from the war-torn Balkans. Finally, and perhaps most vexing, are the global implications of the crisis. Elements beyond Europe include the United Nations, which has committed peacekeeping monitors and observers; Russia, with its strong pro-Serb bias; the Islamic countries (primarily Iraq, Iran, and Saudi Arabia), which support Muslim populations in the Balkans; and the United States, which as a NATO and UN member has committed military forces to enforce Balkan peace agreements.

As is the case with many world regions, the September 11, 2001, terrorist attacks on the United States and the resulting global war on terrorism have affected the European region both internally and internationally. Immediately following the September 11, 2001, attacks, there was a genuine outpouring of sympathy and allegiance with the United States. The day after the attacks, *Le Monde*, a French newspaper that is mildly anti-American, ran an editorial under the headline "Today we are all Americans." Similar sentiments were echoed by most other European newspapers.

Beyond this media response, however, lay complex geopolitical realities that ranged from NATO responsibilities for mutual defense to the fact that Al Qaeda terrorists apparently planned their 9/11 attacks from Europe. Interwoven with these complexities are

- Long colonial histories with Muslim countries;
- Contemporary foreign relations with Muslim countries;
- The large numbers of Muslims living within Germany, France, and England today;
- The festering problem of the Balkans with its ethnic and religious undertones;
- A dynamic relationship between Europe and Russia;
- The relationships between European countries and the United States, both as independent countries, but also as members of supranational organizations such as the EU and NATO.

With the war in Afghanistan, some European countries, namely Britain and Germany, chose to take an active and visible role, while other European countries preferred a less-publicized role or deferred completely from participation. One point of contention that has inhibited and even injured Europe–U.S. relations is a perception of U.S. unilateralism. Put differently, while Europe supports many of the goals of the war on terrorism, contradicting this is a sense that the United States does not take European views seriously and, instead, pushes ahead with its own agenda despite the concerns and suggestions of its European allies. Indeed, relationships between Europe and the United States appear to be suffering as a result.

Europeans are once more talking about an all-Europe security force that would be an alternative to NATO. At the same time, however, the agenda for expanding NATO into eastern Europe moves forward, perhaps even faster than many anticipated since Russia, because of its open support for the U.S. war on terrorism, has dropped some of its resistance to NATO expansion into former Warsaw Pact countries.

Internally, events since the 9/11 attacks have forced Europe to revisit the issues of immigration and unfettered travel between EU countries. While these topics were politically charged before 9/11, they have become even more so in the subsequent months as the specter of terrorists illegally entering Europe—and then traveling about freely—looms large. Of course, nationalistic anti-immigrant forces in countries such as France and Germany are using this fear to their advantage in the context of domestic politics. In the minds of some observers, these intertwined concerns over terrorism and immigration may cause a slowing of the previously accepted agenda of EU expansion into eastern Europe.

Pundits commonly take refuge in Balkan history to explain the current troubles, pointing out centuries of regional rivalries, clan feuds, religious intolerance, and ethnic cleansing. Implied in this perspective is the belief that the Balkans are a region unlike any other part of the world, and thus we should accept as inevitable the violence and terrorism of ethnic separatism. Indeed, the term **balkanization** is used to describe the fragmented geopolitical processes involved with small-scale independence movements and the phenomenon of mini-nationalism as it develops along ethnic fault lines.

While there is some truth to this interpretation of Balkan history, it can also mislead and mask other truths. As a result, some words of caution are needed. First, history is commonly reconstructed and reinvented to serve present-day actions and agendas. Thus, there is rarely agreement about the past by different Balkan groups. How history is used is usually more relevant than the content or facts. Second, historical currents run both deep and shallow. While some groups point to events and misdeeds that took place centuries ago (such as the Serb defeat at the hands of Muslims in 1389), other animosities are driven by more recent events in the Balkans, be they an ethnic group's relationship to Nazi Germany during World War II or ethnic terrorism in the last decade.

That said, a general overview of recent Balkan history is helpful to understanding the current situation. As World War I ended in 1918, so did the Austro-Hungarian Empire's control of the Balkans. The independent kingdom of Serbia became the magnet for the creation of the new state of Yugoslavia. Attached to Serbia were a collection of other Hapsburg lands, including Slovenia, Croatia, Bosnia-Herzegovina, Montenegro, Macedonia, and Kosovo. The result was a new state of three mutually antagonistic religions (Islam, Catholicism, and Eastern Orthodoxy), a handful of languages, and two distinct alphabets. From the outset, a Serbian king ruled over this new country and Serbs dominated political and economic activity. Belgrade and the Serbian core amassed wealth at the expense of the ethnic periphery. As a result, Yugoslavia was marked by internal hostility for two decades until it was invaded by Hitler in 1941.

German troops were welcomed as liberators in Croatia, a region that was particularly restless under Serbian rule. Major resistance to Nazi occupation, however, came from two fronts: Serbian royalists who supported the king and opposed Hitler, and partisans who were pan-Yugoslav communists with a distinct anti-Serb, antiroyalist cast. The latter were led by Marshall Tito, a Croat, and supported by the Allies in their battle against Nazi Germany.

Following World War II, Britain, the United States, and the Soviet Union rewarded Tito for his efforts by backing him as the leader of a new socialist Yugoslavia. Conventional wisdom posits that the three major ethnic groups—the Serbs, Croats, and Muslims—coexisted under Tito's strong leadership by subordinating their separatist agendas to the larger goal of a communist state independent of the Soviet Union. This relatively peaceful coexistence continued for almost a decade after Tito's death in 1980. During that period, leaders of the various Yugoslavian states took turns serving as the country's president.

The region's geopolitical stability began to unravel in 1989 (Figure 8.33). In concert with the lessening of Soviet control in eastern Europe and a renewal of ethnic separatism within Yugoslavia, then President Slobodan Milosevic, an outspoken Serbian nationalist, asked the Yugoslav people to vote on the future direction of the country. At issue was whether it should remain unified and, second, whether it should continue with a communist government. Serbia and Montenegro voted for communist governments and unification, while Croatia, Bosnia-Herzegovina, Slovenia, and Macedonia voted for both independence and noncommunist governments.

Slovenia was the first to secede in 1990, followed by Croatia in 1991, Macedonia in January 1992, and finally Bosnia-Herzegovina in April 1992. A brutal civil war ensued in which Milosevic sought to keep all ethnically Serbian areas under Yugoslav control. The Slovenian army inflicted a humiliating defeat on the Serb-dominated Yugoslavian army early on, and then moved ahead with independence. The Yugoslav army waged a ferocious war of ethnic cleansing to rid Bosnia of Muslims and Croats, which was ended by a cease-fire in 1995. In Croatia fighting continued between Serbs and Croats until September 1995. A fragile peace was achieved with the Dayton Peace Accords of November 1995 in which Serbia agreed to respect a new set of boundaries between Yugoslavia and its newly independent neighbors. Sixty thousand NATO troops, including 20,000 Americans, were brought in to enforce the peace treaty. Bosnia-Herzegovina was carved into two complex political entities, a Serb republic and a Muslim-Croat federation, both ruled by the same legislature and president. This legislature also appointed joint premiers in 1997, one a Serb, the other a Muslim. Despite the Dayton Peace Accords, the ethnic warfare plaguing the region was far from over.

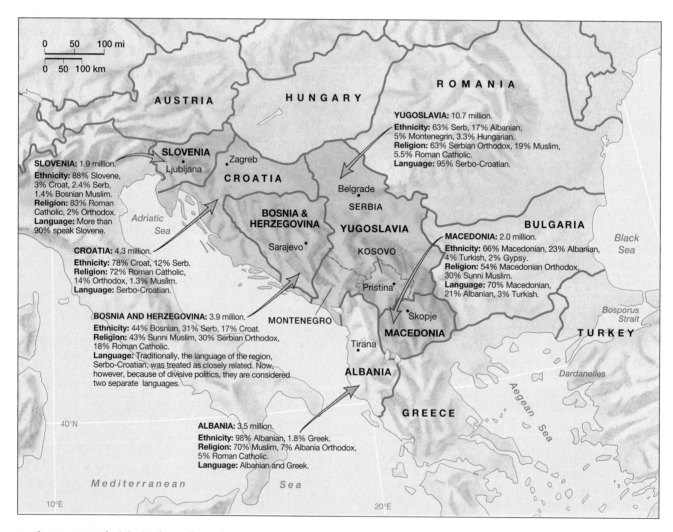

▲ **Figure 8.33 Ethnicity in the Balkans** The diverse and complicated mosaic of ethnic diversity in the Balkans has led to geopolitical fragmentation in recent decades. Not only is the area a meeting ground for Roman Catholicism, Eastern Orthodoxy, and Islam, but linguistic fault lines also complicate ethnic and national identity. Further, a long history of discrimination and retaliation between ethnic groups is embedded in ethnic consciousness.

▲ **Figure 8.34 Kosovar refugees** Thousands of Muslim Kosovars were forced to flee the Yugoslavian province of Kosovo during a period of ethnic cleansing by Serbian forces in the late 1990s. This violence led to aerial bombardment of Yugaslovia by NATO planes to force Serbian leaders to accept UN peacekeepers in the province. *(Buu/Liaison Agency, Inc.)*

Kosovo, to the south of Serbia, is another troublespot (Figure 8.34). This Muslim Albanian-populated province of 2 million lost its autonomous status shortly after the 1989 elections, and Kosovar rebels proclaimed an independent republic in July 1990. While some measures of autonomy were restored with the Dayton Peace Accords, including a return to school instruction in the Albanian language, this was not enough for the rebels. With the effective collapse of the Albanian government in 1997, looting of military bases in that country provided the Kosovo insurgents with a new and enlarged supply of arms, which resulted in escalating attacks on the Serb-dominated Yugoslav army and police. In an attempt to end the conflict, the UN sent 2,000 unarmed monitors into Kosovo in November 1998; however, this did not work, nor did a diplomatic solution that would have guaranteed autonomy for Kosovo within Yugoslavia because of outspoken rejection by former President Milosevic.

As a result of this diplomatic failure, NATO began an intensive air war against Yugoslavia in March 1999. In retaliation, the Serbian army and police killed or exiled hundreds of thousands of ethnic Kosovars in a violent ethnic cleansing campaign designed to create a Serb-only province. After months of warfare, a fragile peace was established in June 1999, enforced by more than 50,000 NATO-led peacekeepers from 30 countries. In October 2000 a popular uprising ousted Milosevic and installed Vojislav Kostunica as president of Yugoslavia. As a result of a new governmental outlook, Yugoslavia was reinstated in the UN and the Council of Europe. In addition, diplomatic relations with the United States and other countries were normalized, ending Yugoslavia's diplomatic isolation and offering hope that the country might somehow peacefully resolve its internal ethnic tensions. That goal, however, may prove elusive as ethnic unrest and terrorism continues in Kosovo and neighboring Macedonia.

Economic and Social Development: Integration and Transition

As the acknowledged birthplace of the Industrial Revolution, Europe in many ways invented the modern economic system. Though Europe was the world's industrial leader in the early twentieth century, it was soon eclipsed by Japan and the United States while Europe struggled to cope with the legacy of two world wars, a decade of global depression, and the Cold War.

In the last 40 years, however, economic integration guided by the European Union has been increasingly successful. In fact, Western Europe's success at blending national economies has given the world a new model for regional cooperation, an approach that could well be imitated in Latin America and Asia in the twenty-first century. Eastern Europe, however, has not fared so well. The results of four decades of Soviet economic planning were, at best, mixed. The total collapse of that system in 1990 cast eastern Europe into a period of chaotic economic, political, and social transition that may well result in a highly differentiated mosaic of rich and poor regions.

Accompanying western Europe's economic boom has been an unprecedented level of social development as measured by worker benefits, health services, education, literacy, and gender equality. Though the improved social services set a laudable standard for the world, they are currently under attack by cost-cutting politicians who argue they add so much to the cost of business that European goods cannot compete in the global marketplace.

The early twenty-first century brings Europe renewed hope that it will once again be a world economic superpower. Yet this region also faces new and difficult challenges. In terms of economic and social development, it is at a major crossroads. Though several different pathways are well marked, the benefits and costs of each are uncertain.

Europe's Industrial Revolution

Europe is the cradle of modern industrialism. Two fundamental changes were associated with this transformation. First, machines replaced human labor in many manufacturing processes, and, second, inanimate energy sources, such as water, steam, electricity, and petroleum, powered the new machines. Though we commonly apply the term *Industrial Revolution* to this transformation, implying rapid change, it took more than a century for the interdependent pieces of the industrial puzzle to come together. This happened first in England between 1730 and 1850. Once it occurred, however, rapid change ensued as industrialization spread to other parts of Europe and the world.

Centers of Change England's textile industry, located on the flanks of the Pennine Mountains, was the center of early industrial innovation. More specifically, innovation and change took place in the small towns and villages of Yorkshire and Lancashire, away from the rigid control of the urban

▲ **Figure 8.35 Water power and textiles** Europe's industrial revolution began on the flanks of England's Pennine Mountains, where swift-running streams and rivers were used to power large cotton and wool looms. Later many of these textile plants switched to coal power. *(James Marshall/Corbis)*

guilds. Yorkshire, on the eastern side of the Pennines, had been a center of woolen textile making since medieval times, drawing raw materials from the extensive sheep herds of that region and using the clean mountain waters to wash the wool before it was spun in rural cottages (Figure 8.35). On the western flank of the mountains, Lancashire workers used cotton from England's overseas colonies (including the United States) as the basis for their newer textile industry. By the 1730s, water wheels drove mechanized looms at the rapids and waterfalls of the Pennine streams.

Locational Factors of Early Industrial Areas

By the 1790s, the steam engine had become the preferred source of energy to drive the new looms because waterpower was hindered by low stream flow during dry weather. However, the steam engine required vast amounts of wood fuel, and England's forests had long been depleted for shipbuilding. While Great Britain possessed large amounts of coal, it was expensive to transport long distances by horse-drawn cart. Only with the advent of the railroad after 1820 could coal be moved long distances at reasonable cost.

Another fundamental innovation was the change from charcoal to coke in the manufacture of iron and steel. Until the early nineteenth century, iron foundries throughout Europe used wood charcoal in the smelting process, which created a pattern of small iron mills dispersed throughout rural forest areas, close to sources of charcoal. In the 1790s, coal-based coke was first substituted for charcoal in the Midlands iron and steel industry. Not only was coke cheaper than charcoal in timber-poor England, but it also permitted the use of lower-grade iron ore. Following this change, iron and steel foundries relocated from rural forest areas to urban clusters adjacent to coalfields. By the 1820s the world's first industrial landscapes had appeared in the English Midlands and Scottish Lowlands. Europe's industrial era had begun.

London was an important exception to the rule that the British Industrial Revolution developed near the coalfields. Despite being far removed from natural resources, this world city developed into the largest industrial center in the United Kingdom because it had access to a large market, a large supply of skilled labor, a port for overseas trade, and a merchant class with access to capital and proximity to banks, insurance firms, investors, and other institutional support systems needed by industry.

Development of Industrial Regions in Continental Europe

By the 1820s, the first industrial districts had begun appearing in continental Europe. These hearth areas were, and still are, near coalfields (Figure 8.36). The first area outside of Britain was the *Sambre-Meuse* region, named for the two river valleys straddling the French-Belgian border. Like the English Midlands, it also had a long history of cottage-based wool textile manufacturing that quickly converted to the new technology of steam-powered mechanized looms. Additionally, a metalworking tradition drew on charcoal-based iron forges in the nearby forests of the Ardenne Mountains. Coal was also found in these mountains; in 1823 the first coke blast furnace outside of Britain started operation in Liege, Belgium.

By the second half of the nineteenth century, the dominant industrial area in all Europe (including England) was the *Ruhr* district in northwestern Germany, near the Rhine River. Rich coal deposits close to the surface fueled the Ruhr's transformation from a small textile region to one oriented around heavy industry, particularly iron and steel manufacturing. By the early 1900s, the Ruhr had used up its modest iron ore deposits and was importing ore from Sweden, Spain, and France. Several decades later the Ruhr industrial region became synonymous with the industrial strength behind Nazi Germany's war machine and thus was bombed heavily in World War II (Figure 8.37).

To the southwest of the Ruhr, the industrial region of *Saar-Lorraine* straddles the border between Germany and France. Though this region contains rich deposits of both coal and iron ore, a much-contested political boundary separated these resources until after World War II. Most of the coal lies in the German Saarland, while iron ore is found primarily in the French Lorraine district. As the industrial strength of this region grew in the late nineteenth century, so did political strife over its control. Twice Germany annexed the iron deposits in France, and twice France attempted unsuccessfully to occupy the coal areas of the Saarland. Only since 1953 have these resources been integrated through trade agreements of the European Coal and Steel Community.

Historical boundary shifts and political tensions between states also plagued the *Upper Silesia* industrial district, which now lies in southern Poland west of Krakow. Historically a center of charcoal-iron smelting, this region was originally part of Prussia before German unification in the 1870s. German and Prussian financial interests developed the region in the late nineteenth century, combining the rich coalfields with imported coke and coal so that by 1900 Upper

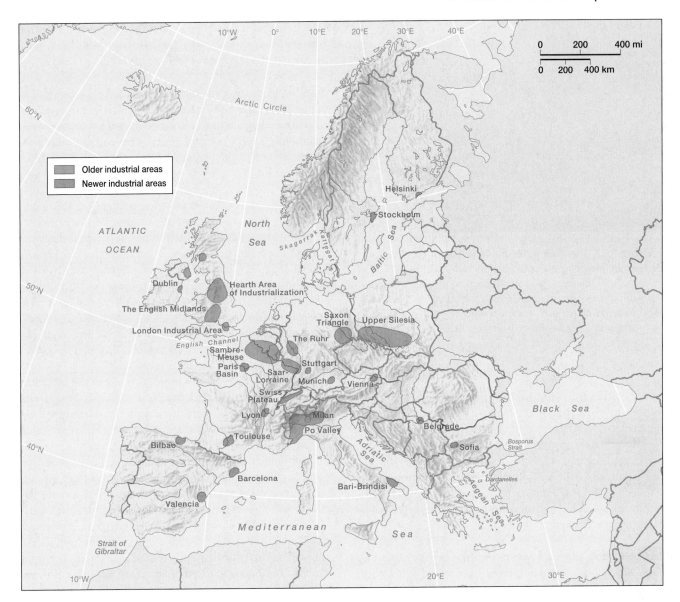

▲ **Figure 8.36 Industrial regions of Europe**   From England, the Industrial Revolution spread to continental Europe, starting with the Sambre-Meuse region on the French-Belgian border, then diffusing to the Ruhr area in Germany. Readily accessible surface coal deposits powered these new industrial areas. Early on, iron ore for steel manufacture came from local deposits, but later it was imported from Sweden and other areas in the shield country of Scandinavia. Most of the newer industrial areas are closely linked to urban areas.

Silesia provided almost 10 percent of Germany's steel. After World War I, however, Germany lost half the region as the Polish border was adjusted westward. The remaining German half was awarded to Poland in 1945 with still another boundary shift.

While most early industrial development took place on or near coalfields, a noteworthy exception is the *Po Plain* in northern Italy, where hydroelectric power generated in the nearby Alps provided energy for industrialization. Because Italy lacked extensive coal deposits, the government subsidized the generation and transmission of electrical energy. Indeed, the northern Italian city of Milan became the first European city to have electric lights in 1883.

Rebuilding Postwar Europe: Economic Integration in the West

Because of its historical momentum, Europe was unquestionably the leader of the industrial world in the early twentieth century. More specifically, before World War I, European industry was estimated to produce 90 percent of the world's manufactured output. However, four decades of political and economic chaos, punctuated by two devastating world wars, left Europe divided and in shambles. By mid-century, its cities were in ruins; industrial areas were destroyed; vast populations were dispirited, hungry, and homeless; and millions of refugees moved about Europe looking for solace and stability.

▲ **Figure 8.37 The Ruhr industrial landscape**   Long the dominant industrial region in Europe, the Ruhr region was bombed heavily during World War II because of its central role in providing heavy arms to the German military. It has been rebuilt, has modernized, and remains competitive with newer industrial regions.   *(Andrej Reiser/Bilderberg Archiv der Fotografen)*

In 1947 the European Recovery Program (ERP) was established to rebuild the region. Through the Marshall Plan, the United States was a major contributor to ERP. In eastern Europe, however, then controlled by the Soviet Union, this aid was rejected because of U.S. involvement. In its place the Soviets created in 1949 the **Council for Mutual Economic Assistance** (CMEA, or, alternatively, Comecon), which linked eastern European aid and recovery to the centralized, command economies of communism.

ECSC and EEC   In 1950 western Europe began discussing a new form of economic integration that would avoid both the historical pattern of nationalistic independence through tariff protection and the economic inefficiencies resulting from the duplication of industrial effort. Robert Schuman, France's foreign minister, proposed that German and French coal and steel production be coordinated by a new supranational authority. In May 1952, France, Germany, Italy, the Netherlands, Belgium, and Luxembourg ratified a treaty that joined them in the European Coal and Steel Community (ECSC). Because of the immediate success of the ECSC, these six states soon agreed to work toward further integration by creating a larger European common market that would foster the free movement of goods, labor, and capital. In March 1957 the Treaty of Rome was signed, establishing the European Economic Community (EEC), popularly called the *Common Market.*

Britain remained outside the new EEC for several reasons, including its economic obligations to former colonies (such as Canada) through Commonwealth agreements and an ambivalence about losing its historic independence from continental Europe, but also because of French objection to English participation. Throughout the 1960s, EEC membership for the United Kingdom was resisted by France for fear the English economy was simply a Trojan horse for American inter-

ests. In response, Britain joined with Switzerland, Austria, Portugal, Denmark, Sweden, and Norway in 1958 to form the short-lived European Free Trade Association (EFTA).

European Community and Union   The EEC reinvented itself in 1965 with the Brussels Treaty, which laid the groundwork for adding a political union to the successful economic community. In this "second Treaty of Rome," aspirations for more than economic integration were clearly stated with the creation of an EEC council, court, parliament, and commission. The EEC also changed its name to the European Community (EC). To be fully successful, though, the EC had to expand its membership, and this occurred eight years later when the French dropped their opposition to the admission of the United Kingdom. In 1973 the EC added not just the United Kingdom, but also Denmark, Ireland, and Norway, though voters in that Scandinavian country subsequently rejected admission in a referendum. The EFTA, as a result, disintegrated. During the 1980s Greece, Portugal, and Spain were added, with Austria, Finland, and Sweden joining in 1995 to round out the current membership of 15 member states (Figure 8.38).

In 1991, the EC again changed its name to the European Union (EU) and expanded its goals once again with the Treaty of Maastricht (named after the town in the Netherlands in which delegates met). While economic integration remains an underlying theme in the new constitution, particularly with its commitment to a single currency through the European Monetary Union, the EU has moved more explicitly into intergovernmental and supranational affairs with its stated goals of common foreign policies and mutual security agreements.

The long list of EU applicants suggests that the advantages of EU membership outweigh any disadvantages from loss of national sovereignty. Currently the EU is entertaining applications from Bulgaria, Cyprus, the Czech Republic, Estonia, Hungary, Malta, Poland, Romania, Slovakia, Slovenia, and Turkey. At least four of these countries may be accepted into the EU in the near future.

Euroland: The European Monetary Union   As any world traveler knows, each state usually has its own monetary system, since coining money has long been a fundamental component of state sovereignty. As a result, crossing a political border usually means changing money and becoming familiar with a new system of bills and coins. In Europe this meant buying pounds sterling in England, marks in Germany, francs in France, lira in Italy, and so on. However, this mosaic of coinage systems is becoming an artifact of the past as Europe moves from individual state monetary systems and toward a common currency. On January 1, 1999, in what many saw as a historical advance for the notion of a united Europe, 11 of the 15 EU member states joined in the European Monetary Union (EMU). As of that day, cross-border business and trade transactions began taking place in the new monetary unit, the *euro.* While the euro remained a fiscal abstraction for three years, on January 1, 2002, new euro coins and bills became available for everyday use. This common coinage comple-

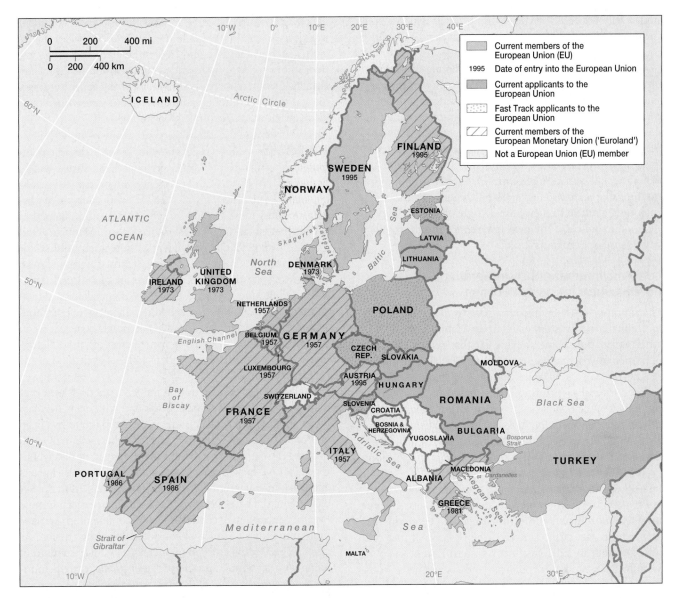

▲ **Figure 8.38 The European Union**   The driving force behind Europe's economic and political integration has been the European Union (EU), which began in the 1950s as an organization focused solely on rebuilding the region's coal and steel industries. After that proved successful, the Common Market (as it was then called) expanded into a wider range of economic activities. Today the EU has 15 members, with a handful of applicants, primarily from eastern Europe, anxiously awaiting membership.

ly replaced the different national currencies of **Euroland** member states in July 2002.

By adopting a common currency, Euroland members expect to increase the efficiency and competitiveness of both domestic and international business. Formerly, when products were traded across borders, there were transaction costs associated with payments made in different currencies; these are now eliminated within Euroland. Germany, for example, exports two-thirds of its products to other EU members. With a common currency, this business now becomes essentially domestic trade, buffered from the fluctuations of different currencies and without transaction costs. Euroland also creates benefits of scale from a mass market of 300 million people and a single economy of $6 billion.

Besides a common currency, EMU member states are also committed to a common set of financial policies designed to control inflation, coordinate economic activity, and standardize wages and benefits. Though attractive in some ways, these common fiscal policies also reduce the traditional flexibility of member states to address the troublesome internal issues of inflation and unemployment. In the past, for example, if a state's economy stagnated, it could always devalue its currency to make its products more competitive in international markets. But under the coordinated fiscal policies of EMU, this will not be an option. Some experts argue that Euroland will succeed only with a strong global economy. If it cools to the point of a global recession, Euroland could be in big trouble. Indeed, uncertainty about the vitality of Euroland, coupled with concerns over loss of sovereignty, led the United Kingdom, Denmark, and Sweden to not join EMU in this first round of currency integration.

Though primarily a tool to generate economic efficiency for Europe, there is little question that a common currency also has strong geopolitical overtones. If European states are bound by a common economic policy and a common currency, fiscal issues should not get in the way of a coordinated and cooperative political interaction. Nationalism may actually diminish. However, should the Euroland economy falter and unemployment remain high, intra-EU competition may return through a rekindled nationalism that finds member states demanding a return to their historic rights of setting national economic policies. The next few years should provide a clearer sense of Euroland's future.

Economic Integration, Disintegration, and Transition in Eastern Europe

Eastern Europe has been historically less well developed economically than its western counterpart (Tables 8.2 and 8.3). Partial explanations come from the fact that eastern Europe is not particularly rich in natural resources. It has only modest amounts of coal, less iron ore, and little to speak of in terms of oil and natural gas. Furthermore, what few resources have been found in the region have been historically exploited by outside interests, including the Ottomans, Hapsburgs, Germans, and, more recently, the Soviet Russians. In many ways eastern Europe has been an economic colony of the more-developed states to its west, south, and east.

The Soviet-dominated economic planning of the postwar period (1945–91) attempted to develop eastern Europe through a coordinated and planned regional economy that also served Soviet goals. However, the collapse of that centralized system in 1991 cast many eastern European countries into deep economic, political, and social chaos. Thus, we cannot understand the present-day problems of the region without some background information on Soviet-bloc economic planning.

The Soviet Plan   In 1949, with the creation of the Council for Mutual Economic Assistance (CMEA, or Comecon), the Soviet Union took control of the region's economies. Its goals were complete economic, political, and social integration through a **command economy,** one that was centrally planned and controlled. The methods were straightforward. In most settings the state owned all means of production, so the economy could be managed for the benefit of the people as a whole, rather than for private interests. As a result, most economic activity was nationalized. Supply and demand were coordinated at a regional level so that states would not compete with each other. Natural resources, such as coal, iron, natural gas, and petroleum, were moved across state boundaries to serve the greater good. The Soviet Union also exported to eastern Europe those resources in short supply and, in return, guaranteed markets for the region's products. Additionally, each individual state subsidized industry to facilitate overall economic development and provide manufacturing jobs for a re-

duced agricultural labor force. This industrial development was decentralized to upgrade the condition of backward regions (such as Slovakia and eastern Poland), as well as implanted in rural areas to provide employment to villagers and farmers. Finally, to increase food supplies, the agricultural sector was reshaped through **collectivization** (state ownership), mechanization, and centralized planning.

The Results   After 40 years of communist economic planning, the results from CMEA were mixed and varied widely within eastern Europe. In Poland and Yugoslavia, for example, many farmers strongly resisted the nationalization of agriculture; thus, most arable and productive land remained in private hands. However, in Romania, Bulgaria, Hungary, and Czechoslovakia, 80 to 90 percent of the agricultural sector was converted to state and collective farms. Across eastern Europe, despite these radical changes in agricultural structure, food production did not increase dramatically, and food shortages, particularly for urban dwellers, were common. Agricultural workers fared somewhat better, being closer to the sources of food production, where most of the rural population tended subsistence gardens to meet their food needs (Figure 8.39).

Perhaps most notable during the Soviet period were dramatic changes to the industrial landscape, with many new factories built in both rural and urban areas and fueled by cheap energy and raw materials imported from the Soviet Union. In defining the pathway to economic development in eastern Europe, CMEA planners chose heavy industry, such as steel plants and truck manufacturing, over consumer goods. As in the Soviet Union, the bare shelves of retail outlets in eastern Europe became both a sign of communist shortcomings and, perhaps more important, a source of considerable public tension. As western Europeans enjoyed an increasingly high standard of living with an abundance of consumer goods, eastern Europeans struggled to make ends meet as the utopian vision promised by Soviet communists became increasingly elusive. Though some good probably resulted from this period of CMEA economic planning, few envisaged the widespread problems that would result with the abrupt end to the economic subsidies provided by the Soviet Union.

Transition and Changes Since 1991   As Soviet political hegemony over eastern Europe disintegrated in the late 1980s, so did the integrated economies of CMEA. In place of the Soviet-dominated era has come a painful period of economic transition bridging on outright chaos in some places as eastern European countries rebuild and reorient their national economies. Most countries are redirecting their economic and political attention to western Europe with the explicit goal of joining the EU sometime in the twenty-first century. To link up with the West, many have moved already from a socialist-based economy of state ownership and control toward a capitalist economy predicated on private ownership and free markets. To achieve this, states such as the Czech Republic, Hungary, and Poland have gone through a period of **privatization,** which is the

TABLE 8.2 *Economic Indicators*

Country	Total GNI (Millions of $U.S., 1999)	GNI per Capita ($U.S., 1999)	GNI per Capita, PPP* ($Intl, 1999)	Average Annual Growth % GDP per Capita, 1990–1999
Western Europe				
Austria	205,743	25,430	24,600	1.4
Belgium	252,051	24,650	25,710	1.4
France	1,453,211	24,170	23,020	1.1
Germany	2,103,804	25,620	23,510	1.0
Liechtenstein	—	—	—	—
Luxembourg	18,545	42,930	41,230	3.8
Netherlands	397,384	25,140	24,410	2.1
Switzerland	273,856	38,380	28,760	−0.1
United Kingdom	1,403,843	23,590	22,220	2.1
Eastern Europe				
Bulgaria	11,572	1,410	5,070	−2.1
Czech Republic	51,623	5,020	12,848	0.9
Hungary	46,751	4,640	11,050	1.4
Moldova	1,481	410	2,100	−10.8
Poland	157,429	4,070	8,390	4.4
Romania	33,034	1,470	5,970	−0.5
Slovakia	20,318	3,770	10,430	1.6
Southern Europe				
Albania	3,146	930	3,240	2.8
Bosnia-Herzegovina	4,706	1,210	—	32.7[†]
Croatia	20,222	4,530	7,260	1.0
Greece	127,648	12,110	15,800	1.8
Italy	1,162,910	20,170	22,000	1.2
Macedonia	3,348	1,660	4,590	−1.5
Malta	3,492	9,210	—	4.2[†]
Portugal	110,175	11,030	15,860	2.3
San Marino	—	—	—	—
Slovenia	19,862	10,000	16,050	2.5
Spain	583,082	14,800	17,850	2.0
Yugoslavia (Serbia)	—	—	—	—
Northern Europe				
Denmark	170,685	32,050	25,600	2.0
Estonia	4,906	3,400	8,190	−0.3
Finland	127,764	24,730	22,600	2.0
Iceland	8,197	29,540	27,210	1.8
Ireland	80,559	21,470	22,460	6.1
Latvia	5,913	2,430	6,220	−3.7
Lithuania	9,751	2,640	6,490	−3.9
Norway	149,280	33,470	28,140	3.2
Sweden	236,940	26,750	22,150	1.2

*Purchasing power parity
[†]Not a 10-year average

Source: *The World Bank Atlas, 2001.*

TABLE 8.3 *Social Indicators and Status of Women*

Country	Life Expectancy at Birth		Under Age 5 Mortality (per 1,000)		Percent Illiteracy (Ages 15 and over)		Female Labor Force Participation (% of total, 1999)
	Male	Female	1980	1999	Male	Female	
Western Europe							
Austria	75	81	17	5	0	0	40
Belgium	75	81	15	6	0	0	41
France	75	83	13	5	0	0	45
Germany	74	81	16	5	0	0	42
Liechtenstein	—	—	—	—	—	—	—
Luxembourg	75	81	—	—	0	0	37
Netherlands	75	81	11	5	0	0	40
Switzerland	77	83	11	5	0	0	40
United Kingdom	75	80	14	6	0	0	44
Eastern Europe							
Bulgaria	68	75	25	17	1	2	48
Czech Republic	71	78	19	5	0	0	47
Hungary	66	75	26	10	1	1	48
Moldova	64	72	—	22	1	2	49
Poland	68	77	—	10	0	0	46
Romania	67	74	36	24	1	3	44
Slovakia	69	77	23	10	0	0	48
Southern Europe							
Albania	69	75	57	—	9	23	41
Bosnia-Herzegovina	65	72	—	18	0	0	38
Croatia	70	77	23	9	1	3	44
Greece	76	81	23	7	2	4	38
Italy	76	82	17	6	1	2	38
Macedonia	70	75	69	17	—	—	42
Malta	74	80	—	—	0	0	28
Portugal	72	79	31	6	6	11	44
San Marino	76	83	—	—	0	0	—
Slovenia	72	79	16	6	0	0	46
Spain	74	82	16	6	2	3	37
Yugoslavia (Serbia)	70	75	—	16	0	0	43
Northern Europe							
Denmark	74	79	10	6	0	0	46
Estonia	65	76	25	12	0	0	49
Finland	74	81	9	5	0	0	48
Iceland	78	81	—	—	0	0	45
Ireland	74	79	14	7	0	0	34
Latvia	65	76	26	18	0	0	50
Lithuania	67	77	24	12	0	1	48
Norway	76	81	11	4	0	0	46
Sweden	77	82	9	4	0	0	48

Sources: Population Reference Bureau, Data Sheet, *2001, Life Expectancy (M/F); The World Bank,* World Development Indicators, *2001, Under Age 5 Mortality Rate (1980/99); The World Bank Atlas, 2001, Female Participation in Labor Force; CIA World Factbook, 2001, Percent Illiteracy (ages 15 and over).*

▲ **Figure 8.40 Post-1989 hardship**   With the fall of communism in eastern Europe after 1989, economic subsidies and support from the Soviet Union ended. Accompanying this transition has been a high unemployment rate resulting from the closure of many industries, such as this plant in Bulgaria.   *(Rob Crandall/ Rob Crandall, Photographer)*

▲ **Figure 8.39 Polish agriculture**   One of the most problematic issues facing eastern European countries is the modernization of the agricultural sector. This is particularly pressing if eastern European countries are to gain membership in the European Union and their farmers are to compete with western European farmers.   *(Hans Madej/Bilderberg Archiv der Fotografen)*

transfer to private ownership of those firms and industries previously owned and run by state governments.

Eastern Europeans have witnessed many changes as a result of this shift in economic policy. Price supports, tariff protection, and subsidies have been removed from consumer goods as countries have transitioned to a free market. For the first time, goods from western Europe and other parts of the world are plentiful and common in eastern European retail stores. The irony, though, is that this prolonged period of economic transition has taken its toll on eastern European consumers, for today few can afford these long-dreamed-of products. Unemployment has often risen to more than 15 percent, and underemployment is common. Financial security is elusive, with irregular paychecks for those with jobs and uncertain welfare benefits for those without. Furthermore, for most people the basic costs of food, rent, and utilities are higher under a free market system than under the subsidized economies of communism. Table 8.2 shows the stark contrast in purchasing power parity between eastern and western Europe

The causes of this economic pain are complex. As the Soviet Union moved to address its own economic and political turmoil, it stopped exporting subsidized natural gas and petroleum to eastern Europe and instead sold it on the open global market to gain hard currency. Without cheap energy, many eastern European industries were unable to operate and shut down operations, laying off thousands of workers. In the first two years of the transition (1990–92), industrial production fell 35 percent in Poland and 45 percent in Bulgaria (Figure 8.40). In addition, the markets guaranteed under a command economy, many of them in the Soviet Union, dissolved overnight with the CMEA's collapse in 1991. Consequently, many factories and services closed because they lacked a market for their goods. In some cases products from

the West, particularly beverages, clothes, and cigarettes, moved in quickly to fill this void (see "From the Field: Traveling in Post-1989 Bulgaria").

With the privatization of industry, services, and property, state subsidies ended and the cost of living increased because of higher prices for goods, services, and rent. Free-market advocates point out that this should be a temporary stage and that stiffer competition should drive prices down. Double- and even triple-digit inflation has also vexed eastern European populations in recent years, aggravating the cost-of-living increases. Food supplies remain problematic since agricultural production fell during the transition. A number of factors explain this decline, including shortages of fuel to run farm equipment, uncertainty about land ownership and access to fields in former state or communal farms, and the inefficiency of operating units where privatization has taken place. In Bulgaria, for example, which previously had 90 percent of its productive land in state or communal farms, most of the new privatized farms are smaller than 2.5 acres (1 hectare).

Regional Disparities Within Eastern Europe   Though certain generalizations hold true throughout eastern Europe, geographers are quick to appreciate significant differences between countries and even subregions. Perhaps more important, there is a troublesome pattern of dual economies emerging in eastern Europe, with some countries speeding ahead to reap the rewards of successful change while others appear to be falling ever farther behind their neighbors (Figure 8.41).

By most accounts, successful economic transitions are taking place in the Czech Republic, Slovenia, Hungary, and Poland. This transformation is supported by the purchasing power parity and per capita GNI data found in Table 8.2. Though far behind affluent western European states, these four countries have all forged strong trade ties with western

FROM THE FIELD   Traveling in Post-1989 Bulgaria

Several of the authors visited Bulgaria at the invitation of the Institute of Geography in Sofia. During our visit, we met a wonderful cross section of people—professionals and workers, students and farmers—and traveled across this intriguing country, staying in large cities and small rural villages alike. It was a remarkable experience, seeing Bulgaria in the throes of change, yet an experience that was not always pleasant because of the many hardships caused by the profound and rapid changes that have affected every aspect of life in the decade since the end of communism.

Unemployment is high, underemployment even more common. Those with jobs do not get regular paychecks. Energetic optimism about the future alternates with a clouded and conditioned uncertainty that often degenerates into outright pessimism. In many ways what we saw in Bulgaria, a country of slightly more than 8 million, is a microcosm for the rest of eastern Europe, though we appreciate the vast differences between places and people in this complex region. Here are a few of our impressions:

- For those of us who knew eastern Europe before 1989, the change in the cities is remarkable. In place of the monotonous, often drab landscape of earlier years, Bulgaria's large cities are now dotted with the garish colors of Western billboards and posters; everywhere are advertisements for soft drinks, cigarettes, and fashionable clothes. Further, the familiar names of western franchises dominate major shopping streets, often displacing traditional retail services, forcing booksellers and hardware stores to operate out of cartons and suitcases from open-air street stands because they cannot pay the high rents asked by the new free-market landlords.

- Traveling in the countryside, we saw many horse- and donkey-drawn carts, not only carrying farm products, but also serving as everyday transportation. This reminded us that even if people own cars, gasoline is scarce and expensive for the rural population. Additionally, little farm machinery is seen in the fields. This is because of uncertain access to formerly state-owned or communal farm equipment, coupled with the problem of expensive fuel. Instead, people have returned to hand tools and various forms of animate power. A rare mechanized wheat combine, for example, was surrounded not by trucks or tractors, but by pony-drawn wooden carts and people carrying heavy sacks of grain.

- Staying with a farm family in a small village of several hundred in northeastern Bulgaria, we saw these changes close up. Farmers are probably the most disgruntled group; although they resent the postcommunist changes, they are reluctant to openly advocate a return to communism before a group of American strangers. Most farmers are essentially unemployed because of the closure of state-owned farms, and few seem anxious to farm the small plots that would result from privatization. Furthermore, markets for their farm products are uncertain. Unlike their urban cousins, at least they have access to market gardens that provide food for the household. Walking about the village, we saw small garden plots, intensively cultivated with vegetables of all sorts, yet usually protected by high fences and locked gates. Even the livestock were under lock and key, with guard dogs everywhere. We could not determine where the threat lay, whether it might come from other village residents or from night-raiding urban dwellers.

- Near this village lay one of the many closed factories we saw throughout Bulgaria, shut for lack of raw materials or cheap fuel or because the organizational structure had simply vanished. This factory formerly repaired large steel shipping containers used to transport goods across the Black Sea. Even though that large water body lies four hours east of this village, the factory was located here as part of the CMEA's decentralization philosophy of bringing work to the countryside. Now, however, there was no longer work in the factory.

  Many of the shipping containers have been "liberated" from the factory and converted into toolsheds, livestock barns, and even small cottages. The most creative use, perhaps demonstrating the fledgling spirit of Bulgarian capitalism, was the conversion of several containers into shelters and food preparation structures for a village beer garden and coffeehouse. Behind the bar was a cheerful electrical engineer from the container factory who had created an attractive social focus for this dispirited village.

As in all of eastern Europe, change has not come easily to Bulgaria since the end of communism in 1990. While many people praise their new personal freedoms, harsh economic realities still cloud their skies and cast uncertainty on their future.

Europe, are receiving investment from the West, and are well along the path to privatization. As a result, a building boom of sorts has recently added a globalized appearance to urban landscapes in these countries. In Prague, for example, seven Western-style megamalls have been built in the last several years; in Warsaw, new high-rise office buildings have transformed the stodgy communist-era central city. As a measure of confidence in the economic future of these countries, 90 percent of new commercial construction in the last years has been financed by foreign developers from western Europe, North America, and Japan. Further, these four countries have applied for membership in the EU, and many expect they will be admitted sometime in this decade.

At the other end of the spectrum, Macedonia, Moldova, and Albania rank lowest in most economic and social development measures. Common to these three countries is widespread political and ethnic strife that has inhibited development. Albania, for example, suffered near-anarchy for

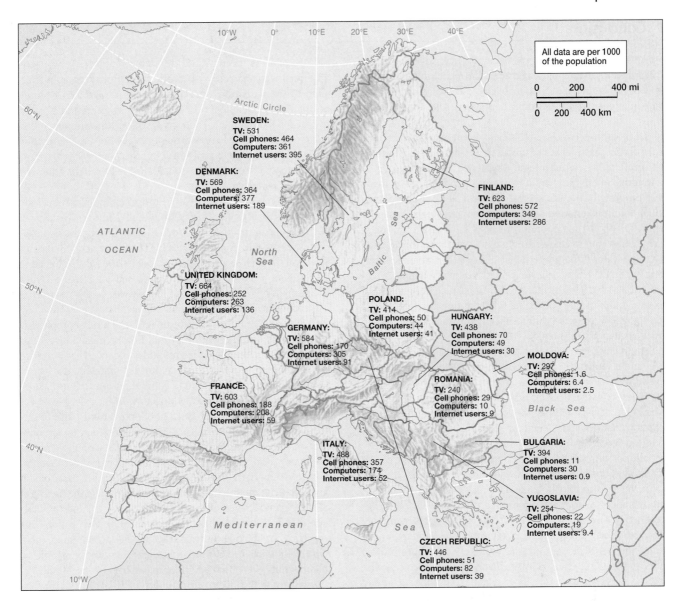

▲ **Figure 8.41 Global linkages and regional disparities**   This map of selected countries in different regions of Europe shows the number of television sets, cell phones, computers, and Internet connections per 1,000 of the population. These data were selected because they are indicators of the differing levels of connectivity to the larger globalizing world. Ponder the implications, for example, of the number of computers and Internet connections available to people in Bulgaria contrasted with those in Germany or France.

several years following the 1997 collapse of pyramid investment plans that destroyed the finances of tens of thousands of families. Violence and looting followed, and a multinational UN force of 7,000 was sent to restore order. Macedonia and Moldova both gained independence in the early 1990s by breaking away from larger states, the former from Yugoslavia and the latter from Ukraine and Romania. However, as with Albania, political and ethnic strife inhibits economic and social development. Additionally, neighboring states and trading partners, such as Romania, Ukraine, Bulgaria, and Yugoslavia, are also struggling with extremely weak economies.

Eastern European countries that make up the middle ground of economic and social change include the three Baltic countries—Latvia, Estonia, and Lithuania—all formerly part of the Soviet Union. Also struggling are Bosnia-Herzegovina, Croatia, and Yugoslavia, weakened and distracted by their mutual animosities and strife; Slovakia, the rural and agricultural tail of the former Czechoslovakia; and, the two Black Sea countries of Bulgaria and Romania, both highly dependent on the Soviet Union for economic stability during the postwar years. Romania's problems, though, have been more internal than external. After a revolving-door series of governments with contrasting views about the country's future, stability appears to have finally set in, which has led to economic aid and loans from the World Bank and International Monetary Fund. This should end triple-digit inflation and help jump-start the Romanian economy.

⊕ Conclusion

The twentieth century brought unparalleled change to Europe. Disruption, even chaos, characterized many decades, with two world wars, a handful of minor ones, the political chaos accompanying a serious economic depression, and almost half a century of political and military division during the Cold War. The scars of these hardships are still apparent, particularly in the east and southeast. Most of Europe, however, meets the twenty-first century with optimism and confidence.

In terms of environmental issues, western Europe has made great progress in the last several decades. Not only have individual countries enacted strong environmental legislation, but the EU has played an important role with its strong commitment to regional solutions that address transboundary concerns, such as air and water pollution or ocean and coastal problems. Additionally, the EU has become an aggressive environmental advocate at the global level, as illustrated by its participation in global climate conferences and treaties where it has demonstrated a willingness to cut back its atmospheric emissions far more than any other major world region.

Serious environmental problems, however, plague eastern Europe, and it remains to be seen whether the EU can solve them. Several variables will shape eastern Europe's environmental future. These new democracies need to enforce existing or new environmental legislation, pay the costs of toxic cleanup and new pollution controls, and cooperate regionally to solve air and water pollution problems. Lastly, eastern Europeans must control the air pollution and surface congestion resulting from an explosive increase in private automobile ownership.

The region also faces ongoing challenges related to population and migration. As for population and settlement issues, the phenomenon of slow and no growth is being appropriately confronted by individual countries as they determine the right amount of natural increase. The most pressing problem is how Europe deals with in-migration from Asia, Africa, Latin America, former Soviet lands, and its own underdeveloped regions. If Europe's economy booms once again, this migration could be the solution to a new labor shortage, but if it stagnates, migration will continue to complicate a large number of issues throughout the region.

Cultural tension will remain one of the problems caused by migration. As the bulk of Europeans meld into a common culture of sorts, the cultural differences between Europeans and outsiders will increase. While some European countries, such as France and even Britain, seem prepared to become multicultural societies, others appear less interested or willing, despite explicit legal actions against ethnic and racial discrimination. While there is little doubt the cultural geography of Europe will change profoundly in the next few years, how Europeans will react to the social changes brought about by migration is unclear.

With the end of the Cold War, Europe's geopolitical issues are continually being redefined. The role and future direction of NATO is one of the important topics facing Europe. Many European states would like to replace NATO with a new military alliance, a European security force without a strong and dominating U.S. presence. Indeed, the United States may not resist this idea because NATO's future could be linked with costly Yugoslavia-like engagements, and the United States might prefer avoiding such actions. The Balkans, unfortunately, will remain problematic for Europe and the world beyond. Europe must address this problem by deciding whether full independence is appropriate for Yugoslavian provinces such as Kosovo and Montenegro. In addition, Yugoslavia will have to continue its precarious balancing act between Russia and the Muslim world.

Issues of economic and social development must also be addressed by twenty-first-century Europeans. Much uncertainty exists as the EU implements its common monetary policy and expands into eastern Europe. Prospective member states, such as Hungary, Poland, the Czech Republic, Slovenia, and Estonia, still have a difficult road to travel before they fully meet the fiscal and political requirements of the EU; should they fail and the invitations be withdrawn, a brighter economic future for eastern Europe could be in jeopardy. Even if these states are admitted, there remains the challenge of alleviating what will surely be significant differences between rich and poor in eastern Europe. While it is easy to envision a positive future for an enlarged EU that includes new member states in eastern Europe, it is much more difficult to think of a positive economic, political, and social future for Moldova, Albania, Bulgaria, and Romania. Sharing the EU's twenty-first-century dream with these countries will constitute the real challenge for this dynamic world region.

⊕ Key Terms

balkanization (page 355)
buffer zone (page 352)
Cold War (page 321)
collectivization (page 362)
command economy (page 362)
continental climate (page 329)
Council for Mutual Economic
   Assistance (CMEA)
   (page 360)
Cyrillic alphabet (page 343)

distributaries (page 332)
Euroland (page 361)
European Union (EU)
   (page 321)
feudalism (page 350)
fjord (page 328)
guest worker (page 335)
hierarchical diffusion
   (page 344)

Iron Curtain (page 325)
irredentism (page 352)
marine west coast climate
   (page 329)
medieval landscape (page 338)
Mediterranean climate
   (page 329)
moraines (page 326)
polders (page 331)

privatization (page 362)
Renaissance–Baroque
   landscape (page 338)
Schengen Agreement
   (page 337)
secularization (page 346)
shield landscape (page 328)

Questions for Summary and Review

1. How does the European Lowland differ north and south of the Rhine River?

2. Describe the different upland and mountain regions of Europe.

3. What are the major factors influencing Europe's weather and climate?

4. Explain why eastern Europe has such serious environmental problems.

5. List those European countries that are below replacement in natural population increase and those that will grow in the next 20 years. Explain factors behind the differences.

6. Explain what is meant by "Schengenland" and "Fortress Europe." What are the geographic advantages and disadvantages of the new arrangement behind these terms?

7. What are the major tongues of the Germanic language family? Which have the most speakers?

8. Map the north and south groups of the Slavic language family. Also map those European countries that do not speak Indo-European languages.

9. Map the interface boundaries, both historical and modern, between Christianity and Islam in different parts of Europe. Where is this interface still a problem? Why?

10. List the new nation-states that appeared on Europe's map in 1919, 1945, and 1992.

11. What were the major geopolitical institutions of the Cold War? Map their areas of influence up to 1990.

12. Describe the major ethnic groups in terms of language and religion in each of the Balkan countries.

13. What are the major components of the Industrial Revolution as seen in Europe? How did the location of industrial areas change through time?

14. Trace the evolution of the EU since 1952, noting how its goals and membership have changed. Be prepared to make a map of its membership at different points in time.

15. Describe the goals and agenda of CMEA, beginning in 1950.

16. Explain what has happened to the eastern European economy since 1990.

Thinking Geographically

1. To compare the scale of Europe to North America, draw a circle 500 miles (800 kilometers) in diameter, which is a day's journey, around Frankfurt, Germany. Then do the same for Chicago. Comment on the similarities and differences in mobility as a result of this exercise.

2. Discuss how the strategic importance of Europe's seas and straits has changed during the twentieth century with changes in naval and shipping technology.

3. Investigate the implications of sea-level rise from global warming for different parts of Europe, including the Dutch coastline. How might this influence the European Union's policy on control of atmospheric emissions?

4. How and why has western Europe become more energy-efficient than North America?

5. Map the different rates of natural increase in Europe, noting which countries will grow and which will decline in the next 20 years. Then link this map to a discussion of migration in Europe. Given these two factors, how might the population map change in 20 years?

6. Find good maps of several mid-sized European cities, and deduce the historical development of the cities based on their street patterns, arrangement of open spaces, boulevards, parks, and so on. Draw on the differences between the medieval and Renaissance–Baroque periods for your interpretation.

7. Look at the ads in a French or German newsmagazine and list the "globalized" English words used. Discuss your findings in terms of what sectors of society seem most open to these foreign terms.

8. Choose an eastern European country and city, and explore how both the rural landscape and the cityscape have changed in the last several decades. One approach would be to set a baseline of, say, 1985, and then compare it to the present.

9. Critically examine the geographic aspects of the Dayton Peace Accords, and explain whether you think the boundaries were drawn wisely.

10. Research the fiscal, political, and social changes that must take place in Hungary and Poland before they are admitted to the EU. Discuss the costs versus the benefits of membership. Within each country, who will profit, and who might suffer?

11. Choose data from Table 8.2 (Economic Indicators) and Table 8.3 (Social Indicators) to map regional disparities in eastern Europe and/or the Balkans. Discuss whether you think these disparities can be reduced.

⊕ Regional Novels and Films

Novels

Jerzy Andrzejewski, *Ashes and Diamonds* (1962, Weidenfeld & Nicholson)

John Berger, *Pig Earth* (1979, Random House)

John Berger, *Once in Europa* (1983, Random House)

Emilie Carles, *A Life of Her Own: A Countrywoman in Twentieth-Century France* (1991, Rutgers University Press)

Natalie Zeamon Davis, *The Return of Martin Guerre* (1983, Harvard University Press)

James Joyce, *The Dubliners* (1967, Viking)

Milan Kundera, *The Unbearable Lightness of Being* (1984, Harper & Row)

W. S. Merwin, *The Lost Upland: Stories of Southwest France* (1992, Alfred Knopf)

Peter Nadas, *The Book of Memories* (1985, Farrar, Straus & Giroux)

V. S. Naipal, *The Enigma of Arrival* (1987, Alfred Knopf)

Charles Power, *In the Memory of the Forest* (1997, Charles Scribner's)

Peter Schneider, *The Wall Jumper: A Berlin Story* (1998, University of Chicago Press)

Dimitur Talev, *The Iron Candlestick* (1964, Foreign Language Press)

Films

Angi Vera (1978, Hungary)

Antonia's Line (1995, The Netherlands)

Au Revoir, Les Enfants (1987, France)

Before the Rain (1995, Macedonia)

The Bicycle Thief (1948, Italy)

The Boat Is Full (1981, Switzerland)

Christ Stopped at Eboli (1979, Italy)

Europa, Europa (1991, Germany)

The Firemen's Ball (1968, Czechoslovakia)

Hey, Babu Riba (1988, Yugoslavia)

Jean de Florette (1987, France)

Man of Iron (1981, Poland)

Manon of the Spring (1987, France)

My Life as a Dog (1985, Sweden)

Playtime (1967, France)

The Postman (1994, Italy)

The Remains of the Day (1994, United Kingdom)

The Stationmaster's Wife (1977, Germany)

Trainspotting (1995, United Kingdom)

Triumph of the Will (1934, Germany)

Underground (1995, Yugoslavia [Bosnia])

Weekend (1967, France)

⊕ Bibliography

Ardaghy, John. 1987. *Germany and the Germans: An Anatomy of Society Today.* New York: Harper & Row.

Barnes, Ian, and Hudson, Robert. 1998. *The Historical Atlas of Europe: From Tribal Societies to a New European Unity.* New York: Macmillan.

Bowler, Ian. 1985. *Agriculture Under the Common Agricultural Policy.* Manchester, England: Manchester University Press.

Carter, Francis, and Turnock, David, eds. 1993. *Environmental Problems in Eastern Europe.* London: Routledge, Chapman & Hall.

Dawson, Andrew. 1993. *A Geography of European Integration: A Common European Home.* New York: John Wiley & Sons.

Denitch, Bogdan. 1996. *Ethnic Nationalism: The Tragic Death of Yugoslavia.* Minneapolis: University of Minnesota Press.

Doherty, Paul, and Poole, Michael. 1997. "Ethnic Residential Segregation in Belfast, Northern Ireland, 1971–1991." *The Geographical Review* 87(4), 520–36.

Engman, Max, ed. 1992. *Ethnic Identity in Urban Europe.* Aldershot, England: Dartmouth Publishing.

Fells, John, and Niznik, Jozef. 1992. "What Is Europe?" *International Journal of Sociology* 22, 201–207.

Glenny, Misha. 1993. *The Fall of Yugoslavia: The Third Balkan War.* New York: Penguin.

Goldfarb, Jeffrey. 1992. *After the Fall: The Pursuit of Democracy in Central Europe.* New York: Basic Books.

Hamilton, Kimberly. 1994. *Migration and the New Europe.* Boulder, CO: Westview Press.

Hoffman, Eva. 1993. *Exit into History: A Journey Through the New Eastern Europe.* New York: Viking.

Houston, James. 1953. *A Social Geography of Europe.* London: Ducksworth.

Jordan, Terry. 1996. *The European Culture Area,* 3rd ed. New York: HarperCollins.

Kaplan, Robert. 1993. *Balkan Ghosts: A Journey Through History.* New York: Vintage Books.

Lewis, Flora. 1992. *Europe: The Road to Unity.* New York: Touchstone.

Liefferink, J.; Lowe, P.; and Mol, A., eds. 1993. *European Integration and Environmental Policy.* New York: Belhaven Press.

Martin, Philip L. 1998. *Germany: Reluctant Land of Immigration.* Washington, DC: American Insitute for Contemporary German Studies.

McDonald, James. 1997. *The European Scene: A Geographical Perspective.* Upper Saddle River, NJ: Prentice Hall.

Netting, Robert. 1981. *Balancing on an Alp: Change and Continuity in a Swiss Mountain Community.* Cambridge, England: Cambridge University Press.

Oberhauser, Ann. 1991. "The International Mobility of Labor: North African Migrant Workers in France." *Professional Geographer* 43, 431–45.

Pells, Richard. 1997. *Not Like Us: How Europeans Have Loved, Hated, and Transformed American Culture Since World War II.* New York: Basic Books.

Renfrew, Colin. 1989. "The Origins of Indo-European Languages." *Scientific American* 261(4), 106–14.

Rosenberg, Tina. 1995. *The Haunted Land: Facing Europe's Ghosts After Communism.* New York: Random House.

Silber, Laura, and Little, Allan. 1995. *Yugoslavia: Death of a Nation.* New York: Penguin Books.

Slavenka, Drakulic. 1993. *The Balkan Express: Fragments from the Other Side of the War.* New York: W. W. Norton and Company.

Smith, A. D. 1986. *The Ethnic Origin of Nations.* Oxford, England: Blackwell.

Watts, Mary. 1971. *Reading the Landscape of Europe.* New York: Harper & Row.

Yarnal, Brent. 1995. "Bulgaria at the Crossroads." *Environment* 37(10), 7–32.

Greenland
(DENMARK)

North Pole +

ARCTIC OCEAN

60°N
20°W

80°N
140°W

160°W

180°

160°E

40°E

60°E

80°E

120°E

140°E

Barents Sea

North Sea

Arctic Circle

EUROPE

Baltic Sea

Murmansk

Kaliningrad

RUSSIA

St. Petersburg

Archangel

Norilsk

SIBERIA

Verkhoyansk

BELARUS

Minsk

Lena River

Yakutsk

UKRAINE

Chernobyl

Moscow

Yaroslavl'
Ivanovo
Nizhniy
Novgorod

Serov

R U S S I A

Kiev
(Kyiv)

Dnieper R.

Kazan

Ob River

Yenisey River

Ob River

Odessa

Kharkov

Dnepropetrovsk

Donetsk

Saratov

Samara

Yekaterinburg

Simferopol'

Don R.

Volga R.

Volgograd

Magnitogorsk

Chelyabinsk

Sevastopol

Black Sea

Omsk

Krasnoyarsk

Baikal-Amur Mainline (BAM) Railr

Amur Ri

GEORGIA

Groznyy

Caspian Sea

Novosibirsk

Novokuznetsk

Lake Baikal

Trans-Siberian Railroad

Transcaucasia

Tbilisi

Irkutsk

ARMENIA

Yerevan

A S I A

RUSSIAN DOMAIN
Political Map

⊕ ● Over 1,000,000

✧ • 500,000–1,000,000
(selected cities)

★ • Selected smaller cities

Elevation in meters

4000+
2000–4000
500–2000
200–500
0–200

Sea Level

Below sea
level

0   250   500 mi

0   250   500 km

CHAPTER 9

The Russian Domain

The Russian domain sprawls across the vast northern half of Eurasia and includes not only Russia itself, but also the nations of Ukraine, Belarus, Georgia, and Armenia (see Figure 9.1; see also "Setting the Boundaries"). The land is rich with superlatives: endless Siberian spaces, unlimited natural resources, legends of ruthless Cossack warriors, and tales of epic wars and revolutions are all part of the region's geographical and historical mythology. Indeed, the rise of Russian civilization remarkably parallels the story of the United States. Both cultures grew from small beginnings to become imperial powers enriched by the fur trade, gold rushes, and transcontinental railroads during the nineteenth century and by dramatic industrialization in the twentieth century. Recently, however, the parallels have ended: incredible political and economic changes have rocked the Russian-dominated region, and its near-term future remains uncertain. The challenges are indeed daunting. Economic collapse across much of the region in the late 1990s produced steep declines in living standards. Political instability includes both tensions between neighboring states as well as divisive pressures within countries. The region also faces some of the most severe environmental challenges anywhere in the developed world.

Globalization is shaping the Russian domain in complex ways. The region's relationship with the rest of the world shifted dramatically during the last 10 years of the twentieth century. Until the end of 1991, all five countries belonged to the Soviet Union, the world's most powerful communist state. Under Soviet control, the region's twentieth-century economy saw large increases in industrial output that made the nation a major global producer of steel, weaponry, and petroleum products. Its communist system offered a powerful ideology that promised economic prosperity and hope to residents within the region and beyond. Indeed, the political and military reach of the Soviet Union spanned the globe, making it a superpower on par with the United States. The Soviet presence dominated many eastern European countries, and nations from Cuba to Vietnam enjoyed close strategic ties with the country.

Suddenly, as the old communist order evaporated early in the 1990s, the now-independent republics of Russia, Ukraine, Belarus, Georgia, and Armenia had to carve out new

◀ **Figure 9.1 The Russian domain** Russia and its neighboring states of Belarus, Ukraine, Georgia, and Armenia make up a dynamic and unpredictable world region that has experienced tremendous change in the past decade. Sprawling from the Baltic Sea to the Pacific, the region includes huge industrial centers, vast farmlands, and almost-empty stretches of tundra. New global connections are on the rise, but it remains unclear how this region will participate in the evolving world economy.

SETTING THE BOUNDARIES

The boundaries of the Russian domain have shifted over time. For decades the regional definitions were relatively easy because the highly centralized Soviet Union (or Union of Soviet Socialist Republics) dominated the region's political geography. The country, born in a communist revolution in 1917, dwarfed all other states in the world and was powerfully united by the Soviet government and largely controlled by ethnic Russians. Most geographers agreed that the Soviet Union could thus be viewed as a single world region. Although the nation contained many cultural minorities, the Soviet Union wielded singular political and economic power from Leningrad on the Baltic Sea to Vladivostok on the Pacific Ocean. After World War II, some geographers even included much of Soviet-dominated eastern Europe within the region in response to the country's expanded military role in nations such as East Germany, Poland, and Hungary.

The maps were suddenly redrawn late in 1991. The once-powerful Soviet state was officially dissolved, and in its place stood 15 former "republics" that had once been united under the Soviet Union. Now independent, each of these republics has tried to make its own way in a post-Soviet world. While some geographers initially treated the region as the "Former Soviet Union," it quickly became clear that diverse cultural forces, economic trends, and political orientations were taking the republics in different directions. Even so, the Russian Republic remained dominant in size and area and thus came to form the nucleus of a new Russian domain that was considerably smaller than the Soviet Union and yet included some characteristics shared by Russia and several of its neighboring states.

The new regional definition reflects the changing political and cultural map since the breakup of the Soviet Union. The term *domain* suggests persisting Russian influence within the four other nations included in the region. Russia, Ukraine, and Belarus make up the core of the new region. Enduring cultural and economic ties closely connect these three countries. In addition, Georgia and Armenia, while more culturally distinctive, are still best classified within a zone of Russian influence. In fact, Armenia and Russia recently signed a treaty of "friendship, cooperation, and mutual understanding" that will maintain Russian military bases in the country for another 25 years. While Georgia has had more stressful relations with its giant neighbor, Russian troops also remain stationed in that former Soviet republic. Two significant areas that were once a part of the Soviet Union have been eliminated from the domain. The mostly Muslim republics of Central Asia and the Caucasus (Kazakstan, Uzbekistan, Kyrgyzstan, Turkmenistan, Tajikistan, and Azerbaijan) have become aligned with a Central Asia world region (Chapter 10), while the Baltic republics (Estonia, Latvia, and Lithuania) and Moldova are best grouped with Europe (Chapter 8).

regional and global relationships. With the breakdown of Soviet control, the region also felt the growing presence of western European and American influences. Westernized popular culture, as well as both modest and radical economic changes, accompanied the political transformations within the realm. These new global relationships have not been easy. Social tensions have risen within the region as people grapple with fundamental economic changes. Political relationships with neighboring regions in Europe and Asia remain uncertain. The region's economic stability has also been threatened by its exposure to the competitive pressures of the global economy. The result is a world region that has seen its global linkages radically redefined in the recent past. Today, fluctuating world oil markets, shifting patterns of foreign investment, new patterns of migration, and the shadowy flows of illegal drugs and Russian mafia money all demonstrate the unpredictable nature of the Russian domain's global connections.

Slavic Russia (population 145 million) dominates the region. Although only about three-quarters the size of the former Soviet Union, Russia's dimensions still make it the largest state on Earth. West of Moscow, the country's European front borders Finland and Poland, while far to the east the nations of Mongolia and China share a thinly peopled boundary with sprawling Russian Siberia. Its area of 6.6 million square miles (17 million square kilometers) dwarfs even Canada, and its 11 time zones are a reminder that dawn in Vladivostok on the Pacific Ocean is still only dinnertime in Moscow. In more ways than one, the Russians sit squarely between sunrise and sunset. With the demise of the Soviet Union, the Russians ended almost 75 years of Marxist rule. What comes next is difficult to predict, and the country currently stands in an economic and political twilight between the old ways and the new. Will Russia, spurred by its numerous ethnic minorities, further fragment into smaller political units, or might it consolidate its power and reassert direct control over neighboring states such as Belarus and Ukraine? Will its economy achieve an appropriate balance between private control and state ownership?

Arguably, stability in the region may be a long way off, but two things are certain. First, by virtue of its size and location, change within Russia will inevitably shape political and economic geographies far beyond its borders. Second, even with its present problems, Russia possesses a long-established and coherent national identity, as well as a rich assemblage of natural and human resources that can act to bind the nation together and provide the foundation for future political stability and economic development.

The bordering states of Ukraine, Belarus, Georgia, and Armenia will inevitably be linked to the evolution of their giant neighbor, even as they attempt to make their own way as newly independent nations. Emerging from the shadows of Soviet dominance has been difficult. Ukraine, in particular, has the size, population, and resource base to become a major European nation, but it has struggled to create real political and economic change since independence. With about 50 million people and a rich storehouse of resources, Ukraine's size of 233,000 square miles (604,000 square kilometers) is similar to

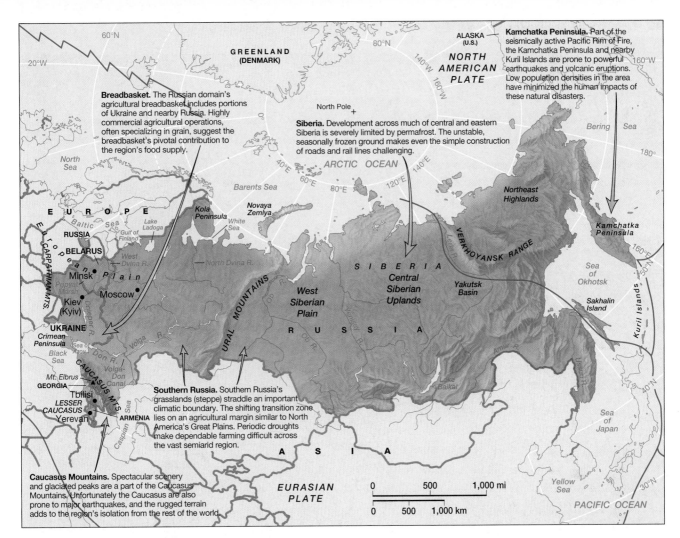

Breadbasket. The Russian domain's agricultural breadbasket includes portions of Ukraine and nearby Russia. Highly commercial agricultural operations, often specializing in grain, suggest the breadbasket's pivotal contribution to the region's food supply.

Siberia. Development across much of central and eastern Siberia is severely limited by permafrost. The unstable, seasonally frozen ground makes even the simple construction of roads and rail lines challenging.

Kamchatka Peninsula. Part of the seismically active Pacific Rim of Fire, the Kamchatka Peninsula and nearby Kuril Islands are prone to powerful earthquakes and volcanic eruptions. Low population densities in the area have minimized the human impacts of these natural disasters.

Southern Russia. Southern Russia's grasslands (steppe) straddle an important climatic boundary. The shifting transition zone lies on an agricultural margin similar to North America's Great Plains. Periodic droughts make dependable farming difficult across the vast semiarid region.

Caucasus Mountains. Spectacular scenery and glaciated peaks are a part of the Caucasus Mountains. Unfortunately the Caucasus are also prone to major earthquakes, and the rugged terrain adds to the region's isolation from the rest of the world.

▲ **Figure 9.2 Physical geography of the Russian domain** In the west, the European Plain stretches across Belarus and the Russian heartland to the Urals. In the far south, the rugged Caucasus Mountains impose significant barriers to movement. East of the Urals, more large plains and plateaus alternate with mountains, particularly in eastern Siberia, where higher peaks and volcanic uplifts further isolate the region from the world beyond.

that of France. Nearby Belarus is smaller (80,000 square miles or 208,000 square kilometers), and its population of 10 million is likely to remain more closely tied economically and politically to Russia. Presently, its strikingly authoritarian and antiforeign leadership echoes remnants of the old Soviet empire. South of Russia and beyond the bordering Caucasus Mountains, the Transcaucasian countries of Armenia and Georgia are smaller still. Their populations depart culturally from their Slavic neighbor to the north. In addition, these two nations face significant political challenges: Armenia shares a hostile border with Azerbaijan (see Chapter 10), and Georgia's ethnic diversity threatens its political stability.

Environmental Geography: A Vast and Challenging Land

The daunting environmental challenges of the Russian domain are poignantly illustrated by the plight of the Caspian and Black Sea sturgeon. Recently the World Wildlife Federation warned that several varieties of this valuable caviar-producing fish were in danger of extinction. Indeed, the annual sturgeon harvest has fallen to barely 3 percent of the levels seen in the late 1970s. The culprit has been the growing global black market for the exotic Russian delicacy, combined with few domestic regulations that effectively protect the caviar beds of the Caspian Sea, the Volga River Delta, and the Sea of Azov (a part of the Black Sea). The Russian mafia, no friend to the environmentalists, has also been a central player in this ecological disaster. Mafia-supported poaching operations and transport connections have resulted in the rapid depletion of the sturgeon as the valuable contraband flows north to Moscow and onward to even more lucrative international markets. The vanishing fish exemplifies the larger plight of the Russian domain's devastated environment. Caught by inadequate legal protections, a global economic setting that encourages its continued harvest, and the powerful presence of the mafia, the luckless fish is being sacrificed to turn a quick profit. Sadly, the same can be said for much of the region's natural resource base.

Still, the Russian domain possesses a vast and diverse physical setting. Size, latitude, and topography create a distinctive physical geography (Figure 9.2). The fact that the

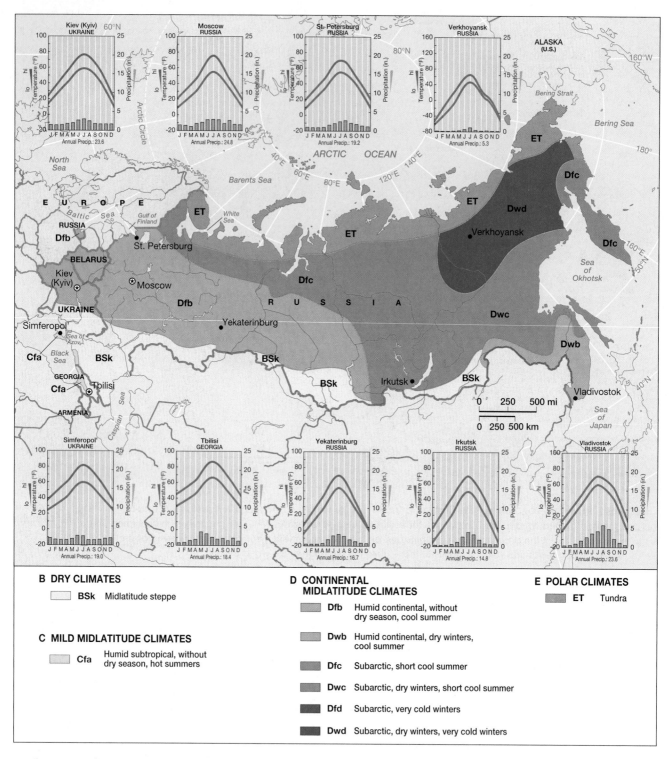

▲ **Figure 9.3 Climate map of the Russian domain** The region's northern latitude and large landmass suggest that continental climates dominate. Indeed, farming is greatly limited by short growing seasons across much of the region. Aridity imposes limits elsewhere. Only a few small zones of mild midlatitude climates are found on the warming shores of the Black Sea in the far southwestern corner of the region, producing subtropical conditions in western Georgia.

region occupies a major portion of the world's largest landmass means that huge distances separate people and resources. Particularly within Russia, thousands of miles and many days of road or rail travel add immeasurable human and economic costs to the movement of people, goods, and services within the country. But size is also a blessing: the domain's varied natural resources are widely distributed across the realm, and the region is still home to some of Eurasia's best farmlands, metals resources, and petroleum reserves.

The region's northern latitudinal position is critical to understanding basic geographies of climate, vegetation, and agriculture. Indeed, the Russian domain provides the world's largest example of a high-latitude continental climate where seasonal temperature extremes and short growing seasons profoundly limit opportunities for human settlement. Large, cold, dry arctic high-pressure systems become anchored over the Russian interior during the winter, providing some of the coldest temperature readings on the planet. Adding to the region's climatic extremes, the sprawling European Plain distances most of Russia from the modifying effects of the North Atlantic and Baltic Sea. Rugged mountains to the south and east further fragment the region and isolate it from other potential maritime influences. In terms of latitude, Moscow is positioned as far north as Ketchikan, Alaska, and even the Ukrainian capital of Kiev (Kyiv) would sit north of the Great Lakes in Canada. Thus, apart from a subtropical zone near the Black Sea, much of the region experiences a classic continental climate with hard, cold winters and marginal agricultural potential (Figure 9.3).

The European West

An airplane flight over the western portions of the Russian domain would reveal a vast, barely changing landscape below. European Russia, Belarus, and Ukraine cover the eastern portions of the vast European Plain, which runs from southwest France to the Ural Mountains. The northern two-thirds of this area was covered by glaciers during the Pleistocene ice age and is characterized by low elevations, subdued topography, and extensive areas of poorly drained land. Europe's largest wetland is the Pripyat marsh of southern Belarus and northern Ukraine. Lands farther south were not glaciated and are somewhat hilly, but these areas also contain little overall physical relief. The smoothly rolling terrain, low levels of evaporation, and moderately abundant precipitation of Ukraine, Belarus, and most of European Russia give rise to broad, gently flowing rivers. These waterways have long been important transportation arteries. One of the major geographical advantages of European Russia is the fact that different river systems, all now linked by canals, flow into four separate drainages. The Dnieper and Don rivers flow into the Black Sea; the West and North Dvina rivers drain into the Baltic and White seas, respectively; and the Volga River runs to the Caspian Sea (Figures 9.2 and 9.4). Russian civilization was historically integrated along these riverine passageways, and much of Russian history can be understood as the struggle to gain control not only of the rivers but also of the seas into which they flow.

Even though European Russia has a milder climate than Siberia, most of it experiences cold winters and cool summers by North American standards. Moscow (Figure 9.3), for example, is about as cold as Minneapolis in January, yet not nearly as warm in July. In Ukraine, Kiev is milder, however, and Simferopol', near the Black Sea, offers wintertime temperatures that average more than 20 °F warmer than those of Moscow.

Three distinctive environments shape agricultural potential in the European West (Figure 9.5). North of Moscow and St.

▲ **Figure 9.4 Volga Valley**   The city of Nizhniy Novgorod lies along the shores of the Volga River east of Moscow. The huge Volga Basin remains a center of Russian settlement and economic development. It drains much of western Russia and empties into the Caspian Sea.   *(Tass/Sovfoto/Eastfoto)*

Petersburg, poor soils and cold temperatures severely limit farming, and much of the land remains in coniferous forest. Belarus and central portions of European Russia possess longer growing seasons, but acidic **podzol soils,** typical of northern forest environments, limit agricultural output and the ability of the region to support a highly productive farm economy. Still, diversified agriculture includes grain (rye, oats, and wheat) and potato cultivation, swine and meat production, and dairying. South of 50° latitude, agricultural conditions improve across much of southern Russia and Ukraine. Forests gradually give way to steppe environments dominated by grasslands and by fertile "black earth" **chernozem soils,** which have proven valuable for commercial wheat, corn, and sugar beet cultivation and for commercial meat production (Figure 9.6). As one approaches the Russian shoreline of the Caspian Sea, however, more desertlike conditions are encountered, limiting agriculture to irrigated tracts and extensive livestock grazing.

The Ural Mountains and Siberia

The Ural Mountains (Figure 9.2) mark European Russia's eastern edge, separating it from Siberia, or Asian Russia. Despite their geographical significance as the traditional division between continents, the Urals are not a particularly impressive range; several of their southern passes are less than 1,000 feet (305 meters) high, and railroad travelers sometimes fail to notice the passage. The setting is even less conducive to farming than farther west: the city of Yekaterinburg is distinctly colder and drier than Moscow, reflecting the increasingly continental climate of the Russian interior. Although not agriculturally productive, the range is still significant for two reasons: its ancient rocks are heavily mineralized, and it once marked Russia's eastern cultural boundary.

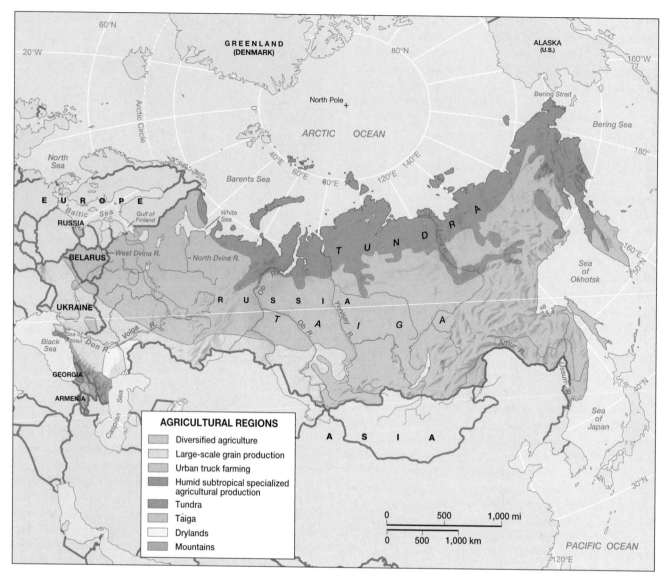

▲ **Figure 9.5 Agricultural regions** Harsh climate and poor soils combine to limit agriculture across much of the Russian domain. Better farmlands are found in Ukraine and in European Russia south of Moscow. Portions of southern Siberia support wheat production, but yield marginal results. In the Russian Far East, warmer climates and better soils translate into higher agricultural productivity. *(Modified from Clawson and Fisher, 1998,* World Regional Geography, *Upper Saddle River, NJ: Prentice Hall)*

East of the Urals, the vastness of Russian Siberia unfolds across the landscape for thousands of miles. The great Arctic-bound Ob, Yenisey, and Lena rivers (Figure 9.5) drain millions of square miles of northern country that includes the level and poorly drained West Siberian Plain, the hills and plateaus of the Central Siberian Uplands, and the rugged and isolated Northeast Highlands. Wintertime climatic conditions vary from legendary Verkhoyansk, with average January minimum temperatures of –58 °F (–50 °C), to the more moderate readings found in southern Siberian settlements, such as Irkutsk (–6 °F/–21 °C) near Lake Baikal. Most precipitation falls in the summer across Siberia, but much of the region receives less than 20 inches (51 centimeters) of moisture annually.

Siberian vegetation and agriculture reflect the climatic setting. The northern portion of the region is too cold for tree growth and instead supports tundra vegetation, which is char-

acterized by mosses, lichens, and a few ground-hugging flowering plants. South of the tundra, the Siberian **taiga,** or coniferous forest zone, dominates a large portion of the Russian interior (Figure 9.7). Indeed, more than 20 percent of the world's forests lie within the region. The trees of the taiga are generally small and slow-growing fir, spruce, and larch, as the climate is harsh and the soils generally acidic and poor in nutrients. Only localized agriculture (marginal wheat farming and potato cultivation) is possible even within favored portions of the zone. In the east, much of the tundra and taiga regions also are associated with **permafrost,** a cold-climate condition of unstable, seasonally frozen ground that limits the growth of vegetation and makes problematic the construction of even simple railroad tracks. Conditions moderate across southwestern Siberia: longer growing seasons and better soils offer more agricultural opportunities, although precipitation

▲ **Figure 9.6 Commercial wheat production**   Ukraine possesses some of the region's best cropland, including a sizable zone of commercial wheat production. Highly mechanized operations improve farm productivity on both state-controlled and privately managed acreage. *(Tass/Sovfoto/Eastfoto)*

▲ **Figure 9.7 Siberian taiga**   Siberia's vast coniferous forests stretch from the Urals to the Pacific. Known as the taiga, these fir, spruce, and larch forests lie in a zone too cold for commercial agriculture. Lumbering and industrial pollution increasingly threaten this national resource.   *(Andrey Zvoznikov/The Hutchison Library)*

decreases and becomes less dependable along the border with Kazakstan. Even with these drier conditions, southwest Siberia offers a narrowed eastward extension of the productive grain-growing belts of the western Russian and Ukrainian steppe.

The Russian Far East

Proximity to the Pacific Ocean, a more southerly latitude, and a pair of fertile river valleys create a distinctive subregion within the Russian Far East. About the same latitude as North America's New England, the region features longer growing seasons and milder climates than those found to the west or north. Here, the continental climates of the Siberian interior meet the seasonal monsoon rains of East Asia. Vladivostok is considerably warmer and wetter than Irkutsk, and the fertile Ussuri and Amur river valleys offer ample opportunities for mixed crop and livestock farming (Figure 9.5). The Amur forms a considerable portion of the Russian-Chinese border in the Far East and is the seventh longest river in the world. It is also a fascinating zone of ecological mixing: conifers of the taiga mingle with Asian hardwoods, and reindeer, Siberian tigers, and leopards find common ground. Natural hazards also haunt the region as part of the Pacific Rim of Fire (Figure 9.2). The Kuril Islands and the Kamchatka Peninsula both experience earthquakes and volcanic eruptions, but low population densities limit the human impact of this tectonic activity.

The Caucasus and Transcaucasia

In European Russia's extreme south, flat terrain gives way first to hills and then to the Caucasus Mountains, a large range stretching between the Black and Caspian seas (Figure 9.2). The highest point in the range, Mt. Elbrus, reaches beyond 18,000 feet (5,486 meters), and many peaks in the central and western Caucasus hold glaciers. Also home to major earthquakes, the Caucasus mark Russia's southern boundary. Farther south lies Transcaucasia and the distinctive natural setting of Georgia and Armenia. Extensive lowlands and low plateaus are found here, and a second, less-formidable moun-

tain range, aptly named the Lesser Caucasus, runs through the southern part of Georgia and along the boundary between Armenia and Azerbaijan.

Patterns of both climate and terrain in the Caucasus and Transcaucasia are tremendously complex. Rainfall is generally high in the western zone, with some slopes supporting dense forests. The area's eastern valleys, on the other hand, are semiarid or arid. In areas of adequate rainfall or where irrigation is possible, agriculture can be quite productive. Georgia in particular has long been a noted producer of fruits, vegetables, flowers, and wines (Figure 9.8). Bordering the Black Sea, western Georgia is dominated by a fertile alluvial

▲ **Figure 9.8 Subtropical Georgia**   The modifying influences of the Black Sea and a more southern latitude produce a small zone of humid subtropical agriculture in Georgia. The verdant landscapes of these tea plantations offer a sharp contrast to the colder country found north of the Caucasus.   *(Sovfoto/Eastfoto)*

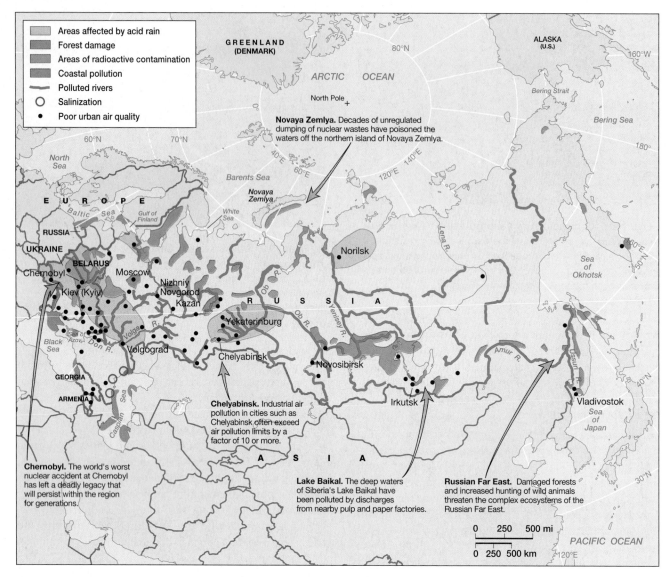

Figure 9.9 Environmental issues in the Russian domain Varied environmental hazards have left a devastating legacy across the region. The landscape has been littered with nuclear waste, heavy metals, and air pollution. Fouled lakes and rivers pose additional problems in many localities. Present economic difficulties and political uncertainties only add to the costly challenge of improving the region's environmental quality in the twenty-first century.

lowland, whereas central and eastern Georgia contain somewhat higher country wedged between the Caucasus and Lesser Caucasus ranges. The Georgian capital of Tbilisi enjoys a rare subtropical climate within the region that averages 30 °F warmer in the winter than does bone-chilling Moscow. Armenia, located mostly to the south of the Lesser Caucasus, is somewhat drier and is dominated by grains, potatoes, and fruits.

A Devastated Environment

As the story of the struggling sturgeon suggests, the Russian domain faces many environmental challenges. The breakup of the Soviet Union and subsequent opening of the region to international public scrutiny revealed some of the world's most

severe environmental degradation (Figure 9.9). The frenetic pace of seven decades of Soviet industrialization took its toll across the region. Even in some of the most remote reaches of Russia, careless mining and oil drilling, the spread of nuclear contamination, and rampant forest cutting have resulted in frightening environmental damage. New Russian environmental and antinuclear movements have protested these ecological disasters, but to date these movements remain a minor political voice in a region dominated by the desire for economic growth. In fact, hostile Russian political leaders eliminated that nation's environmental protection agency in 2000, and green political movements in Belarus and Ukraine have had little success in fostering tougher environmental standards. Indeed, the magnitude of many of these environmental challenges is so great that they have global implications and may affect world climate patterns, water quality, and nuclear safety. For example, since the 1980s, the global environmental costs of Siberian forests lost to lumbering and pollution may have exceeded the more widely publicized destruction of the Brazilian rain forest.

Air and Water Pollution   Poor air quality plagues hundreds of cities and industrial complexes throughout the region. The Soviet penchant for building large clusters of industrial processing and manufacturing plants in concentrated areas, often with minimal environmental controls, has produced an ongoing legacy of fouled air that stretches from Belarus to Russian Siberia. A traditional reliance on abundant, but low-quality coal also contributes to pollution problems. The air quality in dozens of cities within the region typically fails to meet health standards, particularly in the winter when cold-air inversions trap the polluted atmosphere for days on end. In Russia about 45 percent of the population lives in cities that exceed government pollution standards, and large numbers of urban residents suffer from chronic respiratory problems. Siberia's northern mining and smelting city of Norilsk is one of Russia's most polluted urban areas (Figure 9.10). In addition, a large swath of larch-dominated taiga has died in a huge zone of contamination that stretches more than 75 miles southeast of the city. Elsewhere, growing rates of private car ownership have greatly increased automobile-related pollution, especially since many vehicles lack catalytic converters and still use leaded gasoline. Today, 90 percent of Moscow's air pollution has been linked to the city's growing automobile traffic.

Degraded water is another hazard that residents of the region must cope with daily. Municipal water supplies are constantly vulnerable to industrial pollution, flows of raw sewage, and demands that increasingly exceed capacity. The extensive industrialization and dam-building along Russia's Volga Valley has produced a corridor of degraded water that stretches for hundreds of miles. Oil spills and seepage have harmed thousands of square miles in the tundra and taiga of the West Siberian Plain and along the Ob River estuary. Water pollution has also affected much of the northern Black Sea, large portions of the Caspian Sea shoreline, and even Arctic waters off Russia's northern coast.

The pollution of Siberia's Lake Baikal attracted international attention in the 1970s and 1980s. Lake Baikal, the world's largest reserve of freshwater, is a remarkable natural feature. Not only is the lake almost 400 miles (644 kilometers) long, but it is also 5,300 feet (1,615 meters) deep, occupying a structural rift in the continental crust (Figure 9.11). Lake Baikal is home to a large array of unique species, including the world's only freshwater seal. Until recently, Baikal's water was exceptionally pure. In the 1950s and 1960s, however, the Soviet government built pulp and paper factories along its shores, attracted by both the lake's clean water (useful for producing high-quality wood fibers) and the abundant forests in the surrounding uplands. With factory discharges, the lake's purity rapidly declined. By the 1970s Russian ecologists warned that Baikal's entire ecosystem was threatened.

▲ **Figure 9.10 Russian air pollution**   The mineral-processing center of Norilsk achieved a dubious reputation during the Soviet period as one of the dirtiest cities in Siberia, if not the entire Northern Hemisphere. Toxic air, damaged forests, and water pollution still plague the area today.   *(Bilderberg Archiv der Fotografen)*

▲ **Figure 9.11 Lake Baikal**   Southern Siberia's Lake Baikal is one of the world's largest deep-water lakes. Industrialization devastated water quality after 1950 as pulp and paper factories poured wastes into the lake. Recent cleanup efforts have helped, but many environmental threats remain.   *(Digital Image © 1996 Corbis)*

ENVIRONMENTAL HISTORY   The Legacy of Chernobyl

Northern Ukraine's Chernobyl nuclear plant experienced a deadly meltdown on April 25, 1986 (Figure 9.1.1). The reactor burned for 16 days, pouring smoke two miles into the sky and spreading nuclear contaminants from southern Russia to northern Norway. It proved to be the world's worst nuclear accident and one of the greatest environmental disasters of the modern age, and it will continue to impact the ecological health of the region for decades to come. Vladimir Chernousenko, a Ukrainian nuclear physicist, led attempts to clean up the disaster. He describes the process:

As soon as the explosion happened, troops were placed around the area. The government put a lid on the event immediately, and millions were not evacuated in time. . . . I was called in by Mikhail Gorbachev to evaluate what had happened. When I concluded my investigation, I sent a three-volume report to Gorbachev. Immediately, it became a secret document. . . . 65 million people in Russia received a dose, 90 million people north of the Ukraine may have been contaminated, and as many as 7,000 died immediately. . . . A million and a half people in and around Chernobyl (including the people who cleaned up the site) received extremely high doses of radiation, and millions of others still receive internal radiation daily from food contamination. Prior to the Chernobyl disaster, Ukraine had been the breadbasket of Europe; now there is no way to clean up the soil.

Chernousenko reports on the discouraging aftermath of the disaster:

Since Chernobyl, childhood and animal diseases in my country have increased fourfold. . . . In the years to come, many people are going to die. . . . There have been an estimated 200 accidents in nuclear installations in the former USSR, with millions of curies released. In my country not one square inch is free of radioactive fallout. People are losing their hair, and blood is coming out of their mouths.

Nuclear power stations are dangerous . . . even when they do not blow up. My assistants and I researched 10 plants, and we consistently found the water polluted and people around the plants sick.

Chernousenko was dying of cancer as he recalled the event, one of thousands to perish, either directly or indirectly, from one of the greatest environmental disasters ever.

Source: Adapted from "Ten Years Later, Chernobyl Is as Deadly as Ever," *Utne Reader*, May–June, 1996.

▲ **Figure 9.1.1 Aerial view of Chernobyl** One of the world's greatest environmental nightmares unfolded in April 1986 when the Chernobyl nuclear reactor experienced a meltdown. Nearby portions of northern Ukraine, Belarus, and Russia remain a toxic testimony to the disaster. *(Tass/Sovfoto/Eastfoto)*

Owing in part to the resulting international pressure, pollution from the paper mills was reduced. However, the lake is by no means out of danger.

The Nuclear Threat The nuclear era brought its own particularly deadly dangers to the region. The Soviet Union's aggressive nuclear weapons and nuclear energy programs expanded greatly after 1950, and issues of environmental safety were often ignored. Northeast Siberia's Sakha (or Yakutia) region, for example, suffered regular nuclear fallout in the era of aboveground nuclear testing. In other areas, nuclear explosions were widely utilized for simple seismic experiments, oil exploration, and dam-building projects. The once-pristine Russian Arctic has also been poisoned. During the Soviet era, the area around the northern island of Novaya Zemlya served as a huge and unregulated dumping ground for nuclear wastes. Nearby, dozens of atomic submarines have been abandoned to rust away amongst the fjords of the Kola Peninsula. The Navy has no funds to safely remove their nuclear fuel. Aging nuclear reactors also dot the region's landscape, often contami-

nating nearby rivers with plutonium leaks. Nuclear pollution is particularly pronounced in northern Ukraine, where the Chernobyl nuclear power plant suffered a catastrophic meltdown in 1986 (see "Environmental History: The Legacy of Chernobyl"). Large areas of nearby Belarus were also devastated in the Chernobyl disaster, which contaminated soils across much of the southern part of that country.

The region's involvement with the nuclear age continues to take new forms. In 2001 Russia's new Rostov Nuclear Energy Station went on-line, making it the region's first new nuclear power plant since the Soviet era. Growing blackouts and energy shortages have produced a new government drive to revive Russia's nuclear industry, and as many as 10 additional reactors are scheduled for completion by 2007. Ironically, the capital-hungry Russian government has also been exploring the commercial importation of nuclear waste from the rest of the world, not unlike similarly isolated and poverty-stricken communities and Indian reservations in the western United States. This is, however, an example of globalization that is being strongly resisted by Russian environmentalists.

The Post-Soviet Paradox The end of Soviet control has had mixed consequences for the region's environment. The demise of the Soviet Union brought about environmental improvement in some areas. Many factories shut down because they were no longer economically viable, which itself reduced pollution. The huge steel mills in the Russian city of Magnitogorsk, for example, produce less than half the raw steel they did a decade ago, and global competition threatens to reduce demand further. Ironically, cleaner air has been the result. Although costly, advanced pollution control equipment is also beginning to be imported from western Europe. Elsewhere in Russia, nuclear warhead storage facilities have been consolidated, and government authorities are responsible for maintaining control over the nation's 22,000 nuclear weapons. There is a growing environmental consciousness among young, educated Russians and Ukrainians. Increased connections with environmental activists in North America and Europe have also raised public awareness of many environmental problems within the region.

In other areas, however, the breakdown of centralized authority has contributed to more environmental degradation. Waste materials are often handled more casually than in the past, since the government no longer has much effective regulatory power. Given the central government's smaller role in the post-Soviet period, most cities in the region have been left to finance their own cleanup projects. For example, industrial Volgograd (in the southern Volga Valley) needs low-cost antipollution technology that could save hundreds of lives annually, but the city cannot afford the $100,000 price tag. An especially vexing problem is the disposal of nuclear byproducts, now sometimes smuggled out of the country and possibly used in illegal weapons manufacturing. The current economic troubles of the former Soviet world also have brought about an accelerated exploitation of natural resources. Russia must now frantically export oil and other minerals, as well as timber, in order to obtain the cash necessary to operate in the global economy. In the resulting resource rush, few safeguards have been implemented or enforced. Russia also faces an impending crisis of wildlife extinction. The old Soviet regime had some success in protecting both endangered species and sizable areas of natural habitat noted for their biological diversity. Unregulated hunting and trapping are now on the rise, however, and in some areas are virtually uncontrolled. It is questionable whether such animals as the Siberian tiger and the Siberian leopard, both inhabiting the forests of Russia's Far East, will survive the next few decades.

Population and Settlement: An Urban Domain

The five states of the Russian domain are home to more than 200 million residents. While they are widely dispersed across a vast Eurasian landmass, most live in cities. The region's evolving political setting, distinctive distributions of natural resources, and changing migration patterns continue to shape its population geography. The results have been a population strongly concentrated in the European West, periodic impulses to disperse beyond that traditional core, and an overall tendency toward rapid urbanization.

Population Distribution

Striking differences in population densities exist between European and Asian portions of the Russian domain. The more favorable agricultural setting of the European West historically encouraged higher densities of population than did the more inhospitable conditions found across central and northern Siberia. Although Russian efforts over the past century have encouraged a wider dispersal of the population, it remains heavily concentrated in the west (Figure 9.12). European Russia is home to more than 110 million persons, while Siberia, although far larger, holds only some 35 million. When one adds the 60 million inhabitants of Belarus and Ukraine, the imbalance between east and west becomes even more striking.

The European Core The region's largest cities, biggest industrial complexes, and most productive farms are located in the European Core, a subregion that includes Belarus, much of Ukraine, and Russia west of the Urals. The Core supports population densities that greatly exceed those to the east, although they are modest by northwest European or East Asian standards. The sprawling city of Moscow and its nearby satellite centers clearly dominate the settlement landscape with a metropolitan area of more than 8.5 million people (Figure 9.13). Within 250 miles, a series of other major urban centers are closely linked to Moscow. Largest of these is the industrial city of Nizhniy Novgorod (formerly named Gor'kiy in the Soviet period) (1.5 million), traditionally oriented around automobile and heavy equipment manufacturing.

Beyond Moscow's immediate orbit, three other areas of concentrated urban settlement dominate European Russia (Figure 9.12). On the shores of the Baltic Sea, St. Petersburg (Leningrad in the Soviet period) (4.8 million) offers a major trading window to the West. Between 1712 and 1917 it served as the capital of the Russian Empire, and it acquired a rich skyline of baroque architecture and beautiful churches, which gave the city an urban landscape many have compared to the great cities of western Europe. Although industrialization and Soviet-style buildings took their toll on the city during the communist era, St. Petersburg has recently seen an architectural renaissance, interest in preserving its Russian Orthodox churches, and increased tourism (Figure 9.14).

Far southeast of Moscow, a second urban focus is oriented along the lower and middle stretches of the Volga River. Industrialization within the region accelerated greatly during World War II, as the region lay somewhat removed from German advances in the west. Today, the highly commercialized river corridor, also blessed with nearby petroleum reserves, supports a diverse industrial base strategically located to serve the large populations of the European Core. From north to south, the four Volga Valley cities of Kazan (1.1 million), Samara (1.2 million), Saratov (900,000), and Volgograd (1 million; Stalingrad in the Soviet period) are the largest settlements within the region, and each possesses a sizable industrial infrastructure.

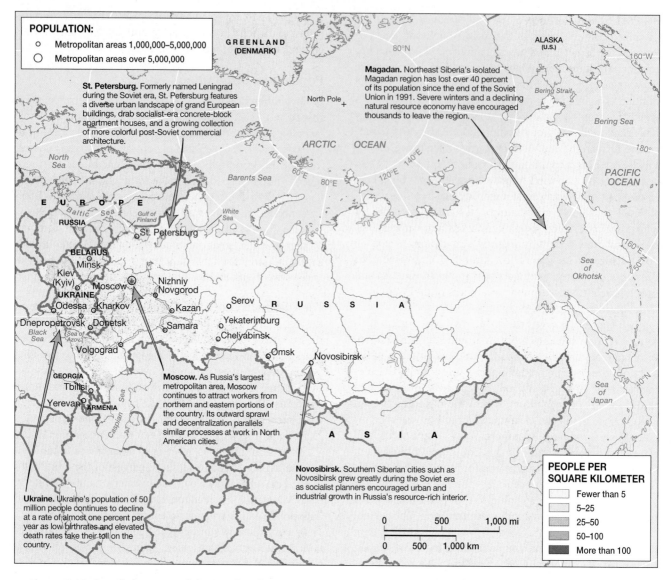

POPULATION:
- ○ Metropolitan areas 1,000,000–5,000,000
- ◯ Metropolitan areas over 5,000,000

St. Petersburg. Formerly named Leningrad during the Soviet era, St. Petersburg features a diverse urban landscape of grand European buildings, drab socialist-era concrete-block apartment houses, and a growing collection of more colorful post-Soviet commercial architecture.

Magadan. Northeast Siberia's isolated Magadan region has lost over 40 percent of its population since the end of the Soviet Union in 1991. Severe winters and a declining natural resource economy have encouraged thousands to leave the region.

Moscow. As Russia's largest metropolitan area, Moscow continues to attract workers from northern and eastern portions of the country. Its outward sprawl and decentralization parallels similar processes at work in North American cities.

Ukraine. Ukraine's population of 50 million people continues to decline at a rate of almost one percent per year as low birthrates and elevated death rates take their toll on the country.

Novosibirsk. Southern Siberian cities such as Novosibirsk grew greatly during the Soviet era as socialist planners encouraged urban and industrial growth in Russia's resource-rich interior.

PEOPLE PER SQUARE KILOMETER
- Fewer than 5
- 5–25
- 25–50
- 50–100
- More than 100

▲ **Figure 9.12 Population map of the Russian domain** Population within the region is strongly clustered west of the Ural Mountains. Dense agricultural settlements, extensive industrialization, and large urban centers are found in Ukraine, much of Belarus, and across western Russia south of St. Petersburg and Moscow. A narrower chain of settlements follows the better lands and transportation corridors of southern Siberia, but most of Russia east of the Urals remains sparsely settled.

A third constellation of key population centers on the eastern edge of the Core is anchored along the resource-rich alignment of the Ural Mountains. From the gritty industrial landscapes of Serov (1.0 million) and Yekaterinburg (1.2 million; Sverdlovsk in the Soviet period) in the north to Chelyabinsk (1.1 million) and Magnitogorsk (425,000) in the south, the Urals region specializes in iron and steel manufacturing, metals smelting and refining, and heavy machinery construction. Outside of these industrial centers, however, the rural population density of the Urals is lower than in the better agricultural lands found farther west within the Core.

Beyond Russia, major population clusters within the European Core are also found in Belarus and Ukraine. The Belorussian capital of Minsk (1.7 million) is the dominant urban center in that country, and its landscape recalls the drab Soviet-style architecture of an earlier era (Figure 9.15). In nearby Ukraine, the capital of Kiev (2.6 million) straddles the Dnieper River, and the city's rich architectural heritage is a reminder of its historic role in the political and economic geography of the European interior. Other large industrial centers, such as Kharkov (1.6 million), Dnepropetrovsk (1.2 million), and Donetsk (1.1 million), are located in resource-rich eastern Ukraine and benefit from their proximity to deposits of coal and iron ore. Smaller population clusters are found in Georgia's fertile Black Sea lowlands and around the capital of Tbilisi (1.3 million). To the southeast, the Armenian capital of Yerevan (1.3 million) contains almost one-third of that nation's population.

Figure 9.14 St. Petersburg Picturesque Canal Street captures some of the architectural character that makes St. Petersburg one of Russia's most beautiful cities. Despite the construction of many concrete buildings in the Soviet period, St. Petersburg retains a charm that echoes the urban landscapes of western Europe. *(Nick Nicholson/The Image Bank)*

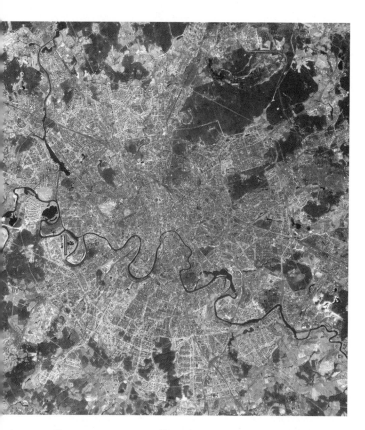

Figure 9.13 Metropolitan Moscow Sprawling Moscow extends more than 50 miles (80 kilometers) beyond the city center. The city is home to more than 8.5 million people, and the relative strength of its urban economy continues to attract migrants from elsewhere in the country, thus putting more pressure on its infrastructure. *(CNES/Spot Image/Photo Researchers, Inc.)*

Siberian Hinterlands Leaving the southern Urals city of Yekaterinburg on a Siberia-bound train, one is acutely aware that the land ahead is ever more sparsely settled (Figure 9.12). The distance between cities grows, and the intervening countryside reveals a landscape shifting gradually from farms to forest. The Siberian hinterland is divided into two characteristic zones of settlement. To the south, an alignment of isolated, but sizable urban centers follows the **Trans-Siberian Railroad** (Figure 9.16), a key railroad corridor to the Pacific completed in 1904. The industrial settlements along the route benefit from good east–west connections as well as from proximity to natural resources such as oil, natural gas, coal, and iron ore. The eastbound traveler encounters Omsk (1.3 million) as the rail line crosses the Irtysh River, Novosibirsk (1.3 million) at its junction with the Ob River, and Irkutsk (600,000) near the southwest corner of Lake Baikal. More than 1,500 miles (2,415 kilometers) beyond, the better agricultural lands and industrial opportunities of the Russian Far East contribute to higher population densities. Khabarovsk (620,000) is the leading settlement of the Amur Valley, while the port city of Vladivostok (730,000) provides the principal Russian access to the Pacific. A thinner sprinkling of settlement appears along the more recently completed (1984) **Baikal-Amur Mainline**

Figure 9.15 Minsk Almost 2 million people reside in the Belorussian capital of Minsk. Its drab apartments and office buildings echo the strong shaping legacy of the Soviet period on the urban scene. Indeed, Belarus retains many Soviet-style characteristics, trailing far behind Russia in the pace of economic reforms. *(Novosti/Sovfoto/Eastfoto)*

(BAM) Railroad, which parallels the older line, but runs north of Lake Baikal to the Amur River. North of the BAM line, however, the seemingly empty vastness of the Siberian hinterland dominates the scene. Settlements are few and far between, and larger urban areas are usually either regional administrative capitals, such as Yakutsk (200,000), or resource extraction and processing centers, such as Norilsk (200,000).

▲ Figure 9.16 Trans-Siberian Railroad   Snaking its way through winter and the vastness of the southern Siberian landscape, the Trans-Siberian Railroad provides a critical connection between European and Asian portions of the Russian domain. *(Tass/Sovfoto/Eastfoto)*

Regional Migration Patterns

Over the past 150 years, millions of people within the Russian domain have been on the move. These major migrations, both forced and voluntary, reveal sweeping examples of human mobility that rival the great movements from Europe and Africa or the transcontinental spread of settlement across North America.

Eastward Movement   Just as settlers of European descent moved west across North America, exploiting natural resources and displacing native peoples, European Russians moved east across the vast Siberian frontier to extend their influence and rework basic population geographies within the Eurasian interior. Although the deeper historical roots of the movement extend back several centuries, the pace and volume of the eastward drift accelerated in the late nineteenth century once the Trans-Siberian Railroad was completed (Figure 9.16). Peasants were attracted to the region by its agricultural opportunities (in the south) and by greater political freedoms than they traditionally enjoyed under the **tsars** (or czars; Russian for *Caesar*), the authoritarian leaders who dominated politics during the pre-1917 Russian Empire. Almost 1 million Russian settlers moved into the Siberian hinterland between 1860 and 1914.

The eastward migration continued during the Soviet period, once communist leaders consolidated power during the late 1920s and saw the economic advantages of developing the region's rich resource base. The German invasion of European Russia during World War II demonstrated that there were also strategic reasons for settling the eastern frontier, and this propelled further migrations during and following the war. Indeed, by the end of the communist era, 95 percent of

Siberia's population was classified as Russian (including Ukrainians and other western immigrants). With the completion of the BAM Railroad in the 1980s, yet another corridor of settlement opened in Siberia, prompting new migrations into a region once remote from the outside world.

Political Imperatives   Political motives have also shaped migration patterns. Particularly in the case of Russia, leaders from both the imperial and Soviet eras saw advantages in moving selective populations to new locations. Clearly, for example, the infilling of the southern Siberian hinterland had a political as well as an economic rationale. Both the tsars and the Soviet leaders saw their geopolitical fortunes rise as Russians moved into the resource-rich Eurasian interior. For some, however, the move to Siberia had a different connotation. The region became a repository for political dissidents and troublemakers. Especially in the Soviet period, uncounted millions were forcibly relocated to the region's infamous **Gulag Archipelago**, a vast collection of political prisons in which inmates often disappeared or spent years far removed from their families and home communities. The communist regimes of Joseph Stalin (1928–1953) and his successors were particularly noted for their forced migrations, including the removal of thousands of Jews to the Russian Far East between 1928 and 1958 and the involuntary relocation of numerous ethnic minorities during World War II.

Russification, the Soviet policy of resettling Russians into non-Russian portions of the Soviet Union, also had profound consequences for the region's human geography. Millions of Russians were given economic and political incentives to move elsewhere in the Soviet Union in order to increase Russian dominance in many of the outlying portions of the country. The migrations were geographically selective in that most of the Russians moved either to administrative centers or to industrial complexes. As a result, by the end of the Soviet period, Russians made up significant minorities within former Soviet republics (now independent nations) such as Kazakstan (38 percent Russian), Latvia (34 percent), and Estonia (30 percent). Among its Slavic neighbors, Belarus remains 13 percent Russian and Ukraine more than 22 percent Russian, with concentrations particularly high in the eastern portions of the countries. Often, these Russian workers were given special employment and housing privileges in their new localities, thus provoking antipathy among the non-Russian population.

New International Movements   In the post-Soviet era, Russification has often been reversed (Figure 9.17). Several of the newly independent non-Russian countries have imposed rigid language and citizenship requirements, which encouraged many Russian residents to leave. In other settings, ethnic Russians simply experienced varied forms of social and economic discrimination. By 2000, about 6 million Russians had left former Soviet republics and had been repatriated in their homeland. As a result, the Central Asia and Baltic regions that were once a part of the Soviet Union saw their Russian populations decline significantly in the 1990s, often by 20 to 35 percent.

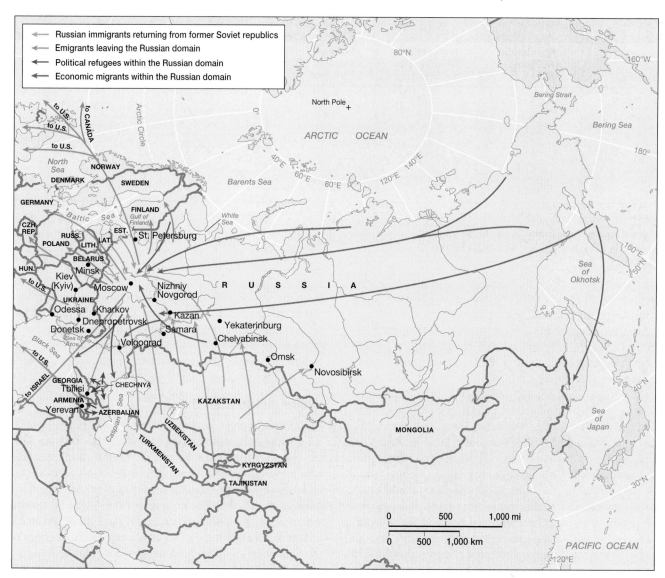

Legend:
- ← Russian immigrants returning from former Soviet republics
- ← Emigrants leaving the Russian domain
- ← Political refugees within the Russian domain
- ← Economic migrants within the Russian domain

▲ **Figure 9.17 Recent migration flows in the Russian domain** Recent events are encouraging the return of ethnic Russians from former Soviet republics, while other Russians are emigrating from the domain for economic, cultural, and political reasons. Within Russia, both political and economic forces are also at work encouraging people to be on the move.

The domain's more open borders have also made it easier for other residents to leave the region (Figure 9.17). Poor economic conditions and the region's unpredictable politics have encouraged many to emigrate. The "brain drain" of young, well-educated, upwardly mobile Russians has been considerable. Sometimes, ethnic links play a part in migration patterns. For example, many Russian-born ethnic Finns have moved to nearby Finland, much to the consternation of the Finnish government. Russia's Jewish population also continues to fall, a pattern begun late in the Soviet period. These emigrants have flocked mostly to Israel or the United States, where they locate in familiar Jewish neighborhoods such as Miami, Los Angeles, and New York City's Brighton Beach district. Many non-Jewish residents have also left: as the title of an old Hollywood movie, *The Russians Are Coming, The Russians Are Coming*, suggests, Rus-

sians have become one of the largest new immigrant groups in the United States. In related fashion, a recent U.S. Immigration Service report suggests that young Ukrainian women have been a favorite choice for American men searching for marriage partners on the Web. Several thousand U.S. marriages annually have resulted, suggesting how the forces and technologies of globalization are playing out in unanticipated ways. Armenians have also joined the exodus, with more than 300,000 now living in Los Angeles County.

The Urban Attraction Regional residents have also been bound for the cities. The Marxist philosophy embraced by Soviet planners encouraged urbanization. In 1917 the Russian Empire was still overwhelmingly rural and agrarian; 50 years later, the Soviet Union was primarily urban. Planners saw

TABLE 9.1 *Demographic Indicators*

Country	Population (Millions, 2001)	Population Density, per square mile	Rate of Natural Increase	TFR[a]	Percent < 15[b]	Percent > 65	Percent Urban
Armenia	3.8	330	0.3	1.1	24	9	67
Belarus	10	125	−0.4	1.3	19	13	70
Georgia	5.5	203	0	1.2	20	13	56
Russia	144.4	22	−0.7	1.2	18	13	73
Ukraine	49.1	211	−0.7	1.1	18	14	68
Total	212.8						

[a]Total fertility rate

[b]Percent of population younger than 15 years of age

Source: Population Reference Bureau. World Population Data Sheet, *2001.*

great economic and political advantages in efficiently clustering the population, and Soviet policies dedicated to large-scale industrialization obviously favored an urban orientation. Today, Russian, Ukrainian, and Belorussian rates of urbanization are comparable to those of the industrialized capitalist countries (Table 9.1).

Soviet cities grew according to strict governmental plans. Planners selected different cities for different purposes. Some were designed for specific industries, while others were allocated primarily administrative roles. All cities were allotted set population levels. A system of internal passports prohibited people from moving freely from city to city. Instead, people generally went where the government assigned them jobs. Moscow, the country's leading administrative city, thrived under the Soviet regime. It formed the undisputed core of Soviet bureaucratic power as well as the center of education, research, and the media. The impact of centralized administrative functions in the Soviet Union was also evident in the rise of the capital cities of its constituent republics. Yerevan, the capital of Armenia, was little more than a town in 1920, yet by the 1990s it had grown into a major city of more than 1 million inhabitants. Minsk, the capital of Belarus, experienced a similar, although not quite as dramatic, expansion. Specialized industrial cities grew at an even faster pace. In the mining and metallurgical zone of the southern Urals, centers such as Yekaterinburg and Chelyabinsk mushroomed into major urban centers. Another cluster of specialized industrial cities, including Kharkov and Donetsk, emerged in the coal districts of eastern Ukraine.

With the end of the Soviet Union, however, people gained basic freedoms of mobility. In addition, with a more open economy, especially in Russia, shifting urban employment opportunities increasingly reflect how effectively local economies could compete in the global market. This new economic reality, for example, has led to the depopulating of many older industrial areas, as they simply cannot produce raw materials or finished industrial goods at competitive global prices (Figure 9.17). Since 1991, for example, over 12 percent of the population of resource-rich northern Russia has emigrated. In many cases, people are freely gravitating toward growth opportunities in locations of new foreign in-

vestment, principally in larger urban areas in western and southern Russia. Still, some persons are reluctant to move because they often still enjoy inexpensive, state-subsidized housing that they would risk losing (many forego paying rent and utilities altogether, yet remain secure in their lodgings).

Inside the Russian City

Large Russian cities possess a core area, or center, that features superior transportation connections; the best-stocked, upscale department stores and shops; the most desirable housing; and the most important offices (both governmental and private) (Figure 9.18). At the top of the urban hierarchy, large cities such as Moscow and St. Petersburg also feature extensive public spaces and examples of monumental architecture at the city center. Inner-city decay, so characteristic of the United States, is not a feature of the Russian central city. Russian cities also lack sprawling decentralized suburbs like those of North America and thus tend to end abruptly. One often passes from a landscape of high-rise apartments into an essentially rural setting without traveling through an extensive zone of single-family dwellings.

Within the city, there is usually a distinctive pattern of concentric land-use zones, each of which was built at a later date as one moves outward from the center. Such a ringlike urban morphology is not a unique phenomenon, but it is probably more highly developed here than in most parts of the world, owing to the extensive power of government planners during the Soviet period. At their very center, the cores of many older cities predate the Soviet Union. Pre-1900 stone buildings often dominate older city centers. Some of these are former private mansions that were turned into government offices or subdivided into apartments during the communist period but are now being reprivatized.

Beyond this pre-communist urban core, one can often find a ring of public housing projects and fully planned *sotzgorods*, or socialist neighborhoods. Sotzgorods are based on a close connection between workplace and home, and typically include spartan dormitory-style housing. These projects became most common in the heavy industrial cities that

▲ Figure 9.18 Downtown Moscow   The bustling traffic
along central Moscow's Novy Arbat parallels urban scenes elsewhere in Europe and North America. The city's landscape remains a complex and fascinating mix of imperial, Soviet, and
post-Soviet influences.   (Itar-Tass/Sovfoto/Eastfoto)

▲ Figure 9.19 Moscow housing   For many residents in larger Russian cities, home is a high-rise apartment house. Most of
these satellite centers were built in the Soviet era. Poor construction and a lack of landscaping often yield a bleak suburban
scene, but nearby stores, entertainment, and public transportation offer important amenities.   (Itar-Tass/Sovfoto/Eastfoto)

mushroomed during the early decades of the Soviet Union.
Their age and their industrialized surroundings now make
them among the least desirable places in which to live.

Another common urban zone, removed some distance
from city cores, is the *chermoyuski*. Chermoyuski are large,
uniform apartment blocks built during the 1950s and 1960s.
At that time a more prosperous Soviet Union could give each
family its own small, private flat. Most of these nearly uniform apartment blocks are five stories or less in height, since
they typically are not equipped with elevators. The main problem with the chermoyuski is shoddy construction. Most were
designed to last only a few decades, since Soviet planners expected a booming economy that would soon allow the construction of better, more permanent, dwellings. Today, many
of these apartment complexes remain, but they are now di-
lapidated, with little money available for renovation.

Farther out from the city centers are the **mikrorayons**, the
much-larger housing projects of the 1970s and 1980s (Figure
9.19). Mikrorayons are typically composed of massed blocks
of standardized apartment buildings, ranging from 9 to 24
stories in height. Each mikrorayon was to form a self-contained community, with grocery stores and other basic serv-

ices located within walking distance of each apartment building. The largest of these supercomplexes contain up to
100,000 residents. While planners hoped that mikrorayons
would foster a sense of community, most now serve largely as
anonymous bedroom communities for larger metropolitan
areas. Although public transportation is well developed in
most cities of the region, commuting time between city centers and these outer apartment complexes is usually 45 minutes or more.

Some of Russia's most rapid urban growth in recent years
has occurred on the metropolitan periphery, paralleling the
North American experience. Moscow, for example, has seen
its urban reach expand far beyond the city center. The surrounding administrative district (the Moscow Oblast) contains about 7 million residents, and local officials estimate
that more than $1 billion was invested in the fast-growing
suburban fringe by foreigners in 2001. Already home to more
than 700 companies, the outer city has been attracting many
global corporate offices, including those of PepsiCo, Procter
and Gamble, and Bristol-Myers. Land prices and tax rates are
lower than in the central city; the bureaucracy is less onerous;
and the transportation and telecommunications infrastructure is relatively new.

Elsewhere on Moscow's urban fringe, elite **dacha,** or cottage communities, appeal to many more well-to-do residents,
particularly during the summer months (Figure 9.20). The
tradition of rural retreats, dating back to the Russian empire,
also thrived during the Soviet era as Communist Party officials
sought an escape from the dreary bureaucratic chores of the
city. Today, there are about 300 cottage settlements on the
Moscow periphery, including many new privatized developments northwest and southeast of the city that cater to the
country's business elite. Some of the new luxury cottages sell
for more than $300,000. The older dacha belt to the west of

▲ **Figure 9.20 Dacha** Moscow's rural fringe remains a popular spot for dachas. The summer homes provide a rural respite for Moscow residents who can afford to rent or purchase these countryside retreats. *(Tass/Sovfoto/Eastfoto)*

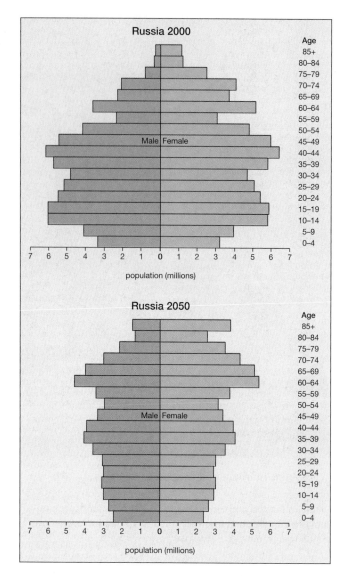

▲ **Figure 9.21 Russia's changing population** These two population pyramids provide a recent glimpse (2000) as well as predicted patterns (2050) of Russia's population structure. Present trends suggest that Russia's population will continue to age, with relatively fewer young people supporting a relatively large elderly population. Also note the impact of earlier wars and higher death rates on older adult Russian males. *(U.S. Census Bureau, International Data Base)*

the city along the Moskva River also offers summer homes, spa retreats, and holiday hotels in a zone dubbed the "Sub-Moscow Switzerland." While such amenities are far beyond the reach of most urbanites, growing dacha communities illustrate how the ongoing evolution of Russian society continues to play out on the everyday scene. They also suggest how modern settlement patterns represent landscape elements inherited from the Soviet period, as well as the imprint of dramatic new forces at work across the Russian domain.

The Demographic Crisis

In early 2001 Russia's Ministry of Labor and Social Development acknowledged that the country faced a crisis of persistently declining populations. Government officials pledged to improve the nation's health-care system, provide more incentives for increasing the birthrate, and foster immigration. The success of those initiatives remains to be seen, however, and government studies in 2000 predicted that Russia's population could fall by 3 million people by 2005 and by a startling 25 million by 2030. Similar conditions are affecting the other countries within the region (Table 9.1). Declining populations, low birthrates, and rising mortality, particularly among middle-aged males, are all symptoms of the grim demographic scenario that has engulfed the region. During World War II, large numbers of deaths combined with low birthrates to produce sizable population losses. While population increases accelerated in the 1950s, growth slowed by 1970, and death rates began exceeding birthrates in the early 1990s. Two population pyramids tell the troubling tale (Figure 9.21). The first shows that, while adult-age populations are prominently represented (with the exception of older males and the birth crash of World War II), few children are being added. Looking ahead to 2050, today's shrinking numbers take an even bigger toll as the pattern of small families is likely to continue.

The region's demographic crisis appears related to the fraying social fabric and uncertain economic times. Very low birthrates within the region may reflect a lack of optimism about the future. In Russia, for example, total live births fell from a peak of 2.5 million in 1987 to fewer than 1.4 million in the late 1990s, a period of tremendous economic and political disruption. Many Russian families report that they simply do not have sufficient incomes to support children. The health of women of childbearing age has also declined, while problem pregnancies, maternal childbirth death rates, and birth defects are on the rise. Similarly, sharp increases in death rates, especially among Russian men, appear related to many stress-related conditions, such as alcoholism and heart disease. Murder and suicide rates have climbed rapidly in the past 15 years. In addition, perhaps 20 to 30 percent of the increase in death rates may be attributed to the region's increasingly toxic environment. Whatever the complex causes, the average life span for Russians has fallen appreciably since the late 1980s, and for men the drop has been truly unprecedented in the developed world.

Cultural Coherence and Diversity: The Legacy of Slavic Dominance

For hundreds of years, Slavic peoples speaking the Russian language expanded their influence from an early homeland in central European Russia. Eventually, this Slavic cultural imprint spread north to the Arctic Sea, south to the Black Sea and Caucasus, west to the shores of the Baltic, and east to the Pacific Ocean. In this process of diffusion, Russian cultural patterns and social institutions spread widely, and they also influenced scores of non-Russian ethnic groups that continued to live under the rule of the Russian empire. The legacy of that Slavic expansion continues today. It offers Russians a rich historical identity and sense of nationhood (see: "Local Voices: Exploring Russian Cultural Identity"). It also provides a meaningful context to understand the way present-day Russians are dealing with forces of globalization and how non-Russian cultures have evolved within the region.

The Heritage of the Russian Empire

The expansion of the Russian Empire paralleled similar events in western Europe. As Spain, Portugal, France, and Britain carved out empires in the Americas, Africa, and Asia, the Russians expanded eastward and southward across Eurasia. Unlike other European empires, however, the Russians formed one single territory, uninterrupted by oceans or seas. Partly because of its contiguous nature, the Russian Empire—after being transformed into the Soviet Union—remained intact while the empires of the other European powers collapsed after World War II. Only with the fall of the Soviet Union in 1991 did this transformed empire finally begin to dissolve.

Origins of the Russian State   The origin of the Russian state lies in the early history of the **Slavic peoples**, defined linguistically as a distinctive northern branch of the Indo-European language family. The Slavs originated in or near the Pripyat marshes of modern Belarus. Some 2,000 years ago they began to migrate to the east, extending as far as modern Moscow by A.D. 200. Slavic political power grew by A.D. 900 as they intermarried with southward-moving warriors from Sweden known as *Varangians*, or *Rus*. Within a century, the state of Rus extended from Kiev (the capital) in modern Ukraine, to Lake Ladoga near St. Petersburg. The new Kiev-Rus state interacted with the rich and powerful Byzantine Empire of the Greeks, and this influence brought Christianity to the Russian realm by A.D. 1000. Along with the new religion came many other aspects of Greek culture, including the Cyrillic alphabet. Even as the Russians converted to **Eastern Orthodox Christianity**, a form of Christianity historically linked to eastern Europe and church leaders in Constantinople (modern Istanbul), their Slavic neighbors to the west (the Poles, Czechs, Slovaks, Slovenians, and Croatians) accepted Catholicism. The resulting religious division split the Slavic-speaking world into two groups, one oriented to the west, the other to the east and south. This early Russian state soon faltered and split into several principalities that were then ruled by invading Mongols and Tatars (a group of Turkish-speaking peoples).

LOCAL VOICES   Exploring Russian Cultural Identity

Anyone who has pored over a novel by Dostoyevsky or Tolstoy, thrilled to the music of Tchaikovsky or Stravinsky, or simply pondered the epic events of Russian history understands that Russians share a passionate and complex cultural identity. There are many elements that contribute to its persistence, including language, religion, the arts, an attachment to special places (such as Moscow, the historic Volga River Valley, or Siberia), and a vividly shared cultural history that has managed to combine dramatic achievements with enduring pain and tragedy.

Russian-born novelist Andrei Makine captures some sense of that traditional identity in *Dreams of My Russian Summers*, a novel set in the Soviet Union of the 1960s and 1970s. At one point in the story, the main character, a young man coming of age, ponders his past and his tangled place within the present:

Yes, I was Russian. Now I understood, in a still confused fashion, what that meant. Carrying within one's soul all those human beings disfigured by grief, those burned villages, those lakes filled with naked corpses. Knowing the resignation of a human herd violated by

a despot. And the horror of feeling oneself participating in this crime. And the wild desire to reenact all these stories from the past—so as to eradicate from them the suffering, injustice, and death . . . Remaking history. Purifying the world . . . Pushing this commitment to the point of delirium, to the point of fainting. Living very mundanely on the edge of the abyss. Yes, that's what Russia is . . .

It is difficult to know whether young twenty-first century Russians will be confronted by such passionate encounters with their past. The breathtaking speed of recent changes suggests that the nation's cultural identity continues to evolve. A Russian cultural historian put it well when he wrote, "We, the Russians, are . . . once more the most interesting phenomenon on earth . . . like a novel whose ending none of us knows." Indeed, the Russians continue to carve a new path, their cultural identity torn between a past that won't let them go and a better future stubbornly slow to appear.

Sources: Andrei Makine, *Dreams of My Russian Summers*. New York: Scribners, 1998, pp. 146–47, and Andrei Sinyavsky, *Soviet Civilization: A Cultural History*. New York: Arcade Publishing, 1988, p. 273.

Growth of the Russian Empire   By the fourteenth century, however, northern Slavic peoples overthrew Tatar rule and established a new and expanding Slavic state (Figure 9.22). The core of the new Russian Empire lay near the eastern fringe of the old state of Rus. The former center around Kiev was now a war-torn borderland (or "Ukraine" in Russian) contested by the Orthodox Russians, the Catholic Poles, and the Muslim Turks. Gradually this area's language diverged from that spoken in the new Russian core, and *Ukrainians* and Russians developed into two separate peoples. Eventually Ukraine was incorporated into the Russian Empire, but the identity of its inhabitants remained distinctive. A similar development took place among the northwestern Russians, who experienced several centuries of Polish rule and over time were transformed into a distinctive group known as the *Belorus-*

sians. Thus the original Russians were divided into three distinctive but closely related peoples: the Russians proper, the Belorussians (of Belarus), and the Ukrainians.

The Russian Empire expanded remarkably in the sixteenth and seventeenth centuries (Figure 9.22). The arctic fringe of European Russia, inhabited primarily by non-Slavic peoples, provided furs as well as a sea route to western Europe (through the White Sea port of Archangel). Former Tatar territories in the Volga Valley (near Kazan) were incorporated into the Russian state in the mid-1500s. The Russians also allied with the seminomadic **Cossacks**, Slavic-speaking Christians who had earlier migrated to the region to seek freedom in the ungoverned steppes (Figure 9.23). The Russian empire granted them considerable privileges in exchange for their military service, an alliance that facilitated Russian expansion into

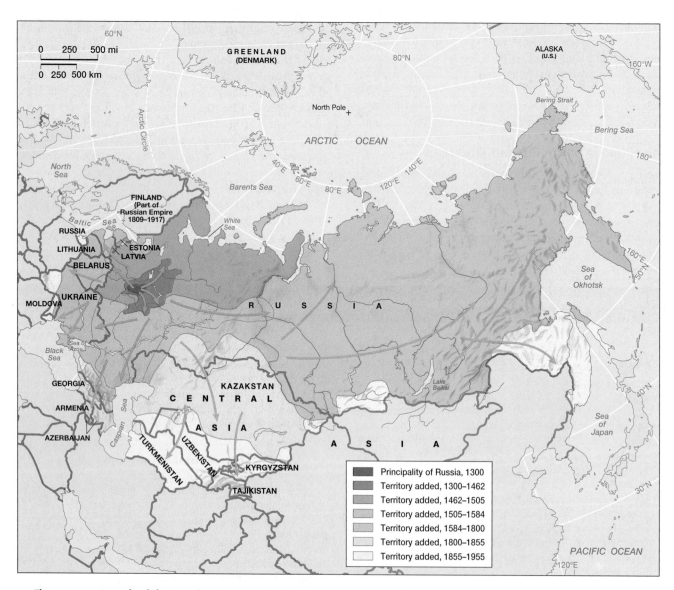

▲ **Figure 9.22 Growth of the Russian Empire**   Beginning as a small principality in the vicinity of modern Moscow, the Russian Empire took shape between the fourteenth and sixteenth centuries. After 1600, Russian influence stretched from eastern Europe to the Pacific Ocean. Later, portions of the empire were added in the Far East, Central Asia, and near the Baltic and Black seas.   *(Modified from Bergman and Renwick, 1999,* Introduction to Geography, *Upper Saddle River, NJ: Prentice Hall)*

▲ **Figure 9.23 Cossack** Natives of the Ukrainian and Russian steppe, the highly mobile Cossacks played a pivotal role in aiding Russian expansion into Siberia during the sixteenth century. Many modern descendents retain their skills of horsemanship and are proud of their distinctive ethnic heritage. *(Julien Chatelin/Getty Images, Inc.)*

Siberia. Furs were the chief lure of this immense northern territory. Since cold climates produce high-quality furs, those of Siberia commanded premium prices in the developing global market for the resource. By the 1630s Russian power was entrenched in central Siberia, and by the end of the century it had reached the Pacific Ocean. Chinese resistance, however, delayed Russian occupation of the Far East region until 1858, and the imperial designs of the Japanese halted further expansion to the southeast when the Russians lost the Russo-Japanese War in 1905.

While the Russian Empire expanded to the east with great rapidity, its westward expansion was slow and halting. In the 1600s Russia still faced formidable enemies in Sweden, Poland, and the Ottoman (Turkish) Empire. By the 1700s, however, all three of these states had weakened, allowing the Russian Empire to gain substantial territories. After defeating Sweden in the early 1700s, Tsar Peter the Great (1682–1725) obtained a foothold on the Baltic, where he built the new capital city of St. Petersburg, designed to be a window on the West. Later in the eighteenth century, Russia defeated both the Poles and the Turks and gained all of modern-day Belarus and Ukraine. Tsarina Catherine the Great (1762–1796) was particularly pivotal in colonizing Ukraine and bringing the Russian Empire to the warm-water shores of the Black Sea.

The nineteenth century witnessed the Russian Empire's final expansion. Large gains were made in Central Asia, where a group of once-powerful Muslim states was no longer able to resist the Russian army. The mountainous Caucasus region proved a greater challenge, as the peoples of this area had the advantage of rugged terrain in defending their lands. South of the Caucasus, however, the Christian Armenians and Georgians accepted Russian power with little struggle, since they found it preferable to rule by the Persian or Ottoman empires.

The Legacy of Empire The expansion of the Russian people was one of the greatest human movements Earth has ever witnessed. By 1900 a traveler going from St. Petersburg on the Baltic to Vladivostok on the Sea of Japan would everywhere encounter Russian peoples speaking the same language, following the same religion, and living under the rule of the same government. Nowhere else in the world did such a tightly integrated cultural region cover such a vast space. Indeed, elements of that imperial era are being resurrected today to foster Russian unity. Recently Russian leaders proposed a return to the tsarist-era coat of arms (the double-headed eagle) to symbolize the ongoing connections between Russian society today and the earlier culture of Dostoyevsky and Pushkin. But despite these unifying elements, the Russian cultural world retains many pockets of linguistic, religious, and national diversity. Even in European Russia, non-Russian peoples hold on to their identity, while in much of the far north, indigenous peoples remain in the majority.

The history of the Russian Empire also reveals points of ongoing tension with the world beyond. One of these tensions centers on Russia's ambivalent relationship with western Europe. Russia shares with the West the historical legacy of Greek culture and Christianity; since the time of Peter the Great, Russia has undergone several waves of intentional Westernization. At the same time, however, Russia has long been suspicious of—even hostile to—European culture and social institutions. Highly nationalistic Russian thinkers have long viewed Western society as decadent and Western Christianity as heretical, and they have seen their country as the unique heir of the Greek and Roman traditions. Although elements of this debate were transformed during the Soviet period, the central tension remained, and it influences Russia to this day.

Another significant source of tension in Russian culture stems from its long history of authoritarianism. From its earliest days, the Russian Empire was highly centralized, with the tsar exercising tremendous power over the entire country. That tradition was certainly reinforced through much of the communist era, when leaders such as Joseph Stalin ruled the Soviet Union with an iron hand. Still, there has also been a countervailing history of resistance against overwhelming state power within the Russian domain, as made evident in the dramatic breakup of the Soviet Union in 1991 and in ongoing debates today about how much power should be centralized in Moscow.

Geographies of Language

Slavic languages dominate the region (Figure 9.24). The distribution of Russian-speaking populations is complicated. Russian, Belorussian, and Ukrainian are closely related languages. Some linguists argue that they ought to be considered separate dialects of a single Russian language, as they are all mutually intelligible. Most Ukrainians, however, insist that Ukrainian is a distinct language in its own right, and there is a well-developed sense of national distinction between Russians and Ukrainians. Belorussians, on the other hand, are more inclined to stress their close kinship with the Russians.

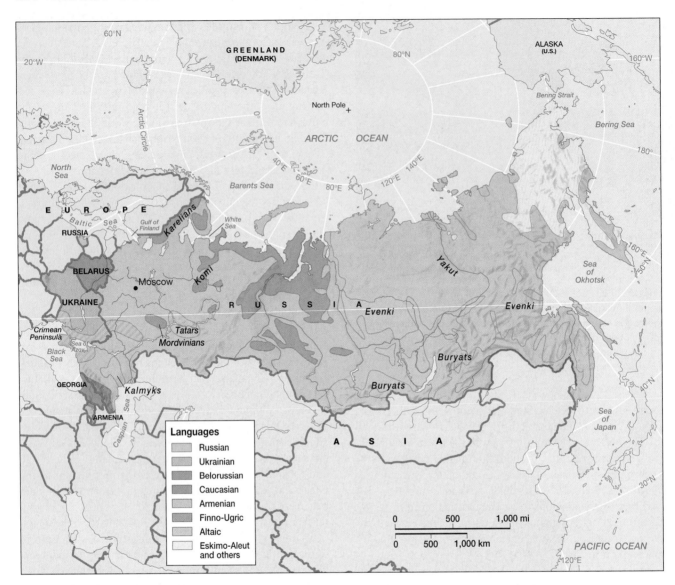

▲ **Figure 9.24 Languages of the Russian domain**   Slavic Russians dominate the realm, although many linguistic minorities are present. Siberia's diverse native peoples add cultural variety in that area. To the southwest, the Caucasus Mountains and the lands beyond contain the region's most complex linguistic geography. Ukrainians and Belorussians, while sharing a Slavic heritage with their Russian neighbors, add further variety in the west.

The Belorussians and Ukrainians   The geographic pattern of the Belorussian people is relatively simple. The vast majority of Belorussians reside in Belarus, and most people in Belarus are Belorussians (Figure 9.24). The country does, however, contain scattered Polish and Russian minorities. For the Russians this presents few problems, since Russians and Belorussians can relatively easily assume each other's ethnic identity.

The Ukrainian situation, however, is more complex. Because of different historical patterns of territorial conquest, Russian speakers dominate large parts of eastern Ukraine, while they make up a much smaller portion of western Ukraine's population. Similarly, the Crimean Peninsula, now a part of Ukraine, has long ethnic and political roots within Russia. Conversely, ethnic Ukrainians live in scattered communities in southern Russia and southwestern Siberia. Dur-

ing the Soviet period, such geographical mixing had little consequence, since the distinction between Russians and Ukrainians was not viewed as important in official circles. Now that Russia and Ukraine are separate countries with a heightened sense of national distinction, this issue has emerged as a significant source of tension across the region.

Patterns Within Russia   Approximately 80 percent of Russia's population claims a Russian linguistic identity. Russians inhabit most of European Russia, but there are large enclaves of other peoples. The Russian zone extends across southern Siberia to the Sea of Japan. In sparsely settled lands of central and northern Siberia, Russians are numerically dominant in many areas, but they share territory with varied indigenous peoples. While the language map suggests that large areas remain beyond the Russian realm, most of these regions are only sparsely populated. Still, their distinctive non-Russian cultural orientation has emboldened many of the groups in these areas to seek more political autonomy since the breakdown of the Soviet Union.

Finno-Ugric (Finnish-speaking peoples), though small in number, dominate sizable portions of the non-Russian north. The Finno-Ugric (or Uralic) speakers comprise an entirely different language family from the Indo-European Russians. While many have been culturally Russified, distinct Finnish-speaking peoples such as the Karelians, Komi, and Mordvinians remain a part of Russia's modern cultural geography. Altaic speakers also complicate the country's linguistic geography. This language family includes the Volga Tatars, whose territory is centered on the city of Kazan in the middle Volga Valley. While retaining their ethnic identity, the Turkish-speaking Tatars have extensively intermarried with and borrowed from their Russian neighbors. Yakut peoples of northeast Siberia also represent Turkish speakers within the Altaic family. Other Altaic speakers in Russia belong to the Mongol group. In the west, examples include the Kalmyk-speaking peoples of the lower Volga Valley who migrated to the region in the 1600s. Far to the east, the Buryats live in the vicinity of Lake Baikal and represent an indigenous Siberian group closely tied to the cultures and history of Central Asia (Figure 9.25).

The plight of many native peoples in central and northern Siberia parallels the situation in the United States, Canada, and Australia. Rural indigenous peoples in each of these settings remain distinct from dominant European cultures. Such groups are also internally diverse and are often divided into a number of unrelated linguistic clusters. One entire linguistic grouping of Eskimo-Aleut speakers is limited to approximately 25,000 people who are widely dispersed through northeast Siberia. Another indigenous Siberian group is the Altaic-speaking Evenki, whose traditional terri-

tory covers a large portion of central and eastern Siberia. Many of these Siberian peoples have seen their traditional ways challenged by the pressures of Russification, just as indigenous peoples elsewhere in the world have been subjected to similar pressures of cultural and political assimilation. Unfortunately, other common traits seen within such settings are low levels of education, high rates of alcoholism, and widespread poverty.

Transcaucasian Languages Although small in size, Transcaucasia offers a bewildering variety of languages (Figure 9.26). From Russia, along the north slopes of the Caucasus, and to Georgia and Armenia east of the Black Sea, a complex history and a fractured physical setting have combined to produce some of the most complicated language patterns in the world. No fewer than three language families are spoken within a region smaller than Ohio, and many individual languages are represented by small, isolated cultural groups. In the Dagestan district of the northeast Caucasus, some 30 distinct languages are still spoken. The Caucasian language family, unrelated to any other on Earth, includes Chechnyan (Chechan) and Dagestani peoples north of the Caucasus and Georgians to the south. Altaic peoples appear on the scene as well, complicating the cultural geographies of the region along the shores of the Caspian Sea and along the northwest border separating Russia and Georgia. Finally, Indo-Europeans dominate Armenia, and other Greek and Iranian peoples are sprinkled through the valleys of the Caucasus in central Georgia and southern Russia. Not surprisingly, language remains a pivotal cultural and political issue within the fragmented Transcaucasian subregion.

Geographies of Religion

Most Russians, Belorussians, and Ukrainians share a religious heritage of Eastern Orthodox Christianity. For hundreds of years Eastern Orthodoxy served as a central cultural presence within the Russian Empire (Figure 9.27). Indeed, church and state were tightly fused, until the demise of the empire in 1917. Under the Soviet Union, however, religion in all forms was severely discouraged and actively persecuted. Most monasteries and many churches were converted into museums or other kinds of public buildings, and schools disseminated the doctrine of atheism. Organized religion enjoyed something of a reprieve during World War II when it was seen as helpful for the war effort, and religion was even more fully tolerated by the 1980s. Still, only a small minority of Soviet people remained actively religious.

With the downfall of the Soviet Union, however, a religious revival has swept much of the Russian domain. In the past 15 years, more than 11,000 Orthodox churches have been returned to religious uses. Now, an estimated 75 million Russians are members of the Orthodox Church, including almost 400 monastic orders dispersed across the country. Within Russia the Orthodox Church appears headed toward a return to its former role as a state church. The government is increasingly using church officials to sanction various state activities, which is particularly ironic since many of Russia's

▲ **Figure 9.25 Minority Buryats** Closely related to residents of Mongolia, Russia's Buryats live in the vicinity of Lake Baikal. They enjoy some degree of political autonomy in recognition of their distinctive ethnic background. *(Hans-Jurgen Burkard/Bilderberg Archiv der Fotografen)*

▶ **Figure 9.26 Languages of the Caucasus region** A bewildering mosaic of Caucasian, Indo-European, and Altaic languages characterizes the Caucasus region of southern Russia and nearby Georgia and Armenia. Persisting political problems have erupted in the region as local populations struggle for more autonomy. Recent examples include independence movements in Chechnya and in nearby Dagestan. *(Modified from Martin Glassner, Political Geography, 2nd ed., 1996, p. 599. Reprinted by permission of John Wiley & Sons, Inc.)*

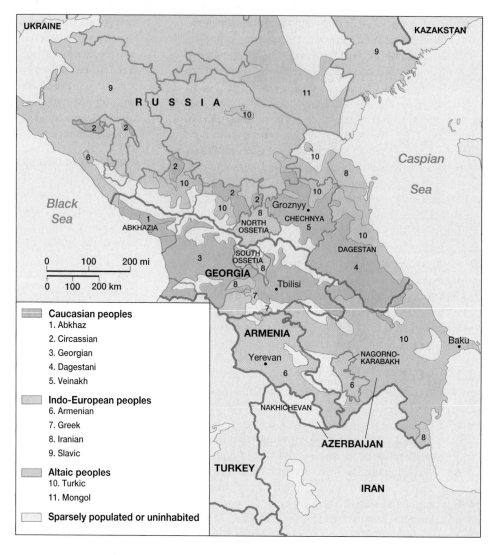

Caucasian peoples
1. Abkhaz
2. Circassian
3. Georgian
4. Dagestani
5. Veinakh

Indo-European peoples
6. Armenian
7. Greek
8. Iranian
9. Slavic

Altaic peoples
10. Turkic
11. Mongol

Sparsely populated or uninhabited

current leaders played roles in the earlier Soviet period when religious observances were banned. For example, when he became president in 2000, former Soviet-era KGB agent Vladimir Putin professed his Orthodox beliefs and claimed that his mother baptized him during the height of Soviet repression. In turn, the Church has supported recent Russian political campaigns, both domestically and internationally.

Other forms of Western Christianity are also present in the region. For example, the people of western Ukraine, who experienced several hundred years of Polish rule, eventually joined the Catholic Church. Eastern Ukraine, on the other hand, remained fully within the Orthodox framework. This religious split reinforces the cultural differences between eastern and western Ukrainians. Western Ukrainians have generally been far more nationalistic, and hence more firmly opposed to Russia, than eastern Ukrainians. Elsewhere, Christianity came early to the Caucasus, but modern Armenian forms—their roots dating to the fourth century A.D.—differ on several doctrinal points from both Eastern Orthodox and Catholic traditions. Georgian Christianity, however, is more closely tied to the Orthodox faith. Evan-

gelical Protestantism has also been on the rise since the demise of the Soviet Union.

Non-Christian religions also appear and, along with language, shape ethnic identities and tensions within the region. Islam is the largest non-Christian religion and claims between 20 and 25 million adherents. Most are Sunni Muslims and they include peoples in the North Caucasus, the Volga Tatars, and Central Asian peoples near the Kazakstan border. To date, Islamic fundamentalism has not become a major cultural or political issue in the region. Still, a growing Islamic political consciousness is present, particularly in Russia. In mid-2001, the first large-scale meeting of the new Muslim Prosperity Party took place. Russia, Belarus, and Ukraine are home to more than 1 million Jews, who are especially numerous in the larger cities of the European West. Jews suffered severe persecution under both the tsars and the communists. Recent out-migrations, prompted by new political freedoms, have further reduced their numbers in Russia, Belarus, and Ukraine. Buddhists are also represented in the region, associated with the Kalmyk and Buryat peoples of the Russian interior. Indeed, Buddhism has witnessed a recent renaissance.

▲ **Figure 9.27 Russian Orthodox Church, Siberia**   A revival of interest in the Russian Orthodox Church followed the collapse of the Soviet Union in the early 1990s. The newly built Znamensky Cathedral in Kemerovo displays many of the faith's characteristic cultural landscape signatures.   *(Novosti/ Sovfoto/Eastfoto)*

Russian Culture in Global Context

Russian culture has interacted in varied ways with the world beyond. Russian cultural norms have for centuries embodied both an inward orientation toward traditional forms of expression and an outward orientation directed primarily to western Europe. By the nineteenth century, even as Russian peasants interacted rarely with the outside world, Russian high culture had become thoroughly Westernized, and Russian composers, novelists, and dramatists gained considerable fame in Europe and the United States. By the turn of the twentieth century, Russian avant-garde artists were among the international leaders in devising the experimental styles of modernism.

Soviet Days   During the Soviet period, a new mixture of cultural relationships unfolded within the socialist state. Initially, European-style modern art flourished in the Soviet Union, encouraged by the radical rhetoric of the new rulers. By the late 1920s, however, Soviet leaders turned against modernism, which they viewed as the decadent expression of

a declining capitalist world. Many Soviet artists fled to the West, and others were exiled to Siberian labor camps. Increasingly, state-sponsored Soviet artistic productions centered on **socialist realism**, a style devoted to the realistic depiction of workers harnessing the forces of nature or struggling against capitalism. Still, traditional high arts, such as classical music and ballet, continued to receive lavish state subsidies, and to this day Russian artists regularly achieve worldwide fame.

Turn to the West   By the 1980s it was clear that the attempt had failed to fashion a new Soviet culture based on working-class solidarity and informed by the aesthetics of socialist realist art. The younger generation instead adopted a rebellious stance, turning for inspiration to fashion and rock music from the West. The mass-consumer culture of the United States proved immensely popular, symbolized above all by brand-name chewing gum, jeans, and cigarettes. The Soviet government attempted to ward off this perceived cultural onslaught, but with little success. Soviet officials could more easily censor books and other forms of written expression, but even here they were increasingly frustrated. By the end of the Soviet period, the persistent secrecy that separated the Soviet Union from the West had broken down, thus enabling more people and information to flow into the region.

After the fall of the Soviet Union in 1991, basic freedoms brought an inrush of global cultural influences, particularly to the region's larger urban areas such as Moscow. Shops were quickly flooded with Western books and magazines; people pondered the financial mysteries of home mortgages and condominium purchases; and they reveled in the newfound pleasures of fake Chanel handbags and McDonald's hamburgers. English-language classes became even more popular in cities such as Moscow, where Russians hurried to embrace the world their former leaders had warned them about for generations. Cultural influences streaming into the country were not all Western in inspiration. Films from Hong Kong and Mumbai (Bombay), as well as the televised romance novels *(telenovelas)* of Latin America, for example, proved far more popular in the Russian domain than in the United States.

Russians and the Net   Russia's enthusiasm for the Internet exemplifies the region's growing links to the rest of the world. Only 7 million people (about 5 percent of the country's population) are estimated to be using the Net in 2003, but that is more than triple the Internet use in 1999, and it represents the fastest growth rate in Europe. While many Russian professionals access the Internet in the workplace, better telecommunications technologies are allowing more home-based users to dial up the world from their own living rooms. Russian use of the Net is also on the rise beyond Moscow, although cities such as St. Petersburg are still struggling to attract foreign investment or government assistance to fund necessary investments in infrastructure. The huge Russian natural gas company, Gazprom, is doing its share. It has been busily laying modern fiber-optic communication cables alongside its natural gas pipelines across much of Russia, Ukraine, and Belarus. While still at an early stage, such initiatives promise to have long-term economic and cultural consequences for millions of residents within the Russian domain.

The Music Scene　Younger residents of the region have also embraced the world of popular music, and their enthusiasm for American and European performers, as well as their support of a budding home-grown music industry, symbolizes the changing values of an increasingly post-Soviet generation. MTV Russia went on the air at midnight on September 26, 1998. A year later the network sponsored a huge open-air concert on Red Square. Headlined by the Red Hot Chili Peppers, the event was a far cry from Soviet-era tank parades in front of Communist Party officials. By 2001 Russian MTV reached more than 60 million viewers. Perhaps even more important, Sony Music Entertainment established its Russian operations in December 1999, followed by several of the world's other major recording companies. Sony, BMG, and other labels have signed multiple Russian acts for domestic markets, helping to boost a native pop-music culture. Universal has also opened the way for Russian performers to go global. For example, it sponsored the English-language debut of Russian pop singer Alsou in 2001 (Figure 9.28). Music piracy is also widespread in the Russian domain, and government officials annually confiscate millions of CDs and cassettes that are part of a large, illegal global-trading network.

The integration of the Russian domain into the global cultural mainstream has provoked strong reactions among the region's more conservative institutions. The rhetoric parallels

▲ **Figure 9.28 Alsou**　Symbol of a new Russia and the globalizing power of popular music, Russian singer Alsou recorded English-language songs in 2001 for young audiences in western Europe and North America.　*(Hy/Newsmakers/Getty Images, Inc.)*

Soviet-era concerns about the damaging effects of new cultural ideas, particularly on the young. Extreme nationalists have resisted foreign influences such as Western-style music, and they have often been supported by more traditional elements of the Russian Orthodox Church.

Geopolitical Framework: The Remnants of a Global Superpower

The geopolitical legacy of the former Soviet Union still weighs profoundly upon the Russian domain. After all, the bold lettering of the "Union of Soviet Socialist Republics" dominated the Eurasian map for much of the twentieth century, and the country's global political reach left no corner of the world untouched. Many of the present political uncertainties that plague the region stem from the Soviet period. Former Soviet republics continue to struggle to define new geopolitical identities for themselves. Neighboring states persist in eyeing the region with trepidation, the legacy of former Soviet political and military power. Present demands for more local political control within countries such as Russia, Ukraine, and Georgia can still be understood in the context of the Soviet era, when a highly centralized political apparatus gave little voice to regional dissent. For all of these reasons, future political geographies in the area remain cartographic uncertainties, more prone than most corners of the world to shift unpredictably as conditions rapidly evolve.

Geopolitical Structure of the Former Soviet Union

The Soviet Union rose from the ashes of the Russian Empire, which collapsed abruptly in 1917. The ultraconservative policies of the Russian tsar generated opposition among businesspeople and workers, while the peasants, who formed the majority of the population, had always resisted the powerful land-owning aristocracy. After the fall of the tsar and the aristocracy, a broad-based coalition government assumed authority. Several months later, however, the **Bolsheviks**, a faction of Russian communists representing the interests of the industrial workers, seized power within the country. The leader of these Russian communists was Vladimir Ilyich Ulyanov, usually known by his self-selected name, Lenin. Lenin was the main architect of the Soviet Union, the Russian Empire's successor state.

The new socialist state reconfigured Eurasian political geography. Although it resembled the territorial contours of the Russian Empire and centralized authority continued to be concentrated in European Russia, the spatial and economic structure of the country was radically transformed. When the Soviet Union emerged in 1917, Lenin and the other communist leaders were aware that they faced a major challenge in organizing the new state. The old empire had contained many distinct nationalities, but Russians had traditionally dominated these peoples. Lenin wanted to end such domination, which he considered contrary to the international and class-based spirit of socialism. Still, the new leaders wanted to retain as much of the imperial territory as possible.

The Soviet Republics and Autonomous Areas   Soviet leaders designed a geopolitical solution that maintained their country's territorial boundaries and acknowledged, at least theoretically, the rights of its non-Russian citizens. Each major nationality was to receive its own "union republic," provided it was situated on one of the nation's external borders (Figure 9.29). Eventually 15 such republics were established, thus creating the Soviet Union. A number of these republics were quite small, while the massive Russian Republic sprawled over roughly three-quarters of the Soviet terrain. Each republic was to be administratively autonomous, vested with the right to secede from the union if it so desired. In practice, however, the Soviet Union remained a centralized state, with important decisions made in the capital of Moscow.

The Soviets came up with different geopolitical solutions to acknowledge smaller ethnic groups and nationalities that were not situated on the country's external borders. Indeed, dozens of significant minority groups pressed for recognition. One solution to this problem was the creation of **autonomous areas** of varying sizes that recognized special ethnic homelands, but did so within the structure of existing republics. Thus, within the Russian Republic the larger nationalities, such as the Yakut and the Volga Tatars, were granted their own "autonomous republics" (not to be confused with the 15 higher-level "union republics," such as Russia or Ukraine). Smaller nationalities, such as the Siberian Evenki, received less-important "autonomous regions." The system's main deficiency was that the autonomy it was designed to provide proved to be more of a charade than a reality.

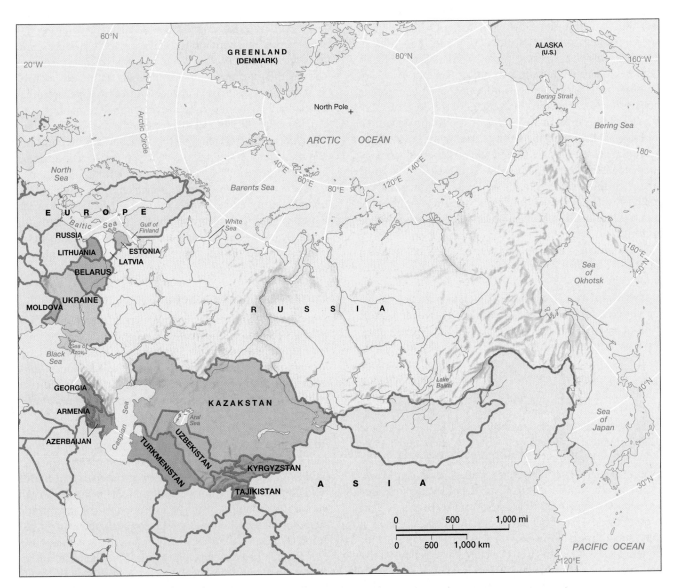

▲ **Figure 9.29 Soviet geopolitical system**   During the Soviet period, the boundaries of the country's 15 internal republics often reflected major ethnic divisions. Ultimately, however, many of the ethnically non-Russian republics pressured the Soviet government for more political power. As the Soviet empire disintegrated, the former republics became politically independent states and now form an uneasy ring of satellite nations around Russia. *(Modified from Rubenstein, 1999, Introduction to Human Geography, Upper Saddle River, NJ: Prentice Hall)*

Centralization and Expansion of the Soviet State

In the early Soviet era, it appeared that the framework of separate republics and autonomous areas might allow non-Russian peoples to protect their own cultures and establish their own social and economic policies (provided, of course, that such policies embodied Marxist principles). From the beginning, however, it was clear that such self-determination would be a temporary measure. According to official ideology, the gradual development of a communist society would see the withering away of all significant ethnic differences and the disappearance of religion. In the future, a new classless Soviet society was supposed to emerge.

By the 1930s it was clear that national autonomy would not have any real significance within an increasingly centralized Soviet state. The chief architect of this political consolidation was Soviet leader Joseph Stalin, who did everything he could to centralize power in Moscow and to assert Russian authority. Stalin launched a ruthless plan of state-controlled agricultural production and industrialization. Although many people initially resisted his policies, Stalin never hesitated to use force to bring about his vision of a purer socialist revolution within the Soviet Union. The Stalin period also saw the enlargement of the Soviet Union. As a victorious power in World War II, the country acquired southern Sakhalin and the Kuril Islands from Japan. It regained the Baltic republics (Lithuania, Latvia, and Estonia—independent between 1917 and 1940), as well as substantial territories formerly belonging to Poland, Romania, and Czechoslovakia. A small but strategic addition on the Baltic Sea was the northern portion of East Prussia (the port of Kaliningrad), previously part of Germany. It still forms a small but strategic Russian **exclave**, which is defined as a portion of a country's territory that lies outside of its contiguous land area.

After World War II, the Soviet Union also gained significant authority, although not actual sovereignty, over a broad swath of eastern Europe. As they pushed the German army west toward the end of the war, Soviet troops advanced across much of the region, actively working to establish communist regimes thereafter. In the words of British leader Winston Churchill, the Soviets extended an **"Iron Curtain"** between their eastern European allies and the more democratic nations of western Europe by restricting the free flow of people and information.

As eastern Europe retreated behind the Iron Curtain, the Soviet Union and the United States, while allies during World War II, became antagonists in a global **Cold War** of escalating military competition that lasted from 1948 to 1991. In its post-World War II heyday, the Soviet Union became a global superpower, one of only two countries equipped with enough nuclear weapons to ensure global destruction. At the height of its power in the late 1970s, it enjoyed close military and economic alliances not only with eastern Europe, but also with certain countries in Asia (Mongolia, North Korea, Vietnam, Laos, and Cambodia), the Caribbean region (Cuba and Nicaragua), and Africa (Angola, Somalia, and several others).

End of the Soviet System

Ironically, Lenin's system of culturally defined republics helped sow the seeds of the Soviet Union's demise. Even though the nationally based republics and autonomous areas of the Soviet Union were never allowed real freedom, they did provide a persisting political framework for the perpetuation of distinct cultural identities. Indeed, contrary to the expectations of Soviet leaders, ethnic nationalism intensified in the post–World War II era as the Soviet system grew less repressive. When Soviet President Mikhail Gorbachev initiated his policy of **glasnost**, or greater openness, during the 1980s, several republics—most notably the Baltic states of Lithuania, Latvia, and Estonia—demanded outright independence.

Meanwhile, other forces worked toward the political end of the Soviet regime. A failed war in Afghanistan in the early 1980s frustrated both the Soviet leaders and population. In eastern Europe, growing protests over Soviet dominance emerged from Czechoslovakia and Poland. Worsening domestic economic conditions, increasing food shortages, and the declining quality of life led to fundamental questions concerning the value of centralized planning within the country. In response, President Gorbachev introduced **perestroika**, or planned economic restructuring, aimed at making production more efficient and more responsive to the needs of Soviet citizens. In 1991, however, Gorbachev saw his authority slip away amid rising pressures for political decentralization and more dramatic economic reforms. During the summer, Gorbachev's regime was further imperiled by the popular election of reform-minded Boris Yeltsin as the head of the Russian Republic and by a failed military coup by communist hardliners. By late December, all of the country's 15 constituent republics had become independent states, and the Soviet Union ceased to exist.

Current Geopolitical Setting

Post-Soviet Russia and the nearby independent republics have radically rearranged their political geographical relationships since the collapse of the Soviet Union in 1991. All of the former republics still struggle to establish stable political relations with their neighbors, and increasing calls for further political decentralization in many settings threaten the internal integrity of Russia and nearby states (Figure 9.30).

Russia and the Former Soviet Republics

For a time it seemed that a looser political union of most of the former republics, called the **Commonwealth of Independent States (CIS)**, would emerge from the ruins of the Soviet Union. All the former republics, with the exception of the three Baltic states, joined the CIS soon after the dismemberment of the old union (Figure 9.30). By the early twenty-first century, however, the CIS had developed into little more than a forum for discussion, without real economic or political power. Newly independent countries, such as Uzbekistan, wanted to keep the CIS ineffectual for fear that Russia would otherwise use the organization to regain authority over the entire area.

Among former Soviet republics, Russia, Belarus, and Ukraine have enjoyed the closest political and economic re-

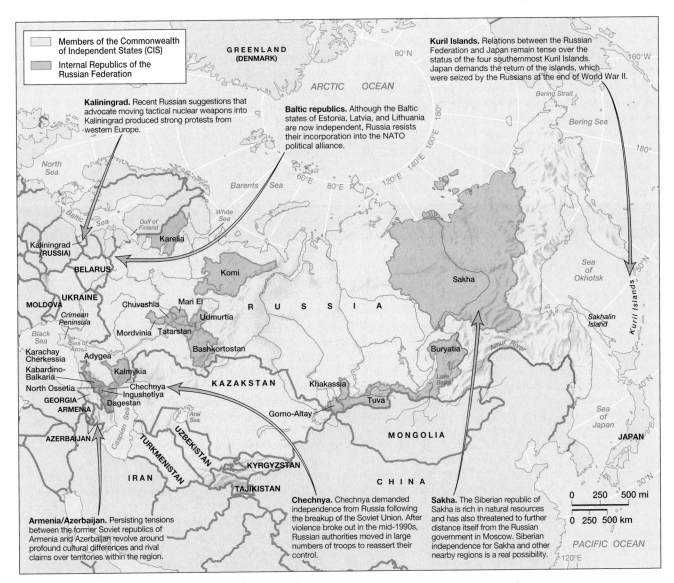

Kuril Islands. Relations between the Russian Federation and Japan remain tense over the status of the four southernmost Kuril Islands. Japan demands the return of the islands, which were seized by the Russians at the end of World War II.

Kaliningrad. Recent Russian suggestions that advocate moving tactical nuclear weapons into Kaliningrad produced strong protests from western Europe.

Baltic republics. Although the Baltic states of Estonia, Latvia, and Lithuania are now independent, Russia resists their incorporation into the NATO political alliance.

Chechnya. Chechnya demanded independence from Russia following the breakup of the Soviet Union. After violence broke out in the mid-1990s, Russian authorities moved in large numbers of troops to reassert their control.

Sakha. The Siberian republic of Sakha is rich in natural resources and has also threatened to further distance itself from the Russian government in Moscow. Siberian independence for Sakha and other nearby regions is a real possibility.

Armenia/Azerbaijan. Persisting tensions between the former Soviet republics of Armenia and Azerbaijan revolve around profound cultural differences and rival claims over territories within the region.

▲ **Figure 9.30 Geopolitical issues in the Russian domain**   The Russian Federation Treaty of 1992 created a new internal political framework that acknowledged many of the country's ethnic minorities. Political stability, however, continues to elude the country as many of these groups press for complete independence. Elsewhere, Armenia struggles with neighboring Azerbaijan, and Russia's relations with nearby states remain in flux.   *(Modified from Bergman and Renwick, 1999,* Introduction to Geography, *Upper Saddle River, NJ: Prentice Hall)*

lations. In 1996 Russia and Belarus declared their membership in a special two-country political and economic union, and they moved even closer to reunification in 1999 with a series of agreements on military cooperation. Still, mutual distrust between the two nation's leaders has slowed progress toward a more comprehensive political union, and Belarus's persistently authoritarian regime has not matched Russia's attempts at greater democracy and economic freedom. Meanwhile, Russia and Ukraine signed a 1998 agreement to develop trade relations and foster economic linkages, and Russian investments in Ukraine grew sharply in 2000 and 2001. Elsewhere, Georgia, Ukraine, Uzbekistan, Azerbaijan, and Moldova formed the GUUAM group, dedicated to facilitating trade through the Caspian–Black Sea corridor.

Another complicating factor in the post-Soviet period has been the ongoing military relationships between Russia and its former republics. **Denuclearization,** the return of nuclear weapons from outlying republics to Russian control and their partial dismantling, was completed during the 1990s. The Soviet-era nuclear arsenals of Kazakstan, Ukraine, and Belarus were removed in the process. Elsewhere, Tajikistan invited the Russian army in during the early 1990s to quell its own internal ethnic struggles, and Armenia has maintained close military ties with Moscow. In other cases, however, Russian involvement has not helped the cause of political cooperation. Georgians, for example, deeply resent Russian assistance to the Abkhazians, a breakaway Muslim group located in the northwestern part of their country. They have also been negotiating with Russia to close all Russian military bases that remain in Georgia. At the same time, Georgia has expressed an interest in joining the NATO alliance, a move sure to annoy Russian leaders in Moscow.

Naval bases in the former Soviet Union present another dilemma. Russia's Baltic fleet is headquartered in the Kaliningrad exclave, now separated from the rest of the country. Reports that the Russians were moving tactical nuclear weapons into the area in 2001 sparked strong protests from the United States and NATO interests in western Europe. To the south,

the Soviet-era Black Sea fleet was traditionally based in the Crimean port of Sevastopol, now in Ukraine. After lengthy negotiations, Russia and Ukraine agreed in 1997 to share the naval base, but the longer-term political sovereignty of the entire Crimean Peninsula remains in doubt (see "Geopolitical Setting: Crimea's Shifting Identity").

The former Transcaucasian republics have witnessed political tensions in the post-Soviet era. The area contains striking ethnic and religious diversity, and rivalries are accentuated by the geopolitical structure inherited from the Soviet period.

GEOPOLITICAL SETTING   Crimea's Shifting Identity

Ukraine's Crimea Peninsula has been contested real estate within the Russian domain for centuries (Figure 9.30, Figure 9.2.1). An exotic land of cherry blossoms, palm trees, hillside farms, and Mediterranean-like vistas, the peninsula juts far into the subtropical waters of the Black Sea. Centuries ago, it caught the eye of Greek colonists and Genoese traders. In the thirteenth century, westward-moving Tatars, a Mongol people from Central Asia, occupied the region and made it their homeland. These Crimean Tatars were later ruled by the Ottoman Turks and, after 1783, were annexed to an expanding Russian Empire (Figure 9.22). In the years that followed, the area was often in the midst of a geopolitical maelstrom that involved claims by Russian, Turkish, British, and French colonial interests. By the late nineteenth century, the region was firmly under Russia's control, and its natural charms attracted the country's wealthy and powerful. Localities such as Yalta became famous summer resorts.

Still, Crimea's political identity did not stabilize during the Soviet era. While many Tatars moved to Turkey at the time of the Russian Revolution, others stayed and pressed for some degree of political autonomy, which was granted in 1921. Nazi Germany invaded the strategic area during World War II, but was later forced out by victorious Russian forces.

Stalin, however, became convinced that the Tatars were Nazi sympathizers and expelled almost the entire population to Central Asia (Uzbekistan). Ethnic Russians flooded into the region thereafter, along with a smaller number of Ukrainians. In 1954 the entire peninsula, while largely Russified culturally, was administratively transferred from the Russian to the Ukrainian Republic as a symbol of Soviet friendship.

Recently, Crimea's convoluted path has taken a few new twists. Now a part of Ukraine, Crimea has attracted the return of hundreds of thousands of Tatars (many illegally) from Central Asia, who advocate independence for their ancient homeland. They still make up only about 10 percent of the population. Most Crimeans (63 percent) today are actually ethnic Russians; spurred by Russian nationalists in their own homeland, many are advocating a return to Russian sovereignty. Not surprisingly, the Ukrainian minority (25 percent) are also edgy, even in their own country, because a Russian military presence (portions of the Black Sea naval fleet) remains in the port city of Sevastopol (Figure 9.2.2). The returning Tatars have been lobbying hard for more land and language rights, while the large Russian population remains unconvinced that its best interests are being served by a weak and bankrupt Ukrainian government. Crimea's geopolitical future is anyone's guess.

▲ **Figure 9.2.1 Crimean Peninsula**   Long an ethnic and political borderland, the subtropical Crimean Peninsula juts far into the Black Sea. Although it is a part of Ukraine, Crimea has a largely Russian population along with an important Tatar minority.

▲ **Figure 9.2.2 Sevastopol**   Located along the subtropical southern coast of the Crimean peninsula, the port city of Sevastopol remains a strategic naval base for the Black Sea fleet. *(Sergei Svetlitsky)*

The territories of the Christian Armenians and the Muslim Azeris, for example, interpenetrate one other in a complex fashion. The far southwestern portion of Azerbaijan (Nakhichevan) is separated from the rest of the country by Armenia, while the important Armenian-speaking district of Nagorno-Karabakh is officially an autonomous portion of Azerbaijan. Almost immediately after independence in 1991, Armenia and Azerbaijan went to war over these territories. After Armenia successfully occupied much of Nagorno-Karabakh, the fighting between the countries diminished. No peace treaty has been signed, however, and Azerbaijan demands the return of the territory. Meanwhile, more than 600,000 Azerbaijani refugees remain homeless. Armenia also has no international support for annexing Nagorno-Karabakh, which contributes to its own sense of isolation.

Devolution and the Russian Federation   Within Russia, further pressures for devolution, or more localized political control, produced the March 1992 signing of a new Russian Federation Treaty. The treaty granted Russia's internal autonomous republics and its lesser administrative units greater political, economic, and cultural freedoms, including more control of their natural resources and foreign trade. Conversely, it weakened Moscow's centralized authority to collect taxes and to shape policies within its varied hinterlands. Defined essentially along ethnic lines, 21 regions possess status as republics within the federation and now have constitutions that often run counter to national mandates. Some republics, such as Tatarstan in the Volga Valley, have even extracted agreements from Moscow to develop their own "foreign economic policy" (Figure 9.31). In addition, dozens of smaller regions and municipalities, including Moscow itself, have declared varying degrees of political autonomy from the federal reins of power.

▲ **Figure 9.31 Kazan**   Capital of the internal Tatarstan Republic, Kazan is home to many Islamic residents who press for even more freedom from distant Moscow.   *(Novosti/Sovfoto/Eastfoto)*

Several clusters of autonomous republics are apparent. One group, composed of a number of particularly small territories, is located along the northern slope of the Caucasus Mountains, an area noted for its extraordinary ethnic diversity. A group of larger autonomous areas is found in the central Volga Valley, homeland of the Volga Tatars and a number of other Turkish-speaking, as well as Finnish-speaking, peoples. A third cluster is located along the southern border of Siberia, homeland of the Mongolian Buryats and several Turkish-speaking ethnic groups. Russia's largest autonomous areas are located in the far north, especially in Siberia. Sakha (or Yakutia), for example, sprawls over 1,200,000 square miles, an area larger than the territory of all but seven of the world's independent countries. But Sakha, like the other autonomous areas of the tundra and taiga zone, is a sparsely inhabited land.

Internal Russian geopolitics remain caught between the forces of devolution, aimed at more local political control, and a renewed effort on the part of the central government in Moscow to assert its national presence. The outcome of that struggle remains uncertain. Local political control has its advantages. After a long era of Soviet dominance, it allows people to have a more direct role in running their own political affairs. They also can claim fuller control of their economic resources and more effectively manage public expenditures on infrastructure and economic development. For example, when the central government slowed progress on land reform, the regional governor in the city of Samara pressed ahead with his own program, privatizing 90 percent of the farmland. Other autonomous units within Russia have also gone ahead with their own development efforts, including offering special tax breaks to foreign companies seeking a foothold within the Russian economy.

Devolution, however, comes with a price. Central leaders in Moscow fear that Russia could fragment, resulting in further political instability. Indeed, many of the country's local and regional governments have been plagued by incompetence and widespread corruption. Political fragmentation makes it more difficult to introduce national reforms, to make large-scale investments in infrastructure, and to engage in comprehensive long-term planning. A more centralized state, advocates argue, can also deal more effectively with global political and economic affairs.

Regional Tensions   Regional geopolitical trouble spots within Russia reveal some of the tensions between local and national demands for power (Figure 9.30). After the Soviet Union disintegrated, several of Russia's internal autonomous areas threatened to secede. Initially, strong challenges came from Tatarstan on the Volga, an oil-rich area with a long history of national identity. For the time being, however, new concessions from Moscow have quelled demands for independence. Nearby Bashkortostan has also been a setting for secession threats.

Elsewhere, outlying regions in Siberia and the Russian Far East have voiced their desire for complete separation from Russia. While Moscow has recently reined in powerful regional leaders in such settings, enduring political and economic forces will continue to encourage more autonomy for these far-flung portions of the country. Siberia's rich natural resource base has encouraged many local politicians and residents to opt for more autonomy. Why should bureaucrats in Moscow benefit from Siberian oil, gas, and timber reserves? Diverse ethnic minorities within the region have increasingly asserted their political rights, further complicating the situation. In the end, political, cultural, and economic self-interest may propel various Siberian regions, such as Sakha, to go their own way.

The Caucasus Mountains have been an even more explosive setting for internal political tensions. In 1994 leaders of the internal Chechnyan Republic vowed to establish a genuinely independent state. Russia responded with a massive military invasion. After intensive fighting, the Russian army took control of Chechnya's capital of Groznyy, located in the plains just to the north of the Caucasus (Figure 9.32). Chechnyan rebels held out in the rugged mountains to the south, and increased their regional and national terrorist attacks in the late 1990s. A new Russian military force of around 100,000 soldiers was sent into the small republic late in 1999 to quell the Chechnyan rebels and rally domestic political support around the struggling central government in Moscow. Since their return, the Russians have been accused of looting Chechnyan natural resources (especially metals and oil) and trying to grab a share of the lucrative global drug trade that flows through the remote mountainous region. Meanwhile, about half the republic's population fled the fighting, causing massive refugee movements and dislocations in neighboring areas, particularly to nearby Ingushetia. No permanent solution appears in sight as Russia asserts its control in the north while rebels remain active across much of the southern and eastern mountains.

The Shifting Global Setting

Much of Russian geopolitical history has been focused on its creation of a vast land-based empire across much of Eurasia. As British geographer Halford Mackinder argued in 1904, Russia's control of the heartland, the huge interior landmass of Eurasia, gave it a global advantage over maritime powers such as Great Britain. The collapse of the Soviet Union, however, resulted in a simultaneous crash of Russian global power. Russia, with only a little more than half of the population of the former Soviet Union and with a teetering economy, could no longer afford to support and subsidize allies such as Cuba. Moreover, the communist ideology that had previously cemented the Soviet alliance rapidly declined in Russia and throughout most of the world. Russia did what it could to maintain the formidable military complex of the Soviet Union, but decay was inevitable. In retrospect, Mackinder's thesis on the dominance of the heartland no longer seems supportable. The Russian example suggests that power in the modern world flows much more from economic productivity and technological prowess than from the mere control of continental territory.

Since the fall of the Soviet Union, regional political tensions continue to challenge the Russians, in both the east

▲ Figure 9.32 Devastated Groznyy　The recent war for independence in the Chechnyan Republic met stiff resistance from Russia's central government. Much of the capital city has been caught in the crossfire and continues to reveal the scars of the conflict. *(Tass/Sovfoto/Eastfoto)*

and west. In East Asia, the boundary between Russia and China was imposed by the Russian Empire in 1858 and has never been fully accepted by Beijing. The two countries battled over the area in 1969, but since then relations have improved. The leaders of both Russia and China now stress the need for cooperation, but the potential for renewed conflict remains. Territorial disagreements also complicate Russia's relationship with Japan. Japan has long demanded the return of the four southernmost islands of the Kuril Archipelago seized by the Soviet Union in 1945. Although Russia could gain major financial benefits in any agreement, it has refused to return any territory to Japan. To the west, Russia worries about the expansion of the North Atlantic Treaty Organization. Although most Russian leaders accept the inclusion of Poland, Hungary, and the Czech Republic into NATO, virtually all strongly oppose the entry of any former Soviet territories.

Today, Russian leaders appear willing to reassert the nation's global political status. Its nuclear arsenal, while reduced in size, remains a powerful counterpoint to American, European, and Chinese interests. While Russia can no longer directly challenge the United States as it did in the days of the Soviet Union, it can act as a partial counterweight to the United States in international maneuverings. Russian leaders, for example, can offer themselves as intermediaries in diplomatic negotiations. Russia also retains a permanent seat on the United Nations Security Council, arguably the world's most important geopolitical body. Russia's recent inclusion in some of the G-7 economic meetings also signifies its remaining international clout. And despite their nominal independence from direct Russian control, the other nations within the Russian domain clearly feel the presence of their mammoth neighbor. Indeed, the long geopolitical history of the domain suggests that, despite the region's recent weakness, it is likely to play a major role in future global political affairs.

Russian leaders, particularly President Vladimir Putin, clearly saw the geopolitical handwriting on the wall after the events of September 11, 2001. The terrorist attacks on the United States and the multinational "War on Terror" which followed gave Russia an opportunity to deftly redefine its domestic and global geopolitical agenda in several important ways.

Within the country, ongoing Russian concerns over its Muslim minority population (about 15 percent of the total), especially its more fundamentalist elements, could now be cast in a new light. Given the West's global rhetoric challenging the Afghan Taliban and other fundamentalist Islamic elements in the Middle East, Russian leaders suddenly had more latitude to declare their own campaigns against their sometimes troublesome Muslim populations and to describe these moves as part of the wider "War on Terror." Indeed, many of the most troublesome Islamic groups are in Chechnya, which now may give the Russians a freer hand to rein in such "ter-

rorists" within their own borders. More broadly, the attacks have encouraged President Putin to continue centralizing power in his once-fragmented country.

On the global stage, the attacks provided Russia with a great opportunity to strengthen its relationships with the United States and with the wealthy nations of western Europe. Suddenly Russia's geopolitical real estate was of greater strategic importance, and Putin allowed for and encouraged overflight rights to the U.S. military. In addition, Russia's close ties with Central Asian countries (parts of the former Soviet Union) such as Uzbekistan and Tajikistan allowed it to play a helpful role in securing airfields and troop bases for incoming Americans as they initiated their war in Afghanistan. What does it all mean? Russian leaders see the U.S. "War on Terror" as an ideal opportunity to rebuild their nation's own global reputation, to cooperate with the West in the hope of receiving increased aid and investment, and to further solidify their own domestic geopolitical agenda in the process.

Economic and Social Development: An Era of Ongoing Adjustment

The economic future of the Russian domain remains difficult to predict. Economic declines devastated the region for much of the 1990s, but the situation stabilized in 2000 and 2001. Russia's gross national product declined by more than 40 percent between 1990 and 1999, and similar economic disasters unfolded in nearby Belarus, Ukraine, Armenia, and Georgia (Table 9.2). These declines marked the most abrupt economic collapse within the industrialized world since the Great Depression of the early 1930s. The causes of the recent economic difficulties have been complex, rooted both in the long-standing policies of the Soviet period as well as in the chaotic nature of economic reforms that have been unevenly carried out since 1991. Recently, at least in the case of Russia, higher oil and gas prices brought selective economic improvement, but significant challenges remain for the region.

The true economic potential of the Russian domain has always been difficult to gauge. Optimists point to the vast size, abundant natural resources, and well-educated, urbanized populations of the region as significant assets. Indeed, in its heyday, the Soviet Union rose to become one of the great industrial powers in the world, and did so in a remarkably short period of time. Skeptics note that size also brings its disadvantages, particularly by raising transportation costs within the region's economy. They also point out that Russia's northern location has always made food production problematic. Most notably, ever since the breakup of the Soviet Union, economies within the region have struggled to evolve in a stable fashion toward greater productivity and output. Instead, much of the 1990s witnessed rising prices and unemployment but declining investment and industrial output, hardly the recipe for economic health. In addition, economic troubles have had profound consequences for the quality of life within the region. The social fabric of families, the quality of health care, and even the

TABLE 9.2 *Economic Indicators*

Country	Total GNI (Millions of $U.S., 1999)	GNI per Capita ($U.S., 1999)	GNI per Capita, PPP* ($Intl, 1999)	Average Annual Growth % GDP per Capita, 1990–1999
Armenia	1,878	490	2,360	−3.9
Belarus	26,299	2,620	6,880	−2.9
Georgia	3,362	620	2,540	
Russia	328,995	2,250	6,990	−5.9
Ukraine	41,991	840	3,360	−10.3

*Purchasing power parity

Source: The World Bank Atlas, 2001.

psychological outlook of the population have all suffered amid the economic uncertainties of the period. The region's future path to prosperity remains clouded as it spends the early years of the twenty-first century reconfiguring an economy that was forged within a radically different political and ideological past.

The Legacy of the Soviet Economy

The birth of the Soviet Union in 1917 initiated a radical change within the region's economy. Under the Russian Empire, most people were peasants, farming the land much as they had done for centuries. Following the revolution, however, the Soviet Union quickly emerged to rival, and even surpass, many of the most powerful economies on Earth. During that era of unparalleled growth, much of the region's present economic infrastructure was established, including new urban centers and industrial developments, as well as a modern network of transportation and communication linkages. Thus, even though the Soviet Union has departed from the scene, much of its economic legacy remains to shape the progress and the challenges of its contemporary economy.

As communist leaders such as Stalin consolidated power in the 1920s and 1930s, they nationalized Russian industries, creating a system of **centralized economic planning** in which the state controlled production targets and industrial output. The Soviets stressed heavy, basic industries (steel, machinery, chemicals, and electricity generation), deferring demand for consumer goods to the future. Huge increases in production were realized, but at a significant human cost. Several mining and industrial areas, most of them in Siberia, were turned into virtual slave-labor camps.

The Soviets also nationalized agriculture. By the late 1920s, communist leader Stalin was shifting agricultural land into large-scale collectives and state farms that organized production around state-mandated production goals. Marxist ideology reconfigured the countryside. As collectivization spread, the agricultural landscape was reorganized. Field boundaries and cropping complexes reflected state-established production goals; rapid mechanization replaced human and animal labor with tractors; and entirely new areas (such as the Virgin and Idle Lands Project east of the Volga River) were put into agricultural production to boost total food output. Still, the agricultural sector struggled. Most peasants initially resisted collectivization, and many slaughtered their animals rather than turn them over to state ownership. During the 1930s, Stalin's troops and crop famines killed millions of farmers, including many of the *kulaks*, or wealthier peasants, who were seen as leaders of the agrarian resistance. Later efforts to put new lands into production also failed to meet the high expectations of Soviet planners. Ironically, the most significant increases in Soviet food production in the 1950s and 1960s did not come from grandiose production schemes such as the Virgin and Idle Lands Project. Rather, the liberalization of rules concerning private plots—small garden-size fields of potatoes, vegetables, and pasturage—succeeded in significantly raising food output across the country. These private plots produced significant quantities of food and set the stage, during the post-Soviet period, for the selective privatization of the agricultural landscape.

Soviet-era industrial expansion was far more successful than any gains seen in agriculture. The nation's development of heavy industries during the 1930s proved vitally important in World War II. Although the German army advanced to within miles of Moscow, new centers of heavy industry in the resource-rich Urals and in Siberia's Kuznetsk Basin in the vicinity of Novosibirsk remained untouched, supplying the Russian army with the materials necessary to conduct the war (Figure 9.33). The industrial economy boomed in the 1950s and 1960s, and Soviet leaders confidently predicted that their country would be the world's most highly developed nation by 1980. It was a pivotal era in establishing the industrial complexes that remain dominant today. Both the Urals and Kuznetsk industrial zones expanded further in the postwar period. While isolated from major population centers, these manufacturing zones enjoyed excellent access to raw energy (coal, oil, and natural gas) and metals (iron ore, bauxite, and copper) resources. Cities such as the steel-production center of Magnitogorsk in the southern Urals appeared almost overnight and became showplaces of Soviet industrial dynamism. Farther east, Novosibirsk and Novokuznetsk boomed in southern Siberia, producing steel, farm machinery, transportation equipment, and chemicals.

Soviet industrial geography also expanded in European portions of the country after 1950. The energy-rich Volga region dominated Soviet oil production until new Siberian fields were opened in the 1970s. Petrochemical manufacturing remains important in the region, along with varied metallurgical and transportation-related industries. Moscow and its nearby cities are less endowed with raw materials, but they benefit from some of the country's best infrastructure, a large skilled labor force, and domestic market demand for industrial goods. As a result, the country's oldest manufacturing region specializes in a variety of high value-added products that continue to be crucial to the nation's industrial output, particularly in consumer-oriented goods. Elsewhere, the eastern Ukraine industrial region remains important to the south, benefiting from nearby access to raw materials. Major Soviet-era developments near the Krivoy Rog iron ore deposits and in the coal-rich Donetsk Basin continue to benefit the region's economy, although a lack of foreign investment and real economic reforms within the country led to declining industrial production in the late 1990s.

Much of the Russian domain's basic infrastructure—its roads, rail lines, canals, dams, and communications networks—also originated during the Soviet period and reflect the planning necessities of that era. Indeed, massive projects to improve Soviet infrastructure were pursued with patriotic fervor. While some proved to be a waste of resources, many have been excellent long-term investments that form the backbone of the post-Soviet Union's economic geography. Dam and canal construction, for example, turned the main rivers of European Russia into a virtual network of interconnected reservoirs. Invaluable links such as the Volga-Don

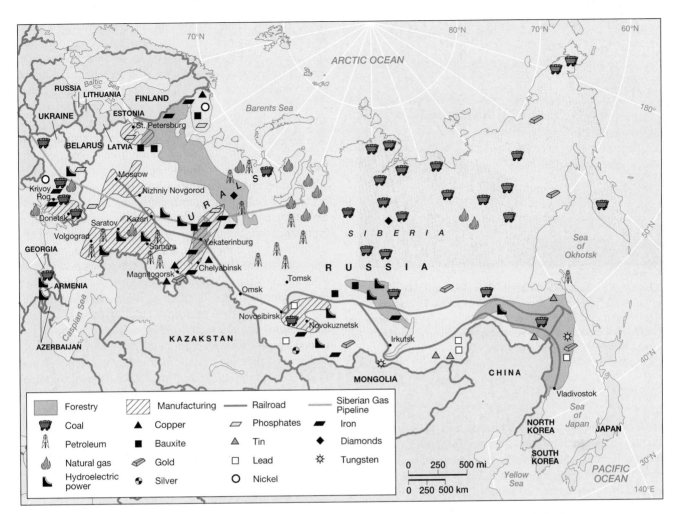

▲ Figure 9.33 Major natural resources and industrial zones   The region's varied natural resources and chief industrial zones are widely distributed. Fossil fuels are in abundance, although their distance from markets often imposes special costs. In southern Siberia, rail corridors offer access to many mineral resources. In the mineral-rich Urals and eastern Ukraine, proximity to natural resources sparked industrial expansion, while Moscow's industrial might is related to its proximity to markets and capital.   (Modified from Bergman and Renwick, 1999, Introduction to Geography, Upper Saddle River, NJ: Prentice Hall, and Rubenstein, 1999, Introduction to Human Geography, Upper Saddle River, NJ: Prentice Hall)

Canal (completed in 1952), which connected those two key river systems, have greatly facilitated the movement of industrial raw materials and manufactured goods within the country (Figure 9.34). The Soviets also improved the country's railroad network. Thousands of miles of new track were added in the European west, and the Trans-Siberian Line was modernized and then complemented by the addition of the BAM link across central Siberia. Farther north, the Siberian Gas Pipeline was built to link the energy-rich fields of the Soviet arctic with growing demand in Europe.

Overall, the postwar period saw real economic improvements for the Soviet people. No longer a peasant nation, Russia, like the other Soviet republics, was now industrialized and largely urbanized. A massive housing campaign in the 1950s and 1960s ensured most families a private flat. Consumer goods, such as televisions, telephones, and even private automobiles, began to spread more widely through society.

The Soviet social system also achieved some notable successes. Literacy was virtually universal, quality health care was readily available, and every capable person was guaranteed employment. In terms of having eliminated dire poverty, the Soviet economy was a success.

Despite these successes, however, economic and social problems increased during the 1970s and 1980s. Soviet agriculture was still inefficient, and the country increasingly relied on grain imports. Manufacturing efficiency and quality failed to match the standards of the West, particularly in regard to consumer goods. Soviet citizens were burdened by recurrent shortages and usually had to spend many hours each week waiting in line to purchase basic necessities. Rigid centralized planning simply did not have the flexibility necessary to provide adequate goods and services. Equally troubling was the fact that the Soviet Union was failing to participate fully in the technological revolutions that were

▲ **Figure 9.34 Volga-Don Canal**   This view near Volgograd suggests the enduring economic importance of the Volga-Don Canal. Built during the Soviet era, the canal remains a key commercial link that facilitates the economic integration of southern Russia.   *(Carl Wolinsky/Stock Boston)*

transforming the United States, Europe, and Japan, a failure that had serious military consequences. Disparities also visibly grew between the Soviet elite and an everyday citizenry that still enjoyed few personal freedoms. Political leaders vacationed in luxurious dachas, shopped in private stores, and avoided the long lines that most people faced to buy shoddier goods in the state-managed retail outlets. Such disparities of wealth, which sometimes seemed to have little to do with skill or education levels, further undermined faith in the socialist system. By the late 1980s, the Soviet Union had reached both an economic and a political impasse.

The Post-Soviet Economy

Fundamental economic changes have shaped the Russian domain since the demise of the Soviet Union. Particularly within Russia itself, much of the highly centralized state-controlled economy has been dismantled, replaced by a mixed economy of state-run operations and private enterprise. The changeover has been very difficult. Fundamental problems of unstable currencies, corruption, and changing government policies plague the system. Indeed, the region experienced an unprecedented economic decline for much of the 1990s (Table 9.2). In Russia, for example, steel output declined from almost 70 million tons in 1992 to less than 50 million tons at the end of the decade. While selective portions of the region's economy stabilized early in the twenty-first century, no one can confidently predict how the economy will evolve in the future.

Redefining Regional Economic Ties   Economic ties still link the five nations of the Russian domain, but the region has seen many of those relationships redefined in the post-

Soviet era. The collapse of the communist state transformed planned internal exchanges between Soviet republics into less-predictable flows of foreign trade. Under the old regime, for example, government officials made certain that Ukraine received adequate supplies of oil from elsewhere in the country, just as Russia obtained cotton for its textile industry from Uzbekistan. Under the new economy, Ukraine must buy oil, or barter other products for it, from Russia, just as Russia suddenly must purchase more cotton from abroad.

Even with these disruptions, Russia retains an overwhelming economic presence within the region. Belarus, for example, is still strongly dependent upon Russian economic assistance, particularly in the form of cheap energy exports. Similarly, Ukraine imports a great deal of its raw materials (particularly oil) from Russia, and its exports (principally metals, food, and machinery) are led by return flows to the north. Recently, cash-starved Ukrainian companies have also looked to Russia for help, and the result has been a Russian-led buying binge of cheap Ukrainian assets. Ukrainian aluminum smelters, television stations, and oil refineries are now controlled by large Russian corporations, certainly a new post-Soviet twist on a long-established Russian presence within the country. Both Georgia and Armenia also retain close trading ties with Russia, although these countries have made efforts to broaden their economic linkages with western Europe, the United States, and Southwest Asia.

Privatization and Economic Uncertainty   Even as hopes initially ran high for a smooth restructuring of the post-Soviet economy, recent years have been extraordinarily difficult for most residents of the region. In 1992 the Russian government suddenly freed prices from state control, encouraging more production and filling store shelves. As prices rose, however, inflation soared, wages failed to keep pace, and savings were wiped out as the value of the Russian currency (ruble) declined. Another dramatic move came in October 1993 when the government initiated a massive program to privatize the Russian economy. Millions of Russians were given the option to buy into newly privatized agricultural lands and industrial companies. These initiatives greatly opened the economy to more private initiative and investment. Unfortunately, however, a lack of legal and financial safeguards also invited many abuses and often resulted in mismanagement and corruption within the new system. Elsewhere in the region, privatization has proceeded much more slowly. Much of the Belorussian and Ukrainian economies remain mired in inflexible, corrupt state-controlled systems.

The agricultural sector continues to struggle. No matter its economic system, much of the region will always be challenged by short growing seasons, poor soils, and moisture deficiencies. Russia's best farmlands remain limited to a small slice of southern territory wedged between the Black and Caspian seas, leaving most of the rest of the country on the agricultural margins. In addition, the recent drastic economic changes have not been easy for the nation's farmers. About two-thirds of the country's farmland was privatized by 2000, with many farmers forming voluntary cooperatives or joint-stock associations to work the same acreage they did under the Soviet system. While crop prices have risen, costs have gone up even faster, and many farmers are not skilled in dealing with the uncertainties of a market-driven agricultural economy. Elsewhere, the government has held some agricultural land in the form of state farms, employing workers less willing to take risks in the fickle farm economy. Overall, agriculture continues to employ about 15 percent of Russia's workforce, and the basic geography of crops remains little changed from the Soviet era. Beyond Russia, much of rural Belarus and Ukraine have seen steeply declining living standards since 1991, and agricultural reforms in these nations have been much slower than in neighboring Russia. In addition, the lingering specter of the Chernobyl disaster casts a shadow across farmlands in northern Ukraine and nearby Belarus.

Russia, in particular, has encouraged the rapid privatization of the service sector. Thousands of privatized retailing establishments have appeared, and they now dominate that portion of the economy. In addition, the long-established "informal economy" continues to flourish. Even during the Soviet era, millions of citizens earned extra money by informally selling Western consumer goods, manufacturing food and vodka, and providing skilled services such as computer and automobile repairs. Today, these barter transactions and informal cash deals form a huge part of the regional economy that is never reported to government authorities.

In Russia, substantial portions of the natural resource and heavy industrial sectors of the economy have also been privatized. By the end of 1994, the government reported that more than 100,000 enterprises were in private hands; three years later, about 70 percent of the nation's wealth was generated in the private sector, a radical departure from the Soviet period. Gazprom, the huge Russian natural gas company, exemplifies the process. Privatized in the spring of 1994, about one-third of the company was auctioned off to Russian citizens; 15 percent went to managers and workers; 10 percent remained in the company's treasury; and most of the rest was kept in government hands. Employing 385,000 people and controlling one-third of Earth's known natural gas reserves, Gazprom has been nicknamed "Russia, Inc."

Privatization, on scales both great and small, has also brought tremendous challenges. The privatization process has often been corrupt, liberally lining the pockets of company insiders and trade union officials who obtained controlling interest of many corporations. Special state decrees permitting privatization also funneled large sums of money to politicians and their Swiss bank accounts. While a small number of Russia's elite have become wealthy through these economic changes (Figure 9.35), most workers have seen little real economic gain. Factory managers still must often depend on barter transactions to obtain basic parts and materials for the products they manufacture. Cumbersome national and regional tax policies also prevent efficient revenue collections from most private companies and individuals involved in the informal economy, and thus the government sees little benefit from such operations.

▲ **Figure 9.35 Living in the lap of luxury**   At one extreme, Russia's new entrepreneurial elite revel in the opportunities that have opened up amid recent economic reforms. Greater regulatory stability and a more stable currency, however, may be needed to offer more sustainable and widely shared paths to economic growth.   *(B. Brecelj/Corbis/Sygma)*

The Russian Mafia   Organized crime is pervasive in Russia and controls many aspects of the economy. The government's own interior ministry estimates that the Russian mafia controls about 40 percent of the private economy and 60 percent of state-run enterprises. Unofficially, sources also suggest that 80 percent of the country's banks are under mafia influence. While organized crime certainly existed in the Soviet era, the more liberal economic and political environment that followed has allowed it to flourish even more widely and openly. The mafia provided critical capital and jobs to many unemployed young men in the unstable months immediately following the collapse of communism. There also remain many close ties between organized crime and Russian intelligence-collecting agencies. Today, bribery and protection are part of the cost of doing business across much of the Russian domain, and many entrepreneurs must turn to criminal elements to obtain loans in an economy where willing banks and foreign lending institutions are in short supply. Different local and regional crime organizations have divided up much of the economy. More than 8,000 syndicates now exist in Russia alone. While one group might control the construction business in a Moscow suburb, another syndicate oversees drug dealing and prostitution, and still another helps to funnel illegal CDs and DVDs to eager consumers.

The Russian mafia has also gone global. In 2001, it was implicated in a huge money-laundering scheme that involved Russian, British, and U.S. banks as well as the flow of International Monetary Fund investments into the region. Banks in Switzerland, Liechtenstein, and Cyprus have been involved in similar schemes. Mafia money has also been invested in every corner of the world economy, including Sri Lankan gambling syndicates, Colombian drug enterprises, and legitimate Israeli high-technology companies. Russian emigrants help diffuse mafia influence, and émigré communities in Tel

Aviv, Paris, London, and New York City act as key local contacts. Not surprisingly, one recent study of global corruption found that Russia ranked as one of the six most corrupt countries in the world.

A Fraying Social Fabric   Tough economic times and political uncertainties have contributed to a fraying social fabric within the Russian domain. Rates of violent crime increased late in the Soviet period, and have risen further since the fall of communism. Organized criminal activities have profited from fewer state restrictions on economic activity, and street crime has escalated with growing urban poverty and insecurity. High unemployment, rising housing costs, and declining social welfare expenditures have hit many families hard. Often, both husband and wife work multiple low-paying jobs with few benefits and long hours. Highlighting the economic and social challenges within the region, one recent study of the Russian middle class suggested that it still included only 10 to 15 percent of the population (compared with 64 percent of the U.S. population).

Women are paying an especially heavy price amid the current economic and social problems. Increasing domestic violence is symptomatic of the turmoil that has particularly impacted women throughout the region. Beatings and rapes are common. A survey in Moscow suggested that one-third of divorced women had experienced domestic violence, while a women's rights group in Ukraine reported that rape was an all-too-common crime in many villages. International organizations have sharply criticized government authorities for doing little to change the situation. Divorce rates also continue to rise as families struggle to make ends meet. Economically, women often have suffered first as unemployment has risen. Forced out of traditional industrial and service-sector jobs, many have been compelled to sell homemade products or services on city streets for extremely low wages (Figure 9.36). Many housewives and schoolteachers are forced into prostitution, simply as a way to survive. Politically, women in Russia, Belarus, and Ukraine serve in fewer governmental positions than their counterparts in western Europe or the United States, suggesting that women are not able to play a high-profile role in shaping major social and economic policies within the re-

▲ **Figure 9.36 Life on the street**   This young street vendor at the Moscow Food Market hopes to sell her pickled vegetables to passing consumers. Economic instability and high unemployment, however, make even modest purchases more difficult for many Russians.   *(John Egan/The Hutchison Library)*

gion. Still, reforms undertaken during the communist era produced high female literacy rates and educational achievements (Table 9.3) that have persisted in all of the countries within the Russian domain.

Russia's health care crisis also illustrates the interplay between the region's troubled economy, limited government expenditures, and larger social issues. The crisis suggests that many of the roots of current social problems predate the fall of communism. For example, Soviet expenditures on health care averaged 6.6 percent of the country's gross national product in 1960 but declined to 2.3 percent by the end of the Soviet period. By the late 1990s, Russia's annual per capita medical expenditures stood at about $20, by far the lowest in the industrial world. Unfortunately, the nation's shrinking commitment to medical care comes at a time when several major health problems are on the rise. A shortage of vaccines and medical services has contributed to the return of such diseases as cholera, typhus, and the bubonic plague. Chronic illnesses, often related to lifestyle, produce the highest mortality

TABLE 9.3 *Social Indicators and Status of Women*

Country	Life Expectancy at Birth		Under Age 5 Mortality (per 1,000)		Percent Illiteracy (Ages 15 and over)		Female Labor Force Participation (% of total, 1999)
	Male	Female	1980	1999	Male	Female	
Armenia	71	76		18	1	3	48
Belarus	62	74		14	0	1	49
Georgia	69	77		20			47
Russia	59	72		20	0	1	49
Ukraine	63	74		17	0	1	48

Sources: Population Reference Bureau, Data Sheet, 2001, Life Expectancy (M/F); The World Bank, World Development Indicators, 2001, Under Age 5 Mortality Rate (1980/99); The World Bank Atlas, 2001, Female Participation in Labor Force; CIA World Factbook, 2001, Percent Illiteracy (ages 15 and over).

rates in the developed world for cardiovascular disease, alcoholism, and smoking, particularly for Russian men. One recent study suggests that half of all Russian men and one-third of Russian women are plagued by long-term drinking problems and that as many as 50 percent of Russian deaths are related to alcohol consumption. Most Russian men (66 percent) also smoke, creating a growth industry for multinational tobacco companies eager to increase sales. AIDS is rapidly on the rise, spreading especially among the region's young drug-using population. In addition, toxic environmental conditions have extracted a huge price from the region, although precise estimates are difficult to make.

Growing Economic Globalization

The relationship between the Russian domain and the world beyond has shifted greatly since the end of communism. During much of the Soviet era, the region was relatively isolated from the world economic system. By the 1970s, however, the Soviet Union began to export large quantities of fossil fuels to the West while importing more food products. Connections with the global economy quickly intensified with the downfall of the Soviet Union. As Russia and its neighbors selectively embraced market economics, the region intersected with the global economy in a variety of ways.

A New Day for the Consumer   Most visibly, a barrage of new consumer imports now reaches many residents of the Russian domain, particularly those living in its larger cities. McDonald's hamburgers, Calvin Klein jeans, and many other symbols of global capitalism are visible in the heart of Moscow and, increasingly, in many other settings throughout the region. Luxury goods from the West have also found a small but enthusiastic market among the newly emergent Russian elite, a group noted for its devotion to BMW automobiles, Rolex watches, and other status emblems. Most Russians find such luxuries far beyond their limited budgets, but they are interested in purchasing basic foods and cheap technology from their Western neighbors. In 2000, for example, frozen chickens were the leading imported product from the United States. In addition, inexpensive consumer goods from East and Southeast Asia and from industrializing neighbors such as Turkey offer affordable alternatives to Russian products. Most important, a consumer consciousness seems more apparent today within the Russian domain than it was during the Soviet era. However, it remains to be seen whether the regional economy will expand rapidly enough to meet the demands of these aspiring consumers.

Attracting Foreign Investment   The Russian domain (particularly Russia itself) has seen some growth in foreign investment in the post-Soviet era, but the region struggles to attract enough outside capital. Still, money has flowed into Russian-controlled enterprises (many Russian companies are now listed on European and American stock exchanges) and has also been directly invested into the region's economy. As measured by total foreign investment, the strongest global ties by far have been with the United States and western Europe, particularly Germany and Great Britain (Figure 9.37). Large investments have been made in Russia's oil and gas economy, as well as in the food, telecommunications, and consumer goods industries. In 2000 foreign investment in Russia totaled more than $9.5 billion and was growing at over 14 percent annually. Progress has also been made in developing more legal frameworks (called production-sharing agreements) that would make it easier and safer for outsiders to make large, long-term investments within the region.

Still, many foreign investors are wary of the region. One recent study suggests that Russian insiders control 80 percent of privatized Russian companies, and many of these corporations have been very slow to adopt open, Western-style accounting practices. The instability of Russian currency, stock, and debt markets has also cooled interest in foreign investment. Potential entrepreneurs are discouraged by the gyrating economy, the ever-present mafia, and government red tape. The entire financial services sector remains poorly developed. Proposed reforms that would modernize Russia's 1,300 banks have also been slow in coming. Although the countries of the Russian domain have grown closer to global institutions such as the World Bank and the International Monetary Fund, the region's unpredictable economy will probably restrain investments from the world beyond. Political uncertainty remains high as well, not only in Russia, but also in remainder of the region.

Globalization and Russia's Petroleum Economy   Russia's oil and gas industry remains one of the strongest economic links between the region and the global economy, and the diverse international connections it has forged suggest the increasing importance of the sector to the region's future. The statistics are impressive: Russia has 35 percent of the world's natural gas reserves (mostly in Siberia), and, even with its recent economic turmoil, it is the world's largest gas exporter. As for oil, Russia is by far the world's largest non-OPEC producer: it outpaces even the United States in annual output (major oilfields are in Siberia, the Volga Valley, the Far East, and the Caspian Sea region), and it possesses more than twice the proven reserves of the United States.

The dynamic geography of the Russian oil and gas business exemplifies the global character of the enterprise and the changing nature of the region's economy. Prior to the breakup of the Soviet Union, about half of Russia's oil and gas exports went to other Soviet republics, such as Ukraine and Belarus. While these two nations still depend on Russian supplies, the primary destination for Russian petroleum products has overwhelming shifted to western Europe. Russia now supplies that region with more than 25 percent of its natural gas and 16 percent of its crude oil, and those linkages are likely to grow even stronger. An agreement between Russia and the European Union in 2000 aimed at the rapid expansion of these East–West linkages. The Siberian Gas Pipeline already weds distant Asian fields with western Europe via Ukraine, and those connections are being supplemented by new lines through Belarus (the Yamal-Europe Pipeline) and Turkey (the Blue Stream Pipeline)(Figure 9.38). Similar expansions are

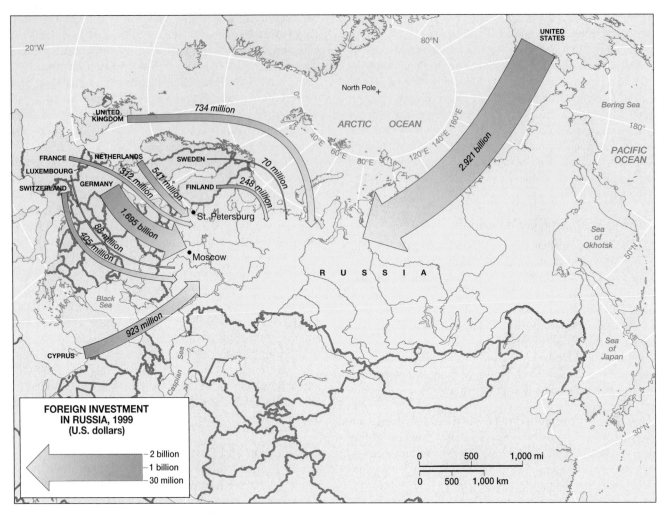

▲ **Figure 9.37 Global links: Leading sources of foreign investment in Russia, 1999** The amount of foreign investment (including direct, portfolio, and long-term credits) into Russia totaled over $9.5 billion in 1999. U.S. investments accounted for more than 30 percent of the total. Several west European nations were also major sources of investment. More than 40 percent of the investment was targeted at the Moscow and St. Petersburg regions, although some capital was funneled to other localities from these centers. *(U.S. Department of Commerce, Russia Commercial Guide 2001)*

dramatically refashioning the geography of oil exports. The huge Druzhba Pipeline already serves western Europe. Large new export terminals are opening on the Baltic and Black seas, including a major new line from the vast Caspian Sea reserves that opened in 2001.

Ponder the larger technological and financial picture as well. British, American, Dutch, and Norwegian oil and gas companies are spearheading many technical aspects of these huge initiatives. Similarly, much of the Blue Stream Pipeline is being financed by Italian and Japanese banks, and a broad assortment of other global financial loan guarantees and assistance are backing other exploration and pipeline projects. Clearly, the world is making a large bet that Russia's petroleum products will help integrate that troubled region with the rest of the global economy. While Russia has much to gain in the process, the region will also have to bear the human and environmental costs associated with that industry.

Local Impacts of Globalization As the geography of the Russian petroleum industry suggests, the local impacts of globalization are highly selective. Obviously, portions of the region that are close to oil and gas wells, pipeline and refinery infrastructure, and key petroleum shipping points are greatly impacted (both economically and environmentally) by the changing global oil economy. The same is true more broadly: globalization has affected different locations within the Russian domain in very distinctive ways. Within Russia, capitalism has brought its most dramatic, though selective, benefits to Moscow. Luxury goods, fancy cars, and upscale restaurants cater to the city's newly rich. Indeed, more than 32 percent of new foreign investment in the country has recently gone to the Moscow area. Aggressive municipal authorities in the city have also managed to become part owners in hundreds of local businesses and foreign joint ventures, thus guaranteeing a continuing flow of urban revenues.

▲ **Figure 9.38 Siberian Gas Pipeline**   Snaking its way across the remote Russian taiga, the Siberian Gas Pipeline connects arctic gas fields to needy consumers in far-off Europe. The costly pipeline requires frequent maintenance in the harsh northern environment.   *(Tass/Sovfoto/Eastfoto)*

▲ **Figure 9.39 Vladivostok**   The busy harbor of Vladivostok remains Russia's leading trade center in the Far East. With easy access to markets in Japan, China, and even the United States, this Pacific port is poised to grow as Russia's economy recovers. *(Itar-Tass/Sovfoto/Eastfoto)*

Beyond Moscow, St. Petersburg and the Siberian city of Omsk have seen growing global investment, while new oil and gas prospects in Siberia, the Far East (Sakhalin Island), and near the Caspian Sea have attracted other global capital. Cities such as Nizhniy Novgorod and Samara have also succeeded in attracting global capital by building a probusiness economic environment. Italy's Fiat Corporation, for example, recently

committed $850 million for a new manufacturing plant in Nizhniy Novgorod, recognizing that the city and its leading private companies could offer a favorable economic setting for investment. In addition, port cities such as Vladivostok are well positioned to take advantage of their accessibility to nearby markets, even though political red tape and corruption have hampered much growth in Pacific Basin trade (Figure 9.39).

Elsewhere, globalization has clearly imposed penalties. Older, less competitive industrial centers in the Urals, Siberia, and eastern Ukraine have been hit hard (see "The Local and the Global: Liquidating Russian Cities"). Aging steel plants, for

---

THE LOCAL AND THE GLOBAL   Liquidating Russian Cities

What happens when a city is no longer deemed necessary? In 2001 Russians living in several northern industrial settlements qualified for a new World Bank program designed to assist governments with "urban liquidation." As the global economy has changed over the past 50 years, many of Russia's mining and manufacturing towns have had to either change or die. In addition, Russia's struggling economy has often meant lower demand for certain natural resources and industrial goods. The result is a growing out-migration from these declining settlements, a process that will be hastened by the World Bank program.

Vorkuta, in northern Russia's Komi Republic, has qualified for the liquidation program. An old coal-mining city of some 170,000 residents, Vorkuta was once at the center of an important Soviet-era industrial complex. Today, many mines have closed and the town has become a major drain on an already-strapped federal budget. At least 50,000 people are being given vouchers for new housing elsewhere in order to encourage them to leave in the first phase of the liquidation process. No one is forced to leave, but authorities say that remaining residents would be "moved to one building,

while the rest are completely shut down and liquidated." One nearby settlement, Zapadny, has already been leveled and returned to tundra.

A northeast Siberian gold-mining center in the Magadan region will also get liquidation assistance. Eleven settlements and 6,500 people will be affected, leaving only four towns occupied within the region. Many residents need little incentive to leave. Indeed, more than 40 percent of Magadan's population has already emigrated from the region since 1991! Declining global gold prices along with escalating costs in the isolated mining region have led to widespread layoffs. The landscape is bleak, winters are long and brutal, and the program will offer a modest beginning to those lucky enough to receive assistance. Whatever its ultimate success, the new World Bank initiative is a recognition that a changing global economy often has dramatic consequences for particular localities, and that one solution is simply to make such unneeded places disappear.

Source: Adapted from "World Bank to Help Liquidate Villages," *The Moscow Times*, March 14, 2001.

example, no longer have guaranteed markets for their high-cost, low-quality products as they did in the days of the planned Soviet economy. Instead, they must compete on the global market, a market that is increasingly prone to lower prices and weakening demand for many of the industrial goods the region produces. Many of the region's extractive centers are similarly vulnerable in a global economy that has recently driven down many commodity prices. Indeed, the time has passed when the region can insulate itself from the larger world.

⊕ Conclusion

Many influences have shaped the human geographies of the Russian domain. Much of its underlying cultural geography was forged centuries ago, the complex product of Slavic languages, Orthodox Christianity, and numerous ethnic minorities that continue to complicate the scene today. Further transforming the contemporary cultural milieu is an onslaught of global influences, a set of products, technologies, and attitudes that increasingly mingle with traditional cultural orientations. Much of the region's political legacy is rooted in the Russian Empire, a land-based system of colonial expansion that greatly enlarged Russian influence after 1600 and was then reasserted in somewhat different form during the Soviet period. Only large remnants of that empire survive on the modern map, yet it has stamped the geopolitical character of the region in lasting ways. Much of the region's economic geography remains defined by its abundant, yet varied natural setting, as well as by the fundamental Marxist-guided changes wrought by twentieth-century industrialization and urbanization. The result is a human scene forged from diverse historical roots and a human geography crafted from forces that range from Tatar nationalism to twenty-first-century consumerism.

Looking toward the future, perhaps no other portion of the world faces uncertainties as deep as those confronting this region. Even its basic geopolitical structure remains in doubt. Will Belarus or eastern Ukraine rejoin Russia? Northern Kazakstan, still home to a large Russian population, is another target for possible Russian expansion. The most nationalistic Russian leaders would like to go several steps further and recreate the Soviet Union, if not the old tsarist empire. While partial reunification of the Russian domain remains a possi-

bility, there is also the potential for further fragmentation. Many non-Russian peoples within Russia might prefer independence. Secession also appeals to ethnic Russians who live in remote but resource-rich portions of the country. The Russian government appears determined to prevent political fragmentation, but its financial resources and even its military credibility remain limited.

Equally uncertain are the region's future political and economic systems, which will significantly influence its international relations. Russia's leaders in the immediate post-Soviet period promised a quick transition to European-style representative democracy and capitalism, as well as rapid integration into the global economic system. Subsequent economic and social difficulties, however, have largely discredited this approach in the eyes of the Russian population and made it possible for more extreme political parties to emerge.

Despair remains the greatest challenge facing the people of the region. The recent disappointments they have faced have been monumental. Communism spectacularly failed in its promise to bring about a social and economic utopia. Unfortunately, with little experience in banking institutions and legal protections, Russia's experimentation with capitalism has brought not the freedom and prosperity its supporters promised, but rather crime, poverty, and uncertainty. Longer term, the Russian domain's links to the world economy will be pivotal, but those fledgling connections have yet to provide many immediate rewards for the majority of the population. More than anything, the people of the Russian domain need a renewal of hope, a prospect that tomorrow's world might be better than today's.

⊕ Key Terms

autonomous areas (page 399)
Baikal-Amur Mainline (BAM)
    Railroad (page 385)
Bolsheviks (page 398)
centralized economic
    planning (page 406)
chernozem soils (page 377)
Cold War (page 400)

Commonwealth of
    Independent States
    (CIS) (page 400)
Cossacks (page 392)
dacha (page 389)
denuclearization (page 401)
Eastern Orthodox Christianity
    (page 391)

exclave (page 400)
glasnost (page 400)
Gulag Archipelago (page 386)
Iron Curtain (page 400)
mikrorayons (page 389)
perestroika (page 400)
permafrost (page 378)
podzol soils (page 377)

Russification (page 386)
Slavic peoples (page 391)
socialist realism (page 397)
taiga (page 378)
Trans-Siberian Railroad
    (page 385)
tsars (page 386)

⊕ Questions for Summary and Review

1. Compare the climate, vegetation, and agricultural conditions of Russia's European west with those of Siberia and the Russian Far East.

2. Describe some of the high environmental costs of industrialization within the Russian domain.

3. Discuss how major river and rail corridors have shaped the geography of population and economic development in the region. Provide specific examples.

4. Contrast Soviet and post-Soviet migration patterns within the Russian domain, and discuss the changing forces at work.

5. Describe some of the major land-use zones in the modern Russian city, and suggest why it is important to understand the impact of Soviet-era planning within such settings.

6. What were the key phases of colonial expansion during the rise of the Russian Empire, and how did each enlarge the reach of the Russian state?

7. What are some of the key ethnic minority groups within Russia and the neighboring states, and how have they been recognized in the region's geopolitical structure?

8. Describe how centralized planning created a new economic geography across the former Soviet Union. How has it had a lasting impact?

9. Briefly summarize the key strengths and weaknesses of the post-Soviet Russian economy, and suggest how globalization has shaped its evolution.

⊕ Thinking Geographically

1. How might it be argued that Russia's natural environment is one of its greatest assets as well as one of its greatest liabilities?

2. In the future, how might the forces of capitalism and free markets reshape the landscapes and land uses of cities within the Russian domain?

3. What options do peoples such as the Volga Tatars or the Siberian Buryats have for preserving their cultural autonomy? What might be some of the advantages and disadvantages of such peoples' pressing for greater political independence?

4. On a base map of the Russian domain, suggest possible political boundaries 20 years from now. What forces will work for a larger Russian state? A smaller Russian state?

5. What were some of the greatest strengths and weaknesses of centralized Soviet-style planning between 1917 and 1991? How were Ukraine and Belarus impacted? Why did the system ultimately fail?

6. Why is organized crime such a critical problem in this part of the world?

7. How have the growing forces of globalization impacted Russian culture?

8. From the perspective of a 22-year-old Russian college student and resident of Moscow, write a short essay that suggests how the economic and political changes over the past 10 to 15 years have changed your life.

⊕ Regional Novels and Films

Novels

Fyodor Dostoyevsky, *The Brothers Karamazov* (orig. pub. 1879; 1995, Bantam Books)

Fyodor Dostoyevsky, *Crime and Punishment* (orig. pub. 1866; 1984, Bantam Classics)

Eugenia Ginzburg, *Journey into the Whirlwind* (orig. pub. 1967; 1975, Harcourt Brace)

Nikolai Gogol, *Dead Souls* (orig. pub. 1842; 1961, Viking Press)

Mikhail Lermantov, *A Hero of Our Time* (orig. pub. 1840; 1966, Penguin)

Andrei Makine, *Dreams of My Russian Summers* (1998, Scribners)

Andrei Makine, *Once Upon the River Love* (1999, Penguin)

Aleksandr Solzhenitsyn, *One Day in the Life of Ivan Denisovich* (orig. pub. 1962; 1998, Signet Classics)

Leo Tolstoy, *War and Peace* (orig. pub. 1869; 1982, Viking)

Irene Zabytko, *The Sky Unwashed: A Novel* (2000, Algonquin Books)

Films

Brother (1997, Russia)

Burnt by the Sun (1994, France/Russia)

Dr. Zhivago (1965, U.S.)

Gorky Park (1983, U.S.)

Leo Tolstoy's Anna Karenina (1997, U.S.)

Little Vera (1988, USSR)

Moscow Does Not Believe in Tears (1980, USSR)

Prisoner of the Mountains (1996, Russia)

Rasputin (1985, USSR)

Reds (1981, U.S.)

Russia House (1990, U.S.)

The Thief (1997, Russia)

War and Peace (1968, USSR)

⊕ Bibliography

Adams, A. E.; Matley, I. M.; and McCagg, W. O. 1966. *An Atlas of Russian and East European History*. New York: Praeger.

Bater, James H. 1996. *Russia and the Post-Soviet Scene: A Geographical Perspective*. London: Arnold.

Brawer, Moshe. 1994. *Atlas of Russia and the Independent Republics*. New York: Simon & Schuster.

"The Caucasus: Where Worlds Collide." 2000. *The Economist*, August 19.

Channon, John (with Robert Hudson). 1995. *The Penguin Historical Atlas of Russia*. New York: Penguin.

Cole, Jonathan P. 1984. *Geography of the Soviet Union*. London: Butterworths.

Colton, Timothy J. 1995. *Moscow: Governing the Socialist Metropolis*. Cambridge: Harvard University Press.

Demko, George J., et al., eds. 1996. *Population Under Duress: The Geo-demography of Post-Soviet Russia*. Boulder, CO: Westview.

Dewdney, John C. 1979. *A Geography of the Soviet Union*. Oxford: Pergamon.

Dukes, Paul. 1990. *A History of Russia: Medieval, Modern, Contemporary*. Durham, NC: Duke University Press.

"Facing Oblivion, Rust-Belt Giants Top Russian List of Vexing Crises." 1998. *New York Times*, November 8.

"Fencing with the Pirates of Music." 2001. *The Moscow Times*, March 26.

Forsyth, James. 1992. *A History of the Peoples of Siberia: Russia's North Asian Colony 1581–1990*. Cambridge: Cambridge University Press.

Gachechiladze, Revaz. 1995. *The New Georgia: Space, Society, Politics*. College Station: Texas A&M University Press.

Gooding, John. 1996. *Rulers and Subjects: Government and People in Russia, 1801–1991*. London: Arnold.

Halperin, Charles J. 1987. *Russia and the Golden Horde: The Mongol Impact on Medieval Russian History*. Bloomington: Indiana University Press.

Hauner, Milan. 1992. *What Is Asia to Us? Russia's Asian Heartland, Yesterday and Today*. London: Routledge.

Heleniak, Timothy. 2001. "Russia's Modest Migration Gains Unlikely to Stop Population Decline." Population Reference Bureau. *Population Today*, May–June issue.

Hosking, Geoffrey. 2001. *Russia and the Russians*. Cambridge: Harvard University Press.

Hutenbach, Henry. 1996. *The Caucasus: A Region in Crisis*. Boulder, CO: Westview.

"In Search of Spring: A Survey of Russia." 1997. *The Economist*, July 12.

ITAR/TASS News Agency. 2001. "Growing Migration in Russia Affects Demographic Situation," February 15.

Kaiser, Robert J. 1994. *The Geography of Nationalism in Russia and the USSR*. Princeton, NJ: Princeton University Press.

Karny, Yo'av. 2000. *Highlanders: A Journey to the Caucasus in Quest of Memory*. New York: Farrar, Straus and Giroux.

Khazanov, Anatoly M. 1996. *After the U.S.S.R.: Ethnicity, Nationalism, and Politics in the Commonwealth of Independent States*. Madison, WI: University of Wisconsin Press.

Lapidus, Gail W., ed. 1994. *The New Russia*. Boulder, CO: Westview.

Lincoln, W. Bruce. 1994. *The Conquest of a Continent: Siberia and the Russians*. New York: Random House.

Lydolph, Paul E. 1990. *Geography of the USSR*. New York: Wiley.

Medvedkov, Yuri; Medvedkov, Olga; Smith, W. Randy; and Krischynas, Raymond. 1993. "Cities of the Former Soviet Union." In Stanley D. Brunn and Jack F. Williams, eds., *Cities of the World: World Regional Urban Development*, pp.150–93. New York: HarperCollins.

Pesmen, Dale. 2000. *Russia and Soul: An Exploration*. Ithaca, NY: Cornell University Press.

Powell, David E. 1998. "The Dismal State of Health Care in Russia." *Current History* 97:335–41.

Pryde, Philip R., ed. 1995. *Environmental Resources and Constraints in the Former Soviet Republics*. Boulder, CO: Westview.

"Putin's Choice: A Survey of Russia." 2001. *The Economist*, July 21.

"The Rise and Rise of the Russian Mafia." 2001. BBC News World Service, Internet Edition, March 21.

Rodgers, Allan, ed. 1990. *The Soviet Far East: Geographical Perspectives on Development*. London: Routledge.

Rorlich, Azade-Ayse. 1986. *The Volga Tatars: A Profile in National Resilience*. Stanford: Stanford University Press.

"Russian Organised Crime: Crime Without Punishment." 1999. *The Economist*, August 28.

"The Russians Are Coming, The Russians Are Here." 2000. *USA Today*, July 11.

"Russia's Orthodox Church Moves to Center Stage in State Affairs." 2001. *Russia Today,* March 6.

Shaw, Denis, ed. 1995. *The Post-Soviet Republics: A Systematic Geography*. Essex: Longman Scientific and Technical.

Shaw, Denis, 1999. *Russia in the Modern World: A New Geography*. Oxford: Blackwell.

Solzhenitsyn, Aleksandr I. 1997. *The Gulag Archipelago 1918–1956: An Experiment in Literary Investigation*. Boulder, CO: Westview.

Stewart, John Massey, ed. 1992. *The Soviet Environment: Problems, Policies and Politics*. Cambridge: Cambridge University Press.

United States Department of Commerce. 2001. "Russia Country Commercial Guide FY2001." Chapter 2: Economic Trends and Outlook; Chapter 10: Economic and Trade Statistics.

United States Department of Energy. 1999. "Russia: Environmental Issues." *Energy Information Administration (EIA) Bulletin,* December.

United States Department of Energy. 2000. "Russia: Oil and Gas Exports." *Energy Information Administration (EIA) Bulletin,* December.

United States Department of State. 2000. "Background Notes: Russia."

Valencia, Mark J., ed. 1995. *The Russian Far East in Transition: Opportunities for Regional Economic Cooperation*. Boulder, CO: Westview.

"Will Russia Hold Together?" 1998. *The Economist,* September 12.

Wilson, Andrew. 2000. *The Ukrainians: Unexpected Nation*. New Haven: Yale University Press.

Wixman, Ronald. 1984. *The Peoples of the USSR: An Ethnographic Handbook*. Armonk, NY: M. E. Sharpe.

Wood, Alan, ed. 1991. *The History of Siberia: From Russian Conquest to Revolution*. London: Routledge.

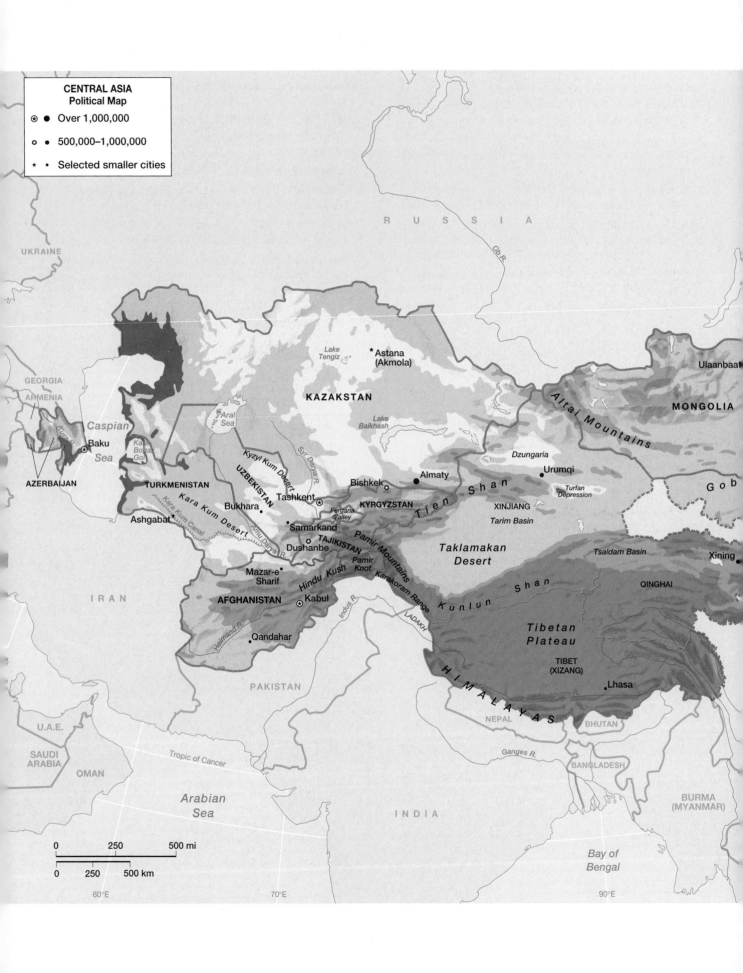

CENTRAL ASIA
Political Map

- ⊛ ● Over 1,000,000
- ⊙ • 500,000–1,000,000
- ★ • Selected smaller cities

UKRAINE

RUSSIA

GEORGIA

ARMENIA

Kura R.

⊛ Baku

AZERBAIJAN

Caspian Sea

Ob R.

ULAANBAAT

MONGOLIA

Altai Mountains

Lake Tengiz

★ Astana (Akmola)

KAZAKSTAN

Lake Balkhash

Aral Sea

Kara Bogaz Gol

TURKMENISTAN

Kyzyl Kum Desert

Syr Darya R.

UZBEKISTAN

Tashkent ⊛

Bishkek ⊙

Almaty ●

Tien Shan

Dzungaria

Urumqi •

Turfan Depression

Gob

Kara Kum Desert

Kara Kum Canal

Bukhara •

Ashgabat ★

Amu Darya R.

Samarkand •

Fergana Valley

KYRGYZSTAN

XINJIANG

Tarim Basin

Dushanbe ⊙

TAJIKISTAN

Pamir Mountains

Pamir Knot

Karakoram Range

Taklamakan Desert

Kunlun Shan

Tsaidam Basin

Xining •

QINGHAI

Mazar-e Sharif •

Hindu Kush

IRAN

AFGHANISTAN

Kabul ⊛

Indus R.

LADAKH

Tibetan Plateau

TIBET (XIZANG)

• Qandahar

Helmand R.

PAKISTAN

HIMALAYAS

• Lhasa

NEPAL

BHUTAN

U.A.E.

SAUDI ARABIA

OMAN

Tropic of Cancer

Ganges R.

BANGLADESH

BURMA (MYANMAR)

Arabian Sea

INDIA

Bay of Bengal

0	250	500 mi

0	250	500 km

60°E   70°E   90°E

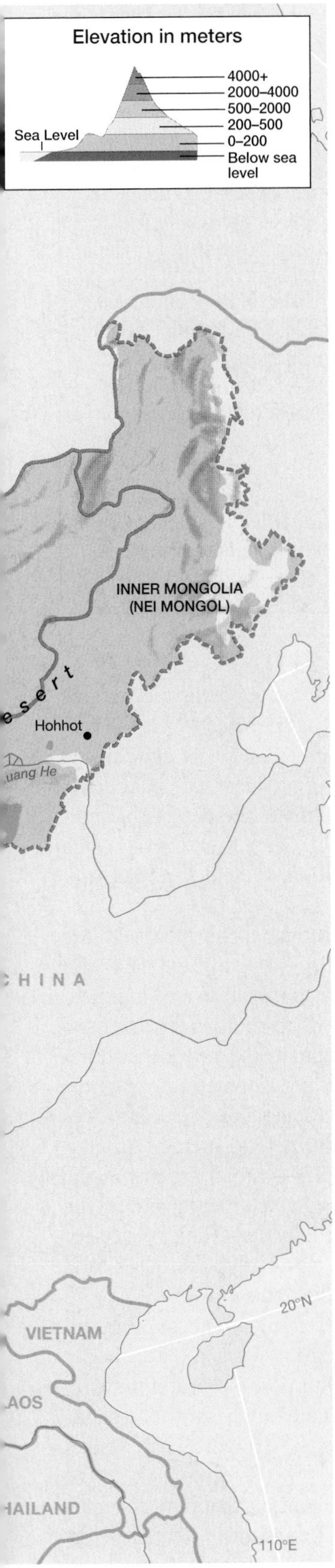

CHAPTER 10

Central Asia

Central Asia does not appear in most books on world regional geography. Although it covers a larger expanse than the United States, it is a remote and lightly populated area dominated by high mountains, barren deserts, and semiarid steppes (grasslands). Central Asia has also been something of a geopolitical void, long dominated by external forces. Until 1991 it contained only two independent countries, Mongolia and Afghanistan. The rest of the region was at that time divided between the Soviet Union and China (Figure 10.1).

Although long absent from world maps, Central Asia began to reappear in discussions of global geography following the breakup of the Soviet Union. Suddenly six new countries appeared on the international scene, prompting scholars to reexamine the position of Central Asia in human affairs. Historians have convincingly shown that the region as a whole has a certain historical coherence and that it once played a central role in the political drama of Eurasia (see "Setting the Boundaries").

Central Asia was more firmly established on the map of the world after September 11, 2001, when it became evident that the attack on the World Trade Center and the Pentagon had been planned and organized by Osama bin Laden and his Al Qaeda organization. Although bin Laden is originally from Saudi Arabia and Al Qaeda operates globally, both had been based for several years in Afghanistan, receiving support from that country's extreme Islamic government known as the Taliban. The Taliban had also been attempting to export its radical version of Islamic politics into neighboring Central Asian countries such as Uzbekistan and Tajikistan. The subsequent war against Al Qaeda and the Taliban, waged both by Afghani rebels and by forces from the United States and Britain, proved to the global community just how geopolitically important Afghanistan and the rest of Central Asia had become.

One reason for Central Asia's lack of prominence until recently is that it was poorly integrated into international trade networks. This began to change, too, in the 1980s and 1990s, however, as large oil and gas reserves were found, especially in Kazakstan, Turkmenistan, and Azerbaijan. Evidence is mounting that larger discoveries await. Estimates of the total oil reserves in the region run between 70 and 200 billion barrels, second only to the 600 billion barrels in the Persian Gulf area. As a result, Western oil companies are

◀ **Figure 10.1 Central Asia** Central Asia, a vast, sprawling region in the center of the Eurasian continent, is dominated by arid plains and basins and lofty mountain ranges and plateaus. Eight independent countries—Kazakstan, Turkmenistan, Uzbekistan, Kyrgyzstan, Tajikistan, Azerbaijan, Afghanistan, and Mongolia—form Central Asia's core. China's lightly populated far west and north are often placed within Central Asia as well, due to patterns of cultural and physical geography.

SETTING THE BOUNDARIES

The term *Central Asia* is defined differently by different writers. Most authorities agree that it includes five newly independent former-Soviet republics: Kazakstan, Kyrgyzstan, Uzbekistan, Tajikistan, and Turkmenistan. This chapter, however, adds another post-Soviet state, Azerbaijan, in addition to Mongolia and Afghanistan, as well as the autonomous regions of western China (Tibet and Xinjiang). Several other provinces and regions of western China, such as Nei Mongol (Inner Mongolia) and Qinghai, are occasionally discussed.

The inclusion of these additional territories within Central Asia is controversial. Azerbaijan is often classified with its neighbors in the Caucasus region (Georgia and Armenia); western China is obviously part of East Asia by political criteria; and Mongolia is also often placed within East Asia because of both its location and its historical connections with China. Afghanistan, for its part, is just as often located within either South Asia or Southwest Asia and the Middle East. Indeed, some writers would include the entire former-Soviet zone within Southwest Asia, although others link it instead with a Russia-centered post-Soviet world region.

But considering Central Asia's historical unity, its common environmental circumstances, and its recent reentry onto the stage of global geopolitics, we think that it deserves consideration in its own right. It also makes sense to define its limits rather broadly. Azerbaijan, for example, is linked by both cultural (language and religion) and economic (oil) factors more to Central Asia than it is to Armenia and Georgia.

At the same time, however, any unity that Central Asia as a whole possesses is far from stable. Continuing Chinese political control over, and Han Chinese migration into, southeastern Central Asia threatens whatever claims may be made for regional coherence. Central Asia itself remains, moreover, deeply divided along cultural lines. Most of the region is Muslim in religious orientation and Turkish in language, but both the northeastern and southeastern sections (Mongolia and Tibet) are firmly Buddhist. Only time will tell whether Central Asia will indeed merit recognition as a distinct world region in its own right.

---

showing increasing interest in Central Asia, although red tape and political complications have prevented the realization of large profits. A number of important countries, moreover, are seeking to exert influence over Central Asia, including Iran, Pakistan, Turkey, the United States, and Russia. China's strict control and periodic repression of its Central Asian lands also highlight the significance of the region.

Central Asia forms a large, compact region in the center of the Eurasian landmass. Alone among all the world regions, it lacks ocean access. Owing to its continental position in the center of the world's largest landmass, Central Asia is noted for its rigorous climate. High mountains, deep basins, and extensive plateaus magnify its climatic extremes. The aridity of the region, as we shall see, has also contributed to some of the most severe environmental problems in the world.

Environmental Geography: Steppes, Deserts, and Threatened Lakes of the Eurasian Heartland

One of the great environmental tragedies of the twentieth century was the virtual destruction of the Aral Sea, a vast lake (until recently, larger than Lake Michigan) located on the boundary of Kazakstan and Uzbekistan in western Central Asia. The Aral's only sources of water are the Amu Darya and Syr Darya rivers, which flow out of the Pamir Mountains, some 600 miles to the southeast. Both of these rivers have been intensively used for irrigation since antiquity, but the scale of diversion vastly expanded after 1950. The valleys of the two rivers formed the southernmost farming districts of the Soviet Union and thus became vital suppliers of warm-season crops. Cotton in particular emerged as an economic mainstay of Central Asia. The acreage devoted to rice—a very water-demanding crop—also increased. Soviet agricultural planners favored huge engineering projects that could deliver water to arid lands and thus "make the deserts bloom." The biggest of these projects was the Kara Kum Canal, which carries water from the Amu Darya across the deserts of southern Turkmenistan.

Unfortunately, the more crops the deserts produced, the less freshwater was available for the Aral Sea. The Aral proved to be particularly sensitive to irrigation development, since it is relatively shallow and its tributaries flow across a vast and arid agricultural landscape. By the 1970s the shoreline had begun to retreat at an unprecedented rate; eventually a number of "seaside" villages found themselves stranded up to 40 miles (64 kilometers) inland. An estimated 135—out of a total of 173—animal species in the lake disappeared. New islands began to emerge, and by the 1990s the Aral Sea had been virtually divided into two separate lakes (Figure 10.2). Sixty percent of the lake's total volume of water is estimated to have disappeared.

The destruction of the Aral Sea resulted in economic and cultural damage, as well as ecological devastation. Fisheries that were once large enough to support a canning industry began to close as the lake grew increasingly salty. Even agriculture has suffered. The retreating lake (Figure 10.3) has left large salt flats on its exposed beds; windstorms pick up the salt, along with the agricultural chemicals that have accumulated in the lake's shallows, and deposit it in nearby fields. Yields have thus declined; desertification has accelerated; and public health has been undermined. Particularly hard hit by this destruction are the Karakalpak people, members of a rel-

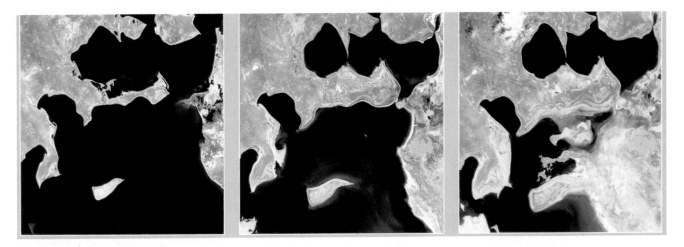

▲ **Figure 10.2 The shrinking of the Aral Sea**   These three satellite images—the first taken in May 1973, the second in August 1987, and the third in July 2000—show the dramatic shrinkage of the Aral Sea. More than 60 percent of the lake's water has been lost since the 1960s, resulting in severe economic damage and environmental degradation.   *(EROS Data Center, U.S. Geological Survey)*

atively powerless ethnic group who have long inhabited the southern shores of the Aral Sea and the formerly rich delta of the Amu Darya River. Recent reports indicate that the Karakalpak have the highest levels of infant and maternal death in the former Soviet Union, and that more than 80 percent of Karakalpak women suffer from anemia. Although the Karakalpak have their own "autonomous republic" within Uzbekistan, any autonomy they might enjoy is of little use in such a devastated landscape.

Since the breakup of the Soviet Union in 1991, government planners have not pushed as hard to increase the cotton crop as they once did. But cotton is still a major foreign exchange earner for Uzbekistan, and the nation cannot afford to abandon all of the lands that were brought under cultivation during the Soviet period. International efforts to conserve water and to slow the ongoing destruction of the Aral Sea are being made, but thus far with relatively little progress.

Other Environmental Issues

Despite the appalling tragedy of the Aral Sea, much of Central Asia has a relatively clean environment, owing largely to its generally low population density. Industrial pollution is a serious problem only in the larger cities, such as Uzbekistan's Tashkent and Azerbaijan's Baku. Some parts of Central Asia, such as northwestern Tibet, remain practically pristine, with little human impact of any kind. Elsewhere, however, the typical environmental dilemmas of arid environments plague the region: desertification (the spread of deserts), salinization (the accumulation of salt in the soil), and desiccation (the drying up of lakes and wetlands). We have already examined the desiccation of the Aral Sea; let us now examine the phenomenon more generally throughout the region, since Central Asia contains some of the world's largest—and most endangered—lakes.

Shrinking and Expanding Lakes   Western Central Asia supports large lakes because it forms a low-lying basin, without drainage to the ocean, that is virtually surrounded by mountains and other more humid areas. The world's largest lake, by a huge margin, is the Caspian Sea, located along the region's western boundary; the fourth largest is (or more precisely, was) the Aral Sea, situated some 300 miles (480 kilometers) to the east, and the fifteenth largest is Lake Balqash, found 600 miles (960 kilometers) farther east. The Caspian Sea is roughly the size of Montana; the Aral Sea was roughly the size of West Virginia; and Lake Balqash is roughly the size of Connecticut and Rhode Island combined. Several other large lakes are also located in the region.

Despite their names, neither the Aral nor the Caspian are true seas, since they are not connected with the ocean. They are incorrectly called seas only because they are extremely large

▲ **Figure 10.3 A dying lake**   Due to the shrinkage of the Aral Sea, former lakeside villages are now located far inland, as evidenced by these two beached ships. Not only have fishing economies been destroyed, but the desiccated lake bed itself is now a source of pollution, as desert winds deposit salt and agricultural chemicals on fields.   *(David Turnley/Corbis)*

ENVIRONMENTAL ISSUES   Saving the Caspian Sea?

Like other lakes without an outlet to the ocean, the Caspian exhibits tremendous natural fluctuations in water level. During moist periods, the level of the lake rises; during drier times, it falls. In the early 1900s Russian geographer L. S. Berg investigated the history of the Caspian to see if there was any natural periodicity to this rise and fall. He discovered maximum levels in 1650, 1770, and 1900, and minimal levels in 1590, 1710, and 1840. Extrapolating into the future, he concluded that maximum levels would again be reached in 2020. Most climatologists today doubt that climatic fluctuations are as regular as Berg supposed. Still, the recent—and quite surprising—rise in the level of the Caspian does conform to his predictions.

Considering the economic importance of the Caspian, Soviet planners were eager to devise methods of stabilization during the earlier period of falling levels. There are two basic approaches that can be used to slow down or reverse the fall of nondrained lakes: increase the flow of freshwater or reduce evaporation.

The influx of freshwater into the Caspian could be significantly increased only by transferring water from another drainage system. This has already been done to a minor extent, as some water is diverted to the Volga from several rivers of northern European Russia. To make up for the irrigation loss in the lower Volga, however, much larger diversions would be required. In the 1960s a truly ambitious scheme was hatched, one that would have entailed tapping the flow of the huge Ob River, which crosses the sparsely populated wetlands of the West Siberian Plain.

Diverting water from the Ob to the Caspian would require building a canal some 1,500 miles (2,400 kilometers) long. Large amounts of water would be lost to evaporation and seepage along the way. Such a massive transfer project might also cause ecological changes in both the Ob Basin and the Arctic Ocean, into which the river drains. Such fears, however, have been largely put to rest; by the 1980s it was obvious that the Soviet Union could not afford the massive price tag that such a project would carry.

By the late 1970s it had become clear that the second and much less expensive alternative, reducing evaporation, was more feasible. Evaporation can be reduced simply by decreasing the surface area of the lake, which in turn can be accomplished by diking off certain sections. Once this is accomplished, the elevation of the water surface can be stabilized even as the lake itself becomes smaller. The most easily diked area is the almost-enclosed gulf known as the Kara-Bogaz Gol in the Caspian's center-east. It loses large amounts of water to evaporation but receives little influx from either rainfall or streams.

In the early 1980s construction commenced on a dike across the 1,800-foot-long (555-meter) mouth of the Kara-Bogaz Gol. By the mid-1990s the gulf was virtually dry—while the lake itself was beginning to cause problems by rising. Engineering projects obviously do not present a perfect solution for the problems of water control in basins characterized by internal drainage.

and somewhat salty. The Caspian is actually less salty than the ocean (particularly in the north), while until the 1970s the Aral was only slightly brackish (salty). Lake Balqash is almost fresh in the west but is quite salty in its long eastern extension. Since none of these lakes is drained by rivers, all naturally fluctuate in size, depending on how much precipitation falls in their drainage basin in a given year. Like the Aral Sea, Balqash and several other lakes have suffered from reduced water flow and hence increasing salinity. The story of the Caspian, however, is more complicated. The Caspian Sea receives most of its water from the large rivers of the north, the Ural and the Volga, which drain much of European Russia. Owing to the construction of large reservoirs and the development of extensive irrigation facilities in the lower Volga basin, the volume of freshwater reaching the Caspian began to decline in the second half of the twentieth century. With a reduced influx of water, the level of the great lake dropped, exposing as much as 15,000 square miles (39,000 square kilometers) of former lake bed. A reduced volume of water resulted in increased salinity levels, undermining fisheries and threatening the entire ecosystem. The Russian caviar industry, centered in the northern Caspian, suffered extensive damage.

The Caspian reached a low point in the late 1970s. At that point it began to rise, presumably because of higher than nor-

mal precipitation in its drainage basin, and by the late 1990s it had risen some 8.2 feet (2.5 meters). This enlargement, too, has caused problems, inundating, for example, some of the newly reclaimed farmlands in the Volga Delta. The most serious current environmental threat to the Caspian, however, is probably pollution from the oil industry, rather than fluctuation in size (see "Environmental Issues: Saving the Caspian Sea?").

Desertification   Desertification is another major concern in Central Asia (Figure 10.4). In the eastern part of the region, the Gobi Desert has gradually spread southward, encroaching on densely settled lands in northeastern China proper. The Chinese have tried to stabilize dune fields and to prevent the march of desert with massive tree-and grass-planting campaigns, but such efforts have been only partially successful. Northern Kazakstan has also seen extensive desertification. This area was one of the main sites of the ambitious Soviet "virgin lands campaign" of the 1950s, in which semiarid grasslands were plowed and planted with wheat. Many of these lands have since returned to native grasses, but not before erosion stripped away much of their productivity.

Deforestation has also harmed the region. Although most of Central Asia is too dry to support forests, many of its mountains were once well wooded. Today extensive forests can be found

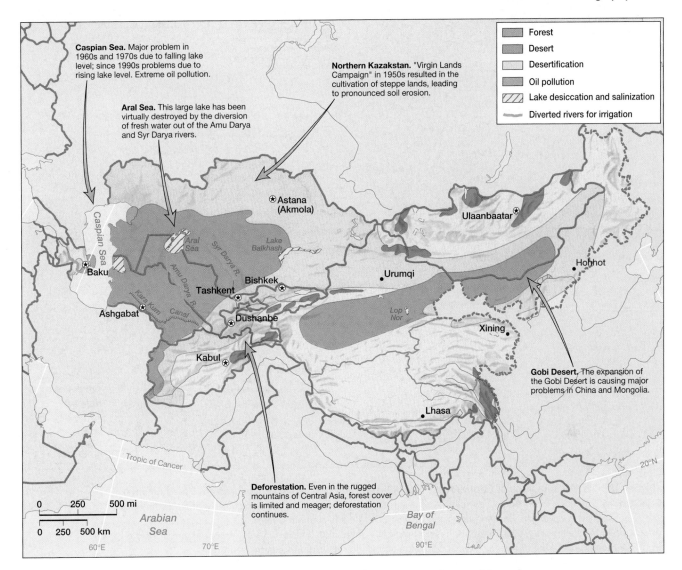

Caspian Sea. Major problem in 1960s and 1970s due to falling lake level; since 1990s problems due to rising lake level. Extreme oil pollution.

Aral Sea. This large lake has been virtually destroyed by the diversion of fresh water out of the Amu Darya and Syr Darya rivers.

Northern Kazakstan. "Virgin Lands Campaign" in 1950s resulted in the cultivation of steppe lands, leading to pronounced soil erosion.

Gobi Desert. The expansion of the Gobi Desert is causing major problems in China and Mongolia.

Deforestation. Even in the rugged mountains of Central Asia, forest cover is limited and meager; deforestation continues.

Legend:
- Forest
- Desert
- Desertification
- Oil pollution
- Lake desiccation and salinization
- Diverted rivers for irrigation

▲ Figure 10.4 Environmental issues in Central Asia   Central Asia has experienced some of the most severe problems associated with desertification in the world. Soil erosion and overgrazing have led to the advance of desertlike conditions in much of western China and Kazakstan. In western Central Asia, the most serious environmental problems are associated with the diversion of rivers for irrigation and the corresponding desiccation of lakes. Oil pollution is a particularly serious issue in the Caspian Sea area.

only in the wild gorge country of the eastern Tibetan Plateau; in some of the more remote slopes of the Tien Shan, Altai, and Pamir mountains; and in the uplands of northern Mongolia.

Central Asia's Physical Regions

To understand why Central Asia suffers from such environmental problems as lake desiccation, it is necessary to examine the region's physical geography in greater detail. In general, Central Asia is dominated by grassland plains (or steppe) in the north, desert basins in the southwestern and central areas, and high plateaus and mountains in the south-center and southeast (Figure 10.5). Lofty mountains do extend, however, into the very heart of the region, dividing the desert zone into a series of separate basins and giving rise to the rivers that flow into the deserts and hence into the imperiled lakes.

The Central Asian Highlands

The highlands of Central Asia originated in one of the great tectonic events of Earth's history: the collision of the Indian subcontinent into the Asian mainland. This ongoing collision has created the highest mountains in the world, the Himalayas, located along the boundary of South Asia and Central Asia. The Himalayas, for all their grandeur and fame, are merely one portion of a much larger network of high mountains and plateaus. To the northwest they merge with the Karakoram Range and then the Pamir Mountains. From the so-called Pamir Knot, a complex tangle of mountains situated where Pakistan, Afghanistan, China, and Tajikistan converge, other towering ranges radiate outward in several directions. The Hindu Kush sweeps to the southwest through central Afghanistan; the Kunlun Shan extends to the east (along the

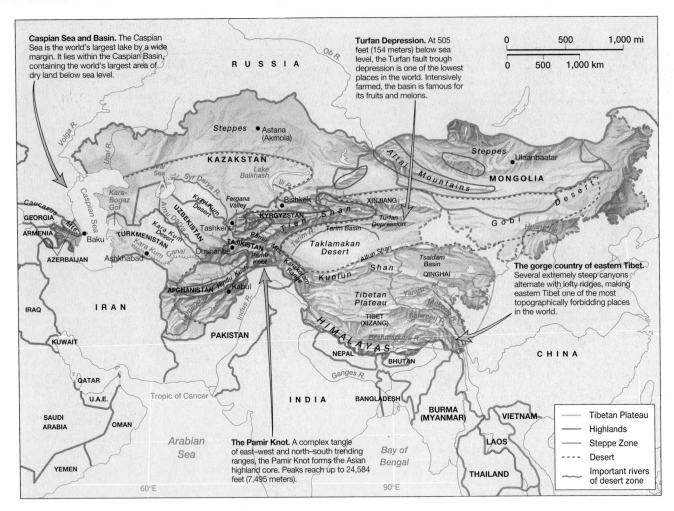

Caspian Sea and Basin. The Caspian Sea is the world's largest lake by a wide margin. It lies within the Caspian Basin, containing the world's largest area of dry land below sea level.

Turfan Depression. At 505 feet (154 meters) below sea level, the Turfan fault trough depression is one of the lowest places in the world. Intensively farmed, the basin is famous for its fruits and melons.

The gorge country of eastern Tibet. Several extremely steep canyons alternate with lofty ridges, making eastern Tibet one of the most topographically forbidding places in the world.

The Pamir Knot. A complex tangle of east–west and north–south trending ranges, the Pamir Knot forms the Asian highland core. Peaks reach up to 24,584 feet (7,495 meters).

Legend:
- Tibetan Plateau
- Highlands
- Steppe Zone
- Desert
- Important rivers of desert zone

▲ **Figure 10.5 Physical regions of Central Asia** Central Asia is divided into three main regions based on physical geography. The north is dominated by relatively flat, grassy plains known as the steppes. Most of the central portion of Central Asia is covered by desert plains and basins. Scattered throughout Central Asia, but particularly pronounced in the south, are the mountain ranges and plateaus of the highland zone. Since most of Central Asia is arid, rivers running out of the highlands have special significance for the region's human geography.

northern border of the Tibetan Plateau); and the Tien Shan swings out to the northeast into China's Xinjiang province. All of these ranges have peaks higher than 20,000 feet (6,000 meters) in elevation. Much lower but still significant ranges are found in the southwestern reaches of the region, along Turkmenistan's boundary with Iran and Azerbaijan's boundaries with Russia, Armenia, and Iran.

Much more extensive than these mountain ranges, however, is the Tibetan Plateau (Figure 10.6). This massive upland extends some 1,250 miles (2,000 kilometers) from east to west and 750 miles (1,200 kilometers) from north to south. More remarkable than its size is its elevation; virtually the entire area is higher than 12,000 feet (3,700 meters) above sea level, and its average height is about 15,000 feet (4,600 meters)—higher, in other words, than the highest mountains in the contiguous United States.

Most of the large rivers of South, Southeast, and East Asia originate in the Tibetan Plateau and adjoining mountains, including the Indus, Ganges, Brahmaputra, Mekong, Yangtze, and Huang He. These rivers pass either through the canyonlands of eastern Tibet or through gaps in the Himalayas to the

▲ **Figure 10.6 Tibetan Plateau** The Tibetan Plateau is dominated by alpine grasslands and tundra interspersed with rugged mountains and saline lakes. In summer the sparse vegetation offers forage for the herds of nomadic Tibetan pastoralists. Much of the northern part of Tibet is too high to sustain such a low-intensity land use. *(Michel Peissel/SIPA Press)*

south. Most of the Tibetan Plateau, however, drains internally. Runoff from mountain snowfields trickles down to evaporate from innumerable lakes and marshes. These lakes and wetlands are generally saline, since salt accumulates wherever bodies of water do not drain to the sea.

The greater part of the Tibetan Plateau, at about 15,000 feet (4,600 meters) of elevation, lies near the maximum elevation at which human life can exist. Rather than forming a flat, tablelike surface, the plateau is punctuated with east–west running ranges alternating with undrained basins. The largest of these basins, the Tsaidam, is also the lowest, situated below 9,000 feet (2,700 meters). Many of Tibet's higher ranges contain glaciers, which owe their existence to cold temperatures more than to heavy snowfall. Although the southeastern sections of the plateau receive ample precipitation, most of Tibet is arid. Cut off by high ranges from any source of moisture, large areas of the plateau receive only a few inches of rain a year (Figure 10.7). Winters on the Tibetan Plateau are cold; and while summer afternoons can be warm, summer nights remain chilly.

The Plains and Basins   Although the mountains of Central Asia are higher and more extensive than those found anywhere else in the world, most of the region is characterized by plains and basins of low and intermediate elevation. This lower-lying zone can be divided into two main areas: a central belt of deserts punctuated by verdant river valleys and a northern swath of semiarid steppe.

Central Asia's desert belt is itself divided into two discontinuous segments by the Tien Shan and Pamir mountains. To the west lie the arid plains of the Caspian and Aral Sea basins, located primarily in Turkmenistan, Uzbekistan, and southern Kazakstan. The most desolate areas, the Kara Kum and Kyzyl Kum deserts, support very meager vegetation and contain extensive sand dunes. Most of this area is relatively flat and very low. The Aral Sea lies only 135 feet (41 meters) above sea level, while the surface of the Caspian Sea is 92 feet (28 meters) below sea level. A significant proportion of the country of Azerbaijan also lies below sea level. The climate of this region is strikingly continental; summers are dry and hot, while winter temperatures average well below freezing. Central Asia's eastern desert belt extends for almost 2,000 miles (3,200 kilometers) from the extreme west of China at the foot of the Pamirs to the southeastern edge of Inner Mongolia. It is conventionally divided into several deserts, even though the belt of aridity is continuous. The most important of these deserts are the Taklamakan, found in the Tarim Basin of Xinjiang, and the Gobi, which runs along the border between Mongolia proper and the Chinese region of Inner Mongolia. Much of the interior portion of the Taklamakan is covered by sand and rock and is virtually devoid of vegetation. The Tarim is a deep basin, about 3,000 feet (900 meters) above sea level, nearly enclosed by some of the world's highest mountains. Several smaller basins in the vicinity are much lower; the Turfan lies at 505 feet (154 meters) below sea level.

The environment of western **Turkestan** (or the former Soviet zone) is distinguished from that of so-called eastern Turkestan (China's Xinjiang) in part by its larger rivers. Much more snow falls on the western than the eastern slopes of the Pamir Mountains, giving rise to more abundant runoff. The largest of these rivers, as we have seen, flow into the Aral Sea. Others, such as the Helmand of Afghanistan, terminate in shallow lakes or extensive marshes and salt flats. In Xinjiang, one substantial river, the Tarim, flows out of the highlands onto the basin floor, where it frequently shifts course across the sandy lowlands. Before the completion of new irrigation projects in the 1960s, it terminated in a salty lake called Lop Nor. Subsequently, the newly dried-out Lop Nor salt flat was used periodically by China for testing nuclear weapons.

North of the desert zone, rainfall gradually increases and desert eventually gives way to the great grasslands, or steppe, of northern Central Asia. Near the region's northern boundary, trees begin to appear in favored locales, outliers of the great Siberian taiga (coniferous forest) of the north. A nearly continuous swath of grasslands extends some 4,000 miles (6,400 kilometers) east to west across the entire region, only partially broken by Mongolia's Altai Mountains. Particularly rich pastures are found in northern and eastern Kazakstan, in the Dzungaria region of northern Xinjiang, and in northern and central Mongolia. Summers on the northern steppe are usually pleasant, but winters are cold, sometimes brutally so. Blizzards and intense cold in the winter of 2000–2001, for example, wiped out roughly half of Mongolia's livestock, resulting in widespread hunger and destitution.

Population and Settlement: Densely Settled Oases amid Vacant Lands

Most of Central Asia is sparsely populated (Figure 10.8). Large areas are essentially uninhabited, either too arid or too high to support human life. Even many of the more favorable areas are populated only by widely scattered groups of nomadic **pastoralists** (people who raise livestock for subsistence purposes). Mongolia, which is more than twice the size of Texas, has only 2.5 million inhabitants—fewer than live in the Dallas metropolitan area. But as is common in arid environments, those few lowland locales with good soil and dependable water supplies can be thickly settled. Despite its overall aridity, Central Asia is well endowed with perennial rivers and fertile oases. While the nomadic pastoralists of the steppe and desert zones have dominated the history of Central Asia, the sedentary peoples of the river valleys have always been more numerous.

Highland Population and Subsistence Patterns

The environment of the Tibetan Plateau is particularly harsh. Not only is the climate cold and water often scarce or brackish, but ultraviolet radiation, owing to the elevation, is always pronounced. Only sparse grasses and herbaceous plants—so-called mountain tundra—can survive such rigors. Human subsistence is obviously difficult under such conditions. The only feasible way of life over most of the Tibetan Plateau is nomadic pastoralism based on the yak, an altitude-adapted relative of

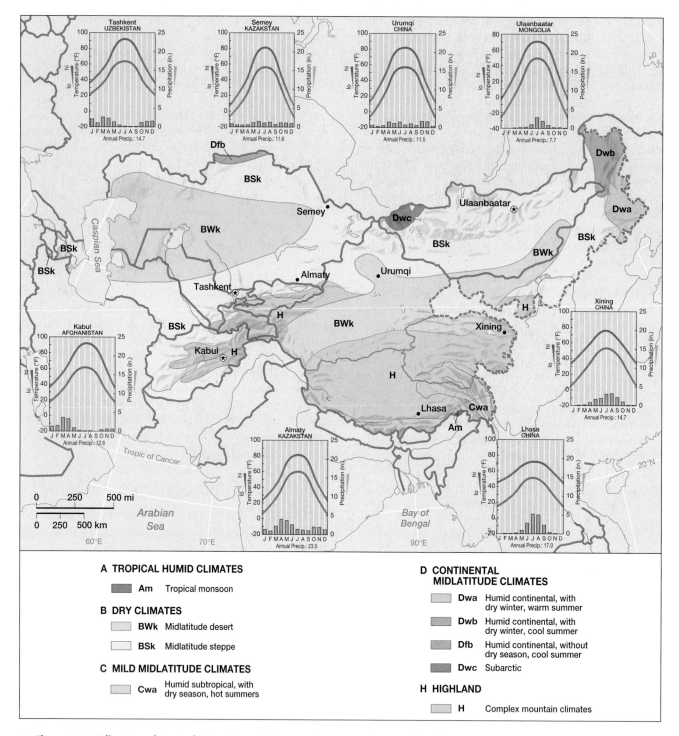

▲ Figure 10.7 Climates of Central Asia   Central Asia is a dry region dominated by desert and steppe climates. Even in most of Central Asia's highlands, marked "H" on this map, arid conditions predominate. Truly humid areas in Central Asia are found only in limited areas of the far north and extreme southeast. As a midlatitude region located in the interior of a vast continent, Central Asia is marked by pronounced continentality, experiencing profound differences between winter and summer temperatures.

the cow. Several hundred thousand people manage to make a living in such a manner, roaming with their herds over vast distances. Much of northwestern Tibet, however, is too high even for yak pastoralism and is thus uninhabited.

Although most of the Tibetan Plateau can support only nomadic pastoralism, most Tibetans are sedentary farmers.

Farming in Tibet is possible only in a few favorable locations, generally those that are *relatively* low in elevation and that have good soils and either adequate rain or a dependable irrigation system. The main zone of sedentary settlement lies in the far south, where protected valleys offer favorable conditions. The population of Tibet proper (the Chinese au-

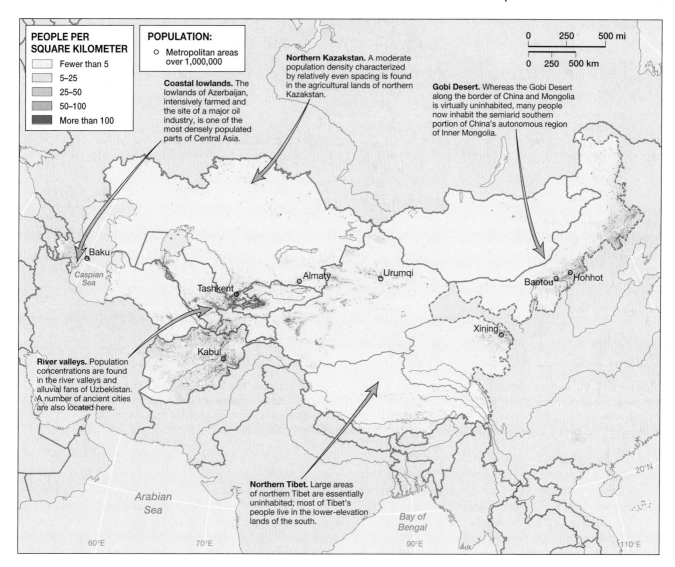

PEOPLE PER SQUARE KILOMETER
- Fewer than 5
- 5–25
- 25–50
- 50–100
- More than 100

POPULATION:
- ○ Metropolitan areas over 1,000,000

Coastal lowlands. The lowlands of Azerbaijan, intensively farmed and the site of a major oil industry, is one of the most densely populated parts of Central Asia.

Northern Kazakhstan. A moderate population density characterized by relatively even spacing is found in the agricultural lands of northern Kazakhstan.

Gobi Desert. Whereas the Gobi Desert along the border of China and Mongolia is virtually uninhabited, many people now inhabit the semiarid southern portion of China's autonomous region of Inner Mongolia.

River valleys. Population concentrations are found in the river valleys and alluvial fans of Uzbekistan. A number of ancient cities are also located here.

Northern Tibet. Large areas of northern Tibet are essentially uninhabited; most of Tibet's people live in the lower-elevation lands of the south.

▲ Figure 10.8 Population density in Central Asia   Central Asia as a whole remains one of the world's least densely populated regions, although it does contain distinct clusters of higher population density. Most of Central Asia's large cities are located near the region's periphery or in its major river valleys.

tonomous region of Xizang) is only 2.5 million, while that of China's Qinghai province (most of which lies on the Tibetan Plateau) is 4.2 million. Considering the vast size of this area, these are small numbers indeed. An area of comparable size in eastern China holds nearly 1 billion human inhabitants.

Population densities are also low in the other highland areas of Central Asia, although settled agricultural communities can be found in the protected valleys of the southern ranges. Owing to its complex topography, the Pamir range in particular offers a large array of small and nearly isolated valleys that are suitable for agriculture and intensive human settlement. Not surprisingly, this area is marked by great cultural and linguistic diversity. Many villages here are noted for their agricultural terraces and well-tended orchards.

Central Asia's mountains are vitally important for people living in the adjacent lowlands, whether they are settled farmers or migratory pastoralists. Many herders use the highlands for summer pasture; when the lowlands are parched, the high meadows provide rich grazing. The Kyrgyz (of Kyrgyzstan) are noted for their traditional economy based on **transhumance,** moving their flocks from lowland pastures in the winter to highland meadows in the summer. The farmers of Central Asia rely on the highlands for their wood supplies (a few of these mountains still contain forests) and, more important, for their water. Settled agricultural life in most of Central Asia is possible only because of the rivers and streams flowing out of the region's mountains.

Lowland Population and Subsistence Patterns

Most of the inhabitants of Central Asian deserts live in the narrow belt where the mountains meet the basins and plains. Here water supplies are adequate and soils are neither salt- nor alkali-impregnated, as is often the case in the basin interiors. The population distribution pattern of China's Tarim Basin

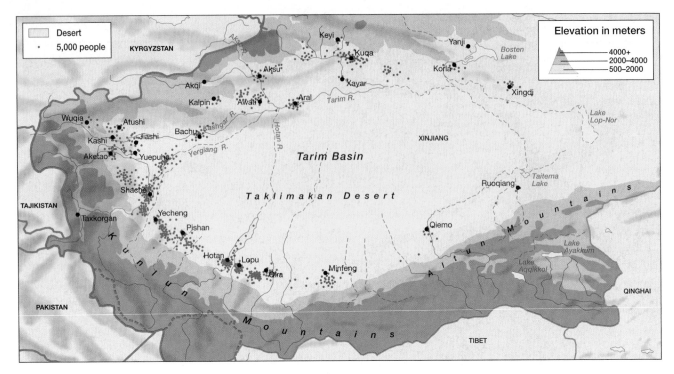

▲ **Figure 10.9 Population patterns in Xinjiang's Tarim Basin**   The central portion of the Tarim Basin is a virtually uninhabited expanse of sand dunes and salt flats. Along the edge of the basin, however, dense agricultural and urban settlements are located where streams running out of the surrounding mountains allow for intensive irrigation. The largest of these oasis communities are found along the southwestern fringe of the basin.

forms an almost perfect ringlike structure (Figure 10.9). Streams flowing out of the mountains are diverted to irrigate fields and orchards in the narrow fertile band situated between the steep slopes of the mountains and the sandy, rocky, or salt-encrusted flats of the central basin.

The population west of the Pamir range, in former Soviet Central Asia, is also concentrated in the transitional zone nestled between the highlands and the plains. A series of **alluvial fans** (fan-shaped deposits of sediments dropped by streams flowing out of the mountains) have long been devoted to intensive cultivation (Figure 10.10). Fertile loess soil abounds (**loess** is a silty soil deposited by the wind), and in a few favored areas winter precipitation is high enough to allow rain-fed agriculture. Several large valleys in this area also offer fertile and easily irrigated farmland. The Fergana Valley of the upper Syr Darya River, which is particularly noted for its productivity, is shared by three countries: Uzbekistan, Kyrgyzstan, and Tajikistan. In the far west of the region, Azerbaijan's Kura River Basin is another area of intensive agriculture (mostly cotton and rice) and concentrated settlement.

Unlike the other deserts of the region, the Gobi has few sources of permanent water. Rivers draining Mongolia's highlands flow to the north or terminate in interior basins, while only a few of the larger streams from the Tibetan Plateau reach the Gobi proper. (The Huang He, however, does swing north to reach the desert edge before turning south to flow through the Loess Plateau.) Owing to this paucity of **exotic rivers** (those originating in more humid

areas) and to its own aridity, the Gobi remains one of Asia's least-populated areas.

The steppes of northern Central Asia are the classical land of nomadic pastoralism. Until the present century, virtually none of this area had ever been plowed and farmed. To this day, pastoralism remains a common way of life across the grasslands, particularly in Mongolia (Figure 10.11). In northwestern China and the former Soviet republics, however,

▲ **Figure 10.10 Farmland in Uzbekistan**   The fertile river valleys of Uzbekistan have been intensely cultivated for many centuries, producing large harvests of fruits and vegetables in addition to cotton and grain. Here tomatoes are harvested—a subtropical crop that has traditionally been exported to Russia and other areas.   *(Jeremy Nicholl/Katz/Corbis/SABA Press Photos, Inc.)*

▲ **Figure 10.11 Steppe pastoralism**   The steppes of northern and central Mongolia offer lush pastures during the summer. Mongolians, some of the world's most skilled horse-riders, have traditionally followed their herds of sheep and cattle, living in collapsible, felt-covered yurts. Many Mongolians still follow this way of life.   (Goussard/SIPA Press)

many pastoral peoples have been forced to adopt sedentary lifestyles. National governments in the region, like those in most other parts of the world, find migratory people hard to control and difficult to provide with medical and other social services. In northern Kazakstan, the Soviet regime converted the most productive pastures into farmland in the mid-1900s in order to increase the country's supply of grain. Some of these lands have since reverted to steppe, but large areas remain under the plow, and Kazakstan is a major producer of spring wheat. Consequently, northern Kazakstan has the highest population density of the steppe belt. Mongolia developed a small agricultural sector after World War II, but since the fall of communism its grain production has decreased by about 50 percent.

Population Issues

Although Central Asia remains a low-density environment, some portions of it are growing at a moderately rapid pace. In western China, much of the population growth over the past 30 years has stemmed from the migration of Han Chinese into the area—an influx much resented by many of the indigenous inhabitants. Population growth in the former Soviet zone, on the other hand, has stemmed not from immigration but rather from relatively high levels of fertility. This area, particularly Kazakstan, has actually witnessed a substantial migration of people out of the region. These emigrants are mostly ethnic Russians returning to the Russian homeland (Figure 10.12).

As Table 10.1 suggests, population statistics are not readily obtainable for the Central Asian portions of China. The numbers available for the rest of the region show that its overall fertility rates are near the middle of those for the developing world as a whole. During the final years of the Soviet Union, Central Asia's birthrates—which were substantially higher than those elsewhere in the country—were a major cause of concern for Russian nationalists and may have contributed to the breakup of the Soviet Union. Some observers attribute Central Asia's somewhat elevated birthrates to Islam, but others think that it rather reflects social and economic factors, including the region's relatively low levels of urbanization. Muslim Azerbaijan, they note, has a lower birthrate than the less-urbanized but equally Muslim countries of Uzbekistan and Tajikistan.

Fertility patterns do vary substantially from one part of Central Asia to another. Afghanistan, the least-developed and most male-dominated country of the region, has the highest birthrate by a substantial margin. Although good data are difficult to find, much evidence would suggest that Tibet's birthrate remains quite low (see "Demographic Issues: Polyandry in Tibet"). Kazakstan's birthrate, just slightly under

---

DEMOGRAPHIC ISSUES   *Polyandry in Tibet*

Early in the twentieth century some outsiders worried that the population of Tibet, never large to begin with, was in some danger of disappearing altogether owing to the extremely low birthrate. Today the main concern is over the influx of Han Chinese, but the Tibetan birthrate remains low. Historically speaking, this is partly the result of widespread monasticism. Another factor is the unusual institution of polyandry—in other words, of a single woman taking more than one husband. The opposite type of plural marriage, polygyny (whereby one man takes more than one wife), is historically common in many parts of the world. Polyandry, however, has almost always been very rare—except in Tibet. It has been especially widespread in western Tibet, that portion of the plateau farthest removed from Chinese cultural and political influences.

In traditional Tibetan society, only the eldest son inherits property, and only those inheriting property are allowed to

marry. Younger brothers thus have had the option of joining a monastery or, in many cases, of sharing the eldest brother's wife. Such a custom not only avoided the endless subdivision of land, which could be a major problem in the harsh environment of Tibet, but also ensured that fertility rates remained low. Since large numbers of men have traditionally lived either in monasteries or in polyandrous households, many women have had no opportunity to marry and thus have remained in their parents' or brother's households. Since polyandrous women are not able to have any more children than monogamous ones (quite unlike polygynous men), a low birthrate is the unavoidable consequence of this custom.

Tibet's unique family structure perhaps functioned in the past to maintain sustainable population levels in a harsh environment. Today, when Tibetans are competing against Han Chinese immigrants—especially in the growing urban sector—polyandry is perhaps no longer so adaptive.

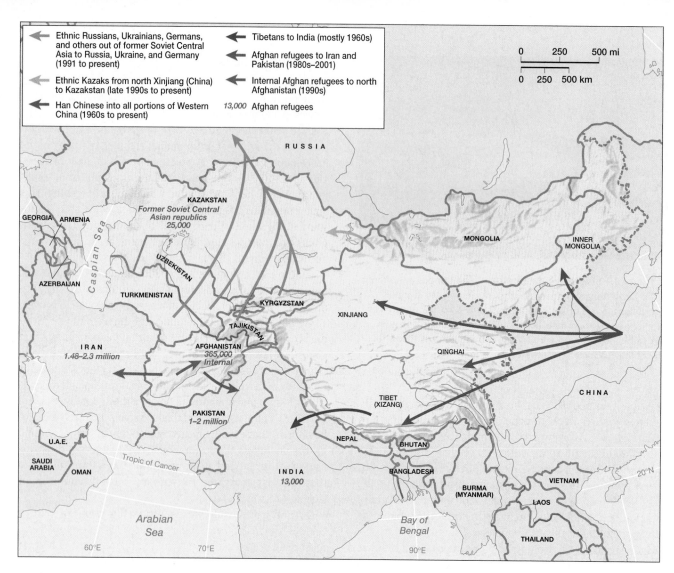

▲ Figure 10.12 **Recent migration and refugee flows** In recent years, large numbers of people have been moving out of the former Soviet Central Asia and into the Chinese portions of Central Asia. Afghanistan, for its part, has one of the largest refugee problems in the world.

TABLE 10.1 *Demographic Indicators*

Country	Population (Millions, 2001)	Population Density, per square mile	Rate of Natural Increase	TFR[a]	Percent < 15[b]	Percent > 65	Percent Urban
Afghanistan	26.8	106	2.4	6.0	43	3	22
Azerbaijan	8.1	243	0.9	2.0	32	6	51
Kazakstan	14.8	14	0.5	1.8	28	7	56
Kyrgyzstan	5.0	65	1.3	2.4	35	5	35
Mongolia	2.4	4	1.4	2.2	34	4	57
Tajikistan	6.2	112	1.4	2.4	42	4	27
Tibet	2.5	4.5	—	—	—	—	—
Turkmenistan	5.5	29	1.3	2.2	38	4	44
Uzbekistan	25.1	145	1.7	2.7	38	4	38
Xinjiang	17.0	24	—	—	—	—	—

[a]Total fertility rate

[b]Percent of population younger than 15 years of age

Source: Population Reference Bureau. World Population Data Sheet, *2001.*

the natural replacement level, is also low for the region, reflecting in part the extremely low fertility level of the Russian speakers in the north of the country.

Urbanization in Central Asia

Although the steppes of northern Central Asia had no real cities before the modern age, the river valleys and oases have been partially urbanized for millennia. Such cities as Samarkand and Bukhara in Uzbekistan were famous even in medieval Europe for their riches and their lavish architecture (Figure 10.13). This early urban fluorescence was built upon the region's economic and political position. The Amu Darya and Syr Darya valleys lay near the midpoint of the trans-Eurasian silk route, and they formed the core of a number of empires based on the cavalry forces of the steppe. The modern era of steamships and oceanic trade brought hardship to the cities of Central Asia. Isolated from maritime routes, these ancient mercantile centers could no longer compete well in world trade and began to diminish.

The conquest of Central Asia by the Russian and Chinese empires resulted in further difficulties, but also ushered in a new wave of urban formation. Cities slowly began to appear on the Kazak steppes, where none had previously existed. The Manchu and Chinese conquerors of Inner Mongolia and Xin-

▲ **Figure 10.13 Traditional architecture in Samarkand** Samarkand, Uzbekistan, is famous for its lavish Islamic architecture, some of it dating back to the 1400s. The city owes part of its rich architectural heritage to the fact that it was the capital of the great medieval conqueror Tamerlane. *(Haley/SIPA Press)*

jiang similarly built new administrative and garrison cities, often placing them only a few miles from indigenous urban sites. As is visible in Figure 10.14, this process created a dualistic urban framework that is still partially visible today. The old indigenous cities of the region are characterized by complex

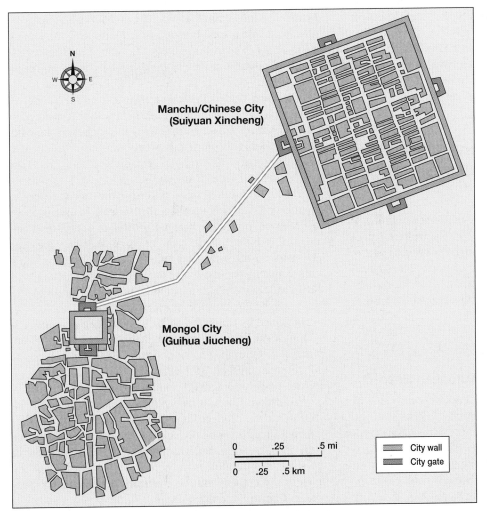

◀ **Figure 10.14 The dual-city plan of traditional Hohot** Hohot, the main city of Inner Mongolia (the Chinese autonomous region of Nei Monggol), exemplifies the dual-city nature of many urban centers of western China. The Manchu/Chinese city shows the clear evidence of traditional Chinese geometrical city planning, while the older Mongol city, at some distance, shows a more haphazard pattern that stemmed from spontaneous growth. Recent developments, however, have tended to fill in the intervening areas and thus obscure these traditional distinctions. *(Source: Piper R. Gaubatz, 1996, Beyond the Great Wall: Urban Form and Transformation on the Chinese Frontiers, p 67, Stanford: Stanford University Press.)*

and almost mazelike networks of streets and alleyways, while the neighboring Manchu cities were constructed according to a strict geometrical order. The recent growth of urban populations in the area has obscured this old dualism, but several new "twin cities" have emerged in Xinjiang, where Han Chinese immigrants have recently built settlements next to indigenous cities.

One may also distinguish Russian/Soviet cities from indigenous cities in the former Soviet zone, but this dichotomy is not so clear-cut. In Uzbekistan, for example, Tashkent is largely a Soviet creation, while many parts of Bukhara still reflect the older urban patterns. Several major cities, such as Kazakstan's former capital of Almaty, did not exist before Russian colonization. In Azerbaijan, Baku emerged as a major city in the early twentieth century as the first Caspian oil fields began to be intensively exploited. Almost everywhere, moreover, one can see the effects of centralized Soviet urban planning and design.

Today well-developed urban networks can be found in both the river and oasis zone of the south and the agricultural areas of northern Kazakstan. North-central Kazakstan is also witnessing the rise of a new major city, Astana, which the government has designated as a new, centrally located capital. Even Mongolia, long a land virtually without permanent settlements, now has more people living in cities than in the countryside. In some parts of the region, however, cities remain relatively few and far between. Only 28 percent of the people of Tajikistan, for example, are urban residents. Tibet similarly remains a predominantly rural society. But wherever war and ethnic strife predominate, as they have in Afghanistan, cities may be expected to swell to uneconomic proportions as refugees seek safety from rural combat zones.

Cultural Coherence and Diversity: A Meeting Ground of Disparate Traditions

Although Central Asia has a certain environmental unity, its cultural coherence is more questionable. The western half of the region is largely Muslim and is often classified as part of Southwest Asia. Northeastern and southeastern Central Asia—Mongolia and Tibet—are characterized by a distinctive form of Buddhism sometimes called Lamaism. Tibet is culturally linked to both South and East Asia, and Mongolia is intimately associated with China, but neither fits easily within any world region.

Historical Overview: An Indo-European Hearth?

The river valleys and oases of Central Asia were early sites of sedentary, agricultural communities. Archaeologists have discovered abundant evidence of farming villages dating back to the Neolithic period (beginning circa 8000 B.C.) in the Amu Darya and Syr Darya valleys and along the rim of the Tarim Basin. After the domestication of the horse around 4000 B.C., nomadic pastoralism emerged in the steppe belt as a new human adaptation. Eventually pastoral peoples gained power over the entire region, transforming not only the history of Central Asia but also that of virtually all of Eurasia.

In the premodern period, pastoral nomads enjoyed profound military advantages over sedentary societies. They had access to large numbers of horses at a time when cavalry almost always held the edge over infantry. Nomads also possessed the benefits of mobility. If pursued by a larger army, they could withdraw to the more inaccessible reaches of the steppe until the danger had passed. Not until the age of gunpowder were the benefits of pastoralism offset by the demographic and economic advantages held by the more populous agriculturally based states.

The earliest recorded languages of Central Asia, spoken both in the oasis communities and among some of the pastoralists, were members of the Indo-European linguistic family. Indeed, Central Asia is often considered to be the birthplace of the Indo-European peoples. In the first millennium B.C., the inhabitants of the Amu Darya and Syr Darya valleys spoke languages closely related to ancient Persian. Many vestiges of this Persian heritage are still found in southwestern Central Asia.

Indo-European languages were replaced on the steppe roughly 2,000 years ago by languages in another major family: Altaic. Three great branches constitute the Altaic family: Tungusic (spoken by most of the indigenous peoples of Manchuria and Siberia), Mongolian, and Turkish. Altaic peoples spread as far west as southeastern Europe by the waning years of the Roman Empire. By the second century B.C., a powerful nomadic empire of Turkish-speaking peoples arose in what today is Mongolia, forcing the Chinese to begin building the Great Wall as a defensive measure. As Turkish power spread through most of Central Asia, Turkish languages gradually began to replace Indo-European tongues in the oasis communities. This process, however, was never completed, and southwestern Central Asia remains a meeting ground of the Persian and Turkish languages.

The Turks were eventually replaced on the eastern steppes by another group of Altaic speakers, the Mongols. In the late 1100s the Mongols united the pastoral peoples of Central Asia and used the resulting force to conquer nearby sedentary societies. By the late 1200s this Mongol Empire had grown into the largest contiguous empire the world had ever seen, stretching from Korea and southern China in the east to the Carpathian Mountains and the Euphrates River in the west (Figure 10.15).

Protected by mountain barriers and by the rigorous conditions of the plateau, Tibet has taken a different course from the rest of Central Asia. Tibet emerged as a strong, unified kingdom around A.D. 700. Tibetan unity and power did not persist, however, and the region reverted to its former state of semi-isolation. Tibet was incorporated for a short period in the 1200s into the Mongol Empire, and in later centuries other Mongol states occasionally enjoyed limited powers over the Tibetans. These interactions resulted in the establishment of Mongolian communities in the northeastern portion of the plateau and in the eventual conversion of the Mongolian people to Tibetan Buddhism.

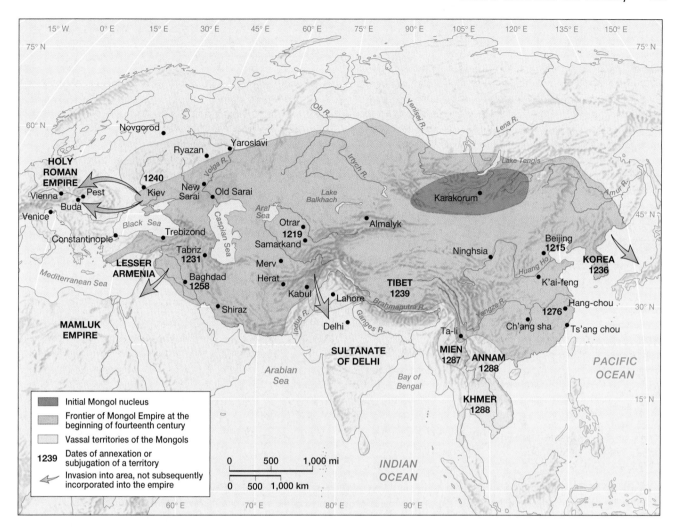

▲ Figure 10.15 The Mongol Empire of the 1200s   In the 1200s the Mongols carved out the largest land-based empire the world has ever seen. From a core area in modern-day Mongolia, the Mongol conquests extended as far as southern China in the southeast, Ukraine in the west, and Iraq in the southwest. Although the empire did not long remain unified, it did have profound effects on the subsequent political and economic history of Eurasia.

Contemporary Linguistic and Ethnic Geography

Today most of Central Asia is inhabited by peoples speaking Mongolian and Turkish languages (Figure 10.16). A few indigenous Indo-European languages are confined to the southwest, while Tibetan remains the main language of the plateau. Russian is also widely spoken in the west, while Chinese is increasingly important in the east. Chinese is perhaps beginning to threaten the long-term survival of several Central Asia languages, particularly Tibetan. One of the major complaints of the indigenous people of western China—Tibet and Xinjiang—is the fact that Mandarin Chinese is the basic language of higher education.

Tibetan   Tibetan is usually placed in the Sino-Tibetan family, implying a shared linguistic ancestry between the Chinese and the Tibetan peoples. Many students of Tibetan, however, argue that no definite relationship between the two has ever been established. Tibetan itself is divided into a number of distinct dialects that are spoken over almost the entire in-

habited portion of the Tibetan Plateau. Only about 1.5 million people in Tibet itself speak it, however, out of a total population of some 2.5 million (most of the rest speak Chinese). Perhaps another 3 million Tibetan speakers live in China's provinces of Qinghai and Sichuan; smaller numbers may be found in the far northern Himalayan reaches of South Asia. Tibetan has an extensive literature written in its own script, most of which is devoted to religious topics.

Mongolian   Mongolian forms a cluster of closely related dialects spoken by approximately 5 million persons. The standard Mongolian of both the independent country of Mongolia and China's Inner Mongolia is called Khalkha; other Mongolian dialects include Buryat (found in southern Siberia) and Kalmyk (found in the extreme southeastern corner of European Russia). Mongolian has its own distinctive script, which dates back some 800 years, but Mongolia itself adopted the Cyrillic alphabet of Russia in 1941. Efforts are now being made to revive the old script.

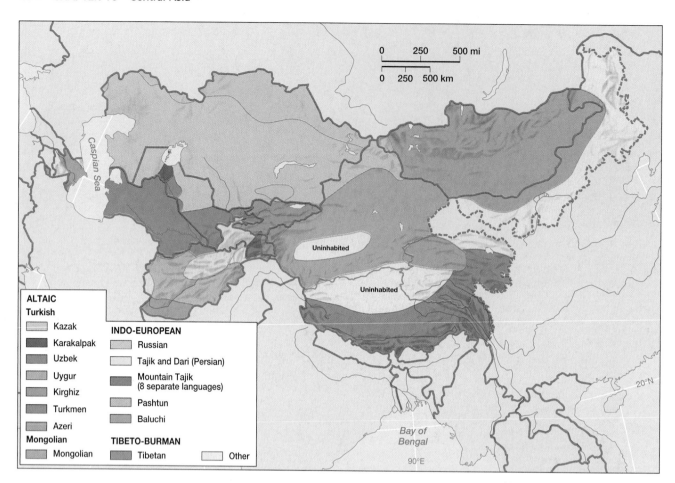

▲ Figure 10.16 Linguistic geography of Central Asia   Most of Central Asia is dominated by languages in the Altaic family, which includes both the Turkish languages (found through most of the center and the west of the region) and Mongolian (found in Central Asia's northeast). Several Indo-European languages, however, are located in both the far northwest and the south-center, while the Tibeto-Burmese language of Tibetan covers most of the Tibetan Plateau in the southeast.

Mongolian speakers form about 90 percent of the population of Mongolia. In China's Inner Mongolian Autonomous Province, however, they have been almost submerged by a wave of Han Chinese migrants over the past 50 years. Today only about 2 million out of 22 million residents of Inner Mongolia speak Mongolian, making some observers wonder whether this area has been permanently lost to the Mongolian cultural sphere.

Turkish Languages   Far more Central Asians speak Turkish languages than Mongolian and Tibetan combined. The Turkish linguistic sphere extends from Azerbaijan in the west through Xinjiang province in the east. The various Turkish languages are not as closely related to each other as are the dialects of Mongolian, but they are still obviously kindred tongues. Six main Turkish languages are found in Central Asia; five are associated with newly independent (former Soviet) republics of the west, while the sixth, Uygur, is the main indigenous language of China's Xinjiang province.

Uygur is an old language, dating back almost 2,000 years. The Uygur number about 8 million, almost all of whom live in Xinjiang (Figure 10.17). As recently as 1953 the Uygur

formed about 80 percent of the population of Xinjiang; now, because of Han Chinese immigration, they actually form a minority in their own homeland. There are also about 1 million Kazak speakers in Xinjiang.

Five of the six countries of the former Soviet Central Asia—Kazakstan, Uzbekistan, Turkmenistan, Kyrgyzstan, and Azerbaijan—are named after the Turkish languages of their dominant populations. In three of these countries the indigenous people still form a clear majority. Some 82 percent of the people of Azerbaijan speak Azeri (there are more Azeris in northern Iran, however, than there are in Azerbaijan); some 70 percent of the people of Uzbekistan speak Uzbek as their native tongue; and some 73 percent of the people in Turkmenistan speak Turkmen. With more than 17 million speakers, Uzbek is the most widely spoken Central Asian language. In the Amu Darya delta in the far north of Uzbekistan, however, most people speak a different Turkish language called Karakalpak, while Kazak speakers are found in sparsely populated Uzbek deserts.

In the two other Turkish republics, the titular nationality forms only about half of the total population. Some 52 percent of the inhabitants of Kyrgyzstan speak Kyrgyz as their

▲ **Figure 10.17 Uygur mosque** A small mosque, illustrating traditional Uygur architecture, survives amid blocks of modern apartments in Urumchi, Xinjiang. Traditional forms of housing and urban design can still be found in Uygur communities in northwestern China, but they are gradually disappearing. *(Chris Stowers/Panos Pictures)*

began to urge ethnic Kazaks living in other countries (particularly those in western China) to return, in part to bolster the sense of Kazak national identity.

Linguistic Complexity in the Former Soviet Zone The sixth republic of the former Soviet Central Asia, Tajikistan, is dominated by people who speak an Indo-European rather than a Turkish language. Tajik is so closely related to Persian that it is often considered to be a Persian (or Farsi) dialect. Iran, the homeland of Persian, is, however, separated from Tajikistan by some 400 miles (640 kilometers) of mostly Turkish-speaking territory. Roughly 3.5 million people in Tajikistan, about 65 percent of the total population, speak Tajik as their main language. The remote mountains of eastern Tajikistan are populated by peoples speaking a variety of distinctive Indo-European languages, sometimes collectively referred to as "Mountain Tajik."

Tajikistan, like much of the rest of former Soviet portion of Central Asia, is noted for its complex mixture of languages. About a quarter of its people, for example, are Uzbeks. The peripheral portions of Azerbaijan—part of the famous Caucasus "mountain of languages"—are also noted for their ethnolinguistic complexity. Such ethnic mixing was actually greater in earlier decades, since Soviet policy resulted in gradual ethnic homogenization. The Soviet authorities also devised complex political boundaries among the different ethnolinguistic groups of Central Asia, partly in order to play one group off against another and thus bolster their own authority.

Language and Ethnicity in Afghanistan The linguistic geography of Afghanistan is even more complex than that of the former Soviet zone (Figure 10.18). Afghanistan was never colonized by outside powers, and it is one of the few

native language, while approximately 42 percent of the people of Kazakstan speak Kazak. The other residents of these countries speak Russian, Uzbek, Ukrainian, German, and a variety of other languages both indigenous and exogenous to Central Asia. In Kazakstan the population is almost evenly split between speakers of Turkish languages, on the one hand, and European languages (Russian, Ukrainian, and German), on the other. In general, the Kazaks and other Turks live in the center and south of the country, while the people of European descent live in the agricultural districts of the north and in the cities of the southeast. In 2001 Kazakstan's government

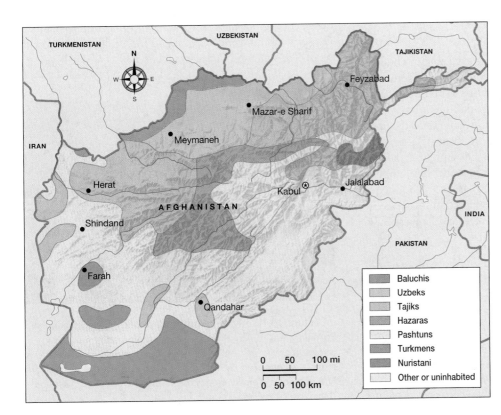

◄ **Figure 10.18 Afghanistan's ethnic patchwork** Afghanistan is one of the world's more ethnically complex countries. Its largest ethnic group is that of the Pashtuns, a people who inhabit most of the southern portion of the country as well as the adjoining borderlands of Pakistan. Northern Afghanistan, however, is mostly inhabited by Uzbeks, Tajiks, and Turkmens— whose main population centers are located in Uzbekistan, Tajikistan, and Turkmenistan, respectively. The Hazaras of Afghanistan's central mountains, like the Tajiks, speak a form of Persian, but are considered to be a separate ethnic group in part because they, unlike other Afghans, follow Shiite rather than Sunni Islam.

Legend:
- Baluchis
- Uzbeks
- Tajiks
- Hazaras
- Pashtuns
- Turkmens
- Nuristani
- Other or uninhabited

countries of the world to have inherited the boundaries of a premodern, indigenous kingdom. This kingdom emerged in the 1700s on traditional dynastic lines that did not reflect ethnic or linguistic divisions ("dynastic" linkages are those based on the family of the monarch). The modern nation-state ideal—that each country should be identified with a single national group—never had any currency in Afghanistan.

The eighteenth-century creators of Afghanistan were mostly members of the Pashtun ethnic group. They did not attempt, however, to build a nation-state around Pashtun identity. Indeed, approximately half of the Pashtun population (whose language is *usually* called Pashtun) live in Pakistan. In Afghanistan itself, estimates of the proportion of the populace speaking Pashtun vary from 40 percent to 60 percent.

Pashtun speakers live primarily to the south of the Hindu Kush (along the Pakistani border) and in the west. Almost as many people in Afghanistan speak Dari, Afghanistan's variant of Persian. Dari speakers are concentrated in the cities of the west, in the central mountains, and near the boundary with Tajikistan. Two separate ethnicities are ascribed to the Dari-speaking people; those in the west and north are considered to be Tajik, whereas those in the central mountains are called Hazaras, reputed to be descendants of Mongol conquerors who arrived in the twelfth century. Another 11 percent of the people of Afghanistan speak Turkish languages, mainly Uzbek.

The ethnic melange present in Afghanistan did not present a problem until the latter part of the twentieth century. As we shall see later in this chapter, the war waged in Afghanistan in the fall of 2001 had a complex ideological and geopolitical background, but it also had roots in the country's linguistic as well as its religious geography.

Geography of Religion

Ancient and medieval Central Asia formed a religious hodge-podge. The major overland trading routes of premodern Eurasia crossed the region, giving easy access to both merchants and missionaries. Several varieties of Buddhism, Islam, Christianity, Judaism, Zoroastrianism, and several minor religions have all thrived at various times and in various places within the region. During the years of Mongol supremacy, adherents of many faiths mingled throughout the empire. Subsequently, however, religious lines hardened, and the region was eventually divided into two opposed spiritual camps: Islam triumphed in the west and center, and Lamaist Buddhism prevailed in Tibet and Mongolia.

Islam in Central Asia   As is true elsewhere in the Muslim world, different Central Asian peoples are known for their different interpretations of Islamic orthodoxy. The Pashtuns of Afghanistan have been noted for their strict Islamic ideals—although critics contend that Pashtun religious strictures, such as never allowing women's faces to be seen in public, are actually based on their pre-Islamic customs (Figure 10.19). The traditionally nomadic groups of the northern steppes, such as the Kazaks, on the other hand, have often been considered lax

in their religious observances. And while most of the region's Muslims are Sunnis, Shiism is dominant among both the Hazaras of central Afghanistan and the Azeris of Azerbaijan.

Under the communist rule of China, the Soviet Union, and Mongolia, all forms of religion were discouraged. Chinese authorities and student radicals attempted to suppress Islam in Xinjiang during the Cultural Revolution of the late 1960s and early 1970s. Mosques were destroyed or converted to museums; religious schools were closed; and people were sometimes forced to eat pork. Chinese Muslims now enjoy basic freedom of worship, but the state still closely monitors religious expression out of fear that it will lead to political separatism. Periodic persecution of Islam also occurred in Soviet Central Asia, and until the 1970s many observers thought that that religion was slowly disappearing from the region.

Religious expression was not, however, so easily repressed. Interest in Islam began to grow in former Soviet Central Asia in the 1970s and 1980s. In the post-Soviet period Islam continues to revive as people reassert their indigenous heritage and identity. Thus far, however, there have been few signs of mass Islamic political fundamentalism. In Xinjiang, Islam does indeed seem to be emerging as a focal point of a political movement among the Uygur people. Most Uygur leaders, however, insist that their beliefs are not fundamentalist.

Only in Afghanistan, parts of Tajikistan, and the Fergana Valley (mostly in Uzbekistan) is Islamic fundamentalism a powerful movement. From the mid-1990s until 2001, most of Afghanistan was controlled by an extremely fundamentalist organization called the Taliban. The Taliban insisted that all aspects of society conform to its own harsh version of Islamic orthodoxy. This commitment was demonstrated in early 2001 when Afghanistan's religious authorities oversaw the destruction of all Buddhist statues in the country—including some of the world's largest and most magnificent works of art—as symbols of ancient idolatry.

▲ **Figure 10.19 Afghan women in public**   Especially in the Pashtun areas of Afghanistan, women have traditionally been forced to cover their entire bodies when in public areas. In areas that were controlled by the Taliban, such dress codes were strictly enforced. Extremely modest dress still prevails in Afghanistan. *(Lemoyne/Getty Images, Inc.)*

Islam is not the only religion represented in former Soviet Central Asia. Many Russians belong to the Russian Orthodox Church, and Uzbekistan has a small Jewish population.

Lamaist Buddhism   Mongolia and Tibet stand apart from the rest of Central Asia—and indeed, from the rest of the world—in the adherence of their people to Lamaist Buddhism. Buddhism entered Tibet from India many centuries ago, where it merged with the indigenous religion of the area, called Bon. The resulting hybrid, Lamaism, is more oriented toward magic than are other forms of Buddhism, and it is far more hierarchically organized. Standing at the apex of Lamaist society is the Dalai Lama, considered to be the reincarnation of the Buddha. Ranking below him is the Panchen Lama, followed by other religious officials. Until the Chinese conquest, Tibet was essentially a **theocracy** (or religious state), with the Dalai Lama enjoying political as well as religious authority.

Lamaism is noted for its dedication to monasticism (Figure 10.20). A substantial proportion of Tibet's male population has long become monks, and monasteries once wielded both economic and political clout. Such widespread monasticism, which demanded that many people remain celibate, ensured that Tibet's population density remained low. Some scholars view Tibetan monasticism almost as an environmental adaptation, arguing that the plateau could not support dense human settlements and thus required some method of limiting fertility.

Lamaism in Tibet suffered particularly brutal persecution after 1959 when China invaded Tibet. The Chinese hold on Tibet has never been as secure as that on Xinjiang, and Tibetan Buddhism is often viewed as a vehicle for political separatism. The Dalai Lama, who fled Tibet for India in 1959, has also long been a powerful advocate for the Tibetan cause in international circles. During the 1960s and 1970s, an estimated 6,000 Tibetan Buddhist monasteries were destroyed and thousands of monks were killed; the number of active monks today is only about 5 percent of what it had been before the Chinese occupation. Many monasteries have, however, been allowed to reopen, but their activities are severely limited. Still, the Lamaist faith continues to provide a bulwark of Tibetan identity, and in so doing helps keep alive the dream of independence.

In Mongolia the downfall of communism and hence of Russian influence has allowed the Lamaist Buddhist faith to experience a renaissance. Several monasteries have been refurbished, and many people are returning to their national religion. The intensity of Buddhist belief, however, is not as strong in Mongolia as it is in Tibet.

Central Asian Culture in International and Global Context

Western Central Asia's closest external cultural relations are with Russia, while those of eastern Central Asia are with China. In the east, the main issue is the migration of Han Chinese, which has resulted in serious ethnic and political tensions. In the west, in contrast, Russian influence is diminishing.

▲ **Figure 10.20 Lamaist Buddhist monastery**   Tibet is well known for its large Buddhist monasteries, buildings that in earlier years served as seats of political as well as religious authority. The Potala Palace in Lhasa, traditional seat of the Dalai Lama, is the largest, most important, and most famous of such monastic establishments.   *(Alain le Garsmeur/Panos Pictures)*

During the Soviet period, the Russian language spread widely through western Central Asia. Russian served both as a lingua franca (or common language) and as a means of instruction in higher education. One had to be fluent in Russian in order to reach any position of responsibility. The Cyrillic (or Russian) script, moreover, replaced the Arabic script that had previously been used for the indigenous languages. Russian speakers settled in all of the major cities, and many became influential. Today, however, many Russian speakers have migrated back to Russia, especially from the region's poorer countries, such as Tajikistan. The use of Russian in education, government, business, and the media is declining in favor of local languages. Efforts are being made to abandon Cyrillic in favor of the Roman alphabet, the script of modern-day Turkey. Russian has long served an important role as the international language of the area, however, and it must still be regarded as the common language of western Central Asia.

Although Central Asia is remote and poorly integrated into global cultural circuits, it is hardly immune to the forces of globalization. The tensions existing throughout the region between religious and secular orientations, and between ethnic nationalism and multiethnic inclusion, are symptoms of the global condition at the start of the third millennium. So, too, the increased usage of English and the influence of U.S. culture throughout Central Asia shows that this part of the world is not cut off from global culture. Such influences have been especially marked in the oil cities of the Caspian Basin, such as Baku (see "Global and Local: Louisiana Oil Culture in Central Asia"). Although the proportion of English speakers in Central Asia is low, most of the region's numerous Web pages are written in English. English-speaking Central Asians with computer skills

GLOBAL AND LOCAL   Louisiana Oil Culture in Central Asia

Oil and gas production are highly specialized businesses, leading to a vigorous international trade in parts, services, and labor. Many U.S. oil workers, petroleum engineers, and geologists have long sought work overseas, a phenomenon that accelerated after falling prices in the late 1980s and 1990s brought recession to the oil fields of Texas, Louisiana, and Oklahoma. Most of the initial movement was to the Persian Gulf area, but after the breakup of the Soviet Union in 1991, the Caspian Sea emerged as a major new target.

Wherever large numbers of U.S. expatriates settle, U.S. culture soon follows. Central Asia, particularly the oil and gas center of Baku in Azerbaijan, is no exception. Most U.S. workers initially arriving in Baku found the lack of familiar foods and amenities discouraging. Soon, however, entrepreneurs began to fill the void. Particularly important have been Charlie and Marie Schroeder, Louisianans who had previously worked in the Middle East. While Mr. Schroeder's company, Caspian Sea Ventures, specializes in importing industrial parts, the couple has found particular success in the restaurant business. By early 1999 they were running four separate establishments, each with a different theme and flavor. Although distinctly Western, these establishments also reflect both the Gulf of Mexico flavor of much of the U.S. oil industry as well as the increasing cosmopolitanism of U.S. foodways. One restaurant, Ragin' Cajun, serves southern Louisiana cuisine (itself heavily influenced by French and African traditions); another, Margaritaville, specializes in Mexican-American food; and a third, Finnegan's, seeks to replicate the atmosphere of an Irish pub.

These "American" restaurants have not yet attracted many Azerbaijani customers, but they do provide employment opportunities. Mr. Schroeder, however, regrets that his employees do not have better options. "These kids are overqualified. Most of them speak anywhere between 4 to 13 languages, which makes me sick to my stomach because I can't even speak English. I speak New Orleans and a little bit of Texan."

Source: Adapted from Stephen Kinzer, "At a Crossroads of an Oil Boom, Everyone Comes to Charlie's." *The New York Times*, May 30, 1999.

---

are increasingly valued as the region strives to find a place and a voice in the global community (see "Local Voices: Virtual Tibet" later in this chapter). Even the Al Qaeda terrorist organization, which was based in Afghanistan from mid-1992 until late 2001, encouraged its agents to learn English.

Geopolitical Framework: Political Reawakening in a Power Void

Central Asia has played a minor role in global political affairs for the past several hundred years. Before 1991 the entire region, except Mongolia and Afghanistan, lay under direct Soviet and Chinese control. Mongolia, moreover, had been a Soviet satellite, and even Afghanistan came under Soviet domination in the late 1970s. The southeastern third of Central Asia, of course, is still an integral part of China. And although the breakup of the Soviet Union saw the emergence of six new Central Asian countries, all of them are economically troubled and geopolitically insecure (Figure 10.21).

Partitioning of the Steppes

Although Central Asia in the twentieth century was contested and controlled by outside powers, this was not always the case. Before 1500 it was a power center, a region whose mobile armies threatened the far more populous, sedentary states of the Eurasian rim. The development of gunpowder and of effective hand weapons changed the balance of power, however, allowing the wealthier agricultural states to vanquish the nomadic pastoralists. By the 1700s their armies had been de-

feated and their lands taken. The winners in this struggle were the two largest states bordering the steppes: Russia and China.

The Manchu conquest of China in 1644 undercut the autonomy of the steppe peoples. The Manchus came from the borderlands of Central Asia (in Manchuria) and were themselves skilled in the arts of cavalry warfare. By the mid-1700s the Chinese empire of the Manchus stood at its greatest territorial extent. Not only Mongolia and Xinjiang but also Tibet and a slice of modern Kazakstan lay within the empire.

From its height in the late 1700s, Manchu-ruled China declined rapidly. Yet it was still able to retain most of its Central Asian dominions through the 1800s, largely because they were so remote. By the early 1900s, however, Chinese authority had begun to diminish in Central Asia as well. When the Manchu (also called Ch'ing or Qing) dynasty fell in 1911, Mongolia became independent, although China did manage to keep the extensive borderlands of Inner Mongolia (Nei Mongol). Tibet had earlier gained de facto independence, and even Xinjiang lay beyond the reach of effective Chinese authority during the 1920s.

Russia began to advance into Central Asia at roughly the same time as China. In the 1700s the Russian Empire undertook the systematic conquest of the Kazak steppes. When China weakened in the 1800s, Russia also advanced into former Chinese territory east of Lake Balqash. Russia's conquest of Kazakstan proceeded fairly easily, since the Kazak cavalry was no match for Russian guns. Expansion farther to the south, however, was blocked by the sedentary states of Uzbekistan. Only in the late 1800s when European military techniques and materials raced ahead of those of Asia was Russia

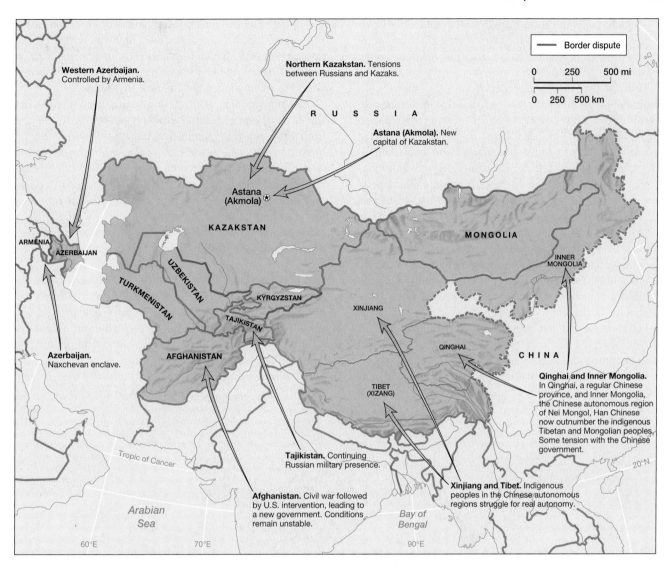

Western Azerbaijan. Controlled by Armenia.

Northern Kazakstan. Tensions between Russians and Kazaks.

Astana (Akmola). New capital of Kazakstan.

Border dispute

Azerbaijan. Naxchevan enclave.

Tajikistan. Continuing Russian military presence.

Afghanistan. Civil war followed by U.S. intervention, leading to a new government. Conditions remain unstable.

Xinjiang and Tibet. Indigenous peoples in the Chinese autonomous regions struggle for real autonomy.

Qinghai and Inner Mongolia. In Qinghai, a regular Chinese province, and Inner Mongolia, the Chinese autonomous region of Nei Mongol, Han Chinese now outnumber the indigenous Tibetan and Mongolian peoples. Some tension with the Chinese government.

▲ Figure 10.21 Central Asian geopolitics Six of the eight independent states of Central Asia came into existence in 1991 with the dissolution of the Soviet Union. A number of border disputes and other geopolitical problems have been inherited from the period of Soviet rule. In eastern Central Asia, the most serious difficulties stem from China's maintenance of control over areas in which the indigenous peoples are not Chinese. Afghanistan, scene of a prolonged and brutal civil war, has experienced the most extreme forms of geopolitical tension in the region.

able to conquer the Amu Darya and Syr Darya valleys. Its conquest of this area was not completed until the early 1900s, just before the Soviet Union replaced the Russian Empire.

One reason for the Russian advance into Central Asia was concern over possible British influence in the area. Britain did indeed attempt to conquer Afghanistan, but was rebuffed by Afghani forces—and by the country's forbidding terrain. Subsequently, Afghanistan's position as an independent "buffer state" between the Russian Empire (later the Soviet Union) and the British Empire in South Asia remained secure. The British also sent a major military expedition to Tibet in the early twentieth century and almost created an autonomous Tibetan state under British "protection." This heightened China's determination to regain control over Tibet.

Central Asia Under Communist Rule

Western Central Asia came under communist rule after the foundation of the Soviet Union; Mongolia followed in 1924. After the Chinese revolution of 1949, the communist system was also imposed on Xinjiang and Tibet. In all of these areas, major changes in the geopolitical order soon followed.

Soviet Central Asia The newly established Soviet Union inherited the Russian imperial domain in Central Asia virtually intact. The policies that it directed toward this region, however, changed. The new regime sought to create a socialist economy and to build a new Soviet society that would eventually knit together all of the massive territories of the

Soviet Union. Central Asia's leaders were replaced by Communist Party officials loyal to the new state; Russian immigration was encouraged; and local languages could no longer be written in Arabic script.

Although the early Soviet leaders foresaw the emergence of a unitary Soviet nationality, they realized that local ethnic diversity would not disappear overnight. Early Soviet leaders such as Vladimir Lenin also hoped to protect non-Russian peoples from Russian domination. They therefore divided the Soviet Union into a series of nationally defined "union republics" in which a certain degree of autonomy would be allowed. They were uncertain, however, about what the relevant units in Central Asia should be. Was there a single Turkish-speaking nationality, or were the Turkish peoples themselves divided into a number of separate nationalities, with the Tajiks forming yet another? For several years boundaries shifted as new "republics" suddenly appeared on the map. Finally, in the 1920s the modern republics of Kazakstan, Kyrgyzstan, Tajikistan, Uzbekistan, Turkmenistan, and Azerbaijan assumed their present configurations. In certain areas, such as the fertile Fergana Valley, the political boundaries so drawn remain extremely complex (Figure 10.22).

Some scholars argue that the Soviet policy backfired severely. Rather than forming a transitional step on the way to a Soviet identity, the constituent republics of the Soviet Union instead nurtured local nationalisms that ultimately undermined the Soviet system. Identities such as Turkmen, Uzbek, and Tajik that had been somewhat vague in the pre-Soviet period were now given real political significance.

Another problem undercutting Soviet unity was the fact the cultural and economic gaps separating Central Asians from Russians did not diminish as much as planned. Islam remained entrenched in some areas and began to revive elsewhere in the 1970s. Central Asia also remained poorer than most other parts of the Soviet Union, and by the 1980s it was becoming something of a burden on the national economy. Equally important were the region's higher birthrates, leading many Russians to fear that the Soviet Union risked being dominated by Turkish-speaking Muslims.

The Chinese Geopolitical Order   After China reemerged as a united country in 1949, it too was able to reclaim most of its old Central Asia territories. China's communist leaders promised the non-Chinese peoples a significant amount of political self-determination as well as cultural autonomy, and thus they found much local support in Xinjiang. Tibet, isolated behind its mountain walls and virtually independent for the previous 150 years, presented a greater obstacle. China

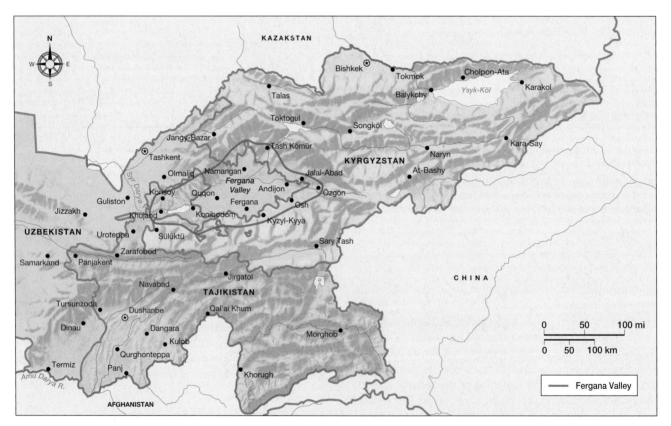

▲ Figure 10.22 **Political boundaries around the Fergana Valley**   Some of the world's most convoluted political boundaries can be found in the vicinity of the Fergana Valley. The central portion of the valley belongs to Uzbekistan, which is otherwise separated from it by high mountains. The lower valley, on the other hand, is part of Tajikistan, the core area of which is likewise separated from the valley by highlands. The Fergana's upper periphery belongs to Kyrgyzstan. Note also the small enclaves of Uzbekistan within Kyrgyzstan.

▲ **Figure 10.23 Chinese invasion of Tibet** In 1959 China launched a massive invasion of Tibet. Pursued by the Chinese army, the 23-year-old Dalai Lama (second from the lead) is shown here escaping over the Zsagola Pass, ultimately to find refuge in India. *(Hg/AP/Wide World Photos)*

occupied Tibet in 1950, but the Tibetans launched a rebellion in 1959. When this was brutally crushed, the Dalai Lama and some 100,000 followers found refuge in India (Figure 10.23).

Loosely following the Soviet nationalities model, China established autonomous regions in areas occupied primarily by non-Han Chinese peoples, including Xinjiang, Tibet proper (called Xizang in Chinese), and Inner Mongolia. Such autonomy, however, often turned out to be more theoretical than real, and it did not prevent the massive immigration of Han Chinese into these areas. Nor were all parts of Chinese Central Asia granted autonomous status. The large and historically Tibetan and Mongolian province of Qinghai, for example, remained an ordinary Chinese province.

Current Geopolitical Tension

The former Soviet portion of Central Asia weathered the post-1991 transition to independence rather smoothly, but the region still suffers from a number of actual and potential ethnic conflicts. Much of China's Central Asian territory is seething, but China retains a firm grip. Afghanistan, unfortunately, suffered from a particularly brutal civil war that was brought to a head by the events and aftermath of Al Qaeda terrorism on September 11, 2001.

Independence in Former Soviet Lands The breakup of the Soviet Union in 1991 generally proceeded peacefully in Central Asia as elsewhere. The six newly independent countries, however, had been dependent on the Soviet system, and it was no simple matter for them to chart their own courses. They still had to cooperate with Russia on security issues, and all opted to remain part of the Commonwealth of Independent States, the rather hollow successor of the Soviet Union. In most cases authoritarian rulers, rooted in the old order, retained power and sought to undermine opposition

groups. All told, democracy made less progress in Central Asia than in other parts of the former Soviet Union.

Kazakstan and Tajikistan have faced a particularly difficult transition. Kazakstan is the largest and most resource-rich Central Asian state, and therefore might seem to have the best chance for success. Potential ethnic strife, however, looms over the country. Many Kazaks want to create a national state centered on Kazak identity, and they resent the presence and power of Russians, Ukrainians, Germans, and others of European background. These Europeans, for their part, fear the imposition of what they consider alien Central Asian cultural standards.

Thus far, tensions in Kazakstan have remained muted, but many fear an unstable future. The predominantly Russian population of northern Kazakstan, the country's breadbasket, could attempt to join their land with Russia. While extreme Russian nationalists would welcome such a move, moderates would reject it as destabilizing. In the late 1990s, Kazakstan moved its capital from Almaty in the south to Astana in the Russian-dominated north, partly to forestall any plans for secession.

In Tajikistan war broke out almost immediately after independence in 1991. Many members of the smaller ethnic groups living in the mountainous east resented the authority of the lowland Tajiks and thus rebelled. They were joined by several radical Islamic groups seeking to overthrow the secular state. Although the civil war officially ended after several years, fighting continues to flare up periodically, and much of the mountainous west remains beyond the control of the central government. New lines of tension are now emerging between the Tajiks and the country's Uzbek minority, a conflict that threatens to involve Uzbekistan as well. The Tajikistan government has been able to remain in power in part because of its use of Russian troops. The presence of Russian military forces, however, opens the question of Russia's long-term intentions in the region. Although the present Russian government does not plan to reabsorb any Central Asian territory, it does regard the entire area as lying within its zone of strategic interest.

Azerbaijan also experienced strife following the breakup of the Soviet Union. Armenia invaded, allowing the Armenian-speaking highlands in the western portion of the country to form a "breakaway republic." Hostile relations with Armenia make it difficult for Azerbaijan to control its **exclave** of Naxcivan, a piece of Azerbaijani territory separated from the rest of the country by Armenia and Iran (Figure 10.21).

Strife in Western China Local opposition to Chinese rule in Central Asia increased during the 1990s, albeit without success. Any form of protest in Tibet has been severely repressed, but the Tibetans have brought the attention of the world to their struggle (see "Local Voices: Virtual Tibet"). China maintains several hundred thousand troops in a region that has only 2.4 million civilian inhabitants. Such an overwhelming military presence is considered necessary because of both Tibetan resistance and the strategic importance of the region. The border between China and India is still contested, with China controlling a small section of the Tibetan Plateau that India claims.

LOCAL VOICES   Virtual Tibet

The Tibetan movement for independence has inspired perhaps the largest number of sites on the World Wide Web of any political resistance movement, creating what is in effect a "virtual Tibet." Not only Tibetan activists and the Tibetan government in exile but also a number of "Tibetan friendship committees" the world over run Web pages dedicated to gaining freedom for Tibet. A good source for locating many of these Web sites is "Tibet Online Resource Gathering" (www.tibet.org/). The political orientation of this site is evident on its home page, which states that "Tibet's ancient and fantastic civilization and ecosystem are faced with extinction due to 48 years of mismanagement and abuse under its colonial ruler, the People's Republic of China."

An equally important site is that of the official Tibetan government in exile, headed by the Dalai Lama and run out of Dharmasala, India (www.tibet.com/). One of the more interesting aspects of this site is its map of Tibet, which includes much more territory than is contained in Xizang, China's autonomous region of Tibet. As the Web page explains, "Tibet is comprised of the three provinces of Amdo (now split by China into the provinces of Qinghai and part of Gansu), Kham (largely incorporated into the Chinese provinces of Sichuan, Gansu, and Yunnan), and U-Tsang (which, together with western Kham, is today referred to by China as the Tibet Autonomous Region)."

Although the vast majority of Tibetan Web sites are devoted to Tibetan independence, the government of China does supply an opposing view. One important site is run by the China News Organization (www.chinanews.org/Tibet/index.html). From this site one can open pages with such titles as "Exposing the Trickery of the Dalai Lama," "Dalai Sabotages Religious Order," and "Historical Records Prove Tibet Is Inseparable Part of China." The last-mentioned page contends that "History has proved time and again that only while in the embrace of the motherland [in other words, China] can Tibet achieve sound and rapid development, an improved living standard, and more freedom."

Until the late 1990s, China's control of Xinjiang seemed much more secure than its hold on Tibet. Although fragmented by deserts and mountains, Xinjiang is less forbidding than Tibet and is now the home of millions of Han Chinese immigrants. Xinjiang is also far more vital to China than is Tibet. It contains a variety of mineral deposits (including oil) essential for Chinese industry, and it has been the site of nuclear weapons tests. Many Uygurs, not surprisingly, oppose such uses of their homeland, and they resent the periodic suppression of their religion and culture by the communist regime (Figure 10.24). In 1990 and again in 1997, antigovernment riots broke out in several major cities. During the 1997 riots, Uygur separatists even planted bombs in Beijing. Immediate reprisals ended this overt display of secessionist sentiment, but the underlying spirit remains unbroken.

China's position is that all of its Central Asian lands are integral positions of its national territory. Those who advocate independence are viewed as traitors and are sometimes considered to be in league with Western political forces that have sought for the past 200 years to keep China weak and divided. Chinese officials have also linked separatist elements in Xinjiang to a global movement of radical Islamic fundamentalists. Leaders of Xinjiang's "Free Eastern Turkestan Movement," however, insist that their movement is founded on territory rather than religion or ethnicity, contending that it is based on an alliance of all non-Han Chinese peoples of Xinjiang. The involvement of a number of Uygur separatists in Osama bin Laden's Al Qaeda organization in Afghanistan, however, did not help their cause in the arena of international public opinion.

War in Afghanistan   None of the conflicts in western China or in the former Soviet republics compares in intensity to the struggle waged in Afghanistan. Afghanistan's troubles began

▲ **Figure 10.24 Ethnic tension in Xinjiang**   Relations between Han Chinese officials and Muslim Central Asians grew increasingly tense in the 1990s. In 1997 severe rioting broke out in several cities in Xinjiang. This photograph shows a Chinese official levying a trading tax on merchants in Kashgar's Sunday market.   *(Chris Stowers/Panos Pictures)*

in 1978 when a Soviet-supported military "revolutionary council" seized power. The new Marxist-oriented government began to suppress religion, which led almost immediately to widespread rebellion (Afghanistan is a very traditional and deeply Islamic country). When the government was about to collapse, the Soviet Union responded with a massive invasion. Despite its power, the Soviet military was never able to gain control of the more rugged parts of the country. Pakistan, Saudi Arabia, and the United States, moreover, ensured that the anti-Soviet forces remained well armed.

The exhausted Soviets finally withdrew their troops in 1989. The puppet government that they installed remained to face the insurgents alone. It managed to hold the country's core around the city of Kabul for a few years, largely because the opposition forces were themselves divided. Local warlords grabbed power in the countryside, destroying any semblance of central authority and perpetrating a number of atrocities.

In 1995–1996, a new movement called the Taliban arrived on the Afghan scene. The Taliban was founded by young Muslim religious students disgusted with the anarchy that was consuming their country. They were convinced that only the firm imposition of Islamic law could end corruption, quell the disputes among the country's different ethnic groups, and ultimately bring peace and unity to Afghanistan. Large numbers of soldiers flocked to the Taliban standard, and by 1997 they had won control of most of Afghanistan. By September 2001 only about a tenth of country, mostly in the far northeast, lay outside of the Taliban's power.

The Taliban acquired strength not only from its religious nature, but also from the ethnic divisions of Afghanistan. It is closely identified with the Pashtun people. The Pashtun are the most populous ethnic group in Afghanistan, and they have a reputation for militarism, as well as a good connection for military supplies among their relatives across the Pakistan border. Pakistan is one of only three countries (the others being Saudi Arabia and the United Arab Emirates) that recognized the Taliban as the legitimate government of Afghanistan.

The main opposition to the Taliban came from the country's other ethnic groups, especially the Uzbeks and Tajiks of the north and the Shiite Hazaras of the central mountains (Figure 10.25). The forces of these anti-Taliban groups then joined together to form the Northern Alliance. It would be a mistake to regard this conflict as reducible to ethnic differences, however, as several notable Uzbek and Tajik commanders joined the Taliban, while a number of Pashtun leaders remained opposed to it.

By the late 1990s, moreover, even most of Afghanistan's Pashtun populace had turned against the Taliban rulers, mostly because of the severe restrictions on daily life imposed by the Taliban's religious authorities in the name of Islamic orthodoxy. These restrictions were most pronounced for women, but even men were compelled, by both whippings and imprisonment, to obey the Taliban's numerous petty decrees. Most forms of recreation were simply outlawed, including television, films, music, and even kite-flying. Men were sometimes severely beaten if their beards were too short or their hair too long. Little escaped the watchful eyes of the Taliban's religious police.

International Dimensions of Central Asian Tension

With the collapse of the Soviet Union in 1991, Central Asia emerged as a key arena of geopolitical tension. A number of important countries, including China, Russia, Pakistan, Iran, Turkey, and the United States, vied for power and influence in the region. The revival of Islam has also generated international geopolitical repercussions.

▲ **Figure 10.25 War in Afghanistan**   Prolonged civil war in Afghanistan created a social disaster. Many Afghans sought refuge from the extreme fundamentalist forces of the Taliban either by heading into other countries or by fleeing to areas of Afghanistan not under Taliban rule. This photo shows refugees who fled Kabul and headed north. Since the fall of the Taliban, many refugees have been heading home.   *(Martin Adler/Panos Pictures)*

Islamic Fundamentalism?   All of the governments of Central Asia except that of Mongolia were concerned in the late 1990s that an Islamic fundamentalist movement similar to that of the Taliban could disrupt their own countries. At that time such worries were seemingly confirmed with the formation of the IMU (Islamic Movement in Uzbekistan), a group dedicated to the creation of a new Islamic state in the Fergana Valley. Based in western Tajikistan and receiving assistance from the Taliban and Al Qaeda, the IMU set off bombs in Tashkent and other cities and engaged the armies of Uzbekistan and Kyrgyzstan. As a result, military budgets increased throughout the region, exacerbating tensions between Central Asian countries.

The aftermath of terrorism on September 11, 2001, however, completely changed the balance of power in the region. The United States, with British assistance, launched a major war against Al Qaeda and the Taliban government that had

In early 2002, Afghan refugees began to stream home from camps in Pakistan and Iran, despite dangerous conditions. From the beginning of March to early April, an estimated 150,000 people returned from Pakistan, a figure that astounded international human rights workers. The UN High Commission for Refugees announced that 800,000 refugees would likely return to Afghanistan in 2002. Even relatively prosperous Afghani merchants began to leave Pakistan for their homeland—a positive sign for the beleaguered Afghan economy.

People are also returning to Afghanistan from Iran, which in early 2002 held some 2.3 million Afghan refugees. The Iranian government predicted that all of the refugees would return home by 2004. Such an occurrence, however, might require some force, as many Afghans have made Iran their new home and have no desire to leave. But the Iranian government resents their presence, accusing them of taking jobs from locals and smuggling drugs. In early 2002, a single mass expulsion forced 1,300 Afghans out of eastern Iran and across the border.

Regardless of whether they come voluntarily or are forced to return, most Afghan refugees will face difficult prospects in their homeland. Food remains scarce, bandits are numerous, and local warlords and the weak central government vie for power. Although foreign aid keeps Afghanistan afloat, it has been wholly inadequate according to most development experts. Some returning Afghans may opt to leave the country yet again. Indeed, in January and February 2002, an estimated 60,000 Afghans lined up at the border trying to flee into Pakistan. Most were Pashtuns, the ethnic group that formed the backbone of the Taliban—but that now faces persecution in many parts of the country.

offered it sanctuary. The U.S. offensive began October 7 with the bombing of Taliban military and governmental targets, and soon focused on supplying air support and other forms of assistance to the Northern Alliance. By mid-November Northern Alliance forces had driven the Taliban out of 80 to 90 percent of the country. Local Taliban fighters began to defect to the Northern Alliance in large numbers, although both the Al Qaeda forces and the foreign troops within the Taliban armies proved more difficult to dislodge. The rapid fall of the Taliban proved that few Afghanis supported its brutal policies. Indeed, the capture of the major cities by the Northern Alliance was marked by spontaneous celebration as music blared, men cut their beards, and carefully wrapped television sets were resurrected. Evidence uncovered early in the war showed that the Taliban and Al Qaeda had been closely intertwined and that a large number of well-equipped internationalist terrorist-training camps had been operating in Afghanistan.

After the fall of the Taliban, military emphasis shifted to the networks of heavily fortified caves and underground bunkers that Al Qaeda had established in various parts of the country, particularly near the Pakistan border. Pashtun tribal authorities, meanwhile, claimed authority in much of southern and eastern Afghanistan. In the western city of Herat, power fell to a local warlord with close ties to Iran.

By the early months of 2002, the remnants of Al Qaeda had been routed from their strongholds in eastern Afghanistan, although reports of regrouping continued to come from various parts of the country. At this time, the emphasis in Afghanistan began to shift to rebuilding the country's institutions and establishing an effective central government. An interim administration was established under Hamid Karzai, pending the meeting of a traditional assembly that would create a more permanent government. But as of April 2002, the Karzai regime had established firm control over only the vicinity of Kabul, aided by a 4,500-strong International Security Assistance Force. In many other areas, local warlords reestablished their power, while bandit groups made travel—and aid deliveries—extremely difficult. Tensions between ethnic Tajiks, who dominated the interim government, and Pashtuns, moreover, made Afghanistan's recovery all the more uncertain.

Although the international struggle against the Al Qaeda terrorist network and its Taliban supporters has taken center stage, a number of other Central Asian geopolitical issues remain troublesome. Several of these concern border conflicts.

Border Conflicts   The border between China and the newly independent Central Asian republics presents a potentially serious international issue. China's government does not formally accept the boundary now separating it from Tajikistan. It also objects to the fact that Kyrgyzstan supports radio broadcasts in Uygur supposedly aimed at Kyrgyzstan's own small Uygur minority. On the other side of the coin, some 1 million Kazaks live in northern Xinjiang, and many would like to be reunited with their conationals on the other side of the border. Leaders in Beijing and in the capitals of Kazakstan, Kyrgyzstan, and Tajikistan, however, do not want to provoke overt conflicts over these unresolved issues. In fact, China recently initiated a series of diplomatic maneuvers designed to quell tensions, fight Islamic separatism, and perhaps counter the influence of the United States. By early 2001 the regional forum thus created—called the "Shanghai Six"—included China, Russia, Kazakstan, Kyrgyzstan, Tajikistan, and Uzbekistan.

The Roles of Russia, Iran, Pakistan, and Turkey   Russia's concerns are directed at its former Central Asian territories. This is not just a matter of potential claims to Russian-speaking northern Kazakstan or even, as extreme nationalists dream, of Russia's reclaiming a Central Asian empire; it is also a question of cultural and economic influence. Russia's economic ties to the region were by no means erased when the Soviet Union collapsed. The main transportation line linking European Russia to central Siberia, moreover, cuts across northern Kazakstan, while former Soviet Central Asia's rail links to the outside world are still largely oriented toward Russian lands.

The Central Asian governments are themselves rather ambivalent about Russian influence. While all are eager to nurture their own sense of nationalism, the Russian connection brings undeniable benefits. Turkmenistan, for example, enacted measures in the late 1990s to discourage Russian speakers from moving back to Russia for fear that the resulting "brain drain" could undermine its economy. Russian influence in the region was also enhanced when Russia gave its full cooperation to the fight against Al Qaeda and the Taliban in late 2001.

Russia is not the only external power interested in the new republics of Central Asia. Iran is now a major trading partner, and it offers the best route to the ocean. Since the completion of a rail link between Iran and Turkmenistan in 1996, part of Central Asia's global trade has been reoriented toward Iran's ports. Iran's cultural ties with the region are old and deep, particularly in Tajikistan and northern Afghanistan, where Persian is the major language. Iran remained a major enemy of Afghanistan's Taliban regime and offered military help to the Northern Alliance. It thus hopes to secure a measure of influence in the country's new, post-Taliban regime.

Pakistan has been keen to gain influence in Central Asia. In earlier centuries close religious and political bonds linked Central Asian and South Asian Muslim communities—bonds that many Pakistanis would like to revive. Extremely rugged topography, however, separates Pakistan from the Amu Darya and Syr Darya valleys; Afghanistan, moreover, occupies the intervening territory. During the Taliban period, Pakistan enjoyed very close relations with Afghanistan. In fact, the Taliban received a tremendous amount of support from Pakistan's army and especially its secret services; some observers even regarded the Taliban as something of a Pakistani creation. The aftermath of September 11 thus put Pakistan in a serious bind. Owing to both U.S pressure and promises, Pakistan joined the anti-Taliban coalition, and it supplied the U.S. military with valuable intelligence. But it also became quickly evident that Pakistan regarded the victorious Northern Alliance with a good deal of hostility. It thus remains to be seen what Pakistan's relations will be with post-Taliban Afghanistan.

Turkey's cultural connections with Central Asia are also close. Most Central Asians speak Turkish languages that are closely related to that of Turkey. In the late 1800s and early 1900s, a political philosophy known as Pan-Turkism advocated that all Turkish-speaking peoples unite to form a single, huge country. This school of thought no longer has many adherents, but a tenuous sense of kinship does link Turkey to the Central Asian countries (including Xinjiang) known collectively as Turkestan.

Turkey offers itself as the model—contrary to those of Iran and Pakistan—of the modern state of Muslim heritage. In this secular model, religion provides social and cultural glue, but politics and economics operate without religious content. Turkey itself, however, is having some difficulty maintaining this balancing act, and it is not obvious that the Turkish system can easily be exported, even to other Turkish-speaking states. Turkey is also separated from Central Asia by Armenia, Russia, and Iran—countries with which it does not enjoy good relations.

Economic and Social Development: Abundant Resources, Devastated Economies

Central Asia is by most conventional measures one of the least prosperous regions of the world. One of its countries, Afghanistan, stands near the bottom of almost every list of economic and social indicators. Central Asia does, however, contain substantial natural resources, particularly oil and natural gas. Much of the region also enjoys relatively high levels of health and education, a legacy of the social programs enacted by the communist regimes. These same regimes, however, built inefficient economic systems, and since the fall of the Soviet Union western Central Asia has experienced a spectacular economic decline.

The Post-Communist Economies

Soviet economic planners sought to spread the benefits of economic development widely across their country. This required building large factories even in such remote areas as Central Asia, regardless of the costs involved. Such Central Asian industries relied heavily on subsidies from the center. When those subsidies ended, the industrial base of the region began to collapse, leading to plummeting living standards. As is true elsewhere in the former Soviet Union, however, certain individuals have grown very wealthy since the fall of communism.

As Table 10.2 shows, no Central Asian country or region could be considered prosperous by any measure. Kazakstan stands as the most developed, and it may have the best prospects. Since Kazakstan's agricultural base is potentially productive and its population density is low, it could emerge as a major food exporter. More important, it has two of the world's largest underutilized deposits of oil and natural gas, the vast Tengiz and Kashagan fields in the northeastern Caspian basin, as well as sizable deposits of other minerals. Kazakstan has signed agreements with Western oil companies to exploit its oil reserves, but work has been slowed by red tape.

Owing to its sizable population, Uzbekistan has total GNI almost as large as that of Kazakstan, giving it the second largest economy in the region. The Uzbek economy, moreover, has not declined as sharply as those of its neighbors. This is largely because it has retained many aspects of the old command economy (one run, in other words, by governmental planners rather than by private firms responding to the market) and has resisted economic liberalization. Critics contend, however, that Uzbekistan's industries will grow increasingly uncompetitive unless they are opened to market forces. At present, Uzbekistan remains a major exporter of cotton. It also has significant gold and natural gas deposits and has inherited a number of chemical and machinery factories from the Soviet period. Cotton production, however, is threatened by environmental degradation, and agriculture in general remains under inefficient state control.

Kyrgyzstan, on the other hand, has moved aggressively to privatize former state-run industries. Its economy, however,

TABLE 10.2 *Economic Indicators*

Country	Total GNI (Millions of $U.S., 1999)	GNI per Capita ($U.S., 1999)	GNI per Capita, PPP* ($Intl, 1999)	Average Annual Growth % GDP per Capita, 1990–1999
Afghanistan	—	—	800**	—
Azerbaijan	3,705	460	2,450	–10.7
Kazakstan	18,732	1,250	4,790	–4.9
Kyrgyzstan	1,465	300	2,420	–6.4
Mongolia	927	390	1,610	–0.6
Tajikistan	1,749	280	1,090**	2.0**
Turkmenistan	3,205	670	3,340	–9.6
Uzbekistan	17,613	720	2,230	–3.1

*Purchasing power parity

Source: The World Bank Atlas, 2001, except data marked **, which are from the CIA World Factbook, 2000.

is largely agricultural, and few of its industries are competitive. Since independence, Kyrgyzstan's economy has shrunk by almost 50 percent. It does, however, have the largest supply of freshwater in the region, a resource that will become increasingly valuable in years to come.

Turkmenistan also has a substantial agricultural base, due mainly to Soviet irrigation projects, and it remains a major cotton exporter. It also retains a state-run economy and has resisted pressure for economic liberalization. Government planners, however, hope that the development of new oil and gas fields will bring prosperity, proclaiming that Turkmenistan will soon become the "Kuwait of natural gas."

The region's best-developed fossil fuel industry is located in Azerbaijan (Figure 10.26). Azerbaijan has attracted a great deal of international interest and investment, promising to revitalize its oil industry. Thus far, however, its economy has largely failed to respond, and Azerbaijan—despite its promise—remains a very poor country.

The most economically troubled of the former Soviet republics is Tajikistan. With a per capita GNI of only some $280, Tajikistan rates as one of the world's poorer countries. It has few natural resources. A prolonged civil war has undercut production, and Tajikistan is burdened by its remote location and rugged topography. Its government has done little to dismantle the old command economy, yet industrial output has declined at an alarming rate.

Tajikistan's recovery probably depends not only on the resumption of peace and the enactment of reforms, but also on the stabilization of its neighbors. Tajikistan is one of the most remote countries of the world, with poor transportation connections to the outside (Figure 10.27). Most rail lines, inherited from the Soviet period, link the country ultimately to Russia. More efficient routes would connect Tajikistan to the world economy through the Caspian and then the Black seas, through Iran to the Arabian Sea, or through Afghanistan to Pakistan. At present, the Iranian route is emerging as a favored alternative, in part because of the completion of a rail line linking Iran to Turkmenistan.

Mongolia, although never part of the Soviet Union, was a close Soviet ally run by a communist party. It too suffered an economic collapse in the 1990s. Mongolia no longer receives Soviet subsidies, and its industries are not competitive in the global market. Its agricultural foundation is meager, and its traditional trade in livestock products is not very profitable. Mongolia thus emerged in the postcommunist period as a poor country, and its economy continued to decline into the late 1990s. Isolation also plagues Mongolia. Mongolia probably has some substantial mineral reserves, however, and its low population density gives it a certain leeway. Subsistence pastoralism helps many of its people maintain their livelihoods regardless of the country's economic conditions. Severe winter storms in 2000–2001, however, destroyed livestock herds in many parts of the country.

▲ Figure 10.26 Oil development in Azerbaijan   Although oil has brought a certain amount of wealth to Azerbaijan, it has also resulted in extensive pollution and visual blight. Since most of the petroleum is located either near or under the Caspian Sea, this sea—actually the world's largest lake—is now ecologically endangered. (John Spaull/Panos Pictures)

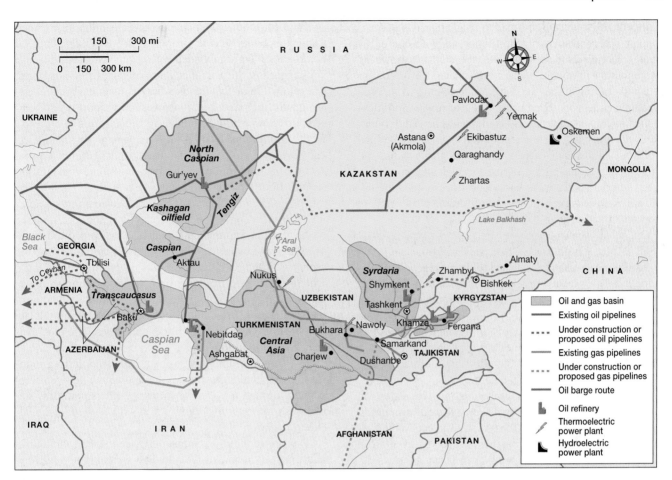

▲ Figure 10.27 Oil and gas pipelines   Central Asia has some of the world's largest oil and gas deposits, and has recently emerged as a major center for drilling and exploration. Due to its landlocked location, Central Asia cannot easily export its petroleum products. Pipelines have been built to solve this problem, and a number of others are currently being planned. Pipeline construction is a contentious issue, however, since several of the potential pathways lie across Iran, a country that remains under U.S. sanctions, and Russia, a politically and economically unstable country.

The Economy of Tibet and Xinjiang in Western China

The Chinese portions of Central Asia have not suffered the same economic crash that has visited the other parts of the region. China as a whole has one of the world's fastest-growing economies, although its centers of dynamism are all located in the coastal zone. Still, the most remote parts of the country have at least maintained their positions. But China as a whole started out much less developed than the Soviet Union, and poverty remains widespread. Although reliable economic figures are difficult to obtain, overall levels of per capita economic production in western China are probably comparable to those of Tajikistan.

Tibet in particular remains one of the world's poorest places. Most of the plateau is relatively isolated from the Chinese economy, much less the global economy. But many Tibetans are at least able to provide for most of their basic needs. Both Tibet and Xinjiang, moreover, do not suffer the overcrowding that strains the poorer parts of China proper. Then again, their harsh environments cannot support high population densi-

ties. This leads critics to contend that Han Chinese immigration threatens to upset the local balance between human numbers and the environment throughout western China.

Xinjiang does have tremendous mineral wealth, including a substantial portion of China's oil reserves. Its agricultural sector is also productive, although limited in extent. Many of the indigenous Muslim peoples believe that the wealth of the region is being monopolized by the Chinese state and the Han Chinese immigrants. Some point to the China Xinjiang Construction Company, a firm that runs more than 340 industrial enterprises as well as 172 giant farms, 500 schools, and 200 hospitals. This huge company was initially formed by the army that rejoined Xinjiang to China in 1949 and is still run almost entirely by Han Chinese. At the turn of the millennium, China announced ambitious new plans to build road and rail links from eastern China to Tibet and Xinjiang. While these lines will bring economic benefits, they may also result in an even larger migration stream, further undermining the indigenous cultures of the region.

Economic Misery in Afghanistan

Afghanistan is unquestionably the poorest country in the region, with one of the weakest economies in the world. Reliable statistics, however, are impossible to obtain. Afghanistan suffered nearly continuous war starting in the late 1970s, undercutting virtually all economic endeavors. Even before the war it was an impoverished country with little industrial or commercial development. Its only significant legitimate exports are animal products, handwoven carpets, and a few fruits, nuts, and semiprecious gemstones. Almost all fuel and consumer goods—and a good deal of basic food stocks—must be imported. Weapons imports have also burdened the Afghani economy in recent years.

One might well wonder how Afghanistan managed to pay for all of these imports and still keep afloat. By the late 1990s Afghanistan had essentially turned to the production of illicit drugs for the global market. According to the CIA, it had emerged by 1999 as the world's largest producer of opium. Most of the opium produced in the country ends up being sold as heroin in the large cities of North America and Europe. In the late 1990s most of the opium was produced in areas under Taliban control, but a significant amount also came form the Northern Alliance's territories.

At the turn of the millennium, economic conditions in Afghanistan were still declining. Several years of drought in the late 1990s, coupled with the tyranny and economic mismanagement of the Taliban regime, reduced many Afghani people to the point of starvation. War had done much to destroy the country's basic economic infrastructure. The Taliban, for example, often demolished houses, irrigation facilities, and orchards in areas that harbored Northern Alliance sympathizers, while the U.S. bombing campaigns also caused widespread damage. The United States, along with a host of other countries, subsequently promised billions of dollars in economic aid to be dedicated first to feeding and clothing the people of Afghanistan and then to rebuilding its economy and infrastructure. Many fear, however, that opium growing will continue to be the economic mainstay of the more remote parts of the country.

Central Asian Economies in Global Context

As the discussion above indicates, despite its poverty and relative isolation, Afghanistan has been thoroughly embedded in the global economy, albeit through illicit products. Illegal drugs are extremely lightweight and can thus be successfully exported from even the most isolated parts of the world.

In the former Soviet area, the most important international connections remain with Russia. Both Russia and the newly independent Central Asian countries are ambivalent about this relationship. Russia has sought to distance itself economically from its former Central Asian territories, most notably by forcing them to develop their own currencies and to quit using the Russian ruble. Turning away from Russia, the former Soviet zone is quickly developing economic as well as political ties to such countries as Iran, Pakistan, Turkey, and China. The last three countries are particularly interested in Central Asia's vast energy reserves.

The United States and other Western countries are drawn to the area by its oil and natural gas deposits. Most large oil companies have established operations in Kazakstan and Turkmenistan, which are thought to contain the world's largest unexploited fossil fuel deposits open to Western firms. Western goods, including luxury items such as sport utility vehicles, are now readily available to those profiting from oil wealth in cities such as Kazakstan's Almaty, Turkmenistan's Ashgabat, and Azerbaijan's Baku. All told, however, Central Asia has yet to experience substantial economic globalization (Figure 10.28).

These same fossil fuel reserves have created a very complex international economic environment. For the region's large oil fields to be economically viable, a vast pipeline system must be constructed. As was shown in Figure 10.27, existing pipelines, which are already inadequate, pass through Russia, which charges exorbitant prices. A large amount of oil is shipped out by rail through Russia and Georgia, an inefficient mode of export. International negotiations began in the late 1990s for the construction of new pipelines. The most efficient route for Turkmenistan would be through Iran, and in 1997 a small gas pipeline was opened between the two countries. The United States, however, opposes any new pipelines passing through Iran. If Afghanistan becomes politically stable, it is quite possible that it will emerge as the favored oil-export corridor.

The United States has lobbied, however, for an alternative, and much more expensive, pipeline system that would extend from Azerbaijan into Georgia and then across Turkey to the Mediterranean port of Ceyhan. In 2001 Azerbaijan, Kazakstan, Georgia, and Turkey signed a memorandum of understanding supporting this route, much to the pleasure of U.S. diplomats.

Social Development in Central Asia

As Table 10.3 indicates, social conditions in Central Asia vary more than economic conditions. In the former Soviet territories, levels of health and education are relatively high, but seem to be declining. Afghanistan, not surprisingly, is at the bottom of the scale by every measure.

Social Conditions and the Status of Women in Afghanistan

Social conditions in Afghanistan are no better than economic circumstances, although again reliable information is difficult to obtain. As can be seen in Table 10.3, the average life expectancy in the country is a mere 45 years, one of the lowest figures in the world. Infant and childhood mortality levels remain extremely high. Not only does Afghanistan suffer from constant warfare, but its rugged topography hinders the provision of basic social and medical services. Illiteracy is commonplace and is notably gender-biased. Afghanistan's 15 percent adult female literacy figure is one of the lowest in the world.

Women in traditional Afghani society—and especially in Pashtun society—lead highly constrained lives. In many areas they must completely conceal their bodies, including their faces, when venturing into public. Such restrictions intensi-

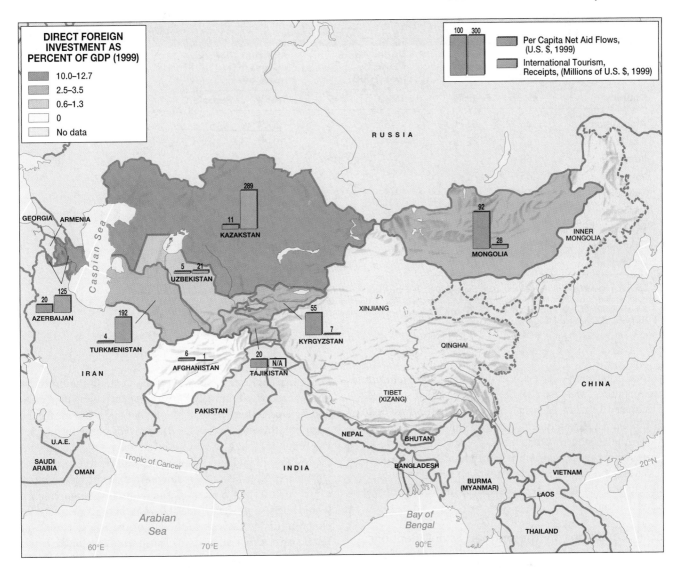

▲ Figure 10.28 Global linkages: direct foreign investment   Kazakstan and Azerbaijan, due to their oil industries, receive the bulk of the direct foreign investment coming into Central Asia. Turkmenistan, also a fossil-fuel exporter, and Mongolia and Kyrgyzstan, both relatively open economies, attract more foreign investment than do Uzbekistan, Tajikistan, or Afghanistan. Similar patterns hold for many other measures of global connectedness, with Kazakstan leading the pack. Overall, however, Central Asia is not closely connected with the global system. This is true even in regard to net aid flows to desperately poor Central Asian countries such as Afghanistan and Tajikistan.

fied in the 1990s. Dress controls were strictly enforced where the Taliban gained authority. In most areas, Taliban forces prevented women from working, attending school, and often even from obtaining medical care. The prohibition on work brought particular hardship to Afghanistan's war widows, leaving many in outright destitution. It apparently forced some younger women—against the Taliban's intentions—into prostitution. Several international aid agencies were forced to scale back their operations in Afghanistan after their female employees began to suffer from constant harassment.

The assault on women by the Taliban did not gone uncontested in Afghanistan. Many women risked their lives to document the horrific abuses that the women of their country were suffering. While Taliban leaders argued that they were upholding Islamic orthodoxy, their opponents con-

tended that they were actually enforcing an extremist version of Pashtun customs. Many of the more prosperous Persian-speaking families in areas under Taliban control therefore sent their daughters to schools in Iran.

The fall of the Taliban brought jubilation to the women of Afghanistan (Figure 10.29). Many promptly uncovered their faces and began working—or seeking work. But Afghani women also remained nervous; most members of the Northern Alliance had not upheld women's rights, and fears that fighting would continue tended to dampen the celebrations. International agencies have promised to fight for women's rights in post-Taliban Afghanistan, but it remains to be seen how successful they will be.

Conditions in Afghanistan have been so miserable over the past decade that some 6 million persons—nearly one-third of

TABLE 10.3 *Social Indicators and Status of Women*

Country	Life Expectancy at Birth		Under Age 5 Mortality (per 1000)		Percent Illiteracy (Ages 15 and over)		Female Labor Force Participation (% of total, 1999)
	Male	Female	1980	1999	Male	Female	
Afghanistan	46	44	—	—	47**	15**	—
Azerbaijan	68	75	—	21	99**	96**	44
Kazakstan	60	71	—	28	99**	96**	47
Kyrgyzstan	65	72	—	38	99**	96**	47
Mongolia	65**	70**	—	73	73	52	47
Tajikistan	66	71	—	34	99	99	45
Tibet	—	—	—	—	—	—	—
Turkmenistan	63	70	—	45	99**	97**	46
Uzbekistan	68	73	—	29	93	84	47

Sources: Population Reference Bureau, Data Sheet, *2001, Life Expectancy (M/F);* The World Bank, World Development Indicators, *2001, data marked ** are from the* CIA World Factbook, *2000, Under Age 5 Mortality Rate (1980/99);* The World Bank Atlas, *2001, Female Participation in Labor Force;* The World Bank, World Development Indicators, *2001, data marked ** are from the* CIA World Factbook, *2000, Percent Illiteracy (ages 15 and over).*

the total population—have at various times sought refuge in other countries. Most have fled to Pakistan, which still holds an estimated 1 million Afghani refugees, and Iran, where up to 2.3 million Afghans remain. Another 1 million have moved within Afghanistan to urban areas. By late 2001, however, refugees were beginning to return to their homelands. Owing to the devastation of war, however, it will be a number of years before they all can return (see "Geography in the Making: Afghan Refugees Return").

Social Conditions in the Former Soviet Republics

Women in the former Soviet portion of Central Asia enjoy a higher social position than those of Afghanistan. Traditional-

▲ Figure 10.29 Afghani women and the fall of the Taliban When the Taliban government fell from power in Kabul on November 13, 2001, many residents of the city erupted into spontaneous celebrations. A number of women demonstrated their new—although still limited—freedom by exposing their faces in public. Such an "offense" would have earned them severe beatings under Taliban rule. *(Scott Peterson/Getty Images, Inc.)*

ly, women had much more autonomy among the northern pastoral peoples (especially the Kazak) than among the Uzbek and Tajik oasis dwellers of the south, but under Soviet rule the position of women everywhere improved. In the former Soviet republics today, women's educational rates are comparable to those of men, and women are well represented in the workplace. Note also that women can expect to live significantly longer than men in these countries. Some evidence would suggest, however, that the position of women has declined in many areas since the fall of the Soviet Union. In Kazakstan several leaders have blamed feminism for the country's low birthrate, and in much of the region newly rich men are increasingly practicing polygamy in order to have large numbers of children.

The figures in Table 10.3 reveal generally favorable levels of social welfare overall. This is especially notable when one considers the dismal state of the region's economy—or contrasts conditions in the former Soviet zone with those of neighboring Afghanistan. Tajikistan, with a per capita GNI of only $280, has a female life expectancy of 71 years as well as almost universal adult literacy. Western Central Asia thus stands with Sri Lanka, Kerala in India, and Cuba as places that enjoy relatively high levels of social development despite economic impoverishment (Figure 10.30).

The social successes of Tajikistan and its northern and western neighbors reflect investments made during the Soviet period. It is unclear, however, whether health and educational facilities can be maintained in the face of economic collapse. Certainly the region's relatively high levels of infant and childhood mortality do not bode well. While Tajikistan might hope that its educated workforce will attract foreign investment, its remote location and continued instability make such a scenario unlikely. Universal education may become a luxury that the country cannot afford.

◀ **Figure 10.30 Tarkent heliostation in Tashkent, Uzbekistan**   Despite its generally poor economic statistics, Uzbekistan boasts a relatively well-developed medical, educational, and technological infrastructure. The Tarkent heliostation focuses the sun's rays to generate the extraordinarily high temperatures needed to produce super-pure materials such as aluminum titanate and magnesium oxide.   *(Novosti/ Getty Images, Inc.)*

Social Conditions in Western China   It is particularly difficult to obtain reliable information about social conditions in the Chinese portion of Central Asia. Certainly China as a whole has made significant progress in health and education, but many reports suggest that the peoples of Tibet and Xinjiang have been left behind. According to the eastern Turkestan Web pages, some 60 percent of the non-Han peoples of Xinjiang are illiterate. China's minority peoples have, however, been granted certain exemptions from the country's strict population control measures. But there are many reports from Xinjiang and elsewhere of forced sterilization and abortions, adding significantly to the tensions in the area.

⊕ Conclusion

Central Asia, long obscured by Russian and Chinese domination, has recently reappeared on the map of the world. Many geographers remain unconvinced that it forms a legitimate world region. The six newly independent former Soviet republics, along with Afghanistan, could just as easily be classified within Southwest Asia. Tibet and Xinjiang, on the other hand, are usually grouped with the rest of China as part of East Asia. Although the Mongol, Uygur, and Tibetan peoples of this area are struggling to maintain their lands and cultural autonomy, they may not be able to withstand the Han Chinese demographic tide. The Chinese government, moreover, shows no indication that it would be willing even to discuss the possibility of genuine autonomy for either Tibet or Xinjiang.

While China maintains a firm grip on Tibet and Xinjiang, the rest of Central Asia has emerged as a key area of geopolitical and economic competition. Russia, the United States, China, Iran, Pakistan, and Turkey all contend for influence. Kazakstan, Uzbekistan, Tajikistan, Kyrgyzstan, Turkmenistan, and Azerbaijan have attempted to play these powers off against each other in order to bolster their own positions. And while political structures throughout the area remain largely authoritarian, the economies of the region are gradually opening up to global connections.

Regardless of what political-economic model (or models) it follows, Central Asia is likely to face serious economic difficulties for some time. The region is not a significant participant in global trade, and it has attracted little foreign investment outside of the oil industry. Some economists argue, moreover, that a continental position by itself is a major economic liability. To thrive in the new economic global order, they contend, a country must have easy access to the sea-lanes that form the main conduits of global exchange. But whether its continental position will undercut Central Asia's economic prosperity in the twenty-first century remains to be seen.

Railroads and even roadways have proved adequate to integrate interior portions of the United States with the global economy.

Three countries in the region, Kazakstan, Azerbaijan, and Turkmenistan, are almost certain to emerge as major players in global trade owing to their vast reserves of oil and natural gas. Whether such natural resources will allow these countries to become prosperous is another matter. Oil wealth—as Nigeria has proved—can be a curse as much as a blessing, since it often results in an inflated currency that undercuts industry and agriculture. Relatively high levels of corruption in much of Central Asia do not bode well for the region's success. Relatively high levels of education, on the other hand, may help these countries use their oil riches to build a sturdy economic base.

Central Asia remains something of a question mark on the map of the world. A large part of the region recently received its independence after more than 100 years of Russian control, and it remains to be seen what economic and political directions it will follow. Much of the rest of the region—Tibet and Xinjiang—remains under the authority of what most indigenous residents consider to be a foreign power. China, however, regards these lands as an integral part of its own territory. Considering the demographic imbalance between the Han Chinese, on the one hand, and the Uygur and Tibetan peoples, on the other, the position of indigenous Central Asian culture in Xinjiang and Tibet must be considered vulnerable.

Afghanistan, of course, remains Central Asia's biggest question mark. It has the region's most serious economic, social, and ethnic problems—problems that now command the attention of the global community. Global efforts will probably be necessary if Afghanistan is to rebuild itself economically and find political stability. If that were to happen, the future of all of Central Asia would be much more secure than it is today.

⊕ Key Terms

alluvial fan (page 428)   exotic river (page 428)   pastoralist (page 425)   transhumance (page 427)
exclave (page 441)   loess (page 428)   theocracy (page 437)   Turkestan (page 425)

⊕ Questions for Summary and Review

1. Describe the three major regions of Central Asia as defined by physical geography.

2. Why is lake level a particularly serious concern in Central Asia?

3. Why is the distribution of population in Central Asia so uneven? Why are certain areas virtually uninhabited?

4. Describe the different agricultural patterns that are found on (a) the plains of northern Kazakstan, (b) the major river valleys of Uzbekistan, and (c) the Tibetan Plateau.

5. Why is much of Central Asia often called "Turkestan"? Why does this label not apply to the entire region?

6. How does religion act as a political force in Central Asia? How does this vary in different parts of the region?

7. Why is the United States concerned about the fossil fuel resources of Central Asia? How do transportation routes play into this concern?

8. What role does Russia currently play in Central Asian geopolitics? How has Russian power historically influenced the region?

9. How do social conditions (health, longevity, literacy, etc.) vary across the border between Afghanistan and Uzbekistan? Why are the disparities so pronounced?

10. If western Central Asia is so rich in natural resources, why did the region experience such pronounced economic decline in the 1990s?

⊕ Thinking Geographically

1. How much of a disadvantage will Central Asia's continental location be in years to come? Is coastal access truly significant in determining a country's competitive position in the global economy?

2. Will the countries of former-Soviet Central Asia be able to maintain their high levels of education now that they are independent? If so, will they be able use education to their own economic advantage?

3. Is Russia's influence in Central Asia bound to decline now that Russia has no direct political authority in the area?

4. What external connection will prove most important for Central Asia in years to come—those based on religion, language, economic ties, or geopolitical connections?

5. Is China's political control over eastern Central Asia justified? Should Han Chinese migration to the area be a concern to the United States?

6. Is Chinese control over Tibet a legitimate concern of U.S. foreign policy? Should the United States use trade sanctions to attempt to influence Chinese policy in the area?

7. How might the desiccation of Central Asia's great lakes best be addressed?

⊕ Regional Novels and Films

Novels

Rinjing Dorje, *Tales of Uncle Tompa, The Legendary Rascal of Tibet* (1997, Barrytown Ltd.)

Mark Frutkin, *Invading Tibet* (1993, Soho Press)

James Hilton, *Lost Horizon* (1996, William Morrow)

Amin Maalouf, *Samarkand: A Novel* (1998, Interlink Publishing)

Hilary Roe Metternich, ed., *Mongolian Folktales* (1996, Avery Press)

Films

Kandahar (2001, Iran)

Kundun (1997, U.S.)

Seven Years in Tibet (1997, U.S.)

Urga: Close to Eden (1991, Mongolia)

Bibliography

Adshead, S. A. M. 1993. *Central Asia in World History*. New York: St. Martin's Press.

Alworth, Edward, ed. 1989. *Central Asia: 120 Years of Russian Rule*. Durham, NC: Duke University Press.

Bacon, Elizabeth E. 1980. *Central Asians Under Russian Rule: A Study in Culture Change*, 2nd ed. Ithaca, NY: Cornell University Press.

Barfield, Thomas J. 1989. *The Perilous Frontier: Nomadic Empires and China from 221 B.C. to A.D. 1757*. Oxford: Basil Blackwell.

Batalden, Stephen K., and Batalden, Sandra L. 1993. *The Newly Independent States of Eurasia: A Handbook of Former Soviet Republics*. Phoenix: Oryx Press.

Beckwith, Christopher I. 1987. *The Tibetan Empire in Central Asia*. Princeton, NJ: Princeton University Press.

Bergholtz, Fred W. 1993. *The Partition of the Steppe: The Struggle of the Russians, Manchus, and the Zunghar Mongols for Empire in Central Asia*. New York: Peter Lang.

Brawer, Moshe. 1994. *Atlas of Russia and the Independent Republics*. New York: Simon & Schuster.

Chin, Jeff, and Kaiser, Robert. 1996. *Russians as the New Minority: Ethnicity and Nationalism in the Soviet Successor States*. Boulder, CO: Westview Press.

Christian, David. 1994. "Inner Eurasia as a Unit of World History." *Journal of World History* 5, 173–211.

Drompp, Michael. 1989. "Centrifugal Forces in the Inner Asian 'Heartland': History *Versus* Geography." *Journal of Asian History* 23, 135–55.

Frank, Andre Gunder. 1992. "The Centrality of Central Asia." *Bulletin of Concerned Asian Scholars* 24, 50–74.

Frye, Richard. 1996. *The Heritage of Central Asia: From Antiquity to the Turkish Expansion*. Princeton, NJ: Markus Wiener.

Fuller, Graham E. 1993. "Turkey's New Eastern Orientation." In Graham E. Fuller and Ian O. Lesser, eds., *Turkey's New Geopolitics: From the Balkans to Western China*, pp. 37–98. Boulder, CO: Westview Press.

Gaubatz, Piper Rae. 1996. *Beyond the Great Wall: Urban Form and Transformation on the Chinese Frontiers*. Stanford: Stanford University Press.

Goodson, Larry. 2001. *Afghanistan's Endless War: State Failure, Regional Politics, and the Rise of the Taliban*. Seattle: University of Washington Press.

Griffin, Michael. 2001. *Reaping the Whirlwind: The Taliban Movement in Afghanistan*. Sterling, VA: Pluto Press.

Gross, Jo-Ann, ed. 1992. *Muslims in Central Asia: Expressions of Identity and Change*. Durham, NC: Duke University Press.

Grousset, Rene. 1970. *The Empire of the Steppes: A History of Central Asia*. New Brunswick, NJ: Rutgers University Press.

Hambly, Gavin. 1969. *Central Asia*. London: Weidenfeld and Nicolson.

Hauner, Milan. 1990. *What Is Asia to Us? Russia's Asian Heartland Yesterday and Today*. London: Routledge.

Hauner, Milan. 1991. "Russia's Geopolitical and Ideological Dilemmas in Central Asia." In Robert Canfield, ed., *Turko-Persia in Historical Perspective*, pp. 189–216. Cambridge: Cambridge University Press.

Humphrey, Caroline, and Sneath, David. 1999. *The End of Nomadism Society? Society, State, and Environment in Inner Asia*. Durham, NC: Duke University Press.

Huntington, Ellsworth. 1907. *The Pulse of Asia*. New York: Houghton Mifflin.

Jagchid, Sechin, and Hyer, Paul. 1979. *Mongolia's Culture and Society*. Boulder, CO: Westview Press.

Kaiser, Robert J. 1994. *The Geography of Nationalism in Russia and the USSR*. Princeton, NJ: Princeton University Press.

Kobori, Iwao, and Glantz, Michael. 1998. *Central Eurasian Water Crisis: Caspian, Aral, and Dead Sea*. Tokyo: United Nations University Press.

Lattimore, Owen. 1988. *Inner Asian Frontiers of China*, 2nd ed. Oxford: Oxford University Press.

Lewis, Robert A., ed. 1992. *Geographical Perspectives on Soviet Central Asia*. London: Routledge.

Morgan, David. 1986. *The Mongols*. Oxford: Basil Blackwell.

Rashid, Ahmed. 2001. *Taliban: Militant Islam, Oil, and Fundamentalism in Central Asia*. New Haven: Yale University Press.

Rossabi, Ralph A. 1990. "The 'Decline' of the Central Asian Caravan Trade." In James D. Tracy, ed., *The Rise of Merchant Empires: Long-Distance Trade in the Early Modern World, 1350–1750*, pp. 351–70. Cambridge: Cambridge University Press.

Rubin, Barnett. 1995. *The Search for Peace in Afghanistan: From Buffer State to Failed State*. New Haven: Yale University Press.

Rumer, Boris, ed. 1998. *Central Asia: The Challenge of Independence*. New York: M. E. Sharpe.

Rumer, Boris, ed. 2000. *Central Asia and the New Global Economy*. New York: M. E. Sharpe.

Shaw, Denis J. B., ed. 1995. *The Post-Soviet Republics: A Systematic Geography*. New York: Wiley/Longman.

Sinor, Denis, ed. 1990. *The Cambridge History of Early Inner Asia*. Cambridge: Cambridge University Press.

Tregear, Thomas R. 1980. *China: A Geographical Survey*. New York: John Wiley & Sons.

EAST ASIA
Political Map

⊕ ● Over 1,000,000

◦ ˙ 500,000–1,000,000
(selected cities)

· Selected smaller cities

(National capitals shown in red)

Elevation in meters

4000+
2000–400
500–2000
200–500
0–200
Below se
level

Sea Level

RUSSIA

KAZAKSTAN

KYRGYZSTAN

MONGOLIA

HEILONGJIANG

MANCHURIA

Harbin

Changchun

JILIN

Shenyang

LIAONING

NORT
KORE

P'yongyang

Ürümqi

XINJIANG
(SINKIANG)

GANSU

Hohhot

INNER MONGOLIA

BEIJING

HEBEI

Tangshan

Beijing

TIANJIN

Tianjin

Seoul

SOUT
KORE

Shijiazhuang

HEBEI

Yinchuan

NINGXIA

Taiyuan

SHANXI

SHANDONG

Jinan

Kwangju

Yellow
Sea

QINGHAI

Xining

GANSU

Lanzhou

Xi'an

SHAANXI

Huang He

Kaifeng

Zhengzhou

HENAN

JIANGSU

Nanjing

ANHUI

Suzhou

SHANGH

TIBET
(XIZANG)

C H I N A

Hefei

Shanghai

Hangzhou

NEPAL

Lhasa

Chengdu

SICHUAN

CHONGQING

Chongqing

HUBEI

Wuhan

ZHEJIANG

Eas
Chin
Sea

BHUTAN

Changsha

JIANGXI

Nanchang

INDIA

BANGLADESH

Guiyang

GUIZHOU

HUNAN

Guilin

FUJIAN

Fuzhou

T'aipei

MYANMAR

YUNNAN

Kunming

GUANGXI

Nanning

Guangzhou

GUANGDONG

MACAU

Shenzhen

Hong Kong

TAIWAN

VIETNAM

Gulf of
Tonkin

Bay of
Bengal

LAOS

THAILAND

HAINAN

Hainan
(CHINA)

South
China
Sea

PHILIPPINE

90°E

110°E

CHAPTER 11

East Asia

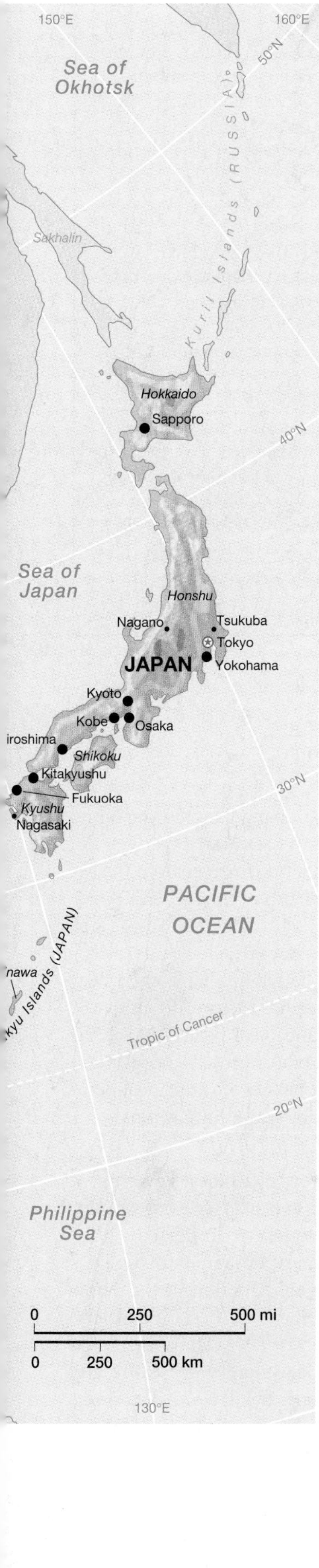
150°E
160°E
Sea of Okhotsk
50°N
Kuril Islands (RUSSIA)
Sakhalin
Hokkaido
Sapporo
40°N
Sea of Japan
Honshu
Nagano
Tsukuba
Tokyo
JAPAN
Yokohama
Kyoto
Kobe
Osaka
iroshima
Shikoku
Kitakyushu
30°N
Fukuoka
Kyushu
Nagasaki
PACIFIC OCEAN
nawa
kyu Islands (JAPAN)
Tropic of Cancer
20°N
Philippine Sea
0
250
500 mi
0
250
500 km
130°E

East Asia, composed of China, Japan, South Korea, North Korea, and Taiwan (Figure 11.1), is the most populous region of the world. China alone is inhabited by more than 1.2 billion persons, more than live in any other region of the world except South Asia. Although East Asia is historically unified by cultural features, in the second half of the twentieth century it was divided ideologically and politically, with the capitalist economies of Japan, South Korea, and Taiwan (as well as the former British colony of Hong Kong) separated from the communist bloc of China and North Korea. Disparities in levels of economic development also remained pronounced. As Japan reached the pinnacle of the global economy, much of China remained mired in extreme poverty.

More recently, however, divisions with East Asia have been muted to a certain extent. While China is still governed by the Communist Party, it has embarked on a path of modified capitalist development. Ties between its booming coastal zone and Japan, South Korea, and Taiwan have quickly strengthened. While mutual animosity still pervades relations between North Korea and South Korea and between China and Taiwan, East Asia as a whole has witnessed a gradual reduction in political tensions (see "Setting the Boundaries").

A thousand years ago, East Asia was not only the most populous but also the most economically advanced part of the world. In the 1800s, however, its fortunes declined precipitously. China suffered repeated famines and revolutions and was exploited by European imperial powers. Japan managed to retain full independence and to build a powerful military, but it remained a poor country until the second half of the twentieth century. In 1945, at the end of World War II, Japan was utterly vanquished, while China remained gripped in a brutal civil war between nationalist and communist forces. The future of the region looked bleak.

Today East Asia must be recognized as one of the core areas of the world economy, and it is emerging as a center of political power as well. Japan, South Korea, and Taiwan are among the world's key trading states. Tokyo, Japan's largest city, stands alongside New York and London as one of the financial centers of the globe. Tokyo is also an increasingly cosmopolitan city, drawing immigrants from around the world (many, if not most of them, illegal) and participating as both a consumer and producer of global culture. China is still more economically and culturally self-contained, but it too is emerging as a global trading power. China's coastal zone, particularly the former British colony of Hong Kong,

◀ **Figure 11.1 East Asia** This region includes China, Japan, North Korea, South Korea, and Taiwan. China, the world's largest country in terms of population, dominates East Asia with more than a billion people. The second largest country is Japan, with 127 million. Japan, South Korea, Taiwan, and Hong Kong (now once again part of China) have long dominated economically; however, China's recent development places that country solidly in the list of world players both political and economically.

SETTING THE BOUNDARIES

East Asia is easily marked off on the map of the world merely by the territorial extent of its constituent countries: China, Taiwan, North Korea, South Korea, and Japan (Figure 11.1). Japan is an island country composed of four main islands, but also including a chain of much smaller islands (the Ryukyus) that extends almost to Taiwan. To the north, Japan claims the four southernmost Kuril Islands, but these are actually controlled by Russia. Taiwan is basically a single-island country, but its political status is ambiguous since China claims it as part of its own territory. China itself is a vast continental state—the third most extensive in the world—that reaches well into Central Asia. It also includes the large island of Hainan in the South China Sea. Korea constitutes a compact peninsula between northern China and Japan, but it is divided into two sovereign states: North Korea and South Korea. South Korea also encompasses the sizable island of Cheju in the Yellow Sea.

In political terms, such a straightforward definition of East Asia is appropriate. If one turns to cultural considerations, however, the issue becomes more complicated. The main cultural problem arises with regard to the western half of China. This is a huge but lightly populated space; some 95 percent of the residents of China live in the east (Figure 11.16). By certain criteria, only the populous eastern half of

China (often referred to as "**China proper**") really fits into the East Asian world region. The indigenous inhabitants of western China are not Chinese by culture and language, and they have never accepted the religious and philosophical beliefs that have given historical unity to East Asian civilization. In the northwestern quadrant of China, called *Xinjiang* in Chinese, most indigenous inhabitants speak Turkish languages and are Muslim in religion. Tibet, in southwestern China, has a highly distinctive culture, and the Tibetans in general resent Chinese authority. Xinjiang and Tibet are thus perhaps more appropriately classified within Central Asia, which is covered in Chapter 10. In this chapter we will examine western China only to the extent that it is politically part of China, the historical core of East Asia.

Vietnam presents another anomaly for the definition of East Asia. By most cultural and historical criteria, Vietnam clearly belongs within East Asia. Vietnam was part of China for almost a thousand years, and many of its cultural patterns are similar to those of China. By modern political measures, however—as well as by its actual location—Vietnam fits much better within Southeast Asia. Vietnam will therefore be discussed occasionally in this chapter, but extended consideration will be reserved for Chapter 13.

is tightly integrated within global networks of information, commerce, and entertainment.

The global political position of East Asia is somewhat ambiguous. Since its defeat in World War II, Japan has maintained a low international profile, pledging itself to keep a relatively small armed force. China, on the other hand, is rising out of poverty and relative isolation to become a major regional, if not global, military power. This development has caused much concern among China's neighbors—and among distant countries, such as the United States.

Environmental Geography: Resource Pressures in a Crowded Land

Environmental problems in East Asia are particularly severe owing to a combination of the region's large population, its rapid industrial development, and its unique physical geography (Figure 11.2). The wealthier countries of the region, particularly Japan, have been able to invest heavily in environmental protection. China, on the other hand, not only has less money available for this purpose, but many of its problems, such as deforestation, soil erosion, and flooding, are more deeply rooted (see "Environment: China's Relationship with Nature"). Let us begin our discussion with one of China's most controversial environmental actions, the building of the Three Gorges Dam on the Yangtze River.

Flooding, Dam-Building, and Related Issues in China

The Yangtze River (also called the Chang Jiang) is one of the most important physical features of East Asia. This river, the third largest (by volume) in the world, emerges from the Tibetan highlands onto the rolling lands of the Sichuan Basin, passes through a magnificent canyon in the Three Gorges area (Figure 11.3), and then meanders across the lowlands of central China before entering into the sea in a large delta near the city of Shanghai. The Yangtze has historically been the main avenue of entry into the interior of China, and it has long been famed in Chinese literature for its beauty and power. More recently, however, it has become the focal point of an environmental controversy of global proportions.

The Three Gorges Controversy The Chinese government wants to control the Yangtze for two main reasons: to prevent flooding and to generate electricity. To do so, it must build a series of large dams, the most prominent of which is a massive structure now under construction in the Three Gorges area. This $25 billion structure—600 feet (180 meters) high and 1.2 miles (1.9 kilometers) long—will be the largest hydroelectric dam in the world when completed in 2009, and will form a reservoir 350 miles long. It will jeopardize several endangered species (including the Yangtze River dolphin), inundate a major scenic attraction, and displace an estimated 1.2 to 1.9 million persons—entailing the largest forced re-

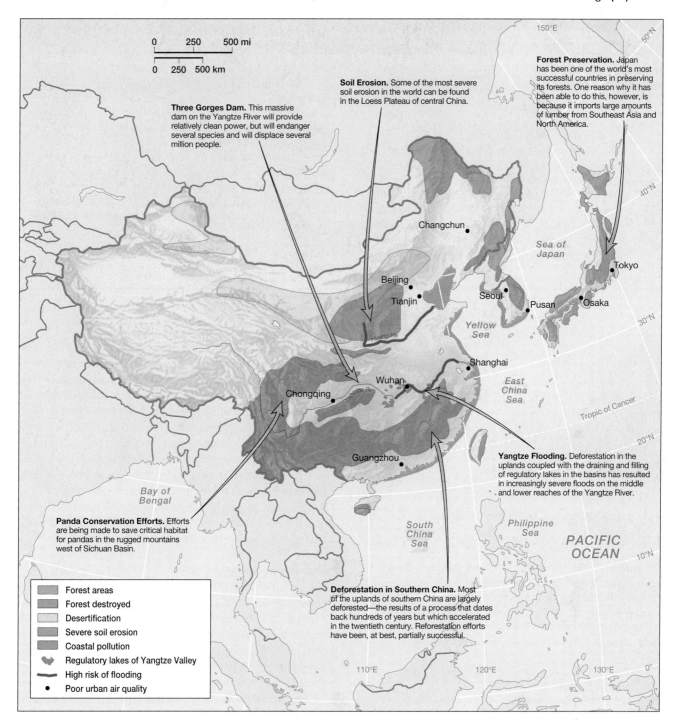

Three Gorges Dam. This massive dam on the Yangtze River will provide relatively clean power, but will endanger several species and will displace several million people.

Soil Erosion. Some of the most severe soil erosion in the world can be found in the Loess Plateau of central China.

Forest Preservation. Japan has been one of the world's most successful countries in preserving its forests. One reason why it has been able to do this, however, is because it imports large amounts of lumber from Southeast Asia and North America.

Yangtze Flooding. Deforestation in the uplands coupled with the draining and filling of regulatory lakes in the basins has resulted in increasingly severe floods on the middle and lower reaches of the Yangtze River.

Panda Conservation Efforts. Efforts are being made to save critical habitat for pandas in the rugged mountains west of Sichuan Basin.

Deforestation in Southern China. Most of the uplands of southern China are largely deforested—the results of a process that dates back hundreds of years but which accelerated in the twentieth century. Reforestation efforts have been, at best, partially successful.

Forest areas
Forest destroyed
Desertification
Severe soil erosion
Coastal pollution
Regulatory lakes of Yangtze Valley
High risk of flooding
Poor urban air quality

▲ **Figure 11.2 Environmental issues in East Asia**   This vast world region has been almost completely transformed from its natural state and continues to have serious environmental problems. In China, some of the more pressing environmental issues involve deforestation, flooding, water control, and soil erosion.

settlement for a single project in history. Few experts expect that the resettlement program will be completed in time. As of early 2001, reports of violence and repression were already coming from the resettlement camps. The ecological and human rights consequences of the dam are so negative that the World Bank, which initially funded the project, withdrew support. The Chinese government, however, decided that the dam was so important that it would finance it by itself with

support from private firms, most notably the U.S. investment bank Morgan Stanley Dean Witter.

Other Yangtze dams will also generate substantial amounts of electricity, meeting up to 11 percent of China's energy requirement. As China industrializes, its demand for power is skyrocketing. At present, nearly four-fifths of China's total energy supply comes from burning coal, which results in horrendous air pollution. Smog is particularly severe in the winter

ENVIRONMENT   China's Relationship with Nature

Vaclav Smil, a Czech-born geographer currently teaching in Canada, has written extensively on environmental degradation and energy use in China. The following passages are excerpted from his book, *The Bad Earth* (1984).

A reverence for nature runs unmistakably through the long span of Chinese history. The poet . . . found the mountains his most faithful companions; emperors . . . painted finches in bamboo groves and ascended sacred mountains; Buddhist monks sought their dhyana "midst fir and beech"; craftsmen located their buildings to "harmonize with the local currents of the cosmic breath"; painters were put through the rigors of mastering smooth, natural, tapering bamboo leaves and plum branches; and who couldn't admire the symphony of plants, rocks, and water in countless gardens.

To stop here, however, . . . would be telling only the more appealing half of the story. There was also a clearly discernible current of destruction and subjugation: the burning of forests just to drive away dangerous animals; massive, total, and truly ruthless deforestation to create new fields, to get fuel and charcoal, and to obtain timber for fabulous palaces and ordinary houses, wood for cremation of the dead and (to no small effect) for making ink from the soot of burnt pines . . . the erection of sprawling rectilinear cities (fires would rage for days to consume the vast areas of wooden buildings) eliminating any trace of nature, save for some artificial gardens.

Source: From *The Bad Earth* by Vaclav Smil. Armonk, NY: M. E. Sharp, 1984, pp. 6–7.

▲ **Figure 11.3 The Three Gorges of the Yangzte**   The spectacular Three Gorges landscape of the Yangzte River is the site of a controversial flood control dam that will displace several million people. Not only are the human costs high to those displaced, but there may also be significant ecological costs to endangered aquatic species.   *(Greg Baker/AP/Wide World Photos)*

months, when stagnant air masses often sit over northern and central China. Coal burning also forms acid rain, which is becoming a serious problem throughout the region. But while dam-building may reduce air pollution somewhat, it will not be adequate to supply China's vast energy needs. Most environmentalists therefore argue that the damage caused by the dams will be greater than the benefits they confer. Most industrialists and government planners disagree.

Equally important, argue Chinese planners, are the water-control benefits that the Three Gorges Dam will confer. The lower and middle stretches of the Yangtze periodically suffer from devastating flooding. Flooding has been exacerbated over the past several decades by the draining and filling of floodplain lakes and by deforestation in the river's headwaters. East of Sichuan, the Yangtze passes through several large hill-

and mountain-rimmed basins (in Hunan, Hubei, and Jiangxi provinces), each containing a group of lakes. During flood periods, water flows from the river into these **regulatory lakes,** reducing the flood crest downstream (they are called *regulatory* because they reduce the flow of water during floods). As the lakes have been gradually converted to farmland, flooding downstream has intensified.

Damming the Yangtze is thus considered necessary in part to compensate for environmental changes made by human activities elsewhere in the river's course. Experiences in other parts of the world, however, suggest that no amount of dam-building can prevent the largest, and therefore most destructive, floods. Certainly that has been the case in regard to the Huang He, China's second most important river.

Flooding in Northern China   The North China Plain, which has been deforested for thousands of years, has long been plagued by both drought and flood. This area is dry most of the year, yet often experiences heavy downpours in summer. Since ancient times, large-scale works in hydraulic engineering have both controlled floods and allowed irrigation. But no matter how much effort has been put into water control, disastrous flooding has never been completely prevented (Figure 11.4).

The worst floods in northern China are caused by the Huang He, or Yellow River, which cuts across the North China Plain. Owing to upstream erosion, the Huang He carries a huge **sediment load** (the amount of suspended clay, silt, and sand in the water), giving it the dubious distinction of being the muddiest major river in the world. When the river enters the low-lying plain, its velocity slows and its sediments begin to settle and accumulate in the riverbed. As a result, the level of the river gradually rises above that of the surrounding lands. Eventually it must break free of its course to find a new route to the sea over lower-lying ground. Twenty-six such course changes have been recorded in Chinese history.

Through the process of sediment deposition and periodic course changes, the Huang He has actually created the vast North China Plain. In prehistoric times the Yellow Sea ex-

▲ **Figure 11.4 Flooding on the North China Plain** Major flooding has been a historical problem with the Huang He River, sometimes inundating large sections of the North China Plain. At other times, severe droughts can plague the same region. Extensive dikes have been built along much of the river to protect the countryside from flooding, as seen in this photo taken near the historical city of Kaifeng. *(Yang Xiuyun/ChinaStock Photo Library)*

tended far inland. Even today the sea is retreating as the Huang He's delta expands; one study revealed a six-mile advance of the land in a three-year period. Such a process occurs in other alluvial plains, but nowhere else is it so pronounced.

Nowhere else, moreover, is it so destructive. The North China Plain has been densely populated for millennia and is now home to more than 300 million persons. Since ancient times, the Chinese have attempted to keep the river within its banks by building progressively larger dikes. Eventually, however, the riverbed rises so high that the flow can no longer be contained. When this occurs, catastrophic flooding results. Such floods have been known to kill several million persons in a single episode. For this reason, the Huang He has been aptly called "the river of China's sorrow." While the river has not changed its course since the 1930s, most geographers agree that another correction is inevitable.

Erosion on the Loess Plateau

The Huang He's sediment burden is derived from the eroding soils of the Loess Plateau, located to the west of the North China Plain. **Loess** is a fine, wind-blown material that was deposited on this upland area during the last Ice Age. The dust storms of the period must have been overwhelming, for in some places several hundred feet of loess accumulated.

Loess forms fertile soil, but it washes away easily when exposed to running water. At the dawn of Chinese civilization, the semiarid Loess Plateau was covered with tough grasses and scrubby forests that helped retain the soil. Chinese farmers, however, began to clear the land for the abundant crops that it yields when rainfall is adequate. Cultivation required plowing, which, by exposing the soil to runoff, exacerbated erosion. As the population of the region gradually increased,

the remaining areas of woodland diminished, leading to ever-greater rates of soil loss. As the erosion process continued, great gullies—some of them hundreds of feet deep—cut across the plateau, steadily reducing the extent of arable land.

Today the Loess Plateau is one of the poorest parts of China, and thus one of the poorer parts of the world. Population is only moderately dense by Chinese standards, but good farmland is limited and drought is common. The population continues to expand, moreover, even while the extent of arable land declines. The Chinese government encourages the construction of terraces to conserve the soil, but such efforts have not been effective everywhere. Campaigns to plant woody vegetation on the most severely eroded lands have been even less successful. Seedlings require careful attention if they are to survive the harsh climate, but local farmers do not always have the necessary equipment or knowledge to care for them, nor are they often provided with adequate incentives.

China's government has also attempted to reduce the flow of silt out of the Loess Plateau by building small check-dams on the region's streams. Most of these dams were soon rendered ineffective by rapid siltation. A more ambitious flood-control project entailed the construction of a major dam on the Huang He itself. Within a few years, however, the reservoir behind this Sanmenxia Dam had simply filled in with silt, destroying its flood-control capability. Partly because of these environmental disasters, large numbers of people have been migrating out of the Loess Plateau, seeking work in the burgeoning cities of coastal China.

Other East Asian Environmental Problems

Although the problems associated with the Yangtze and Huang He rivers are among the most spectacular environmental issues in East Asia, they are hardly the only ones. Pollution, deforestation, and the loss of wildlife plague much of the region.

Forests and Deforestation

Most the uplands of China and South Korea support only grass, meager scrub, and stunted trees (Figure 11.5). China lacks the historical tradition of forest conservation that, as we shall see, characterizes Japan. In earlier periods, hillsides were often cleared for fuel-wood, and in some instances entire forests were burned for ash that could be used as fertilizer in rice fields. In much of southern China, sweet potatoes, maize, and other crops have been grown on steep and easily eroded hillsides for several hundred years. After centuries of exploitation, many uplands have been so degraded that they cannot easily regenerate forests.

Although the Chinese government has initiated large-scale reforestation programs, few have been very successful. As it now stands, substantial forests are found only in China's far north, where a cool climate prevents fast growth, and along the eastern slopes of the Tibetan Plateau, where rugged terrain restricts commercial forestry. As a result, China suffers a severe shortage of forest resources. If its economy continues to boom, China will likely become a major importer of lumber, pulp, and paper in the near future. One must wonder which part of the world could supply China's potentially vast demand for forest products.

▲ **Figure 11.5 Denuded hillslopes in China**   Because of the need to clear forests for wood products and agricultural lands, China's mountain slopes have long been problem areas. Without forest cover, soil erosion is a serious issue.   *(Andrew Wong/ Reuters/Getty Images, Inc.)*

Mounting Pollution

As China's industrial base expands, other environmental problems, such as water pollution and toxic-waste dumping, are growing more acute, particularly in the booming coastal areas. The burning of high-sulfur coal has resulted in particularly acute air pollution, a problem exacerbated by the increasing number of automobiles being driven by the Chinese. Such problems emerged earlier in other portions of East Asia. But Taiwan and South Korea, which have large chemical, steel, and other heavy industries, responded by imposing more stringent environmental controls as they grew wealthier. These two countries are also following in Japan's footsteps by setting up new factories in poorer countries—including China—that have less exacting environmental standards. (Taiwan has even negotiated with North Korea, the poorest state in the region, for a nuclear-waste dumping site.) Such an option is not open to a country as poor as China. It thus remains to be seen whether China can successfully respond to the environmental degradation generated by its rapidly industrializing economy.

Environmental Issues in Japan

Considering its large population and intensive industrialization, Japan's environment is relatively clean. Actually, the very density of its population gives certain environmental advantages, allowing, for example, a very efficient public transportation system. In the 1950s and 1960s, Japan's most intensive period of industrial growth, the country did suffer from some of the world's worst water and air pollution, and several infamous toxic-waste disasters killed and maimed thousands of persons. Soon afterward the Japanese government passed stringent air and water pollution laws.

Japan's cleanup was aided by its insular location, since winds usually carry smog-forming chemicals out to sea. Equally important has been the phenomenon of **pollution exporting**. Because of Japan's high cost of production and its

strict environmental laws, many Japanese companies have relocated their dirtier factories to other countries, particularly those of Southeast Asia. In effect, Japan's pollution has been partially displaced to poorer countries. Of course, the same argument can also be made about the United States and western Europe, but to a somewhat lesser extent.

Endangered Species

Another environmental question of global dimensions has been linked to the economic rise of East Asia. This region is one of the main centers of trade in endangered species. Many forms of traditional Chinese medicine are based on products derived from rare and exotic animals. Deer antlers, bear gallbladders, snake blood, tiger penises, and rhinoceros horns are believed by many to have medical effectiveness, and certain individuals will pay fantastic sums of money to purchase them. As wealth has accumulated, trade in such substances has expanded. China itself has relatively little remaining wildlife, but virtually all other areas of the world help supply its demand for wildlife products. China has, however, made good efforts to protect some of its remaining areas of habitat. Among the most important of these are the high-altitude forests and bamboo thickets of western Sichuan province, home of the panda bear (Figure 11.6). Efforts are also being made to preserve wildlands in northern Manchuria, where evidence of a surviving population of Siberian tigers was discovered in 1997, and in the canyonlands of northwestern Yunnan, which were declared off-limits to loggers in 1999.

▲ **Figure 11.6 The endangered panda bear**   One of the more successful Chinese environmental solutions has protected high-altitude forests and bamboo thickets in Sichuan province as panda bear habitat. Efforts are also being made elsewhere in China to protect endangered species and habitats.   *(Keren Su/Pacific Stock)*

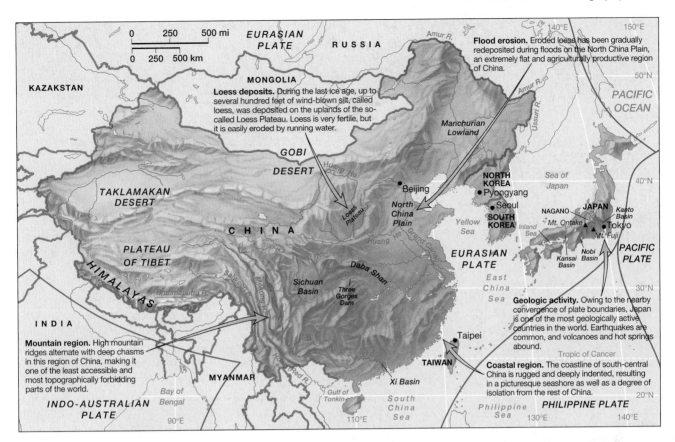

▲ **Figure 11.7 Physical geography of East Asia**   The physical geography of mainland East Asia varies widely from the high plateaus and deserts basins of western China to the broad river valleys and vast plains of eastern China. In the island region, landscapes are shaped by the convergence of three major tectonic plates: the Eurasian, Philippine, and Pacific.

East Asia's Physical Geography

To understand more fully the environmental problems being faced in East Asia, it is necessary to examine the region's physical geography, including its climates and landforms (Figure 11.7), in greater detail.

East Asia is situated in the same general latitudinal range as the United States, although it extends considerably farther north and south. The northernmost tip of China lies as far north as central Quebec, while China's southernmost point is at the same latitude as Mexico City. The climate of southern China is thus roughly comparable to southern Florida and the Caribbean, while that of northern China is similar to south-central Canada (Figure 11.8).

The insular belt of East Asia, extending from northern Japan through Taiwan, is situated at the intersection of three **tectonic plates** (the basic building blocks of Earth's crust): the Eurasian, the Pacific, and the Philippine. This area is therefore geologically active, experiencing numerous earthquakes and dotted with volcanoes (Figure 11.9). The mainland is more geologically stable than the islands, although many parts of China are earthquake-prone. East Asia is not a particularly mineral-rich region. (Although the far western region of China has a number of important mineral deposits, this region is discussed in Chapter 10, Central Asia, rather than as part of East Asia proper [see "Setting the Boundaries"].) East Asia's

modest petroleum reserves are concentrated in northern and northeastern China, and perhaps offshore in the South China Sea. China does have vast deposits of coal.

Japan's Physical Environment   Although slightly smaller than California, Japan is more elongated, extending farther north and south. As a result, Japan's extreme south, in southern Kyushu and the Ryukyu Archipelago, is subtropical, while northern Hokkaido is almost subarctic. Most of the country, however, is distinctly temperate. The climate of Tokyo is not unlike that of Washington, D.C.—although Tokyo is distinctly rainier.

Japan's climate varies not only from north to south, but also from southeast to northwest across the main axis of the archipelago. In the winter, the area facing the Sea of Japan is cloudier and receives much more snow than the Pacific Ocean coastline. During this time of the year, cold winds from the Asian mainland blow across the relatively warm waters of the Sea of Japan. The air picks up moisture over the sea and deposits it, usually as snow, when it hits land (Figure 11.10). This often cloud-shrouded northwestern coastline, sometimes called "the dark side of Japan," is well noted in Japanese literature. The Pacific coast of Japan, on the other hand, is far more vulnerable to the typhoons (hurricanes) that frequently strike the country.

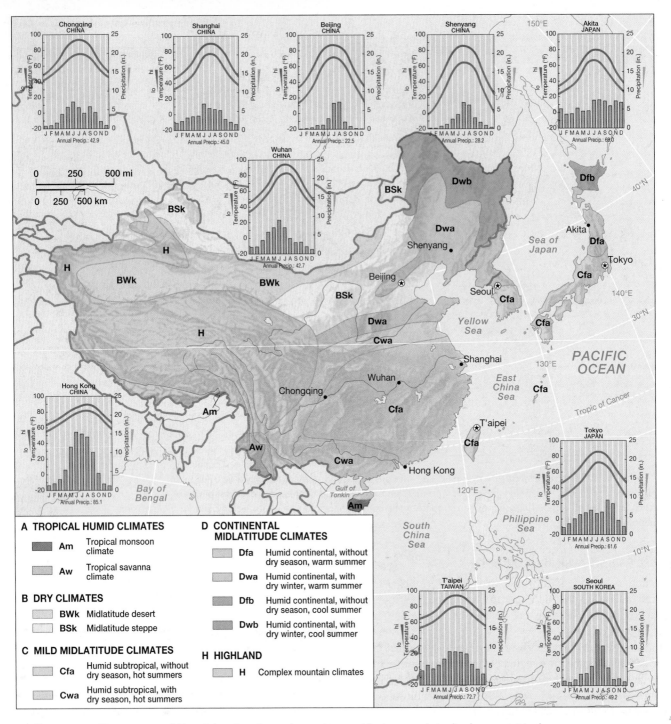

▲ **Figure 11.8 Climate map of East Asia** East Asia is located in roughly the same latitudinal zone as North America, so there are climatic parallels between the two world regions. The northernmost tip of China lies at about the same latitude as Quebec and shares a similar climate, whereas southern China approximates the climate of Florida. In Japan, maritime influences produce a milder climate.

The Pacific coast of Japan is separated from the Sea of Japan coast by a series of mountain ranges. Japan is one of the world's most rugged countries, with mountainous terrain covering some 85 percent of its territory. As Figure 11.11 shows, most of these uplands are thickly wooded. By some measures, Japan is the world's second most heavily forested industrialized country (after Finland). Japan owes its lush forests both to its mild, rainy climate and to its long history of forest conservation. For hundreds of years, both the Japanese state and

its village communities have enforced strict conservation rules, ensuring that timber and firewood extraction would be balanced by tree growth.

Along Japan's coastline and interspersed among its mountains are limited areas of alluvial plains (Figure 11.9). In prehistoric times, these lowlands were covered by forests and wetlands. They have long since been cleared and drained for intensive agriculture. The largest Japanese lowland is the Kanto Plain to the north of Tokyo, but even it is only some 80 miles

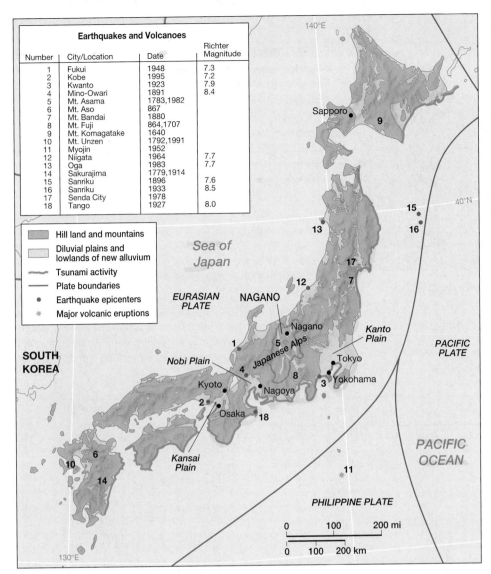

Figure 11.9 Japan's physical geography Japan has several sizable lowland plains, primarily along the coastline, but they are interspersed among rugged mountains and uplands. Because of its location at the convergence of three major tectonic plates, Japan experiences numerous earthquakes. Volcanic eruptions can also be hazardous and are linked directly to Japan's location on tectonic plate boundaries. Additionally, much of Japan's coast is vulnerable to devastating tsunamis (tidal waves) caused by earthquakes in the Pacific basin.

Earthquakes and Volcanoes

Number	City/Location	Date	Richter Magnitude
1	Fukui	1948	7.3
2	Kobe	1995	7.2
3	Kwanto	1923	7.9
4	Mino-Owari	1891	8.4
5	Mt. Asama	1783,1982	
6	Mt. Aso	867	
7	Mt. Bandai	1880	
8	Mt. Fuji	864,1707	
9	Mt. Komagatake	1640	
10	Mt. Unzen	1792,1991	
11	Myojin	1952	
12	Niigata	1964	7.7
13	Oga	1983	7.7
14	Sakurajima	1779,1914	
15	Sanriku	1896	7.6
16	Sanriku	1933	8.5
17	Senda City	1978	
18	Tango	1927	8.0

Legend:
- Hill land and mountains
- Diluvial plains and lowlands of new alluvium
- Tsunami activity
- Plate boundaries
- • Earthquake epicenters
- * Major volcanic eruptions

wide and 100 miles long (130 by 160 kilometers). The country's other main lowland basins are the Kansai, located around Osaka, and the Nobi, centered on Nagoya. In mountainous Nagano prefecture, smaller basins are sandwiched between the imposing peaks of the Japanese Alps.

Taiwan's Environment Taiwan, an island about the size of Maryland, sits at the edge of the continental landmass. To the west, the Taiwan Strait is only about 200 feet (60 meters) deep; to the east, ocean depths of many thousands of feet are found 10 to 20 miles (16 to 32 kilometers) offshore.

Taiwan itself forms a large tilted block. Its central and eastern regions are rugged and mountainous, while the west is dominated by an alluvial plain. Bisected by the Tropic of Cancer, Taiwan has a mild winter climate, but it is not infrequently battered by typhoons in the early autumn. Unlike nearby areas of China proper, Taiwan still has extensive forests. These are concentrated in its remote central and eastern upland areas.

Chinese Environments China, even if one excludes its Central Asian provinces of Tibet and Xinjiang, is a vast country with diverse environmental regions. For the sake of convenience, it

▲ Figure 11.10 Heavy snow in Japan's mountains Cold, moist air moving off the Sea of Japan produces heavy snows in northwestern Japan and along the country's mountainous spine. Numerous major ski areas dot the Japanese Alps, several of which have hosted world-class sports competitions. *(George Mobley/NGS Image Collection)*

▲ **Figure 11.11 Forested landscapes of Japan** Although much of Japan is heavily forested and supports a viable wood products industry, the yield is not large enough to satisfy demand. As a result, Japan imports timber extensively from North America, Southeast Asia, and Latin America. *(Robert Holmes/Corbis)*

can be divided into two main areas, one lying to the north of the Yangtze River Valley, the other including the Yangtze and all areas to the south. As Figure 11.12 shows, each of these can be subdivided into a number of distinctive regions.

Southern China is a land of rugged mountains and hills interspersed with lowland basins. The lowlands of southern China are far larger than those of Japan. One of the most distinctive is the former lake bed of central Sichuan, often called the Red Basin (in the central Upper Yangtze region of Figure 11.12). Protected by imposing mountains, it has a unique climate. Passing south over the Daba Mountains in the winter, a traveler might experience a major surprise; to the north, conditions are generally frigid and dry, whereas to the south, in the Red Basin, the land is green and wet. Eastward from Sichuan, the Yangtze River passes through several other broad basins (in the Middle Yangtze region on Figure 11.12), partially separated from each other by hills and low mountains, before flowing into a large delta in the vicinity of Shanghai.

South of the Yangtze Valley, the mountains are higher (up to 7,000 feet, or 2,150 meters, in elevation), but they are still

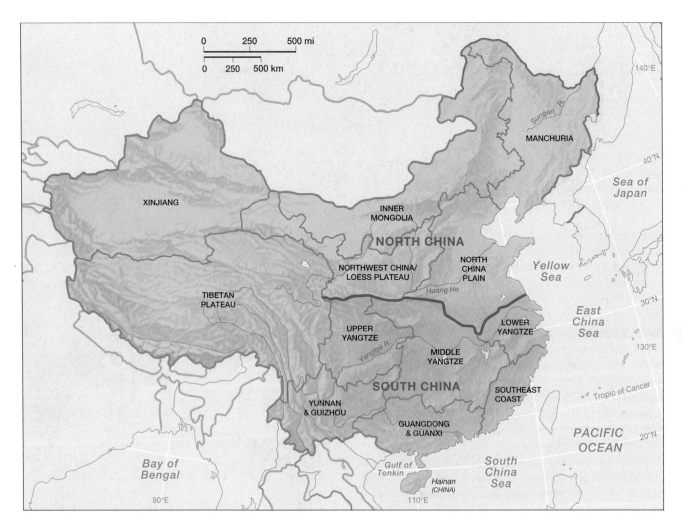

▲ **Figure 11.12 Landscape regions of China** The Yangtze Valley divides China into two general areas. Immediately to the north is the large fertile area of the North China Plain, bisected by the Huang He (or Yellow) River. To the west is the Loess Plateau, an upland area of soil derived from wind-deposited silt after the prehistoric glacial period, about 15,000 years ago.

▲ **Figure 11.13 The Fujian coast of China**   In southeast China lies the rugged coastal province of Fujian. Here the coastal plain is narrow and the shoreline deeply indented, producing a picturesque landscape. Because of limited agricultural opportunities along this rugged coastline, many Fujianese people work in maritime activities.   *(Dean Conger/Corbis)*

interspersed with alluvial lowlands. Larger valleys are found in the far south, the Xi Basin in Guangdong province being the largest and most densely populated. Here the climate is truly tropical, free of frost. West of Guangdong lie the moderate-elevation plateaus of Yunnan and Guizhou, the former noted for its perennial springlike weather. Finally, to the northeast of Guangdong lies the rugged coastal province of Fujian. Fujian's coastal plain is narrow and its coastline deeply indented, features that have long encouraged the Fujianese people to seek a maritime way of life (Figure 11.13).

North of the Yangtze Valley, the climate is both colder and drier than it is to the south. Summer rainfall is generally abundant except along the edge of the Gobi Desert (Figure 11.7), but the other seasons are often dry. Immediately north of the lower course of the Yangtze lies the North China Plain, a large area of fertile soil crossed by the Huang He (or Yellow) River. With the exception of a few low mountains in Shandong province, the entire area is a virtually flat plain. The North China plain is cold and dry in winter and hot and humid in the summer. Overall precipitation is not high, however, and some areas are in danger of **desertification** (or the spread of desert conditions). Seasonal water shortages are growing increasingly severe through much of the region as withdrawals for irrigation and industry increase, leading to some concern about an impending water crisis.

West of the North China Plain sits the Loess Plateau. This is a fairly rough upland of moderate elevation and uncertain precipitation. It does, however, have fertile soil—as well as huge coal deposits. West of the Loess region lie the semiarid plains and uplands of Gansu province, situated at the foot of the great Tibetan Plateau.

China's far northeastern region is called *Dongbei* in Chinese and *Manchuria* in English. Manchuria is dominated by

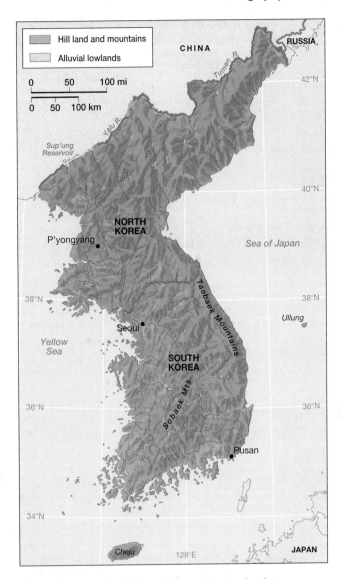

▲ **Figure 11.14 The mountains and lowlands of Korea** Like Japan, Korea has extensive uplands interspersed with lowland plains. The country's highest mountains are in the north, whereas its most extensive alluvial plains are in the south.

a broad, fertile lowland sandwiched between mountains and uplands stretching along China's borders with North Korea, Russia, and Mongolia. Although winters here can be brutal, summers are usually warm and moist. Manchuria's peripheral uplands have some of China's best-preserved forests and wildlife refuges. In the extreme northeast, extensive wetlands can still be found in the valleys of the Ussuri and Amur rivers.

Korean Landscapes   Korea forms a well-demarcated peninsula, partially cut off from Manchuria by rugged mountains and sizable rivers (Figure 11.14). Like Japan, its latitudinal range is pronounced. The far north, which just touches Russia's Far East, has a climate not unlike that of Maine, whereas the southern tip is more reminiscent of the Carolinas. Korea, again like Japan, is a mountainous country with scattered alluvial basins. The lowlands of the southern portion of the peninsula

TABLE 11.1 *Demographic Indicators*

Country	Population (Millions, 2001)	Population Density, per square mile	Rate of Natural Increase	TFR[a]	Percent < 15[b]	Percent > 65	Percent Urban
China	1,273.3	344	0.9	1.8	23	7	36
Hong Kong[c]	6.9	16,743	0.3	1.0	17	11	100
Japan	127.1	872	0.2	1.3	15	17	78
South Korea	48.8	1,274	0.9	1.5	22	7	79
North Korea	22.2	472	1.5	2.3	26	6	59
Taiwan	22.5	1,608	0.8	1.7	21	9	77

[a]Total fertility rate

[b]Percent of population younger than 15 years of age

[c]Although still a part of China, Hong Kong's statistics are often compiled separately.

Source: Population Reference Bureau. World Population Data Sheet, 2001.

are more extensive than those of the north, giving South Korea a distinct agricultural advantage over North Korea. The north, however, has much more abundant mineral deposits, as well as forest and hydroelectric resources. Much of the upland area of South Korea has long suffered from extensive deforestation.

Population and Settlement: A Realm of Crowded Lowland Basins

East Asia, along with South Asia, is the most densely populated region of the world. The lowlands of Japan, Korea, and China are among the most intensely used portions of Earth, containing not only the major cities, but also most of the agricultural lands of these countries.

Although the density of East Asia is extremely high, the region's population growth rate has declined dramatically since the 1970s. In Japan, the current concern is impending population loss, and correspondingly an aging populace that will need to be supported by a shrinking cohort of younger workers. While China's population is still expanding, its rate of growth is by global standards relatively low (Table 11.1).

Japanese Settlement and Agricultural Patterns

Japan is a highly urbanized country, supporting two of the largest urban agglomerations in the world. Yet it is also one of the world's most mountainous countries, and its uplands are lightly inhabited. Agriculture must therefore share the limited lowlands with cities and suburbs, resulting in extremely intensive farming practices.

Japan's Agriculture Lands
Japanese agriculture is largely limited to the country's coastal plains and interior basins. Rice is Japan's major crop, and irrigated rice demands flat land. Japanese rice farming has long been one of the most productive forms of agriculture in the world, helping support a large population—some 127 million persons—on a relatively small and rugged land. Although rice is grown in almost all Japan-

ese lowlands, the country's premier rice-growing districts lie along the Sea of Japan coast of central and northern Honshu.

Vegetables are also grown intensively in all of the lowland basins, even on tiny patches within urban neighborhoods (Figure 11.15). The valleys of central and northern Honshu are famous for their temperate-climate fruit, while citrus comes from the milder southwestern reaches of the country. Crops that thrive in a cooler climate, such as potatoes, are produced mainly in Hokkaido and northern Honshu, as are dairy products.

Settlement Patterns
All Japanese cities—and the vast majority of the Japanese people—are located in the same lowlands that support the country's agriculture. Not surprisingly, the three largest metropolitan areas—Tokyo, Osaka, and Nagoya—sit near the centers of the three largest plains.

▲ **Figure 11.15 Japanese urban farm** Japanese landscapes often combine dense urban settlement with small patches of intensively farmed agriculture. Here a cabbage farm coexists with an urban neighborhood. *(Kyodo News International, Inc.)*

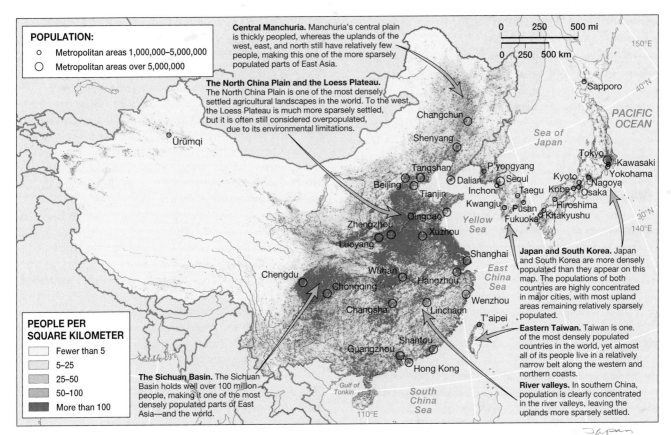

▲ **Figure 11.16 Population map of East Asia** Parts of East Asia are extraordinarily densely settled, particularly in the coastal lowlands of China and Japan. This contrasts with the sparsely settled lands of western China, North Korea, and northern Japan. Although the total population of this world region is high, as is the overall density, the rate of natural population increase has slowed rather dramatically in the last several decades because of several factors, primarily China's well-known "one-child" policy.

[handwritten margin notes: Japan · Highly Urbanized · Types of crops · Location of cities]

The overall population density of Japan is high (Figure 11.16). Japan as a whole contains some 870 persons per square mile (2,137 per square kilometer), a figure substantially higher than the 72 persons per square mile found in the United States. The fact that its settlements are largely restricted to roughly 15 percent of the country's land area means that Japan's effective population density—the actual crowding that the country experiences—is one of the highest in the world. This is especially true in the main industrial belt, which extends from Tokyo through Nagoya and Osaka and hence along the Inland Sea (the maritime region sandwiched between Shikoku, western Honshu, and Kyushu) to the northern coast of Kyushu. The lowland portion of this zone is perhaps the most intensively used part of the world for human habitation, industry, and agriculture.

Japan's Urban-Agricultural Dilemma Due to such space limitations, all Japanese cities are characterized by dense settlement patterns. In the major urban areas, the amount of available living space is highly restricted for all but the most affluent families. In Tokyo, an apartment the size of a large U.S. living room often shelters an entire middle-class family. Many observers—especially American ones—argue that Japan should allow its cities and suburbs to expand into nearby rural areas. However, since most uplands are too steep for residential use, such expansion would have to come at the expense of agricultural land.

As it now stands, the wholesale conversion of farmland to neighborhoods would be difficult. Although most farms are extremely small (the average is only several acres) and only marginally viable, farmers are politically powerful, and croplands are often protected by the tax code. Moreover, most Japanese citizens believe that it is vitally important for their country to remain self-sufficient at least in rice. Rice imports are therefore all but banned. Memories of wartime shortages linger, and many people consider rice cultivation to be an essential cultural or even spiritual aspect of the Japanese nation.

U.S. economic interests, along with certain Japanese consumer advocates, stress the fact that the present system forces Japanese consumers to pay up to six times the world market price for rice, their staple food. They also point out that Japan relies on imports for many of its other food needs as well as for the energy that is required to grow rice. But whatever arguments are made, change will probably come gradually, if at all. For the moment, Japan must live with the tensions resulting from the fact that its affluent population is forced to live in tight proximity and to pay high prices for basic staples.

Settlement and Agricultural Patterns in China, Taiwan, and Korea

Like Japan, Taiwan and Korea are essentially urban. China, however, remains largely rural, with only some 36 percent of its population living in cities. Chinese cities are rather evenly distributed across the plains and valleys of China proper. As a result, the overall pattern of population distribution in China closely follows the geography of agricultural productivity. Almost all Chinese cities are growing at a rapid pace, despite governmental efforts to keep them under control.

China's Agricultural Regions

A line drawn just to the north of the Yangtze Valley divides China into two main agricultural regions. To the south, rice is the dominant crop. To the north, wheat, millet, and sorghum prevail.

In southern and central China, population is highly concentrated in the broad lowlands, which are famous for their fertile soil and intensive agriculture. More than 100 million persons live in the Sichuan Basin, while more than 70 million reside in deltaic Jiangsu, a province smaller than Ohio. (With 11 million people, Ohio is one of the most densely populated U.S. states.) Cropping occurs year-round in most of southern and central China; summer rice alternates with winter barley or vegetables in the north, while two rice crops can be harvested in the far south. In some of the most fertile and densely populated areas, as many as three crops can be produced in a single year. Southern China also produces a wide variety of tropical and subtropical crops, and moderate slopes throughout the area supply sweet potatoes, corn, and other upland crops.

The North China Plain is one of the most thoroughly **anthropogenic landscapes** in the world (an anthropogenic landscape is one that has been heavily transformed by human activities). Virtually its entire extent is either cultivated or occupied by houses, factories, and other structures of human society. Too dry in most areas for rice, the North China Plain specializes in wheat, millet, and sorghum.

Manchuria was a lightly populated frontier zone as recently as the mid-1800s. Today, with a population of more than 100 million, its central plain is thoroughly settled. Still, Manchuria remains less crowded than many other parts of China, and it is one of the few parts of China to produce a consistent food surplus.

The Loess Plateau is more thinly settled yet, supporting only some 70 million inhabitants. But considering its aridity and widespread soil erosion, this is a high figure indeed. Like other portions of northern China, the Loess Plateau produces wheat and millet, largely for subsistence purposes. One of its unique settlement features is the prevalence of subterranean housing. Although loess erodes quickly under the impact of running water, it coheres well in other circumstances. For millennia, villagers have excavated pits on the surface of the plateau, from which they have tunneled into the earth to form underground houses (Figure 11.17). These subterranean dwellings are cool in the summer and warm in the winter. Unfortunately, they tend to collapse during earthquakes, leading to extremely high mortality rates. One quake in 1920 killed an estimated 100,000 persons, while another in 1932 killed some 70,000.

Settlement and Agricultural Patterns in Korea and Taiwan

Like China and Japan, Korea is a densely populated country. It contains some 70 million persons (22 million in the north and 48 million in the south) in an area smaller than Minnesota. South Korea's population density is 1,150 per square mile (2,980 per square kilometer), a significantly higher figure than that of Japan. Most of these people are crowded into the alluvial plains and basins of the west and south. The highland spine, extending from the far north to northeastern South Korea, remains relatively sparsely settled. South Korean agriculture is dominated by rice. North Korea, in contrast, relies heavily on corn and other upland crops that do not require irrigation.

Taiwan is the most densely populated state in East Asia. Roughly the size of the Netherlands, it contains more than 22 million inhabitants (7 million more than the Dutch homeland, one of the most crowded countries of Europe). Its overall population density is some 1,500 persons per square mile (3,885 per square kilometer), one of the highest figures in the world. Since mountains cover most of central and eastern Taiwan, virtually the entire population is concentrated in the narrow lowland belt in the north and west. Large cities and numerous factories are scattered here amid lush farmlands.

East Asian Agriculture and Resource Procurement in Global Context

Although East Asian agriculture is highly productive, it is not productive enough to feed the huge number of people who live in the region. Japan, Taiwan, and South Korea are major food importers, and China is moving in the same direction. Other resources are also being drawn in from all quarters of the world by the powerful economies of East Asia.

The Global Dimensions of Japanese Agriculture and Forestry

Japan may be self-sufficient in rice, but it is still one of the world's largest food importers. As the Japanese have

▲ **Figure 11.17 Loess settlement** A typical subterranean dwelling carved out the soft loess sediment in central China. Approximately 70 million people live similarly in the Loess Plateau region. Unfortunately, this region is also prone to major earthquakes that take a high toll on the local population because of dwelling collapse. *(Christopher Liu/ChinaStock Photo Library)*

grown more prosperous over the past 50 years, their diet has grown more diverse. That diversity is made possible largely by importing food from elsewhere.

Japan imports food from a wide array of other countries. It procures both meat and the feed used in its domestic livestock industry from the United States, Canada, and Australia. These same countries supply wheat needed to produce bread and noodles. Even soybeans, long a staple of the Japanese diet, must be purchased from Brazil and the United States. Japan has one of the highest rates of fish consumption in the world, and the Japanese fishing fleet scours the world's oceans to supply the demand. Japan also purchases prawns and other seafood, much of it farm-raised in former mangrove swamps, from Southeast Asia and Latin America.

Japan depends on imports to supply its demand for forest resources. While its own forests produce high-quality cedar and cypress logs, it obtains most of its construction lumber and pulp (for papermaking) from western North America and Southeast Asia. Almost half of Southeast Asia's forest-product exports are shipped to Japan. As the rain forests of Malaysia, Indonesia, and the Philippines diminish, Japanese interests are beginning to turn to Latin America and Africa as sources of tropical hardwoods. Japanese and South Korean firms are also looking to Siberia (eastern Russia), a nearby and previously little-exploited forest zone.

Japan is able to support its large and prosperous population on such a restricted land base because it can purchase resources from abroad. Certainly all countries engage in trade, but Japan's basic resource dependence is particularly pronounced. Almost all of the oil, coal, and other minerals that it consumes are imported. If forced to rely on its own resources, Japan would find itself in an economic and ecological bind. But because its industries are successful in the global market, Japan has avoided such a predicament.

Although geographers once looked at resource endowments as a main support of each country's economic performance, such a view is no longer tenable in the increasingly interconnected global economy. As long as a country can export items of value, it can obtain whatever imports it requires. This also means, however, that the environmental degradation generated by a successful economy like that of Japan is no longer limited to its home territory. It, too, has been globalized.

The Global Dimensions of Chinese Agriculture

Through the 1980s, China was essentially self-sufficient in food, despite its huge population and crowded lands. But rapid economic growth, combined with changing diets, has brought about an increased consumption of meat, which requires large amounts of feed-grain. It has also resulted in the loss of agricultural lands to residential and industrial development. In the mid-1990s, China had to import large amounts of grain for several years.

Optimists argue that China could produce much more food than it now does by increasing its use of fertilizers and by converting its "wastelands" of grass and scrub into agricultural fields. Chinese leaders are well aware of the current grain shortage, but they promise that agricultural production will soon increase. Many concede, however, that by the year 2030 China will have to import 5 to 10 percent of its total grain demand.

Korean Agriculture in a Global Context   South Korea has already made the transition, as Japan and Taiwan have, to a global resource procurement pattern. In the mid-1990s, South Korea was the world's fifth leading importer of wheat (after China, Egypt, Japan, and Brazil) and the second leading importer of corn (after Japan). North Korea, on the other hand, has pursued a goal of strict self-sufficiency. While relatively successful for a number of years, in the late 1990s this policy resulted in widespread famine after a series of floods, followed by drought, destroyed most of the country's rice and corn crops. Undernutrition remains rampant in North Korea.

Urbanization in East Asia

China has one of the world's oldest urban foundations, dating back more than 3,500 years. In medieval and early modern times, East Asia as a whole possessed a well-developed system of cities that included some of the largest settlements on the planet. In the early 1700s, Tokyo, then called Edo, probably overshadowed all other cities, with a population of more than 1 million.

But despite this early start, East Asia was overwhelmingly rural at the end of World War II. Some 90 percent of China's people then lived in the countryside, and even Japan was only about 50 percent urbanized. China had failed to urbanize largely because its economy had failed to develop. But as the region's economy began to grow after the war, so did its cities. Japan, Taiwan, and South Korea are now between 70 and 80 percent urban, which is typical for advanced industrial countries. Some 64 percent of China's people still live in rural areas, but this figure is decreasing steadily. And because China's population is immense, its urban foundation is still sizable. Twelve of the world's 100 largest cities are in China, 2 of which, Beijing and Shanghai, rank among the top 15.

Chinese Cities   Traditional Chinese cities were clearly demarcated from the countryside by defensive walls. Most were planned in accordance with strict geometrical principles that were thought to reflect the cosmic order. The old-style Chinese city was horizontal, dominated by low buildings and characterized by straight streets. Houses were typically built around courtyards, and narrow alleyways served both commercial and residential functions.

China's urban fabric began to change during the colonial period. A group of port cities was taken over by European interests, which proceeded to build Western-style buildings and modern business districts. By far the most important of these semicolonial cities was Shanghai, built near the mouth of the Yangtze River, the main gateway to interior China.

When the communists came to power in 1949, Shanghai, with a population of more than 10 million persons, was the second-largest city in the world. The new authorities viewed it as a decadent foreign creation. They therefore milked it for taxes, which they invested elsewhere. Most of the old civic elite fled to Hong Kong, and relatively few migrants were allowed in. As a result, much of the city, and especially the old business district, began to decay.

Since the late 1980s, Shanghai has experienced a major revival and is again in many respects China's premier city. Migrants are now pouring into Shanghai, even through the state still tries to restrict the flow, and building cranes crowd the skyline. Official statistics now put the population of the metropolitan area at some 14 million, but the actual number, including temporary migrants, may well be significantly larger. The new Shanghai is a city of massive high-rise apartments and concentrated industrial developments (Figure 11.18). Shanghai also sits at the core of an old, and now revitalizing, network of urban places situated around the greater Yangtze Delta. But despite Shanghai's revived economic fortunes, the city remains politically secondary to Beijing, China's capital.

Beijing was China's capital during the Manchu period (1644–1912), a status it regained in 1949. Under communist rule, Beijing was radically transformed; old buildings were razed and broad avenues were plowed through old neighborhoods (Figure 11.19). Crowded residential districts gave way to large blocks of apartment buildings and massive government offices. Some historically significant buildings were saved; the buildings of the Forbidden City, for example, where the Manchu rulers once lived, survived as a complex of museums. The area immediately in front of the Palace Museum, however, was cleared. The resulting plaza, Tienanmen Square, is reputed to be the largest open square in any city of the world. It makes a convenient display area for public spectacles and government-sponsored political rallies. Crit-

▲ Figure 11.18 Contemporary Shanghai   This vibrant city of more than 14 million embodies the new China with its massive high-rise apartments, industrial developments, and office towers. This photo shows the Nan Pu bridge and high-rise buildings in the newly developed Pu Dong area.   *(Keren Su/Pacific Stock)*

ics regard Beijing, a city of some 13 million persons, as a place designed primarily to express state power.

China's urban system as a whole is fairly well balanced, with sizable cities relatively evenly spaced across the landscape and with no city overshadowing all others. This balance stems from China's indigenous heritage of urbanism, its vast size and division into separate physical-geographical regions, and its legacy of socialist planning. In the early 1960s,

▶ Figure 11.19 Beijing, China   The historic capital during the Manchu period (1644–1912), Beijing regained its status as capital city in 1949. Under communist rule, much of Beijing's historical landscape was razed and replaced with large blocks of government offices and massive apartment buildings. Only the historically significant Forbidden City, home to the Manchu rulers, was saved from this transformation of the urban landscape. Tienanmen Square, reputed to be the largest open square in any city, was created by clearing buildings away from the area in front of the Palace Museum.

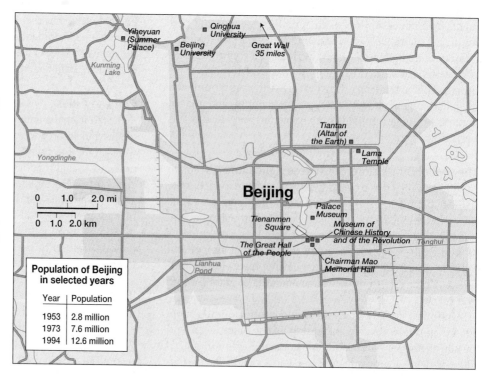

Population of Beijing in selected years	
Year	Population
1953	2.8 million
1973	7.6 million
1994	12.6 million

LOCAL VOICES   China's Rural Poor and the Lure of the City

One of the largest migrations in China's history is under way as at least 100 million migrant laborers roam the country looking for work; most are rural poor who have left their farms and villages to seek a better life in the city. The story of Sanzi, one family's 16-year-old third son who just graduated from junior high school, is typical.

My parents work a tiny plot of land that cannot support the costs of sending me to high school, so if I stay here in Qingdong I'm just another mouth to feed. There is no future other than working in the fields. Most of my friends have already left for the city; I feel like I'm choking to death in this village. I'm the only one in the family who has not tried the city.

My father went to Shanghai to work in a lumber business, but now he's back in the village working in the fields and the rest of us have to help pay off the debt from his failed attempt. My mother and two older brothers then went to Shanghai to work in a hotel kitchen but that lasted only a year because of mother's poor health.

One of the older brother's bosses has sent word to the village that there might be some sort of job for Sanzi in Guangzou, at the mouth of the Pearl River. The boss promised to pick up Sanzi at the train station, instructing him: "Wear your brother's white windbreaker and wait at the main exit."

The government has concentrated economic reform efforts in Guangzou, and every day thousands of peasants pour into the city hoping to cash in on the action. Greeting them at the train station, however, are an equal number of peasants on their way back to the countryside, having tired of—or failed at—their stint in the city.

Sanzi stumbles off of the train after a 26-hour ride in the crowded fourth-class car. He puts on the white windbreaker and waits. "I sat down in the shade to look at my new home. It was unbelievable. Blaring taxis, women in spiked heels and miniskirts, men with cell phones; all of it was so new."

After Sanzi endures almost two hours of anxious waiting, the boss pulls up in an early-1990s Cadillac. He motions for Sanzi, who grabs his bag and jumps in the car—the first time he's been in an automobile. He asks about the job, but the boss, seeing Sanzi's youth and slight build, is noncommittal. All he says is that it will be with a construction crew working for a friend of his. "Can you handle hard labor?" asks the boss.

"Of course I can. I grew up in the village doing hard labor, working all day in the fields," Sanzi answers. The boss closes the door and the Caddy joins the stream of traffic heading into the city. Another of China's rural poor begins a new life in the city.

Source: Adapted from "Desperate Journeys: China's Rural Poor Take Flight" by Tim Kao, *San Francisco Chronicle*, February 2, 1999.

noted China scholar G. William Skinner argued that the distribution of Chinese cities is best explained through the use of **central place theory.** Central place theory, developed by German geographer Walter Christaller in the 1930s, holds that an evenly distributed rural population will give rise to a regular hierarchy of urban places, with uniformly spaced larger cities surrounded by constellations of smaller cities, each of which, in turn, will be surrounded by smaller towns. Such a regular pattern, central place theorists contend, is generated by retail marketing; every family must have ready access to a nearby town to procure basic necessities, but larger urban areas, where more expensive items are obtainable, can be situated at a more distant location. Critics of the theory, however, contend that the even distribution of cities and towns found in much of China and many other parts of the world stems more from political administration and other kinds of economic activities than from retail marketing.

In the 1990s Beijing and Shanghai vied for the first position among Chinese cities, with Tianjin, serving as Beijing's port, coming in a close third. All three of these cities have long been removed from the regular provincial structure of the country and granted their own metropolitan governments. In 1997 Chongqing in Sichuan was added to this list of province-level municipalities. In the same year another major city, Hong Kong, passed from British to Chinese control and was granted a unique status as an autonomous "special administrative region." While not as populous as Beijing or Shanghai, Hong Kong is far wealthier. The emerging greater metropolitan area of the Xi Delta, composed of Hong Kong, Shenzhen, and Guangzhou (called *Canton* in the West), may now perhaps be regarded as China's premier urban area (see "Local Voices: China's Rural Poor and the Lure of the City").

Below these four primary cities are several dozen other large and growing urban centers. Some 32 of them had more than a million inhabitants each in the late 1990s, and in the near future many more will pass that milestone. It is still questionable, however, whether China will become a predominantly urban society. Industrial development is now moving into rural areas, as factories sprout in a seemingly haphazard fashion amid rice fields and vegetable plots.

City Systems of Japan, Taiwan, and South Korea

The urban structures of East Asia's capitalist fringe are quite different from those of China. Both Taiwan and South Korea are noted for their pronounced **urban primacy** (the concentration of urban population in a single city), whereas Japan is the center of a new urban phenomenon, that of the **superconurbation** (a superconurbation, or megalopolis, is a huge zone of coalesced metropolitan areas).

Seoul, the capital of South Korea, overwhelms all other cities in the country. Seoul itself is home to more than 10 million persons, and its greater metropolitan area contains some 40 percent of South Korea's total population. All of South Korea's major governmental, economic, and cultural institutions are concentrated there. Seoul's explosive and generally unplanned growth has resulted in serious congestion. The

South Korean government is promoting industrial growth in other cities, but none of its actions has yet challenged the primacy of the capital.

Taiwan is similarly characterized by a high degree of urban primacy. The capital city of Taipei, located in the far north, mushroomed from some 300,000 people during the Japanese colonial period (from 1895 to 1945) to some 6.2 million (in the metropolitan area) in the late 1990s.

Japan has traditionally been characterized by urban "bipolarity" rather than urban primacy. Until the 1960s, Tokyo, the capital and main business and educational center, together with the neighboring port of Yokohama, was balanced by the mercantile center of Osaka and its port of Kobe. Kyoto, the former imperial capital and the traditional center of elite culture, is also situated in the Osaka region. A host of secondary and tertiary cities serve to balance Japan's urban structure. Nagoya, with a metropolitan area of 4.8 million persons, remains the center of the automobile industry and is one of the few large Japanese cities in which it is more efficient to travel by car than by public transportation. Many other sizable and tightly packed cities dot the basins and coastal lowlands, most of which are little known outside of the country. For example, Kitakyushu, in far western Japan, is home to 1.5 million persons and is a center of the Japanese chemical industry.

As Japan's economy boomed in 1960s, 1970s, and 1980s, so did Tokyo. The capital city then outpaced all other urban areas in almost every urban function. The Greater Tokyo metropolitan area contains 25 to 28 million persons, depending on how it is defined. The Osaka–Kobe metropolitan area stands at a distant second, with 10 to 14 million inhabitants. Concerned about the increasing primacy of Tokyo, the Japanese government has been steering new developments to other parts of the country. Such efforts, however, have been only moderately successful. Most firms and government agencies prefer to be near the center of power and wealth. The newly built "science city" of Tsukuba, for example, has failed to attract much private investment. Tsukuba often seems almost deserted on weekends, when many of its residents seek the stimulation of Tokyo.

But Japan's other cities have not withered away to support Tokyo's growth. Most of them have expanded as well, just not at so rapid a pace. In general, urban growth has been supported by rural depopulation. Metropolitan expansion has been particularly pronounced in the cities linking Tokyo to Osaka, an area known as the Tokkaido corridor. Transportation connections are superb along this route, and proximity to Tokyo, Osaka, and Nagoya encourages development. As can be seen in Figure 11.20, the result has been the creation of a superconurbation (explained above). One can travel from

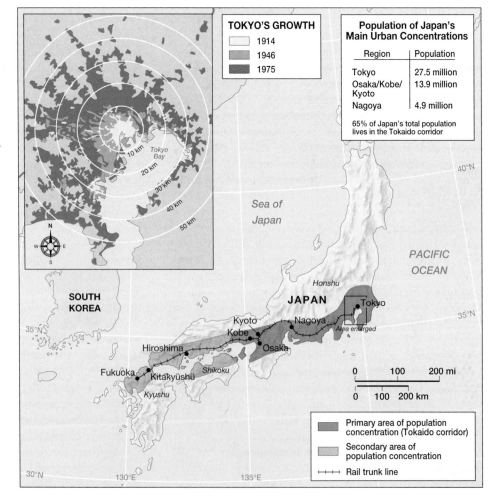

▶ Figure 11.20 Urban concentration in Japan The inset map shows the rapid expansion of Tokyo in the postwar decades. Today the Greater Tokyo metropolitan area is home to almost 30 million people. The larger map shows the cluster of urban settlements along Japan's southeastern coast. The major area of urban concentration is between Tokyo and Osaka, a distance of some 300 miles, known as the Tokkaido corridor. By some accounts, 65 percent of Japan's population lives in this area. The Osaka–Kobe metropolitan area ranks second to Tokyo, with approximately 14 million inhabitants.

TOKYO'S GROWTH
- 1914
- 1946
- 1975

Population of Japan's Main Urban Concentrations

Region	Population
Tokyo	27.5 million
Osaka/Kobe/Kyoto	13.9 million
Nagoya	4.9 million

65% of Japan's total population lives in the Tokaido corridor

Primary area of population concentration (Tokaido corridor)

Secondary area of population concentration

Rail trunk line

Tokyo to Osaka on the main rail line, a distance of almost 300 miles (480 kilometers), and never leave the urbanized area. By some accounts, 65 percent of Japan's 127 million persons are crowded into this narrow supercity.

Japanese cities sometimes strike foreign visitors as rather gray and monotonous places, lacking historical interest. Little of the country's premodern architecture remains intact. Traditional Japanese buildings were made of wood, which survives earthquakes much better than stone or brick. Fires have therefore been a long-standing hazard, and in World War II the U.S. Air Force fire-bombed most Japanese cities, virtually obliterating them (Hiroshima and Nagasaki were, on the other hand, completely destroyed by atomic bombs). The one exception was Kyoto, the old imperial capital, which was spared devastation. As a result, Kyoto is famous for its beautiful (wooden) Buddhist monasteries and Shinto temples, which ring the basin in which central Kyoto lies. Other Japanese cities were largely reconstructed in the late 1940s and 1950s, a period when Japan was still relatively poor and could afford only inexpensive concrete buildings. In the boom years of the 1980s, however, modern skyscrapers rose in many of the larger cities and postmodernist architecture began to lend variety, especially to the wealthier urban neighborhoods.

Cultural Coherence and Diversity: A Confucian Realm?

East Asia is in some respects one of the world's more unified cultural regions. Although different East Asian countries, as well as the regions within them, have their own unique cultural features (see "Food and Culture: East Asian Cuisine"), the entire region shares certain historically rooted ways of life and systems of ideas.

Most of these East Asian commonalties can be traced back to ancient Chinese civilization. Chinese civilization emerged roughly 4,000 years ago, largely in isolation from the Eastern Hemisphere's other early centers of civilization in the valleys of the Indus, Tigris-Euphrates, and Nile rivers. As a result, East Asian civilization developed along several unique lines. Before 1800, the entire region remained somewhat self-contained, with only secondary cultural and economic connections extending to the rest of Eurasia. The most prominent intellectual traditions of the region regarded China not only as the Middle Kingdom—or center of the world—but also as the world's only significant civilization. Such an inward focus was characteristic of other civilizations, but not to the same extent as in East Asia.

FOOD AND CULTURE   East Asian Cuisine

Many Americans tend to lump the various cuisines of East Asia into single category, based, perhaps, on their common focus on rice. Such a perception, however, is inaccurate, and East Asia as a whole cannot really be considered a coherent "food region." In most of northern China, for example, rice is a luxury food, and most people subsist largely on grains such as wheat and sorghum. Rice, moreover, is just as important in Southeast Asia and eastern India as it is in East Asia. Within East Asia there are also marked differences in seasoning and styles of preparation. Korean food, for example, is noted for its heavy use of hot (red) pepper and garlic, whereas Japanese food by comparison sometimes seems either bland or merely salty. In much of China, traditional cuisine revolves around vegetables and sometimes meat cooked in a sauce, which is then placed on rice; in Japan rice is usually served plain, offered alongside but not in combination with various side dishes.

Japanese cooking is, however, remarkably complex. The country has long borrowed liberally from other traditions (Chinese, European, and North American especially), while maintaining a distinct sense of what constitutes Japanese food. Its own cuisine is noted for the variety of its seafoods, pickles, and vegetables. Fruit is also popular in Japan, but is prohibitively expensive by American standards. Melons, for example, often cost more than $20 apiece and are considered appropriate gifts. One of the most popular Japanese foods is *sushi,* which literally means "vinegared rice." Sushi is often, but by no means always, eaten with *sashimi,* or raw fish. Before the late 1800s the Japanese ate virtually no red meat, but that is no longer true. The consumption of beef especially has risen dramatically since the 1960s.

China has no single cuisine, but rather a variety of food styles associated with different provinces. The food of Guangdong, for example—the first Chinese food introduced to America—is seldom heavily spiced, while that of Sichuan, which has more recently gained favor in the United States, is noted for its spiciness. China has a long history of elite gastronomy, with many expensive delicacies based on rare and exotic plants and animals. A classic example is "bird's-nest" soup, the main constituent of which is the dried spittle that certain swallows use to glue their nests together. After the coming of communist rule, however, elite cooking traditions nearly vanished from mainland China. During this period, little was available other than basic grains, vegetables, and small quantities of pork. The more elaborate forms of cuisine survived, however, in Taiwan, Hong Kong, and overseas Chinese communities, and they are now slowly returning to the Chinese mainland.

Despite the disparities in cooking traditions across East Asia, the region does have a few dietary commonalties. For example, tofu, or soybean curd, is a traditional source of protein throughout the region, which has historically been noted for its low levels of meat consumption. East Asia can also be considered the "chopstick region." In Japan, China, Korea, and Vietnam, food is generally eaten with chopsticks, and is cut and prepared to be handled with these implements. Chopsticks are not traditionally used in other parts of the world. Americans often ask for them in Thai restaurants, but in doing so they commit a minor error in cultural geographical categorization.

Unifying Cultural Characteristics

The most important unifying cultural characteristics of East Asia are related to religious and philosophical beliefs. Throughout the region, Buddhism and especially Confucianism have shaped not only individual beliefs, but also social and political structures. Although the role of traditional belief systems has been seriously challenged in recent decades, especially in China, historically grounded cultural patterns never disappear overnight. Even something as basic as written communication reveals a distinctly East Asian cultural background.

The Chinese Writing System
The clearest distinction between East Asia and the world's other cultural regions is found in written language. Existing writing systems elsewhere in the world are based on the alphabetic principle, in which each symbol represents a distinct sound. All of humankind's varied alphabetical scripts evidently can be traced back to a single point of origin in Syria and Lebanon. From there, alphabetical writing spread westward to Europe, eastward to Indonesia, and southward to Ethiopia. East Asia, on the other hand, evolved an entirely different system of **ideographic writing**. In ideographic writing, each symbol (or ideograph—more commonly called "character") represents primarily an idea rather than a sound (although the symbols can denote sounds in certain circumstances). As a result, ideographic writing requires the use of a large number of distinct symbols.

The East Asian writing system can be traced to the dawn of Chinese civilization. The first symbols were essentially pictures representing different words, but over time they grew more abstract. As the Chinese Empire expanded and the prestige of Chinese civilization carried its culture to other lands, the Chinese writing system spread. Japan, Korea, and Vietnam all came to use the same system, although in Japan it was substantially modified, while in Korea and Vietnam it was much later largely replaced by alphabetic systems.

The Chinese ideographic writing system has one major disadvantage and one major advantage when compared with alphabetic systems, both of which stem from the fact that it is largely divorced from spoken language. The disadvantage is that it is difficult to learn; to be literate, a person must memorize thousands of characters. The main benefit is that two literate persons do not have to speak the same language to be able to communicate, since the written symbols that they use to express their ideas are the same.

This advantage was tremendously important for the creation of a unified Chinese culture. When the Chinese Empire expanded south of the Yangtze River beginning in about 200 B.C., ethnic groups speaking a variety of languages were suddenly brought into the same political and cultural system. Because they could adopt the Chinese writing system, the peoples of southern China were able to integrate fully into the Middle Kingdom without adopting its spoken language. To this day, the residents of Beijing in the north and Guangzhou in the south speak entirely different languages, yet write them in identical form.

Korean Modifications
In Korea, Chinese characters were adopted at an early date and were used exclusively for hundreds of years. In the 1400s, however, Korean officials decided that Korea needed its own alphabet. They wanted to allow more widespread literacy, hoping also to more clearly differentiate Korean culture from that of China. The use of the new script spread quickly through the country. Korean scholars and officials, however, continued to use Chinese characters, regarding their own script as suitable only for popular writings. Today the Korean script is used for almost all purposes, but scholarly works still contain interspersed Chinese characters.

Japanese Modifications
The writing system of Japan is even more complex (Figure 11.21). Initially the Japanese simply borrowed Chinese characters, referred to in Japanese as **kanji**. Owing to the profound grammatical differences between Japanese and Chinese, the exclusive use of kanji resulted in awkward sentence construction. The Japanese solved this quandary by developing a quasi-alphabet, or more precisely a syllabary, known as **hiragana**, that allowed the expression of words and parts of speech not easily represented by Chinese characters. In hiragana, each symbol represents a distinct syllable, or combination of a consonant and a vowel

▲ **Figure 11.21 Japanese writing** The writing system of Japan was originally based on Chinese characters, known in Japan as *kanji*. Because of grammatical differences, however, the Japanese developed two unique "alphabets" of syllables, known as *katakana* and *hiragana*. Here *kanji* and *katakana* symbols are visible. *(Hiroshi Harada/DUNQ/Photo Researchers, Inc.)*

sound. Because of the restricted sound system of Japanese, only 51 hiragana symbols are necessary. A different but essentially parallel system, called *katakana*, is used in Japan for spelling words of foreign origin. Use of the hiragana resulted in increased literacy in medieval Japan, especially for women. The greatest early works in Japanese literature were written by aristocratic women in the hiragana script. Men with official duties and scholarly ambitions, however, continued to write in kanji. Eventually the two styles of writing merged, and written Japanese came to employ a complex mixture of symbols. In general, the more advanced a text is, the more kanji it contains. Japanese kanji today differ slightly from Chinese characters, as the latter have been simplified to some extent. Some Japanese characters, moreover, have evolved different meanings. Even so, it is still relatively easy for anyone literate in Japanese to learn how to read Chinese—but not to speak it!

Unlike Chinese, Japanese is easily written in the Roman alphabet. The resulting *romanji* style of writing is especially important in advertisements and for computer use. Computer keyboards using roman letters are not only much less cumbersome than Japanese keyboards, but are also a component of modern global culture. Computer technology, however, allows romanji entries to be easily converted into the traditional Japanese mixture of kanji, hiragana, and katakana symbols. Modern technology in this case helps spread a global attribute while allowing the continued use of local cultural forms.

The Confucian Legacy

Just as the use of a common writing system helped forge cultural linkages throughout East Asia, so too the idea system of **Confucianism** (the philosophy developed by Confucius) came to occupy a significant position in all of the societies of the region. Indeed, so strong is the heritage of Confucius that some writers refer to East Asia as the "Confucian world." In Japan, however, Confucianism never had the influence that it did in China and Korea.

The premier philosopher of Chinese history, Confucius (or Kung Fu Zi, in Mandarin Chinese) lived during the sixth century B.C., a period of marked political instability. The North China Plain and Loess Plateau, then the main areas of Chinese culture, were divided into a number of hostile states. Confucius's goal was to create a philosophy that could generate social stability. While Confucianism is often considered to be a religion, Confucius himself was far more interested in the "here and now," focusing his attention on how to lead a correct life and organize a proper society. Confucian thought does not deny the existence of deity or of an afterlife, but neither does it give them much consideration.

Confucius stressed deference to the properly constituted authority figures, but he thought that authority has a responsibility to act in a benevolent manner. The most basic level of the traditional Confucian moral order is the family unit, considered the bedrock of society. The ideal family structure is patriarchal, and children are told to obey and respect their parents—especially their fathers—as well as their elder brothers. At the highest level of the traditional moral order sat the emperor of China, who was regarded as an almost godlike father figure for the entire country.

Confucian philosophy also stresses the need for a well-rounded and broadly humanistic education. To a certain extent, Confucianism advocates a kind of meritocracy, holding that an individual should be judged on the basis of behavior and education, rather than on family background. The high officials of Imperial China (pre-1912)—the powerful **Mandarins**—were thus selected by competitive examinations. Only wealthy families, however, could afford to give their sons the education that was needed for success on those grueling tests.

Confucianism in Japan

In Japan, Confucianism was never as important as it was on the mainland. Japanese officials were actually able to exclude certain Confucian beliefs that they considered dangerous. The most important of these was the revocable "mandate of heaven." According to this notion, the emperor derived his authority from the principle of cosmic harmony, but such a mandate could be withdrawn if the emperor failed to fulfill his duties. This idea was used both to explain and to legitimize the rebellions that occasionally resulted in a change of China's ruling dynasty. In Japan, on the other hand, a single imperial dynasty has persisted throughout the entire period of written history. Although the emperor of Japan has had little real power for more than a thousand years, the absolute sanctity of his family lineage continues to form a basic principle of Japanese society.

The Modern Role of Confucian Ideology

The significance of Confucianism in East Asian development has been hotly debated for most of this century. In the early 1900s many observers believed that the conservatism of the philosophy, derived from its respect for tradition and authority, was responsible for the economically backward position of China and Korea. But since East Asia has more recently enjoyed the world's fastest rates of economic growth, such a position is no longer supportable. New voices now argue that Confucianism's respect for education and the social stability that it generates give East Asia a tremendous advantage in international competition.

Other scholars remain skeptical of both views, preferring to credit such factors as economic policy for the region's rapid economic growth. They also note that most interior portions of China have not participated much in the current economic boom, even though they share the Confucian legacy. Confucianism has, moreover, recently lost much of the hold that it once had on public morality throughout the entire region.

Religious Unity and Diversity in East Asia

Certain explicitly religious beliefs have worked alongside Confucianism to cement the East Asian region. The most important culturally unifying beliefs are associated with Mahayana Buddhism. Other religious practices, however, have had a more disunifying role.

Mahayana Buddhism

Buddhism, a religion that stresses the human soul's quest to escape an endless cycle of rebirths and reach union with the divine cosmic principle (or nirvana), originated in India in the sixth century B.C. By the second century A.D., Buddhism had reached China, and within a few hundred years it had spread throughout East Asia. Today Buddhism remains widespread everywhere the region, although it is far less significant here than it is in mainland Southeast Asia, Sri Lanka, and Tibet.

The variety of Buddhism (Figure 11.22) practiced in East Asia—Mahayana, or Greater Vehicle—is distinct from the Therevada Buddhism of South and Southeast Asia. Most important, Mahayana Buddhism simplifies the quest for nirvana (total enlightenment), in part by positing the existence of souls (*boddhisatvas*) who refuse divine union for themselves in order to help others spiritually. Mahayana Buddhism, unlike other forms, is nonexclusive; in other words, one may follow it while simultaneously professing the beliefs of other faiths. Thus, many Chinese consider themselves to be both Buddhists and Taoists (as well as Confucianists), while most Japanese are at some level both Buddhists and followers of Shinto.

As Mahayana Buddhism spread through East Asia in the medieval period, many different sects emerged. Probably the best known of these is Japanese Zen, which demands that its followers engage in the rigorous practice of "mind emptying." At one time, Buddhist monasteries associated with Zen and other sects were rich and powerful. In all East Asian countries, however, periodic reactions against Buddhism resulted in the persecution of monks and the suppression of monasteries. One reason was that government officials often viewed Buddhism as a foreign religion that placed India—rather than China—at the center of the world. Despite such hardships, East Asian Buddhism was never extinguished. But it also never became the focal point of society as it did in mainland Southeast Asia—or as Islam did in Southwest Asia and Christianity in Europe. In Japan, that position was partially captured by a different religion altogether, Shinto.

Shinto
Shinto is so closely bound to the idea of Japanese nationality that it is questionable whether a non-Japanese person can truly follow the religion. Shinto began as the animistic worship of nature spirits, but it was gradually refined into a subtle set of beliefs about the harmony of nature and its connections with human existence. Until the late 1800s, Buddhism and Shinto were complexly intertwined. Subsequently, the Japanese government began to disentangle them while elevating Shinto into a nationalistic cult focused on the divinity of the Japanese imperial family. After World War II, the more excessive aspects of nationalism were removed from the religion.

Shinto is still a place- and nature-centered religion. Certain mountains, particularly two volcanoes—Fuji and Ontake—are considered sacred and are thus climbed by large numbers of people (Figure 11.23). Major Shinto shrines, often located in picturesque places, attract numerous pilgrims; the most notable of these is the Ise Shrine south of Nagoya, which is the central site of the cult of the emperor. Small local shrines, like Buddhist temples, offer verdant oases in otherwise largely treeless Japanese urban neighborhoods.

▲ Figure 11.22 The Buddhist landscape   This Buddhist temple is located in Yunnan Province, China, and was used as the headquarters of the Yunnan provincial Buddhist association before the communist revolution of the middle twentieth century. Today it is preserved as a historical museum.   *(Brian Vikander/Corbis)*

Taoism and Other Chinese Belief Systems   The Chinese religion of Taoism (or Daosim) is similarly rooted in nature worship. Like Shinto, it stresses the acquisition of spiritual harmony and the pursuit of a balanced life. Taoism is indirectly associated with feng shui, commonly called **geomancy** in English, the Chinese and Korean practice of designing buildings in accordance with the spiritual powers that supposedly course through the local topography. Even in hypermodern Hong Kong, skyscrapers worth millions of dollars have occasionally gone unoccupied because their construction

▲ Figure 11.23 Mt. Fuji, Japan   This picturesque volcanic mountain, sacred to Japan's Shinto religion, is climbed by large numbers of religious pilgrims each year. In the foreground are tea fields.   *(Pacific Stock)*

failed to accord with feng shui principles. While Taoism is traditionally limited to China, many of its ideas have been embraced by New Age faiths in North America and Europe.

Despite the fact that both Taoism and Buddhism were historically followed throughout the country, traditional religious practice in China has always embraced local **particularism** (in other words, focusing on the unique attributes of particular places). Many minor gods were traditionally associated with single cities or other specific areas. Such beliefs never had much significance for the country's educated elite, but in premodern China rituals associated with them were sometimes carried out by state officials. In modern-day rural China, village gods are often still honored.

Minority Religions
Small numbers of adherents of virtually all world religions can be found in the increasingly cosmopolitan cities of East Asia. Millions of Chinese and Japanese belong to Christian churches (Henan province in China alone has as many as 3 million Protestants), even though they constitute less than 1 percent of the population of either country. Far more South Koreans, some 6 million in all, are Christian, mostly Protestant. Some reports indicate that Christianity is growing rapidly in China—despite persecution—but reliable information is scarce.

Much larger than China's Christian population is its Muslim community. Several tens of millions of Chinese-speaking Muslims, called *Hui,* are concentrated in Gansu and Ningxia in the northwest and in Yunnan province along the south-central border. Smaller clusters of Hui, often segregated in their own villages, live in almost every province of China. The only Muslim congregations in Japan, on the other hand, are associated with recent and probably temporary immigrants from South, Southeast, and Southwest Asia.

Secularism in East Asia
For all of these varied forms of religious expression, East Asia is still one of the most secular regions of the world. In Japan, most people occasionally observe Shinto or Buddhist rituals and maintain a small shrine for their own ancestors, and a small segment of the populace is devout. Japan also has a number of "new religions," sometimes called cults, a few of which are noted for their fanaticism. But for Japanese society as a whole, religion is simply not very important.

Elite culture in China was formerly dominated by Confucianism, which is more of a philosophy than a faith. After the communist regime took power in 1949, all forms of religion and traditional philosophy—including Confucianism—were discouraged and sometimes severely repressed. Under the new regime, atheistic **Marxist** philosophy (the communistic belief system developed by Karl Marx) became the official ideology. In the 1960s many observers thought that the traditional Chinese religious complex would survive only in overseas Chinese communities. With the easing of Marxist orthodoxy during 1980s and 1990s, however, many forms of religious expression began to return. This seems to be especially true in China's more prosperous coastal areas. In North Korea, on the other hand, Marxist orthodoxy is still rigidly enforced.

Linguistic and Ethnic Diversity in East Asia
While written languages may have helped unify East Asia, the same cannot be said for spoken languages (Figure 11.24). Japanese and Mandarin Chinese may partially share a system of writing, but the two languages bear no direct relationship. In their grammatical structures, Chinese and Japanese are more different from each other than are Chinese and English. Like Korean, however, Japanese has adopted many words of Chinese origin, just as English has borrowed heavily from Greek and Latin.

Language and National Identity in Japan
Japanese, by most accounts, is not related to any other language. Korean is also usually classified as the only member of its language family. Many linguists, however, think that Japanese and Korean should be classified together because they share many basic grammatical features. Only the mutual distrust of the Japanese and the Koreans, some suggest, has prevented this linguistic relationship from being acknowledged. A few linguists would go a step further to argue that Japanese and Korean are distantly related to the Altaic languages of Mongolia and Turkey.

From many perspectives, the Japanese form one of the world's more homogeneous peoples, and they tend to regard themselves in such a manner. To be sure, minor cultural and linguistic distinctions are noted between the people of western Japan (centered on Osaka) and eastern Japan (centered on Tokyo), and many individual prefectures (particularly Nagano, in the mountainous heart of Honshu) have distinctive customs and lifeways. Overall, however, such differences are of little significance.

In earlier centuries, however, the Japanese archipelago had been divided between two very different peoples: the Japanese living to the south, and the Ainu inhabiting the north. The Ainu are completely distinct from the Japanese. They possess their own language and have a different physical appearance. Unlike Japanese men, Ainu men usually have heavy beards, and were at one time disparaged by their southern neighbors as "hairy barbarians." Owing to their general facial features, the Ainu were once categorized as members of the "Caucasian race." Few scholars, however, now believe that humankind is divided into separate races, and Ainu do not appear to be closely related to Europeans by genetic criteria.

For centuries the Japanese and the Ainu competed for land, and by the tenth century A.D. the Ainu had been largely driven off the main island of Honshu. Until the 1800s, however, Hokkaido, as well as Sakhalin (now part of Russia), largely remained Ainu territory. The Japanese people subsequently began to colonize Hokkaido, putting renewed pressure on the Ainu. Today only about 24,000 Ainu remain, and most of them have mixed Japanese-Ainu ancestry.

Minority Groups in Japan
While the Japanese are relatively homogeneous, their language is divided into several dialects. But dialectal differences in the main islands are minor, and only in the Ryukyu Islands does one encounter a variant of Japanese so distinct that it might be considered a separate

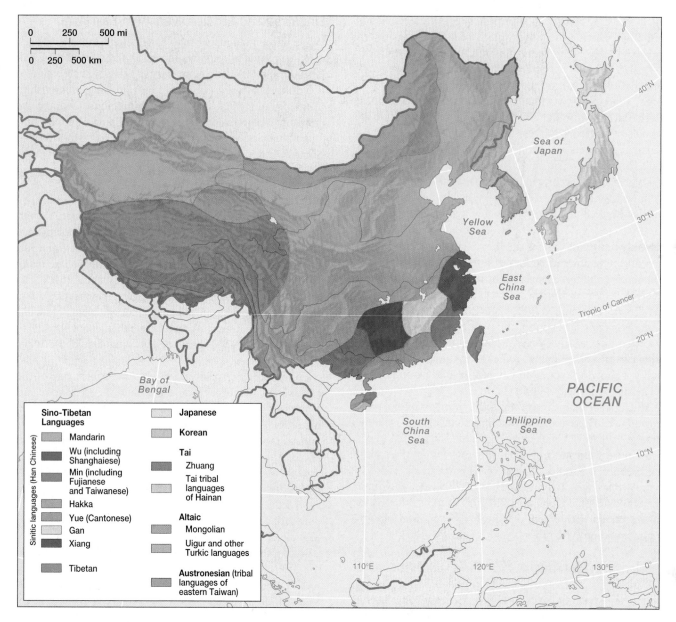

▲ Figure 11.24 The language geography of East Asia   The linguistic geography of Korea and Japan is very
straightforward, as the vast majority of people in those countries speak Korean and Japanese, respectively. In China,
the dominant Han Chinese speak a variety of closely related *sinitic* languages, the most important of which is Mandarin Chinese. In the peripheral regions of China, a large number of languages—belonging to several different linguistic families—can be found.

language. Many Ryukyu people believe that they have not
been considered full members of the Japanese nation, and
they have suffered certain forms of discrimination.

Discrimination has also been felt by approximately
700,000 persons of Korean descent living in Japan today.
Many of them were born in Japan (their parents and grandparents having left Korea early in the century) and speak
Japanese rather than Korean as their primary language. But
despite their deep bonds to Japan, such individuals are rarely
able to obtain Japanese citizenship. Perhaps as a result of such
treatment, many Japanese Koreans hold radical political views.

Starting in the 1980s, other immigrants began to arrive in
Japan, mostly from the poorer countries of Asia. Most do not
have legal status. Men from China and southern Asia typically work in the construction industry and in other dirty and
dangerous jobs; women from Thailand and the Philippines
often work as entertainers or prostitutes. Almost 200,000
Brazilians of Japanese ancestry have returned to Japan for the
relatively high wages they can earn. Immigration is, however, less pronounced in Japan than in most other wealthy countries, and relatively few migrants acquire permanent residency,
let alone citizenship.

The most victimized people in Japan are probably not foreigners but rather the **Burakumin,** or *Eta,* an outcast group of Japanese whose ancestors worked in "polluting" industries such as leather-craft. While discrimination is now illegal, the Burakumin are still among the poorest and least-educated people in Japan. Private detective agencies do a brisk business checking prospective marriage partners and employees for possible Burakumin ancestries. The Burakumin, however, have banded together to demand their rights; the Buraku Liberation League is politically powerful and is reputed to have close connections with the Yakuza (the Japanese mafia). The Burakumin usually live in separate neighborhoods and are concentrated in the Osaka region of western Japan.

Language and Identity in Korea
The Koreans are also a relatively homogeneous people. The vast majority of residents in both North and South Korea speak Korean and unquestioningly consider themselves to be members of the Korean nation. There is, however, a strong sense of regional identity, some of which can be traced back to the medieval period when the peninsula was divided into three separate kingdoms. The people of southwestern South Korea, especially those living in Kwangju and its environs, tend to be viewed as distinctive. Many southwesterners believe that they have suffered periodic discrimination.

Not all Koreans live in Korea. Several hundred thousand reside directly across the border in northern China. Substantial Korean communities can also be found in Kazakstan in Central Asia, owing to the deportations ordered by the Soviet Union in the mid-1900s. (At that time, the Maritime Province of Russia's extreme southeast was largely cleared of its Korean residents, as the Soviet government feared that they might be disloyal.) A more recent Korean **diaspora** has brought hundreds of thousands of people to the United States (a diaspora is a scattering of a particular group of people over a vast geographical area). As recently as 2001 some 15,000 Koreans were leaving annually, mostly moving to Canada, the United States, Australia, and New Zealand.

Language and Ethnicity Among the Han Chinese
The geography of language and ethnicity in China is far more complex than that of Korea or Japan. This is true even if one considers only the eastern half of the country, so-called China proper. The most important distinction is that separating the Han Chinese from the non-Chinese peoples. The Han, who form the vast majority, are those people who have long been incorporated into the Chinese cultural and political systems and whose languages are expressed in the Chinese writing system. They do not, however, all speak the same language.

Northern, central, and southwestern China—a vast area extending from Manchuria through the middle and upper Yangtze Valley to the valleys of Yunnan in the far south—constitutes a single linguistic zone. The spoken language here is generally called *Mandarin Chinese.* In China today Mandarin— often called simply "the common language"—is the national tongue. Mandarin is divided into a number of closely related and mutually intelligible dialects. The official Mandarin dialect of the Beijing area, however, is gradually spreading.

In southeastern China, from the Yangtze Delta to China's border with Vietnam, a number of separate but related languages are spoken. Peoples speaking these languages are Han Chinese, but they are not native Mandarin speakers. The most important of these languages are located along the coast. Traveling from south to north, one encounters Cantonese (or Yue) spoken in Guangdong, Fujianese (alternatively Hokkienese, or Min locally) spoken in Fujian, and Shanghaiese (or Wu), spoken in and around the city of Shanghai and in Zhejiang province. These are true languages, not dialects, since they are not mutually intelligible. They are usually called dialects, however, because they have no distinct written form.

One group of people speaking a southern Chinese language, the Hakka, is occasionally considered by others not to be true Han Chinese. The Hakka are sometimes disdained by others as rootless wanderers. Evidently their ancestors fled northern China roughly a thousand years ago to settle in the rough upland area where Guangdong, Fujian, and Jiangxi provinces meet. As is evident in Figure 11.25, later migrations took them throughout southern China, where they typically settled in hilly wastelands. The Hakka traditionally made their living by growing upland crops such as sweet potatoes and by working as loggers, stonecutters, and metalworkers. They today form one of the poorest communities of southern China.

Ironically, however, a significant number of Hakkas have reached high positions in the Chinese government (where they often conceal their Hakka origins). The Hakka have a tradition of rebellion against lowland landowners, so they quickly lent their support to the communist movement that eventually gained national power. Also paradoxically, the Hakka fiercely proclaim their own Han identity and often consider themselves exemplars of Chinese culture. As such, they may disdain the sentiments of local particularism among their neighbors—just as they are disdained by others for their lack of local rootedness.

Despite their many differences, all of the languages of the Han Chinese (including Hakka) are closely related to each other, belonging to the same Sinitic language subfamily. Since their basic grammars and sound systems are similar, it is not difficult for a person speaking one of these languages to learn another. It is usually difficult, however, for speakers of European languages—or of Japanese or Korean—to gain fluency in these tongues. All Sinitic languages are **tonal** and monosyllabic; their words are all composed of a single syllable (although compound words can be formed from several syllables), and the meaning of each basic syllable changes completely according to the pitch in which it is uttered.

The Non-Han Peoples
Many of the more remote upland districts of China proper are inhabited by various groups of non-Han peoples speaking non-Sinitic languages. Such peoples are usually classified as **tribal,** implying that they have a traditional social order based on autonomous village communities. Such a view is not entirely accurate, however, since

▶ **Figure 11.25 The Hakka diaspora** The Hakka form an important and distinctive subgroup of the Han Chinese. They seem to have originated in north-central China but long ago migrated to the area where Fujian, Guangdong, and Jiangxi provinces converge. Here they developed distinctive agricultural patterns well suited to the rough uplands of the region. Later movements took Hakka communities to upland areas throughout much of southern China. In the 1800s and early 1900s, many Hakka moved to Southeast Asia, where they still form important communities.

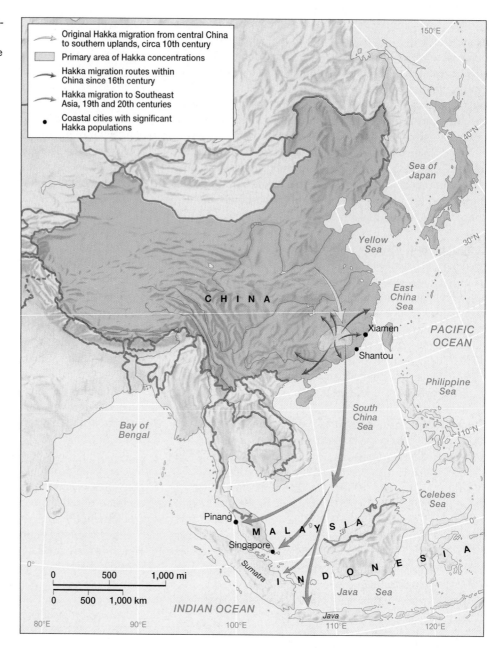

some of these groups once had their own kingdoms and all are now subject to the Chinese state. What they do have in common is a heritage of cultural and sometimes political struggle against the Han Chinese (Figure 11.26).

Over the course of many centuries, the territory occupied by these non-Han communities has been steadily reduced, both because of the continued expansion of the Han and because of their own emigration. Acculturation into Chinese society and intermarriage with the Han have also reduced many non-Han groups. Their main concentrations today are in the rougher lands of the far north and the far south.

As many as 11 million Manchus live in the more remote portions of Manchuria. The Manchu languages are related to those of the tribal peoples of central and southeastern Siberia. Few Manchus, however, still speak their own languages, hav-

ing abandoned them for Mandarin Chinese, and the community is in danger of linguistic extinction. This is an ironic situation, since the Manchus ruled the entire Chinese Empire from 1644 to 1912. Until the end of this period, the Manchus prevented the Han from settling in central and northern Manchuria, which they hoped to keep as their homeland. Once Chinese were allowed to settle in Manchuria in the 1800s—in part to prevent Russian expansion—the Manchus soon found themselves vastly outnumbered. As they began to intermarry with and adopt the language and lifeways of the newcomers, their own culture began to disappear.

Much larger communities of non-Han peoples are found in the far south, especially in Guangxi. Most of the inhabitants of Guangxi's uplands and remote valleys speak languages of the Tai family, closely related to those of Thailand. Since there

▲ Figure 11.26 Tribal villages in south China   Non-Han people are usually classified as "tribal" in China, which assumes they have a traditional social order based upon autonomous village communities. Shown are Yi people at an open-air market in the village of Xhanghe in Yunnan province.   *(Michael S. Yamashita/NGS Image Collection)*

are as many as 18 million non-Han people in Guangxi, it has been designated an **autonomous region.** Such autonomy was designed to allow non-Han peoples to experience "socialist modernization" at a different pace from that expected of the rest of the country. Critics contend that very little real autonomy has ever existed, despite the official designation. (In addition to Guangxi, there are four other autonomous regions in China. Three of these, Xizang [Tibet], Nei Monggol [Inner Mongolia], and Xinjiang, are located in Central Asia and are thus discussed at length in Chapter 10. The final autonomous region, Ningxia, located in northwestern China, is distinguished by its large concentration of Hui [Mandarin-speaking Muslims].)

Other areas with sizable numbers of non-Han peoples are Yunnan and Guizhou, in southwestern China, and western Sichuan. Most tribal peoples here practice **swidden agriculture** (also called "slash and burn"; see Chapter 13) on rough slopes; lowlands are generally occupied by rice-growing Han Chinese. A wide variety of separate languages, falling into several linguistic families, are found among the scattered ethnic groups living in the uplands. Figure 11.27 shows that in Yunnan the resulting ethnic mosaic is staggeringly complex.

Language and Ethnicity in Taiwan

Taiwan is also noted for its linguistic and ethnic complexity. In the island's mountainous eastern region, a few small groups of "tribal" peoples speak languages related to those of Indonesia (belonging to the Austronesian language family). These peoples resided throughout Taiwan before the sixteenth century. At that time, however, Han migrants began to arrive in large numbers. Most of the newcomers spoke Fujianese dialects, which eventually evolved into the distinctive language of Taiwanese.

Taiwan was transformed almost overnight in 1949, when China's nationalist forces, defeated by the communists, sought refuge on the island. Most of the nationalist leaders spoke Mandarin, which they made the official language. Taiwan's new leadership discouraged Taiwanese, viewing it as a mere local dialect. As a result, considerable tension developed between the Taiwanese and the Mandarin communities. Only in the 1990s did Taiwanese speakers begin to reassert their language rights.

East Asian Cultures in Global Context

East Asia, like most other parts of the world, has long exhibited tensions between an internal orientation and tendencies toward cosmopolitanism. This dichotomy has both a cultural and, as we shall later see, an economic dimension. Until the mid-1800s, all East Asian countries attempted to insulate themselves from Western cultural influences. Japan subsequently opened its doors, but remained highly ambivalent about foreign ideas. Only after its defeat in 1945 did Japan really opt for a globalist orientation. It was followed in this regard by South Korea, Taiwan, and Hong Kong (then a British colony). The Chinese and North Korean governments decided during the early Cold War decades of the 1950s and 1960s to isolate themselves as much as possible from Western and global culture.

The Cosmopolitan Fringe   The capitalist countries of East Asia are characterized by a vibrant cosmopolitan internationalism, especially in the large cities, which coexists with strong national and local cultural variation. Virtually all Japanese, for example, study English for 6 to 10 years, and although relatively few learn to speak it fluently, most can read and understand a good deal. Business meetings among Japanese, Chinese, and Korean firms are more often than not conducted in English. Relatively large numbers of advanced students, especially from Taiwan, study in the United States and other English-speaking countries, and thus bring home a kind of cultural bilingualism. Internet usage, with its implicit globalism, is also widespread in East Asia's cosmopolitan fringe.

The current cultural flow is not merely from a globalist West to a previously isolated East Asia. Instead, the exchange is growing more reciprocal. Hong Kong's action films are popular throughout most of the world, and with the success of director John Woo in the United States, they are beginning to influence filmmaking techniques in Hollywood. The recent international blockbuster *Crouching Tiger, Hidden Dragon* was filmed in China, in Mandarin, by a Taiwanese director who had several Hollywood hits to his credit. Japan, on the other hand, virtually dominates the world market in video games, and its ubiquitous comic-book culture and animation techniques are now following karaoke bars in their overseas march.

Cultural globalization is, of course, just as controversial in Japan as it is elsewhere. Japanese ultranationalists are few but vocal, calling their fellow citizens to resist the decadence of the West and to retain the military traditions of the **samurai** (the warrior class of premodern Japan). Many other Japanese people, however, worry that their country is too insular, and that they do not possess the English-language and global cultural skills necessary to operate effectively in the world economy.

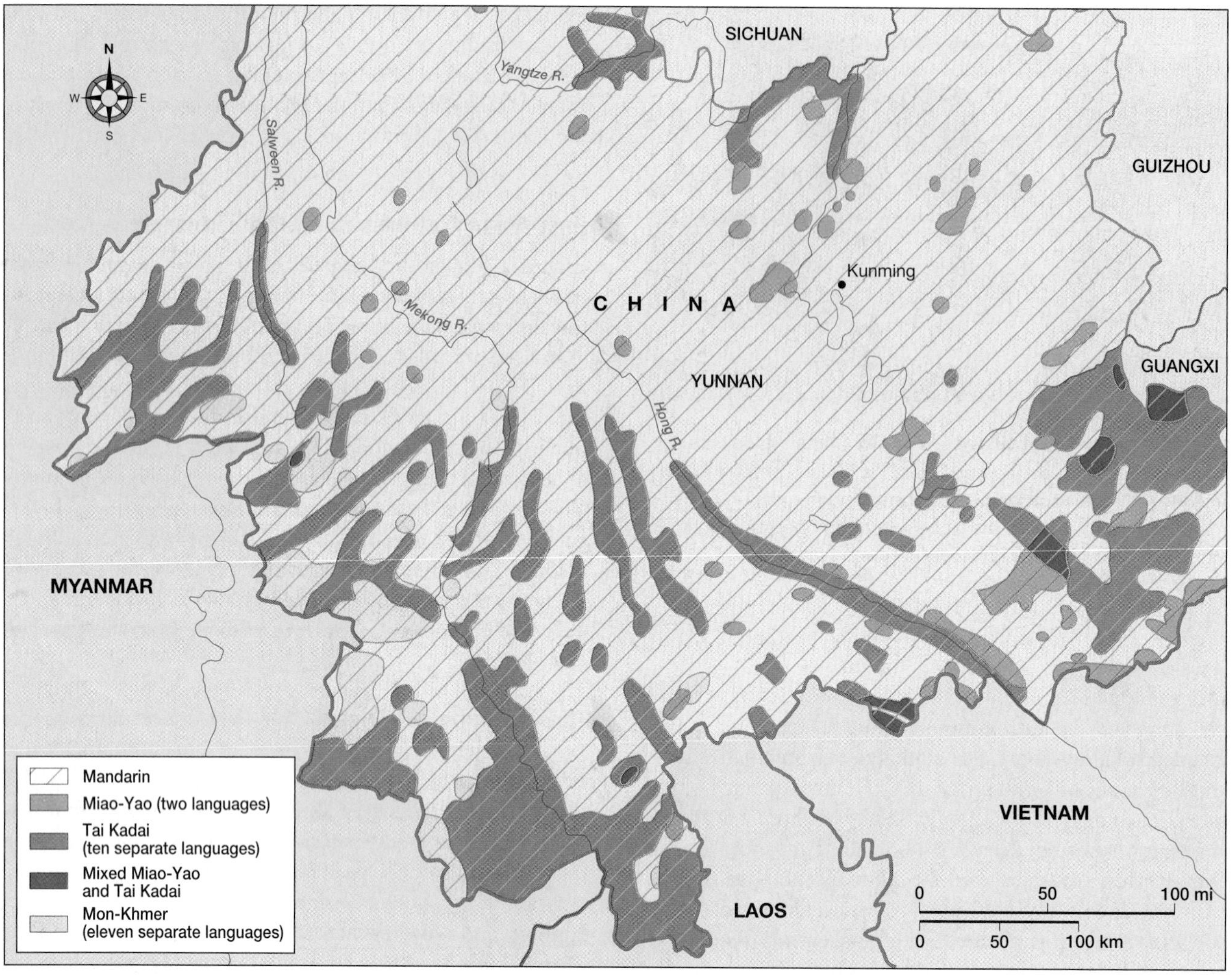

▲ **Figure 11.27 Language groups in Yunnan** China's Yunnan Province is the most linguistically complex area in East Asia. In Yunnan's broad valleys and relatively level plateau areas and in its cities, most people speak Mandarin Chinese. In the hills, mountains, and steep-sided valleys, however, a wide variety of tribal languages, falling into several linguistic families, are spoken. In certain areas, several different languages can be found in very close proximity.

The Chinese Heartland In one sense, Japan is more culturally predisposed to cosmopolitanism than is China. The Japanese have always borrowed heavily from other cultures (particularly from China itself), whereas the Chinese have historically been more self-sufficient. For most of Chinese history, cosmopolitanism has implied an orientation to the norms established by the Mandarin class in the core of the vast Chinese Empire. The southern coastal Chinese have, however, often embraced a different version of cosmopolitanism, one linked to the Chinese diaspora communities of Southeast Asia and the Pacific, and ultimately to maritime trading circuits extending over much of the globe.

In most periods of Chinese history, the interior orientation of the center prevailed over the external orientation of the southern coast. After the communist victory of 1949, only the small British enclave of Hong Kong was able to pursue international cultural connections. In the rest of the country,

a dour and puritanical cultural order was rigidly enforced. While this culture was largely founded on the norms of Chinese peasant society, it was also influenced by the socialist system that had emerged in the Soviet Union.

After China began to liberalize its economy and open its doors to foreign influences in the late 1970s and early 1980s, the southern coastal region suddenly assumed a new prominence. Through its doors global cultural patterns began to penetrate the rest of the country. The result has been the emergence of a vibrant but somewhat gaudy urban popular culture in China that is replete with such global features as nightclubs, karaoke bars, fast-food franchises, and theme parks.

The recent liberalization of the Chinese cultural order has also allowed the reemergence of regional identities. From the late 1960s to the late 1970s, all forms of localism were rigorously suppressed as China's leaders sought to build a nationally uniform, working-class culture. Today markers of local ethnicity are on the rise. This is again especially true in south-eastern China, where most people do not speak Mandarin and have remained culturally distinctive. This resurgence of local identity, ironically, is also tied up with the process of globalization. Southern Chinese culture, particularly Cantonese culture, is now widely considered to be attractive precisely because it is identified as internationalist. The Guangdong region, including Hong Kong, has been the main gateway for foreign culture to enter China, and is thus viewed as being at the forefront of globalization. Its Cantonese food, music, and films are now becoming popular in cities throughout China.

The Geopolitical Framework and Its Evolution: The Imperial Legacies of China and Japan

The political history of East Asia revolves around the centrality of China and the ability of Japan to remain outside of China's grasp. The traditional Chinese conception of geopolitics was based on the idea of a universal empire: all territories were supposed to be a part of the Chinese Empire, pay tribute to it and acknowledge its supremacy, or stand outside the system altogether. Until the 1800s, the Chinese government would not recognize any other as its diplomatic equal. When China could no longer maintain its power in the face of European aggression, the East Asian political system fell into disarray. As European power declined in the 1900s, China and Japan again contended for regional leadership. After World War II, East Asia was split by larger **Cold War** rivalries. The resulting tension between China and Japan persists to this day (Figure 11.28), although it is increasingly softened by economic ties.

The Evolution of China

The original core of Chinese civilization (dating back at least to 1800 B.C.) was the North China Plain and the Loess Plateau. For many centuries, periods of unification alternated with times of division into competing states. The most important episode of unification occurred in the third century B.C. Once political unity was achieved, the Chinese Empire began to expand vigorously to the south of the Yangtze Valley. Subsequently, the ideal of the imperial unity of China triumphed, with periods of division being seen as indicating profound disorder. This ideology helped cement the Han Chinese into a single people. The potential for disunification, however, was always present.

Several Chinese dynasties rose and fell between 219 B.C. and 1912, most of them, as visible in Figure 11.29, controlling roughly the same territory. The core of the Chinese Empire remained China proper, excluding Manchuria. Other lands, however, were sometimes ruled as well. The most important of these was a western projection extending north of the Tibetan Plateau into the desert basins of Central Asia (modern Xinjiang). China valued this area because the vital trading route to western Eurasia (the "silk road") passed through it. But while China often ruled this territory (and indeed, does so today), it was never directly incorporated into the Chinese social and cultural systems.

Various Chinese dynasties attempted to conquer Korea, but the Koreans resisted. Eventually China and Korea worked out an arrangement whereby Korea paid token tribute and acknowledged the supremacy of the Chinese Empire, and in return received trading privileges and retained independence. When foreign armies invaded Korea—as did those of Japan in the late 1500s—China sent troops to support its "vassal kingdom."

For most of the past 2,000 years, the Chinese Empire was Earth's wealthiest and most powerful state. Its only real threat came from the pastoral peoples of Mongolia and Manchuria. Although vastly outnumbered by China, these societies were organized on a highly effective military basis. Usually the Chinese and the Mongols enjoyed a mutually beneficial trading relationship. Periodically, however, they waged war, and on several occasions the northern nomads conquered China (see Chapter 10). (The Great Wall along China's border did not, in other words, provide adequate defense; Figure 11.30.) In time the conquering armies adopted Chinese customs in order to govern the far more numerous Han people. In the long run they had a relatively small impact on Chinese society.

The Manchu Ch'ing Dynasty
The final and most significant conquest of China occurred in 1644, when the Manchus toppled the Ming Dynasty and replaced it with the Ch'ing (also spelled Qing) Dynasty. As earlier conquerors did, the Manchus retained the Chinese bureaucracy and made few institutional changes. Their strategy was to adapt themselves to Chinese culture, yet at the same time to preserve their own identity as an elite military group. Their system functioned well until the mid-nineteenth century, when the empire began to crumble before the onslaught of European and, later, Japanese power.

China's most significant legacy from the Manchu Ch'ing Dynasty was the extension of its territory to include much of Central Asia. The Manchus subdued the Mongols and eventually established control over eastern Central Asia, including Tibet. Even the states of mainland Southeast Asia sent tribute and acknowledged Chinese supremacy. Never before had the Chinese Empire been so extensive or so powerful.

The Modern Era
From its height in the 1700s, the Chinese Empire descended rapidly in the 1800s as it failed to keep pace with the technological progress of Europe. Threats to the empire had always come from the north, and Chinese and Manchu officials saw little peril from European merchants operating along their coastline. But the Europeans were distressed by the amount of silver needed to obtain Chinese silk, tea, and other products, and by the fact that the Chinese disdained the manufactured goods that they offered. In response, the British began to sell opium, which Chinese authorities viewed, probably correctly, as a threat to their nation. When the imperial government tried to suppress the opium trade in the 1840s, the British attacked and quickly prevailed.

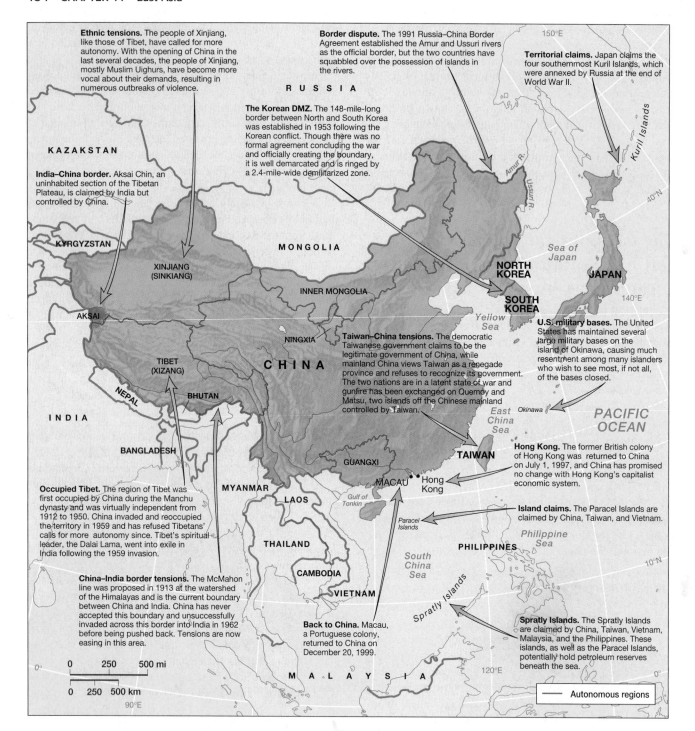

Ethnic tensions. The people of Xinjiang, like those of Tibet, have called for more autonomy. With the opening of China in the last several decades, the people of Xinjiang, mostly Muslim Uighurs, have become more vocal about their demands, resulting in numerous outbreaks of violence.

Border dispute. The 1991 Russia–China Border Agreement established the Amur and Ussuri rivers as the official border, but the two countries have squabbled over the possession of islands in the rivers.

Territorial claims. Japan claims the four southernmost Kuril Islands, which were annexed by Russia at the end of World War II.

India–China border. Aksai Chin, an uninhabited section of the Tibetan Plateau, is claimed by India but controlled by China.

The Korean DMZ. The 148-mile-long border between North and South Korea was established in 1953 following the Korean conflict. Though there was no formal agreement concluding the war and officially creating the boundary, it is well demarcated and is ringed by a 2.4-mile-wide demilitarized zone.

Taiwan–China tensions. The democratic Taiwanese government claims to be the legitimate government of China, while mainland China views Taiwan as a renegade province and refuses to recognize its government. The two nations are in a latent state of war and gunfire has been exchanged on Quemoy and Matsu, two islands off the Chinese mainland controlled by Taiwan.

U.S. military bases. The United States has maintained several large military bases on the island of Okinawa, causing much resentment among many islanders who wish to see most, if not all, of the bases closed.

Occupied Tibet. The region of Tibet was first occupied by China during the Manchu dynasty and was virtually independent from 1912 to 1950. China invaded and reoccupied the territory in 1959 and has refused Tibetans' calls for more autonomy since. Tibet's spiritual leader, the Dalai Lama, went into exile in India following the 1959 invasion.

Hong Kong. The former British colony of Hong Kong was returned to China on July 1, 1997, and China has promised no change with Hong Kong's capitalist economic system.

Island claims. The Paracel Islands are claimed by China, Taiwan, and Vietnam.

China–India border tensions. The McMahon line was proposed in 1913 at the watershed of the Himalayas and is the current boundary between China and India. China has never accepted this boundary and unsuccessfully invaded across this border into India in 1962 before being pushed back. Tensions are now easing in this area.

Back to China. Macau, a Portuguese colony, returned to China on December 20, 1999.

Spratly Islands. The Spratly Islands are claimed by China, Taiwan, Vietnam, Malaysia, and the Philippines. These islands, as well as the Paracel Islands, potentially hold petroleum reserves beneath the sea.

— Autonomous regions

▲ **Figure 11.28 Geopolitical issues in East Asia** East Asia remains one of the world's geopolitical hot spots. Tensions are particularly severe between capitalist, democratic South Korea and the isolated communist regime of North Korea, and between China and Taiwan. China has had several border disputes, which include a number of small islands in the South China Sea. Japan and Russia have not been able to resolve their quarrel over the southern Kuril Islands.

This first "opium war" ushered in a century of political and economic chaos in China. The British demanded free trade in selected Chinese ports, and in the process overturned the traditional policy of managed exchange based on the acknowledgment of Chinese supremacy. As European enterprises penetrated China and undermined local economic interests, anti-Manchu rebellions began to break out. At first, all such uprisings were crushed, but not before causing tremendous destruction. Meanwhile, European power continued to advance. In 1858 Russia annexed the northernmost reaches of Manchuria, and by 1900 China had been divided—as shown in Figure 11.31—into separate "**spheres of influence**" in which different European economic interests prevailed. (In a "sphere of influence," the colonial power had

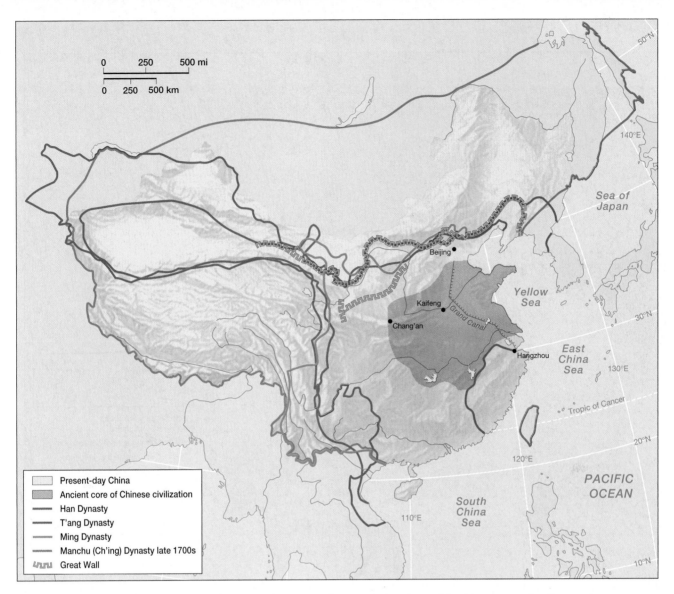

Present-day China
Ancient core of Chinese civilization
—— Han Dynasty
—— T'ang Dynasty
—— Ming Dynasty
---- Manchu (Ch'ing) Dynasty late 1700s
ᴨᴨ Great Wall

▲ Figure 11.29 The historical extent of China   China is usually regarded as the world's oldest country, but the territorial extent of the Chinese state has varied greatly over the centuries. The earliest states were limited to the Loess Plateau and North China Plain, but most historical Chinese dynasties controlled the entire core area of modern China as well as the Tarim Basin in Xinjiang. Before the 1600s, however, China seldom held control of Tibet, Inner Mongolia, or central and northern Manchuria.

◀ Figure 11.30 The Great Wall of China   The Great Wall of China runs 1,500 miles (2,400 kilometers) east to west from the Yellow Sea to deep within Central Asia. The first parts of the wall were built in the fourth century B.C.; however, most of the wall was either rebuilt or finished in later times, mainly in the fifteenth and sixteenth centuries.   (Michael Howell/Pacific Stock)

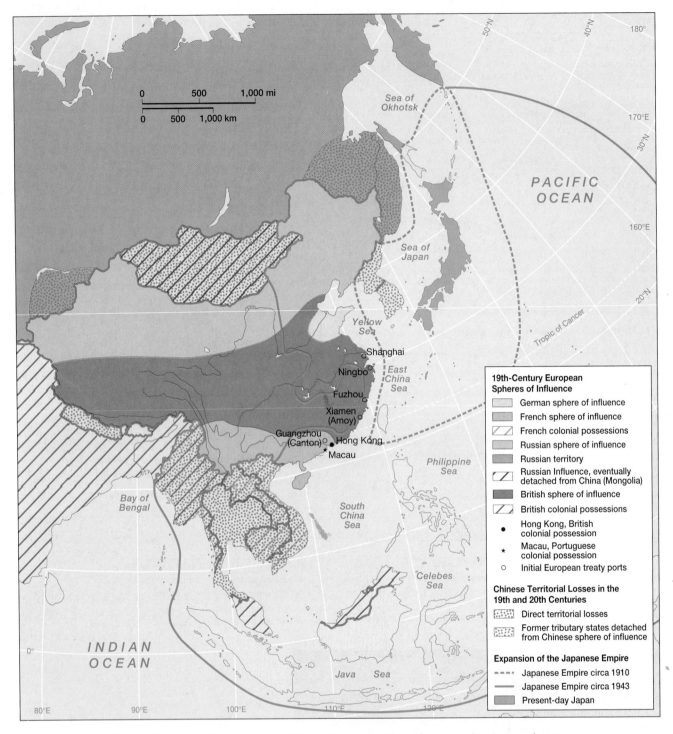

▲ **Figure 11.31 Nineteenth-century European colonialism**   The Chinese lost influence and territory in the nineteenth century as European power expanded. Although China regained its autonomy and most of its territory in the 1900s, Russia retained large areas that were formerly under Chinese control. The first half of the twentieth century saw the rapid expansion of the Japanese Empire, which ended with the defeat of Japan in World War II.

no formal political authority, but it did have informal influence and tremendous economic clout.)

A successful rebellion in 1911 finally toppled the Manchus and destroyed the empire, but subsequent efforts to establish a unified Chinese Republic were not successful. In many parts

of the country, local military leaders ("warlords") grabbed power for themselves. By the 1920s it appeared that China might be completely dismembered. The Tibetans had gained autonomy; Xinjiang was under Russian influence; and in China proper Europeans and local warlords vied with the weak Chinese Republic for power. Japan was also increasing its demands and seeking to expand its territory. To understand this part of the story, it is necessary to examine the history of Japan.

The Rise of Japan

Japan did not emerge as a unified state until the seventh century, more than 2,000 years later than China. From its earliest days, Japan looked to China (and, at first, to Korea as well) for intellectual and political models. Its offshore location insulated Japan from the threat of rule by the Chinese Empire, but the Japanese state did periodically acknowledge Chinese supremacy. At the same time, the Japanese conceptualized their islands as a separate empire, equivalent in certain respects to China. Between 1000 and 1580, however, Japan had no real unity, being divided into a number of mutually antagonistic feudal realms.

The Closing and Opening of Japan

In the early 1600s, Japan was reunited by the armies of the Tokugawa **Shogunate** (a shogun is a military leader who theoretically remains under the emperor but who actually holds power). At this time, Japan asserted its total autonomy from China and attempted to isolate itself from the rest of the world. Until the 1850s, Japan traded with China mostly through the Ryukyu islanders (who paid tribute to both China and Japan) and with Russia only through Ainu intermediaries. The only Westerners allowed to trade in Japan were the Dutch, and their activities were strictly limited.

Japan remained largely closed to foreign commerce and influence until U.S. gunboats sailed into Tokyo Bay to demand access in 1853. Aware that China was losing power and that they could no longer keep the Westerners out, Japanese leaders set about modernizing their economic, administrative, and military systems. This effort accelerated when the Tokugawa Shogunate was toppled in 1868 by the Meiji Restoration. (It is called a *restoration* because it was carried out in the emperor's name, but it did not give the emperor any real power.) Unlike China, Japan successfully accomplished most of its reform efforts.

The Japanese Empire

Japan's new rulers realized that their country remained threatened by European imperial power. They therefore nurtured the development of a silk export industry, which gave them the funds needed to buy modern equipment. They also decided that the only way to meet the challenge was to become expansionistic themselves. Japan therefore took control over Hokkaido and began to move farther north into the Kuril Islands and Sakhalin.

In 1895 the Japanese government tested its newly modernized army against China, winning a quick and profitable victory that gave it control of Taiwan. Tensions then mounted with Russia as the two countries vied for power in Manchuria and Korea. The Japanese defeated the Russians in 1905, giving them considerable influence in northern China. With no strong rival in the area, Japan annexed Korea in 1910. Alliance with Britain, France, and the United States during World War I brought further gains, as Japan was awarded Germany's island colonies in Micronesia.

The 1930s brought a global depression, putting a resource-dependent Japan in a difficult situation. The country's leaders sought a military solution, and in 1931 Japan conquered Manchuria. In 1937 Japanese armies moved south, occupying the North China Plain and the coastal cities of southern China. The Chinese government withdrew to the relatively inaccessible Sichuan Basin to continue the struggle. During this period, Japan's relations with the United States steadily deteriorated. When the United States cut off the export of scrap iron, Japan began to experience a resource crunch.

In 1941 Japan's leaders decided to destroy the American Pacific fleet in order to clear the way for the conquest of resource-rich Southeast Asia. Their grand strategy was to unite East and Southeast Asia into a "Greater East Asia Co-Prosperity Sphere." This "sphere" was to be ruled by Japan, however, and was designed to keep the Americans and Europeans out. Japanese forces sometimes engaged in brutal acts; in the infamous "rape of Nanjing," Japanese troops slaughtered up to several hundred thousand Chinese civilians. In Korea, the colonial government evidently planned to extinguish the Korean language in favor of Japanese.

Postwar Geopolitics

With the defeat of Japan at the end of World War II, East Asia became a power vacuum, and hence an arena of rivalry between the United States and the Soviet Union. Initially, American interests prevailed in the maritime fringe, while Soviet interests advanced on the mainland. Soon, however, East Asia began to experience its own revival.

Japan's Revival

Japan lost its colonial empire when it lost the war. Its territory was reduced to the four main islands plus the Ryukyu Archipelago and a few minor outliers. In general, the Japanese government acquiesced to this loss of land. The only remaining territorial conflict concerns the four southernmost islands of the Kuril chain, which were taken by the Soviet Union in 1945. Although Japan still claims these islands, Russia refuses even to discuss relinquishing control, causing considerable strains in Russo-Japanese relations.

After losing its overseas possessions, Japan was forced to rely on trade to obtain the resources needed for its economy. Here it proved remarkably successful. Japan's military power, on the other hand, was limited by the constitution imposed on it by the United States. Because of these restrictions, Japan has relied in part on the U.S. military for its defense needs. The U.S. Navy patrols many of its vital sea-lanes, and U.S. armed forces maintain several bases within Japan. This U.S. military presence, however, is becoming increasingly controversial. Many Japanese citizens believe that their country ought to provide its own defense, and they resent the presence of U.S. troops. Slowly but steadily, meanwhile, Japan's own military has emerged as a strong regional force despite the constitutional limits imposed on it (see "Geopolitical Tensions: Military Bases in Okinawa").

GEOPOLITICAL TENSIONS   Military Bases in Okinawa

Okinawa, the largest and most important of Japan's Ryukyu Islands, was occupied by the United States from 1945 to 1972. Although Okinawa was returned to Japan in 1972, the United States retained control over several large military bases concentrated in the northern half of the island. Some 20 percent of Okinawa's land area remains devoted to military purposes. The U.S. bases are economically significant. In 1972 military spending accounted for almost half of the island's economic output. This figure had declined to less than 10 percent by 1997, but it still represented an important source of income for one of Japan's poorest areas.

Local opposition to the military bases, however, began to mount in the early 1990s. Several highly publicized crimes by U.S. servicemen, including one case of child rape, turned many Japanese against the U.S. military presence altogether.

High levels of noise from military jets, as well as periodic aircraft accidents, further upset the Okinawans. In October 1995 mass demonstrations broke out in Okinawa, leading the island's governor to demand that all bases close by 2015. Okinawans later voted by a wide margin to substantially reduce the land devoted to military operations.

While most Okinawans clearly want U.S. military activities—and personnel—to become less disruptive, they do not necessarily want the bases to be closed. Public polling, in fact, has revealed that three-quarters of Okinawa's residents support some form of U.S. military activity, largely for the economic advantages that it brings. The U.S. military presence in the far south of Japan thus remains secure for the moment, but its long-term future remains in doubt.

The Division of Korea   The end of World War II brought much greater changes to Korea than to Japan. As the end of the war approached, the Soviet Union and the United States agreed to divide the country; Soviet forces were to occupy the area north of the 38th parallel, whereas U.S. troops would occupy the south. This soon resulted in the establishment of two separate regimes. In 1950 North Korea invaded South Korea, seeking to reunify the country. The United States, with support from the United Nations, supported the south, while China aided the north. The war ended in a stalemate, and Korea remained a divided country, its two governments locked in a bitter cold war.

Large numbers of U.S. troops remained in the south after the war, numbering some 37,000 in 2001. South Korea in the 1960s was a poor agrarian country that could not defend itself. Over the past 30 years, however, the south has emerged as a wealthy trading nation while the fortunes of the north have plummeted. Many South Korean students resent the presence of U.S. forces and seek rapprochement with the north. Their periodic and sometimes violent demonstrations have been a significant force in South Korean politics. The South Korean government long regarded North Korea as a desperate rogue state that could attack at any time (Figure 11.32). Many U.S. policymakers concurred, and some feared that North Korea was on the verge of constructing nuclear arms. A crisis was reached in 1994 when North Korea refused to allow international inspection of its nuclear power plants. Soon afterward, however, it relented in exchange for help in meeting its energy needs.

A major change occurred in 1998 when the South Korean people elected Kim Dae Jung president. Kim had been a political prisoner during South Korea's period of authoritarian government (South Korea had been under a virtual dictatorship in the 1960s and 1970s), and he represents the discriminated-against southwestern region. He immediately began to pursue better relations with North Korea, hoping for eventual reconciliation. In 2001 the government of the south promised $18 million in food aid to the north. Much evidence indicates, however, that North Korea has failed to live up to its agreements, and many outside experts suspect that it is still supporting terrorism and developing weapons of mass destruction.

The Division of China   World War II brought tremendous destruction and loss of life to China. Before the war began, China had already been engaged in a civil conflict between nationalists (who favored an authoritarian capitalist economy) and communists, both of whom hoped to unify the country. The communists had originally been based in the middle Yangtze region, but in 1934 nationalist pressure forced them out. Under the leadership of Mao Zedong, they retreated in the "Long March," which took them to the Loess Plateau, an area close to both the traditional power center of northern China and the industrialized zones of Manchuria. After the Japanese invaded China proper in 1937, the two camps cooperated, but as soon as Japan was defeated, China again found itself embroiled in civil war. In 1949 the communists proved victorious, forcing the nationalists to retreat to Taiwan.

A latent state of war has persisted ever since between China and Taiwan. Although no battles have been fought, gunfire has periodically been exchanged over Kinmen and Matsu, two small Taiwanese islands just off the mainland. The Beijing government still claims Taiwan as an integral part of China and vows eventually to redeem it. The Taiwanese nationalists, for their part, maintain that they represent the true government of China. It was actually made a crime in Taiwan to advocate Taiwanese independence, since the fiction had to be maintained that Taiwan was merely one province of a temporarily divided China. Taiwan is thus equipped with a provincial government that governs exactly the same territory that its national government does.

▲ Figure 11.32 The demilitarized zone in Korea   North and South Korea were divided along the 38th parallel after World War II. Today, even after the conflict of the early 1950s, the demilitarized zone, or DMZ (which runs near the parallel), separates these two states. U.S. armed forces are active in patrolling the demilitarized zone.   *(Yonhap/AP/Wide World Photos)*

▲ Figure 11.33 Chinese soldiers in Tibet   Following a full-scale invasion of Tibet in 1959, China continues to increase its presence through its military forces, the relocation of migrants into the area from other parts of China, and rebuilding programs that mask the traditional Tibetan landscape. Here Chinese soldiers observe a praying Tibetan while eating ice cream bars. *(Galen Rowell/Corbis)*

The idea of the intrinsic unity of China continues to be influential both in China and abroad. In the 1950s and 1960s, the United States recognized Taiwan as the only legitimate government of China, but its policy changed after U.S. leaders decided that it would be more useful—and more realistic—to recognize mainland China. Soon China entered the United Nations, and Taiwan found itself diplomatically isolated, virtually without international recognition. (A number of African and Caribbean countries, however, recognize Taiwan and in return receive Taiwanese aid.) In reality, however, Taiwan is a separate country, and increasing numbers of its citizens would like to proclaim it an independent republic. In 2000 Taiwan elected a former advocate of Taiwanese independence. China, however, threatens to invade if Taiwan declares independence. Some observers think that China might try to reclaim Taiwan, pointing to its overwhelming military advantages; others view such a scenario as highly unlikely, pointing instead to the $20 billion annual trade flow, routed through other countries, between the island and the mainland, and to the international complications that such an action would cause. In 2001 tensions eased slightly after the two countries agreed to allow limited direct links between them. The same year also saw the entry of both China and Tai-wan into the WTO (World Trade Organization). Taiwan was forced to join this organization under the awkward name of "Chinese Taipei," however, in order to avoid the suggestion that it is in actuality a separate sovereign country.

The Chinese Territorial Domain   Despite the fact that it has been unable to regain Taiwan, China has been successful in retaining the Manchu territorial legacy. In the case of Tibet, this has required considerable force; resistance by the Tibetans compelled China to launch a full-scale invasion in 1959. The Tibetans, however, have continued to struggle for real autonomy if not actual independence, as they fear that the Han Chinese now moving to Tibet will eventually outnumber them and undermine their culture (Figure 11.33). Tibet proper (or *Xizang* in Chinese) is an autonomous region by virtue of its non-Han indigenous society, but true autonomy has never been granted (see Chapter 10).

The postwar Chinese government also retained control over Xinjiang in the northwest, as well as Inner Mongolia (or Nei Monggol), a vast territory stretching along the Mongolian border. The indigenous peoples of Xinjiang prefer to call the region "Eastern Turkestan" to emphasize its Turkish heritage. The Han Chinese, however, reject this term because it challenges the unity of China. Like Tibet, Nei Monggol and Xinjiang are classified as autonomous regions. The peoples of Xinjiang are increasingly asserting their religious and ethnic identities, and separatist sentiments are growing. Most Han Chinese, however, regard Nei Monggol and Xinjiang as integral parts of their country, and they regard any talk of succession as treasonous. In the case of Xinjiang, they cite the precedence of Chinese control dating back to the Han Dynasty some 2,000 years ago.

A few ardent Chinese nationalists dream of reclaiming Manchu territories that were taken by other powers. While any actual gains are unlikely, such claims complicate China's international relations. The most important potential conflict is with Russia, which controls a sizable territory north of Manchuria that was grabbed from China in the 1850s. China also claims that some of its former territories in the Himalayas were illegally annexed by Britain when it controlled South Asia, resulting in several unresolved border disputes with India. The two countries went to war in 1962 when China occupied an uninhabited highland district in northeastern Kashmir. Tensions between India and China eased in the 1990s, but their territorial disputes remain unresolved. Owing to this disagreement with India, China has maintained an informal alliance with Pakistan.

China also claims a group of tiny islands—most of them submerged at high tide—in the South China Sea. The Paracel Islands, however, are also claimed by Taiwan and Vietnam, while the Spratly Islands are claimed by Taiwan, Vietnam, Malaysia, and the Philippines. While the islands themselves have little importance, they may overlay considerable petroleum reserves. China has recently constructed buildings on some of these islands that it calls fishing platforms but that the Philippines maintains are military installations. The struggle over these small islands has complicated China's relations with the ASEAN countries of Southeast Asia (see Chapter 13).

One territorial issue was finally resolved in 1997 when China reclaimed Hong Kong. In the isolationist 1950s, 1960s, and 1970s, Hong Kong acted as China's window on the outside world, and it grew prosperous as a capitalist enclave and refuge for wealthy Chinese industrialists. As Chinese relations with the outer world opened in the 1980s, Britain decided to honor its treaty provisions and return Hong Kong to China. China in turn promised that Hong Kong would retain its fully capitalist economic system for at least 50 years. Civil liberties not enjoyed in China itself also remained protected in Hong Kong. Wealthy citizens of Hong Kong, however, are nervous about the situation, and many have acquired residency rights in Canada and Australia as a fallback (Figure 11.34).

In 1999 Macao, the last colonial territory in East Asia, was returned to China. This small Portuguese enclave, located across the estuary from Hong Kong, has functioned largely as a gambling refuge. It remains to be seen whether the unusual function of this incongruous little piece of territory will survive in the new millennium.

The Global Dimension of East Asian Geopolitics

In the early 1950s East Asia was divided into two hostile Cold War camps: China and North Korea were allied with the Soviet Union, while Japan, Taiwan, and South Korea were linked to the United States. The Chinese–Soviet alliance soon deteriorated into mutual hostility, however, and in the 1970s China and the United States found that they could accommodate each other, sharing as they did an enemy in the Soviet Union.

The end of the Cold War, coupled with the rapid economic growth of China, again reconfigured the balance of power in East Asia. The United States no longer needs China to offset the Soviet Union, and the U.S. military has become increasingly worried about the growing power of the Chinese army. China's neighbors have also become more concerned. China now has the largest army in the world, nuclear capability, and sophisticated missile technology. From 1994 to 2001, China's official military budget grew from roughly $7 billion to more than $16 billion. China also has, by some measures, the world's third-largest economy.

China is thus coming of age as a major force in global politics. Whether it is a force to be feared by other countries is a matter of considerable debate. Chinese leaders insist that they have no expansionistic designs and no intention of interfering in the internal affairs of other countries. They do, however, regard concerns expressed by the United States and other countries about their human rights record, as well as their activities in Tibet, as undue meddling in their internal affairs, reminiscent of the European imperialism of the 1800s and early 1900s. Tension between the United States and China reached a peak in April 2001, when a Chinese jet fighter collided with an U.S. spy plane that was monitoring the Chinese navy some 65 miles south of Hainan. After the U.S. plane was forced to land on the island, the Chinese authorities detained its crew and asked for an official apology from the American government. After 11 days an extremely vague

▲ Figure 11.34 China reclaims Hong Kong Fireworks celebrate the return of Hong Kong to China in 1997 after a long period under British colonial rule. Britain agreed to honor its treaty provisions and return Hong Kong to China under the promise that Hong Kong's capitalistic system would remain intact for at least 50 years. (D. Groshong/Corbis/Sygma)

apology was issued and the crew was returned, but U.S.-Chinese relations remained tense.

Relations between China and the United States, however, improved significantly after the September 11, 2001, terrorist attack on the World Trade Center in New York City. China's government generally endorsed the U.S-led antiterrorism campaign and cooperated with the Americans over actions in Afghanistan designed to defeat the terrorists. China has its own concerns about Islamic separatism in Xinjiang, and has long feared infiltration by militants from neighboring Afghanistan. Critics contended, however, that the U.S.-China rapprochement would allow increased repression of Muslims, Tibetans, and other restive Chinese minority groups.

Opinions on China in the United States vary tremendously. Many American leaders, particularly those in the business community, contend that the two countries should ignore their political differences and develop closer economic and cultural ties. Some East Asian experts similarly argue that the United States must respect China's sovereignty more carefully—or risk a future devastating war in the region. Critics, on the other hand, think that China's trade practices are unfair, its labor and human-rights record appalling, and its actions in Tibet unsupportable. Noted political scientist Samuel Huntington goes a step further to view China as the core of a Confucian power bloc locked in dire competition with the West for global influence.

Huntington's critics see more paranoia than clairvoyance in his vision of Confucian unity. They point to the deep internal rifts within the Confucian world. Korea is bitterly divided, and South Korea is wary of Chinese power. China itself remains split, and although Taiwan is small and diplomatically isolated, its economic and even its military power are not insignificant. Japan (which Huntington admittedly sees as more Western than Confucian) is nervous about China and has a long history of animosity toward Korea. As of 2001, Japan's relations with China were deteriorating, largely because of a growing trade imbalance. Japanese nationalists also blame Chinese immigrants for taking local jobs and for joining crime syndicates. They call for a stronger Japanese military and a more assertive attitude toward China.

China and Korea, on the other hand, remain angry at Japan for downplaying the atrocities that its army committed in World War II. This controversy intensified in 2001, leading South Korea to recall its ambassador from Tokyo when Japan issued new textbooks that minimized Japan's aggression in the early 1900s.

Regardless of what one thinks about East Asia's international relations, its recent economic ascent is undeniable. Yet as we shall see in the next section, East Asia does have a number of serious economic problems.

Economic and Social Development: An Emerging Core of the Global Economy

East Asia exhibits extreme disparities of economic and social development. Japan's urban belt contains one of the world's greatest concentrations of wealth, whereas many interior districts of China remain among the world's poorest places. Overall, however, East Asia experienced rapid economic growth in the 1970s, 1980s, and 1990s, and two of its economies, those of Taiwan and South Korea, jumped from the ranks of underdeveloped to developed (Table 11.2). Increasingly, East Asia functions as a global economic core (Figure 11.35). But again, growth has not been evenly distributed. North Korea, for example, has during the same period experienced a rather desperate decline.

In regard to social development, the picture is brighter. Even in the poorer parts of China, most people are reasonably healthy and well educated. As China moves to a market economy, however, such "modern" problems as unemployment and homelessness are beginning to appear.

Japan's Economy and Society

Japan was the pacesetter of the world economy in the 1960s, 1970s, and 1980s. In the early 1990s, however, the Japanese economy experienced a major setback, and growth has been slow ever since. But despite its current problems, Japan is still the world's second largest economic power.

TABLE 11.2 *Economic Indicators*				
Country	Total GNI (Millions of $U.S., 1999)	GNI per Capita ($U.S., 1999)	GNI per Capita, PPP* ($Intl, 1999)	Average Annual Growth % GDP per Capita, 1990–1999
China	979,894	780	3,550	9.5
Hong Kong	165,122	24,570	22,570	1.9
Japan	4,054,545	32,030	25,170	1.1
South Korea	397,910	8,490	15,530	4.7
North Korea	—	—	1,000**	1.0**
Taiwan	—	—	16,100**	5.5**

*Purchasing power parity

Source: The World Bank Atlas, 2001, except those marked **, which note data from the CIA World Factbook, 2000.

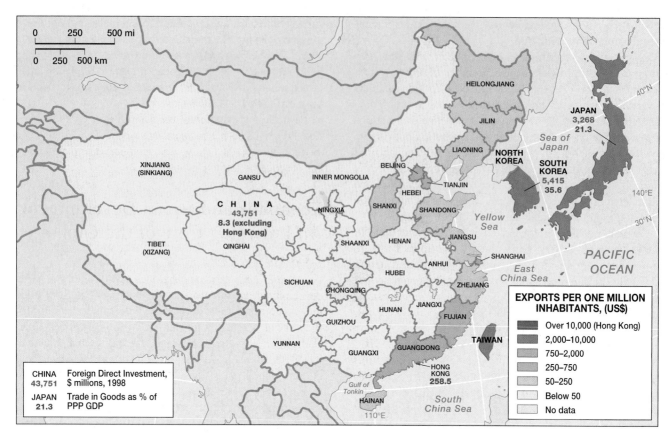

▲ Figure 11.35 East Asia's global ties As this map shows, Hong Kong is one of the world's most important trading ports, while Taiwan, Japan, and South Korea are also highly integrated into global economic networks. In the rest of China, Guangdong and Shanghai are relatively well connected to the global economy through trade, as are Beijing, Tianjin, and Fujian. Interior portions of China, on the other hand, are still largely isolated from the world economy.

Japan's Boom and Bust Although Japan's heavy industrialization began in the late 1800s, most of its people remained poor. The 1950s, however, saw the beginnings of the Japanese "economic miracle." Shorn of its empire, Japan was forced to export manufactured materials. Beginning with inexpensive consumer goods such as clothing and toys, Japanese industry moved to more sophisticated products, including automobiles, cameras, electronics, machine tools, and computer equipment (Figure 11.36). By the 1980s it was the leader in many segments of the global high-tech economy, and its currency, the yen, emerged as one of the world's strongest.

The early 1990s saw the collapse of Japan's inflated real estate market, leading to a banking crisis. At the same time, many Japanese companies discovered that producing labor-intensive goods at home had become too expensive. They therefore began to relocate factories to Southeast Asia and China. Because of these and related difficulties, Japan's economy slumped through the 1990s and into the first years of the new millennium. The Japanese government made several attempts to revitalize the economy through massive state spending, resulting in large public deficits.

Despite its downturn in the 1990s, Japan remains a core country of the global economic system. Its economic system increasingly spans the globe, as Japanese multinational firms invest heavily in production facilities in North America and Europe, as well as in Third World countries. Japan remains a world leader in a large array of high-tech fields, including robotics, optics, and machine tools for the semiconductor industry. It is also the world's largest creditor nation, owning a large percentage of U.S. government bonds.

The Japanese Economic System The business environments in Japan and the United States are distinctive, reflecting different versions of the capitalist economic system. In Japan the bureaucracy maintains far greater control over the economy than it does in the United States. In particular, the Ministry of International Trade and Industry (MITI) has sometimes been viewed as the main engine behind Japan's industrial expansion. Other scholars, however, argue that MITI has been as much an obstacle as an initiator of business success.

Japanese corporations are also structured differently from those of the United States. In Japan large groups of companies (called *keiretsu*) are complexly intertwined, owning each other's stock and buying products and services from each other. Because of these interconnections, Japanese firms are

▲ **Figure 11.36 Automated Japanese auto factory**   Part of Japan's economic success has resulted from automation of its factory assembly lines. Here Mazda automobiles are assembled in a Hiroshima plant. These cars are destined for the east coast of the United States.   *(Jodi Cobbings/NGS Image Collection)*

much less influenced by investors and stockbrokers than are those of the United States. The connections between employers and employees are also tighter in Japan. Japanese workers seldom switch companies, and the core workforce of each corporation is rarely subjected to layoffs.

Proponents of the Japanese system argue that it creates business and social stability and encourages long-term planning. Opponents argue that it reduces flexibility, results in high prices and low profits, and will ultimately prove too costly to maintain. They also point out that Japanese agriculture, wholesaling, and distribution remain rather inefficient. Japan is only now beginning to see the rise of large-scale discount stores, which offer lower prices but also threaten the viability of small-scale merchants as well as downtown shopping districts in small cities.

Living Standards and Social Conditions in Japan
Despite its difficulties in the 1990s, Japan still had a higher per capita **gross national income (GNI)** than the United States in 1999 when calculated on the basis of currency equivalents. Living standards are, however, somewhat lower in Japan, and America's per capita GNI remains larger when calculated on the basis of purchasing power parity. Housing, food, transportation, and services are particularly expensive in Japan. Certain amenities that are almost standard in the United States, such as central heating, remain relatively rare.

Although the Japanese may live in cramped quarters and pay high prices for basic products, they also enjoy many benefits unknown in the United States. Unemployment remains lower than in the United States; health care is provided by the government; and crime rates are extremely low. By such social measures as literacy, infant mortality, and average longevity, Japan surpasses the United States by a comfortable margin. Japan also lacks the extreme poverty found in certain pockets of American society. The disparities of wealth between the haves and the have-nots, while still substantial, are not nearly as great as in the United States. Furthermore, while wealth is concentrated in Tokyo and a few other large cities, regional income disparities are minor.

Japan, of course, has its share of social problems. Koreans and alien residents from other Asian countries suffer discrimination, as do members of the indigenous Japanese underclass, the Burakumin. Japan's more remote rural areas have few jobs, and many have seen population decreases. In small villages it often seems that most of the remaining people are elderly. Farming is an increasingly marginal occupation, and many farm families survive only because one family member works in a factory or office. Professional and managerial occupations in Japan's cities are noted for their long hours and high levels of stress. Overall, social regimentation is greater and civil liberties fewer in Japan than in the United States.

Women in Japanese Society
Critics often contend that Japanese women have not shared the benefits of their country's success. Advanced career opportunities remain limited for women, especially those who marry and have children. The expectation remains that mothers will devote themselves to their families and to their children's education. (The Japanese educational system is hierarchically organized, and poor performance in secondary school almost precludes career success.) Japanese businessmen often work, or socialize with their co-workers, until late every evening, and thus contribute little to child care. Japan's recession of the 1990s seems to have resulted in further reductions in career opportunities for women.

One response to the arduous conditions faced by Japanese women has been a drop in the marriage rate. Many young Japanese women are delaying marriage, and a sizable number may be abandoning it altogether. Japan has seen an even more dramatic decline in its fertility rate. Whether this is due to the domestic difficulties faced by Japanese women or merely the result of the pressures of a postindustrial society is an open question. Fertility rates have, after all, dropped even lower in many parts of Europe.

But regardless of the cause, Japanese women are now bearing so few children that the country's population will begin to decline if the fertility rate remains constant. Economic planners are concerned about the increasing dependency burden that this will cause. A shrinking population means an aging population, and increasing numbers of retirees will have to be supported by shrinking numbers of workers.

The Newly Industrialized Countries

The Japanese path to development was successfully followed by its former colonies: South Korea and Taiwan. Hong Kong also emerged as a newly industrialized economy, although its economic and political systems remained distinctive.

The Rise of South Korea
The postwar rise of South Korea was even more remarkable than that of Japan. During the period of Japanese occupation, Korean industrial development

was concentrated in the north, which is rich in natural resources. The south, in contrast, remained a densely populated, poor agrarian region. South Korea emerged from the bloody Korean War as one of the world's least-developed countries.

In the 1960s the South Korean government initiated a program of export-led economic growth. It guided the economy with a heavy hand and denied basic political freedom to the Korean people. By the 1970s such policies had proved highly successful in the economic realm. Huge Korean industrial conglomerates, known as *chaebol,* moved from exporting inexpensive consumer goods to heavy industrial products and then to high-tech equipment.

At first South Korean firms remained dependent on the United States and Japan for basic technology. By the 1990s, however, this was no longer the case, as South Korea emerged as one of the world's main producers of semiconductors. South Korean wages have also risen at a rapid clip. The country has invested heavily in education (by some measures it has the world's most intensive educational system), which has served it well in the global high-tech economy. Increasingly, South Korean companies are themselves becoming **multinational,** building new factories in the low-wage countries of Southeast Asia and Latin America, as well as in the United States and Europe (a multinational firm operates and manufactures in more than one country).

Contemporary South Korea The political and social development of South Korea has not been nearly as smooth as its economic progress. Throughout the 1960s and 1970s, student-led protests against the dictatorial government were brutally repressed. Dissension was particularly acute in the country's southwest region, an area that had suffered some discrimination. As the South Korean middle class expanded and prospered, pressure for democratization grew, and by the late 1980s it could no longer be denied. But even though open elections are now held and basic freedoms allowed, political tension has not disappeared. In the late 1990s several scandals erupted, revealing substantial corruption at high levels of government and business. South Korea's transition from an underdeveloped country ruled by a dictator to a prosperous democracy has, in other words, been both rapid and troubled.

The South Korean political crisis of the late 1990s was accompanied by economic difficulties. By 1997 the country's banking system was in chaos, and its economy entered a deep recession (Figure 11.37). Although the economy soon recovered, it remained unstable through 2002, racked by bankruptcies and suffering from relatively high unemployment. Critics contend that the South Korean economy needs substantial reforms, in particular the breaking up of the large conglomerates (*chaebol*).

Taiwan and Hong Kong Taiwan and Hong Kong have also experienced rapid economic growth since the 1960s. Both, in fact, have substantially higher per capita **gross domestic product (GDP)** levels than South Korea. The Taiwanese

government, like that of South Korea and Japan, has guided the economic development of the country. Taiwan's economy, however, is organized not around large conglomerates and linked business firms, but rather around small to mid-size family firms. This characteristically Chinese form of business organization is sometimes said to give Taiwan greater economic flexibility than its northern neighbors, but it has prevented it from entering certain industries that require huge concentrations of capital. Unlike the other capitalistic economies of the region, Taiwan suffered little from the "Asian crisis" of the late 1990s. In 2001, however, Taiwan experienced a significant recession, owing largely to the global downturn in the high-tech sector.

Hong Kong, unlike its neighbors, has been characterized by one of the most **laissez-faire** economic systems in the world (*laissez-faire* refers to market freedom, with little governmental control). State involvement has been minimal, which is one reason why the city's business elite was nervous about the transition to Chinese rule. Hong Kong traditional-

▲ **Figure 11.37 Protests in South Korea** During the economic crisis of 1997, workers and students demonstrated against government financial and trade policies that had—in their minds, at least—led to the sudden downturn in the country's economic state. This protest rally is in the capital city, Seoul. *(Yun Jai-hyoung/AP/Wide World Photos)*

ly functioned as a trading center, but in the 1960s and 1970s it emerged as a major producer of textiles, toys, and other consumer goods. By the 1980s, however, such cheap products could no longer be made in such an expensive city. Hong Kong industrialists subsequently began to move their plants to southern China, while Hong Kong itself increasingly specialized in business services, banking, telecommunications, and entertainment. Fears that its economy would falter after the Chinese takeover in 1997 were not substantiated. But by late 2001, Hong Kong's economy was also in recession.

Both Taiwan and Hong Kong have close overseas economic connections. Linkages are particularly tight with Chinese-owned firms located in Southeast Asia and North America. Taiwan's high-technology businesses are also intertwined with those of the United States; there is a constant back-and-forth flow of talent, technology, and money between Taipei and Silicon Valley. Hong Kong's economy is also closely bound with that of the United States (as well as those of Canada and Britain), but its closest connections, not surprisingly, are with the rest of China.

Chinese Development

China dwarfs all of the rest of East Asia in both physical size and population. Its economic takeoff is thus reconfiguring the economy of the entire region. But despite its recent growth, China's economy has a number of serious weaknesses. The vast interior remains trapped in poverty, and many of its largest industries are not competitive. The future of the Chinese economy is thus one of the biggest uncertainties facing both East Asia and the world economy as a whole.

China Under Communism   More than a century of war, invasion, and near-chaos in China ended in 1949 when the communist forces led by Mao Zedong seized power. The new government, inheriting a weak economy, set about nationalizing private firms and building heavy industries. Certain successes were realized, especially in Manchuria, where a large amount of heavy industrial equipment was inherited from the Japanese.

In the late 1950s, however, China experienced an economic disaster ironically called the "Great Leap Forward." One of the main ideas behind this scheme was that small-scale village workshops could produce the large quantities of iron needed for sustained industrial growth. Communist Party officials demanded that these inefficient workshops meet unreasonably high production quotas. In some cases the only way they could do so was to melt down peasants' agricultural tools. Peasants were also forced to contribute such a large percentage of their crops to the state that many went hungry. The result was a horrific famine that may have killed 20 million persons.

The early 1960s saw a return to more pragmatic policies, but toward the end of the decade a new wave of radicalism swept through China. This "Cultural Revolution" aimed at mobilizing young people to stamp out the remaining vestiges of capitalism. Thousands of experienced industrial managers and college professors were expelled from their positions. Many were sent to villages to be "reeducated" through hard physical labor; others were simply killed. The economic consequences of such policies were devastating.

Toward a Postcommunist Economy   When Mao Zedong, who had been revered as an almost superhuman being, died in 1976, China faced a crucial turning point. Its economy was nearly stagnant and its people desperately poor, but the economy of Taiwan, its rival, was booming. A political struggle ensued between pragmatists hoping for change and dedicated communists. The pragmatists emerged victorious, and by the late 1970s it was clear that China would embark on a different economic path. The new China would seek closer connections with the world economy and take a modified capitalist road to development (Figure 11.38; see also "Economic Growth: Housing in China").

China did not, however, transform itself into a fully capitalist country. The state continued to run most heavy industries, and the Communist Party retained a monopoly on political power. Instead of suddenly abandoning the communist model, as the former Soviet Union did, China allowed cracks to appear in which capitalist ventures could take root and thrive.

One of China's first capitalist openings, in the late 1970s, was in agriculture, which had previously been dominated by large-scale communal farms. Individuals were suddenly allowed to act as agricultural entrepreneurs, selling produce in the open market. Owing to this change, the income of many farmers rose dramatically. By the late 1980s, however, the focus of growth had shifted to the urban-industrial sector. As the government became concerned about inflation, it placed price caps on agricultural products and increased taxes on farmers. By 2000 many rural areas in China's poorer interior provinces were experiencing economic distress.

▲ **Figure 11.38 Industrial expansion in coastal China**   One of the important economic reforms that has led to China's recent development was the creation of Special Economic Zones (SEZs) along its eastern coast. Here workers in a coastal automobile plant assemble cars for the rapidly expanding domestic market. *(Serge Attal/Getty Images, Inc.)*

ECONOMIC GROWTH   Housing in China

With the imposition of communist rule in 1949, the residents of Chinese cities were given a housing tradeoff. The state offered them virtually free housing, but it used its monopoly to enforce social control. Those in charge of apartment buildings, for example, were expected to report all foreign visitors and suspicious activities to the local police. Standards, moreover, remained minimal. In 1990 the average person in Shanghai enjoyed only some 50 square feet (4.6 square meters) of living space, and as of 1998 approximately half of the households in the city—one of China's most prosperous—still used communal kitchens and bathrooms.

As China gradually moved from a socialistic to a capitalistic system, pressure began to mount on its housing system. In 1998 the government announced a major liberalization program; rents throughout the country would rise, and housing stock would begin to revert to private control. Similar reform efforts had been implemented earlier in cities such as Shanghai, which are not only the pacesetters in the transition to market economics, but which have also been burdened by high levels of immigration and crowding. By 1998 almost half of Shanghai's housing units were under private ownership, and by 2001, 25 percent of the city's residents actually owned their own homes.

The privatization of housing in China will present tremendous opportunities, but it will also come at a cost. The poorest people may be forced out of the market altogether, leading to the development of shantytown slums. More technical problems may emerge as well. China's banks, which are already burdened by a high level of bad loans, have little experience in mortgages. Some banks may well prosper in this new market, but others could easily fail.

Industrial Reform   Another important early industrial reform involved opening **Special Economic Zones (SEZs)** in which foreign investment was welcome and state interference minimal. The Shenzhen SEZ, adjacent to Hong Kong, proved particularly successful after Hong Kong manufacturers found it a convenient source of cheap land and labor. Additional SEZs were soon opened, mostly in the coastal region. The basic strategy was to attract foreign investment that could generate exports, the income from which could supply China with the capital that it needed to build its infrastructure and thus achieve conditions for sustained economic growth.

Other market-oriented reforms followed. Former agricultural cooperatives were allowed to transform themselves into quasi-capitalist entities. Many of these "township and village enterprises" proved highly successful. By the early 1990s, the Chinese economy was growing at some 8 to 15 percent a year, perhaps the fastest rate of expansion the world has ever seen. China emerged as a major trading nation, and by the mid-1990s it had amassed huge trade surpluses, especially with the United States. Seeking to strengthen its connections with the global economic system, China joined the World Trade Organization (WTO), a body designed to facilitate free trade and provide ground rules for international economic exchange in November 2001. The East Asian economic crisis of the late 1990s had a relatively minor effect on China. Its annual economic growth rate slowed to roughly 6 to 7 percent, still high by global standards. Some evidence indicates, however, that China's banking system has serious problems, making the country vulnerable to recession. Critics contend that China must fully abandon centralized planning if its economic expansion is to continue. China's leadership, however, has made it clear that economic reform will be a gradual process.

Social and Regional Differentiation   The Chinese economic surge unleashed by the reforms of the late 1970s and 1980s resulted in growing **social and regional differentiation.** In other words, certain groups of people—and certain portions of the country—prospered, while others faltered. Despite its official socialism, the Chinese state encouraged the formation of an economic elite, having concluded that only wealthy individuals can adequately transform the economy. The least-fortunate Chinese citizens were sometimes left without work, and many millions migrated from rural villages to seek employment in the booming coastal cities. The government attempted to control the transfer of population, but with only partial success. Shantytowns, as well as homeless populations, began to emerge around some of China's cities. Many migrant children, owing to their lack of official residence, have trouble enrolling in school. If they fail to gain education, they may end up forming an enduring underclass.

China, like all other countries, has always had relatively rich and relatively poor areas. Before the communist period, the Yangtze Delta was the most prosperous part of China, while the flood-prone area immediately to its north (in northern Jiangsu province) was one of the poorest. The communist government attempted to equalize the fortunes of the different regions, giving special privileges to individuals from poor places, such as northern Jiangsu. Such efforts were not wholly successful, and some provinces continued to be deprived. Since the coming of market reforms, moreover, the process of regional economic differentiation has accelerated.

The Booming Coastal Region   Most of the benefits from China's economic transformation have flowed to the coastal region and to the capital city of Beijing. The first beneficiaries were the southern provinces of Guangdong and Fujian. This region was perhaps predisposed to the new economy, since the southern Chinese have long been noted for their mercantile orientation. Guangdong and Fujian have also benefited from their close connections with the overseas Chinese communities of Southeast Asia and North America. (The vast majority of overseas Chinese emigrants came from these provinces.) Proximity

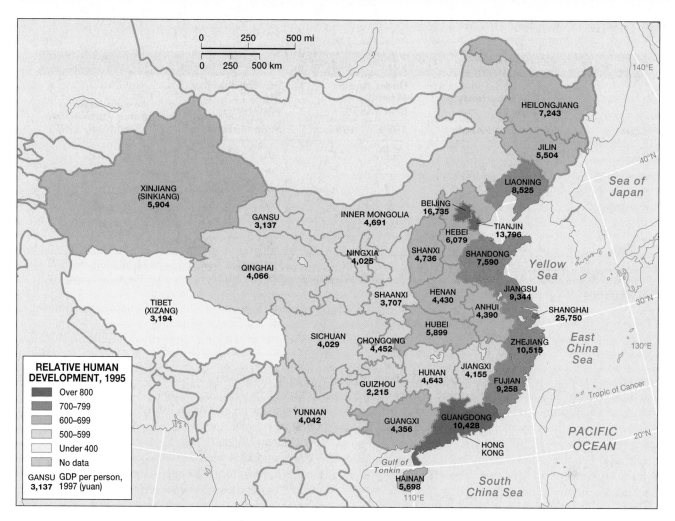

▲ Figure 11.39 Economic and social differentiation in China Although China has seen rapid economic expansion in recent years, the benefits of growth have not been evenly distributed throughout the country. Economic prosperity and social development are concentrated on the coast, especially in Shanghai, Guangdong, Beijing, and Tianjin. Most of the interior remains mired in poverty. The poorest part of China is the upland region of Guizhou in the south-central part of the country. (Source: Robert Benewick and Stephanie Donald, 1999, The State of China Atlas, p. 35, New York: Penguin Reference)

to Taiwan and especially Hong Kong also proved helpful. Vast amounts of capital have flowed to the south coastal region since the 1980s from foreign (and Hong Kong-based) Chinese business networks. American, Japanese, and European firms have also invested heavily in the region.

By the 1990s the Yangtze Delta, centered on the city of Shanghai, reemerged as the most dynamic region of China. The delta was the traditional economic (and intellectual) core of China, and before the communist takeover, Shanghai had been its premier industrial and financial center. The Chinese government, moreover, has encouraged the development of huge industrial, commercial, and residential complexes, hoping to take advantage of the region's dynamism. The Suzhou Industrial Park in Jiangsu is emerging as a hypermodern city of more than a half million persons, thanks largely to a $20 billion investment, most of it Singaporean. Shanghai's Pudong industrial development zone has attracted $10 billion, much of it going to the construction of a new airport and subway system.

The Beijing–Tianjin region has also played a major role in China's economic boom. Its main advantage is its proximity

to political power and its position as the gateway to northern China. The other coastal provinces of northern China have also done relatively well.

Interior and Northern China Most of the other parts of China, in contrast, have seen relatively little economic expansion. Central and northern Manchuria remain relatively prosperous, owing to fertile soils and early industrialization, but have not participated much in the recent boom. Many of the state-owned heavy industries of the Manchurian "**rust belt**," or zone of decaying factories, are relatively inefficient.

Most of the interior provinces of China have likewise missed the wave of growth that struck China in the 1980s and 1990s. In many areas, rural populations continue to grow while the natural environment deteriorates. One consequence is high levels of underemployment and out-migration. By most measures, relative deprivation increases with distance from the coast (Figure 11.39). In 1997 per capita GDP in Guizhou in the south-central area stood at 2,215 yuan (China's currency), whereas in Shanghai it had reached 25,750 yuan. As a result

TABLE 11.3 *Social Indicators and Status of Women*

Country	Life Expectancy at Birth		Under Age 5 Mortality (per 1,000)		Percent Illiteracy (Ages 15 and over)		Female Labor Force Participation (% of total, 1999)
	Male	Female	1980	1999	Male	Female	
China	69	73	65	37	91	75	45
Hong Kong	77	82	—	5	96	90	37
Japan	77	84	11	4	—	—	41
South Korea	71	78	27	9	99	96	43
North Korea	67	73	43	93	99**	99**	41
Taiwan	72	78	—	—	93**	79**	—

*Sources: Population Reference Bureau Data Sheet, 2001, Life Expectancy (M/F); World Develoment Indicators, 2001, Under Age 5 Mortality Rate and Percent Illiteracy, except those marked **, which note data from the CIA World Factbook, 2000; The World Bank Atlas, 2001, Female Participation in Labor Force.*

of such discrepancies, China is encouraging development in the west (including Tibet and Xinjiang as well as the western provinces of China proper), but thus far its efforts have been largely limited to transportation improvement and natural resource extraction.

Scholars debate about how bad conditions really are in the poorer parts of interior China. Some believe that China's official statistics are too positive, hiding a significant amount of hunger and destitution. Others think that the economic boom of the coastal zone is helping raise living standards even in the poorest districts.

Rising Tensions　China's explosive but uneven economic growth has generated a number of problems. Inflation ran as high as 30 percent per year in the late 1980s and early 1990s, making planning difficult and creating hardship for those on fixed incomes. Corruption by state officials is by some accounts rampant; success often seems to depend on knowing the right people and having the proper connections. China's crime rate, moreover, which had been extremely low, began to rise rapidly. Organized crime in particular is emerging as a major problem in many areas.

A more momentous issue has been the struggle for free expression and democracy. The desire for democratic reform has been enhanced by rising incomes and the development of a sizable middle class. In 1989, however, the state crushed a movement for government accountability and democracy and forced opposition to go underground. Whether most Chinese people really want democracy is a controversial issue. Some scholars argue that a Confucian heritage predisposes China toward authoritarian government; others regard such a notion as little more than an apology for tyranny. If the latter camp is correct, tensions will probably mount as long as China's economy prospers while its ruling class denies basic freedoms. Such a warning seemed to many to be born out in 2000 and 2001 after the Chinese government out-

lawed the Falun Gong, a quasi-Buddhist group devoted to meditation and exercise, claiming that it was an evil cult threatening the nation. When Falun Gong members began to protest, they were severely repressed, and many hundreds were evidently killed.

China's political and human-rights policies have complicated its international relations. Sources of tension with the United States and other wealthy countries are also economic in nature. China's large and growing trade surplus and its reluctance to enforce copyright and patent law irritate many of its trading partners. Several U.S. firms have accused Chinese concerns of pirating music, software, and brand names. China has made efforts to stop such activities, but critics contend that its actions have been minimal. As the global market grows and as popular culture becomes globalized, copyright and trademark infringement are becoming increasingly lucrative, and hence increasingly difficult to control.

Social Conditions in China

Despite its pockets of persistent poverty, China has achieved significant progress in social development. Since coming to power in 1949, the communist government has made large investments in medical care and education, and today China boasts impressive health and longevity figures (Table 11.3). The illiteracy rate remains fairly high, but since 97 percent of children supposedly attend elementary school, it will probably drop substantially in the coming years.

Human well-being in China is also geographically structured. The literacy rate, for example, remains relatively low in many of the poorer parts of China, including the uplands of Yunnan and Guizhou and the interior portions of the North China Plain. Literacy is much more widespread in Manchuria, the Yangtze Delta, and most major urban areas. Such regional disparities may increase as the gap between the wealthy and the poor grows.

China's Population Quandary   Population policy also remains an unsettling issue for China. With more than 1.2 billion persons highly concentrated in less than half of its territory, China has one of the world's highest effective population densities. By the 1980s its government had become so concerned that it instituted the famous "one-child policy." Under this plan, couples in normal circumstances are expected to have only a single offspring and can suffer financial and other penalties if they do not comply (Figure 11.40). This strategy has been successful; the average fertility level is now only 1.8, and the population is growing at the relatively slow rate of 0.9 percent a year. Still, when one considers how large the population already is, even a 0.9 percent annual growth rate is worrisome. China will likely reach much more than 1.5 billion persons before stabilization occurs.

Fertility levels in China, as might be expected, vary from province to province. Birthrates are relatively low in most large cities, the Yangtze Delta, parts of the Sichuan Basin, and parts of the North China Plain, and are relatively high in many upland areas of south China, the Loess Plateau, and northwestern China (Figure 11.41). Higher levels of fertility in these poorer and more rural areas will probably lead to increased migration to the booming coastal cities, and perhaps also to heightened regional differentiation.

While China's population policy has reduced its growth rate, it has also generated social tensions and human-rights abuses. Particularly troubling is the growing gender imbalance in the Chinese population. Baby boys in the country now far outnumber baby girls. This asymmetry reflects the practice of honoring one's ancestors; since family lines are traced through male offspring, one must produce a male heir to maintain one's lineage. Many couples are therefore desperate to produce a son. Some opt to bear more than one child, regardless of the penalties they may face. A few even to turn to kidnappers; childless couples reportedly pay up to $3,500 for a young boy, many of whom are nabbed in impoverished Guizhou province. Another option is gender-selective abortion; if ultrasound reveals a female fetus, the pregnancy is sometimes terminated. Baby girls are also commonly abandoned, and well-substantiated rumors of female infanticide circulate. International women's organizations, as well as anti-abortion groups, are concerned about these effects of China's population policy. Many environmentalists, however, applaud China for lowering its birthrate.

The Position of Women   Women have historically had a relatively low position in Chinese society, as is true in most other civilizations. One particularly blatant traditional manifestation of this was the practice of foot binding: the feet of elite girls were usually deformed by breaking and binding them in order to produce a dainty appearance. This crippling and painful practice was eliminated only in the twentieth century. In certain areas of southern China it was also common in traditional times for girls to be married, and

▲ **Figure 11.40 China's population policies**   One aspect of China's population policy is the expansion of child-care facilities so that mothers can be near their children while at work. This enables women to resume participating in the workforce soon after giving birth. This photo shows a typical day-care center attached to an industrial plant in Guangdon province in coastal China. *(Xinhua/Getty Images, Inc.)*

hence to leave their own families, when they were mere toddlers (such marriages, of course, would not be consummated for many years).

Not all women suffered such disabilities in premodern China. Some individuals achieved fame and fortune—a few even through military service. Among the Hakka, women enjoyed a relatively high social position and were seldom subjected to foot binding. Both the nationalist and communist governments have, moreover, sought to begin equalizing the relations between the sexes. Many of their measures have been successful, and women now have a relatively high level of participation in the Chinese workforce (Table 11.3). But it is still true that throughout East Asia—in Japan no less than in China—few women have achieved positions of power in either business or government. As China modernizes and its urban economy grows, the position of its women will probably improve.

▶ Figure 11.41 Demographic change in China In this cartogram, the relative size of each of China's provinces reflects the size of its population. The color coding shows that recent population growth has been fastest in the major urban-industrial areas (Shanghai, Beijing, Tianjin, and Fujian), due to in-migration, and in the poorer areas of the northwest (Gansu, Ningxia) and southwest (Yunnan and Guizhou), due to high fertility. (Source: Robert Benewick and Stephanie Donald, 1999, The State of China Atlas, p. 15, New York: Penguin Reference)

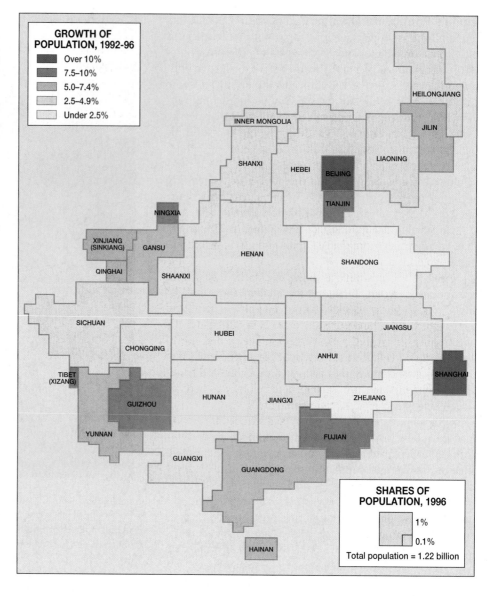

Conclusion

East Asia is united by deep cultural and historical bonds. Particularly important has been the role of the Chinese state, which at various times has encompassed virtually the entire region. Japan is an exception, yet even it has profound historical connections to Chinese civilization. Because of these deep transregional ties, some observers predict the formation of a "Confucian bloc" that will challenge the West for global primacy. Others, however, point to the tensions and mutual animosities that pervade the region, and to the economic and cultural linkages that increasingly connect all of East Asia, including China, to the global system.

A few observers not only doubt the formation of a Confucian bloc, but even question whether China itself will persist as a unified state. The Tibetans and other non-Han peoples of western China long for independence, although this probably remains an unrealistic goal. Internal tensions also seem to be mounting within China proper. The relatively wealthy coastal provinces increasingly resent the control of Beijing and the flow of their tax receipts to the national coffers. The old split between China's more mercantile south and its more bureaucratic north may also be reemerging. Chinese provinces are starting to set their own economic policies, rather than waiting for orders from the center. Contributing to this centrifugal process (one leading to a spreading out or a breaking apart) is the explosive growth of the private economy coupled with the near-stagnation of the state-owned sector. Some China watchers believe that this growing imbalance may eventually sap the strength of the Chinese Communist Party, undermining the country's central authority.

It is, however, difficult to discern present conditions in China, let alone to predict the political future. Just how strong is popular sentiment for democratization in China? How intense are feelings of provincial, as opposed to national, loyalty? Such questions are not easy to address in a country in which freedom of expression is limited. China does, however, have a several-thousand-year legacy of unity under a single government. And if China's spy-plane conflict with the United States in 2001 revealed anything, it was that nationalism is extremely strong in China. It is therefore reasonable to expect China to remain united. If it does, and if its economy continues to grow at current rates, China will clearly be one of the world's leading countries later in the twenty-first century.

Other important questions hinge on environmental issues. Will China's booming economy lead to the emergence of a mass consumer culture like that of Japan? If so, would China's own territory be able to support the massive levels of consumption that this would generate? Since the answer to this query would likely be "no," one must then ask what the ramifications on the global environment and economy would be if China were to follow Japan in provisioning its new needs largely from abroad. Perhaps more important, would global ecological systems be able to handle the increased output of carbon dioxide that full-scale Chinese industrialization, especially if based on China's abundant coal resources, would produce? These are some of the most significant, but ultimately unanswerable, questions of global geography today.

The most important questions about East Asia's future center on China, but significant issues face the region's other countries. It remains to be seen, for example, whether South Korea can successfully manage the transition from being a low-wage exporter to becoming a high-wage technological powerhouse. South Korea also faces a daunting challenge to the north. As long as the present regime remains in power in Pyongyang, South Korea will feel threatened, despite the peace overtures that its government is now making. But other dilemmas would arise if the North Korean government were to fail and the North Korean people were to seek unification with the south. The German experience shows how complicated and how expensive such reunification can be.

In the case of Korea, where the two halves of the country are more evenly balanced in land area and population, reunification would be much more difficult still.

Japan, for its part, now stands at a crossroads brought on by the waning of its long-lasting economic miracle. For all of its economic power and prestige, Japan has yet to find its place in the world. Will Japan's political influence ever begin to match its economic power? This question is hotly debated both in Japan and abroad. Equally important is whether Japan will be able to retain its unique economic and social institutions in the face of mounting global competition. There is a profound sense of uncertainty in the country today; young people no longer feel assured about their futures, and leaders debate whether significant institutional changes are required.

One of the most intensive debates in Japan today centers around the educational system. Japanese education, beginning in junior high school, is noted for its severity and intense pressure. Standardized testing and rote memorization are emphasized, and if one does poorly on tests, one's fate is virtually sealed. This system has served Japan well, resulting in high levels of basic knowledge and competence. Critics contend, however, that it stifles creativity—and that creativity will be the ultimate secret to success in the coming postindustrial global economy. The education ministry is responding to such criticisms. As of 2002, Saturday morning classes will be eliminated, and bureaucratic guidelines for teachers will be reduced.

In an important sense, the education question cuts to the heart of Japan's current dilemma. One must wonder whether Japan's basic social and economic institutions will continue to be markedly different from those of the United States and other Western countries. Those who believe in enduring Confucian values usually answer in the affirmative. Others, however, believe that the process of globalization will force a certain degree of global economic and social convergence. Some thus predict that Japanese women will eventually enter the professions in large numbers; rigid hierarchies at work and school will begin to break down; and Japanese workers will find themselves increasingly threatened with layoffs and downsizing, just as U.S. workers are. Only the future, of course, can tell us who is right.

⊕ Key Terms

anthropogenic landscape (page 468)

autonomous region (page 481)

Burakumin (page 479)

central place theory (page 471)

China proper (page 456)

Cold War (page 483)

Confucianism (page 475)

desertification (page 465)

diaspora (page 479)

geomancy (page 476)

gross domestic product (GDP) (page 494)

gross national income (GNI) (page 493)

hiragana (page 474)

ideographic writing (page 474)

kanji (page 474)

laissez-faire (page 494)

loess (page 459)

Mandarin (page 475)

Marxism (page 477)

multinational corporation (page 494)

particularism (page 477)

pollution exporting (page 460)

regulatory lakes (page 458)

rust belt (page 497)

samurai (page 481)

sediment load (page 458)

Shogun, Shogunate (page 487)

social and regional differentiation (page 496)

Special Economic Zones (SEZs) (page 496)

spheres of influence (page 484)

superconurbation (page 471)

swidden agriculture (page 481)

tectonic plates (page 461)

tonal language (page 479)

tribal peoples (page 479)

urban primacy (page 471)

⊕ Questions for Summary and Review

1. What is the major climatological difference between the Pacific coast of Japan and the coast facing the Sea of Japan?

2. How does the physiography of the Huang He River valley differ from that of the valley of the Yangtze?

3. Why is Japan so much more heavily forested than China?

4. How does the settlement pattern of North Korea differ from that of South Korea?

5. Why has Shanghai emerged as the most populous urban area in China?

6. What are the major ways in which Japanese cities differ from those of the United States?

7. What major cultural features are common to the entire East Asian world region?

8. Where are the non-Han peoples of China concentrated? Why are they concentrated in these areas?

9. In what ways have China and Japan reacted differently—or similarly—to the forces of global culture?

10. What have been the main consequences of the geographical division of Korea into two states?

11. Historically speaking, how did China and Japan act differently as imperial powers?

12. What role does the United States play in the contemporary geopolitics of East Asia?

13. How have the different countries of East Asia followed different paths to economic development?

14. Where in China would one find the most rapid economic development, and why would one find it there?

15. How does the position of women in Japan compare to the position of women in other wealthy, industrialized countries?

⊕ Thinking Geographically

1. Discuss the advantages and disadvantages of China's proceeding to build major dams.

2. Discuss the ramifications, both positive and negative, of Japan's allowing the importation of rice and opening its agricultural lands to urban development.

3. What would be the consequences of China's granting true autonomy to the Tibetans and other non-Han peoples? Independence?

4. What are the potential implications of Taiwan's declaring itself an independent country?

5. Discuss the potential ramifications of the United States' restricting the importation of Chinese goods in order to put pressure on the Chinese government for human-rights reforms.

6. Discuss the advantages and disadvantages of China's current population policies.

7. Do you think that East Asia will emerge as the center of the world economy in the next century?

⊕ Regional Novels and Films

Novels

Sawako Ariyoshi, *The River Ki* (1982, Kodansha)

Xingjian Gao, *Soul Mountain* (2000, Flamingo)

Kazuo Ishigura, *An Artist of the Floating World* (1989, Vintage Books)

Peter H. Lee, *Flowers of Fire* (1986, University of Hawaii Press)

Wang Shuo, *Playing for Thrills* (1998, Penguin)

Hsueh-Chin Tsao, *The Dream of the Red Chamber* (1958, Doubleday)

Films

Fallen Angel (1995, Hong Kong)

Farewell, My Concubine (1993, China)

The Gate of Heavenly Peace (1995, China)

Nomugi Pass (1979, Japan)

Seven Samurai (1954, Japan)

Shall We Dance? (1996, Japan)

The Story of Qiu Ju (1992, China)

A Taxing Woman (1987, Japan)

301,302 (1995, South Korea)

Why Has Bodi-Darma Left for the East? (1989, Korea)

Yellow Earth (1984, China)

⊕ Bibliography

Adshead, Samuel A. M. 1988. *China in World History*. New York: St. Martin's Press.

Boserup, Ester. 1965. *The Conditions of Agricultural Growth: The Economics of Agrarian Change Under Population Pressure.* London: Allen & Unwin.

Cannon, Terry, and Jenkins, Alan, eds. 1990. *The Geography of Contemporary China: The Impact of Deng Xiaoping's Decade.* London: Routledge.

Chapman, Graham P., and Baker, Kathleen M., eds. *The Changing Geography of Asia*. London: Routledge.

Chiu, T. N. 1986. *A Geography of Hong Kong*. Oxford: Oxford University Press.

The Contemporary Atlas of China. 1988. Boston: Houghton Mifflin.

Cotterell, Arthur. 1993. *East Asia: From Chinese Predominance to the Rise of the Pacific Rim*. Oxford: Oxford University Press.

Cybriwsky, Roman A. 1991. *Tokyo: The Changing Profile of an Urban Giant*. Boston: G. K. Hall.

Foret, Philippe. 2000. *Mapping Chengde: The Qing Landscape Enterprise*. Honolulu: University of Hawaii Press.

Gaubatz, Piper Rae. 1996. *Beyond the Great Wall: Urban Form and Transformations on the Chinese Frontiers*. Palo Alto, CA: Stanford University Press.

Gernet, Jacques. 1982. *A History of Chinese Civilization*. Cambridge: Cambridge University Press.

Hanley, Susan B., and Wolf, Arthur P. 1985. *Family and Population in East Asian History*. Palo Alto, CA: Stanford University Press.

Hoare, James, and Pares, Susan. 1988. *Korea: An Introduction*. London: Routledge.

Knapp, Ronald G. 2000. *China's Old Dwellings*. Honolulu: University of Hawaii Press.

Knapp, Ronald G., ed. 1992. *Chinese Landscapes: The Village as Place*. Honolulu: University of Hawaii Press.

Kolb, A. 1971. *East Asia, China, Japan, Korea, Vietnam: Geography of a Culture Region*. London: Methuen.

Kornhauser, David H. 1982. *Japan: Geographical Background to Urban-Industrial Development*. London: Longman.

Lee, Ki-baik. 1984. *A New History of Korea*. Cambridge, MA: Harvard University Press.

Leeming, Frank. 1993. *The Changing Geography of China*. Cambridge, MA: Blackwell.

Litzinger, Ralph. 2000. *Other Chinas: The Yao and the Politics of National Belonging*. Durham, NC: Duke University Press.

McCune, Shannon. 1956. *Korea's Heritage: A Regional and Social Geography*. Rutland, VT: Charles E. Tuttle.

Myers, Ramon H., and Peattie, Mark R. 1984. *The Japanese Colonial Empire, 1895–1945*. Princeton, NJ: Princeton University Press.

Perdue, Peter C. 1987. *Exhausting the Earth: State and Peasant in Hunan, 1500–1850*. Cambridge, MA: Harvard University Press.

Pomeranz, Kenneth. 1993. *The Making of a Hinterland: State, Society, and Economy in Inland North China, 1853–1937*. Berkeley, CA: University of California Press.

Rowe, William T. 1984. *Hankow: Commerce and Society in a Chinese City, 1796–1889*. Palo Alto, CA: Stanford University Press.

Rozman, Gilbert, ed. 1991. *The East Asian Region: Confucian Heritage and Its Modern Adaptation*. Princeton, NJ: Princeton University Press.

Skinner, G. William. 1964. "Marketing and Social Structure in Rural China." *Journal of Asian Studies* 24(1), Part I: 1–43; Part II, 195–228.

Skinner, G. William, ed. 1977. *The City in Late Imperial China*. Palo Alto, CA: Stanford University Press.

Smil, Vaclav. 1984. *The Bad Earth: Environmental Degradation in China*. New York: M. E. Sharpe.

Smith, Christopher J. 1991. *China: People and Places in the Land of One Billion*. Boulder, CO: Westview.

Songqiao, Zhao. 1986. *Physical Geography of China*. New York: John Wiley & Sons.

Spence, Jonathan D. 1990. *The Search for Modern China*. New York: W. W. Norton.

Sun, Jingzhi, ed. 1988. *The Economic Geography of China*. Oxford: Oxford University Press.

Totman, Conrad. 1989. *The Green Archipelago: Forestry in Pre-industrial Japan*. Berkeley, CA: University of California Press.

Tregear, T. R. 1965. *A Geography of China*. Chicago: Aldine.

Trewartha, Glenn T. 1965. *Japan: A Geography*. Madison: University of Wisconsin Press.

Veeck, Gregory, ed. 1991. *The Uneven Landscape: Geographical Studies in Post-Reform China*. Baton Rouge, LA: Geoscience Publications.

Wigen, Karen. 1995. *The Making of a Japanese Periphery, 1750–1920*. Berkeley, CA: University of California Press.

TURKMENISTAN

TAJIKISTAN

AFGHANISTAN

CHINA

Karakoram Range

K a s h m i r

Islamabad ★
Srinagar • JAMMU AND
KASHMIR

Indus R.

HIMACHAL
PRADESH

BHUTAN

Lahore •
Amritsar •
PUNJAB

CHANDIGARH

H I M A L A Y A S

ARUNACHAL
PRADESH

IRAN

PAKISTAN

HARYANA

UTTARANCHAL

Thimphu ★

Delhi •
New Delhi ⊛

DELHI

NEPAL

SIKKIM

Brahmaputra R.

ASSAM

NAGALAND

Kathmandu ★

Indus R.

RAJASTHAN

UTTAR PRADESH

MEGHALAYA

MANIPUR

Karachi •

S i n d

Ganges R.

BIHAR

BANGLADESH

Tropic of Cancer

Dhaka ⊛

TRIPURA

MIZORAM

Arabian
Sea

GUJARAT

MADHYA
PRADESH

JHARKHAND

WEST
BENGAL

MYANMAR
(BURMA)

20°N

DAMAN
AND DIU

Narmada R.

CHHATTISGARH

Calcutta •
Sundarbans

DADRA AND
NAGAR HAVELI

I N D I A

ORISSA

Bay of
Bengal

MAHARASHTRA

Mumbai
(Bombay) •

Godavari R.

D e c c a n

Eastern

P l a t e a u

Krishna R.

Ghats

Goa
GOA •

KARNATAKA

ANDHRA
PRADESH

PONDICHERRY

Andaman
Islands
(INDIA)

Western Ghats

Bangalore •

Madras •

Andaman
Sea

10°N

Lakshadweep
(INDIA)

PONDICHERRY

PONDICHERRY

Malabar Coast

LAKSHADWEEP

TAMIL NADU

ANDAMAN AND
NICOBAR ISLANDS

KERALA

Jaffna •

Nicobar
Islands
(INDIA)

SRI LANKA

Colombo ⊛

Male ★

MALDIVES

INDIAN OCEAN

0°

Equator

70°E

80°E

90°E

CHAPTER 12

South Asia

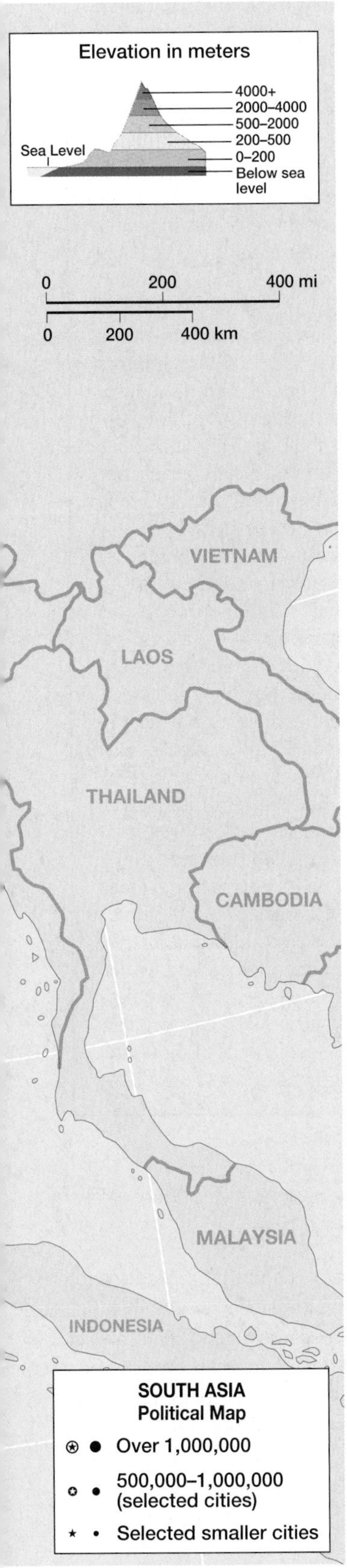

Elevation in meters

4000+
2000–4000
500–2000
200–500
0–200
Below sea level

Sea Level

0   200   400 mi

0   200   400 km

VIETNAM

LAOS

THAILAND

CAMBODIA

MALAYSIA

INDONESIA

SOUTH ASIA
Political Map

⊛ ● Over 1,000,000

○ • 500,000–1,000,000
(selected cities)

★ • Selected smaller cities

Southern Asia, a land of deep historical and cultural commonalities, has recently experienced intense political conflict. Since independence from British colonialism in 1947, the two largest countries, India and Pakistan, have fought several wars and remain locked in bitter animosity. This political tension reaches such heights that many arms control experts maintain that South Asia is the leading candidate for a nuclear war. Religious divisions underpin this geopolitical turmoil, for India is primarily a Hindu country (with a large Muslim minority), while neighboring Pakistan and Bangladesh are both predominantly Muslim (Figure 12.1; see "Setting the Boundaries"). Even within India, cultural and political tensions, inflamed by the rise of Hindu nationalism, jeopardize the future of this huge federal state.

Parallel to these geopolitical tensions are demographic concerns. Given its current rate of growth, South Asia could soon surpass East Asia as the world's most populous region. The underlying issue, however, is not simply the gigantic population of this region, but whether it can support these people, given its economy and resource base. Although agricultural production has increased slightly faster than population over the past three decades, many experts think that these improvements are approaching their limit and that population may once again outpace food resources. Compounding this serious situation is the widespread poverty of South Asia; it is, along with Sub-Saharan Africa, the poorest part of the world. Roughly half of India's people subsist on less than one dollar a day.

South Asia is far less connected to the contemporary globalized world than are East or Southeast Asia. Few would use the term *economic tiger* to describe any South Asia country because of their historically slow rates of economic growth and inward orientation designed to meet internal needs rather than produce export goods. South Asia may soon, however, have a significant global impact, given the impressive levels of scientific and technical skills found within the labor force, the international links that the region has established through migration, and the enormous size of its local markets. Parts of India, for example, have recently emerged as major players in the global software industry, and are now tightly linked to California's Silicon Valley.

◀ Figure 12.1 South Asia   This region is the second most populous in the world, primarily because of India's population of more than 1 billion people. Bracketing India on the west and east are Pakistan and Bangladesh, two large countries with predominantly Muslim populations. The two Himalayan countries of Nepal and Bhutan, along with the island nations of Sri Lanka and the Maldives, round out the region.

South Asia forms a distinct landmass separated from the rest of the Eurasian continent by a series of sweeping mountain ranges, including the Himalayas—the highest in the world. For this reason it is often called the Indian **subcontinent**, in reference to its largest country. The boundaries of the region, however, do not exactly coincide with the mountain crest, which to the northwest lies in central Afghanistan. At one time Afghanistan was often classified within South Asia, but at present it is better placed in Central Asia. To the south, South Asia includes a number of islands in the Indian Ocean, including the countries of Sri Lanka and the Maldives, as well as the Indian territories of the Lakshadweep and the Andaman and Nicobar Islands. Farther west, the islands of Reunion and Mauritius are occasionally placed in South Asia, but for our purposes are classified with Sub-Saharan Africa.

The unity of this region comes more from a shared history than from any contemporary political or cultural coherence. In the past, the region could be characterized as a meeting ground between the major religions of Hinduism, Islam, and, to a lesser degree, Buddhism. Additionally, South Asia shares a colonial history dominated by British administration. Today, however, the region is torn by ethnic and political tensions that have surfaced more intensely since independence in 1947. As often as not, such tensions have religious underpinnings.

India is by far the largest South Asian country, both in size and in population. Covering more than 1 million square miles from the Himalayan crest to the southern tip of the peninsula at Cape Comorin, India is the world's seventh largest country in terms of area and, with more than 1 billion inhabitants, second only to China in population. Although ostensibly a Hindu country, India contains a tremendous amount of religious, ethnic, linguistic, and political diversity. It is important to realize that India, as a result of both its size and its diversity, is much more comparable to Europe as a whole than to any particular European country. Correspondingly, each Indian state is roughly comparable to any given European country. Gujarat, for example, is roughly of the same size and population as Italy, and its culture and language are equally distinctive.

Pakistan, the next largest country, is less than one-third the size of India. Stretching from the lofty northern mountains to the arid coastline on the Arabian Sea, its population of 145 million is only about 15 percent of India's. Despite this imbalance, these two countries—both with nuclear weapon systems—are locked in a tense power struggle, especially over the disputed territory of Kashmir. Until independence in 1947, Pakistan was one portion of a larger undivided British colonial realm simply called *India*. Because of its strong ties to Islam, however, some Pakistanis argue that their country is now more closely connected to its Muslim neighbors in Southwest Asia than it is to India and the rest of South Asia.

Bangladesh, on India's eastern shoulder, is also a Muslim country. Originally created as East Pakistan in the hurried partition of India in 1947, it achieved independence after a brief civil war in 1971. Although a small country in area (54,000 square miles), Bangladesh is one of the most densely populated in the world—and also one of the poorest—with a population of 133.5 million in an area about the size of Wisconsin. Bangladesh has a short border with Burma, but it is otherwise virtually engulfed by India, which wraps around the country to the north and the northeast.

Nepal and Bhutan are both located in the Himalayan Mountains, sandwiched between India and the Tibetan plateau of China. Nepal, with some 23.5 million people, is much the larger of the two in both area and population, and is far more open to and engaged with the contemporary world. Bhutan, on the other hand, has purposely remained disconnected from the global system, remaining a relatively isolated Buddhist kingdom of less than 1 million inhabitants.

The two island countries of Sri Lanka (formerly Ceylon) and the Maldives round out South Asia. Each of these countries has its own problems that cloud the future. Sri Lanka (population of 19.5 million) has since 1983 been mired in a civil war that has taken a high toll on its potentially vibrant economy. The predicament facing the small island nation of the Maldives, in contrast, is not of its own making, but instead speaks to the linkages of all countries in our globalized world. If global warming continues and if sea levels rise as predicted, this island nation—where the highest point is only 6 feet above sea level—will be entirely flooded and its 300,000 inhabitants will have to seek higher ground somewhere else.

Environmental Geography: Diverse Landscapes, from Tropical Islands to Mountain Rim

South Asia's environmental geography covers a wide spectrum that ranges from the highest mountains in the world to densely populated delta islands barely above sea level; from one of the wettest places on Earth to dry, scorching deserts; from tropical rain forests to degraded scrublands to coral reefs (Figure 12.2). All of these ecological zones have their own distinct and complex environmental problems. To illustrate the complexity of South Asian environmental issues, let us begin with the gripping saga of Rajkumar and Veerappan— the film star and the poacher king.

The Film Star and the Poacher King

In southern India, where the states of Kerala, Tamil Nadu, and Karnataka converge, lies a large area of steep mountains and wild forests containing elephants, leopards, tigers, bears, and a huge array of other rare species. Although several national parks have been established, access is still difficult and perilous, largely because these forests are also the haunts of one of India's most notorious outlaw gangs, let by the famed Koose Veerappan. Due to Veerappan's poaching, ivory-bearing male elephants are extremely rare; herds are largely composed of tuskless females. By the late 1990s, as ivory grew scarce and the Indian government clamped down on exports, Veerappan turned increasingly to the profitable venture of poaching sandalwood, a rare tree that yields a fragrant oil highly esteemed in much of Asia.

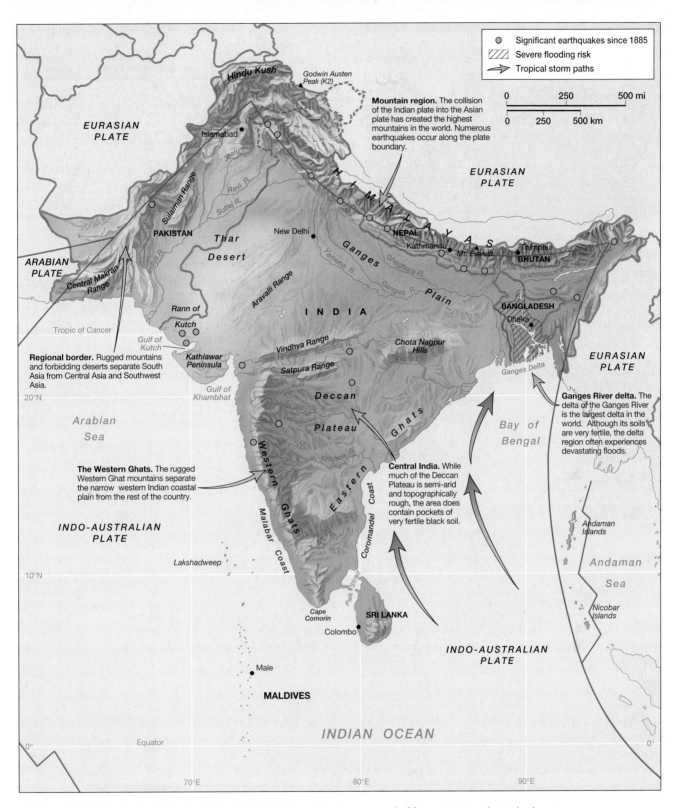

Legend:
- ● Significant earthquakes since 1885
- ▨ Severe flooding risk
- → Tropical storm paths

Mountain region. The collision of the Indian plate into the Asian plate has created the highest mountains in the world. Numerous earthquakes occur along the plate boundary.

Regional border. Rugged mountains and forbidding deserts separate South Asia from Central Asia and Southwest Asia.

The Western Ghats. The rugged Western Ghat mountains separate the narrow western Indian coastal plain from the rest of the country.

Central India. While much of the Deccan Plateau is semi-arid and topographically rough, the area does contain pockets of very fertile black soil.

Ganges River delta. The delta of the Ganges River is the largest delta in the world. Although its soils are very fertile, the delta region often experiences devastating floods.

▲ **Figure 12.2 Physical geography of South Asia** This region is composed of four extensive physical subregions: the high Himalayan mountains in the north; the expansive Indus-Ganges lowland that reaches from Pakistan in the west to the delta lands of Bangladesh; peninsular India, dominated by the Deccan Plateau; and the island realm that includes Sri Lanka and the Maldives. Many of this seismically active region's landscapes are products of the slow northward movement of the Indo-Australian tectonic plate against the Eurasian plate. Typhoons from the Bay of Bengal pose a significant risk to parts of eastern South Asia.

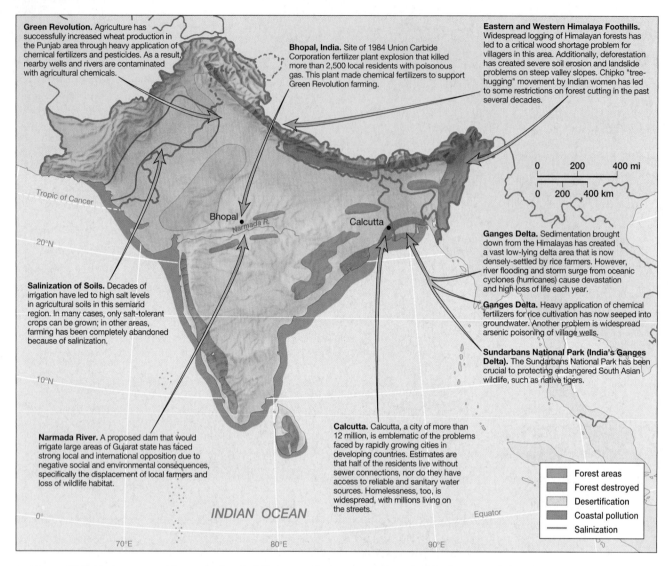

Green Revolution. Agriculture has successfully increased wheat production in the Punjab area through heavy application of chemical fertilizers and pesticides. As a result, nearby wells and rivers are contaminated with agricultural chemicals.

Bhopal, India. Site of 1984 Union Carbide Corporation fertilizer plant explosion that killed more than 2,500 local residents with poisonous gas. This plant made chemical fertilizers to support Green Revolution farming.

Eastern and Western Himalaya Foothills. Widespread logging of Himalayan forests has led to a critical wood shortage problem for villagers in this area. Additionally, deforestation has created severe soil erosion and landslide problems on steep valley slopes. Chipko "tree-hugging" movement by Indian women has led to some restrictions on forest cutting in the past several decades.

Salinization of Soils. Decades of irrigation have led to high salt levels in agricultural soils in this semiarid region. In many cases, only salt-tolerant crops can be grown; in other areas, farming has been completely abandoned because of salinization.

Ganges Delta. Sedimentation brought down from the Himalayas has created a vast low-lying delta area that is now densely-settled by rice farmers. However, river flooding and storm surge from oceanic cyclones (hurricanes) cause devastation and high loss of life each year.

Ganges Delta. Heavy application of chemical fertilizers for rice cultivation has now seeped into groundwater. Another problem is widespread arsenic poisoning of village wells.

Sundarbans National Park (India's Ganges Delta). The Sundarbans National Park has been crucial to protecting endangered South Asian wildlife, such as native tigers.

Narmada River. A proposed dam that would irrigate large areas of Gujarat state has faced strong local and international opposition due to negative social and environmental consequences, specifically the displacement of local farmers and loss of wildlife habitat.

Calcutta. Calcutta, a city of more than 12 million, is emblematic of the problems faced by rapidly growing cities in developing countries. Estimates are that half of the residents live without sewer connections, nor do they have access to reliable and sanitary water sources. Homelessness, too, is widespread, with millions living on the streets.

Tropic of Cancer
20°N
10°N
0°
Equator
70°E   80°E   90°E
Bhopal   Narmada R.   Calcutta
INDIAN OCEAN

0   200   400 mi
0   200   400 km

Forest areas
Forest destroyed
Desertification
Coastal pollution
Salinization

▲ **Figure 12.3 Environmental issues in South Asia** As might be expected in a highly diverse and densely populated region, there are a wide range of environmental problems. These range from salinization of irrigated lands in the dry lands of Pakistan and western India to groundwater pollution from Green Revolution fertilizers and pesticides. Additionally, deforestation and erosion are widespread in upland areas.

On July 30, 2000, Veerappan—who is viewed as a Robin Hood figure in some quarters—shocked India by kidnapping one of the country's most noted actors, Rajkumar. Although nationally famous, Rajkumar is especially beloved in his home state of Karnataka, where he has been an ardent champion of the local language and culture. Such activities did not escape the attention of Veerappan, whose initial ransom demands were basically political. The bandit chief is an equally passionate defender of his native state of Tamil Nadu, and he holds a serious grudge against Karnataka. Not only did he demand that Karnataka release larger amounts of water from the Cauvery River to downstream areas in Tamil Nadu, but he also asked for Tamil to be made an official language in Karnataka and for a statue of the Tamil poet Thiuruvalluvar to be erected in Bangalore, Karnataka's capital and the center of India's internationally powerful software industry.

Rajkumar was released four months later without any of these demands having been met. The evidence now suggests that Veerappan's political claims were at least in part a smokescreen. Rajkumar seems to have run afoul of Veerappan because of his son's shady dealings in the high-quality granite business on lands controlled by the bandit group. The woman who negotiated the actor's release, moreover, had recently been accused of cheating an Italian granite importer out of a large amount of money. Although the details of the case will probably remain hidden, it is likely that Veerappan was mainly seeking a larger cut of this lucrative internationally oriented business.

The story of Rajkumar and Veerappan tells a lot about contemporary India, with its fantastic but threatened biological treasures, serious water shortages, and rapidly increasing global economic connections. It also reveals the intensity of local ethnic sentiments in a country where each state prizes its own

◀ **Figure 12.4 Flooding in Bangladesh** Devastating floods are common in the low-lying delta lands of Bangladesh. Heavy rains come with the southwest monsoon, especially to the Himalayas, and powerful cyclones often develop over the Bay of Bengal. *(Baldev/Corbis/Sygma)*

language and distinctive cultural heritage. That, however, is a story for later in this chapter. For the moment, let us examine in detail a few key environmental issues faced by contemporary South Asia.

Environmental Issues in South Asia

As is true in other poor and densely settled regions of the world, a raft of serious ecological issues plague South Asia. The region also suffers from the usual environmental problems of water and air pollution that accompany early industrialization and manufacturing. South Asia has, in fact, suffered from some of the world's worst environmental disasters. The 1984 explosion of a fertilizer plant in Bhopal, India, for example, killed more than 2,500 persons and maimed many more. Compounding all of these problems are the immense numbers of new people added each year through natural population growth. Behind most environmental issues lies the shadow of South Asia's rapidly mounting population (Figure 12.3).

Natural Hazards in Bangladesh   The link between population pressure and environmental problems is nowhere clearer than in the delta area of Bangladesh, where the search for fertile land has driven people into hazardous areas, putting millions at risk from seasonal flooding as well as from the powerful *cyclones* that form over the Bay of Bengal. For millennia, drenching monsoon rains have eroded and transported huge quantities of sediment from the Himalayan slopes to the Bay of Bengal by the Ganges and Brahmaputra rivers, gradually building this low-lying and fertile deltaic environment. Until a few hundred years ago, most of the delta was lightly inhabited and heavily forested. But with continual population growth, people gradually moved into the swamps to transform them into highly productive rice fields. While this agricultural activity has supported Bangladesh's vast population, it has exacerbated the area's natural hazards.

Although periodic flooding is a natural, even beneficial, phenomenon that enlarges deltas by depositing fertile riverborne sediment, flooding has become a serious problem for people inhabiting these low-lying areas. In September 1998, for example, more than 22 million Bangladeshis were made homeless when water covered two-thirds of the country (Figure 12.4). Flood levels were the highest on record, even though the death toll of 700 was much reduced from the 1988 disaster that killed more than 3,000.

With the populations of both Bangladesh and northern India growing rapidly, there is a strong possibility that flooding will take even higher tolls in the next decade as desperate farmers relocate into the hazardous lower floodplains. Deforestation of the Ganges and Brahmaputra headwaters magnifies the problem. Since forest cover and ground vegetation intercept rainfall and slow runoff, deforestation in the river headwaters results in increased flooding.

Forests and Deforestation   Forests and woodlands once covered most of the region, except for the desert areas in the northeast, but in most areas tree cover has vanished as a result of human activities. The Ganges Valley and coastal plains of India, for example, were largely deforested thousands of years ago to make room for agriculture. Elsewhere, forests were cleared more gradually for agricultural, urban, and industrial expansion. Railroad construction in the nineteenth century was especially destructive. More recently, hill slopes in the Himalayas and elsewhere have been logged for commercial purposes, serving both internal and export needs. Extensive forests can still be found, however, in the far northern, the southwestern, and the east-central areas.

As a result of deforestation, many villages of South Asia suffer from a shortage of fuel-wood for household cooking, forcing people to burn dung cakes from cattle. While this low-grade fuel provides adequate heat, it also diverts nutrients

▲ **Figure 12.5 Chipko tree-huggers** As a protest against deforestation in their countryside, women have resorted to the ancient practice of hugging trees to prevent them from being cut. Because of these protests, the state of Uttar Pradesh recently banned commercial logging. *(Rod Johnson/Panos Pictures)*

that could be used as fertilizers to household fires. Where wood is available, collecting it may involve many hours of female labor because the remaining sources of wood are often far from the villages. In many areas, extensive eucalyptus stands have been planted to supply fuel-wood and timber. Unfortunately, these Australian trees support very little wildlife, resulting in virtual biological deserts.

Concern about India's fast-disappearing forests has led women in northern India to engage in the historical practice of "tree-hugging" to protect ancient groves from logging. The modern **Chipko movement**, named after an Indian word for "hug," started in 1973 as a women's protest movement against deforestation and has spread throughout many Himalayan villages (Figure 12.5). Because of this movement, the Indian state of Uttar Pradesh recently banned commercial cutting in its uplands. Other states have also recently made significant strides in forest protection.

Wildlife: Extinction and Protection
Although the overall environmental situation in South Asia might be grim, there are a few issues that inspire optimism. The region has, for example, managed to retain a diverse assemblage of wildlife despite population pressure and intense poverty. The only remaining Asiatic lions live in India's Gujarat state, and even Bangladesh has retained a viable population of tigers in the Sundarbans, the mangrove forests of the Ganges delta. Wild elephants still roam several large reserves in India, Sri Lanka, and Nepal. This protection of wildlife in South Asia far exceeds that in other Asia regions; in China, for example, elephants and many other large mammal species have been extinct for centuries.

Although the current state of wildlife protection may be positive, the future is uncertain as pressure mounts to convert

wildlands to farmlands. The best remaining zone of extensive wildlife habitat is in India's far northeast, an area subjected to rapid immigration. Moreover, wild animals, particularly tigers and elephants, threaten crops, livestock, and even people living near the reserves. Adjacent to the Sundarbans reserve, almost 200 people were injured or killed by tigers in a recent five-year period. When a rogue elephant herd ruins a crop or a tiger kills livestock, government agents are usually forced to destroy the animal. As result, several key species now seem to be in decline, although conservation groups struggle valiantly on their behalf.

The Four Subregions of South Asia

To better understand environmental conditions in this vast region, South Asia can be broken down into four physical subregions, starting with the high mountain ranges of its northern fringe and extending to the tropical islands of the far south. Lying south of the mountains are the extensive river lowlands that form the heartland of both India and Pakistan. Between river lowlands and the island countries is the vast area of peninsular India, extending more than 1,000 miles (1,600 kilometers) from north to south (Figure 12.6).

Mountains of the North South Asia's northern rim of mountains is dominated by the great Himalayan Range, forming the northern borders of India, Nepal, and Bhutan. These mountains are linked to the equally high Karakoram Range to the west, extending through northern Pakistan. More than two dozen peaks exceed 25,000 feet (7,620 meters), including the world's highest mountain, Everest, on the Nepal–China (Tibet) border at 29,028 feet (8,848 meters). To the east are the lower Arakan Yoma Mountains, forming the border between India and Myanmar (Burma) and separating South Asia from Southeast Asia.

These formidable mountain ranges were produced by tectonic activity caused by peninsular India pushing northward into the larger Eurasia Continental Plate; as a result of the collision between these two tectonic plates, great mountain ranges have been folded and upthrust. The entire region is seismically active, putting all of northern South Asia in serious earthquake danger (Figure 12.2). While most of South Asia's northern mountains are too rugged and high to support dense human settlement, there are major population clusters in the Katmandu Valley of Nepal, situated at 4,400 feet (1,341 meters), and the Valley, or Vale, of Kashmir in northern India, at 5,200 feet (1,585 meters).

Indus-Ganges-Brahmaputra Lowlands South of the northern mountains lie large lowlands created by three major river systems that have carried sediments eroded off of the mountains through millions of years, building vast alluvial plains of fertile and easily farmed soils. These river lowlands are densely settled and constitute the population core areas of Pakistan, India, and Bangladesh.

Of these three rivers the Indus is the longest, covering more than 1,800 miles (2,880 kilometers) as it flows southward from the Himalayas through Pakistan to the Arabian

ters), joining the Ganges in central Bangladesh and spreading out over the vast delta, the largest in the world. Unlike the sparsely populated Indus delta in Pakistan, the Ganges-Brahmaputra delta is very densely settled, with more than 3,000 people per square mile (or 1,200 per square kilometer).

Peninsular India   Jutting southward is the familiar shape of peninsular India, made up primarily of the Deccan Plateau, which is bordered on each coast by narrow coastal plains backed by elongated north–south mountain ranges. On the west are the higher Western Ghats, which are generally about 5,000 feet (1,524 meters) but reach higher than 8,000 feet (2,438 meters) near the peninsula's southern tip; to the east, the Eastern Ghats are lower and discontinuous, thus forming less of a transportation barrier to the broader eastern coastal plain of peninsular India. On both coastal plains, fertile soils and an adequate water supply support population densities comparable to the Ganges lowland to the north.

Soils are poor or middling over much of the Deccan, but in Maharashtra state basaltic lava flows have produced particularly fertile black soils. A reliable water supply for agriculture is a major problem in most areas. Much of the western plateau lies in the rain shadow of the Western Ghats, giving it a somewhat dry climate. Small reservoirs or tanks have been the traditional method for collecting monsoon rainfall for use during the dry season. More recently, deep wells and powerful pumps have mined groundwater to support irrigated crops and village water needs.

Partly because of the overuse of these aquifers, the Indian government plans a series of large dams to provide for irrigation. These plans, however, are controversial because—like dams in many parts of the world—the reservoirs will dislocate hundreds of thousands of rural residents. A case in point is the Sardar Sarovar Dam project on the Narmada River (Figure 12.3) in the state of Madhya Pradesh, which alone will displace more than 100,000 people. Local residents and activists throughout India have joined forces in opposition, but farmers in neighboring Gujarat, the main beneficiaries, strongly support the project.

The Southern Islands   At the southern tip of peninsular India lies the island country of Sri Lanka, which is almost linked to India by a series of small islands called Adam's Bridge. Sri Lanka is ringed by extensive coastal plains and low hills, but mountains reaching over 8,000 feet (2,438 meters) occupy the southern interior, providing a cool, moist climate. Because the main monsoon winds arrive from the southwest, that portion of the island is much wetter than the rain shadow area of the north and east.

Forming a separate country are the Maldives, a chain of more than 1,200 islands stretching south to the equator some 400 miles (640 kilometers) off the southwestern tip of India. The combined land area of these islands is only about 116 square miles (290 square kilometers), and only a quarter of the islands are actually inhabited. The islands of the Maldives are flat, low coral atolls. With its highest elevation just over 6 feet (2 meters) above sea level, the country plays

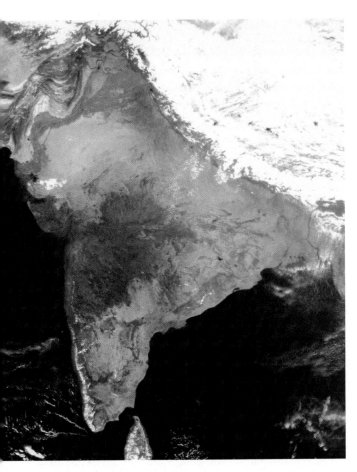

▲ **Figure 12.6 South Asia from space**   The four physical subregions of South Asia are clearly seen in this satellite photograph, from the snow-clad Himalayan mountains in the north to the islands of the south. The Deccan Plateau is dark, fringed by white clouds as moist air is lifted over the uplands of the Western Ghats.   *(Earth Satellite Corporation/Science Photo Library/Photo Researchers, Inc.)*

Sea, providing much-needed irrigation waters to the desert areas in the southern portions of that country. It was along the Indus Valley that one of the world's earliest civilizations arose some 5,000 years ago.

Even more densely settled is the vast lowland of the Ganges, which, after flowing out of the Himalayas, travels southeasterly some 1,500 miles (2,400 kilometers) to empty into the Bay of Bengal. The Ganges has not only provided the fertile alluvial soil that has made northern India one of the world's most densely settled areas, but has also long served as a transportation corridor. Given the central role of this important river in South Asia's past and present, it is understandable why Hindus consider the Ganges sacred. As with the Indus Valley, the Ganges basin was a hearth area of early civilization, and the river is the lifeblood of the most populous area of contemporary India.

Although this large South Asian lowland is often referred to as the Indus-Ganges Plain, this term neglects the Brahmaputra River. This river rises on the Tibetan Plateau and flows easterly, then southward and westerly over 1,700 miles (2,720 kilome-

ENVIRONMENT   The Monsoon Arrives in Southern India

In 1987 journalist Alexander Frater realized a lifelong dream of witnessing one of the most dramatic of meteorological events—the arrival and progress of the South Asia monsoon. He followed it from its first "burst" on the southern tip of India at Cape Comorin in June, northward to Mumbai and Delhi, then, to pick up the eastern arm of the monsoon, to Calcutta, and finally into the Himalayan foothills to experience the monsoon in Cherrapunji, one of the world's wettest places. Here he writes about the arrival of the monsoon in Cochin (Kochi) in southwestern India.

At l P.M. the serious cloud build-up started. Two hours fifty minutes later racing cumulus extinguished the sun and left everything washed in an inky violet light. At 4:40, announced by deafening ground-level thunderclaps, the monsoon finally rode into Cochin. The cloud-base blew through the trees like smoke; rain foamed on the hotel's harbourside lawn and produced a bank of hanging mist opaque as hill fog. In the coffee shop the waiters rushed to the windows, clapping and yelling, their customers forgotten.

Heaving a door open I stepped outside. Soaked to the skin within seconds I felt a wonderful sense of flooding warmth and invigoration; it was, indubitably, a little bit like being born again.

Raindrops rang like coins on the flagstoned path and the air was filled with fusillades of crimson flowers from the flamboyant trees; they went arcing by like tracers and raked by an especially mean burst, I can testify that flamboyant blossoms hitting you at 60 k.p.h. cause pain and temporary loss of vision. At Fort Cochin they were ringing the bells in St. Francis Church.

Then, from the corner of an eye still watering from the flower strike, I witnessed an astonishing scene. Two straining waiters held the coffee-shop door open while a party of men and women filed into the storm. The men wore button-down shirts and smart business suits, the women best-quality silk saris and high-heeled shoes; as they emerged, they opened their arms and lifted their faces to the rain.

The Spices Board had come out to greet the monsoon. . . .

Buffeted by the gusts, unbalanced by the waves, the Spices Board executives clung to each other with water in their eyes and looks of sublime happiness on their faces. A young woman in a soaked and flapping gold-coloured sari laughed at me and clapped her hands. "Paradise will be like this!" she shouted.

Source: Adapted from Alexander Frater, 1990. *Chasing the Monsoon: A Modern Pilgrimage Through India.* New York: Henry Holt & Company.

---

a prominent role in the international debate about global warming and the accompanying rise in sea level (see Chapter 2). Should the worst-case sea-level-rise scenario come to pass, the Maldives would be a seriously endangered country.

South Asia's Monsoon Climates

The dominant climatic factor for most of South Asia is the **monsoon,** the distinct seasonal change of wind direction, which corresponds to wet and dry periods. Most of South Asia has three distinct seasons. First is the warm and rainy season of the southwest monsoon from June through October. This is followed by a relatively cool and dry season, extending from November until February, when the dominant winds are from the northeast. Only a few areas in the far northwest and southeast get rainfall during this otherwise dry time. The third season is the hot period from March to late May, which builds up with great heat and humidity until the monsoon's much-anticipated and rather sudden "burst" in early June (see "Environment: The Monsoon Arrives in Southern India").

This monsoon pattern is caused by large-scale meteorological processes that affect much of Asia (Figure 12.7). During the Northern Hemisphere's winter, a large high-pressure system forms over the cold Asian landmass. Cold, dry winds flow outward from the interior of this high-pressure cell, over the Himalayas and down across South Asia. As winter turns to spring, these winds diminish, resulting in the hot, dry season of March through May. Eventually this buildup of heat over South and Southwest Asia produces a large thermal low-pressure cell. By early June this low-pressure cell is strong enough to draw in warm, moist air from the Indian Ocean.

Usually the first monsoon rains arrive at the southern tip of peninsular India in late May; after this first "burst," it takes about six weeks for monsoon rains to travel northward to the Himalayas.

Orographic rainfall results from the uplifting and cooling of moist monsoon winds over the Western Ghats. As a result, some stations receive more than 200 inches (508 centimeters) of rain during the four-month wet season. On the climate map (Figure 12.8), these are the areas of Am, or tropical monsoon, climate. Inland, however, a strong rain-shadow effect dramatically reduces rainfall on the Deccan plateau.

Farther north, as the monsoon winds are forced up and over the Himalayan foothills, copious amounts of rainfall are characteristic. Simla, India, at 7,000 feet (2,134 meters), averages more than 16 inches (40 centimeters) in July alone. To the east, directly in line for moist air off the Bay of Bengal, Cherrapunji, India, at 4,000 feet (1,220 meters) is a strong contender for the title of world's wettest place, with an average rainfall of 451 inches (1,128 centimeters).

Not all of South Asia, however, receives high rainfall totals from the southwest monsoon. In Pakistan and the Indian state of Rajasthan, precipitation is low enough to result in steppe and desert climates, and is highly uncertain as well. In Karachi the annual total is less than 10 inches (25 centimeters). Only in northeastern Pakistan does the monsoon bring abundant rainfall. Here it is often July before the rains arrives.

Regardless of whether rainfall is heavy or light, the monsoon rhythm affects all of South Asia in many different ways, from the delivery of much-needed water for crops and villages, to the mood of millions of people as they eagerly await relief from oppressive heat (Figure 12.9). Some years the monsoon

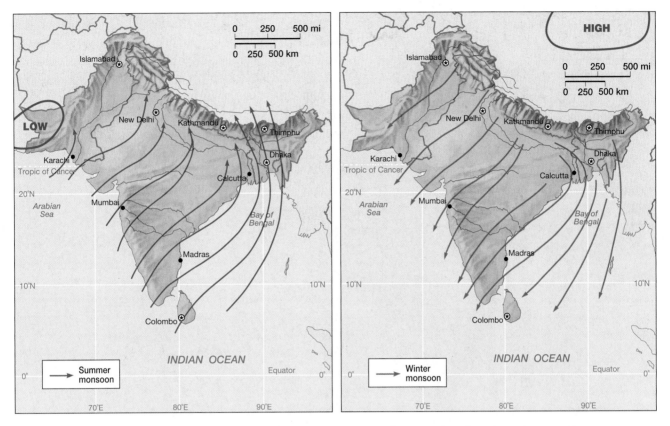

▲ Figure 12.7 The summer and winter monsoons   Low pressure centered over South and Southwest Asia draws in warm, moist air masses during the summer that bring heavy monsoon rains to most of the region. Usually these rains begin in June and last for several months. During the winter, high pressure forms over northern Asia. As a result, winds are reversed from those of the summer. During this season only a few coastal locations along India's east coast and in eastern Sri Lanka receive substantial rain.

delivers its promise with abundant moisture; in other years, though, it brings only scant rainfall to much of South Asia, resulting in crop failure, famine, and hardship.

Population and Settlement: The Demographic Dilemma

South Asia could soon surpass East Asia as the world's most populous region. India alone is home to more than 1 billion people, second only to China in population, while Pakistan and Bangladesh, with 145 and 133.5 million inhabitants, respectively, rank among the world's 10 most populous countries (Table 12.1). Furthermore, South Asia is growing more than twice as fast as East Asia. India alone adds some 18 million people each year.

This rapid population growth takes a high toll on South Asia, for it has more undernourished and malnourished people than any other world region. About a third of India's population lives below the country's official poverty line; Bangladesh is poorer still, with two-thirds of its children classified by the World Bank as underweight.

Although South Asia has made remarkable agricultural gains over the last several decades, there is still widespread concern over its ability to feed itself. The threat of crop fail-

ure, although much diminished, remains, in part because much South Asian farming is vulnerable to the fickle monsoon rains. Although recent gains in economic development have been heartening, they may be neutralized by rapid population growth; whether a vibrant industrial and export sector can provide the necessary bootstraps to pull up the whole region remains to be seen.

The Geography of Family Planning

While all South Asian countries have family planning programs, the commitment to these policies—along with the results—varies widely from place to place. A brief overview of population policies in the three largest South Asian countries provides important background for further discussion.

India   Widespread concern over India's population growth began in the 1960s. To some extent, the measures taken over the last 40 years have been successful; the total fertility rate (TFR) dropped from 6 in the 1950s to the current rate of 3.2. Fertility rates vary widely within India, from lows of 1.9 and 2.0 in the states of Goa and Kerala, to a problematic high of 4.8 in Uttar Pradesh, which, with more than 155 million people, would be the sixth largest country in the world were it independent. Because of the size and high fertility of this Indian

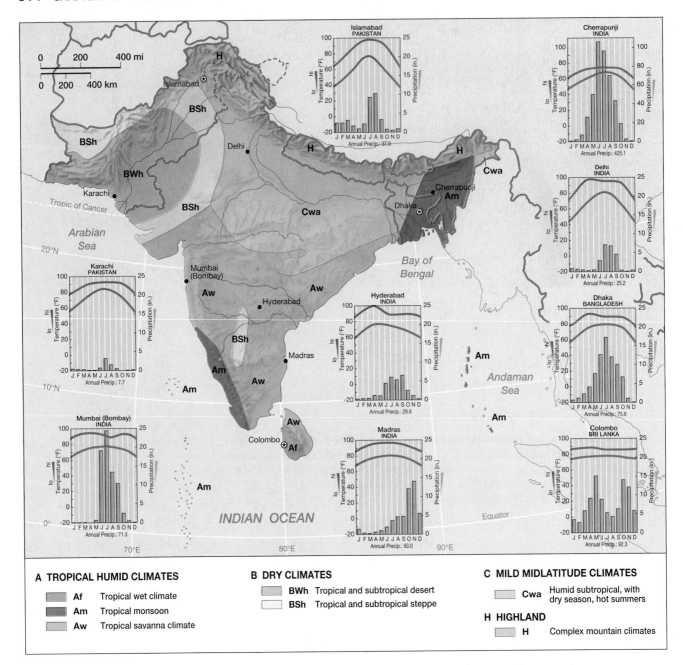

▲ Figure 12.8 Climates of South Asia   Except for the extensive Himalayas, South Asia is dominated by tropical and subtropical climates. Many of these climates show a distinct summer rainfall season that is associated with the southwest monsoon. The climographs for Mumbai (Bombay) and Delhi are excellent illustrations. However, the climographs for east-coast locations such as Madras, India, and Colombo, Sri Lanka, show how some locations also receive rains from the northeast monsoon of the winter.

state, much of the progress made in other areas of the country has been effectively canceled. A strong relationship is evident between women's education and family planning; where women's literacy has increased most dramatically, fertility levels have plummeted.

Sterilization has been an important method of family planning. Twenty-seven percent of India's married women are now sterilized, the highest rate in the world. Male sterilization has also been common. In both cases, however, criticism has been directed at government agencies for setting up rigid quotas

and for failing to deliver on promises of agricultural loans in return for sterilization. A high rate of infection, even death, moreover, has resulted from hurried operations. Further criticism implies that widespread sterilization was possible only because of the low rate of women's literacy, and that government family planners took advantage of illiterate rural women to achieve their quotas.

As is the case in China, a distinct cultural preference for male children is found in most of South Asia, a tradition that further complicates family planning. Where allowed (and they

▲ **Figure 12.9 Monsoon rain** During the summer monsoon, some Indian cities such as Mumbai (Bombay) receive more than 70 inches of rain in just three months. These daily torrents cause floods, power outages, and daily inconvenience. However, these monsoon rains are crucial to India's agriculture. If the rains are late or abnormally weak, crop failure often results. *(Sharad J. Devare/Dinodia Picture Agency)*

are now banned in some Indian states), sex determination clinics provide couples with information about the sex of the fetus, resulting in a higher rate of abortion for female fetuses. Consequently, especially in northern India, a lower TFR is accompanied by a higher ratio of male to female infants. In southern South Asia (particularly Sri Lanka and the Indian state of Kerala), where women have a much higher social position, sex ratios are balanced and birthrates are much lower. Southern India, at least, can now look forward to demographic stabilization.

Pakistan This large country of more than 145 million appears to have an ambivalent attitude toward family planning. While the government's official position is that the birthrate is excessive, the country still lacks an effective, coordinated family planning program. As a result, the TFR remains very high at 5.6, and the current rate of natural increase of 2.8 percent adds about 4 million children each year. A partial explanation for the high birthrate may come from the fact that Pakistan still experiences massive early childhood mortality: 126 per 1,000 children born, as compared to 90 per 1,000 in neighboring India.

Another contributing factor to high fertility is Pakistan's low rate of female contraceptive usage; less than one woman in five uses any sort of birth control. While some attribute Pakistan's ambivalence toward family planning to its strong Muslim culture, the two are not necessarily linked, as Bangladesh demonstrates.

Bangladesh This country has one of the highest settlement densities in the world, with a population about half that of the United States packed into an area smaller than the state of Wisconsin. Although Bangladesh is predominantly Muslim, it has made significant strides in family planning. As recently as 1975 the TFR was 6.3, but it had dropped to 3.3 by the late 1990s. Unlike India, most family planning comes from oral contraception, which is used by more than 50 percent of the women in Bangladesh (Figure 12.10). The success of family planning can be attributed to strong support from the Bangladesh government, advertised through radio and billboards. Also important are the more than 35,000 women fieldworkers who take information about family planning into every village in the country.

Migration and the Settlement Landscape

South Asia is one of the least urbanized regions in the world, with only a quarter of its huge population living in settlements classified as cities. The majority of its people live in compact rural villages and small towns. Still, one of the important themes in South Asia's settlement and population geography is the rapid migration from villages to large cities. This often results as much from desperate conditions in the countryside as from the attraction of employment in the city.

TABLE 12.1 *Demographic Indicators*							
Country	Population (Millions, 2001)	Population Density, per square mile	Rate of Natural Increase	TFR[a]	Percent < 15[b]	Percent > 65	Percent Urban
India	1,033.0	814	1.7	3.2	36	4	28
Pakistan	145.0	472	2.8	5.6	42	4	33
Bangladesh	133.5	2,401	2.0	3.3	40	3	21
Nepal	23.5	413	2.4	4.8	41	3	11
Bhutan	0.9	50	3.1	5.6	42	4	15
Sri Lanka	19.5	771	1.2	2.1	28	6	22
Maldives	0.3	2,495	3.2	5.8	46	3	25

[a]Total fertility rate

[b]Percent of population younger than 15 years of age

Source: Population Reference Bureau. World Population Data Sheet, 2001.

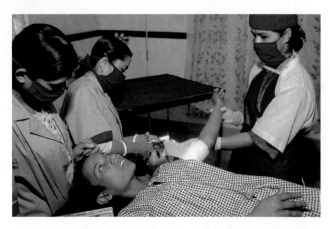

▲ **Figure 12.10 Family planning in Bangladesh** Bangladesh has been one of the most successful nations in South Asia in reducing its fertility rate through family planning. Many women in Bangladesh use oral contraceptives. This photo shows a woman receiving a contraceptive implant. *(Peter Barker/Panos Pictures)*

A primary cause of dislocation is changes in agriculture: increased mechanization with resulting under- or unemployment; the higher costs of farming modern crops; expansion of large estate farms at the expense of subsistence farming; uncertain access to irrigation water; and the environmental deterioration of exploited lands.

The most densely settled areas of South Asia still coincide with areas of fertile soils and dependable water supplies (Fig-ure 12.11). The largest rural populations are found in the core area of the Ganges and Indus river valleys and on the coastal plains of India. Settlement is less dense on the Deccan Plateau, and is relatively sparse in the highlands of the far north and the arid lands of the northwest.

As is true in most other parts of the world, South Asians have long migrated from poor and densely populated areas to places that are either less densely populated or wealthier. This process is ubiquitous in South Asia today, but four areas stand out as zones of intensive out-migration: Bangladesh, the Indian states of Bihar and Rajasthan, and the northern portion of India's Andhra Pradesh (Figure 12.12). Migrants are often attracted to large cities such as Mumbai (Bombay), but those from Bangladesh are settling in large numbers in rural portions of adjacent Indian states, exacerbating ethnic and religious tensions. In Nepal, migrants have long been moving from crowded mountain valleys to formerly malaria-infested lowlands along the Indian border. Sometimes migrants are forced out by war; sizable streams of Tamils from Sri Lanka and Hindus from Kashmir have recently sought security away from their battle-scarred homelands.

Agricultural Regions and Activities

South Asian agriculture has historically been relatively unproductive, especially when compared with that of East Asia. Although the reasons behind such poor production are complex, many experts cite the relatively low social status of most

▶ **Figure 12.11 Population map of South Asia** Except for the desert areas of the west and the high mountains of the north, South Asia is a densely populated region. Particularly high densities are found on the fertile plains along the Indus and Ganges rivers and in India's coastal lowlands. In rural areas the population is typically clustered in villages, often located near water sources, such as streams, wells, canals, or small tanks that store water between monsoon rains.

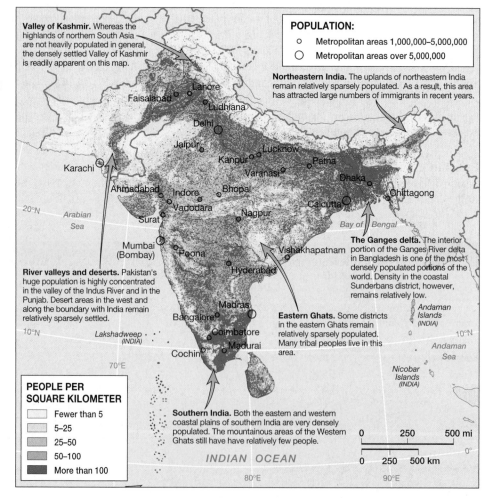

Valley of Kashmir. Whereas the highlands of northern South Asia are not heavily populated in general, the densely settled Valley of Kashmir is readily apparent on this map.

Northeastern India. The uplands of northeastern India remain relatively sparsely populated. As a result, this area has attracted large numbers of immigrants in recent years.

POPULATION:
○ Metropolitan areas 1,000,000–5,000,000
◯ Metropolitan areas over 5,000,000

River valleys and deserts. Pakistan's huge population is highly concentrated in the valley of the Indus River and in the Punjab. Desert areas in the west and along the boundary with India remain relatively sparsely settled.

The Ganges delta. The interior portion of the Ganges River delta in Bangladesh is one of the most densely populated portions of the world. Density in the coastal Sunderbans district, however, remains relatively low.

Eastern Ghats. Some districts in the eastern Ghats remain relatively sparsely populated. Many tribal peoples live in this area.

Southern India. Both the eastern and western coastal plains of southern India are very densely populated. The mountainous areas of the Western Ghats still have have relatively few people.

PEOPLE PER SQUARE KILOMETER
Fewer than 5
5–25
25–50
50–100
More than 100

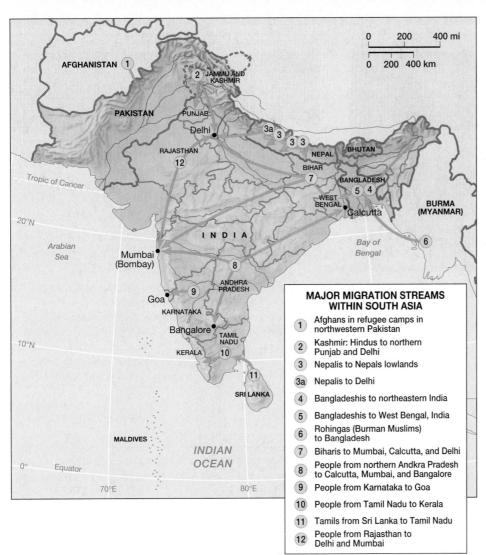

◀ **Figure 12.12 Major migration streams in South Asia** Large-scale movements of people occur in all parts of South Asia. Despite the region's poverty, it has still attracted large numbers of refugees from war and oppression in Afghanistan and Burma. Within the region, most movement directs people away from particularly poor and overcrowded areas to large cities, more prosperous areas, and less densely populated districts.

MAJOR MIGRATION STREAMS WITHIN SOUTH ASIA

1. Afghans in refugee camps in northwestern Pakistan
2. Kashmir: Hindus to northern Punjab and Delhi
3. Nepalis to Nepals lowlands
3a. Nepalis to Delhi
4. Bangladeshis to northeastern India
5. Bangladeshis to West Bengal, India
6. Rohingas (Burman Muslims) to Bangladesh
7. Biharis to Mumbai, Calcutta, and Delhi
8. People from northern Andkra Pradesh to Calcutta, Mumbai, and Bangalore
9. People from Karnataka to Goa
10. People from Tamil Nadu to Kerala
11. Tamils from Sri Lanka to Tamil Nadu
12. People from Rajasthan to Delhi and Mumbai

cultivators and the fact that much farmland has long been controlled by absentee elites. Others place the blame on the legacy of British colonialism, which emphasized export crops for European markets.

Regardless of the causes, low agricultural yields in the context of a huge, hungry, and rapidly growing population constitute a pressing problem. Since the 1970s, however, agricultural production has grown faster than the population, primarily because of the **Green Revolution,** agricultural cultivation techniques based on hybrid crop strains and the heavy use of industrial fertilizers and chemical pesticides. Because the Green Revolution also carries significant social and environmental costs, it has been highly controversial, as will be discussed shortly.

Crop Zones South Asia can be divided into several distinct agricultural regions, all with different problems and potentials. The most fundamental division is between the three primary subsistence crops of rice, wheat, and millet.

Rice is the main crop and foodstuff in the lower Ganges Valley, along the lowlands of India's eastern and western coasts, in the delta lands of Bangladesh, along Pakistan's lower Indus Valley, and in Sri Lanka. This distribution reflects the

large volume of irrigation water needed to grow rice. The sheer amount of rice grown in South Asia is impressive: India ranks behind only China in world rice production, and Bangladesh is the fourth largest producer (Figure 12.13).

Wheat is the principal crop of the northern Indus Valley and in the western half of India's Ganges Valley. South Asia's "breadbasket" is the northwestern Indian state of Punjab and adjacent areas in Pakistan. Here the Green Revolution has been particularly successful in increasing grain yields. In the less fertile areas of central India, millet and sorghum are the main crops, along with root crops such as manioc. In general, wheat and rice are the preferred staples throughout South Asia (and, indeed, most of the world), and it is generally poorer people who consume "rough" grains such as the various millets.

Many other crops are also widely cultivated in South Asia, some commercially, others for local subsistence. Oil seeds, such as sesame and peanuts, for example, are grown in semiarid districts, while the humid southwestern state of Kerala state (along with Sri Lanka) is noted for its coconut groves, spice gardens, and tea plantations. In both Pakistan and west central India, cotton is widely grown, while Bangladesh has long supplied most of the world's jute, a tough fiber used in the manufacture of rope. In general,

▲ **Figure 12.13 Rice cultivation** A large amount of irrigation water is needed to grow rice, as is apparent from this photo from Sri Lanka. Rice is also the main crop in the lower Ganges Valley and delta, along the lower Indus River of Pakistan, and in India's coastal plains. *(Mahaux Photography/The Image Bank)*

▲ **Figure 12.14 Green Revolution farming** Because of "miracle" wheat strains that have increased yields in the Punjab area, this region has become the breadbasket of South Asia. India more than doubled its wheat production in the last 25 years and has moved from chronic food shortages to self-sufficiency. Increased production, however, has led to both social and environmental problems. *(Earl Kowall/Corbis)*

though, the cultivation of staple foods for the local population has been of higher priority than export crops since the end of British colonialism.

Livestock Many if not most South Asians receive inadequate protein, and meat consumption is extremely low. Part of this is simply a reflection of poverty, since meat is expensive to produce. In India, religion is equally important, since most Hindus are vegetarians.

Despite this prohibition against eating meat, animal husbandry is vitally important throughout South Asia. India has the world's largest cattle population, partly because cattle are sacred in Hinduism but also because milk is one of South Asia's main sources of protein. While India's cattle have traditionally produced little milk, a so-called white revolution has increased dairy efficiency in recent decades. Cattle are also widely used for plowing and pulling carts, but they are gradually being replaced in many areas by tractors. In most of South Asia, the main meat-producing animals are sheep, goats, and chickens.

The Green Revolution The only reason South Asian agriculture has been able to keep up with population growth is the Green Revolution, which originated during the 1960s in agricultural research stations established by international development agencies. One of the major problems researchers faced was the fact that higher yields could not be attained by simply fertilizing local seed strains, since the plants would simply grow taller and then fall to the ground before the grain could mature. The solution was to cross-breed new "dwarf" crop strains that would respond to heavy chemical fertilization by producing extra grain rather than longer stems (Figure 12.14).

By the 1970s it was clear that these efforts had succeeded in reaching their initial goals. The more prosperous farmers of the Punjab quickly adopted the new "miracle wheat" varieties, so-

lidifying the Punjab's position as the region's breadbasket. Green Revolution rice strains were also adopted in the more humid areas. As a result, South Asia was transformed from a region of chronic food deficiency to one of self-sufficiency. India more than doubled its annual grain production between 1970 and the mid-1990s, from 80 to 191 million tons.

While the Green Revolution was clearly an agricultural success, many argue that it has been an ecological and social disaster. Serious environmental problems result from the chemical dependency of the new crop strains (Figure 12.3). Not only do they typically need large quantities of industrial fertilizer, which is both expensive and polluting, but they also require frequent pesticide applications since they lack natural resistance to local plant diseases and insects.

Social problems have also followed the Green Revolution. In many areas, particularly where wheat is cultivated, only the more prosperous farmers are able to afford the new seed strains, irrigation equipment, farm machinery, fertilizers, and pesticides necessary to support this new high-technology agriculture. As a result, poorer farmers have sometimes been forced from their lands. Many of these people either become wage laborers for their more successful neighbors or migrate to the region's already crowded cities.

Although the Green Revolution has been subjected to serious criticism, it also has been staunchly defended. Some advocates contend that its environmental dangers have been reduced as farmers learn to grow new crop varieties using traditional methods of fertilization and pest control. Others contend that overall poverty has decreased markedly in the Punjab. On the regional scale, the more important issue is whether South Asia could have fed its large and growing population during the last three decades without the Green Revolution.

Future Food Supply   While the Green Revolution has fed South Asia's expanding population over the past several decades, it remains unclear whether it will be able to continue doing so in the near future. Many of the crop improvements have seemingly exhausted their potential, although optimists believe that genetic engineering might provide another breakthrough, ushering in a second wave of increased crop yields. Most environmentalists, however, see serious dangers in this new technology.

Another option is expanded water delivery, since many fields remain unirrigated in South Asia's semiarid areas, and even in the humid zone dry-season fallow is often the norm. Irrigation, however, brings its own problems. In much of Pakistan and northwestern India, where irrigation has been practiced for generations, soil **salinization** (Figure 12.3), or the buildup of salt in agricultural fields, is already a major constraint. Additionally, water tables are falling in the Punjab, India's breadbasket, because double-cropping has pushed water use beyond the sustainable yield of the underlying aquifers. Villagers and farmers without capital to deepen their wells are left high and dry, forcing them to abandon irrigated agriculture. Because of this, some experts worry that future food supplies could actually decrease.

Urban South Asia

Although South Asia is one of the least urbanized regions of the world, with about a quarter of its population living in cities (see Table 12.1), this does not mean that cities are a minor part of the settlement landscape; South Asia actually has some of the largest urban areas in the world. India alone lists more than 30 cities with populations greater than a million, most of which are growing rapidly. Greater Mumbai (Bombay), with more than 16 million persons, may become the world's largest urban agglomeration by the year 2025.

Because of this rapid growth, most South Asian cities have staggering problems with homelessness, poverty, congestion, water shortages, air pollution, and sewage disposal. Calcutta's homeless are legend, with perhaps half a million sleeping on the streets each night. In that city and others, sprawling squatter settlements, or **bustees,** mushroom in and around urban areas, providing temporary shelter for many urban migrants. A brief survey of the region's major cities gives clues to the problems and prospects of the region's urban areas.

Mumbai (Bombay)   The largest city in South Asia, Mumbai (Bombay) is India's financial, industrial, and commercial center. The major port on the Arabian Sea, Mumbai is responsible for half the country's foreign trade (Figure 12.15). Long noted as the center of India's textile manufacturing, the city is also the hub of its film industry, the largest in the world.

Mumbai's economic vitality draws people from all over South Asia, resulting in ethnic tensions and strife in the sprawling suburbs. Partly in reaction to this ethnic incursion, the nationalist party that governs the city asserted its local ethnic identity by officially changing the city's name from Bombay—a colonial name—to Mumbai, after the Hindu goddess Mumba. Although the political implications of the name change are highly charged, residents seem to take it in stride, using use the two names interchangeably.

Mumbai (Bombay) occupies a peninsular site originally composed of seven small islands. Since the seventeenth century, drainage and reclamation projects have joined the islands into a larger body known as Bombay Island; two parallel ridges of 100 feet (30 meters) form the spines of the inner city, one occupied by the historic colonial fort, now the commercial center, the other, Malabar Hill, an expensive residential area. Because of restricted space, most growth has taken place to the north and east of the historic peninsula so that now the metropolitan region occupies an area 10 times the size of the original city. Spatial restrictions in downtown Mumbai have also resulted in skyrocketing commercial and residential rents, which are now some of the highest in the world. Even members of the city's thriving middle class find it difficult to find adequate housing. Hundreds of thousands of less-fortunate immigrants, eager for work in central Mumbai, live in "hutments," crude shelters built on formerly busy sidewalks (Figure 12.16). The least fortunate sleep on the street or in simple plastic tents, often placed along busy roadways.

Despite Mumbai's extraordinary contrasts of wealth and poverty, it remains in many ways an orderly and relatively crime-free city. It is less dangerous to walk the streets of central Mumbai than those of many major North American cities. Organized crime is a problem, especially in the massive film industry, but it has few effects on the lives of the average people.

Delhi   Delhi, the sprawling capital of India, has more than 11 million people in its greater urban area. It consists of two contrasting landscapes expressing its past: Delhi (or old Delhi), a former Muslim capital, is a congested town of tight neighborhoods; New Delhi, in contrast, is a city of wide boulevards, monuments, parks, and expansive residential

▲ **Figure 12.15 Mumbai (Bombay) central city**   The heart of this metropolitan area of 16 million is Bombay Peninsula. Because space is limited and building restrictions are severe, most recent growth has been in the east and to the north of the city center.   *(Rob Crandall/Rob Crandall, Photographer)*

▲ **Figure 12.16 Mumbai hutments**   Hundreds of thousands of people in Mumbai live in crude hutments, with no sanitary facilities, built on formerly busy sidewalks. Hutment construction is forbidden in many areas, but wherever it is allowed, sidewalks quickly disappear.   *(Rob Crandall/Rob Crandall, Photographer)*

areas. It was born as a planned city when the British moved their colonial capital from Calcutta in 1911. Located here are the embassies, luxury hotels, government office buildings, and airline offices necessary for a vibrant political capital. South of the government area are more expensive residential areas, some of which are focused on expansive parks and ornamental gardens (see "The Urban Scene: Round-up Time in Delhi").

Rapid growth, along with the government's inability to control auto and industrial emissions, has given Delhi some of the worst air pollution in the world. In a largely symbolic move, the city government addressed the problem in 1997 by banning smoking in public places. Delhi, like most large northern cities, also has a much higher crime rate than does Mumbai.

Calcutta   To many, Calcutta is emblematic of the problems faced by rapidly growing cities in developing countries. Not only is homelessness rife, but this city of more than 12 million falls far short of supplying its residents with water, power, or sewage treatment (Figure 12.3). Electrical power is so inadequate that every hotel, restaurant, shop, and small business has to have some sort of standby power system. During the wet season, many streets are routinely flooded. No other Indian city has problems of this magnitude.

Aggravating the poverty, pollution, and congestion of Calcutta is chronic labor unrest resulting from the decline of its suffering industries. As a British creation, Calcutta found its primary role as a trading center on the Hugli River. The river, however, is silting up, impeding navigation from Calcutta. Furthermore, the independence of South Asia hurt Calcutta more than any other Indian city. Until partition, Calcutta was the main export center of northeastern India, but with the drawing of India's border just to the east, Calcutta was cut off from most of its trade area. This economic blow was accompanied by a massive influx of Hindu refugees from the eastern areas, first in 1947, and then again with Bangladeshi independence in 1971.

With continued rapid growth as migrants drain in from the countryside, a mixed Hindu-Muslim population that generates ethnic rivalry, a declining economic base, and an overloaded infrastructure, Calcutta faces a problematic future. Yet it remains a culturally vibrant city noted for its fine educational institutions, its theaters, and its publishing firms.

THE URBAN SCENE   Round-up Time in Delhi

About 40,000 cows wander the streets of Delhi, taking full advantage of their sacred status by slowly crossing busy highways and snacking at an open-air vegetable stand. Revered as they may be, most live the life of the homeless, for they are typically turned loose because they are old and milkless. With little grass to graze in this city of 13 million people, these cows often scavenge through household trash looking for table scraps. Unfortunately, most garbage is packed into indigestible plastic bags. While animal rights groups have asked that garbage bags be banned, that is not likely in the near future.

Dr. Vijay Chaudry, the veterinarian who runs a shelter that cares for Delhi's sacred animals, says that he has found glass, iron, wire, electrical cords, shoes, shirts, and razor blades inside cows. But the real killer is the plastic garbage bag. "We lose two or three cows a day," he says, "and when we cut them open, it is terrible what we find. For an animal so sacred, they die a bad death."

Delhi employs about 100 cow-catchers, whose job it is to keep the animals out of trouble by herding them away from busy streets and the many trash piles found throughout the city. Usually it takes eight men to capture a street-smart cow; getting the beast into the small truck they use is a real challenge, aggravated by the fact that citizens who gather to watch the contest usually root for the cow. "It is necessary, if misunderstood work," says Raman Kumar Sharma, a crew chief for the cowcatchers. "Sometimes people do not realize we have the cow's best interests at heart. We've had violence with the crowds."

Hindus venerate the cow as a symbol of motherhood and a source of life. Killing cows is banned in most of India's 27 states, though there seems to be no shortage of steaks for those who can afford them. Beef is sold on the black market by butchers who deliver prime cuts door to door, yet never discuss how the meat was obtained.

Source: Adapted from Barry Bearak, 1998. "In New Delhi, Cows and Cow Catchers Run Risks," *International Herald Tribune*, October 22, 1998, page 12.

▲ **Figure 12.17 Dhaka street scene** This busy capital of Bangladesh was a former colonial river port but is now a global center for the manufacturing of clothing, shoes, and sports equipment. As seen in this photo, a common mode of transportation is the human-powered pedicab. *(Dirk R. Frans/The Hutchison Library)*

Dhaka As the capital and major city of Bangladesh, Dhaka (also spelled Dacca) has experienced rapid growth due to migration from the surrounding countryside. In 1971, when the country gained independence from Pakistan, Dhaka had about 1 million inhabitants; today it numbers close to 8 million. Like Calcutta, Dhaka originated as a trade center and river port (Figure 12.17). Unlike Calcutta, however, its economic vitality has increased since independence, since it combines the administrative functions of government with the largest industrial concentration in Bangladesh. Cheap and abundant labor in Dhaka has made the city a global center for clothing, shoe, and sports equipment manufacturing. North American shoppers are readily reminded of this new role when looking through clothing tags at any department store.

Karachi Karachi, a rapidly growing port city of more than 7 million people, is Pakistan's largest urban area and its commercial core. It also served as the country's capital until 1963, when the new city of Islamabad was created in the northeast. Karachi, however, has suffered little from the exodus of government functions; it is the most cosmopolitan city in Pakistan, with its checkerboard street pattern lined with high-rise buildings, hotels, banks, and travel agencies. In many ways, its landscape conveys the sense that Karachi is a model metropolis for a developing country.

Karachi, however, suffers from serious political and ethnic tensions that have turned parts of the city into armed camps if not battlegrounds. During the worst of the violence in 1995, more than 200 people were killed on the city's streets each month, and army bunkers were lodged at major urban crossroads. The major problems are between the Sindis, who are the region's indigenous inhabitants, and the Muhajirs, the Muslim refugees from India who settled in and around the city

▲ **Figure 12.18 Lahore landscape** Lahore, Pakistan's cultural, educational, and artistic center, plays a central role in the country's Islamic national identity. It is noted for its rich architectural heritage. *(Nik Wheeler)*

after independence and partition in 1947. There are also clashes between Sunni and Shiite Muslims. As a result, travel both within and beyond the city is often difficult, particularly for international travelers, inhibiting Karachi's commercial activity.

Lahore Although considerably smaller than Karachi with its five million inhabitants, Lahore is Pakistan's cultural, educational, and artistic center. Its landscape expresses its central role in the region's Muslim history with striking mosques, palaces, and monuments (Figure 12.18). After the British captured the city in 1846, they added their stamp with parks, monuments, and Victorian buildings. Migration to this city has been substantial in the last decades, much of it linked to Pakistan's ongoing geopolitical problems, but the city has not suffered from the same level of violence as Karachi.

Islamabad Upon independence, Pakistan's leaders determined that Karachi was too far from the center of the country and thereafter decided that an entirely new capital was necessary. This planned city would make a statement through its name—Islamabad—about the religious foundation of Pakistan. Located close to the contested region of Kashmir, it would also make a geopolitical statement. In geographic terms, such a city is referred to as a **forward capital** since it signals—both symbolically and geographically—the intentions of the country. By building its new capital in the north, Pakistan sent a clear message that it was not giving up its claims to the portion of Kashmir controlled by India. Islamabad is closely linked to the historic city of Rawalpindi, once a major British encampment, which is about 8 miles (13 kilometers) away. While these two cities form a single metropolitan region of about 1 million people, they are completely different in appearance and character. To avoid congestion, planners designed Islamabad around self-sufficient sectors, each with its own government buildings, residences, and shops. At this point six sectors are complete, offering a rather bland suburban landscape.

Cultural Coherence and Diversity: A Common Heritage Rent by Religious Rivalries

Historically, South Asia is a well-defined cultural region. A thousand years ago, virtually the entire area was united by ideas and social institutions associated with Hinduism. The subsequent arrival of Islam added a new religious dimension without fundamentally undercutting the region's cultural unity. British imperialism subsequently grafted a number of cultural features over the entire region, from the widespread use of English to a common passion for cricket. Since the mid-twentieth century, however, religious strife has intensified, leading some to question whether South Asia can still be conceptualized as a culturally coherent world region.

India has been a resolutely secular state since its inception, the Congress Party, its guiding political organization until the 1980s, struggling to keep politics and religion separate. In doing so, it relied heavily on support from Muslims as well as the lower "untouchable" castes. Since the 1980s, this secular political tradition has come under increasing pressure from the growth of Hindu "fundamentalism," which is probably better referred to as **Hindu nationalism.** These nationalists promote Hindu values as the essential and exclusive fabric of Indian society. Not only have Hindu nationalists gained considerable political power both at the federal level and in many Indian states through the Bharatiya Janata Party (BJP), but open agitation against the country's Muslim minority became rampant in the mid-1990s. In several high-profile instances, Hindu mobs demolished Muslim mosques that had allegedly been built on the sites of ancient Hindu temples; the destruction of a mosque in Ayodhya in the Ganges Valley galvanized the nationalistic BJP membership in 1992 (Figure 12.19). Hindu-Muslim strife, although it has lessened since the mid-1990s, remains one of the tragedies of modern South Asia.

Fundamentalism, Islamic in this case, has also been a divisive issue in Pakistan. Powerful fundamentalists leaders want to make Pakistan a religious state under Islamic law, a plan virulently rejected by the country's secular intellectuals and international businesspeople. The government has attempted to intercede between the two groups, but it has often been viewed as tilting toward the fundamentalists. Anti-blasphemy laws, for example, have been used to persecute members of the country's small Hindu and Christian communities, as well as liberal Muslim writers. Large amounts of money from Saudi Arabia, moreover, has gone into the development of super-fundamentalist religious schools that have fanned the flames of Islamic extremism.

Origins of South Asian Civilizations

Many scholars think that the roots of South Asian culture extend back to the Indus Valley civilization, which flourished more than 5,000 years ago in what is now Pakistan. This remarkable urban-oriented society vanished almost entirely in the second millennium B.C., after which the record grows dim. By 800 B.C., however, a new urban focus had emerged in the middle Ganges Valley (Figure 12.20). The social, religious, and intellectual norms associated with this civilization eventually spread throughout the lowlands of South Asia. Although pronounced local diversity continued to exist, by the first millennium A.D. a traveler would have observed similar political institutions and religious practices throughout South Asia.

Hindu Civilization   The religion, or perhaps more precisely, religious complex, of this early South Asian civilization was Hinduism, a tremendously complicated faith that incorporates diverse forms of worship and that lacks any orthodox creed. Certain deities are recognized, however, by all believers, as is the notion that these various gods are all manifestations of a single divine entity (Figure 12.21). All Hindus, moreover, share a common set of epic stories, usually written in **Sanskrit**, the sacred language of this religion. Hin-

▲ **Figure 12.19 Destruction of the Ayodhya Mosque**   A group of Hindu nationalists are seen listening to speeches urging them to demolish the mosque at Ayodhya, allegedly built on the site of a former Hindu temple. The mosque was subsequently destroyed by Hindu fundamentalists.   *(Sunil Malhotra/Reuters/Getty Images, Inc.)*

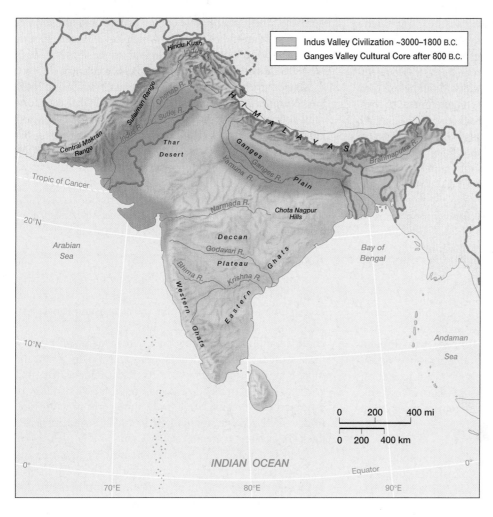

◀ **Figure 12.20 Early South Asian civilizations** The roots of South Asian culture may extend back 5,000 years to an Indus Valley civilization based on irrigated agriculture and vibrant urban centers. What happened to that civilization remains a topic of conjecture, because the archaeological record grows dim by 1800 B.C. Later, a new urban focus emerged in the Ganges Valley, from which social, religious, and intellectual influences spread throughout lowland South Asia.

duism is noted for its mystical tendencies, which have long inspired many men (and few women) to seek an ascetic lifestyle, renouncing property and sometimes all regular human relations. One of its hallmarks is a belief in the transmigration of souls from being to being through reincarnation; the nature of one's acts in the physical world influences the course of these future lives.

Scholars once confidently argued that Hinduism originated from the fusion of two distinct religious traditions: the mystical beliefs of the subcontinent's indigenous inhabitants (including the people of the ill-fated Indus Valley civilization) and the sky-god religion of the Indo-European invaders who swept down into the region sometime in the second millennium B.C. from Central Asia. Such a scenario also proved convenient for explaining India's **caste system**, the strict division of society into different hierarchically ranked hereditary groups. The elite invaders, according to this theory, wished to remain separate from the people they had vanquished, resulting in an elaborate system of social division. Recent research, however, has called this theory into question, and it now seems more likely that the caste system, as well as Hinduism itself, emerged through more gradual social and cultural processes.

▲ **Figure 12.21 Hindu temple** Hindu temples dot India's landscape, taking many different forms, and are devoted to a wide range of deities. This is the Shantadurga temple in Ponda, Goa. *(Rob Crandall/Rob Crandall, Photographer)*

Buddhism While a caste system of some sort seems to have existed in the early Ganges Valley civilization, it was soon challenged from within by Buddhism. Siddhartha Gautama, the Buddha, was born in 563 B.C. in an elite caste. He rejected the life of wealth and power that was laid out before him, however, and sought instead to attain enlightenment, or mystical union with the cosmos. He preached that the path to such nirvana was open to all, regardless of social position. His followers eventually established Buddhism as a new religion. Buddhism spread through most of South Asia, becoming something of an official faith under the Mauryan Empire, which ruled much of the subcontinent in the third century B.C. Later centuries saw Buddhism expand through most of East, Southeast, and Central Asia.

But for all of its successes abroad, Buddhism never replaced Hinduism in India. It remained focused on monasteries rather than permeating the wider society. Many Hindu priests, moreover, struggled against the new faith. One of their techniques was to embrace many of Buddhism's philosophical ideas and enfold them within the intellectual system of Hinduism. By A.D. 500 Buddhism was on the retreat throughout South Asia, and within another 500 years it had virtually disappeared from the region. The only major exceptions were certain peripheral areas, notably the island of Sri Lanka and the high Himalayas, both of which remain mostly Buddhist to this day.

Arrival of Islam The next major challenge to Hindu society—Islam—came from the outside. Arab armies conquered the lower Indus Valley (an area called *Sind*) around A.D. 700, but advanced no farther. Then around the year 1000, Turkish-speaking Muslims began to invade from Central Asia. At first they merely raided, but eventually they began to settle and rule on a permanent basis. By the 1300s, most of South Asia lay under Muslim power, although Hindu kingdoms persisted in southern India and in the arid lands of Rajasthan. Later, during the sixteenth and seventeenth centuries, the **Mughal Empire,** the most powerful of the Muslim states, dominated much the region from its power center in the upper Indus-Ganges basin.

At first Muslims formed a small ruling elite, but over time increasing numbers of Hindus converted to the new religion, particularly those from lower castes who sought freedom from their rigid social order. Conversions were most pronounced in the northwest and northeast, and eventually the areas now known as Pakistan and Bangladesh became predominantly Muslim.

At first glance, Islam and Hinduism are strikingly divergent faiths. Islam is resolutely monotheistic, austere in its ceremonies, and spiritually egalitarian (all believers stand in the same relationship to God). Hinduism, by contrast, is at least superficially polytheistic, lavish in its rituals, and caste-structured. Because of these profound differences, Hindu and Muslim communities in South Asia are sometimes viewed as utterly distinct, living in the same region, but not sharing the same culture or civilization. Increasingly, such a view is expressed in South Asia itself. Many residents of modern Pakistan stress the Islamic nature of their country and its complete separation from India. In India, which has a Muslim minority some 150 million strong, religious animosity sometimes seems to threaten the political fabric of the country.

Yet by overemphasizing the separation of Hindu and Muslim communities, one risks missing much of what is historically distinctive about South Asia. Until the twentieth century, Hindus and Muslims usually coexisted on amicable terms; the two faiths stood side by side for hundreds of years, during which time they came to influence each other in many ways. In earlier generations, especially in rural areas, many people actually participated in both Hindu and Muslim ceremonies. Moreover, aspects of caste organization have persisted even among Muslims, and among Indian Christians as well.

The Caste System Although caste is one of the historically unifying features of South Asia, the system is not uniformly distributed across the subcontinent. It has never been significant in India's tribal areas; in modern Pakistan and Bangladesh its role is fading; and in the Buddhist Singhalese society of Sri Lanka its influence has long been rather marginal. Even in India, caste is now being de-emphasized, especially among the more educated segments of society. But it remains true that caste has been of tremendous social and historical significance over most of the region, and that it continues to structure day-to-day social existence for many Indians. Marriage across caste lines, for example, remains relatively rare.

Caste is actually a rather clumsy term for denoting the complex social order of the Hindu world. The word itself is of Portuguese origin; more problematic, it combines two distinct local concepts: *varna* and *jati*. Varna refers to the ancient fourfold social hierarchy of the Hindu world, whereas *jati* refers to the hundreds of local endogamous ("marrying within") groups that exist at each varna level (different jati groups are thus usually called *subcastes*). Jati, like varna, are hierarchically arranged, although the exact order of precedence is not so clear-cut (see "Cultural Tensions: More About the Caste System in India").

Contemporary Geographies of Religion

South Asia thus has a predominantly Hindu heritage overlain by a substantial Muslim imprint. Such a picture fails, however, to capture the enormous diversity of modern religious expression in contemporary South Asia. The following discussion looks specifically at the geographical patterns of the region's main faiths (Figure 12.22).

Hinduism Fewer than 1 percent of the people of Pakistan are Hindu, and in Bangladesh and Sri Lanka Hinduism is a distinctly minority religion. Almost everywhere in India, however—and in Nepal, as well—Hinduism is very much the faith of the majority. In east-central India, more than 95 percent of the population is Hindu. Hinduism is itself a geographically complicated religion, with different aspects of faith varying between different parts of India. Even within a given Indian state, forms of worship differ from region to region and, more importantly, from caste to caste.

Islam   Islam may be considered a "minority" religion for the region as a whole, but such a designation obscures the tremendous importance of this religion in South Asia. With some 400 million members, the South Asian Muslim community is one of the largest in the world. Bangladesh and especially Pakistan are overwhelmingly Muslim. India's Islamic community, although constituting only some 15 percent of the country's population, is still roughly 150 million strong—a figure substantially larger than the total population of any country in the Muslim heartland of Southwest Asia and North Africa.

As can be seen on Figure 12.22, Muslims live in almost every part of India. They are, however, concentrated in four main areas: in most of India's cities; in Kashmir, in the far north, particularly in the densely populated Vale of Kashmir (more than 80 percent of the population here follows Islam); in the upper and central Ganges plain, where Muslims constitute 15 to 20 percent of the population; and in the southwestern state of Kerala, which is approximately 25 percent Muslim (parts of northern Kerala are more than 50 percent Muslim).

Interestingly, Kerala was one of the few parts of India that never experienced prolonged Muslim rule. Islam in Kerala is historically connected not to Central Asia, but rather to trade across the Arabian Sea. Kerala's Malabar Coast long supplied spices and other luxury products to Southwest Asia, enticing many Arabian traders to settle. Gradually many of Kerala's indigenous inhabitants converted to the new religion as well. Sri Lanka is approximately 9 percent Muslim, and the Maldives is virtually entirely Muslim. Like Kerala in India, Islam in these two island countries is rooted in the trade networks of the Arabian Sea.

Sikhism   The tension between Hinduism and Islam in medieval northern South Asia gave rise to a new religion called **Sikhism.** Sikhism originated in the late 1400s in the Punjab, near the modern boundary between India and Pakistan. The Punjab was the site of religious fervor at the time; Islam was gaining converts and Hinduism was increasingly on the defensive. The new faith combined elements of both religions, and thus appealed to many who felt trapped between the competing claims of faith. Many orthodox Muslims, however, viewed Sikhism as a dangerous heresy precisely because it incorporated elements of their own religion. Periodic bouts of persecution led the Sikhs to adopt a militantly defensive stance, and in the political chaos of the early 1800s they were able to carve out a large kingdom for themselves. Even today, Sikh men are noted for their work as soldiers and bodyguards.

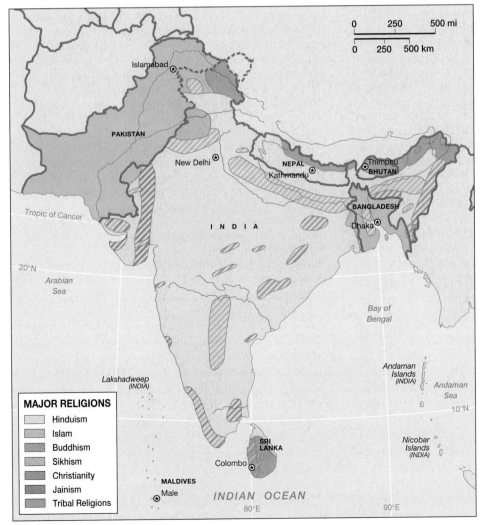

◄ **Figure 12.22 Religious geography of South Asia**   Hindu-dominated India is bracketed by the two important Muslim countries of Pakistan and Bangladesh. Some 150 million Muslims, however, live within India, comprising roughly 15 percent of the total population. Of particular note are the Muslims in northwest Kashmir and in the Ganges Valley. Sikhs form the majority population in India's state of Punjab. Also note the Buddhist populations in Sri Lanka, Bhutan, and northern Nepal, the areas of tribal religion in the east, and the centers of Christianity in the southwest.

CULTURAL TENSIONS   More About the Caste System in India

It has often been argued that the essence of the caste system is the notion of social pollution. The lower one's position in the hierarchy, the more potentially polluting one's body supposedly is. Members of higher castes were traditionally not supposed to eat or drink with, or even use the same utensils as, members of lower castes.

At the apex of the *varna* system sit the Brahmins, members of the traditional priestly caste. Brahmins perform the high rituals of Hinduism, and they form the traditional intellectual elite of India. Almost all Brahmin groups value education highly, and today they are disproportionately represented among India's professional classes. In many areas of India, Brahmins have historically controlled large expanses of land, providing them with substantial wealth.

Below the Brahmins in the traditional hierarchy are the Kshatriyas, members of the warrior or princely caste. In premodern India this group actually had far more power and wealth than did the Brahmins; it was they who ruled the old Hindu kingdoms. Belonging to an expressly military group, Kshatriyas are exempted from many of the restrictions encountered by the other high-status groups. Whereas Brahmins are prohibited from handling weapons, for example, Kshatriyas considered this to be an occupational duty. Unlike members of other high castes, they are seldom vegetarians and do not traditionally shun alcohol.

Next stand the Vaishyas, members of the traditional merchant caste. Although ranked below the Brahmins and the Kshatriyas, this is still an elite group. In earlier centuries, a near monopolization of long-distance trade and moneylending in northern India gave many Vaishyas ample opportunities to accumulate wealth. Like the Brahmins, Vaishyas are noted for their exacting standards of personal behavior. The precepts of vegetarianism and nonviolence, for example, are most highly developed among certain merchant subcastes of western India. One prominent representative of this tradition was Mohandas Gandhi, the founder of modern India and one of the twentieth century's greatest religious and political leaders.

The majority of India's population fits into the fourth varna category, that of the Sudras. The Sudra caste is composed of an especially large array of subcastes (*jati*), most of which originally reflected occupational groupings. These varied endogamous groups are themselves ranked in a hierar-

chy of purity, although the distinctions between them are not always clear-cut. Most Sudra subcastes were traditionally associated with peasant farming, but others were based on craft occupations, including those of barbers, smiths, and potters.

While the Brahmins, Kshatriyas, Vaishyas, and Sudras form the basic fourfold scheme of caste society, another sizable group stands outside of the varna system altogether. These are the so-called **scheduled castes** of the *untouchables* (or **dalits,** as they are now preferably called), people considered so polluting that they were until recently not allowed to enter Hindu temples. Untouchables are traditionally subdivided into more than 1,000 jati, some of which are considered more defiling than others. Such low status positions seem to have been derived in most instances from "unclean" occupations, such as those of leather workers (who dispose of dead animals), scavengers, latrine cleaners, and swine herders. In the southern Indian state of Kerala, some groups were until the nineteenth century actually considered "unseeable"; these unfortunate persons had to hide in bushes and avoid walking on the main roads to protect members of higher castes from "visual pollution." The traditional social codes of many dalit groups allowed behaviors considered repulsive by members of higher castes, such as eating pork or even beef. Not surprisingly, many dalits have converted to Islam, Christianity, and Buddhism in an attempt to escape from the caste system. Even so, they continue to suffer from discrimination.

The caste system is clearly in a state of flux in India today. Its original occupational structure has long been undermined by the necessities of a modern economy, and various social reforms have chipped away at the discrimination that it embodies. Gandhi fought strenuously against caste prejudice, and consistently championed the cause of the untouchables. The dalit community itself has produced several notable national leaders who have waged partially successful political struggles. Owing to such efforts, the very concept of "untouchability" is now technically illegal in India, and a variety of controversial affirmative-action plans have been enacted to give members of scheduled castes special consideration in university admissions and government-sector employment. In several Indian states, moreover, successful alliances of low-caste groups have virtually excluded Brahmins from high-level government positions.

At present the Indian state of Punjab is approximately 60 percent Sikh. Small but often influential groups of Sikhs are scattered across the rest of India. Devout Sikh men are immediately visible, since they do not cut their hair or their beards. Instead, they wear their hair wrapped in a turban and often tie their beards close to their faces.

Buddhism and Jainism   Although Buddhism virtually disappeared from India in medieval times, it persisted in Sri Lanka. Among the island's dominant Singhalese people, Theravada Buddhism developed into a virtual national religion, fostering close connections between Sri Lanka and mainland

Southeast Asia. In the high valleys of the Himalayas, Buddhism also survived as the majority religion (Figure 12.23). Here one finds Lamaism, or Tibetan Buddhism. Tibetan Buddhism, with its esoteric beliefs and huge monasteries, has been better preserved in the small and isolationist country of Bhutan and in the Ladakh region of northeastern Kashmir than in Chinese-controlled Tibet itself. The small town of Dharmsala in the northern Indian state of Himachal Pradesh is the seat of Tibet's government-in-exile and of its spiritual leader, the Dalai Lama, who fled Tibet in 1959 after an unsuccessful revolt.

At roughly the same time as the birth of Buddhism (circa 500 B.C.), another religion emerged in northern India as a protest against orthodox Hinduism: **Jainism.** This religion similarly stressed nonviolence, taking this creed to its ultimate extreme. Jains are forbidden to kill any living creatures, and as a result the most devout adherents wear gauze masks to prevent them from inhaling small insects.

Agriculture is forbidden to Jains, since plowing can kill small creatures. As a result, most members of the faith have looked to trade for their livelihoods. Many have prospered, aided no doubt by the frugal lifestyles required by their religion. The small Jain community is now one of the wealthiest in India, although it includes many poor individuals. Prosperous and devout Jains often contribute substantial sums of money to *pinjrapoles,* institutions devoted to caring for sick and injured animals, especially cattle. Today Jains are concentrated in northwestern India, particularly Gujarat.

Other Religious Groups   Even more prosperous than the Jains are the Parsis, or Zoroastrians, concentrated in the Mumbai area. The Parsis arrived as refugees, fleeing from Iran after the arrival of Islam in the seventh century. Zoroastrianism is an ancient religion that focuses on the cosmic struggle between good and evil. Although numbering only a few hundred thousand, the Parsi community has had a major impact on the Indian economy. Several of the country's largest industrial firms are still controlled by Parsi families. Intermarriage and low fertility, however, now threaten the survival of the community.

Indian Christians are more numerous than either Parsis or Jains. Their religion arrived some 1,800 years ago; early contact between the Malabar Coast and Southwest Asia brought Christian as well as Muslim traders. A Jewish population also established itself but later languished, today numbering only a few hundred. Kerala's Christians, by contrast, are counted

in the millions, constituting some 20 percent of the state's population. Several Christian sects are represented, but the largest are historically affiliated with the Syrian Christian Church of Southwest Asia. Another stronghold of Christianity is the small Indian state of Goa, a former Portuguese colony, where Roman Catholics and Hindus each comprise roughly half of the population.

During the colonial period, British missionaries went to great efforts to convert South Asians to Christianity. They had very little success, however, in Hindu, Muslim, and Buddhist communities, where people were largely content with their own faiths and were wary of foreign missionaries associated with the ruling power. The remote tribal districts of British India, on the other hand, proved to be more receptive to missionary activity. In the uplands of India's extreme northeast, entire communities abandoned their traditional animist faith in favor of Protestant Christianity. Christian missionaries are still active in many parts of India, but during the 1990s they began to experience severe pressure in many areas from Hindu nationalists.

Geographies of Language

South Asia's linguistic diversity matches its religious diversity. In fact, one of the world's most important linguistic boundaries runs directly across India (Figure 12.24). North of the line, languages belong to the Indo-European family, the world's largest linguistic family. The languages of southern India, on the other hand, belong to the **Dravidian** family, a linguistic group that is unique to South Asia. Along the mountainous northern rim of the region a third linguistic family, Tibeto-Burman, prevails, but this area is marginal to the South Asian cultural sphere. Scattered tribal groups in eastern India speak Austro-Asiatic languages related to those of mainland Southeast Asia, but their populations are small. South Asia can thus be divided into two major linguistic zones, the Indo-European north and the Dravidian south. But within these broad divisions there are many different languages, each associated with a distinct culture. In many parts of South Asia, several languages are spoken within the same region or even city, and multilingualism is common everywhere.

How or when Indo-European languages came to South Asia is uncertain, but it has long been argued that they arrived with pastoral peoples from Central Asia who invaded the subcontinent in the second millennium B.C., largely replacing the indigenous Dravidian peoples. According to this hypothesis, offshoots of the same original cattle-herding people also swept across both Iran and Europe, bringing their language to all three places. Supposedly, the ancestral Indo-European tongue introduced to India was similar to Sanskrit. This rather simplistic scenario, however, is now regarded with some suspicion, and many scholars argue for a more gradual infiltration of Indo-European speakers from the northwest.

Any modern Indo-European language of India, such as Hindi or Bengali, is more closely related to English than it is to any Dravidian language of southern India, such as Tamil. That said, it should also be noted that South Asian languages on both sides of this linguistic divide do share a number of

▲ **Figure 12.23 Buddhist monastic landscape**   Buddhism is the dominant faith in most of the Himalayan districts of the far north and in central and southern Sri Lanka. Monasteries are particularly important in the north, where Tibetan Buddhism predominates.   *(Linde Waidhofer/Getty Images, Inc.)*

▶ Figure 12.24 Language map of South Asia A major linguistic divide separates the Indo-European languages of the north from the Dravidian languages of the south. In the Himalayan areas, most languages instead belong to the Tibeto-Burmese family. Of the Indo-European family, Hindi is the most widely spoken, with some 480 million speakers, which makes it the second-most widely spoken language in the world. Most other major languages are closely associated with states in India.

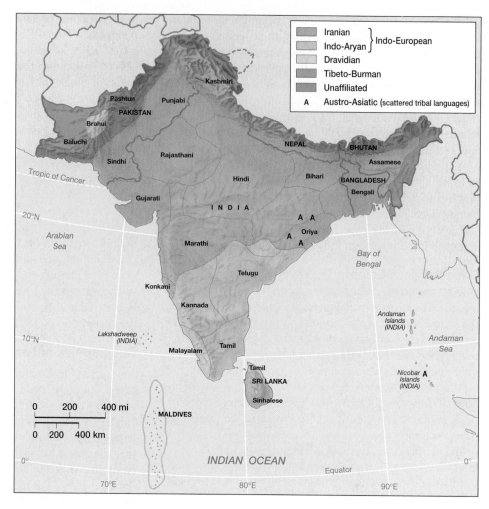

superficial features. Dravidian languages, for example, have borrowed many words from Sanskrit, particularly those associated with religion and scholarship.

The Indo-European North

As can be seen in Figure 12.24, South Asia's Indo-European languages are themselves divided into two subfamilies: Iranian and Indo-Aryan. As might be expected, Iranian languages, such as Baluchi and Pashtun, are found in western Pakistan, near the border with Iran and Afghanistan. Languages of the much larger Indo-Aryan groups are all closely related to each other, yet are still quite distinctive, in some cases even to the extent of being written in different scripts. Each of the major languages of India is associated with an Indian state, since the country deliberately remapped its political subdivisions along linguistic lines a decade after attaining independence in 1947. As a result, one finds Gujarati in Gujarat, Marathi in Maharashtra, Oriya in Orissa, and so on. Two of these languages, Punjabi and Bengali, span India's international boundaries to extend into Pakistan and Bangladesh, respectively, since these borders were established on religious rather than linguistic lines. Nepali, the national language of Nepal, is largely limited to that country. Minor dialects and languages, most of which are neither written nor standardized, abound in many parts of the region.

The most widely spoken language of South Asia is **Hindi**—not to be confused with the Hindu religion. With some 480 million speakers, Hindi is the second-most widely spoken language in the world. It occupies a prominent role in contemporary India, both because so many people speak it and because it is the main language of the Ganges Valley, the historical and demographic core of India. More specifically, it is the main tongue of the populous states of Uttar Pradesh and Madhya Pradesh in central northern India. Many northerners would like Hindi to become the country's common language, but many other Indians, particularly those living in the south, resist the idea. Still, almost all Indian students learn some Hindi, often as their second or third language.

Bengali is the second-most widely spoken language in South Asia. It is the national language of Bangladesh and the main language of the Indian state of West Bengal. Spoken by almost 200 million persons, Bengali is the world's ninth-most widely spoken language. Its significance extends beyond its official status in Bangladesh and its total numerical strength. Equally important is its extensive literature. West Bengal, particularly its capital of Calcutta, has long been one of South Asia's main literary and intellectual centers. Calcutta may be noted for its appalling poverty, but it also has one of the highest levels of cultural production in the world, as measured by the output of drama, poetry, fiction, and film.

The Punjabi-speaking zone in the west was similarly split at the time of independence, in this case between Pakistan and the Indian state of Punjab. While an estimated 100 million persons speak Punjabi, it does not have the significance of Bengali. Although Punjabi is the main vehicle of Sikh religious writings, it lacks an extensive literary tradition. More importantly, Punjabi did not become the national language of Pakistan, even though it is the day-to-day language of some two-thirds of the country's inhabitants. Instead, that position was accorded to Urdu.

Urdu, like Hindi, originated on the plains of northern India. The difference between the two was largely one of religion: Hindi was the language of the Hindu majority, Urdu that of the Muslim minority, including the former ruling class. Owing to this distinction, Hindi and Urdu were from the start written differently—the former in the Devanagari script (derived from Sanskrit) and the latter in the Arabic script. Although Urdu contains many words borrowed from Persian, its basic grammar and vocabulary are almost identical to those of Hindi.

With independence in 1947, millions of Urdu-speaking Muslims from the Ganges Valley fled to the new state of Pakistan. Since Urdu had a higher status than Pakistan's indigenous tongues, it was quickly established as the new country's official language. Karachi, Pakistan's largest city, is now largely Urdu-speaking, but elsewhere other languages, such as Punjabi and Sindhi, remain primary. Most Pakistanis, however, speak Urdu as their second language.

Languages of the South

Four thousand years ago, Dravidian languages were probably spoken across most of South Asia, even in the north. As Figure 12.24 indicates, a Dravidian tongue called Brahui is still found in the uplands of western Pakistan. The four main Dravidian languages, however, are confined to southern India and northern Sri Lanka. As in the north, each language is closely associated with an Indian state: Kannada in Karnataka, Malayalam in Kerala, Telugu in Andhra Pradesh, and Tamil in Tamil Nadu. Tamil is usually considered the most important member of the family because it has the longest history and the largest literature. Tamil poetry dates back to the first century A.D., making it one of the world's oldest written languages.

Although Tamil is spoken in northern Sri Lanka, the country's majority population, the Singhalese, speak an Indo-European language. Apparently the Singhalese migrated from northern South Asia several thousand years ago. Although this movement is lost to history, the migrants evidently settled on the island's fertile and moist southwestern coastal and central highland areas, which formed the core of a succession of Singhalese kingdoms. These same people also migrated to the Maldives, where the national language, Divehi, is essentially a Singhalese dialect. The drier north and east of Sri Lanka, on the other hand, were settled many hundreds of years ago by Tamils. In the 1800s British landowners imported Tamil peasants from the mainland to work on their tea plantations in the central highlands, giving rise to a second population of Tamil-speakers in Sri Lanka.

Linguistic Dilemmas   The multilingual countries Sri Lanka, Pakistan, and India are all troubled by linguistic conflicts. Such problems are most complex in India, simply because India is so large and has so many different languages. India's linguistic environment is changing in complicated ways, pushed along by modern economic and political forces.

Indian nationalists have long dreamed of a national language, one that could help forge the disparate communities of the country into a more unified nation. But this **linguistic nationalism,** or the linking of a specific language with nationalistic goals, meets the stiff resistance of provincial loyalty, which itself is intertwined with local languages. The obvious choice for a national language would be Hindi, and Hindi was indeed declared as such in 1947. Raising Hindi to this position, however, alienated speakers of important northern languages such as Bengali and Marathi, and even more so the speakers of the Dravidian tongues. As a result of this cultural tension, in the 1950 Indian constitution Hindi was demoted to sharing the position of "official language" of India with 14 other languages.

Regardless of opposition, the role of Hindi is expanding, especially in the Indo-European-speaking north. Here local languages are fairly closely related to Hindi, which can therefore be learned without too much difficulty. Hindi is spreading through education, but even more significantly through popular media, especially television and motion pictures. Films and television programs are made in several Indo-Aryan languages, but Hindi remains the primary vehicle. In a poor but modernizing country such as India, where many people experience the wider world largely through moving images, the influence of a national film and television culture can be tremendous.

Even if Hindi is spreading, it still cannot be considered anything like a common national language, even in northern India. In the Dravidian south, more importantly, its role remains strictly secondary. National-level political, journalistic, and academic communication thus cannot be conducted in Hindi, or in any other indigenous language. Only English, an "associate official language" of contemporary India and the language of administration across the subcontinent during the colonial period, serves this function.

Before independence, many educated South Asians learned English for its political and economic benefits under colonialism. It therefore emerged as the de facto common tongue, albeit one largely limited to the upper and middle classes. Today many nationalists wish to de-emphasize English, considering it to be the language of imperial oppression. Others, however, and particularly southerners, advocate English as a neutral national language, since all parts of the country have an equal stake in it. Furthermore, English confers substantial international benefits.

▲ **Figure 12.25 Multilingualism** This four-language sign, in Malayalam, Tamil, Kannada, and English, shows the multilingual nature of contemporary South Asia. In many ways, English, the colonial language of British rule, still serves to bridge the gap between the many different languages of the region. *(Rob Crandall/Rob Crandall, Photographer)*

English is thus the main integrating language of India, and it remains widely used elsewhere in South Asia. Indeed, India is sometimes said to be the most populous English-speaking country in the world (Figure 12.25). English-medium schools abound in all parts of the region, and many children of the elite learn this global language well before they begin their schooling.

South Asians in a Global Cultural Context

The widespread use of English in South Asia has not only facilitated the spread of global culture into the region, but has also helped South Asians' cultural production to reach a global audience. The global spread of South Asian literature, however, is nothing new. As early as the turn of the twentieth century, Rabindranath Tagore gained international acclaim for his poetry and fiction, earning the Nobel Prize for Literature in 1913. In the 1980s and 1990s, such Indian novelists as Salman Rushdie and Vikram Seth became major literary figures in Europe and North America.

The spread of South Asian culture abroad has been accompanied by the spread of South Asians themselves. Migration from South Asia during the time of the British Empire led to the establishment of large communities in such far-flung places as eastern Africa, Fiji, and the southern Caribbean (Figure 12.26). Subsequent migration targeted the developed world; there are now several million persons of South Asian descent living in Britain (mostly Pakistani), and a similar number in North America (mostly Indian). Many contemporary migrants to the United States are doctors, software engineers, and other professionals, making Indian-Americans one of the country's wealthiest ethnic groups.

In South Asia itself, the globalization of culture has brought tensions as severe as those felt anywhere in the world. Traditional Hindu and Muslim religious norms frown on any overt display of sexuality—a staple feature of global popular culture. While romance is a recurrent theme in the often melodramatic

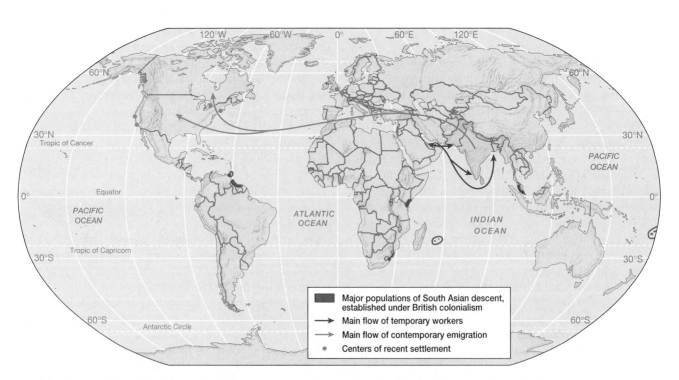

▲ **Figure 12.26 The South Asian global diaspora** During the British imperial period, large numbers of South Asian workers settled in other colonies. Today, roughly 50 percent of the population of such places as Fiji and Mauritius are of South Asian descent. More recently, large numbers have settled, and are still settling, in Europe (particularly Britain) and North America. Large numbers of temporary workers, both laborers and professionals, are employed in the wealthy oil-producing countries of the Persian Gulf.

films of Mumbai (or "Bollywood," as it is commonly called), even kissing is considered risqué. Religious leaders often denounce Western films and videos as immoral. Still, the pressures of internationalization are hard to resist. Many Indians were surprised in the mid-1990s when the fundamentalist Hindu government in Mumbai relented and allowed Michael Jackson to perform in public. In the tourism-oriented Indian state of Goa, such cultural tensions are on full display. There, German and British sun-worshipers often wear nothing but thong bikini-bottoms, whereas Indian women tourists go into the ocean fully clad. Young Indian men, for their part, often simply walk the beach and gawk, good-naturedly, at the outlandish foreigners (Figure 12.27).

Cultural globalization is beginning to allow South Asian novelists, particularly women, to explore themes that would otherwise be difficult for them to address. For example, Arundhati Roy, author of the 1997 international best-seller *The God of Small Things*, examined a tragic romance between a high-caste woman (a Syrian Christian rather than a Hindu) and an untouchable man in her native state, Kerala. Such a topic is virtually forbidden in the local cultural context, and Roy's novel was initially ignored in her homeland. Only its international acclaim allowed it widespread recognition in Kerala and elsewhere in India.

Geopolitical Framework: A Deeply Divided Region

Before the coming of British imperial rule, South Asia had never been politically united. While a few empires at various times ruled most of the subcontinent, none spanned its entire extent. Whatever unity the region had was cultural, not political. The British, however, brought the entire region into a single political system by the middle of the nineteenth cen-

▲ Figure 12.27 Goa beach scene   The liberal Indian state Goa, formerly a Portuguese colony, is now a major destination for tourists, both from within India and from Europe and Israel. European tourists come in the winter for sunbathing and for "Goan Rave parties," where ecstasy and other drugs are widely available. Indian tourists typically find the scantily clad foreigners outlandish if not bizarre.   *(Rob Crandall/Rob Crandall, Photographer)*

tury. A hundred years later, independence in 1947 witnessed the traumatic separation of Pakistan from India; in 1971 Pakistan itself was divided with the independence of Bangladesh, formerly East Pakistan. Serious internal tensions began to rise in India itself in the 1980s, perhaps threatening another round of political division (Figure 12.28).

The major geopolitical issue with global implications, however, is the continuing tension between Pakistan and India, which is aggravated by the fact that both countries are now nuclear powers capable of delivering weapons of mass destruction to neighboring territory. Furthermore, certain Indian and Pakistani leaders seem infatuated with their recent nuclear status, inflaming the passions of nationalism. Some experts think that the highest possibility for a nuclear war now lies embedded in the geopolitical anxieties of South Asia.

South Asia Before and After Independence in 1947

During the 1500s, when Europeans began to arrive on the coasts of South Asia, most of the northern subcontinent came under the power of the Mughal Empire, a mighty Muslim state ruled by people of Central Asian descent (Figure 12.29). Southern India remained under the control of a powerful Hindu kingdom called *Vijayanagara*. European merchants, keen to obtain spices, textiles, and other Indian products, established a series of coastal trading posts. The Mughals and other rulers were little concerned with the growing European naval power, as their own focus was the control of land. The Portuguese carved out an enclave in Goa on the west coast, while the Dutch gained control over much of Sri Lanka in the 1600s, but neither was a significant threat to the Mughals.

The Mughal Empire grew stronger in the 1600s, while Hindu power declined until it was limited to the peninsula's far south. In the early 1700s, however, the Mughal Empire weakened rapidly. A number of contending states, some ruled by Muslims, others by Hindus, and a few by Sikhs, emerged in former Mughal territories. The largest of these was the Hindu state of the Marathas, centered in modern Maharashtra state. The Maratha rulers hoped to unify India, but failed after several key defeats on the Ganges Plain. As a result, the 1700s was a century of political and military turmoil, often bordering on chaos.

The British Conquest   These unsettled conditions provided an opening for European imperialism. The British and French, having largely displaced the Dutch and Portuguese, competed for trading posts. Before the Industrial Revolution, Indian cotton textiles were considered the best in the world, and British and French merchants needed to obtain huge quantities for their global trading networks. With Britain's overwhelming victory over France in the Seven Years' War (1756–1763), France was reduced to a few marginal coastal possessions. Britain, or more specifically the **British East India Company**, the private organization that acted as an arm of the British government, now monopolized trade in the area and was free to exploit the political chaos of the interior and stake out a South Asian empire of its own.

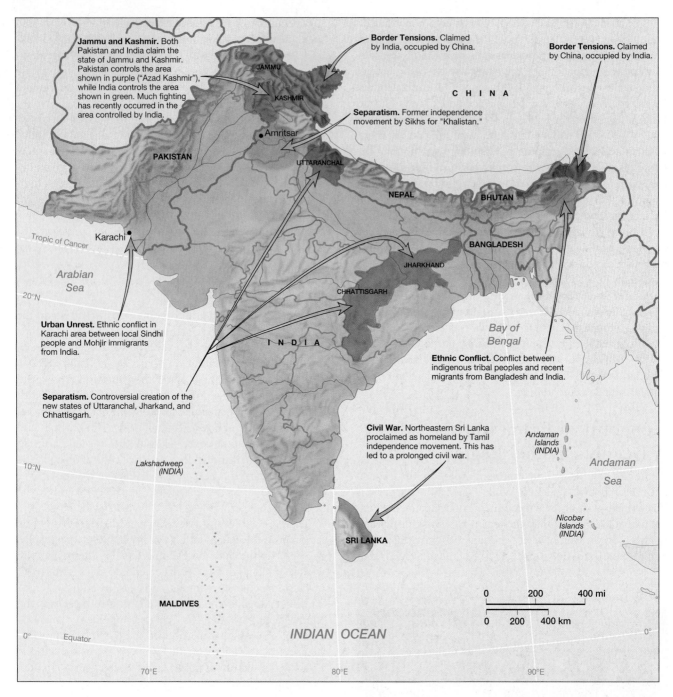

Jammu and Kashmir. Both Pakistan and India claim the state of Jammu and Kashmir. Pakistan controls the area shown in purple ("Azad Kashmir"), while India controls the area shown in green. Much fighting has recently occurred in the area controlled by India.

Border Tensions. Claimed by India, occupied by China.

Border Tensions. Claimed by China, occupied by India.

Separatism. Former independence movement by Sikhs for "Khalistan."

Urban Unrest. Ethnic conflict in Karachi area between local Sindhi people and Mohjir immigrants from India.

Separatism. Controversial creation of the new states of Uttaranchal, Jharkand, and Chhattisgarh.

Ethnic Conflict. Conflict between indigenous tribal peoples and recent migrants from Bangladesh and India.

Civil War. Northeastern Sri Lanka proclaimed as homeland by Tamil independence movement. This has led to a prolonged civil war.

▲ Figure 12.28 Geopolitical issues in South Asia Given the cultural mosaic of South Asia, it is not surprising that ethnic tensions have created numerous geopolitical problems in the region. Particularly vexing are ethnic tensions in Sri Lanka, Kashmir, and the Punjab.

The company's usual method was to make strategic alliances with Indian states in order to defeat the latter's enemies, most of whose territories it would then grab for itself. As time passed, its army, largely composed of South Asian mercenaries, grew ever more powerful. Several Indian states put up heroic resistance, but none could ultimately resist the immense resources of the East India Company. With the defeat of the Sikh state in the Punjab in the 1840s, British control of South Asia was essentially completed. Valuable local allies, as well as a few former enemies, were allowed to remain in power, provided that they no longer threatened British interests. The territories of these indigenous states, however, were gradually whittled back, and British advisors increasingly dictated their policies.

From Company Control to British Colony The continuing reduction in size of the indigenous states, coupled with the growing arrogance of British officials, led to a rebellion in 1856

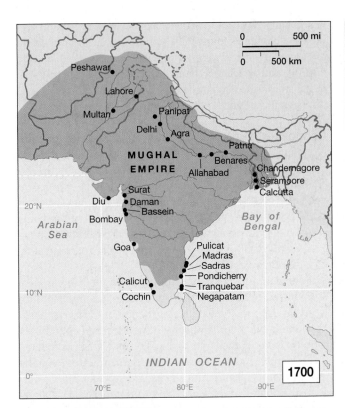

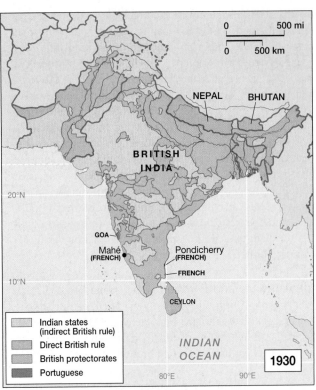

▲ **Figure 12.29 Geopolitical change** At the onset of European colonialism before 1700, much of South Asia was dominated by the powerful Mughal Empire. Under Britain, the wealthiest parts of the region were ruled directly, but other lands remained under the partial authority of indigenous rulers. Independence for the region came after 1947, when the British abandoned their extensive colonial territory. Bangladesh, which was formerly East Pakistan, gained its independence in 1971 after a short struggle against centralized Pakistani rule from the west.

The British enjoyed direct control over the region's most productive and most densely populated areas, including virtually the entire Indus-Ganges Valley and most of the coastal plains. The British also ruled Sri Lanka directly, having supplanted the Dutch in the 1700s. The major areas of indirect rule, where Hindu, Muslim, and Sikh rulers retained their princely states under British advisors, were in Rajasthan, the uplands of central India, southern Kerala, and along the frontiers. The British administered this vast empire through three coastal cities that were largely their own creation: Bombay (Mumbai), Madras (now Chenai), and, above all, Calcutta. In 1911 they established a new capital in New Delhi, near the strategic divide between the Indus and Ganges drainage systems.

While the political geography of British India stabilized after 1856, the empire's frontiers remained unsettled. British officials continually worried about threats to their immensely profitable colony, particularly from the Russians advancing across Central Asia. In response, they attempted to expand as far to the north, and especially the northwest, as possible.

across much of South Asia. When this uprising (called the *Sepoy Mutiny* by the British) was finally crushed, a new political order was implemented. South Asia was now to be ruled directly by the British government, with the Queen of England as its head of state. The continuing usurpation of the indigenous states ended, and a stable political map of British India emerged.

In some cases this merely entailed making strategic alliances with local rulers. In such a manner Nepal and Bhutan retained their independence, although they would no longer be free of British interference. In the extreme northeast, a number of small states and tribal territories, most of which had never been part of the South Asian cultural sphere, were more directly brought into the British Empire, hence into India. A similar policy was conducted on the vulnerable northwestern frontier. Here, however, local resistance was much more effective, and the British-Indian army suffered defeat at the hands of the Afghans. Afghanistan thus retained its independence, forming an effective buffer between the British and Russian empires. The British also allowed the tribal Pashtun-speaking areas of what is now northwestern Pakistan to retain almost all of their autonomy, thus forming a secondary buffer.

Independence and Partition

The framework of British India began to unravel in the early twentieth century as the people of South Asia increasingly demanded independence. The British, however, were equally determined to stay, and by the 1920s South Asia was embroiled in massive political protests.

The rising nationalist movement's leaders faced a major dilemma in attempting to organize a potentially independent regime. Many leaders, including Mohandas Gandhi—the father-figure of Indian independence—favored a unified state that would encompass all British territories in mainland South Asia. Most Muslim leaders, however, feared that a unified India would leave their people in a vulnerable position. They therefore argued for the division of British India into two new countries: a Hindu-majority India and a Muslim-majority Pakistan. In several parts of northern South Asia, however, Muslims and Hindus were settled in roughly equal proportions. A more significant obstacle was the fact that the areas of clear Muslim majority were located on opposite sides of the subcontinent, in present-day Pakistan and Bangladesh.

No longer able to maintain their world empire after World War II, the British withdrew from South Asia in 1947. As this occurred, the region was indeed divided into two countries: India and Pakistan. Partition itself was a horrific event; not only were millions of persons displaced, but tens of thousands were massacred in the process. Millions of Hindus and Sikhs fled from Pakistan, to be replaced by millions of Muslims fleeing India, especially from India's allotted portion of the Punjab (Figure 12.30).

The Pakistan that emerged from partition was for several decades a clumsy two-part country, its western section in the Indus Valley, its eastern portion in the Ganges Delta. This situation did not bode well for the country. The Bengalis, occupying the poorer eastern section, complained that they were treated as second-class citizens, with political and economic power remaining entrenched in the west. In 1971 they launched a rebellion, and, with the help of India, quickly prevailed. Bangladesh then emerged as a new country. This second partition did not solve Pakistan's problems, however, as it remained politically unstable and prone to military rule. It

is also important to note that Pakistan retained the British policy of allowing almost full autonomy to the Pashtun tribes of the northwest, a relatively lawless area marked by clan fighting and vengeance feuds. This area subsequently lent much support to Afghanistan's Taliban regime and to Osama bin Laden's Al Qaeda organization.

Geopolitical Structure of India   The leaders of newly independent India, committed to democracy, faced a major challenge in organizing such a large and culturally diverse country. They decided to chart a middle ground between centralization and local autonomy. India itself was thus organized as a **federal state**, with a significant amount of power being vested in its individual states. The national government, however, retained full control over foreign affairs and a large degree of economic authority.

Following independence, India's constituent states were reorganized to accord with linguistic geography. The idea was that each major language group should have its own state and hence a certain degree of political and cultural autonomy. Yet only the largest groups received their own territories, which has led to recurring demands from smaller groups that have felt politically excluded. Over time, several smaller states have been added to the map. Goa, following the forcible expulsions of the Portuguese in 1961, became a separate state in 1987 over the objection of its large northern neighbor of Maharashtra. In 2000 three new states were added, Jharkand, Uttaranchal, and Chhattisgarh. Two of these involved major controversies. Jharkand emerged in the largely tribal areas of southern Bihar against the protests of the dominant Biharis of the northern lowlands. In contrast, local tribal and dalit ("untouchable") groups opposed the creation

▲ **Figure 12.30 Partition, 1947**   Following Britain's decision to leave South Asia, violence and bloodshed broke out between Hindus and Muslims in much of the region. With the creation of Pakistan (originally in two different sectors, west and east), millions of people relocated both to and from the new states. Many were killed in the process, embittering relations between India and Pakistan.   *(Bettmann/Corbis)*

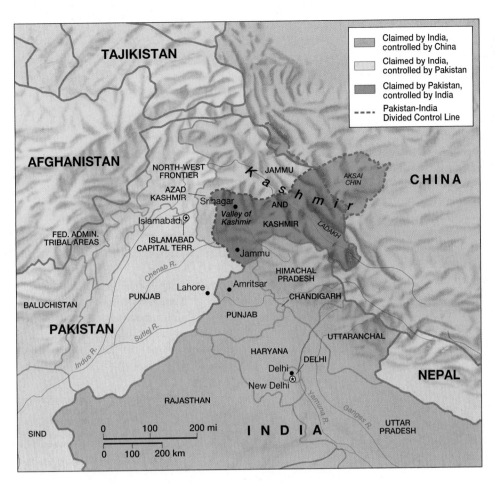

◀ **Figure 12.31 Conflict in Kashmir** Unrest in Kashmir inflames the continually hostile relationships between the two nuclear powers of India and Pakistan. Under the British, this region of predominantly Muslim population was ruled by a Hindu maharaja, who managed to join the province to India upon partition. Today many Kashmirs wish to join Pakistan, while many others argue for an independent state.

of Uttaranchal, formerly the Himalayan districts of Uttar Pradesh, concerned about the fact that the new state, unlike all others, has a substantial majority of upper-caste people.

Ethnic Conflicts in South Asia

The movement for new states in India is rooted in ethnic tensions. Unfortunately, similar ethnic conflicts have emerged in many different portions of South Asia. Several of these are far more extreme than those of Jharkand and Uttaranchal. Of these conflicts, the most complex—and perilous—is that in Kashmir.

Kashmir Relations between India and Pakistan were hostile from the start, and the situation in Kashmir has kept the conflict burning (Figure 12.31). During the British period, Kashmir was a large princely state with a primarily Muslim core joined to a Hindu district in the south (Jammu) and a Tibetan Buddhist district in the far northeast (Ladakh). Kashmir was then ruled by a Hindu **maharaja,** or a king subject to British advisors, whose ancestors had gained control during the chaotic waning years of the Mughal Empire. During partition, Kashmir came under severe pressure from both India and Pakistan. Troops from Pakistan gained control of western and much of northern Kashmir, at which point the maharaja opted for union with India. India thus retained the core areas of the state. But neither Pakistan nor

India would accept the other's control over any portion of Kashmir, and they subsequently fought several inconclusive wars over the issue.

Although the Indo-Pakistani boundary has remained fixed, fighting in Kashmir intensified, reaching a peak in the 1990s. Many Muslim Kashmiris hope to join their homeland to Pakistan; others would rather see it emerge as an independent country. Hindu militants and other Indian nationalists, on the other hand, are adamant that Kashmir remain part of India, and the Indian government agrees (although it may be willing to grant substantial autonomy). The result has been a low-level but periodically brutal war involving several different factions. Efforts have been made to reach a peaceful settlement, but they seem unlikely to succeed. India refuses to talk directly with Pakistan, accusing the Pakistani government of supporting terrorism. The Vale of Kashmir, with its lush fields and orchards nestled among some of the world's most spectacular mountains, was once one of South Asia's premier tourist destinations; now, however, it is a battle-scarred war zone. The war has also unleashed an internal diaspora, as Kashmiris (especially Hindu ones) increasingly flee their homeland to find peace elsewhere in northern India (Figure 12.12).

The Punjab Religious conflict also lies at the root of political violence in India's Punjab. The original Punjab, an area of intermixed Hindu, Muslim, and Sikh communities, was divided between India and Pakistan in 1947. During partition, virtually all Hindus and Sikhs fled from Pakistan's allotted portion, just as Muslims left the zone awarded to India. Relations between Hindus and Sikhs had previously been harmonious, but they too began to deteriorate, and India's Punjab was itself later divided into two states: Hindu-majority Haryana and Sikh-majority Punjab.

But the division of the Indian state of Punjab did not solve the area's problems. Most Sikhs resented the fact that the Indian government refused to recognize Sikhism as a separate religion, instead classifying it as a sect of Hinduism. In the 1970s and 1980s, as the area grew increasingly prosperous due to the Green Revolution, Sikh leaders began to strive for autonomy and Sikh radicals began to press for outright independence. In their vision, the Punjab should separate from India, renaming itself Khalistan, the "land of the pure."

Since many Sikh men had maintained the military traditions of their ancestors, this secession movement soon became a formidable fighting force. The Indian government reacted firmly, and tensions mounted. An open rupture occurred in 1984 when the Indian army raided the main Sikh temple at Amritsar, in which a group of militants had barricaded themselves (Figure 12.32). Shortly thereafter, the president of India, Indira Gandhi, was assassinated by her own Sikh bodyguards. As hostility escalated, the Indian government placed the Punjab under martial law, instituting heavy repression. Such a policy eventually proved relatively successful in quelling political violence, at least in the short run. Renewed strife in the area, however, remains a distinct possibility.

The Northeast Fringe   A more complicated ethnic conflict emerged in the 1980s in the uplands of India's extreme northeast, particularly in the states of Arunachal Pradesh, Nagaland, Manipur, and portions of Assam. The underlying problem is rooted in demographic change and cultural collision. Much of this area is still relatively lightly populated, and as a result has attracted hundreds of thousands of migrants from Bangladesh and adjacent provinces of India. Many local inhabitants consider this movement a threat to both their lands and their cultural integrity. On several occasions indigenous war-bands have attacked newcomer villagers, and, in turn, have suffered reprisals from the Indian military. This is a remote area, however, and relatively little information from it reaches the outside world. The Indian government has placed travel restrictions over much of the northeastern frontier, making access by journalists difficult.

Sri Lanka   Interethnic violence in Sri Lanka is especially severe. Here the conflict stems from both religious and linguistic differences. Northern Sri Lanka and parts of its eastern coast are dominated by Hindu Tamils, while the island's majority group is Buddhist in religion and Singhalese in language. Relations between the two communities have historically been fairly good, but tensions mounted soon after independence (Figure 12.33).

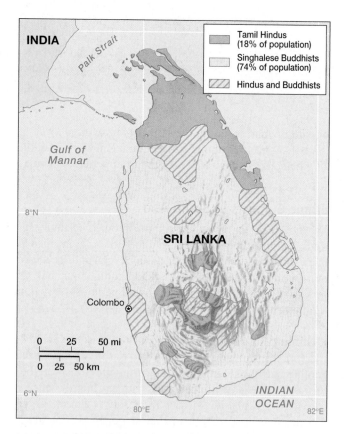

▲ **Figure 12.33 Civil war in Sri Lanka**   The majority of Sri Lankans are Singhalese Buddhist, many of whom maintain that their country should be a Buddhist state. A Tamil-speaking Hindu minority in the northeast strenuously resists this idea. Tamil militants, who have waged war against the Sri Lankan government for several decades, hope to create an independent country in their northern homeland.

▲ **Figure 12.32 Sikh temple at Amritsar**   Separatism in India's Punjab region is strong because of hostility between the Sikh majority and the Indian government. These tensions were magnified in 1984 when the Indian army raided the Amritsar temple to dislodge Sikh militants.   *(Daniel O'Leary/Panos Pictures)*

The basic problem is that Singhalese nationalists favor a unitary government, some of them going so far as to argue that Sri Lanka ought to be a Buddhist state. Most Tamils, on the other hand, support political and cultural autonomy, and they have accused the government of discriminating against them. In 1983 war erupted when the rebel force known as the *Tamil Tigers* attacked the Sri Lankan army. Northern Sri Lanka has been embroiled in brutal fighting ever since. Both extreme Tamil and extreme Singhalese nationalists remain unwilling to compromise, making a near-term solution unlikely.

International and Global Geopolitics

South Asia's major international geopolitical problem is the continuing cold war between India and Pakistan (Figure 12.34). Since independence, these two countries have regarded each other as implacable enemies, and both maintain large military forces to neutralize the other. Today, however, the stakes are considerably higher because both India and Pakistan have nuclear capabilities and, more troubling, leaders in both countries sometimes fuel nationalistic pride with nuclear testing and atomic saber-rattling.

During the global Cold War, Pakistan allied itself with the United States and India remained neutral while leaning slightly toward the Soviet Union. Such entanglements fell apart with the end of the superpower conflict in the early 1990s. Since then Pakistan has forged an informal alliance with China, from which it has obtained sophisticated military equipment. China's military connection to Pakistan is rooted in its own animosity toward India; it claims a large territory within India's northeastern boundary, and after a brief war in 1962, it occupied a section of the Tibetan Plateau that for-

▲ **Figure 12.34 Border tensions** Relationships between India and Pakistan have been extremely tense since independence in 1947. Moreover, with both countries now nuclear powers, the fear that border hostilities will escalate into wider warfare has become a nightmarish possibility. *(Deepak Sharma/AP/Wide World Photos)*

merly belonged to India. As a result, relations between China and India have remained icy, although in the late 1990s a slight thaw was detected. Still, India's development of a nuclear arsenal was designed to send a message not just to Pakistan, but to China as well.

The conflict between India and Pakistan became more complex in the aftermath of the terrorists attacks in the United States on September 11, 2001. Until that time, Pakistan had been strongly supporting Afghanistan's Taliban regime (see Chapter 10), in part because Pakistan fears having a hostile neighbor across its northwestern frontier. India correspondingly offered some assistance to the Northern Alliance that was struggling against the Taliban. After the attack on the World Trade Center and Pentagon (orchestrated from Afghanistan by Osama bin Laden), the United States gave Pakistan a stark choice: either it would assist the United States in its fight against the Taliban and receive in return debt reductions and other forms of aid, or it would completely lose favor with the United States. Pakistan's president Pervez Musharraf quickly agreed to help, and Pakistan offered both military bases and valuable intelligence to the U.S. military. Pakistan cooperated in part to prevent U.S. foreign policy from tilting too far in the direction of its archrival, India.

The decision to help the United States was not without risks. Osama bin Laden had gained much popularity among Pakistan's more extreme Islamic fundamentalists. Both he and the Taliban, moreover, enjoyed significant support among the Pashtun populace of Pakistan's virtually ungoverned North-West Frontier province. Many observers predicted that Pakistan's alliance with the United States would lead to a massive uprising against its government. But although a few large demonstrations were held—and a number of Pakistanis did go to Afghanistan to fight for the Taliban—the mood of the country remained generally quiet.

But while Musharraf weathered the initial storm easily, it also true that there is a great deal of anti-American sentiment in Pakistan. This anti-Americanism has been encouraged by a network of politically active, and extremist, Islamic fundamentalists. Due in part to the struggle in Kashmir, the government of Pakistan has generally promoted the growth of this network. It has been forced to reassess that policy, however, in the wake of the events following September 11.

The tension between India and Pakistan, and the triangular relationships that this entails with China and United States, overshadows all other international geopolitical issues in South Asia. Elsewhere in the region, the power of India is simply overwhelming, although often resented. In the late 1990s, a Marxist rebel movement emerged in Nepal, aimed in part at dissolving the country's close ties to India. Bangladesh, on the other hand, owes its very existence to Indian support, and the two countries have enjoyed relatively cordial relations. In other parts of the region, India has used its power more forcefully. It simply annexed the formerly semi-independent country of Sikkim in the Himalayas in 1975, and later sent its army into

GEOGRAPHY IN THE MAKING   Pakistan and India Since September 11, 2001

In the wake of the terrorist attacks of September 11, 2001, religious and geopolitical tensions mounted in South Asia. The situation became particularly perilous in Pakistan, as Islamic extremists began to target representatives of the United States.

Pressured by the United States, Pakistan cracked down on Jaish-e-Muhammad, a militant group fighting against India's control of Kashmir that previously had been tolerated, if not supported, by Pakistan's government. In retaliation, Jaish-e-Muhammad kidnapped and murdered Daniel Pearl, an American journalist. Pakistani authorities subsequently kidnapped the wife and other family members of Ahmad Omar Saeed Sheikh, the man believed to be responsible for Pearl's death. Although Sheikh then gave himself up, other members of his organization vowed to continue fighting against both India and the United States. On March 17, 2002, Islamic militants attacked a Christian church in Islamabad, killing five people, two of them Americans. Meanwhile, Pakistani authorities vowed to uproot all terrorist cells. Collaborating with the FBI, they quickly arrested several key Al Qaeda operatives working in their country.

Although the decision by Pakistan's President Pervez Musharraf to align with the United States has generated outrage on the part of radical fundamentalists, it has undoubtedly brought the country economic benefits, and thus receives widespread support. International banks rescheduled Pakistan's crushing $12.5 billion debt burden, while the United States lifted all sanctions previously imposed to protest Pakistan's nuclear weapons program. Confident of his popularity with the majority—and perhaps with his ability to pull political strings—Musharraf held a general referendum in April 2002 that confirmed his presidency for the next 5 years. Skeptics, however, viewed the election as basically fraudulent.

Although not directly related to the attacks of September 11, 2001, religious tensions in India also escalated in early 2002. On February 27, 2002, a mob of Muslim extremists attacked a train carrying Hindu pilgrims from the contested site of Ayodhya, killing 58 people. In retaliation, Hindu mobs killed more than 600 Muslims in several Indian states. Violence was most pronounced in Ahmadabad, capital of the prosperous state of Gujarat. In many other Gujarati cities, however, such as the textile center of Surat, relations between Hindus and Muslims remained peaceful.

---

northern Sri Lanka to try to quell the fighting there. India's military leaders, however, quickly discovered how intractable Sri Lanka's problems are, and they withdrew their forces before their own losses mounted.

India's ambition has long been to become the dominant power in South Asia and, ultimately, in the Indian Ocean basin. Its army is large and capable, and its military technology is well advanced. It has been thwarted in these goals, however, not only by its internal conflicts and struggle with Pakistan, but also by its economic problems.

Economic and Social Development: Burdened by Poverty

South Asia is a land of developmental paradoxes. It is, along with Sub-Saharan Africa, the poorest world region, yet it is also the site of some immense fortunes. It has a sizable and growing middle class, yet large areas, and many social groups, remain virtually cut off from the processes of development. Many of South Asia's scientific and technological accomplishments are world-class, but it also has some of the world's highest illiteracy rates. While South Asia's high-tech businesses are closely integrated with Silicon Valley and other centers of the global information economy, the South Asian economy as a whole has been one of the world's most self-contained and inward-looking. Predictions about South Asia's economic future vary accordingly. Some observers foresee intensified misery as the region's large and growing population undercuts natural systems and exhausts its resource base; others predict a regional renaissance as modern, globally interconnected economic enterprises replace age-old subsistence systems.

South Asian Poverty

One of the clearest measures of human well-being is nutrition, and by this score South Asia ranks very low indeed. Probably nowhere else can one find so many undernourished and malnourished people. More than 300 million Indian citizens live below their country's official poverty line, which is set at a very meager level. Bangladesh is poorer still. This country is so impoverished, and so beset by environmental problems, that some critics recently regarded it as almost hopeless. According to World Bank statistics, 67 percent of Bangladeshi children are underweight and therefore malnourished—the highest figure in the world. By measures such as infant mortality and average longevity, Nepal and Bhutan are in even worse condition, matched only by the poorer countries of Sub-Saharan Africa and, in Asia, by Laos, Cambodia, and Afghanistan (Figure 12.35). In urban slums throughout South Asia, rapidly growing populations have little chance of finding housing or basic social services. Moreover, it is estimated that roughly half a million South Asian children work as virtual slaves in carpet-weaving workshops and other small-scale factories.

Despite such deep and widespread poverty, South Asia should not be regarded as a zone of uniform misery. India especially has a large and growing middle class, as well as a small but wealthy upper class. Roughly 100 million Indians are able to purchase such modern consumer articles as

▲ **Figure 12.35 Food aid** Partly because of the large population, food problems have long plagued South Asia. In this photo, military personnel dispense wheat mix and dried skimmed milk to people in Nepal as aid against a famine. *(Jeremy Hartley/Panos Pictures)*

televisions, motor scooters, and washing machines. This is a large market by any definition, and it has begun to excite the imaginations of corporate executives worldwide. India's economy has grown since the 1950s at a moderate but accelerating pace, and certain places and sectors are now virtually booming. But if several Indian states have shown marked economic vitality (Figure 12.36), others have seen only stagnation or even deterioration. Similarly, some parts of South Asia have made impressive gains in social well-being, far above what might be predicted on the basis of their economic production, but others have made only modest improvements, remaining burdened by high levels of illiteracy and poor health.

South Asia's developmental contradictions are even more profound when considered in light of economic history.

From ancient times to the early modern period (1500–1700s), India was famed throughout the globe as a land of great riches. It was, of course, always afflicted with poverty—as were all civilizations—but it may have held more raw accumulated wealth (especially in the form of gold and jewels) than any other world region. Not surprisingly, South Asia has long been the target of outsiders' greed. Central Asians began to pillage the region a thousand years ago, but the British did a much more thorough job. Not only did they send much of India's hoarded wealth back to Europe, but they also systematically stifled Indian industrialization, since they wanted the colony to form a captive market for British exports. The British did, however, build an extensive railroad system that still forms the foundation of South Asia's transportation network.

Geographies of Economic Development

Since independence, the governments of South Asia have attempted to create new economic systems that would benefit their own people rather than foreign countries or corporations. As in most other parts of the world, planners stressed heavy industry and economic autonomy. While some major gains were realized, the overall pace of development remained slow. As is true of most developing regions, core areas of development and social progress emerged, surrounded by large peripheral areas that lagged behind, creating landscapes of striking economic disparity (Table 12.2).

The Himalayan Countries Both Nepal and Bhutan are disadvantaged by their rugged terrain and remote locations and by the fact that they have been relatively isolated from modern technology and infrastructure. But such measurements are somewhat misleading, especially for Bhutan, since they fail to convey the fact that many areas in the Himalayas are still largely subsistence-oriented and are not full participants in market economies.

Bhutan has purposely remained virtually disconnected from the modern world economy, and its small population lives in a relatively pristine natural environment. Indeed, Bhutan is so isolationist that it has only recently allowed tourists to enter—provided that they pay $1,000 to obtain a visa. Nepal, on the other hand, is more heavily populated and suffers much more severe environmental degradation. It is also more closely integrated with the Indian, and ultimately the world, economy. Nepal relies heavily on international tourism. Tourism has brought some prosperity to a few favored locales, but often at the cost of heightened ecological damage.

Bangladesh The economic figures for Bangladesh are not quite as low as those of the Himalayan countries, but they are more indicative of widespread hardship, since most people there require cash to meet their basic needs. Most Bangladeshis are involved with the market, often through the production of commercial crops, such as rice and jute. Partly because of the country's massive population, poverty is extreme and widespread.

▶ Figure 12.36 South Asia's manufacturing and high-tech areas   This map shows the region's major industrial areas. Of these, perhaps the most vibrant are those of greater Mumbai (Bombay), central Gujarat (around Ahmadabad), and the so-called Silicon Plateau of the south, near Bangalore. Global niche manufacturers include Sialkot (surgical implements and soccer balls), Landi Kotal (arms and armaments), and Moradabad (brass faux-antiques).

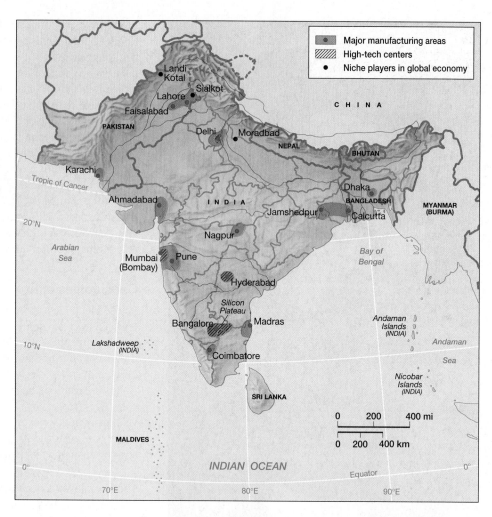

TABLE 12.2 *Economic Indicators*

Country	Total GNI (Millions of $U.S., 1999)	GNI per Capita ($U.S., 1999)	GNI per Capita, PPP* ($Intl, 1999)	Average Annual Growth % GDP per Capita, 1990–1999
India	441,834	440	2,230	4.1
Pakistan	62,915	470	1,860	1.3
Bangladesh	47,071	370	1,530	3.1
Nepal	5,173	220	1,280	2.3
Bhutan	399	510	1,260	3.4
Sri Lanka	15,578	820	3,230	4.0
Maldives	322	1,200	1,800**	3.9

*Purchasing power parity

Source: The World Bank Atlas, 2001, except those marked **, which note data from the CIA World Factbook, 2000.

Environmental degradation and the colonial legacy have contributed to Bangladesh's impoverishment, as did the partition of 1947. Most of pre-partition Bengal's businesses were located in the western area, which went to India. The division of Bengal tore apart an integrated economic region, much to the detriment of the poorer and mainly rural eastern section. Bangladesh has also suffered because of its agricultural em-

phasis on jute, a plant that yields tough fibers useful for making ropes and burlap bags. Bangladesh failed to discover any major alternative export crops as synthetic materials undercut the global jute market.

Not all of the economic news coming from Bangladesh is negative. The country is internationally competitive in textile and clothing manufacture, in part because its wage rate is

▲ **Figure 12.37 Grameen Bank**   This innovative institution loans money to rural women so they can buy land, purchase homes, or start cottage industries. In this photo, taken in Bangladesh, women proudly repay their loans to a bank official as testimony to their success.   *(John Van Hasselt/Corbis/Sygma)*

so abysmally low. Low-interest credit provided by the internationally acclaimed Grameen Bank has given hope to many poor women in Bangladesh, allowing the emergence of a number of vibrant small-scale enterprises (Figure 12.37). By 2000, with the country's birthrate steadily falling, the Bangladeshi economy was finally beginning to grow substantially faster than its population.

Pakistan   Pakistan also suffered the effects of partition in 1947. But unlike Bangladesh, Pakistan at least inherited a reasonably well-developed urban infrastructure. By the measure of per capita GNI, Pakistan today has a more productive economy than either Bangladesh or India. The country has a productive agricultural sector, as it shares the fertile Punjab with India. Pakistan also boasts a large textile industry, based in part on its huge cotton crop. Several Pakistani cities, moreover, have important niches in the global economy. Sialkot, for example, is noted for its export of surgical implements, a legacy of its days as a sword-making center, and Landi Kotal, near the unstable border with Afghanistan, is famous through much of the world for its weapons, mostly inexpensive imitations of "global brands" (such as Russian Kalishnikovs) (Figure 12.36).

Many experts argue, however, that Pakistan's economy is less dynamic than that of India, with a lower potential for growth. Part of the problem is that Pakistan is burdened by especially high levels of defense spending. Additionally, a small but powerful landlord class that pays virtually no taxes to the central government controls its best agricultural lands. Unlike India, moreover, Pakistan has not been able to develop a successful information-technology industry.

Sri Lanka and the Maldives   As can be seen in Table 12.2, Sri Lanka's economy is the second most highly developed in South Asia by conventional criteria. Its exports are concen-

trated in textiles and agricultural products such as rubber and tea. By global standards, however, it is still a very poor country. Its progress, moreover, has been undercut by its seemingly perennial civil war. Were it not for this conflict, Sri Lanka would benefit more from the prime location of the port of Colombo and from its high levels of education. The Maldives is the most prosperous South Asian country based on per capita GNI, but its total economy, like its population, is very small. Most of its revenues are gained from fishing and international tourism.

India's Lesser Developed Areas   India's economy, like its population, dwarfs those of the other South Asian countries. While India's per capita GNI is lower than that of Pakistan, its total economy is more than five times larger. As the region's largest country, India also exhibits far more internal variation in economic development. The most basic economic division is that between India's more prosperous west and its poorer districts in the east. A less clear-cut rift is that between India's more dynamic south and its relatively stagnant north.

The so-called tribal states of the northeastern fringe generally form the bottom economic rungs as measured by per capita GNI, but the prevalence of subsistence economies makes such statistics misleading. More extreme deprivation is found in the lower Ganges Valley, where cash economies generally prevail. Bihar is India's poorest state by virtually all economic indicators, as well as its most politically corrupt.

Neighboring Uttar Pradesh, India's most populous state, is also extremely poor. Like Bihar, it is densely populated and has experienced little industrial development. While both Bihar and Uttar Pradesh have fertile soils, their agricultural systems have not profited as much from the Green Revolution as have those of the Punjab. Both states are also noted for their social conservatism; the caste system is deeply entrenched; tensions between Hindus and Muslims are bitter; and opportunities for most peasants are extremely limited. Ironically, South Asia's wealth was historically concentrated in the fertile lowlands of the Ganges Valley, yet today the area ranks among the poorest parts of this impoverished region.

Other relatively poor states in eastern India include Orissa and West Bengal. The worst slums in India—and perhaps the world—are located in West Bengal's Calcutta. But Calcutta also supports a substantial and well-educated middle class, and it is the site of a large industrial complex. For most of the period of Indian independence, West Bengal has been governed by a socialist party that has fostered, without great success, heavy, state-led industry. In a dramatic turnaround in the 1990s, West Bengal's Marxist leaders began to advocate internationalization, encouraging large multinational firms to build new factories in the state.

Although western India is in general much more prosperous than eastern India, the large western state of Rajasthan still ranks among the country's poorest areas. Rajasthan suffers from an arid and drought-prone climate; nowhere else in the world are deserts and semideserts so densely populated. It is also noted for its social conservatism. During the British period, almost all of this large state remained outside of the sphere of direct imperial power. Here, in the courts of maharajas, the military and political traditions of Hindu India persisted up until recent times. Rajasthan's rulers not only maintained elaborate courts and fortifications, but also supported many traditional Indian arts. Because of this political and cultural legacy, Rajasthan, despite its poverty, is one of India's most important destinations for international tourists.

India's Centers of Economic Growth   North of Rajasthan lie the Indian states of Punjab and Haryana, showcases of the Green Revolution. Their economies rest largely on agriculture, but substantial investments have been recently made in food processing and other industries. On Haryana's eastern border lies the capital district of New Delhi. India's political power and much of its wealth are concentrated here.

The west-central states of Gujarat and Maharashtra are noted for their industrial and financial clout, as well as for their agricultural productivity. Gujarat was one of the first parts of South Asia to experience substantial industrialization, and its textile mills are still among the most productive in the region. Gujaratis have long been famed as merchants and overseas traders, and they are disproportionately represented in the **Indian diaspora,** the migration of large numbers of Indians to foreign countries. As a result, cash remittances from these emigrants help to bolster the state's economy.

The state of Maharashtra is usually viewed as India's economic pacesetter. The huge city of Mumbai (Bombay) has long been the financial center and media capital of India; Mumbai's port, moreover, carries some 25 percent of India's total foreign trade. Major industrial zones are located around Mumbai and in several other parts of Maharashtra. In recent years, Maharashtra's economy has grown more quickly than those of most other Indian states, reinforcing its primacy. Mumbai now has some of the highest office rents in the world, a distinction earned, admittedly, as much by stringent building regulations as by economic vitality (Figure 12.38).

The center of India's fast-growing high-technology sector lies farther to the south, especially in Karnataka's capital of Bangalore. The Indian government selected the upland Bangalore area, which is noted for its pleasant climate, for its fledgling aviation industry in the 1950s. Other technologically sophisticated ventures soon followed. In the 1980s and 1990s, a quickly growing computer software and hardware industry emerged, earning Bangalore the label of "Silicon Plateau" (Figure 12.39). In the 1980s growth was spurred by the investments of U.S. and other foreign corporations eager to hire relatively inexpensive Indian technical talent. Since the 1990s, these multinational companies have been joined by an expanding group of locally owned firms concentrated in software production. More recently, the city of Hyderabad in the reform-oriented southern state of Andhra Pradesh has emerged as a secondary focus of high-tech growth. Small but vibrant zones of information technology are now emerging in other parts of southern and western India.

▲ **Figure 12.38 Mumbai (Bombay) stock exchange**   Evidence of this city's central role in the globalizing South Asia economy is the stock exchange, upper right, located in the heart of the city.   *(Rob Crandall/Rob Crandall, Photographer)*

India has proved especially competitive in software because software development does not require a sophisticated infrastructure; computer code can be exported via wireless telecommunication systems without the use of modern roads or port facilities. What is necessary, of course, is technical talent, and this India has in great abundance. Many Indian social groups have long been highly committed to education, and India has been a major scientific power for decades. With the growth of the software industry, India's brain power has finally begun to translate into economic gains. Whether such developments can spread benefits beyond the rather small

▲ **Figure 12.39 India's Silicon Plateau**   Since a significant proportion of its population is extremely well educated, India is suited for high-tech jobs both in computer assembly and manufacture and software development. Many of California's Silicon Valley firms draw upon facilities in India, which is 12 hours away in time zones, to run nonstop operations.   *(Chris Stowers/Panos Pictures)*

high-tech enclaves they presently occupy remains to be seen. What is certain is that information technology has tightly linked certain parts of India to the global economy, and especially to its nerve center in California's Silicon Valley, where up to 25 percent of engineers are Indian or Indian-American. An increasing flow of both information and people between Silicon Valley and Silicon Plateau indicates that this avenue of globalization is only in its infancy.

Globalization and India's Economic Future

As Figure 12.40 indicates, South Asia is one of the world's least globalized regions by conventional economic criteria. The volume of foreign trade is not large; foreign direct investment is minor; and (with the exception of the Maldives) international tourists are few. But globalization is advancing rapidly, especially in India but also in much of the rest of the region. Direct foreign investment in India, for example, went from $156 million in 1990 to $2,635 million in 1998.

To understand the low globalization indicators for the region, it is necessary to look at its recent economic history. India's postindependence economic policy, like those of other South Asian countries, was based on widespread private ownership combined with governmental control of planning, re-

source allocation, and certain heavy industrial sectors. Independent India established high trade barriers to protect its economy from global competition, which was seen as biased in favor of the wealthy countries. This mixed socialist-capitalist system brought a fairly rapid development of heavy industry and allowed India to become virtually self-sufficient, even in the most technologically sophisticated goods. It also resulted in slow but steady economic growth and, through the 1970s, a low level of foreign indebtedness.

By the 1980s, however, problems with this model were becoming apparent, and frustration with India's economic progress was mounting among the business and political elite. While growth was persistent, it remained in most years only a percentage point or two above the rate of population expansion. The percentage of Indians living below the poverty line, moreover, remained virtually constant. At the same time, countries such as China and Thailand were experiencing far faster development after opening their economies to global forces. Many Indian businesspeople chafed at the governmental regulations that they believed thwarted their ability to expand. Under the so-called license raj, entrepreneurs had to struggle to get official permission to operate, after which they were largely protected from competition. In the 1980s, foreign indebtedness began to mushroom, putting further

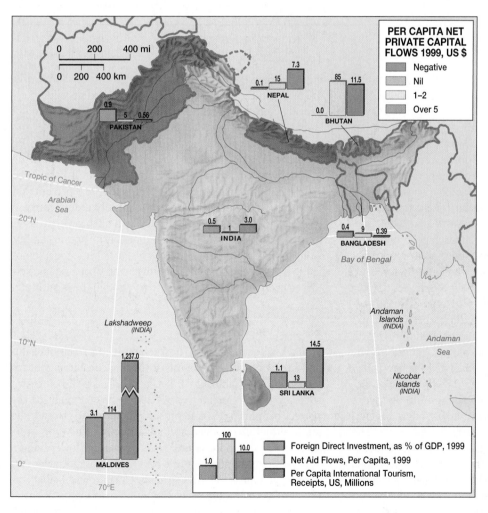

◀ Figure 12.40 South Asia's global linkages Despite South Asia's growing global connections, the region as a whole is still relatively self-contained, especially in regard to finance. In 1999, for example, more private investment capital flowed out of Pakistan and Nepal than flowed in; India and Bangladesh, at the same time, received modest capital inflows. Direct foreign investment was also modest across the region. Note also that the large countries of South Asia (especially India) receive very little foreign aid, whereas the small countries (Bhutan and the Maldives) receive substantial amounts. In regard to international tourism, the Maldives is simply off the chart, whereas Bangladesh and Pakistan hardly participate at all.

THE LOCAL AND THE GLOBAL   Supermarkets Versus Corner Grocery Stores in Indian Cities

One common aspect of economic globalization is the appearance of the Western-style supermarket in the cities of developing countries. These international food chains hope to profit by selling food to the billions of potential customers, but some international development workers maintain that supermarkets can actually provide fresher, lower-cost food than traditional corner stores.

Not so in India, based upon the recent experience of several supermarket chains that have been attempting to change the shopping habits of people in Mumbai and Delhi. Instead, the local corner groceries—the "mom-and-pops," as they're called—are holding their own by providing shoppers with convenience and good customer service. Home delivery for groceries is still common and free, as are zero-interest charge accounts. Additionally, these small markets cater to shoppers by stocking food local people prefer, unlike the supermarkets, which offer only standard brands in international packaging. "I never go to the supermarket," says a Delhi shopper in a cramped corner grocery. "Here they have whatever I want. Further, if I ring him up, he'll send things over. He's very obliging." While neighborhood stores don't provide the air-conditioned comfort or wide aisles of the supermarkets, they underscore once again the importance of local tastes, neighborhood traditions, and resistance to one-size-fits-all international marketing.

Supermarket chains in Indian cities also face problems of high land costs and an unreliable infrastructure. Because of old laws limiting real estate development in Delhi and Mumbai, land prices for space-extensive shopping centers are sky high. In contrast, because family-owned grocery stores are passed on from one generation to another, land prices are not an issue. Additionally, since electrical service is unreliable in Indian cities, large supermarkets must have expensive back-up generators to keep their extensive food freezers going during outages. With smaller stocks on hand, corner stores are not bothered by this problem.

Consequently, the familiar neighborhood grocery stores that dot Indian cities will persist, at least in the near term. International food chains, however, have made it quite clear they are not about to give up—not when there is a potential market of more than a billion shoppers. This illustrates once again that tensions between the local and the global are everywhere, even at the neighborhood level.

Source: Adapted from Miriam Jordan, 1988. "India's 'Mom-and-Pops' Hold Off Supers," *International Herald-Tribune,* October 12, 1998, p. 14.

---

pressure on the economy. Frustration also grew over India's insufficient infrastructure, particularly its roads and electrical systems.

In response to these difficulties, the Indian government earnestly began to liberalize its economy in the early 1990s. Many regulations were modified and some were eliminated, and the economy was gradually opened to imports and multinational businesses. Other South Asian countries have followed a somewhat similar path. Pakistan, for example, began to privatize many of its state-owned industries in 1994.

This gradual internationalization and deregulation of the Indian economy has generated substantial opposition. Foreign competitors are now seriously challenging domestic firms, threatening the livelihoods of their employees. Cheap manufactured goods from China are seen as an especially serious threat. But India has a strong heritage of economic nationalism stemming from the colonial exploitation it long suffered. Several growing political parties, including the Hindu-nationalistic BJP, have thus challenged the new internationalist policies. On gaining power, however, it soon became apparent that the BJP did not really want to disengage India from the global economy; rather, it has sought to renegotiate with foreign firms hoping to operate in India (see "The Local and the Global: Supermarkets Versus Corner Grocery Stores in Indian Cities"). Some of the most visible reactions against internationalization have been largely symbolic in nature, such as the closing down of a Kentucky Fried Chicken outlet in which inspectors had discovered several flies. Agri-

cultural liberalization, a central agenda of both the World Trade Organization and the United States, had become an especially contentious issue by the early 2000s, since much of India's huge farming sector is not globally competitive.

India's future economic policies thus remain uncertain. By the turn of the millennium, the Indian economy was growing at roughly 6 percent a year, much faster than it had in the 1950s, 1960s, and 1970s, but still short of the 8 percent that is generally considered necessary to rapidly reduce poverty. And while globalization and liberalization often generate faster growth, they also bring heightened insecurity and instability. Considering as well the country's history of fractious democratic politics, the current trend toward an open and market-driven economy could be quickly reversed in the near future.

Social Development

South Asia's social indices show relatively low levels of health and education, which is hardly surprising considering the region's poverty (Table 12.3). Levels of social well-being, not surprisingly, vary greatly across the region. As might be expected, people in the more prosperous areas of western India are healthier, live longer, and are better educated, on average, than people in the poorer areas, such as the lower Ganges Valley. Bihar thus stands at the bottom of most social as well as economic indices, while Punjab, Gujarat, and Maharashtra stand near the top. Several key measurements of social welfare are slightly higher in India than in Pakistan, despite

TABLE 12.3 *Social Indicators and Status of Women*

Country	Life Expectancy at Birth Male	Life Expectancy at Birth Female	Under Age 5 Mortality (per 1,000) 1980	Under Age 5 Mortality (per 1,000) 1999	Percent Illiteracy (Ages 15 and over) Male	Percent Illiteracy (Ages 15 and over) Female	Female Labor Force Participation (% of total, 1999)
India	60	61	177	90	68	44	32
Pakistan	60	61	161	126	59	30	28
Bangladesh	59	59	211	89	52	29	42
Nepal	58	57	180	109	58	23	40
Bhutan	66	66	—	—	56**	28**	40
Sri Lanka	70	74	48	19	94	89	36
Maldives	60	63	—	—	93**	93**	43

Sources: Population Reference Bureau Data Sheet, 2001, Life Expectancy (M/F); World Develoment Indicators, 2001, Under Age 5 Mortality Rate and Percent Illiteracy, except those marked **, which note data from the CIA World Factbook, 2000; The World Bank Atlas, 2001, Female Participation in Labor Force.

Pakistan's slightly superior per capita economic output. Pakistan has done a particularly poor job of educating its people, which is one reason why extreme fundamentalist Islamic organizations have been able to recruit so effectively in much of the country. Particularly in the areas near the Afghan border, the only educational option is often the extremist Islamic schools lavishly funded from Saudi Arabia.

Several discrepancies stand out when one compares South Asia's map of economic development with its map of social well-being. Punjab's birthrate, for example, is higher than might be expected, considering its general prosperity. Portions of India's extreme northeast show relatively high literacy rates despite their poverty; in this case, the pattern is explained by the educational efforts of Christian missionaries. Calcutta and its immediate environs clearly stand out as a relatively well-educated area, despite the general distress of the lower Ganges basin. The most pronounced discrepancies, however, strike the eye when one examines the southern reaches of South Asia. In regard to health, longevity, and education, the extreme south far outpaces the rest of the region.

The Educated South   Southern South Asia's relatively high levels of social welfare are clearly visible when one examines Sri Lanka. Considering its meager economic resources and interminable civil war, Sri Lanka must be considered one of the world's great success stories of social development. It demonstrates that a country can achieve significant health and educational gains even in the context of an "undeveloped" economy. Sri Lanka's average longevity of 72 years stands in favorable comparison with many of the world's industrialized countries, as does its literacy rate. Equally impressive is the fact that Sri Lanka's fertility rate has been reduced essentially to the replacement level. The Sri Lankan government has achieved these results by funding universal primary education and inexpensive medical clinics. With a well-educated, relatively healthy population growing at a slow rate, Sri Lanka could probably make rapid economic gains as well if only it could solve its political problems.

On the mainland, Kerala in southwestern India has achieved even more impressive results. Kerala is not a prosperous state; it is extremely crowded and has long had some difficulty feeding its population. Its overall economic figures are only average for India. Kerala's indices of social development, however, most notably in regard to longevity, literacy, and fertility, are the best in India, comparable to those of Sri Lanka (Figure 12.41). Since Kerala is poorer than Sri Lanka, its social accomplishments must be considered all the more impressive (see the *Author Field Trip: Kerala in Global Context*).

Some observers attribute Kerala's social successes to its state policies. Kerala has long been led by a socialist party that has stressed mass education and community health care. While this has no doubt been an important factor, it does not seem to offer a complete explanation. West Bengal, for example, also has a socialist political heritage, but it has not

▲ **Figure 12.41 Education in Kerala**   India's southwestern state of Kerala, which has virtually eliminated illiteracy, is South Asia's most highly educated region. It also has the lowest fertility rate in South Asia. Because of this, many argue that women's education and empowerment is the best and most enduring form of contraception.   *(Rob Crandall/Rob Crandall, Photographer)*

been nearly as successful in its social welfare programs. Kerala's neighboring state of Tamil Nadu, on the other hand, has made significant social progress despite having a different political environment. Some researchers suggest that one of the key variables for explaining the success of the far south is the relatively high social position of its women.

The Status of Women

It is often said that South Asian women are accorded a very low social position in both the Hindu and Muslim traditions. In upper-class families of both religions throughout the Indus-Ganges basin, women were traditionally secluded to a large degree, their social relations with men outside of the family being severely restricted.

Some scholars have argued that the Hindu tradition is more limiting to women than the Muslim tradition. Hindu women are expressly forbidden from engaging in certain economic activities (such as plowing), and in some areas they are excluded from inheriting land. Throughout northern India, women traditionally leave their own families shortly after puberty to join those of their husbands. As outsiders, often in distant villages, young brides have few opportunities. Widows in the higher castes, moreover, are not supposed to remarry, and instead are encouraged to go into permanent mourning.

Several social indices show that women in the Indus-Ganges basin suffer much discrimination (Table 12.3). In Pakistan, Bangladesh, and such Indian states as Rajasthan, Bihar, and Uttar Pradesh, female levels of literacy are far lower than those of males. An even more telling statistic is that of gender ratios, the relative proportion of males and females in the population. All things being equal, there should be slightly more women than men in any population, since women generally have a longer life expectancy. Northern South Asia, however, contains many more men than women. Western Uttar Pradesh probably has the most male-biased population in the world.

An imbalance of males over females often results from what is known as "differential neglect." In poor families of the region, boys typically receive better nutrition and medical care than do girls, which results in higher rates of survival. Economics play a major role. In rural households, boys are usually viewed as a blessing, since each typically remains with his family and works for its well-being. In the poorest groups, elderly people (especially widows) subsist largely on what their sons can provide. Girls, on the other hand, marry out of their families at an early age and must be provided with a dowry. They are thus seen as a net economic liability. Considering the economic desperation of the South Asian poor, it perhaps not surprising to see such unbalanced gender ratios.

Much evidence suggests that the social position of women is improving, especially in the more prosperous parts of northwestern India, where employment opportunities outside the family context are emerging. But even in many of the region's middle-class households, women still suffer major disabilities. Indeed, dowry demands seem to be increasing in some areas, and there have been a number of well-publicized murders of young brides whose families failed to deliver an adequate dowry. In some areas, gender ratios may be growing even more male-biased now that technology allows the possibility of gender-selective abortion.

While the social bias against women across northern South Asia is striking, it is much less evident in southern India and Sri Lanka. Geographic location, in other words, seems to play a much greater role than religion in determining the social position of women. In Kerala especially, women have relatively high status, regardless of whether they are Hindus, Muslims, or Christians. Here the gender ratio shows the normal pattern, with a slight predominance of females over males. Female literacy is very high in Kerala, which is one reason why the state's overall illiteracy rate is so low. Kerala's fertility rate is the lowest in India, which is another sign of women's social power. Not surprisingly, the high social position of women in southwestern India has deep historical roots. Among the Nairs—Kerala's traditional military and land-holding caste—all inheritance up to the 1920s had to pass through the female line. In earlier times, moreover, Nair women not only could divorce with impunity, but were even allowed more than one husband at a time.

⊕ Conclusion

South Asia, a large and complex area of more than a billion people, has in many ways been overshadowed by neighboring world regions: by the economic ascendance and volatility of Southeast Asia, the size and political weight of East Asia, and the geopolitical tensions of Southwest Asia. In some ways, South Asian leaders seemed willing to let other parts of the world dominate global headlines, following, perhaps, the nonalignment model set by India during the Cold War.

Much of that is changing, however, for South Asia now figures prominently in discussions of world problems and issues. Not only is it quite possible that South Asia will surpass East Asia as the most populous world region, but South Asia also demands attention over the daunting question of whether its population can be supported by its resource base. While most South Asian countries have shown remarkable progress in reducing fertility, population growth will remain relatively rapid over the next several decades because of the large number of young people in the region. With a huge cohort group entering its childbearing years, South Asian countries can ill afford to relax their family planning agendas.

Geopolitical tensions, both between countries and within them, also remain worrisome. Because of South Asia's complex cultural geography shaped by peoples speaking several dozen languages and following four major religions, and inflamed by several centuries of British colonial domination, cultural differences are often translated into geopolitical animosity. The long-standing feud between Pakistan and India escalated dangerously in the 1990s with reciprocal threats to use weapons of mass destruction. Some international arms experts now rank South Asia as the world region most likely to initiate nuclear war.

Internal geopolitics, too, are a cause for concern. Sri Lanka's civil war continues well into its second decade, with no end in sight. Ethnic and religious fighting also rages in the streets of Pakistan's largest city, Karachi, making it one of the most dangerous cities in the world. More than in most Muslim countries, in Pakistan an extremely fundamentalist Islamic movement challenges all aspects of the state and secular society. In giant India internal politics also demand world attention. As with other parts of this complicated region, cultural differences sometimes flare into violence. Religious strife between Muslims and Hindus continues; Hindu radicals attack Christian missionaries; Sikhs clash with Hindus in the Punjab. Off the street and in the statehouse, fundamentalist politicians—sometimes armed with funds raised by the Indian diaspora in the United States—work to impose their agenda on education, the arts, and even on India's industry and economics. While successful to some degree, their influence may be waning in the face of a more moderate multicultural backlash. As in other parts of the world, most people shun violence and desire peace and prosperity.

Globalization affects South Asia in many different ways, reminding us of the many facets of this world phenomenon. Nepal, for example, has found a lucrative link to international tourism by putting its spectacular mountain scenery on sale to trekkers from the wealthy parts of the world. Although the cultural and environmental costs of this enterprise are sometimes questionable, Nepal seems committed to this open-door policy for visitors. In contrast, Bhutan is much more cautious about international visitors, using quotas, strict guidelines, and high fees to profit from, yet limit the impact of, the outside world.

Bangladesh, a country that international aid workers recently viewed with profound pessimism because of its huge population, hazardous lowland environment, and limited resources, may actually have found some relief through economic globalization. As has been demonstrated countless times in the past decade, international industry will locate in countries with a large, low-paid labor force and a stable political environment. Currently, Bangladesh meets those needs. How long this will last and how beneficial it may actually prove remain to be seen.

Many argue that India is perfectly positioned to take advantage of economic globalization. Sizable segments of its vast labor force are well educated and speak excellent English, the major language of global commerce. An illustration of how these skills connect with a globalized world comes from the increasing number of North American firms developing "back-office" activities in India. These are labor-intensive record-keeping tasks that connect electronically to business headquarters. The credit records for millions of American consumers, for example, are kept in computer files that demand constant updating and cross-referencing. Today these electronic files are often maintained in India, where they can be processed daily by skilled low-wage workers, yet accessed instantaneously by electronic connections to North America. Often as not, when you ask a customer service representative about a bank card transaction, the answer comes from India's "back offices."

But can these global connections make the difference in saving South Asia from food shortages and ecological degradation as population growth threatens to outrun the resource base? As with all complicated world problems, the answers vary from the resounding optimism of economic free-trade advocates to the gloomy pessimism of environmentalists. Perhaps it is fitting to conclude with the Hindu proverb: To make the gods laugh, attempt to predict the future.

⊕ Key Terms

British East India Company (page 531)
bustees (page 519)
caste system (page 523)
Chipko movement (page 510)
dalit (page 526)
Dravidian language (page 527)

federal state (page 534)
forward capital (page 522)
Green Revolution (page 517)
Hindi (page 528)
Hindu nationalism (page 522)
Indian diaspora (page 542)
Jainism (page 527)

linguistic nationalism (page 529)
maharaja (page 535)
monsoon (page 512)
Mughal Empire (also spelled Mogul) (page 524)
orographic rainfall (page 512)

salinization (page 519)
Sanskrit (page 522)
scheduled castes (page 526)
Sikhism (page 525)
subcontinent (page 506)
Urdu (page 529)

⊕ Questions for Summary and Review

1. What are the four subregions of South Asia? Describe their similarities and differences.

2. What causes the South Asian monsoon? How does it affect different parts of South Asia?

3. What is orographic rainfall? Where is it important for South Asian agriculture?

4. How and why does the birthrate differ geographically within South Asia?

5. What are the similarities and differences between family planning programs in the major South Asian countries? With what results?

6. What are the considerations and variables when looking at the relationship between population growth and agricultural production? What is the outlook for the next several decades?

7. What are some of the "infrastructural" problems faced by South Asian cities? Give some specific examples.

8. Describe the geography of Islam within India. That is, where are the significant Muslim minorities located in India?

9. Describe the areas of origin in South Asia for Hinduism, Islam, Sikhism, and Buddhism.

10. What are the main features of the caste system? How has this system changed in India over the last several decades?

11. What are the major Indo-European languages in South Asia? Where are they located? Where are the non-Indo-European languages located?

12. How has the political geography of South Asia changed in the postcolonial era?

13. Describe three different regions of geopolitical and ethnic tension within South Asia.

14. Where are the centers or core areas of economic development within the different South Asian countries? Conversely, what regions would be considered marginal or peripheral to these cores?

15. What kinds of relationships are seen between women's literacy and different aspects of economic and social development?

16. How does the geography of gender ratios (the number of males and females within a population) differ within South Asia? What does this tell us?

⊕ Thinking Geographically

1. As a geographer, suggest different strategies for solving (or at least lessening) the serious flooding problems in Bangladesh. Consider the fact that this crowded country must maximize most of its area for agricultural production.

2. What are the advantages and disadvantages of expanding irrigated agriculture in India and Pakistan? How can the disadvantages be reduced to acceptable levels?

3. What are the pros and cons of the Green Revolution as a means of increasing South Asia's food supplies? What is the outlook for the next decade?

4. Discuss the conflict between wildlife protection and rural villages in India by evaluating the tension between preserving habitat and the needs of the rural poor.

5. What are the drawbacks and benefits of using English as a national language in India? Might it help or hinder unity? Would this increase or decrease India's links to the contemporary world?

6. Choose one of the areas of geopolitical tension (Kashmir, Punjab, Sri Lanka, etc.) and, after becoming better acquainted with the complex issues that underlie this conflict, evaluate the different proposals currently offered for solving (or at least ameliorating) the problem.

7. Conventional wisdom maintains that India will overtake China as the world's most populous country in the next several decades. First, using current rate of natural increase (RNI), calculate when this might be if rates of growth in each country do not change. Then explicate the different variables in both China and India that might change this outcome.

8. As a geographer, you work for an international arms reduction agency that is working to reduce tensions that could lead to nuclear war between Pakistan and India. What would you suggest?

9. Examine both the positive and negative aspects of the Indian diaspora. Begin with the less-than-obvious positive aspects of this migration by working through how it might benefit India in today's globalized context. In doing this exercise, pay attention to what has been written about the personal characteristics and talents of those who migrate.

10. From a geographical point of view, what is the best course of action for near-term future economic development in Pakistan, India, and Bangladesh?

11. Acquaint yourself with the environmental, social, and economic implications of Nepal's open-door policy toward trekking and other forms of tourism. Do the benefits seem to outweigh the costs?

12. Is the state of Kerala a good model for economic and social development in other parts of India? Why?

⊕ Regional Novels and Films

Novels

Anita Desai; *Fasting, Feasting* (1999, Houghton Mifflin)

Kamala Markandaya, *Nectar in a Sieve* (1954, J. Day)

Arundhati Roy, *The God of Small Things* (1997, Random House)

Salman Rushdie, *Midnight's Children* (1981, Alfred Knopf)

Paul Scott, *The Raj Quartet* (1980, Morrow)

Vikram Seth, *A Suitable Boy* (1993, HarperCollins)

Bapsi Sidhwa, *Cracking India: A Novel* (1991, Milkweed Editions)

Kushwant Singh, *Train to Pakistan* (1956, Chatto and Windus)

Films

Bandit Queen (1994, India)

City of Joy (1992, Great Britain)

Earth (1998, India)

East Is East (1999, Great Britain)

Monsoon Wedding (2001, India)

Phantom India (1969, France)

Saalam Bombay (1988, India)

Surjo Dighal Bari (The House Along the Sun (1979, Bangladesh)

⊕ Bibliography

Ahmad, Kazi. S. 1964. *A Geography of Pakistan.* Karachi: Oxford University Press.

Bayly, C. A. 1988. *Indian Society and the Making of the British Empire.* Cambridge: Cambridge University Press.

Bhardwaj, Surinder M. 1973. *Hindu Places of Pilgrimage in India: A Study in Cultural Geography.* Berkeley: University of California Press.

Carstairs, G. Morris. 1975. *The Twice-Born: A Study of a Community of High-Caste Hindus.* Bloomington: Indiana University Press.

Crossette, Barbara. 1995. *So Close to Heaven: The Vanishing Buddhist Kingdoms of the Himalayas.* New York: Knopf.

Dutt, Ashok K., and Geib, Margaret. 1987. *An Atlas of South Asia.* Boulder, CO: Westview.

Eaton, Richard M. 1993. *The Rise of Islam and the Bengal Frontier, 1204–1760.* Berkeley: University of California Press.

Er-Rashid, Haroun. 1977. *Geography of Bangladesh.* Boulder, CO: Westview.

Fox, Richard G., ed. 1977. *Realm and Region in Traditional India.* Durham, NC: Duke University Program in Comparative Studies on Southern Asia.

Frater, Alexander. 1990. *Chasing the Monsoon.* New York: Henry Holt.

Guha, Ramachandra. 1989. *The Unquiet Woods: Ecological Change and Peasant Resistance in the Himalaya.* Berkeley: University of California Press.

Inden, Ronald. 1990. *Imagining India.* Oxford: Blackwell.

Johnson, Basil L. 1983. *Development in South Asia.* New York: Penguin.

Kolanad, Gitanjali. 1994. *Culture Shock! A Guide to Customs and Etiquette in India.* Portland, OR: Graphic Arts Center Publishing Company.

Kothrai, Ashis, et al. 1995. "People and Protected Areas: Rethinking Conservation in India." *The Ecologist* 25(5), 188–194.

Lodrick, Deryck O. 1981. *Sacred Cows, Sacred Places: Origins and Survivals of Animal Homes in India.* Berkeley: University of California Press.

Malik, Yogendra K., and Singh, V. B. 1994. *Hindu Nationalists in India: The Rise of the Bharatiya Janata Party.* Boulder, CO: Westview.

Masica, Colin P. 1976. *Defining a Linguistic Area: South Asia.* Chicago: University of Chicago Press.

McGowan, William. 1992. *Only Man Is Vile: The Tragedy of Sri Lanka.* New York: Farrar, Straus & Giroux.

Naipaul, V. S. 1991. *India: A Million Mutinies Now.* New York: Penguin Books.

Rothermund, Dietmar. 1993. *An Economic History of India.* London: Routledge.

Schwartzberg, Joseph E. 1992. *A Historical Atlas of South Asia.* Oxford: Oxford University Press.

Singh, R. L., ed. 1968. *India: Regional Studies.* Calcutta: Indian National Committee for Geography.

Sopher, David E., ed. 1980. *An Exploration of India: Geographical Perspectives on Society and Culture.* Ithaca, NY: Cornell University Press.

Spate, O. H. K., and Learmonth, A. T. A. 1967. *India and Pakistan: A General and Regional Geography.* London: Methuen.

Steven, Stanley F. 1993. *Claiming the High Ground: Sherpas, Subsistence, and Environmental Change in the Highest Himalaya.* Berkeley: University of California Press.

Tandon, Prakash. 1968. *Punjabi Century 1857–1947.* Berkeley: University of California Press.

Verghese, Abraham. 2001. "The Bandit King and the Movie Star." *The Atlantic* 287(2), 71–78.

von Furer-Haimendorf. 1982. *Tribes of India: The Struggle for Survival.* Berkeley: University of California Press.

Wolpert, Stanley. 1991. *India.* Berkeley: University of California Press.

Woodcock, George. 1967. *Kerala: A Portrait of the Malabar Coast.* London: Faber and Faber.

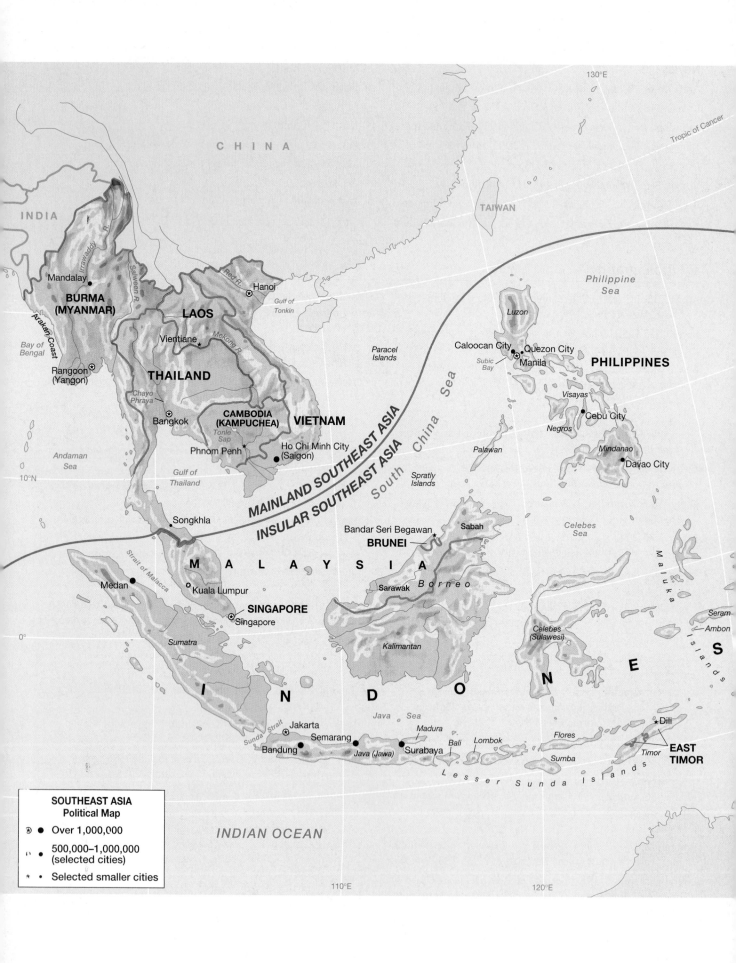

CHINA

Tropic of Cancer

INDIA

Mandalay

BURMA
(MYANMAR)

Irrawaddy R.

Salween R.

Arakan Coast

Bay of
Bengal

Rangoon
(Yangon)

LAOS

Red R.

Hanoi

Gulf of
Tonkin

TAIWAN

Philippine
Sea

Luzon

Caloocan City   Quezon City

Subic
Bay

Manila

PHILIPPINES

Paracel
Islands

Vientiane

Mekong R.

THAILAND

Chayo
Phraya

Bangkok

CAMBODIA
(KAMPUCHEA)

VIETNAM

Tonle
Sap

Phnom Penh

Ho Chi Minh City
(Saigon)

Andaman
Sea

10°N

Gulf of
Thailand

MAINLAND SOUTHEAST ASIA

INSULAR SOUTHEAST ASIA

South China Sea

Spratly
Islands

Palawan

Visayas

Negros

Cebu City

Mindanao

Davao City

Celebes
Sea

Songkhla

Bandar Seri Begawan

BRUNEI

Sabah

Sarawak

Borneo

M A L A Y S I A

Strait of Malacca

Medan

Kuala Lumpur

SINGAPORE

Singapore

Sumatra

I   N

Kalimantan

D   O

Celebes
(Sulawesi)

N   E

Maluka Islands

Seram

Ambon

S

0°

Java   Sea

Jakarta

Semarang

Madura

Bandung

Java (Jawa)

Surabaya

Bali

Lombok

Sunda Strait

Flores

Sumba

Dili

Timor

EAST
TIMOR

L e s s e r   S u n d a   I s l a n d s

INDIAN OCEAN

SOUTHEAST ASIA
Political Map

● Over 1,000,000

• 500,000–1,000,000
(selected cities)

★ Selected smaller cities

130°E

110°E

120°E

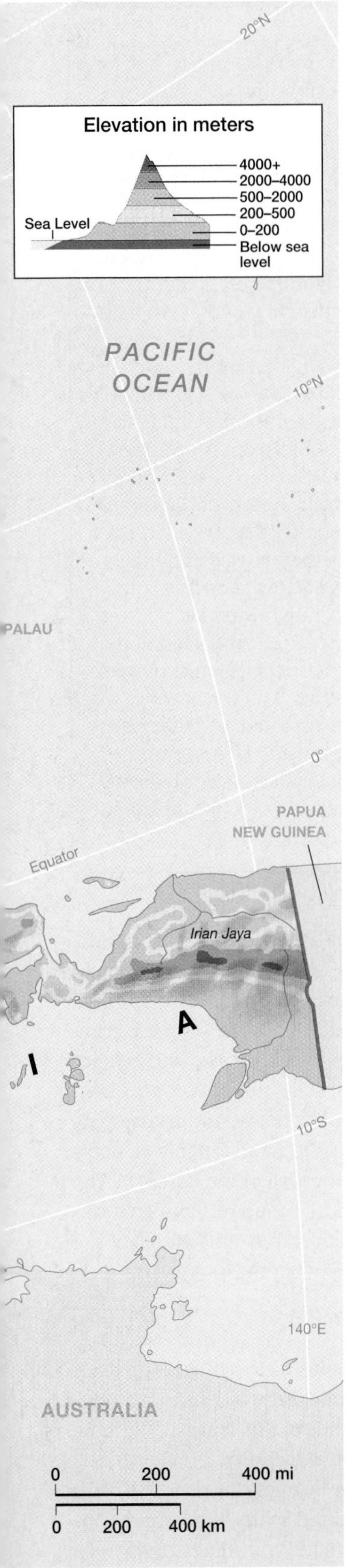

13

Southeast Asia

Elevation in meters

4000+	
2000–4000	
500–2000	
200–500	
Sea Level	0–200
	Below sea level

PACIFIC
OCEAN

20°N

10°N

PALAU

0°

PAPUA
NEW GUINEA

Equator

Irian Jaya

A

I

10°S

140°E

AUSTRALIA

0	200	400 mi

0	200	400 km

Southeast Asia, perhaps more than any other world region, illustrates both the promises and the perils of globalization. In the 1990s, much of Southeast Asia experienced a roller-coaster ride of economic boom and bust, descending almost overnight from the giddy heights of fast-growing "tiger economies" to a severe recession. In the 1980s Southeast Asia had often been seen as a model for world economic development. By the late 1990s, however, that model demanded revision as repercussions from Southeast Asia's economic woes affected countries across the globe. As of 2002, most Southeast Asian economies were again expanding, but their revival seemed unstable. Others, most notably Indonesia, remained mired in economic difficulties and political instability. While some argue that Southeast Asian countries need substantial economic reform if they are to thrive once more, others blame globalization itself, with its emphasis on free trade, global capital investment, and low-wage assembly lines.

Southeast Asia's involvement with the larger world is not new. The region (Figure 13.1) has long been heavily influenced by external forces. Chinese and especially Indian influences date back many centuries. Later, commercial ties with the Middle East opened the doors to Islam, and today Indonesia ranks as the most populous Muslim country in the world. More recently came the heavy-handed imprint of Western colonialism, as Britain, France, the Netherlands, and the United States administered large Southeast Asian colonies. During this period, national territories were rearranged, populations relocated, and new cities built to serve trade and military needs.

Southeast Asia's resources and its strategic location made it a major battlefield during World War II. Yet long after world peace was restored in 1945, warfare of a different sort continued in this region. As colonial powers withdrew and were replaced by newly independent countries, Southeast Asia became a battleground for world powers and their global ideologies. In Vietnam, Laos, and Cambodia, communist forces, tacitly supported by China and the Soviet Union, waged a fierce and determined struggle for control of local territory and people. Resisting this were the United States and several of its allies, equally determined to defeat communism out of fear that it would rapidly spread to the whole of Southeast Asia.

Ironically, while the communist forces did prevail in Vietnam, Laos, and Cambodia, they subsequently opened their economies to capitalist influences. Today with the end of the Cold War, competing ideologies have taken a back seat to other problems vexing Southeast Asia.

◀ **Figure 13.1 Southeast Asia** This region includes the large peninsula in the southeastern corner of Asia as well as a vast number of islands scattered to the south and east. It is conventionally divided into two subregions: mainland Southeast Asia, which includes Burma (Myanmar), Thailand, Laos, Cambodia, and Vietnam, and insular (or island) Southeast Asia, which includes Indonesia, the Philippines, Malaysia, Brunei, Singapore, and East Timor. Malaysia includes the tip of the mainland peninsula and most of the northern part of the island of Borneo.

SETTING THE BOUNDARIES

The region of Southeast Asia traditionally consists of 10 countries that vary widely in spatial extent, population, cultural attributes, and levels of economic and social development. With the recent emergence of East Timor as an independent state, the list must now be expanded to 11. Geographically, these countries are commonly divided into those on the Asian mainland and those on islands, or the insular realm. The mainland includes Burma (called Myanmar by the current government), Thailand, Cambodia, Laos, and Vietnam. Of these countries, Burma is the largest in area, about the same size as Texas. Thailand is next in size, covering an area somewhat larger than California. Although Burma is the largest in territory, Vietnam has the largest population of the mainland states, with almost 80 million people, about the same as Germany.

Insular Southeast Asia includes the sizable countries of Indonesia, the Philippines, and Malaysia, as well as the very small countries of Singapore, Brunei, and East Timor. Although classified as part of the insular realm because of its cultural and historical background, Malaysia actually splits the difference between mainland and islands. Part of its national territory is on the mainland's Malay Peninsula and part is on the large island of Borneo, some 300 miles distant. Borneo also includes the Muslim sultanate of Brunei, a small but oil-rich island country of 300,000 people covering an area slightly larger than Rhode Island. Singapore is essentially a city-state, occupying a small island just to the south of the Malay Peninsula.

Indonesia is the quintessential island nation, stretching 3,000 miles (4,800 kilometers, or about the same distance as from New York to San Francisco) from the large island of Sumatra in the west to New Guinea in the east, and containing more than 13,000 separate islands. Not only does it dwarf all other Southeast Asian states in size, but it is by far the largest in population. With more than 200 million people, it is ranked as the world's fourth most populous country. Additionally, because most of its inhabitants are Muslim, Indonesia is also counted as the world's largest Islamic nation. Lying north of the equator is the Philippines, a country of 77 million people spread over a number of islands, both large and small.

Until the second half of the twentieth century, much of this world region was sometimes referred to as "Indochina," a term that accurately reflects the strong historical influences of the large neighboring countries of India and China. Western colonial powers, including France, Britain, the Netherlands, and the United States, controlled most of the region until World War II, when Japan temporarily expanded its empire into Southeast Asia. Because of the strategic importance of this region, many heated battles were fought on its territory during World War II, and it was during this period that the geographic term "Southeast Asia" replaced "Indochina" and similar terms. After World War II, with the colonial powers gradually and often reluctantly withdrawing their hold on territory, newly independent states appeared as the modern geopolitical map emerged. Today, because of its continued strategic value, coupled with its close linkages to the dynamic world economy, Southeast Asia occupies a prominent place in the list of world regions.

Accompanying the recent economic turmoil has been an increase in ethnic and social tensions within many countries, most notably Burma (called Myanmar by the current military government), Indonesia, and, to a lesser degree, the Philippines. Many argue that these countries must reinvent not only their economies but also their political systems before they can participate fully and productively in the twenty-first century. Geopolitically, the **Association of Southeast Asian Nations (ASEAN)** has brought a new level of regional cooperation to the area, nurtured in part by the desire of Southeast Asia's countries to control—rather than be controlled by—external global forces. In many ways, the ASEAN agenda captures the problematic geography of this diverse region as Southeast Asia forges its own identity within the context of world globalization (see "Setting the Boundaries").

Environmental Geography: A Once-Forested Region

The mountainous area along the border between northern Thailand from Burma is a rugged place. Slopes are steep and thickly wooded in most places; rainfall is often torrential; leeches abound; and several strains of medicine-resistant malaria are prevalent. These same uplands are also, however, home to one of the most distinctive Southeast Asian ethnic groups: the Karen, a people numbering roughly 7 million. Unfortunately, the story of the Karen and their homeland is not a happy one. Much of their territory has been overrun by the Burmese army, and many of the Karen have been forced into squalid refugee camps in Thailand, where they still suffer periodic attacks from Burma. Furthermore, the Karen's dispossession has been accompanied by the loss of the magnificent and highly valuable teak forests of upland Burma. In the saga of the Karen we can see the meshing of cultural, political, economic, and environmental forces, resulting in a tragedy that is unfortunately not atypical for the more remote areas of Southeast Asia.

The Tragedy of the Karen

The Karen, like other tribal peoples of the upland areas of Burma, were never fully incorporated into the Burmese kingdom, which long ruled the lowlands of the country. With the imposition of British colonial rule in the 1800s, however, the Karen territory was joined to the Burmese lowlands. British and American missionaries educated many Karen and converted roughly 30 percent of them to Protestant Christianity. A number of Karen Christians obtained positions in Burma's colonial government. The Burmans of the lowlands, a strongly Buddhist people, resented this deeply, for they had long viewed the Karen as culturally inferior (according to conventional but confusing terminology, the term "Burmese" refers to all of the inhabitants of Burma, whereas "Burmans" refers only to the country's dominant, Burmese-speaking ethnic group). After independence, the Karen lost their favored position and soon grew to resent what they saw as Burman cultural and economic domination.

By the 1970s the Karen were in open rebellion and soon managed to establish an insurgent state of their own (Figure 13.2). They supported their rebellion by smuggling goods between Thailand and Burma and by mining the gemstones of their territory. Since Burma has one of the world's most protected economies, with a full range of governmental controls inhibiting commerce, smuggling is an especially lucrative occupation. Some estimates in the 1980s ranked the Karen economy as almost as large as the official Burmese economy.

The Burmese army, however, began to make headway against the insurgents in the early 1990s, and by the end of the decade had overrun most of the Karen territory. Although some Karen continue to fight, as a whole they are demoralized and divided. Crucial to Burma's success was an agreement made with Thailand in which the Thai government agreed to prevent Karen soldiers from finding sanctuary on its side of the border. This agreement was made in part in exchange for access by Thai timber interests to Burma's valuable teak forests.

Thailand's interests in Burma's forests stems from its own environmental problems. Thailand was once a major exporter of teak and other tropical hardwoods, but by the 1990s it had been largely deforested. The Thai government responded by banning commercial logging within the country. The only way Thai logging firms could stay in business was to move into the extensive forests of Burma and Laos, countries that had previously been relatively isolated from the world economy, and therefore little-touched by deforestation. The price of admission in the case of Burma was for Thailand to crack down on Karen guerrillas on its side of the border. Logging in Burma, however, has been reduced somewhat by a recent U.S. ban on importing teak products from both Thailand and Burma. To get around this restriction, Thai loggers have moved into the forests of Laos in search of valuable teak trees.

The Deforestation of Southeast Asia

The story of the Karen and Burma's teak forests is unique, but unfortunately deforestation and related environmental problems are major issues throughout most of Southeast Asia (Figure 13.3). Globalization has had a particularly profound effect on the Southeast Asian environment. Export-oriented logging companies have reached deep into the region's forests, cutting trees, damaging watersheds, and dispersing vast quantities of atmospheric pollutants as cutover lands are burned. This forest burning aggravates often intolerable air pollution within the rapidly growing major cities of Southeast Asia. Pollution levels in Jakarta, for example, are almost three times higher than World Health Organization guidelines specify.

Although colonial powers cut Southeast Asian forests for tropical hardwoods and naval supplies, and indigenous peoples have long cleared small areas of forest for agricultural use, rampant deforestation has come only in the last several decades with large-scale international commercial logging. This activity is largely driven by Asia's seemingly insatiable appetite for wood products such as plywood and paper pulp. The demand for paper pulp alone has increased at about 6 percent per year over the last decade throughout Asia.

While most forestry experts agree that Japan first globalized world forestry in the 1960s, other Asian countries, such as Taiwan, South Korea, Malaysia, and Indonesia, have followed suit with their own wood-products firms. Although the headquarters, boardrooms, and profits of these companies may be global, the damaging effects of commercial logging are both local and global. Landscapes are denuded, watersheds destroyed, wildlife habitat devastated. Additionally, the costs are high for rural peoples who rely on forest resources for their traditional way of life. The global effects of tropical forest clearing are also becoming increasingly problematic as more evidence links forest cutting to atmospheric warming and air pollution.

Two points are important for understanding forestry problems in Southeast Asia. First, although countries such as Indonesia look to their forest lands for increasing food supplies through expanded agriculture and for relieving population pressure in their more densely settled areas, agriculture and population growth are usually not the main cause of deforestation. Most forests are cut so that the wood products can be exported to other parts of the world. Subsequently, some of the

▲ **Figure 13.2 Karen rebels**   The Karen people have been in rebellion against Burma (Myanmar) since the 1970s. For several years they maintained a separate insurgent state, with its own capital city and regular army. The Burmese military advanced in the 1990s, however, forcing the Karen back to a guerrilla-style war.   *(Dean Chapman/Panos Pictures)*

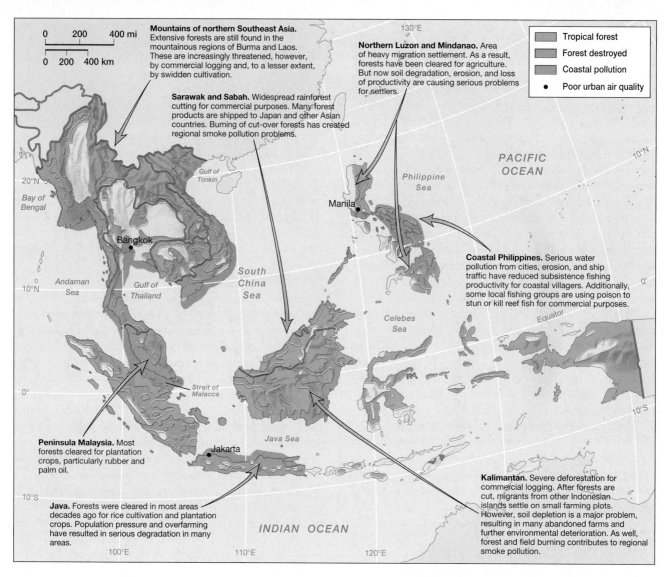

Figure 13.3 Environmental issues in Southeast Asia Southeast Asia was once one of the most heavily forested regions of the world. Most of the tropical forests of Thailand, the Philippines, peninsular Malaysia, Sumatra, and Java, however, have been destroyed by a combination of commercial logging and agricultural settlement. The forests of Kalimantan (Borneo), Burma (Myanmar), Laos, and Vietnam, moreover, are now being rapidly cleared. Water and urban air pollution, as well as soil erosion, are also widespread in Southeast Asia.

Map labels:

Mountains of northern Southeast Asia. Extensive forests are still found in the mountainous regions of Burma and Laos. These are increasingly threatened, however, by commercial logging and, to a lesser extent, by swidden cultivation.

Sarawak and Sabah. Widespread rainforest cutting for commercial purposes. Many forest products are shipped to Japan and other Asian countries. Burning of cut-over forests has created regional smoke pollution problems.

Northern Luzon and Mindanao. Area of heavy migration settlement. As a result, forests have been cleared for agriculture. But now soil degradation, erosion, and loss of productivity are causing serious problems for settlers.

Coastal Philippines. Serious water pollution from cities, erosion, and ship traffic have reduced subsistence fishing productivity for coastal villagers. Additionally, some local fishing groups are using poison to stun or kill reef fish for commercial purposes.

Peninsula Malaysia. Most forests cleared for plantation crops, particularly rubber and palm oil.

Java. Forests were cleared in most areas decades ago for rice cultivation and plantation crops. Population pressure and overfarming have resulted in serious degradation in many areas.

Kalimantan. Severe deforestation for commercial logging. After forests are cut, migrants from other Indonesian islands settle on small farming plots. However, soil depletion is a major problem, resulting in many abandoned farms and further environmental deterioration. As well, forest and field burning contributes to regional smoke pollution.

Legend: Tropical forest / Forest destroyed / Coastal pollution / • Poor urban air quality

logged-off lands are replanted (often with fast-growing, "weedy" tree species) while others are opened for agricultural settlement.

Second, most Southeast Asian countries now have widely publicized bans on the export of raw logs. Some (Thailand is an example) have even gone so far as to ban forest cutting altogether. Log export bans are enacted so that Southeast Asian countries can maximize their income from forest resources by milling and processing logs internally. In the past, logs were usually shipped to intermediary processing plants in Singapore or South Korea before entering the world market. Today most countries in the region require international companies to coinvest in building internal processing plants. Although there can be considerable economic benefits from this arrangement, a log ban does not necessarily mean either that the rate of forest cutting declines or that the environmental issues have been resolved. As for those countries with outright bans on forest cutting, either the bans are not generally enforced, or—as in certain parts of Malaysia—there are simply no trees left to cut (Figure 13.4).

Malaysia has been the leading exporter of tropical hardwoods from Southeast Asia. In recent decades, 60 percent of these log exports went to Japan. Income from these exports puts Malaysian forest products on par with petroleum exports and plantation products such as rubber. The cost to the environment, however, has been high. Peninsular Malaysia was largely denuded by 1985 when a cutting ban was imposed. Since then, forest cutting has been concentrated in the states of Sarawak and Sabah on the island of Borneo, where the granting of logging concessions to Malaysian and foreign firms has caused considerable problems with local tribal people by disrupting their traditional resource base. International forestry experts estimate that at current rates of cutting, Malaysia will be almost completely deforested in the near future.

Thailand cut more than 50 percent of its forests between 1960 and 1980. This loss was followed by a series of logging

▲ **Figure 13.4 Commercial logging** Southeast Asia has long been the world's most important supplier of tropical hardwoods. Unfortunately, most of the tropical forests of the Philippines and Thailand, as well as the Indonesian islands of Java and Sumatra, have been destroyed by the logging process. *(Jean-Leo Dugast/ Panos Pictures)*

▲ **Figure 13.5 Urban air pollution** Air pollution has reached a crisis stage in the rapidly industrializing cities of Southeast Asia, particularly in Bangkok, Manila, and Jakarta. People sometimes resort to using face masks to filter out soot and other forms of particulate matter. Forest fires, which often follow logging, greatly exacerbate the problem. *(V. Miladinovic/ Corbis/Sygma)*

bans so that by 1995 forest cutting was no longer legal. Damage to the landscape, however, had been severe, with heavy silting of irrigation works and hydroelectric facilities, increased flooding in lowlands areas, and severe erosion on hill slopes. Increasingly, these cutover lands are being reforested with fast-growing Australian eucalyptus trees. Eucalyptus forests, however, cannot support the local wildlife, and are thus biologically impoverished. The spread of eucalyptus in Thailand and much of the rest of the tropical world thus represents a kind of botanical globalization that has troubling implications for the preservation of biological diversity.

Indonesia, the largest country in Southeast Asia, has fully two-thirds of the region's forest area, including about 10 percent of the world's true tropical rain forests. Most of Sumatra's forests have been denuded, however, and those of Kalimantan (Borneo) will not last long at present rates of cutting. Indonesia's last forestry frontier is on the island of New Guinea.

Smoke and Air Pollution

Until recently, most of Southeast Asia's residents seemed oblivious to the widespread pall of air pollution created by a combination of urban smog and the smoke from tropical forest clearing. Then, late in the 1990s, the region suffered from two consecutive years of disastrous air pollution that served

as a wake-up call. During that period, a commercial airliner crashed because of poor visibility; countless road accidents resulted; two ferries collided in smoke-laden conditions; and hundreds of thousands of people were admitted to hospitals with life-threatening respiratory problems (Figure 13.5). Additionally, because of global publicity, billions of tourism dollars were lost when reservations were cancelled.

Several factors, both natural and economic, combined to produce the region's horrendous air pollution problems. First, large portions of insular (or island) Southeast Asia suffered a severe drought caused by El Niño, turning the normally wet tropical forests into tinder boxes. This drought also dried out the widespread peat bogs of coastal Kalimantan, which, once fired, continued to burn for months. Second, commercial forest cutting was—and is—responsible for most forest burning, even though the large logging firms commonly blame small farmers. In Sumatra's forest, 80 percent of the fires were from commercial forest plantations, with only 20 percent from small slash-and-burn agriculture. As commercial lumbering has increased over the region, so has air pollution from logging-related fires.

The third part of the equation is Southeast Asia's burgeoning cities, where cars, trucks, and factories emit huge quantities of pollutants. Regardless of forest burning, many cities, such as Bangkok, Jakarta, and Manila, already have unhealthy levels of pollution. In Bangkok, rolling down a window during a taxi ride is considered a very serious breach of etiquette, since people protect themselves from street-level pollution by sequestering themselves in air-conditioned cars. As a general rule, lung cancer and pulmonary disease resulting from urban air pollution kills people at five times the rate in the United States. Given the increase in auto traffic and other sources of urban pollution coupled with the increasing presence of smoke from forest fires, this unhealthy situation is likely to grow worse before any decisive action is taken.

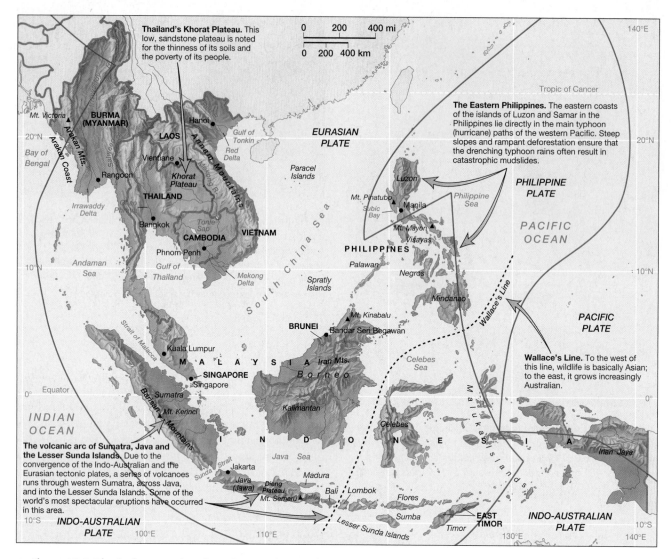

▲ Figure 13.6 Physical geography of Southeast Asia   Southeast Asia is one of the world's most geologically active regions owing to the intersection of several tectonic plates. As a result, earthquakes are a frequent occurrence in many areas, and volcanism and other forms of active mountain-building abound. Several mountain ranges radiate out from the rugged uplands of northern Southeast Asia, dividing the area's broad river valleys and deltas into several distinct physical regions.

Patterns of Physical Geography

To understand why forestry issues are so important in Southeast Asia, it is necessary to look in more detail at its physical geography. The southern portion of the region—known as insular Southeast Asia—is one of the world's three main zones of tropical rain forest. The northern part of the region, known as mainland Southeast Asia, is located in the tropical wet-and-dry zone that is noted for particularly valuable timber species, such as teak. Distinguishing between the mainland and the islands is thus one of the main keys for understanding the geography of Southeast Asia.

Island Southeast Asia is less geologically stable than the mainland, and thus it is a region of more diverse landforms. Four of Earth's tectonic plates converge here: the Pacific, the Philippine, the Indo-Australian, and the Eurasian (Figure 13.6). As a result of this tectonic structure, earthquakes occur frequently near the plate boundaries. Large, often explosive vol-

canoes are another consequent feature of the insular Southeast Asian landscape (Figure 13.7). A string of active volcanoes extends the length of eastern Sumatra (in Indonesia) across Java and into the Lesser Sunda Islands, the small islands east of Java. One of the world's most famous volcanic explosions occurred in this area, that of Krakatau in 1883, which killed more than 30,000 people and sent clouds of ash around the world. Numerous volcanoes also dot the Philippine archipelago, including Mt. Pinatubo, which erupted in 1991, causing 800 deaths and temporarily cooling Earth's atmosphere.

Mainland Environments   Mainland Southeast Asia is an area of rugged uplands interspersed with broad lowlands associated with large rivers. The region's northern boundary lies in a cluster of mountains connected to the highlands of western Tibet and south-central China. In the far north of Burma (Myanmar), peaks reach 18,000 feet (5,500 meters).

▲ **Figure 13.7 Volcanic eruption in the Philippines** The Philippines and Indonesia, sitting astride plate-tectonic boundaries, have many volcanoes. Although several areas of insular Southeast Asia—most notably central and eastern Java—owe their fertile soils to past eruptions, volcanism remains a major threat to human life. This was demonstrated in the Philippines by the catastrophic eruption of Mt. Pinatubo in 1991. *(Durieux/ SIPA Press)*

▲ **Figure 13.8 Delta landscape** Southeast Asia has some of the world's largest delta landscapes. Deltaic environments are used for intensive irrigated rice cultivation, allowing very high rural population densities. Delta wetlands are also used for aquaculture (fish farming) and other forms of intensive food production. Most of mainland Southeast Asia's large cities are located in delta areas, resulting in periodic flooding and other environmental problems. *(John Elk III/Stock Boston)*

From this point, a series of distinct mountain ranges radiates out, extending through western Burma, along the Burma–Thailand border, and through Laos into southern Vietnam. Most of these ranges are less than 10,000 feet (3,500 meters) in height. Population is relatively sparse throughout these mountains. The mountainous country of Laos, for example, has only 5.4 million people in an area as large as the United Kingdom in western Europe, which has almost 60 million people.

Several large rivers flow southward out of Tibet and its adjacent highlands into mainland Southeast Asia. The longest is the Mekong, which is about as long as North America's Missouri River at 2,600 miles (4,190 kilometers). It flows through Laos and Thailand, then across Cambodia before entering the South China Sea through an extensive delta in southern Vietnam. Second longest is the Irrawaddy at about 1,300 miles (2,100 kilometers), which flows through Burma's central plain before reaching the Bay of Bengal. This river also has a large delta. Two smaller rivers are equally significant: the Red River, which forms a sizable and heavily settled delta in northern Vietnam, and the Chao Phraya, which has created the fertile alluvial plain of central Thailand (Figure 13.8).

The centermost area of mainland Southeast Asia is Thailand's Khorat Plateau, which is neither a rugged upland nor a fertile river valley. This low sandstone plateau averages about 500 feet (175 meters) in height and is noted for its thin, poor soils. Water shortages and droughts make this extensive area difficult for settlement and, as we shall see, relatively poor.

Monsoon Climates Almost all of mainland Southeast Asia lies in the tropical monsoon zone, characterized by a distinct hot and rainy season from May to October. This is followed by dry but still generally hot conditions from November to April (Figure 13.9). Only the central highlands of Vietnam and a few coastal areas receive significant rainfall during this autumn and winter period. In the far north, the winter months bring mild and sometimes rather cool weather.

Figure 13.9 shows two tropical climate regions in mainland Southeast Asia. While both are connected to the monsoon regime (see the discussion in Chapter 12), the difference between the two is in the total amount of precipitation received during the year. Along the coasts and in the highlands, the Am, or tropical monsoon, climate dominates. Rainfall totals for the Am climate usually register more than 100 inches (254 centimeters) each year. In the uplands, this figure can easily reach more than 200 inches (508 centimeters).

The second climate region, the Aw climate, covers most of mainland Southeast Asia. Here annual rainfall totals are about half those of the Am rainforest region. In most cases this can be explained by interior locations removed from the oceanic source of moisture, along with the lack of orographic lifting in these areas of more subdued topography. A good portion of Thailand, for example, receives only about 50 inches (127 centimeters) of rain during the year. Much of Burma's central Irrawaddy Valley is almost semiarid, with rainfall totals below 30 inches.

The Forest Landscape The original vegetation of most of the mainland was tropical monsoon forest. Because of the dry winters, this forest was not as dense or diverse as the equatorial rain forests of insular Southeast Asia, which receive copious rainfall year-round. But until the late nineteenth century, this was one of the more heavily wooded portions of the world. Even the great deltas of the Irrawaddy and Mekong rivers remained largely uncultivated and covered with forests. Massive forest cutting began in the late 1800s in mainland Southeast Asia. Lowland forests were gradually converted to farmland both to feed the expanding local population and to supply rice for the growing world economy. Under British and French colonial influence, forests were cleared to create

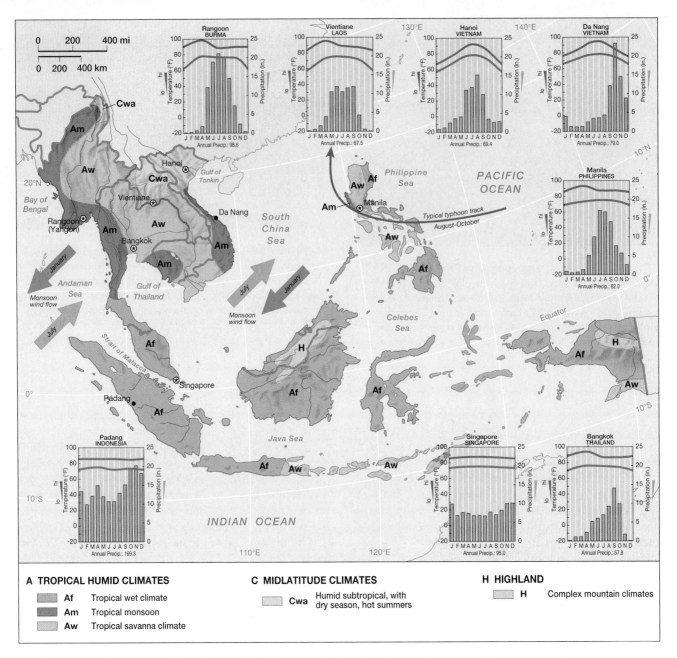

▲ Figure 13.9 **Climate map of Southeast Asia** Most of insular Southeast Asia is characterized by the constantly hot and humid climates of the equatorial zone. Mainland Southeast Asia, on the other hand, has the seasonally wet and dry climates of the tropical monsoon and tropical savanna types. Only in the far north are subtropical climates, with relatively cool winters, encountered. The northern half of the region is strongly influenced by the seasonally shifting monsoon winds. Northeastern Southeast Asia—and especially the Philippines—often experiences typhoons from August to October.

an export economy based on rice cultivation. Within 30 years in the late nineteenth century, the Irrawaddy delta was deforested and converted into commercial rice paddies as the British relocated settlers from India and central Burma to work the plantations.

A second round of forest loss began in the mid-twentieth century, after World War II. This was focused on the forests of the upland areas. Monsoon forests in mainland Southeast Asia contain several valuable tree species, most notably teak. It is said that today one prime teak tree can be worth as much as $40,000.

Insular Environments The signal feature of insular Southeast Asia is its archipelagic environment. Indeed, this is a region of countless islands of all sizes and shapes. Borneo and Sumatra are the third and sixth largest islands in the world, respectively, while many thousands of others are little more than specks of land rising at low tide from a shallow sea.

Indonesia alone is said to contain more than 13,000 islands, dominated by the four great landmasses of Sumatra, Borneo (the Indonesian portion is called Kalimantan), Java, and the oddly shaped Sulawesi. This island nation also includes the western half of New Guinea and the Lesser Sunda Islands, which extend to the east of Java. A prominent mountain spine runs through these islands as a result of tectonic forces. In the western Indonesian islands, volcanic peaks of 10,000 feet (3,500 meters) are common. The highest mountains, however, are on the island of New Guinea, where a handful of spectacular peaks reach around 15,000 feet (5,250 meters) in elevation.

The Philippines, another insular country, includes more than 7,000 islands. The two largest and most important are Luzon (about the size of Ohio) in the north and Mindanao (the size of South Carolina) in the south. Sandwiched between them are the Visayan Islands, which number roughly a dozen. Again, because of tectonic forces that generate active volcanoes, the topography of the Philippines includes mountainous landscapes that reach elevations of 10,000 feet (3,500 meters).

Closely related to this impressive collection of islands is the world's largest expanse of shallow ocean. These waters cover the **Sunda Shelf,** which is an extension of the continental shelf extending from the mainland through the Java Sea between Java and Kalimantan. Here waters are generally less than 200 feet (70 meters) deep. Because of the rich marine life, some insular Southeast Asian peoples have adopted maritime ways of life, essentially living on their boats and setting foot on land only on a temporary basis.

Equatorial Island Climates

The climates of insular Southeast Asia are somewhat more complex than those of the mainland because of three factors: a more complicated monsoon effect, the stronger influence of Pacific typhoons, and the equatorial location of the Indonesian islands.

Most of insular Southeast Asia, unlike the mainland, receives rain during the Northern Hemisphere's winter because the winter monsoon winds cross large areas of warm equatorial ocean, where they absorb moisture. As a result, these winds can bring heavy rains as the saturated air masses are lifted over island uplands. On Sumatra and Java, for example, where north winds blow between November and March, these air masses produce heavy rains on the northern side of these east–west running islands. As an illustration, the climograph for Jakarta, in northern Java, shows the heaviest rainfall during January and February, with the lowest monthly rainfall coming during the May to September period (see Figure 13.9). But during the May–September period, the heaviest rains are found on the southern flanks of these same islands because of the south winds associated with the Asian summer monsoon.

A second factor in the climate pattern of insular Southeast Asia are the tropical hurricanes, or **typhoons** as they are called in the western Pacific, that bring heavy rainfall to the northeastern reaches of insular Asia during the period of August to October. These strong storms bring devastating winds and torrential rain. They develop east of the Philippines and then move westward into the South China Sea, where they often intensify. Each year a number of typhoons affect parts of the Philippines with heavy damage and loss of life through flooding and landslides (Figure 13.10). These strong storms also commonly strike the eastern coast of Vietnam.

The third climate control, the equatorial influence, results from the area's low latitude. More so than mainland Southeast Asia, the islands experience very little seasonality. Temperatures remain high throughout the year, with very little variation. In Jakarta, for example, the average daily high temperature varies only 4 °F (2.2 °C) during the year, from 84 °F (29 °C) in January and February (during the winter monsoon) to 88 °F (31 °C) in September. The average low temperature varies even less because high humidity retains heat during the night.

Also associated with the equatorial influence is the fact that rainfall is both higher and more evenly distributed during the year than on the mainland. Although it is common to have a period of heavier rain linked to one of the two monsoon patterns, island climates close to the equator do not experience a distinct dry season, as do those on the mainland. Instead, rain falls throughout the year from tropical thunderstorms and squalls common to equatorial climates. As a result of this year-round precipitation, most of island Southeast Asia is placed into the Af, or tropical rainforest, climate category, as can be seen in Figure 13.9. The southeastern islands of Indonesia, however, do experience a distinct dry season from May to October, whereas parts of the western Philippines are typically dry from November to April.

Wallace's Line and Island Biogeography   Within insular Southeast Asia there is a dramatic difference in animal and plant life between western and eastern islands that has long fascinated scientists (Figure 13.6). On the western islands of Sumatra, Java, and Borneo, one finds large Asian mammals such as tigers, bears, elephants, and rhinoceros, as well as apes—orangutans and gibbons. But most large mammals are missing completely in the eastern islands. Instead, one finds mammal and bird species usually associated with Australia—most notably marsupials such as opossums, wallabies, and arboreal kangaroos.

▲ **Figure 13.10 Typhoon damage**   Typhoons are a major environmental threat in the Philippines and parts of mainland Southeast Asia. Typhoons often result in wind damage, but flooding and mudslides cause the most destruction. Deforestation over the past 30 years has greatly increased the magnitude of floods and mudslides. *(Vogel/Getty Images, Inc.)*

This striking difference in flora and fauna was first documented by British naturalist Alfred Wallace, the cofounder with Charles Darwin of the theory of natural selection, who traveled extensively in these islands during the second half of the nineteenth century. Wallace was one of the first to examine **island biogeography,** the distribution and ecology of plant and animal life unique to islands.

Wallace correctly surmised that the differences in plant and animal life can be explained by the last global ice age, which ended about 12,000 years ago. At that time, the world's oceans were about 300 feet (105 meters) lower because extensive glaciers covered much of the Northern Hemisphere. With a lower ocean level, the shallow Sunda Shelf was dry land, thus forming an extension of mainland Asia. In the east, this lower sea level connected New Guinea and the other islands to the Australian continent, thereby allowing animals from that landmass to colonize new territory.

The advance of these species was halted by open water between the islands of Borneo and Celebes, as it was in the narrow channel between Bali and Lombok in the Lesser Sunda Islands. As the world's ice sheets melted, sea levels rose, reaching their current level about 5,000 years ago. With these higher levels, the islands of Southeast Asia were once again cut off from the mainland; thus animals were stranded in their modern locations. Because of Alfred Wallace's work on this fascinating ecological issue, the division between the two natural realms is called **Wallace's Line,** separating plants and animals from Asia from those originating in Australia.

Population and Settlement: Subsistence, Migration, Cities

The scale of Southeast Asia's population issue is quite different from those of its giant neighbors, China and India. With just over 500 million people, Southeast Asia is still *relatively* sparsely settled. Part of the reason can be attributed to extensive tracts of infertile soil and rugged mountains. Such areas generally remain thinly inhabited. In contrast, relatively dense populations are found in the region's deltas, coastal areas, and zones of fertile volcanic soil (Figure 13.11).

Many of the favored lowlands of Southeast Asia have experienced striking population growth over the past several decades. Demographic growth and family planning have thus become increasingly important issues through much of the region. Different Southeast Asian countries have responded to their demographic situations in very different ways. While Indonesia promotes migration to its outer islands as a way to relieve population pressure in the core areas, Thailand places considerable emphasis on family planning. Malaysia's government, on the other hand, has concluded that its current population is too small, and thus advocates population growth.

As is the case throughout the world, one of Southeast Asia's major changes in settlement geography derives from massive migration to cities. This has been particularly true in countries that experienced rapid economic growth during the boom years of the 1980s and 1990s. Now, with an economic slowdown in the region, Southeast Asian cities may have a respite of sorts to address the environmental and infrastructural problems generated during the boom years.

Settlement and Agriculture

Part of the reason for the historical paucity of population in much of Southeast Asia is the infertile soil, which is unable to support intensive agriculture and high population densities. The island rain forests, though lush and biologically rich, grow on poor soils. Plant nutrients are locked up in the vegetation itself, rather than being stored in the soil where they would easily benefit agriculture. Furthermore, the incessant rain of the equatorial zone tends to wash nutrients out of the soil. Agriculture must be carefully adapted to this limited soil fertility by constant field rotation or the application of heavy amounts of fertilizer.

There are, however, some notable exceptions to this generalization about soil fertility and settlement density in equatorial Southeast Asia. Unusually rich soils connected to volcanic activity are scattered through much of the region, but are particularly prevalent on the island of Java. Java, with more than 50 volcanoes, is blessed with rich soils that support a large array of tropical crops and a very high population density. It has well over 100 million people—more than half the total population of Indonesia—in an area smaller than the state of Iowa. Dense populations are also found in pockets of fertile alluvial soils along the coasts of insular Southeast Asia, where people augment land-based farming with marine resources and trade activities.

The demographic patterns in mainland Southeast Asia are less complicated than those of the island realm. In all the mainland countries, population is concentrated in the agriculturally intensive valleys and deltas of the large rivers, whereas the uplands remain relatively lightly settled (Figure 13.11). The population core of Thailand, for example, is formed by the valley and delta of the Chao Phraya River, just as Burma's is focused on the Irrawaddy. In Vietnam there are two distinct foci: the Red River delta in the far north and the Mekong delta in the far south. In contrast to these densely settled areas, the middle reaches of the Mekong River provide only limited lowland areas in Laos, which is one of the reasons that country has a much smaller population than its neighbors. In Cambodia the largest population historically has clustered around Tonle Sap, a large lake with a very unusual seasonal flow reversal. During the rainy summer months the lake receives water from the Mekong drainage, but during the drier winter months it contributes to the river's flow. Cambodia's modern population core and its capital city of Phnom Penh are situated where the Tonle Sap drainage meets the Mekong River.

Agricultural practices and settlement forms obviously vary widely across the complex environments of Southeast Asia. Generally speaking, however, three farming and settlement patterns are apparent.

Swidden in the Uplands   Also known as shifting cultivation or "slash-and-burn" agriculture, swidden is practiced throughout the rugged uplands of both mainland and island

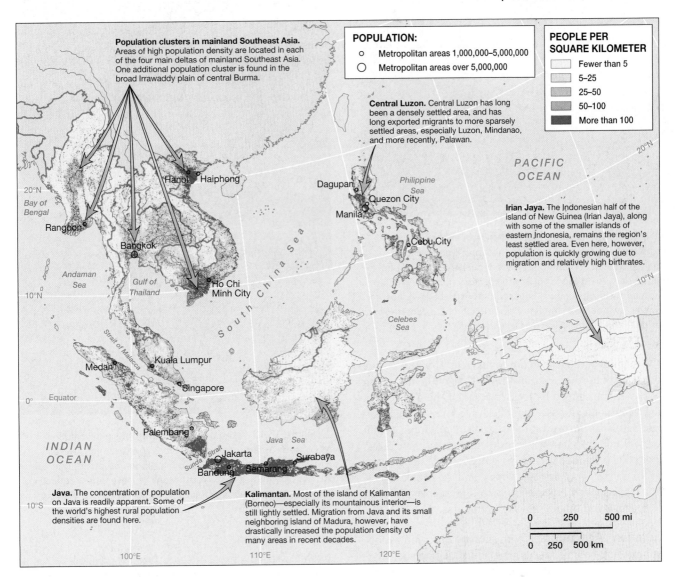

Population clusters in mainland Southeast Asia.
Areas of high population density are located in each of the four main deltas of mainland Southeast Asia. One additional population cluster is found in the broad Irrawaddy plain of central Burma.

POPULATION:
○ Metropolitan areas 1,000,000–5,000,000
◯ Metropolitan areas over 5,000,000

PEOPLE PER SQUARE KILOMETER
- Fewer than 5
- 5–25
- 25–50
- 50–100
- More than 100

Central Luzon. Central Luzon has long been a densely settled area, and has long exported migrants to more sparsely settled areas, especially Luzon, Mindanao, and more recently, Palawan.

Irian Jaya. The Indonesian half of the island of New Guinea (Irian Jaya), along with some of the smaller islands of eastern Indonesia, remains the region's least settled area. Even here, however, population is quickly growing due to migration and relatively high birthrates.

Java. The concentration of population on Java is readily apparent. Some of the world's highest rural population densities are found here.

Kalimantan. Most of the island of Kalimantan (Borneo)—especially its mountainous interior—is still lightly settled. Migration from Java and its small neighboring island of Madura, however, have drastically increased the population density of many areas in recent decades.

▲ **Figure 13.11 Population map of Southeast Asia**   In mainland Southeast Asia, population is concentrated in the valleys and deltas of the region's large rivers. In the intervening uplands, population density remains relatively low. In Indonesia, density is extremely high on Java, an island noted for its fertile soil and large cities. Some of Indonesia's outer islands, especially those of the east, remain lightly settled. Overall, population density is high in the Philippines, especially in central Luzon.

Southeast Asia (Figure 13.12). In the **swidden** system, small plots of several acres of dense tropical forest or brush are periodically cut or "slashed" by hand. Then the fallen vegetation is burned to transfer nutrients to the soil before subsistence crops are planted. Yields remain high for several years, then drop off dramatically as the soil nutrients from burned vegetation are exhausted and insect pests and plant diseases multiply. These plots are abandoned completely after a few years and allowed to revert to woody vegetation. The cycle of cutting, burning, and planting is moved to another small plot not far away—thus the term *shifting cultivation*. Families and small villages generally control a large amount of territory so they can rotate their fields on a regular basis. After a period of 10 to 75 years, the farmers return to the original plot, which once again has nutrients in the dense vegetation. The cycle then continues with another round of slashing and burning.

Swidden is a sustainable form of agriculture when population densities remain relatively low and stable and when upland people control enough territory. Today, however, the swidden system is threatened in Southeast Asia for two reasons. First, it cannot easily support the increasing population that has resulted from relatively high human fertility and, in some cases, migration. With a higher population density, the swidden rotation period must be shortened, undercutting soil resources. Second, the upland swidden system is often a casualty of commercial forest logging because of the devastation done to forest environments. Perhaps nothing illustrates the long arms of globalization better than the impact of international commercial forest cutting on the traditional agricultural system of upland tribal peoples.

When swidden can no longer support the population, upland people often adapt by switching to a cash crop that will

▲ Figure 13.12 Swidden agriculture   In the uplands of Southeast Asia, swidden (or "slash-and-burn") agriculture is widely practiced. When done by tribal peoples with low population densities, swidden is not environmentally harmful. When practiced by large numbers of immigrants from the lowlands, however, swidden can result in deforestation and extensive soil erosion.   (Paula Bronstein/Getty Images, Inc.)

allow them to participate in the commercial economy. In the mountains of northern Southeast Asia, one of the main cash crops is opium, grown by local farmers for the global drug trade. This mountainous area is often called the "**Golden Triangle**." In 2000 Burma was the world's second largest opium producer (after Afghanistan), and it may regain first place. Reportedly, 100 tons of opium leave the Golden Triangle each year for markets in Europe, Australia, and the United States. Many reports imply linkages between local military officials, teak logging, and opium growing in this area.

Plantation Agriculture   With European colonization, Southeast Asia became a focus for commercial plantation agriculture, growing high-value specialty crops ranging from rice to rubber. Even in the nineteenth century, Southeast Asia was linked to a globalized economy through the plantation system. Forests were cleared and swamps drained to make room for these plantations; labor was supplied (often unwillingly) by indigenous people or by contract laborers brought in from India or China. These plantations were usually in the coastal lowlands, from which products could be easily shipped to Europe and North America. Over the years, with fluctuating world demand, competition from other world regions, and changing political conditions within the region, the array of plantation crops has changed (Figure 13.13).

Plantations are still an important part of Southeast Asia's geography and continue to play a major role in today's global economy. Most of the world's natural rubber, for example, is produced in Malaysia, Indonesia, and Thailand. Natural rubber, however, constitutes a small fraction of the total world rubber consumption since most rubber is now made synthetically from petroleum. Cane sugar has long been a plantation crop of the Philippines and parts of Indonesia. More recently, pineapple plantations have appeared in both the Philippines and Thailand, which are now the world's leading exporters. Indonesia is the region's leading producer of tea,

▲ Figure 13.13 Tea harvesting in Indonesia   Plantation crops, such as tea, are major sources of exports for several Southeast Asian countries. Coconut, rubber, oil palms, and coffee are other major cash crops. Many of these crops require large amounts of labor, particularly at harvest time.   (Dermot Tatlow/Panos Pictures)

and Malaysia dominates the production of palm oil. Coconut oil and **copra** (dried coconut meat) are widely produced in the Philippines, Indonesia, and elsewhere.

Although plantations are still widespread, there is a general tendency for Southeast Asian countries to depend less on agricultural exports and more on other forms of economic enterprise, particularly industry. One partial exception is Malaysia, a former British colony with long experience with plantation agriculture. Malaysia focuses on plantation crops in part because it has de-emphasized food crops for its local markets. It can easily afford to buy food staples on the world market because of its successful export economy.

Rice in the Lowlands   The lowland basins of mainland Southeast Asia are largely devoted to intensive rice cultivation. Throughout almost all of Southeast Asia, rice is the preferred staple food. In fact, in several local languages "to eat rice" means "to eat a meal." Traditionally, rice was mainly cultivated on a subsistence basis by rural farmers. But as the number of wage laborers in Southeast Asia has grown due to economic development, so has the demand for commercial rice cultivation. Rice harvests are increasingly traded to fulfill the food needs of the region's expanding urban markets, and a large amount of rice is exported to the global system from Thailand. Three delta areas have been the focus for commercial rice cultivation: the Irrawaddy in Burma, the Chao Praya in Thailand, and the Mekong in Vietnam and Cambodia. The use of agricultural chemicals and high-yield crop varieties, along with improved water control and the expansion of irrigation facilities to facilitate dry-season cropping, have allowed production to keep pace with population growth. Such techniques have, however, also resulted in significant environmental degradation.

In those areas without irrigation and water control, yields remain relatively low. Rice growing on the Khorat Plateau, for example, depends largely on the uncertain rainfall, without the benefits of the more sophisticated water control methods available elsewhere in Thailand. In some lowland districts lacking irrigation, dry-field crops, especially sweet potatoes and manioc, form the staple foods of people too poor to buy market rice on a regular basis.

Recent Demographic Change

Because Southeast Asia is not facing the same kind of population pressure as East or South Asia, a wide range of government population policies is found. While several countries show concern about rapid growth and thus have strong family planning programs, others believe that their populations are too small. In those countries with rapid demographic expansion, internal relocation away from densely populated areas to outlying districts is a common policy.

Population Contrasts   The Philippines, the third most populous country in Southeast Asia, has a relatively high growth rate (Table 13.1). Complicated internal politics in the Philippines tend to impede effective family planning. When a popular democratic government replaced a dictatorship in the 1980s, the Philippine Roman Catholic Church, which played an active role in the peaceful revolution, pressured the new government to cut funding for family planning programs. As

a result, many clinics and centers that had dispensed family planning information were closed. Although high birthrates are not always associated with Catholicism, the Church's outspoken stand on birth control seems to inhibit the dispersal of family planning information.

The highest total fertility rate (TFR) in Southeast Asia (5.4 children) is found in Laos, a country of Buddhist religious tradition. Here the high birthrate is best explained by the country's low level of economic and social development. Thailand, which shares cultural traditions with Laos yet is considerably more developed, demonstrates the other end of that spectrum. Here the TFR has dropped dramatically within the last 30 years from 5.4 (in 1970) to 1.8, a figure that will soon bring population stability. But while economic growth may explain some of this decrease, it is also important to note that the Thai government has promoted family planning for both population and health reasons, including the high incidence of AIDS in the country.

Indonesia, with the region's largest population at more than 206 million, has also seen a dramatic decline in fertility in recent decades, although its fertility rate remains above the replacement level. If the present trend continues, however, Indonesia will reach population stability well before most other large developing countries. As with Thailand, this drop in fertility seems to have resulted from a strong government family planning effort, coupled with improvements in education.

Cambodia, like Laos, has a persistently high fertility rate. A partial explanation is the fact that Cambodia has one of the lowest life expectancy rates and one of the highest infant mortality rates in Southeast Asia. As has been seen in other world regions, high birthrates are often associated with high mortality. Cambodia's high mortality rates are linked to its recent history of civil strife and internal violence, factors that have also inhibited economic and social development. Now that some semblance of stability has returned to Cambodia, it will be interesting to see whether its fertility rate declines.

TABLE 13.1 *Demographic Indicators*

Country	Population (Millions, 2001)	Population Density, per square mile	Rate of Natural Increase	TFR[a]	Percent < 15[b]	Percent > 65	Percent Urban
Burma (Myanmar)	47.8	183	1.6	3.3	33	5	27
Brunei	0.3	156	2.0	2.7	32	3	67
Cambodia	13.1	187	1.7	4.0	43	4	16
Indonesia	206.1	280	1.7	2.7	31	4	39
Laos	5.4	59	2.5	5.4	44	4	17
Malaysia	22.7	178	2.0	3.2	33	4	57
Philippines	77.2	666	2.2	3.5	37	4	47
Singapore	4.1	17,320	0.9	1.6	17	6	100
Thailand	62.4	315	0.8	1.8	24	6	30
Vietnam	78.7	623	1.4	2.3	33	6	24

[a]Total fertility rate

[b]Percent of population younger than 15 years of age

Source: Population Reference Bureau. World Population Data Sheet, 2001.

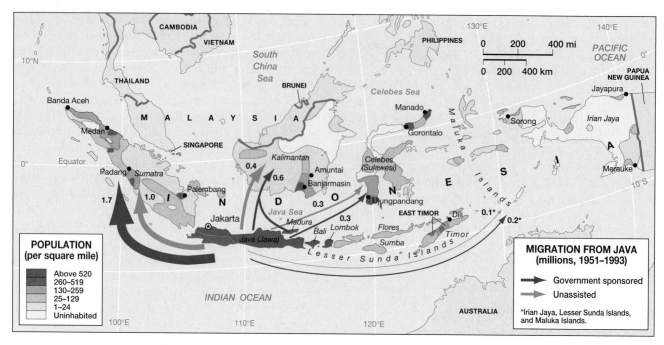

▲ **Figure 13.14 Indonesian transmigration**   The distribution of population in Indonesia shows a marked imbalance; Java, along with the neighboring island of Madura, is one of the world's most densely settled places, whereas most of the country's other islands remain rather lightly populated. As a result, the Indonesian government has encouraged the resettlement of Javanese and Madurese people to the outer islands, often paying the costs of relocation. This transmigration scheme has resulted in a somewhat more balanced population distribution pattern, but has also caused substantial environmental degradation and intensified several ethnic conflicts.

Growth and Migration   Indonesia has the most explicit policy of **transmigration,** or relocation of its population from one region to another within its national territory (Figure 13.14). Primarily because of migration from densely populated Java, the population of the outer islands of Indonesia has grown rapidly since the 1970s. The province of East Kalimantan, for example, experienced an astronomical growth rate of 30 percent per year during the last two decades of the 1900s. As a result of this shift in population, many parts of Indonesia outside of Java now have moderately high population densities, although many of the more remote districts remain lightly inhabited.

As is the case worldwide, high social and environmental costs often accompany these relocation schemes. Javanese peasants, accustomed to working the highly fertile soils of their home island, often fail in their attempts to grow rice in the former rain forest of Kalimantan. Field abandonment is high after repeated crop failures. In some areas farmers have little choice but to adopt a semi-swidden form of cultivation, moving to new sites once the old ones have been exhausted, a process associated with further deforestation and soil degradation. The term **shifted cultivators** is sometimes used for these rural migrants who are moved from one area to another through relocation schemes (Figure 13.15; see "Local Voices: Land Rights and Indigenous Peoples").

The Philippines have also used internal migration as a means of alleviating population pressure on the main islands. Beginning in the late nineteenth century, Philippine society has responded to increasing population pressure in the core areas of central Luzon by sending colonists to frontier zones. In the early twentieth century this settlement frontier still lay in Luzon; by the postwar years it had moved south to the island of Mindanao. That frontier is currently saturated, however, so the emphasis is now on international migration. One unfortunate aspect of this new globalized migration is the thousands of young Philippine men who work as low-paid deckhands on ill-fitted merchant ships plying the oceans.

Slow Growth and Pro-Growth   The city-state of Singapore stands out on the demographic charts with a fertility rate well below replacement levels. Unless the deficit is offset by immigration or a dramatic turnabout in the birthrate, population will soon begin to decline. The government is concerned about this situation and is actively promoting marriage and childbearing, particularly among the most highly educated segment of its population. Ironically, this is the very segment that has undergone the most pronounced fertility decline.

The Malaysian government also supports population growth, even though its fertility rate is still well above replacement levels. Larger families are encouraged, as is immigration to the country. Additionally, the government is particularly interested in increasing settlement in its Borneo island states of Sabah and Sarawak, which it considers underpopulated. In fact, several government officials have recently suggested that Malaysia should double its population to around 50 million.

▲ **Figure 13.15 Migrant settlement in Indonesia** Migration from densely settled to sparsely settled areas of Southeast Asia has resulted in the creation of thousands of new communities. Many of these communities have minimal infrastructure and services, and some of them struggle to survive. *(Charly Flyn/ Panos Pictures)*

Urban Settlement

Despite the relatively high level of economic development found in Southeast Asia, the region is not heavily urbanized; less than 30 percent of its population lives in cities. Even Thailand retains a rural flavor, which is somewhat unusual for a country that has experienced so much recent industrialization. Cities, however, are growing rapidly throughout the region, so the rate of urbanization will undoubtedly increase significantly over the next decade.

Many Southeast Asian countries have **primate cities**, single, large urban settlements that overshadow all others. Thailand's urban system, for example, is dominated by Bangkok, just as Manila far surpasses all other cities in the Philippines. Both have grown recently into mega-cities with more than 10 million residents. More than half of all city-dwellers in Thailand live in the Bangkok metropolitan area (Figure 13.16), which is 25 times larger than the country's second city, Nakhon Ratchasima. In both Manila and Bangkok, as is the case in most developing countries, explosive urban growth has led to housing problems, congestion, and pollution. With the recent arrival of mass automobile culture, Bangkok may now suffer from the worst traffic congestion in the world. In Manila, it is estimated that more than half of the city's population lives in squatter settlements, usually without basic water and electricity service.

Thailand, the Philippines, and Indonesia are all making efforts to encourage growth of secondary cities by decentralizing economic functions. The goal is to stabilize the population of the primate cities. In the case of the Philippines, the city of Cebu has emerged in recent years as a more dynamic economic center than Manila, leading to hopes that a more balanced urban system may soon emerge.

Urban primacy is considerably less pronounced in other Southeast Asian countries. Vietnam, for example, has two main cities, Ho Chi Minh City (formerly Saigon) in the south,

---

LOCAL VOICES   Land Rights and Indigenous Peoples

In the mountains of northern Luzon, as in many other parts of Southeast Asia, indigenous small-scale societies often have no legal rights to their ancestral lands. This lack of basic land rights makes them vulnerable to outside interests eager to gain control of their resources. The following dialogue, written by Filipino author Pedro Bundok, illustrates the tensions generated by the contradictions between two legal systems of land rights—one customary and indigenous, the other codified and identified with the modern state. Although the text was written 30 years ago, the issues that it brings up are still current. Here we find Pedro Bundok trying to explain the official land system to an elderly gentleman named Bugtong:

> "Pedro, I have a problem. . . . They tell me that I should not farm my land anymore because I am a squatter. What do they mean by a squatter?"
>
> "A squatter," I said, . . . "is someone who does not get permission to use land that belongs to someone else."
>
> "How can the government say that I am a squatter? My father's bones and my grandfather's bones are buried on the land where my camote [sweet potatoes] are planted!"
>
> "The government man says that you should have a piece of paper called a 'title,' Bugtong. If you have a paper it means that you own the land."
>
> "Pedro, you act like you don't know very much. . . . We all know that the one from Heaven owns the land. We can only bor-

row the land. . . . How can that young boy from the government think that a paper is going to make me own my land?"

> "Bugtong, . . . the government wants every person to decide which land he is borrowing and then get a paper that they call a 'title' which stops the arguments about who is allowed to farm each part of the land."
>
> "But Pedro, our ancestors have been here since Balitok and Bogan and they have not had any papers and they have not had any trouble. If anyone had an argument, they just called for a conference and the elders helped them to remember and they agreed what to do. Then they ate a pig and everyone went home happy. . . ."
>
> "Bugtong, . . . the elders in the lowlands have forgotten how to do that, so now they have papers to help them remember. . . ."
>
> "Pedro, I understand about the papers, but why did that man say that I am a squatter? This I cannot understand. . . ."
>
> "The government says that any land which does not have papers now belongs to the government. You do not have papers for the land yet, so he thinks that you are farming the land which belongs to the government. . . ."
>
> "Now I have two problems instead of one," he said. "First, why should I ask permission to farm the lands where the bones of my grandfather and brother and father are buried? Second, who is the government?"

Source: From *Democracy Among the Mountaineers* by Pedro Bundok. Quezon City, Philippines: New Day Publishers, 1973.

▲ **Figure 13.16 Bangkok** Bangkok saw the development of an impressive skyline during its boom years from the late 1970s through the late 1990s. Unfortunately, infrastructure did not keep pace with population and commercial growth, resulting in one of the most congested and polluted urban landscapes in the world. *(Robert Holmes/Robert Holmes Photography)*

▲ **Figure 13.17 Singapore** Singapore remains the economic and technological hub of Southeast Asia. It is famous for its clean, efficiently run, and hypermodern urban environment. Some residents complain, however, that Singapore has lost much of its charm as it has developed. *(Adina Tovy/Lonely Planet Images/Photo 20-20)*

with more than 3 million people, and the capital city of Hanoi in the north, with slightly more than 1 million residents. Jakarta, a vast city, is the largest urban area in Indonesia, but the country has a host of other large and growing cities, including Bandung and Surabaya. Yangon (formerly Rangoon) is the capital and primate city of Burma. This city has doubled its population in the last two decades to more than 4 million residents. In Cambodia, the capital city of Phnom Penh has less than 1 million; the same is true of Vientiane, the capital and primate city of Laos.

Kuala Lumpur, the largest city in Malaysia, has received heavy investments from both the national government and the global business community in recent decades. This has produced a modern city of grandiose ambitions that is free of most infrastructural problems plaguing other Southeast Asian cities. As testimonial to its outlook, the Petronas Towers, owned by the country's national oil company, were the world's tallest buildings at almost 1,500 feet (450 meters) when completed in 1996. However, Shanghai, China, is currently building a taller structure, reminding us that skyscrapers are common symbols of national prestige and economic power.

The independent republic of Singapore is essentially a city-state of 3 million people on an island of 240 square miles (600 square kilometers), about three times the size of Washington, D.C. While space is at a premium, Singapore has been very successful at developing high-tech industries that have brought it great prosperity. Unlike other Southeast Asian cities, Singapore has no squatter settlements or slums. Only in the fast-disappearing Chinatown and historic colonial district does one find older buildings. Otherwise, Singapore is an extremely clean but rather sterile city of modern high-rise skyscrapers and space-intensive industry (Figure 13.17).

Cultural Coherence and Diversity: A Meeting Ground of World Cultures

Unlike many other world regions, Southeast Asia lacks the historical dominance of a single civilization. Instead, the region has been a meeting ground for cultural diffusion from South Asia, China, the Middle East, Europe, and even North America. Abundant natural resources, along with the region's strategic location on oceanic highways connecting major continents, have long made Southeast Asia attractive to outsiders. As a result, the contemporary cultural geography of this diverse region is in part a product of borrowing and amalgamation from external influences.

The Introduction and Spread of Major Cultural Traditions

In Southeast Asia contemporary cultural diversity is embedded in the historical influences connected to the major religions of Hinduism, Buddhism, Islam, and Christianity (Figure 13.18).

South Asian Influences The first major external influence arrived from South Asia some 2,000 years ago when migrants from what is now India helped establish Hindu kingdoms in coastal locations in Burma, Thailand, Cambodia, central and southern Vietnam, Malaysia, and western Indonesia. Although Hinduism faded away in most locations, this tradition is still found on the Indonesian islands of Bali and Lombok, and vestiges remain in many other areas. The ancient Indian script formed the basis for many Southeast Asian writing systems, and in Muslim Java (Indonesia) the Hindu epic called the **Ramayana** remains a central cultural feature today.

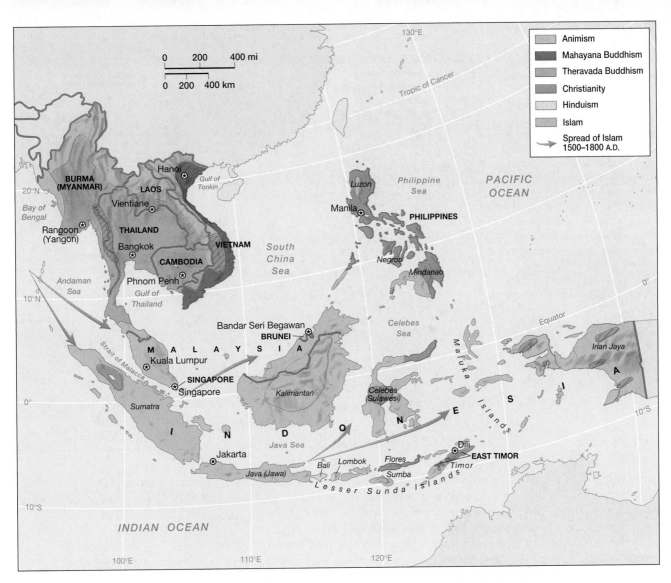

▲ **Figure 13.18 Religion in Southeast Asia** Southeast Asia is one of the world's most religiously diverse regions. Most of the mainland is predominantly Buddhist, with Theravada Buddhism prevailing in Burma (Myanmar), Thailand, Laos, and Cambodia, and Mahayana Buddhism (combined with other elements of the so-called Chinese religious complex) prevailing in Vietnam. The Philippines is primarily Christian (Roman Catholic), but the rest of insular Southeast Asia is primarily Muslim. Substantial Muslim minorities are found in the Philippines, Thailand, and Burma. Animist and Christian minorities can be found in remote areas throughout Southeast Asia, especially in Indonesia's province of Irian Jaya.

A second wave of South Asian religious influence reached mainland Southeast Asia in the thirteenth century in the form of Theravada Buddhism, which is closely associated with Sri Lanka. Virtually all of the people in lowland Burma, Thailand, Laos, and Cambodia converted to Buddhism at that time, and today it still forms the foundation for their social institutions. Saffron-robed monks, for example, are a common sight in Thailand and Burma, where Buddhist temples abound. In Thailand, the country's revered constitutional monarchy remains closely connected with Buddhism. While Southeast Asian Theravada Buddhism shares many traits with the Mahayana Buddhism of East Asia, there are enough cultural and religious differences that the two are mapped separately in Figure 13.18.

Chinese Influences Unlike most other mainland peoples, the Vietnamese were not heavily influenced by South Asian civilization. Instead, their early connections were to East Asia. Vietnam was actually a province of China until about A.D. 1000, when the Vietnamese established a kingdom of their own. But while the Vietnamese rejected China's political rule, they retained many attributes of Chinese culture. The traditional religious and philosophical beliefs of Vietnam are, for example, centered around Mahayana Buddhism and Confucianism. Furthermore, until the French colonial period of the nineteenth century, the Vietnamese used Chinese characters for their written communication.

East Asian cultural influences in many other parts of Southeast Asia are directly linked to more recent immigration of

▶ **Figure 13.19 Chinese in Southeast Asia** People from the southern coastal region of China have been migrating to Southeast Asia for hundreds of years, a process that reached a peak in the late 1800s and early 1900s. Most Chinese migrants settled in the major urban areas, but in peninsular Malaysia sizable numbers were drawn to the countryside to work in the mining industry and in plantation agriculture. Today Malaysia has the largest number of people of Chinese ancestry in the region. Singapore, however, is the only Southeast Asian country with a Chinese majority.

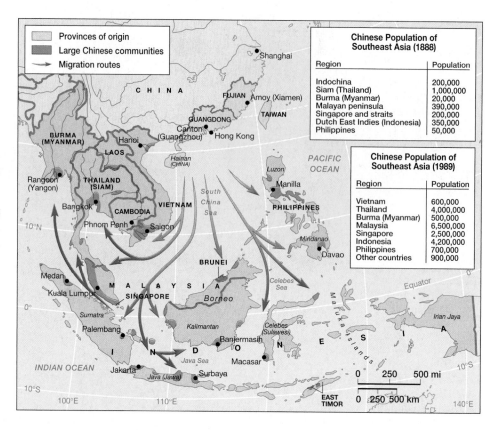

Chinese Population of Southeast Asia (1888)

Region	Population
Indochina	200,000
Siam (Thailand)	1,000,000
Burma (Myanmar)	20,000
Malayan peninsula	390,000
Singapore and straits	200,000
Dutch East Indies (Indonesia)	350,000
Philippines	50,000

Chinese Population of Southeast Asia (1989)

Region	Population
Vietnam	600,000
Thailand	4,000,000
Burma (Myanmar)	500,000
Malaysia	6,500,000
Singapore	2,500,000
Indonesia	4,200,000
Philippines	700,000
Other countries	900,000

southern Chinese. While this migration dates back hundreds of years, it reached a peak in the nineteenth and early twentieth centuries (Figure 13.19). China was then a poor and crowded country, which made sparsely populated Southeast Asia appear to be a place of great opportunity. At first most migrants were single men. Many returned to China after accumulating money, but others married local women and established mixed communities. This is especially true in the Philippines, where the elite population is often described as "Chinese Mestizo," or people of mixed Chinese and Filipino descent. In the nineteenth century Chinese women begin to migrate in large numbers, allowing the creation of ethnically distinct Chinese settlements. Urban areas throughout Southeast Asia are still characterized by large and cohesive Chinese communities. In Malaysia the Chinese minority constitutes roughly one-third of the population, whereas in the city-state of Singapore some three-quarters of the people are of Chinese ancestry.

In many places in Southeast Asia, relationships between the Chinese minority and the indigenous majority are strained. Even though their ancestors arrived many generations ago, many Chinese are still considered resident aliens because they maintain their Chinese citizenship. Probably a more significant source of tension is the fact that most overseas Chinese communities are relatively prosperous. Many Chinese emigrants prospered because they were able to find a mercantile niche that was unfilled by the local people. As a result, they often exert tremendous economic influence on local affairs—influence that is sometimes deeply resented by others. Anti-Chinese uprisings have occurred periodically throughout Southeast Asia. In 2000 and 2001, especially se-

vere rioting broke out in several Indonesian cities. Since many Indonesian Chinese had recently converted to Christianity, rioters burned a number of Christian churches.

The Arrival of Islam Muslim merchants from South and Southwest Asia arrived in Southeast Asia more than a thousand years ago, and by the 1200s their religion began to spread. From an initial focus in northern Sumatra, Islam diffused into the Malay Peninsula, through the main population centers in the Indonesian islands, and east to the southern Philippines. By 1650 Islam had largely replaced Hinduism and Buddhism throughout Malaysia and Indonesia. The only significant holdout was the small but fertile island of Bali, where thousands of Hindu musicians and artists fled from the courts of Java, giving the island a strong tradition of arts and crafts that has been maintained to the present day. Partly because of this artistic legacy, Bali is today one of the premier destinations of international tourism.

Today the world's most populous Muslim country is Indonesia, where some 87 percent of the nation's 206 million inhabitants follow Islam. This figure, however, masks a significant amount of internal diversity. In some parts of Indonesia, such as in northern Sumatra (Aceh), orthodox forms of Islam took root (Figure 13.20). In others, such as central and eastern Java, a more lax form of worship emerged in which certain Hindu and even animistic beliefs have been maintained. Islamic reformers, however, have long been striving to instill more orthodox forms of worship among the Javanese.

In Malaysia and especially in northern Sumatra, on the other hand, Islamic fundamentalism has gained ground. While the current Malaysian government has generally supported

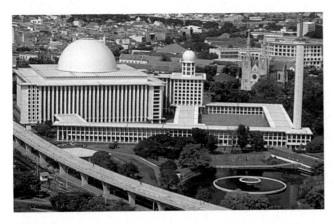

▲ **Figure 13.20 Indonesia's largest mosque**   Indonesia is often said to be the world's largest Muslim nation, since more Muslims reside here than in any other country. Islamic architecture in Indonesia is often somewhat modernistic, especially when contrasted with the more traditional styles found in Southwest Asia and North Africa.   *(Dana Downie/Lonely Planet Images/ Photo 20-20)*

the revitalization of Islam, it is wary of the growing power of the fundamentalist movement. Some Malaysian Muslims, moreover, view the fundamentalists as advocating social practices derived from the Arabian Peninsula that are not necessarily religious in origin. These include women covering their heads or even veiling themselves. As a result of these tensions, religious controversies abound in modern Malaysia. In the late 1990s a national debate was waged after several contestants in a beauty pageant were arrested for having violated a religious ban on such exhibitions. Subsequently, the national government ordered a review of all of Malaysia's Islamic laws, which had been enforced by local officials.

Islam was still spreading eastward through insular Southeast Asia when the Europeans arrived in the sixteenth century. When the Spanish claimed the Philippine Islands in the 1570s, they found the southwestern portion of the archipelago to be thoroughly Islamic. The Muslims resisted the Christian religion introduced by the Spanish and derived their cultural identity from their Islamic ties. To this day, most of the far southwest Philippines is thoroughly Muslim. In the northern and central Philippines, Islam had arrived only a few decades before the Europeans and was extinguished rather quickly by Spanish priests and soldiers. Today the Philippines is roughly 85 percent Roman Catholic, making it the only predominantly Christian country in all of Asia.

Christianity and Indigenous Cultures   Christian missions
spread throughout Southeast Asia in the late nineteenth and early twentieth century when European colonial powers controlled the region. While French priests converted many people in southern Vietnam to Catholicism, they had little influence elsewhere. Beyond Vietnam, missions failed to make headway in areas of Hindu, Buddhist, or Islamic heritage. Missionaries were, however, far more successful in Southeast Asia's highland areas, where they found a wide array of hill tribes who had never accepted the major lowland religions.

Instead, these people retained their indigenous belief systems, which generally focus worship on nature spirits and ancestors. The general name for such religions is **animism.** While some modern hill tribes remain animist today, others were converted to Christianity. As a result, notable Christian concentrations are found in the Lake Batak area of north-central Sumatra, the mountainous borderlands between southern Burma and Thailand, the northern peninsula of Sulawesi, and the highlands of southern Vietnam. Animism retains strongholds in the mountains of northern Southeast Asia, central Borneo, far eastern Indonesia, and the highlands of northern Luzon in the Philippines.

Indonesia, more than anywhere else in the region, has experienced religious strife in recent years, especially between its Muslim majority and Christian minority. Relations between Muslims and Christians in Indonesia were generally quite good until the late 1990s. Indonesia, despite its strong Muslim majority, is a resolutely secular state that has always emphasized religious tolerance (such toleration, however, has not always been extended to animist communities). With the Indonesian economic disaster of the late 1990s, however, relations quickly deteriorated, and religious fighting began to break out, especially in the Maluku Islands of eastern Indonesia. Transmigration is also implicated, as in many instances indigenous Christian and animist groups are now struggling against Muslim immigrants from Java and Madura.

Religion and Communism   By 1975 communism had triumphed in Vietnam, Cambodia, and Laos. In all three countries, religious practices were then strongly discouraged. At present, Vietnam's communist government is struggling against a revival of faith among the country's Buddhist majority and its 8 million Christians. Buddhist monks are frequently detained, and the government reserves for itself the right to appoint all religious leaders.

Geography of Language and Ethnicity

As with religion, language in Southeast Asia expresses the long history of external cultural influences and migration. The linguistic geography of the region is extremely complicated—far too complicated, in fact, to be adequately conveyed in a single linguistic map (Figure 13.21). The several hundred distinct languages of the region can, however, all be placed into five major linguistic families. These are *Austronesian,* which covers most of the islands from the Philippines to Indonesia along with the Malay Peninsula; *Tibeto-Burman,* which includes the languages of Burma; *Tai-Kadai,* centered on Thailand and Laos; *Mon-Khmer* encompassing most of the languages of Vietnam and Cambodia; and *Papuan,* found in eastern Indonesia.

The Austronesian Languages   One of the world's most widespread language families is Austronesian, extending from Madagascar, off the coast of Africa, to Easter Island in the eastern Pacific. Linguistic geographers suggest that this family originated prehistorically in Taiwan and adjacent areas of East Asia, then spread widely across the Indian and

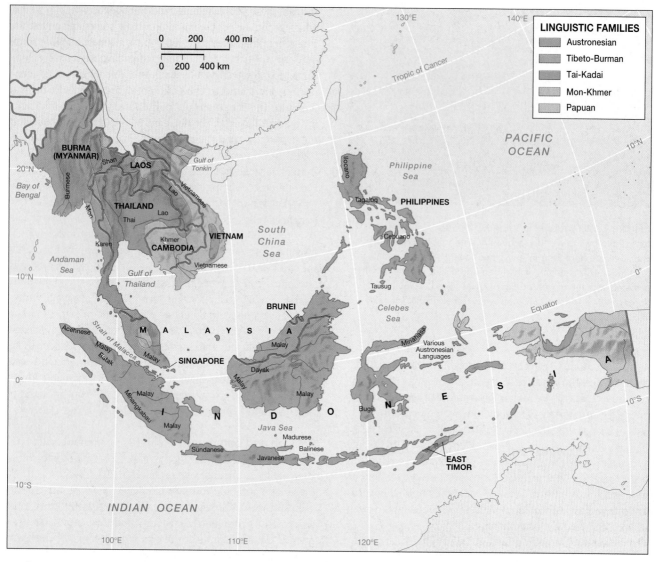

▲ Figure 13.21 Language map of Southeast Asia A vast number of languages are found in Southeast Asia, but most are tribal tongues spoken by only a few thousand persons. In mainland Southeast Asia—the site of three major language families—the central lowlands of each country are dominated by people speaking the national languages: Burmese in Burma, Thai in Thailand, Lao in Laos, and Vietnamese in Vietnam. Almost all languages in insular Southeast Asia belong to the Austronesian linguistic family. There were no dominant languages here before the creation of such national tongues as Filipino and Bahasa Indonesia in the mid-twentieth century.

Pacific oceans by seafaring people who migrated from island to island.

Today almost all insular Southeast Asian languages are within the Austronesian family. This means that elements of both grammar and vocabulary are widely shared across the insular realm; thus, it is relatively easy for a person who speaks one of these languages to learn any other. But despite this linguistic commonality at the family level, more than 50 distinct languages are spoken in Indonesia alone. And in far eastern Indonesia, a variety of languages fall into the completely separate family of Papuan, closely associated with New Guinea.

One language, however, overshadows all others in insular Southeast Asia: Malay. Malay is native to the Malay Peninsula, eastern Sumatra, and coastal Borneo, yet was spread historically throughout the regions by merchants and seafarers.

As a result, it became a common trade language, or **lingua franca**, understood and used by people of different languages throughout much of the insular realm. Dutch colonists in Indonesia eventually employed Malay as an administrative language, although they wrote it with the Roman alphabet, rather than in the Arabic-derived script used by native speakers. When Indonesia became an independent country in 1949, its leaders elected to use the lingua franca version of Malay as the basis for a new national language called "Bahasa Indonesia" (or more simply, "Indonesian"). Although Indonesian is slightly different from the Malaysian spoken in Malaysia, they essentially form a single, mutually intelligible language.

The goal of the new Indonesian government was to offer a common language that would overcome ethnic differences throughout the far-flung state. In general, this policy has been successful in that more than 80 percent of all Indonesians (over 160 million persons) now understand the language. Indonesian is also widely used in government, education, and entertainment. However, regionally based languages, such as Javanese, Balinese, and Sundanese, continue to be primary languages in most homes. More than 75 million people speak Javanese, which makes it one of the world's major tongues.

The Philippines is not as linguistically cohesive as either Malaysia or Indonesia, even though the eight major languages spoken in the archipelago are all Austronesian. Despite more than 300 years of Spanish colonialism, the Spanish language never became a unifying force for the islands. During the American period (1898–1946), English served as the language of administration and education. After independence following World War II, Philippine nationalists decided to forge a national language that could replace English and help unify the new country. They selected Tagalog, the tongue of Manila and a language with a fairly well-developed literary tradition. The first task was to standardize and modernize Tagalog, which had many distinct dialects. After this was accomplished, it was renamed "Pilipino" (or "Filipino," although there is no "f" sound in Tagalog), and today, mainly because of its use in education, television, and movies, it is gradually becoming a unifying national language.

Tibeto-Burman Languages   Each country of mainland Southeast Asia is closely identified with the national language spoken in its core territory. This does not mean, however, that all the inhabitants of these countries speak these official languages on a daily basis. In the mountains and other remote districts, nonnational languages are common. This linguistic diversity reinforces ethnic diversity even in the face of national educational programs designed to foster integration and unity.

A good example comes from Burma. There the national language is Burmese, a language that is closely related to Tibetan and perhaps more distantly to Chinese. Based upon linguistic evidence, the ancestors of the Burmans (as the Burmese speakers are properly called) probably lived in either Tibet or southern China centuries ago. Today some 22 million people speak Burmese (Figure 13.22). Although the nationalistic military government of Burma is determined to force its version of unity on the population, a major schism has developed with several non-Burman "hill tribes" that inhabit the rough uplands that flank the Burmese-speaking Irrawaddy Valley. Although most of these tribal groups speak languages in the Tibeto-Burman family, they are quite distinctive from Burmese.

Tai-Kadai Languages   The Tai-Kadai linguistic family seems to have originated in southern China and then to have spread into Southeast Asia starting around 1200. Today closely related languages within the Tai subfamily are found through most of Thailand and Laos, in the uplands of northern Vietnam, in Burma's Shan Plateau, and in parts of southern China. Most Tai languages are quite localized, and a number are spoken by small tribal groups. But two of them, Thai and Lao, are important national languages of Thailand and Laos, respectively.

Linguistic terminology in this part of the world is admittedly quite complicated. Historically, the main language of the Kingdom of Thailand, called Siamese (just as the kingdom was called Siam), was restricted to the lower Chao Phraya valley, which formed the national core. In the 1930s, however, the country changed its name to Thailand to emphasize the unity of all the peoples speaking the closely related Tai languages within its territory. Siamese was similarly renamed Thai, and it has gradually become the unifying language for the country. There are still substantial variations in dialect, however, with those of the north often still considered separate languages. Even more distinctive is Lao, the Tai language that became the national tongue of Laos. More Lao speakers reside in Thailand than in Laos, however, where they form the majority population of the impoverished Khorat Plateau.

Mon-Khmer Languages   This language family probably covered virtually all of mainland Southeast Asia prior to the arrival of the Tibeto-Burmans and the Tai. It contains two major languages, Vietnamese (the national tongue of Vietnam) and Khmer (the national language of Cambodia), as well as a host of minor languages spoken by hill peoples and a few lowland groups scattered throughout the area but concentrated in the uplands of Vietnam, Thailand, and Cambodia. Because of the historic Chinese influence in Vietnam, the Vietnamese language was written with Chinese characters until the French imposed the Roman alphabet during their colonial reign, which remains in use. Khmer, like the other national languages of mainland Southeast Asia, is written in its own script, ultimately derived from India.

The most important aspect of linguistic geography in mainland Southeast Asia is probably the fact that in each country the national language is limited to the core area of densely populated lowlands, whereas the peripheral uplands are populated by tribal peoples speaking separate languages. In Vietnam, for example, less than half of the national territory is

▲ **Figure 13.22 Burmese road signs**   The Burmese language is written in a unique script, ultimately derived from South Asia. Roman letters and the English language are still used for some purposes. Burma has, however, discouraged the use of English, viewing it as the language of colonial oppression.   *(Alain Evrard/Getty Images, Inc.)*

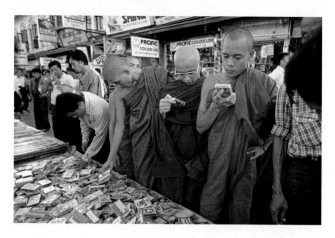

▲ **Figure 13.23 Monks in a Burmese market** Buddhism remains important in mainland Southeast Asia, where many men become Buddhist monks. Religious traditionalism does not preclude engagement with modern global culture, as is evident in this photograph of monks in a Burmese market. *(Andres Hernandez/Getty Images, Inc.)*

▲ **Figure 13.24 Sex tourism in Thailand** Thailand has one of the highest rates of prostitution in the world. While most prostitutes cater to a local clientele, those working in Bangkok's infamous Patpong district are usually hired by foreign men. Prostitution in Thailand is associated with high rates of HIV infection and with the brutal exploitation of women and girls. *(AFP/Corbis)*

occupied by Vietnamese speakers, even though they constitute a sizable majority of the country's population. This linguistic contrast between the lowlands and the uplands poses an obvious problem for national integration. Such problems are most extreme in Burma, where the upland peoples are numerous and well organized—and resist strenuously the domination of the lowland Burmans.

Southeast Asian Culture in Global Context

The imposition of European colonial rule ushered in a new era of globalization to Southeast Asia, bringing with it European languages, Christianity, and novel governmental, economic, and educational systems. This period also deprived most Southeast Asians of their cultural autonomy. As a result, with decolonialization and political independence after World War II, many countries attempted to isolate themselves from the cultural and economic influences of the emerging global system. Burma, for example, retreated into its own form of Buddhist socialism, placing strict limits on foreign tourism, which the government viewed a source of cultural contamination. Although the door has opened a bit wider since the 1980s, the government of Burma remains wary of foreign practices and influences, an attitude that has become a major source of tension within that country. Ironically, Burma's 500,000 Buddhist monks are among the strongest opponents of the country's isolationist as well as repressive government (Figure 13.23).

Other Southeast Asian countries, however, have been receptive to foreign cultural influences. This is particularly true in the case of the Philippines, where American colonialism may have predisposed the country to many of the more popular forms of Western culture. As a result, Filipino musicians and other entertainers are much in demand throughout East and Southeast Asia, in part because their performance styles echo those of the United States. Thailand, which was never subjected to colonial rule, appears to be the most receptive mainland Southeast Asian country to global culture, with its open policy toward tourism (including a notorious sex trade;

Figure 13.24), mass media, and economic interdependence. But cultural globalization has been challenged in some Southeast Asian countries, most decisively in Malaysia and Singapore. There one finds outspoken criticism of American films and satellite television, with government campaigns against the more explicit and crass expressions of Western entertainment. Censorship of global TV, newspapers, and magazines, especially of anything critical of the Malaysian or Singaporean governments, also occurs.

English, as the global language, also causes ambivalence in many countries. On one hand, it is the language of questionable popular culture, yet on the other, citizens need proficiency in English if they are to participate in global business and politics. In Malaysia the widespread use of English grew increasingly controversial in the 1980s as nationalists stressed the importance of their native tongue. This distressed the business community, which considers English vital to Malaysia's competitive position, as well as the influential Chinese communities, for which Malay is not a native language.

In Singapore, the situation is more complex. Mandarin Chinese, English, Malay, and Tamil (from southeastern India) are all official languages of this country. Furthermore, the local languages of southern China are common in home environments, since 75 percent of Singapore's population is of southern Chinese ancestry. In recent years the Singapore government has encouraged the use of Mandarin Chinese, in large part because it wishes to instill the traditional Confucian cultural values supposedly associated with this language, and has discouraged the use of southern Chinese dialects.

In the Philippines, nationalists have long decried the common use of English, even though widespread fluency has proved beneficial to the millions of Filipinos who have emigrated for better economic conditions. The Philippine government is now gradually replacing English with Filipino and, as a result, competency in English is slowly declining. At the

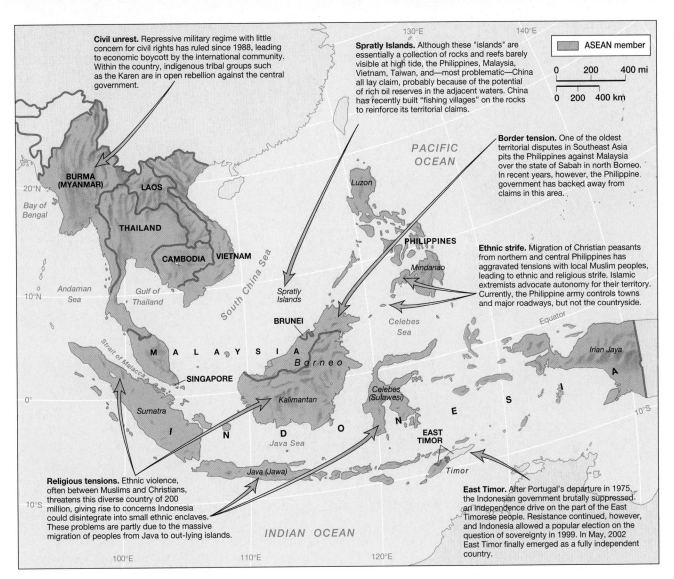

Civil unrest. Repressive military regime with little concern for civil rights has ruled since 1988, leading to economic boycott by the international community. Within the country, indigenous tribal groups such as the Karen are in open rebellion against the central government.

Spratly Islands. Although these "islands" are essentially a collection of rocks and reefs barely visible at high tide, the Philippines, Malaysia, Vietnam, Taiwan, and—most problematic—China all lay claim, probably because of the potential of rich oil reserves in the adjacent waters. China has recently built "fishing villages" on the rocks to reinforce its territorial claims.

Border tension. One of the oldest territorial disputes in Southeast Asia pits the Philippines against Malaysia over the state of Sabah in north Borneo. In recent years, however, the Philippine government has backed away from claims in this area.

Ethnic strife. Migration of Christian peasants from northern and central Philippines has aggravated tensions with local Muslim peoples, leading to ethnic and religious strife. Islamic extremists advocate autonomy for their territory. Currently, the Philippine army controls towns and major roadways, but not the countryside.

Religious tensions. Ethnic violence, often between Muslims and Christians, threatens this diverse country of 200 million, giving rise to concerns Indonesia could disintegrate into small ethnic enclaves. These problems are partly due to the massive migration of peoples from Java to out-lying islands.

East Timor. After Portugal's departure in 1975, the Indonesian government brutally suppressed an independence drive on the part of the East Timorese people. Resistance continued, however, and Indonesia allowed a popular election on the question of sovereignty in 1999. In May, 2002 East Timor finally emerged as a fully independent country.

▲ **Figure 13.25 Geopolitical issues in Southeast Asia** The countries of Southeast Asia have managed to solve most of their border disputes and other sources of potential conflicts through ASEAN (the Association of Southeast Asia Nations). Internal disputes, however, mostly focused on issues of religious and ethnic diversity, continue to plague several of the region's states, particularly Indonesia and Burma (Myanmar). ASEAN also experiences tension with China over the Spratly Islands of the South China Sea.

same time, as is the case with many other languages, the official language, based on Tagalog, is increasingly incorporating words and phrases from English, giving rise to a hybrid dialect known as "Taglish." Similarly, Singapore has recently seen the emergence of an especially complex informal hybrid tongue called "Singlish."

Geopolitical Framework: War, Ethnic Strife, and Regional Cooperation

Southeast Asia is readily defined as a geopolitical entity of 10 different states that, after a long period of colonialism shaped their current territories, have joined together under the umbrella organization of the Association of Southeast Asian Nations, or ASEAN. Today ASEAN, more than anything else, gives

Southeast Asia a geopolitical regional coherence. Within this framework, however, many states are still struggling with serious ethnic and regional tension. In some areas, these struggles could actually force a reshaping of the geopolitical landscape (Figure 13.25). In fact, in 1999 Indonesia surprised much of the world by relinquishing its control of the eastern portion of the island of Timor. Now under the control of the UNTAET (United Nations Transitional Authority for East Timor), East Timor is gradually emerging as an independent country.

Before European Colonialism

The modern countries of mainland Southeast Asia all existed in one form or another as indigenous kingdoms before the onset of European colonialism. Cambodia emerged earliest, reaching its height in the twelfth century, when it controlled much of what is now Thailand and southern

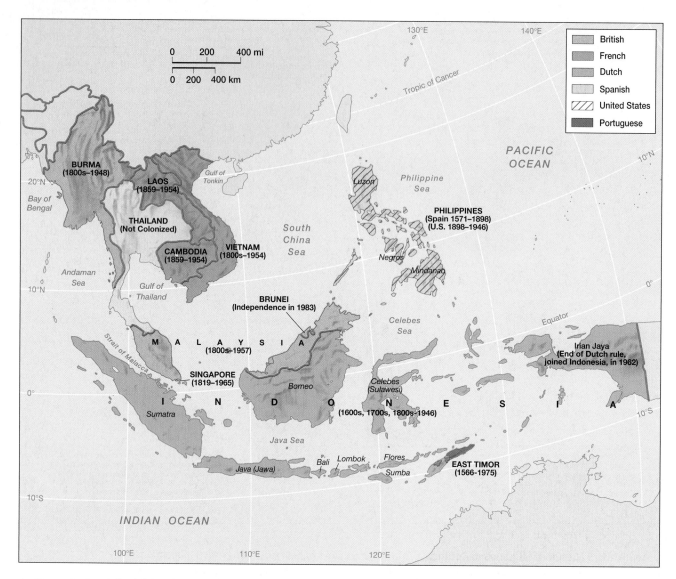

▲ **Figure 13.26 Colonial Southeast Asia** With the exception of Thailand, all of Southeast Asia was under Western colonial rule by the early 1900s. The Netherlands had the largest empire in the region, covering the territory that was later to become Indonesia. France maintained a substantial imperial realm in Vietnam, Laos, and Cambodia, as did Britain in Burma and Malaysia (including Singapore and Brunei). The Philippines had been colonized by Spain, but passed to the control of the United States in 1898.

Vietnam. By the 1300s, independent kingdoms had been established by the Burmese, Siamese, Lao, and Vietnamese people who were centered on the major river valleys and deltas. These kingdoms were in a nearly constant state of war with one another, fighting more for labor than for territory. Victors would typically take home thousands of prisoners to settle their lands, which led to considerable ethnic mixing during the precolonial period.

The situation in insular Southeast Asia was quite different from that of the mainland, with the premodern map bearing no resemblance to that of the modern nation-states. Many kingdoms existed on the Malay Peninsula and on the large islands of Sumatra and Java, but few were territorially stable. In the Philippines, eastern Indonesia, and central Borneo, most societies were organized at the village level. The countries of Indonesia, the Philippines, and Malaysia thus owe their territorial configuration almost wholly to the European colonial powers (Figure 13.26).

The Colonial Era

The Portuguese were the first Europeans to arrive (around 1500), lured by the cloves and nutmeg of the Maluku Islands (formerly the Spice Islands) in what is now eastern Indonesia. In the late 1500s, the Spanish conquered most of the Philippines, which they used as a base for their silver trade between China and the Americas. By the 1600s, the Dutch had started staking out territory, followed by the British. With superior naval weaponry, the Europeans were quickly able to conquer key ports and strategic trade locales. Yet for the first

200 years of colonialism, except in the Philippines, the Europeans made no major geopolitical changes. During this early period, trade in exotic items was more important than consolidated political power.

By the 1700s, the Dutch had become the most powerful force in the region. As a result, a Dutch Empire in the "East Indies" (or Indonesia) began appearing on world maps. This empire continued to grow into the early twentieth century, when the Dutch vanquished their last main adversary, the powerful Islamic sultanate of Aceh in northern Sumatra. Later, the Dutch invaded the western portion of New Guinea in response to German and British advances in the eastern half. In a subsequent treaty, these imperial powers sliced New Guinea down the middle, with the Dutch taking the western half.

There was, however, resistance to the expanding colonial empire of the Netherlands. In the 1920s, several anticolonial groups began struggling for independence. During World War II, Indonesia was occupied by the Japanese, and upon their surrender, Indonesian independence was proclaimed. Although the Dutch attempted to reestablish colonial rule, they were forced to acknowledge Indonesia's independence in 1949. Decolonization was finalized in 1963, when Indonesia gained control of the last Dutch outpost of Irian Jaya, or western New Guinea.

The British, preoccupied with their empire in India, concentrated their attention on the sea-lanes linking South Asia to China. As a result, they established several fortified trading outposts along the vital Strait of Malacca, the most notable being on the island of Singapore. To avoid conflict, the British and Dutch agreed that the British would limit their attention to the Malay Peninsula and the northern portion of Borneo. The British allowed Muslim sultans to retain limited powers, much as they had done in parts of India. When the British left this area in 1963, the country of Malaysia emerged in their wake. Two small portions of the former British sphere did not join the new country. In northern Borneo, the Sultanate of Brunei became an independent country in its own right, backed by riches from substantial oil reserves. Singapore briefly joined Malaysia, but then withdrew and became fully independent in 1965. This divorce was carried out partly for ethnic reasons. Malaysia was to be a primarily Malay state, but with Singapore its population would have been almost 50 percent Chinese. Partly to avoid this, the two states agreed to an amicable separation.

In the 1800s European colonial power spread through most of mainland Southeast Asia. The British, seeking to safeguard their South Asian empire, fought several wars against the indigenous kingdom of Burma before annexing the entire area in 1885, including considerable upland territory that had never been under Burmese rule. During the same period, the French moved into Vietnam's Mekong Delta, gradually expanding their territorial control to the west into Cambodia and north to China's border. This colonial empire was formally consolidated as the Union of French Indo-China in 1887. Thailand was the only country to avoid colonial rule, although it did lose substantial territories to the British in Malaysia and the French in Laos. This independence was maintained partly because it served both British and French interests to have a buffer state between their colonial empires.

The final colonial power to enter the area was the United States, which took the Philippines first from Spain and then from Filipino nationalists between 1898 and 1900. The U.S. army subsequently conquered the Muslim areas of the southwest, which had never been fully under Spanish authority.

Organized resistance to European rule began in the 1920s in mainland countries, but it took the Japanese occupation of World War II to shatter the myth of European colonial invincibility. After Japan's surrender in 1945, agitation for independence was renewed throughout Southeast Asia. As Britain realized that it could no longer control its South Asian empire, it also withdrew from adjacent Burma, which achieved independence in 1948. In the Philippines, the United States granted long-promised independence on July 4, 1946, although it retained military bases, as well as considerable economic influence, for several decades.

The Vietnam War and Its Aftermath

Unlike the United States, France was determined to maintain full control of its Southeast Asian colonies. Resistance to French rule was organized primarily by communist groups that were deeply rooted in northern Vietnam. As the French reoccupied Indo-China in 1946 after Japan's defeat, the leader of this resistance movement, Ho Chi Minh, became president of a separatist government based in Hanoi. France ceded local autonomy to this northern region but not to the south. Open warfare between French soldiers and the communist forces of the north went on for almost a decade, with the colonial French army suffering demoralizing losses against the guerrilla troops. After the French suffered a decisive defeat at Dienbienphu in 1954, they agreed to withdraw. An international peace council in Geneva then determined that Vietnam would, like Korea, be divided into two countries. As a result, the leaders of the communist rebellion came to power in North Vietnam, and, not surprisingly, allied themselves with the Soviet Union and China. South Vietnam emerged as an independent capitalist-oriented state with close political ties to the United States.

The Geneva peace accord did not, however, end the fighting. Communist guerrillas in South Vietnam fought to overthrow the new government and unite it with the north. For its part, North Vietnam sent troops and war materials across the border to aid the rebels. Most of these supplies reached the south over the Ho Chi Minh Trail, an ill-defined network of forest passages through Laos and Cambodia, thus steadily drawing these two countries into the conflict. In Laos the communist Pathet Lao forces challenged the government, while in Cambodia the **Khmer Rouge** guerrillas gained considerable power. In South Vietnam the government gradually lost control of key areas, including much of the Mekong Delta, a region perilously close to the capital city of Saigon.

By 1962 the United States was sending large numbers of military advisors to South Vietnam to train and organize resistance to the communist guerrillas. In Washington, D.C., the **domino theory** became accepted foreign policy. According to

War tourism is a booming business in Vietnam. As a result, the landscapes and experiences that were nightmares for hundreds of thousands during the 1960s and 1970s are now attracting foreign tourists and U.S. veterans alike. For example, outside of Ho Chi Minh City (formerly Saigon), one can visit the Cu Chi tunnels, one of the most famous battlegrounds of the Vietnam War. There visitors can tour the 75 miles of tunnels that once offered haven for Vietcong fighters. Underground, visitors crawl through narrow tunnels connecting living quarters, storerooms, kitchens, and hospitals. Above ground, young women dressed in the characteristic black pajamas of the Vietcong guerrillas sell T-shirts, pens made from bullets, and replicas of the coveted Zippo lighters carried by American GIs. Nearby, on a big screen TV, news footage from the war plays endlessly: B-52 bombers drop strings of bombs; villagers flee U.S. soldiers; and Vietcong guerrillas fight heroically.

While foreign currency is the main goal of war tourism, few opportunities are missed for propaganda. Every now and then, however, sensibility rules. In Ho Chi Minh City, the Museum of War Remnants attracts many visitors. Business seems to have improved since its name was changed from the "Museum of American War Crimes." Hue, the scene of heavy fighting during the Tet Offensive of 1968, has become a hub for war tours that list such attractions as "Khe Sanh, Dong Ha, Marble Mountain, China Beach, bombed-out church, and the DMZ." Even My Lai, the site of an infamous U.S. troop massacre of villagers, has been turned into something of a theme park. Guides escort tourists to a cemetery museum, and storytellers re-create the horrible events that destroyed the village and its people. Farther east, the government plans to open parts of the Ho Chi Minh Trail.

In the last few years, an increasing number of U.S. veterans have come back to Vietnam in search of remembered landscapes and battlefield experiences. Why? One tour leader put it this way: "They get a sense of closure; that's the big benefit of going back as a veteran. We [the Americans] left suddenly. Now you know how the story has ended. All the Vietnamese are very friendly. It's a different country now." Often, however, the veterans are disappointed because the landscape has changed so dramatically. Forests have reclaimed the countryside, obscuring small hills that were previously major battlefield landmarks. Was the fire-base here, or over there? Many veterans can't tell because the countryside looks so different. Instead of former military bases, one often finds coffee plantations and foreign trade warehouses. One enterprising tour operator uses GPS (global positioning system) coordinates and old military maps from the 1960s to guide U.S. veterans to specific spots as they revisit remembered landscapes from an earlier, more troubled era.

Source: Adapted from Seth Mydans, "Visit the Vietcong's World: Americans Welcome," *New York Times*, July 7, 1999, p. A4.

this notion, if Vietnam fell to the communists, then so would Laos and Cambodia; once those countries were lost, Burma and Thailand, and perhaps even Malaysia and Indonesia, would surely become members of the Soviet-dominated communist bloc. Fearing such an outcome, the United States was drawn ever deeper into the war. By 1965 thousands of U.S. troops had begun a ferocious land war against the communist guerrillas. By 1969 there were more than half a million U.S. troops in the region (Figure 13.27). However, despite superiority in arms and troops—and total domination of the air—U.S. forces gradually lost control over much of the countryside. As casualties mounted and the antiwar movement back home strengthened, the United States initiated secret talks in search of a negotiated settlement. U.S. troop withdrawals began in earnest by the early 1970s (see "Remembered Landscapes: War Tourism in Vietnam").

With the withdrawal of U.S. forces, the noncommunist governments began to collapse. Saigon fell in 1975, and in the following year Vietnam was officially reunited under the government of the north. Reunification was a traumatic event in southern Vietnam. Hundreds of thousands of people fled from the new regime to other countries, especially to the United States. The first wave of refugees consisted primarily of wealthy professionals and businesspeople, but later migrants included many relatively poor ethnic Chinese. Most of these refugees fled on small, rickety boats; large numbers suffered shipwreck or pirate attack. Most

found their way to squalid refugee camps elsewhere in Asia, where some still languish.

Vietnam proved fortunate compared to Cambodia. There the Khmer Rouge installed one of the most brutal regimes the world has ever seen. City-dwellers were forced into the countryside to become peasants, and most wealthy and highly educated persons were summarily executed. The Khmer Rouge's goal was to create a wholly new agrarian society by returning to what they called "year zero." After several years of this bloodshed, neighboring Vietnam invaded Cambodia and installed a far less brutal, but still repressive, communist regime. Fighting between different factions continued for more than a decade. Most Vietnamese troops had left Cambodia by 1989 as the United Nations worked to broker a settlement to this civil war. Since that time, several coalition governments have brought a tenuous peace to the shattered country.

Vietnam stationed significant numbers of troops in Laos after 1975. Large numbers of Hmong and other tribal peoples—many of whom had fought on behalf of the United States—fled to Thailand and the United States (Figure 13.28). Elections were not held in Laos until 1989, when a people's assembly was formed to approve a new constitution. Single-party elections were held in 1992 but did little to alter the country's political landscape. Although Laos remains an impoverished and repressive country, in 1997 it was admitted to the Association of Southeast Asian Nations (ASEAN), giving its government substantial regional credibility.

▲ Figure 13.27 U.S. soldier and Vietcong prisoners   The United States maintained a substantial military presence in Vietnam in the 1960s and early 1970s. Although U.S. forces claimed many victories, they were ultimately forced to withdraw, leading to the victory of North Vietnam and the reunification of the country.   (Getty Images, Inc.)

Geopolitical Tensions in Contemporary Southeast Asia

A number of contemporary conflicts in Southeast Asia are rooted in the region's colonial past. In several instances, locally based ethnic groups are struggling against the centralized national governments that inherited territory from former colonial powers. Often, in such transfers the rights of smaller groups are sacrificed for larger national goals. Tension also results when tribal groups attempt to preserve their homeland from the forces of globalization, be it international logging, mining, or interregional migration. Both forms of geopolitical conflict are pronounced in the large, multiethnic country of Indonesia.

Conflicts in Indonesia   When Indonesia gained independence in 1949, it encompassed all of the former Dutch possessions in the region except Irian Jaya, or western New Guinea. The Netherlands retained this territory, arguing that its cultural background distinguished it from Indonesia. In 1962 the Dutch organized a plebiscite to see whether the people of Irian Jaya wished to join Indonesia or form an independent country. The vote went for union, but many observers believe that the election was rigged by the Indonesian government. Tensions increased in the following decades as Javanese immigrants, along with mining and lumber firms, arrived in western New Guinea. Faced with the loss of their land and the degradation of their environment, the indigenous inhabitants began to rebel. Rebel leaders demanded independence, or at least autonomy, but they faced a far stronger force in the Indonesian army. The war in Irian Jaya is still a smoldering, sporadic, and occasionally bloody guerrilla affair. Indonesia is determined to maintain control of the region in part because it is home to the country's largest taxpayer, the Freeport-McMoRan copper and gold mine.

An even more brutal war erupted in 1975 on the island of Timor, in southeastern Indonesia (Figure 13.29). The eastern half of this poor and rather dry island had been a Portuguese

▲ Figure 13.28 Hmong people in the United States   The tribal Hmong of Laos sided with the United States during the Vietnam War. As a result, many were forced to flee the country after the victory of the Pathet Lao communist forces. Many languished for years in squalid refugee camps; many others found refuge in the United States. The Hmong have had a much more difficult time adjusting to life in the United States than have other Southeast Asian refugee groups.   (Syracuse Newspapers/The Image Works)

colony (the only survivor of Portugal's sixteenth-century foray into the region), and had therefore evolved into a largely Christian society. The East Timorese expected independence when the Portuguese finally withdrew. Indonesia, however, viewed the area as rightfully its own, largely by virtue of its geographical position, and immediately invaded. A ferocious war ensued, which the Indonesian army won in part by starving the people of East Timor into submission.

After the economic crisis of 1997, Indonesia's power in the region slipped. A new Indonesian government promised an election in 1999 to see whether the East Timorese still wanted independence. At the same time, however, the Indonesian army began to organize loyal militias in an attempt to intimidate the people of East Timor into voting to remain within the country. When it was clear that the vote would be strongly for independence, the militias began rioting, looting, and slaughtering civilians. Under international pressure, Indonesia finally withdrew its armed forces, and the East Timorese began to build a new country—the eleventh sovereign state in Southeast Asia. Considering the devastation caused by the militias, the process has not been easy. The Timorese, moreover, are not even sure what their official language should be. Should it be Portuguese, spoken mainly by the elderly elite? Indonesian, spoken widely by the young but politically suspect? Or one of the 11 indigenous languages? Some have gone so far as to suggest moving directly to English, the main global tongue, but that seems unlikely.

▲ **Figure 13.29 East Timorese demonstration in Jakarta**
Indonesia invaded and claimed East Timor in 1975 when the
Portuguese left their last colonial outpost in Southeast Asia. The
East Timorese resisted the Indonesian takeover, and violence
reached a climax in 1999 when the East Timorese voted for in-
dependence, after which pro-Indonesian militias attacked the
Timorese leaders and large segments of Timorese society.
(V. Miladinovic/Getty Images, Inc.)

Secession struggles have occurred elsewhere in Indonesia
as well. In the late 1950s central Sumatra and the northern
portion of Sulawesi rebelled, but they were quickly defeated
and eventually reconciled to Indonesian rule. Another small-
scale war was initiated in 1997, this one in western and southern Kalimantan (Borneo). Here the indigenous Dayaks, a tribal
people now partly converted to Christianity, began to clash
with Muslim migrants from Madura, a poor and densely inhabited island north of Java. By 2001 Dayak warriors, some of
whom had revived the head-hunting practices of their ancestors, had driven the Madurese out of much of their territory.
The Indonesian military has restored order in Kalimantan's
cities, but has intervened relatively little in the countryside.
An equally violent struggle erupted in the late 1990s in the
Aceh region of northern Sumatra. Many of the Acehnese, the
most orthodox Muslim people of Indonesia, are demanding
the creation of an independent Islamic state. While the Indonesian government has promised Aceh "special autonomy,"
it is determined to do whatever is necessary to prevent actual

independence. Finally, the southern Maluku Islands, especially
Ambon and Seram, are now a zone of war between Muslims
and Christians. Even the area's cities are divided into Muslim
and Christian sectors, and crossing to the wrong side can be
quite dangerous.

Indonesia has obviously had difficulties creating a unified
nation over its vast, sprawling extent and across its myriad
cultural and religious communities. As a creation of the colonial period, Indonesia has a weak historical foundation. Some
critics have viewed it as something of a Javanese empire, but
actually many non-Javanese individuals have risen to high
governmental positions. It is undeniable, however, that the Indonesian state has relied on repression. While it is now basically democratic, the military remains politically powerful.
Most Indonesian voters, moreover, live in Java and other central portions of the country, and most have little sympathy
for secessionist movements in Irian Jaya, Aceh, and other peripheral areas.

But even in Java itself, political tensions run high. Two
huge rival Muslim organizations contend for influence, one of
which rejects all pre-Islamic traditions (the Muhammadiya),
the other far more accommodating (the N.U., or Nahdlatul
Ulama, which is rooted in eastern and central Java). Until the
crisis of 1997, these groups largely stayed out of politics; since
then they have been centrally involved. Leaders of the
Muhammadiya accuse both the government and the N.U. of
being soft on separatists, especially Christians in Irian Jaya. As
of 2001, the two groups were threatening violence against
each other. In addition, fundamentalist Muslims organized
massive demonstrations in the fall of 2001 against the United States and its bombing campaign in Afghanistan.

Regional Tensions in the Philippines   The Philippines
has also suffered from regional secession movements, although not to the same extent as Indonesia. Its most persistent problem area is the Islamic southwest. In the 1990s the
government was apparently making some headway in talks
with the main rebel group by promising greater autonomy,
but the more extreme Muslim factions continued to demand
total independence. The migration of Christian peasants from
northern and central Philippines to Mindanao has exacerbated local tensions. The Philippine army generally maintains
control over the area's main cities and roadways, but it has
little power in the more remote villages. Extremists in the
Abu Sayyaf terrorist group have recently taken to kidnapping
wealthy tourists staying in resorts on the relatively pristine
island of Palawan, and in some cases executing them when
their demands are not met. As Abu Sayyaf reportedly has close
ties with Osama bin Laden and his Al Qaeda network, the
Philippines quickly became a key site in the U.S-led struggle
against global terrorism. In October 2001 some 30 American
counterterrorism experts arrived in the southern Philippines
to aid the Philippine army in its struggle against Muslim separatists.

The Philippines has also faced a communist-oriented na-
tionwide rebellion. In the mid-1980s the NPA (New People's
Army) controlled one-quarter of the country's villages scat-

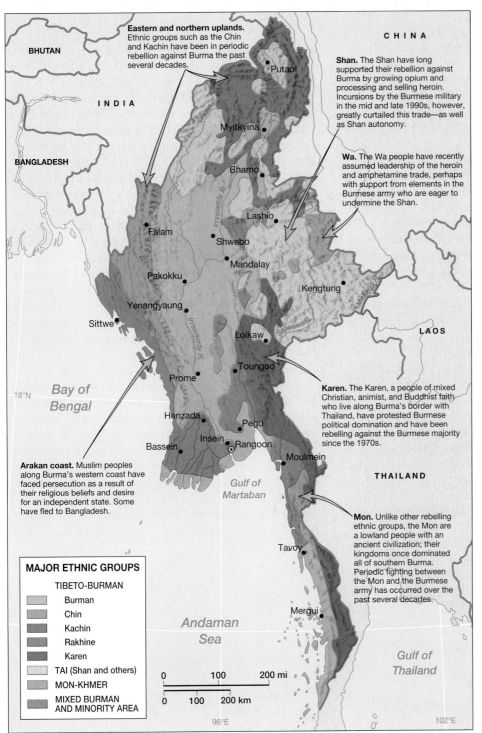

Although the central lowlands of Burma are primarily populated by people of the dominant Burman ethnic group, the peripheral highlands and the southeastern lowlands are the home of numerous non-Burman peoples. Most of these peoples have been in periodic rebellion against Burma since the 1970s, owing to their perception that the national government has been attempting to impose Burman cultural norms. Economic stagnation and political repression by the central government have intensified these conflicts.

Eastern and northern uplands. Ethnic groups such as the Chin and Kachin have been in periodic rebellion against Burma the past several decades.

Shan. The Shan have long supported their rebellion against Burma by growing opium and processing and selling heroin. Incursions by the Burmese military in the mid and late 1990s, however, greatly curtailed this trade—as well as Shan autonomy.

Wa. The Wa people have recently assumed leadership of the heroin and amphetamine trade, perhaps with support from elements in the Burmese army who are eager to undermine the Shan.

Karen. The Karen, a people of mixed Christian, animist, and Buddhist faith who live along Burma's border with Thailand, have protested Burmese political domination and have been rebelling against the Burmese majority since the 1970s.

Arakan coast. Muslim peoples along Burma's western coast have faced persecution as a result of their religious beliefs and desire for an independent state. Some have fled to Bangladesh.

Mon. Unlike other rebelling ethnic groups, the Mon are a lowland people with an ancient civilization; their kingdoms once dominated all of southern Burma. Periodic fighting between the Mon and the Burmese army has occurred over the past several decades.

MAJOR ETHNIC GROUPS

TIBETO-BURMAN

- Burman
- Chin
- Kachin
- Rakhine
- Karen
- TAI (Shan and others)
- MON-KHMER
- MIXED BURMAN AND MINORITY AREA

tered across all the major islands. Since the mid-1990s, however, the strength of this insurgent group has declined considerably. The 1980s also saw the rebellion of tribal groups in northern Luzon, peoples whose lands and livelihoods had been threatened by dam-building and forestry projects. This movement also diminished in the 1990s after the government offered concessions.

The Quagmire of Burma

Burma (Myanmar) is perhaps the most war-ravaged country of Southeast Asia. Burma's simultaneous wars have pitted the central government, dominated by the Burmese-speaking ethnic group (the Burmans), against the country's varied non-Burman societies. Fighting intensified gradually after independence in 1948, and by the 1980s almost half of the country's territory had become a combat zone (Figure 13.30). Burma's troubles, moreover, are by no means limited to the country's ethnic minorities. Since 1988 Burma has been ruled by a very repressive military regime, one that has little popular support even among the majority Burmans.

The rebelling peoples of Burma want to maintain their cultural traditions, lands, and resources, and they generally see the national government as something of a Burman empire seeking to impose its language and, in some cases, religion,

Theravada Buddhism, upon them. Most of the insurgent ethnic groups live in rugged and inaccessible terrain, and many of them are animists or Christians. But even some lowland groups have rebelled. The Muslim peoples of the Arakan coast in far western Burma have long faced discrimination because of their religious convictions. In the 1980s many were forced to flee to neighboring Bangladesh.

Several of Burma's ethnic insurgencies have been financed largely by opium-growing and heroin manufacture. This is especially true in the case of the Shan rebellion. The Shan are a Tai-speaking people inhabiting a plateau area within the famous Golden Triangle of drug production. In the late 1980s and early 1990s, a breakaway Shan state gained a firm financial basis through the narcotics trade and a strong political foundation through ethnic solidarity. For a number of years, this strategy proved to be successful, but by the mid-1990s it began to falter. Burmese agreements with Thailand reduced the Shan heroin trade, while military operations cut off the supply of raw opium reaching the Shan factories. But while the government apparently triumphed over the Shan, other ethnic groups have been able to move into the void. By 2001 the United Wa State Army (UWSA), having largely abandoned rebellion for the drug trade, emerged as the major supplier of both opium/heroin and amphetamines. Evidently Burma's military leaders support the USWA because it helps them in their struggle against the Shan. Thailand, which now has a growing amphetamine problem on top of its heroin problem, is not pleased with the situation and has threatened to cut off its export of electricity to eastern Burma.

As long as Burma retains its repressive Burman-dominated state structure, social unrest and ethnic turmoil are not likely to diminish. Protests against the government are mounting even in the country's core. Additionally, the international community is putting some pressure for democratic change on Burma's government with trade boycotts and other sanctions. The combination of internal and external pressures makes the future of this volatile state especially difficult to predict.

International Dimensions of Southeast Asian Geopolitics

Geopolitical conflicts in Southeast Asia have occurred not only within countries, but also between countries. Tensions have typically arisen when two countries claimed the same territory.

One of the oldest territorial debates in Southeast Asia pits the Philippines against Malaysia over the Malaysian state of Sabah in north Borneo. The Philippines bases its claim on the fact that the Islamic Sultanate of the Sulu Archipelago, which was incorporated into the Philippines during the American colonial period, at one time controlled part of Sabah. Malaysia, not surprisingly, has steadfastly rejected the Philippine claim. Malaysia also has historical quarrels with Thailand over Thailand's Malay-speaking Muslim region in its far south, and with Indonesia over the border in Borneo.

In recent years, however, the various countries of Southeast Asia have largely agreed to drop these and other border disputes. With the rise of the Association of Southeast Asian Nations (ASEAN) and the growth of trade relations, national leaders have concluded that amicable relations with one's neighbors are more important in the long run than the possible gain of additional territory. Militant Malay separatists in southern Thailand were thoroughly demoralized in 1998, for example, when Malaysia offered to help the Thai government arrest them.

More intractable has been the dispute over the Spratly and Paracel islands in the South China Sea, two groups of insignificant rocks and reefs that are virtually submerged at high tide (Figure 13.31). The Philippines, Malaysia, and Vietnam have all advanced territorial claims there, as have China and Taiwan. In the mid-1990s China began to actively assert its claims by building ambiguous structures that it calls fishing shelters, but which its neighbors refer to as military posts. Increasingly, the dispute in the South China Sea pits Southeast Asia as a whole against China, clearly a far more powerful geopolitical and military entity. One reason for the keen display of interest in these islands is that there is some evidence of substantial undersea oil reserves in their vicinity.

It is partly because of mutual concern over growing Chinese power that the countries of Southeast Asia are banding together. This is especially visible in the development and enlargement of ASEAN. In its earlier days, ASEAN was an alliance of nonsocialist countries fearful of the communist regimes that had come to power in Vietnam, Cambodia, and Laos. Through the 1980s, the United States maintained a strong military presence in the region through its control of naval and air bases in the Philippines, and U.S. military force helped bolster the anticommunist coalition. In the early

▲ Figure 13.31 The Spratly Islands   The Spratly Islands are small and barely above water at high tide, but they are geopolitically important. Oil may exist in large quantities in the vicinity, heightening the competition over the islands. Southeast Asian countries are especially concerned about China's military activities in the Spratlys.   *(Nouvelle Chine/Getty Images, Inc.)*

1990s, however, the United States, under mounting pressure from Filipino nationalists, withdrew its military forces from the Philippines. By this time the struggle between communism and capitalism had become an increasingly moot issue, and it was apparent that Vietnam offered no threat to the ASEAN states. In 1995 Vietnam itself gained membership in the organization, followed by Laos and Cambodia. While ASEAN attempts to address economic issues, its main role has been to facilitate political cooperation and to present something of a united front against the rest of the world. Increasingly, the foreign power most at issue in Southeast Asia is China. Although Vietnam and China have similar political and economic systems—as well as cultural backgrounds—they also have a long history of antagonism. It is therefore not surprising that Vietnam would opt to join ASEAN.

While ASEAN is on friendly terms with the United States, one purpose of the organization is to prevent the United States or any other country from gaining undue influence in the region. For that reason, ASEAN leaders are keen to include all Southeast Asian countries within the association. In 1997 Burma gained membership, a move that disappointed the United States and other Western countries concerned about Burma's human-rights record. The ASEAN states, however, were more concerned about Burma's increasing ties with China, which they hoped to forestall by granting Burma admission into their organization. ASEAN's ultimate international policy is negotiation rather than confrontation. For this reason, the ASEAN Regional Forum (ARF) was established in 1994 as an annual conference in which ASEAN leaders could meet with the leaders of both the East Asian and Western powers to attempt to ease tensions within the region.

ASEAN has not defused all of the tensions existing between the countries of Southeast Asia. Minor disagreements within the ASEAN bloc have also surfaced over relations with the outside world. Malaysia in particular has pushed an anti-Western line on cultural and political (but not economic) issues, and it has strongly opposed Australia's desire to gain closer association with its Asian neighbors. Powerful Islamic fundamentalist groups in both Malaysia and Indonesia, moreover, are becoming increasingly hostile to the United States, a tendency much exacerbated by U.S. military operations in Afghanistan. Thailand and the Philippines have been more willing to accommodate the West.

While ASEAN is concerned about Chinese power in the South China Sea, one must not overstate the potential for conflict between China and Southeast Asia. Economic ties between these two parts of the world are growing quickly. Especially important are the networks of family business groups that link the various Chinese communities of Southeast Asia with Taiwan and China proper. Such networks, which ultimately extend throughout the so-called Pacific Rim (which includes all countries bordering on the Pacific, whether in the Americas, Asia, or Oceania), have been an important component of Southeast Asia's economic development.

Economic and Social Development: The Roller-Coaster Ride of Tiger Economies

Until the downturn of the late 1990s, economic development in Southeast Asia was often held up to the world as a model for a new globalized capitalism. With investment capital flowing first from Japan and the United States, then from international investment portfolios, countries such as Thailand, Malaysia, and Indonesia moved quickly into the ranks of booming "tiger" economies characterized by high annual growth rates. Because of the successful economic development of these countries, the notion of an "Asian model" evolved as a strategy that might also work in other world regions, such as Africa or South Asia.

More recently, however, Southeast Asian countries have suffered from the roller-coaster ride of globalized economic development. Within a decade most economies have experienced both unforeseen highs and lows (Figure 13.32). Since the summer of 1997, regional economies have been in a state of periodic crisis. At that time, currencies in both Thailand and Indonesia were devalued almost 50 percent, and other countries have suffered similarly. Several of the economies of the region subsequently rebounded, but they remain insecure

▲ **Figure 13.32 Bank closings** The economic crisis of 1997–1998 hit Southeast Asia's banking sector particularly hard. Many banks have been closed, and many others may soon be closed, since their liabilities are much greater than their assets. The economic crisis has brought devastation to the poor and lower middle class. *(Marcus Rose/Panos Pictures)*

TABLE 13.2 *Economic Indicators*

Country	Total GNI (Millions of $U.S., 1999)	GNI per Capita ($U.S., 1999)	GNI per Capita, PPP* ($Intl, 1999)	Average Annual Growth % GDP per Capita, 1990–1999
Burma (Myanmar)	—	—	1,200**	—
Brunei	—	—	17,400**	–0.5
Cambodia	3,023	260	1,350	1.9
Indonesia	125,043	600	2,660	3.0
Laos	1,476	290	1,430	3.8
Malaysia	76,944	3,390	7,640	4.7
Philippines	77,967	1,050	3,990	0.9
Singapore	95,429	24,150	22,310	4.7
Thailand	121,051	2,010	5,950	3.8
Vietnam	28,733	370	1,860	6.2

*Purchasing power parity

Source: The World Bank Atlas, *2001, except those marked **, which note data from the CIA World Factbook, 2000.*

and have not recaptured the giddy optimism of the boom years. Others are still characterized by high unemployment, social unrest, truncated industrial development, and lost fortunes. Many argue that the region's spectacular economic growth contained the seeds of its own destruction because of its marked dependence on foreign investment.

Economic Leaders and Laggards

Southeast Asia today is a region of very uneven economic and social development. While some countries, such as Indonesia, have experienced both boom and bust, others, such as Burma and Cambodia, either hesitated or were not able to join the global economic bandwagon in the first wave. Singapore, on the other hand, managed to avoid the economic crisis of the late 1990s almost completely. As a result, the regional economic geography remains highly uneven and fluid (Table 13.2).

The Philippine Decline Forty years ago, the Philippines was the most highly developed Southeast Asian country and was considered by many to have the brightest prospects in all of Asia. It boasted the best-educated populace in the region, and it seemed to be on the verge of sustained industrialization and economic development. Per capita GNI in 1960 was higher in the Philippines than in South Korea. By the late 1960s, however, Philippine development had been derailed. Through the 1980s and early 1990s, the country's economy failed to outpace its population growth, resulting in declining living standards for both the poor and the middle class. The Philippine people are still well educated and reasonably healthy by world standards (see Table 13.3), but even the country's educational and health systems declined during the dismal decades of the 1980s and 1990s.

TABLE 13.3 *Social Indicators and Status of Women*

Country	Life Expectancy at Birth Male	Female	Under Age 5 Mortality (per 1,000) 1980	1999	Percent Illiteracy (Ages 15 and over) Male	Female	Female Labor Force Participation (% of total, 1999)
Burma (Myanmar)	54	57	134	120	89	80	43
Brunei	71	76	—	—	93**	83**	35
Cambodia	54	58	330	143	59	21	52
Indonesia	65	70	125	52	91	81	41
Laos	51	54	200	143	63	32	—
Malaysia	70	75	42	10	91	83	38
Philippines	64	70	81	41	95	95	38
Singapore	76	80	13	4	96	88	39
Thailand	70	75	58	33	97	93	46
Vietnam	63	69	105	42	95	91	49

Sources: Population Reference Bureau Data Sheet, *2001, Life Expectancy (M/F);* World Development Indicators, *2001, Under Age 5 Mortality Rate and Percent Illiteracy, except those marked **, which note data from the CIA World Factbook, 2000;* The World Bank Atlas, *2001, Female Participation in Labor Force.*

Why did the Philippines fail so spectacularly despite its earlier promise? While there are no simple answers, it is clear that dictator Ferdinand Marcos (who ruled from 1968 to 1986) squandered—and perhaps even stole—hundreds of millions of dollars while failing to enact programs conducive to genuine development. The Marcos regime instituted a kind of **crony capitalism** in which the president's many friends were granted huge sectors of the economy, while those perceived to be enemies had their properties expropriated. After Marcos declared martial law in 1972 and suspended Philippine democracy, revolutionary activity intensified and the country began to fall into a downward spiral. Wealthy Filipinos increasingly lost faith in their country and began to send their money to investment havens in the United States and Europe.

While it is tempting to blame the failure of the Philippines on the policies of the Marcos regime, such an explanation must be regarded as only partially adequate. Indonesia has seen the development of a similar kind of crony capitalism, yet in this case the companies in question generally proved competitive, at least until the financial collapse of 1997. In the Philippines, it seems that more money was actually stolen from the state by corrupt officials and well-connected businesspeople. Unfortunately, this kind of **kleptocracy** (or "government by thieves") is not uncommon in the poorer parts of the world.

The Marcos dictatorship was finally replaced by an elected democratic government in 1986, but the Philippine economy continued to languish through the 1990s. Many Filipinos have responded to the economic crisis by working overseas, either in the oil-rich countries of Southwest Asia, the wealthy cities of North America and Europe, or the newly industrialized nations of East and Southeast Asia. Men primarily work in the construction industry, women as nurses or domestic servants. Although foreign remittances have kept the Philippine economy afloat, this exodus of labor represents in many respects a tragedy for the country as a whole. Tens of thousands of Filipina teachers, no longer able to support themselves on their meager salaries, now work as maids and nannies in Singapore, Hong Kong, and the United Arab Emirates. With the loss of so many teachers, educational standards at home continue to decline (Figure 13.33).

By the mid-1990s, the Philippine economy showed some signs of revival. The government turned its attention to infrastructural problems, such as electricity provision, and foreign investments began to flow into the country. A vibrant local economy emerged in the former U.S. naval base of Subic Bay. Boasting both world-class facilities and a highly competent local government, Subic Bay has emerged as a major export-processing center. Cebu City, on the central Visayan island of Cebu, has also expanded quickly, giving rise to its local nickname of "Ceboom." The economic crisis of the late 1990s, however, again undercut the Philippine economy, as did the election of Joseph Estrada, a corrupt former actor, to the country's presidency. Estrada was forced out of office in 2001, and the new leader, Gloria Arroyo, is again trying to enact economic reforms.

Despite his clear corruption, Estrada continued to enjoy widespread support among the poor even after he was removed from office. Most Filipinos did not realize any personal gains from the upturn of the mid-1990s. The Philippines still has one of the least equitable distributions of wealth in the region, if not the world. While many members of the elite are fantastically wealthy by any measure, many of the poor are simply destitute, often subsisting on less than $1 a day. Although some efforts have been made toward land reform, vast estates of sugarcane and other plantation crops remain in the hands of a few super-elite—and politically powerful—families.

The Regional Hub: Singapore   If the Philippines has been the biggest economic disappointment in Southeast Asia, Singapore and Malaysia have surely been the region's greatest developmental successes. Singapore has transformed itself from an **entrepôt** port city, a place where goods are imported, stored, and then transshipped, to one of the world's most prosperous and modern states. Singapore is now the communications and financial hub of Southeast Asia, as well as a thriving high-tech manufacturing center. The Singaporean government has played an active role in the development process, but has also allowed market forces freedom to operate. Singapore has encouraged investment by multinational companies (especially those involved in technology), and has itself invested heavily in housing, education, and some social services. This government, however, is also repressive and generally undemocratic, allowing little freedom of the press or public expression. Chewing gum in public is illegal in Singapore, for example, and even minor acts of civil disobedience are severely punished. One is allowed to make money, but not to express one's political opinions. While many Singaporeans chafe under such repressive policies, others counter that they have brought fast growth as well as a clean, orderly, and safe society (Figure 13.34).

▲ **Figure 13.33 Filipina migrant workers in Kuwait**   The long period of economic stagnation in the Philippines has resulted in an outflow of workers from the country. Women from the Philippines often work as domestic servants in the Persian Gulf region and in Singapore and Hong Kong.   *(Penny Tweedie/ Panos Pictures)*

▲ **Figure 13.34 Social order in Singapore**   Singapore has one of the world's most regulated societies. Freedom of expression is highly limited, and severe fines are imposed on activities that are considered to be disruptive or dirty, such as public eating, smoking, and gum chewing.   *(Jean-Leo Dugast/Panos Pictures)*

Although the Singaporean government has thus far been able to repress dissent, its authoritarian form of capitalism confronts a new technological challenge in the Internet. National leaders want the communication services that the Net provides, but they are worried about the free expression that it allows, fearing that it will lead to excessive individualism. It will be interesting to see how Singapore responds to this challenge, in part because the governments of China and Vietnam are impressed by, and may be seeking to emulate, the technocratic, authoritarian capitalism pioneered by Singapore.

The Malaysian Boom   Although not nearly as prosperous as Singapore, Malaysia has recently experienced very rapid economic growth and has now broken into the ranks of the middle-income countries. Development was initially concentrated in agriculture and natural resource extraction, focused on tropical hardwoods, plantation products (mainly palm oil and rubber), and tin. More recently, manufacturing, especially in labor-intensive high-tech sectors, has become the main engine of growth. As Singapore prospers, moreover, many of its enterprises are spilling over into neighboring Malaysia. Increasingly, Malaysia's economy is multinational; many Western high-tech firms operate in the country, while several Malaysian companies are themselves establishing subsidiaries in foreign lands.

Unfortunately, Malaysia was hard hit by the Asian economic crisis of the late 1990s. In 1998 alone its economy declined by 8 percent. The Malaysian government's response to the crisis was different from that of its neighbors. Spurning the advice of the International Monetary Fund, Malaysia instituted currency controls, regulating the flow of international funds in and out of the country. Although it is too soon to tell, there are some indications that the Malaysian response is working better than the more financially orthodox responses of other Southeast Asian countries, such as Thailand. In

early 2001 Malaysia's economy was growing at the rapid annual rate of 6.5 percent, but by the end of the year it was again threatened by recession, this time by problems confronting the global high-tech industry.

The modern economy of modern Malaysia is not uniformly distributed across the country. One disparity is geographical: most industrial development has occurred on the west side of peninsular Malaysia, with most of the rest of the country remaining largely dependent on agriculture and resource extraction. More important, however, have been disparities based on ethnicity. The industrial wealth generated in Malaysia has been concentrated in the Chinese community. Ethnic Malays remain less prosperous than Chinese-Malaysians, and those of Tamil descent are poorer still. Many tribal peoples, moreover, have suffered as development proceeds and their lands are expropriated. Not surprisingly, such disparities of wealth have generated considerable ethnic tensions within the country.

To be sure, the disproportionate prosperity of the local Chinese community is a feature of most Southeast Asian countries. The problem is particularly acute in Malaysia, however, simply because its Chinese minority is so large (some 30 percent of the country's total population). The government's response has been one of aggressive "affirmative action," by which economic clout is transferred to the numerically dominant Malay, or **Bumiputra** ("sons of the soil"), community. This policy has been reasonably successful. Since the economy as a whole grew from 1980 to 1997 at 8 percent a year, the Chinese community thrived even as its relative share of the country's wealth declined. Considerable resentment, however, is still felt by the Chinese. Especially galling is the fact that most seats in the country's universities are reserved for students of Malay background. Partly for this reason, Malaysia sends a larger proportion of its student population abroad for education than any other Southeast Asian county.

Thailand: An Emerging Tiger?   Thailand, like Malaysia, climbed rapidly during the 1980s and 1990s into the ranks of the world's newly industrialized countries. Yet it also experienced a major downturn in the late 1990s that negated much of this development through a devalued currency and the exodus of investment capital. The Asian crisis of 1997 actually began in the overheated Thai economy and real estate market. Subsequent recovery has been slow; as of April 2002 Thailand's economy was growing by the low annual rate of 2.1 percent.

Japanese companies were leading players in the earlier Thai boom. As Japan itself became too expensive for assembly and other manufacturing processes, Japanese firms began to relocate factories to such places as Thailand. They were particularly attracted by the country's low-wage, yet reasonably well-educated, workforce. Thailand is politically stable, lacking the severe ethnic tensions found in many other parts of Asia. Although Thailand's Chinese population is large and economically powerful, relations between the Thai and the Chinese have generally been good, marked by much intermarriage and cultural synthesis. Thailand also has a demo-

SOCIAL GEOGRAPHY   Heroin, AIDS, and Prostitution in Thailand

Although Thailand has made considerable progress in providing health services to its citizens, the country faces one of the worst AIDS epidemics in the world. The prevalence of AIDS in Thailand stems from the combination of a high level of intravenous drug use and rampant prostitution (both female and male). Large quantities of opium are processed into heroin in the Golden Triangle, which includes parts of northern Thailand as well as Burma and Laos. Although opium smoking is an old cultural practice among the tribal peoples of the area, injecting heroin is new. Unfortunately, the use of heroin is now spreading rapidly through both the hills and adjacent lowlands. When narcotics are widely injected and needles shared, AIDS spreads silently but rapidly.

As southern Thailand is beginning to prosper, its prostitution industry is being increasingly staffed by women from poorer areas. The Khorat Plateau supplies many prostitutes, but recruitment is also focused on the Golden Triangle drug zone, where the HIV virus is widespread. Many of the women are forced into prostitution, and it is not unknown for addicted parents to sell their daughters, and sometimes their sons, into virtual sex slavery. Prostitutes from Burma, having been smuggled illegally into Thailand, are especially vulnerable to exploitation. Child prostitution is also widespread. Children and adolescents seem to be especially susceptible to infection, hastening the spread of the disease.

Thailand's government has worked hard to encourage the use of condoms. It has done very little, however, to suppress—or even carefully regulate—prostitution. Profiting from a huge international business, prostitution ringleaders, many of whom have important positions in the Thai military, have found it easy to circumvent the relatively weak Thai laws and regulation. Much evidence suggests that the AIDS crisis in Burma is almost as severe as that in Thailand. The government of Burma, however, has done very little to stop the epidemic, in part because it maintains—contrary to the facts—that a commercial sex industry cannot thrive in a Buddhist country such as Burma.

---

cratic government and a thriving free press, although it did have a legacy of military coups followed by periods of authoritarian rule. In recent years, however, Thailand's revered constitutional monarchy has emerged as a democratic bulwark. Proponents of democracy can thus point to Thailand as a counterexample to the Singaporean notion that rapid growth requires authoritarian rule.

To repeat a familiar story, however, Thailand's economic boom has by no means benefited the entire country to an equal extent. Most industrial development has occurred in the historical core, especially in the city of Bangkok itself. Yet even in Bangkok the blessings of progress have been decidedly mixed. As the city begins to choke on its own growth, industrial growth has begun to spread outward. The entire Chao Phraya lowland area shares to some extent in the general prosperity because of both its proximity to Bangkok and its rich agricultural resources. In northern Thailand, the Chiang Mai area has benefited from the fact that it attracts many international tourists.

Thailand's Lao-speaking northeast (the Khorat Plateau) is the country's poorest region. Soils here are too thin to support intensive agriculture, yet the population is sizable. Because of the poverty of their homeland, northeasterners are often forced to seek employment in Bangkok. As Lao-speakers, they often experience ethnic discrimination. Men typically find work in the construction industry; northeastern women more often make their living as prostitutes.

Thailand has actually earned the dubious distinction of being one of the world's prostitution centers. Thai prostitutes cater largely to a domestic clientele, but they also attract many foreign customers. Most of these "sex tourists" come from wealthy countries such as Japan and Germany, but a significant number are Malaysian. Prostitutes, especially those who are underage, are frequently coerced into their jobs. The "sex industry" of Thailand presents an ironic and tragic situation; in general, the social position of Thai women is relatively high, yet the exploitation of women in Thai brothels can be extraordinarily severe (see "Social Geography: Heroin, AIDS, and Prostitution in Thailand").

Recent Economic Expansion in Indonesia
At the time of independence (1949), Indonesia was one of the poorest countries in the world. The Dutch had used their colony largely as an extraction zone for tropical crops and other resources and had invested little in infrastructure or education. The population of Java mushroomed in the nineteenth and early twentieth centuries, and serious land shortages burdened peasant communities. Political instability, moreover, inhibited economic development during the first decades of independence.

The Indonesian economy finally began to expand in the 1970s. Oil exports fueled the early growth, as did the logging of tropical forests. But unlike most other oil exporters, Indonesia continued to grow even after oil prices plummeted in the 1980s. Today production is relatively low, and Indonesia may have to import oil in the near future. Like Thailand and Malaysia, Indonesia proved attractive to multinational companies eager to export from a low-wage economy. Large Indonesian firms, some three-quarters of them owned by local Chinese families, have also capitalized on the country's low wages and abundant resources. The national government has nurtured technologically oriented businesses, although their success remains uncertain.

But despite rapid growth in the 1980s and 1990s, Indonesia remains a poor country. Its pace of economic expansion never matched that of Thailand, Singapore, and Malaysia, and it has remained much more dependent on the unsustainable exploitation of natural resources. The financial crisis of the late 1990s, moreover, hurt Indonesia more severely than any other country. In 1998 alone, industrial production declined by some 15 percent. Millions of Indonesians suddenly found themselves so destitute that they could no longer afford rice, the country's basic food staple. Political instability is also a continuing concern. Economic recovery began by 2000, but thus far has remained quite tenuous.

As one would expect, development in Indonesia exhibits pronounced geographical disparities. Northwest Java, close to the capital city of Jakarta, has boomed, and much of the resource-rich and moderately populated island of Sumatra has long been relatively prosperous. In the overcrowded rural districts of central and eastern Java, however, many peasants have little if any land, and thus remain on the margins of subsistence. Far eastern Indonesia has not experienced much economic or social development, and throughout the remote areas of all the "outer islands," tribal peoples have suffered terribly as their resources have been taken and their lands lost to outsiders.

Persistent Poverty in Vietnam, Laos, and Cambodia

The three countries of former French Indochina—Vietnam, Cambodia, and Laos—experienced modest economic expansion during the boom years of the 1980s and 1990s. This area endured almost continual warfare between 1941 and 1975, and fighting persisted until the late 1990s in Cambodia. Critics contend that the socialist economic system adopted by these countries forestalled economic growth. This debate is now moot, however, since a partially capitalist model of development is gradually supplanting socialism in all three countries.

Of these three countries, Vietnam is by most measures the most prosperous. Its per capita GNI of $370, however, still places it among the world's poorest economies. Postwar reunification in 1975 did not bring the anticipated growth, and economic stagnation ensued. Conditions grew worse in the early 1990s after the fall of the Soviet Union, Vietnam's main supporter and trading partner. Until the mid-1990s, Vietnam remained under embargo by the United States and thus somewhat isolated from the world economy. Frustrated with their country's economic performance, Vietnam's leaders began to embrace the market while retaining the political forms of a communist state. They have, in other words, followed the Chinese model. Vietnam now welcomes multinational corporations such as Nike, which are attracted by the extremely low wages received by its relatively well-educated workforce. The government also hopes that a market economy will unleash the entrepreneurial potential of the Vietnamese people, thus initiating a phase of rapid economic expansion (Figure 13.35).

These reforms did bring on an upsurge in economic activity in Vietnam, particularly in the formerly capitalist south. In 2000, a stock market was even allowed to open in Ho Chi Minh City. The world market, however, has been somewhat slow to invest, and exports remain dominated by primary products, especially rice and crude oil. U.S. firms were finally allowed into the country in 1994, but many U.S. companies fear that the country's political climate remains too

▲ **Figure 13.35 Storefront capitalism in Vietnam** Although Vietnam is a communist state, it has—like China—embraced many forms of capitalism. Private shops are now allowed, and foreign investment is welcome. In general, the market economy is much more highly developed in the south than in the north. *(Liba Taylor/Panos Pictures)*

unstable. South Korean corporations have been more eager to build factories. Labor tensions have arisen, however, owing to the rigid discipline standards imposed by Korean managers. Noncompetitive, state-owned industries also remain a burden on the Vietnamese economy, and the country's leaders are determined to move slowly with further economic reforms. Merchants still complain of harassment by state officials, and one recent survey listed Vietnam as the most corrupt country in Asia.

Laos and Cambodia face more serious problems than Vietnam. In Cambodia, the ravages of war exacerbated an already difficult situation, while Laos faces special difficulties owing to its rough terrain and relative isolation. Both countries are also hampered by a lack of infrastructure; outside the few cities, paved roads and reliable electricity are rarities. As a result, the economies of both Cambodia and Laos remain largely agricultural in orientation. The Laotian government is pinning its economic hopes on hydropower development. The country is mountainous and has many large rivers, and could therefore generate large quantities of electricity, which is in high demand in neighboring Thailand. The Laotian government, however, remains extremely repressive, discouraging economic development. The Cambodian economy, for its part, is being increasingly integrated with that of Thailand. Several Cambodian border towns recently set themselves up as gambling havens, attracting major investments from Thai prostitution ringleaders frustrated with the fact that gambling is not allowed in Thailand.

Despite their basic lack of development, Cambodia and Laos are perhaps not as miserable as one might expect from the official economic statistics. Both countries have relatively low population densities and abundant resources. Their low per capita GNI figures, more importantly, partially reflect the fact that many of their people remain in a subsistence economy. While the highland peoples of Laos make few contributions to GNI, most of them have adequate shelter and food. Although Laos, Cambodia, and Vietnam have lower per capita GNI figures than India, a much smaller percentage of their people are seriously malnourished.

Burma's Troubled Economy

Burma (Myanmar) stands with Cambodia and Laos at the bottom of the scale of Southeast Asian economic development. For all of its many problems, however, Burma is a land of great potential. It has abundant natural resources (including oil and other minerals, water, and timber), as well as a large expanse of fertile farmland. Its population density is moderate, and its people are reasonably well educated. Burma's economy, however, has remained virtually stagnant since independence in 1948.

Burma's economic woes can be traced in part to the continual warfare the country has experienced. Most observers also blame economic policy. Beginning in earnest in 1962, Burma attempted to isolate its economic system from global forces in order to achieve self-sufficiency under a system of Buddhist socialism. While its intentions may have been admirable, the experiment was not successful; instead of a creating a self-contained economy, Burma found itself burdened by smuggling and black-market activities. Certainly its very lack of economic dynamism has made Burma's capital of Yangon (Rangoon) a much more pleasant place than its ancient rival Bangkok, since it is not nearly as congested and polluted. Some critics of industrial development, moreover, argue that the Burmese people are wary of economic progress owing to their religious beliefs, and that they prefer the slow pace and reduced tensions of an agrarian economy. Other scholars counter that the Burmese people are thoroughly dissatisfied with their limited economic opportunities and that the government has been able to maintain its power only through repression.

Regardless of the debates over Buddhist socialism, the government of Burma has since the early 1990s been attempting to revitalize its economy by tolerating market forces and gradually opening to the global system. By the mid-1990s Burma's economy was supposedly growing at an annual pace of 6 to 9 percent, but per capita income was still lower than it had been in 1980 (reliable statistics, however, are impossible to obtain). Burma remains a notoriously difficult place in which to conduct business, and its political future is cloudy. Several U.S. companies suspended their initial forays into Burma owing to protests over the country's abysmal human-rights record. Other companies, many of them American and Japanese, continue to operate in Burma despite the fact that several human-rights groups have organized boycotts against them. Burma's trading connections with China, moreover, are growing.

Globalization and the Southeast Asian Economy

As the discussion above shows, Southeast Asia as a whole has undergone rapid integration into the global economy (Figure 13.36). Singapore has thoroughly staked its future to the success of multinational capitalism, and several neighboring countries seem to be following suit. Every year Malaysia exports goods—mostly electronics—to the United States that are worth 23 percent of its total gross domestic product, one of the highest figures in the world. Even Marxist Vietnam and once-isolationist Burma are opening their doors to the global system, although with much hesitation in the case of Burma.

Much debate has arisen among scholars over the roots of Southeast Asia's economic gains, as well as its more recent economic problems. Those who credit primarily the diligence, discipline, and entrepreneurial skills of the Southeast Asian peoples are optimistic that the economies will again rebound. Some skeptics argue, however, that most of the region's growth has come from the application of large quantities of labor and capital unsupported by real advances in productivity. Others contend that Southeast Asia moved too quickly from its traditional state-run economies to embrace free-market capitalism. The tiger boom, they argue, was an artificial blossoming resulting from external international capital that ultimately skimmed off the cream of regional economies. But regardless of how this debate turns out, it is clear that Southeast Asia's globalized economies are heavily dependent upon exports, especially to the United States. When the U.S. economy undergoes a serious downturn, they too suffer.

Regardless of what happens in coming years, global economic integration has already brought about a significant amount of development in Singapore, Malaysia, Thailand, and even Indonesia. Over Southeast Asia as a whole, however, economic development has also resulted in accelerating environmental degradation and growing social inequality. Outside of Singapore and Malaysia, moreover, successful development has generally been based on labor-intensive manufacturing in which workers are paid miserably low wages (often less than $1 a day) and subjected to harsh discipline. Consumers in the world's wealthy countries are increasingly aware that many of their basic purchases, such as sneakers and clothing, are produced under exploitative conditions. Movements have thus begun in Europe, the United States, and elsewhere to pressure both multinational corporations and Southeast Asian governments to improve the working conditions of laborers in the export industries. Some Southeast Asian leaders object. Malaysia's Prime Minister Mahathir, for example, accuses Western activists of wanting to prevent Southeast Asian development, and protect their own home markets, under the pretext of concern over worker rights.

Issues of Social Development

As might be expected, several key indicators of social development in Southeast Asia correlate well with levels of economic development. Singapore thus ranks among the world leaders in regard to health and education, as does the small, yet oil-rich country of Brunei. Laos and Cambodia, not surprisingly, come out near the bottom of the chart (Table 13.3). Note, however, that the people of Vietnam are healthier and better educated than might be expected on the basis of their country's overall economic performance. Although Vietnam's Marxist government has not yet been

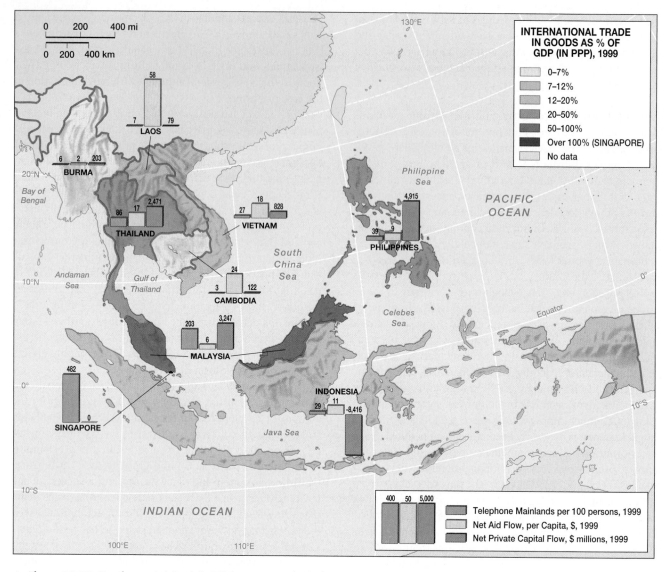

▲ Figure 13.36 Southeast Asia's global linkages   Much of Southeast Asia is tightly integrated into global trade networks, exporting large amounts of both primary products and manufactured goods. This is particularly true of Malaysia and Singapore. Laos, Cambodia, and especially Burma, on the other hand, remain relatively isolated from the global system. In the 1980s and early 1990s, Indonesia's ties with the global economy were rapidly strengthening. Since that time, however, economic collapse, political instability and ethnic violence have resulted in a tremendous flow of capital out of the country.

able to generate sustained economic growth, it has provided reasonably good social services.

With the exception of Laos, Cambodia, and Burma, Southeast Asia has achieved relatively high levels of social welfare. In Laos and Cambodia, however, life expectancy at birth hovers around a miserable 55 years (as compared to Thailand's 72 years), and literacy rates remain below 50 percent, largely because of the extremely low rates of female literacy. But even the poorest countries of the region have made some improvements in mortality under the age of 5, as is evident from the figures in Table 13.3. Note, however, that progress has been more pronounced in prosperous countries such as Malaysia and Singapore, whereas war-torn Cambodia has made relatively small gains. Cambodia, critics note, devotes

roughly 40 percent of its budget to its armed forces and only about 10 percent to social welfare programs.

Most of the governments of Southeast Asia have placed a high priority on basic education. Literacy rates are relatively high in most countries of the region. Much less success, however, has been realized in university and technical education. As Southeast Asian economies continue to grow, this educational gap is beginning to have negative consequences, forcing many students to study abroad. In 1997, for example, more than 50,000 Malaysians were attending colleges in other countries. Economic and social development in Southeast Asia have also led to reduced birthrates. With population growing much more slowly now than before, economic gains are more easily translated into improved living standards.

⊕ Conclusion

In many ways, Southeast Asia is the world region most emblematic of this book's focus on diversity amid globalization. Unarguably, diversity within the region is pronounced, from the array of different cultural traditions to persistent and problematic disparities in economic and social development. Given this wide diversity, one might argue that modern globalization, for better or worse, has helped Southeast Asia find a new sense of regional identity and cohesiveness through common struggles with economic development and revival, as well as geopolitics. Southeast Asia has gained an unusual degree of solidarity in recent years through the regional forum of ASEAN, while shared concerns about economic restructuring and global trading have also nurtured regional identity. As a result, the historical and cultural unity that Southeast Asia has lacked may be forced upon it by twenty-first-century globalization.

That said, the role of ASEAN and the economic coherence of Southeast Asia should not be exaggerated. Most of the region's trade is still directed outward toward the traditional cores of the global economy—North America, Europe, and especially Japan. This external orientation is not surprising, considering the export-focused economic policies embraced by most Southeast Asian countries. A significant question for Southeast Asia's future is whether the region will develop an integrated regional economy, or whether linkages with Japan and the United States will remain more important. Certainly the phenomenal success of Singapore has led to some regional integration. Labor migration is also intensifying local connections as workers from low-wage countries such as the Philippines and Indonesia stream into higher-wage countries, such as Malaysia.

An equally important question concerns the future of the region's environment. Given the emphasis placed by global trade ties on exporting resources such as wood products, it is perhaps understandable that Southeast Asia has sacrificed so many of its forests to support other kinds of economic development. But in most of the region, forests are now seriously depleted. One of the questions Southeast Asia must address is whether to cooperate on forest conservation or, instead, to cave in to special interests promoting continued forest destruction. If the larger Asian economy rebounds, market demand for Southeast Asian forest products will also increase. It is yet unclear whether the region can resist this temptation. Southeast Asia must also meet the needs of its burgeoning cities by making significant infrastructural improvements, specifically those that will improve traffic flow, curtail water pollution, and somehow reduce air pollution.

Finally, some degree of geopolitical change is needed to bring stability to the region. Specifically, the national governments of Indonesia, Burma, and Cambodia may need to reinvent themselves, seeking greater congruence and more cooperation between their people and their leaders. Indonesia has yet to adequately address internal ethnic and cultural tensions. Cambodia must move beyond gratuitous recognition of Khmer Rouge atrocities to genuine efforts at rebuilding and reconciliation. Last, Burma must achieve some sort of stability, coming to terms with both its separatist tribal peoples and the urban population that opposes the current military regime.

Southeast Asia will continue to be a region heavily influenced by globalization, yet there will be more scrutiny and criticism than in the past. This is because its recent economic downturn had widespread repercussions throughout the world. No one, be they in Southeast Asia, Latin America, Europe, or North America, wants to see another such crisis.

⊕ Key Terms

animism (page 569)
Association of Southeast Asia Nations (ASEAN) (page 552)
Bumiputra (page 584)
copra (page 562)
crony capitalism (page 583)
domino theory (page 575)
entrepôt (page 583)
Golden Triangle (page 562)
island biogeography (page 560)
Khmer Rouge (page 575)
kleptocracy (page 583)
lingua franca (page 570)
primate cities (page 565)
Ramayana (page 566)
shifted cultivators (page 564)
Sunda Shelf (page 559)
swidden (page 561)
typhoons (page 559)
transmigration (page 564)
Wallace's Line (page 560)

⊕ Questions for Summary and Review

1. Why are river deltas so important in the settlement pattern of Southeast Asia?
2. Explain how and why the monsoon climates of Southeast Asia differ between mainland and islands.
3. What is Wallace's Line? What is its significance, both prehistorically and today?
4. What do we mean when we refer to the "globalization of world forestry"?
5. Explain why smoke and air pollution are so pronounced in Southeast Asia.
6. Compare and contrast the three major types of agriculture in the region—swidden, plantation, and rice cultivation.
7. How might "transmigration" solve and also aggravate a country's population problems?
8. What major religions are found in Southeast Asia? Describe the historical and contemporary patterns for each religion.
9. Why is there ambivalence about the English language in some Southeast Asian countries?
10. What are the ethnic tensions facing Indonesia? Locate the different problem areas on a map.

11. What are the goals of ASEAN, and how have those goals changed in the last several decades?

12. What caused the recent financial downturn in Southeast Asian economies?

13. Why is Singapore an *entrepôt?*

14. Explain how Malaysia's financial reforms differ from those of Thailand.

15. Why does the "Asian model" for economic development need rethinking?

⊕ Thinking Geographically

1. Discuss the ramifications of state-sponsored migration in Indonesia from areas of high population density to areas of low population density in both the "sending" and the "receiving" areas.

2. Why should—or should not—citizens of the United States be concerned about deforestation in Southeast Asia? If they should be, what would be the proper ways they might show such concern?

3. What might the fate of animism be in the new millennium? Consider whether it might be doomed to extinction before the forces of modern economics and national integration, or whether it may persist as tribal peoples struggle to retain their cultural identities.

4. What should be the position of the English language in the educational systems of Southeast Asia? What should be the position of each country's national language? What about local languages?

5. How might ethnic tensions in countries such as Burma and Indonesia be reduced? Does Indonesia have a reasonable claim to such areas as Irian Jaya?

6. What roles might ASEAN play in the coming years? Should it concentrate on economic or political issues?

7. How could Malaysia successfully integrate its economy into the global system and at the same time regulate the flow of global (or Western) culture? Evaluate whether Singapore can continue to experience economic growth while severely limiting basic freedoms.

8. Is the Southeast Asian economic path of integration into the global economy, which is marked by an openness to multinational corporations, going to prove wise in the long run, or do its potential hazards outweigh its benefits?

⊕ Regional Novels and Films

Novels

Anthony Burgess, *The Long Day Wanes: A Malaysian Trilogy* (1965, Norton)

Joseph Conrad, *Victory* (1936, Doubleday)

Graham Greene, *The Quiet American* (1956, Viking)

Mochtar Lubis, *A Road with No End* (1982, Graham Bush)

W. Somerset Maugham, *Borneo Stories* (1976, Heinemann)

Paul Theroux, *The Consul's Fire* (1977, Houghton Mifflin)

Films

The Deer Hunter (1978, U.S.)

Full Metal Jacket (1987, U.S.)

Heaven and Earth (1993, U.S.)

Indochine (1992, France)

Platoon (1987, U.S.)

The Scent of Green Papaya (1993, Vietnam)

Sunset at Chaopraya (1996, Thailand)

The Ugly American (1962, U.S.)

⊕ Bibliography

Bello, Walden. 1998. "The Rise and Fall of South-east Asia's Economy." *The Ecologist* 28(1), 9–17.

Broad, Robin, and Cavanagh, John. 1993. *Plundering Paradise: The Struggle for the Environment in the Philippines*. Berkeley: University of California Press.

Broek, Jan O. 1944. "Diversity and Unity in Southeast Asia." *Geographical Review* 34, 175–195.

Dixon, Chris. 1991. *South East Asia in the World Economy: A Regional Geography*. Cambridge: Cambridge University Press.

Dobby, E. G. H. 1958. *Southeast Asia*. London: University of London Press.

Dutt, Ashok. K, ed. 1985. *Southeast Asia: Realm of Contrasts*. Boulder, CO: Westview Press.

Freyer, Donald W. 1979. *Emerging Southeast Asia: A Study in Growth and Stagnation*. New York: John Wiley & Sons.

Hall, D. G. E. 1955. *A History of South-East Asia*. New York: St. Martin's Press.

Hanks, Lucien. 1972. *Rice and Man: Agricultural Ecology in Southeast Asia*. Chicago: Aldine Atherton.

Hardjono, Joan, ed. 1991. *Indonesia: Resources, Ecology, and Environment*. Singapore: Oxford University Press.

Hart, Gillian; Turton, Andrew; and White, Benjamin, eds. 1989. *Agrarian Transformations: Local Processes and the State in Southeast Asia*. Berkeley: University of California Press.

Herring, George. 1995. *America's Longest War: The U.S. and Vietnam 1950–1975*. New York: McGraw-Hill.

Hussey, Antonia. 1993. "Rapid Industrialization in Thailand." *Geographical Review* 83, 14–28.

Keyes, Charles F. 1977. *The Golden Peninsula: Culture and Adaptation in Mainland Southeast Asia*. New York: Macmillan.

Leach, E. R. 1954. *Political Systems of Highland Burma*. Boston: Beacon.

Leinbach, Thomas R., and Sien, Chia Lin. 1989. *South-East Asian Transport*. New York: Oxford University Press.

Leinbach, Thomas R., and Ulack, Richard. 1993. "Cities of Southeast Asia." In Brunn, Stanley D., and Williams, Jack F., eds., *Cities of the World: World Regional Urban Development*, pp. 389–429. New York: HarperCollins.

Lewis, Martin W. 1992. *Wagering the Land: Ritual, Capital, and Environmental: Degradation in the Cordillera of Northern Luzon, 1900–1986*. Berkeley: University of California Press.

McGee, Terence G. 1967. *The Southeast Asian City: A Social Geography*. New York: Praeger.

McLennan, Marshall S. 1980. *The Central Luzon Plain: Land and Society on the Inland Frontier*. Quezon City, Philippines: Alemar-Phoenix.

Peluso, Nancy Lee. 1992. *Rich Forests, Poor People: Resource Control and Resistance in Java*. Berkeley: University of California Press.

Pelzer, Karl J. 1945. *Pioneer Settlement in the Asiatic Tropics*. New York: American Geographical Society.

Pye, Lucian W. 1967. *Southeast Asia's Political Systems*. Upper Saddle River, NJ: Prentice Hall.

Rambo, Terry, and Sajise, Percy, eds. 1984. *An Introduction to Human Ecological Research on Agricultural Systems in Southeast Asia*. Los Banos, Philippines: University of the Philippines.

Reid, Anthony. 1988. *Southeast Asia in the Age of Commerce*. Volume I. *The Lands Below the Winds*. New Haven, CT: Yale University Press.

Reid, Anthony. 2000. *Charting the Shape of Early Modern Southeast Asia*. Seattle: University of Washington Press.

Rigg, Jonathan. 1991. *Southeast Asia—A Region in Transition: A Thematic Geography of the ASEAN Region*. New York: HarperCollins.

Steinberg, David Joel. 2001. *Burma: The State of Myanmar*. Washington, DC: Georgetown University Press.

Steinberg, David Joel, ed. 1985. *In Search of Modern Southeast Asia: A Modern History*. Honolulu: University of Hawaii Press.

Taylor, Alice, ed. 1972. *Focus on Southeast Asia*. New York: Praeger.

Taylor, John, and Turton, Andrew, eds. 1988. *Sociology of "Developing Societies": Southeast Asia*. New York: Monthly Review Press.

Tucker, Shelby. 2000. *Among Insurgent: Walking Through Burma*. Houndsmills, Basingstoke, UK: Palgrave.

Ulack, Richard, and Pauer, Gyula. 1989. *Atlas of Southeast Asia*. New York: Macmillan.

von der Mehden, Fred. 1986. *Religion and Modernization in Southeast Asia*. Syracuse, NY: Syracuse University Press.

Wernstedt, Frederick, and Spencer, Joseph. 1967. *The Philippine Island World: A Physical, Cultural, and Regional Geography*. Berkeley: University of California Press.

Wyatt, David K. 1984. *Thailand: A Short History*. New Haven, CT: Yale University Press.

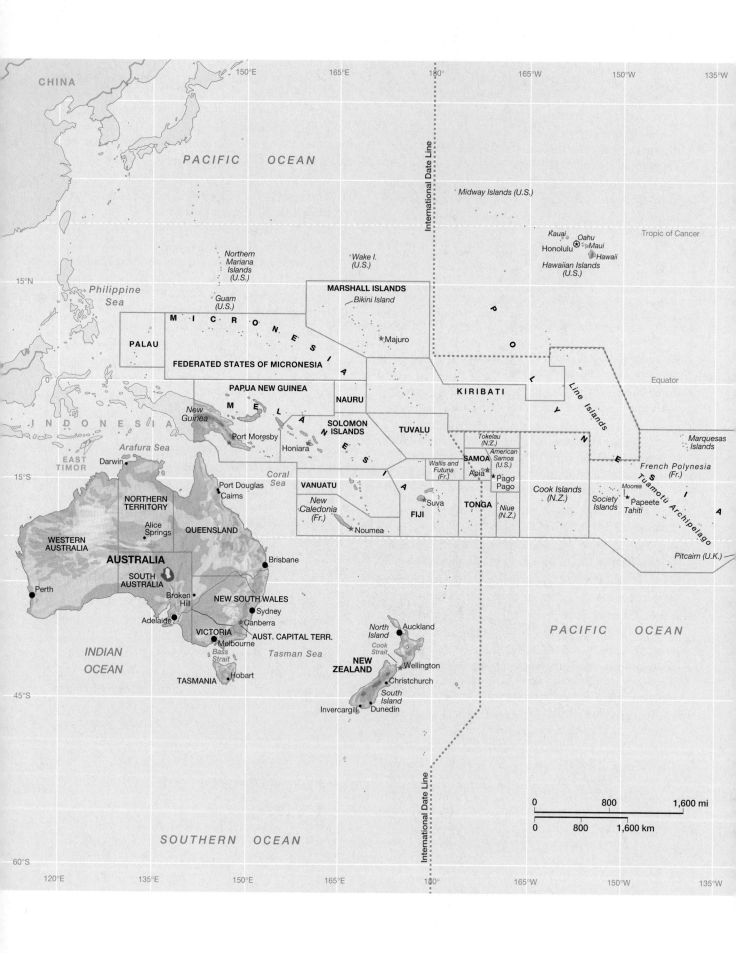

CHINA

PACIFIC OCEAN

International Date Line

Midway Islands (U.S.)

Tropic of Cancer

Kauai
Oahu
Honolulu ⊗ Maui
Hawaii
Hawaiian Islands
(U.S.)

Philippine Sea

15°N

Northern
Mariana
Islands
(U.S.)

Wake I.
(U.S.)

MARSHALL ISLANDS

Bikini Island

M I C R O N E S I A

Guam
(U.S.)

PALAU

FEDERATED STATES OF MICRONESIA

Majuro

P
O
L
Y
N
E
S
I
A

0°

Equator

I N D O N E S I A

PAPUA NEW GUINEA

New
Guinea

M E L A

N

NAURU

KIRIBATI

Line Islands

Marquesas
Islands

EAST
TIMOR

Darwin

Arafura Sea

Port Moresby

Honiara

SOLOMON
ISLANDS

TUVALU

E

Tokelau
(N.Z.)

American
Samoa
(U.S.)

SAMOA

French Polynesia
(Fr.)

S
I
A

15°S

Port Douglas
Cairns

*Coral
Sea*

S
I
A

Wallis and
Futuna
(Fr.)

Apia ★
Pago
Pago

Tuamotu Archipelago

NORTHERN
TERRITORY

VANUATU

New
Caledonia
(Fr.)

Suva

Cook Islands
(N.Z.)

Moorea
★ Papeete
Tahiti

Society
Islands

Alice
Springs

QUEENSLAND

FIJI

TONGA

Niue
(N.Z.)

WESTERN
AUSTRALIA

AUSTRALIA

Brisbane

Noumea

Pitcairn (U.K.)

SOUTH
AUSTRALIA

Perth

Broken
Hill

NEW SOUTH WALES

PACIFIC OCEAN

Adelaide

Sydney
★ Canberra
AUST. CAPITAL TERR.

North
Island

Auckland

INDIAN
OCEAN

VICTORIA
Melbourne

Bass
Strait

Tasman Sea

Cook
Strait

Wellington

NEW
ZEALAND

Christchurch

TASMANIA

Hobart

South
Island

Invercargill

Dunedin

45°S

International Date Line

SOUTHERN OCEAN

60°S

0   800   1,600 mi
0   800   1,600 km

120°E   135°E   150°E   165°E   180°   165°W   150°W   135°W

14

Australia
and Oceania

This vast world region, dominated mostly by water, includes the island continent of Australia as well as **Oceania**, a sweeping collection of islands that reach from New Guinea and New Zealand to the U.S. state of Hawaii in the mid-Pacific (Figure 14.1; see "Setting the Boundaries"). While the area was settled by indigenous peoples long ago, more recent European and North American colonization began the process of globalization that now characterizes the region. From the perspective of its European colonizers, the region was enlivened by its mythical character: exotic and otherworldly, its "lost islands" and "bush" were populated by strange plants and animals as well as by peoples free from the conventional cultural mores of European society. From the perspective of its indigenous inhabitants, the region was a world rich in natural and cultural complexity, a way of life rudely interrupted and changed forever by the arrival of European peoples. In the early twenty-first century, the region is caught up in powerful processes of globalization that are transforming rural and urban landscapes and producing new and sometimes unsettled cultural and political geographies.

Ongoing political and ethnic unrest in Fiji illustrates how the heat of twenty-first-century globalization has fired the cauldron of Pacific globalization (Figure 14.2). Currently, the country struggles to address new political and economic relationships between the two dominant ethnic groups, indigenous Fijians and the descendants of South Asian sugarcane workers (called Indo-Fijians) who were brought to the islands in the nineteenth century as a solution to labor shortages in the cane fields. Cultural and religious barriers between the two groups have always been high, resulting in extremely low rates of intermarriage. Generally speaking, the Indo-Fijians dominate the country's commercial and urban sectors, which has caused resentment in the largely rural and village-based indigenous Fijian community. While tensions have long been precarious between these two groups, this unrest has been accentuated by contemporary globalization as foreign investors push for economic restructuring and market development. Achieving these economic goals requires a radical change to the traditional land tenure system, which is primarily controlled by a confederacy of indigenous Fijian groups. As a result, island governance has been unsettled by a series of coups and counter-coups over the last two decades. In May 2000, armed indigenous Fijians took hostage the nation's first prime minister of Indian descent. Large numbers of Indo-Fijians fled the islands, leaving the commercial sector in shambles. Despite calls from

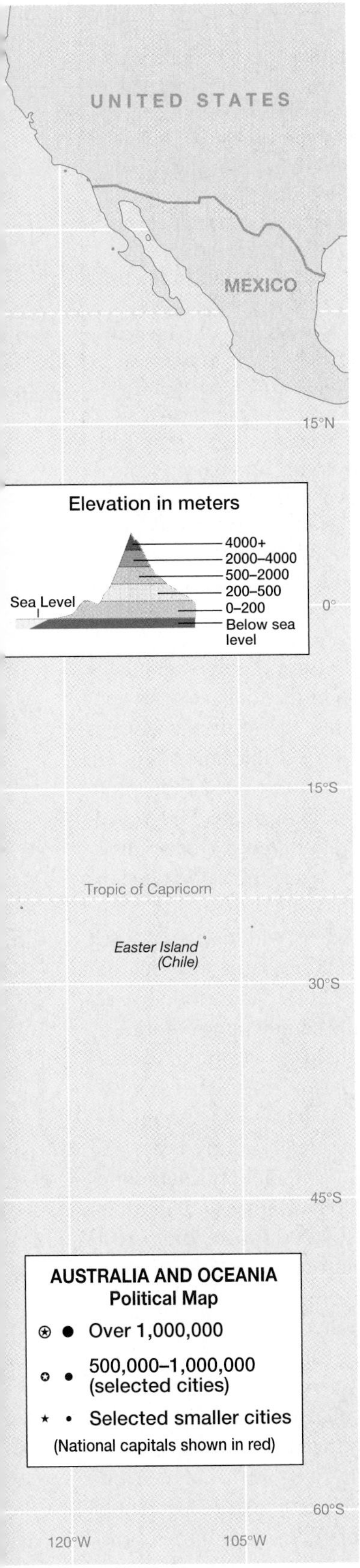

◀ **Figure 14.1 Australia and Oceania** More water than land, the Australia and Oceania region sprawls across the vast reaches of the western Pacific Ocean. Australia dominates the region, both in its physical size and in its economic and political clout. Along with New Zealand, Australia represents largely Europeanized settlement in the South Pacific. Elsewhere, however, the island subregions of Melanesia, Micronesia, and Polynesia contain important native populations that have mixed in varied ways with later European and American arrivals.

UNITED STATES

MEXICO

15°N

Elevation in meters

4000+
2000–4000
500–2000
200–500
0–200
Sea Level
Below sea
level

0°

15°S

15°S

Tropic of Capricorn

Easter Island
(Chile)

30°S

45°S

AUSTRALIA AND OCEANIA
Political Map

⊛ ● Over 1,000,000

○ ● 500,000–1,000,000
(selected cities)

★ • Selected smaller cities
(National capitals shown in red)

60°S

120°W   105°W

SETTING THE BOUNDARIES

The vast, watery distances of the Pacific, as well as shared elements of indigenous and colonial history that stretch from the South Pacific to Hawaii, help to define the boundaries of the Australia and Oceania region. Still, some of its regional boundaries are born from convenience, while others remain ill-defined. Australia (or "southern land") forms a coherent political unit and subregion and clearly dominates the region in both area and population. Across the Tasman Sea, New Zealand, while usually considered part of Oceania, is easily linked with Australia, particularly in their parallel histories of European—mostly British—colonization and development. Both countries also have significant native populations, although Australia's Aborigines are culturally and ethnically distinct from New Zealand's Maori peoples.

More challenging questions arise when pondering how those larger landmasses (often termed "Meganesia") relate to the island worlds beyond, as well as to nearby portions of Southeast Asia. Increasing economic ties and the common borders of the Pacific Ocean link Australia and New Zealand with the smaller islands of Oceania. Still, the area's environmental and cultural diversity does not offer any easy defining characteristics. More broadly, however, traditional island cultures often intermingled, and they shared relationships with the sea and the varied fruits of the tropical world. Later, Europeans and Americans converged upon the scene. As they rushed to carve up the realm into colonial possessions, their strategic maneuverings often had lasting consequences. Originally Polynesian, Hawaii, once it was securely in the U.S. sphere, evolved into a multicultural and political borderland that expresses the complex globalization of the Pacific region. To the south, the French produced their own distinctive island subregion, mostly scattered across the eastern reaches of Polynesia. On the island of New Guinea, however, complex colonial-era boundaries produced a persisting and problematic regional border. Today, an arbitrary boundary line bisects the island, and while Papua New Guinea (eastern half) is usually considered a part of Oceania, neighboring Irian Jaya (western half) is a part of Indonesia and thus typically grouped with Southeast Asia.

▲ **Figure 14.2 Unrest in Fiji** Ethnic tensions between Indo-Fijians, the descendants of South Asian sugar cane workers brought to the islands in the nineteenth century but who are now the dominant economic class, and the indigenous Fijians often take a violent turn. Here, an Indo-Fijian family examines the ruins of their house after it was torched by Fijian rebels during the military coup of August 2000. *(AFP/Corbis)*

the international community to restore the democratically elected Fijian government, leaders of the military coup continue to hold power. By all accounts, these current tensions are more than a continuation of historic ethnic unrest between the two groups. Instead, observers attribute much of the problem to unrest within the traditional indigenous Fijian community as new players become empowered by the spoils of economic globalization. In many ways, the same tensions found in Fiji are also expressed in many other island countries of Oceania and, to a lesser degree, within Australia and New Zealand, which serves to remind us of the way environmental, settlement, cultural, geopolitical, and economic activities are inseparably linked in the contemporary world.

Australia and New Zealand share many geographical characteristics and dominate the regional setting. Major population clusters in both countries are located in the middle latitudes. Australia's 19.4 million residents occupy a vast land area of 2.97 million square miles (7.69 million square kilometers), while New Zealand's combined North and South Islands (104,000 square miles, or 269,000 square kilometers) are home to 3.9 million people. Most residents of both countries live in urban settlements near the coasts (Figure 14.3). Australia's huge and dry interior, often termed the **outback,** is as thinly settled as North Africa's Sahara Desert, and much of the New Zealand hinterland is a visually spectacular but sparsely occupied collection of volcanic peaks and rugged, glaciated mountain ranges. Taken together, the land areas of these two South Pacific nations almost equal that of the United States, but their populations total less than 10 percent of their distant North Pacific neighbor. All three countries, however, share a pervasive European cultural heritage, the product of common global-scale processes that sent Europeans far from their homelands over the past several centuries. The

▲ **Figure 14.3 Sydney, Australia** Most Australians live in cities, and the country's urban landscapes often resemble their North American counterparts. This view of Sydney features its world-famous harbor and displays the dramatic interplay of land and water. *(Rob Crandall/Rob Crandall, Photographer)*

highly Europeanized populations of both Australia and New Zealand also retain particularly close cultural links with Britain, and they maintain relatively high levels of income and economic development.

Punctuated with isolated chains of sand-fringed and sometimes mountainous islands, the blue waters of the tropical Pacific dominate much of the rest of the region. Three major subregions of Oceania each contain a surprising variety of human settlements and political entities, although overall land areas and populations are small compared with Australia. Farthest west, **Melanesia** (meaning "dark islands") contains the culturally complex, generally darker-skinned peoples of New Guinea, the Solomon Islands, Vanuatu, and Fiji. The largest of these countries, Papua New Guinea (179,000 square miles, or 463,000 square kilometers), includes the eastern half of the island of New Guinea (the western half is part of Indonesia), as well as nearby portions of the northern Solomon Islands. Its population of 5 million people is slightly higher than that of New Zealand. To the east, the small island groups, or **archipelagos,** of the central South Pacific are called **Polynesia** (meaning "many islands"), and this linguistically unified subregion includes French-controlled Tahiti in the Society Islands, the Hawaiian Islands, and smaller political states such as Tonga, Tuvalu, and Samoa. New Zealand is also often considered a part of Polynesia since its native peoples, known collectively as the **Maori,** share many cultural and physical characteristics with the somewhat lighter-skinned peoples of the mid-Pacific region. Finally, the more culturally diverse region of **Micronesia** (meaning "small islands") is north of Melanesia and west of Polynesia and includes microstates such as Nauru and the Marshall Islands, as well as the U.S. territory of Guam.

Three characteristics unite the Australia and Oceania region. First, the realm's unique physical setting has historically isolated it from much of the more densely settled parts of the world. This basic geographical fact contributed to the distinctive evolution of the region's exotic plants and animals, affected the timing of its human occupation, shaped later processes of European colonization, and impacted the economic potential for trade with much of the rest of the modern world.

Second, the region reveals a fascinating set of responses between indigenous peoples and varied European cultures. It is a vast laboratory of cultural adaptations, assimilation, and conflicts that displays the unpredictable outcomes that result when one culture comes in contact with another. Thus, from the largely Anglo and wealthy Sydney suburbanites to the still-elusive peoples of the remote New Guinea Highlands, the region offers a wide range of examples of the many-layered legacies of the region's multicultural roots.

Third, the region is united by its youthful political geography. Essentially a product of the twentieth century, the current political map includes a mosaic of enduring colonies, as well as a varied collection of states that are still emerging from their political infancy. Thus, fluidity and uncertainty characterize the region's geopolitical identity. Even Australians are still torn about their ties to Britain, and many of the region's new states have been marked by internal political disputes, violent conflicts, and open debates about the future of their economic base and the nature of their civic society. The result is a world region still ill defined in terms of its internal self-identity as well as its way of relating to the rest of the globe during the twenty-first century.

Environmental Geography: A Varied Natural and Human Habitat

The region's physical setting is dramatic testimony to the power of space: the geology and climate of the seemingly limitless Pacific Ocean define much of the physical geography of Oceania, and the almost equally expansive interior of Australia shapes the basic physical setting of that island continent (Figure 14.4). Not coincidentally, the greatest human impacts upon the region and the densest clusters of population exist at the meeting place of land and sea: coastal strips and island shores share environmental settings where residents have opted to look inward as well as outward.

Environments at Risk

Even with relatively modest populations, many settings in Australia and Oceania face significant human-induced environmental problems. Some challenges are simply the outcome of natural events that increasingly impact larger and more widely distributed human populations. For instance, Pacific Rim-related seismic hazards, periodic Australian droughts, and unwelcome tropical cyclones now pose greater threats than they once did as new settlements have made increasing populations vulnerable to these problems. Other environmental issues, however, are even more directly related to

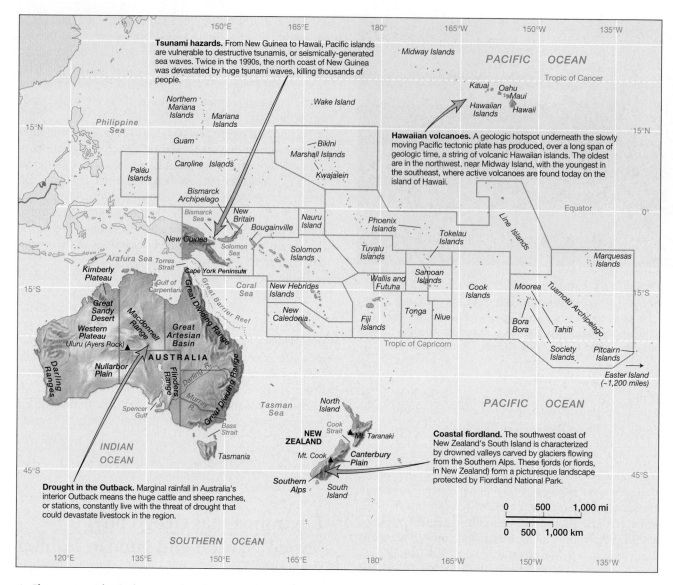

Tsunami hazards. From New Guinea to Hawaii, Pacific islands are vulnerable to destructive tsunamis, or seismically-generated sea waves. Twice in the 1990s, the north coast of New Guinea was devastated by huge tsunami waves, killing thousands of people.

Hawaiian volcanoes. A geologic hotspot underneath the slowly moving Pacific tectonic plate has produced, over a long span of geologic time, a string of volcanic Hawaiian islands. The oldest are in the northwest, near Midway Island, with the youngest in the southeast, where active volcanoes are found today on the island of Hawaii.

Coastal fiordland. The southwest coast of New Zealand's South Island is characterized by drowned valleys carved by glaciers flowing from the Southern Alps. These fjords (or fiords, in New Zealand) form a picturesque landscape protected by Fiordland National Park.

Drought in the Outback. Marginal rainfall in Australia's interior Outback means the huge cattle and sheep ranches, or stations, constantly live with the threat of drought that could devastate livestock in the region.

▲ **Figure 14.4 Physical geography of Australia and Oceania**   Varied physical processes have shaped the Australia and Oceania region. Ancient shield rocks form the geological core of Australia, while active volcanoes from Hawaii to Papua New Guinea literally produce examples of Earth's newest landscapes. New Zealand is home to some of the region's most complex and varied physical settings. Dominant almost everywhere in the region, the omnipresent waters of the blue Pacific shape land and life in fundamental ways.

human causes (Figure 14.5). Specifically, European colonization introduced many environmental threats, and recent economic globalization has further pressured the region's natural resource base.

Global Resource Pressures   Processes of globalization have exacted an environmental toll upon Australia and Oceania. Specifically, the region's considerable base of natural resources has been opened to development, much of it by outside interests. While gaining from the benefits of global investment, the region has also paid a considerable price for encouraging development, and the result is an increasingly threatened environment. Major mining operations have profoundly impacted Australia, Papua New Guinea, New Caledonia, and Nauru. Some of Australia's largest gold, silver,

copper, and lead mines are located in sparsely settled portions of Queensland and New South Wales, but watersheds in these semiarid regions are highly susceptible to metals pollution. In Western Australia, huge open-pit iron mines dot the landscape, unearthing ore that is often bound for global markets, particularly Japan. To the north, Papua New Guinea's Bougainville copper mine has transformed the Solomon Islands, while even larger gold mining ventures have raised increasing environmental concerns on the island of New Guinea (Figure 14.6). Elsewhere, Micronesia's tiny Nauru has been virtually turned inside out as much of the island's jungle cover was removed to get at some of the world's richest phosphate deposits. Former Australian and New Zealand mine owners have already paid millions of dollars to settle environmental damage claims.

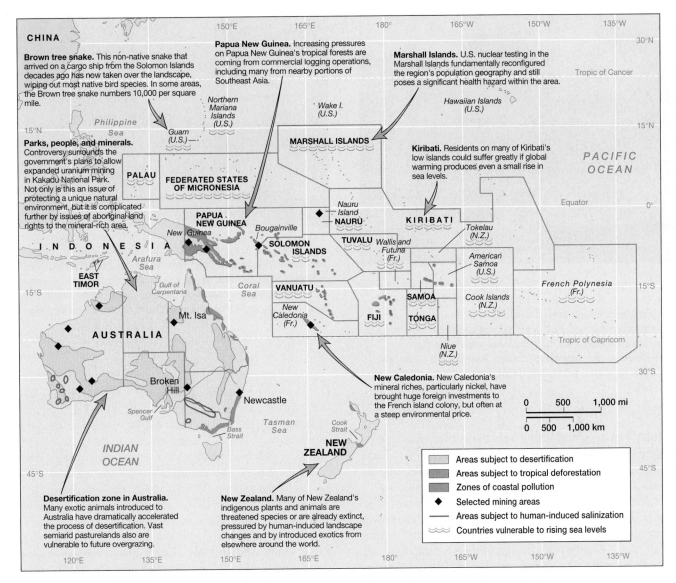

Figure 14.5 Environmental issues in Australia and Oceania Modern environmental problems belie the region's myth as an earthly paradise. Tropical deforestation, extensive mining, and a long record of nuclear testing by colonial powers have brought varied challenges to the region. Human settlements have also extensively modified the pattern of natural vegetation. Future environmental threats loom on low-lying Pacific Islands if global sea levels rise.

Deforestation is another major environmental threat across the region. Vast stretches of Australia's eucalyptus woodlands, for example, have been destroyed to produce better pastures. In addition, coastal rain forests in Queensland are only a fraction of their original area, although a growing environmental movement in the region is fighting to save the remaining forest tracts. Tasmania has also been an environmental battleground, particularly given the diversity of its midlatitude forest landscapes. While the island's earlier European and Australian development featured many logging and pulp mill operations, more than 20 percent of the island is now protected by national parks. The Australian Green Party is pushing hard for further preservation on the island. Deforestation is also a major problem in many parts of Oceania. With limited land areas, smaller islands are subject to rapid tree loss, which in turn often leads to soil erosion. Sizable expanses of rain forest are still found on most of the larger islands of Melanesia. Although rain forests still cover 70 percent of Papua New Guinea, more than 37 million acres (15 million hectares) have been identified as suitable for logging (Figure 14.7). Some of the world's most biologically diverse environments are being threatened in these operations, but landowners see the quick cash sales to loggers as attractive, even though the practice is a nonsustainable alternative to their traditional lifestyles. The global expansion of commercial logging also threatens nearby portions of the Solomon Islands, as well as many unique environmental settings in Polynesia (see "Environment: Saving the Samoan Rain Forest"). Increasingly, residents question unbridled development: recently Solomon Islanders resisted a government resettlement plan that threw them off their land so that a Malaysian lumber company could clear the area of trees. Logging proceeded, despite continued protests.

▲ **Figure 14.6 Mining in Papua New Guinea** Open-pit mining for gold, silver, copper, and lead mark the landscapes of Papua New Guinea, New Caledonia, and Nauru. Although bringing some economic benefit to local peoples, these activities also cause immense environmental damage to the region. In New Guinea, for example, sediments from upland mines have severely damaged the Fly River ecosystem. *(Michael Freeman/ Corbis)*

▲ **Figure 14.7 Logging in Oceania** Foreign logging companies have made large investments in tropical Pacific settings such as Papua New Guinea. While bringing new jobs, these ventures dramatically alter local environments as precious hardwood forests are harvested for export. *(David Austen/Woodfin Camp & Associates)*

Global Warming and Rising Sea Levels Oceania's greatest future environmental threat may be global warming. If Earth grows significantly warmer in the near future, as many climate models predict, the impact on many low-lying atolls will be devastating. Higher global temperatures will partially melt polar ice caps, and resulting rises in sea levels could literally drown the region's low islands. Countries such as Kiribati and the Marshall Islands could simply disappear, forcing residents to flee elsewhere, probably to other overcrowded islands. Not surprisingly, the Pacific states have been major supporters of global conventions to limit the production of greenhouse gases. Nauru's leaders recently reported that their low island is seeing increased coastal erosion and that storm surges, elevated sea levels, and unusual drought conditions have damaged the island's supply of fresh water. Island residents, already displaced to the coast by inland mining operations, are being squeezed out of their homes from all directions. Unfortunately, these tiny island nations have little international clout compared to developed countries, such as the United States, which contribute most global warming pollution.

Nuclear Testing The region's peripheral economic and political status have often been costly. When the United States and France required atomic testing grounds for their nuclear weapons programs, they chose the South Pacific as an ideal location because of its remote location and sparse population. Local residents had little voice in such matters, although French tests in the 1990s provoked more widespread and sustained negative responses from the regional populations than had earlier tests. The environmental consequences of the tests have been long lasting. In 1968 President Johnson confident-

ly told former residents of Bikini and Rongelap in the Marshall Islands that they could return to their bomb-transfigured homelands. The area had been used for 67 nuclear weapons explosions between 1946 and 1958. Tragically, the natives returned, rebuilt their houses, and resettled their villages, only to find that radioactive cesium 137 was still concentrated in the soil and soon entered the food chain. Residents were later evacuated to other islands, but the health and environmental costs in such settings have been monumental.

Exotic Plants and Animals Exotic (nonnative) species pose a major problem in many settings because they often threaten the viability of indigenous life-forms. In Australia, some native fauna could not compete with nonnative animals. On the other hand, exotic rabbits successfully multiplied in an environment that lacked the diseases and predators that elsewhere kept their numbers in check. Before long, rabbit populations had reached plague proportions, and large sections of land were virtually stripped of vegetation. The animals were brought under control only through the purposeful introduction of the rabbit disease myxomatosis. Introduced sheep and cattle populations have also stressed the region's environment by accelerating soil erosion and contributing to desertification. In addition, feral livestock have spread. In various parts of Australia, one can find large numbers of once-domestic but now wild goats, horses, pigs, cattle, sheep, water buffalo, and even camels.

Island settings have also witnessed the dramatic consequences of plant and animal invasion. For example, many small islands possessed no native land mammals, and their native bird and plant species proved vulnerable to the ravages of introduced rats, pigs, and other animals. The larger islands of the region, such as those of New Zealand, originally supported several species of large, flightless birds that filled some

ENVIRONMENT   Saving the Samoan Rain Forest

Ethnobotanist Paul Alan Cox has spent years in the Samoan rain forest. His encounter with this remote Polynesian setting began in 1973 as a young Mormon missionary. He became fascinated with how native peoples utilized rainforest plants to treat disease. Eleven years later, he returned as a trained botanist, hoping to decipher the secrets held by native female healers as they worked among the local population. In his travels to remote villages far beyond the reach of Western-style cures, Cox discovered that these Samoan "medicine women" possessed an intimate knowledge of how dozens of plant species could be used to treat human ailments. Passed down from mother to daughter, this folk medicine included using the polo leaf for fighting infections, the ulu ma'afala root for diarrhea, and the bark of a local tree to combat hepatitis. Healers introduced Cox to 70 of these remedies, including one that may be effective against AIDS. Indeed, these are not magical native cures: 25 percent of prescription drugs made in the United States are extracted from flowering plants, many of which grow in tropical forest settings.

This Samoan treasure trove of botanical riches is currently under assault, however. Already, more than 80 percent of Samoa's coastal rain forest has been logged, much of it in just the past 25 years. Cox sees this as a tragic waste, both environmentally and economically. He points out that research institutes and drug companies can pay handsome royalties to people who assist in the search for health remedies. In other words, the forests are worth much more standing than they are cut down. Awarded the Goldman Environmental Prize in 1997 for his forest-saving efforts, Cox fears that the island's many sawmills may complete their work before he and other researchers can even catalogue the area's vanishing riches.

Such environmental races with time are commonplace elsewhere in Oceania and across the broader tropical world. No wonder the islanders have given Cox the name Nafanuna, after one of their traditional protective gods.

Source: Adapted from "Rainforest Pharmacist," *Audubon,* January 1999.

▲ **Figure 14.1.1 Samoan rain forest**   Logging threatens Samoa's lush and diverse tropical vegetation. In addition to posing increased local environmental hazards, these operations may eliminate important medicinal plants useful for treating many diseases.   *(Jean-Marc La Roque/Auscape International Pty. Ltd.)*

of the ecological niches held by mammals on the continents. The largest of these, the moas, were substantially larger than ostriches. In the first wave of human settlement in New Zealand some 1,500 years ago, moa numbers fell rapidly as they were hunted, their habitat burned, and their eggs consumed by invading rats. By 1800, the moas had been completely exterminated.

A second wave of extinctions occurred in Polynesia and other portions of Oceania soon after the arrival of Europeans. As in Australia, introduced plants and animals often competed successfully with native species and threatened to eliminate very valuable species that occur nowhere else on the planet. In particular, the Hawaiian Islands suffered devastating extinctions following the arrival of European and American influences. The spread of nonnative species continues today, perhaps even at an accelerated pace. In Guam, for example, the brown tree snake, which arrived accidentally by cargo ship from the Solomon Islands in the 1950s, has taken over the landscape (Figure 14.8). In some forest areas, with over 10,000 snakes per square mile, they have virtually wiped

▲ **Figure 14.8 Island pest**   The brown tree snake, which arrived in Guam accidentally in the 1950s, has now taken over large parts of the island's forest lands and killed off most native bird species. Further, because these snakes, which reach 10 feet in length, climb along electrical wires, they frequently cause power outages throughout Guam.   *(John Mitchell/Photo Researchers, Inc.)*

out all native bird species. Additionally, the snakes cause frequent power outages as they crawl along electrical wires. While the brown tree snake has already done its damage to Guam, it threatens other islands as well, since it readily hides in cargo containers awaiting air or water shipment to island destinations. Further, because the snake is an agile climber, it has even hitched rides to other islands on airplane landing gear. Cargo handlers and custom agents throughout Oceania are on alert to keep this globalized pest from getting established beyond Guam.

Australian Environments

Curiously, Australia is one of the world's most urbanized societies, yet most people associate the country with its vast and arid outback, a sparsely settled land of sweeping distances, scrubby vegetation, and exotic animals (Figure 14.9). Much of the Australian continent conforms to the stereotype: placenames such as the "Nullarbor [no trees] Plain," "Sandland," and the "Great Sandy Desert" suggest that Europeans exploring the region encountered a hostile and unfamiliar land, particularly as they probed its huge and arid interior. Indeed, distance and dryness have left a lasting mark on Australia's physical setting and shaped the ways in which people have subsequently settled its more-favored corners.

Regional Landforms Journeying across Australia from west to east, three major landform regions dominate the continent's physical geography (Figure 14.4). The Western Plateau occupies more than half of the continent and geologically represents an ancient shield landmass that once joined Antarctica. Most of the region is a vast, irregular plateau that averages only 1,000 to 1,800 feet in height (305 to 550 meters). As one leaves the plateau, the Interior Lowland Basins stretch

north to south for more than 1,000 miles from the swampy coastlands of the Gulf of Carpentaria to the Murray and Darling valleys, Australia's largest river system. Most of the region is a flat, featureless plain, punctuated occasionally by dry lake beds and by stream valleys where water is only a rare visitor. The Great Artesian Basin of western Queensland is included in this lowland region. Filled with sedimentary rocks, this vast basin, mostly below 500 feet (150 meters) in elevation, possesses a rich supply of underground water that has drained there from the higher country to the east. Finally, more forested and mountainous country defines the horizon as one arrives in the Eastern Highlands along Australia's Pacific Ocean rim. The Great Dividing Range extends from the Cape York Peninsula in northern Queensland to southern Victoria. Nearby Tasmania is also dominated by mountains. These ancient eroded highlands include rugged tablelands but only a few peaks above 5,000 feet (1,525 meters). The east-facing edges of the mountains often feature steep escarpments and rugged, dissected canyons that slow easy interaction between the interior and the narrow, often densely settled coastal plains on the Pacific. Nearby, off the eastern coast of Queensland, the Great Barrier Reef offers a final dramatic subsurface feature: over the past 10,000 years, one of the world's most spectacular examples of coral reef-building has produced a living legacy now protected by the Great Barrier Reef Marine Park (Figure 14.10).

Climate and Vegetation Australia's varied climates strongly shape patterns of natural vegetation across the country (Figure 14.11). Generally, zones of somewhat higher precipitation encircle Australia's arid center. In the tropical low-latitude north, seasonal changes are dramatic and unpredictable. Localities such as Darwin can experience drenching monsoon-

▲ **Figure 14.9 The Australian Outback** Arid and generally treeless, the vast lands of the Australian outback resemble some of the dry landscapes of the U.S. West. In this photo, wildflowers blossom along a dirt road near Tom Price, in the Pilbara region of Western Australia. *(Rob Crandall/Rob Crandall, Photographer)*

▲ **Figure 14.10 The Great Barrier Reef** Stretching along the eastern Queensland coast, the famed Great Barrier Reef is one of the world's most spectacular examples of coral reef-building. Threatened by varied forms of coastal pollution, much of the reef is now protected in a national marine park. *(Hilarie Kavanagh/Getty Images, Inc.)*

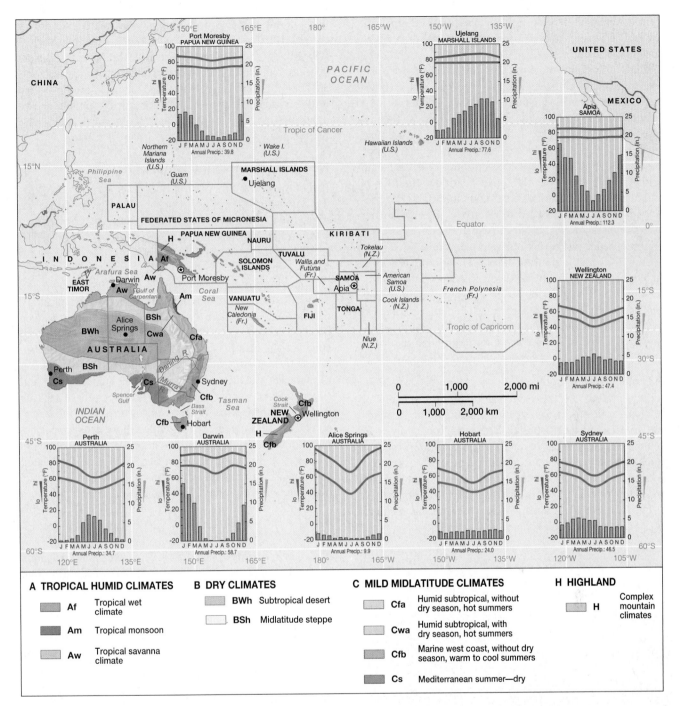

A TROPICAL HUMID CLIMATES

	Af	Tropical wet climate
	Am	Tropical monsoon
	Aw	Tropical savanna climate

B DRY CLIMATES

| | BWh | Subtropical desert |
| | BSh | Midlatitude steppe |

C MILD MIDLATITUDE CLIMATES

	Cfa	Humid subtropical, without dry season, hot summers
	Cwa	Humid subtropical, with dry season, hot summers
	Cfb	Marine west coast, without dry season, warm to cool summers
	Cs	Mediterranean summer—dry

H HIGHLAND

| | H | Complex mountain climates |

▲ Figure 14.11 **Climate map of Australia and Oceania**   Latitude and altitude shape the climatic patterns of the region. Equatorial portions of the Pacific bask in all-year warmth and humidity, while the Australian interior is predictably dry and under the dominance of subtropical high pressure. Cool and moisture-bearing storms of the southern Pacific Ocean provide midlatitude conditions across New Zealand and portions of Australia. More locally, mountain ranges dramatically raise precipitation totals in many highland zones.

al rains in the summer (December to March), followed by bone-dry winters (June to September). Indeed, life across the region is shaped by this annual rhythm of "wet" and "dry." Much of the north is clothed in a mix of tropical woodlands and thorn forests interspersed with open grasslands. As one ventures south along the east coast of Queensland, precipitation remains high (60 to 100 inches, or 153 to 254 cen-

timeters), but it diminishes rapidly as one moves into the interior. Vegetation follows the pattern: coastal tropical rain forests give way to eucalyptus and acacia woodlands in the hill country, and eventually the trees thin out west of the mountains, replaced by the short grass and desert scrub vegetation of southwest Queensland's arid outback. Indeed, some of the country's dry heartland is almost devoid of vegetation, and precipitation at interior locations such as the Northern Territory's Alice Springs averages less than 10 inches (25 centimeters) annually.

South of Brisbane, more midlatitude influences dominate eastern Australia's climate. Coastal New South Wales, southeastern Victoria, and Tasmania experience the country's most dependable year-round rainfall in a climatic regime that averages 40 to 60 inches (102 to 152 centimeters) of precipitation per year. Nearby mountains see frequent winter snows. Forests clothe most of the wetter southern coastlands and highlands, but tracts of more open heath dot the hills of southern Victoria, and drier conditions in the interior produce grasslands often suited to extensive grazing. Farther west, summers are hot and dry in much of South Australia and in the southwest corner of Western Australia, producing a distinctively Mediterranean climate. These zones of southern Mediterranean climate produce a mix of vegetation types. In moister hills near Perth, acacia, eucalyptus, and conifer forests cover the slopes, while portions of South Australia feature abundant **mallee** vegetation, a tough and scrubby eucalyptus woodland of limited economic value that has often been stripped away by determined farmers (Figure 14.12).

An Exotic Zoogeography   Isolation and genetics have combined to produce Australia's animal kingdom, surely one of the most exotic on the planet. Long separated from Asia, animal species on the Australian landmass have evolved in unique ways. Marsupials, or animals that carry their young in pouches, never faced much competition from other invading species and adapted well to the continent's varied natural settings. More than 120 species of marsupials inhabit the country. Europeans such as Captain Cook expressed amazement as they gazed upon leaping kangaroos, duck-billed platypuses,

▲ **Figure 14.12 Mallee scrub vegetation**   The interior of South Australia and New South Wales features thousands of square miles of a tough, scrubby eucalyptus vegetation known as mallee. Farmers often remove this native cover in order to plant exotic grass and grain crops.   *(Jaime Plaza Van Roon/Auscape International Pty. Ltd.)*

hairy-nosed wombats, and snarling Tasmanian devils during their early exploratory visits. Many of these animals, such as the red and gray kangaroo, thrive on the drought-resistant vegetation of the Australian outback. Bird life is similarly varied. Flightless emus, shrieking kookaburras, rainbow-hued lorikeets, and dozens of other parrot species celebrate the biological fruits of spatial isolation in an extensive and diverse natural environment.

New Zealand's Varied Landscape

Part of the Pacific Rim of Fire, New Zealand owes its geological origins to undersea mountain-building that produced two rugged and spectacular islands in the South Pacific more than 1,000 miles (1,600 kilometers) southeast of Australia. The active volcanic peaks and geothermal features of the North Island, in particular, reveal the country's fiery origins. Blessed with such indigenous names as Ruapehu, Ngauruhoe, and Tongariro, these volcanic peaks tower over nearby tablelands, reaching heights of more than 9,100 feet (2,775 meters). Even higher and more rugged mountains run down the western spine of the South Island. Mt. Cook (or Aoraki) is New Zealand's highest peak, cresting at more than 12,000 feet (3,660 meters). Often mantled by high mountain glaciers and surrounded by steeply sloping valleys, the Southern Alps are known to the Maori as *Te Tapu Nui* (the Peaks of Intense Sacredness). They are one of the world's most visually spectacular mountain ranges, complete with narrow, fjordlike valleys that intricately indent much of the South Island's isolated western coast. East of the mountains, small patches of coastal plain are found between Christchurch and Invercargill.

As in Australia, New Zealand's isolation offered opportunities for the development of unique plant and animal species. Eighty-five percent of the country's native trees and seed plants are found nowhere else on Earth. Bats are the region's only native mammals, while ancient tuatara reptiles, kiwi birds, and the flightless and now-extinct moas illustrate the country's special biological legacy.

North Island Environments   Most of New Zealand's North Island is distinctly subtropical. The coastal lowlands near Auckland are mild and wet year-round. Still, local variations can be striking, as the area's volcanic peaks create their own microclimates. For example, 8,000-foot (2,440-meter) Mt. Taranaki, a volcano on the southwest side of the island, includes fern-enshrouded lowland forests, sagebrush-like subalpine slopes speckled with wildflowers, and the snow-clad peak itself, which is the haunt of high-country climbers and adventurous skiers (Figure 14.13). While much of the North Island's natural vegetation has been replaced with human-introduced species, portions of the region's ancient and towering kauri forests remain in government preserves. Elsewhere, European forest and grassland species have profoundly altered the look of the North Island landscape.

South Island Environments   Across the narrow Cook Strait, New Zealand's South Island offers an equally complex natural setting. Although the encircling waters of the Pacific moder-

▲ Figure 14.13 Mt. Taranaki New Zealand's North Island contains several volcanic peaks, including Mt. Taranaki. The 8,000-foot (2,440-meter) peak offers everything from subtropical forests to challenging ski slopes, and attracts both local and international tourists. *(Ken Graham/Ken Graham Agency)*

▲ Figure 14.14 Central Otago, South Island On New Zealand's South Island, the Southern Alps capture rainfall on the west coast but leave areas to the east in a drier rain shadow. As a result, the Central Otago region has a semiarid landscape resembling portions of the U.S. West. *(John Lamb/Getty Images, Inc.)*

ate the climate, conditions become distinctly cooler as one moves poleward south of the Cook Strait. Indeed, South Island's southern edge feels the seasonal breath of Antarctic chill, as it lies more than 46° south of the equator (and more than 700 miles, or 1,120 kilometers, south of Auckland). It is no accident that the southern towns of Dunedin and Invercargill have a strong Scottish flavor, a function of both settlement history and climatic setting. To the north, milder conditions prevail on the Canterbury Plain in Christchurch, but nearby snow-covered mountains suggest the region's midlatitude character. Mountain ranges on New Zealand's South Island also display incredible local variations in precipitation: west-facing slopes are drenched with more than 100 inches (254 centimeters) of precipitation annually, while lowlands to the east (such as the Canterbury Plain) average only 25 inches (64 centimeters) per year. The damp, west-facing or windward side of the Southern Alps, for instance, supports a dense, sometime impenetrable rain forest of misty, fern-covered canyons and forested slopes. East of the divide, however, much drier and less extreme conditions prevail. The Otago region, inland from Dunedin, sits partially in the rain shadow of the Southern Alps, and its rolling, open landscapes resemble the semiarid expanses of North America's Intermountain West (Figure 14.14).

The Oceanic Realm

Leaving the region's large, midlatitude islands behind, the vast expanse of Pacific waters reveals another set of environmental settings that are as rich and complex as they are fragile and threatened with human alteration. Literally born from the sea, these thousands of mostly tiny, scattered landmasses are everywhere, shaped by their surrounding maritime envi-

ronments. Oceanic currents and wind patterns have historically served as natural highways of movement between these island worlds. Those same forces define the rhythm of weather patterns across the region in a broad band that extends more than 20° north (Hawaiian Islands) and south (New Caledonia) of the equator.

Creating Island Landforms Much of Melanesia and Polynesia is part of the seismically active Pacific Rim. As a result, volcanic eruptions, major earthquakes, and **tsunamis,** or seismically induced sea waves, are not uncommon across the region, and they impose major environmental hazards upon inhabitants. For example, volcanic eruptions and earthquakes on the island of New Britain (Papua New Guinea) forced more than 100,000 people from their homes in 1994. Only four years later, a massive tsunami triggered by an offshore earthquake swept across the north coast of New Guinea, killing 3,000 residents and destroying numerous villages. Such events are unfortunately a part of life in this geologically active part of the world.

Island geology varies across the realm. The large islands of Melanesia, as well as New Zealand, are composed of fragments of continental rock and are thus geologically quite complex. New Guinea, for example, is dominated by multiple, generally east-west trending mountain ranges separated by rugged, elevated plateaus that remain difficult to traverse. Extensive coastal lowlands flank these intricate highland regions, particularly in the southern portion of the island. Indeed, the nearby islands of the Bismark Archipelago are an eastward extension of this continental mountain-building zone.

Most of the islands of Polynesia and Micronesia, however, are truly oceanic, having originated from volcanic activity on

▶ **Figure 14.15 Bora Bora**
Jewel of French Polynesia, Bora
Bora displays many of the classic
features of Pacific high islands. As
the island's central volcanic core retreats, surrounding coral reefs produce a mix of wave-washed sandy
shores and shallow lagoons. *(Paul
Chesley/Getty Images, Inc.)*

the ocean floor without any geological connection to larger
landmasses. The larger active and recently active volcanoes
form **high islands,** which often rise to a considerable elevation and cover a substantial area. The island of Hawaii, the
largest and youngest of the Pacific's high islands, is more than
80 miles (128 kilometers) across and rises to a height of more
than 13,000 feet (3,980 meters). Indeed, the entire Hawaiian
archipelago exemplifies a geological **hot spot** where moving
oceanic crust passes over a supply of magma, thus creating a
chain of volcanic uplifts. Many of the islands of French Polynesia, including Bora Bora, are smaller examples of high islands (Figure 14.15). Indeed, high islands are widely scattered
throughout Micronesia and Polynesia. In tropical latitudes,
most high islands are ringed by coral reefs, which quickly establish themselves in the shallow waters near the shore.

High islands have a limited lifespan in geological terms.
Gradually, the processes of volcanism that created them diminish; the islands subside; and erosion steadily wears them
away. After a few hundred thousand years, only a few low
peaks may rise out of a shallow lagoon surrounded by coral
reefs. Indeed, nineteenth-century scientist Charles Darwin
selected the island of Bora Bora to exemplify the geological
evolution of these island settings (Figure 14.16). Eventually,
even remnant peaks erode away, leaving only a coral reef surrounding a shallow lagoon. Reefs tend to persist, however,
because they are composed of living organisms that create
new coral even as the (former) island base continues to subside. The top of the reef, therefore, tends to maintain its position at or just below sea level. **Low islands** are formed as
large waves periodically break off and pulverize large pieces
of coral, which are then deposited on adjacent sections of the
reef to form narrow, wave-washed, and sandy islands.

The combination of narrow sandy islands, barrier coral
reefs, and shallow central lagoons is also known as an **atoll.**

The islands and reefs of the atoll characteristically form a circular or oval shape, although some are quite irregular. The
world's largest atoll, Kwajalein in Micronesia's Marshall Islands,
is 75 miles (120 kilometers) long and 15 miles (24 kilometers)
wide. Polynesia and Micronesia are dotted with extensive atoll
systems, and a number are found in Melanesia as well. Some
extensive archipelagos, such as the Marshall Islands in Micronesia and the Tuamotus in Polynesia, are composed entirely of atolls.

Patterns of Climate   Many Pacific islands receive abundant precipitation, and high islands in particular are often
noted for their heavy rainfall and dense tropical forests (Figure 14.11). Much of the zone is located in the rainy tropics or in a tropical wet-dry climate region where abundant
summer rains and even tropical cyclones can bring heavy
seasonal precipitation. Apia, Samoa, for example, has a dry
fall and winter (April to August), but the town receives almost 18 inches (46 centimeters) of precipitation in the rainy
summer month of January. In Melanesia's Papua New
Guinea, Port Moresby is distinctly drier but still averages
almost 50 inches (127 centimeters) of rain per year, most
of it coming in the Southern Hemisphere summer (December to March). In the nearby highlands, however, conditions are often wetter, and local microclimates provide
varied precipitation patterns. Frequent snows even visit the
higher 13,000-foot (3,980-meter) peaks. Some oceanic locales, however, experience significantly less precipitation.
Low-lying atolls usually receive less precipitation than high
islands and very often experience water shortages. This is
partly because they have limited water-storage capacity,
where a small "lens" of fresh water often "floats" above the
salt water in the center of each sandy island. In dry periods,
however, such stores are quickly depleted.

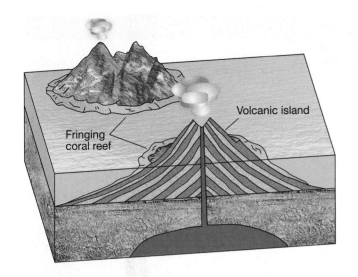

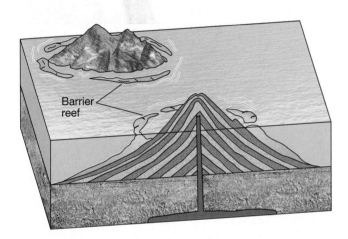

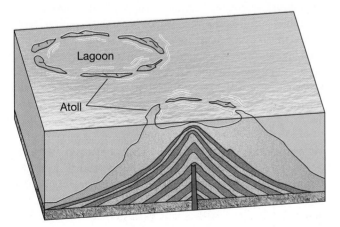

▲ **Figure 14.16 Evolution of an atoll** *Many Pacific islands begin as rugged volcanoes (top) with fringing coral reefs. However, as the extinct volcano subsides and erodes away, the coral reef expands, becoming a larger barrier reef (middle). The term "barrier reef" comes from the hazards these features post to navigation when approaching the island from sea. Finally, all that remains (bottom) is a coral atoll surrounding a shallow lagoon.*

Population and Settlement: A Diverse Cultural Landscape

Modern population patterns across the region reflect the combined influences of indigenous and European settlement. In settings such as Australia, New Zealand, and the Hawaiian Islands, European migrations have structured the distribution and concentration of contemporary populations, while island settings elsewhere in Oceania display population geographies essentially determined by the movement and adaptations of native peoples. The cultural landscape also reflects these diverse influences: more urban and commercial agricultural settings are largely transplanted from Europe and North America, but traditional societies still leave their mark upon the scene in many localities, ranging from the New Guinea Highlands to Polynesia's Society Islands.

Contemporary Population Patterns

Despite the popular stereotypes of life in the outback, modern Australia has one of the most highly urbanized populations in the world (Table 14.1). Indeed, 40 percent of the country's residents live within either the Sydney or Melbourne metropolitan areas (Figure 14.17). Australia's eastern and southern rimland is home to the overwhelming majority of its almost 20 million people. Most residents of Queensland, for example, live along the well-watered, amenity-rich coast, culminating in the state's capital city of Brisbane (1.5 million). Inland, population densities decline as rapidly as the rainfall: semiarid hills west of the Great Dividing Range still contain significant rural settlement, but the state's southwestern periphery remains sparsely peopled. New South Wales is the country's most populous state, and its sprawling capital city of Sydney (4 million), focused around one of the world's most magnificent natural harbors, is the largest metropolitan area in the entire South Pacific. In the nearby state of Victoria, Melbourne's 3.3 million residents have long vied with Sydney for urban status, claiming cultural and architectural supremacy over their slightly larger neighbor. In between these two metropolitan giants, the much smaller federal capital of Canberra (346,000) represents a classic geopolitical compromise in the same spirit that created Washington, D.C., midway between the populous southern and northern portions of the United States.

Smaller clusters of population are found beyond Australia's eastern rim. More favored farmlands in the interior of New South Wales and Victoria feature higher population densities than are found in most of the nation's arid heartland. Inland Aboriginal populations are widely but thinly scattered across districts such as northern Western Australia and South Australia, as well as in the Northern Territory. In addition to these rural settlements, urban clusters are focused around other Australian state capitals. Western Australia's growing metropolis of Perth (1.3 million) is now the largest of these peripheral cities, with Adelaide (South Australia, 1 million), Hobart (Tasmania, 195,000), and Darwin (Northern Territory, 85,000) accounting for smaller but regionally important centers of settlement.

TABLE 14.1 *Demographic Indicators*

Country	Population (Millions, 2001)	Population Density, per square mile	Rate of Natural Increase	TFR[a]	Percent < 15[b]	Percent > 65	Percent Urban
Australia	19.4	6	0.6	1.7	20	12	85
Fed. States of Micronesia	0.1	444	2.5	4.6	44	4	27
Fiji	0.8	119	1.9	3.3	33	4	46
French Polynesia	0.2	153	1.6	2.6	30	5	54
Guam	0.2	744	2.5	4.2	32	5	38
Kiribati	0.1	337	2.4	4.5	40	3	37
Marshall Islands	0.1	1,007	2.2	6.6	49	3	65
Nauru	0.01	1,412	1.4	3.7	43	1	100
New Caledonia	0.2	30	1.7	2.6	30	6	71
New Zealand	3.9	37	0.8	2.0	23	12	77
Palau	0.02	107	1.0	2.5	28	6	71
Papua New Guinea	5	28	2.3	4.8	39	4	15
Samoa	0.2	155	2.4	4.5	32	6	33
Solomon Islands	0.5	41	3.4	5.7	43	3	13
Tonga	0.1	349	2.1	4.2	41	4	32
Tuvalu	0.01	1,100	2.1	3.1	34	3	18
Vanuatu	0.2	44	3.0	4.6	42	3	21

[a]Total fertility rate

[b]Percent of population younger than 15 years of age

Source: Population Reference Bureau. World Population Data Sheet, 2001.

The population geography of the rest of Oceania reflects a broad sprinkling of peoples, both native and European, who have clustered near favorable resource opportunities. In New Zealand, more than 70 percent of the country's 3.9 million residents live on the North Island, with Auckland (1 million) dominating the metropolitan scene in the north and the capital city of Wellington (165,000) anchoring settlement along the Cook Strait in the south. Settlement on the South Island is strikingly clustered in the somewhat drier lowlands and coastal districts east of the mountains, with Christchurch (320,000) serving as the largest urban center. Elsewhere, rugged and mountainous terrain on both the North and South Islands feature much lower population densities. Such is not the case in Papua New Guinea: less than 15 percent of the country's population is urban, and many people live in the isolated, interior highlands. The nation's largest city is the capital of Port Moresby (200,000), located along the narrow coastal lowland in the far southeastern corner of the country. The largest city on the northern periphery of Oceania is Honolulu (1 million), on the island of Oahu, where rapid metropolitan growth since World War II has occurred because of U.S. statehood and the scenic attractions of its mid-Pacific setting.

Legacies of Human Occupancy

The historical settlement of the Pacific realm can never be precisely reconstructed, but humans in several major migrations succeeded in occupying the region over time. The region's remoteness from many of the world's early population centers meant that it often lay beyond the dominant migratory paths of earlier peoples. Even so, settlers found their way to the isolated Australian interior and the far reaches of the Pacific. The pace of new in-migrations accelerated greatly once Europeans identified the region and its resource potential.

Peopling the Pacific The large islands of New Guinea and Australia, given their proximity to the Asian landmass, were settled much earlier than the more distant islands of the Pacific, which could not be reached until the invention of sophisticated watercraft. Around 40,000 years ago, the ancestors of today's native Australian or **Aborigine** populations were making their way out of Southeast Asia and into Australia (Figure 14.18). The first Australians most likely arrived via some kind of watercraft. However, since such boats were probably not very seaworthy, the more distant islands remained inaccessible to humankind for tens of thousands of years. During glacial periods, however, sea levels were much lower than they are now, which would have allowed easier movement to Australia across relatively narrow spans of water. It is not known whether the original Australians came in one wave or in many, but the available evidence suggests that they soon occupied large portions of the continent, including Tasmania (which was then connected to the mainland by a land bridge).

Eastern Melanesia was settled much later than Australia and New Guinea. Distant islands could not be reached until better sailing craft were developed. By approximately 3,500 years ago,

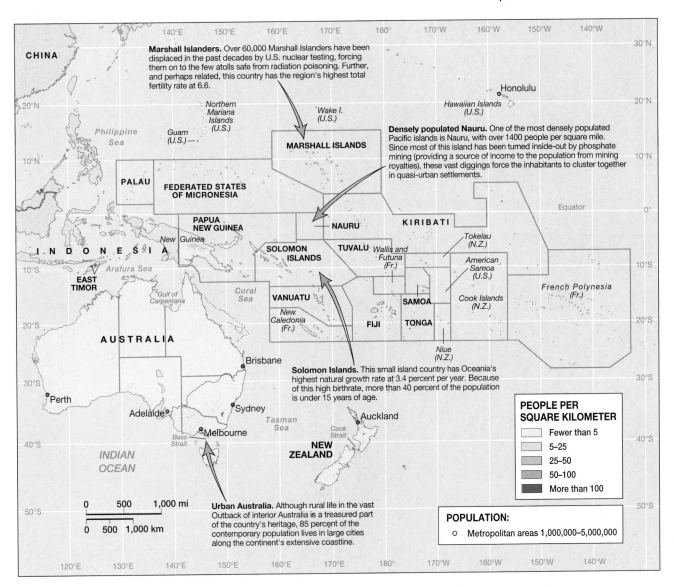

▲ **Figure 14.17 Population map of Australia and Oceania** Fewer than 30 million people occupy this world region. While Papua New Guinea and many Pacific islands feature mostly rural settlements, most regional residents live in the large urban areas of Australia and New Zealand. Sydney and Melbourne account for almost half of Australia's population, and most New Zealand residents live on the North Island, home to the cities of Auckland and Wellington.

however, certain Pacific peoples had mastered long-distance sailing and navigation, which eventually opened the entire oceanic realm to human habitation. In that era, people gradually moved east to occupy New Caledonia, the Fiji Islands, and Samoa. From there, later movements took seafaring folk north into Micronesia, with areas such as the Marshall Islands occupied around 2,000 years ago. Continuing movements from Asia further complicated the story of these migrating Melanesians. Some of the migrants mixed culturally and eventually reached western Polynesia, where they formed the nucleus of the Polynesian people. By A.D. 800, they had reached such distant oceanic outposts as New Zealand, Hawaii, and Easter Island. Debate has centered on whether these Polynesians migrants purposefully set out to colonize new lands or whether they were blown off course during routine voyages, only to end up on new islands. Certainly, population pressures could quickly

reach a crisis stage on relatively small islands, encouraging people to make spectacularly dangerous voyages. Equipped with sturdy outrigger sailing vessels and ample supplies of food, the Polynesians were quickly able to colonize most of the islands they discovered.

European Colonization About six centuries after the Maori brought New Zealand into the Polynesian realm, Dutch navigator Abel Tasman spotted the islands on his global reconnaissance of 1642. Tasman's initial sighting marked the beginning of a new chapter in the human occupation of the South Pacific. Late in the following century, more lasting European contacts cemented ties to the region. British sea captain James Cook surveyed the shorelines of both New Zealand and Australia between 1768 and 1780. Cook and others believed that these distant lands might be worthy of European

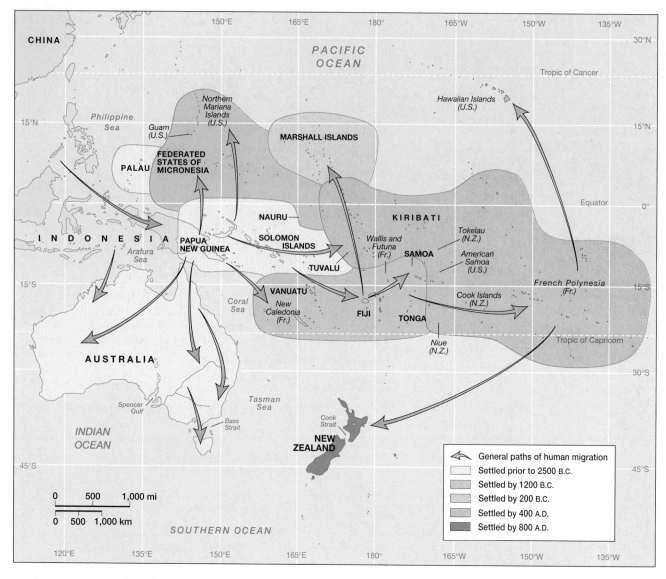

▲ **Figure 14.18 Peopling the Pacific**   Ancestors of Australia's aboriginal population may have made their way into the island continent more than 40,000 years ago. Much more recent settlement of Pacific islands by Austronesian peoples from Southeast Asia shaped cultural patterns across the oceanic portions of the realm. Eastward migrations through the Solomon, Fiji, and Cook islands were followed by late movements to the north and south.

development. In addition, other expeditions were probing the Pacific, and most of Oceania's major island groups were assuming a place on European maps by the end of the eighteenth century.

Actual European colonization of the region began in Australia. The British needed a remote penal colony to which convicts could be exiled. The southeastern coast of Australia was selected as an appropriate site, and in 1788 the First Fleet arrived with 750 prisoners in Botany Bay near what is now modern Sydney. Other fleets and more convicts soon followed. Before long, however, free settlers outnumbered the convicts, who were themselves gradually gaining their freedom. The growing population of English-speaking people soon moved inland and also settled other favorable coastal locales. British and Irish settlers were attracted by

the agricultural and stock-raising potential of the distant colony and by the lure of gold and other minerals (a major gold rush occurred in the 1850s). The British government also encouraged the emigration of its own citizens, often paying the transportation fare of those too poor to afford it themselves.

The new settlers came into conflict with the Aborigines almost immediately after arriving. No treaties were signed, however, and in most cases Aborigines were simply expelled from their lands. In some places, most notably Tasmania, they were hunted down and killed. In mainland Australia, the Aborigines were greatly reduced in numbers by disease, dispossession, and exploitation, but they survived in substantial numbers. By the mid-nineteenth century, Australia was primarily an English-speaking land.

British settlers were also attracted to the lush and fertile lands of New Zealand. European whalers and sealers arrived shortly before 1800, but more permanent agricultural settlement took shape after 1840 as the British formally declared sovereignty over the region. As new arrivals grew in numbers and the scope of planned settlement colonies on the North and South islands expanded, tensions increased with the native Maori population. Organized in small kingdoms or chiefdoms, the Maori were formidable fighters. In 1845 one indigenous group decided to resist further encroachments, leading to the more generalized Maori wars that engulfed New Zealand until 1870. The British eventually prevailed, however, and the Maori lost most of their land as well as control of their country.

The native Hawaiians also lost control of their lands to more numerous immigrants. Hawaii emerged as a united and powerful kingdom in the early 1800s, and for many years its native rulers limited U.S. and European claims to their islands. Increasing numbers of missionaries and settlers from the United States were allowed in, however, and by the late nineteenth century, control of the Hawaiian economy had largely passed to foreign plantation owners. By 1898, U.S. forces were strong enough to overthrow the Hawaiian monarchy and to annex the islands to the United States.

Elsewhere in Oceania, actual European or U.S. settlement was much less significant. While European and U.S. powers eventually gained political control over the entire area, few opportunities awaited prospective settlers. Whaling, timber harvesting, and opportunities for plantation agriculture drew Europeans to some islands, while mineral resources attracted selected colonization to others. Native populations often dropped as European diseases arrived, but in most cases recovery eventually occurred. Some islands, such as the Marquesas in Polynesia, however, have yet to rebound to the population levels they had prior to contact with the Europeans.

Modern Settlement Landscapes

The settlement geography of Australia and Oceania offers a fascinating juxtaposition of local and exotic influences. The contemporary cultural landscape still reflects the imprint of indigenous peoples in many settings where native populations have remained numerically dominant. Elsewhere, patterns of recent colonization have produced a modern scene profoundly shaped by processes of Europeanization. The result includes everything from German-owned vineyards in South Australia to houses on New Zealand's South Island that appear to be plucked directly from the British Isles. Further, processes of economic and cultural globalization have reshaped the settlement landscape, particularly in the twentieth century, resulting in urban forms that make cities such as Perth or Auckland look strikingly similar to such places as San Diego or Seattle.

The Urban Transformation Both Australia and New Zealand are highly urbanized, Westernized societies, and thus the vast majority of their populations live in urban and suburban environments. As in Europe and North America, much of this urban transformation came during the twentieth century as the rural economy became less labor-intensive and as opportunities for urban manufacturing and service employment grew. As urban landscapes evolved, they took on many of the characteristics of their largely European populations, but blended these with a strong dose of North American influences, as well as with the unique settings native to each urban place. The result is an urban landscape in which many North Americans, for example, are quite comfortable, even though the varied local accents heard on the street and many features of the metropolitan scene are reminders of the strong and enduring attachments to British traditions.

All of the major cities of Australia and New Zealand are focused around vibrant and dynamic downtown areas that often resemble those of metropolitan North America. Coastal features, port settings, and waterfront districts shape the central city landscapes of every major urban area in the region and give each center a unique identity. Perth, Melbourne, Sydney, Brisbane, Wellington, and Auckland are anchored on the sea, a function of their important commercial connections to the global economy. Thus, central city skylines, complete with growing numbers of North American-style high-rises, are never far from their harbor settings (Figure 14.19). In addition, these urban places are often lauded for their modest crime rates, lack of slums, relatively clean streets, spacious parks and open spaces, and efficient public transportation. Sydney and Melbourne, in particular, are receiving millennial face-lifts. Many new hotels, sports facilities, and entertainment complexes were built in Sydney in association with the Summer 2000 Olympics, and Melbourne's ambitious Federation Square Redevelopment Project is dramatically transforming 7.9 acres (3.2 hectares) of aging downtown buildings

▲ **Figure 14.19 Downtown Melbourne** Metropolitan Melbourne lies along the Yarra River. Capital of the Australian state of Victoria, Melbourne resembles many growing North American cities with its high-rise office buildings, entertainment districts, and downtown urban redevelopment. *(Fritz Prenzel/Peter Arnold, Inc.)*

▲ **Figure 14.20 Sydney's suburbs** Like North American cities, urban settlements in both New Zealand and Australian now sprawl far beyond their traditional centers as inhabitants seek larger lots, open space, and other amenities of the country-side. Urban planners note how this suburban sprawl complements the outdoor orientation of Australians and New Zealanders. *(Patrick Ward/Corbis)*

▲ **Figure 14.21 Port Moresby, Papua New Guinea** Urban poverty and high crime haunt the city of Port Moresby, the capital of Papua New Guinea. The city's slums, many built out on the water, reflect stresses of recent urban growth as rural residents emigrate from nearby highlands. *(Chris Rainier/Corbis)*

and railyards into a new entertainment, museum, and business district.

Suburbanization and peripheral commercial development are also changing these large cities of the Southern Hemisphere (Figure 14.20). For example, Sydney's metropolitan reach now extends west for more than 20 miles from the downtown area, and Melbourne's urban grasp is expanding inland toward the hills as well as farther east along Port Phillip Bay. Amenity-rich Perth (Western Australia) and Brisbane (Queensland) have seen some of the most rapid recent population gains, particularly in their expanding suburban fringe. Similarly, New Zealand's Auckland, given that city's water-bound limits between Manukau and Waitemata harbors, is spreading well beyond the old boundaries of the traditional city. Land uses and settlement landscapes in these dynamic urban peripheries parallel patterns across North America: low-density residential areas are linked by modern commercial highways and punctuated by "edge city" collections of office centers, industrial parks, and entertainment complexes.

The affluent Western-style urban settings in Australia and New Zealand offer a stunning contrast with the urban landscapes found in less-developed settings in the region. Walk the streets of Port Moresby in Papua New Guinea, and a very different urban landscape attests to the yawning gap between rich and poor within Oceania (Figure 14.21). Rapid growth in Port Moresby, the country's political capital and largest commercial center, has produced many of the classic challenges of urban underdevelopment: there is a shortage of adequate housing; the creation of basic infrastructure lags far behind the need; and street crime and alcoholism are on the rise. Elsewhere, urban centers such as Suva (Fiji), Noumea (New Caledonia), and Apia (Samoa) also reflect the econom-

ic and cultural tensions generated as indigenous populations are exposed to Western influences. Rapid growth is a common problem in these smaller urban Pacific settings because many native peoples from neighboring rural areas and nearby islands gravitate toward the job opportunities available in these areas. In the past 50 years, the spectacular global growth of tourism in places such as Fiji and Samoa has also transformed the urban scene: nineteenth-century village life has often been replaced by a landscape dominated by souvenir shops, honking taxicabs, and crowded seaside resorts.

The Rural Scene Rural landscapes across Australia and the Pacific region also express a complex mosaic of cultural and economic influences. In some settings, Australian Aborigines or native Papua New Guinea Highlanders can still be found in their familiar homelands, their traditional lifeways and settlements barely changed from pre-European times. Yet such settlement landscapes are becoming increasingly rare. Global influences penetrate the scene as the cash economy, foreign tourism and investment, and the accoutrements of popular culture work their way into the hinterlands. That still leaves relatively large areas of rural settlement across many areas of the Australian and New Zealand interiors, as well as smaller examples sprinkled among the islands of the Pacific. Even here, however, land uses are often shaped by the regional or global commercial agricultural economy or by the recreational values associated with such settings.

Much of rural Australia is too dry for farming or serves as only marginally valuable agricultural land. Although modest in size, the area in crops has doubled since 1960 as increased use of fertilizers, more widespread irrigation, and more aggressive rabbit eradication efforts have opened up new areas for development. Much of the remainder of the interior, however, features range-fed livestock, areas beyond the pale of

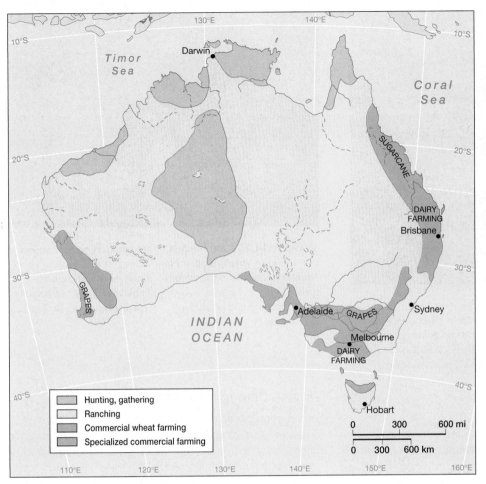

◀ **Figure 14.22 Australian agriculture** Australia's best farmlands are found along its eastern and southern rim, where specialty crops such as sugarcane and grapes thrive in selected locations. Commercial wheat farming remains important in slightly drier districts. In the interior are large tracts where Aboriginals still practice foraging and hunting. *(Modified from Clawson and Fisher, 1998,* World Regional Geography, *Upper Saddle River, NJ: Prentice Hall)*

any agricultural potential, and isolated areas where Aboriginal peoples still pursue their traditional forms of hunting and gathering (Figure 14.22).

Sheep and cattle dominate the livestock economy. Many rural landscapes in the interior of New South Wales, Western Australia, and Victoria, for example, are oriented around isolated sheep stations, or ranch operations that move the flocks from one vast pasture to the next. Cattle can sometimes be found in these same areas, although many of the more extensive, range-fed cattle operations are concentrated farther north in Queensland. Cattle are even grazed on the tough tropical grasslands and woodlands of northern Australia, although environmental conditions in these settings are often marginal for any agricultural activity. More specialized dairy cattle operations, often on much smaller family-owned farms, are found in moister settings from the Queensland coast to Victoria.

Croplands also vary across the region. Sometimes mingling with the sheep country, a band of commercial wheat farming includes southern Queensland; the moister interiors of New South Wales, Victoria, and South Australia; and a swath of more-favored land east and north of Perth. Elsewhere, specialized sugarcane operations thrive along the narrow, warm, and humid coastal strip of Queensland. To the south and west, productive irrigated agriculture has devel-

oped in localities such as the Murray River Basin, allowing for the production of orchard crops and vegetables. **Viticulture,** or grape cultivation (and other specialized horticulture), increasingly shapes the rural scene in places such as South Australia's Barossa Valley, New South Wales' Riverina district, and Western Australia's Swan Valley. Indeed, the area under grape cultivation grew by 50 percent between 1991 and 1998 as the popular Chardonnay, Cabernet Sauvignon, and Shiraz varieties propelled wine production to revenues of more than $540 million per year.

Although much smaller in area, New Zealand's rural settlement landscape includes a variety of agricultural activities. Pastoral pursuits clearly dominate the New Zealand scene, with the vast majority of agricultural land devoted to livestock production, particularly sheep grazing and dairying. Commercial livestock outnumber people in New Zealand by a ratio of more than 20 to 1, and this is apparent everywhere on the rural scene. Sheep gnaw on largely imported European grasses from the isolated hinterlands of the South Island to the volcanic slopes and shorelines of the North Island. In addition, dairy operations dot the rural scene, mostly in the humid lowlands of the subtropical north, where they sometimes mingle with suburban landscapes in the vicinity of Auckland. One of the largest zones of more specialized cropping spreads across the fertile Canterbury Plain near Christchurch (Figure 14.23).

▲ **Figure 14.23 Canterbury Plain** The varied agricultural landscape of South Island's Canterbury Plain offers a mix of grain fields, livestock, orchard crops, and vegetable gardens. The rugged Southern Alps are a dramatic backdrop to this productive region. *(Robert Frerck/Woodfin Camp & Associates)*

▲ **Figure 14.24 Yam harvest** These farmers on the Melanesian island of Vakuta (Papua New Guinea) are harvesting yams. Traditional tropical agriculture features a mix of crops, often grown in the same field. Where possible, fields are periodically rotated to maintain productivity. *(Peter Essick/Aurora & Quanta Productions)*

This spectacular South Island setting, spotted early by British colonizer Edward Gibbon Wakefield, proved fertile ground for English settlement and continues to feature a varied landscape of pastures, grain fields, orchards, and vegetable gardens, all spread beneath the towering peaks of the Southern Alps.

Elsewhere in Oceania, varied influences shape the rural scene. On better-watered high islands, denser populations take advantage of more diverse agricultural opportunities than are usually found on the more-barren low islands, where fishing is often more important. Several types of rural settlement can be identified across the island realm. In rural New Guinea, village-centered shifting cultivation dominates: farmers clear a patch of forest and then, after a few years, shift to another patch, thus practicing a form of land rotation. Subsistence foods such as sweet potatoes, taro (another starchy root crop), coconut palms, bananas, and other garden crops often are intercropped in the same field, and growing numbers of planters also include commercial crops such as coffee. In other parts of Oceania, traditional agricultural patterns are similar: most rural settlements are organized around village-based societies surrounded by nearby fields in which varied crops are produced (Figure 14.24). Commercial plantation agriculture has also made its mark in many more accessible rural settings. In these localities, settlements consist of worker housing near crops that are typically controlled by absentee landowners. For example, copra (coconut), cocoa, and coffee operations have transformed many agricultural settings in places such as the Solomon Islands and Vanuatu. Sugarcane plantations have reshaped other island settings, particularly in Fiji and Hawaii.

Diverse Demographic Paths

Varied population-related issues face residents of the region today. In Australia and New Zealand, while populations grew rapidly (mostly from natural increases) in the twentieth century, today's low birthrates parallel the pattern in North America. Just as in the United States and Canada, however, significant population shifts within these countries continue to impose challenges. For example, the exodus of farmers from Australia's wheat-growing and sheep-raising interior mirrors similar processes at work in the rural Midwest of the United States and in the Canadian prairies. Communities see many of their productive young people and professionals leave for the better employment opportunities of the city. Older urban and industrial areas, particularly near Sydney and Melbourne, have also lost population. Conversely, other portions of Australia and New Zealand face rapid growth. As in North America, amenity-rich settings, retirement communities, and new upscale suburbs draw ever more people, thus pressuring infrastructure and public services. Some people in these two countries also complain about liberal national immigration policies allowing in too many new residents and workers, although such frustrations are often more rooted in cultural fears than they are in concerns about overall population numbers.

Different demographic challenges grip many of the less-developed island nations of Oceania. Population growth rates are often above 2 percent per year and are even higher in countries such as Vanuatu and the Solomon Islands. While the larger islands of Melanesia contain some room for settlement expansion, competitive pressures from commercial mining and logging operations limit the amount of new agricultural land that will probably be available in the future. On some of the smaller island groups in Micronesia and Polynesia, population growth is even more pressing. Tuvalu (north of Fiji), for example, has just over 10,000 inhabitants, but they are crowded onto a land area of about 10 square miles (26 square kilometers), making it one of the world's more thickly populated countries. In atoll environments, a single island might be considered overpopulated if it supports only 100 people. Thus, even relatively small population centers in the Pacific Islands have little internal flexibility in coping with higher birthrates. Making matters worse, many young people are migrating to already crowded urban centers in these island nations. Reflecting the region's bizarre entanglements with nuclear testing, Marshall Islanders face the additional daunting task of crowding their country's 100,000 people onto those atolls that were lucky enough to escape extensive U.S. bomb blasts. Today, more than half the nation's residents reside on Majuro Atoll, a narrow strip of sand crammed with tin shacks and heaps of nondisposable trash.

Cultural Coherence and Diversity: A Global Crossroads

Australia and Oceania offer superb laboratories to examine how cultural geographies are transformed as different groups migrate to a region, interact with one another, and evolve over time. Many general processes of cultural change are exemplified. For example, some portions of the region saw multiple waves of implantation of cultural groups from other islands. Once these groups were in place, their relative isolation often facilitated cultural differentiation in which originally similar cultures evolved in dissimilar ways. In some cases, when different cultures later came in fresh contact with one another, assimilation took place, and one culture absorbed important components of the other. As Europeans and other outsiders arrived in the region, colonization compelled cultural accommodation as native peoples adjusted to externally imposed forces of change. Finally, worldwide processes of globalization have also redefined the region's cultural geography, provoking fears of homogenization while at the same time promoting more coordinated cultural preservation efforts as threatened groups attempt to protect their cultural heritage. A contemporary snapshot of the region reveals that all of these processes remain at work today and help account for the diverse cultural signatures found in Australia, New Zealand, and the other Pacific islands.

Multicultural Australia

Australia's cultural patterns illustrate many of these fundamental processes at work. Today, while still dominated by its colonial European roots, the country's multicultural character is becoming increasingly visible as native inhabitants assert their cultural identity and as varied immigrant populations play larger roles in society, particularly within major metropolitan areas.

Aboriginal Imprints   For thousands of years, Australia's Aborigines dominated the cultural geography of the continent. These indigenous peoples never practiced agriculture,

opting instead for a hunting-gathering way of life that persisted up to the time of the European conquest. As the population consisted of foragers and hunters living in a relatively dry land, settlement densities remained low; tribal groups were often isolated from one another; and overall populations probably never numbered more than 300,000 inhabitants. To survive, Aborigines developed great adaptive skills, often subsisting in harsh environments that Europeans avoided. Their low densities meant linguistic fragmentation. Although precise counts vary, there were probably 250 languages spoken at the time of European contact, and almost 50 indigenous languages can still be found.

Radical cultural and geographical changes accompanied the arrival of Europeans, and Aboriginal populations were decimated in the process. The geographical results of colonization were striking as Aboriginal settlements were relegated to the sparsely settled interior, particularly in northern and central Australia, where fewer Europeans competed for land. In most cases, the European attitude toward the Aboriginal population was even more prejudicial than it was toward native peoples of the Americas. As hunter-gatherers lacking centralized political organization, Aborigines were usually considered less than human and more often than not simply as another strange animal species. No treaties were signed and no concessions were made. Typically, Aborigines were simply run off any lands desired by Europeans. Although decimated by these disruptions, as well as by European diseases, the Aborigines survived, particularly in the isolated outback, where they were far from entanglements with encroaching Europeans.

Today Aboriginal cultures persevere in Australia, and a growing native peoples movement parallels similar activities in the Americas. Indigenous peoples comprise approximately 2 percent (or 400,000) of Australia's population but their geographical distribution changed dramatically in the last century. Aborigines account for almost 30 percent of the Northern Territory's population (many of these in Arnhemland near Darwin), and other sizable native reserves are located in northern Queensland and Western Australia. Most native peoples, however, live in the same large urban areas that dominate the country's overall population geography. Indeed, more than 70 percent of Aborigines live in cities, and very few of them still practice traditional hunting-and-gathering lifestyles. Processes of cultural assimilation are clearly at work: urban Aborigines are frequently employed in service occupations; Christianity has often replaced traditional animist religions; and only 13 percent of the native population still speaks an indigenous language (Figure 14.25). Recently, native-language bilingual education programs have lost funding in the Northern Territory, mirroring the rejection of similar efforts in multicultural California.

Still, forces of diversity are at work, suggesting a growing Aboriginal interest in preserving traditional cultural values. Particularly in the outback, a handful of Aboriginal languages retain their vitality and have growing numbers of speakers. In addition, cultural leaders are selectively preserving Aboriginal spiritualism, and these religious practices often link local

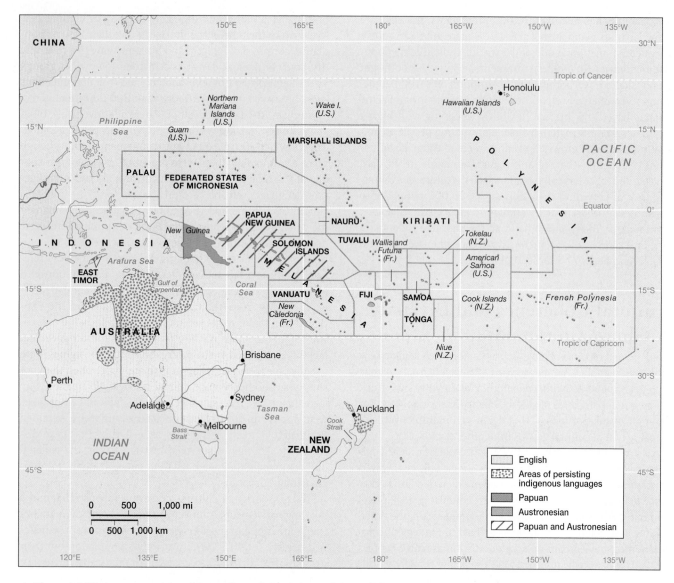

▲ **Figure 14.25 Language map of Australia and Oceania**   While English is spoken by most residents, native peoples and their linguistic traditions remain an important cultural and political force in both Australia and New Zealand. Elsewhere, traditional Papuan and Austronesian languages dominate Oceania. The French colonial legacy also persists in select Pacific locations. Tremendous linguistic diversity has shaped the cultural geography of Melanesia, and more than 1,000 languages have been identified in Papua New Guinea.

populations to surrounding places and natural features that are considered sacred. In fact, a growing number of these sacred locations are at the center of land-use and resource controversies between Aboriginal populations and Australia's European majority. The future for Aboriginal cultures remains unclear: pressures for cultural assimilation will be intense as many native peoples relocate to more Western-oriented urban settlements and lifestyles. At the same time, more rapid rates of natural increase (almost twice the national average) and a growing cultural awareness of Aboriginal traditions will work to preserve elements of the country's indigenous cultures.

A Land of Immigrants   Most Australians reflect the continent's more recent European-dominated migration history, but even these patterns have become more complex as a ris-

ing tide of Asian cultures becomes important. Overall, more than 70 percent of Australia's population continues to reflect a British or Irish cultural heritage. These groups dominated many of the nineteenth- and early twentieth-century migrations into the country, and the close cultural ties to the British Isles remain fundamental. Sizable numbers of Italians, Greeks, and Germans have added diversity to the European mix. Some of these migrants became involved in specialized agricultural activities, particularly in southeastern Australia, while others moved to the cities, where they added ethnic variety to the mostly Anglo mix. A need for laborers along the fertile Queensland coast also prompted European plantation owners to import inexpensive workers from the Solomons and New Hebrides. These Pacific Island laborers, known as **kanakas,** were spatially and socially segregated from their

Anglo employers, but further diversified the cultural mix of Queensland's "sugar coast." Historically, however, nonwhite migrations to the country were strictly limited by what is often termed a **White Australia Policy,** in which governmental guidelines promoted European and North American immigration at the expense of other groups. This remained national policy until 1973.

Recent migration trends have reversed this historical bias, and more diverse inflows of new workers and residents are adding to the country's multicultural character. Since the 1970s, the Migration Program has been dominated by a varied people chosen on the basis of their educational background and potential for succeeding economically in Australian society. For example, a growing number of families have come from places such as China, India, Malaysia, and the Philippines. Smaller numbers have qualified as migrants through their New Zealand citizenship, while others have arrived as refugees from troubled parts of the world, such as Vietnam and Yugoslavia. The result is a much more varied foreign-born population. Indeed, 22 percent of Australia's people are now immigrants, reflecting the country's global popularity as a migration destination. In the early twenty-first century, almost 40 percent of the settlers arriving in the country have been from Asia. Major cities offer particularly attractive possibilities: Sydney's Asian population already exceeds 10 percent and is growing rapidly, while Perth's urban landscape increasingly displays cultural and economic links to its Asian neighbors.

Australian society, while enduringly Anglo, has been changed forever by its varied immigrant mix. For example, although the country is still largely Christian (27 percent Catholic and 22 percent Anglican), some of the nation's fastest growing religions are Islam, Buddhism, and Hinduism. In similar fashion, while English is Australia's national language, more than 2.5 million residents now speak another language, including more than 100,000 speakers each of Italian, Greek, Cantonese, Arabic, and Vietnamese. Economic ties with Japan are also reshaping the country's cultural geography: more Japanese travel to Australia for business and pleasure, and a growing number of Australians are learning Japanese. A vocal minority of native-born whites resist the current cultural mosaic and advocate much more restrictive immigration policies on both racial and economic grounds. To illustrate, the recent resurgence of Pauline Hanson's One Nation Party demonstrates the tensions between whites and the Aboriginal and immigrant populations some see as a threat to traditional Anglo cultural values and economic opportunities. Though failing to dominate the national political agenda, this anti-immigrant political activity is a reminder of the tense ethnic politics underlying Australian society.

Australian society has also been shaped by global popular culture, and the relationship has often been a two-way street. The country's largely English-speaking population has embraced North American movies, music, and television. Many American television programs appear down under, although the Australian media are mandated to show at least 50 percent local programming from 6 A.M. to midnight in order to "keep the Australia in Australian television." Australian influences have also shaped global media. Sydney's Summer Olympic Games in 2000 provided Australians with an ideal opportunity to share their culture and country with the world. The popular film *Crocodile Dundee* provided a humorous mix of fact and fancy in its portrayal of an Australian maverick in the outback. More significantly, Australian ownership (News Corporation) of many British newspapers as well as the Fox television network in the United States is a reminder that global cultural influences move both in and out of the region. In addition to the cultural influences of the mass media, Australians also give leisure time a high priority. Their recreational habits parallel North American passions for outdoor activities and for competitive and spectator sports, in this case, cricket, rugby, and horse racing.

Patterns in New Zealand

New Zealand's cultural geography broadly reflects the patterns seen in Australia, although the precise cultural mix differs slightly. Native Maori populations are more numerically important and culturally visible in New Zealand than their Aboriginal counterparts in Australia. While British colonization clearly mandated the dominance of Anglo cultural traditions by the late nineteenth century, Maori populations survived, although they lost most of their land in the process. After declining with initial European contacts and conflicts, native populations began rebounding in the twentieth century, and today the Maori account for more than 15 percent of the country's 3.9 million residents. Geographically, the Maori remain most numerous on the North Island, including a sizable concentration in metropolitan Auckland. While urban living is on the rise, many Maori, akin to their Aboriginal counterparts, are also committed to preserving their religion, traditional arts, and Polynesian lifeways (Figure 14.26; see "Local Voices: Land and Life in Maori Religion"). In addition, Maori is now an official language in the country, along with English.

▲ **Figure 14.26 Maori artisans**   New Zealand's native Maori population actively preserves its cultural traditions and has recently increased its political role in national affairs. These artisans are carving decorations for a traditional Maori canoe. *(Arno Gasteiger/Bilderberg/Aurora & Quanta Productions)*

LOCAL VOICES   Land and Life in the Maori Religion

The traditional Maori religion is closely bound to ideas about the natural world, and the Maori's environmental philosophy has attracted growing interest not only among Maori keen on preserving their culture, but also among Westerners sympathetic to its ideals. The Maori envision each of the components of the natural world, including people, as possessing a harmonious and interrelated life force or essence, known as *mauri*. People must respect the integrity and vitality of this force, including its manifestation in mountains, rivers, trees, animals, and humans themselves. While understanding that one cannot live in the world without transforming it, the Maori believe that the natural world must be damaged as little as possible and that interfering with rivers or other animals must be done respectfully and for justified reasons. As Maori writer Rangimarie Rose Pere suggests, the *mauri* of each creature interacts with Earth, and if people respect Earth, all of its inhabitants will prosper.

This Maori perspective on the environment has many consequences. Recently, a local Maori leader protested the discharge of sewage into a river, claiming that it upset the harmony, or *mauri*, of the stream. Indeed, such an attitude has often defined the rift separating Maori environmentalists from more development-minded New Zealanders. On the other hand, the Maori find it hard to conceive of the Western idea of wilderness, since it involves a fundamental split between people and nature. How can one segregate a wilderness space in the midst of one's forest home? Even Maori artwork is shaped by their environmental ethics. When artists such as wood carver Rangi Hetet work with their materials, they typically choose raw slabs rather than milled boards, suggesting that the natural wood better retains its original *mauri*, a quality that survives within the properly crafted carving. Europeans, conversely, are guilty of destroying the forest's *mauri* when they carelessly fell the timber and send it to the sawmill.

Source: Adapted from John Patterson, "Respecting Nature: The Maori Way," *The Ecologist* 29 (January–February 1999).

While many New Zealanders still identify with their largely British heritage, the country's twentieth century's cultural identity matured with an increasing sense of separateness from its British roots. Several processes have forged New Zealand's special cultural character. As Britain tightened its own links with the European continent after World War II, New Zealanders increasingly forged a more independent and eclectic identity. In many ways, popular culture ties the country ever more closely to Australia, the United States, and continental Europe, a function of increasingly global mass media. For example, the most famous film ever made in New Zealand, *The Piano,* was produced by an Australian, financed with French capital, and reached its largest audiences in the United States. Indeed, by the late 1990s, locally produced programming made up less than 20 percent of New Zealand television.

Diversity also continues to shape the cultural setting. The nation's unique Polynesian roots impart a special regional character: in addition to its Maori population, more than 5 percent of its people are Pacific Islanders, and Auckland has the largest Polynesian population of any city in the world. Adding further complexity to the cultural mix are growing numbers of Asians who now make up another 5 percent of New Zealand's residents and who are making their own dynamic contribution to the country's culture and economy, particularly in its larger urban centers. The end result is a national character truly forged at a Pacific crossroads, an accumulation of varied and vastly different cultural influences that have been amalgamated into a uniquely New Zealand identity that resists easy definition.

The Mosaic of Pacific Cultures

Native and exotic cultural influences produce a varied mosaic across the islands of the South Pacific. In more isolated locales, traditional cultures maintain their integrity largely insulated from outside influences. In most cases, however, modern life in the islands revolves around an intricate cultural and economic interplay of local and Western influences. One thing is certain: the relative cultural insularity of the past is gone forever, and in its place is a Pacific realm rapidly adjusting to powerful forces of colonization, global capitalism, and popular culture.

Traditional Culture Worlds   Defining the pre-European cultural setting is no simple task. Anthropologists were once confident that the division of Oceania into Melanesia, Micronesia, and Polynesian reflected clear cultural and racial distinctions. In the racist thinking of the early twentieth century, Polynesians were usually considered to be superior to the other peoples of the Pacific, and Melanesians inferior. According to the once-standard view, Polynesians were brown-skinned peoples (perhaps distantly related to Europeans) who possessed advanced political structures (chiefdoms and kingdoms) and complex systems of social stratification; Melanesians, on the other hand, were black-skinned peoples (probably related to Africans) living in simple, rather egalitarian village communities. Melanesians, moreover, were often regarded as savage cannibals, while Polynesians were sometimes considered to be noble exemplars of "natural" existence. Micronesians, by the same line of thinking, were usually placed in an intermediate position. European sailors in the 1800s were often known to jump ship in order to marry Polynesian women and remain in these idyllic islands. Indeed, many European and American artists and writers viewed Polynesia as a kind of natural and social paradise that had remained uncorrupted by the repressive social system of European society.

Today such notions of cultural superiority and inferiority have been abandoned by serious scholars, and the division between Polynesian and Melanesian cultures is no longer as clear as it once was. Certainly most Polynesian societies were more politically centralized and class-based than those of Melanesia, but this was not always the case. Small-scale Melanesian societies, moreover, were sometimes linked to one another through very complex systems of trade and social exchange. Equally significant, upon close geographical inspection, the boundary of the two regions turns out to be a zone of pronounced cultural melding. The Melanesians of Fiji and the Polynesians of Tonga, for example, not only traded with each other, but often intermarried, and thus came to share a number of political and cultural institutions. Even in the heart of Melanesia, there are a number of atolls that are actually Polynesian in culture and language.

The modern language map reveals some significant cultural patterns that both unite and divide the region (Figure 14.25). Most of the indigenous languages of Oceania belong to the Austronesian language family, which encompasses wide expanses of the Pacific, much of insular Southeast Asia, and Madagascar. Linguists hypothesize that the first great oceanic mariners spoke Austronesian languages, and thus disseminated them throughout this vast realm of islands and oceans. Within the broad Austronesian family, the Malayo-Polynesian subfamily includes most of the related languages of Micronesia and Polynesia, suggesting a common cultural and migratory history for these far-flung peoples.

Melanesia's language geography is more complex and still incompletely understood by outside experts: while coastal peoples often speak languages brought to the region by the seafaring Austronesians, more isolated highlander cultures, particularly on the island of New Guinea, speak varied Papuan languages. Indeed, the linguistic complexity of that island is so daunting—more than 1,000 languages have been identified—that many experts question whether they even comprise a unified "Papuan family" of related languages. Some scholars estimate that half of New Guinea's languages are spoken by fewer than 500 persons, suggesting the cultural role played by the region's rugged topography in isolating cultural groups. These New Guinea highlands may hold some of the world's few remaining **uncontacted peoples,** cultural groups that have yet to be "discovered" by the Western world.

Traditional patterns of social life are as complex and varied as the language map. In many cases, however, life revolves around predictable settings. For example, across much of Melanesia, including Papua New Guinea, most people live in small villages often occupied by a single clan or family group. Many of these traditional villages contain fewer than 500 residents, although some larger communities may house more than 1,000 people. Life often revolves around the gathering and growing of food, an annual round of rituals and festivals,

▲ **Figure 14.27 Tonga village**   Although the economic and technological effects of globalization have arrived in Tonga, village life remains important in many Polynesian settings. Most village housing in these tropical environments reflects the use of locally available construction materials.   *(Ted Streshinsky/Corbis)*

and complex networks of kin-based social interactions, often including the potential for conflict with other nearby groups.

Traditional Polynesian settings also feature village life (Figure 14.27), although there are often strong class-based relationships between local elites (often religious leaders) and ordinary residents. Polynesian villages are also more likely linked to other islands by wider cultural and political ties. Despite the Western penchant for depicting Polynesian communities in idyllic terms, violent warfare was actually quite common across much of the region prior to European contact. Another cultural myth is the notion that Polynesians led a natural sex life uncomplicated by Western constraints. The truth of the matter seems to be that lower-class women were often ordered by Polynesian political elites to sleep with Western men so that the elites might obtain steel, knives, guns, and other goods unavailable in their own economy. Not surprisingly, such actions tended to be seriously misinterpreted by European men.

External Cultural Influences   While traditional culture worlds persist in some settings, most Pacific islands have witnessed tremendous cultural transformations in the past 150 years. Outsiders from Europe, the United States, and Asia brought new settlers, values, and technological innovations that have forever changed Oceania's cultural geography and its place in the larger world. The result is a modern setting where Pidgin English has broadly supplanted native languages, Hinduism is practiced on remote Pacific Islands, and traditional fishing peoples now work at resort hotels and golf course complexes.

European colonialism transformed the cultural geography of the Pacific world by broadly, if loosely, incorporating it into new political and economic systems and by bringing in new peoples to the region who directly reconfigured its cultural makeup. Hawaii illustrates the pattern. By the mid-nineteenth century, Hawaii's King Kamehameha was already entertaining

▲ Figure 14.28 Multicultural Hawaiians   Many residents of the Hawaiian Islands represent a blend of Pacific Island, Asian, and European influences. These young women express a mixture of Polynesian and Asian ancestors. *(Porterfield/Chickering/Photo Researchers, Inc.)*

▲ Figure 14.29 South Asians in Fiji   British sugar plantation owners imported thousands of South Asian workers to Fiji during the colonial era. Today almost half of Fiji's population is South Asian, including many urban residents. This group is often in conflict with indigenous Fijians.   *(Frank Fournier/Woodfin Camp & Associates)*

a varied assortment of whalers, Christian missionaries, traders, and navy officers from Europe and the United States. A small elite group of **haoles,** or light-skinned European and American foreigners, were successfully profiting from commercial sugarcane plantations and Pacific shipping contracts. Labor shortages on the islands, however, prompted the importation of Chinese, Portuguese, and Japanese workers who further complicated the region's cultural geography. By 1900, the Japanese had become a dominant part of the island workforce. The United States formally annexed the islands in 1898. The cultural mosaic revealed in the Hawaiian census of 1910 suggests the magnitude of change: more than 55 percent of the population was Asian (mostly Japanese and Chinese), native peoples made up another 20 percent, and about 15 percent (mostly imported European workers) were white. By the end of the twentieth century, the Asian population was less dominant but more ethnically varied, about 40 percent of Hawaii's residents were white, and the small number of remaining native Hawaiians had been joined by an increasingly diverse group of other Pacific Islanders. In addition, ethnic mixing has produced a rich mosaic of Hawaiian creole cultures that offer a unique blend of North American, Asian, Pacific Island, and European influences (Figure 14.28).

Hawaii's story has been played out in many other Pacific island settings. In the Mariana Islands, Guam was absorbed into America's Pacific empire as part of the Spanish-American War in 1898. Thereafter, not only did native peoples feel the effects of Americanization (the island remains a self-governing U.S. territory today), but thousands of Filipinos were moved there to supplement its modest labor force. To the southeast, the British-controlled Fiji Islands offered similar opportunities for fundamentally redefining Oceania's cultural mix. The same sugar plantation economy that spurred changes in Hawaii prompted the British to import thousands of South Asian laborers to Fiji. The descendents of these Indians (most

practice Hinduism) now comprise almost half the island country's population and often come into sharp conflict with the native Fijians (Figure 14.29). In French-controlled portions of the realm, small groups of traders and plantation owners filtered into the Society Islands (Tahiti), but a larger contingent of French colonial settlers (many originally a part of a penal colony) had a major impact on the cultural makeup of New Caledonia. Still a French colony, New Caledonia's population is more than one-third French, and its capital city of Noumea reveals a cultural setting curiously forged from French society, Melanesian traditions, and balmy South Pacific breezes.

Given the frequency of contact between different island cultures, it is no surprise that people have generated shared forms of intercultural communication. For example, several forms of **Pidgin English** (also known simply as "Pijin") are found in the Solomons, Vanuatu, and New Guinea, where it is the major language used between ethnic groups. In Pijin, a largely English vocabulary is reworked and blended with Melanesian grammar. Conventional wisdom traces its origin to nineteenth-century Chinese sandalwood traders ("pijin" is the Chinese pronunciation of the word for "business"). While of historical origin, Pijin is now becoming a "globalized" language of sorts in Oceania as trade and even political ties develop between different indigenous island groups.

In many Pacific settings, Christianity has also been absorbed into the indigenous cultural mix. Catholicism is more prevalent in areas of French colonization, while Protestant groups are heavily concentrated on islands where their missionaries were particularly active. Many Tongans, for example, were converted to Methodism in the nineteenth century. More recently, Mormon missionaries have been active through much of Oceania and have been particularly successful in parts of Polynesia. Several settings have also seen the emergence of indigenous Christian churches that often blend Western beliefs with traditional animism. One distinctive form of quasi-animist religion in lowland Melanesia was stimulated by outside contacts and became known as the **"cargo cult."** Many areas of the western Pacific witnessed intense fighting during World War II, and Allied forces received airdrops of war materials and other supplies. Many of these goods came into the hands of the local people, who were mystified by their sudden appearance. A complex set of ritual practices quickly emerged, designed to entice the cargo-dropping planes and ships to return. The cargo cult peaked in the immediate postwar period, but its remnants persist to this day in some areas.

A tidal wave of outside influences has engulfed the Pacific realm since World War II, producing further cultural changes as well as growing indigenous responses designed to preserve traditional values. Some groups, particularly in Melanesia, remain more isolated from the outside world, although even there growing demands for natural resources offer an avenue for increasing Western or Asian contacts. In many settings, however, the global growth of tourism has brought Oceania into the relatively easy reach of wealthier Europeans, North Americans, Asians, and Australians. The Hawaiian Islands, Fiji, French Polynesia, and American Samoa are being joined by an increasing number of other island tourist destinations. While offering tremendous economic benefits to certain localities, the onrush of tourists and their consumer-driven values has often come into sharp conflict with native cultures. Change is often seen in generational terms: grandparents may recall simpler times of more limited outside contact; parents might work long, hard hours at resort facilities or souvenir shops; and youngsters find themselves surrounded by fast-food restaurants and the glittery attractions of television. Increasingly, however, selected native groups resist these pressures of cultural globalization. In Melanesia, a growing number of countries have joined the "Spearhead Group" of nations dedicated to preserving cultural traditions. For example, to keep native dialects from dying, many primary schools in Papua New Guinea now teach local languages to students in early grades before switching to Pidgin English thereafter. Even more dramatically, some native Hawaiians, paralleling indigenous movements elsewhere in the world, are advocating the return of one or more of the major islands to Polynesian control, thus preserving their cultural heritage.

Geopolitical Framework: A Land of Fluid Boundaries

Pacific geopolitics reflect a complex interplay of local, colonial-era, and global-scale forces. Each political unit within the region reflects how these three types of forces have played out in particular places. The complexities become apparent in the story of Micronesia's Marshall Islands. This sprinkling of islands and atolls (covering 70 square miles, or 180 square kilometers, of land) historically held varied ethnic groups in essentially local political units before they were loosely incorporated into the Spanish and then German empires by the end of the nineteenth century. In 1914, the Japanese moved

into the islands, and the area remained under their control until 1944, when U.S. troops occupied the region. Following World War II, a United Nations trust territory (administered by the United States) was created across a wide swath of Micronesia, including the Marshall group. Demands for local autonomy grew during the 1960s and 1970s, resulting in a new constitution and independence for the Marshall Islanders by the early 1990s. Today, still benefiting from U.S. aid, government officials in the modest capital city on Majuro Atoll struggle to unite island populations, protect sprawling maritime sea claims, and adjudicate a generation of legal and medical problems that grew from U.S. nuclear bomb testing in the region (Figure 14.30). Such narratives are typical across the realm, suggesting a twenty-first-century political geography that is still very much in the making.

Creating Geopolitical Space

Geopolitical space across Australia and Oceania has been reconfigured many times as different cultural groups and political powers have asserted themselves across the region.

▲ **Figure 14.30 Marshall Islands** The political control of Micronesia has shifted numerous times during the last two centuries. While now part of the independent Marshall Islands, these palm-fringed beaches were once claimed by Spanish, German, Japanese, and U.S. interests. *(Douglas Peebles Photography)*

Present patterns of political organization are merely a snapshot in time and perhaps more prone than in most areas of the world to be redefined on future maps.

Indigenous Patterns

Prior to European contact, the region's political geography was an intricate and ever-changing mosaic of indigenous territories. The indigenous political forms of Australia and Oceania varied widely. Aboriginal Australian society was organized around fluid bands of 30 to 60 people, most of whom were related by blood or marriage. Thus, the periodic movements of hunters and gatherers defined the shifting political space of the Aboriginal world. Melanesia was oriented around more sizable and sedentary communities, also based largely on kinship ties. In Micronesia, Polynesia, and some coastal sections of Melanesia, however, larger and more highly structured chiefdoms evolved. Centralized kingdoms in which power was concentrated in the hands of a single monarch existed on several of the larger volcanic islands of Polynesia. The most powerful of these kingdoms was based in Hawaii, which eventually encompassed all of the islands of the archipelago. While the days of ruling Hawaiian royalty have long since passed, these indigenous, frequently tribal affiliations still help native peoples define their political worlds, an orientation that sometimes runs counter to the boundaries of the region's modern states.

An Imposed Colonial Framework

The European world turned the native political pattern on its head, replacing the nuanced, often fluid territorial boundaries of the indigenous world with more precise yet unstable colonial borders. Great Britain claimed and colonized both Australia and New Zealand. By the 1860s, the various colonial footholds in Australia, focused on widely separated coastal settlements, were formally delineated by the simple geometrical boundaries still apparent today. These separate administrative units functioned as distinctive colonial entities until they joined together in the formal Confederation of Australia in 1901 as an essentially independent commonwealth. New Zealand followed a similar path. Early on, isolated settlement clusters dotted the shoreline, marking the first tentative imprints of European political authority upon the Maori-controlled islands. After the Maori were conquered in the late nineteenth century, both islands were united in a formal and independent dominion under the British Crown in 1907.

The global reach of European colonialism also extended deep into the South Pacific during the nineteenth century. Interest in the area began to grow as early as 1842, when France claimed a "protectorate" over Tahiti. Most of Oceania was not formally colonized, however, until the end of the century, the same period that witnessed the scramble for African territory. By then, relatively few areas of the world were still open for annexation, and colonial powers were eager to acquire the parts that remained before their rivals had a chance to do so. Certain oceanic islands were also seen as particularly desirable because they could serve as coaling supply points for steamships and act as relay stations for submarine telegraph cables.

Germany played a major role in the division of Oceania, just as it did in the partition of Africa. German imperialists viewed New Guinea and the larger islands of the nearby Bismarck Archipelago as the main prize. Other colonial powers opposed the German initiatives, however. By 1885, New Guinea was divided into three territories: the Dutch got the western half (near their empire in Indonesia); the British received the southeastern quarter (near Australia); and the Germans acquired the northeastern quarter, along with the Bismarck Archipelago and a portion of the nearby Solomon Islands chain. Portions of Micronesia and Polynesian Samoa rounded out Germany's colonial holdings.

France, Great Britain, and the United States also had major territorial ambitions in Oceania. By the end of the nineteenth century, virtually all of eastern Polynesia lay under French control. Their domain included the Society Islands, the Marquesas Islands, and the Tuamotu Archipelago. Farther west, France colonized New Caledonia in Melanesia and began to move into the nearby New Hebrides Islands (now Vanuatu). The British, however, having just acquired nearby territories to the north, were also interested in the archipelago. Rather than dividing it or fighting over it, the French and British simply agreed to an unprecedented "condominium" in the New Hebrides under which they shared power. Britain gained additional islands in the region, including the southern Solomon chain, the Fiji Islands, many of the smaller islands of western Polynesia, and a few outposts in eastern Polynesia (Pitcairn). Although most of these acquisitions became direct territories of the British Crown, Tonga (in western Polynesia) remained a protectorate with considerable local autonomy. Indeed, Tonga was the only indigenous state of Oceania to retain its own political structure, and it remains to this day a (constitutional) monarchy. For the United States, Hawaii was the main acquisition, but eastern Samoa also came under U.S. rule in 1900 as a result of an agreement signed with Britain and Germany. The United States also gained Guam, the largest island in Micronesia, as a result of its victory over Spain in the Spanish-American War.

Wars and treaty agreements among colonizing powers further changed Oceania's patterns of political geography. For example, Germany lost all of its foreign territories after World War I: those in Micronesia passed to Japan; those in Melanesia were ceded to Australia; and those in Polynesia went to New Zealand. Japan's gains proved crucial during World War II, as they provided the Japanese military with naval and air bases over a huge swath of the Pacific. After World War II, however, Japan's Micronesian empire evaporated, passing to the United States under UN auspices as the Trust Territory of the Pacific. Meanwhile, following World War I, Australia inherited the British territory of southeastern New Guinea, to which it added the German territories in northeastern New Guinea and the Bismarck Archipelago. At this time the interior reaches of New Guinea were essentially unexplored, and the Australian government had little idea of what its large new colony actually contained. In the 1920s, however, gold was discovered, and by the 1930s Australian prospectors had penetrated the highlands. There they were shocked to discover broad highland plateaus inhabited by roughly a million people. The New Guinea Highlanders, for their part, were probably even more shocked at the intrusive habits and new technologies of the light-skinned interlopers who set about establishing Australian rule over the entire area.

Roads to Independence   The modern political states of the region have arrived at independence along many different paths. Other political units remain colonial entities to this day. The newness and fluidity of the political boundaries are remarkable: the region's oldest independent states are Australia and New Zealand, and both were twentieth-century creations that are only now pondering the desirability of completing their formal political separation from the British Crown. Elsewhere, political ties between colony and mother country are even more intimate and enduring. Even many of the newly independent Pacific **microstates**, with their tiny overall land areas, retain special political and economic ties to countries such as the United States.

Independent Australia (1901) and New Zealand (1907) only gradually created their own political identities and still struggle with their ultimate political configuration. Although Australia became a commonwealth in 1901, it still acknowledges the British Crown as its sovereign (Figure 14.31). The Crown's role is strictly symbolic, and a national referendum in 1999 forced Australians to ponder whether they would like their country to drop this remaining tie to Britain and instead become a genuine republic with its own president replacing the British queen as head of state. However, a majority of 55 percent voted to retain Australia's ties to the Crown. Australia, like the United States, is a federal country, with each of its six states retaining substantial powers. The Northern Territory, however, remains directly under the authority of the central government, although it does enjoy limited self-rule. In New Zealand, formal legislative links with Great Britain were not severed until 1947, and in 1994 some New Zealand officials began discussing the same formal break with the British Crown being debated by the Australians.

Elsewhere in the Pacific, colonial ties were severed even more slowly, and the process has not yet been completed. In the 1970s, Britain and Australia began relinquishing their colonial empires in the Pacific. Fiji (Great Britain) gained independence in 1970, followed by Papua New Guinea (Australia) in 1975 and the Solomon Islands (Great Britain) in 1978. The small island nations of Kiribati and Tuvalu (Great Britain) also became independent in the late 1970s, even though some observers argued that they did not have adequate populations or territorial bases to be viable states.

The United States has recently turned over most of its Micronesian territories to local governments, while retaining considerable influence in the area. After gaining these islands from Japan in the 1940s, the U.S. government supplied large subsidies to islanders but also utilized a number of islands for military purposes. Bikini Atoll was obliterated by nuclear tests, and the large lagoon of Kwajalein Atoll was used as a giant missile target. A major naval base, moreover, was established in Palau, the westernmost archipelago of Oceania. By the early 1990s, both the Marshall Islands and the Federated States of Micronesia (comprising the Caroline Islands) had gained independence. Their ties to the United States, however, remain close. A number of other Pacific islands remain under loose U.S. sovereignty. Palau is an American "trust territory," although the Palauans have some local autonomy. The people of the Northern Marianas chose to become a "self-governing commonwealth in association with the United States," a rather ambiguous political position that allows them to become U.S. citizens. The residents of self-governing Guam and American Samoa are also U.S. citizens. Hawaii became a full-fledged U.S. state in 1959 and thus an integral part of the United States.

Other colonial powers were less inclined to relinquish their oceanic possessions. New Zealand still controls substantial territories in Polynesia, including the Cook Islands, Tokelau, and the island of Niue. France has even more extensive holdings in the region. Its largest maritime possession is French Polynesia, encompassing a vast expanse of mid-Pacific territory. To the west, France retains the much smaller territory of Wallis and Futuna in Polynesia and the larger island of New Caledonia in Melanesia.

▲ **Figure 14.31 Britain's Queen in Australia**   By the voters' choice, Australia retains its links to the British Crown as their sovereign. Despite governmental action to make Australia a fully independent republic with its own president—instead of the Queen—as head of state, in a national referendum in 1999, voters chose to retain traditional ties to the British Crown. Here, Queen Elizabeth II visits Aboriginal dancers at the Muda Aboriginal Language and Culture Centre in the Australian outback. *(AP/Wide World Photos)*

Persisting Geopolitical Tensions

Cultural diversity, colonial legacy, youthful states, and a rapidly changing political map contribute to ongoing geopolitical tensions within the Pacific world (Figure 14.32). Indeed, some of these conflicts have consequences that extend far beyond the boundaries of the region. Others are more locally based but are still vivid reminders of the complexities involved as political space is radically redefined and refashioned across varied natural and cultural settings.

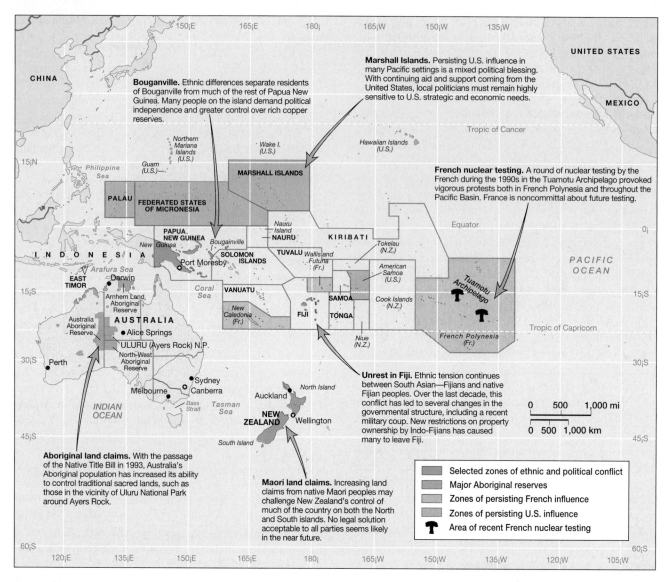

▲ **Figure 14.32 Geopolitical issues in Australia and Oceania** Native land claim issues increasingly shape domestic politics in Australia and New Zealand. Elsewhere, ethnic conflicts have raised political tensions in settings such as Fiji and Papua New Guinea. Colonialism's impact endures as well: American and French interests remain particularly visible in the region, including a legacy of nuclear testing that continues to impact selected Pacific Island populations.

Native Rights in Australia and New Zealand   Indigenous peoples in both Australia and New Zealand have used the political process to gain more control over land and resources in their two countries. Indeed, the strategies these native groups have used parallel efforts in the Americas and elsewhere. In Australia, Aboriginal groups are discovering newfound political power from both more effective lobbying efforts by native groups and a more sympathetic federal government (see "Global and the Local: Aboriginal Videoconferencing in the Outback"). Since land treaties were generally not signed with Aborigines as whites conquered the continent, native peoples originally had no legal land rights whatsoever. More recently, the Australian government established a number of Aboriginal reserves, particularly in the Northern Territory, and expanded Aboriginal control over sacred national parklands

such as Uluru (Ayers Rock; Figure 14.33). Further concessions to indigenous groups were made in 1993 as the government passed the **Native Title Bill,** which compensated Aborigines for already ceded lands, gave them the right to gain title to unclaimed lands they still occupied, and legally empowered them in dealings with mining companies in native-settled areas. Efforts to expand Aboriginal land rights have met concerted opposition. In 1996 an Australian court ruled that pastoral leases (the form of land tenure held by the cattle and sheep ranchers who control most of the outback) do not necessarily extinguish Aboriginal land rights. Grazing interests were infuriated, which led the government to respond that Aboriginal claims might allow the visiting of sacred sites and some hunting and gathering, but not substantial economic control.

GLOBAL AND THE LOCAL   Aboriginal Videoconferencing in the Outback

Cultivating indigenous political consciousness, particularly in a sparsely settled land, is not an easy task. Australian Aborigines, increasingly interested in asserting their rights to tribal lands and local political power, have made ingenious use of twenty-first-century technologies in their quest for tribal unity and respect. Since 1993, the Northern Territory's Warlpiri Aborigines have made extensive use of a sophisticated video-conferencing system to more effectively unite the widely dispersed tribe and to link with Warlpiri expatriates who now live in cities such as Darwin, Alice Springs, and even far-off Sydney. Massachusetts-based PictureTel Corporation helped set up the system, networking a maze of television monitors, video cameras, and satellite dish connections. The result has been the Tanami Network, a series of videoconferencing sites that can link up to 16 participants in a single conversation.

The network has had a major personal as well as political impact upon many residents of the region. Since Aboriginal peoples depend heavily on extensive hand gestures in their communication, the interactive video component of the system has been essential. Family members can converse with distant relatives, increasing social cohesion among a group often fractured by distance and dislocation. Just as important, Aboriginal political activists have effectively used the system to foster a common political consciousness among their members and to provide more responsive government services to those in need. The Aborigines are also going global: they have had a series of video-conferences with other indigenous peoples, from the Saami in northern Scandinavia to the Little Cree Nation in Alberta, Canada. Land rights and language preservation have been common topics of discussion, all made possible by a $1.5 million system largely financed by tribal mineral royalties and community funds.

Source: Adapted from Mark Hodges, "Online in the Outback," *Technology Review* 99 (April 1996).

In New Zealand, Maori land claims have generated similar controversies in recent years. The Maori constitute a far larger proportion of the overall population, and the lands that they claim tend to be much more valuable, further complicating the issue. Recent protests include periodic civic disobedience, growing Maori land claims over much of North and South islands, and a call to return the country's name to the indigenous **Aotearoa**, "Land of the Long White Cloud." The government response, complete with a 1995 visit from Queen Elizabeth, has been to acknowledge increased Maori land and fishing rights as well as to propose a series of financial and land settlements that have yet to be agreed to by the Maoris.

▲ **Figure 14.33 Aborigines at Uluru National Park**   Australian Aborigines gathered recently at Uluru National Park to celebrate their increased political control over the region. Since then, the Native Title Bill has promoted numerous land cessions and further legal settlements.   *(Michael Jensen/Auscape International Pty. Ltd.)*

Conflicts in Oceania   Other geopolitical issues simmer elsewhere in the Pacific, periodically threatening to further redefine the region's fluid territorial boundaries. For example, ethnic differences in Fiji have threatened to tear apart that small island nation. The populations of indigenous Fijians and South Asian immigrants (from the British colonial period) are roughly equal. South Asians dominate most of the country's businesses and commercial settlements, but the Fijians maintain control of virtually all of the land (by law, 83 percent of Fiji's territory is reserved for indigenous village communities) on both main islands of Viti Levu and Vanua Levu. Ethnic strife reached a boiling point in the late 1980s when Fijian military leaders staged a coup and took control of the government after an election threatened to increase Indian control over the country. That coup was followed by a new and controversial constitution in 1990 that angered many Indians, furthered ethnic violence, and prompted a severe scolding from Fiji's former British rulers that included a 10-year exile from the British Commonwealth. (Fiji was readmitted in 1997.) As noted earlier, still another coup took place in May 2000 when indigenous Fijians arrested a democratically elected prime minister of Indian descent. This flagrant violation of the democratic process has chilled diplomatic relations with Fiji's Pacific neighbors.

Papua New Guinea (PNG) must also contend with ethnic tensions. The country is composed of different cultural groups, many of which have a long history of mutual animosity. Most of these peoples now get along with each other reasonably well, although tribal skirmishes occasionally break out in highland market towns. A much bigger problem for the national government has been the rebellion on Bougainville. This sizable island, which has large reserves of copper and other minerals, is located in the Solomon archipelago but belongs to PNG because Germany colonized it in the late 1800s and it thus became politically attached to the eastern portion of New Guinea. Many of Bougainville's indigenous inhabitants believe that their resources are being exploited by foreign interests and by an unsympathetic national government, and they demand local control. Papua New Guinea has reacted with military force; about 5 percent of the island's entire population has already been killed in the conflict; and recent efforts have failed to create a more stable regional government.

The continued French colonial presence within the Pacific region has also created political uncertainties, both in relations with native peoples and between the French and other independent states within the region. Continued French rule in New Caledonia has provoked much local opposition. This large island has substantial mineral reserves (especially of nickel) and sizable numbers of French colonists. French settlement and exploitation of mineral resources angered many indigenous inhabitants. By the 1980s a local independence movement was gaining strength, but in 1987 and 1998 the island's inhabitants (indigenous and French immigrants alike) voted to remain under French rule, at least until 2018. Still, an underground independence movement continues to operate. Troubles have also reverberated in French Polynesia. Although the region receives substantial subsidies from the French and residents have voted to remain a colony, a large minority of the population opposes French control and demands independence. The independence movement was greatly strengthened in 1995 when France decided to resume nuclear testing in the Tuamotu Archipelago. Those activities angered not only the people of French Polynesia (there was antigovernment rioting in Tahiti), but also those of other Pacific countries and territories. Australian leaders called the tests "an act of stupidity"; New Zealand recalled its ambassador in protest; and other Pacific nations from Japan to Chile registered their disapproval. Future French nuclear policies in the Pacific will undoubtedly continue to impact geopolitical relations far beyond Polynesia.

The Imprint of Maritime Claims Global policies on the legal status of maritime boundaries have fundamentally changed the political and economic importance of many small Pacific states. Under the International Law of the Sea, most nations now agree on an **Exclusive Economic Zone (EEZ)**, an area extending seaward for 200 miles (320 kilometers) within which a coastal or island state has exclusive rights to develop or lease potential natural resources such as fishing and mining. The new global geography of EEZs has great importance even for Oceania's microstates, since clusters of small but far-flung atolls can effectively claim huge maritime domains for exclusive resource development. For example, tiny Tonga has a land area of only 270 square miles (700 square kilometers), but its wide areal coverage secures it a large oceanic zone of exclusive economic development rights that it can use or sell to others (Figure 14.34). Nearby Fiji, Samoa, and other states that are within the 200-mile limit must divide the intervening waters equally, thus all but eliminating legal-

ly open high seas in large portions of the island-studded South Pacific. The result has frustrated many development interests globally, and has caused friction and confusion among the island nations themselves, but it has also stimulated interest in the enormous potential value of these newly defined maritime resources.

A Regional and Global Identity?

Australia and New Zealand have emerged to play key political roles in the South Pacific. Although these two countries sometimes disagree on strategic and military matters, their size, wealth, and collective political clout in the region make them important forces for political stability. Special colonial relationships still connect these nations with present and former Pacific holdings. Australia maintains close political ties with its former colony of Papua New Guinea, and New Zealand's continuing control over Niue, Tokelau, and the Cook Islands in Polynesia suggests that its persisting political sphere of influence extends well beyond its borders. When political and ethnic conflicts arise elsewhere in Oceania, Australia and New Zealand are often involved in negotiating peace settlements. Recently, for example, both nations assisted in mediating ongoing disputes on Papua New Guinea's island of Bougainville. In general, the two countries enjoy close political and strategic relations and participate in joint military and reconnaissance efforts in the region. Given the other global interests in the region, however, it remains unclear whether these nations can or wish to assert their political dominance across the entire South Pacific.

Many other political connections tie the region to the world beyond. In 1951, Australia, New Zealand, and the United States forged the ANZUS strategic alliance, which provided a series of mutual security agreements between the three countries. New Zealand sorely tested those links in the 1980s with its outspoken opposition to nuclear weapons and the use of nuclear-powered warships and submarines. Since 1992, however, relations have warmed as the United States has modified its policy toward dispatching nuclear-armed warships to the region. Other strategic links are strengthening geopolitical ties with Southeast Asia. Australia, in particular, has played a growing role in the ARF (Association of South-East Asian Nations Regional Forum). This group is developing multilateral plans to deal with future defense issues, such as the possibility of nuclear conflict in nearby South Asia. In addition, colonial links to the region tie various settings to distant lands: New Caledonia and French Polynesia remain parts of France, while the Marshall Islands, Marianas, and Federated States of Micronesia are still connected politically and economically to Washington, D.C. Indeed, the persistence of French and U.S. interests in the region is not always welcomed by Australia and New Zealand, most notably with France's round of Polynesian nuclear tests in the 1990s. Whatever geopolitical balance of power exists in the future, one thing is certain: the region will remain on the global political periphery.

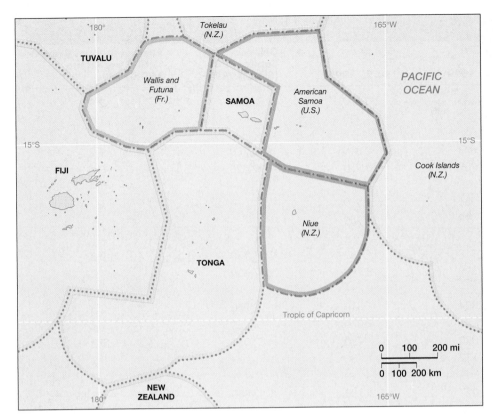

◀ **Figure 14.34 Tonga's EEZ**
Tonga's Exclusive Economic Zone (EEZ) sprawls across vast stretches of the South Pacific. The 200-mile (320-kilometer) limit defines an area where Tonga has rights to lease or develop potential natural resources, such as fishing and mining. Where EEZs overlap, boundary lines are drawn to equally divide the intervening territory.

Economic and Social Development: A Hard Path to Paradise

Wealth and poverty coexist in the Pacific realm, but regional patterns are complex (Table 14.2). Affluent Australia and New Zealand, for example, also contain pockets of pronounced poverty. On the other hand, Oceania offers varied settings that include well-fed subsistence-based populations, relatively prosperous and more commercialized economies, and truly malnourished and impoverished peoples highly dependent on limited government assistance. The twenty-first century poses significant economic challenges for the entire region. Its nations have small populations and domestic markets; the realm retains an enduringly peripheral position in the global economy; and its diminishing resource base is set amid a vulnerable natural environment.

Uncertain Avenues to Affluence

The per capita gross national incomes of both Australia ($20,950) and New Zealand ($13,990) exemplify living standards well above the global norm. Indeed, these two nations are generally grouped among the world's developed countries, along with much of Europe, North America, and Japan. Their many economic assets include highly educated populations, a diverse base of natural resources, and a modern, highly integrated urban and industrial infrastructure. Even so, the relative economic affluence of these two South Pacific nations declined late in the twentieth century due to slower economic growth that was still heavily dependent on the extraction and export of raw materials. Can the trend be reversed? Both nations are currently exploring ways to make the most of their natural resources while at the same time creating a more diversified economy integrated with the global economy.

The Australian Economy   Much of Australia's past economic affluence has been built upon the cheap extraction and export of abundant raw materials. Export-oriented agriculture has long been one of the key supports of Australia's economy. Australian agriculture is highly productive in terms of labor input, and it produces a wide variety of both temperate and tropical crops, as well as huge quantities of beef and wool for world markets. While farm exports are still important to the economy, the mining sector has grown much more rapidly since 1970. Today, Australia is one of the world's mining superpowers. The years since the 1850s gold rush in Victoria have seen a huge expansion in the nation's mineral output, a pattern that accelerated in the 1960s. Among Australia's many assets are coal; rich reserves of iron ore, particularly in Western Australia; and an assortment of other metals, such as bauxite (for aluminum), copper, gold, nickel, lead, and zinc. Indeed, the New South Wales-based Broken Hill Proprietary Company (BHP) is one of the world's largest mining corporations. Given the country's small population, much of that mineral base is consumed elsewhere, making Australia a huge source for raw materials for developed nations, particularly Japan.

TABLE 14.2 *Economic Indicators*

Country	Total GNI (Millions of $U.S., 1999)	GNI per Capita ($U.S., 1999)	GNI per Capita, PPP* ($Intl, 1999)	Average Annual Growth % GDP per Capita, 1990–1999
Australia	397,345	20,950	23,850	2.9
Fed. States of Micronesia	212	1,830	—	−1.8
Fiji	1,849	2,310	4,780	1.2
French Polynesia	3,909	16,930	22,220	−0.1
Guam	—	—	—	—
Kiribati	81	910	—	1
Marshall Islands	99	1950	—	—
Nauru	—	—	—	—
New Caledonia	3,169	15,160	21,130	−0.8
New Zealand	53,299	13,990	17,630	1.8
Palau	—	—	—	—
Papua New Guinea	3,834	810	2,260	2.3
Samoa	181	1,070	4,070	1.4
Solomon Islands	320	750	2,050	0.3
Tonga	172	1,730	—	0.7
Tuvalu	—	—	—	—
Vanuatu	227	1,180	2,880	−0.8

*Purchasing power parity

Source: The World Bank Atlas, 2001.

Although the nation's mineral wealth served it well in the commodity boom years of the 1970s, recent global economic trends have been less favorable for its economic fortunes. Australia's manufacturing sector continues to be a concern. Many of its industries are oriented around the simple processing of raw materials, but often these are not highly profitable, value-added operations. At the same time, small domestic markets discourage the blossoming of more specialized industries. What has been lacking is an entrepreneurial edge in the fast-growing high-technology and information-based industries that have powered recent economic expansion in North America, Japan, and Europe. That may be changing: recent changes in government policy have encouraged more domestic investment, higher savings rates, and more rapid economic growth. Growing numbers of Asian immigrants and economic links with potential Asian markets also bode well for the future. In addition, an expanding tourism industry is helping to diversify the economy. More than 7 percent of the nation's workforce is now devoted to serving the needs of more than 4 million visitors annually. Popular destinations include Melbourne and Sydney, as well as recreational settings such as Queensland's resort-filled Gold Coast, the Great Barrier Reef, and the vast arid outback. Along the Gold Coast, most luxury hotels are owned by Japanese firms and provide a bilingual resort experience for their Asian clientele.

Australia's affluence is distributed widely but unevenly around the country. As is typical in the developed world, much of the wealth is concentrated in the major cities, especially those of the southeast. The high-income suburbs of Melbourne and Sydney resemble those of Boston and Vancouver. Overall, however, no Australian state is significantly richer or poorer than any other, and the disparity of income between the most prosperous fifth of the population and the least prosperous fifth is relatively small in global terms. The Aborigine population, however, is often prone to poverty. Whether living in cities or in the outback, most Aborigines have low standards of living, and their incomes average only 65 percent of the national average. Although a small number of Aborigines still subsist off the land in the large reserves, most occupy a marginal place in the cash economy of the cities.

New Zealand's Economic Challenge

New Zealand is also a wealthy country, but it is somewhat less prosperous than Australia. Before 1970, New Zealand relied heavily on exports to Great Britain, mostly agricultural products such as wool and butter. The strategy faltered, however, once Britain joined the European Union, which then adopted stringent agricultural protection policies. Unlike Australia, however, New Zealand lacked a rich base of mineral resources to export to global markets. By the 1980s, the country had slipped into a serious recession. Eventually, the New Zealand government enacted drastic reforms. The country had previously been noted for its lofty taxes, high levels of social welfare, and state ownership of large economic concerns. Suddenly "privatization" became the watchword, and most state industries were sold off to private parties. As a result, New Zealand has been transformed into one of the most market-oriented countries of the world. To diversify from its traditional export base, the nation has also promoted more aggressive development of its timber resources, fisheries, and tourist industry.

New Zealand's economic prospects remain uncertain. Whether its growth-oriented strategy will pay off in the long run remains to be seen, particularly as the country struggled again with economic stagnation in the late 1990s. Critics also worry that growing income disparities will undermine New Zealand's social foundation. While suburbanites in the north benefit from some of the country's economic reforms, the country's hinterland populations, still often bound to the land, continue to struggle as global prices for their products decline. In addition, the Maori population suffers some of the same economic problems as Australia's Aborigines.

Oceania's Economic Diversity   Varied economic activities shape the Pacific island nations. One way of life is oriented around subsistence-based economies, such as shifting cultivation or fishing. In other settings, the commercial extractive economy dominates, with large-scale plantations, mines, and timber activities often competing for land and labor with the traditional subsistence sector. Elsewhere, the tremendous growth in global tourism has fundamentally transformed the economic geographies of many island settings, forever changing the way that people make a living across much of the vast oceanic realm. Many island nations also benefit from direct subsidies and economic assistance that flow from present and former colonial powers, all designed to promote development and stimulate employment.

Melanesia is, by most measures, the least-developed and poorest part of Oceania. Melanesian countries have benefited less from tourism or from subsidies from wealthy colonial and ex-colonial powers. Most Melanesians live in remote villages that remain somewhat isolated from the modern economy. The Solomon Islands, for example, with few industries other than fish canning and coconut processing, has a per capita GNI of only $750 per year. Papua New Guinea's economy is somewhat more commercially oriented, supporting a per capita GNI of $810. Outbound shipments of coconut products and coffee have increasingly been supplemented with the rapid development of tropical hardwoods in the forest products sector. Gold- and copper-mining ventures have also dramatically transformed the landscape, although political instability has intermittently suspended mineral production in settings such as Bougainville. In New Guinea's interior highlands, however, many villages are entirely oriented around subsistence activities. Fiji remains the most prosperous Melanesian country with a per capita GNI of almost $2,310. Fiji is a major sugar producer, and it has developed a tourist economy popular with North Americans and Japanese.

Among the smaller islands of Melanesia and Micronesia, mining economies dominate New Caledonia and Nauru. New Caledonia's nickel reserves, the world's second largest, are both a blessing and a curse: they sustain much of the island's export economy, yet they inevitably will dwindle in the future. Dramatic price fluctuations for the industrial economy also hamper economic planning for the French colony. Other activities include coffee growing, cattle grazing, and tourism.

To the north, the tiny, phosphate-rich island of Nauru also depends on mining. The citizens of Nauru, for the most part, live directly off the royalties they receive from the mines. Much of the money gained in mining has been invested in a massive trust fund. The Nauruan government has assured its citizens that they will be prosperous even after the phosphate deposits have been exhausted. Much of the trust fund money, however, recently found its way into the Asian real estate market, and these risky investments have proved less than stellar. It now seems possible that the Nauruans will end up with little more than an environmentally devastated island.

Elsewhere in Micronesia and Polynesia, conditions depend upon the viability of local subsistence economies or economic linkages to the world beyond. Many archipelagos export a few food products, but native populations survive mainly on fish, coconuts, bananas, and yams. Some island groups, however, enjoy substantial subsidies from either France or the United States, although such support often comes with a political price. Change is also afoot in some of these island settings: in 1999, Japan agreed to build a spaceport for its future shuttlecraft on Micronesia's Christmas Island (Kiribati), and the Marshall Islands are the site of a planned industrial park financed by mainland Chinese.

Other island groups have been fundamentally transformed by tourism. In Hawaii, more than one-third of the state's economy flows directly from tourist dollars. With almost 5 million visitors annually (including more than 1.25 million from Asia), Hawaii represents all of the classic benefits and risks of the tourist economy. While job creation and economic growth have reshaped the island realm, congested highways, high prices, and the unpredictable spending habits of tourists have left the region vulnerable to future problems. Elsewhere, French Polynesia has long been a favored destination of the international jet set (Figure 14.35). More than 20 percent of

▲ **Figure 14.35 Tahitian resort**   Luxury resort settings in Tahiti (near Papeete) exemplify a growth industry in the South Pacific. While bringing important investment capital, such ventures reorder the region's social and economic structure as well as refashion its cultural landscape. *(Bob & Suzanne Clemenz)*

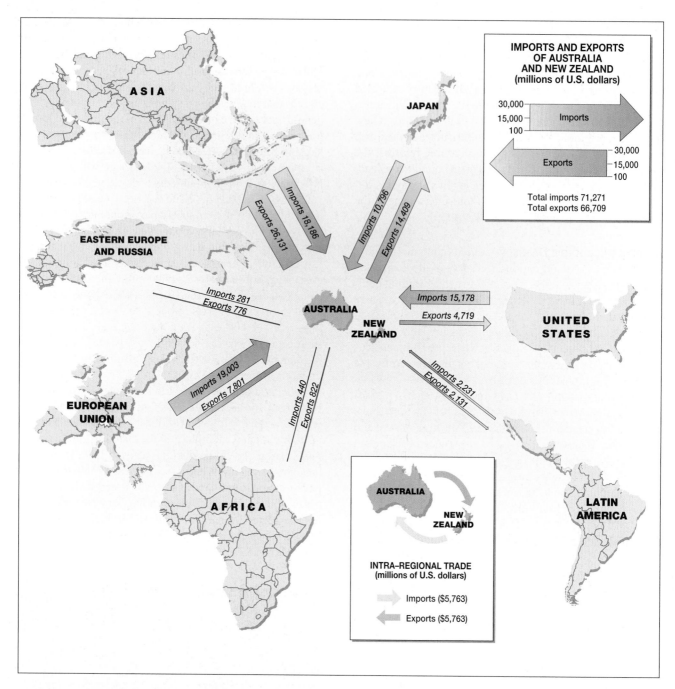

▲ **Figure 14.36 Global trade in Australia and Oceania** Trade flows from Australia and New Zealand demonstrate growing economic ties with nearby Asia. Persisting links with the United States and Europe (particularly Britain) also remain important, especially importing needed manufactured goods. Important intraregional linkages also cement connections between Australia and New Zealand. *(Data from Euromonitor, International Marketing Data and Statistics, 1997)*

French Polynesia's GNI is derived from tourism, making it one of the most prosperous parts of the Pacific. More recently, Guam has emerged as a favorite destination of Japanese and Korean tourists, especially those on honeymoons. Indeed, on a smaller scale, tourism is on the rise across much of the island realm, and many economic planners see it as the avenue to future prosperity. Critics, however, warn that tourist jobs tend to be low-paying (outsiders are often imported for managerial positions); the local quality of life may in fact decline with the presence of tourists; and the natural environments of small islands can quickly be overwhelmed by an influx of demanding visitors.

The Global Economic Setting

Even as Australia and Oceania remain in the global economic hinterlands, their relationships with the world economy will increasingly shape the quality of life and the prospects for development within the region. Several intriguing questions remain. Will future trade patterns in the region shift away from Europe and North America in favor of closer links with Asia? Will there be a move away from the traditional extractive economies that have shaped economic development in the region everywhere from giant Australia to tiny Nauru? Can growing economic linkages within the region make a difference in a part of the world whose total population is less than that of California? None of these questions have ready answers, but they suggest how residents of the region will need to ponder their entry into the twenty-first century's global economy.

Many international trade flows link the area to the far reaches of the Pacific and beyond. Australia and New Zealand dominate global trade patterns in the region (Figure 14.36). In the past 30 years, ties to Great Britain, the British Commonwealth, and Europe have weakened in comparison with growing trade links with Japan, East Asia, the Middle East, and the United States. Australia, for example, now imports more manufactured goods from Japan and the United States than it does from Britain and Europe. Similarly, Australian exports, principally raw materials, increasingly flow toward destinations in Asia and North America, although Europe remains an important trading partner. Other global economic ties have come in the form of capital investment in the region. U.S. and Japanese banks and other financial institutions now dot the South Pacific landscape from Sydney to Suva. Both Australia and New Zealand also participate in the **Asia-Pacific Economic Co-operation Group (APEC),** an organization designed to foster economic development in Southeast Asia and the Pacific Basin. The region's economic ties with Asia also carry risks, however; the Asian downturn in the late 1990s, for example, dampened the popular Korean tourist trade in New Zealand and slowed Asian demand for a variety of Australian raw material exports.

Economic integration within the region has also been promoted. In 1982, Australia and New Zealand signed the **Closer Economic Relationship (CER) Agreement,** which successfully slashed trade barriers between the two countries. New Zealand benefited from the opening of larger Australian markets to New Zealand exports, and Australian corporate and financial interests gained new access to New Zealand business opportunities (see "Economic Change: The Awkward 'Australianization' of the New Zealand Economy"). Today, more than 20 percent of New Zealand's imports and exports come from Australia, and the pattern of regional free trade is likely to strengthen in the future. Smaller nations of Oceania, while often closely tied to countries such as Japan, the United States, and France, also benefit from their proximity to Australia and New Zealand. More than half of Fiji's imports come from those two nearby nations, and other countries, such as Papua New Guinea, Vanuatu, and the Solomon Islands, enjoy a similarly close trading relationship with their more-developed Pacific neighbors.

Enduring Social Challenges

Australians and New Zealanders enjoy high levels of social welfare but face some of the same challenges evident elsewhere in the developed world (Table 14.3). Lifespans average more than 75 years in both countries, and rates of child mortality have fallen substantially since 1960. Paralleling patterns in North America and Europe, cancer and heart disease are leading causes of death, and alcoholism is a persisting social problem, particularly in Australia. Unfortunately, Australia's rate of skin cancer is among the world's highest, the result of having a largely fair-skinned, outdoors-oriented population from northwest Europe in a sunny, low-latitude setting. Overall, Australia's Medicare program (initiated in 1984) and New

---

ECONOMIC CHANGE   The Awkward "Australianization" of the New Zealand Economy

New Zealanders are nervous. Casting a glance across the Tasman Sea, they see an encroaching onslaught of Australian money and financial power muscling its way into the country on the back of the Closer Economic Relationship between the two nations sanctioned in 1982. About half of all overseas companies doing business in New Zealand originate in Australia. Indeed, almost 20 percent of New Zealand's largest 100 corporations are actually controlled by Australians. The trend is particularly pervasive in the all-important financial sector: 70 percent of New Zealand's banking assets are now managed by Australian institutions, and recent economic reforms have even allowed the Bank of New Zealand to be bought out by the National Australia Bank. As one business manager from Victoria put it, "I basically treat New Zealand as another state of Australia." Meanwhile, Australian mass-media magazines, movies, and television programs also flood the New Zealand market. Naturally, many New Zealanders are more than a little uneasy with this coercive brand of Australian "mateship"!

The current economic trend may have many long-lasting consequences. One outcome of the corporate coziness is especially unsettling: when Australians buy out their New Zealand compatriots, corporate home offices often move to Australia, thus triggering job losses and distancing New Zealanders from important business decisions. Indeed, as larger multinational corporations move into the region, they increasingly avoid setting up shop in New Zealand, preferring instead to establish a single South Pacific office in Sydney, Melbourne, or Brisbane. Some New Zealand investment bankers predict that the New Zealand stock market will merge with its Australian counterpart and that a single Australian-dominated currency will follow. As in Europe, such moves make economic sense even as they ruffle this South Pacific nation's cultural and political feathers.

Source: Adapted from Simon Robinson, "Calling Australia Home . . . Control of the New Zealand Economy Is Shifting Across the Tasman," *Time International,* November 9, 1998.

TABLE 14.3 *Social Indicators and Status of Women*

Country	Life Expectancy at Birth		Under Age 5 Mortality (per 1,000)		Percent Illiteracy (Ages 15 and over)		Female Labor Force Participation (% of total, 1999)
	Male	Female	1980	1999	Male	Female	
Australia	76	82	13	5			40
Fed. States of Micronesia	65	67					30
Fiji	65	69					
French Polynesia	69	74					
Guam	72	77					
Kiribati	59	65					
Marshall Islands	63	67					
Nauru	57	65					
New Caledonia	70	76					
New Zealand	74	80	16	6			45
Palau	64	71					
Papua New Guinea	56	55			20	30	42
Samoa	65	72					
Solomon Islands	67	68					47
Tonga	70	72					34
Tuvalu	64	70					
Vanuatu	64	67					

Sources: Population Reference Bureau, Data Sheet, 2001, Life Expectancy (M/F); The World Bank, World Development Indicators, 2001, Under Age 5 Mortality Rate (1980/99) and Percent Illiteracy (ages 15 and over); The World Bank Atlas, 2001, Female Participation in Labor Force.

Zealand's targeted system of social services provide high-quality health care to their populations. The position of women is also high in both countries, including participation in the workforce. Women have recently played key political roles in New Zealand, in particular.

Not surprisingly, the social conditions of the Aborigines and Maoris are much less favorable than those of the population overall. Schooling is irregular for many indigenous peoples, and levels of advanced postsecondary education for Aborigines (12 percent) and Maoris (14 percent) remain far below national averages (32 to 34 percent) in the two countries. Many other social measures reflect the pattern, as well. For example, less than one-third of Aboriginal households own their own homes, while more than 70 percent of white Australian households are homeowners. Furthermore, considerable discrimination against native peoples persists in both countries, a situation that has been aggravated and more publicized with the recent assertion of indigenous political rights and land claims. As with North American African-Americans, Hispanic, and Native American populations, no simple social policies offer solutions to these enduring disparities.

Levels of social welfare in Oceania are higher than one might expect, based on the region's economic circumstances.

Many of its countries and colonies have invested heavily in health and education services and have achieved considerable success. For example, the average life expectancy in the Solomon Islands, one of the world's poorer countries as measured by per capita GNI figures, is a respectable 67 years. By other social measures as well, the Solomon Islands and a number of other Oceania states have reached higher levels of human well-being than is the case in most Asian and African countries with similar levels of economic output. This is partly a result of successful policies, but it also reflects the relatively healthy natural environment of Oceania. Many of the tropical diseases that are so debilitating in Africa simply do not exist within the region.

Papua New Guinea (PNG) is the major exception to these relatively high levels of social welfare. Here the average life expectancy is only 56 years, and a recent study suggests that 34 percent of its young people suffer from malnutrition, particularly protein deficiencies. Adult illiteracy (25 percent) is also more prevalent in PNG than elsewhere in the region. Unfortunately, PNG has found it difficult to provide even basic educational and health services to its people. The country possesses the largest expanse of land in Melanesia, and much of the population lives in relatively isolated villages in the rugged central highlands.

Conclusion

Relative location and small populations contribute to the enduringly peripheral position of Australia and Oceania on the global stage. One of the last habitable portions of the planet to be occupied, the region was also a late chapter in the story of Europe's global colonial expansion. Similarly, its contemporary political geography reveals its still fluid character as countries struggle to disentangle themselves from colonial ties and assert their own political identities. Globalization complicates the process both culturally and politically: new people, many from Asia, are adding ethnic variety as well as providing flows of investment capital that cut across old colonial relationships. In addition, the spatial isolation that water and distance once offered is fast disappearing, casting even the most insulated societies of the Australian outback and the Polynesian periphery rapidly into the postindustrial world of the twenty-first century. The transformation has not been easy or predictable: some native peoples have embraced the attractions of modern life, while others resist the cultural, economic, and political imperatives of the world beyond. Meanwhile, the natural environment, already broadly if unevenly transformed by earlier indigenous peoples and European colonists, has witnessed accelerating changes in the past 50 years as urbanization, extractive economic activities, and recent amenity-oriented investments reconfigure the landscape. Today, sprawling suburbs, open-pit copper and gold mines, and thatched-hut resorts are the dynamic, sometimes intrusive signatures of a once-distant world now brought near by modern communications, improved air travel, and international trade flows.

Australia remains dominant in the greater Pacific realm and should continue to play a pivotal role as a principal economic entry point into the region. Its large land area, resource base, and population are complemented by its increasing emergence as a regional finance center for the South Pacific, as well as nearby portions of Southeast and East Asia. Indeed, the Asian connection seems destined to play a key part in Australia's future economic and even cultural identity. Clearly, the country is no longer simply a distant outpost of Europe. Asian migrants, many of them skilled and affluent, add a dynamic and creative component to Australia's largely European population. In addition, intermingled flows of money, raw materials, and manufactured goods also bind Australia to its Asian neighbors. Will "Asianization" increasingly make Australia a part of that continent, or will political, economic, and cultural resistance to such linkages once again assert themselves? These important questions remain to be answered.

Even while Australia is the giant of the Pacific, it is a secondary player in the global arena. The entire population of the country is about half that of California, and its total economic output is only about as large as that of Texas. Australian optimists, however, have long contended that their country will achieve a more important position in the global economy in decades to come. While it remains to be seen whether Australia's current economic vigor will persist and whether it can develop more high-growth industries, it seems likely that the nation can remain prosperous with enviable social and environmental conditions.

If Australia sits at the edge of Asia, New Zealand might be viewed as sitting on the threshold of Polynesia. New Zealand, of course, was once an integral part of Polynesia, but by the early 1900s it seemed to many to have been transformed into a little England exiled to the South Pacific. Today, however, its Maori population is growing quickly, joined by immigrants from other parts of Polynesia. New Zealand has taken an active role in the political affairs of the entire Pacific basin, and it is in many respects Oceania's leading state. Overall, its multicultural identity and its postindustrial economy continue to evolve. Along with the Polynesian links, closer ties with Australia and Asia will also shape its future cultural milieu, economic base, and political agenda.

The remaining islands of Oceania, while always wedded to the waters that surround and sustain them, have become ever more closely tied to the world beyond. Colonial links often persist. The French presence in the region is dominant, and sometimes unwelcome, particularly when it comes to the issue of nuclear testing. U.S. connections to present possessions and former territories attest to its influence across the region. In addition, globalization has brought fresh interconnections to these island worlds: Japanese-financed golf courses pop up along tropical shores; Canadian and Australian mining companies invest in the New Guinea Highlands; and Korean newlyweds honeymoon beneath the coconut palms. How are native cultures being transformed in this process, and what new cultural hybrids will emerge as Pacific, European, North American, and Asian peoples mingle among the atolls and archipelagos of the South Pacific? Will integration with the global economy irreparably harm the oceanic environment and diminish its appeal as a tourist destination? Can expanded Exclusive Economic Zones allow these island states to assert their broader economic and political role within the Pacific realm? These pivotal issues remain unresolved, but they share a common attribute. As quintessentially geographical concerns, they remind us that the South Pacific realm is a fascinating laboratory in which to study the interplay between land and life, and the often startling outcomes that result when one portion of the world encounters another in fresh and unanticipated ways.

Key Terms

Aborigine (page 606)

Aotearoa (page 623)

archipelagos (page 595)

Asia-Pacific Economic Co-operation Group (APEC) (page 629)

atoll (page 604)

cargo cult (page 619)

Closer Economic Relationship (CER) Agreement (page 629)

Questions for Summary and Review

1. Why does it make sense to group together Australia, New Zealand, and the other Pacific islands as a region? What complexities are involved in doing so?

2. Briefly contrast the topographic settings and climates of Australia's Murray River Valley with that of New Zealand's North Island. What geological and climatic factors account for the similarities and differences?

3. For the following localities, identify a similar North American environment and support your answer: Alice Springs, Adelaide, Nelson (New Zealand), and Brisbane.

4. Describe how a Pacific high island is created and then transformed into a low island. How do these environments produce distinctive settings for human settlement?

5. What might be some notable similarities between Australian cities and North American cities? Notable differences?

6. Describe the types of farming you might observe in (a) South Australia, (b) the New Guinea Highlands, and (c) Fiji. Explain the differences, using both environmental and human variables in your discussion.

7. How do cargo cults and Pidgin English represent similar cultural processes at work in the Pacific world?

8. Discuss some of the key geopolitical issues faced by Pacific Islanders in the early twenty-first century.

9. Why is there such a wide economic gap separating affluent Australia and New Zealand from the rest of Oceania?

10. Why might it be argued that the social and environmental consequences of tourism are even more important than the economic changes it brings?

Thinking Geographically

1. Should New Zealand welcome or be wary of Australian and Japanese investments in the country? Defend your answer.

2. As a local economic development official in Papua New Guinea, argue both for and against a large new Malaysian-financed pulp mill being planned for your district.

3. Select a small Pacific island, and briefly research its political history since 1800. What do your findings illustrate about more general processes of European colonization and political independence movements over the last 200 years?

4. Identify what you see as the three principal economic challenges facing Papua New Guinea. Discuss how the country's physical and human geography may affect its future prospects.

5. Does the continuing French colonial presence in the South Pacific help or hinder economic development within the region? Why?

6. As an Australian Aborigine in touch with your North American counterparts, what are some of the principal similarities and differences in the political and economic plight of your peoples?

7. It is 2042, and global warming has accelerated in the past 30 years. You are the ruler of a small Pacific island nation now being inundated by rising sea levels and altered climatic patterns. What is your strategy to survive, considering these new geographical circumstances?

8. As a maker of educational films, you have been hired to film a three-part series titled *The Essential South Pacific*. The episodes are to be set in (a) Australia, (b) New Zealand, and (c) Polynesia. Budgets are limited, and only a single location in each of those settings can be used to communicate the personality of the region. Identify three specific spots for filming, and explain your answers to the producer of the series.

⊕ Regional Novels and Films

Novels

Thea Astley, *The Beachmasters* (1986, Viking)

Eleanor Dark, *The Timeless Land* (1941, Collins)

Derek Hansen, *Sole Survivor* (1999, Simon and Schuster)

Xavier Herbert, *Capricornia* (1938, Publicist)

Keri Hulme, *The Bone People* (1985, Hodder and Stoughton)

Charlotte Jay, *Beat Not the Bones* (1952, Avon)

Herman Melville, *Typee* (1846, Wiley and Putnam)

James Michener, *Hawaii* (1959, Random House)

Arthur Upfield, *The Sands of Windee* (1958, Angus and Robertson)

Silvia Watanabe, *Talking to the Dead* (1992, Doubleday)

Patrick White, *Riders in the Chariot* (1961, Viking) and *Voss* (1957, Viking)

B. Wongar, *Walg: A Novel of Australia* (1983, Dodd, Mead)

Films

The Adventures of Priscilla, Queen of the Desert (1994, Australia)

Crocodile Dundee (1986, U.S.)

A Cry in the Dark (1988, Australia)

Gallipoli (1981, Australia)

Goodbye Pork Pie (1981, New Zealand)

The Man from Snowy River (1982, Australia)

Mauri (1988, New Zealand)

On the Beach (1959, U.S.)

Once Were Warriors (1994, New Zealand)

Oscar and Lucinda (1997, U.S.)

The Piano (1993, Australia, filmed in New Zealand)

Picture Bride (1993, Hawaii, Japanese)

South Pacific (1958, U.S.)

Thin Red Line (1998, U.S.)

Walkabout (1971, Australia)

⊕ Bibliography

Australia Department of Immigration and Multicultural Affairs. 1998. "Birthplace and Related Data from the 1996 Census, Australia." Report C96.2.0.

Bambrick, Susan, ed. 1994. *The Cambridge Encyclopedia of Australia*. New York: Cambridge University Press.

Bellwood, Peter. 1979. *Man's Conquest of the Pacific: The Prehistory of Southeast Asia and Oceania*. New York: Oxford University Press.

Brookfield, Harold C., ed. 1973. *The Pacific in Transition: Geographical Perspectives on Adaptation and Change*. New York: St. Martin's Press.

Buchholz, Hanns J. 1987. *Law of the Sea Zones in the Pacific Ocean*. Singapore: Institute of Southeast Asian Studies.

Campbell, Ian C. 1990. *A History of the Pacific Islands*. Berkeley: University of California Press.

Cumberland, Kenneth B. 1968. *Southwest Pacific: A Geography of Australia, New Zealand and Their Pacific Island Neighbourhoods*. Christchurch: Whitcombe and Tombs.

Cumberland, Kenneth B., and Whitelaw, James S. 1970. *New Zealand*. Chicago: Aldine Publishing.

Forster, Clive A. 1995. *Australian Cities: Continuity and Change*. Melbourne: Oxford University Press.

Hall, Colin M. 1994. *Tourism in the Pacific Rim*. New York: John Wiley & Sons.

Head, Lesley. 2000. *Second Nature: The History and Implications of Australia as Aboriginal Landscape*. Syracuse, NY: Syracuse University Press.

Head, Lesley. 2000. *Cultural Landscapes and Environmental Change*. London: Arnold.

Heathcote, R. L. 1994. *Australia*. New York: John Wiley & Sons.

Hughes, Robert. 1986. *The Fatal Shore: The Epic of Australia's Founding*. New York: Alfred A. Knopf.

Kirch, Patrick V. 2000. *On the Road of the Winds: An Archaeological History of the Pacific Islands Before European Contact*. Berkeley: University of California Press.

Kirch, Patrick V., and Hunt, Terry. 1997. *Historical Ecology in the Pacific Islands: Prehistoric Environmental and Landscape Change*. New Haven, CT: Yale University Press.

Kluge, P. F. 1991. *The Edge of Paradise: America in Micronesia*. New York: Random House.

Lines, William. 1992. *Taming the Great South Land: A History of the Conquest of Nature in Australia*. Berkeley: University of California Press.

Marriott, Edward. 1998. "All Is (Most Definitely) Not Lost." *Geographical Magazine* 70(11), 10–16.

McKnight, Tom L. 1995. *Oceania: The Geography of Australia, New Zealand, and the Pacific Islands*. Englewood Cliffs, NJ: Prentice Hall.

Meinig, D. W. 1962. *On the Margins of the Good Earth: The South Australian Wheat Frontier, 1869–1884*. Chicago: Rand McNally.

———. 1998. *The Shaping of America: Transcontinental America, 1850–1915*. Volume 3. New Haven and London: Yale University Press.

Mitchell, Andrew. 1991. *The Fragile South Pacific: An Ecological Odyssey.* Austin: University of Texas Press.

Oliver, Douglas L. 1989. *The Pacific Islands.* Honolulu: University of Hawaii Press.

Pilger, John. 1991. *A Secret Country: The Hidden Australia.* New York: Alfred A. Knopf.

Powell, J. M. 1988. *An Historical Geography of Modern Australia: The Restive Fringe.* New York: Cambridge University Press.

"Reborn: Melbourne." 1997. *The Economist,* August 23.

Rose, Deborah Bird. 1996. *Nourish Terrains: Australian Aboriginal Views of Landscape and Wilderness.* Canberra: Australian Heritage Commission.

Ross, Robert, ed. 1998. *Australia: A Traveler's Literary Companion.* San Francisco: Wherabouts Press.

Sekhran, Nik. 1997. "Green or Greed." *Geographical Magazine* 69(10), 75–81.

Shadbolt, Maurice. 1988. *Reader's Digest Guide to New Zealand.* Sydney: Reader's Digest.

Spate, O. H. K. 1968. *Australia.* New York: Praeger.

Stanley, David. 2000. *South Pacific Handbook.* 7th ed. Emeryville, CA: Avalon Travel Publishing.

Terrill, Ross. 1987. *The Australians.* New York: Simon & Schuster.

Ward, Gerard, ed. 1972. *Man in the Pacific Islands: Essays on Geographical Change in the Pacific.* New York: Oxford University Press.

Woodard, Colin. 1998. "Marshall Islands: You Can't Go Home Again." *Bulletin of the Atomic Scientists* 54(5), 10–12.

Glossary

Aborigine An indigenous inhabitant of Australia.

acid rain Harmful form of precipitation high in sulfur and nitrogen oxides. Caused by industrial and auto emissions, acid rain damages aquatic and forest ecosystems in regions such as eastern North America and Europe.

African diaspora The forced dispersion of African peoples from their native lands to various parts of the Americas and the Middle East through the slave trade.

agrarian reform A popular but controversial strategy to redistribute land to peasant farmers. Throughout the twentieth century, various states redistributed land from large estates or granted title from vast public lands in order to reallocate resources to the poor and stimulate development. Agrarian reform occurred in various forms, from awarding individual plots or communally held land to creating state-run collective farms.

agricultural density The number of farmers per unit of arable land. This figure indicates the number of people who directly depend upon agriculture, and it is an important indicator of population pressure in places where rural subsistence dominates.

alluvial fan A fan-shaped deposit of sediments dropped by a river or stream flowing out of a mountain range.

Altiplano The largest intermontane plateau in the Andes, which straddles Peru and Bolivia and ranges in elevation from 10,000 to 13,000 feet (3,000 to 4,000 meters).

altitudinal zonation The relationship between higher elevations, cooler temperatures, and changes in vegetation that result from the environmental lapse rate (averaging 3.5°F for every 1,000 feet). In Latin America, four general altitudinal zones exist: tierra caliente, tierra templada, tierra fria, and tierra helada.

animism A wide variety of tribal religions based on the worship of nature's spirits and human ancestors.

anthropogenic An adjective for human-caused change to a natural system, such as the atmospheric emissions from cars, industry, and agriculture that are causing global warming.

anthropogenic landscape A landscape heavily transformed by human agency.

Aotearoa Maori name for New Zealand, meaning "Land of the Long White Cloud."

apartheid The policy of racial separateness that directed the separate residential and work spaces for white, blacks, coloureds, and Indians in South Africa for nearly 50 years. It was abolished when the African National Congress came to power in 1994.

archipelagos Island groups, often oriented in an elongated pattern.

areal differentiation The geographic description and explanation of spatial differences on Earth's surface; this includes physical as well as human patterns.

areal integration The geographic description and explanation of how places, landscapes, and regions are connected, interactive, and integrated with each other.

Asia-Pacific Economic Co-operation Group (APEC) An international group of Asian and Pacific Basin nations that fosters coordinated economic development within the region.

Association of Southeast Asia Nations (ASEAN) A supranational geopolitical group linking together the 10 different states of Southeast Asia.

atoll Low, sandy islands made from coral, often oriented around a central lagoon.

autonomous areas Minor political subunits created in the former Soviet Union and designed to recognize the special status of minority groups within existing republics.

autonomous region In the context of China, provinces that have been granted a certain degree of political and cultural autonomy, or freedom from centralized authority, owing to the fact that they contain large numbers of non-Han Chinese people. Critics contend that they have little true autonomy.

Baikal-Amur Mainline (BAM) Railroad Key central Siberian railroad connection completed in the Soviet era (1984), which links the Yenisey and Amur rivers and parallels the Trans-Siberian Railroad.

Balfour Declaration Statement issued by Great Britain in 1917 pledging its support for establishing a home for the Jewish people in Palestine.

balkanization Geopolitical process of fragmentation of larger states into smaller ones through independence of smaller regions and ethnic groups. The term takes its name from the geopolitical fabric of the Balkan region.

barrios Urban Hispanic neighborhoods, often associated with low-income groups in North America.

Berlin Conference The 1884 conference that divided Africa into European colonial territories. The boundaries created in Berlin satisfied European ambition but ignored indigenous cultural affiliations. Many of Africa's civil conflicts can be traced to ill-conceived territorial divisions crafted in 1884.

biofuels Energy sources derived from plants or animals. Throughout the developing world, wood, charcoal, and dung are primary energy sources for cooking and heating.

biome Ecologically interactive flora and fauna adapted to a specific environment. Examples are deserts or tropical rain forests.

bioregion A spatial unit or region of local plants and animals adapted to a specific environment, such as a tropical savanna.

Bolsheviks A faction within the Russian Communist movement led by Lenin that successfully took control of the country in 1917.

boreal forest Coniferous forest found in high-latitude or mountainous environments of the Northern Hemisphere.

brain drain Migration of the best-educated people from developing countries to developed nations where economic opportunities are greater.

British East India Company Private trade organization that acted as arm of colonial Britain—backed by the British army—in monopolizing trade in South Asia until 1857, when it was abolished and replaced by full governmental control.

bubble economy A highly inflated economy that cannot be sustained. Bubble economies usually result from rapid influx of international capital into a developing country.

buffer zone An array of nonaligned or friendly states that "buffer" a larger country from invasion. In Europe, keeping a buffer zone has been a long-term policy of Russia (and also of the former Soviet Union) to protect its western borders from European invasion.

Bumiputra The name given to native Malay (literally, "sons of the soil"), who are given preference for jobs and schooling by the Malaysian government.

Burakumin The indigenous outcast group of Japan, a people whose ancestors reputedly worked in leather-craft and other "polluting" industries.

bustees Settlements of temporary and often illegal housing in Indian cities, caused by rapid urban migration of poorer rural people and the inability of the cities to provide housing for this rapidly expanding population.

capital leakage The gap between the gross receipts an industry (such as tourism) brings into a developing area and the amount of capital retained.

cargo cult Quasi-animist Melanesian religious practice, originating with military cargo supply drops during World War II.

Caribbean Community and Common Market (CARICOM) A regional trade organization established in 1972 that includes former English colonies as its members.

Caribbean diaspora The out-migration of Caribbean peoples to various destinations in North America and Europe. This livelihood strategy has been widely practiced since the 1950s, when many people began leaving due to the region's limited economic opportunities. Although difficult for migrants and their families, the Caribbean diaspora has become a way of life and explains the region's dependence on foreign remittances.

caste system Complex division of South Asian society into different hierarchically ranked hereditary groups. Most explicit in Hindu society, but also found in other cultures to a lesser degree.

Central Place theory A theory used to explain the distribution of cities, and the relationships between different cities, based on retail marketing.

centralized economic planning An economic system in which the state sets production targets and controls the means of production.

centrifugal forces Those cultural and political forces, such as linguistic minorities, separatists, and fringe groups, that pull away from and weaken an existing nation-state.

centripetal forces Those cultural and political forces, such as a shared sense of history, a centralized economic structure, and the need for military security, that promote political unity in a nation-state.

chain migration A pattern of migration in which a sending area becomes linked to a particular destination, such as Dominicans with Queens, New York.

chernozem soils A Russian term for dark, fertile soil, often associated with grassland settings in southern Russia and Ukraine.

China proper The eastern half of the country of China where the Han Chinese form the dominant ethnic group. The vast majority of China's population is located in China proper.

Chipko movement The "tree-hugging" movement of northern India in which women, drawing upon Hindu tradition, attempt to save forests from destruction by embracing the trees as loggers approach.

circular migration Temporary labor migration in which an individual seeks short-term employment overseas, saves money, and then returns home.

clan A social unit that is typically smaller than a tribe or ethnic group but larger than a family, based on supposed descent from a common ancestor.

climate region A region of similar climatic conditions. An example would be the marine west coast climate regions found on the west coasts of North America and Europe.

climograph Graph of average annual temperature and precipitation data by month and season.

Closer Economic Relationship (CER) Agreement An agreement signed in 1982 between Australia and New Zealand designed to eliminate all economic and trade barriers between the two countries.

Cold War The ideological struggle between the United States and the Soviet Union that was conducted between 1946 and 1991.

collective farms Group-farmed agricultural units organized around state-mandated production goals.

collectivization The agglomeration of small, privately owned agricultural parcels into larger, state-owned farms. This was a central component of communism in eastern Europe and the Soviet Union.

colonialism The formal, established (mainly historical) rule over local peoples by a larger imperialist government for the expansion of political and economic empire.

Coloureds A racial category used throughout South Africa to define people of mixed European and African ancestry.

Columbian Exchange An exchange of people, diseases, plants, and animals between the Americas (New World) and Europe/Africa (Old World) initiated by the arrival of Christopher Columbus in 1492.

command economy Centrally planned and controlled economies, generally associated with socialist or communist countries, in which all goods and services, along with agricultural and industrial products, are strictly regulated. This was done during the Soviet era in both the Soviet Union and its eastern European satellites.

Commonwealth of Independent States (CIS) A loose political union of former Soviet republics (without the Baltic states) established in 1992 after the dissolution of the Soviet Union.

concentric zone model A simplified description of urban land use: a well-defined central business district (CBD) is surrounded by concentric zones of residential activity, with higher-income groups living on the urban periphery.

Confucianism The philosophical system developed by Confucius in the sixth century B.C.

connectivity The degree to which different locations are linked with one another through transportation and communication infrastructure.

continental climate Climate regions in continental interiors, removed from moderating oceanic influences, that are characterized by hot summers and cold winters. At least one month must average below freezing.

convection cells Large areas of slow-moving molten rock in Earth's interior that are responsible for moving tectonic plates.

copra Dried coconut meat.

Cossacks Highly mobile Slavic-speaking Christians of the southern Russian steppe who were pivotal in expanding Russian influence in sixteenth- and seventeenth-century Siberia.

Council of Mutual Economic Assistance (CMEA) The communist agency that coordinated economic planning and development between the Soviet Union and satellite countries in eastern Europe.

counterurbanization The movement of people out of metropolitan areas toward smaller towns and rural areas.

creolization The blending of African, European, and even some Amerindian cultural elements into the unique sociocultural systems found in the Caribbean.

crony capitalism A system in which close friends of a political leader are either legally or illegally given business advantages in return for their political support.

cultural assimilation The process in which immigrants are culturally absorbed into the larger host society.

cultural imperialism The active promotion of one cultural system over another, such as the implantation of a new language, school system, or bureaucracy. Historically, this has been primarily associated with European colonialism.

cultural landscape Primarily the visible and tangible expression of human settlement (house architecture, street patterns, field form, etc.), but also includes the intangible, value-laden aspects of a particular place and its association with a group of people.

cultural nationalism A process of protecting, either formally (with laws) or informally (with social values), the primacy of a certain cultural system against influences (real or imagined) from another culture.

cultural syncretism The blending of two or more cultures, which produces a synergistic third culture that exhibits traits from all cultural parents.

culture Learned and shared behavior by a group of people empowering them with a distinct "way of life"; it includes both material (technology, tools, etc.) and immaterial (speech, religion, values, etc.) components.

culture hearth An area of historical cultural innovation.

Cyrillic alphabet Based on the Greek alphabet and used by Slavic languages heavily influenced by the Eastern Orthodox Church. Attributed to the missionary work of St. Cyril in the ninth century.

dacha A Russian country cottage used especially in the summer.

Dalit The currently preferred term used to denote the members of India's most discriminated-against ("lowest") caste groups, those people previously deemed "untouchables."

decolonialization The process of a former colony's gaining (or regaining) independence over its territory and establishing (or reestablishing) an independent government.

demographic transition A four-stage model of population change derived from the historical decline of the natural rate of increase as a population becomes increasingly urbanized through industrialization and economic development.

denuclearization The process whereby nuclear weapons are removed from an area and dismantled or taken elsewhere.

dependency theory A popular theory to explain patterns of economic development in Latin America. Its central premise is that underdevelopment was created by the expansion of European capitalism into the region that served to develop "core" countries in Europe and to impoverish and make dependent peripheral areas such as Latin America.

desertification The spread of desert conditions into semiarid areas owing to improper management of the land.

diaspora The scattering of a particular group of people over a vast geographical area. Originally, the term referred to the migration of Jews out of their original homeland, but now it has been generalized to refer to any ethnic dispersion.

digital divide The uneven and inequitable access to the Internet and to computer technology based on wealth and education.

distributaries The different channels or branches of a river in its delta area.

dollarization An economic strategy in which a country adopts the U.S. dollar as its official currency. A country can be partially dollarized, using U.S. dollars alongside its national currency, or fully dollarized, when the U.S. dollar becomes the only medium of exchange and a country gives up its own national currency. Panama fully dollarized in 1904; more recently, Ecuador fully dollarized in 2000.

domestication The purposeful selection and breeding of wild plants and animals for cultural purposes.

domino theory A U.S. geopolitical policy of the 1970s that stemmed from the assumption that if Vietnam fell to the communists, the rest of Southeast Asia would soon follow.

Dravidian language One of the earliest (from perhaps 4,000 years ago) language families and, unlike Hindi, not Indo-European. Once spoken throughout South Asia, Dravidian languages are now found only in southern India and part of Sri Lanka.

Eastern Orthodox Christianity A loose confederation of self-governing churches in eastern Europe and Russia that are historically linked to Byzantine traditions and to the primacy of the patriarch of Constantinople (Istanbul).

economic convergence The notion that globalization will result in the world's poorer countries gradually catching up with more advanced economies.

economic growth rate the annual rate of expansion for GNI.

El Niño An abnormally large warm current that appears off the coast of Ecuador and Peru in December. During an El Niño year, torrential rains can bring devastating floods along the Pacific coast and drought conditions in the interior continents of the Americas.

encomienda A social and economic system used in the early Spanish colonies where there were large native populations. Groups of Indians would be "commended" to a Spaniard, who would exact tribute from them in the form of labor or products. In return, the Spaniard was obligated to educate the Indians in the Spanish language and the Catholic faith.

entrepôt A city and port that specializes in transshipment of goods.

environmental lapse rate The decline in temperature as one ascends higher in the atmosphere. On average, the temperature declines 3.5°F for every 1,000 feet ascended or 6.5°C for every 1,000 meters.

ethnic religion A religion closely identified with a specific ethnic or tribal group, often to the point of assuming the role of the major defining characteristic of that group. Normally, ethnic religions do not actively seek new converts.

ethnicity A shared cultural identity held by a group of people with a common background or history, often as a minority group within a larger society.

ethnographic boundaries State and national boundaries that are drawn to follow distinct differences in cultural traits, such as religion, language, or ethnic identity.

Euroland The 11 states that form the European Monetary Union, with its common currency, the euro. This monetary unit completely replaced national currencies in July 2002.

European Union (EU) The current association of 15 European countries that are joined together in an agenda of economic, political, and cultural integration.

exclave A portion of a country's territory that lies outside of its contiguous land area.

Exclusive Economic Zone An agreement under the International Law of the Sea that provides for a 200-mile zone of exclusive legal offshore development rights that can be utilized or sold by coastal or island nations.

exotic river A river that issues from a humid area and flows into a dry area otherwise lacking streams.

federal state Political system in which a significant amount of power is given to individual states; in India, these states were created upon independence in 1947 and were drawn primarily along linguistic lines, so that today state power is often associated with specific ethnic groups within the nation.

Fertile Crescent An ecologically diverse zone of lands in Southwest Asia that extends from Lebanon eastward to Iraq and that is often associated with early forms of agricultural domestication.

feudalism The formal power relationship that consists of well-defined responsibilities and obligations between a superior (such as an aristocrat) and those lower in social status (such as serfs or vassals).

fjords Flooded, glacially carved valleys; in Europe, found primarily along Norway's western coast.

forward capital A capital city deliberately positioned near the international border of a contested territory, signifying the state's interest—and presence—in this zone of conflict.

fossil water Water supplies that were stored underground during wetter climatic periods.

Free Trade Area of the Americas (FTAA) A proposed hemispheric trade association that will include Latin America, North America, and the Caribbean (minus Cuba). If the treaty is successfully negotiated, it will integrate some 800 million people in the largest trade association in the world. The FTAA is scheduled to go into effect in 2005.

free trade zone (FTZ) A duty-free and tax-exempt industrial park created to attract foreign corporations and create industrial jobs.

gentrification A process of urban revitalization in which higher-income residents displace lower-income residents in central city neighborhoods.

geomancy The traditional Chinese and Korean practice of designing buildings in accordance with the principles of cosmic harmony and discord that supposedly course through the local topography.

geometric boundaries Boundaries of convenience drawn along lines of latitude or longitude without consideration for cultural or ethnic differences in an area.

ghettos Urban ethnic neighborhoods often associated with low-income groups in North America.

glasnost A policy of greater political openness initiated during the 1980s by Soviet President Mikhail Gorbachev.

globalization The increasing interconnectedness of people and places throughout the world through converging processes of economic, political, and cultural change.

Golden Triangle An area of northern Thailand, Burma, and Laos that is known as a major source region for heroin and is plugged into the global drug trade.

Gondwanaland The ancient mega-continent that included Africa, South America, Antarctica, Australia, Madagascar, and Saudi Arabia. Some 250 million years ago, it began to split apart due to plate tectonics.

Gran Colombia A short-lived Latin American state during the early years of independence (1822–1830). A dream of Simon Bolívar, it included Colombia, Venezuela, Ecuador, and Panama.

grassification The conversion of tropical forest into pasture for cattle ranching. Typically, this processes involves introducing species of grasses and cattle, mostly from Africa.

Great Escarpment A landform that rims southern Africa from Angola to South Africa. It forms where the narrow coastal plains meet the elevated plateaus in an abrupt break in elevation.

Greater Antilles The four large Caribbean islands of Cuba, Jamaica, Hispaniola, and Puerto Rico.

Green Line United Nations-patrolled dividing line separating Greek (southern) and Turkish (northern) portions of Cyprus.

Green Revolution Term applied to the development of agricultural techniques used in developing countries that usually combine new, genetically altered seeds that provide higher yields than native seeds when combined with high inputs of chemical fertilizer, irrigation, and pesticides.

greenhouse effect The natural process of lower atmosphere heating that results from the trapping of incoming and reradiated solar energy by water moisture, clouds, and other atmospheric gases.

gross domestic product (GDP)—gross national product (GNP) GDP is the total value of goods and services produced within a given country (or other geographical unit) in a single year. GNP is a somewhat broader measure that includes the inflow of money from other countries in the form of the repatriation of profits and other returns on investments, as well as the outflow to other countries for the same purposes.

gross national income (GNI) The value of all final goods and services produced within a country's borders (gross domestic product, or GDP) plus the net income from abroad (formerly referred to as gross national product, or GNP).

gross national income, per capita The figure that results from dividing a country's GNI by the total population.

Group of 7 (G-7) A collection of powerful countries that confers regularly on key global economic and political issues. It includes the United States, Canada, Japan, Great Britain, Germany, France, and Italy; sometimes Russia also takes part.

growth poles Planned industrial centers developed by states in order to increase manufacturing and stimulate economic growth in underdeveloped areas.

guest workers Workers from Europe's agricultural periphery—primarily Greece, Turkey, southern Italy, and the former Yugoslavia—solicited to work in Germany, France, Sweden, and Switzerland during chronic labor shortages in Europe's boom years (1950s to 1970s).

Gulag Archipelago A collection of Soviet-era labor camps for political prisoners, made famous by writer Aleksandr Solzhenitsyn.

Hajj An Islamic religious pilgrimage to Makkah. One of the five essential pillars of the Muslim creed to be undertaken once in life, if an individual is physically and financially able to do it.

haoles Light-skinned Europeans or U.S. citizens in the Hawaiian Islands.

hierarchical diffusion The spread of an idea or cultural trait through adoption by leaders and other elite at the top of the social structure or hierarchy. In the case of religion, it was common for all members of a clan or tribe to convert if the leader adopted a new religion.

high islands Larger, more elevated islands, often focused around recent volcanic activity.

Hindi An Indo-European language with more than 480 million speakers, making it the second-largest language group in the world. In India, it is the dominant language of the heavily populated north, specifically the core area of the Ganges Plain.

Hindu nationalism A contemporary "fundamental" religious and political movement that promotes Hindu values as the essential—and exclusive—fabric of Indian society. As a political movement, it appears to have less tolerance of India's large Muslim minority than other political movements.

hiragana The main Japanese syllabary, used for writing indigenous words. Each symbol stands for a particular vowel-consonant combination.

homelands Nominally independent ethnic territories created for blacks under the grand apartheid scheme. Homelands were on marginal land, overcrowded, and poorly serviced. In the post-apartheid era, they were eliminated.

Horn of Africa The northeastern corner of Sub-Saharan Africa that includes the states of Somalia, Ethiopia, Eritrea, and Djibouti. Drought, famine, and ethnic warfare in the 1980s and 1990s resulted in political turmoil in this area.

hot spot A supply of magma that produces a chain of mid-ocean volcanoes atop a zone of moving oceanic crust.

houseyard A rural subsistence property that is matriarchal in organization.

hurricanes Storm systems with an abnormally low-pressure center sustaining winds of 75 mph or higher. Each year during hurricane season (July–October), a half dozen to a dozen hurricanes form in the warm waters of the Atlantic and Caribbean, bringing destructive winds and heavy rain.

hydropolitics The interplay of water resource issues and politics.

ideographic writing A writing system in which each symbol represents not a sound but rather a concept.

import substitution A development policy that discourages imports (via high tariffs) and encourages substituting domestically produced manufactured goods.

indentured labor Foreign workers (usually South Asians) contracted to labor on Caribbean agricultural estates for a set period of time, often several years. Usually the contract stipulated paying off the travel debt incurred by the laborers. Similar indentured labor arrangements have existed in most world regions.

Indian diaspora The historical and contemporary propensity of Indians to migrate to other countries in search of better opportunities. This has led to large Indian populations in South Africa, the Caribbean, and the Pacific islands, along with western Europe and North America.

informal sector A much-debated concept that presupposes a dual economic system consisting of formal and informal sectors. The informal sector includes self-employed, low-wage jobs that are usually unregulated and untaxed. Street vending, shoe shining, artisan manufacturing, and even self-built housing are considered part of the informal sector. Some scholars include illegal activities such as drug smuggling and prostitution in the informal economy.

insolation Incoming solar energy that enters the atmosphere adjacent to Earth.

internally displaced persons Groups and individuals who flee an area due to conflict or famine but still remain in their country of origin. These populations often live in refugee-like conditions but are harder to assist because they technically do not qualify as refugees.

Iron Curtain A term coined by British leader Winston Churchill during the Cold War that defined the western border of Soviet power in Europe. The notorious Berlin Wall was a concrete manifestation of the Iron Curtain.

irredentism A state or national policy of reclaiming lost lands or those inhabited by people of the same ethnicity in another nation-state.

Islamic fundamentalism A movement within both the Shiite and Sunni Muslim traditions to return to a more conservative, religious-based society and state. Often associated with a rejection of Western culture and with a political aim to merge civic and religious authority.

island biogeography The study of the ecology of plants and animals unique to island environments.

isolated proximity A concept that explores the contradictory position of the Caribbean states, which are physically close to North America and economically dependent upon that region. At the same time, Caribbean isolation fosters strong loyalties to locality and limited economic opportunity.

Jainism A religious group in South Asia that emerged as a protest against orthodox Hinduism about the sixth century B.C. Its ethical core is the doctrine of noninjury to all living creatures. Today, Jains are noted for their nonviolence, which prohibits them from taking the life of any animal.

Kanakas Melanesian workers imported to Australia, historically often concentrated along Queensland's "sugar coast."

kanji The Chinese characters, or ideographs, used in Japanese writing.

Khmer Rouge Literally, "Red (or communist) Cambodians." The left-wing insurgent group led by French-educated Marxists rebelled against the royal Cambodian government in the early 1960s and again in a peasants' revolt in 1967.

kibbutzes Collective farms in Israel.

kleptocracy A state where corruption is so institutionalized that politicians and bureaucrats siphon off a huge percentage of a country's wealth.

laissez-faire An economic system in which the state has minimal involvement and in which market forces largely guide economic activity.

latifundia A large estate or landholding.

Lesser Antilles The arc of small Caribbean islands from St. Maarten to Trinidad.

Levant The eastern Mediterranean region.

lingua franca An agreed-upon common language to facilitate communication on specific topics such as international business, politics, sports, or entertainment.

linguistic nationalism The promotion of one language over others that is, in turn, linked to shared notions of nationalism. In India, some Hindu nationalists promote Hindi as the national language, yet this is resisted by many other groups in which that language is either not spoken or does not have the same central cultural role, as in the Ganges Valley. The lack of a national language in India remains problematic.

location factors The various influences that explain why an economic activity takes place where it does.

loess A fine, wind-deposited sediment that makes fertile soil but is very vulnerable to water erosion.

low islands Low, small, sandy islands formed from eroding coral reefs. Generally less fertile and populated than high islands.

machismo A stereotypical cultural trait of male dominance.

Maghreb A region in northwestern Africa, including portions of Morocco, Algeria, and Tunisia.

maharaja Regional Hindu royalty, usually a king or prince, who ruled specific areas of South Asia before independence, but who was usually subject to overrule by British colonial advisers.

mallee A tough and scrubby eucalyptus woodland of limited economic value that is common across portions of interior Australia.

Mandarin A member of the high-level bureaucracy of Imperial China (before 1911). Mandarin Chinese is the official spoken language of the country and is the native tongue of the vast majority of people living in north, central, and southwestern China.

Maori Indigenous Polynesian people of New Zealand.

maquiladora Assembly plants on the Mexican border built by foreign capital. Most of their products are exported to the United States.

marianismo An idealized model for women that stresses the virtues of patience, deference, and working in the home.

marine west coast climate Moderate climate with cool summers and mild winters that is heavily influenced by maritime conditions. Such climates are usually found on the west coasts of continents between latitudes of 45 to 50 degrees.

maritime climate Climate moderated by proximity to oceans or large seas. It is usually cool, cloudy, and wet, and lacks the temperature extremes of continental climates.

maroons Runaway slaves who established communities rich in African traditions throughout the Caribbean and Brazil.

Marxism The philosophy developed by Karl Marx, the most important historical proponent of communism. Marxism, which has many variants, presumes the desirability and, indeed, the necessity of a socialist economic system run through a central planning agency.

Medieval landscape Urban landscapes from A.D. 900 to 1500 characterized by narrow, winding streets, three- or four-story structures (usually in stone, but sometimes wooden), with little open space except for the market square. These landscapes are still found in the centers of many European cities.

medina The original urban core of a traditional Islamic city.

Mediterranean climate A unique climate, found in only five locations in the world, that is characterized by hot, dry summers with very little rainfall. These climates are located on the west side of continents, between 30 and 40 degrees latitude.

mega-city Urban conglomerations of more than 10 million people.

megalopolis A large urban region formed as multiple cities grow and merge with one another. The term is often applied to the string of cities in eastern North America that includes Washington, D.C.; Baltimore; Philadelphia; New York City; and Boston.

Melanesia Pacific Ocean region that includes the culturally complex, generally darker-skinned peoples of New Guinea, the Solomon Islands, Vanuatu, New Caledonia, and Fiji.

Mercosur The Southern Common Market established in 1991 that calls for free trade among member states and common external tariffs for nonmember states. Argentina, Paraguay, Brazil, and Uruguay are members; Chile is an associate member.

mestizo A person of mixed European and Indian ancestry.

Micronesia Pacific Ocean region that includes the culturally diverse, generally small islands north of Melanesia. Includes the Mariana Islands, Marshall Islands, and Federated States of Micronesia.

microstates Usually independent states that are small in both area and population.

mikrorayons Large, state-constructed urban housing projects built during the Soviet period in the 1970s and 1980s.

minifundia A small landholding farmed by peasants or tenants who produce food for subsistence and the market.

mono-crop production Agriculture based upon a single crop.

monotheism A religious belief in a single God.

Monroe Doctrine A proclamation issued by U.S. President James Monroe in 1823 that the United States would not tolerate European military action in the Western Hemisphere. Focused on the Caribbean as a strategic area, the doctrine was repeatedly invoked to justify U.S. political and military intervention in the region.

monsoon The seasonal pattern of changes in winds, heat, and moisture in South Asia and other regions of the world that is a product of larger meteorological forces of land and water heating, the resultant pressure gradients, and jet-stream dynamics. The monsoon produces distinct wet and dry seasons.

moraines Hilly topographic features that mark the path of Pleistocene glaciers. They are composed of material eroded and carried by glaciers and ice sheets.

Mughal Empire (also spelled Mogul) The preeminent Islamic period of rule that covered most of South Asia during the early sixteenth to late-seventeenth centuries and attempted to unify both Muslims and Hindus into a large South Asian state. The capital of this empire was Lahore, in what is now Pakistan. The last vestiges of the Mughal dynasty were dissolved by the British following the uprisings of 1857.

multinational corporation A corporation that produces goods and services in a wide range of different countries. The classical multinational corporation, in contrast to the truly transnational corporation, remains solidly based in a single country.

nation-state A relatively homogeneous cultural group (a nation) with its own political territory (the state).

Native Title Bill Australian legislation signed in 1993 that provides Aborigines with enhanced legal rights over land and resources within the country.

neocolonialism Economic and political strategies by which powerful states indirectly (and sometimes directly) extend their influence over other, weaker states.

neoliberal policies Economic policies widely adopted in the 1990s that stress privatization, export production, and few restrictions on imports.

neotropics Tropical ecosystems of the Americas that evolved in relative isolation and support diverse and unique flora and fauna.

North America Free Trade Agreement (NAFTA) An agreement made in 1994 between Canada, the United States, and Mexico that established a 15-year plan for reducing all barriers to trade among the three countries.

Oceania A major world subregion that usually includes New Zealand and the major islands of Melanesia, Micronesia, and Polynesia.

offshore banking Islands or microstates that offer financial services that are typically confidential and tax-exempt. As part of a global financial system, offshore banks have developed a unique niche, offering their services to individual and corporate clients for set fees. The Bahamas and Cayman Islands are leaders in this sector.

Organization of African Unity (OAU) Founded in 1963, the organization grew to include all the states of the continent except South Africa, which finally was asked to join in 1994. It is mostly a political body that has tried to resolve regional conflicts.

Organization of American States (OAS) Founded in 1948 and headquartered in Washington, D.C., the organization advocates hemispheric cooperation and dialog. Most states in the Americas belong except Cuba.

Organization of Petroleum Exporting Countries (OPEC) An international organization of 12 oil-producing nations (formed in 1960) that attempts to influence global prices and supplies of oil. Algeria, Gabon, Indonesia, Iran, Iraq, Kuwait, Libya, Nigeria, Qatar, Saudi Arabia, UAE, and Venezuela are members.

orographic rainfall Enhanced precipitation over uplands that results from lifting (and cooling) of air masses as they are forced over mountains.

Ottoman Empire A large, Turkish-based empire (named for Osman, one of its founders) that dominated large portions of southeastern Europe, North Africa, and Southwest Asia between the sixteenth and nineteenth centuries.

outback Australia's large, generally dry, and thinly settled interior.

overurbanization A process in which the rapid growth of a city, most often because of in-migration, exceeds the city's ability to provide jobs, housing, water, sewers, and transportation.

oxisols Reddish or yellowish soils found in the tropical shields of Latin America. They are formed over long periods of time and accumulate rusted or oxidized iron. When these fine soils are disturbed or compacted, a hard-pan surface results that is virtually impossible to farm.

Palestinian Authority (PA) A quasi-governmental body that represents Palestinian interests in the West Bank and Gaza.

Pan-African Movement Founded in 1900 by U.S. intellectuals W. E. B. Du Bois and Marcus Garvey, this movement's slogan was "Africa for Africans" and its influence extended across the Atlantic.

particularism A mode of thought that emphasizes the uniqueness of different places and different phenomena. Particularism is opposed to universalism, which emphasizes locality-transcending commonalties.

pastoral nomadism A traditional subsistence agricultural system in which practitioners depend on the seasonal movements of livestock within marginal natural environments.

pastoralism A way of life and form of livelihood centered around raising large animals, usually cattle, sheep, goats, and horses, but also sometimes including camels, yaks, and other species.

pastoralists Nomadic and sedentary peoples who rely upon livestock (especially cattle, camels, sheep, and goats) for their sustenance and livelihood.

perestroika A program of partially implemented, planned economic reforms (or restructuring) undertaken during the Gorbachev years in the Soviet Union designed to make the Soviet economy more efficient and responsive to consumer needs.

permafrost A cold-climate condition in which the ground remains permanently frozen.

physiological densities A population statistic that relates the number of people in a country to the amount of arable land.

Pidgin English A version of English that also incorporates elements of other local languages, often utilized to foster trade and basic communication between different culture groups.

plantation America A cultural region that extends from midway up the coast of Brazil, through the Guianas and the Caribbean, and into the southeastern United States. In this coastal zone, European-owned plantations, worked by African laborers, produced agricultural products for export.

plate tectonics The theory that explains the gradual movement of large geological platforms (or plates) along Earth's surface.

podzol soils A Russian term for an acidic soil of limited fertility, typically found in northern forest environments.

polders Reclaimed agricultural areas along the Dutch coast that have been diked and drained. Many are at or below sea level.

pollution exporting The process of exporting industrial pollution and other waste material to other countries. Pollution exporting can be direct, as when waste is simply shipped abroad for disposal, or indirect, as when highly polluting factories are constructed abroad.

Polynesia Pacific Ocean region, broadly unified by language and cultural traditions, that includes the Hawaiian Islands, Marquesas Islands, Society Islands, Tuamotu Archipelago, Cook Islands, American Samoa, Samoa, Tonga, and Kiribati.

postindustrial economy An economy in which the tertiary and quaternary sectors dominate employment and expansion.

prairie An extensive area of grassland in North America. In the more humid eastern portions, grasses are usually longer than in the drier western areas, which are in the rain shadow of the Rocky Mountain range.

primate city The largest urban settlement in a country that dominates all other urban places, economically and politically. Often—yet not always—the primate city is also the country's capital.

privatization The process of moving formerly state-owned firms into the contemporary capitalist private sector.

protectorate During the period of global Western imperialism, a state or other political entity that remained autonomous but sacrificed its foreign affairs to an imperial power in exchange for "protection" from other imperial powers.

purchasing power parity (PPP) A method of reducing the influence of inflated currency rates by adjusting a local currency to a composite baseline of one U.S. dollar based upon its ability to purchase a standardized "market basket" of goods.

qanat system A traditional system of gravity-fed irrigation that uses gently sloping tunnels to capture groundwater and direct it to needed fields.

Quran (also spelled Koran) A book of divine revelations received by the prophet Muhammad that serves as a holy text in the religion of Islam.

rain shadow effect A weather phenomenon in which mountains block moisture, producing an area of lower precipitation on the leeward side of the uplift.

rate of natural increase (RNI) The standard statistic used to express natural population growth per year for a country, region, or the world based upon the difference between birth and death rates. RNI does not consider population change from migration. Though most often a positive figure (such as 1.7 percent), RNI can also be expressed as a negative (–.08) for no-growth countries.

refugee A person who flees his or her country because of a well-founded fear of persecution based on race, ethnicity, religion, ideology, or political affiliation.

remittances Money sent by immigrants to their country of origin to support family members left behind. For many countries, remittances are a principal source of foreign exchange.

Renaissance-Baroque landscape Urban landscapes generally constructed during the period from 1500 to 1800 that are characterized by wide, ceremonial boulevards, large monumental structures (palaces, public squares, churches), and ostentatious housing for the urban elite. A common landscape feature in European cities.

rift valley A surface landscape feature formed where two tectonic plates are diverging or moving apart. Usually, this forms a depression or large valley.

rimland The mainland coastal zone of the Caribbean, beginning with Belize and extending along the coast of Central America to northern South America.

rural-to-urban migration The flow of internal migrants from rural areas to cities that began in the 1950s and intensified in the 1960s and 1970s.

Russification A policy of the Soviet Union designed to spread Russian settlers and influences to non-Russian areas of the country.

rust belt Regions of heavy industry that experience marked economic decline after their factories cease to be competitive.

Sahel The semidesert region at the southern fringe of the Sahara, and the countries that fall within this region, which extends from Senegal to Sudan. Droughts in the 1970s and early 1980s caused widespread famine and dislocation of population.

salinization The accumulation of salts in the upper layers of soil, often causing a reduction in crop yields, resulting from irrigation with water of high natural salt content and/or irrigation of soils that contain a high level of mineral salts.

samurai The warrior class of traditional Japan. After 1600, the military role of the samurai declined as they assumed administrative positions, but their military ethos remained alive until the class was abolished in 1868.

Sanskrit The original Indo-European language of South Asia, introduced into northwestern India perhaps 4,000 years ago, from which modern Indo-Aryan languages evolved. Over the centuries, it has become the classical literary language of the Hindus and is widely used as a scholarly second language, much like Latin in medieval Europe.

scheduled castes Part of the contemporary reform movement in India to lessen the social stigma of lower castes, particularly of the dalits (formerly untouchables), by reserving a certain number of places in universities, governmental offices, and seats in national and state legislatures for them.

Schengen Agreement The 1985 agreement between some—but not all—European Union member countries to reduce border formalities in order to facilitate free movement of citizens between member countries of this new "Schengenland." For example, today there are no border controls between France and Germany, or between France and Italy.

sectoral transformation The evolution of a labor force from being highly dependent on the primary sector to being oriented around more employment in the secondary, tertiary, and quaternary sectors.

secularization The widespread movement in western Europe away from regular participation and engagement with traditional organized religions such as Protestantism or Catholicism.

sediment load The amount of sand, silt, and clay carried by a river.

shatterbelt A geopolitically unstable area where the superpowers vie for power. Eastern Europe was the classic shatterbelt until World War II.

shield landscape Barren, mostly flat lands of southern Scandinavia that were heavily eroded by Pleistocene ice sheets. In many places, this landscape is characterized by large expanses of bedrock with little or no soil that resulted from glacial erosion.

shields Large upland areas of very old exposed rocks that range in elevation from 600 to 5,000 feet (200 to 1,500 meters). The three major shields in South America are the Guiana, Brazilian, and Patagonian.

shifted cultivators Migrants, with or without agricultural experience, who are transplanted by government relocation schemes.

shifting cultivation An agricultural system in which plots of land are farmed and then abandoned for a number of years until fertility is restored, at which point they are again brought under cultivation. See also *swidden*.

Shiites Muslims who practice one of the two main branches of Islam; especially dominant in Iran and nearby southern Iraq.

Shogun, Shogunate The true ruler of Japan before 1868, as opposed to the emperor, whose power was merely symbolic.

Sikhism An Indian religion combining Islamic and Hindu elements, founded in the Punjab region in the late fifteenth century. A long tradition of militarism continues today, and a large proportion of Sikh men are in the Indian armed forces.

Slavic peoples A group of peoples in eastern Europe and Russia who speak Slavic languages, a distinctive branch of the Indo-European language family.

social and regional differentiation "Social differentiation" refers to a process by which certain classes of people grow richer when others grow poorer; "regional differentiation" refers to a process by which certain places grow more prosperous while others become less prosperous.

socialist realism An artistic style once popular in the Soviet Union that was associated with realistic depictions of workers in their patriotic struggles against capitalism.

Special Economic Zones (SEZ) Relatively small districts in China that have been fully opened to global capitalism.

spheres of influence In countries not formally colonized in the nineteenth and early twentieth centuries (particularly China and Iran), limited areas called "spheres of influence" were gained by particular European countries for trade purposes and more generally for economic exploitation and political manipulation.

squatter settlements Makeshift housing on land not legally owned or rented by urban migrants, usually in unoccupied open spaces within or on the outskirts of a rapidly growing city.

steppe Semiarid grasslands found in many parts of the world. Grasses are usually shorter and less dense than in prairies.

structural adjustment programs Controversial yet widely implemented programs used to reduce government spending, encourage the private sector, and refinance foreign debt. Typically, these IMF and World Bank policies trigger drastic cutbacks in government-supported services and food subsidies, which disproportionately affect the poor.

subcontinent A large segment of land separated from the main landmass on which it sits by lofty mountains or other geographical barriers. South Asia, separated from the rest of Eurasia by the Himalayas, is often called the "Indian subcontinent."

subduction zones Areas where two tectonic plates are converging or colliding. In these areas, one plate usually sinks below another. They are characterized by earthquakes, volcanoes, and deep oceanic trenches.

subnational organizations Groups that form along ethnic, ideological, or territorial lines that can induce serious internal divisions within a state.

subsistence agriculture Farming that produces only enough crops or animal products to support a farm family's needs. Usually, little is sold at local or regional markets.

Suez Canal Pivotal waterway connecting the Red Sea and the Mediterranean opened by the British in 1869.

Sunda Shelf An extension of the continental shelf from the Southeast Asia mainland to the outlying islands. Because of the shelf, the overlying sea is generally shallow (less than 200 feet deep).

Sunnis Muslims who practice the dominant branch of Islam.

superconurbation A massive urban agglomeration that results from the coalescing of two or more formerly separate metropolitan areas.

supranational organizations Governing bodies that include several states, such as trade organizations, and often involve a loss of some state powers to achieve the organization's goals.

sustainable development A vision of economic change and growth seeking a balance with environmental protection and social equity so that the short-term needs of contemporary society do not compromise needs of future generations. The operational scale of sustainable development is local rather than global.

swidden agriculture Also called "slash-and-burn agriculture." A form of cultivation in which forested or brushy plots are cleared of vegetation, burned, and then planted to crops, only to be abandoned a few years later as soil fertility declines.

syncretic religions The blending of different belief systems. In Latin America, many animist practices were folded into Christian worship.

tectonic plates The basic building blocks of Earth's crust; large blocks of solid rock that very slowly move over the underlying semimolten material.

terraces Flat areas carved across the face of slopes, usually in a steplike fashion.

theocratic state A political state led by religious authorities.

tonal language Language in which the same set of phonemes (or basic sounds) may have very different meanings depending on the pitch in which they are uttered.

total fertility rate (TFR) The average number of children who will be borne by women of a hypothetical, yet statistically valid, population, such as that of a specific cultural group or within a particular country. Demographers consider TFR a more reliable indicator of population change than the crude birthrate.

township Racially segregated neighborhoods created for non-white groups under apartheid in South Africa. They are usually found on the outskirts of cities and classified as black, coloured, or South Asian.

transhumance A form of pastoralism in which animals are taken to high-altitude pastures during the summer months and returned to low-altitude pastures during the winter.

transmigration The planned, government-sponsored relocation of people from one area to another within a state territory.

transnational corporation Firms and corporations that, although they may be chartered and have headquarters in one specific country, do international business through an array of global subsidiaries.

transnationalism Complex social and economic linkages that form between home and host countries through international migration. Unlike earlier generations of migrants, early twenty-first-century immigrants can maintain more enduring and complex ties to their home countries as a result of technological advances.

Trans-Siberian Railroad Key southern Siberian railroad connection completed during the Russian empire (1904) that links European Russia with the Russian Far East terminus of Vladivostok.

Treaty of Tordesillas A treaty signed in 1494 between Spain and Portugal that drew a north-south line some 300 leagues west of the Azores and Cape Verde islands. Spain received the land to the west of the line and Portugal the land to the east.

tribal peoples Peoples who were traditionally organized at the village or clan level, without broader-scale political organization.

tribalism Allegiance to a particular tribe or ethnic group rather than to the nation-state. Tribalism is often blamed for internal conflict within Sub-Saharan states.

tribe A group of families or clans with a common kinship, language, and definable territory but not an organized state.

tsars A Russian term (also spelled *czar*) for "Caesar," or ruler; the authoritarian rulers of the Russian empire before its collapse in the 1917 revolution.

tsetse fly A fly that is a vector for a parasite that causes sleeping sickness (typanosomiasis), a disease that especially affects humans and livestock. Livestock is rarely found in those areas of Sub-Saharan Africa where the tsetse fly is common.

tsunamis Very large sea waves induced by earthquakes.

tundra Arctic region with a short growing season in which vegetation is limited to low shrubs, grasses, and flowering herbs.

Turkestan That portion of Central Asia populated primarily by Turkish-speaking peoples; "eastern Turkestan" comprises those areas presently controlled by China, and "western Turkestan" comprises areas formerly held by the Soviet Union.

typhoons Large tropical storms, similar to hurricanes, that form in the western Pacific Ocean in tropical latitudes and cause widespread damage to the Philippines and coastal Southeast and East Asia.

uncontacted peoples Cultures that have yet to be contacted and influenced by the Western world.

unitary state A political system in which power is centralized at the national level.

United Provinces of Central America Formed in 1823 to avoid annexation by Mexico, this union collapsed in the 1830s, yielding the independent states of Guatemala, Honduras, El Salvador, Nicaragua, and Costa Rica.

universalizing religion A religion, usually with an active missionary program, that appeals to a large group of people regardless of local culture and conditions. Christianity and Islam both have strong universalizing components. This contrasts with ethnic religions.

urban decentralization The process in which cities spread out over a larger geographical area.

urban form The physical arrangement or landscape of the city, made up of building architecture and style, street patterns, open spaces, housing types, etc.

urban primacy A state in which a disproportionately large city, such as London, New York, or Bangkok, dominates the urban system and is the center of economic, political, and cultural life.

urban realms model A simplified description of urban land use, especially descriptive of the modern North American city. It features a number of dispersed, peripheral centers of dynamic commercial and industrial activity linked by sophisticated urban transportation networks.

urban structure The distribution and pattern of land use, such as commercial, residential, or manufacturing, within the city. Often, commonalties give rise to models of urban structure characteristic of the cities of a certain region or of a shared history, such as cities shaped by European colonialism.

urbanized population That percentage of a country's population living in settlements characterized as cities. Usually, high rates of urbanization are associated with higher levels of industrialization and economic development, since these activities are usually found in and around cities. Conversely, lower urbanized populations (less than 50 percent) are characteristic of developing countries.

Urdu Although Urdu originated in the region of northern India and arose from a similar colloquial base as Hindi, it is one of the official languages of Pakistan today because of its long association with Muslim culture. Urdu borrows heavily from Persian vocabulary and Arabic grammatical construction and, further, is written in a modified form of the Persian Arabic alphabet.

viticulture Grape cultivation.

Wallace's Line A line in Southeast Asia delineating the abrupt difference in flora and fauna from that found on the Asian mainland to plants and animals more common to Australia.

water stress An environmental planning tool used to predict areas that have—or will have—serious water problems based upon the per capital demand and supply of fresh water.

White Australia Policy Before 1975, a set of stringent Australian limitations on nonwhite immigration to the country. Largely replaced by a more flexible policy today.

World Trade Organization (WTO) Formed as an outgrowth of the General Agreement on Tariffs and Trade (GATT) in 1995, a large collection of member states dedicated to reducing global barriers to trade.

Index

Page numbers in *italics* indicate figures.

Installation Instructions
The Prentice Hall CD can be run directly from the CD or copied to the hard drive. If you are copying to the hard drive you need 640 megabytes of space on your hard drive.

Program Instructions
When the CD is inserted it should run at startup. If not, double click on the DAG2e.exe icon to launch the program on a PC or click the MAC-START icon to launch the program on a Macintosh.

Minimum System Requirements
Windows/PC
Processor
K6-2 300mhz or Intel Celeron 300mhz processor
RAM
32 megs of RAM for Windows 98/ME (Windows system minimum)
64 megs of RAM for Windows NT (Windows system minimum)
64 megs of RAM for Windows 2000 (Windows system minimum)
128 megs or RAM for Windows XP (Windows system minimum)
Operating System
Windows 98/ME, ,NT, 2000, XP
Monitor Resolution
800x600
Software Required
Adobe PDF reader is required to view and print the maps.
Required Hardware
8X CD-ROM
Graphic card capable of 800x600 resolution and 8bit color
Sound Card
Printer required to print maps
Internet Connection
Not Required

Macintosh
PowerMac 9500 series processor or greater

RAM
64 megs of RAM
Operating System
OS 8.6 - 9.2
Note: As of Director 8.5 There Is No Projector Support For OSX, please use one of the above operating systems.
Monitor Resolution
800x600
Software Required
Adobe PDF reader is required to view and print the maps.
Required Hardware
8X CD-ROM
Graphic card capable of 800x600 resolution and 8bit color
Sound Card
Printer required to print maps
Internet Connection
Not Required

Adobe Acrobat Reader Copyright (c) 2002 is required to print the exercises from the CD. Installers for both Mac and PC systems are included on the CD.